3rd International Conference on Material Engineering and Advanced Manufacturing Technology (MEAMT 2019)

IOP Conference Series: Materials Science and Engineering
Volume 715

Shanghai, China
26 - 28 April 2019

Editor:

Nader Asnafi

ISBN: 978-1-7138-2574-6
ISSN: 1757-8981

Printed from e-media with permission by:

Curran Associates, Inc.
57 Morehouse Lane
Red Hook, NY 12571

Some format issues inherent in the e-media version may also appear in this print version.

Institute of Physics
Dirac House, Temple Back
Bristol BS1 6BE UK

Phone: 44 1 17 929 7481
Fax: 44 1 17 920 0979

techtracking@iop.org

Additional copies of this publication are available from:

Curran Associates, Inc.
57 Morehouse Lane
Red Hook, NY 12571 USA
Phone: 845-758-0400
Fax: 845-758-2633
Email: curran@proceedings.com
Web: www.proceedings.com

TABLE OF CONTENTS

DATA MINING, ALGORITHMS AND INTERNET OF THINGS

FAILURE AND OUTLIERS DETECTION AND ANALYSIS

FORECASTING, PREDICTIONS, ANALYSIS AND PLANNING

NANOFABRICATION, NANOMETROLOGY AND APPLICATIONS

NEURAL NETWORKS

PERFORMANCE AND BEHAVIOUR IN DIFFERENT APPLICATIONS

PRECISION ENGINEERING, INSPECTION, MEASUREMENT AND METROLOGY

ROBOTICS, MECHATRONICS AND MANUFACTURING AUTOMATION

SENSORS, CONTROLLERS, ACTUATORS AND ADAPTIVITY

MICROSTRUCTURAL CHARACTERIZATION AND MECHANICAL PROPERTIES

OTHERS

Preface

We are very pleased to present the proceedings of the 3rd International Conference on Material Engineering and Advanced Manufacturing Technology (MEAMT 2019) that was held successfully under the auspices of IASED in Shanghai, China, April 26-28, 2019.

The theme of the conference - Material Engineering and Advanced Manufacturing Technology – is of key importance to the science and technology and a sustainable industrial development. This conference promotes valuable contacts between academia and industry and addresses both basic research and the societal/industrial technological needs within Material Engineering and Advanced Manufacturing Technology.

In this era of Industry 4.0, we hope that the conference provided applicable data/information, addressed the issues in the societal/industrial transformation, and facilitated exchange within and between academia and industry.

We want to express our gratitude to the program chairs and all members of the advisory, publicity, and technical committees for their valuable time and advices. We are grateful to the world renowned scientists who acted as keynote speakers at the conference.

After a rigorous review process, where each paper was reviewed by at least two reviewers, high quality papers were accepted for presentation at this conference. We would like to thank all the reviewers for their time, effort, and for completing their assignments on time albeit tight deadlines.

Many thanks to the authors for their valuable contributions and to the attendees for their active participation.

Finally, a big thank to the organizers who made this conference to yet another unforgettable experience.

Nader Asnafi
Professor of Mechanical Engineering, Örebro University, Sweden
Editor of the proceedings of MEAMT 2019
One behalf of the conference committee

IOP Conf. Series: Materials Science and Engineering **715** (2020) 011002 doi:10.1088/1757-899X/715/1/011002

Peer review statement

All papers published in this volume of *IOP Conference Series: Materials Science and Engineering* have been peer reviewed through processes administered by the proceedings Editors. Reviews were conducted by expert referees to the professional and scientific standards expected of a proceedings journal published by IOP Publishing.

Development of microwave processed Ni + 20% SiC based composite clads on AISI-304 steel

Sarbjeet Kaushal[1,*], Dheeraj Gupta[2] and Hiralal Bhowmick[2]

[1]Mechanical Engineering Department, Gulzar Group of Institutes, Ludhiana, India
[2]Mechanical Engineering Department, Thapar Institue of Engineering and Technology, Patiala, India

*E-mail: sarbjeet.kaushal1988@gmail.com

Abstract. In the present investigation microwave hybrid heating technique was used to develop the Ni + 20% SiC based composite clads on AISI-304 substrate. The pilot experiments were carried out inside a domestic microwave oven working at 900W and 2.45GHz. The composite clads were successfully formed within 360 s of microwave exposure time. The so developed microwave processed clads were characterized through relevant techniques like XRD, SEM, EDS and Vicker's microhardness measurement. The XRD study reveals the presence of various carbides and silicides in the composite clad region. Microstructure study results reveal the presence of hard carbide particles in the soft Ni matrix phases. The average value of microhardness in the clad region was observed as 763 ± 24HV, which was approximately 3 times that of the substrate microhardness.

1. Introduction

Metal ceramic based composite claddings/coatings are widely used on the surface of materials vulnerable to corrosion, oxidation and wear losses. These types of composite claddings/coatings constitute both tough and hard phases, which can enhance the corrosion and wear resistance of the material. Ni-based metallic materials along with various ceramic materials like tungsten carbide (WC), chromium carbide (Cr_3C_2), Silicon carbide etc. are mainly used for producing composite claddings [1-3]. These claddings are developed through various techniques like thermal spraying, plasma spraying, laser cladding etc. [4-6]. Laser claddings technique is widely used for developing composite claddings of various materials due to its unique properties such as the formation of an excellent metallurgical bond between the substrate and clad materials, higher cooling rate etc. [7]. However, in case of laser cladding, the heat is directly supplied on the surface of the material to be processed and hence results in large thermal gradient between the surface and the core of the material. This large thermal gradient causes the large thermal distortion and poor microstructure of the clad region. Apart from this, higher installation and running costs are associated with this process. In the recent years, the microwave cladding process is emerging as a noble and versatile surface modification process. Microwave cladding process involves some unique features such as less processing time, lower energy consumptions and volumetric heating. The volumetric heating characteristic results in uniform thermal gradient throughout the material during microwave cladding process and hence excellent microstructure is achieved. Gupta and Sharma (2012) [3] and Kaushal et al. (2017) [2] reported the composite claddings of Ni-WC8Co and Ni-Cr_3C_2 based materials using microwave heating at 2.45

GHz. Ni-SiC based claddings have been developed through various techniques such as thermal spraying, laser cladding etc. Less literature has been reported on the development of Ni-SiC clads through microwave cladding process. Kaushal et al. (2018) [8] successfully developed the microwave cladding of Ni + 10% SiC based material on martensitic steel and reported that the clads were metallurgically bonded to the substrate material. It was also reported that composite claddings can be useful for anti-wear applications. In the present work, Ni-based EWAC + 20% SiC material was developed on austenitic stainless of AISI-304 grade using a domestic microwave applicator at 2.45 GHz frequency and variable power of 180-900 W. The so developed clads were further characterized using XRD, SEM/EDS, and Vicker's micro-hardness measurements.

2. Materials and methods

2.1. Materials selection

Austenitic stainless steel (AISI-304) was used as the substrate material and was machined to the dimensions $10\ mm \times 10\ mm \times 6\ mm$. The commercial available EWAC powder (~98% Ni; Maker: Larsen and Toubro) of average grain size 42 ± 8 μm was mixed with 20% SiC powder (by weight%) of average grain size 30 ± 5 μm. The SEM images (Figure 1(a, b)) reveal the sharp edge morphology of SiC particles and round edge morphology of Ni based particles. The XRD spectra of both the powders are shown in figure 2 (a, b), which reveal the presence of majority SiC particles and small traces of SiO_2 in SiC based powder and enriched Ni particles in EWAC powder. Further, the chemical compositions of raw powders and substrate materials are shown in table 1.

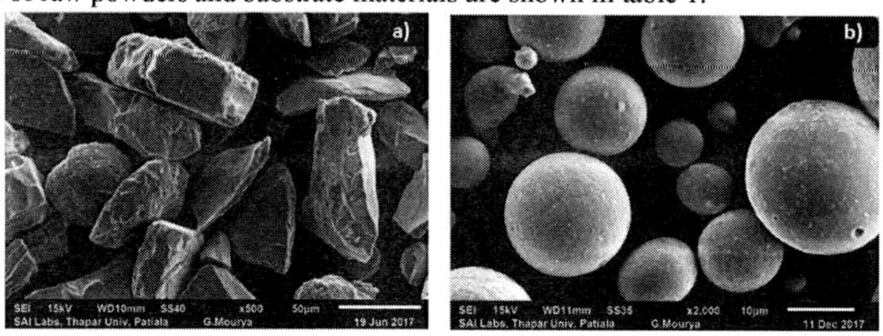

Figure 1. SEM morphology of (a) SiC powder; (b) EWAC powder.

Table 1. Chemical composition of raw materials.

Material	Elements							
	Cr	Ni	Fe	P	C	Mn	Si	O
SiC powder	-	0.003	0.091	-	32.13	-	Bal.	8.11
EWAC powder	0.17	Bal.	-	-	0.2	-	2.8	-
AISI-304 substrate	18.7	10	Bal.	0.45	0.08	2	1	-

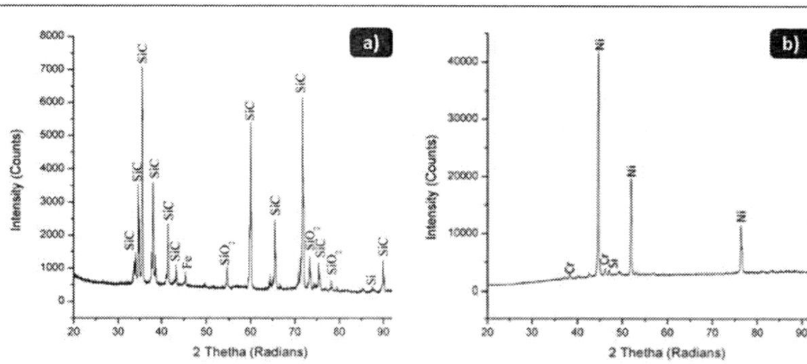

Figure 2. XRD spectra of (a) SiC powder; (b) EWAC powder.

2.2. Development and characterizations of claddings

Cladding development through microwave heating is a very complex phenomenon. Prior to the deposition of the composite clad, the Ni + 20% SiC powder layer was preheated at 180°C in the microwave oven under convection mode. The preheated powder layer was placed manually on AISI-304 substrate by maintaining the approximate thickness of 1.1 mm. The substrate was placed on the refractory brick. The interaction behavior of microwaves is not same for each material. Microwave material interaction depends upon the skin depth of the material, which is defined as the depth from the surface into the material up to which the microwave power intensity falls by 36.8% of the surface value. In other words, the skin depth of the material is directly proportional to its microwave absorption capacity. The materials with lower skin depth have lower microwave absorbing capacity. However, the skin depth of the given material can be raised by increasing its initial temperature up to a critical value. The concept of microwave hybrid heating (MHH) was used by Gupta and Sharma (2014) [9] to achieve this critical temperature value. In case of MHH the susceptor particles (have good microwave absorbing characteristics) are placed on composite powder layer. An alumina separator sheet is placed between powder layer and susceptor particles to avoid any contamination. The whole arrangement was then placed inside the domestic microwave oven and the setup was turned on. Initially, the powder layer does not absorb the microwave radiations due to the lesser skin depths of the constituent powders. However, the susceptor particles start interacting with microwaves instantaneously and start converting microwave energy into heat energy. This heat energy is transferred to the composite powder layer through conventional modes of heat transfer. After getting heat from the susceptor particles, the skin depth of the composite powder layers rises up and the powder layers start absorbing microwave energy. Hence the melting of the powder layer is started and the cladding is achieved within 360 s of exposure time. The detailed principle of MHH is illustrated in figure 3. The different microwave cladding process parameters are tabulated in table 2.

To determine the behavior and various properties of the so developed composite clads, the characterizations of these clads were carried out using relevant techniques. Prior to the characterization, the microwave processed Ni + 20% SiC clad was sectioned across its thickness using the low-speed diamond saw. Polishing of the sectioned specimens was carried out through standard metallographic techniques. The X-ray diffraction spectrometer was used to analyze the different phases present in the composite clad region at the scan rate of 1° min^{-1} and a scan range of 20^{0} to 100^{0}. The Microstructural characterization of the clad region was carried out using scanning electron microscopy (SEM) equipped with energy dispersive spectroscopy (EDS). The microhardness of the clad region was evaluated using Vicker's microhardness measurement at 300 g of load and 20 s of dwell time.

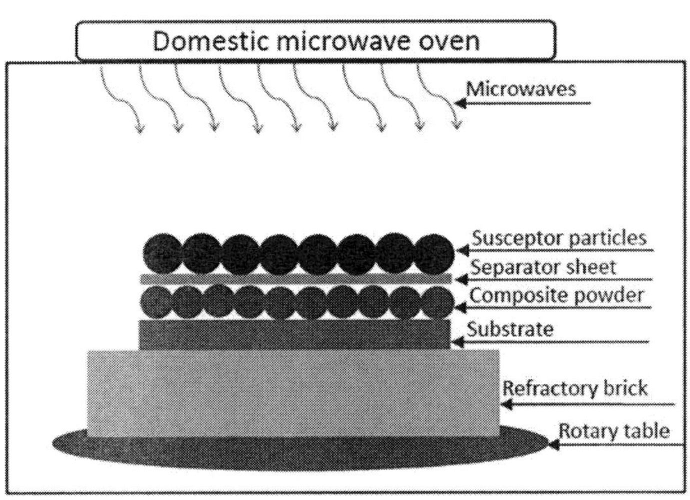

Figure 3. The principle of microwave hybrid heating.

Table 2. Process parameters used for development of microwave composite cladding.

Parameters	Description
Applicator	Multimode (Maker: LG, Model: Charcoal)
Exposure power and Frequency	900 W and 2.45 GHz
Exposure time	360 s
Powder	Ni based + 20% SiC
Preheating powder temperature	180° C
Separator	99.1% pure alumina sheet
Susceptor	Charcoal powder

3. Results and discussion

3.1. Phase study analysis

A typical XRD spectrum of microwave processed Ni + 20% SiC clad is shown in figure 4. The presence of various phases such as Ni_3Si_2 (2Θ: 29.49°, 44.68°), SiC (2Θ: 35.74°, 75.45°), Ni (2Θ: 39.66°, 45.06°), $Cr_{23}C_6$ (2Θ: 48.30°, 92.12°) and NiSi (2Θ: 52.07°) were observed in the composite clad region during microwave hybrid heating. During microwave hybrid heating, the SiC dissociates into Si and C at a higher temperature [8]. This free Si reacts with Ni element present in the EWAC powder to form Ni_3Si_2 and NiSi. The presence of $Cr_{23}C_6$ in the clad region may be due to the partial dilution of Cr elements from the substrate to the clad region and hence confirms the metallurgical bonding between the substrate and clad region. Some of the SiC is present in its original form corresponding to (2Θ: 35.74°, 75.45°). The presence of SiC, $Cr_{23}C_6$, Ni_3Si_2 in the clad region results in higher microhardness of the composite clad.

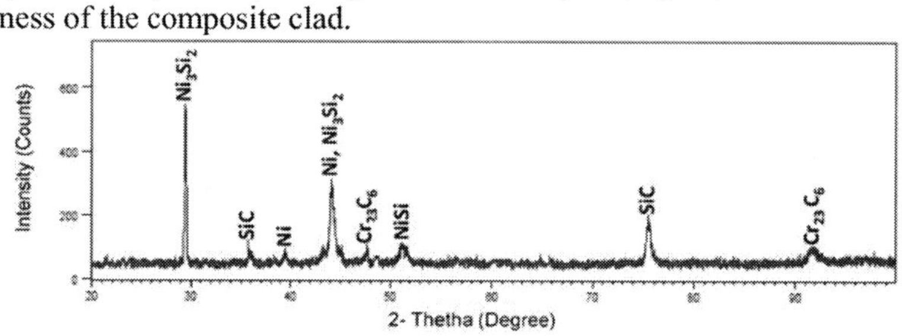

Figure 4. XRD spectrum of microwave processed Ni + 20% SiC clad.

3.2. Microstructural study

To know the quality of the microwave processed composite clad, the microstructural analysis was carried out. It was observed in figure 5 (a), that the composite clad of approximate 1.02 mm thickness was developed using MHH technique. The clad was free from any type of visible cracks and porosity. The wavy interface between the substrate and clad region can be clearly seen. This wavy interface formation was due to the different thermal properties of constituent powders and substrate material, which results into the setup of convection currents at the interface region during microwave heating and hence causes the wavy interface. The same wavy interface was reported by Gupta and Sharma (2010). Further, the zoomed image of the clad region was shown in figure 5 (b), which reveals the randomly dispersed carbide particles in the form of flakes inside the soft Ni matrix. To determine the presence of different elements in the composite clad regions, the EDS analysis study was carried out at point X and Y corresponding to figure 5 (b). The presence of majority Ni element with small traces of C and oxygen was observed at point X (Figure 6(a)), while the presence of enriched carbon and Si elements was found at point Y as shown in figure 6(b).

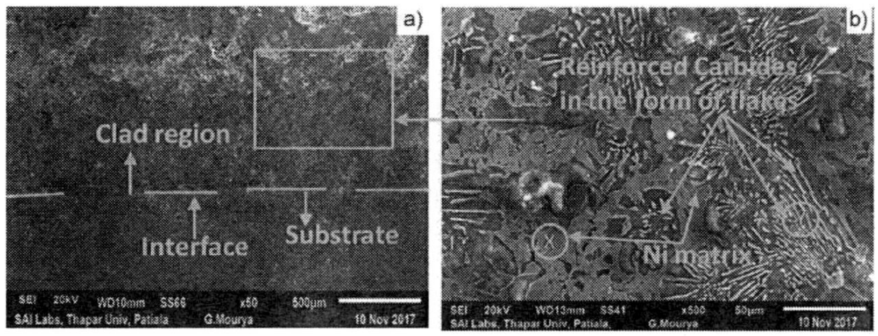

Figure 5. SEM image showing (a) transverse section of the composite clad; (b) zoomed image of the enclosed region in figure 5 (a).

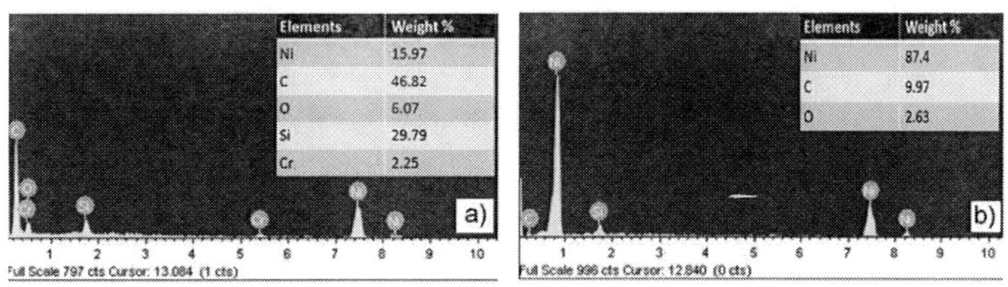

Figure 6. EDS analysis (a) at point X corresponding to figure 5 (b); (b) at point Y corresponding to figure 5 (b).

3.3. Vicker's microhardness study

The presence of various carbides and silicides phases in the composite clad region could attribute to increasing its micro-hardness value. To verify this fact, the Vicker's microhardness indentations were taken across the composite clad region starting from the top clad surface towards the substrate region at the interval of 170 μm. Three indentations were taken at each level and the average value of three readings was taken into consideration. The average value of the microhardness in the clad region was observed 763 ± 24 HV (Figure 7). The value of microhardness at the interface region was 420 ± 104 HV. The large standard deviation at the interface region was due to the dilution of different elements from the substrate and clad region.

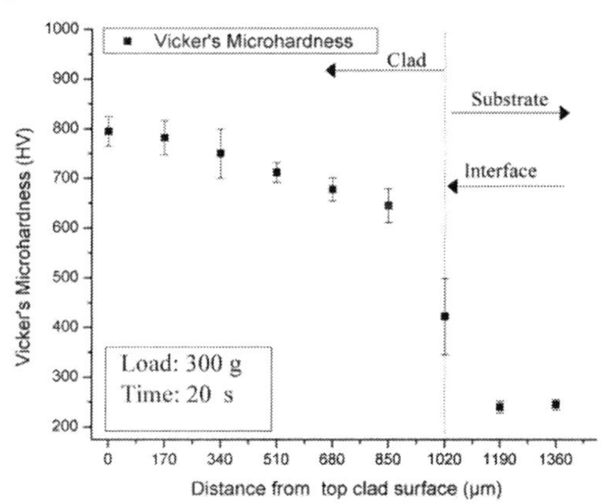

Figure 7. Vicker's microhardness profile of the microwave processed Ni + 20% SiC clad.

4. Conclusions

Microwave hybrid heating technique was successfully employed for the development of Ni + 20% SiC clads on AISI-304 substrate. The various conclusions drawn from the study are as following.

- The microstructure analysis of the composite clad reveals that the composite clads of 1.02 mm thickness were free from any type of visible porosity and cracks.
- The Carbide particles were randomly dispersed in the Ni matrix in form of flakes like structure.
- The presence of Ni_3Si_2, SiC, $Cr_{23}C_6$, NiSi phases were observed in the composite clad region during XRD study.
- The average value of the microhardness in the clad region was observed as 763 ± 24 HV which was almost 2.5 times that of substrate microhardness.

References

[1] A. Pathania, S. Singh, D. Gupta and V. Jain: J. Manuf. Processes Vol. **20** (2015), p. 79-87.
[2] S. Kaushal, D. Gupta and H.L. Bhowmick: J. Tribol- T. ASME.Vol. **139** (2017), p. 061602-1-8.
[3] D. Gupta, P.M. Bhovi, A.K. Sharma and S. Dutta: J. Manuf. Processes Vol. **14** (2012), p. 243-249.
[4] H. S. Grewal, H. Singh and A. Agrawal: Wear Vol. 301 (2013), p. 424-433.
[5] G. Xie, X. Lin, K. Wang, X. Mo. Z. Zhang and P. Lin: Corros. Sci. Vol. **49** (2007), p. 662-671.
[6] L. St-Georges: Wear Vol. 263 (2007), p. 562-566.
[7] E. M. Birger, G.V. Moskvitin, A.N. Polyakov and V.E. Arkhipov: Weld. Int. Vol. **25** (2011), p. 234-243.
[8] S. Kaushal, V. Sirohi, D. Gupta, H.L. Bhowmick and S. Singh: P. I. Mech. Eng. L-J. Mat. Vol. **232** (2019), p. 80-86.
[9] D. Gupta and A.K. Sharma: J. Manuf. Processes Vol. **16** (2014), p. 176-182.

IOP Conf. Series: Materials Science and Engineering **715** (2020) 012002 doi:10.1088/1757-899X/715/1/012002

Structure and properties of graphene reinforced Ti-based composite coatings on TC4 alloy by argon arc cladding

J S Meng*, X P Shi, S J Zhang, M Y Wang, B B Liu, F Xue and WC Cui

College of Naval architecture and Marine Engineering, Shandong Jiaotong University, 1508 Hexing Road, 264310, Weihai, P. R. China

*E-mail: mengjs2008@163.com

Abstract. The TiC coating was prepared on TC4 alloy using argon arc cladding method of graphene and titanium powder to improve wear resistance. The microstructure and phase of coating was observed by XRD, OM and SEM. The properties of coating were tested by microhardness-tester and dry slipping friction tester. The results show that the coating consists of TiC and Ti, and the composite coatings have good bonding with substrate. The in-situ synthesis of TiC particles are uniformly distributed in the composite coatings, and the morphology of TiC particles are composed of particle-shaped, rod-like and flower-like morphologies. The average microhardness of composite coating can reach 835 HV, about 2.8 times of the substrate. The composite coating has excellent tribological properties.

1. Introduction

The TC4 is a kind of titanium alloy which have the characteristics of light quality, high strength, corrosion resistance and easy to processing, and it can be applied to aviation, aerospace, petroleum, transportation and so on [1-3]. However, the application of TC4 alloy is limited due to low hardness, and low wear resistance [4,5]. Therefore, surface modification of TC4 alloy is necessary. The mainly surface modification methods of the TC4 alloy include ion implantation [6,7],spraying technology[8,9], sol-gel method[10,11] and surface cladding technology [12,13], especially the surface cladding technology has become one of the most widely studied at present. Ceramic composites exhibits high microhardness, excellent wear resistance can be employed as cladding materials. Zhang et al[14] have studied the microstructrue of TiC/Ti composite coating by laser cladding technology. The results showed that the highest haredness of coating is HV1400, about 4.5 times of the substrate. Li, et al[15] have studied SiC reinforced Ti-based composite coatings produced by in laser cladding technique using titanium and silicon carbide powder. The result that the maximum hardness of cladding layer can reached 1607.9HV. Yang et al[16] evaluated the microstructure and properties of TiCp/Ti composite coating by tungsten inert gas cladding.

Graphene is a two-dimensional carbon nano-material composed of carbon atoms and sp^2 hybrid orbital with hexagonal honeycomb lattice. Graphene has excellent optical, strength and modulus, electrical and mechanical properties [17-19]. A small amount of graphene can significantly improve the tensile strength and wear resistance of the pure Ti composites [20]. Compare to other cladding technology, the argon arc cladding features easy operation, energy concentration, low cost and non-oxidation and so on other advantages [16,21,22]. However, there was little research about graphene reinforced Ti-based composite coatings by argon arc cladding. So the present study employed the argon arc cladding to deposit graphene and titanium powder on the surface of TC4 to synthesize coating. The phase, microstructures and components of the composite coating was analyzed.

2. Experimental metal

2.1. Materials and cladding processing

The TC4 alloy was used as the substrate material with a dimension of 50 mm × 10mm × 10 mm. The surface of TC4 alloy need be polished with 320 grit silicon carbide sandpaper and washed with acetone and alcohol. The Ti powder (99.5% purity, ~20μm diameter) and graphene powder (99.0% purity; Chengdu Organic Chemicals Co. Ltd., Chinese Academy of Sciences, Chengdu, China) initial materials were used for synthesizing materials. The total mass of powder is 20 g, and the mass fraction of the graphene in the powder is 3%. Figure 1 shows that the SEM images of graphene powder and the pre-processed titanium powder. Next, the pre-processed Ti powder (97% by mass) and graphene (3% by mass) were mixed in the ball milling (F-P400H) for 4 h under argon (99.99% Ar) atmosphere. The EDS of mixed powders show that the powder did not occur reaction by ball milling(figure 2). Then coatings were carried out by MW3000 welding machine. The optimum cladding parameters are the cladding current (90A), cladding voltage (14.5V), cladding speed (140mm/min) , respectively. argon flow rate: 10L/min. Furthermore, argon gas (99.99%) was used as shielding gas in the cladding processing and the argon gas flow rate 10 L/min.

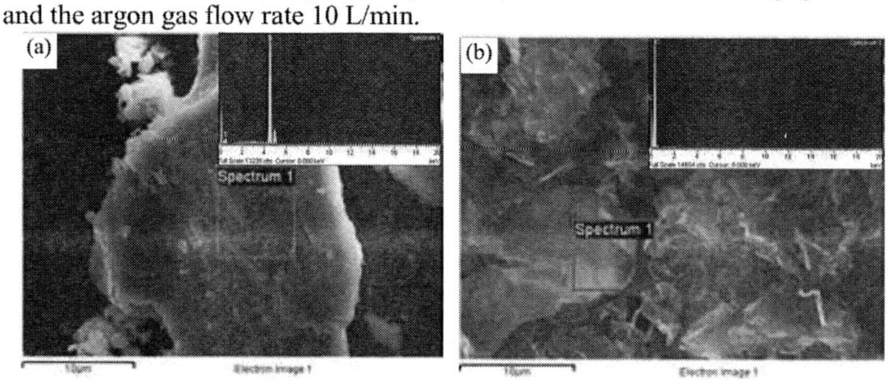

Figure 1. SEM images of the pre-processed titanium powder (a) and graphene powder (b).

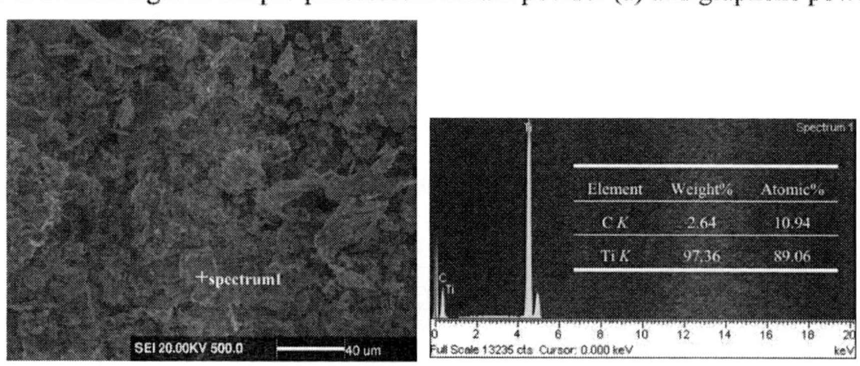

Figure 2. The SEM and EDS patterns of graphene and titanium powders after the ball milling.

2.2. Microanalysis and performance testing

The phase structure of the cladding coatings were analyze by DX-2700B X-ray diffraction with Cu Kα irradiation (λ = 0.1540nm). The scanning speed was 4°/min and the step size was 0.02°. The microstructure and components of the cladding coatings were observed by MX-2600 SEM and EDS.

The microhardness tester (type: MHV-2000) used to measure the microhardness of cladding coatings. The loading force is 2.942 N and the time is 15 s. The wear properties of cladding coatings were measured by wear machine (type: MMS-2A). The friction pair was GCr15 steel with a rock well hardness of 58HRC. The wear conditions were a normal load of 200 N, a sliding speed of 200 r/min, and the time of abrasion of 30 min.

3. Results and discussion

3.1. Microstructure and phase of the composite coatings

Figure 3 shows the XRD diagram of the composite coating. It reveals that the composite coatings comprised of Ti and TiC phases. There is no graphene in the cladding coating, because of carbon titanium in the coatings, in which the graphene and titanium have happened in-situ reaction. The result of the XRD analysis confirms that the TiC reinforced titanium matrix composite coatings have been produced by argon arc cladding technique.

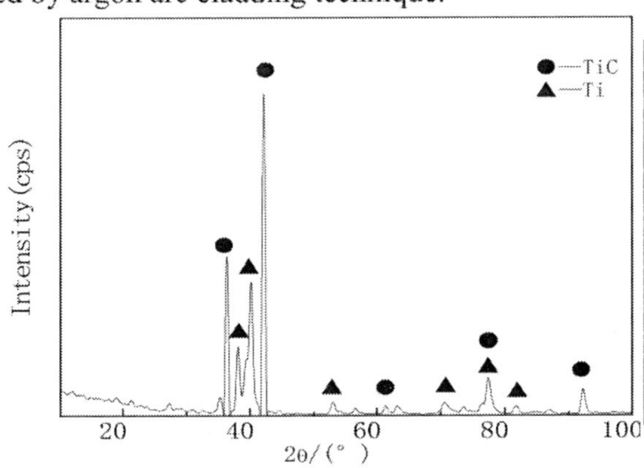

Figure 3. X-ray diffraction patterns of the cladding coating.

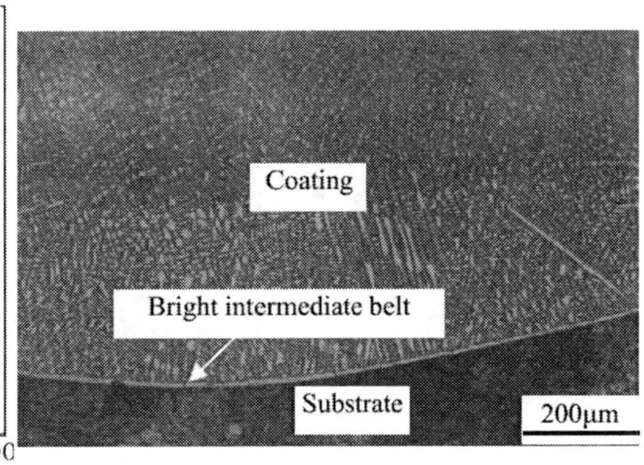

Figure 4. Cross-section morphologies of the composite coating by argon arc cladding.

The cross-section morphologies of the composite coating by argon arc cladding is given in figure 4. It reveals that the surface of the composite coating is relatively smooth, free from pores and cracks, and has metallurgical bonding with matrix. There is an obvious bright intermediate belt between cladding coating and the substrate, forming good metallurgic bonding[23].

Figure 5 shows the upper surface SEM images of the composite coating's cross-section. For the top surface of the coating, as shown in figure 5(a), the amount of particles is smaller and the size of the in-situ formed particles is bigger. For the coating in which depth from surface is equal to 300 μm, as we can see in figure 5(b), it can be observed clearly that the microstructure in the clad zone exists in the form of fine dendrites. The particles are formed by direct metallurgical reaction and distributes evenly in the coating. Argon arc cladding is an integrated physical and metallurgical process of the quick heating and quick cooling far away from the equilibrium state, which determines the particles synthesized in-situ were fine. SEM micrographs of composite coating, as shown in figure 6, clearly reveal the morphology of each phase. It can be found that, the composite coating are composed of particle-shaped and rod-like morphologies, and around the rod-like structures, where there are lots of particle-shaped phases (figure 4(a)). As shown in figure 4(b), the microstructure is composed mainly of flower-like structures. From the component content and element in table 1, the particle-shaped, rod-like and flower-like morphologies were found to be Ti and C by EDX analysis, and so they are presumably the TiC phase identified by XRD. The spectrum 2 were found to be rich in Ti, Al, and V, and so we can infer that they are α-Ti phase. The TiC particles are dominated by fine particles with an average grain size of less than 1 μm and the rod-like structures with average particle size of4 μm×500 nm. The results show that graphene and titanium powder generate a large number of TiC particles by in-situ reaction in the composite coating. These TiC particles play the role of particle reinforced.

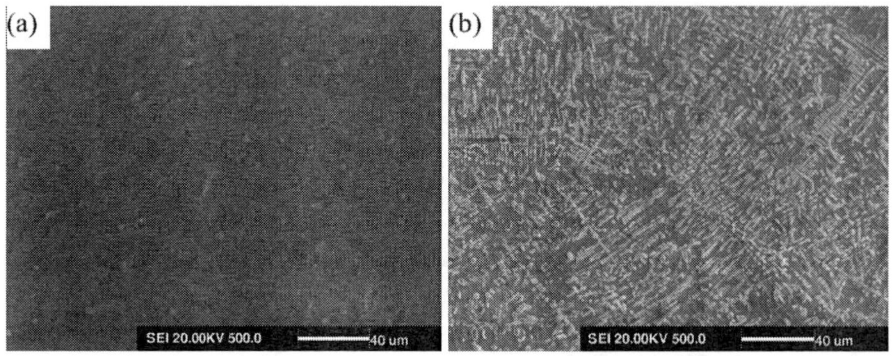

Figure 5. SEM images of the composite coating (a) Top surface; (b) Surface at about 300 μm.

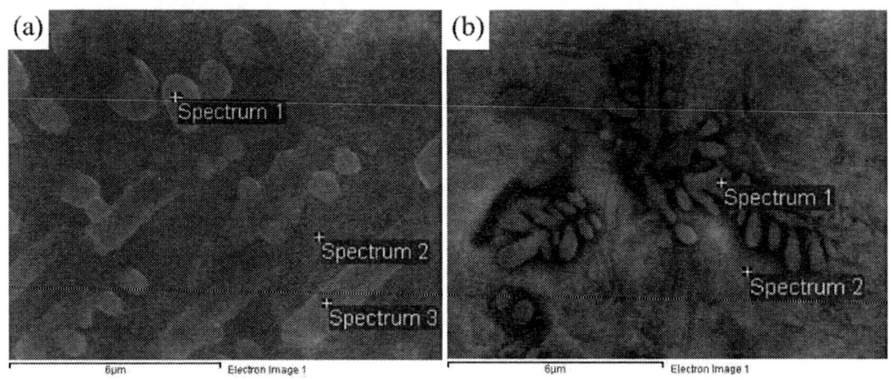

Figure 6. The high magnification morphology of different reinforcement shape.

Table 1. Component content and element of different zone in figure 6.

Position	Element and content (Wt. %)			
	C	Al	Ti	V
Spectrum1(a)	17.7	-	82.3	14.2
Spectrum2(a)	1.32	5.70	89.73	3.25
Spectrum3(a)	20.36	-	79.64	-
Spectrum1(b)	15.61	-	84.39	-
Spectrum2(b)	1.47	5.58	89.86	3.09

3.2. Microhardness

Figure 7 shows the microhardness of the composite coating. As seen from figure 7, the composite coating presents higher microhardness. From distribution of the microhardness curves, there are three main regions: the coating zone, heat-affected zone and substrate. The microhardness of the composite coating is decreased gradually from the coating through the interface to substrate, and the microhardness of top surface of composite coating is relatively lower than that of central region of the coating. The average microhardness of composite coating is 835HV, which is about 2.8 times of the initial hardness. This is mainly attributed to the fine hard particles on the near surface region of the cladding coating. The TiC particles have higher hardness(3400HV)[24, 25]. Therefore, the hardness of the cladding coating can be improved.

3.3. The wear weight loss

Figure 8 shows the wear weight loss of the composite coating surface and TC4 alloy. The argon arc cladding coatings has better wear resistance under dry sliding wear condition in comparison to the TC4 alloy substrate. Wear mass loss of the composite coating is 1/90 of that of the TC4 alloy. The main strengthening mechanism of the composite coating is fine-grain strengthening and dispersion strengthening. These in-situ synthesized TiC reinforcements inhibit the dislocation motion and result in the strengthening effect. The high hardness TiC particles which is formed at the substrate will improve the hardness and wearable ability of the coating, which decrease the mass-loss wear of composite coating.

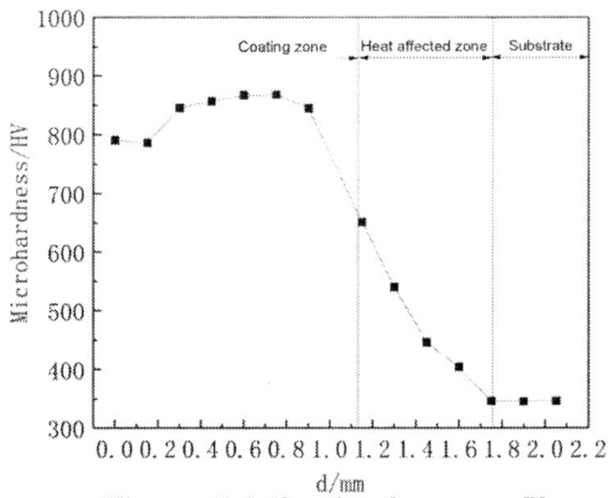

Figure 7. Microhardness profiles across the depth of composite coating.

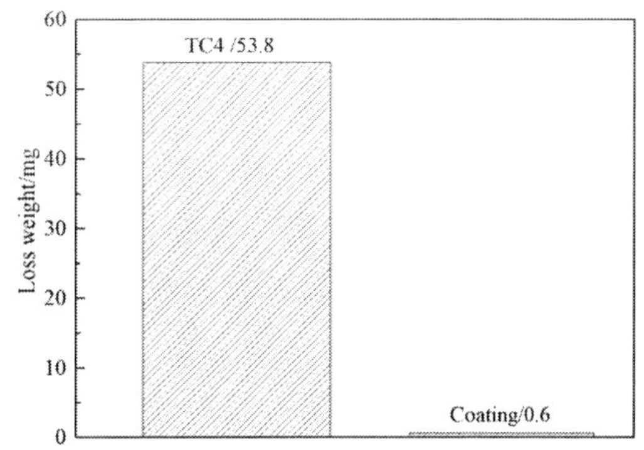

Figure 8. Comparison of wear loss weight for TC4 alloy substrate and the composite coating.

4. Conclusions

(1) The titanium matrix composite coating reinforced by TiC particles being prepared by argon arc cladding with reactions of graphene and titanium powder.

(2) The main phase of the composite coating was TiC phase and α-Ti phase. The morphology of TiC particles are composed of particle-shaped, rod-like and flower-like morphologies.

(3) The microhardness of the composite coating from the surface to the substrate is gradually decreased and the microhardness was up to 835 HV. The argon arc cladding coating exhibits excellent wear resistance under dry sliding wear test conditions due to in situ formation of high hardness of TiC ceramic reinforcements.

Acknowledgments

This research was supported by the General Financial Grant from the China Postdoctoral Science Foundation (Nos.2017M621245), Harbin Municipal Science and Technology (Nos.2016RAQXJ049), Scientific Research Fund of Heilongjiang Provincial Education Department (Nos. 2017-KYYWF-0512), Ph. D. Scientific Research Foundation of Shandong Jiaotong University (Nos.BS2018005), Shandong Provincial Key R&D plan (Nos. 2018GGX105002), Shandong Jiaotong university "Climbing" Research Innovation Team program (Nos. SDJTUC1802) and Natural Science Foundation of Shandong Province (Nos. ZR2019MEE107).

References

[1] Brandl E, Baufeld B, Leyens C and R.Gault 2010 Phys. Procedia 5 495
[2] Luo Y and Ge S R 2009 Tribol Int 42 1373
[3] Zhang K M, Zou J X, Li J, Yu Z S and Wang H P 2012 Trans. Nonferrous Met. Soc. China 22 1817
[4] Hager C H J, Sanders J H and Sharma S 2008 Wear 265 439
[5] Pang W, Man H C and Yue T M 2005 Mater. Sci. Eng. A 390 144

[6] Feng C F, Khor K A, Liu E J and Cheang P 1999 Scripta Mater. 42 103
[7] Wang B Yand Li Z X 2008 Rare Metal Mater. Eng. 37 1407
[8] Luo X T, Wei Y K, Wang Y and Li C J 2015Mater. Design, 85 527
[9] Huang C J, Yan X C, Chen C Y, Xie Y C, Liu M, Kuang M and Liao H L 2018 Vacuum 151 275
[10] Zhang W G, Wang C T and Liu W M 2006 Wear 260 379
[11] Peter B Kirk, Mark J Filiaggi, Rana N S Sodhi and Robert M Pilliar J 1999 Biomed Mater. Res. A 48 424
[12] Cheng X, He K and He J 2010 Tribology 30 250
[13] Wang H, Li J, Zhang G, Dai J and Xi W 2010 Applied Laser 30 199
[14] Zhan K, Zou G J, Li J, Yu Z and Wang H 2010 Trans. Nonferrous Met. Soc. China 20 2192
[15] Li S N , Xiong H P, Li N, Chen B Q, Gao C, Zou W J and Ren H S 2017 Ceram. Int. 43 961
[16] Yang R, Liu Z, Yang G and Wang Y 2012 Procedia Eng. 36 349
[17] Bolotin K I, Sikes K J, Jiang Z, Klima M, Fudenberg G, Hone J, Kim P and Stormer H L2008 Solid State Commun 146 351
[18] Balandin AA, Ghosh S, Bao W, Calizo I, Teweldebrhan D, Miao F and Lau C N 2008 Nano. Lett. 8 902
[19] Stoller MD, Park S, Zhu Y, An J and Ruoff RS 2008 Nano. Lett. 8 3498
[20] Mu X N, Cai H N, Zhang H M, Fan Q B, Zhang Z H, Wu Y, Ge Y X and Wang D D 2018 Mater. Design 140 431
[21] Meng J S and Ji Z S 2018 Rare Metal Mater. Eng. 47 13
[22] Meng J S, Jin G, Shi X P 2018 Applied Surface Science 431 135
[23] Hu Z and Li W 2017 Surface Technology 46 1
[24] Pang W, Man H C and Yue T M 2005 Mater. Sci. Eng. A 390 144
[25] Yang Y, Zhang D, Yan W, Zheng Y 2010 Opt Laser Eng 48 119

IOP Conf. Series: Materials Science and Engineering **715** (2020) 012003 doi:10.1088/1757-899X/715/1/012003

Novel tubular graphene synthesized via chemical vapor deposition process

Ming Yang[1,2], Yanyan Wang[1,2,*], Lei Dong[1,2], Zhekun Wu[1,2], Yanhua Liu[1,2] and Changsi Peng[1,2]

[1]School of Optoelectronic Science and Engineering & Collaborative Innovation Center of Suzhou Nano Science and Technology, Soochow University, Suzhou 215006, People's Republic of China
[2]Key Lab of Advanced Optical Manufacturing Technologies of Jiangsu Province & Key Lab of Modern Optical Technologies of Education Ministry of China, Soochow University, Suzhou 215006, People's Republic of China

*Email: yywang@suda.edu.cn

Abstract. In this work, we prepared a novel nanoscale tubular graphene (TG) by depositing it on the surface of nickel nanowires (Ni-NWs). This method allowed us to synthetize Ni-NWs by a chemical reduction process carried out under water-bath heat and magnetic field. The TG was subsequently deposited on the as-prepared Ni-NWs through a chemical vapor deposition (CVD) process under atmospheric pressure. Scanning electron microscopy (SEM) and Raman spectroscopy revealed the presence of Ni-NWs and TG/Ni-NW composites containing a few layers of graphene. The diameters of the Ni-NWs and TG/Ni-NW composites were ca. 300–500 and 500–1000 nm, respectively. The novel tubular structure of graphene provided this material with large specific surface area and excellent electrical conductivity, making it an ideal for sensor applications.

1. Introduction

Graphene isa flat monolayer of carbon atoms packed in a two-dimensional (2D) honeycomb lattice. This material was prepared by Andre Geim and Konstantin Novoselov bymechanical exfoliation in 2004 [1]. Graphene has receivedsignificant research attentionowing to its excellent performance. In subsequent years, a whole range of carbon materials including 2D graphene, zero-dimensional (0D) fullerene, one-dimensional (1D) carbon nanotubes, and three-dimensional (3D) graphite and diamond materials have been developed. As revealed in the past decade, single-layer graphene possesses a number of interesting characteristics including extremely high charge (electrons and holes) mobility (230000 cm^2/Vs), superior thermal conductivity (3000 W/mK), a visible light absorption as low as 2.3, and ultrahigh specific surface area (2600 m^2/g) [2–4]. Owing to these excellent properties, graphene has been widely used in many areas including electronic devices, information, energy, materials, and biomedicine, among others [5].

Following the pioneer production of graphene by mechanical stripping of graphite crystals, manyother preparation methodshave been developed including epitaxial growth on SiC [6], chemical reduction of graphite oxide [7], chemical vapor deposition(CVD) growth on copper or nickel substrates [8], and unzipping carbon nanotubes (CNTs) [9]. Chemically derived graphene is widely used in combination with other materials (e.g., metal oxides) and self-assembled into 3D macroporous structures owing to its convenient preparation and processing characteristics [10]. However, as a result of the redox process, a large number of defects appears on the surface graphene, which alters greatly its conductivity. Large-area and high-quality graphene materials can be prepared by CVD [11]. CVD achieves materials with significantly higher charge mobility compared to chemically reduced graphene sheets. A large number of studies have recently focused on growing 2D graphene sheets and macroscopic 3D graphene foams by CVD [12–15]. However, despite these advances, the preparation of high-quality graphene with microscopic tubular structure by CVD has not been targeted yet.

Herein, we report a simple method for synthesizing microscopic 3D foam-like tubular graphene (TG). The simple flowchart of this experiment was described in figure 1. Nickel nanowires (Ni-NWs) prepared by chemical reduction under water-bath heat and magnetic field conditions were used as substrates for depositing this TG. The TG was grown by ambient-pressure CVD on Ni-NWs. As revealed by scanning electron microscopy (SEM) and Raman spectroscopy, the Ni-NWs and TG/Ni-NW composites contained graphene with a few layers. The resulting TG showed excellent specific surface area, making it a promising sensor material.

2. Experimental

2.1. Materials

Nickel chloride hexahydrate ($NiCl_2{\cdot}6H_2O$) and hydrazine hydrate ($N_2H_4{\cdot}H_2O$, 85 wt%) were purchased from Sinopharm Chemical Reagents Co.Ltd (China). Ethylene glycol (EG) and sodium hydroxide (NaOH) were obtained from J&K Scientific Ltd (Shanghai). All organic solvents were analytical grade and were used without further purification.

2.2. Preparation of Ni-NWs

Ni-NWs were synthesized by chemical reduction under a magnetic field [16]. Briefly, 0.7 g of $NiCl_2{\cdot}6H_2O$ and 100 mL of EG were added to a 200 mL beaker, and the mixture was magnetically stirred for 1h at room temperature to ensure uniform dispersion of Ni^{2+} ions in the solution. 2 g of NaOH were dissolved in 100 mL of EG to obtain a NaOH solution in EG. After adding 20 mL of a $N_2H_4{\cdot}H_2O$ solution to theabove mixture solutions, the reaction mixture was continually sonicated until a homogeneous blue solution was obtained. This solution was subsequently poured into a 500 mL flask. The flask was placed in a water bath under a 0.5 T magnetic field and subsequently heated at 70 °C for 45 min, after which fluffy black products were formed. The solid product was separated from the solution by a magnet and cleaned three times with deionized water, ethanol, and acetone (one time each). Finally, the as-cleaned Ni-NWs were placed in a vacuum oven and dried at 60 °C.

2.3. Deposition of the TG

The TG was grown on Ni-NWs by CVD to obtain TG/Ni-NW composites. Specifically, the prepared Ni-NWs were placed in a tubular furnace and the pressure was decreased to 8 Pa. Ar (300 sccm) was then introduced into the tube furnace until the pressure in the furnace reached 1 atm. Then, the temperature inside the furnace was increased to 850 °C under flowing Ar (300 sccm) and the material was annealed for 5 min to clean the Ni-NWs surface. After that, H_2(50 sccm) and CH_4(3–15 sccm) were introduced into the tube furnace for 10 min. The number of graphene layers of the TG increased with the concentration of CH_4. Finally, the samples were cooled down to room temperature at a rate of 10 °C/min under flowing Ar (300 sccm) and H_2(50 sccm) after 10 min of the reaction-gas mixture flow. Thus, TGdeposited on Ni-NWs substrate was successfully synthesized.

2.4. Characterization

Field-emission SEM (FE-SEM, Sigma 300, Carl Zeiss, Germany) was employed to study the morphologies of the samples. The structure of the samples was investigated by Raman spectroscopy (HR-800, HORIBA JobinYvon, France) with a 632.8 nm He-Ne laser. X-ray diffraction (XRD, X 'Pert-Pro MPD, Panalytical, Holland) was used to determine the crystalline structure of the samples.

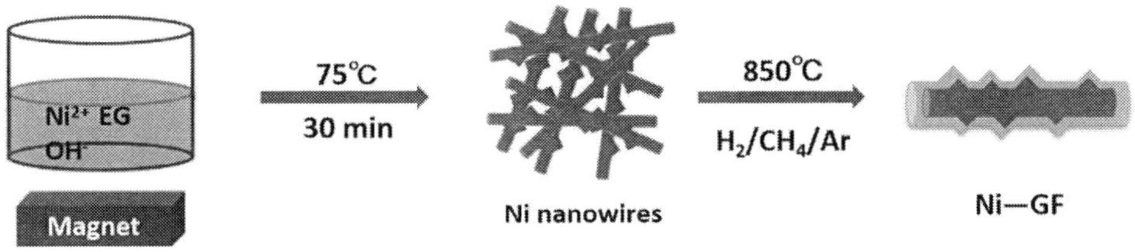

Figure 1. Simple flow chart showing the methodused for preparing Ni-NWs and TG.

3. Results and discussion

The morphologies and microstructures of the Ni-NWs and the TG/Ni-NW composites are shown in figure 2. Low- and high-magnification images revealedthatNi-NWswith a diameter of 300 nm were successfully prepared (figures 2a and b). The Ni-NWs grown undera magnetic field contained numeroussurface spikes. After depositing graphene on Ni-NWs by CVD, the surface protrusion disappeared and a fraction of the NWsmelted at high temperature, which resulted in the diameter of the NWs increasing to 500 nm (figures 2c and d). By comparing pictures before and after the CVD process, the successful synthesis of graphene was clearly revealed by the surface morphology and color change of the material.

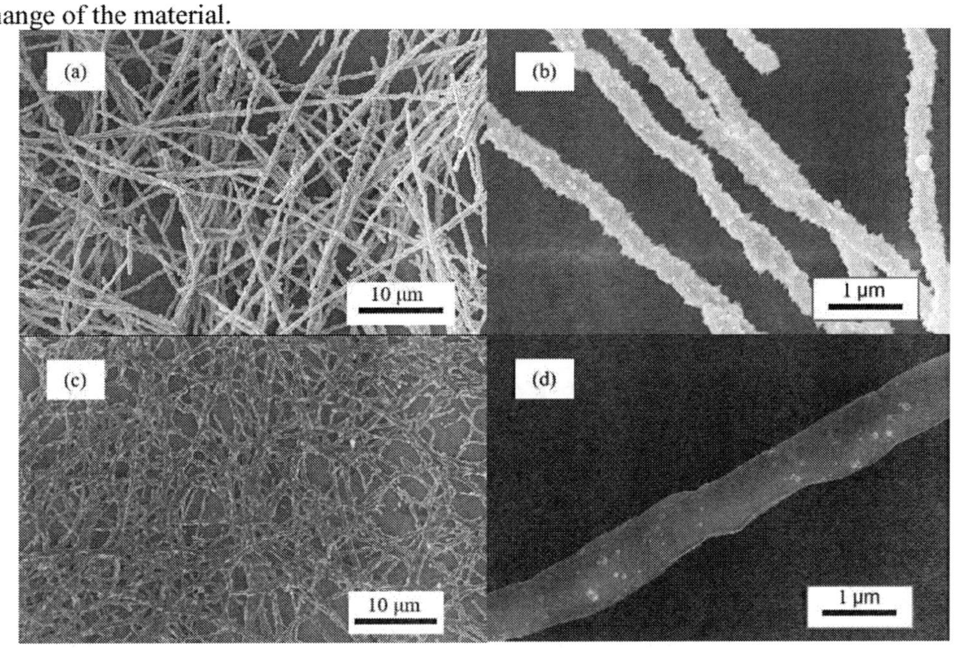

Figure 2.SEM images of: (a) Ni-NWs, (b) high-magnification SEM images of: Ni-NWs and (c) TG/Ni-NW composites, (d) high-magnification SEM image of the TG/Ni-NW composites.

Raman spectroscopy further revealed the presence of graphene on the Ni-NWs. As shown in figure 3, the Raman spectrum of Ni-NWs contained three characteristic peaks at 1323, 1579, and ca. 2670 cm^{-1} [17]. The G band is produced by the first-order scattering of the E_{2g} phonon plane vibration on the sp^2 domain. Therefore, this bandprovides an indication of the degree of crystallization of graphene. The D band is produced by the discontinuity of the crystal plane of the graphene or the doping of foreign atoms. Theintensity of the D band reflects the defect density of graphene. The most intense peak for graphene is the G' peak produced by the double phonon resonance Raman peak of graphene. The appearance of this G' peak is the main indicative of the presence of graphene. The intensity ofthis peakindicates the degree of stacking of the material. Figure 3 shows the Raman spectra of few-layers and multi-layers graphene materials. The ratio of G'/G decreases as the number of graphenelayers increases.As can be seen in figure 3, the ratio of G'/G decreased from 0.94 to 0.42, indicating that the number of graphene layers changed from afew layer (2~3) to multilayer (>3). The peak of D is very low, which indicates that the synthesized graphene has a very high quality, which is consistent with previous reports[18, 19].

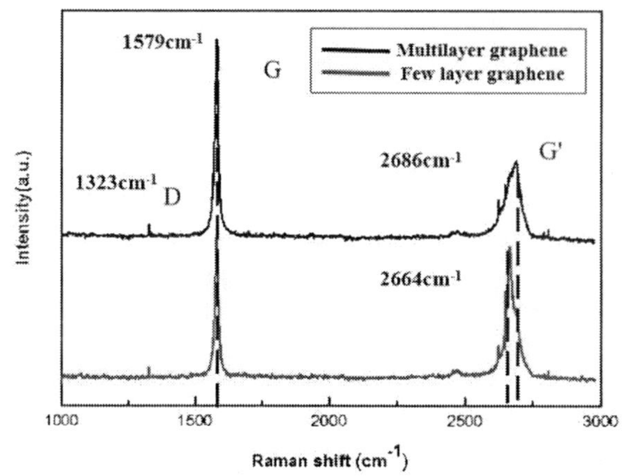

Figure 3. Raman spectra of the TG deposited on Ni-NWs.

4. Conclusions

In conclusion, graphene with a novel tubular structure was easily synthesized by CVD over Ni-NW substrates. Ni-NWs were prepared by chemical reduction under water-bath heat and magnetic field. Graphene was deposited on the surface of Ni-NWs by CVD using H_2 and Ar as carrier gases and CH_4 as a carbon source. The number of graphene layers was found to depend on the concentration of the carbon source, the growth time, and the cooling rate. This particular tubular structure of graphene provided this material with higher specific surface area compared to ordinary 2D graphene materials. As a result, the resulting material can have potential applications in the sensor and energy-storage fields.

Acknowledgments

The authors gratefully acknowledge financial supports by the National Natural Science Foundation of China (no. 61871281 and 51302179), the Natural Science Foundation of Jiangsu Province (no. BK2012184 and BK20181166), the Natural Science Foundation of the Jiangsu Higher Education Institutions of China (no. 18KJB510040), the Priority Academic Program Development (PAPD) of Jiangsu Higher Education Institutions.

References

[1] K.S. Novoselvo, A. K. Geim, S.V.Morozov, D.Jiang, Y. Zhang, S.V. Dubonos, I.V. Grigorieva, A.A. Firsov, Science. **306**, 666-669 (2004)

[2] K.S. Novoselov, A.K. Geim, S.V.Morozov, D.Jiang, M.I. Katsnelson, I.V. Grigorieva, S.V. Dubonos,A.A. Firsov, Nature.**438**,197-200(2005)

[3] Y. Zhang, Y-W. Tan, H.L. Stormer, P. Kim, Nature.**438**, 201-204(2005)

[4] A.A. Balandin, S. Ghosh, W. Bao, I. Calizo, D. Teweldebrhan, F. Miao, C.N. Lau, Nano Lett. **8**,902-07(2008)

[5] K.S. Novoselov, V.I. Fal'ko, L. Colombo, P.R. Gellert, M.G. Schwab, K. Kim, Nature.**490**, 192-200(2012)

[6] J. Penuelas, A. Ouerghi, D. Lucot, C. David, J. Gierak, H. Estrade-Szwarckopt, C. Andreazza-Vignolle Phys. Rev. B. **79**, 033408 (2009)

[7] S. Stankovich, D.A. Dikin, G.H.B. Dommett, K.M. Kohlhaas, E.J. Zimney, E.A. Stach, R.D. Piner, S.T. Nguyen,R.S. Ruoff, Nature.**442**, 282-86 (2006)

[8] K.S. Kim, Y. Zhao, H. Jang, S.Y. Lee, J.M. Kim, K.S. Kim, J-H. Ahn, P. Kim, J-Y. Choi, B.H.

Hong, Nature.**457**, 706-710 (2009)
[9] V. Singh, D. Joung, L. Zhai, S. Das, S.I. Khondaker, S. Seal, Prog. Mater. Sci. **56,** 1178-1271 (2011)
[10] Y. Xu, K. Sheng, C. Li, G. Shi, ACS. Nano. 4, 4324-4330 (2010)
[11] S. Bae, H. Kim, Y. Lee, X. Xu, J-S. Park, Y. Zheng, J. Balakrishnan, T. Lee, H.R. Kim, Y. Song, Y-J. Kim, K.S. Kim, B. Özyilmaz, J-H. Ahn, B.H. Hong, S. Ligima, Nature Nanotech. **5**, 574-578 (2010)
[12] J. Lv, M. Yang, S. Ken, M. Hideo, Mater. Lett. **196**, 369-372 (2017)
[13] Y. Liu, M. Hu, M. Zhang, P. Lang, H. Wei, Y. Gao, J. Alloy. Compd. **696**, 1159-1167 (2017)
[14] Z. Chen, W. Ren, L. Gao, B. Liu, S. Pei, H-M. Cheng, Nat. Mater. **10**, 424-428 (2011)
[15] G. Hu, C. Xu, Z. Sun, S. Wang, H-M. Cheng, F. Li, W. Ren, Adv. Mater. **28**, 1603-1609 (2016)
[16] J. Wang, L.M. Wei, L.Y. Zhang, C.H. Jiang, E.S.W. Kong, Y.F. Zhang, J. Mate. Chem. **22**, 8327-8335 (2012)
[17] A.C. Ferrari, J. C. Meyer, V. Scardaci, C. Casiraghi, M. Lazzeri, F. Mauri, S. Piscanec, D. Jiang, K.S. Novoselvo, S. Roth, A.K. Geim, Phys. Rev. Lett. **97**, 187401 (2006)
[18] X. Li,W. Cai,J.An, S.Kim, J.Nah, D.Yang, R.Piner, A.Velamakanni, I.Jung, E.Tutuc, S.K.Banerjee, L. Colombo, R.S.Ruoff, Science **324**,1312-1314 (2009)
[19] K.S. Kim, Y. Zhao, H. Jang, S.Y. Lee, J.M. Kim, K.S, Kim, J-H Ahn, P. Kim, J-Y Choi, B. H. Hong, Nature **457**, 706-710 (2009)

IOP Conf. Series: Materials Science and Engineering **715** (2020) 012004 doi:10.1088/1757-899X/715/1/012004

Fire evacuation and patient assistance simulation in a large hospital building

Hattaya Boonngam[1,*] and Supat Patvichaichod[2]

[1]Safety Engineering and Environmental Management Program Kasetsart University Sriracha Campus Chonburi, Thailand
[2]Department of Mechanical Engineering, Safety Engineering Research Group Kasetsart University Sriracha Campus Chonburi, Thailand

*E-mail: hattaya22.b@gmail.com

Abstract. This research studied the fire evacuation simulation and patient assistance who were unable to help themselves out a hospital building. The studied building is a high and large building as Thai law which was 10th floors on 90,842.18 square meters and able to accommodate 4,017 occupants that calculated by the NFPA 101 standard which simulated from 3evacuation cases. The experiment result found that all people can be moved out from the building within 3,048 seconds, 2,931 seconds, and 2,958.3 seconds respectively which didn't over than the standard as High Building Controlled under Thai Law. The reason that time-limited needed as 1 hour when transferred the people out of the building because enhance the safety of people, especially critically ill patients. Thus, the researchers suggested more channels of evacuation through the elevators by testing for 2 situations. The result of 2situations showed that this method can be decreased the fire evacuation frill time as 2,073 seconds and 1,979.5 seconds respectively.

1. Introduction

The conflagration is the unexpected event which causes to the loss of life and asset. Especially, the burning of high or large building [1] that there are high, large area, used by a lot of users, and complex building pattern [2] are the factors cause difficult and slow in the way of moving people out of the building.

Thus, the researcher has studied the fire evacuation of the large hospital to investigate the escape behavior of people who stuck in the building and analyze the capability of fire exit route in the building. The reasons of this study are the fire evacuation from the hospital building does not control by factors about the size of building and number of people in the building, but also related to the category and character of building users [3]. Moreover, the building users contain general users who can move out of the building by themselves and building users who need someone to move them out. However, the fire evacuation from hospital building might still be delay because of the way of transferring patients out from the building need to use the time to prepare [4].

2. Model building

The simulation of building and fire evacuation from the large hospital building is created by Pathfinder program which is the simulation program about the human movement to find the exit that can define the evacuation behavior of building occupants by showing the result on both 2 and 3 dimensions [5].

And in this study, the study of fire evacuation time out of the building compared with the law governing large buildings and high-rise buildings in Thailand [6] that determine buildings with sizes from 10,000 square meters or more is a large building And buildings with a height from the ground ranging from 23 meters or more is a high-rise building. The building that is considered a high-rise building must be able to evacuate the building user within 1 hour.

2.1. Model building

The researcher has created the hospital building simulation by importing the file of 2dimensions building (.dwg file) of Auto CAD program to Pathfinder program. The hospital building that we study is the high and large building as Thai law [6] and just finished constructing as 10th floors building (including the basement and deck), height from the ground as 43.49 meters, total area as 90,842.16 square meters, and also consist of 6 fire exit stair and 1 ramp fire escape as shown in figure 1 and figure 2.

Figure 1. Model of a large hospital.

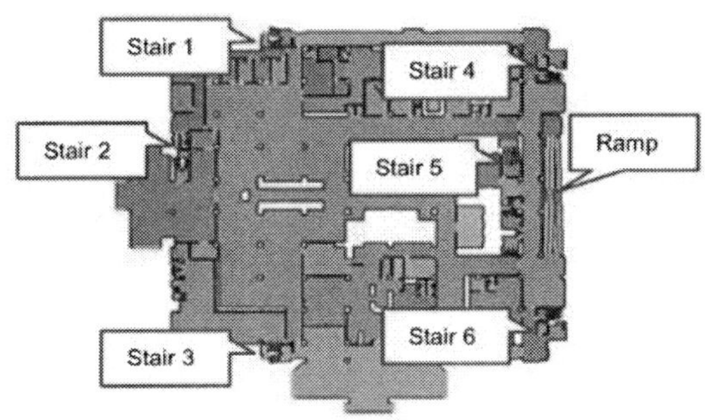

Figure 2. Fire exit stair and ramp of hospital building.

2.2. Simulation settings

Set the building occupants as 4,017 occupants which are calculated through NFPA 101 [7] standard and define the category of user by speediness, evacuation behavior, and the preparing time of evacuation [4] as shown in table 1.

Table 1. Details of each profile of Occupant.

Profile	Occupant (person)	Speed (m/s)	Behaviors
Assistant team	236	1.35	-Assist -Wait1 10s (Preparation time) -Go to exit
General staff	464	1.35	-Go to exit
Visitor	1428	1.20	-Go to exit
Out-Patient	1429	0.95	-Go to exit
In-Patient	325	0.95	-Go to exit
Critically ill patient	135	0.4	-Wait for Assistant -Wait1 10s (Preparation time) -Go to exit

The critically ill patients are defined to transfer by hospital bed which support by assistance evacuation team and can be divided the type of hospital bed as table 2.

Table 2. Type of bed for Critically ill patient.

Type of bed	Size of bed (m.)	Assistant Request / bed (person)
Adult Bed	0.9×2×1	2
Baby Bed	0.5×0.9×0.88	1

IOP Conf. Series: Materials Science and Engineering **715** (2020) 012004 doi:10.1088/1757-899X/715/1/012004

101 adult hospital beds which contain by emergency department's bed at 1st floor as 13 beds, operating and childbirth department's bed at 4th floor as 45 beds, ICU's bed at 5th floor as 43 beds, newborn baby's bed as 34 beds which consist by 10 baby's beds at childbirth department and other 24 beds at ICU department.

Occupants of the building that can take care of themselves are designated to can be evacuated through every exit routes. The critically ill patients who need help from assistant team to move them from 4th and 5th floor are set to evacuate through ramp fire escape only, except the critically ill patients at emergency department at 1st floor that can be evacuated through every exit routes on 1st floor.

Simulation of 3 scenario using steering mode [5] as follows.

Scenario 1: Evacuate every people out from the building through any existing fire exit routes as shown in figure 3.

Figure 3. Situation 1, Every fire exit routes are useable.

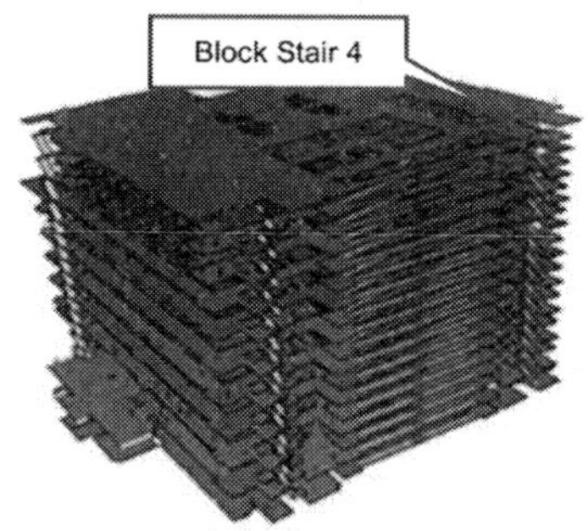

Figure 4. Situation 2, Fire exit stair no. 4 is unusable.

Scenario 2: The fire exit stair no. 4 is unusable because there are being on fire at fire pump room which located on the basement and causes the smoke to spread into fire exit stair no. 4 as shown in figure 4.

Scenario 3: 5 fire exit doors near emergency department on 1st floor is unusable because the car which delivered the patient is burned that make people in the building cannot use such fire exit as shown in figure 5.

Figure 5. Situation 3; 5 fire exit doors which located near emergency department on 1st floor is unusable.

3. Simulation result and analysis

3.1. Simulation result

From these 3 simulations, we found that the time of fire evacuation from the building is 3,048 seconds, 2,931 seconds, and 2,958.3 seconds respectively which does not over the limit as controlled by High Building Controlled under Thai Law that fix the time of evacuation as lower than 1 hours [6].

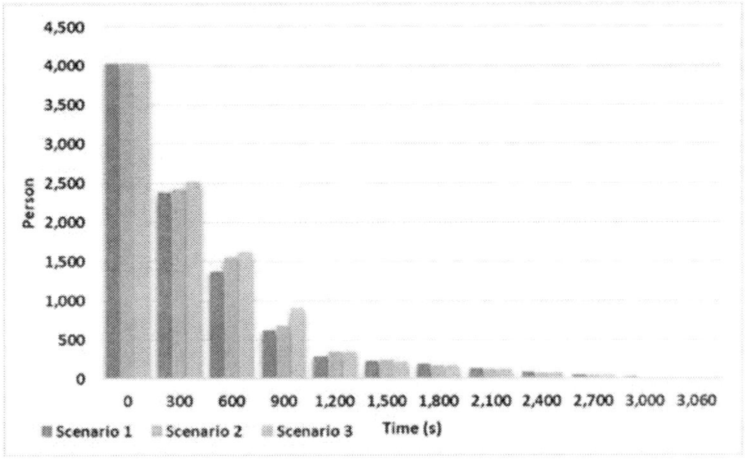

Figure 6. The number of building users in each period.

Furthermore, the researcher also found that the remaining building occupants who still stuck in the building are similar number among 3 simulations when the fire evacuation starts for 1,200 seconds as shown in figure 6. The last groups which move out from the building are critically ill patients and their assistance team which are aided from 4th floor and 5th floor as shown in figure 7. Since the building occupants can evacuate only one route that is ramp fire accompany with the physical factor which needs to be moved through the bed that makes the speed decrease. Moreover, the high of building also cause to slow movement of critical illness patients evacuation.

Figure 7. The last group of building occupants who remain in the building.

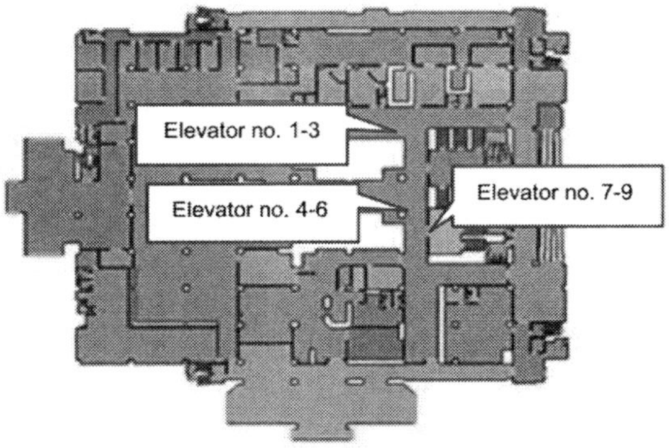

Figure 8. Increase the fire exit channel by using elevator to evacuate the critically ill patients.

3.2. Analysis of simulation result
Although these 3 simulations use the time to evacuate less than 1 hour as Thai Law controlled, we still need to concern about the safety of evacuation from the building, especially the critically ill patients and assistance team which have limitation about movement time. Besides, it can be an alternative channel when the ramp fire exit is unusable. Thus, the researcher has suggested the channel to evacuate from the building by adjusting the passenger elevator to be used as the fire escape elevator. Which according to the original drawing of this building, there are 9 passenger elevators that can be renovated into a fire escape elevator in the middle of the building as shown in figure 8.

In case of increasing the elevator channel to evacuate critically ill patients and assistance team by simulating as 2 Alternative as following;

Alternative 1: Set the assistance team to help the 4th floor critically ill patients by using the ramp fire escape and 5th floor critically ill patients by using elevator as shown in figure 9.

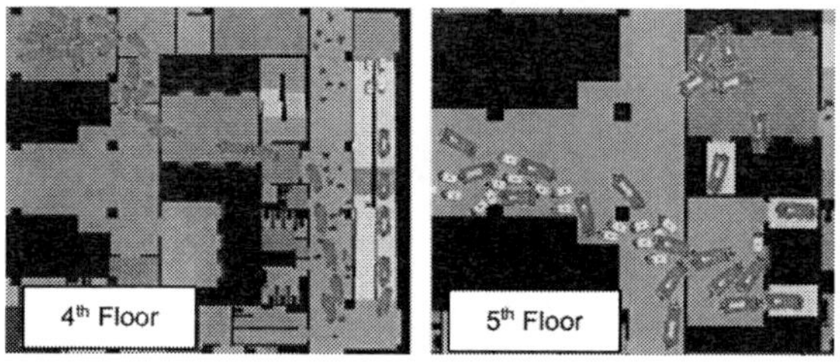

Figure 9. Evacuate the 4th floor critically ill patients by using the ramp fire escape and 5th floor critically ill patients by using elevator.

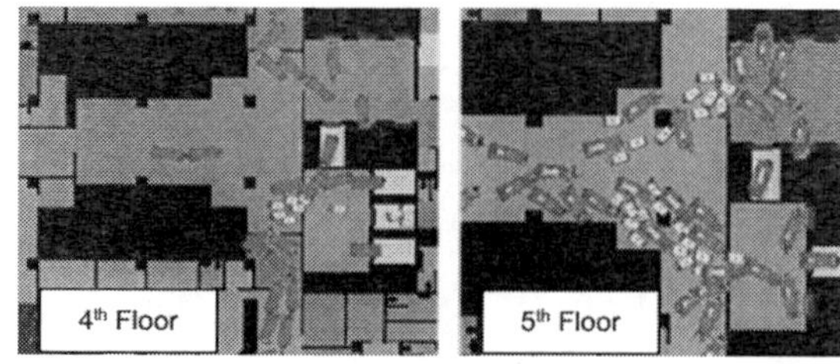

Figure 10. Evacuate the 4th and 5th floor critically ill patients by using elevator.

Alternative 2: Set the assistance team to help the 4th and 5th floor critically ill patients by using elevator as shown in figure 10.

The simulation result found that after increasing the fire exit by adjusting the passenger elevator to fire escape elevator for critically ill patients and assistance team, the time of evacuation from the building of 2 simulations are decreased to 2,073 seconds and 1,979.5 seconds respectively as shown in figure 11.

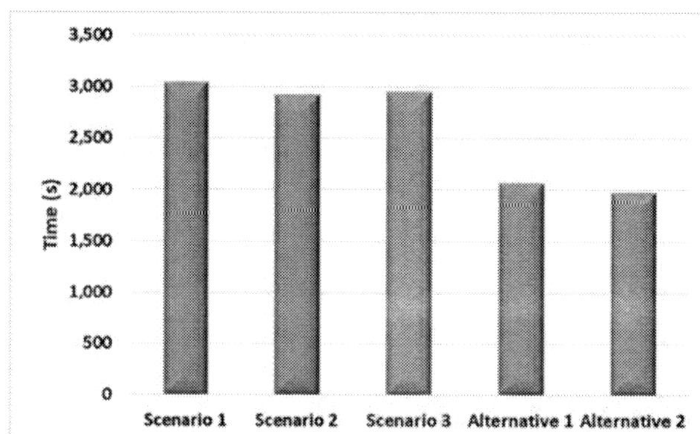

Figure 11. The fire evacuation time in each situation.

4. Results and discussion

From the fire escape route adjusting of large hospital building by improving the passenger elevator to be fire escape elevator, the experiment of 2 simulations found that they can decrease the time to evacuation from the building when compared with the simulation in Scenario 1, 2, and 3 as followings;

Alternative 1: Able to decrease the time of evacuation from Scenario 1, 2, and 3 as 31.98%, 29.27%, and 29.92% respectively.

Alternative 2: Able to decrease the time of evacuation from Scenario 1, 2, and 3 as 35.05%, 32.46%, and 33.08% respectively.

To sum up, the Alternative 2 use the least time to move out all occupants of the building.

5. Conclusion

The fire evacuation simulation from the large hospital building result found that we can move all people out from the building within 1 hour as the High Building Controlled under Thai Law.

The passenger elevator needs to be adjusted to apply as fire escape elevator to increase more evacuation channel for critically ill patients who cannot use the fire exit because they must be moved together with patient's bed.

In case of using the elevator to escape the fire, users need to make sure that such elevators can be used safely when emergency events occur.

The fire emergency evacuation needs to be set clearly, prepare, train as well as the assistance team also need to be ready all the time to be more effective fire evacuation.

IOP Conf. Series: Materials Science and Engineering **715** (2020) 012004 doi:10.1088/1757-899X/715/1/012004

Acknowledgments
The author is grateful to Thunderhead Engineering for Pathfinder Program academic license and Burapha University Hospital for Hospital building Layout and others data.

References

[1] Narin Boonprasert. 2016. *"Fire Evacuation Simulations of High-Rise Building Using PathfiderProgram Case Study of Multiple Occupancy Building"*. Master Engineering thesis, Kasetsart University, Thailand.

[2] Ze-min Jiang, Pei-hong Zhang , Rong-xue Shang, Xiang-liang Tian. 2014. *"Investigation and Simulation on Human Evacuation Behaviour in Large Hospital Building in Shenyang"*. School of Resources & Civil Engineering, Northeastern University, Shenyang 110819, China.

[3] Virginia Alonso-Gutierrez, Enrico Ronchi. 2018. "*The simulation of assisted evacuation in hospitals*". Ashes Fire Consulting, Madrid, Spain and Department of Fire Safety Engineering, Lund University, Sweden

[4] A. Tinaburri, F.A. Ponziani, V. Ricci. 2018. "*Agent Based Modeling of Meta-Communication with Assisted People during Emergency Egress*". Department of firefighters, of the public rescue and civil defence – Ministry of the Interior – Italy University of Rome

[5] Thunderhead Engineering Consultants. "*Pathfinder 2018 User Manual*". https://www.thunderheadeng.com/pathfinder/, USA

[6] Ministerial Regulation No. 33 (B.E. 2535) issued by virtue of the Building Control Act, B.E. 2522

[7] National Fire Protection Association. 2018. "*NFPA 101 Life Safety Code 2018 edition*", July 2018.

IOP Conf. Series: Materials Science and Engineering **715** (2020) 012005 doi:10.1088/1757-899X/715/1/012005

Performance analysis of automatic sprinkler systems in warehouses using fire dynamic simulation

Pennurat Nonsawat[1,]*and Supat Patvichaichod[2]
[1]Safety Engineering and Environmental Management Program Kasetsart University Sriracha Campus, Chonburi, Thailand.
[2]Department of Mechanical Engineering, Safety Engineering Research Group Kasetsart University Sriracha Campus, Chonburi, Thailand.

*E-mail: pennurat@gmail.com

Abstract. This research is based on studying the capability of automatic sprinkler systems by applying fire dynamics simulation. The selection of scenarios based on NFPA 101. A warehouse containing diving instrument in carton placed on wood pallets, which can cause class A fires. To prevent potential damage to the goods, the conflagration in this specific scenarios is simulated to study the effects of temperature on the roof deterioration, the proficiency of the automatic sprinkler systems to extinguish the fire. Researchers found that the sprinklers located on the ceiling of the warehouse are not able to put out the fire but are able to control further spreading of the fire. Researchers, then, install more fire sprinklers in the racks, following the standard from NFPA 13. Results have shown that the second set of sprinklers is not capable of extinguishing the fire but controlling the spread and significantly reducing the temperature. Furthermore, installing both sets of fire sprinklers can prevent the collapse of the roof.

1. Introduction

Conflagration is one of the most serious disasters, capable of inflicting significant damage to lives and properties. Especially in industries in which manufacture and storage are present, the fire can spread to the nearby rack, allowing the damage to expand easily [1]. Nowadays, warehouses are designed with greater height and size of the racks to maximize the efficiency of its use, causing the goods to be more crowded on the rack [2]. Hence, the extinguishing becomes more difficult. In Thailand, most warehouses are only installed with fire sprinkler on the ceiling. Only few are installed with in-rack sprinklers.

Thus, researchers, seeing that sprinklers on the ceiling alone are not enough to put out the fire in warehouses, design the simulation of installation of both sets of sprinklers to follow the standard from NFPA 13 Standard for the Installation of Sprinkler Systems [3]. This simulation is created to let researchers study the temperature inside the warehouse and analyze the effectiveness of the automatic sprinkler system.

2. Theories

This research is used for studying the proficiency of automatic sprinkler system using Fire Dynamics Simulator and Smoke View

2.1. Fire dynamics simulator

The National Institute of Standard and Technology (NIST) has created this software to simulate fire in specific situation. This software is capable of predicting motion of the smoke, movement of the hot air from the fire, affected by wind, ventilation system, and other factors [4].

The software can assist in installing devices to prevent the disaster, using Mesh. Mesh, a certain area used in the simulation, is divided into sub-area, called Grid, to calculate the flow of thermal energy. To process the result, the equation from UT Fire research group [5].

$$D^{*} = \left(\frac{\dot{Q}}{\rho_{\infty} c_{p} T_{\infty} \sqrt{g}}\right)^{2/5} \tag{1}$$

when Q = Heat Release Rate (kW/m^2), $p\infty$ = Density (m^2), cp = Specific Heat (Kw), $T\infty$ = Ambient Temperature (K) and g = Gravity (m/s^2) Furthermore, simulation from CAD Models can be imported into Pyroism.

2.2. Automatic sprinkler systems

Automatic sprinkler systems are unanimously the most effective fire extinguisher. The fire can be put out even when the spread has not begun. The systems are being installed worldwide, as the water from the system has no effect on the ecosystem [6].

For the automatic sprinkler systems in warehouses, NFPA13 suggested that the sprinklers are installed both on the ceiling and in the racks, as this was mentioned: "In-Rack Sprinklers. Minimum K-8.0 quick response sprinklers.

Sprinklers shall be installed at each rack upright within the longitudinal flue space and at the face of the rack. The maximum linear spacing between sprinklers shall not exceed 5 ft (1.5 m) at the rack face and 10 ft (3.0 m) within the longitudinal flue space [3].

3. Methods

The simulation of the warehouses has been made by FDS, in which the occupancy is selected as extra hazard as in NFPA 13 [3], and details about the cargo are filled in the software.

3.1. Creating the simulation

This simulation is designed as shown in figure 1, using FDS, as a 252 m^2 room, chosen from the area of sprinkler operator from the real 3,000 m^2 construction. Inside the room, diving instrument is stored on 6 m high metal racks. The arrangement of the goods is double row racks. The merchandise is stored inside 2 layers of corrugated carton, with the maximum thickness of 7 mm, placed on wood pallets, which can cause class A fire.

Figure 1. Model for Simulation.

The material in table 1 derived out of Maximum Heat Release Rates – Warehouse Materials from NFPA 72 National Fire Alarm and Signaling Code 2013 Edition the materials can release heat as calculated from this equation:

$$Qm = qA \tag{2}$$

IOP Conf. Series: Materials Science and Engineering **715** (2020) 012005 doi:10.1088/1757-899X/715/1/012005

When Qm = maximum heat release rate (kW),

q = heat release rate density per unit of surface area kW/m^2), A = surface area of the fuel (m^2), and the merchandise is as packed as possible [7].

Table 1: Parts of the building and materials.

No.	Parts of the building and Materials	Materials in FDS
1	Floor	Concrete
2	Wall	Concrete
3	Roof	Steel
4	Racks	Steel
5	Pallets	Wood
6	Boxes	Wood
7	Diving instrument	PVC

3.2. Size of the mesh
Evaluation of the fire dynamics could be dramatically affected by the size of the Mesh. Researchers have calculated according to the UT Fire research group. The resolution is 0.15 m [5].

3.3. Location of the sprinklers and thermocouples
The researcher created a model by installing automatic sprinkler systems only at the ceiling. In this simulation, there are 26 upright sprinklers, arranged in branch line with 3 m space between each sprinkler and 3 m space between each branch, with red glass bulb, K-factor 8.0 3/4 in. orifice, which will be activated when the temperature reach 68 °C. The highest flow rate of the sprinkler is 73.323 L/m, calculated from this equation:

$$Q = As * Ds \tag{3}$$

when Q = flow (gpm), As = distance between each branch (m)*distance between each sprinkler (m), and Ds = density of storage (m^2) [3].

Besides, thermocouples are also installed on the ceiling to determine if the roof can withstand the fire, as ASTM E119 has limited the temperature at 538 °C [8], as shown in figure 2.

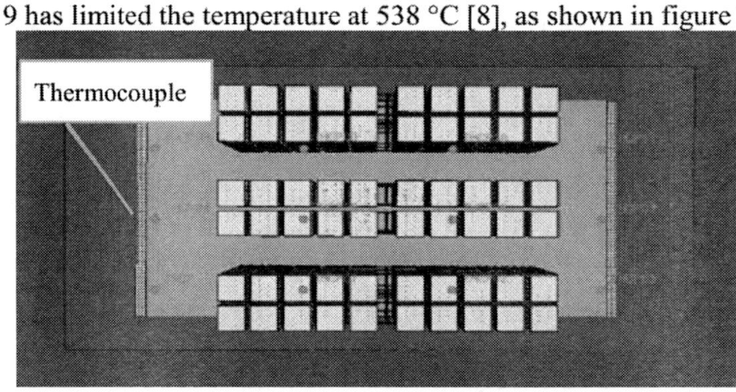

Figure 2. Thermocouple installation point.

3.4. Setting the scenarios
As referred in NFPA 101 Life Safety Code [9], there are 3 possible scenarios:

1) The fire from cigarette stub lighting the carton and the instrument, heat release rate = 3,405 kW/m^2 [7], as shown in figure 3.

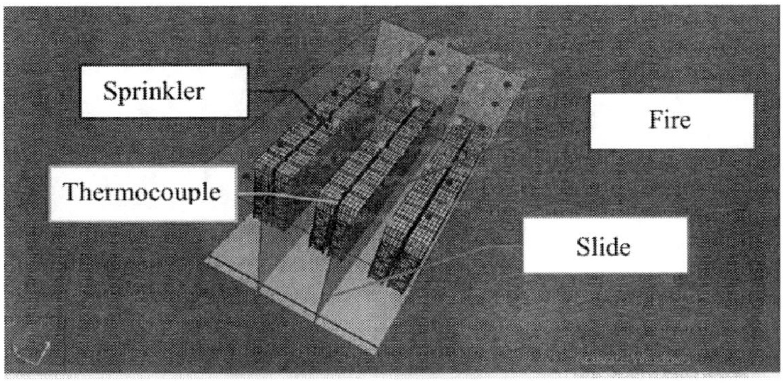

Figure 3. Scenario 1 PVC Fuel.

2) The fire from diesel, leaked from car crash, ignited by a spark, heat release rate = 2,043 kW/m^2 [7], as shown in figure 4.

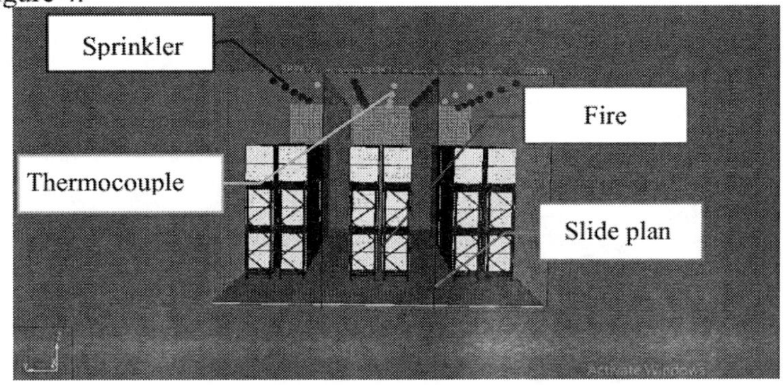

Figure 4. Scenario 2 Diesel oil Fuel.

3) The fire from a short circuit creating a spark lighting the carton, heat release rate = 1,248 kW/m^2 [7], as shown in figure 5.

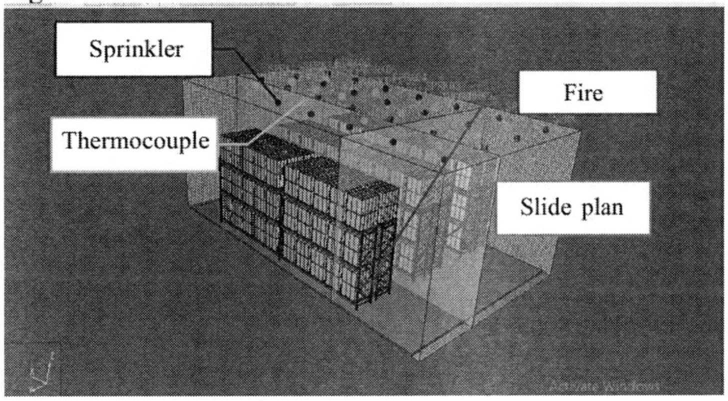

Figure 5. Scenario 3 carton fuel.

4. Results

The results from each scenario are as written below:

Scenario 1: after 5 seconds of fire, the sprinklers are activated but the fire cannot be extinguished. The lowest temperature measured on the ceiling level is 120 °C, while the highest is 180 °C, and the average is 165 °C, as shown in figure 6.

IOP Conf. Series: Materials Science and Engineering **715** (2020) 012005 doi:10.1088/1757-899X/715/1/012005

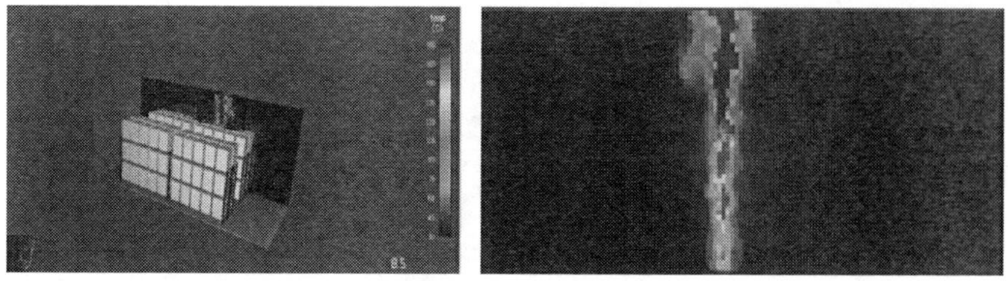

Figure 6. Shows temperature from Scenario 1 – Ceiling.

With two sets of sprinklers (ceiling and in-rack), activated after 3 seconds, the fire also cannot be put out, but the lowest temperature is lowered to 90 °C, while the highest stays the same, making the average temperature 137.5 °C. as shown in figure 7.

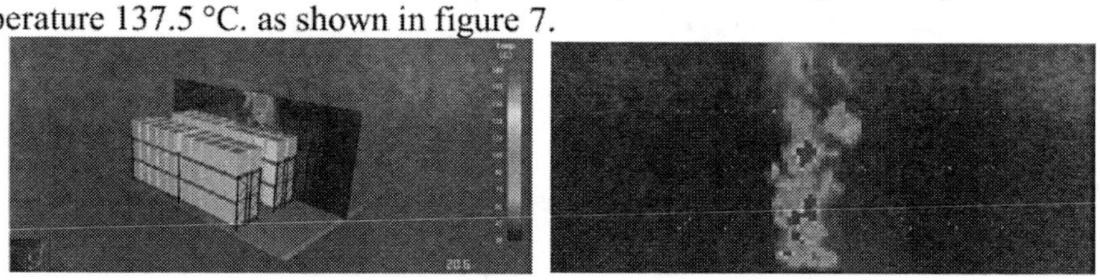

Figure 7. Shows temperature from Scenario 1 – Ceiling and In rack.

Scenario 2: after 10 seconds, the sprinklers are activated but the fire cannot be extinguished. The lowest temperature measured on the ceiling level is 34 °C, while the highest is 47.5 °C, and the average is 43.75 °C, as shown in figure 8.

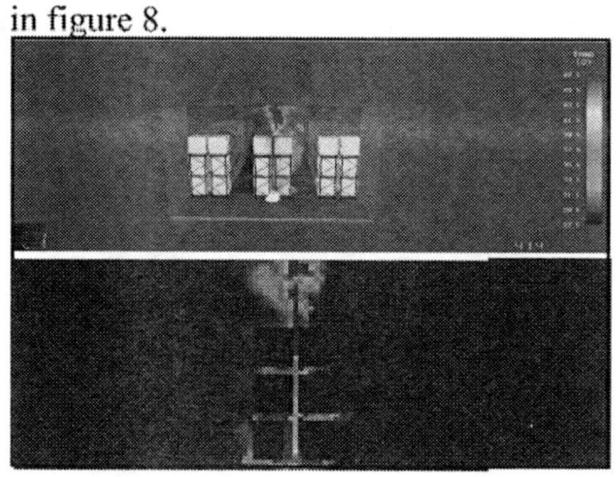

Figure 8. Shows temperature from Scenario 2 – Ceiling.

With two sets of sprinklers (ceiling and in-rack), activated after 9 seconds, the fire also cannot be put out. The lowest temperature is lowered to 33 °C, while the highest is 47 °C, making the average temperature 40 °C, as shown in figure 9.

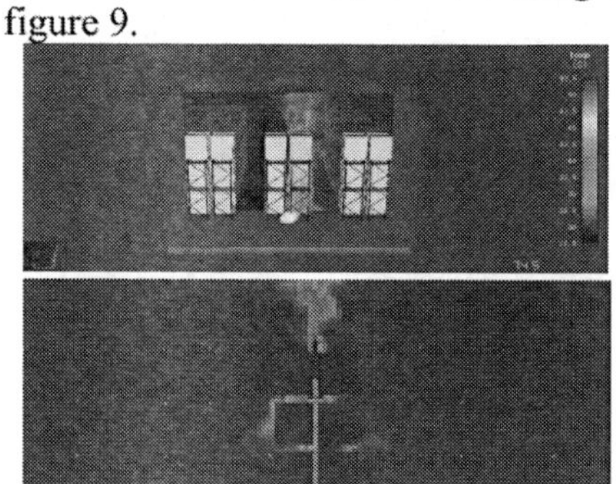

Figure 9. Shows temperature from Scenario 2 – Ceiling and In rack.

IOP Conf. Series: Materials Science and Engineering **715** (2020) 012005 doi:10.1088/1757-899X/715/1/012005

Figure 10. Shows temperature from Scenario 3 – Ceiling.

Scenario 3: after 12 seconds, the sprinklers are activated but the fire cannot be extinguished. The lowest temperature measured on the ceiling level is 43 °C, while the highest is 63 °C, and the average is 51.58 °C, as shown in figure 10.

With two sets of sprinklers (ceiling and in-rack), activated after 12 seconds, the fire also cannot be put out. The lowest temperature is lowered to 30 °C, whereas the highest is also lowered to 49 °C, making the average temperature 38.92 °C, as shown in figure 11.

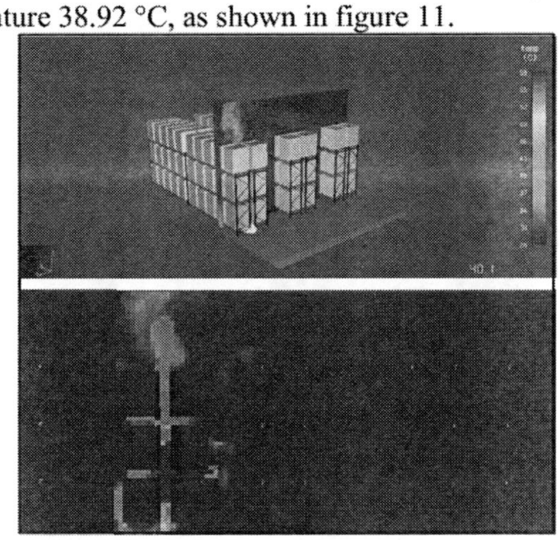

Figure 11. Shows temperature from Scenario 3 – Ceiling and In rack.

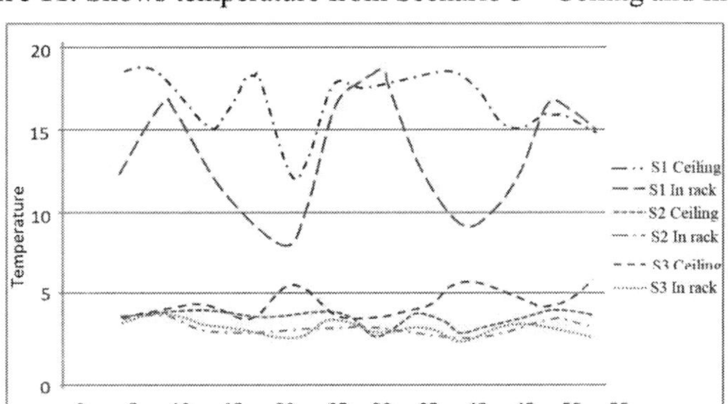

Figure 12. Comparison of heat at the ceiling.

As shown in figure 12, it can be seen that the scenarios 1, the ceiling temperature is 165 °C, when the fire sprinkler head is added to the rack can be reduced to 137.5 °C (16.67%). The second scenario, the average ceiling temperature is 43.75 °C. The fire extinguishing water added to the rack can be reduced to 40 °C (8.57%). Scenario 3 The ceiling temperature is an average of 51.58 0C. When the fire sprinkler head is added to the rack can be reduced to 38.92 °C (24.56%), Average temperature drop rate as shown in figure 13

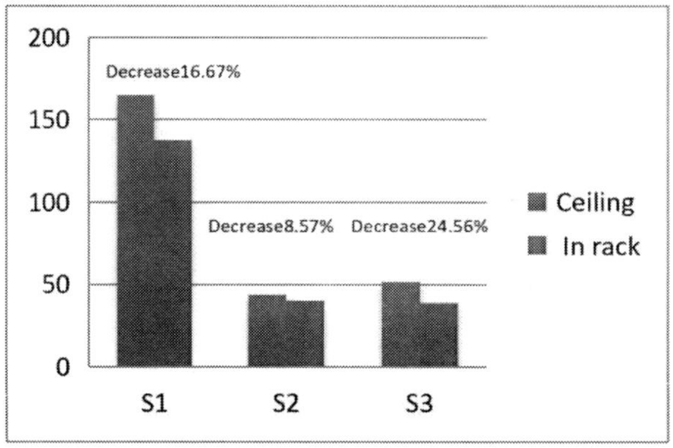

Figure 13. Average temperature drop.

When increasing the flow rate of fire water to 20%, 40%, 60%, 80% and 100% will see that can not be extinguished But the average temperature is reduced to 148.75 °C, 139.58 °C, 130.83 °C, 124.58 °C and 118.33 °C, respectively, as shown in figure 14.

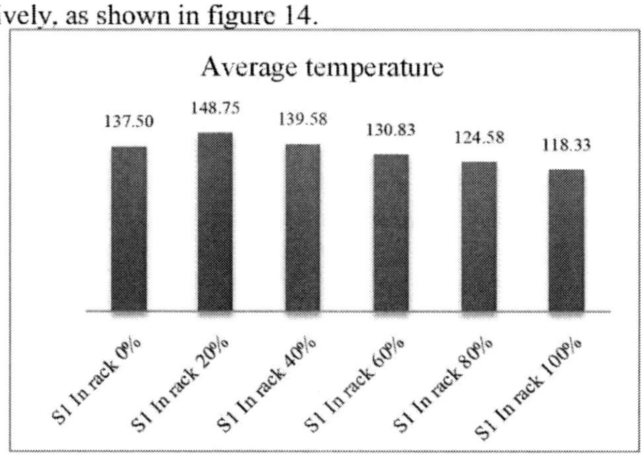

Figure 14. Increasing the flow rate of the atomatic sprinkler system.

5. Conclusion

From the simulation and its evaluation of all scenarios, results are, both systems (ceiling-top alone and with in-rack sprinklers) can decrease the temperature from the fire, stopping the further spread. As analyzed from the measured temperature, installation of the in-rack sprinklers, in addition to the first system, is able to considerably reduce the temperature. Additionally, both systems can prevent roof collapse as the temperature did not exceed 538 °C

In this research, it can be concluded that the automatic sprinkler system in warehouse can control the temperature from the fire and when the flow rate of the fire extinguishing water is increased, the fire spread head is found the fire can control the temperature more effectively.

In addition, this research can be further developed. To find suitable methods for automatic fire extinguishing in the warehouse

Acknowledgments

The author is grateful to Thunderhead Engineering for FDS Program academic license and Unique Sea Products Co., Ltd for warehouse data.

References

[1] Ning Ren, Jaap de Vries, Xiangyang Zhou, Marcos Chaos1, Karl V. Meredith, Yi Wang. 2017. *"Large-scale fire suppression modeling of corrugated cardboard boxes on wood pallets in rack-storage configurations"*. FM Global, Research Division, USA.
[2] Andrew C. Trapp, Ali S. Rangwala. 2015. "Analyzing the impact of in-rack sprinklers in a warehouse fire: A demonstration of the role optimization has in mitigating damage". School of Business, Worcester Polytechnic Institute, USA.
[3] National Fire Protection Association. 2016. "NFPA 13 Standard for the Installation of Sprinkler Systems 2016 Edition. National Fire Protection Association", Quincy, Massachusetts, USA.
[4] Thunderhead Engineering Consultants. *"Fire Dynamics and Smoke Control"*. https://www.thunderheadeng.com/pyrosim/, USA
[5] UT Fire Research Group. *"FDS Mesh Size Calculator"*. 2017. https://www.utfireresearch.com/fds-mesh
[6] Apinun Leenothai. 2002 . *"Designed Program for Automatic Sprinkler Systems"*. Master's thesis, Kasetsart University, Thailand.
[7] National Fire Protection Association. 2013. "NFPA 72 National Fire Alarm and Signaling Code 2013 Edition. National Fire Protection Association", Quincy, Massachusetts. USA.
[8] ASTM-119-98, *"Standard Test Methods for fire Test of Buiding Consruction and Material"*, American Sosiety for Testing and Material, Philadelphia, 1998.
[9] National Fire Protection Association . 2018. *"NFPA 101 Life Safety Code 2018 edition"*, July, 2018.

IOP Conf. Series: Materials Science and Engineering **715** (2020) 012006 doi:10.1088/1757-899X/715/1/012006

Performance analysis of inert gas systems in main distribution board room using fire dynamics simulation

Nummon Rattananon[1,*] and Supat Patvichaichod[2]

[1]Safety Engineering and Environmental Management Program, Kasetsart University Sriracha Campus, Chonburi, Thailand.
[2]Department of Mechanical Engineering, Safety Engineering Research Group, Kasetsart University Sriracha Campus, Chonburi, Thailand.

*E-mail: blue_biot@hotmail.com

Abstract. This study focused on the fire-extinguishing performance of inert gas in main distribution board room using fire dynamics simulator. The study included 4 types of inert gas: Nitrogen, Argon, Argonite and Inergen. The simulation was done in a 1 meters wide x 2.1 meters height x 1 meters deep main distribution board. In simulation, according to each type of inert gas, 3405.0 kW/m^2 heat release rate of electrical wires with ramp-up time at 492.0 second using concentration of inert gases based on NFPA 2001 standard. With initial concentration of Nitrogen at 41.9%, Argon 52.9%, Argonite 42.7% and Inergen 38.5% and also simulated the incidents when increased safety factor of Class C fire to 50% and 100% respectively in all 4 experimental gases which divided into 12 cases.
The results showed that different types of inert gases did not affect the fire-extinguishing performance caused by the same fuel in the same environment. And the increasing of safety factor of Class C fire by 50% and 100% respectively in all 4 types of inert gas did not improve their fire-extinguishing performance.

1. Introduction

Fire accident is one of a disaster causing severe damage to lives and properties. It can happen anywhere and anytime, unavoidable. Because fire or heating is a part of our day to day life and this can be a source of fire hazard [1]. Most of fire accidents caused by electrical systems, some examples of those accidents described as follows: A fire accident in electrical control room in Patong's water quality improvement plant. Before the incident a loud explosion occurred, then the flame was spotted in the room. After the situation was controlled, the inspector found electrical devices in the control room received heavy damage, assuming a short circuit. Another example from a fire accident occurred in electrical control room in Khon Kaen airport; a thousand of passengers riotously attempted to flee from the building. After the incident they found major damage only in the control room area causing airport shutdown. As we can see that electrical production and control system not only improves everyday life, industry and economy but also can cause severe damage to lives and properties as well. As the system has to operate all the time; which can result in heat accumulation, damaging the equipment and lead to fire hazard.

From the above accidents, this should be taken into account, fire detection and suppression systems should be installed in the electrical control room. As a fire-control system, inert gas fire-extinguishing is one of an interesting option. Since inert gases are in the atmosphere, no production is needed, do not break down to toxic gas, Not conductive and do not have environmental impacts. Plus, it does not damage electrical equipment and also save cleaning time (after used) due to no chemical residue [2].

2. Theories

2.1. Inert gas fire-extinguishing system

Inert gas fire extinguisher or specifically referred as clean agents (from NFPA 2001 standard). It is one of the gas-type fire extinguishers that was invented to replace Halon. As it is environmentally friendly, inert gas has 0 ODP and GWP values. It is also low toxicity, non-corrosive and naturally available. The gases which are utilized as a fire extinguisher are nitrogen and argon, available in both single and mixed compound as show in table 1. The main fire extinguishing mechanism is to reduce oxygen content; by releasing the gas to the area, making the oxygen content below 12%; then the fire will be extinguished [3].

Table 1. Show Inert gas components.

Type of Inert gas	Components
IG-100 (Nitrogen)	N_2 100%
IG-01 (Argon)	Ar 100%
IG-541 (Inergen)	N_2 52% Ar 40% and CO_2 8%
IG-55 (Argonite)	N_2 50% and Ar 50%

2.2. Fire dynamics simulator (FDS)

FDS is a study of the model of fire with a computer, developed by the NIST (National Institute of Standard and Technology) in the United States to illustrate flame's behavior in a simple way. The system will predict movements of smoke or the flow of hot air caused by fire, also including ventilation system, wind force and other factors, by using fire and combustion mechanisms tools. With the smoke view added to the system, make it easier for demonstration [4].

The idea of FDS to divide buildings or rooms into small sections called computational cell in order to calculate density, pressure, temperature, speed and concentration of the gas in each cell using law of conservation of energy and mass. Various properties of materials such as decoration materials, walls, floors and ceilings will be taken into account in order to simulate fire-accident scenarios. Number of the cells is major factor to determine the time used in data processing of the scenario and simulation details [5]. And it can be derived from the optimum mesh matching with the simulation details and preferred processing time according to UT Fire research group's recommendations.

$$D^* = \left(\frac{\dot{Q}}{\rho_\infty c_\rho T_\infty \sqrt{g}}\right)^{2/5} \quad (1)$$

when D*=Characteristic Fire Diameter, Q= Heat Release Rate, p∞=Density, cp=Specific Heat, T∞= Ambient Temperature and g=Gravity. To create the most realistic fire simulation, t-squared Fire Ramp can be calculated according to UT Fire research group's guidelines [6].

$$\dot{Q} = \alpha t^2 \quad (2)$$

when $\dot{Q}$ = Heat Release Rate (kW), α = Fire Growth Coefficient (kW/s^2), t = Time (s).

3. Methods

Create a model of main distribution board room (MDB) with MDBs cabinet located in the middle of the room. Then installed 4 types of inert gas fire-extinguishing systems, i.e., IG-100, IG-01, IG-541, and IG-55. Finally, simulate the scenario using FDS with concentration and volume of substances based on NFPA 2001 standard [3].

3.1. Create a modeling

We have prepared a questionnaire in order to gather detailed information about MDB room, including size of the room and MDB, location of MDB, thickness and materials used in making the MDB, floors, walls, ceilings, ventilation system etc. Then analyzed the obtained data to determine optimum parameters in order to create a model as an example of modern MDB room and its equipment. MDB in this study is studied is a 3-section interior MDB with 3 meters wide, 1 meter deep, 2.1 meters high, with side vents. MDB's frame and floor are made of 3 mm thick steel, side-cover and doors are made of 1.5 mm thick steel. It is located in the middle of the 3.2 meters wide x 5.2 meters length x 3 meters high room, derived from the size of the MDB plus working space according to the EIT standards [7].

We determined contacting surface of the room model as follows: concrete for the room, steel for the MDB and PVC for electrical cord. The electrical cord is the source of fire with the maximum heat release rates 3,405 kW/m^2 according to Warehouse Materials table from NFPA 72 standard[8]. And we used 492.0 s for the ramp-up time according to UT Fire research group's calculation [6] as shown in figure 1.

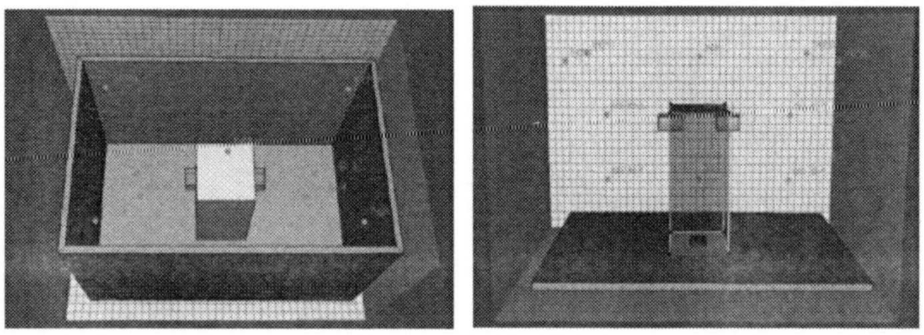

Figure 1. Model of main distribution board.

3.2. Determine the size of Fire Mesh

Fire mesh's size, 3.2 meters wide x 5.2 meters length x 3 meters high, was calculated according to UT Fire research group. The grid's size used in the model was 0.15 x 0.15 x 0.15 m which is a moderate.

3.3. Installing a gas nozzle and determining the concentration of inert gas

Create one nozzle was installed in the middle of the room above the MDB. The required nozzle must be able to inject solid particles for 60 s in a circular shape with the volume of inert gas derived from equation as follows [3] :

$$X = 2.303 \left\{ \frac{V_s}{S} \right\} \text{Log}_{10} \left\{ \frac{100}{100 - C} \right\} \quad (3)$$

when X = Volume of inert gas added at standard conditions per volume of hazard [ft^3/ft^3(m^3/m^3)], V_s = Specific volume of inert gas at 70°F (21°C) [ft^3/ib(m^3/kg)], S = Specific volume of the inert gas at 1 atmosphere and temperature [ft^3/ib(m^3/kg)], C = Agent design concentration. The gas's flow rate can be calculated from equation as follows:

$$M = \frac{\rho V}{S} \quad (4)$$

when M = Mass flow rate (kg/(m^2*s)), ρ = Density (kg/m^3), V=Volume of agent (m^3), S=Time of injection agent (s).The simulation was divided into 12 cases according to the type and concentration of inert gas which NFPA 2001 recommended. And we also simulated scenario when the safety factor of Class C fire increased by 50% and 100% respectively as shown in table 2.

Table 2. Show all case for simulation.

Case	Agent	Design Concentration	Flow rate (kg/(m^2*s))
1	IG-100	41.9 %	0.527
2	IG-100	46.5 %	0.606
3	IG-100	62 %	0.938
4	IG-01	52.9 %	1.036
5	IG-01	58.8 %	1.21
6	IG-01	78.4 %	2.109
7	IG-541	38.5 %	0.471
8	IG-541	42.75 %	0.541
9	IG-541	57 %	0.819
10	IG-55	42.7 %	0.767
11	IG-55	47.4 %	0.884
12	IG-55	63.2 %	1.376

5 Thermocouples were installed over the ceiling area for monitoring of temperature change in each section of the room as shown in figure 2.

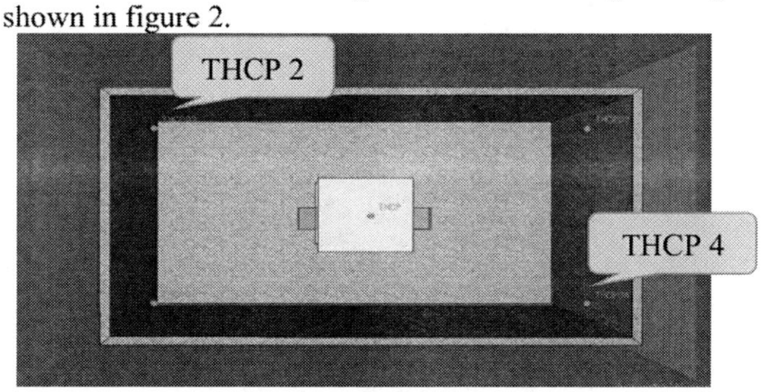

Figure 2. Show the location of the Thermocouples.

We simulated the fire scenario by setting the fire at power line below, inside the MDB. The flame has 0.3 x 0.3 x 0.05 meters in dimension.

4. Results

The results from each scenario are as written below:

4.1. First, second and third cases

In these cases, we used IG-100 as a fire suppressor with percentage of concentration from NFPA 2001 recommendation and also varied according to the safety factor of Class C fire which increased by 50% and 100% respectively. The results from all 3 cases showed that when IG-100 was released, it cannot suppress the fire within 60 seconds. Plus, the increased concentration did not help reducing (or just insignificantly reducing) the temperature inside the room as shown in table 3. The temperature inside the MDB remained at 42.0 °C as shown in figure 3.

Table 3. The results of the simulation of IG-100.

Case	Safety Factor	Concentration	Temperature room
1	NFPA recommended 35 %	41.9 %	23.70
2	increase 50 %	46.5 %	23.70
3	increase 100 %	62 %	23.60

Figure 3. Show the temperature in room & cabinet (IG-100).

4.2. 4th, 5th and 6th cases

IG-01 was used for these cases, with the concentration from NFPA 2001 recommendation and also varied according to the safety factor of Class C fire which increased by 50% and 100% respectively. The results from all 3 cases showed that when IG-01 was released, it cannot suppress the fire within 60 seconds. Plus, the increased concentration did not help reducing (or just insignificantly reducing) the temperature inside the room as shown in table 4. The temperature inside the MDB remained at 42.0 °C as shown in figure 4.

Table 4. The results of the simulation of IG-01.

Case	Safety Factor	Concentration	Temperature room
4	NFPA recommended 35 %	52.9 %	23.50
5	increase 50 %	58.8 %	23.60
6	increase 100 %	78.4 %	23.10

Figure 4. Show the temperature in room & cabinet (IG-01).

4.3. $7^{th}8$,th and 9th cases

In these cases, we used IG-541 as a fire suppressor with percentage of concentration from NFPA 2001 recommendation and also varied according to the safety factor of Class C fire which increased by 50% and 100% respectively. The results from all 3 cases showed that when IG-541 was released, it cannot suppress the fire within 60 seconds. Plus, the increased concentration did not help reducing (or just insignificantly reducing) the temperature inside the room as shown in table 5. The temperature inside the MDB remained at 42.0 °C as shown in figure 5.

Table 5. The results of the simulation of IG-541.

Case	Safety Factor	Concentration	Temperature room
7	NFPA recommended 35 %	38.5 %	23.80
8	increase 50 %	42.75 %	23.80
9	increase 100 %	57 %	23.70

Figure 5. Show the temperature in room & cabinet (IG-541).

4.4. 10th11 ,th and 12th cases

IG-55 was used for these cases, with the concentration from NFPA 2001 recommendation and also varied according to the safety factor of Class C fire which increased by 50% and 100% respectively. The results from all 3 cases showed that when IG-55 was released, it cannot suppress the fire within 60 seconds. Plus, the increased concentration did not help reducing (or just insignificantly reducing)the temperature inside the room as shown in table 6. The temperature inside the MDB remained at 42.0 °C as shown in figure 6.

Table 6. The results of the simulation of IG-55.

Case	Safety Factor	Concentration	Temperature room
10	NFPA recommended 35 %	42.7 %	23.70
11	increase 50 %	47.4 %	23.70
12	increase 100 %	63.2 %	23.50

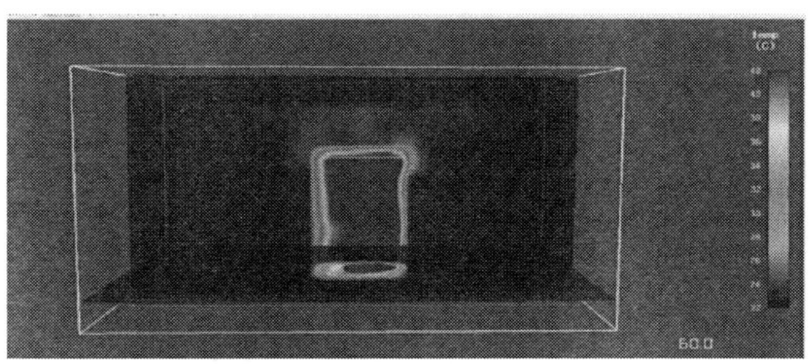

Figure 6. Show the temperature in room & cabinet (IG-55).

5. Conclusion

From the study of fire-extinguishing model with 4 types of inert gas in MDB room. It was found that the fire-extinguishing performance of inert gas: Nitrogen, Argon, Inergen and Argonite. The experimental gases had initial concentration using NFPA 2001 recommendation and varied concentration according to the safety factor of Class C fire which increased by 50% and 100%. The results showed that the increased concentration did not (or insignificantly) help reducing or extinguishing the fire caused by the same fuel in the same environment. In addition, when compared the fire-extinguishing performance between all 4 types of inert gas with the concentration according to NFPA 2001, the results showed that all 4 types of inert gas had no significantly difference in fire-extinguishing performance.

Therefore, we can conclude that in order to install a fire-extinguishing system with inert gas in the model environment; regardless of which type of inert gas is used, it will yield the same fire-extinguishing performance. And the concentration according to NFPA 2001 is recommended for all 4 types of the inert gas. The recommended concentration is the most cost-effective because it requires lesser amount of substances, though giving the same fire-extinguishing performance.

Acknowledgments

We would like to express our sincere thanks to Thunderhead Engineering for granting us a permission to use FDS program. Also, thanks to Assistant Professor Supat Phatwichaichote for the continuous support of this study. Besides my advisor, we would like to thank everyone who gave us their invaluable time to provide information in every questionnaire.

Reference

[1] Sompoth Thongchaaum and Kringchai Pungprawat. 2016. Understanding of Fire Prevention and Extinction:Prince Palace Hotel Case Study. http://grad.kbu.ac.th/wp-content/uploads/2016/08/mpacontent/uploads/2016/08/mpa_2558-15.pdf. : 20/3/2016.

[2] Sathit Witoonchatree andKiatiyuth Kveeyarn. 2011. Fire Dynamic Simulation for Fire Protection Design of Main Distribution Board. KUEngineering Journal 24 (77): 58-68.

[3] National Fire Protection Association. 2018. NFPA 2001 Standard on Clean Agent Fire Extinguishing Systems. National Fire Protection Association, Quincy, Massachusetts.

[4] Ee H. Yii, Andrew H. Buchanan, Charles M. Fleischmann. 2003. Simulating the effects of fuel type and geometry on post-flashover fire temperatures. Arup Fire, Australia.

[5] McGrattan, K. 2004. Fire Dynamics Simulator (Version 4) Technical Reference Guide. National Institute of Standards and Technology, Gaithersburg, MD.

[6] UT Fire Research Group. FDS Mesh Size Calculator.2017. https://www.utfireresearch.com/fds-mesh

[7] The Engineering Institute of Thailand under H.M.The King's Patronage. 2013. Thai Electrical Code 2013. The Engineering Institute of Thailand under H.M.The King's Patronage, Thailand.

[8] National Fire Protection Association. 2013. NFPA 72 National Fire Alarm and Signaling Code 2013 Edition. National Fire Protection Association, Quincy, Massachusetts. USA.

IOP Conf. Series: Materials Science and Engineering **715** (2020) 012007 doi:10.1088/1757-899X/715/1/012007

Evacuation time analysis of high-rise building by Using Pathfinder case study: Residential Occupancy

Jeeraphon Sudte[1,*] and Supat Patvichaichod[2]

[1]Fire Protection Engineering Program, Graduate School, Kasetsart University, Bangkok, Thailand
[2]Department of Mechanical Engineering, Faculty of Engineering at Sriracha, Kasetsart University Sriracha Campus, Chonburi, Thailand

Email: jeeraporn.yyst@gmail.com

Abstract. This research involves the study of evacuation time for a high-rise and extra-large building in Bangkok, which has 41 stories. The height of building is 143.7 meters and the building have usage area of 31,898 square meters. The objectives of this research are the estimate time of the building evacuation. The building occupant load of 2,461 persons was estimated according to NFPA 101 Life Safety Code. The simulation via Pathfinder was carried out for 4 cases; Case 1 fire in resident room on the 13th floor is to ensure that the type fires most likely to occur, Case 2 fire in lounge at the 35th floor that a large number of occupants in a large area and Case 3,4 from case 1 and case 2 the exit stair at the inside is not available. The building evacuation times are 2,321 seconds, 2,410.8 seconds, 2,736.3 seconds, and 2,841.8 seconds respectively. In the case where only one exit stair is available, the building evacuation time may be increase because of the congestions at the exit stair.

1. Introduction

At present, Residence in Bangkok (Thailand) is a high-rise building, due to the land prices is increase and to make it worth the investment, according to National Fire Protection Association 101 (NFPA 101, Life Safety Code), The most important thing about High-rise building with sleeping area is life safety of the residents. The analysed of evacuation time should be the first priority in building design. In the past, Evacuation time was calculated by hydraulic calculation method refer to International standard (NFPA 101) but there was some problems. The Hydraulic calculation had many human errors and not accurate with real situation. Now Pathfinder Program, is developed for simulate evacuation time, is lower human error and more accurate than hydraulic calculation.

Definition of high-rise buildings that a person may enter or use, with a height of 23 meters or more. Measuring the height of the building is measured from the ground level of construction to the deck floor. For a gable or hip building, measure from the ground level that is constructed to the top of the top wall [1].

2. Model capabilities and modelling studies

2.1. Model building

Pathfinder is an emulation simulation program. The display is graphical. This allows users to design both in 2 and 3 dimensions for results analysis with motion simulation divided into 2 operating modes:

IOP Conf. Series: Materials Science and Engineering **715** (2020) 012007 doi:10.1088/1757-899X/715/1/012007

Steering Mode and SFPE Mode, the use of Steering Mode will be the most practical use that is, imitating human behaviour and movement. In SFPE mode as use based on the theory, assumptions, and calculations by hand as defined in the SFPE manual, this mode is controlled by the flow rate through the gate or through the density of the population. Which is allowed to infiltrate and do not have to try to evade. [2]. A Condominium building has 41 floors. The height from the ground is 143.7 meters, with a total area of 31,898 square meters, show on figure 1.

Figure 1. High-rise condominium building model.

2.2. Simulation setting

1. This simulation has 2,461 people in the building. Occupant load factor is based on the NFPA 101, Life Safety Code [3].
2. The user profile of the building is divided into equal proportions, such as children, teenagers, middle-aged, elderly including male and female, etc. [4]
3. Behaviour model has been set to "Steering" mode; jam velocity is 0.25 m/s; control the speed threshold at which occupants are recorded as being jammed at the exit [5].
4. According to NFPA 101 the simulated scenarios are as follows:

- Case 1, Fire in resident room on the 13th floor is to ensure that the type fires most likely to occur.
- Case 2, Fire in the lounge at the 35th floor that a large number of occupants in a large area.
- Case 3 from case 1 the exit stair 2 is not available.
- Case 4 from case 2 the exit stair 2 is not available.

3. Simulation result

3.1. Simulation result

Case 1: Evacuation times 2,321.0 seconds.

IOP Conf. Series: Materials Science and Engineering **715** (2020) 012007 doi:10.1088/1757-899X/715/1/012007

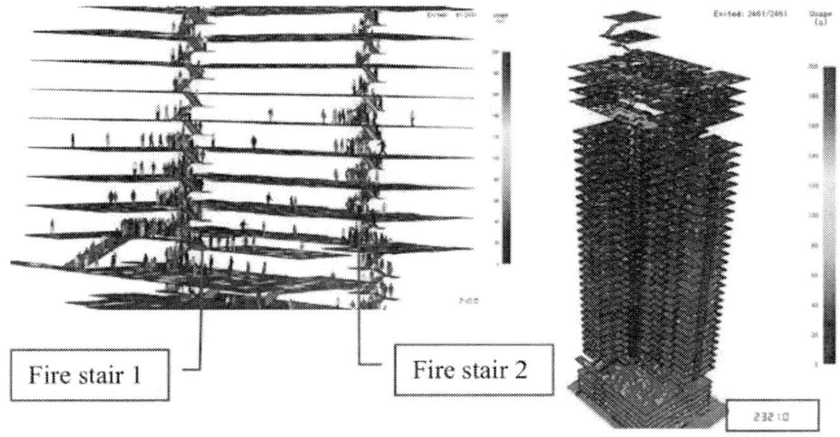

Figure 2. Simulation result of case 1.

Case 2: Evacuation times 2,410.8 seconds.

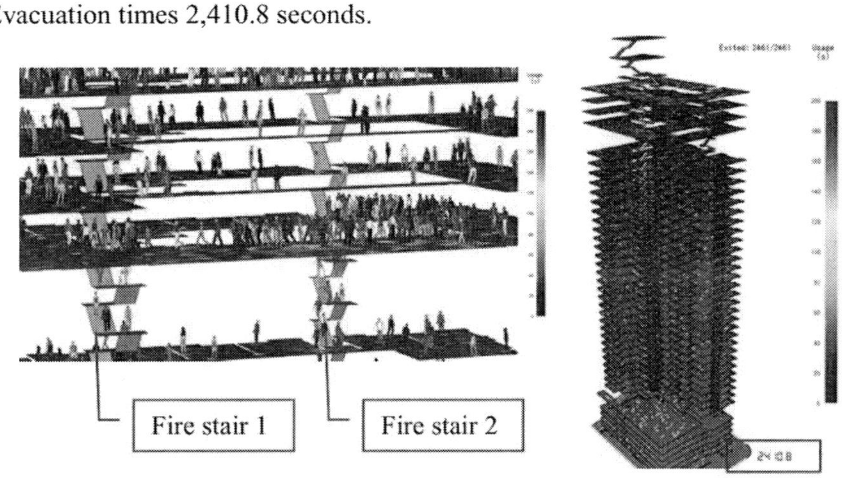

Figure 3. Simulation result of case 2.

Case 3: Evacuation times 2,736.3 seconds.

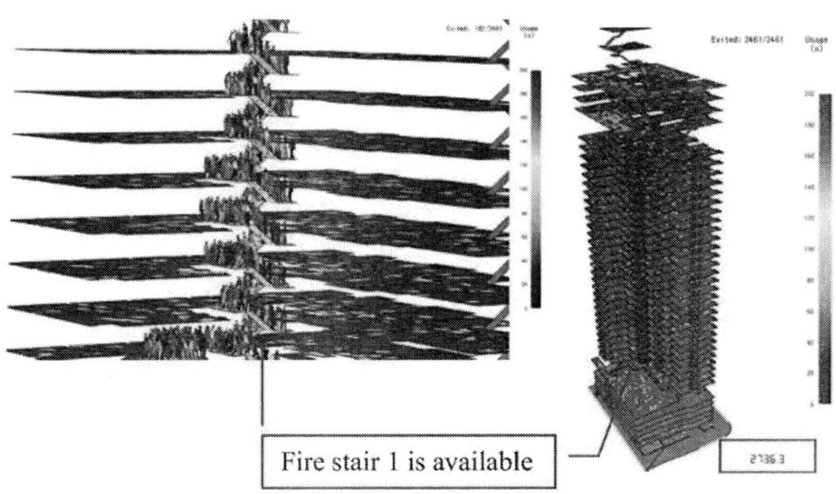

Figure 4. Simulation result of case 3.

Case 4: Evacuation times 2,841.8 seconds.

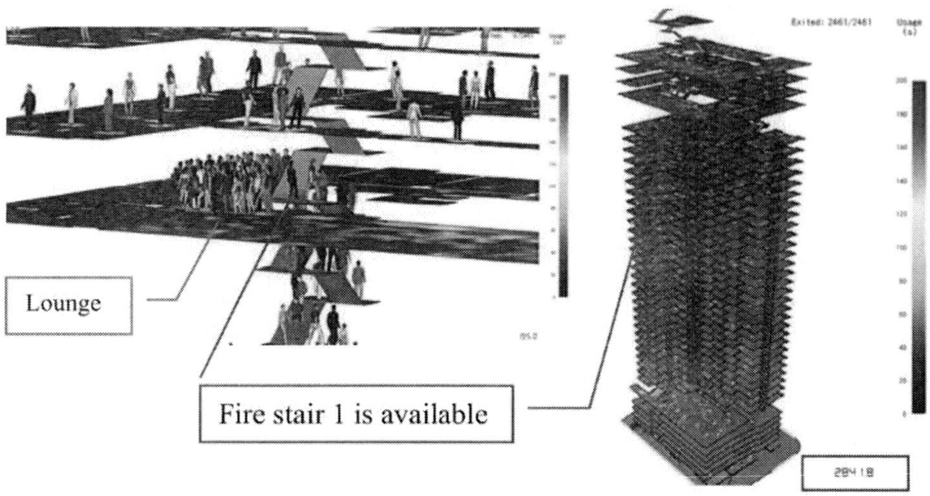

Figure 5. Simulation result of case 4.

4. Results and discussions

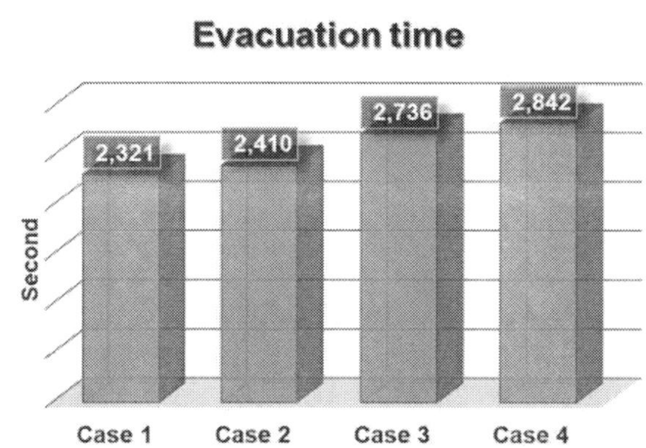

Figure 6. Summary simulation result.

The evacuation time consists of 4 parts; Time from fire ignition to detection (Time to Notification), Time from detection to notification of occupants of a fire emergency (Reaction Time), Time from notification (or cue reception) until evacuation commences (Pre-evacuation Time) and Time from the start of purposive evacuation movement until safety is reached (Travel or Movement Time). The first 3 parts are Delay time or Pre-evacuation Time.

Case 1 Fire in resident room on the 13th floor is to ensure that the type fires most likely to occur. When a fire occurs on the 13th floor, a smoke detector detects the density of smoke (Time to Notification) and sends a signal to the Fire alarm. Until all people react (Reaction Time) takes about 60 seconds. After that, people on the 13th floor begin to evacuate (Pre-evacuation Time). After 120 seconds, the fire alarm will send a sandwich alarm (The floor where the fire occurred start alarm and setting time delay above 1 floor and below 1 floor, time not more than 180 seconds) to floor 14 and floor 12, and people from floor 14 and floor 12 start to evacuate, which the occupant inside the fire escape is not overcrowd. And evacuated quickly. After 180 seconds, fire alarms from level 15 upwards and level 11 downward start to evacuate and panic. From 240 seconds onwards, the people on the low level ran into the fire escape stairs impact into someone who had escaped from the upper level causing inside the fire escape stair 1 and 2 have overcrowd people and slower movement, show on figure 2.

From the incident, it takes time to evacuate to a place of safety. (Travel or Movement Time) 2321.0 seconds, which take the least evacuation time from a total of 4 cases.

Case 2 fire in lounge at the 35th floor that a large number of occupants in a large area. Delay Time is like Case 1. After 120 seconds, the fire alarm will send signals to the 36th and 34th floors from 36th and 34th floor. Start to evacuate, but due to the large number of people on the 35th floor, there was a bottleneck at the entrance to the fire escape causing the evacuation to be slow. After 180 seconds, fire alarms from Level 37 upwards and Level 34 downward start to evacuate, the people on the low level ran into the fire escape stairs impact into someone who had escaped from the upper level causing inside the fire escape stair 1 and 2 have overcrowd people and slower movement, show on figure 3, so it has Travel or Movement Time 2410.8 seconds, which is more than the case 1; 89.8 seconds

According to NFPA 101, the worst-case scenarios will be studied: at least 1 fire escape path is blocked or not available. Case 3 from case 1 the exit stair at the inside is not available. There will be bottlenecks at the entrance to every fire escape door and inside the fire escape stair will be overcrowded causing moves slowly, show on figure 4. Have the evacuation time of 2736.3 seconds, which is more than case 1; 415.3 seconds and more than case 2; 325.5 seconds.

Case 4 from case 2 the exit stair at the inside is not available, Like the case 3, the evacuation time is 2841.8 seconds, show on figure 5, which is more than the case 2; 431.0 seconds, which takes the most evacuation time from a total of 4 cases, show on figure 6.

5. Conclusion

Research studies on fire scenarios simulation and evacuation times. To sum up, Case 1 fire in resident room on the 13th floor is to ensure that the type fires most likely to occur, the people on the low level ran into the fire escape stairs impact into someone who had escaped from the upper level causing inside the fire escape stair 1 and 2 have overcrowd people and slower movement, the building evacuation times are 2,321 seconds. Case 2 fire in lounge at the 35th floor that a large number of occupants in a large area, due to the large number of people on the 35th floor, there was a bottleneck at the entrance to the fire escape causing the evacuation to be slow, the building evacuation times are 2,410.8 seconds. Case 3 from case 1 the exit stair 2 is not available, there will be bottlenecks at the entrance to every fire escape door and inside the fire escape staircase will be overcrowded causing moves slowly. Have the evacuation time of 2736.3 seconds, which is more than case 1; 415.3 seconds and more than case 2; 325.5 seconds. And Case 4 from case 2 the exit stair 2 is not available, like the case 3, the evacuation time is 2841.8 seconds, which is more than the case 2; 431.0 seconds, which takes the most evacuation time from a total of 4 cases. From the simulation case 3 and 4, can be said that if at least one fire escape route is blocked or not available, the evacuation time will increase by approximately 420.0 seconds. The building evacuation time may be increase because of the congestions at the exit stair. In conclusion, scenarios and cause of the fire effect to evacuation time. However, we can use Pathfinder Program for simulation other scenarios to find evacuation time accord to local law and develop performance of mean of egress. For future work can use Pathfinder Program to simulation other building such as Hospital, Extra hazard factory and Office building etc.

Acknowledgments

This is a project completed by Pathfinder with a license to study from Thunderhead Engineering. The high-rise building plan is supported by an architect company design.

References

[1] E. Ronchi, D. Nilsson, Fire evacuation in high-rise buildings: a review of human behavior and modelling research (Fire Science Reviews, licensee Springer, 2013)
[2] Thunderhead Engineering, User Manual, Pathfinder, 2014
[3] NFPA, NFPA 101: Life Safety Code, National Fire Protection Association, 2018
[4] G. Grewolls, K. Grewolls, Sensitivity Analysis of Evacuation Simulations (SIMTEGO GmbH, Dynardo, 2014)

IOP Conf. Series: Materials Science and Engineering **715** (2020) 012007 doi:10.1088/1757-899X/715/1/012007

[5] Thunderhead Engineering, Technical Reference, Pathfinder, 2013

IOP Conf. Series: Materials Science and Engineering **715** (2020) 012008 doi:10.1088/1757-899X/715/1/012008

Performance – based life safety analysis of the hospital building

Natnaree Chiangaek [1,*] and Supat Patvichaichod[2]

[1]Fire Protection Engineering Program, Graduate School, Kasetsart University, Bangkok, Thailand
[2]Department of Mechanical Engineering, Faculty of Engineering at Sriracha, Kasetsart University Sriracha Campus, Chonburi, Thailand

Email: natnaree.chi@gmail.com

Abstract. This study Performance – Based Life Safety Analysis of the Hospital Building reference NFPA 101, Life Safety Code. Four simulation scenarios were selected for this simulation. First, fire in a patient room; evacuation time is 2,847.3 seconds after the announcement. Second, fire in a storage room near the canteen; evacuation time is 3,014.5 seconds after the announcement. Third, from the first and second scenarios have a nonoperative fire alarm system which increases evacuation time that is 3,122.0 seconds and 3,235.8 seconds respectively. From simulation results shows that the first problem is the bottleneck at the fire escape door. It must be modified by expanding the door. And the second problem is the slow movement of the patient that has to solve the problem by having a staff to help evacuate.

1. Introduction

There are more than one thousand fires in global hospitals each year, and quite a lot of injured people in hospitals and health care facilities. Require to effective strategies for the safe evacuation of all hospital occupants in fire situations. This strategic planning for the migration process in the health care environment needs to be thoroughly considered. Most hospitals have large and complex building structures, including furniture and various portable equipment; therefore, there is great variability and dynamism in mean of egress. Wards have different uses (e.g. intensive care, outpatient clinics, etc.) and differences in the duties of staff to support these applications [1]. Inpatient wards in the high-rise hospital building, have the risk of fire occur. Because many occupants are in the building for a long time and different patient classifications (self-care to intensive care) [2]. Hospital emergency evacuation requires assistance for patients with reduced mobility. The extensive set of experiments to assess evacuation that has been helped using a stretcher, portable chair, evacuation chair, and life-saving pad. Then, the experimental data were used to predict the evacuation time for the entire hospital under the circumstances of personnel management and equipment assistance. In this research, study and analysed the fire escape route performance of the hospital building. Simulate the hospital building by Pathfinder program to simulate a fire situation and find the required safe egress time (RSET) by using the principles and theories in the setting program is based on the NFPA 101 (Life Safety Code) and actual suitability [3].

2. Model buildings and calculations

2.1. Model building

Pathfinder is an agent-based egress simulator that uses steering behaviors to model occupant motion. It consists of three modules: a graphical user interface, the simulator, and a 3D results viewer. Pathfinder provides two primary options for occupant motion: an SFPE mode and a steering mode [4]. The SFPE mode implements the concepts in the SFPE Handbook of Fire Protection Engineering [Nelson and Mowrer, 2002]. This is a flow model, where walking speeds are determined by occupant density within each room and flow through doors is controlled by door width. The steering mode is based on the idea of inverse steering behaviors. Steering behaviors were first presented in Craig Reynolds' paper "Steering Behaviors for Autonomous Characters" [Reynolds, 1999] and later refined into inverse steering behaviors in a paper by Heni Ben Amor [Amor et. al., 2006]. Pathfinder's steering mode allows more complex behaviors to naturally emerge as a by-product of the movement algorithms - eliminating the need for explicit door queues and density calculations [5]. The hospital building has a total area of 42,379 square meters, 72.2 height meters, 15 floors and 2 basements consisting of several business segments, including the SRT station, the Outpatient Department including I.C.U. and Inpatient Department on the upper floor of the building as shown in figure 1.

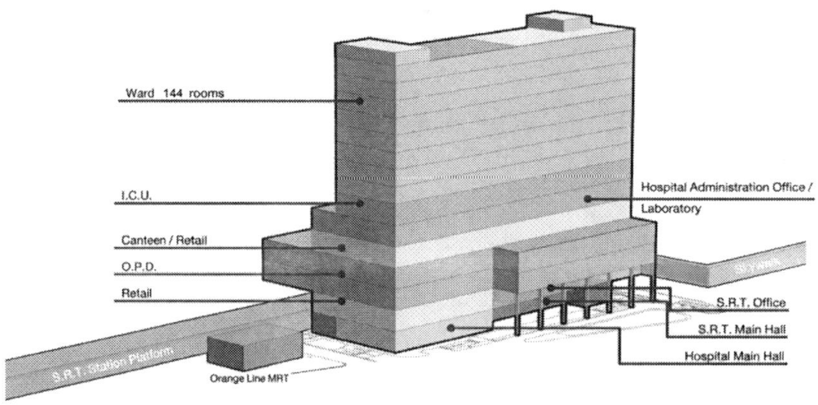

Figure 1. Showing the usable area of the hospital building.

2.2. Simulation setting

1. In this simulation, there are 4,100 Persons are involved. This number derived from the occupant load factor (m2/person) as defined in NFPA 101 [3]. Persons divided into three groups that staff, inpatient and outpatient.
2. Evacuation patterns of various groups.
   - The staff is designated as patient assistance for evacuation.
   - Outpatients can evacuate on their normal conditions.
   - Patients must be rescued by the staff with the bed or wheelchairs, etc.
3. The behaviors model has been set to "Steering" mode; jam velocity is 0.25 m/s; control the speed threshold at which occupants are recorded as being jammed at the exit [3].
4. Design fire scenarios base on NFPA 101. Four scenarios were selected for this simulation.

-The idea behind design fire scenario 1 is to ensure that the types of fires most likely to occur in a given occupancy type that is, the statistically most significant scenarios are considered in the design analysis. Therefore, this building has 15 floors and 2 basements. From all of building have 8-story ward or 45 percent of the building, therefore, set a fire origin that is short circuit in ward at the 8th floor which is the middle floor of the building and the first floor of ward as shown in figure 2. To analyse that many patients and staff on the 9th floor and above will evacuate safely or not.

-Design fire scenario 2, It addresses the concern regarding a fire starting in a normally unoccupied room and migrating into the space that potentially holds the greatest number of occupants in the

IOP Conf. Series: Materials Science and Engineering **715** (2020) 012008 doi:10.1088/1757-899X/715/1/012008

building. The fire origin is on the 5th floor, which had a storage room next to canteen. There was a lot of people which has both normal people and patients as shown in figure 3.

- Design fire scenario 3 and 4 addresses a set of conditions with a typical fire originating in the building with any one passive or active fire protection system or feature being ineffective.

Examples include

- Unprotected openings between floors or between fire walls or fire barrier walls
- Failure of rated fire doors to close automatically
- Shut-off of sprinkler system water supply
- Non-operative fire alarm system
- Inoperable smoke management system
- Automatic smoke dampers blocked open.

Therefore, Scenario 3 and 4, from case 1 and 2, will require a delay of 300 seconds before starting to evacuate cause non-operative fire alarm system.

3. Simulation result and analysis

3.1. Simulation result

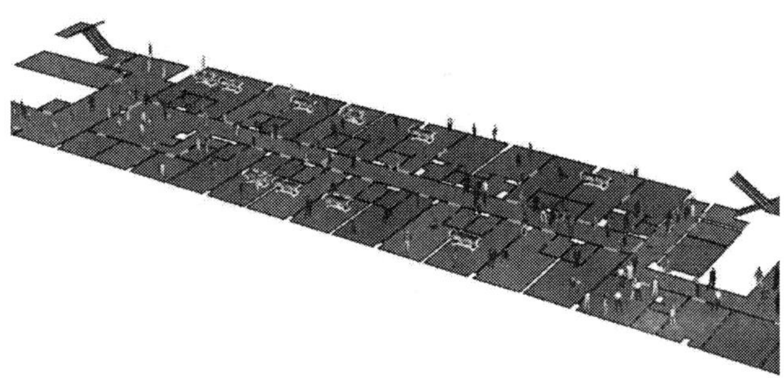

Figure 2. Enlarged picture of ward on the 8th floor.

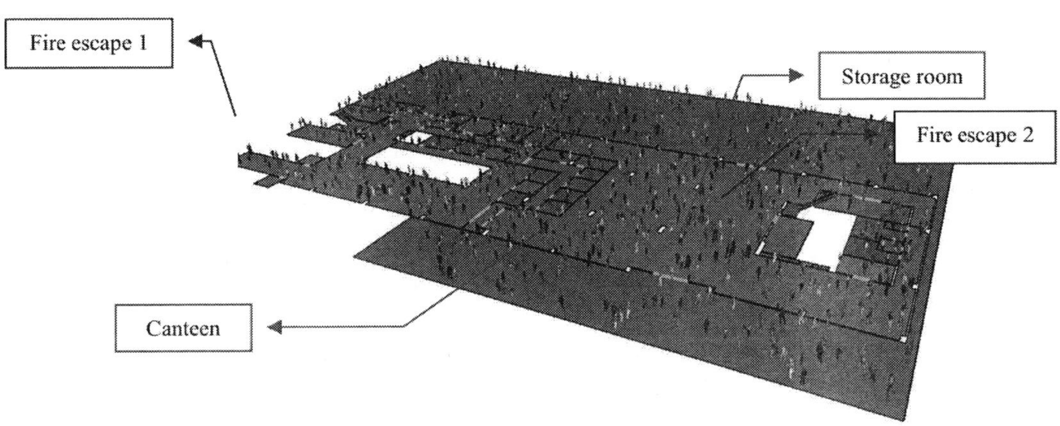

Figure 3. Enlarged picture of canteen and storage room on the 5th floor.

IOP Conf. Series: Materials Science and Engineering **715** (2020) 012008 doi:10.1088/1757-899X/715/1/012008

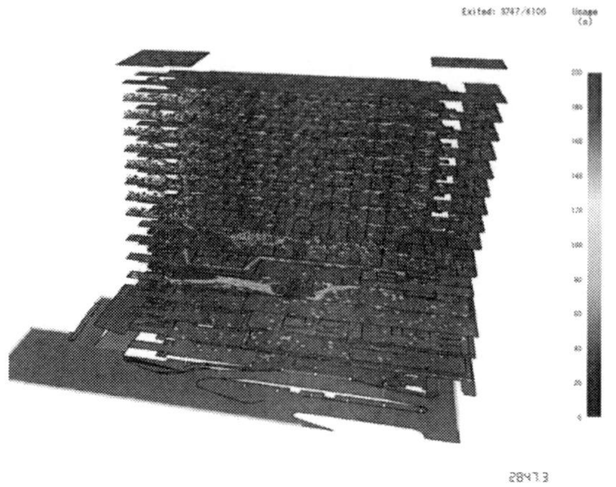

Figure 4. Simulation results for case 1.

Figure 5. Simulation results for case 2.

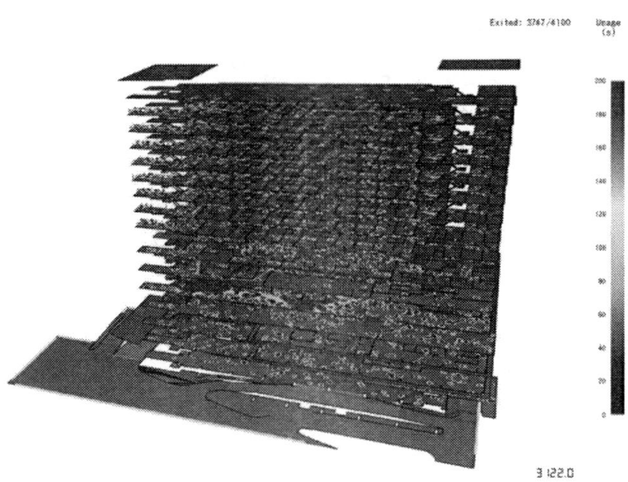

Figure 6. Simulation results for case 3.

IOP Conf. Series: Materials Science and Engineering **715** (2020) 012008 doi:10.1088/1757-899X/715/1/012008

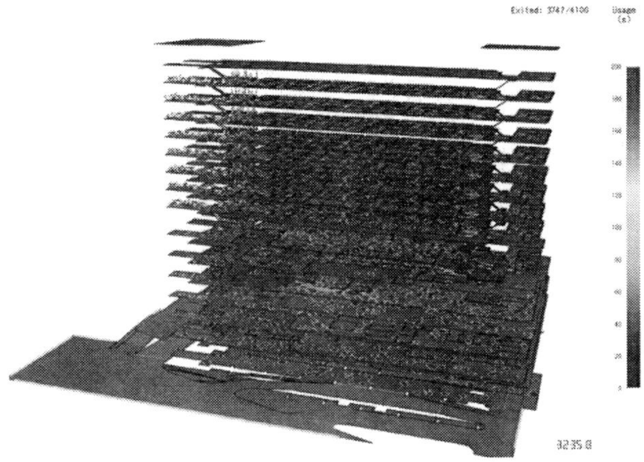

Figure 7. Simulation results for case 4.

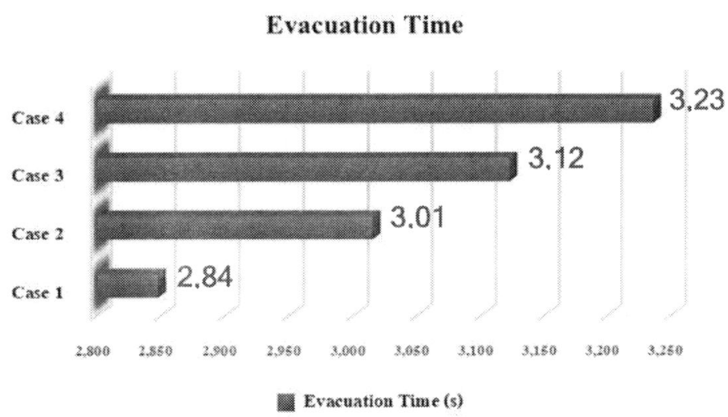

Figure 8. Evacuation time for both 4 cases.

4. Results and discussion

Scenario 1, when a fire occurs at the ward, the assistant will evacuate the patient to the refuge area in phase evacuation. Normally, the assistant is less than the patient. Resulting in a lot of evacuation time. In the part of ICU room, it is made into a fire compartment and does not evacuate critically patients. Allowing time to evacuate 2,847.3 seconds as shown in figure 4.

Scenarios 2, fire occur near the fire escape 2. Therefore, the people on the 5th floor have to escape by using fire escape 1, causing the fire escape 1 to get a bottleneck at the door. And crowded inside the stairs. Allowing time to evacuate 3,014.5 seconds as shown in figure 5, evacuation time is greater than the first case 167.2 seconds.

Scenario 3, when the fire escape 2 was unavailable, allowing time to evacuate 3,122.0 seconds, the evacuation time increases from the first case 274.7 seconds. Because of a bottleneck at the fire door as shown in figure 6.

Scenario 4, when the fire escape 2 was unavailable, allowing time to evacuate 3,235.8 seconds as shown in figure 7, the evacuation time increases from the second case 221.3 seconds as shown in figure 8. Because of a bottleneck at the fire door.

5. Conclusions

Even though the fire escape has the correct width according to the standard but in the real evacuation, cannot support all the building user if evacuated simultaneously, should develop evacuation plans such as phase evacuation.

The problem of the width of the exit door of the area that has many building users, there should be more than 1 exit.

Comparison of evacuation time for four scenarios was shown. Maximum evacuation time is scenario 4, this means that we may use this result to design available safe egress time (ASET). This simulation result can be used to find the fault location of the fire escape route to makes changes in the future.

Use the evacuation elevator to increase the evacuation method. And reduce evacuation time.

Acknowledgments

This is a project accomplished by Pathfinder with an education license from Thunder Engineering. Layout supported by high-rise hospital building.

References

[1] A. Hunt, Simulating Hospital Evacuation, Thesis for Doctor of Philosophy, University of Greenwich, UK., 2016
[2] A. Kajornsiririerk, *Evacuation routes of inpatient wards case study: high rise building of government hospital*, Thesis for Master of Architecture Program in Architecture, Chulalongkorn University, Thailand, 2017
[3] Y. Ding, L. Yang, F. Weng, Z. Fu, P. Rao, *Investigation of combined stairs elevators evacuation strategies for high rise buildings based on simulation*, Simulation Modelling Practice and Theory, ScienceDirect, 2015
[4] Thunderhead Engineering, *User Manual*, Pathfinder, 2014
[5] Thunderhead Engineering, *Technical Reference*, Pathfinder, 2013

IOP Conf. Series: Materials Science and Engineering **715** (2020) 012009 doi:10.1088/1757-899X/715/1/012009

Optimal model of locating charging stations with massive urban trajectories

Danhui Huang*, Yuan Chen and Xiaochen Pan

School of Big Data and Software Engineering, Chongqing University, Chongqing, China

*E-mail: hdh@cqu.edu.cn

Abstract. Popularization of electric vehicles (EVs) is a promising approach towards realizing environmentally friendly transportation. While government has implemented advantageous policy efforts to boost EVs, how to effectively build electric charging stations becomes a key link. A reasonable distribution should not only conform to user habits and potential behaviour patterns, but also satisfying the charging reliability. In this paper, vehicle track data and POI data are used to cluster with road network information. A clustering algorithm based on R serving radius is proposed to improve the results and make them accommodate the rationality of the real world. Finally, suggestions are made for each charging station power supply scale according to regional difference. Experiments show that the proposed model has practical guiding significance.

1. Introduction

In recent years, the aggravation of environmental pollution and the development of battery technology have promoted the development of electric vehicle industry. As the promotion of EVs is expected to reduce unhealthy pollution in the future, electric vehicle has drawn attention of automobile manufacturers and the government [1].

In view of the charging station location problem of EVs, many research institutions and researchers have done a lot of research works. From macroscopic aspect. the most k-location set was proposed, within an interactive mining processes involving users' input [2]. From microscopic aspect, all sorts of methods had been tried to sitting these k points such as theoretical framework [3] and GIS-based [4]; In terms of distance and waiting time, OCSP and OCPA optimization components were designed to minimize the average time respectively [5]. Different from previous works, we take user's potential charging behaviour into consideration, analyses their movement rule hide in trajectories to carry out charging station layout.

The rest of this paper is organized as follows. Section 2 gives a general description of the model. Section 3 describes charging station location optimization algorithm and scale calculation rule. Section 4 presents experimental results. Section 5 concludes the paper

2. Model formulation

From the perspective of discovering people's potential charging habits, and meeting more charging needs, we apply the method of clustering POI and trajectory data to Chongqing taxi trajectory data [6].

IOP Conf. Series: Materials Science and Engineering **715** (2020) 012009 doi:10.1088/1757-899X/715/1/012009

Based on fusion data, a charging station location model for site selection, optimization and scale recommendation is proposed.

The main solutions of charging station location problem are as follows. The first step is data processing, the staying points are extracted from the taxi GPS track data, and the POI data is obtained from the social network at the same time. The second step is fusion data clustering, the center of cluster can be more convenient for users to reach After fusion, site selection recommendation will be more close to business districts, office buildings, government agencies and so on, these locations are more user-friendly also. In the clustering process, density-based DBSCAN algorithm has been used [7].

The third step is to improve the algorithm. Due to the original result may produce a phenomenon of candidate sites gathering in one region, it's unreasonable for the construction of the charging station. R-DBSCAN could find a compromise point to replace these gathering sites and optimize the model. The fourth step is scale recommendation. A charging station includes a variety of attributes, such as location, floor space, power supply capacity, scale, construction costs and so on. According to the research on regional charging demands, this model will recommend the number of charging piles in various charging stations by using scale conversion rule.

2.1. *Data description*

The dataset used in this paper is the GPS trajectory of more than 12,000 taxis in Chongqing(2017.03.01-2017.03.31), the frequency is 4 records per minute, the original data set is about 132 GB. This paper selects six attributes, a GPS point is represented as a six-tuple: P<date, time, id, lng, lat, speed>, details showed in table 1.

Table 1. Original GPS trajectory data.

Date	Time	Id	Lng	Lat	Speed
20170301	2001	0600307115	106.4851	29.53619	0
20170301	2435	0100504146	119.0178	33.59274	13.2
20170301	2249	0600305444	106.4894	29.60768	39.8

More than 20,000 POI data from Baidu Map was obtained, covering the nine districts of Chongqing City. There are eight categories of POI data: restaurants, hotels, companies, hospitals, parking lots, office buildings, government agencies, shopping centers. Each piece of data contains five attributes: ID, name, longitude, latitude, and POI category.

2.2. *Scale definition and grid partition*

Charging stations in different locations should have different scales according to charging needs. In prosperous areas, such as business gathering areas, the demand for EVs charging is far greater than that in dispersed residential areas [8]. For this reason, we propose the definition of charging station scale, and incorporates the construction scale of each charging station into the charging station selection planning.

Consider the heat value HValue and the threshold e as two factors that influence the scale S, set the threshold e to 10, the relationship between them is shown with Eq.(1).

$$S = \lfloor HValue/\, e \rfloor \tag{1}$$

Moreover, if a charger is available near the destination of the EV driver, the user may be willing to park the charging station and walk for a few minutes to the destination. Therefore, instead of using exact geographical locations, each staying point is allocated to a grid. In this paper, nine districts in the main city of Chongqing are divided into grids according to latitude and longitude. The size of grids is about 1.1Km*1.0Km, and the total number of grids is nearly 10,000.

2.3. *Staying points identification*

Assuming that a taxi has n GPS track sampling points, the taxi GPS track P=p1 p2...pn, is a continuous point sequence, composed of GPS sampling data points in time sequence[9], There are many methods for GPS staying point detection, ranging from the method based on missing point [10], direction feature [11], to merge trajectory points below a certain velocity threshold into candidate residence positions [12]. In this paper, the following definition of GPS staying point are selected [13].

A sub-trajectory is represented as:

$$P_{k,m} = \{[(t_j,loc_j)], j \in [k,k+m-1]\} \tag{2}$$

The length of sub-trajectory Pk, m is m, m $\leq$ n and k $\in$ (0, n-m). If there is a data point loci satisfied:

$$loc_i = loc_{i+1}, \forall i \in [k,k+m], when\, t_{i+1} - t_i \leq \Delta t\, and\, t_{m-1} - t_0 \geq \Delta T \tag{3}$$

If two points have same latitude and longitude, and the time interval is less than Δt, it indicates these two points are continuous points. Moreover. if the stay time at this point is greater than ΔT, then this position can be considered as a staying point. Because of the current technical limitations of EVs, in general, the charging time is more than one hour, so the stay time is set to 60 minutes.

3. Locating algorithm

DBSCAN is a density-based clustering algorithm which divides high density regions into clusters, and can find clusters of arbitrary shape in noisy spatial data. In this paper, the sample set is too large converge, so the KD tree is used to search for the nearest neighbour [14].

3.1. R-DBSCAN

In this model, the central point of cluster is selected as a candidate location in the final charging station layout. The results of the original DBSCAN algorithm may make the two cluster points too close to each other. To solve this problem and improve the adaptability in reality, we introduced R service radius, which can control the final site selection points won't appear in the same area.

Each charging station will cover the serving area with a radius R, and if multiple site selection points are in the R scope, they will be merged into a new one.

As shown in figure 1(a), the radius ranges of points p1, p2 and p3 cover each other, while p4 only belong to its R range. In this paper, p1, p2 and p3 are merged into one, and their average coordinates are taken as the new point P1 in figure 1(b), as a result, four site selection points are finally optimized to two site selection points, and the location of P2 remains unchanged. Calculate the distance between two points as:

$$d_{12} = \sqrt{(x_1 - x_2)^2 + (y_1 + y_2)^2} \tag{4}$$

R-DBSCAN can be summarized as the following four steps:

-Save the original site selection points to set S, randomly select a point p and create another set T1 for p1.

-Calculate the distance d between other points in S and point p. If point d1m < R, save pm to T1 until there is no point in S.

-Remove the point that has been stores in T1 from S, and then repeat steps 1-3

-While S is empty, the average coordinates of the points in each set T1T2...Tn are the final site selection point. The algorithm ends.

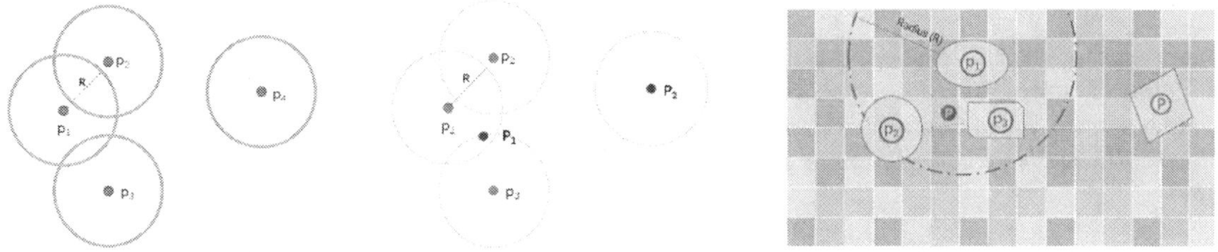

(a) Original results (b) Final results (c) Example of scale calculation

Figure 1. (a) And (b) shows the process of merging, (c) give an example of scale calculation.

3.2. HValue calculation

In the previous preparation work, we divided the nine districts in Chongqing into grids. Each cluster will occupy one or more grids, the sum of the HValue of these grids is also the HValue of the cluster. In figure 1(c), the red dot is the final site selection point generated by the three original points, add up the three HValue and the result is the HValue of red final site. This relationship can be expressed as:

$$HValue_P = \sum_{j\in P}\sum_{i\in p_j} HValue(grid_i) \tag{5}$$

For all original sites j belong to P, and all grids i belong to pj, calculate the HValue of each original site and add them up, then we can get the *H*Valup of the final site selection point P.

After we get the HValue, the scale and the number of charging piles for each charging station can be calculated to an exact value respectively.

4. Experiments

4.1. Visualization of experiment data R-DBSCAN validation

We extract the road network of Chongqing, which contains 11,311 roads and 1479 vertices, as shown in figure 2(a). figure 2(b) shows taxi trajectory spatial distribution after coordinate transformation. Figure 2(c) shows charging demands using a heat map, the lighter the denser. Figure 2(d) shows POI distribution on the map, by scaling the map, we can see the number and exact location in the region.

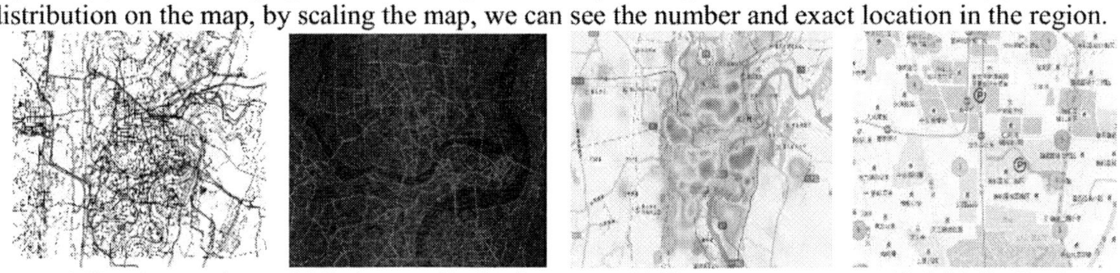

(a)Road network (b) Trajectory distribution (c) Heat map (d) POI distribution

Figure 2. Visualization of road network, Trajectory, heat map and POI in Chongqing urban area.

4.2. RDBSCAN validation

The original DBSCAN algorithm mainly contains two parameters, Eps and Minpts. These two parameters affect the distribution of charging stations by controlling the distance between two points in the cluster and the number of clustering points.

Figure 3 present the comparison between original DBSCAN and R-DBSCAN, as we can see in the red circle of figure 3(a), figure.3(c), the original DBSCAN algorithm regards the multiple clusters in adjacent areas as multiple site selection points, and suggests to build the same number of charging stations. However, considering the realistic factors such as land use restriction, capital restriction, business planning, the more reasonable way is to build only one charging station in this area. R-DBSCAN integrates the charging stations in the R radius region by setting R constraints, and finally gives a recommendation point that can cover all the charging requirements in the original region, as shown in figure.3 (b), figure.3 (d).

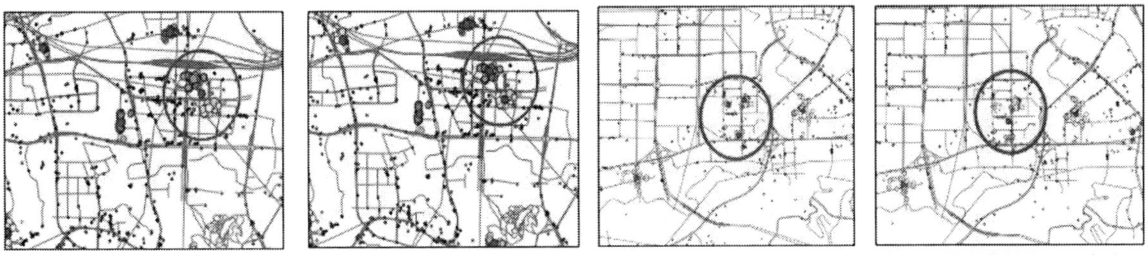

(a) Original result 1 (b) R-DBSCAN result 1 (c) Original result 2 (d) R-DBSCAN result 2

Figure 3. Comparison between original DBSCAN and R-DBSCAN (Eps=0.0015, Minpts=40).

The heat value of the final recommendation point is the sum heat value of all the original cluster grids, that is to say, the scale of the final recommendation point will be larger and there will be more charging piles.

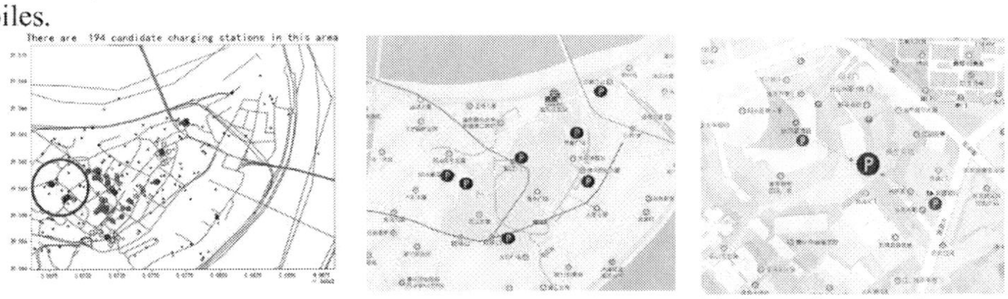

(a) R-DBSCAN results (b) Results on real map (c) Two adjacent points

Figure 4. Locating charge stations in Jiefangbei district, Chongqing(Eps = 0.0008, Minpts = 10, R = 0.001).

Figure 4 shows an example on real map in Jiefangbei Business Circle, Chongqing. The original DBSCAN system recommends 7 charging stations in this area as shown in figure 4 (a). figure 4 (b) projected these 7 points onto the real map, 5 recommended points outside the red circle cover several commercial locations such as Jie Fangbei Pedestrian Street, Dayuan Square and Hongyadong with reasonable distance. However, as shown in figure 4(c), the other two recommended points in the red circle, one is near Marriott hotel and the other is near internet cafe, it's not necessary to build two charging stations on the same street. The R-DBSCAN algorithm re-integrates the two clusters and assigns them to a new recommendation point, which can cover the original two clusters. R-DBSCAN algorithm improves the adaptability of the reality, not only ensures convenience and power supply demand, but also contributes to avoid unnecessary waste of social resources.

4.3. HValue matrix and scale recommendation

The grid area in figure 5 (a) is part of the nine districts in the main city of Chongqing. The grid size is 16*16, HValue matrix calculates parking demand in each area, which is the numerical basis for making scale recommendations for charging station.

As shown in figure 5 (b), each charging station recommendation point has scale information, including the number of recommended charging piles, the heat value of the grid occupied by the charging station, the threshold and the number of charging piles per scale. The charging station recommendation point in Chongqing North Railway Station area has a high heat value , this is Chongqing's largest passenger station for high-speed train and bus traffic with a large number of parking bus and private car. Compared with other areas with business circle or park, its charging demand is larger than others, which is more in line with the reality.

(a)HValue matrix (b)Scale recommendation

Figure 5. Heat grids distribution and scale recommendation in Chongqing.

5. Conclusion

Based on the user's behavioural habits, this paper analyses the trajectory data of the vehicle, and combines various facilities layout and road network distribution in the city to improve the clustering algorithm. The coverage radius R is used to control the number of charging stations in the area, and the final charging station location is optimized and adjusted. According to the difference in charging demand of each area, the proposal for the construction scale of each charging station is given. While guaranteeing convenience and satisfying charging demand, resources are reasonably allocated. Combined with the real map, we validate the recommendation site of nine districts charging station in the main city. The results can cover the crowded area of traffic vehicles in various streets, shopping districts, entertainment areas, etc. The distance between charging stations is much more reasonable, and the scale recommendation is also positively correlated with the density of local traffic flow, which fully illustrates the effectiveness of this model.

References

[1] Lu C, Rong K, You J, et al. Business ecosystem and stakeholders' role transformation: Evidence from Chinese emerging electric vehicle industry[J]. Expert Systems with applications, 2014, 41(10): 4579-95.

[2] Li Y, Bao J, Li Y, et al. Mining the Most Influential k-Location Set from Massive Trajectories[J]. IEEE Transactions on Big Data, 2018, 4(4): 556-70.

[3] Xiong Y, Gan J, An B, et al. Optimal electric vehicle fast charging station placement based on game theoretical framework[J]. IEEE Transactions on Intelligent Transportation Systems, 2018, 19(8): 2493-2504.

[4] Erbaş M, Kabak M, Özceylan E, et al. Optimal siting of electric vehicle charging stations: A GIS-based fuzzy Multi-Criteria Decision Analysis[J]. Energy, 2018, 163: 1017-31.

[5] Li Y, Luo J, Chow C Y, et al. Growing the charging station network for electric vehicles with trajectory data analytics[C]//2015 IEEE 31st International Conference on Data Engineering. IEEE, 2015: 1376-87.

[6] Morris B T, Trivedi M M. Trajectory learning for activity understanding: Unsupervised, multilevel, and long-term adaptive approach[J]. IEEE transactions on pattern analysis and machine intelligence, 2011, 33(11): 2287-2301.

[7] Ester M, Kriegel H P, Sander J, et al. A density-based algorithm for discovering clusters in large spatial databases with noise[C]//Kdd. 1996, 96(34): 226-31.

[8] Lingfeng K, Zifa L, Huan Z. Modeling algorithm of charging station planning for regional electric vehicle[J]. Modern Electric Power, 2010,

[9] Yan Z, Chakraborty D, Parent C, et al. SeMiTri: a framework for semantic annotation of heterogeneous trajectories[C], Proceedings of the 14th international conference on extending database technology. ACM, 2011: 259-70.

[10] Deng K, Xie K, Zheng K, et al. Trajectory indexing and retrieval[M]//Computing with spatial

trajectories. Springer, New York, NY, 2011: 35-60.

[11] Du J, Aultman-Hall L. Increasing the accuracy of trip rate information from passive multi-day GPS travel datasets: Automatic trip end identification issues[J]. Transportation Research Part A: Policy and Practice, 2007, 41(3): 220-32.

[12] Wolf J, Guensler R, Bachman W. Elimination of the travel diary: Experiment to derive trip purpose from global positioning system travel data[J]. Transportation Research Record: Journal of the Transportation Research Board, 2001 (1768): 125-34.

[13] Kong Y, Wu J, Xu M, et al. Charging pile siting recommendations via the fusion of points of interest and vehicle trajectories[J]. China Communications, 2017, 14(11): 29-38.

[14] Bentley J L. Multidimensional binary search trees used for associative searching[J]. Communications of the ACM, 1975, 18(9): 509-17.

IOP Conf. Series: Materials Science and Engineering **715** (2020) 012010 doi:10.1088/1757-899X/715/1/012010

Thermal design and analysis of payload cabin of solar UAV in near space

X Li[1,*], F Lu[2], F H Peng[1], W C Fu[1], F K Meng[1] and G L Man[1]

[1]Beijing Key Laboratory of Space Thermal Control Technology, Beijing Institute of Spacecraft System Engineering, 100094 Beijing, China
[2]China Academy of Aerospace Aerodynamic, 100074 Beijing, China

*E-mail: lix501rk@163.com

Abstract. Aiming at the temperature control demand of the near space solar unmanned aerial vehicle (UAV) payload cabin, this paper establishes the thermodynamic model of the payload cabin through thermal characteristic analysis, and proposes a system design method of structural thermal control integration. The thermal design of the payload cabin is carried out by means of thermal control such as uniform temperature, heat insulation, coating and active temperature control. The design weight of the whole payload cabin is only 7kg. The results of thermal analysis show that the temperature of the equipment in the cabin meets the requirement of the target during the whole UAV mission period, which verifies the correctness of the thermal design method. Compared with the traditional active temperature control scheme, the low temperature level increases by 3°C under the same heating power. The heat dissipation of the payload cabin is mainly convective heat transfer, but with the increase of altitude, the proportion of radiation heat dissipation is gradually increased. The research results can provide reference for the thermal design of similar aircraft.

1. Introduction

Near space refers to the space area with altitude between 20 km and 100 km. It is neither in the field of aviation nor in the field of space [1]. General Rod, commander of the U.S. Air Force Space Command, first proposed the concept of near space at the 20th Pacific Space Summit in Hawaii, and stressed that near space vehicles would have the characteristics of 'long-term, low-cost, high survival and rapid response'. With the continuous improvement of understanding of the near space field, the huge military and civil value of the near space has become increasingly prominent, and has become a space field that countries are scrambling to enter.

Due to the drastic changes in the near space environment conditions, the space thermal environment at different altitudes and in different regions has significant differences. Different from aviation and aerospace, the external environmental conditions to be considered include pressure, temperature, humidity, ozone, ultraviolet and space particle radiation, etc., which is far beyond people's imagination. Thermodynamic analysis and thermal control of aircraft interior and exterior are very important problems in the field of near space vehicle design. The thermal design of the loading cabin as the platform is the key to ensure the temperature of the equipment in the aircraft at a reasonable level.

At present, Japan, Korea, Germany and the United States are the countries that have studied the thermal analysis and thermal control of aircraft. Due to the technical blockade, there is no public report on the successful thermal design methods and cases of the payload cabin of near space vehicles. In China, from the published papers [2-4], it can be seen that the research of near space thermal control technology is still in its infancy, and many engineering application problems have not been solved. However, with the continuous development of the near space vehicle technology, the flight climbing height and flight time of the aircraft are required to continue to improve, which leads to the higher requirements of the near space vehicle thermal control technology. It is urgent to put forward and apply the targeted and practical thermal control scheme.

In this paper, the thermal design of the payload cabin of a solar UAV with altitude range of 0-30km is carried out to ensure the temperature of the equipment in the cabin. In view of the harsh environmental conditions and the restrictions of weight and energy, a thermal design scheme of the structure thermal control integrated payload cabin system is put forward by analysing the thermal characteristics of the UAV in the near space environment, and is verified by thermal analysis.

2. Thermal characteristic analysis of payload cabin

2.1. Mission characteristic

Solar UAV can be left vacant for tens of days or even months from takeoff to cruise to landing. There are 16 equipment in the UAV payload cabin, the total weight is 17kg, the total heat consumption is 270W, and the working temperature is between - 30 ~ + 55°C (the temperature of the equipment shell is taken when UAV is working), due to the limited resources, only two active temperature control loops are allocated to the thermal control system, totalling 100W.

The UAV is designed to fly at a maximum altitude of 27,000m. As shown in figure 1, typical mission profiles can be divided into three stages: first day climb, circular flight and landing. Gliding landing is mostly in the night, especially in the daytime. During normal flight, UAV's pitch angle is -5 ~8 degree, and the rolling range is small. The take-off time is 5~10 am and the climb / glide speed is 0~2m/s. The whole mission profile has a flying time of 30 days.

2.2. Physical model

The payload cabin is located in the center of UAV, and it is the core compartment of the whole UAV, which is equipped with flight control, navigation, measurement and control equipment. As an independent cabin, the entire payload cabin is located in the outer skin of the UAV; only the bottom surface is exposed as a heat dissipation surface, and contacts with the external atmospheric environment.

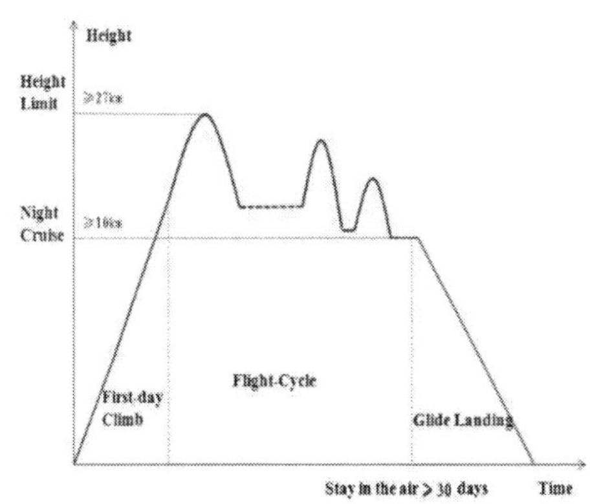

Figure 1. Typical mission profile of UAV.

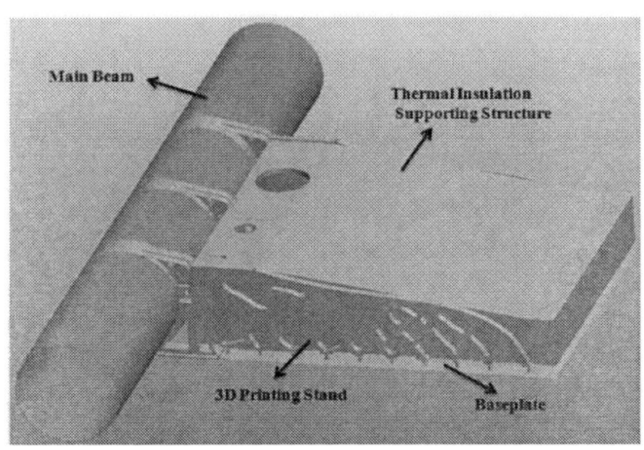

Figure 2. Structure of payload cabin.

As shown in figure 2, to maximize mechanical decoupling with the UAV, the entire payload cabin is connected only to the main girder of the UAV; to maintain good aerodynamic characteristics and appearance, the outer surface of the payload cabin bottom is in conformity with the local wing, and the inner surface needs to ensure the contact area between the equipment and the mounting surface due to the installation equipment and the contact heat exchange, so the inner surface is still level; because the solar unmanned aerial vehicle is extremely demanding on weight, in order to minimize weight loss, the bottom plate uses 0.3 mm thickness of aluminium skin aluminium honeycomb sandwich structure, the area is 0.79 m^2. The main bearing structural members on both sides of the load compartment are based on the additive manufacturing 3D printing independently developed by the mechanical division of the CAST. Dot matrix structure, each design weight is only 750g. The rest of the tie rods and brackets are made of ultra-light material magnesium-lithium alloy, which is nearly 500g less than the traditional aluminium alloy structural parts; the thermal insulation support structure is based on PMI foam board Hollow sandwich structure. The weight of the entire load compartment is estimated to be around 7kg.

2.3. *Analysis of thermal environment characteristics*
The adjacent stratosphere is 12~50km away from the ground. The flying height of the solar UAV is at the bottom of the stratosphere. According to the vertical distribution curve of the atmospheric properties of the 1976 version, as shown in figure 3, the atmospheric temperature, density and pressure vary with height. In the ground preparation stage before flight, it takes 1-2 hours, the extreme minimum temperature is -40°C, the extreme maximum temperature is 42.9 °C, and a suitable takeoff time should be selected according to the weather conditions. In the climbing stage, when the altitude is less than 11 km, the average temperature decreases about 6.5 °C/Km, 11-20 Km, the temperature is about -56.5 °C, the air speed increases with the altitude, and the air pressure increases. And the temperature decreases with the altitude rising; during the mission flight, according to the national military standard, the maximum temperature is -36.2°C, and the minimum temperature is -82.4°C.

2.4. *Thermodynamic model*
The thermal environment of solar UAV in near space is rather complicated. From take-off to mission, the external thermal environment is constantly changing, and the total influencing factors are the sun, the earth, cold space and the surrounding atmosphere. As shown in figure 4, the payload cabin includes direct solar radiation Q_{SUN} , ground (or cloud) reflected solar radiation $Q_{Reflect}$, infrared radiation Q_{E-air} between the earth surface (or cloud) and the payload cabin, infrared radiation Q_{Rad} from the payload cabin to the outside, convective heat transfer Q_{Conv} from the bottom of the payload cabin to the outside atmosphere, and natural heat transfer inside the payload cabin Q_{N-Conv} and heat source in the payload cabin, the heat consumption of the equipment $\overline{Q}$.

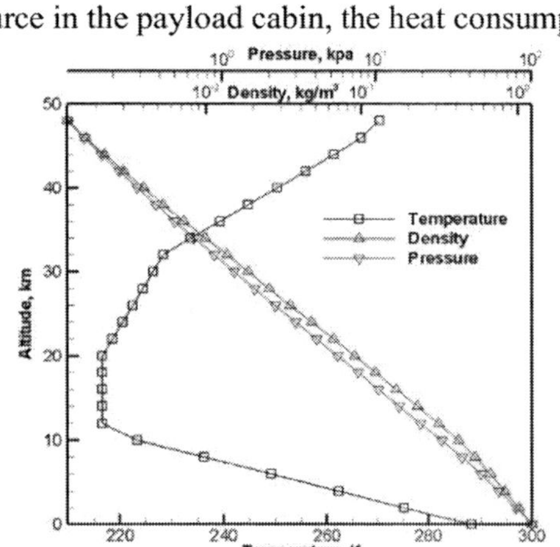

Figure 3. 1976 vertical distribution of atmospheric physical properties[5].

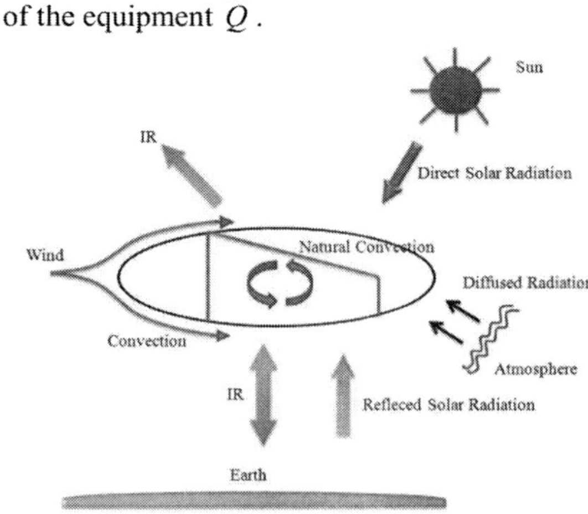

Figure 4. Schematic diagram of thermal environment for near space solar UAV.

For the entire UAV, the transient energy control equation[6] is:

$$\frac{\partial(\rho c_P T)}{\partial t}+div(\rho c_P \mathbf{u} T)=div(k \cdot gradT)+S_T \quad (1)$$

In the equation, ρ-density, T-temperature, t-time, k-thermal conductivity, div-divergence, $\mathbf{u}$ -velocity vector, c_P-constant pressure specific heat, $gradT$-temperature gradient, S_T-source term. The source term includes the sum of the factors affecting the thermal environment of the UAV described above.

$$S_T=\dot{q}_{SUN}+\dot{q}_{Reflect}+\dot{q}_{E-air}+\dot{q}_{Rad}+\dot{q}_{Conv}+\dot{q}_{N-Conv}+\dot{\bar{q}} \quad (2)$$

Assuming that the outer skin is isothermal, equation (1) can be simplified as:

$$\rho c_P \frac{\partial(T)}{\partial t}=\dot{q}_{SUN}+\dot{q}_{Reflect}+\dot{q}_{E-air}+\dot{q}_{Rad}+\dot{q}_{Conv}+\dot{q}_{N-Conv}+\dot{\bar{q}} \quad (3)$$

For the payload cabin, since the bottom only faces the atmosphere, and which is facing the ground and is not directly exposed to the sun, the steady-state energy balance equation of the payload cabin is:

$$Q_{Reflect}+Q_{E-air}+\bar{Q}=Q_{Rad}+Q_{Conv}+Q_{N-Conv} \quad (4)$$

Among them, the atmosphere albedo heat flux of the earth[7] is:

$$Q_{Reflect}=\alpha AS \quad (5)$$

A is the area of the bottom plate, S is the intensity of solar radiation, and α is the albedo. Generally, it takes α=0.30~0.35, which is closely related to the surface state, latitude and even time of the earth. The corresponding albedo at different latitudes of the earth is zonal distribution[8].

Infrared radiation heat flux of the earth is:

$$Q_{E-air}=\varepsilon_E \sigma A T_E^4 \quad (6)$$

Among them, the earth's surface temperature is related to latitude, geographical location, weather and other conditions. Generally, it decreases with the increase of latitude. According to the season statistics, the average earth surface temperature is approximately 288K.

The infrared radiation heat flux from the payload cabin to the outside is:

$$Q_{Rad}=\varepsilon_{ext}\sigma A(T_\infty^4-T_w^4) \quad (7)$$

T_w—Outside surface temperature of payload cabin bottom.

T_∞—External radiation source temperature, take T_∞ as the effective temperature of the sky.

Sky effective temperature[9] is a function of atmospheric water vapor content, cloud amount (or sunshine percentage), atmospheric temperature and surface temperature defined by absolute blackbody radiation flux according to Stephen-Boltzmann law. Because these-meteorological factors have obvious regional and temporal variations, the sky effective temperature is also a function of geographical location (space) and seasonal (time). Under the clear sky condition, the Brunt equation is used to estimate.

$$T_\infty=(0.51+0.208\sqrt{e_a})^{0.25}\times T_a \quad (8)$$

e_a—Partial pressure of water vapor in the atmosphere.

In the stratosphere of 20 km, there is very little water vapor in the atmosphere, and the partial pressure of water vapor is considered to be zero, thus the effective temperature of the sky can be expressed as $0.51^{0.25}T_a$.

For convective heat transfer:

$$Q=hA(T_w-T_f) \quad (9)$$

Available from $Nu=\frac{hl}{\lambda}$:

$$h = Nu\frac{\lambda}{l} \tag{10}$$

The convective heat transfer between the bottom plate of the payload cabin and the outer atmosphere is correlated by the mean surface heat transfer coefficient of the external forced convection tube.

$$Nu_F = 0.0266\,\mathrm{Re}^{0.08}\,\mathrm{Pr}^{1/3} \tag{11}$$

Experimental correlation of natural convection heat transfer in large space is used in natural convection inside the payload cabin:

$$Nu_N = 0.11(\mathrm{Gr}\cdot \mathrm{Pr})^{1/3} \tag{12}$$

2.5. Brief summary

From the above analysis, we can see that:

(1)The thermal environment of the solar UAV is similar to that of the satellite. We can draw lessons from the mature thermal design method of the satellite, such as temperature equalization design of heat pipe and active temperature control.

(2)Compared with the satellite, due to the influence of rarefied air, the thermal environment in the near space is more complex and changes more dramatically. Natural convection and forced convection heat transfer exist in the payload cabin under sealed or unsealed conditions, so the thermal design needs to consider the corresponding insulation requirements.

(3)Due to the UAV cruising day and night, the flight altitude is constantly changing, so it is necessary to analyse the influence of convective heat transfer and radiative heat transfer at different altitudes on the UAV, and to reasonably design the heat dissipation surface and heating power to adapt to the influence of thermal environment at different heights.

Thermal design should take into account the above factors, so as to put forward a reasonable and systematic thermal design scheme.

3. Thermal design

3.1. Temperature equalization design

A total of 16 equipments in the payload cabin, the maximum heat consumption is 120W, and the minimum heat consumption is 1W. In order to reduce the temperature non-uniformity between the equipment in the payload cabin, two I-shaped single-hole aluminum-ammonia channel heat pipes with a cross-section of 30×11.4 were semi-embedded in the honeycomb panel, as shown in figure 5, has a length of 1.3m and 1.4m respectively, and is designed as a U-shaped structure. The thickness of the honeycomb panel is 25.6mm, and the upper surface of the heat pipe is attached to the upper surface of the honeycomb skin. To reduce weight, the fins on the lower surface of the heat pipe are cut off. The temperature difference between the devices is reduced by the uniform temperature of the heat pipes and reasonable layout of the devices.

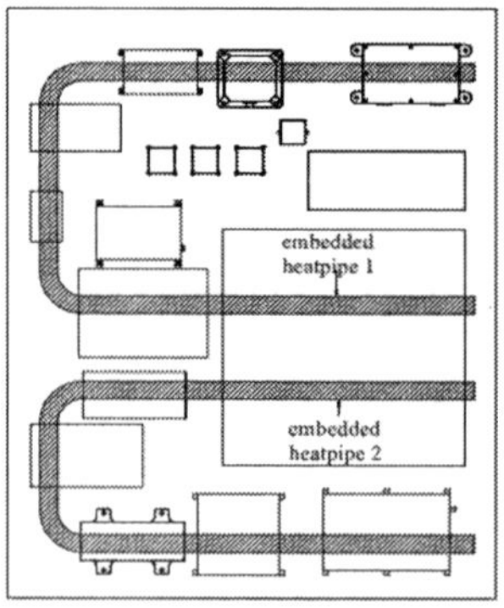

Figure 5. Schematic diagram of thermal design of heat sink.

3.2. Heat insulation design

According to the standard atmospheric physical properties, the lowest temperature of the UAV cruising altitude in the vicinity of 15-27 km is - 56.5°C. According to the national military marking, the extremely low temperature can reach - 82.4, far exceeding the operating temperature limit of the equipment. Therefore, insulation design is required. Due to the existence of thin gas in the adjacent space, this paper proposes a hollow sandwich insulation structure based on PMI foam board. As shown in figure 6, the selected PMI foam density is $32kg/m^3$, corresponding to the environment of -100~80°C, the thermal conductivity is 0.02~0.04W/m·K[10], which increases with the increase of temperature.

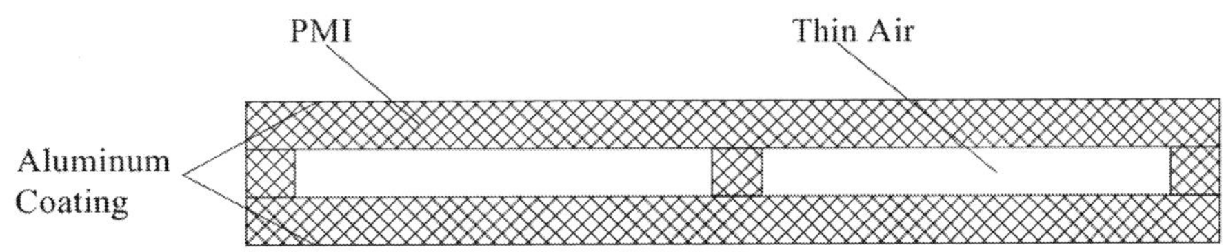

Figure 6. Hollow sandwich insulation structure based on PMI foam board.

The PMI insulation structure serves as both a support structure and an insulation material to keep the entire payload cabin warm. The single-layer PMI has a thickness of 5 mm and the heat-insulating structure has a total thickness of 15 mm. Aluminized film is adhered to both inner and outer surfaces to further reduce heat transfer of radiation between the bulkhead and the external environment as well as the equipment inside the cabin.

3.3. Heat dissipation design

According to the evaluation of the thermal characteristics of the payload cabin, this paper only designs passive heat dissipation to meet the heat dissipation requirements of the payload cabin. The honeycomb panel is exposed in the atmosphere, and is the main channel of heat dissipation through convection and radiation. As can be seen from figure 4, when the UAV flying altitude exceeds 20 km, on the one hand, the ambient temperature begins to rise continuously, but the air density and pressure are still decreasing. From equation (9) ~ (12), the convective heat transfer is decreasing continuously. When the flying altitude is 27 km, it is known from equation (4) that, without considering radiation heat dissipation, convective heat transfer alone cannot meet the requirements of heat dissipation; if radiation heat dissipation is considered, but the aluminum alloy material which keeps the original surface of honeycomb panel, its emissivity is about 0.1, and the calculation cannot meet the load cabin dispersion. Therefore, it is designed to spray a certain area of low-absorption and high-emission thermal control coating SR107 white paint on the outer surface of the honeycomb panel with a solar absorption ratio of 0.2 and an emissivity of 0.88.

According to the calculation boundary conditions of external heat flow under extreme high temperature conditions given by CSBF[11], equation (5) ~ (8) is calculated according to the parameters of high temperature conditions corresponding to UAV flying to 27 km and 40 degrees north latitude. Assuming that the temperature of payload cabin bottom plate is 40°C, the area of white paint, which can be calculated by equation (4), is about 0.325 m^2, and the maximum external heat flux at the bottom plate is 168.4W.

3.4. Active temperature control design

Based on the limitation of temperature control loop and power, taking into account the bottom honeycomb panel as the main heat dissipation channel, which is the cause of strong convection heat transfer, in order to improve the ability of active temperature control and reduce the heat leakage during heat transfer, instead of the traditional method of pasting heating sheet on the heat pipe on the satellite, the heating sheet is directly pasted on the equipment. The power of the heater applied to each

IOP Conf. Series: Materials Science and Engineering **715** (2020) 012010 doi:10.1088/1757-899X/715/1/012010

device is determined based on the results of the thermal analysis. Each heater is 50W, a total of 2 channels, and the temperature control threshold is [-20,0] °C.

4. Thermal analysis verification

Through the analysis of the external environment, the high and low temperature working conditions are established. The thermal analysis modeling calculation is carried out by using NX 8.5 software. Three steady-state working conditions are calculated. The detailed analysis results are shown in table 1.

Among them, the low temperature working condition is used to evaluate the temperature control ability of the heater under the extremely low temperature in the adjacent space, and the standard condition is used to verify the temperature level of the equipment in the cabin if the heater is not used, so as to judge whether the design can achieve the purpose of saving energy.

Table 1. Thermal Analysis Calculation Conditions.

Condition	Height (m)	Flight speed (m/s)	Ambient temperature (K)	Explain
High temperature	27000	46	236.8	Considering the external heat flux of the sun and the earth's infrared, the heater closes
Low temperature	18000	24.3	190.8	No solar outside heat flux, heaters on
Standard working	18000	24.3	216.5	No solar outside heat flux, heaters off

The calculated temperature results of each equipment in the payload cabin are shown in table 2, all within the operating temperature range. Figure 7 shows the temperature cloud chart of the payload cabin under different conditions.

Table 2. Temperature level of equipments.

Condition	High temperature	Low temperature	Standard working
Temperature level /°C	37.3~50.0	-19.5~-5.5	-21.1~0.0

From the calculation results, it can be concluded that: under high temperature conditions, the area of the heat dissipation surface is suitable to meet the heat dissipation demand; under low temperature conditions, the power of the heater is enough to ensure that the equipment does not exceed the lower limit of the working temperature; under standard operating conditions, the heater can be closed to save UAV energy.

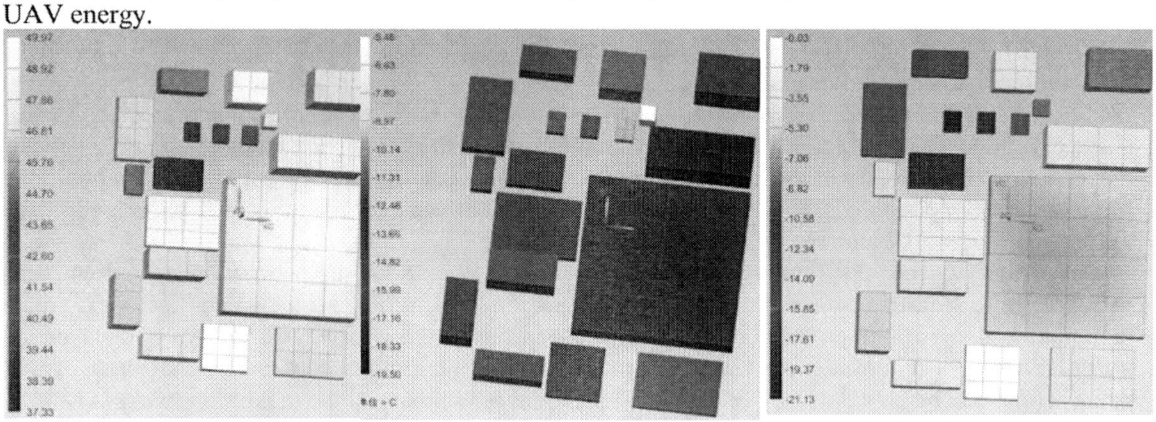

High temperature condition. Low temperature condition. Standard working condition.

Figure 7. Temperature map of equipment under various working conditions.

4.1. Comparison of temperature levels in low temperature working conditions under different active temperature control modes

For the active temperature control design, compared with the traditional scheme of attaching the heating sheet to the heat pipe. As shown in figure 8, the temperature level of the device is -22.4~-

10.5°C under low temperature conditions with the heaters long-passed, the minimum temperature is reduced by 3°C compared with the heating plate directly pasted on the equipments, and the overall temperature level of the equipments is much lower than the direct attachment of the heating sheet to the equipments.

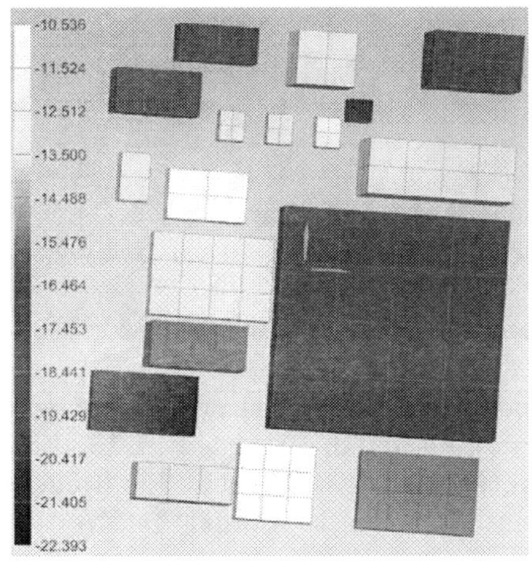

Figure 8. Temperature chart of low temperature condition (heat sheet pasted on heat pipe).

4.2. Analysis of the effects of radiative heat dissipation and convection heat dissipation at different heights

According to the results of the thermal analysis, the radiative heat dissipation and the convection heat dissipation at the heights of 18 km and 27 km are respectively counted as shown in table 3. Combined with the thermodynamic model of the UAV payload cabin in section 2.4, the convective heat transfer accounted for a large proportion, but with the increase of the height, the convective heat transfer was significantly weakened, and the proportion of radiative heat dissipation increased gradually. Therefore, in thermal design, it is necessary to combine the specific level of heat consumption, thermal characteristics and flight altitude and other factors to determine whether it is necessary to strengthen radiation and cooling surface size.

Table 3. The proportion of radiation and convective heat dissipation at different heights.

Height(km)	Q_{Radi}(%)	Q_{Conv} (%)
18	15.3	84.7
27	40.4	59.6

5. Conclusion

Aiming at the thermal control requirements of the payload cabin of the near space solar UAV, this paper analyzes the thermal environment of the payload cabin in the near space, establishes the thermodynamic model, puts forward a thermal design method of the integrated structural thermal control system which is easy to be realized in engineering, and determines the area of the heat dissipation surface. The thermal analysis proves that the temperature of the equipment meets the requirements of the index, which shows that the thermal design method is reasonable and feasible. The main conclusions are as follows:

(1)The payload cabin of the structure integrated thermal control design, weighing only 7kg, meets the requirements of thermal control performance and saves weight on the premise of meeting the structural strength.

(2)Compared with the traditional heating sheet sticking on the heat pipe, the active temperature control method with direct heating equipment can raise the low temperature level by 3 °C at the same heating power, and the temperature control effect is better,and is more suitable for the thermal environment in the near space.

(3)The heat dissipation of the payload cabin is mainly convective heat transfer, but as the altitude increases, the proportion of radiation heat dissipation increases. For specific thermal design, the size of the heat dissipation surface should be determined according to the heat consumption level and the flying height.

The thermal design method proposed in this paper has certain engineering application value and can be used as a reference for similar aircraft thermal design.

References

[1] Xia Chun-Li. On the legal status of the near space: The demarcation of air space and outer space[J]. Journal of Beijing University of Aeronautics and Astronautics (Social Sciences Edition), 2009, 22(4):47-51.
[2] Deng Li-jun. Research on a thermal control system for near space air conditioner[D]. Nanjing University of Sciences & Technology, 2009.
[3] MA Wei,XUAN Yi-min,HAN Yu-ge. Thermal Management and Design Principles of Stratospheric Vehicles[J]. Journal of Astronautics, 2009(5).
[4] MA Wei,XUAN Yi-min,HAN Yu-ge. Thermal Environment Analysis and Thermal Design Principle of Stratospheric Vehicles[J]. Journal of Astronautics, 2010(5).
[5] U.S. Standard Atmosphere, 1976, U.S. Government Printing Office, Washington D. C., 1976.
[6] Tao Wen-quan. Numerical heat transfer[M]. Xian: Xian Jiaotong University Press, 2001.
[7] Hou Zengqi, Hu Jingang. Foundation and application of spacecraft thermal control technology[M].Beijing:China Science and Technology Press,2007:273(in Chinese).
[8] Kang Cao , John Baker. Application of Transient Network Models for Near Space Thermal Management[C]. 46th AIAA Aerospace Sciences Meeting and Exhibit 7 - 10 January 2008, Reno, Nevada.
[9] LV Chen, JIANG Lu-hua,CAI Jing-jing.The Innuence of Thin Film Solar Cells on the Temperature of Helium Inside the Airship[J]. Computer Simulation,2016, 33(7):104~110.
[10] Yang Ruping, Ma Xiaojing, You Jie, et al. Development of usage for polymethacrylimide(PMI)foam in thermal insulation system of reusable launching vehicle cryogenic tank[J]. Cryogenics, 2010,1:54~59.
[11] Brian O'Connor. Thermal Control of the Balloon-Borne HEROES Telescope[R]. NASA, 2014.

Numerical simulation of flow control in lift-increase effect using air-blowing

Xu Kao, Zhu Xiaojun*, Zhu Jianghong and Li Feng

China Academy of Aerospace Aerodynamics, Beijing, 100074, China

*E-mail: zxj011010826@163.com

Abstract. In this study, numerical simulation method is applied to analyze the aerodynamic characteristics of the airfoil influenced by the parameters of the jet port. The NACA0012 airfoil is studied by using the incompressible Navier-Stokes equations with the structural grid. The influence of air-blowing speed, width, position of the jet and angle of attack on the aerodynamic characteristics of the airfoil is presented. The results show that, for the airfoil, the lift coefficient can be impacted by blowing speed, width and position of the jet to a certain extent. Jet control can reduce the stall angle of attack of the airfoil, but before the stall angle of attack, the effect of lift-increase is almost same at different angle of attack. A basis for the jet design and the selection of jet position along the wing is provide.

1. Introduction

In terms of reducing drag and increasing lift, the application of flow control have attracted the attention of a large number of scholars [1]. Jet blowing vertically or aslant from the lower surface of the wing, can change the pressure distribution of wing, increase the pressure difference of upper and lower surfaces, resulting in the increase of lift. At present, experiment and numerical simulation are two main analysis methods for research of jet control [2]. To accurately simulate the effect of jet control on lift, the first thing is to be able to simulate the flow field with a jet control accurately and meticulously [3]. As the flow field with a jet control is quiet complicated and disturbing, the numerical simulation is confronted with many problems. Most of the research are based on wind tunnel experiment so far. With the fast development of the computer technology, the numerical simulation of the flow field around airfoil with high angle of attack become possible [4]. Based on this issue, numerical simulation can be conducted on the research of jet effect on the wing, and the mechanism of the lift-increase.

This kind of jet control has been applied in AV-8 Harrier aircraft preliminarily. It can help aircraft achieve short-range vertical takeoff and landing, so the length of track is reduced. There are two factors influencing the application of jet control method so far. The first is to get the steady gas source. The second is that full wingspan jet can bring big a significant impact on the safety of the wing structure [5]. The first question can be solved by entraining gas from the engine. But the second question haven't been solved well yet. Based on the above, only a part of wingspan jet control have been put forward in this paper. Local jet control can avoid problems of structural safety caused by the full wingspan jet control. What's more, the best location of the jet port on the wingspan can be found and the effects of different jet velocity on the aerodynamic characteristics of the wing can be studied through the simulation of the flow field with jet control.

IOP Conf. Series: Materials Science and Engineering **715** (2020) 012011 doi:10.1088/1757-899X/715/1/012011

2. Jet control numerical simulation method

2.1. Control equation and turbulent model

In this paper, the impressible RANS Naveir-Stokes equation is adopted as control equation to the flow field numerical simulation [6], it can be expressed as

$$\text{Continuous equation } \frac{\partial u_i}{\partial x_i} = 0 \tag{1}$$

$$\text{Momentum equation } \frac{\partial u_i}{\partial t} + u_j \frac{\partial u_i}{\partial x_j} = -\frac{1}{\rho}\frac{\partial p}{\partial x_i} + \upsilon \frac{\partial^2 u_i}{\partial x_j \partial x_i} + \frac{\partial \tau_{ij}}{\partial x_j} \tag{2}$$

The finite volume method is used to carry out the space discretization, and the convection terms adopt two order upwind schemes, diffusion terms use central difference scheme. Pressure based simple algorithm is used for pressure-velocity coupling, the discrete algebraic equations are solved by Gauss-Seidel algebraic method [7]. Spalart-Allmaras (SA) turbulent model is used as turbulent model. SA turbulent model is a relatively simple one equation model which solved a transport equation for the eddy viscosity. This model is more suitable for the flow problem with a wall limit [8]. It can give a good result to the boundary layer problem with the inverse pressure gradient. It often used in aerodynamic problems, such as the flow field analysis to aircraft, airfoil and so on.

2.2. Boundary conditions

Airfoil and wing surface adopt wall boundary condition, meet the conditions of no-slip, insulation, surface normal gradient is zero. Jet port adopt the mass flow inlet as the boundary condition. Mass flow is defined as, $\dot{m} = \rho s v$. Among them ρ is the density of the fluid of jet, which is equal to the air flow, s is the area of the jet, v is the velocity of the jet. For wing, the plane of symmetry adopt symmetric boundary condition. In addition to airfoil surface, symmetric plane and jet port, the other boundaries of flow field all be set as pressure far field.

2.3. Physical model and grid division

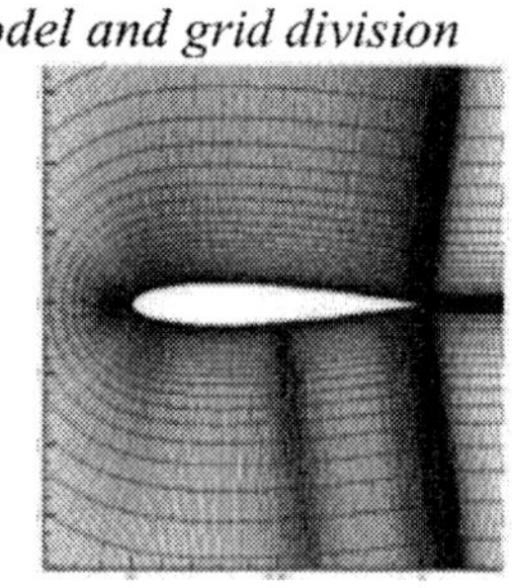

Figure 1. Local refinement of 2D grids.

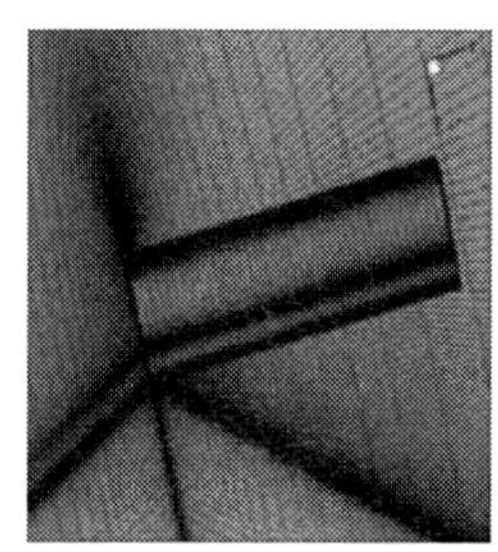

Figure 2. 3D mesh model.

Two-dimensional model adopts the NACA0012 airfoil. The chord length of the airfoil is one meter. Computational grid adopt the C-H type grid, which is shown in figure 1 Around the boundaries of the computational domain have a distance of 20 times the chord length from the airfoil surface, airfoil downstream direction is taken as 20 times the chord length also. The total number of the grid is 43586 after the encryption of jet. The height of the most inner mesh is 10^{-5} m, $y^+ \approx 1$.

Three-dimensional model adopts finite wingspan model based on the NACA0012 airfoil. As is shown in the figure 2, wing root tangential section(z=0) is the symmetric plane, the reference area of the semispan is 3 square meters. Half length of the computational domain is 12 meters. So the wing tip(z=3m) is in the flow field, the effect of the wing tip vortex is taken in to the consideration.

3. 2D model numerical simulation result and analysis

3.1. Numerical simulation algorithm validation

When $Ma_\infty = 0.2, \mathrm{Re} = 4.6\times10^6$ and there is no jet control on NACA0012 airfoil, the lift coefficient(Cl) change for variation of angle of attack gotten by numerical simulation and experiment is shown in the figure 3. From the picture, it shows that before stall angle of attack, the results of numerical simulation are in agreement with the experimental results. The maximum deviation between two methods is 3.25%. Another very important parameter to validate the accuracy is the pressure distribution. When $Ma_\infty = 0.2, \mathrm{Re} = 4.6\times10^6$, and the angle of attack $\alpha = 10^0$, the pressure coefficient distribution got by numerical simulation and experiment is shown is figure 4. From the picture, it shows that the pressure coefficient of numerical simulation is also in agreement with the experimental results. Thus conclusion can be drew that this numerical simulation algorithm meets to the requirement of engineering application.

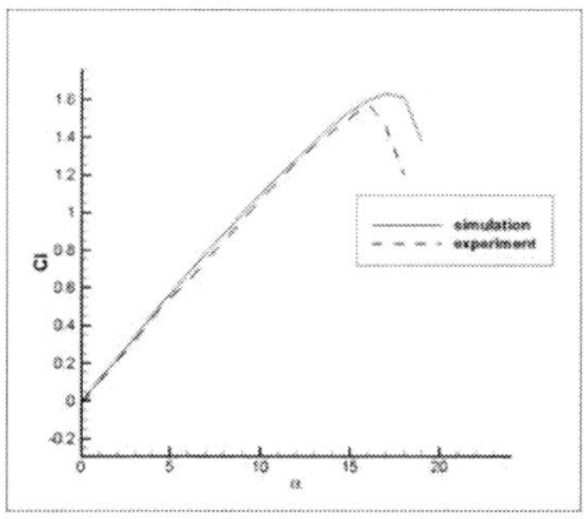

Figure 3. The comparison of lift coefficients.

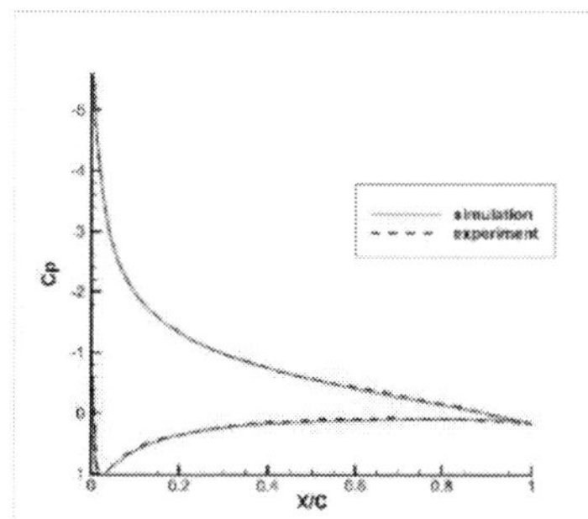

Figure 4. The comparison of pressure coefficients.

3.2. The mechanism of jet's effect on lift-increase

The following study compares the airfoil pressure distribution between whether have jet or not. The basic airfoil is NACA0012. The angle of attack $\alpha = 0^0$, Mach number $Ma_\infty = 0.2$. Reynolds $\mathrm{Re} = 4.6\times10^6$. The location of jet $\overline{x}_j = x_j / c = 0.5$, The width of jet $\Delta\overline{x}_j = 0.04$. The jet vertically blowing from the lower surface of the airfoil, the velocity of jet $\overline{v}_j = v_j / v_\infty = 1$, the pressure around jet is $\overline{p}_j = p_j / p_\infty = 1$. Figure 5 shows the pressure contours plot around the airfoil with jet or not. Figure 6 shows the pressure coefficient distribution around the airfoil with jet or not. From the picture, it shows that there is a positive value of the integral of the pressure coefficient of the airfoil surface when there is a jet, and the pressure of the lower surface of the airfoil is bigger than the upper surface. Figure 7 shows the velocity vector diagram near the jet port and the streamline of the flow near the airfoil. From the picture, a separation zone is formed in front of the jet port and behind the jet port respectively can be seen by us. The separation zone in front of the jet is very small while the separation zone behind the jet is relatively large.

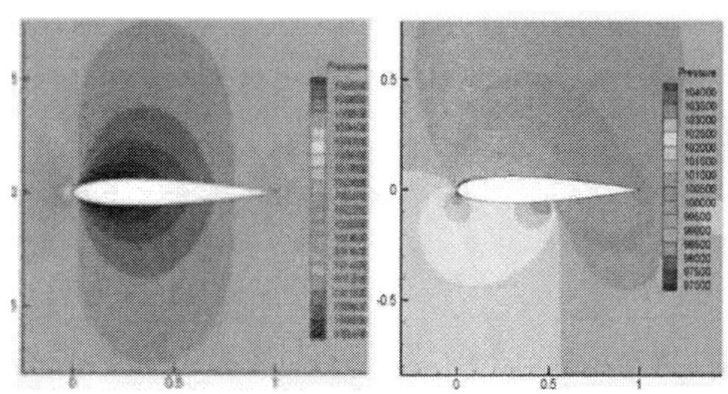

Figure 5. The pressure contours plot around the airfoil with jet or not.

IOP Conf. Series: Materials Science and Engineering **715** (2020) 012011 doi:10.1088/1757-899X/715/1/012011

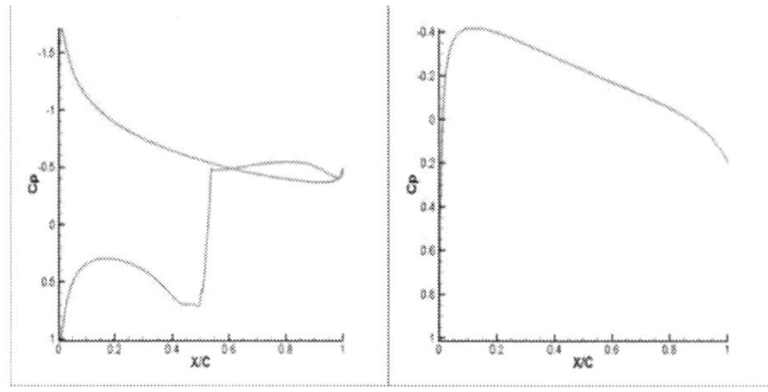

Figure 6. The pressure coefficients distribution of the airfoil.

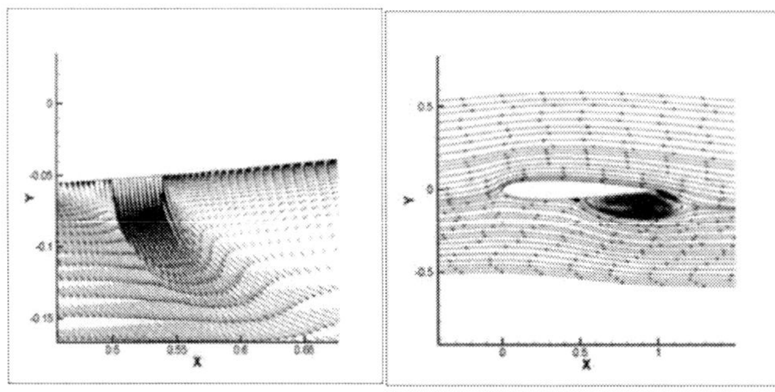

Figure 7. The velocity vector diagram near the jet port and the streamline of the flow near the airfoil.

3.3. Influence of jet velocity in the flow field

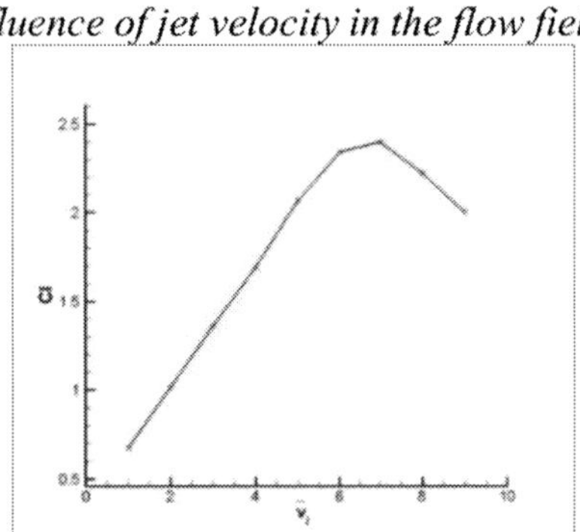

Figure 8. The effect of jet velocity on the Cl.

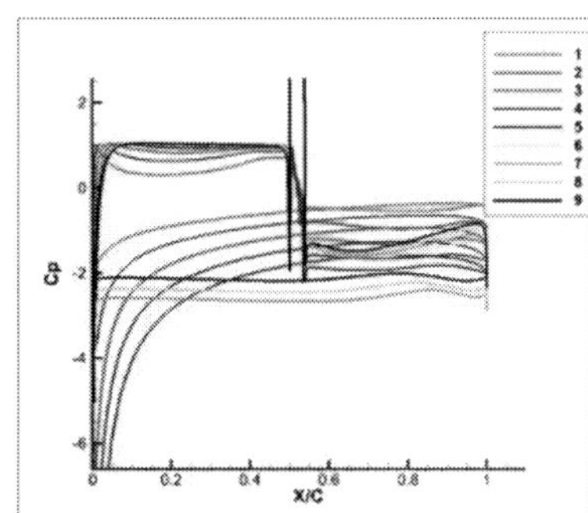

Figure 9. The pressure coefficients distribution of different jet velocity.

Based on the calculation conditions above, the following research only change the magnitude of $\bar{v}_j$. Figure 8 shows the effect of jet velocity on the Cl. Figure 9 shows the pressure coefficients distribution of different jet velocity. As is shown in the picture, when $\bar{v}_j$ is less than a fixed value, the lift increases with the raising of jet velocity. But beyond this value the lift decreases with the raising of jet velocity. The reason is that when $\bar{v}_j$ <7, as the jet velocity raising, the pressure differences before and behind the jet port are all climbing up, the pressure difference on the jet port only reduce a little; when $\bar{v}_j$ >7, the pressure difference on the jet port changes drastically and drops a little before and behind the jet port, which causes the lift coefficient to reduce.

3.4. Influence of jet location in the flow field

Similarly, the following research only change the chordwise location of the jet. All other parameters are equal to case B. The effect of jet location $\bar{x}_j = 0.5, 0.6, 0.7, 0.8$ on the Cl is shown in figure 10. The

Cp difference between different jet location is shown in figure 11.

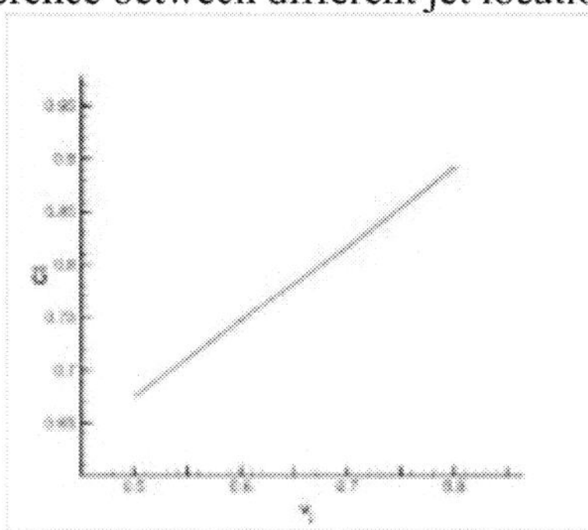

Figure 10. The effect of jet velocity on the Cl.

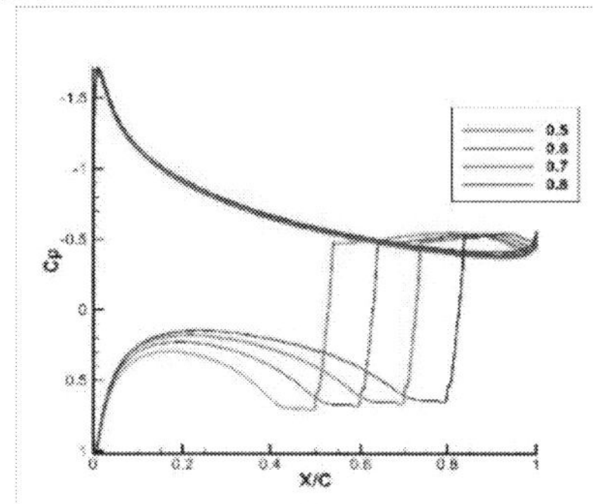

Figure 11. The pressure coefficients distribution of different jet location.

From the picture, it shows that due to the influence of the jet, a large positive pressure is generated in front of the jet, while the negative pressure is formed after the jet, and the lift increases with the jet moving near to the trailing edge. The reason is that while the jet moving near to the trailing edge, the pressure value of the lower surface of the airfoil is more positive, resulting in the increase of the lift.

3.5. Influence of jet width in the flow field

The location of jet is fixed in $\bar{x}_j = 0.5$. Only change the width of jet, and all other parameters are the same with case B. The lift coefficient and pressure coefficient difference between different jet width $\Delta\bar{x}_j = 0.005, 0.01, 0.02, 0.03, 0.04, 0.05$ are shown in figure 12 and figure 13.

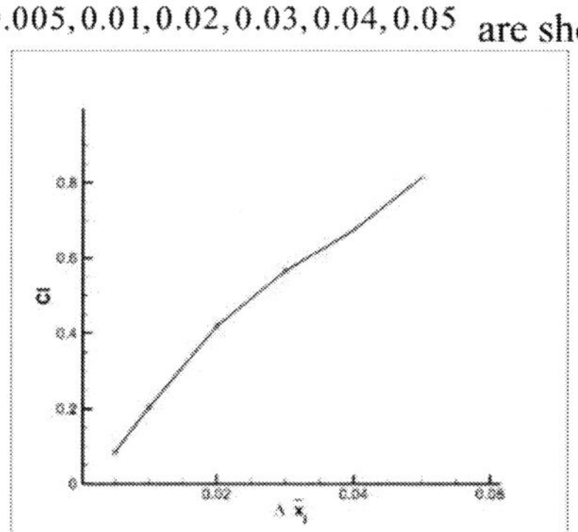

Figure 12. The effect of jet width on the Cl.

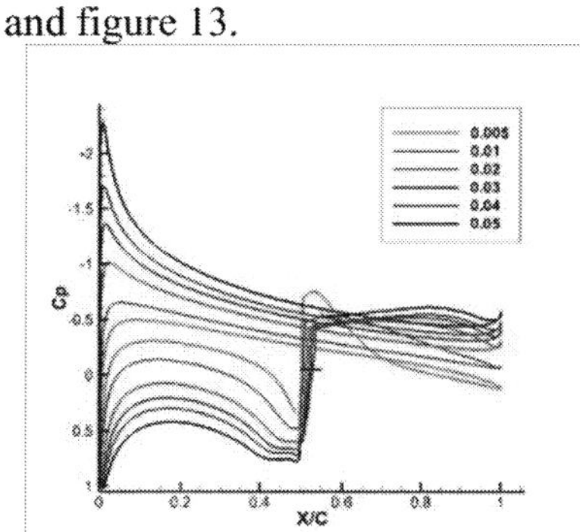

Figure 13. The pressure coefficients distribution of different jet width.

As is shown in the picture, when the width of jet is in a certain range, the lift coefficient increases with the increase of the jet width. The reason can be said in two aspects. With the increase of jet with, on the one hand, the mass flow is increased also, so the reaction of the gas to airfoil is strengthened; one the other hand, the positive pressure of the lower surface before jet port is increased, while the negative pressure of the upper surface before jet port is increased also, so the pressure difference between lower surface and upper surface before jet port is increased, while the pressure difference between lower surface and upper surface behind jet port only change a little, resulting in the increase of the lift.

3.6. Jet effect on the lift at different angle of attack

Based on case B, the following research merely change the airfoil angle of attack. With the change of the angle of attack, the lift coefficient differences between using jet control or not are shown in figure 14. As is shown is the picture, the stall angle of attack is obviously reduced, but before stall angle of attack, jet control can make the lift coefficient increase at any angle of attack, and the effect of lift-increase is almost same. In comparison, the maximum lift coefficient can increase by 0.411 approximately. Conclusion can be drew that the effect of lift-increase is very obvious when the angle of attack is not very big.

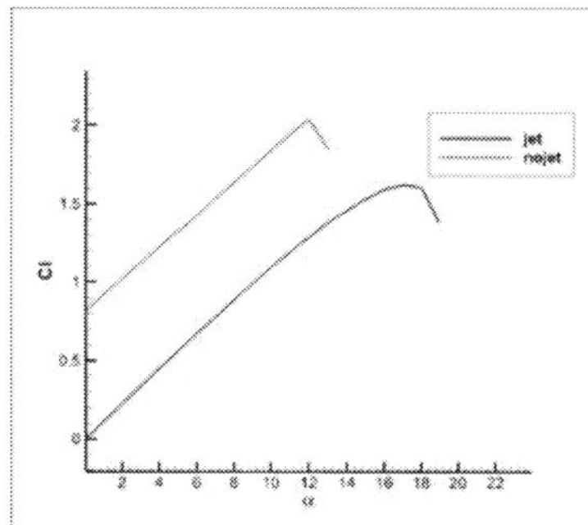

Figure 14. The lift coefficients of different angle of attack.

4. Conclusion

Based on the NACA0012 airfoil, a research is conducted on jet control on the lower surface of the airfoil and finite span model in order to increase the lift. The main conclusions are as follows:

(1) With appropriate settings, the aerodynamic characteristics of the airfoil can be accurately simulated. The method is satisfied with the requirement of engineering application;

(2) Jet control on the lower surface of NACA0012 airfoil can increase the lift coefficient of the airfoil by changing the airfoil surface pressure distribution. The effect of jet control have a certain relationship with the jet velocity, jet width and jet location, and the stall angle of attack is reduced, but before the stall angle of attack, the effect of lift-increase is almost the same at different angle of attack, while the maximum lift coefficient is increased by 0.411.

References

[1] Tong Zengjun, Liu Peiqing, Duan Huishen. Numerical simulation of the flap blowing and suction control for two dimension multi-element airfoil[J].Aircraft Design, 2010, 30(4):10-15.

[2] Zhao Wanli, Liu Peiqing, Zhu Jianyong. Numerical simulation of the flow control using gurney flaps[J]. Power system and clean energy, 2011, 27(9).

[3] Li Feng, Deng Ningfeng, Zhouweijiang, Ma Handong. The Numerical simulation of lift-increase effect on an airfoil with a jet from its surface.[J]. Acta Aeronautica et Astronautica Sinica.1994, 15(5).

[4] Guo Tongqing. Transonic unsteady aerodynamics and flutter computations for complex assemblies[D].Nanjing: Nanjing University of Aeronautics and Astronautics, 2008, 48-59.

[5] Li Dawei, Qiu Jinhao, Ji Hongli, Nie Rui. Numerical simulation of blowing control based on limited wingspan model [J]. Chinese journal of applied mechanics, 2014, 31(4).

[6] Di Yachao, Gao Ge, Dong He, Xu Jinglei, Tang Yangyang, Xu Huan. Numerical investigation of impressible turbulent flows past airfoils[J].Journal of Aerospace power,2013, 28(12).

[7] Krathapalli A, Leopold D. An experiment investigation of flow around an airfoil with a jet exhausting from lower surface.[J] NASA-166131, 1981.

[8] Zhu Yixi, Lu Zhiliang, Guo Tongqing. Numerical simulation of multi-element airfoil in unsteady ground effect[J].Acta Aeronautica et Astronautica Sinica, 2014.

IOP Conf. Series: Materials Science and Engineering **715** (2020) 012012 doi:10.1088/1757-899X/715/1/012012

Numerical simulation of ultrasonic testing reliability of civil aircraft considering the influence of the angle between the sound beam axis and the crack orientation

PeiQiang Tian[1,*], YunWen Feng[1], ShaoZheng Zhan[2], XueLiang Zhang[1] and XiaoFeng Xue[1]

[1]Northwestern Polytechnical University, Xi'an710072, China
[2]Aviation Industry Aircraft Strength Research Institute, Xi'an710072, China

*E-mail: peiqiangtian@163.com

Abstract. In order to quantify the influence of the angle between the axis of the sound beam and the crack orientation on the detection reliability during the ultrasonic testing, this paper uses COMSOL Multiphysics to simulate the numerical simulation of ultrasonic testing and explores the variation of the ultrasonic detection amplitude with the angle. The function of fitting the function is given, and then the numerical simulation method of ultrasonic detection reliability considering the influence of the angle between the beam axis and the crack orientation is proposed. According to the numerical simulation case of ultrasonic detection reliability, the following conclusions are obtained. When the detecting device detects a crack of less than 2.23 mm and the detection probability is greater than 90%, the angle between the axis of the sound beam emitted by the detecting device and the direction of the crack must be greater than 70°; If the detecting device detects a crack greater than 0.7 mm and the angle is greater than 86°, the probability of detecting the crack is greater than 61.35%.

1. Introduction

Ultrasonic testing is one of the important techniques in modern industrial in-situ non-destructive testing. It has a series of advantages such as high sensitivity, strong penetrability, good direction, fast detection speed, low cost, relatively simple equipment and harmless to the human body [1]. In the non-destructive testing process, not all defects existing in the object to be inspected can be accurately found in all cases. There are many factors affecting the reliability of the non-destructive testing technology, such as the performance of the equipment, the material of the test block, the internal structure, and the levels of inspectors [2]. These uncertainties have been dominated. For a detection system, what are the main influencing factors? What are the secondary factors? How to correct and compensate for the uncertainty caused by these factors ? These problems should be studied and quantified one by one so that the test results meet the specified requirements. It then provides an effective basis for product designed, material selection and quality assessment.

At present, the reliability of non-destructive testing is the basis of non-destructive testing technology, and it is the focus of research and technical research by domestic and foreign scholars. The breakthrough of this technology will make the application prospect of non-destructive testing technology more broad. In order to quantify the ability of the detection system to detect cracks and determine the applicability of the detection scheme, "the probability of detection (POD)" can be introduced to characterize the reliability of ultrasonic detection [3], and the POD curve is used as a criterion for quantifying the capability evaluation of the non-destructive testing system.

2. Research content

Practical engineering experience is shown below. When the angle between the axis of the sound beam and the orientation of the crack (hereinafter referred to as "the angle") is 90°, the crack detection amplitude is often also the largest. When the angle is greater than or less than 90°, the crack detection amplitude will decrease. As the angle is furthering away from 90°, the crack detection amplitude will be smaller. When the angle is too large or too small, the probe may not receive crack echoes, making the crack undetectable. In this paper, the angle between the beam axis and the crack orientation is taken as the research object. By exploring the variation law of the angle and the detection amplitude, the function relationship is fitted and the variation amplitude of the detection amplitude under different angles is quantified. Applying the above results to the calculation of the POD curve, the POD curves at different angles and the crack size with a detection probability of 90% can be obtained. This provides a scientific approach to the research and evaluation of ultrasonic testing reliability.

3. Reliability theory of non-destructive testing

3.1. Theoretical overview of POD

Non-destructive testing reliability analysis is based on statistical and mathematical statistical methods to calculate and analyze the test results. It determines the functional relationship between POD and crack size, fits the POD(a) curves, and calculates the minimum crack size for a specific probability of detection [4]. The crack size under the condition of 95% confidence and 90% detection probability is often taken as an evaluation index for non-destructive testing reliability.

The common ultrasonic test result data is divided into two types: "Hit/Miss" type and "Signal Response" type [5]. Among them, "Signal Response" type data contains more information about cracks. It not only reflects that the signal response amplitude exceeds the detection threshold, but also quantifies the extent of exceeding the detection threshold and is ultimately reflected on the POD. For "Signal Response" type data, a represents the crack size, and $\hat{a}$ represents the amplitude of the detection signal, and there is a log-linear relationship between the two and obeys a log-normal distribution [6], which is called "Log- Logistic" model.

3.2. POD calculation of Log-logistic model

The linear relationship between $\ln\hat{a}$ and lna in the Log-logistic model is as follows:

$$ln\hat{a} = A + B \cdot lna + \delta \tag{1}$$

where A and B are the intercept and the slope respectively. δ is the error term and is generally considered to be δ~N(0, σ_δ^2). The POD(a) based on this linear relationship can be represented by a probability density function:

$$POD(a) = \int_{\hat{a}_{th}}^{\infty} g(a)(\hat{a})d\hat{a} \tag{2}$$

where g(a)($\hat{a}$)—probability density function; $\hat{a}_{th}$ —detection threshold. When $\hat{a}$ > $\hat{a}_{th}$, the crack is detectable and POD(a) can be expressed as:

$$\mathrm{POD}(a) = \mathrm{Prob}\left[\delta > \ln(\hat{a}_{th}) - (A + B \times \ln(a))\right] \tag{3}$$

Because of δ~N(0, σ^2), the equation (3) can be expressed as:

$$\mathrm{POD}(a) = 1 - \Phi\left[\frac{\ln(\hat{a}_{th}) - (A + B \times \ln(a))}{\sigma_\delta}\right] \tag{4}$$

IOP Conf. Series: Materials Science and Engineering **715** (2020) 012012 doi:10.1088/1757-899X/715/1/012012

where $\Phi(z)$ is the distribution function of the standard normal distribution. Using the characteristics of the standard normal distribution, the above equation can be simplified to:

$$POD(a) = \Phi\left[\frac{\ln a - \frac{\ln(\hat{a}_{th}) - A}{B}}{\frac{\sigma_\delta}{B}}\right] \quad (5)$$

The probability integral lower limit μ and the standard deviation σ of the log size logarithm are:

$$\mu = \frac{\ln(\hat{a}_{th}) - A}{B}; \sigma = \frac{\sigma_\delta}{B} \quad (6)$$

Therefore POD(a) can be written as follows:

$$POD(a) = \Phi\left[\frac{\ln(a) - \mu}{\sigma}\right] \quad (7)$$

Using equation (6), we can also estimate the lower confidence limit of the log-mean of the crack detection amplitude for a given confidence level, so that the curve can be plotted under the confidence, and the logarithm of the crack detection amplitude is given at a given confidence. The lower confidence limit for the mean is calculated as follows:

$$\ln\hat{a}_{CL} = A + B \times \ln a - u_{\frac{1-\gamma}{2}} \times \frac{\sigma}{\sqrt{n}} + \delta \quad (8)$$

Substituting equation (8) into equations (3) to (5), the lower limit of the probability integral of the log size logarithm (lna) at a given confidence is:

$$\mu_{CL} = \mu + u_{\frac{1-\gamma}{2}} \times \frac{\sigma}{\sqrt{n}} \quad (9)$$

Therefore, the POD(a) is written as follows under the given confidence:

$$POD(a)_{CL} = \Phi\left[\frac{\ln(a) - \mu_{CL}}{\sigma}\right] \quad (10)$$

4. Establishment and verification of ultrasonic testing simulation model

In order to study the influence of the angle between the beam axis and the crack orientation on the reliability of ultrasonic testing, this paper uses COMSOL Multiphysics to carry out ultrasonic testing simulation analysis. Ultrasonic detection amplitude which changed with the angle of inclusion was studied under different crack sizes.

This paper uses the two-dimensional axisymmetric model in COMSOL Multiphysics software to model. The numerical simulation verification model of ultrasonic testing is a crack-free cylinder with a diameter of 50 mm and a height of 20 mm. The material is selected as the built-in material Ti-6Al-4V, and the wave velocity is set at 6000 m/s. The model is shown in figure 1.

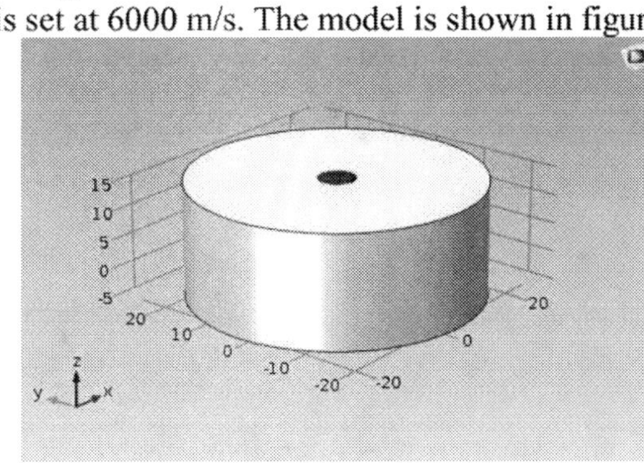

Figure 1. Simulation model of crack-free cylinder.

Set the probe diameter to 6mm, the frequency is 5MHz, and the simulation obtains 4 bottom echoes. The actual echo detection of the four echo waveforms given by Vamsi is shown in figure 2(a) [7], and the four echo waveforms obtained by the simulation are shown in figure 2(b). After the simulation is completed, the amplitude and propagation time of the bottom surface echo is extracted from four consecutive times, and the attenuation ratio of the adjacent amplitude (the ratio of the previous echo amplitude to the amplitude of the next echo) is compared with the experimental data of Vamsi. The result is as table 1.

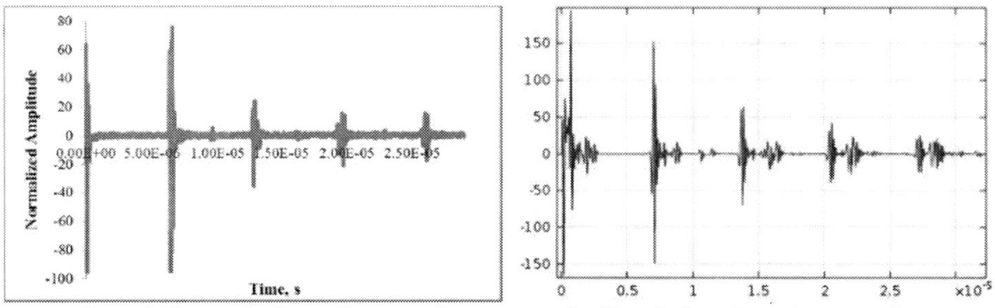

a. Actual ultrasonic detection waveform. **b.** Simulation detection waveform.

Figure 2. Echo waveforms obtained by actual detection and simulation.

Table 1. Comparison of actual ultrasonic testing and simulation results.

Number	Echo amplitude		Echo time		Attenuation ratio	
	Test value /%	Simulation value /Pa	Test value /μs	Simulation value /μs	Test value	Simulati on value
1	170	153	6.94	6.99	\	\
2	59	71	13.92	13.8	2.86	2.16
3	36	42	20.88	20.6	1.66	1.69
4	21	26	27.33	27.4	1.69	1.62

It can be seen from the table that the error of the experimental values and the simulated values of the three eigenvalues are 4.45%, 1.81% and 4.14%, respectively. The error is small and the attenuation of the ultrasonic echo is consistent. In addition, the propagation time between consecutive backwall echoes of the A-scan signal obtained by the experimental and numerical simulations is substantially the same. Therefore, it can be considered that the software simulation model is reliable in ultrasonic propagation and attenuation, and can be further used to simulate the ultrasonic response of cracks.

Therefore, take the 2mm flat bottom hole crack as an example. A two-dimensional model was established for simulation research, in which the model length was 50 mm and the height was 30 mm. The flat bottom hole model was 2 mm long and 10 mm high. The two-dimensional model ultrasonic detection instantaneous sound field slice diagram and waveform diagram is shown in figure 3.

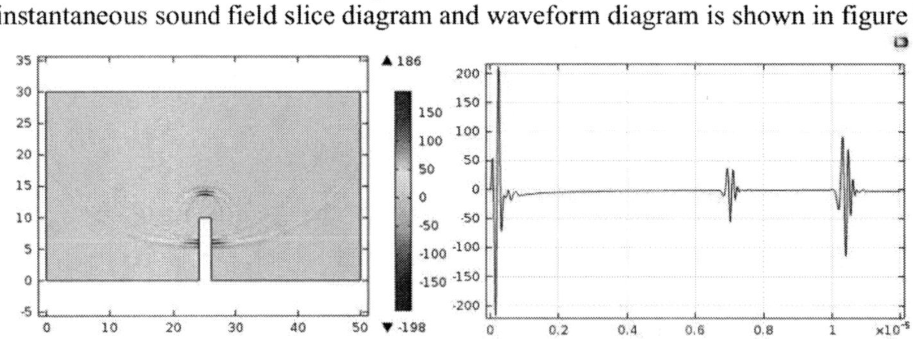

a. Instantaneous sound field slice. **b.** Ultrasonic detection waveform.

Figure 3. Ultrasonic testing of two-dimensional model flat-bottom hole crack.

5. Study on the influence of the angle on crack detection

In order to quantify the influence degree of the angle on the ultrasonic testing results, the 5mm flat bottom hole crack is taken as the research object, and an angle value is taken at intervals of 2° in the $[70^\circ, 90^\circ]$ interval for simulation study. In the simulation study, the incident direction of the acoustic wave is kept unchanged, and the crack orientation is changed to change the angle. Taking the angle of 82° as an example, the instantaneous sound field slice diagram and the ultrasonic detection waveform diagram are shown in figure 4 and figure 5.

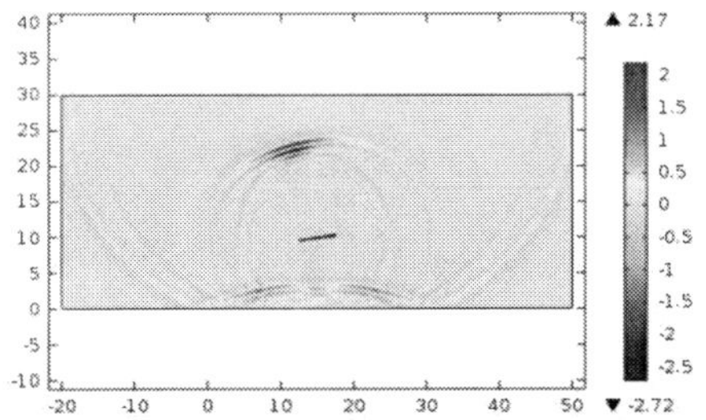

Figure 4. Instantaneous sound field reflection slice.

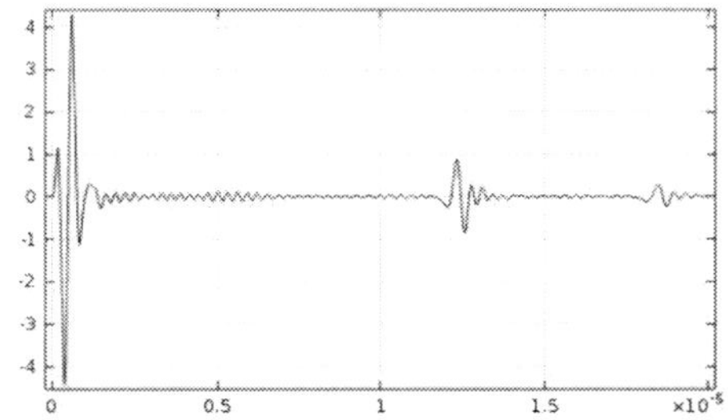

Figure 5. Ultrasonic detection waveform.

The simulation results with the angle $[70^\circ, 90^\circ]$ are as follows:

Table 2. Simulation results of detecting amplitude changes with angle.

Parameter	Simulation results										
The angle/°	70	72	74	76	78	80	82	84	86	88	90
Echo sound pressure /Pa	25	30	36	40	51	64	90	130	178	218	240
Reduction /%	89.66	87.31	84.98	83.24	78.72	73.43	62.24	45.87	25.59	8.97	0

The detection amplitude at 90° is regarded as 1. The simulation results under other angles are converted. Matlab is used to fit the simulation results in table 2, and then the function relationship is as follows:

$$K_\theta = \frac{0.176 \cdot \theta^2 - 4.2 \cdot \theta + 57.73}{\theta^2 - 3.68 \cdot \theta + 57.89} \tag{11}$$

where θ is the angle value at which the angle is offset by 90°, and K_θ is the proportional coefficient of the amplitude of the echo. According to formula (11), the variation of the ultrasonic detection echo amplitude of the crack with the angle is shown in figure 6:

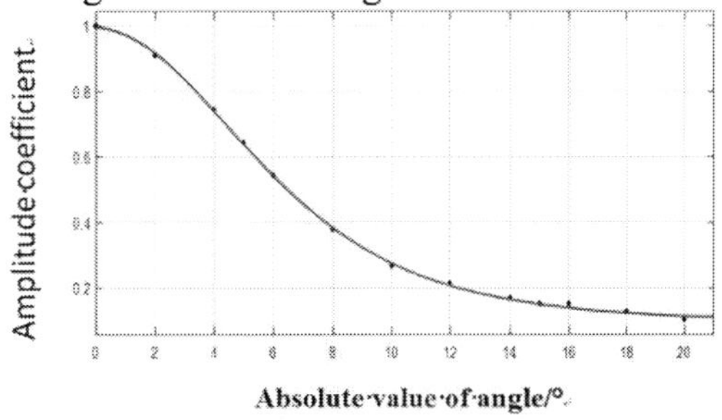

Figure 6. Fitting the amplitude coefficient of the echo amplitude under each angle.

It can be seen that after the angle changes, we can multiply the corresponding value by the crack detection amplitude at 90° to obtain the detection amplitude at the corresponding angle. Since the Log-logistic model is based on the log-linear relationship between the crack size and the detected amplitude, the POD is calculated. Therefore, when the detected amplitude is multiplied by a certain proportional coefficient and then the logarithm is obtained, there is still a log-linear relationship between the crack size and the crack size. That means the Log-logistic model can still be used to calculate the POD curve affected by the angle.

6. Reliability of ultrasonic testing considering the influence of angle

The simulation results of ultrasonic testing under different crack sizes are given. The corresponding detection amplitudes of cracks of different sizes at angles of 86°, 82°, 78°, 74° and 70° were calculated using the contents of the fourth section, as shown in table 3.

Table 3. Ultrasonic simulation detection amplitude at different angles.

The angle/°	Detecting amplitude								
90	14.77	19.55	27.93	28.70	36.56	55.59	75.56	81.62	130.33
86	10.92	14.45	20.65	21.22	27.03	41.10	55.86	60.34	96.36
82	5.65	7.48	10.69	10.99	14.00	21.28	28.93	31.25	49.90
78	3.06	4.05	5.79	5.95	7.57	11.52	15.65	16.91	27.00
74	2.06	2.73	3.90	4.01	5.10	7.76	10.54	11.39	18.19
70	1.70	2.25	3.21	3.30	4.20	6.38	8.68	9.37	14.97

According to the crack detection amplitude under different angles, the POD calculation method of "Signal Response" type data is adopted. The Matlab is used to calculate and fit the POD of a certain crack size at a specific angle. The minimum crack size corresponding to the curve at different angles and the 90% probability of detection is shown in figure 7.

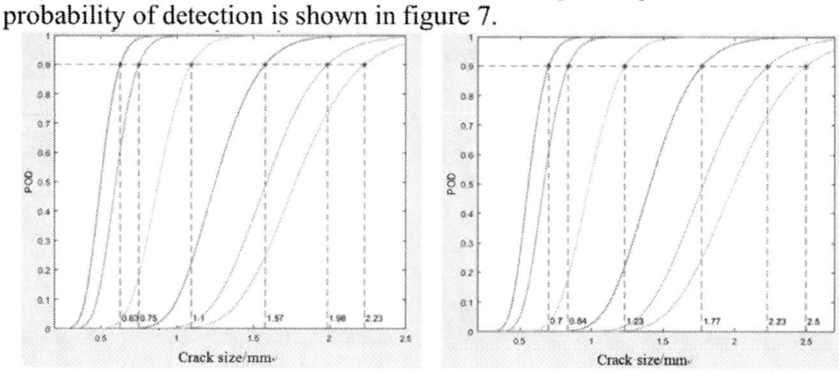

a. POD curve and a_{90} size corresponding to different angles.
b. The POD curve and $a_{90/95}$ size corresponding to different angles when the confidence is 95%.

Figure 7. POD curves at different angles and their corresponding a_{90} and $a_{90/95}$ crack sizes (The angles from left to right are 90°, 86°, 82°, 78°, 74°, and 70°, respectively).

Under the same detection conditions, when the ultrasonic detecting sound beam axis is perpendicular to the crack orientation, the detection reliability is the highest and the corresponding a_{90} and $a_{90/95}$ detectable crack sizes are also the smallest. As the angle θ deviates from 90°, the crack detection amplitude gradually decreases, and the detection reliability decreases in turn. The corresponding a_{90} and $a_{90/95}$ crack sizes also gradually increase. The minimum detectable minimum crack size is shown in table 4.

Table 4. Detectable minimum crack size at different angles.

The angle/°	Detectable minimum crack size	
	a_{90}/mm	$a_{90/95}$/mm
90	0.63	0.70
86	0.75	0.84
82	1.10	1.23
78	1.57	1.77
74	1.98	2.23
70	2.23	2.50

7. Conclusions

In order to quantify the influence of the angle between the beam axis and the crack orientation on the detection reliability, the crack detection probability (POD) under different conditions was obtained by numerical simulation of in-situ ultrasonic inspection. The main conclusions are summarized as follows:

1) Considering the influence of the angle, the numerical simulation of COMSOL Multiphysics can be used to obtain the ultrasonic detection amplitude under different angles. After substituting into the Log-logistic model, the POD curve of the ultrasonic detection affected by the angle of inclusion can be calculated. The minimum crack size can be calculated, which the detection device can detect with a probability of 90%. This forms a numerical simulation method for the reliability of in-situ ultrasonic testing.

2) As the angle is offset by 90°, the crack detection amplitude decreases rapidly. When the angle is changed from 85° to 80°, the detection amplitude is changed from 64.17% to 26.67% of the highest level, which is reduced by 37.5%.

3) In the case of a 90% probability of detection, when the angle is changed from 90° to 70°, the minimum detectable crack size is reduced from 0.63 mm to 2.23 mm, which is 71.75% lowered. If the detecting device detects a crack of less than 2.23 mm with a 90% detection probability, the angle between the axis of the sound beam and the orientation of the crack must be greater than 70°. If the detecting device detects a crack greater than 0.7 mm and the angle is greater than 86°, the probability of detecting the crack is greater than 61.35%. If the detecting device detects a crack greater than 0.7 mm and the angle is greater than 86°, the probability of detecting the crack is greater than 61.35%.

8. Prospect

In this paper, the reliability of in-situ ultrasonic testing is carried out for the edge crack, and many numerical simulations and calculations are carried out. However, there is still much room for development.

1) The aircraft has many machine structures and many forms of damage. In addition to ultrasonic testing, there are various non-destructive testing techniques such as eddy current testing, radiation detection and magnetic particle testing. Therefore, various non-destructive testing reliability studies can be carried out for different defects in other structures. In addition, the probability of detecting defects in the structure can be more effectively improved, and the reliability of the aircraft structure can be improved.

2) In this paper, the influence of two influencing factors on the reliability of in-situ ultrasonic testing is studied in the form of two-dimensional curve. In order to analyze the influence of multi-factors more directly, the three-dimensional form can be used to study the reliability of other structures.

Acknowledgements

Thanks for the project support from the Aviation Industry Strength Institute.

References

[1] Z. Li. National University of Defense Technology. D. 5 (2010)
[2] Y. Y. Li. Dalian University of Technology. D. 8 (2014)
[3] V. K. Rentala, P. M. Mylavarapu, J. P. Gautam. Ultrasonics. J. 59 (2018)
[4] R. B. Thompson, L. Brasche. E-Journal of Advanced Mainteance. J. 25(2012)
[5] J. R. Davis, D. Banjevic. ASM International. M. 94(2009)
[6] A. P. Berens, P. M. Hovey. Philadephia: Astm Special Technical Publication, C, 70 (2011)
[7] T. S. Jarsveld, R. Dekker. Reliability Engineering & System Safety. J. 1576(2011)

IOP Conf. Series: Materials Science and Engineering **715** (2020) 012013 doi:10.1088/1757-899X/715/1/012013

Co-simulation of transfer alignment and CAN busbased on CANoe-MATLAB

Wang Yongsheng*, Luo Wei and Zha Fan

Design Institute of Hubei Space Technology Academy, Wuhan 430040, China

*E-mail:wysh320@163.com

Abstract. The research on rapid transfer alignment for INS on moving base is very important in the field of the inertia technology. In order to build the real-time CAN bus network simulation system with the velocity matching model of the transfer alignmentalgorithm, CANoe(CAN open environment)-MATLAB interface is applied. The CANoe-MATLAB interface is connected with MATLAB/Simulink which has powerful modeling functions and CANoe which has perfect function of bus simulation. Modeling the transfer alignmentalgorithm by MATLAB/Simulink and providing the real-time CAN bus network communications in HIL(hardware-in-loop) simulation by CANoe. The simulations tests show that co-simulation system built by the CANoe and the MATLAB/Simulink performs favorably with rapid convergence and validity.

1. Introduction

CANoe is a total network simulation tool designed to both model nodes on a network and the network that binds them[1]. Typically the functionality of these network nodes are modelled using CAPL(Communication Access Programming Language) which is a C style programming languageprovided with CANoe. However, due to the limited ability of this language, it is difficult to design nodes with complex control algorithms or control functions.

TheVector_AddOn_Matlab_Interfaceextends CANoe's node modeling capability by adding the strength of MATLAB/Simulink environment. This interface provides data exchange with CANoe for simulations running inside Simulinkand assures time synchronization between both tools[2].

The transfer alignment process is made to the non-aligned slave INS by use of information of the aligned master INS when the master carrier is moving[3]. When the system is built in the real-time CAN bus network communications, the The CANoe-MATLAB interface is used necessarily.

2. Transfer alignment modelling

The process of transfer alignment is based on the system error model, establishing system equation and observation equation, and estimating the misalignment angle by means of various estimation methods of the stochastic system. The most commonly used estimation method in transfer alignment is the Kalman filtering; meanwhile the velocity matching error model is used as the observability analysis[3][4][5].

2.1. Kalman filter initialization

The following matrix need to be initialized: the identity matrixI_k, the state transition matrix A_k, the mean square error matrix P_k, the process noise covariance matrix Q_k, the measurements noise covariance matrixR_k, the relation matrix H_k and the filter state matrix X_k.

Table 1. Initial value.

Symbol	Dimensionality	Initial Value
I_k	5×5	The elements on the diagonal line are 1.0 and the remaining elements are 0
A_k	5 ×5	The value is the same as I_k
P_k	5×5	The diagonal elements are 3.0e-4, 1.0e-4, 3.0e-4, 0.25,0.25 successively and the remaining elements are 0
Q_k	5×5	The diagonal elements are 8.350e-13, 8.350e-13, 8.350e-13, 1.0e-6, 1.0e-6 successively and the remaining elements are 0
R_k	2×2	$R_k = \begin{bmatrix} 0.01 & 0 \\ 0 & 0.01 \end{bmatrix}$
H_k	2×5	$H_k = \begin{bmatrix} 0 & 0 & 0 & 1 & 0 \\ 0 & 0 & 0 & 0 & 1 \end{bmatrix}$
X_k	5×1	$X_k = [0 \quad 0 \quad 0 \quad 0.5 \quad 0.5]$

2.2. Attitude matrix of slave INS and master INS calculation
Calculate the slave INS attitude matrix $\mathbf{C}_\mathrm{b}^\mathrm{n}$ and the master INS attitude matrix $\mathbf{C}_\mathrm{a}^\mathrm{n}$ as follows.

$$C_b^n = \begin{pmatrix} \cos\theta_s\cos\varphi_s & -\sin\theta_s\cos\varphi_s\cos\gamma_s + \sin\varphi_s\sin\gamma_s & \sin\theta_s\cos\varphi_s\sin\gamma_s + \sin\varphi_s\cos\gamma_s \\ \sin\theta_s & \cos\theta_s\cos\gamma_s & -\cos\theta_s\sin\gamma_s \\ -\cos\theta_s\sin\varphi_s & \sin\theta_s\sin\varphi_s\cos\gamma_s + \cos\varphi_s\sin\gamma_s & -\sin\theta_s\sin\varphi_s\sin\gamma_s + \cos\varphi_s\cos\gamma_s \end{pmatrix} \quad (1)$$

$$C_a^n = \begin{pmatrix} \cos\theta_m\cos\varphi_m & -\sin\theta_m\cos\varphi_m\cos\gamma_m + \sin\varphi_m\sin\gamma_m & \sin\theta_m\cos\varphi_m\sin\gamma_m + \sin\varphi_m\cos\gamma_m \\ \sin\theta_m & \cos\theta_m\cos\gamma_m & -\cos\theta_m\sin\gamma_m \\ -\cos\theta_m\sin\varphi_m & \sin\theta_m\sin\varphi_m\cos\gamma_m + \cos\varphi_m\sin\gamma_m & -\sin\theta_m\sin\varphi_m\sin\gamma_m + \cos\varphi_m\cos\gamma_m \end{pmatrix} \quad (2)$$

2.3. State transition matrix updating
After the traditional transfer alignment model has been initialized, improve the robustness of the system by updating the state transition matrix as follows.

$$A_k(4,2) = \Delta V_x^n A_k(5,2) = -\Delta V_z^n \quad (3)$$

2.4. Observation vector updating
The observation vector Z_k is the velocity difference between master INS and slave INS.

$$Z_k(1) = V_E^s - V_E^m Z_k(2) = V_N^s - V_N^m \quad (4)$$

2.5. Kalman filter calculating
The Kalman filter state vectorX_k contains the attitude misalignment angle in 3 directions and the velocity error in northern and eastern directions.

$$X = [\varphi_x, \varphi_y, \varphi_z, \delta V_E, \delta V_E]^T \quad (5)$$

The recursive formula of Kalman filter is generalized as follows. Update the filter state matrix X_k and the mean square error matrix P_k, and the result of transfer alignment is $X[0]\,X[1]\,X[2]$ in 3 directions.

$$X_{k,k+1} = A_k X_k \quad (6)$$

$$P_{k,k+1} = A_k P_k A_k^T + Q_k \quad (7)$$

$$K_k = P_{k,k+1}H_k^T(H_kP_{k,k+1}H_k^T + R_k)^{-1} \tag{8}$$

$$P_{k+1} = (I_k - K_kH_k)P_{k,k+1} \tag{9}$$

$$X_{k+1} = X_{k,k+1} + K_k(Z_k - H_kX_{k,k+1}) \tag{10}$$

$$X_k = X_{k+1} \tag{11}$$

$$P_k = P_{k+1} \tag{12}$$

3. CAN bus co-simulation

There are three different approaches of time synchronization [1]: (1) The Simulinktime base is used for the CANoe simulation. CANoe is operated in Slave mode whereas MTALAB Simulink is the simulation Master. This will be referred to as the Offline Mode. (2) In contrast to the Offline Mode the CANoe time base is used in MATLAB Simulink. Therefore the simulation is run in almost real-time, which is calledthe Synchronized Mode. (3) In the HIL Mode, a Windows DLL is producedwith the Simulink which can be loaded in CANoe'ssimulation environment.

Create a new database with CANdb++ in the CANoe environment, which contains environment variables,network nodes,messages and signals. In the CAN database, the following objects need to be linked: signals with messages,messages with network nodes and message signals with network nodes.

Open the model in Simulink. The model contains 3 CANoe specific blocks: Simulation step, Signal Input and Signal Output. The first block-Simulation step is only relevant to the simulation in Offline and Synchronized Mode. The others are signal input and output blocks respectively which are linked to the CAN database. Design the transfer alignment filter model by Chapter 2, which is shown as figure 1.

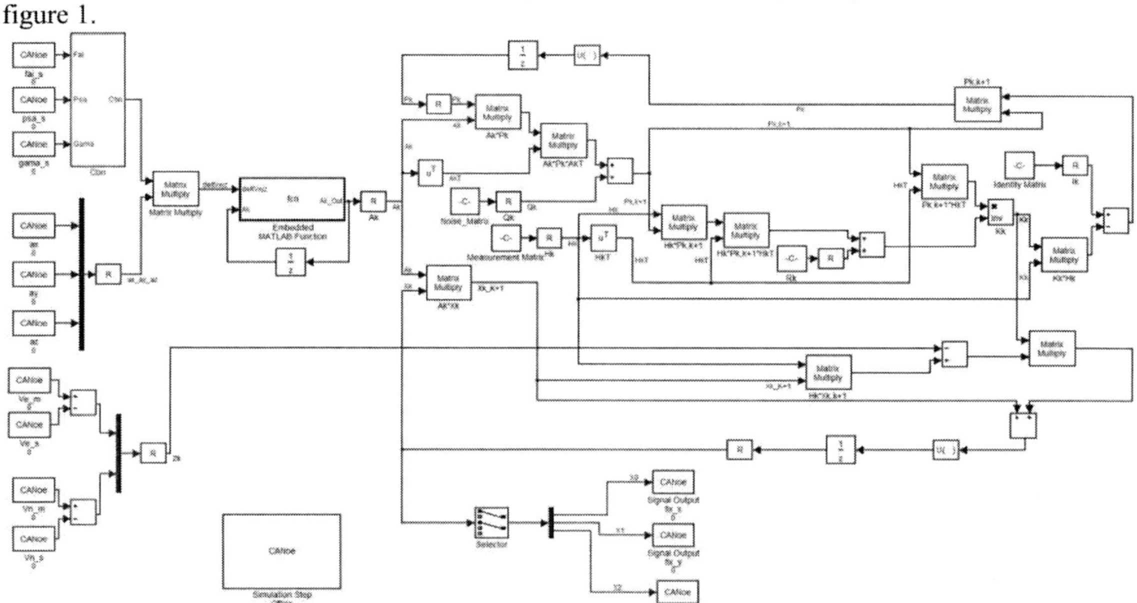

Figure 1. Simulink model.

Open the simulink setup and import the database into CANoe.The signals can be displayed in data and graphics windows. The whole system simulation process is based on the offline mode of the CANoe-MATLAB interface, which is used as a slave mode. When the simulink mode is started, the CANoe automatically opens the bus messages and signals, which can be used to feed back the Simulink model accuracy and real-time.

4. Analysis of simulation results

The purpose of the simulation consists of two parts. On one hand, verify the correctness of the transfer alignment model. On the other hand, verify the accuracy and real-time between CANoe and MATLAB.

Figure 2 shows the velocity in northern and eastern directions of master INS and slave INS.

Figure 3 shows the attitude of slave INS. What is clear from the data is that the master INS and slave INS are both on the moving base which is rolling, as shown in the figure 4. Through simulation the attitude misalignment angles in 3 directions are convergent using Kalman filter, which indicate the precision of model is good and is agree with actual stages, as shown in the figure 5.

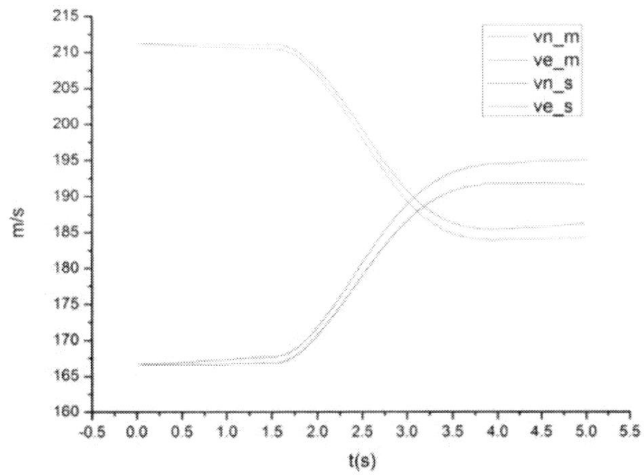

Figure 2. The velocity of master INS and slave INS.

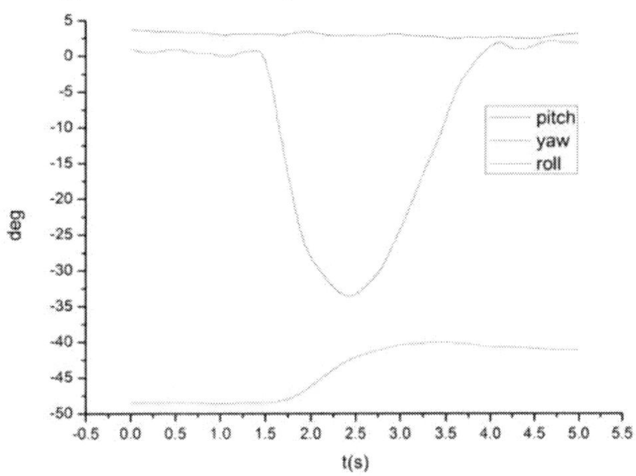

Figure 3. The attitude of slave INS.

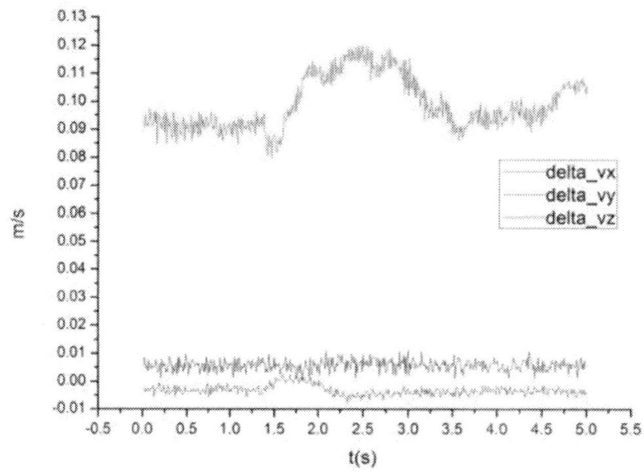

Figure 4. The variation of master INS in velocity directions.

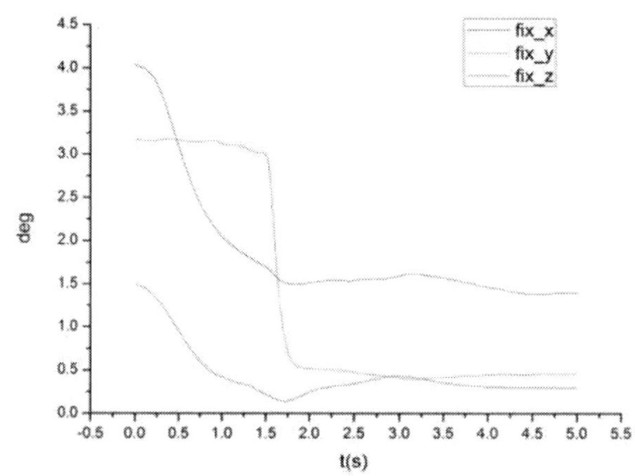

Figure 5. The result of attitude misalignment angles.

5. Conclusion

The CANoe addon MATLAB co-simulation has a wide range of application, with this approach it is possible to test and verify the transfer alignment model designs with real CAN hardware in a real-time networked environment.The method covers the shortage of MATLAB's capability to the CAN bus, meanwhile improves the ability to check out the defects of the model. The result of co-simulation based on CANoe and MATLAB/Simulink performs favorably with rapid convergence and validity. The simulation method will have more extensive applications.

References

[1] Vector Informatik GmbH. CANoe AddOn MATLAB Interface User Guide[Z]. 2008
[2] VectorInformatik GmbH. Using_MATLAB_with_CANoe[Z]. 2008
[3] WANG Si, DENG zheng-long. Technique Review of Transfer Alignment for InertialNavigation Systems on Moving Base[J]. Journal of Chinese Inertial Technology.2003,11(2): 61-67
[4] ZHOU Feng, MENG Xiu. Research on Simulation Environment of Transfer Alignment for Airborne Weapon SINS[J]. Journal of Projectiles,Rockets,Missles and Guidance.2009,29(1): 35~42
[5] Li Hongcheng. Study on Kalman Filter techniques of Initial Alignment of Strapdwn Inertial Navigation System[D].Jiangxi University of Technology. 2016: 50-58

IOP Conf. Series: Materials Science and Engineering **715** (2020) 012014 doi:10.1088/1757-899X/715/1/012014

Effect of twist angle on starting capability of a Savonius rotor – CFD analysis

Ahmad Zakaria* and M S N Ibrahim

Numerical Simulation Lab, Universiti Kuala Lumpur Malaysia Italy Design Institute, 56100 Kuala Lumpur, Wilayah Persekutuan Kuala Lumpur, Malaysia

*E-mail: dzakaria@unikl.edu.my

Abstract. Starting capability is an important feature of any high performance wind turbine. This parameter is usually referred to the rotor static performance. In this paper, the results of CFD analysis on four Savonius rotor designs are presented with the objective of investigating the effect of twist angle on the starting capability. Essentially it is a comparison between a conventional Savonius rotor and three helical rotors with 0°, 90° and 180° twist angle. The proposed methodology was first validated against the published wind tunnel test data. The best parameters were then used to evaluate the static performance of the four Savonius models. It is observed that helical Savonius rotor contributes to positive static torque coefficient for all rotor angles. The rotor with 90° twist angle gives the highest average static torque coefficient of 0.442 followed by 0° twist angle rotor and 180° twist angle with static torque coefficient of 0.434 and 0.385 respectively. This improvement in static torque coefficient will not only contribute to the better starting capability but also to the overall performance of Savonius rotor.

1. Introduction

Despite having a low power efficiency of less than 0.15, Savonius rotor has proven to be well suited for low speeds or urban applications. In many Southeast Asian counties like Malaysia, where the average wind speed is rather low (about 2 m/s) Savonius-solar system may be adopted in order to generate a useful and reliable electric power. In other parts of the world, many studies have been carried out in an attempt to increase the performance of a Savonius rotor. These include the addition of end plates, incorporation of offset, aspect ratio, bucket angle and twist angle [1]. To this end, power coefficient of exceeding 0.5 has been achieved with the addition augmentation device such as deflecting plates [2]. In addition to this, power enhancement has also been made possible via multiple rotor clusters approach [3].

Irrespective of turbine design either Horizontal Axis Wind Turbine (HAWT) or Vertical Axis Wind Turbine (VAWT), starting capability is one of the important criteria for designing high performance Savonius rotors. It is the ability of a wind turbine to start generating power at any initial azimuthal position for given wind speed. This definition may, however, differ from others [4]. A turbine with this capability should, therefore, be able to self-start at relatively lower wind speed as it has lower cut-in wind speed [5]. Ease of starting is highly dependent on its static torque. Higher average static torque means the better is its starting capability. Savonius rotor has the reputation of having such a good self-starting capability. An introduction of twist angle and gap ratio to Savonius blade design resulted in a positive coefficient of static torque for all rotor angle and thus improving the starting capability. Wind tunnel tests performed by [6, 7] revealed that helical Savonius with twist of 90° and 180° can overcome the drawback of negative torque. Augmentation techniques such as guide box tunnel, guide vanes, obstacle plate, frontal nozzle, and frontal guiding plates can improve the static torque by 35% and up to 50% but only suitable for a certain condition such as small operating range and at specific orientation for wind direction. Other factors influencing the starting torque include blade shape optimization [8]. Starting capability of Savonius is generally evaluated by CFD simulation and followed by validation via wind tunnel experiment. The accuracy of such an analysis completely dependent on parameters such as grid size, domain size, turbulence model, azimuthal increment, number of revolution to reach convergence [9].

The main objective of the present study was to study the effect of twist angle on the Savonius rotor starting capability through CFD analysis. The proposed methodology was first validated against the published wind tunnel test data. Torque coefficients for different Savonius rotor models were then compared. This study used a commercial computational fluid dynamics solver called AcuSolve.

2. Validation of the methodology and Savonius rotor model

The reference model for the validation is obtained from the published experimental data [10]. The main objective of this exercise was to compare the predicted static torque coefficient results with the former. Minimum error is achieved when the selected parameters are optimized.

2.1. Rotor validation model

The design parameter was graphically shown in figure 1 and summarized in table 1.

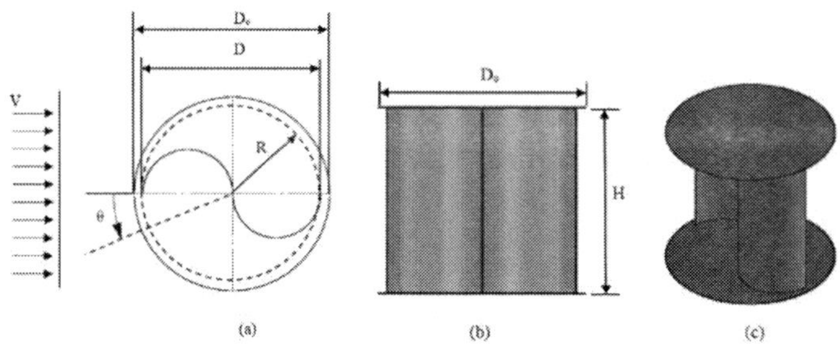

Figure 1. The reference Savonius rotor model showing different views: (a) Top (b) Front (c) Isometric.

Table 1. Design parameter for validation model.

Parameter	Value	Unit
Height, H	1	m
Diameter, D	1	m
End plate diameter, D_o	1.1D	m
Number of buckets	2	-
Aspect ratio, H/D	1	-
Overlap ratio, δ	0	-
Twist angle, φ	0°	Degree

2.2. Helical Savonius rotor model

Three helical Savonius rotors with different twist angles used in this study are as shown in figure 2. Each rotor has common dimensions: H = 1 m H, D = 0.5 m with δ (e/d) = 0.242 and D_o = 1.1D.

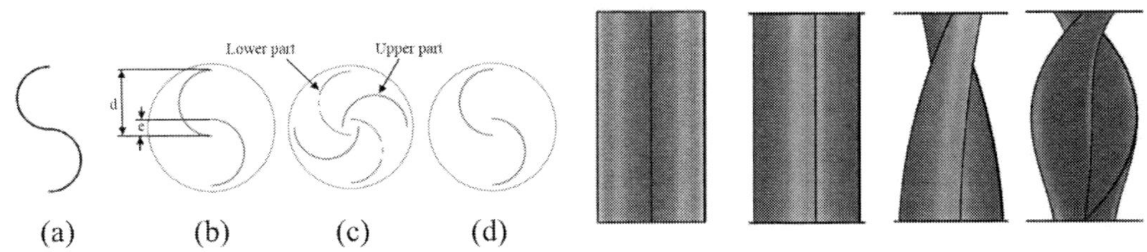

Figure 2. Different design models of Savonius rotors(a) Conventional 'S' shape (b) Straight blade (c, d) Helical Savonius with different twist 90° and 180° respectively. Relative position of lower part and upper part of the blade indicates the twist angle.

3. Numerical analysis

The 3D computational domain with the dimension of 30D x 16D x 16D was used in this study [11]. The static Savonius rotor was placed in the center of the domain. A steady flow of 7 m/s was set at the inlet. The experiment was repeated using different turbulence models namely Spalart-Allmaras (S-A), SST and K-ω. The boundary conditions (BC) are shown in figure 3.

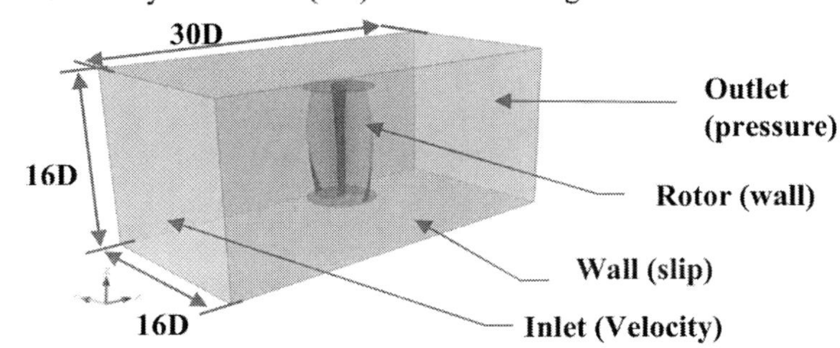

Figure 3. BCs assigned to 3D-computational domain.

The static analysis was carried out in order to evaluate the rotor performance at still condition focusing on torque generated by the rotor [12]. The static position of the Savonius rotor was initially placed as in figure 1 (a) position, θ. The rotor position was changed at 10° interval [10]. 19 models of Savonius static position were required in order to complete one half-rotation. The static torque was calculated at every position and compared finally with the available wind tunnel data. The static torque coefficient, C_{ts} was calculated by using equation (1).

$$C_{ts} = \frac{4Ts}{\rho V^2 D^2 H} \tag{1}$$

where Ts is the rotor static torque at θ rotor position (N.m), ρ is the air density (kg/m3), V is flow stream velocity (m/s), D is rotor diameter (m) and H is rotor height (m).

Grid or mesh sensitivity study was conducted to evaluate the effects of the mesh size on the analysis result. The mesh was refined to a certain boundary, and result from the analysis was observed. Each mesh size was solved using Acusolve based on the same input parameter. Further refinement will not give a significant effect on the analysis result [13]. Several meshing sizes of tetrahedral shape were utilized to increase the mesh density and mesh quality. Sufficient mesh density was selected and generated in the region of high-pressure gradient around the rotor profile. In this study, three mesh sizes were selected. Mesh #1 setting consists of elements 20164721. Whereas Mesh #2 and mesh #3 generated 5229713 and 253365 number of elements respectively The boundary layer was set around the rotor and adaption of y+ value of 1 for enhanced wall treatment [14].

The selected parameter setting from CFD validation was then used to evaluate the proposed helical Savonius rotor designs. The input wind velocity is set as Cartesian velocity with a uniform value of 3 m/s.

IOP Conf. Series: Materials Science and Engineering **715** (2020) 012014 doi:10.1088/1757-899X/715/1/012014

4. Results and discussion

The static analysis used to explore the Savonius rotor static performance which contributes to the explanation of rotor self-starting term [5].

4.1. Mesh sensitivity study

Figure 4 shows the static torque coefficient at variation of rotor angle for tested mesh sizes. The mesh density increases with the mesh setting from mesh #1 to mesh #3. As expected an increase in the number of elements increases the computational time.

The results show similar patterns as observed by the wind tunnel test data [10]. Coarse mesh denoted by mesh #1 shows predicted torque coefficients are close to the experiment for rotor angle between -35° to 65° but starts to deviate after rotor angle 75°. Medium mesh represented by mesh #2 shows a good agreement with the wind tunnel data. However fine mesh #3 shows a large deviation from the experimental data.

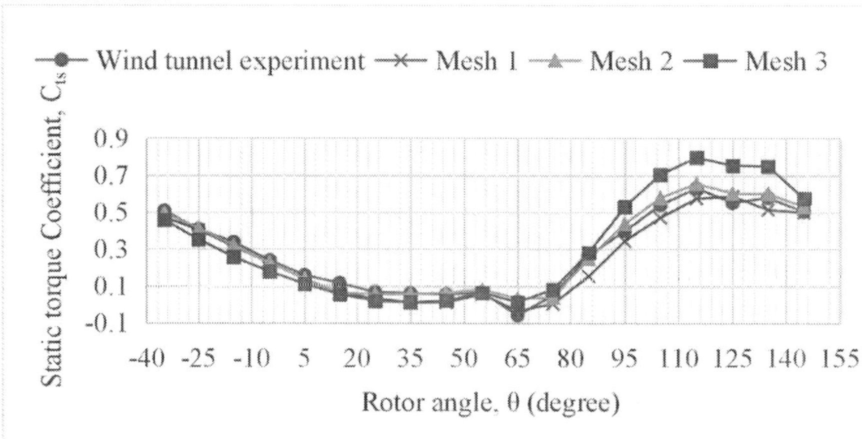

Figure 4. Mesh sensitivity results.

4.2. Turbulence model sensitivity study

The numerical study results of turbulence model investigation are presented in figure 5. The SA turbulence model provides a fairly good estimate to those data obtained by the experiment. However, the SST and k-ω turbulence models are seemed to have overestimated and underestimated the static toque at certain rotor angles. Therefore, all CFD analysis for the four proposed models were performed based on the SA turbulence model.

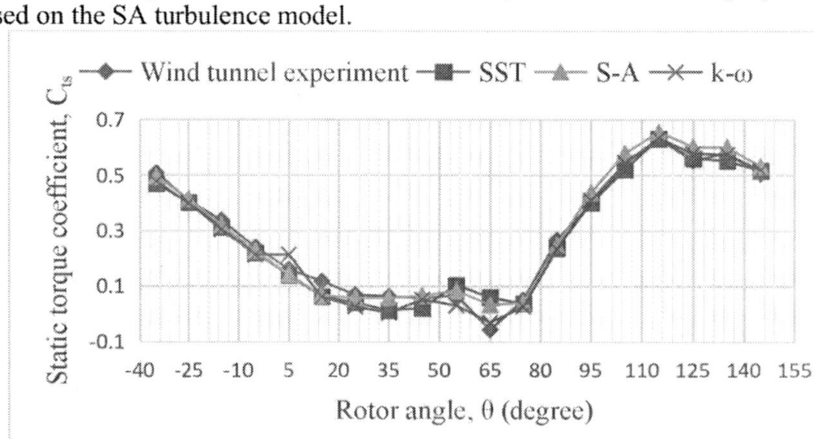

Figure 5. Sensitivity analysis of turbulence models.

4.3. Effect of twist angle on static torque coefficient

Figure 6 shows static torque coefficients generated by the reference(conventional) rotor and three other rotors with twist angle at different rotor angles.

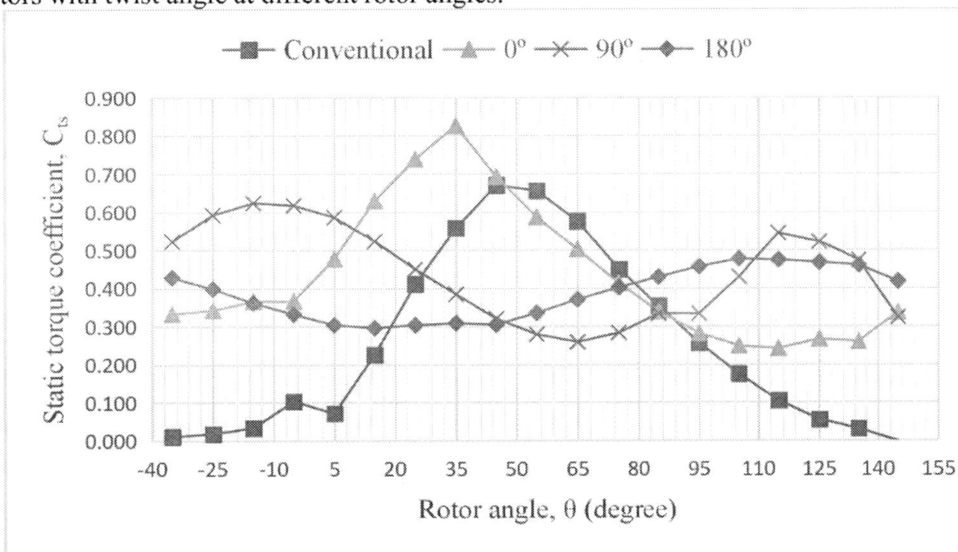

Figure 6. Static torque coefficient for different rotor design.

The conventional Savonius rotor experiences a low and nearly zero torque coefficient at certain rotor angles and thus revealing its unlikelihood to rotate at those rotor angles. Additional end plate and adaption of overlap ratio as depicted by 0° twist angle seems to slightly increase the static torque coefficient. The helical Savonius rotor with twist angle 90° and 180° show a great improvement in the static torque coefficient variation. It is also observed that the average static torque coefficient for conventional Savonius rotor and proposed Savonius rotor with additional of end plate, adaption of overlap ratio with twist angle of 0°, 90° and 180° are 0.25, 0.434, 0.442 and 0.385 respectively.

5. Conclusions

Conclusions derived from this study are as follows:

a.The validation process initially taken in this study shows that the proposed evaluation methodology is acceptable.

b. The Spalart-Allmaras (SA) turbulence model adopted in superior in terms of accuracy when compared to other models.

c. The helical Savonius rotor with endplate and overlap ratio shows a low variation of static torque coefficient at various rotor angles. In fact, it can overcome the drawback of negative and nearly zero static torque coefficient experienced by a conventional Savonius rotor.

d. The best static torque performer of this study is the helical Savonius with 90° twist angle where the static torque coefficient obtained is 0.442. This finding is fairly similar to [6].

Acknowledgments

The authors would like to acknowledge the Malaysian Electricity Supply Industries Trust Account (MESITA) through the Ministry of Energy, Science, Technology, Environment & Climate Change (MESTECC) for funding this research.

References

[1] Lee JH, Lee YT and Lim HC 2016 Effect of twist angle on the performance of Savonius wind turbine. *Renewable Energy*. **89** 231-244

[2] Alom N and Saha U 2018 Four Decades of Research into the Augmentation Techniques of Savonius Wind Turbine Rotor. *Journal of Energy Resources Technology.* **140**(5) 050801

[3] Shaheen M, El-Sayed M and Abdallah S 2015 Numerical study of two-bucket Savonius wind turbine cluster. *Journal of Eind Engineering and Industrial Aerodynamics.* **139** 8-15

[4] Mahmoud N, El-Haroun A, Wahba E and Nasef M 2012 An experiment study of improvement of Savonius rotor performance. *Alexandria Engineering Journal.* **51**19-25

[5] Han N and Li X 2017 Study of Static and Dynamic Performance of Miniature Savonius-type Wind Energy Harvesters. *Energy Procedia.* **105** 119-124

[6] Kamoji MA, Kedare SB and Prabhu SV 2009 Performance tests on helical Savonius rotors. *Renewable Energy.* **34**(3) 521-529

[7] Damak A, Driss Z and Abid M 2013 Experimental Investigation of helical Savonius rotor with a twist of 180°. *Renewable Energy.* **52** 136-142

[8] Mohamed M, Janiga G, Pap E and Thévenin D 2011 Optimal blade shape of a modified Savonius turbine using an obstacle shielding the returning blade. *Energy Conversion and Management.* **52**(1) 236-242

[9] Rezaeiha A, Montazeri H and Blocken B 2018 Towards accurate CFD simulations of vertical axis wind turbines at different tip speed ratios and solidities: Guidelines for azimuthal increment, domain size and convergence. *Energy Conversion and Management.* **156** 301-316

[10] Sheldahl R, Feltz L and Blackwell B 1978 Wind tunnel performance data for two- and three-bucket Savonius rotors. *Journal of Energy.* **2**(3) 160–164

[11] Zakaria A and Ibrahim M 2018 Analysis of Savonius rotor performance at low speeds using numerical study. *International Journal of Engineering & Technology.* **7**(4.36) 1549-52

[12] Alaimo A, Esposito A, Messineo A, Orlando C and Tumino D 2015 3D CFD analysis of a vertical axis wind turbine. *Energies.* **8**(4) 3013-33

[13] Debnath P, Gupta R and Pandey K 2014 Performance Analysis of the Helical Savonius Rotor Using Computational Fluid Dynamics. *ISESCO Journal of Science and Technology.* **10**(18) 17-28

[14] Roy S and Saha U 2013 Computational study to Assess the Influence of Overlap Ratio on Static Torque Characteristics of a Vertical Axis Wind Turbine. *Procedia Engineering.* **51** 694-702

IOP Conf. Series: Materials Science and Engineering **715** (2020) 012015 doi:10.1088/1757-899X/715/1/012015

Aerodynamic optimization of unmanned aerial vehicle for offshore search and rescue (SAR) operation

Sharul Sham Dol

Department of Mechanical Engineering, Abu Dhabi University, Abu Dhabi, United Arab Emirates

*E-mail: sharulshambin.dol@adu.ac.ae

Abstract. Workers on offshore oils rigs are constantly risking their lives due to the location of these rigs in the middle of the ocean. One of the biggest risks to these workers are falling from the rigs and drowning especially during bad weather. Operating UAVs can reduce the risks of injury, human errors and enhances inspections, surveying and emergency operations. The applications of UAVs in offshore tasks are discussed in this paper as well as the aerodynamic optimization of UAV's propellers. Computational Fluid Dynamics (CFD) analysis and experimental procedures were conducted on multiple propeller alternatives to determine the drag and lift properties and selecting the optimal option based on the lift to drag ratio and desired output. The CFD analysis and the experimental test found that the medium sized (9.0 x 4.5 in) propellers were the optimal option with a drag to lift ratio of 1.75.

1. Introduction

An offshore platform requires a myriad of safety guidelines to prevent disasters. Nowadays, some companies demand the presence of an unmanned aerial vehicle (UAV) at their sites for multiple purposes including security and inspection. UAVs can be operated by trained personnel or pre-programmed to fly a certain path. Autonomous UAVs minimize the human operational role and offer quicker response time to emergency SAR operations.

During bad weather, offshore workers could be falling off oil rigs and means of search and rescue are costly and time ineffective. Technologies used for rescue rely on onshore rescue centers and at times colleagues who throw a floatation device [1]. UAV technology can greatly assist in delivering both cost and time effective solutions and will increase the rate of rescue success. It is intended to develop a small rotary UAV able to hover and deliver floatation device.

One of the major drawbacks of quadrotor is its high energy consumption. The matter becomes worst during windy weather especially on the oil rig offshore platform. Therefore, they are not recommended to operate in windy conditions as the battery power drains very quickly and eventually results in poor flight time. To extend the operation time without increasing battery capacity, the aerodynamic performance of a drone becomes very important [2-3]. The power consumption can be reduced by improving the designs of the propeller of the drone to increase the overall thrust. By modelling the propeller design of drone, its aerodynamics performance can be evaluated for further improvement through simulations and experiments. Factors including speed, versatility, fly-ability and design of the drone body as well as the propeller can affect the efficiency of the drone and the SAR operation [3-4]. A concept that is widely thought of is attaching a life buoy to the base of a drone and dropping at near a swimmer in distress, however, the extra weight of the payload will affect the performance of the drone battery thus requiring extra optimization.

The goal of this work is to improve the hovering capabilities and increase the flight time of a UAV by modifying the propellers as well as determining the best alternative that will produce the highest lift to drag ratio.

2. Methodology

The generation of lift is mainly due to the diversion of air molecules by a solid body (i.e. propellers) in a certain direction. In drone, the propellers push the air downwards and consequently the air pushes the drone upwards, creating lift force. In a vertical plane, the drone can hover as well as changing altitude. Hovering is a result of creating lift force that is equal to the gravitational force downwards (i.e. weight). Turning the drone to a certain direction requires manipulating the amount of thrust the rotor can produce.

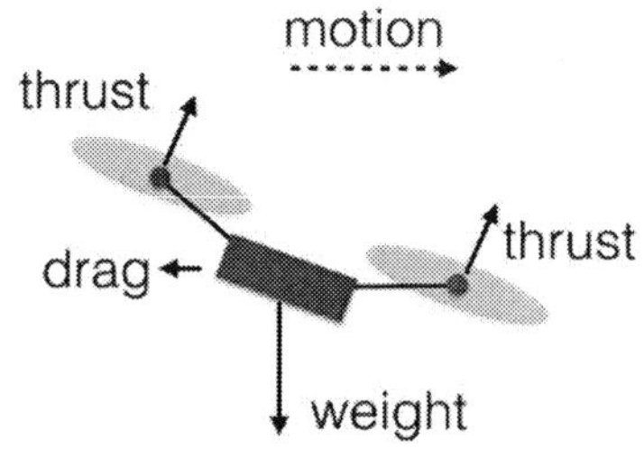

Figure 1. Direction change in a drone.

Figure 1 shows a drone turning to the right side and the red rotor indicates lower thrust production than the opposing green rotor which will lead to motion in that direction. The rotors are the components responsible for creating lift as well as turning the drone around and moving it forward.

Thrust to weight ratio (TWR) is an important parameter to take into consideration to calculate the efficiency of drone and can aid in preventing excessive thrust that will consume more energy. The TWR will indicate if a drone can lift a specific weight and the required power to do so, the equation can be simply stated as,

$$TWR = \frac{T}{W} = \frac{Thrust}{Weight} = \frac{ma}{mg} = \frac{a}{g} \tag{1}$$

where m is mass of the drone and the payload, a is the drone acceleration and g is a gravitational acceleration. TWR decides if the drone will be able to perform as desired. For a drone to maintain flight steadily the TWR must be equal to unity and to produce turning effects with respect to the angle of attack, a TWR of minimum 1.3 must be achieved in order to perform the vertical acceleration (take-off) the TWR must be greater than 1. Drone manufacturers aim to produce drones that can create their products with a minimum TWR of 2, which means each rotor in a hexa-copter (6 rotors) must be able to produce approximately TWR of 0.33.

Wind is considered the toughest opponent to drone performance especially for offshore application. Wind blowing at high speeds will make it difficult for the drone to hold its position or maneuver as desired. Drag force is amplified if the wind is blowing in an opposite direction of the drone's motion. The windy weather requires more power from the motors that can drain the battery faster. Adding an extra battery would increase the weight of the drone (i.e. additional payload). The maneuverability and flight time can be enhanced by modifying the propellers properties (size, shape, material) as well as streamlining the frame of the drone to reduce pressure and friction drags. Computational Fluid Dynamics simulation requires adding the wind speed of the region at which the drone will operate. The wind conditions at Upper Zakum Offshore Oil Field will be used for example.

Upper Zakum Offshore Oil Field is an oil field located 84 kilometers offshore of the capital Abu Dhabi (Figure 2). The field is producing 640,000 barrels of oil per day through a chain of oil wells and

2150 people working on 90 different platforms that increase the possibilities of danger occurring [5]. For the wind speed data for a day starting from 6:00 AM to 6:00 PM, the maximum wind speed reached was 17 km/h while the lowest was 8 km/h. As the season changes and is approaching summer, the wind speed decreases and fluctuates at a range between 10-11 km/h.

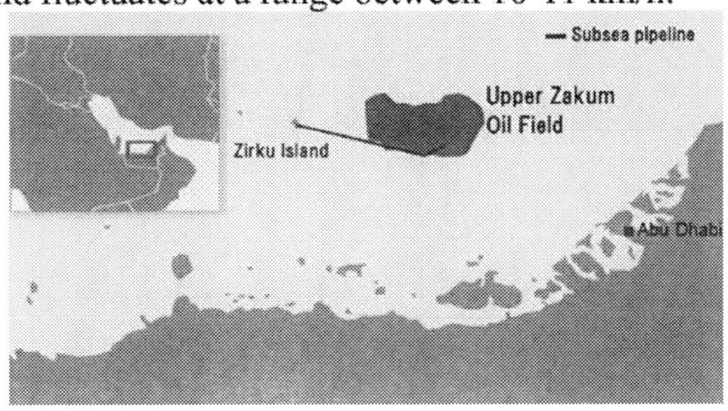

Figure 2. Upper Zakum offshore oil field.

The drone frame's size and type must be able to withstand rough winds and high temperatures as well as being powerful enough to carry loads for survivors in distress. The frame used for this project is the DJI F550 frame, which is commercially available as shown in figure 3.

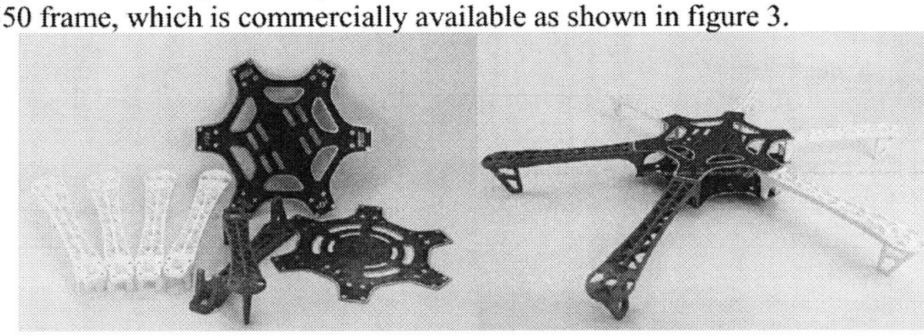

Figure 3. DJI F550 frame kit.

The program used for the CFD analysis is ANSYS (Fluent). The drawings of the chosen alternatives for propellers were exported as IGS format files from SolidWorks to ANSYS as shown in figure 4. Details of the CFD equations and turbulence models can be referred in [6-7].

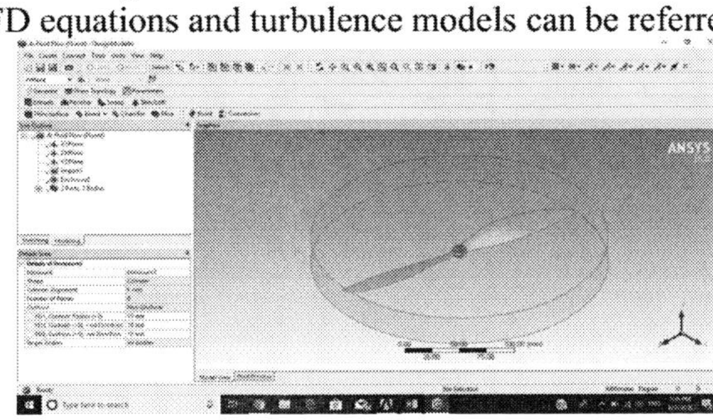

Figure 4. Imported propeller drawing with enclosure.

3. Results and discussion

Multi-rotor drones provide the necessary functions for SAR operations. There are ranges of multi-rotor drones that can be used for the objective however some designs are more suitable than others after taking into considerations the conditions of operation and the added modifications that will result in the increase of weight. Multi-rotor drones are classified by the number of rotors attached and their configuration. A powerful drone with high stability and lifting capability is needed. The most suitable choices are the quad-, hexa- and octo-copters. Table 1 shows the decision matrix for initial evaluation.

Table 1. Decision matrix showing Hexa-copter as the preferred option.

	Quad-copter	Hexa-copter	Octo-copter
Power	3	5	5
Loading capacity	1	5	5
Flight time	5	3	1
Price	5	3	1
Total score	14	16	12

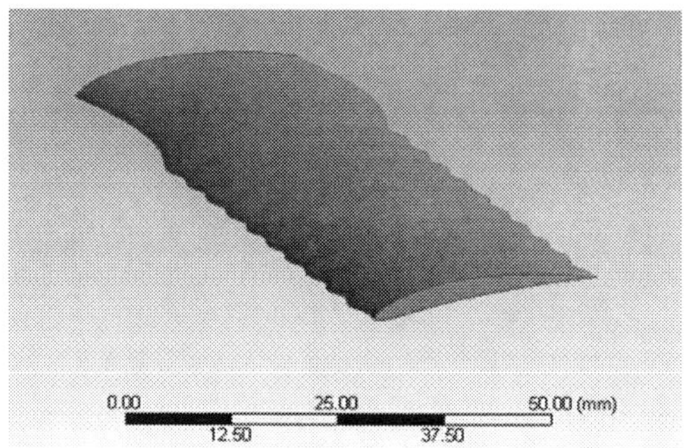

Figure 5. Bumps on airfoil.

Numerous improvements on current mechanical designs are derived through biomimicry. Biomimicry is an approach to innovation that seeks sustainable solutions to human challenged by emulating nature's time-tested patterns and strategies [8]. The idea of whale fin blade originate from whale flippers. Bumps on an airfoil are called tubercles and are said to consume 25% less energy than normal blades spinning at the same speed. An airfoil with tubercles technology is said to produce 32% less in drag and 8% increase in lift compared to a smooth end airfoil. The main idea behind the tubercles technology is to have steeper angles to enhance lift by pushing more air and prevent stalling. Wind tunnel testing showed that the bumps at the edge of an airfoil created vortices that were deflected to the valleys between the hills and had opposite spins making sure that the airflow did not separate until the end of the airfoil which led to higher lift and less drag [9-10]. The work on this type of propeller (Figure 5) is currently in progress and the results will be presented in the subsequent publication.

Using equation (1) for TWR of 2.5 (taken into consideration of rough offshore winds as discussed in previous for Upper Zakum Offshore Oil Field) and the propeller chord length of 0.75 of its radius, the Reynolds number that is calculated from $Re = \rho VL/\mu$ is found to be around 20,000, which justifies the full turbulence condition in our model. The maximum coefficient of lift achieved throughout the 1000 iterations was 12.2 and a minimum of 3. The lift to drag ratio will be calculated on the numbers supplied by this test. It can be seen in the graph (Figure 6) that the coefficient of lift is higher than the coefficient of drag meaning that the drone lift force will be able to overcome the drag forces. The calculation resulted in a hovering thrust of 12.6 N.

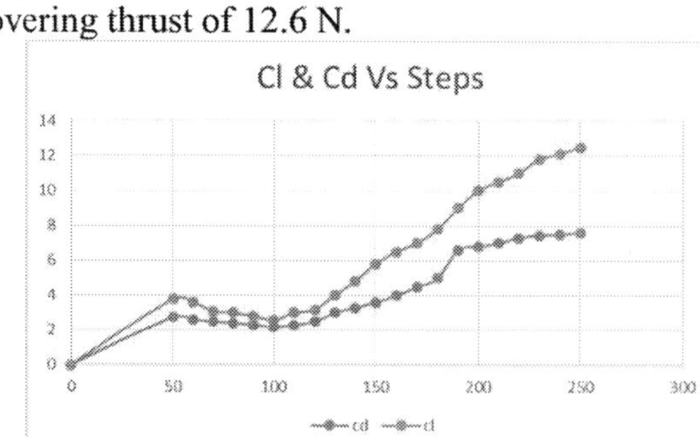

Figure 6. C_L and C_D vs. steps.

IOP Conf. Series: Materials Science and Engineering **715** (2020) 012015 doi:10.1088/1757-899X/715/1/012015

Figure 7 demonstrates the outlet velocity of the propeller that is the region where the air is accelerated into higher velocity. Figure 8 shows the streamline of the air particles as they come in contact with the rotating propeller. The figure illustrates the velocity at which the propeller is rotated, created sufficient lift forces during take-off and maneuvers. The red region is the region with the highest velocity opposed to the blue region that rotates at a lower velocity compared to the tip.

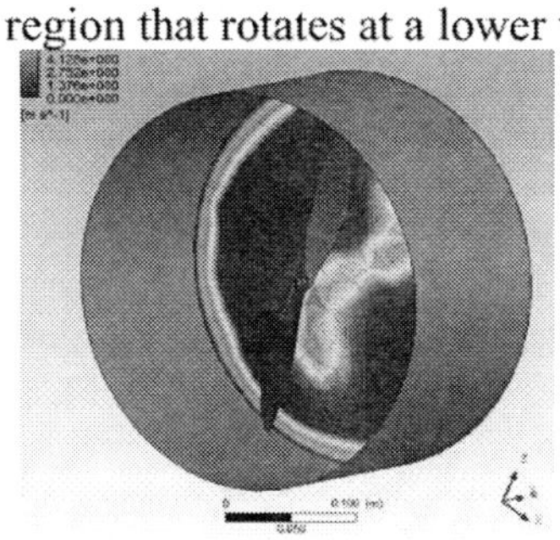

Figure 7. Outlet velocity gradient.

Figure 9 shows the complete drone prototype and its hovering capability, justified by all the parameters used in the previously mentioned simulations and analyses.

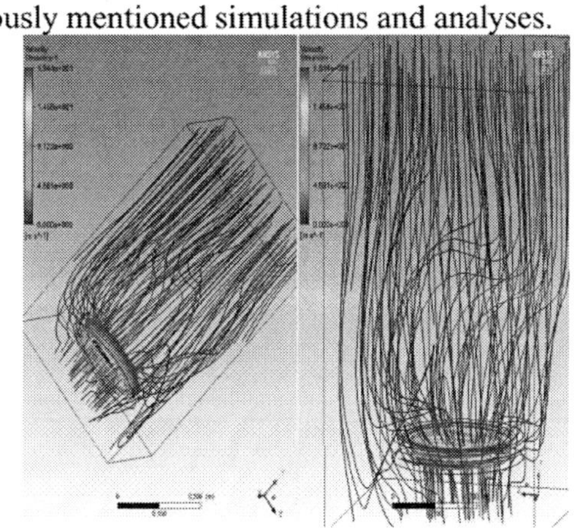

Figure 8. Velocity streamlines.

Figure 9. Drone prototype.

4. Conclusion

The objective of the project was to optimize the performance of drone propellers by increasing its flight time and enhancing its ability to combat harsh weather conditions for SAR. A hexa-drone frame type was chosen due to its design, power output potential and price. The 9.0 x 4.5 in propellers were the final choice due to their output aerodynamics performance. The advantage of this prototype is it is faster than its counterparts and has a greater flight time for longer operations, making it suitable for offshore SAR applications.

References

[1] Bureau of Labor Statistics. Oil and Gas Industry Fatal and Nonfatal Occupational Injuries. U.S. Bureau of Labor Statistics, Washington, DC (2010).

[2] Yeong, S. P., and S. S. Dol. Aerodynamic Optimization of Micro Aerial Vehicle. Journal of Applied Fluid Mechanics **9,** no. 5, 2111-2121 (2016).

[3] Eid, Saif Eldin, and Sharul Sham Dol. "Design and Development of Lightweight-High Endurance Unmanned Aerial Vehicle for Offshore Search and Rescue Operation." In 2019 Advances in Science and Engineering Technology International Conferences (ASET), pp. 1-5. IEEE, (2019).

[4] Yeong, S. P., L. M. King, and S. S. Dol. A Review on Marine Search and Rescue Operations Using Unmanned Aerial Vehicles." World Academy of Science, Engineering and Technology, International Journal of Mechanical, Aerospace, Industrial, Mechatronic and Manufacturing Engineering **9** no. 2, 396-399 (2015).

[5] NA, "Upper Zakum Offshore Oil Field Development, Abu Dhabi," N.d N.d N.d. [Online]. Available: https://www.offshore-technology.com/projects/upper-zakum-offshore-uae/. [Accessed 1 6 2018].

[6] Chan, H. B., T. H. Yong, P. Kumar, S. K. Wee, and S. S. Dol. The numerical investigation on the effects of aspect ratio and cross-sectional shape on the wake structure behind a cantilever. ARPN J. of Eng. and App. Sci. **11** (2016).

[7] Azeez, Abid Abdul, Sharul Sham Dol, and Mohammad S. Khan. "Effects of Cylinder Shape on the Performance of Vortex Induced Vibration for Aquatic Renewable Energy." In 2019 Advances in Science and Engineering Technology International Conferences (ASET), pp. 1-4. IEEE, (2019).

[8] Chear, C. K., and S. S. Dol. "Vehicle aerodynamics: drag reduction by surface dimples." International Journal of Mechanical, Aerospace, Industrial, Mechatronic and Manufacturing Engineering **9**, no. 1 (2015).

[9] Dol, S. S., and M. A. M. Nor. "Flow visualization of the vortex shedding of a stationary circular cylinder by an improved smoke-wire technique." WSEAS transactions on fluid mechanics **1**, no. 6 (2006): 745.

[10] Yong, T. H., and Sharul Sham Dol. "Design and development of low-cost wind tunnel for educational purpose." In IOP Conference Series: Materials Science and Engineering, vol. **78**, no. 1, p. 012039. IOP Publishing, (2015).

A fast mesh deformation method for fluid–structure interaction problems

A. Jendoubi

PMU, Mechanical Engineering Department, Prince Mohammad Bin Fahd University, P.O. Box 1664 Al Khobar 31952 KSA

*E-mail: ajendoubi@pmu.edu.sa

Abstract. Previous work by the author [1] has developed a very simple moving mesh method, called parabolic extension, allowing for very large deformation in static problems and to very large time steps for time-dependent problems. In this paper, we significantly accelerate the previous method by using a forward Euler scheme with a lumped mass matrix to solve the parabolic extension and we applied it for simulation in fluid-structure interaction. The use of a lumped mass matrix in the forward scheme leads to an explicit algorithm (no linear system to solve) which make the new technique presented here very fast and less computational cost. Numerical benchmarks including complex boundary with irregular deformation are carried out. The results demonstrate that the new method generates the deformed mesh very fast and with high quality even in large deformations, especially concerning the boundary meshes.

1. Introduction

Fluid-structure interaction (FSI) problems occur in various engineering applications and their numerical simulations is still challenging nowadays. This phenomenon occur when a potentially deformable solid interacts with a surrounding fluid. The flow of the fluid deforms the solid and/or changes its position thus modifying the geometry of the fluid domain. Many difficulties arise and we will briefly present some of them in this work. Arbitrary Lagrangian Eulerian (ALE) formulations (see [2], [3]) are frequently used to solve such problems where the solid and the fluid are respectively in Lagrangian and Eulerian coordinates. The ALE approach is based on an arbitrary motion of the reference frame, which is continuously rezoned in order to allow a precise representation of moving interfaces. Original developments were made, among others, by Noh [4], Franck and Lazarus [5] and Hirt et al. [6]. Early applications are to be found in the work of Donea et al., Belytschko and Kennedy [7] and Hughes et al. [8] for free surface flows and FSI problems.

The computer implementation of the ALE technique requires the introduction of a mesh-update technique that assigns mesh-node velocities or displacements at each time step, or load step of a calculation.

Knowing the node displacements on the solid boundary (obtained by solving the structure equations), various techniques have been proposed to determine how to update the fluid mesh. In a finite element method (or any other technique based on a discrete mesh representation of the domain), the ability to modify the mesh without a complete reconstruction is central to an efficient implementation of the method. The most frequently encountered problem being the excessive deformation of the elements adjacent to the boundary, leading to computational problems and eventually to the divergence of the solution process. In order word, If care is not taken, elements adjacent to the solid boundary can degenerate leading to computational problems and eventually to the divergence of the solution process.

We consider the general case where the flow of the fluid deforms the solid and changes its position thus modifying the geometry of the fluid domain. We present briefly in this work different moving mesh methods such as kriging, harmonic extension, displacement methods based on an elasticity problem, the parabolic extension (which are presented clearly in [1] and we propose a fast method called the modified parabolic extension to accelerate the step of updating the mesh in the ALE method to solve the FSI problems. The outline of this paper is as follows. As a model for problem with moving boundary, and a typical example of usage of mesh-update techniques, we present, in Section 2, the general context of fluid-structure interactions in ALE formulation. In Section 3, we remind briefly some mesh-update methods presented [1] and we present the new accelerated parabolic extension proposed in this paper. The efficiency of the new parabolic extension algorithm is illustrated with numerical FSI examples in Section 4.

2. The coupled FSI problem: governing equations

In order to introduce the general non-linear fluid-structure problem, let us consider a time-dependent domain $\Omega(t) \subset R^d$, d = 2 or 3. We assume, for all time t, that $\Omega(t) = \Omega_f(t) \cup \Omega_s(t)$ and $\Omega_f(t) \cap \Omega_s(t) = \Phi$, where $\Omega_f(t)$ is occupied by an incompressible viscous fluid and $\Omega_f(t)$ by an elastic solid. The reference (initial) configuration of the system is defined $\Omega_0 = \Omega_{f0} \cup \Omega_{s0}$. The fluid-structure interface at time t is denoted by $\sum(t) = \overline{\Omega_f(t)} \cap \overline{\Omega_s(t)}$ and $\sum_0 = \overline{\Omega_{f0}} \cap \overline{\Omega_{s0}}$ on the initial configuration.

For the fluid, since we are dealing with a moving interface, we consider the incompressible Navier-Stokes equations in ALE formulation (see [9]):

$$\begin{cases} \rho_f\left(\frac{\partial \boldsymbol{v}_f}{\partial t} + (\boldsymbol{v}_f - \boldsymbol{v}_s^e)\cdot\nabla\boldsymbol{v}_f\right) + \nabla p - \nabla\cdot(2\eta\dot{\gamma}(\boldsymbol{v}_f)) &= \boldsymbol{r}_f \text{ in } \Omega_f(t) \\ \nabla\cdot\boldsymbol{v}_f &= 0 \text{ in } \Omega_f(t) \\ \boldsymbol{v}_f(\boldsymbol{x},t) &= \boldsymbol{v}_s(\boldsymbol{x},t) \text{ on } \Sigma(t), \end{cases} \tag{1}$$

where v_f (x, t) is the Eulerian velocity field of the fluid. The vector v_s^e (x, t) is the Eulerian domain (mesh) velocity that will be described later on. As we will see, the domain velocity is naturally defined in the solid domain and must somehow be extended to the fluid domain thus the subscript s → f. Note that in general v_s^e (x, t) ≠ v_f (x, t). The continuity of the fluid and solid velocities is imposed on the interface Σ(t). The system is completed by initial and boundary conditions on the other parts of the boundary $\Gamma_f^D(t)$ (Dirichlet) and $\Gamma_f^N(t)$ (Neumann). The complete problem is thus defined (and solved) on the current geometry Ω_f (t).

The motion of the structure is described by the following equations:

$$\begin{aligned} p_{s0}\dot{V}_s - \nabla\cdot(\Pi_s(U_s)) &= p_{s0}r_s \text{ in } \Omega_{s0} \\ \dot{U} &= V_s \quad \text{in } \Omega_{s0} \end{aligned} \tag{2}$$

where $\mathbf{U}_s(\mathbf{X}, t)$ and $\mathbf{V}_s(\mathbf{X}, t)$ are respectively the Lagrangian displacement and velocity fields of the solid, ρ_{s0} its density in the undeformed geometry and $\mathbf{r}_s$ a given body force (usually vanishing). The problem is thus written in a total Lagrangian formulation on Ω_{s0}.

The tensor $\mathbf{\Pi}_s = \mathbf{\Pi}_s(\mathbf{U}_s)$ is the fifirst Piola-Kirchhoffff tensor (see Bonet [10]) and is related to the Cauchy stress tensor $\boldsymbol{\sigma}_s$ by the relation:

$$\Pi_s = J_s\sigma_s \cdot F_s^{-T} \tag{3}$$

where $\boldsymbol{F}_s$ is the gradient of deformation tensor and J_s is the Jacobian of the transformation from the initial geometry Ω_{s0} to the current configuration $\Omega_s(t)$. These quantities come out naturally from the solution of problem (2). This problem is also completed with a proper set of initial and boundary conditions (Dirichlet on Γ_s^D and Neumann on Γ_s^N).

From a mechanical point of view, the coupling between the systems (1) and (2) is realized by imposing the equilibrium of the stresses at the interface:

$$\sigma_s \cdot n_s = -\sigma_f \cdot n_f \quad on \sum(t),$$

The Cauchy tensor $\boldsymbol{\sigma}_f$ comes out of the solution of system (1) and is defined on the current configuration. System (2) is however solved on the reference configuration. $\boldsymbol{\sigma}_f$ must therefore be transferred on the reference geometry. It is easily seen from (3) and from the transformation of the normal vectors that the above equilibrium condition on the deformes geometry becomes:

$$\prod_s \cdot N_s = -J_{s\mapsto f\sigma f} \cdot F_{s\mapsto f}^{-T} \cdot N_f \quad \text{on} \sum_{0.} \tag{4}$$

on the reference geometry. In the above equation, **N** and **n** stand for the normal vectors to the initial and deformed geometries respectively. The equilibrium condition (4) is added to system (2) as a Neumann boundary condition. Note that $J_{s\mapsto f}$ and $F_{s\mapsto f}$ are extensions to Ω_{f0} of similar quantities already defined in Ω_{s0}. Their computation will require the definition of a Lagrangian displacement in Ω_{f0}. In practice, any reasonable extension (denoted U_s^e) of $U_s\big|_{\Sigma_0}$ over Ω_{f0} can be used and we will see in Section 3 a number of ways to achieve that goal. It is however quite natural to impose the continuity of the displacements at the interface:

$$U_s^e = U_s \quad \text{on} \sum_0 \tag{5}$$

The Lagrangian mesh velocity $V_{s\mapsto f}$ in the fluid domain can then be easily obtained from a finite difference in time. For instance, if an Euler scheme is used, then:

$$V_{s\mapsto f}(X,t) = \frac{U_s^e(X,t+\Delta t) - U_s^e(X,t)}{\Delta t} \tag{6}$$

Which, together with (5), imposes the continuity of the velocities on the interface.

We therefore suppose that we have defined an extension U_s^e of the mesh displacement in Ω_{f0} from which we can evaluate $J_{s\mapsto f}$ and $F_{s\mapsto f}$. As a consequence, the current configuration of the fluid domain, $\Omega_f(t)$, is parametrized by:

$$\Omega_f(t) = \Omega_{f0} + U_s^e(\Omega_{f0}) \tag{7}$$

In other terms, each node X of the initial fluid domain has a corresponding node x in the deformed configuration satisfying:

$$x = X + U_s^e(X,t)$$

In this way, we can follow each node of the fluid mesh. Consequently, the Eulerian domain velocity v_s^e at node x satisfies:

$$v_{s\mapsto f}(x,t) = V_{s\mapsto f}(X,t) \tag{8}$$

and will be used in System (1).

3. Some Mesh-update techniques for FSI problems

We propose in this section different constructions of the extension U_s^e. The main difficulty is to avoid the occurrence of overly distorted elements. Once U_s^e has been constructed, quantities such as $J_{s\mapsto f}$ and $F_{s\mapsto f}$ are easily computed. The mesh velocity is then obtained from (6) and (8). The update of the mesh is thus a crucial step. In the following, we summarize briefly some methods but more details regarding these different methods are presented in details in [1].

Let us emphasize that we are interested in comparing those methods in the context of time-dependent (or quasi-static) phenomena. In this work, we want to accelerate the parabolic extension presented in [1] and thus not all mesh-update procedures are presented here. For example, the evaluation of the computational performance of Radial Basis Functions (RBF) methods, described in de Boer et al. [11], Gong et al. [12], in Rendall and Allen [13, 14] and in Romani et al. [15], is not straightforward and is outside the scope of this paper. The only methods considered in this paper are Laplacian smoothing, elasticity model and the proposed new method which is accelerated parabolic extension.

3.1. Laplacian smoothing

Laplacian smoothing is a classical method for updating meshes (see [16]). It consists in solving a Laplace (or Poisson) equation for each component of the node position. The following problem is solved separately for each component of the displacement.

$$\begin{aligned} -\Delta U_s^e &= 0 \text{ in } \Omega_{f0}, \\ U_s^e &= \boldsymbol{Us} \text{ on } \Sigma_0, \\ U_s^e &= 0 \text{ on } \partial\Omega_{f0} \backslash \Sigma_0, \end{aligned} \tag{9}$$

The idea is to harmonically expand the evolution of the boundary onto the whole of Ω_{f0}. This technique has the merit of simplicity and lead to a low computational cost.

3.2. Elasticity model

Another possibility (see [17]) is to solve the small deformation elasticity system:

$$\begin{cases} -\nabla\cdot\left[2\mu\gamma(U_s^e)-\dfrac{2\mu}{3}(\nabla\cdot U_s^e)I\right]+\nabla p=0 & \\ -\dfrac{1}{k}p-\nabla\cdot U_s^e & =0 \\ \mathrm{U}_s^e & =\mathrm{U}_s \text{ on } \sum_{0,} \\ U_s^e\cdot n & =0 \quad \text{on } \partial\Omega_{f0}\backslash\sum_{0,} \end{cases} \tag{10}$$

where λ and k (bulk modulus) are obtained from the Poisson coefficient ν and Young's modulus E as

$$k=\frac{E}{3(1-2\nu)} \; and \; \mu=\frac{E}{2(1+\nu)}$$

This problem cannot be solved component per component. It is thus more expensive that the Laplacian smoothing. The value of the Poisson coefficient ν can be chosen close to 0.5 in order to preserve the area (volume) of the elements. The computational cost of such an approach would however be prohibitive as shown in [18]. Notice that in this elasticity approach, we cannot construct the extension component wise. This method is clearly more expensive (from a computational point of view) then the harmonic extension.

3.3. Accelerated parabolic extension

We propose an extension method based on the heat equation using an artificial and arbitrary time.

$$\begin{cases} \dfrac{\partial U_s^e}{\partial t}-\nabla\cdot(k(x)\nabla U_s^e)=0 \quad \text{in } \Omega_{f0,} t\geq 0 \\ U_s^e\cdot n \quad =0 \quad \text{on } \partial\Omega_{f0}\backslash\sum_{0,} \\ \mathrm{U}_s^e \quad =\mathrm{U}_s \text{ on } \sum_{0,} \end{cases} \tag{11}$$

The basic idea and starting point of this method was to keep the simplicity of the harmonic extension while improving its behavior by avoiding the generation of invalid elements near the moving boundary.

3.3.1. Solving the parabolic extension

The accuracy of the approximate solution of (11) is not essential and a first order time scheme (Euler) is largely sufficient and will lead to a low computational cost. Two methods were explored, a backward Euler scheme ([1]) and a forward Euler scheme with a lumped mass matrix. The first method was clearly studied in [1]. For this paper we retain the last method since the purpose is to accelerate the first method.

This method is extremely efficient: the use of a lumped mass matrix in the forward scheme leads to an explicit algorithm (no linear system to solve). However the conditional stability of this scheme makes it difficult to use. From our experience, only a few very small time steps can be used. A strict observance of a the CFL stability condition is not necessary. Monitoring the mesh validity and quality is more relevant.

3.3.2. Artifificial time step strategy

Solving (11) over a long period of time (eventually with many time steps) would lead to meshes very close to those obtained with the harmonic extension. It would be indeed a very inefficient way to obtain an harmonic extension.

In the FSI numerical examples presented in the paper, to limit the number of time steps (for computational reason), we need to use a step length sufficient for the diffusion effect to take place while avoiding the steady state. From the definition of the diffusion coefficient, we have

$$\kappa = 1 = \frac{D^2}{\tau}$$

where τ is the time needed for the diffusion to spread over a distance D. In practice, D can be chosen as the maximal value of the imposed displacement thus giving τ .

4. Numerical results

4.1. Case #1: 2D Turek et Hron problem

We present a numerical example proposed by Turek et Hron [19]. An elastic bar is attached behind a circular cylinder. The vortex shedding behind the cylinder will provoque oscillations of the flexible structure. The geometry is described as below.

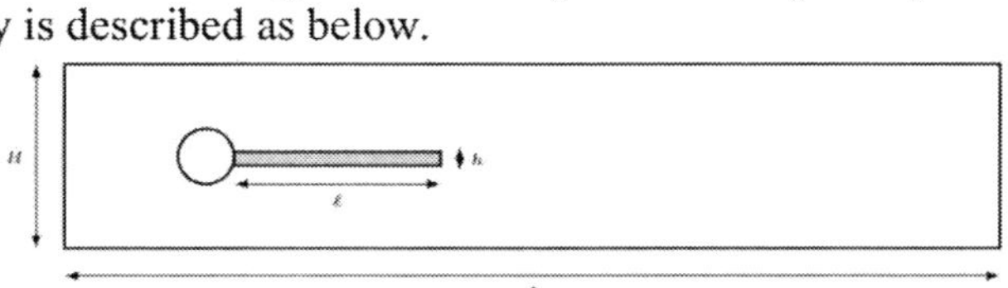

The same dimensions as in [19] were used: $H = 0.41$m, $L = 2.5$m, $h = 0.02$m, and $l = 0.35$m. The cylinder (with radius 0.05m) is centered at (0.2, 0.2). A parabolic velocity profifile is imposed at the inlet of the fluid domain as in [19]. The fluid is Newtonian and the solid is modelled using a Saint-Venant-Kirchhoff hyperelastic model:

$$\prod^{s} = F_S(\lambda_s(tr\ \mathrm{E})\mathrm{I} + 2\mu_s E)$$

where $E = \frac{F^t F - I}{2}$ is the Green-Lagrange tensor. The parameters for the problem are:

Parameters for the solid	Values	Parameters for the fluid	Valeurs
Density ρ_s [kg · m^{-3}]	10^4	Density ρ_f $[kg \cdot m^{-3}]$	10^3
Lamé coefficient λ_S [Pa]	2.0×10^6	Kinematic viscosity ν_f $[m^2 \cdot s^{-1}]$	10^{-3}
Shear modulus μ_s [Pa]	0.5×10^6	Reynok's number Re	100

IOP Conf. Series: Materials Science and Engineering **715** (2020) 012016 doi:10.1088/1757-899X/715/1/012016

As an illustration of the displacement methods described in section 3, we present the fluid meshes obtained after a rigid rotation of $\pi/8$ of the structure. Figure 1 presents the different meshes near the tip of the structure where the displacement is the largest. As can be easily observed, all meshes contain reversed and unusable elements except for the accelerated parabolic extension method.

Using thus the accelerated parabolic extension method with lumped matrix mass, we have successfully solved Turek's benchmark problem. We have been able to reproduce accurately the amplitude end frequency of the displacement of the tip of the structure. Figure 2 shows the velocity field and the position of the structure at three time steps. Note that the kriging method can also be used for the update od the fluid domain.

Let us finish this first numerical example with a simple look at the clock time and memory requirements for the harmonic, elastic and accelerated parabolic extensions. This is done in table 1 We have not indicated the computing time for the kriging method since our implementation is not at all optimal and the comparison would be misleading. The parabolic extension produces valid meshes at a very low cost (comparable to the harmonic extension) and appears as one of the best method for all cases.

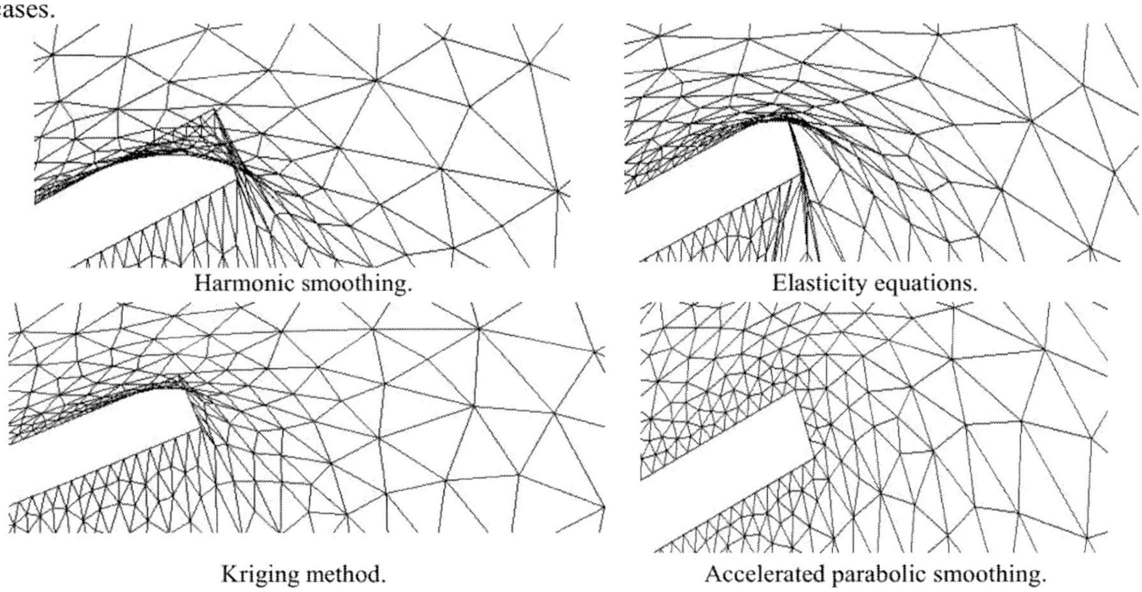

Figure 1. Mesh displacement.

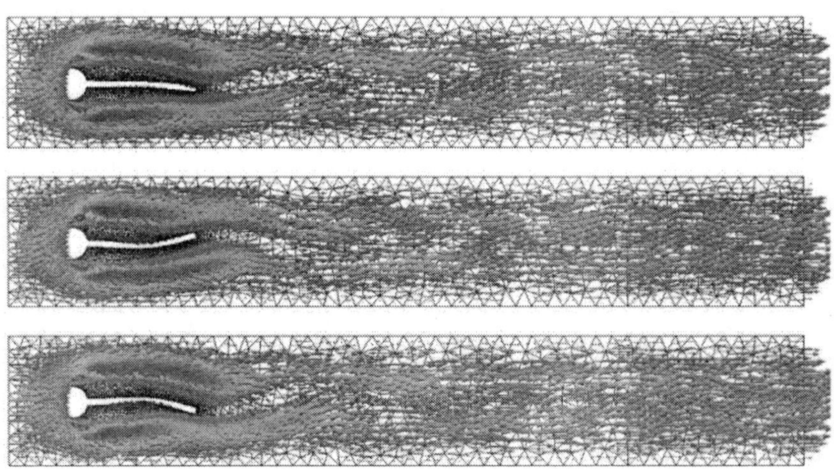

Figure 2. Velocity field and position of the structure.

Table 1. Case #1: Computing times for one iteration. All computations performed on a typical 3GHz 64 bit Intel processor.

Case #1: Turek et Hron		
Methods	Clock time (s)	Memory (KB)
Harmonic extension	16	1 186 508
Elastic extension	86	4 157 202
Parabolic extension	18	1 190 042

4.2. Case #2: 3D Tire interaction with Turbulent flow

We present a second 3D numerical example which a tire interacted with a flow. This problem can be seen as a flow of water around a rolling tire for evaluating the aquaplaning of the vehicle. The purpose of the paper here again is to study the aquaplanning pheniomeno but to show that the accelerated parabolic extension works well even in a very complicated problem. The problem simulation is given in figure 3. Using thus the accelerated parabolic extension method with lumped matrix mass , we have successfully solve the problem. Table 2 presents the clock time and memory requirements for the harmonic, elastic and accelerated parabolic extensions.

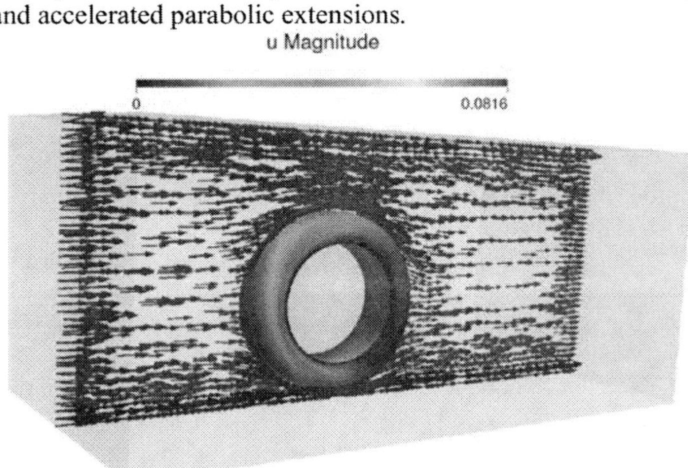

Figure 3. Velocity field and position of tire.

Table 2. Case #2: Computing times for one iteration. All computations performed on a typical 3GHz 64 bit Intel processor.

Case #2: 3D Tire interaction with Turbulent flow.		
Methods	Clock time (s)	Memory (KB)
Harmonic extension	89	10 186 508
Elastic extension	786	40 157 202
Parabolic extension	90	10 190 042

5. Conclusions

In this work, we have introduced a new mesh-update technique called the accelerated parabolic extension with lumped mass matrix. The new method performs very well in all the tests cases considered. It can sustain very large deformations thus avoiding frequent remeshing of the whole domain for fluidstructure interactions problems. The main advantages of the accelerated parabolic extension are its simplicity, its robustness and its low computational cost.

Acknowledgments

The author would like to thank the Mechanical Engineering Department at Prince Mohammad Bin Fahd University, Al Khobar City, Kingdom of Saudi Arabia for all supports into pursuing this work.

References

[1] A. Jendoubi, J. Deteix, and A. Fortin. A simple mesh-update procedure for fluid–structure interaction problems. *Computers & Structures*, 169:13 – 23, 2016.

[2] M.A. Fernández and J.-F Gerbeau. Algorithms for fluid-structure interaction problems. In L. Formaggia, A. Quarteroni, and A. Veneziani, editors, *Cardiovascular Mathematics*, volume 1 of MS&A, pages 307–346. 2009.

[3] M. Astorino. *Interaction fluide-structure dans le système cardiovasculaire. Analyse numérique et simulation*. PhD thesis, Département de mathématiques appliquées, Université Pierre et Marie Curie, Paris VI, France, 2010.

[4] W.F. Noh. CEL : A time-dependent two-space dimensional coupled Eulerian-Lagrangian code. In Fernbach S Alder B and Rotenberg M, editors, *Methods in Computational Physics*, volume 3, pages 117–179, New York, 1964. Academic Press.

[5] R.M. Franck and R.B. Lazarus. Mixed Eulerian-Lagrangian method. In Fernbach S Alder B and Rotenberg M, editors, *Fundamental methods in Hydrodynamics*, volume 3, pages 47–67, New York, 1964. Academic Press.

[6] C.W. Hirt, A.A. Amsden, and J.L. Cook. An arbitrary Lagrangian-Eulerian computing method for all flow speeds. J. *Comput. Phys*., 135:203–216, 1997.

[7] T. Belytschko and J.M. Kennedy. Computer methods for subassembly simulation. *Nucl. Eng. Des*., 49:17–38, 1978.

[8] T.J.R. Hughes, W.K. Liu, and T.K. Zimmermann. Lagrangian-Eulerian finite element formulation for incompressible viscous flows. *Comput. Methods Appl. Mech. Eng*., 29(3):329–349, 1981.

[9] M.A. Fernández. *Modèles simplififiés d'interaction fluide-structure*. PhD thesis, University of Paris IX, 2001.

[10] J. Bonet and R. D. Wood. *Nonlinear Continuum Mechanics for Finite Element Analysis*. Cambridge University Press, Cambridge (UK), New York, 2008.

[11] A. De Boer, M.S. Van der Schoot, and H. Bijl. Mesh deformation based on radial basis function interpolation. *Comp. & Struct*., 85:784–795, 2007.

[12] D. Gong, J. Chang, and C. Wei. An adaptive method for choosing center sets of RBF interpolation. *Journal of Computers*, 6(10):2112–2120, 2011.

[13] T.C.S. Rendall and C.B. Allen. Unified fluid–structure interpolation and mesh motion using radial basis functions. *International Journal for Numerical Methods in Engineering*, 74(10):1519–1559, 2008

[14] T. C. S. Rendall and C. B. Allen. Efficient mesh motion using radial basis functions with data reduction algorithms. *Journal of Computational Physics*, 228:6231–6249, 2009.

[15] L. Romani, M. Rossini, and D. Schenone. Edge detection methods based on rbf interpolation. *Journal of Computational and Applied Mathematics*, 2018.

[16] J.U. Brackbill and J.S. Saltzman. Adaptive zoning for singular problems in two dimensions. J. *Comput. Phys*., 46(3):342–368, 1982.

[17] J.S. Chen, W.K. Liu, and T. Belytschko. Arbitrary Lagrangian–Eulerian methods for materials with memory and friction. In Hughes TJR Tezduyar TE, editor, *Recent developments in computational fluid dynamics*, volume 95, San Francisco, 1988.

[18] T. Wick. Fluid-structure interactions using different mesh motion techniques. *Comput. Struct*., 89(13-14):1456–1467, July 2011.

[19] S. Turek and J. Hron. Proposal for numerical benchmarking of fluidstructure interaction between an elastic object and laminar incompressible flow. In H.-J. Bungartz and M. Schäfer, editors, *Fluid-Structure Interaction*, volume 53 of *Lecture Notes in Computational Science and Engineering*, pages 371–385. Springer Berlin Heidelberg, 2006.

IOP Conf. Series: Materials Science and Engineering **715** (2020) 012017 doi:10.1088/1757-899X/715/1/012017

Modelling research on oxidation of CH4 & C7H16 from small to large scale of temperature at high pressures

Xuejian Cui*

Balwyn High School, Melbourne, 3104, Australia

*E-mail: 1735822788@qq.com

Abstract. The change of CH4 oxidation cause by adding C7H16 from small to large scale of temperature at high pressures were precisely examined under laboratory condition with the temperatures span from 400K to 1200K and the pressures are regulated at 1atm and 10atm at a reaction container with the equivalence ratio of 0.5. The C7H16 proportion in the CH4/C7H16 mixed substances was varies from zero to one hundred percent. The molar ratios for CH4, C7H16, O2, CO, CO2, formaldehyde, ethene were investigated during several trial runs, and the mechanism is selected to establish the research results over the identical terms with the ideally stirred reaction simulation. It is revealed by the resultant that initial T for the oxidation of CH4 becomes smaller as the C7H16 proportion from the mixed substances increases, and a minor portion of C7H16 leads to a significant shift of the oxidation of CH4 from high to low temperature. The result also reflects that under the higher pressure, the CH4 oxidation is significantly promoted, and the oxidation of CH4 appears to have two reaction intervals which have different speed of reaction. According to the reaction pathway analysis at 100 atm, in lower temperature, the CH4 oxidation is enhanced by adding C7H16.

1. Introduction

Natural gas and oxygenated hydrocarbon fuels derived from biomass have become the efficient substituted fuels, because of the high performance on the price and environmental impact contrast to traditional fuel [1-20]. Due to the high proportion of H to C, Natural Gas is supposed to dramatically decrease the CO2 emissions. Moreover, Natural Gas possesses a larger #octane than conventional fuel, and this indicates it is easy to achieve a bigger compression ratio and more efficient thermal reaction inside the engine [3-6]. Regrettably, because of the unpromising reaction rate of CH4 in smaller T, it is challenging to activate natural gas by compressing it in the machines. In order to conquer this trouble, the Natural Gas–Diesel Dual-Fuel engines are established [7, 10-11]. In this model, Natural Gas is activated by a minor portion of diesel and the activation and combusting procedures will be influenced by the Diesel, and this leads to an effect on the activation performance and emissions behaviours of the machines. As a result, it is significant to do research on how the reactions of Natural Gas is influenced by adding more Diesel into the reaction, and this will help us to design and optimize better and more efficient Dual-Fuel engine.

Both of Diesel and Natural Gas are complicated mixtures consist of various substances, causing it challenging to operate the scientific researches for the combustion fuels. Therefore, it is applicable to use surrogate fuel simulation to study the chemical reaction for the complicated mixtures [20-22]. While the majority of natural gas is CH4, C7H16 is regarded as surrogate of diesel due to the likewise low-temperature activation behaviour and cetane numbers. Consequently, in the researches, the CH4/C7H16

IOP Conf. Series: Materials Science and Engineering **715** (2020) 012017 doi:10.1088/1757-899X/715/1/012017

mixture is chosen to be the surrogate simulation of natural gas–diesel dual-fuel to interpret the combination of diesel and natural gas from low to high temperature.

As a stable substance, CH4 is formed because of the large C-H bonds energy in molecular formation, yet C7H16 displays larger reaction rate at lower T by the low-temperature-chemistry [23-27]. The variant reaction behaviour among CH4 and C7H16 and the significant non-linearity of the reaction procedure conduct into bigger challenge at related research, and consequently, just a small amount of information about the impact of adding C7H16 to CH4 is provided in the available resources. Aggarwal Et Al. [28] mathematically researched the activation features for the CH4/C7H16 mixed substances, and figured out that the delayed activation of the CH4/C7H16 mixed substances could be significantly lowered by C7H16, indicating that the activation behaviour of the substance was basically influenced by the oxidation reaction of C7H16. Li Et Al. [29] conducted a scientific experiment on Laminar Burning Speed and Markstein Lengths of premixed CH4/C7H16 mixed substances inside a combusting container, whose outcome revealed that flame instabilities and Laminar Burning Speed of the mixed substances are sensitive to the C7H16 portion less than twenty five percent. Furthermore, Liang Et Al. [30] from the crew recorded the activation delay of CH4/C7H16 mixed substances at various CH4 portion and different starting temperatures through a Shock Tube setting, and the contrast among the experimental and mathematical units of activation delay of the mixed substances are finished, by applying the reacting theories in the modelling of the 0-dimensional Consistent-Volume adiabatic activation procedures of the mixtures was activated. Nevertheless, the substances information of CH4 oxidating reaction with adding up C7H16 are barely covered, and the kinetic impact of C7H16 on the oxidating reaction of CH4 is not researched detailly either.

As a result, the purpose of the presented research is discovering the oxidation behaviour of CH4/C7H16 mixture from small to large scale of temperatures and from small to large scale of pressures. Throughout this experiment, the oxidation of CH4/C7H16 is set at 0.5 equivalent ratio, in the Atmospheric Jet-Stirred Reactor, and is recorded among the temperatures from 400 to 1200K. The pressure is regulated on 1atm and 10atm. The C7H16 portion from the CH4/C7H16 substances varies from zero to one hundred percent. The molar proportion of methane, n-heptane, oxygen, carbon monoxide, carbon dioxide, formaldehyde, and Ethene are detected in the presented research. Moreover, the reaction pathway interpretation of the CH4/ C7H16 mixed substances as well as the sensitivity interpretation of the molar proportion of CH4 are displayed in the experimental environment in order to study the reinforcement of CH4 oxidation by adding C7H16 content.

2. Simulation

A specific scientific mechanism of C7H16 suggested by Zhang Et Al. [31] is used for the simulation, and that is early mentioned in the paper. Every experiment was established inside the ideally stirred reaction container with "chemkin-pro" program applying a transitional solver [32], and the outcomes were contrasted at the following simulation conditions as shown in the table.

Table 1. Simulation environment.

Mixture	methane (%)	n-heptane (%)	Oxygen (%)	Argon (%)	Nitrogen (%)	Flow speed (mL/min)	Temperature (K)
M0	1.0	0.0	4.0	2.0	93.0	1800	400-1200
M10	0.9	0.1	5.8	2.0	91.2	1800	400-1200
M50	0.5	0.5	13.0	2.0	84.0	1800	400-1200
M100	0.0	0.5	11.0	2.0	86.5	1800	400-1200

IOP Conf. Series: Materials Science and Engineering **715** (2020) 012017 doi:10.1088/1757-899X/715/1/012017

3. Experiment Outcome and Interpretation

3.1. Substances temperature variation

The molar proportion of CH4, nC7H16, CO, CO2, O_2, C2H4 and CH2O throughout the oxidation of CH4, C7H16, and CH4/C7H16 were detected by employing the experiment devices introduced above, and the outcomes of the contrast between different experiment conditions were shown in Figure 1. As shown in Figure 1, this model is able to estimate the molar proportion of essential substances in the oxidation at various C7H16 content proportion efficiently.

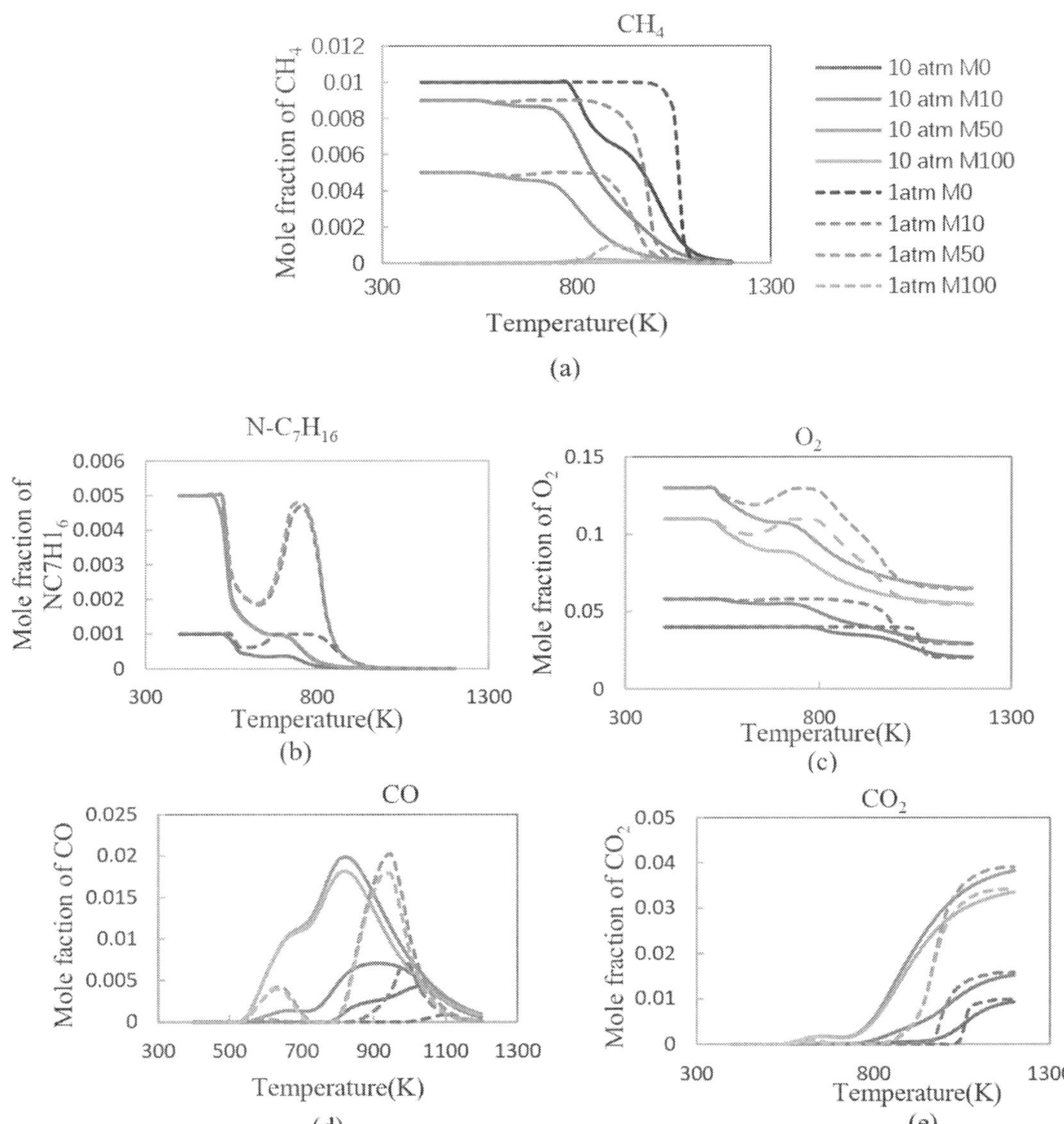

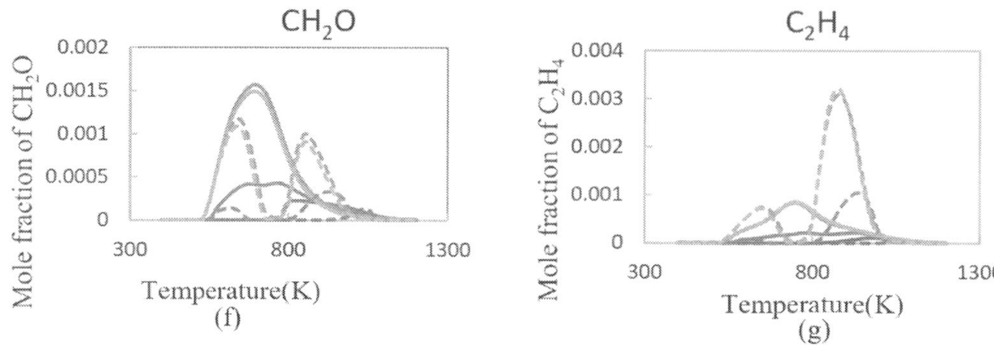

Figure 1. Changes of the molar proportions of main substances with temperatures at various C7H16 portion (P = 1atm, φ= 0.5 and 10atm).

Figure 1 (a-e) presents the changes of the molar proportion of reacting substances (methane, n-heptane and oxygen) and production (carbon monoxide and carbon dioxide) with temperatures variation. Figure 1 (f-g) also displays the changes of the molar proportion of main intermediates (CH_2O and C_2H_4) among different temperatures. Under 1atm condition, the beginning of the oxidation of C7H16 is 550K and low-temperature-chemistry occurs, whilst the onset temperature of CH4 reaction begins around 850K-900K. As seen in Figure 1, between 1000 K and 1100 K, with oxygen relatively reacted and a minor portion of CH_2O discovered at the same time, the molar proportion of CH4 quickly decreases, and CH4 was completely converted at about 1200 K. Under low pressure, CO_2 is produced around the temperature of 880K. The production of CO occurs at 550K, following by the low-temperature-chemistry behaviour. While the C7H16 portion were 10 percent and 50 percent in CH4-C7H16 mixtures, the beginning temperature of CH4 reactions were approximately 900 K and 850 K, and this indicated that as the C7H16 proportion becomes larger, the beginning temperature for the oxidation of CH4 decreases. Furthermore, the outcomes of the molar proportion information throughout the oxidation of CH4–C7H16 mixtures demonstrated that the beginning temperature for CH4 oxidating reaction vary as the C7H16 increases. It is detected that a minor content of C7H16 in the CH4–C7H16 mixtures makes a dramatic promotion on the oxidating reaction of CH4, whilst adding more C7H16 has an ineligible effect. A similar trend was observed from production of CO and CO_2. A small addition (from M0 to M10) makes a dramatic change on the production of CO_2, whereas adding more C7H16 (from M50 to M100) has a marginal impact on the production of CO_2. Similarly, a minor content of C7H16 addition promotes the oxidation of O_2, as the onset temperature becomes much smaller (from M10 to M50), whereas further increase on C7H16 content did not assist in promoting the O_2 oxidation. From the mole fraction-temperature graph of C7H16, it is detected that before the temperature reaches 650K, the temperature is positively proportional to the reactivity. The negative temperature coefficient takes place between 650K-800K. As the temperature exceeds 800K, the high temperature chemistry occurs.

In contrast, under the high pressure (10atm), the onset temperature of CH4 oxidation at M10 (720K) is lower than at 1atm M10 (900K), indicating the CH4 oxidation is promoted. In addition, the C7H16 addition does not have a significant impact on the promotion of CH4 oxidation compared to the 1atm condition, because the onset temperatures of reaction are 800K, 790K, 770K respectively when the amount of C7H16 from the reactant increases. Observed from the graph, the oxidation of CH4 has two reaction intervals at 10atm condition, as the molar proportion of CH4 slowly lowers down at the beginning, its reduction becomes faster in the following interval. Regarding C7H16, the efficiency of low-temperature-chemistry and high-temperature-chemistry is significantly promoted, and the difference between them is reduced. For CO and CO_2, the appearance of high temperature chemistry moves towards the direction of lower temperature. In addition, it is observed that CO and CO_2 are produced in the interval of 500-700K under M10 condition, whereas no CO and CO_2 is produced under M0 condition. Since CH4 does not have low temperature oxidation, the production of CO and CO_2 during 500-700K is mainly due to the low-temperature-oxidation of C7H16. Under low pressure, the difference among low-temperature-chemistry and high-temperature-chemistry was significant for CH_2O and C_2H_4. Nevertheless, there is no apparent distinguish among low-temperature- chemistry and high-temperature-chemistry under high pressure, and only one peak occurs on the graph. Observation from the graph of O_2 under 10atm indicates that onset temperature is slightly reduced in M0 and M10 condition compared to 1atm condition.

3.2. Reaction pathway interpretation

To better explain the process for oxidation of CH4 with adding C7H16, the reaction pathway analysis for the mixed substances at the C7H16 content of 10 percent were conducted under the experiment environment, as displayed in Figure 2.

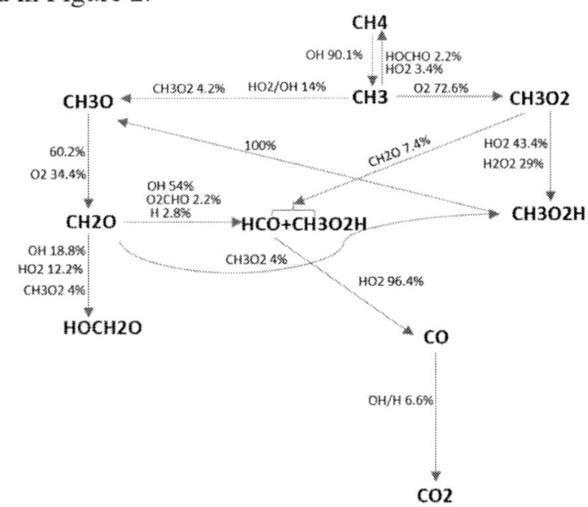

(a)

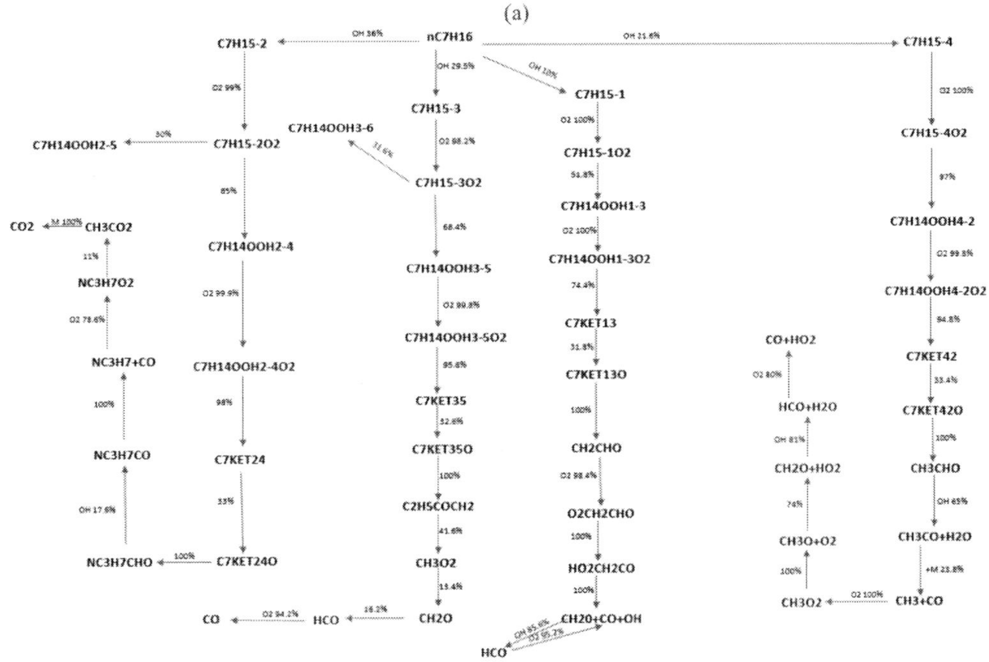

(b)

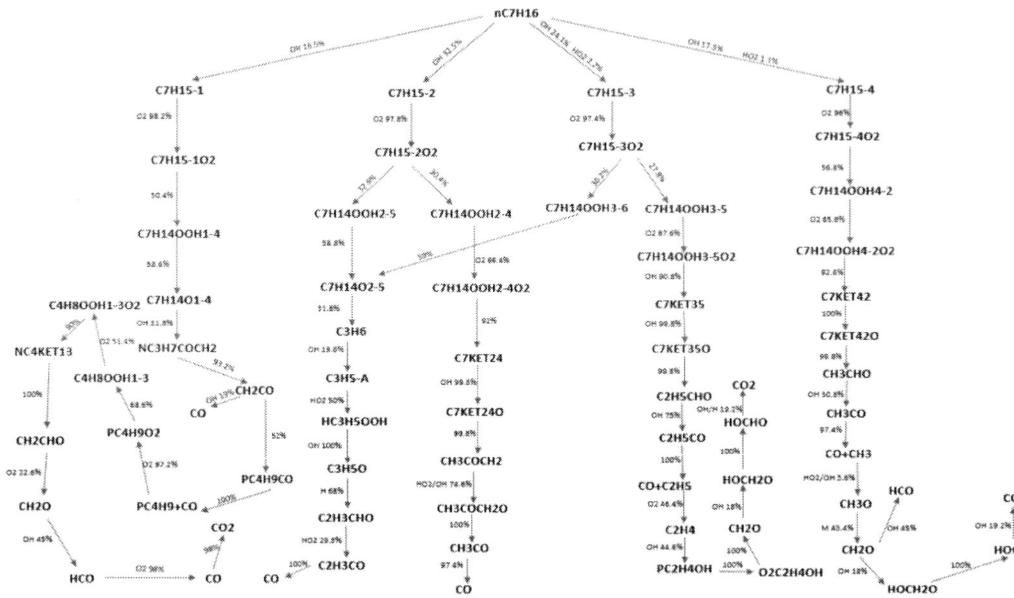

(c)

Figure 2. (a-c) Reaction pathway analysis for CH4 in the reaction with adding ten percent C7H16 at 800 K, 10atm and 0.5 equivalence ratio, and C7H16 in the mixed substances with 10% C7H16 addition at 550K and 750K, 10atm and 0.5 equivalence ratio.

The reaction path analysis for CH4 at the C7H16 content of 10% are conducted under the experiment environment of 10atm at the temperatures of 800K, and the reaction pathway interpretation for C7H16 at the C7H16 content of 10 percent are also simulated under the experiment environment of 10atm at the temperatures of 550K, 750K, respectively, as shown in Figure 2.

With regard to the CH4/C7H16 mixture at C7H16 proportion of 10% at 800K, CH4 was mostly transferred to CH_3 radicals by the H-abstractions at the start of the reactions, and the CH4 reactions with OH through the reaction CH_4 + OH <=> CH_3 + H_2O made the most significant impact to the CH4 consume, accounting for 90.10%. Three channels for the conversion of CH_3 were detected. The majority of CH_3 reacts with O_2 to form CH_3O_2, and is transferred into CH_3O_2H later, and 14.00% of CH_3 was oxidized by oxygen to compose CH_3O straightforward. Then, the formed CH_3O are oxidized by O_2 to form CH_2O, which completely experience the H abstraction with Hydrogen, Oxygen, and Hydroxide radicals, and results in HCO which then is transferred into Carbon Monoxide and Carbon Dioxide as the final production from methane oxidation. The remaining CH_3 mostly reacts with HO_2 and HOCHO to compose CH_4.

Under the temperature of 550K, the C7H16 oxidation starts with the fuel particles completely proceeding the hydrogen abstractions and yielding C7H15 in 4 kinds of isomer, mainly through the radicals of hydroxide in the process C_7H_{15} + H_2O <=> nC_7H_{16} + OH. After that, the substances react with O_2 to form RO_2 (e.g.C_7H_{15}-4 <=> C_7H_{15}-4O_2). The products RO_2 experience the isomerization to form QOOH (e.g.C_7H_{15}-4 <=> C_7H_{14}OOH4-2). Due to the low temperature, the products then experience two types of reactions. The chain branching reaction with the addition of O_2 produces O_2QOOH (e.g. C_7H_{14}OOH4-2O_2), which finally ends up with OQ'O + OH + OH. In addition, the remaining QOOH is transferred into QO + OH through chain propagating reaction.

Under the pressure of 10atm, 750K is already in the high temperature chemistry temperature range, which is because the low temperature chemistry has shifted to the left and happens at lower temperature

IOP Conf. Series: Materials Science and Engineering **715** (2020) 012017 doi:10.1088/1757-899X/715/1/012017

due to the pressure. As the temperature is modified to 750K, the reaction pathway for C7H16 is slightly different from that at 500K. Similarly, C7H16 is resolved into isomers of C_7H_{15}. Moreover, transferred from $C_7H_{14}OOH2$-5, $C_7H_{14}O_2$-5 does not transfer to any substance but is decomposed into C_3H_6.

3.3. Sensitivity interpretation

The interpretation of the sensitivity of the molar proportion of CH4 at different C7H16 ratio is conducted at 800 K along with the sensitivity of C7H16 at various C7H16 proportion at 550K and 750K, under the pressure of 10atm and the 0.5 equivalence ratio to discover the most significant procedure impacting CH4 oxidation and C7H16, as displayed in Figure 3. The sensitivity ratio is determined to be $S = \partial \ln c_j / \partial \ln k_i$, where c_j is the molar proportion, and k_i is the specific rate ratio. A sensitivity ratio number that is bigger than zero demonstrates that the reaction prohibits reaction rate and vice versa.

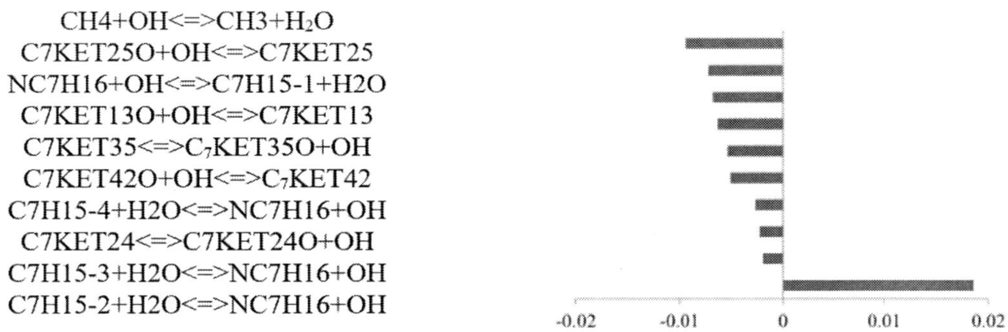

(a) C7H16 sensitivity at 550K.

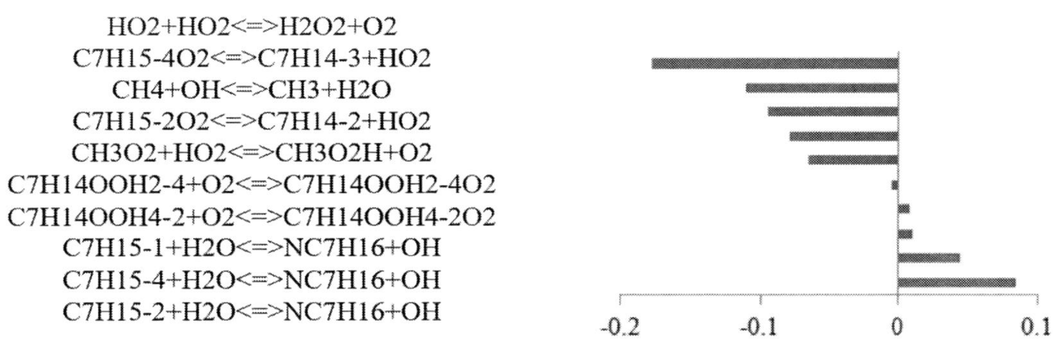

(b) C7H16 sensitivity at 750K.

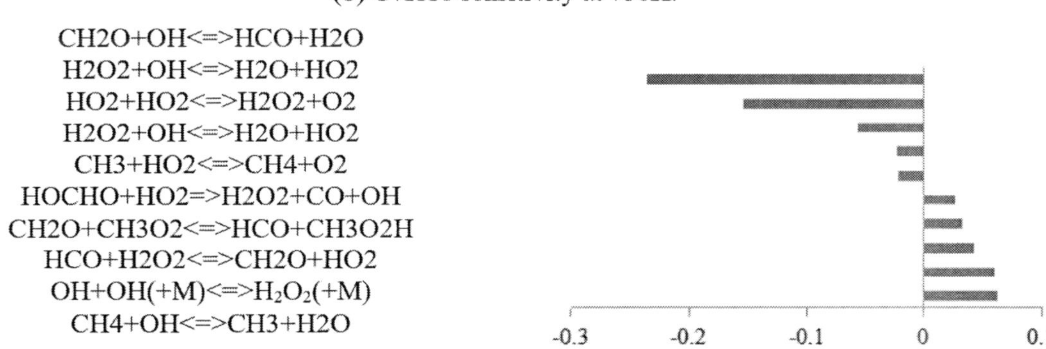

(c) CH_4 sensitivity at 800K.

IOP Conf. Series: Materials Science and Engineering **715** (2020) 012017 doi:10.1088/1757-899X/715/1/012017

Figure 3. Sensitivity interpretation of the molar proportion of C7H16 at 550K and 750K and CH4 at 800 K, 10atm and at 0.5 equivalence ratio.

It is seen from the graphs (a) and (b) that due to the CH4 addition, the reaction of CH4 with OH (CH_4 + OH <=> CH_3 + H_2O) is important at both 500K and 750K. In addition, the RO_2 decomposition reaction (C_7H_{15}-2O_2 <=> C_7H_{14}-2 + HO_2) is also significant at graph (b), even though 750K is the intermediate temperature. Compared to the lower pressure condition, the concentration of HO_2 is relatively high, so the sensitivity of the reaction HO_2 + HO_2 <=> H_2O_2 + O_2 becomes more significant at high pressure.

According to graph (c), because of the influence of C7H16, the reaction system has produced a large amount of CH_2O and HO_2, and those intermediates products impacts the oxidation of CH4 through reactions such as CH_2O + OH <=> HCO + H_2O and H_2O_2 + OH <=> H_2O + HO_2. Therefore, the extra production of CH_2O, H_2O and H_2O_2 leads to the CH4 oxidation at a lower temperature.

4. Conclusions

The impact of adding C7H16 on the CH4 and C7H16 mutual oxidation at 400-1200 K, 10atm and at 0.5 equivalence ratio was mathematically and experimentally tested with the C7H16 increment from 0 to 100%. The conclusion of the presented research is summarized below:

1. When the proportion of C7H16 in the CH4/C7H16 mixed substances becomes bigger, the beginning temperature for the CH4 oxidation decreases and it is observed that there is a nonlinear relationship between the C7H16 portion and the beginning temperature. The observation is that a minor content of C7H16 makes a dramatic promotion on CH4 oxidation, while adding more C7H16 makes minor kinetic reinforcement effect. Under the high pressure, the beginning temperature of the oxidation of CH4 becomes lower, indicating that CH4 oxidation is promoted. Regarding C7H16, the efficiency of low-temperature-chemistry and high-temperature-chemistry is significantly promoted under high pressure.
2. The reaction pathway interpretation of the CH4/C7H16 substances demonstrates that the addition of C7H16 has a vital impact on oxidizing CH4. By increasing the C7H16 content, the H-abstraction of CH_4 by hydroxide and CH_3 oxidation by HO_2 are promoted. After that, the beginning kinetic reinforcement reactions of CH4 oxidation by adding C7H16 content were identified, and the rates of reactions appeared to become faster at low temperature.
3. Based on the outcomes results of the sensitivity explanation and the net rates of reaction of the most vital reactions, it is seen that due to the CH4 addition, the RO_2 decomposition reaction is significant at intermediate temperature (750K). Compared to the lower pressure condition, the sensitivity of the reaction HO_2 + HO_2 <=> H_2O_2 + O_2 becomes more significant at high pressure. Moreover, because of the influence of C7H16, the reaction system has produced a large amount of intermediate products, which impact the oxidation of CH4 to appear at a lower temperature.

Therefore, the presented research demonstrates the ability of C7H16 addition promoted CH4 combustion in diesel and marine engines.

References

[1] Cho HM and He BQ 2007 Energy Convers. Manage. 48 608-618.
[2] Papagiannakis RG and Hountalas DT 2004 Energy Convers. Manage. 45 2971-2987.
[3] Aslam MU, Masjuki HH, Kalam MA, Abdesselam H and Mahlia 2006 TMI Fuel. 85 717-724.
[4] Zhao H, Dana A, Zhang Z, Green W and Ju Y 2018 Energy. 165 727-738.
[5] Zhao H, Fu J, Haas FM and Ju Y 2017 Combust. Flame 183 253-260.
[6] Kalam MA and Masjuki HH 2011 Energy. 36 3563-3571.
[7] Carlucci AP, Risi AD, Laforgia D and Naccarato F 2008 Energy. 33 256-263.
[8] Tran LS, Pieper J, Carstensen HH, Zhao H, Isabelle G, Ju Y and etc. 2017 Proc. Combust. Inst. 36 1165-1173.
[9] Zhao H, Yang X and Ju Y 2016 Combust. Flame. 173 187-194.
[10] Cordiner S, Gambino M, Iannaccone S, Rocco V and Scarcelli R 2008 Energy Fuels. 22 193-197.
[11] Mansour C, Bounif A and Aris A 2001 Int. J. Therm. Sci. 40 409-424.

[12] Zhang T, Zhao H, Ju Y 2018 AIAA. J. 1-5.
[13] Felsmann D, Zhao H, Wang Q, Graf I and etc. 2017 Proc. Combust. Inst. 36 543-551.
[14] Wei L and Peng G 2016 Fuel Proc. Technol. 142 264-278.
[15] Zhang Q, Li M and Shao, S. 2015 Appl. Energy. 157 217-228.
[16] Zhao H, Fu J and Ju Y 2017 In: 55th AIAA Aerospace Sciences Meeting.1964.
[17] Liu J, Yang F, Wang H, Ouyang M and Hao S. 2013 Appl. Energy. 110 201-206.
[18] Papagiannakis RG, Rakopoulos CD and Hountalas DT 2010 Fuel. 89 1397-1406.
[19] Papagiannakis RG and Hountalas DT 2003 Appl. Therm. Eng. 23 353-365.
[20] Westbrook CK, Pitz WJ, Mehl M and Curran HJ 2011 Proc. Combust. Inst. 33 185-192.
[21] Mehl M, Pitz WJ, Westbrook CK and Curran HJ 2011 Proc. Combust. Inst. 33 193-200.
[22] Mueller CJ, Cannella WJ, Bruno TJ, Bunting B, Dettman HD, Franz JA and etc. 2012 Energy Fuels. 26 3284-3303.
[23] Tang C, Wei L, Zhang J, Man X and Huang, Z. 2012 Energy Fuels. 26 6720-6728.
[24] Chen Z, Qin X, Ju Y, Zhao Z, Chaos M and Dryer FL 2007 Proc. Combust. Inst. 31 1215-1222.
[25] Burke U, Somers KP, O'Toole P, Zinner CM, Marquet N, Bourque G and etc. 2015 Combust. Flame. 162 315-330.
[26] Zhao H, Wu L, Patrick C, Zhang Z, Rezgui Y, Yang X and etc. 2018 Combust. Flame. 197 78-87.
[27] Zhang Z, Zhao H, Cao L, Li G and Ju Y 2018 Kinetic effects of C7H16 addition on low and high temperature oxidation of CH4 in a jet-stirred reactor, Energy. Fuels. 32 11970–11978
[28] Aggarwal SK, Awomolo O and Akber K 2011 Int. J. Hydrogen Energy. 36 15392-15402.
[29] Li G, Liang J, Zhang Z, Tian L, Cai Y and Tian L 2015 Energy Fuels. 29 4549-4556.
[30] Liang J, Zhang Z, Li G, Wan Q, Xu L and Fan S 2019 Fuel. 235 522-529.
[31] Zhang K, Banyon C, Bugler J, Curran HJ, Rodriguez A, Herbinet O and etc. 2016 Combust. Flame. 172 116-135.
[32] <http://www.reactiondesign.com/products/chemkin/chemkin-pro/>

Constitutive modelling of high strength low alloy steel under hot compression test

Wilasinee Kingkam[1], Ning Li[1], Hong Pang[1], Cheng-zhi Zhao[1,2], He-xin Zhang[1,2,*] and Zhi-ming Li[3]

[1]College of Materials Science and Chemical Engineering, Harbin Engineering University, Harbin 150001, China
[2]Key Laboratory of Superlight Materials and Surface Technology, Ministry of Education, Harbin Engineering University, Harbin 150001, China
[3]College of Power and Energy Engineering, Harbin Engineering University, Harbin 150001, China

*E-mail: zhanghx@hrbeu.edu.cn

Abstract. The HSLA steel was studied on an MMS-200 thermal mechanical simulation in the temperature range of 800-1100 °C and a strain rate range of 0.1-1 s^{-1}. The isothermal hot compression tests were used to investigate the flow behaviour of HSLA steel and develop a constitutive equation based on an Arrhenius equation. A modification of Zener–Hollomon parameter was considering the strain rate compensation after hot compression for improve the accuracy of the developed constitutive equation. The comparison of the results shows that the experimental flow stress data are in good agreement with the predicted flow stress data. The average absolute relative error with a correlation coefficient was found to be 6.81% and 0.96, respectively.

1. Introduction

The hot working processes at a high-temperature deformation and material flow behavior during hot deformation of metals are mostly complex [1-3]. Hot working processes such as forging, rolling and so on are important to explain the changed in the mechanical response of steel under external loadings. The deformation temperature and strain rate are both parameters significantly affect in the softening and dynamic recrystallization mechanisms. The constitutive equation for the hot deformation is the mathematical representation of the flow stress behavior of metals and alloys can be introduced into the finite element formula for simulating the response of materials under specified loading condition [4-5]. Therefore, many researchers have been attempted to develop the constitutive equation to describe the hot deformation behavior of the materials

High strength low alloy steel (HSLA) is a type of steel widely used for ship construction, gas pipeline and the part of transportation industries [6]. Nowadays, HSLA steel has been improving with suitable chemical composition and thermomechanical processes in order to enhance application in heavy industries and automotive. The flow stress characterizes of HSLA steels is necessary to improve in industrial of the hot forming process. During hot deformation processes, the flow properties can determine the power requirements of industrial equipment manufacturers including the mechanical properties and dimensional accuracy of finished products. During hot deformation processes, the flow properties can determine the power requirements of industrial equipment manufacturers including the mechanical properties and dimensional accuracy of finished products. The relationships between the constitutive equation and the relating process variables such as flow stress of the deforming material are required and it is important to calculate the flow stress [7].

2. Experimental procedures

The materials studied in the experimental was HSLA steels. The chemical composition of HSLA casting consists of low carbon content was listed in table 1. The specimens were cut into a small size with a diameter of 8 mm and height 15 mm. Uniaxial compression tests were performed by MMS-200 thermal-mechanical simulator. The specimens were heated from room temperature to 1250 °C with a heating rate of 20 °C/s, and cooled to a deformation temperature. The hot compression test was operated at temperature of 800-1100 °C with strain rates 0.1, 1 and 10 s^{-1}. The total true strain for each specimen was 0.55. After compression testing finished, all specimens were quenched immediately in water.

Table 1. Chemical compositions of HSLA steel (wt%).

C	Si	Mn	P	S	Cr	Mo	Ni	Al	Cu	Nb	Ti	V
0.01	0.14	1.38	0.01	0.008	0.21	0.13	0.51	0.019	0.39	0.027	0.006	0.009

3. Results and discussion

3.1. Stress-strain curves

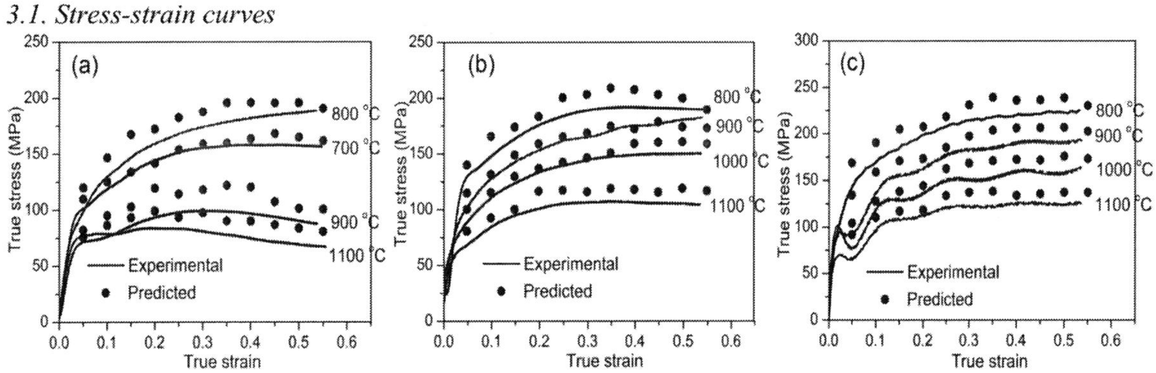

Figure 1. True stress-strain of the experiment and predicted HSLA steel at (a) 0.1 s^{-1}, (b) 1 s^{-1} and (c) 10 s^{-1}.

True stress-strain curves of HSLA steel obtained from the hot compression test are shown in figure 1. It was found that the peak stress is sensitive to temperature and strain rate. When the flow stress being shifted to lower stresses and lower strains with the deformation temperature are increased [8-9]. The results also show that the dynamic recrystallization takes place in the samples at the conditions of high temperature and low strain rate. Besides, the flow stress rising to a peak and followed by softening state until to a stable state.

3.2. Deformation constitutive equations

The Arrhenius equation is often explained the relationship between flow stress, strain rate, and temperature [10]. Moreover, the effects of temperature and strain rate on the deformation behaviour can be shown clearly by Zener Hollomon parameter (Z) in an exponential equation [11-12]. Both equations are given by:

$$Z = \dot{\varepsilon} \exp\left(\frac{Q}{RT}\right) \tag{1}$$

$$\dot{\varepsilon} = Af(\sigma)\exp\left(-\frac{Q}{RT}\right) \tag{2}$$

where Z is the Zener-Hollomon parameter, $\dot{\varepsilon}$ is the strain rate (s^{-1}), R is the universal gas constant and Q is the activation energy of hot deformation (kJ mol^{-1}) with α, β, n_1 and n are materials constant by $\alpha = \beta / n_1$.

Substituting (the power law and exponential law of f(x)) equation (1) into equation (2) with the suitable function and taking the natural logarithm of both sides leads to equations (3)-(4) as follows:

$$\ln \dot{\varepsilon} = n_1 \ln \sigma + \ln A - Q / RT \tag{3}$$

$$\ln \dot{\varepsilon} = \beta\sigma + \ln A - Q / RT \tag{4}$$

Consideration any flow stress at a strain rate of 0.55 s^{-1} are used to analyzed equations. (1), (3) and (4) to determine material constants. The value of α , β and n_1 can be obtained by the average slopes of linear fitting of $\ln \dot{\varepsilon} - \sigma$ and the slopes of the plots of $\ln \dot{\varepsilon} - \ln \sigma$ are shown in figure 2 (a-b). The values of β and n_1 can be calculated as 14.527 and 0.0899 MPa^{-1} and the material constant α (β / n_1) was calculated as 0.00618 MPa^{-1}

The relationship between $\ln \dot{\varepsilon} - \ln[\sinh(\alpha\sigma)]$ obtained from substituting the values of the strain rate and flow stress for all temperature conditions are shown in figure 2c. The value of the material constant (n) obtained from the average slope of the lines in the $\ln \dot{\varepsilon} - \ln[\sinh(\alpha\sigma)]$ plots were 10.594. For the fixed strain rate, by taking the partial derivative of equation (5), the activation energy (Q) can be calculated as:

$$Q = R_n \left[\frac{\partial \ln(\sinh(\alpha\sigma))}{\partial(1/T)} \right]_{\dot{\varepsilon}} \tag{5}$$

The value of Q can be calculated from the slope of every plots in the lines of $\ln \dot{\varepsilon} - 1000/T$ plots (figure 2d). Thus, activation energy Q in the studied HSLA steel was determined to be 361.138 $kJ.mol^{-1}$. These Q values obtained from the experiment fall in the range that reported by previous researchers. For comparison, the activation energy evaluated by Zhang et al. [13] was 368.56 kJ/mol and Wei and liu [14] was 360.063 kJ/mol. Figure 3 shows the relationship between $\ln[\sinh(\alpha\sigma)]$ -ln Z. From the slope of every plot, the straight line can be used to determine the value of ln A. Thus, the exact values of A can be calculated as 3.256×10^{15} s^{-1}. In a similar way, the material constants values (n_1, β, α, n and ln A) as well as activation energy (Q) can be calculated for all condition of HSLA steel. The presents the values of constitutive constants and deformation activation energy of the HSLA steel are given in table 2.

Table 2. Values of constitutive constants and deformation activation energy of HSLA steel.

n_1	β	α	n	ln A	Q (kJ/mol)
14.527	0.0899	0.00618	10.594	35.723	361.138

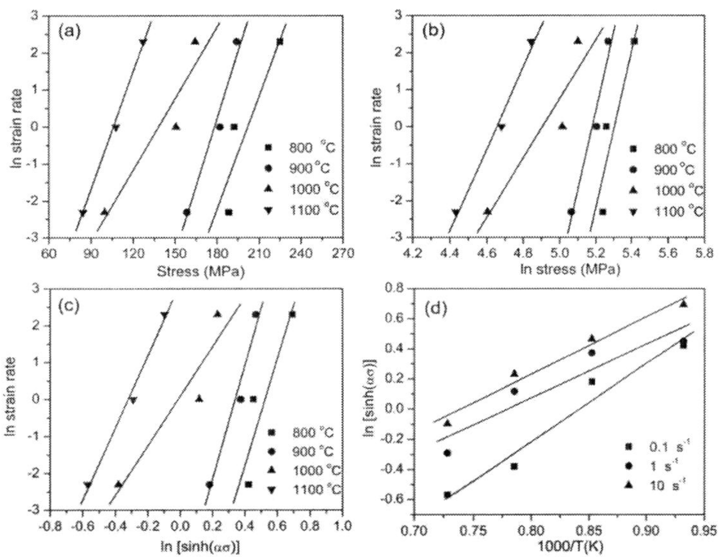

Figure 2. Relationship between (a) $\ln \dot{\varepsilon} - \sigma$, (b) $\ln \dot{\varepsilon} - \ln \sigma$ (c) $\ln \dot{\varepsilon} - \ln[\sinh(\alpha\sigma)]$ and (d) $\ln \dot{\varepsilon} - 1000/T$.

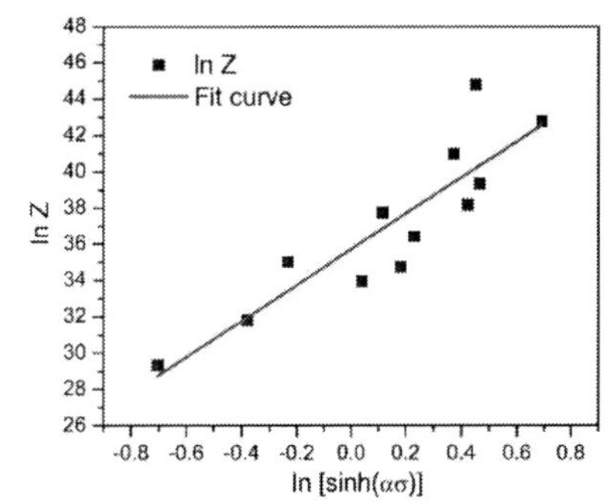

Figure 3. Relationship between $\ln[\sinh(\alpha\sigma)]$ and ln Z.

3.3. Compensation of strain

It is well known that the strain is a significant effect with the deformation activation energy (Q) and material constants (α, β, n and ln A) [15]. Thus, the compensation of the strain can take effect on the prediction of the flow stress accuracy to take into account in order to derive the proper constitutive equations. In this work, the material constant values were determined at different strains in the range of 0.1-0.55 at the intervals of 0.05, the corresponding fitting curves of the variation of material constants with a function of strain shown in figure 4. These material constants values were the polynomial function of the strain. A third order polynomial are shown in equation (6) and the coefficients of polynomial fitting curves of HSLA steel are listed in table 3.

$$\begin{aligned} \alpha &= \alpha_0 + \alpha_1\varepsilon + \alpha_2\varepsilon^2 + \alpha_3\varepsilon^3 \\ n &= n_0 + n_1\varepsilon + n_2\varepsilon^2 + n_3\varepsilon^3 \\ Q &= Q_0 + Q_1\varepsilon + Q_2\varepsilon^2 + Q_3\varepsilon^3 \\ \ln A &= A_0 + A_1\varepsilon + A_2\varepsilon^2 + A_3\varepsilon^3 \end{aligned} \tag{6}$$

In this study, the materials constants are evaluated at a particular strain in the range of 0-0.55. Therefore, the constitutive equation relating to flow stress and Zener-Holloman parameter as following:

$$\sigma = \frac{1}{\alpha}\ln\left\{\left(\frac{Z}{A}\right)^{1/n} + \left[\left(\frac{Z}{A}\right)^{2/n} + 1\right]^{1/2}\right\} \tag{7}$$

Table 3. Coefficients of polynomial fitting curves for material coefficients of HSLA steel.

α		n		Q		ln A	
α_0	0.014	n_0	11.81	Q_0	432.82	A_0	37.94
α_1	-0.063	n_1	38.31	Q_1	392.38	A_1	76.29
α_2	0.174	n_2	-152.34	Q_2	-1277.1	A_2	-235.7
α_3	-0.154	n_3	140.79	Q_3	654.73	A_3	163.43

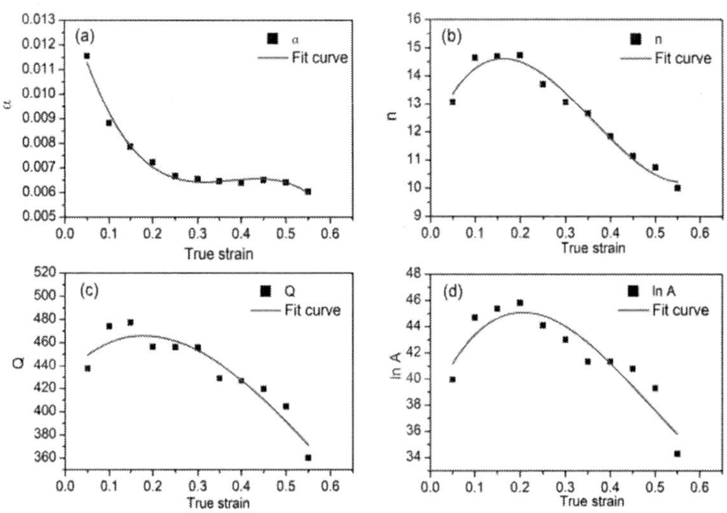

Figure 4. The variation of material constants with a function of strain (a) α , (b) n , (c) Q and (e) $\ln A$.

3.4. *Verification of the developed constitutive equations*

A comparative study between the experimental and predicted hot deformation behaviour of HSLA steel results was carried out to verify the constitutive equations for HSLA steel after hot compression test is shown in figure 5.

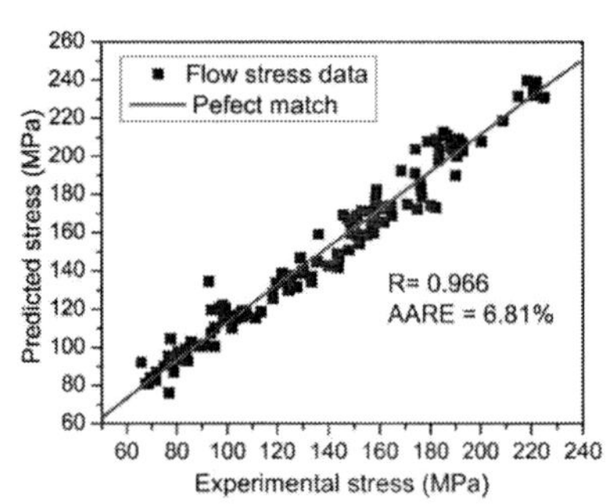

Figure 5. Comparison between the experimental and predicted flow stress from the developed constitutive equation.

The prediction of the constitutive equation was also quantified using standard statistical parameters such as the correlation coefficient (R^2) and average absolute relative error (AARE). The experimental of flow stress were plotted versus flow stress prediction base on develop constitutive equation as shown in figure 5. The correlation coefficient exists between experimental data and predicted data, and the R is 0.966. The AARE is calculated by equation (8):

$$AARE = \frac{1}{N}\sum_{i=1}^{n}\left|\frac{E_i - P_i}{E_i}\right| \times 100 \tag{8}$$

where E is the experimental value and P is the predicted value obtained from the constitutive equation and N is the total number of data, respectively. Therefore, the AARE of experimental and predicted data is 6.81%. These results also confirm the accuracy of the developed constitutive equations with HSLA steel strain compensation and the proposed constitutive equations can effectively predict the flow behaviour steel used to simulate the process of hot forming using a finite element method.

4. Conclusions

The hot deformation behaviour of HSLA steel was studied using hot compression tests. Base on the experimental data, the constitutive analysis of HSLA steel was carried out. The conclusions can be demonstrated as follows:

1. The true stress-strain curves of the HSLA steel shows the characteristics of dynamic recrystallization or dynamic recovery at different temperatures and strain rate. The flow stress increases with the increasing of strain rate and decreasing of deformation temperature.
2. The dependence of flow stress on strain and temperature can be calculated using the constitutive equation and the material constants were determined. The activation energy of this experimental for HSLA steel was determined as 361.138 kJ mol^{-1}.
3. The influence of the strain was incorporated in the constitutive equation by examining the effect of strain on material constants (i.e. α, β, n and ln A) base on the experimental data. The third order polynomial was used to represent the strain in these material constants with having a good correlation. The average absolute relative error associated (AARE) related to total temperature prediction and strain rate was 6.81% and the correlation coefficient (R) was 0.966.

Acknowledgements

This work was supported financially by the Fundamental Research Funds for the Central Universities (Nos. HEUCFP201719 and HEUCFP201731).

References

[1] Y. C. Lin, X.M. Chen, Mater. Des. **32** 1733 (2011)
[2] S.M. Abbasi, A. Momeni, Mater. Sci. Eng. A. **552** 330 (2012)
[3] C.S. Zhang, J. Ding, Y.Y. Dong, G.Q. Zhao, Int. J. Mech. Sci **98** 195 (2015)
[4] S.K. Singh, K. Mahesh, A. Kumar, M. Swathi, Mater. Des **31** 4478 (2010)
[5] B. Guo, S.C. Zhang, C.C. Ge, J. Iron Steel Res. Int **21** 389 (2014)
[6] W. Kingkam, C.Z. Zhao, H. Li, H.X. Zhang, Z.M. Li, Int. J. Electrochem. Sci **13** 2766 (2018)
[7] L.C. Yang, Y.T. Pan, I.G. Chen, D.Y. Lin, Metals **5** 1717 (2015)
[8] C.Z. Zhao, W. Kingkam, N. Li, H.X. Zhang, Z.M. Li J. Mater. Eng. Perform **27** 4129 (2018)
[9] Nourollahi G A, Farahani M, Babakhani A, S S Mirjavadi 2013 Mat. Res. **16** 1309
[10] C.M. Sellars and McTegart W J Acta. Metall **14** 1136 (1966)
[11] H. Farnoush, A. Momeni, K. Dehghani, J. Aghazadeh Mohandesi, H. Keshmiri, Mater. Des. **31** 220 (2010)
[12] A. Momeni, H. Arabi, A. Rezaei, H. Badri, S.M. Abbasi, Mater. Sci. Eng. A **528** 2158 (2011)
[13] J. Zhang, H. Di, X. Wang, Y. Cao, J. Zhang, T. Ma, Mater. Des. **44** 354 (2013)
[14] H.L. Wei, G.Q. Liu, Mater. Des. **56** 437 (2014)
[15] Y.C. Lin, M.S. Chen, J.Zhang, Mater. Sci. Eng. A **499** 88 (2009)

IOP Conf. Series: Materials Science and Engineering **715** (2020) 012019 doi:10.1088/1757-899X/715/1/012019

Modelling of a tibial bone-knee implant assembly in order to analyse its mechanical behaviour

Mihai-Constantin Balaşa[1], Ştefan Cuculici[2] and Viviana Filip[1,*]

[1]Valahia University of Targoviste, Doctoral School of Engineering Sciences, Targoviste, Romania
[2]Emergency Hospital St Pantelimon – Orthopedic Department – Bucharest, Romania

*E-mail: v_filip@yahoo.com

Abstract. Orthopaedic implants have a relatively long yet limited lifespan, but sometimes problems occur after a relatively short time, making it necessary to have the original implant replaced by a new one [1-3]. This intervention, which is called revision surgery, is more difficult than the initial surgical procedure, as the prosthesis is attached to the bone tissue, and by removing the initial implant, the surgeon will also remove part of the bone it is attached to [4]. This is why it would be really helpful for the surgeon to be familiar, before performing the first surgery, with how the implanted bone [5] behaves in the long run, depending on the type of implant, the patients characteristics and the illnesses they may suffer from [6], in order to customise the choice of the implant as best as they can so as to avoid or delay as much as possible the revision surgery [7-9]. For this purpose, the virtual model of the tibial bone and the virtual model of the implant are performed, the bone-implant assembly is created, and, finally, the finite element analysis is performed on the resulting assembly in order to determine its behaviour over time. The purpose of this paper is to show how the cemented tibial bone – knee implant assembly is made.

1. Introduction

In order to provide medical staff with information on the long-term behaviour of an implanted bone, the virtual model of the tibial bone-knee implant assembly is made with a view to being subjected to a finite element analysis.

Since the bone has a complex and irregular shape and is difficult to model, first we performed a 3D bone scanning using the 3D ExaScan portable scanner and then we processed the resulting model using the CATIA software, as shown in figure 1. However, we could not use this virtual bone model to perform the finite element analysis as, because of the large number of surfaces it comprised, it would have been necessary for us to divide it into several parts and assign different material properties to each one. Nevertheless, we used the scan model to create the virtual model through classic modelling techniques.

Thus, we sequentially sectioned the model resulted from scanning, took the curves from each section and imported them into a new drawing, took three representative bone points and we used them as guiding curves and, with the Loft command in SolidWorks, we developed the virtual model of the full body tibial bone [9]. Since the bone does not have identical material properties over its entire length, it is necessary for us to model all the regions that make up the bone, in order to assign distinct material properties to each one. A long bone (such as the tibia) consists of a compact bone tissue

IOP Conf. Series: Materials Science and Engineering **715** (2020) 012019 doi:10.1088/1757-899X/715/1/012019

(outer shell), a medullary cavity in the middle, and a block of spongy substance (spongy bone tissue) surrounded by a layer of compact substance at each of the two ends (proximal epiphysis and distal epiphysis) [10, 11].

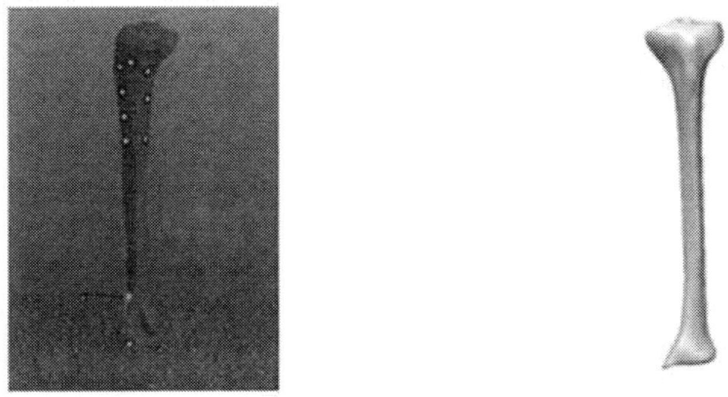

(a) (b)

Figure 1. The 3D surface of the tibial bone, obtained by scanning (a) raw scanning, (b) processed scan in CATIA [9].

In view of the above, we will divide the resulting 3D model in 3 pieces, as seen in figure 2:

- the outer shell (to which we will assign compact / cortical bone-specific properties)
- the proximal epiphysis (to which we will assign spongy bone-specific properties)
- the distal epiphysis (to which we will assign same spongy bone-specific properties)

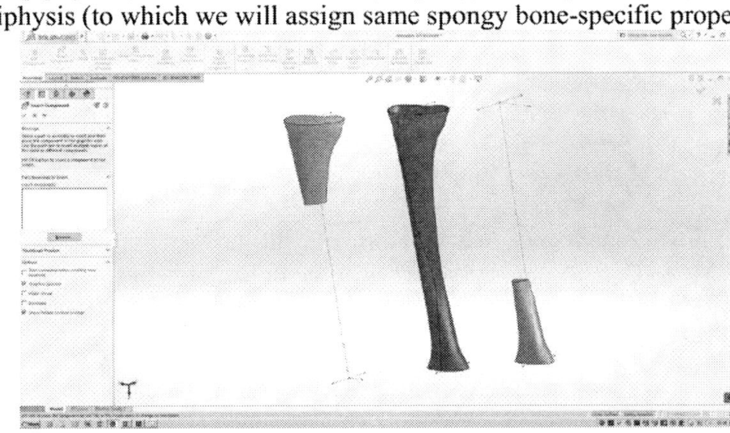

Figure 2. Proximal epiphysis, outer shell, and distal epiphysis.

The knee implant may be fastened to the tibial bone either by pressing (press fit method) or by cementation. We decided to use cementation as a fastening method. For this purpose, it was necessary for us to model the material used for fastening, which is the cement. The cement must be placed between the metallic tibial component of the knee implant and the proximal epiphysis of the tibial bone.

2. Modeling the cement cylinder between the proximal epiphysis and the metallic tibial component

In order to build the tibia-knee implant assembly, the proximal epiphysis must be processed (by drilling a hole into it) to enable the insertion of the metallic tibial component. Since we will model the cemented version, we will place the bonding cement between the Proximal Epiphysis and the Metallic Tibial Component. The bonding cement is shaped as a cylinder that fits on the metallic tibial component and takes its shape. The cement and the tibial component are inserted into the proximal epiphysis. To do this, we will proceed as follows:

Step I - we will create a part named Full Body Cement (which follows the shape of the tibial component) and we will use it to drill (process) the proximal epiphysis.

Step II - we will drill the Full Body Cement part (the hole will follow the shape of the tibial component) in order to create a part named Hollow Cement, into which we will later insert the tibial component.

Step I: Creating the Full Body Cement part and Processing the Proximal Epiphysis

To create the virtual model of the Full Body Cement, we extracted the sketches from the metallic tibial component.

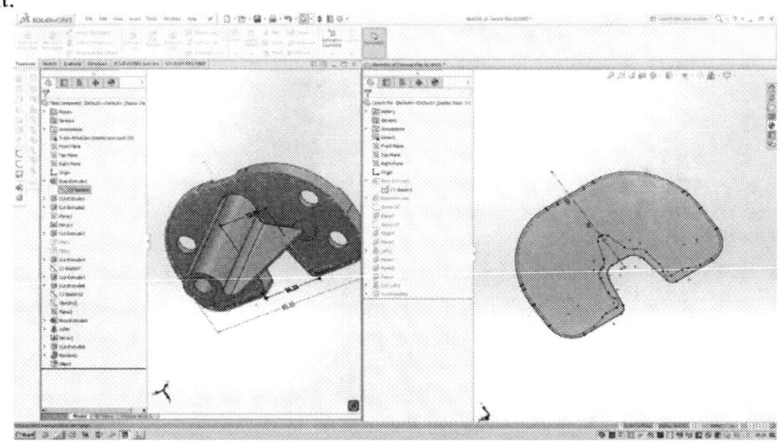

Figure 3. Modelling the Full Body Cement part. Metallic tibial component to the left, cement to the right.

Using the Offset Entities command and setting the offset distance to 1 mm, we doubled the shape to get a larger extraction of the material, in order to fill it with the hollow cement part that will be created later, figure 3. We created guide curves and fixed them using the "Pierce" command. With the "Loft" command, we selected the two sketches and the three guide curves, and created the Full Body Cement part.

With the Full Body Cement part, we will process the proximal epiphysis part. Activating the "Insert - Features - Move/Copy" command, we select the object to be moved (Full Body Cement) and, using the constraints as shown in figure 4, we assemble it into the Proximate epiphysis part for processing.

The next step consists in activating and using the "Insert-Features-Intersect" command to extract the material (so as to be able to perform the processing). With the "Insert-Features-Intersect" command, we select the two parts, and check the first option – "Create intersecting regions". Clicking the "Intersect" button and selecting "Cap planar openings on surfaces", the program enables the selection of the regions we want to exclude. We unselect the region we want to exclude (region 1) and keep only the part that will be saved as Processed Proximity Epiphysis, figure 5.

Step II. Processing (drilling) the Full Body Cement part so it can accommodate the metallic tibial component. We will name the resulting drilled component Hollow Cement part.

With the "Insert – Part" command, we placed the Metallic Tibial Component part into the Full Body Cement part. We will use the "Insert - Features - Move / Copy" command, which at a later moment will enable us to select the constraints or mates.

By activating the "Move / Copy" command, we select the object to be moved (Tibial Component), and with the help of the constraints, we fit it into the Full Body Cement part in order to be processed, with the "Insert-Features-Combine" command. With this command activated, we selected "Subtract" as type of operation, we selected the main body (the Full Body Cement part), then we selected the body to be combined into it and extracted, namely the Tibial Component. Using the "Subtract" command, we obtained the processed Cement (Hollow Cement) part, figure 6.

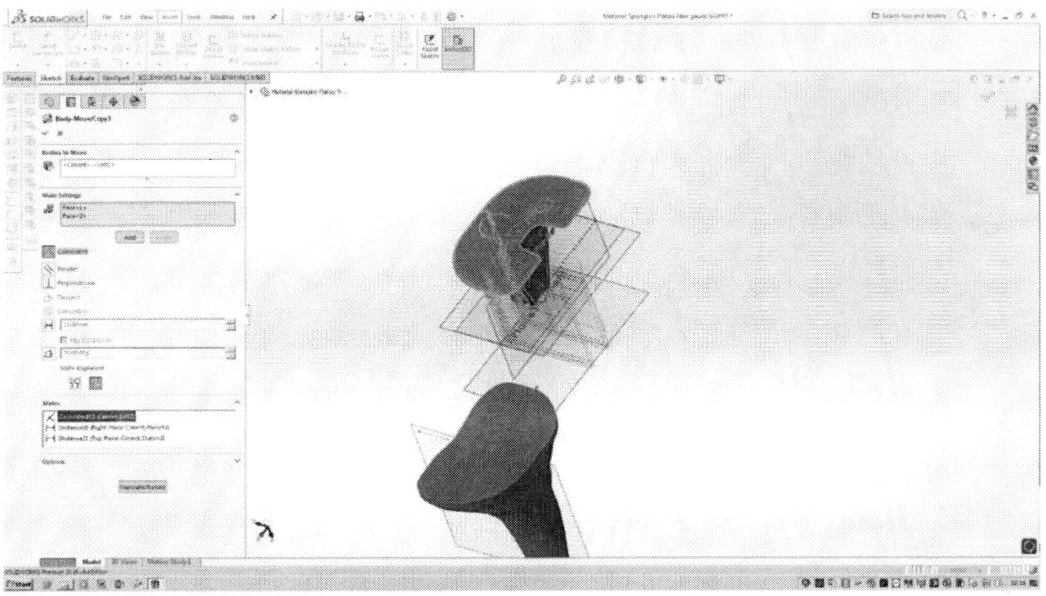

Figure 4. Selecting constraints.

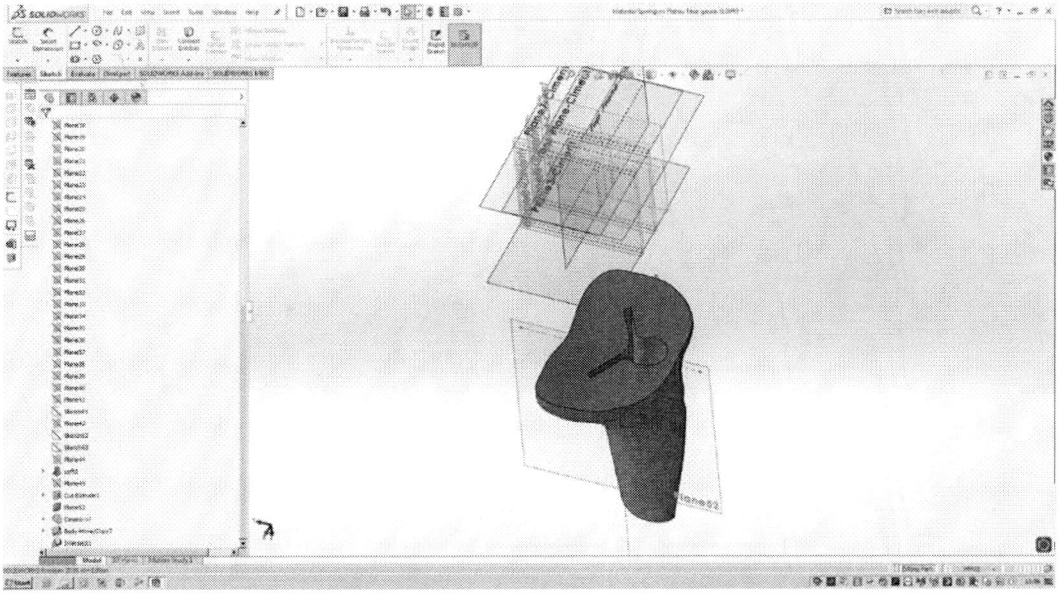

Figure 5. Result of Intersect command.

IOP Conf. Series: Materials Science and Engineering **715** (2020) 012019 doi:10.1088/1757-899X/715/1/012019

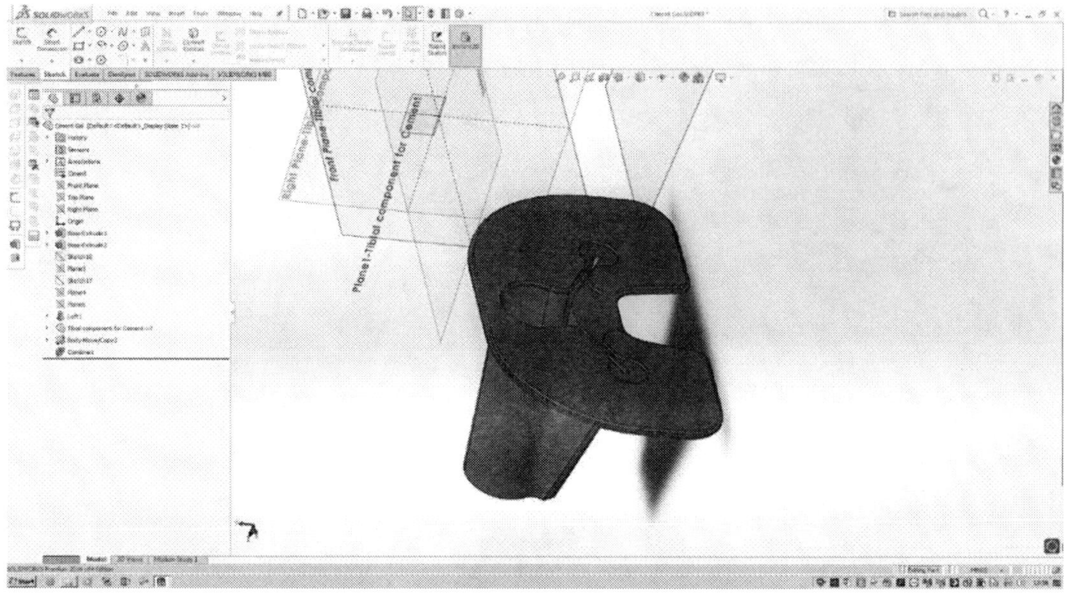

Figure 6. Result of the Subtract command.

3. Making the tibia-knee implant assembly

In order to obtain the virtual model of the implant-fitted tibial bone, we open a new file for the entire assembly and use the "Insert component" command and the 'axis-axis', 'plane-plane', 'plane-axis', 'coincidence', 'tangency', and 'distance' constraints. First we bring in the Outer shell, then we add the Distal epiphysis, then the processed Proximal epiphysis, then the Cement, and in the end the Metallic Tibial Component. The polyethylene insert and the metallic femoral component are placed over it (figure 7).

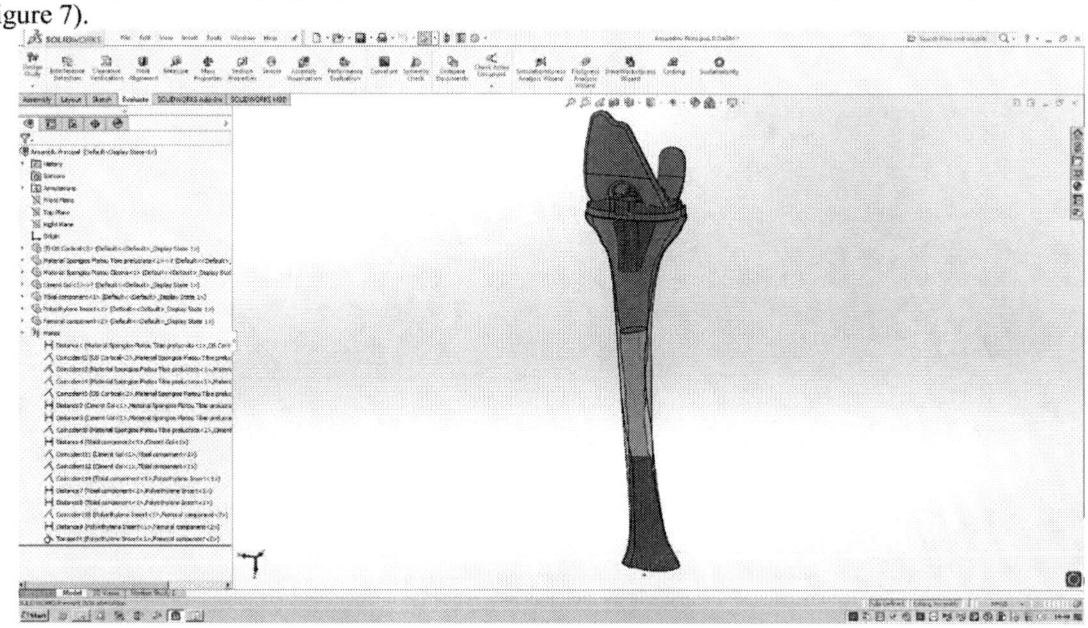

Figure 7. Tibial bone-knee implant assembly.

4. Conclusions, future research directions

Once the virtual model is created in Solid Works, it can be assigned customized parameters (sizes, material properties, varus - valgus deformities), depending on the patient [12]. We assembled the knee implant onto the tibial bone and we fastened it with the cement we modelled, with a view to performing the finite element analysis of the assembly, which will be subjected to various static and dynamic loads (human body weight when standing, walking, running etc.). This numerical simulation of the bone-implant behaviour is helpful for the optimal choice of the shape and size of the implant, depending on the patient's characteristics [9, 13].

References

[1] M. Ben 1. I.I. Cofaru, Summary of the PhD thesis, Researches regarding the biomechanics of the axial deviations of the human lower member and the development of the correspondent surgical devices. (2013).

[2] Şt. Cristea, A. Prundeanu, Fl. Groseanu and D. Gârtone, The Role of Arthroscopy in Mini-Invasive Treatment of Tibial Plateau Fractures, Chapter from the book Modern Arthroscopy, pp. 225-236 (2011).

[3] S. Mihai and V. Filip, "New design concept for reducing torque wear on implant", Applied Mechanics and Materials, pp. 453-458, **658** (2014).

[4] S. Mihai, V. Filip, M. Vladescu, Contributions to the improvement of the tribological behaviour of hip implant joints, Journal of Science and Arts, Year 16, pp. 177-184, **2(35)** (2016).

[5] C.M. Nicolescu, M. Bumbac, S. Mihai, A.I. Gheboianu, M.C. Balasa, V. Filip, S. Cuculici, S. Cristea, C. Pantu, X-RAY diffraction and nanoindentation characterization of bone tissue affected by severe osteoarthritis, Journal of Science and Arts, ISSN: 1844 – 9581, Year 18, pp. 265-274, **1(42)** (2018).

[6] S. Mihai, V. Filip, Studies on explanted orthopedic implants, The Scientific Bulletin of VALAHIA University – MATERIALS and MECHANICS –, ISSN: 1844-1076, pp. 149-152, **8** (year 11) (2013).

[7] 7. S. Mihai, Constructive concepts for reducing wear caused by the friction torque of hip implants, International Journal of Research in Engineering and Technology, pp. 289-295, **2** Issue 10 (2013).

[8] S. Mihai, Orthopaedic implants and prostheses. Conventional and unconventional processing technologies, International Conference on Architecture, Civil & Mechanical Engineering, pp.132-136 (2012)

[9] MC Balaşa, S Mihai, V Filip, AD Negrea, G Tomescu, "Modelling the tibial bone using cad techniques, starting from the 3D scan model", International Journal of Mechatronics and Applied Mechanics, Issue 3, pp.213-219 (2018).

[10] F. F. Cintra, A. K. Yepéz, M. G. S. Rasga, M. Abagge, P. G. C. Alencar, Tibial component in revision of total knee arthroplasty: Compararison between cement and hybrid fixation, Sociedade Brasileira de Ortopedia e Traumatologia, pp. 585-590, **46(5)** (2011).

[11] A.Williams, M. Logan, Understanding Tibio-Femoral motion, The Knee, pp. 81–88, **11** (2004).

[12] M.C. Balaşa, Ş. Cuculici, C. Panţu, S. Mihai, A.D. Negrea, M.O. Zdrafcu, D.D. Leţ, V. Filip, Ş. Cristea, Using 3D scanning techniques in orthopedic systems modeling, The Scientific Bulletin of VALAHIA University MATERIALS and MECHANICS, DOI 10.1515/bsmm-2017-0016, pp. 41-47, **15** No. 13 (2017).

[13] S. Mihai, V. Filip, 3D modeling and performing of orthopedic implants by material deposition rapid prototyping, Romanian Review Precision Mechanics, Optics & Mechatronics, pp.128-131, **41** (2012).

IOP Conf. Series: Materials Science and Engineering **715** (2020) 012020 doi:10.1088/1757-899X/715/1/012020

Finite element analysis of automotive arm brake pedal for rapid manufacturing

MIM Sargini[1,3], SH Masood[2], Suresh Palanisamy[2,*], Elammaran Jayamani[1] and Ajay Kapoor[2]

[1]Faculty of Engineering, Computing and Science, Swinburne University of Technology (Sarawak), Malaysia
[2]Dept. of Mechanical and Product Design Engineering, Swinburne University of Technology, Hawthorn, VIC, Australia
[3]Malaysia Automotive, Robotics and IoT Institute, Cyberjaya, Malaysia

*E-mail: spalanisamy@swin.edu.au

Abstract. Current automotive arm brake pedal uses conventional manufacturing method such as metal press and CO_2 welding processes, which are known to be heavy, costly and demand higher quality control. This paper will discuss how finite element analysis can be applied to optimise the design of a brake pedal for Rapid Manufacturing application based on current additive manufacturing process to produce automotive brake pedals with reduced material weight. The performance of the new design is compared with the current design of arm brake pedal to realise the potential of rapid manufacturing to save time and cost in automotive parts.

1. Introduction

There are many components that control the movement and dynamics of a vehicle. However, there are not many components that are in direct contact with the driver and one of these key components is a pedal brake. A pedal brake act as a switch for the driver to stop the vehicle and it is a critical part of the vehicle. The driver pedals the pedal brake to commence the car to stop. The pedal force is then transferred to the hydraulics, which then triggers the braking sequence. During the emergency braking, a huge force from the driver is transferred into the brake system within the least time, making the pedal brake to provide great mechanical strength and resistance to hold the force. Automotive engineers are continually exploring new manufacturing technologies like Additive Manufacturing to reduce the cost of designing and manufacturing key automotive components like brake pedals.

According to ISO/ASTM 52900:2015 [1], Additive Manufacturing (AM) is the process of joining materials to make objects directly from 3D model data, usually layer upon layer, as opposed to subtractive manufacturing methodologies. Synonyms are additive fabrication, additive techniques, additive layer manufacturing, layered manufacturing and solid freeform fabrication [2]. Rapid Manufacturing (RM) is a mass manufacturing technique based on AM technologies. The future of RM is very promising since its demand among the manufacturers continues to expand. We have yet to see the dream of manufacturing engineers today to extend the total automation to functional and full-life components [3] via RM.

RM technologies are creating a world of possibilities that can lead an organization in new directions and help launch new businesses and business models. Wohler's Report [4] has highlighted that the final part production has been the highest growing application of RM in recent years and would be the largest and most significant application of RM technology. RM also is used to produce precision parts and final parts for automotive, aerospace, medical and healthcare, and consumer durables industries [5]. In other projects, the researchers believe that as for real direct RM of products, its application is still in its infancy, even though very promising. The real breakthrough of RM will mainly depend on cost and productivity improvements, which have to be accompanied by further technical progress in material properties and most of all in accuracy and reliability [6].

In the automotive industry, Prada [7] successfully applied RM in improving an accelerator pedal for his Formula Student racing car. He came up with two prototypes of an accelerator pedal, both made by using RM technologies which are Fullcore 720 with poly-jet technology, and other prototype uses ABS-plus (P430) with FDM technology. The newly design accelerator pedal resulted in 38% of weight reduction compared to its original design made by 7075 aluminium.

In this paper, a new optimised design of an arm brake pedal, which can be produced by RM technologies, material and process, will be introduced using a commercial finite element analysis (FEA) software. The new design will be evaluated to ensure it reliability for the market. In this study, the current design of an arm brake pedal, which uses conventional manufacturing technologies will also be analysed by FEA to compare with the performance of the new design. Several structural analysis tests will be conducted for both arm brake pedal design to study its performance. Both results will be discussed and compared to verify whether the new design is good enough for the market or not.

2. FEA of arm brake pedal: current design

For this study, the arm brake pedal of a five-door hatchback car will be investigated. Figure 1 shows the computer aided design (CAD) model of the brake pedal. The arm brake pedal generally consists of four child parts, which are boss (1), arm (2), pad (3) and stopper (4) as numbered in figure 1. The boss is made of cold drawn steel and acts as a connection between the arm brake pedal and its mounting. The arm is the main body of the arm brake pedal, made of hot rolled steel, which holds and transfer the braking force accordingly. The pad, made of hot rolled steel, is where the driver rests his/her foot to transfer the braking force during the braking event. The stopper function is to maintain the arm brake pedal location during the non-braking event, thus this child part will be ignored in this study since it is not involved during the braking event. Table 1 shows the mechanical properties of the three main child parts used in the analysis. All three child parts then joined by a CO_2 welding process at two different areas which are boss and arm, and arm with the pad.

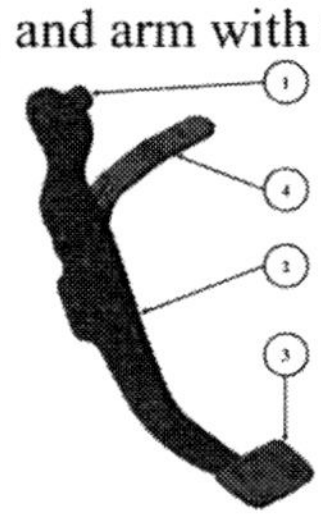

Figure 1. The CAD data of the arm brake pedal.

Table 1. The mechanical properties of each child parts of the arm brake pedal.

#	Child Parts	Thickness mm	Young Modulus (E), MPa	Poisson's Ratio, ν	Tensile Strength (TS), MPa	Yield Strength (YS), MPa	Density (ρ), Ton/m³	Manufacturing Process
1	Boss	ID 15.1 OD 19.0	200,000	0.290	290	290	7.87	Metal extrusion
2	Arm	7.0	201,000	0.306	440	295	7.87	Metal stamping
3	Pad	2.3	201,000	0.306	440	295	7.87	Metal stamping

Basically, the arm brake pedal is a simple machine of a lever with a Mechanical Advantage (MA). Figure 2 below shows the arm brake pedal modelled as a simple machine; a lever with MA. Mathematically, we can model the MA to be:

$$MA = \frac{F_p}{L_1} = \frac{F_d}{L_2} \tag{1}$$

The brake force requirement to test the performance of the arm brake pedal has been determined based on Limpert's study [8], which suggested that the maximum force exerted with right foot for man is 823N and women is 445N [8]. For maximum performance evaluation, the maximum force of 823N is used in this study. To evaluate the performance of the design, the Finite Element Analysis (FEA) software Altair HyperWorks has ben used in this study. 2D meshing (PSHELL) mixed of quad and tria elements have been used with a target elements size of 1mm. For the mechanical connection (CO_2 welding), a 1-D rigid element has been used to represent the mechanical connection. This is to ensure no loss of force transfer during the analysis.

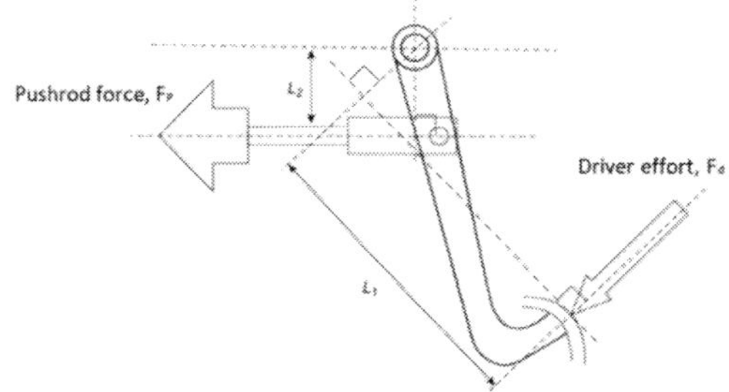

Figure 2. A model of a lever with MA, image courtesy of [8].

Figure 3 shows the boundaries conditions for the arm brake pedal model. The load is applied on one end of the arm. The load will be nodally distributed to the area of pad and arm join. The model is constraint at two areas – the boss and clevis hole of the arm. Rotational of Y-axis is the only degree of freedom enabled for both constraints.

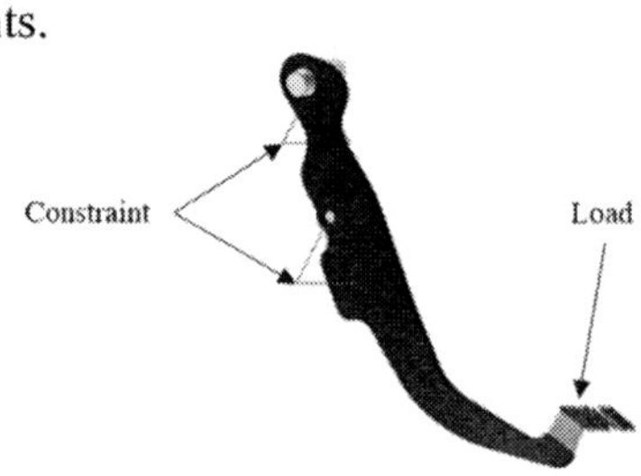

Figure 3. The Boundaries Condition of the arm brake pedal FE model.

To evaluate the performance of the current design of the arm brake pedal, three common structural analysis has been performed in the Altair HyperWorks environment, which includes linear static analysis, non-linear static analysis and fatigue analysis. The linear static analysis will be used to determine the maximum stress the arm brake pedal absorb during the analysis. The non-linear static analysis will be used to determine the maximum deflection/displacement of the subject under the stress. For this study, the geometric non-linearity parameter will be enabled. For evaluation, any displacement/deflection more than 10mm will be considered as a failure. The fatigue test will be used to evaluate the design durability for 1 million test cycle. All the parameters for these tests were set to the default value as suggested by the software developer.

Figure 4 shows the results of the linear static analysis for the current design of the arm brake pedal. Since the arm and the boss use different type of steel materials, they have different limits. For the arm, there is a maximum stress concentration valued at 375.59 MPa located at the clevis hole area while for the boss, maximum stress concentration valued at 174.805 MPa is located at the welding connection area. The maximum displacement for the current design of an arm brake pedal is 1.482 mm at the one end of the arm. Table 2 summarises the overall linear analysis result.

Table 2. The overall result of linear static analysis.

Child parts	Tensile Strength, MPa	Maximum element stress, MPa	Displacement, mm	Safety factor
Boss	290.000	174.805	[redacted]	0.60
Arm	440.000	375.590	[redacted]	0.85
Arm brake pedal	[redacted]		1.482	0.85

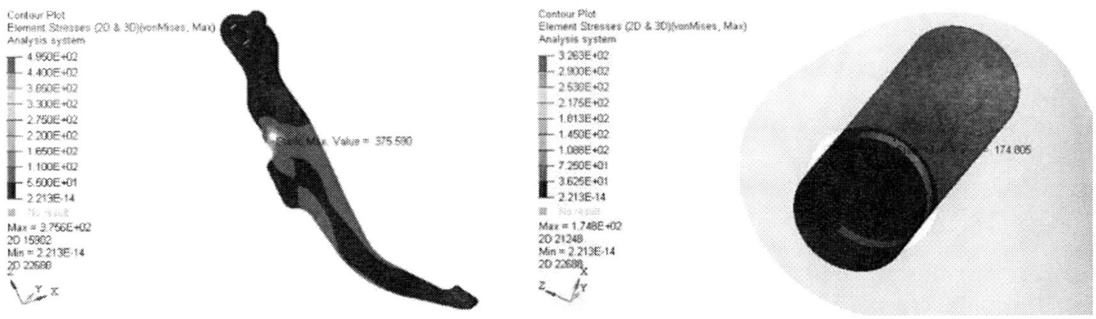

Figure 4. The overall result of the linear static analysis; stress concentration of arm (left) and boss (right).

Figure 5 shows the results of non-linear static analysis of the current design. For the arm, there is a maximum stress concentration valued at 294.497 MPa located at the clevis hole area while for the boss, maximum stress concentration valued at 175.950 MPa located at the welding connection area. The maximum stress concentration for non-linear static analysis for the arm is lower than linear static analysis. This is true since, for the non-linear static analysis, the geometric non-linearity property of the material was taken into account. The maximum displacement for the current design of the arm brake pedal is 1.482 mm located at the one end of the arm. Table 3 summarizes the overall non-linear analysis results.

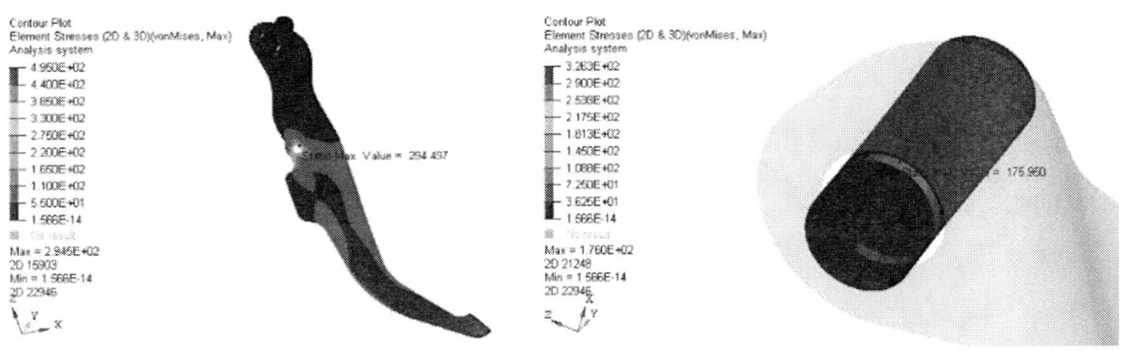

Figure 5. The overall result of the non-linear static analysis; stress concentration of arm (left) and boss (right).

Figure 6 shows the result of the fatigue analysis towards the arm brake pedal. This result shows that the design is reliable even after 10^{20} cycles. As a conclusion, the current design of the arm brake pedal passes all three structural analysis.

IOP Conf. Series: Materials Science and Engineering **715** (2020) 012020 doi:10.1088/1757-899X/715/1/012020

Table 3. The overall result of non-linear static analysis.

Child parts	Tensile Strength, MPa	Maximum element stress, MPa	Displacement, mm	Safety factor
Boss	290.000	175.950	[redacted]	0.60
Arm	440.000	294.497	[redacted]	0.67
Arm brake pedal	[redacted]		1.484	0.85

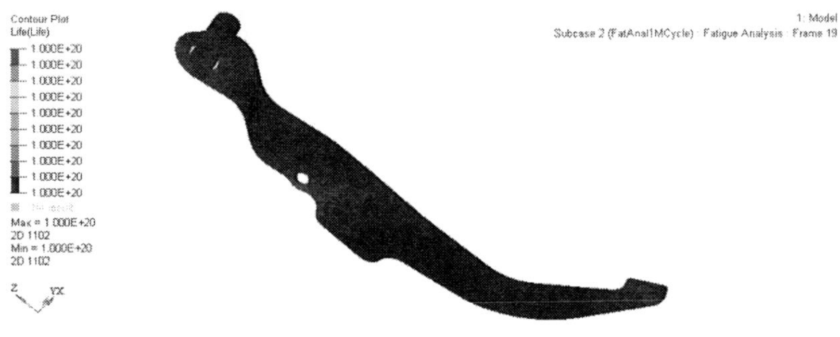

Figure 6. Fatigue analysis result.

3. FEA of arm brake pedal: optimized design for RM

Altair Inspire software has been used to come up with a new optimized design of the arm brake pedal that can be manufactured by the rapid manufacturing (RM) process using an appropriate RM material matching the properties of the conventional pedal material. The RM material selected for this study is a metal-polymer composite filament, Ultrafuse 316L by BASF, which is used by the Fused Filament Fabrication (FFF) technology, also known as Fused Deposition Modelling (FDM). Table 4 shows the mechanical properties of Ultrafuse 316L used in the analysis [9].

Table 4. The mechanical properties of Ultrafuse 316L [9].

Material	Young Modulus (E), MPa	Tensile Strength (TS), MPa	Yield Strength (YS), MPa	Density (ρ), Ton/m³	Percentage of Elongation, %
Ulrafuse 316L (FDM)	152,000	465	167	7.88	31

A design space with the same boundary conditions has been used as the input for the Altair Inspire, which is then processed to generate a topologically optimized model, eliminating the non-necessary mass in the part, hence reducing the mass of the arm brake pedal without affecting its structural integrity. Figure 7 shows the input data with boundary conditions for the Altair Inspire environment.

After several iterations, Altair Inspire comes up with an organic optimized design of the arm brake pedal. Figure 8 shows the final result of the topology optimized design of the arm brake pedal. The new topology optimized design will be analyzed and compared to the performance of the current arm brake pedal. All three structural analyses have been performed with this new design in the Altair HyperWorks environment, identically as applied with the current design.

Figure 9 shows the results of the linear static analysis of the optimized design. Identical to current design results, it is noted that the maximum stress concentration is located at clevis hole area, with a value of 410.830 MPa. The maximum deflection of the new arm brake pedal is 1.76 mm, located at the one end of the pad.

Figure 10 shows the results of the non-linear static analysis for the optimized design. For the non-linear static analysis, the results are also found to be identical with the current design results. The maximum stress concentration is located at clevis hole area, value at 168.943 MPa. The geometric non-linearity property gives an identical result for both designs and the maximum stress is lower. The maximum displacement is found to be 1.84 mm located at one end of the pad.

IOP Conf. Series: Materials Science and Engineering **715** (2020) 012020 doi:10.1088/1757-899X/715/1/012020

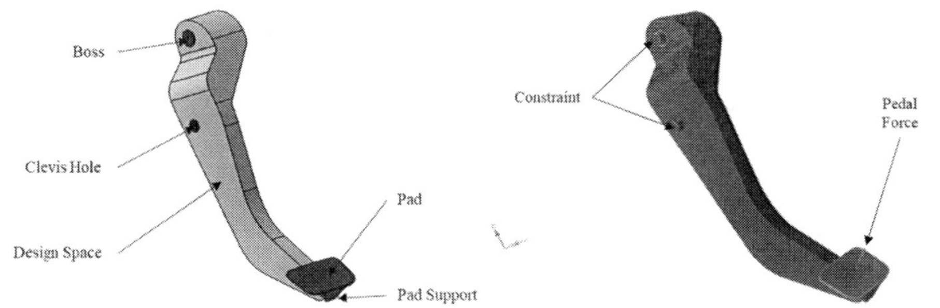

Figure 7. Input data with identical boundaries conditions.

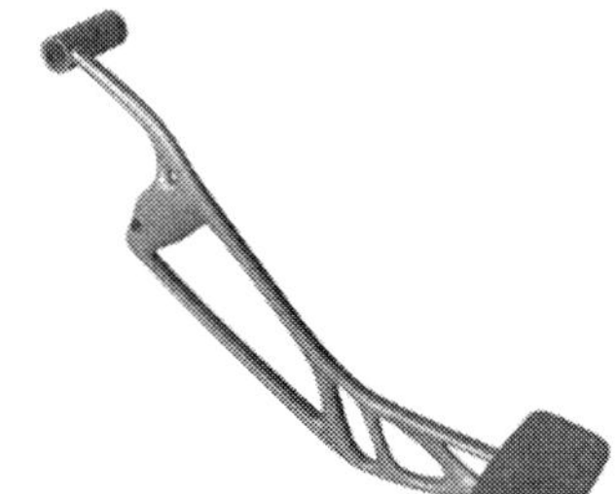

Figure 8. The topology optimized design of the arm brake pedal, utilizing selected RM material.

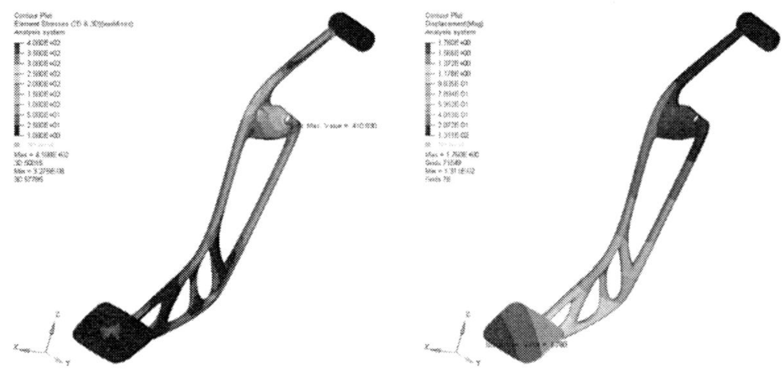

Figure 9. The result of linear static analysis for stress concentration (left) and displacement (right).

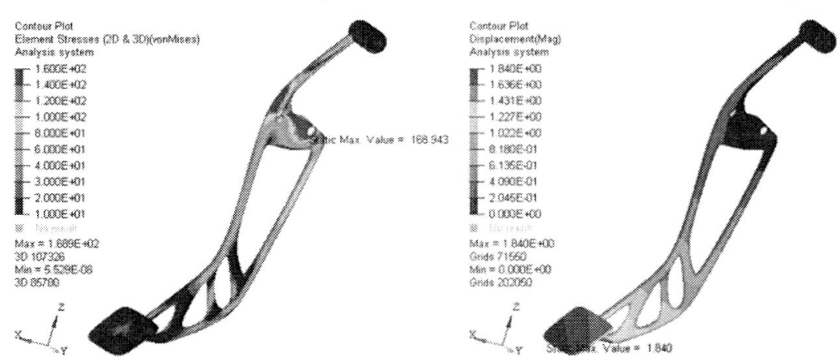

Figure 10. The result of non-linear static analysis for stress concentration (left) and displacement (right).

Figure 11 shows the fatigue analysis result of the new optimized design. Interestingly, it is noted that the fatigue analysis result for both designs are not identical. The result for new design shows the variation of the damage result, instead of current design, where the damage result is equal within the pedal brake. This shows that the mass of the new arm brake design has been topology optimized; maximizing the most important area without scarifying the structural integrity. The minimum damage fatigue life for the new arm brake design is 1.27×10^6 cycles located at clevis hole area. The minimum limit for this test is 1.0×10^6 cycles, hence the new arm brake pedal is reliable and passes the fatigue test. Thus, as a conclusion, the new arm brake design has passed all three structural integrity tests.

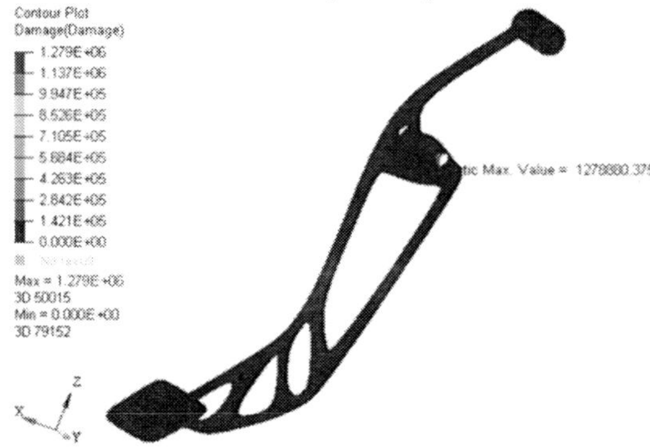

Figure 11. The result of fatigue analysis.

Table 5 summaries the overall performance for the new arm brake pedal design. The result shows a marked reduction of 54% in overall weight of the new optimized design compared to the current design.

Table 5. The overall result of the new arm brake FE verification.

Analysis		Limit	Result	Safety Factor
Linear static	Stress	<465 MPa	411 MPa	0.88
	Displacement	< 10mm	1.76 mm	0.18
Non-linear static	Stress	<465 MPa	169 MPa	0.36
	Displacement	< 10mm	1.84 mm	0.18
Fatigue		> 10^6	1.27×10^6	1.27
Weight		1 Kg	0.46 Kg	54%

4. Conclusions

The results of this study have shown that the current design of the arm brake pedal, manufactured by conventional techniques, has passed all three structural analysis integrity tests, since this design is already in the market. The new optimized design of the arm brake pedal, based on RM material, also passes all three structural analysis integrity tests. This shows that the new pedal design, which uses the RM material and technologies, is also capable to be used for the market. The new design was not only success in passing all three structural integrity test but it also came out to be 54% less in weight.

Based on the result, we can observe that the new arm brake pedal design is applicable to Rapid Manufacturing technologies and can be considered to replace current design based on costly conventional manufacturing technology. The structural analysis results proved that the RM technologies can be applied for the mass market application. A significant weight reduction over the current design without affecting structural performance should open an opportunity for the industries to further understand the potential benefit of RM technologies.

Acknowledgement

This project is sponsored by the Malaysia Automotive, Robotics and IoT Institute (MARii), previously known as Malaysian Automotive Institute. MARii is an agency under Ministry of International Trade and Industry of Malaysia. This project is a part of Transportation Innovation Centre (TIC), a joint venture project between MARii and Swinburne University of Technology (SUT).

References

[1] ASTM International, *Standard Terminology for Additive Manufacturing – General Principles – Terminology*, in *ISO / ASTM52900-15*, ASTM International: West Conshohocken, PA, (2015)

[2] M. Aliakbari, *Additive Manufacturing State-of-the-Art, Capabilities, and Sample Applications with Cost Analysis*, in *Department of Industrial Production*, KTH Royal Institute of Technology: Sweeden, (2012)

[3] K.P. Karunakaran et al., *Rapid manufacturing of metallic objects.* Rapid Prototyping Journal, **18**(4): p. 264-280, (2012)

[4] T. Wohlers, *Wohlers Report 2012 : Additive Manufacturing and 3D Printing State of the Industry Annual Worldwide Progress Report*, Wohlers Associates, Inc, (2012)

[5] G. Dwivedi, S.K. Srivastava, and R.K. Srivastava, *Analysis of barriers to implement additive manufacturing technology in the Indian automotive sector.* International Journal of Physical Distribution & Logistics Management, **47**(10): p. 972-991, (2017)

[6] G.N. Levy, R. Schindel, and J.P. Kruth, *Rapid Manufacturing and Rapid Tooling with Layer Manufacturing (Lm) Technologies, State of the Art and Future Perspectives.* CIRP Annals. **52**(2): p. 589-609, (2003)

[7] J.G. Prada et al., *Direct digital manufacturing of an accelerator pedal for a Formula Student racing car.* Rapid Prototyping Journal, **22**(2): p. 311-321, (2016)

[8] R. Limpert, *Brake Design and Safety*, SAE International, (2011)

[9] H. Gong et al., *Comparison of Stainless Steel 316L Parts Made by FDM- and SLM-Based Additive Manufacturing Processes.* JOM, (2018)

IOP Conf. Series: Materials Science and Engineering **715** (2020) 012021 doi:10.1088/1757-899X/715/1/012021

Applying data mining techniques with data of campus card system

Yufan Li*, Huifu Zhang and Shangli Liu

School of Computer Science and Engineering, Hunan University of Science and Technology, 411201 Xiangtan, China

*E-mail: 494958939@qq.com

Abstract. In recent years, campus card system (CCS) has been widely used in universities. At the same time, the scale of the data stored in the database of the CCS is also increasing rapidly. Therefore, it would be better to adopt data mining techniques to reveal the status, causes and trends of students' consumption data. In this research, the K-Means clustering algorithm is applied to the students' consumption behavior.The Apriori algorithm is used to study the students' consumption behavior and performance. The results not only improve the efficiency of student work but also help to build smart campus for our university.

1. Introduction

Data mining is the process which reveals relationships, trends and patterns by analyzing large amounts of data [1]. It has been widely used in banking, marketing, insurance, education and other domains. And nowadays, college students usually have various consumption types, such as dining, shopping, checking out books, purchasing water, surfing on the net [2], it will generate magnanimous data. Therefore, data mining is an effective approach, which can provide a great help in finding hidden relationships behind the data of CCS.

Jiang et al. [3] analyzed all the consumer records of masters, it is concluded that the boys overall consumption is higher than the girls, but there is no difference between their session and major. Xue et al. [4] used cross tabulations and Kohonen neural network to the data of shop consumption and they found that the types of consumer have a greater influence on consumption, while gender is nearly have no impact on it. By using OLAP(On-line Analytical Processing), Zhang [5] proposed an effective data mining model of the campus card system to classify the students' consumption data and it helps the university leadership to make appropriate decisions.

In this research, we study the CCS data of Hunan University of Science and Technology in China. The paper is organized as follows: First, data mining process is briefly described in Section 2.Then the results and discussion are shown in Section 3. Finally, conclusions and future work are outlined.

2. Data mining process

2.1. Data set

The original data of CCS is stored in Oracle database. The users include undergraduates, graduate students, teachers, etc. Since freshmen came to school for a short time, they have not yet formed a stable habit of consumption, while senior students often have their internships for a long time out of school. Consequently, we collected the consumption data from session 2015 to 2016. Initially size of the data is 380723 and it contains 960 students. Table 1 shows the structure of table stored in transaction database:

Table 1. Structure of table in database.

SNO	MERCNAME	TRANAMT	EFFECTDATA
15xxxxxx01	4	1900	2018-3-27 11:03:34
15xxxxxx01	2	550	2018-6-5 11:55:28
...	...	...	...
16xxxxxx30	1	200	2018-4-10 7:13:42
16xxxxxx30	6	900	2018-4-11 12:08:33

In table 1, SNO (student number) is the primary key, MERCHAME is the ID of canteen, TRANAMT means the expense of every record, and EFFECTDATA is the transaction time. For example, the first record means a student with a student number is 15xxxxxx01 and expenses ¥19 at No.4 canteen on March 27th, 2018, at 11:03:34.

2.1.1. Analysis of student consumption records

We created a new field named DT which means the results that we divided transaction time into time section by an hour. The time before 7am is value T1 in DT, the time between 7:00 and 8:00 is T2 and the time after 10:00 is T17. Figure 1 shows the student records of consumption based on time.

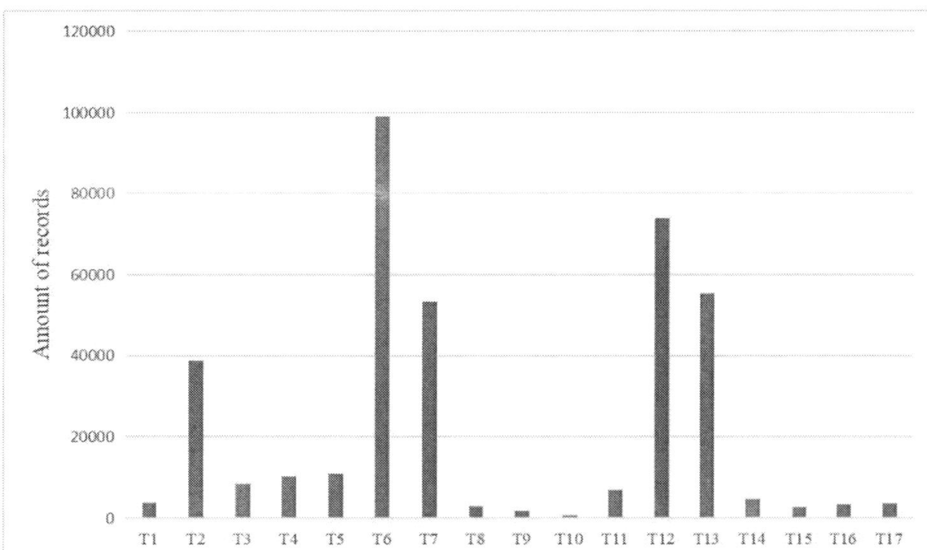

Figure 1. Based on time.

In figure 1, we can see that the amount of student consumption records during 7:00 to 8:00, 11:00 to 13:00, 17:00 to 19:00 are lager than others, and T6 (11:00 to 12:00) has the biggest amount of records while T10 (15:00 to 16:00) is the smallest in all time sections. And we argue that if a student who has breakfast before 8:00 or dinner between 17:00 to 19:00, then he or she is consider to have a regular consumption behavior.

IOP Conf. Series: Materials Science and Engineering **715** (2020) 012021 doi:10.1088/1757-899X/715/1/012021

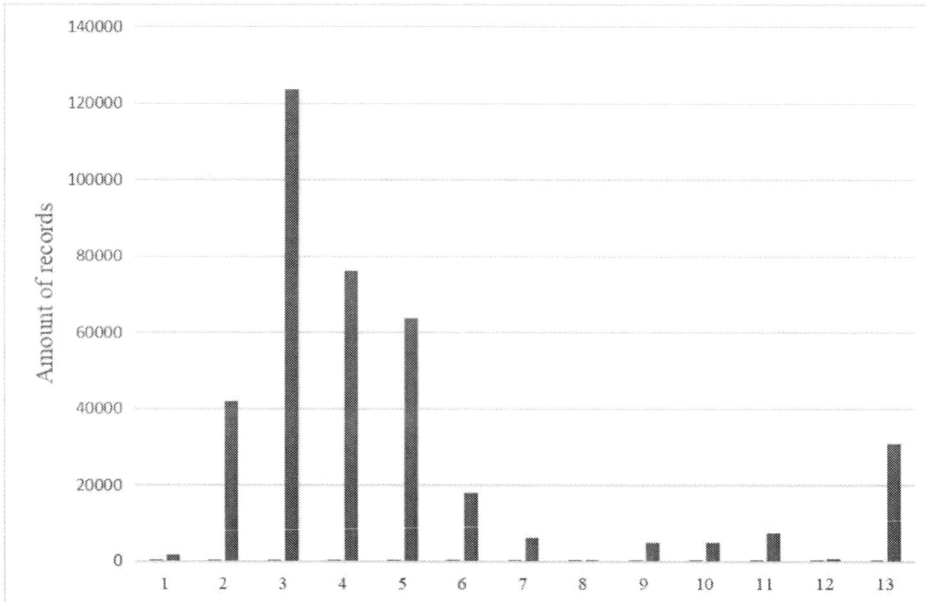

Figure 2. Based on location.

Figure 2 shows the relationship between amount of records and canteen's location. We can see that No.3, No.4, No.5 have more records than other canteen. Those canteens which have low amount of records may far away from residence halls, but they can attract students by offering discount or handing out leaflets.

2.2. Data selection and transformation

In this step, only those fields were selected which were necessary for applying data mining algorithms. We chose a few derived variables, while some variables were extracted from the database. All the response variables which were derived from the Oracle database will be elaborated in section 3.

2.3. Data mining methodologies

A. K-Means Clustering Algorithm:

Clustering is a method which can divide data into classes of similar objects. K-Means clustering algorithm is based on a simple iterative method which is partitioning a dataset into a user-specified number of clusters, k [6].The following figure 3 shows the steps of K-Means clustering algorithm.

Input: Number of desired clusters K
D= {x_i| 1, 2…N}

Methods:
(1) Select k data objects from dataset D as initial centers randomly.
(2) Repeat;
(3) Assign each data object to its closest centriod after calculate the distance between each data object x_i.
(4) Recalculate new mean for each cluster.
(5) Until no change in the center of cluster.

Output:A set of k clusters

Figure 3. Steps of K-Means clustering algorithm.

B. Apriori Algorithm:

Apriori algorithm is one of the most popular association rule mining algorithm to find frequent item set from large data sets and hence to get association rules A→B contains which meet the following conditions [7]:

$$Support(A \rightarrow B) = P(AB) \geq Min\_sup \quad (1)$$

$$Confidence(A \rightarrow B) = P(B \mid A) \geq Min\_conf \quad (2)$$

The Min_Sup and Min_Conf are threshold values defined by users.

3. Results and discussion

To determine the best number of desired K for K-Means algorithm we use the measure called Silhouette Coefficient. The silhouette coefficient, relative to mean intra-cluster distance (a) and mean nearest-cluster distance (b) for each sample, is defined as:

$$\frac{b-a}{\max(a,b)} \quad (3)$$

Table 2. Student related variables.

Variables	Description	Possible values
AC	Amount of Consumption Per Month	{High:,Medium,Low}
SC	Number of Times for Swiping Cards Per Month	{High:,Medium,Low}
RB	Number of Times for Regular Breakfast Per Month	{High:,Medium,Low}
RD	Number of Times for Regular Dinner Per Month	{High:,Medium,Low}
GCS	Grades of Compulsory Subject	{Poor:,Average,Good,Outstanding}
GES	Grades of Elective Subject	{Poor:,Average,Good,Outstanding}
LC	Library Consumption	{Yes,No}
RC	Retaking Course	{Yes,No}

Table 2 explains all the related variables in this research.

In table 3, the comparison of silhouette coefficient with K is illustrated. It shows that clustering effect is the best when K=2. Table 4 shows the results after the application of K-Means algorithm.

Table 3. Comparison of silhouette coefficient with K.

K	Silhouette coefficient
2	0.549
3	0.511
4	0.505
5	0.508
6	0.490
7	0.479

Table 4. Clustering results.

Cluster center	AC	SC	RB	RD	Students
I	369.002	141.241	14.148	47.634	434
II	160.852	64.414	6.835	22.105	526

Students in cluster I consume more in canteen and appear to have a regular eating time. While in cluster II, students exhibit a low interest in canteen, which means they may have a take-away for their meal. Since the students in cluster 2 have a higher proportion of complete 960 students, the canteen should update their menu or offer discount to attract students to consume.

IOP Conf. Series: Materials Science and Engineering **715** (2020) 012021 doi:10.1088/1757-899X/715/1/012021

Table 5. Results of Apriori algorithm.

Min_sup:0.1 Min_conf:0.8 Valuable rules found: 1.GCS=Outstanding,LC=No,RD=Medium,AC=Medium==>SC=Medium(conf=1) 2.AC=Low,RB=Low,RC=Yes==>RD=Low(conf=0.915) 3.SC=High,RD=High,GES=Outstanding ==> AC=High(conf=0.843) 4.SC=High,RC=No==>GES=Outstanding(conf=0.841) 5.RD=Medium,AC=Medium,SC=Medium,GES=Outstanding==>GCS=Good(conf=0.800)

Table 5 shows the results after adopted Apriori algorithm. From the results of aproiri algorithm as shown in row numbered four of table 5, we can clearly see that students who have high SC and without retaking courses appear to have outstanding GES. Moreover, from rows numbered one, two, three and five, we find that students with good performance usually have a regular consumption behavior, habit of checking out books and daily routine.

4. Conclusion

This paper aims to study the data of CCS and analyzes it with data mining techniques like K-Means clustering algorithm and Apriori association rule algorithm. The results we got can help relevant departments in decision-making and improve the efficiency of education management. Data is the core part in CCS and also has great potentialities. In the future, we hope that data mining techniques such as classification algorithms can be widespread used in CCS and thus we could build an effective model to support the construction and application on smart campus.

References

[1] J. W. Han, M. Kamber. Data Mining: Concepts and Techniques, Morgan Kaufmann Publishers Inc. (2005)

[2] W. C. Lv. Design of Campus Smart Card System. Applied Mechanics and Materials, 347-50, 3915-3918 (2013)

[3] T. J. Jiang, J. M. Cao, S. Dan, X. L. Yang. Analysis and Data Mining of Students Consumption Behavior Based on a Campus Card System. International Conference on Smart City and Systems Engineering (2017)

[4] L. M. Xue, W. X. Luan. Data Analysis and Consumption Model Research of E-card system. International Industrial Informatics and Computer Engineering Conference. (2015)

[5] Y. Zhang. The Research of Data Mining Technology in Campus Card Consumption System. International Conference on Electronics & Optoelectronics. (2011)

[6] X. D. Wu, V. Kumar, JR. Quinlan, et al. Top 10 algorithms in data mining. Knowledge & Information Systems, 14(1),1-37(2008)

[7] S. Rathee, A. Kashyap. Adaptive-Miner: an efficient distributed association rule mining algorithm on Spark. Journal of Big Data, 5(1), 1-17(2018)

IOP Conf. Series: Materials Science and Engineering **715** (2020) 012022 doi:10.1088/1757-899X/715/1/012022

Research on improvement of high utility pattern mining algorithm over data streams

Feng Guo*, Yuqiang Li and Lin Li

Wuhan University of Technology, Computer Science and Technology, 430061 Wuhan, China

*E-mail: 2284387841@qq.com

Abstract. Aiming at the problem that the existing algorithms for high utility pattern mining over data streams based on sliding window have multiple datasets scans or redundant items, an efficient HUIGRT algorithm for mining high utility patterns over data streams based on global revision header table is proposed in this paper. First, the global revision header table and the utility tree are constructed. The global revision header table is used to store the items and transaction utility of the current data domain that need to be processed, and the utility tree is used to store all of the utility information on the item sets in the transactions to avoid multiple datasets scans. Then, this algorithm can mine all high utility patterns using the global revision header table and the utility tree. Finally, the redundant items are deleted by revising the global revision header table, meanwhile the utility tree is updated to fill in new data. This paper compares the algorithm with the existing high efficiency algorithm HUPMS and HUM-UT on the three datasets with different sparse: Mushroom、T10.I4.D100K and Retail. The results show that the space-time performance of HUIGRT algorithm is better than the two other algorithms.

1. Introduction

With the explosive growth of data, how to extract hidden useful information from massive data has become a major problem in the field of data mining [1]. Frequent pattern mining [2-4] is one of the main technologies in the field of data mining. It can mine patterns from all the datasets that exceed the user-specified threshold. However, frequent pattern mining only considers whether an item in the transaction item set appears, but ignores the nature of the item itself (profit, value, quantity, etc.), resulting in items that don't appear many times but have high utility may be discarded.

Unlike frequent pattern mining, high utility pattern mining [5] solves the problem that high utility items may be discarded. Each data item has its own internal utility (quantity) and external utility (profit, value, etc.). Yao et al. [6] defined the high utility pattern mining for the first time and gave a specific mathematical calculation model. Liu et al. [7] proposed the two-phase algorithm. However, this algorithm generates a large number of candidates in the first phase of mining, and needs to multiple datasets scans in the second phase, which reduces the efficiency of the algorithm. Subsequently, Wang et al. [8] proposed an algorithm called HUPM-FP and Wu et al. [9] proposed an algorithm called TKU, all does not need to generate candidates, so that all high utility patterns can be mined. The former uses the pattern-growth method to mine high utility patterns, while the latter is used to solve the problem that the user is difficult to set minimum utility threshold in the actual

IOP Conf. Series: Materials Science and Engineering **715** (2020) 012022 doi:10.1088/1757-899X/715/1/012022

application process. Although these algorithms are already very efficient, they are only suitable for static datasets. With the emergence of a large number of data streams [10-11] in many fields, traditional high utility mining algorithms based on static datasets are no longer adapt to the processing of data streams. In order to solve this problem, Jin et al. [12] proposed a model for processing data streams, which can be divided into landmark model, sliding window model [13-14] and damped window model. Among them, the sliding window model is a common model for processing data streams. Based on this model, many scholars have carried out in-depth research on high utility pattern mining over data streams.

Ahmed et al. [15] proposed an algorithm over data streams based on sliding window called HUPMS. In the first phase, a large number of candidates are generated and in the second phase, it is necessary to re-verify the candidates, the space-time performance is poor. Subsequently, Wang [16] proposed an algorithm called HUM-UT. This algorithm uses the global header table (Global-HT) to store the current pending data and uses the utility Tree (UT-Tree) to store data transaction information. Through the Global-HT simulation sliding window in the excavation process, all high utility patterns over data streams can be mined through UT-Tree, which avoids the generation of candidates, and its space-time performance is obviously superior to HUPMS algorithm.

Although the efficiency of the high utility mining of the HUM-UT algorithm is already very high, the Global-HT established by the algorithm contains redundant items that are independent of the current data domain to be processed, resulting in a lot of useless time cost to traverses these redundant items for mining process. Aiming at the above problems, this paper proposes an improved high utility pattern mining based on global revision header table algorithm (HUIGRT), which uses a new global revision header table (GRHT-Table) to delete redundant items that are independent of the current data domain to be processed, thereby reducing the extra space-time consumption generated by the algorithm during the mining process. Experimental results on three different datasets show that the HUIGRT algorithm is superior to the HUPMS algorithm and the HUM-UT algorithm in both time and space efficiency. The HUIGRT algorithm will be further explained below.

2. HUIGRT algorithm

For the two-phase algorithm HUPMS, there are multiple datasets scans, and the one-phase algorithm HUM-UT has redundant items in the global header table. This paper proposes the HUIGRT algorithm, whose GRHT-Table contains a window size data, after adding a new batch of data, the redundant items in the GRHT-Table that are independent of the current data domain to be processed are removed, thereby reducing the extra space-time consumption of the algorithm in the mining process. In addition, UT-Tree is used to avoid scanning datasets multiple times and improve the space-time performance of the algorithm.

The basic process of this algorithm is shown in Algorithm 1.

Algorithm 1 HUIGRT
Input Data streams(*DS*)
Output high utility patterns of the data streams

```
create a null GRHT-Table and a null UT-Tree
while(DS isn't finish)
  read a batch of data from DS
  if(sliding window is filled)
    if(sliding window is filled for the first time)
      call procedure Mining high utility patterns of DS
    else
      call procedure revision of GRTH-Table and update of UT-Tree
      call procedure Mining high utility patterns of DS
    end if
  else
    call procedure Construction of GRTH-Table and UT-Tree
```

```
    end if
end while
```

The HUIGRT algorithm consists of three processes: 2.1. Construction of GRHT-Table and UT-Tree; 2.2. Mining high utility pattern based on GRHT-Table and UT-Tree; 2.3. Revision of GRHT-Table and update of UT-Tree. The three processes will be further explained below.

2.1. Construction of GRTH-Table and UT-Tree

The function of this process is to ensure that the data in sliding window is filled for the first time, so that the current data domain to be processed contains the window size data, so as to realize the first time mining high utility patterns. The entire mining process uses GRHT-Table to store each data item in each batch of transactions and its corresponding transaction weighted utilization. The new batch of transactions is mapped to the tree structure of TN-Tree (Tail Node on Tree), and then multiple TN-Tree is merged into UT-Tree to store the batch utility corresponding to all batch transactions.

GRHT-Table contains different data items and each data item corresponds to the transaction weighted utilization in different batches of data and a list of link pointers, which sw represent the window size, that is, the number of batches contained in the window; the list of link pointers is created during the mining process and points to the node with the same name in the UT-Tree. The TN-Tree consists of only one root node, multiple general nodes and multiple tail nodes. The general nodes stores each data item in the transaction, and the tail nodes not only stores its corresponding one data item, but also stores the utility information of all the data items of the data item to the root node path. When a TN-Tree corresponding to a new batch of data is created, if there have the same prefix data items between different transactions, the node corresponding to the prefix data item is shared on the TN-Tree. When multiple TN-Trees are merged into UT-Tree, the utility information of the batch of data saved by the tail node is merged into the node corresponding to the UT-Tree. At the same time, the tail node on table (TN-Table) is used to store the tail node information of each transaction in each batch of data.

In order to explain the construction process of GRTH-Table and UT-Tree to better, the concepts of utility of data item in transaction, transaction utility and transaction weighted utilization of pattern need to be defined as follows.

Let $I=\{i_1,i_2,i_3,\cdots,i_m\}$ be a collection of all the different data items, where each data item $i_r(1\le r\le m)$ has an external utility, denoted as $p(i_r)$. $D=\{T_1,T_2,T_3,\cdots,T_n\}$ represents a dataset consisting of n transactions, each transaction $T_d\in D(1\le d\le n)$ is a subset of I, and has a unique identifier TID. Each i_r in T_d has an internal utility, denoted as $q(i_r,T_d)$.

Definition 1 (utility of data item in transaction) The utility of data item i_r in transaction T_d is the product of the internal utility and the external utility of the data item in T_d, denoted as $u(i_r,T_d)$, which is defined as follows.

$$u(i_r,T_d)=q(i_r,T_d)\times p(i_r) \tag{1}$$

Definition 2 (transaction utility) The utility of a transaction T_d refers to the sum of the utility of all data items in T_d, denoted as $tu(T_d)$, which is defined as follows.

$$tu(T_d)=\sum_{i_r\in T_d}u(i_r,T_d) \tag{2}$$

Definition 3 (transaction weighted utilization of pattern) The transaction weighted utilization (twu) of pattern X refers to the sum of transaction utility of all transactions containing pattern X in a dataset D, denoted as $twu(X)$, which is defined as follows.

$$twu(X)=\sum_{(X\subseteq T_d)\cap(T_d\in D)}tu(T_d) \tag{3}$$

Combined with the definitions (1) to (3), and using the transaction weighted utilization (twu) as the estimated-utility, the construction process of GRTH-Table and UT-Tree is shown in Procedure 1.

IOP Conf. Series: Materials Science and Engineering **715** (2020) 012022 doi:10.1088/1757-899X/715/1/012022

Procedure 1 Construction of *GRTH-Table* and *UT-Tree*
Input a *GRHT-Table*,a *UT-Tree*, a *TN-Table*, a batch of data(*BD*)
Output a *GRHT-Table*,a *UT-Tree*, a *TN-Table*

```
if(GRHT-Table has included sw batches of data) //sw is the size of window
   call procedure revision of GRTH-Table and update of UT-Tree
else
   for each(item Qi in BD)
      calculate the estimated-utility (twu) of Qi, twu(Qi)
      if(Qi doesn't exist in GRTH-Table)
         create a new item for Qi into  the GRHT-Table
      end if
      set twu(Qi) as the estimated-utility of Qi into the GRHT-Table
   end for
   create a TN-Tree for BD, then merge TN-Tree to UT-Tree
   for each(tail nodes TN on TN-Tree)
      put TN into the TN-Table
   end for
end if
```

This paper takes the data in table 1 and table 2 as an example to illustrate the specific execution process of the algorithm in Procedure 1. Table I is a pending instance of a dataset containing 8 transactions, and table 2 is the external utility corresponding to the data items in table 1, and all have been sorted according to the dictionary table order.

Table 1. Pending instance of dataset.

TID	Transaction item set
T1	(a,6)(b,1)
T2	(b,4)(c,1)(e,1)(f,1)
T3	(b,3)(c,6)(d,3)
T4	(a,1)(b,1)(c,8)
T5	(b,2)(e,3)
T6	(a,5)(b,3)(d,4)
T7	(c,2)
T8	(a,2)(d,3)

Table 2. External utility of each data item.

Item	External utility
a	3
b	8
c	5
d	6
e	10
f	1

Set the size of window $sw=3$, the size of batch data $b=2$, that is, the window contains 3 batches of data, and each batch of data contains 2 transactions. Initialize GRTH-Table, UT-Tree, and TN-Table, and read in the first batch of data (composed of transactions T1 and T2). At this time, the GRTH-Table is empty, and the data in the sliding window is not filled. Calculate the twu of each data item, add non-existent data items to the GRTH-Table in the order of the dictionary table, fill the twu corresponding to each data item into the corresponding position, and fill the GRTH-Table as table 3 shows

Table 3. GRTH-Table after reading the first batch of data.

Item	T[1]	T[2]	T[3]
a	26	null	null
b	74	null	null
c	48	null	null
e	48	null	null
f	48	null	null

Since the first batch of data is read in, the other batch utility in the tail node after the TN-Tree merge are empty, and are marked as empty "{}". The results as shown in figure 1.

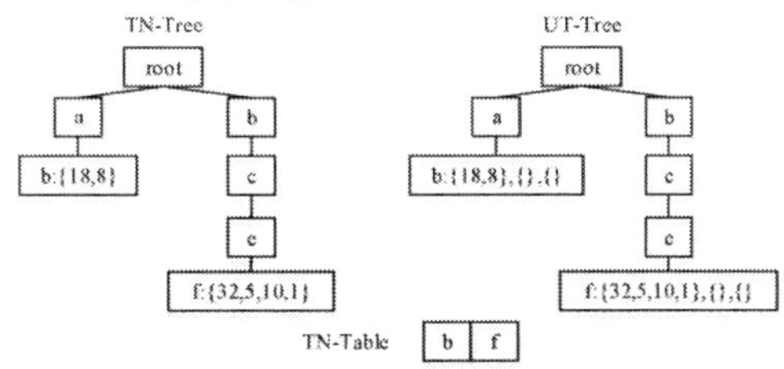

Figure 1. TN-Tree, UT-Tree, and TN-Table after reading the first batch of data.

Next, when read the second batch of data, it has the same prefix data items "ab" and "bc" as the first batch of data; the third batch of data is read, and the other two batches of data have the same prefix data items "ab" and "b", so UT- Tree shares these nodes. Since TN-Tree is created based on the current batch data, it is not related to the specific mining process, and it will not be separately described later. The GRTH-Table is shown in table 4.

Table 4. GRTH-Table after reading the third batch of data.

Item	T[1]	T[2]	T[3]
a	26	51	63
b	74	123	109
c	48	123	0
d	0	72	63
e	48	0	46
f	48	0	0

The UT-Tree and TN-Table corresponding to table 4 are shown in figure 2.

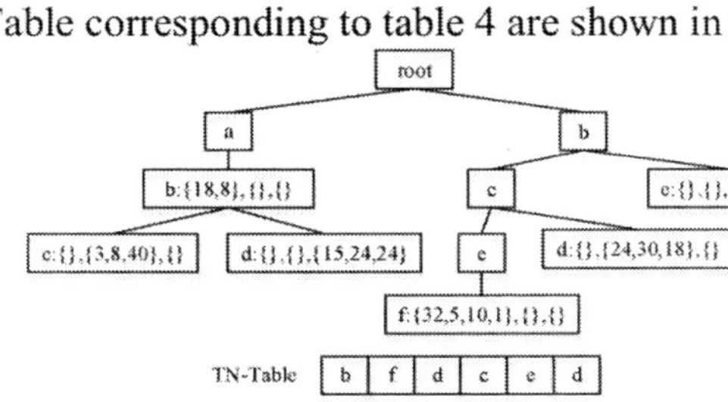

Figure 2. TN-Tree, UT-Tree, and TN-Table after reading the third batch of data.

Since the third batch of data is filled, and the size of window is 3, the condition that the data in sliding window is filled is satisfied, and the process ends, and then the process of mining high utility patterns is performed.

2.2. Mining high utility patterns based on GRHT-Table and UT-Tree

When the data in sliding window is filled, the mining high utility patterns is started until the data in the window is completely traversed.

In order to explain the process of high utility pattern mining to better, the utility of the pattern in the transaction, the utility of the pattern in a dataset, the total utility of the dataset and the minimum utility need to be defined as follows.

Definition 4 (the utility of the pattern in the transaction) The utility of the pattern X in the transaction T_d refers to the utility sum of each data item in X in the transaction T_d, denoted as $u(X,T_d)$, which is defined as follows.

$$u(X,T_d)=\begin{cases}0 & X\not\subseteq T_d\\ \sum_{i_r\in X}u(i_r,T_d) & X\subseteq T_d\end{cases} \tag{4}$$

Definition 5 (the utility of the pattern in a dataset) The utility of the pattern X in a dataset D is the utility sum of X in all transactions containing X in D, denoted as $uti(X)$, which is defined as follows.

$$uti(X)=\sum_{T_d\in D}u(X,T_d) \tag{5}$$

Definition 6 (the total utility of the dataset) The total utility of a dataset D is the sum of the utilities of all the transactions contained in D, denoted as $Ttu(D)$, which is defined as follows.

$$Ttu(D)=\sum_{T_d\in D}tu(T_d) \tag{6}$$

Definition 7 (minimum utility) The minimum utility is the product of a user-specified minimum utility threshold ($minUT$) more than 0 and smaller than 1 and the total utility of dataset D, denoted as $\min Uti$, which is defined as follows.

$$minUti = minUT\times Ttu(D) \tag{7}$$

High utility pattern mining refers to: for a dataset D, calculate the minimum utility ($minUti$), and mine all the patterns whose utility (uti) is no smaller than $minUti$ from D.

The whole mining process is divided into main mining and sub-mining. The mining process is shown in Procedure 2.

Procedure 2 Mining high utility patterns
Input a *GRHT-Table*,a *UT-Tree*, $minUT$
Output high utility patterns(*HUPs*)

```
calculate minUti, then create a link list pointer in GRHT-Table for UT-Tree
add an attached list (Utility-cache) to each leaf node on UT-Tree
for each(item Qi in GRHT-Table)
  calculate the estimated-utility (twu) of Qi,twu(Qi) from GRHT-Table
  calculate the real utility(uti) of Qi, uti(Qi) from Utility-cache
  if(twu(Qi)≥minUti)
    generate a data pattern, X={ Qi}
    if(uti(Qi) ≥minUti)
      attach X to HUPs
    end if
    create a sub global revision header table(SGRHT-TableX) for X
    create a sub utility on tail tree (SUT-TreeX) for X
    call sub procedure SubMining(SGRHT-TableX, SUT-TreeX, X)
  end if
  for each(node Ni corresponding to Qi)
    pass the Utility-cache of Ni to its parent node on UT-Tree
  end for
end for
Sub Procedure: SubMining(SGRHT-TableX, SUT-TreeX, X)
for each(item Qj in SGRHT-TableX)
  calculate the estimated-utility(twu) of {Qj}∪X,twu({Qj}∪X) from
```

```
SGRHT-TableX
    calculate the real utility(uti) of{Qj} ∪ X, uti({Qj} ∪ X),from Utility-Information
    if (twu({Qj} ∪ X) ≥minUti)
       generate a pattern, Y= {Qj} ∪ X
       if(uti({Qj} ∪ X) ≥minUti)
          attach Y to HUPs
       end if
       create a sub global revision header table(SGRHT-TableY) for Y
       create a sub utility on tail tree (SUT-TreeY) for Y
       call procedure SubMining(SGRHT-TableY , SUT-TreeY, Y)
    end if
    for each(node Nj corresponding to Qj)
       pass the Utility-Information of Nj to its parent node on SUT-TreeY
    end for
end for
```

The link pointer list exists in the last column added to the GRTH-Table, pointing to all nodes of the same name in the UT-Tree. The element of Utility_cache is the sum of the utility of the corresponding data items of each batch of data of the node, that is, the sum of the corresponding values in the three "{}", which is recorded as "[]". Since the TN-Table is not used in the current high utility pattern mining process, it will not be separately described later. The UT-Tree that contains the Utility_cache is shown in figure 3.

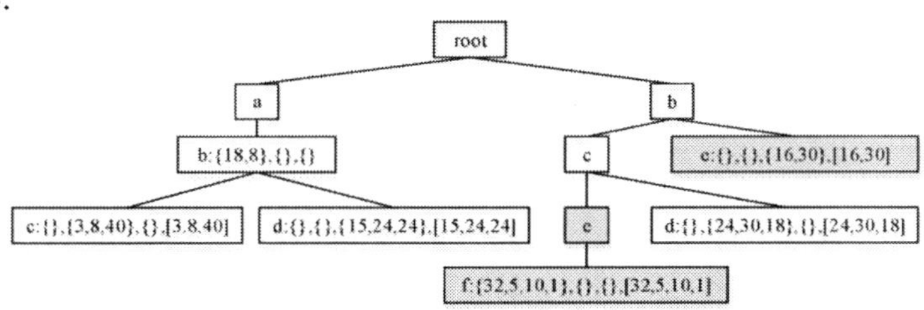

Figure 3. UT-Tree with Utility_cache attached.

Each data item is processed in turn from the last item in the created GRTH-Table. For example, to process data items "f" and "e" (the grayscale fill has been marked in figure 3), set the minimum utility $minUti$ =80 and assume that the currently processed data item is the last item "f". After calculation, $twu(\mathrm{f})$ =32+5+10+1=48, $uti(\mathrm{f})$ =1, at this time $twu(\mathrm{f}) < minUti$, the condition is not satisfied. According to the transaction weight downward closure property [7], all the collections containing the data item "f" are not high utility pattern.

Next, the data item "f" corresponding to the Utility_cache on the UT-Tree is passed to the corresponding parent node "e" (the grayscale fill has been marked in figure 4). Before this, you need to remove the utility information of the current processing node. If the parent node does not contain utility information, the Utility_cache is directly passed to the parent node; otherwise, the utility information of the parent node is added to the utility information of the corresponding data item in the Utility_cache. Each time one node is processed, the Utility_cache is passed to the parent node, ensuring that each item in the processing header table has its corresponding Utility_cache, and the resulting UT-Tree is shown in figure 4.

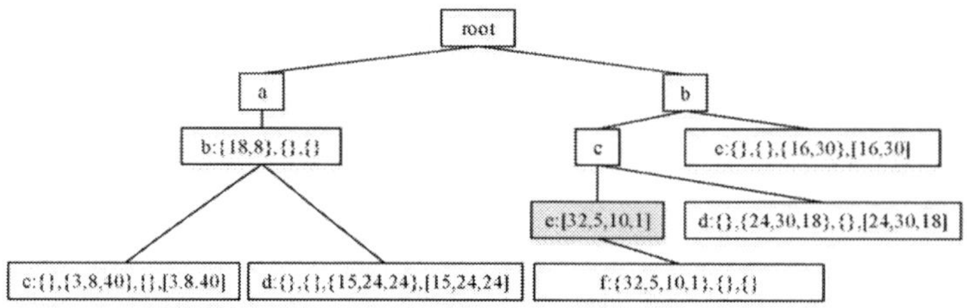

Figure 4. The Utility_cache on the f node is passed to the parent node.

As can be seen from figure 4, there are two paths contains node "e". For the path root-b-c-e, the transaction weighted utilization (twu) is the sum of all the values in the Utility_cache on node "e", which is 47. Similarly, for the path root-b-e, the value of twu is 46. It is calculated that for the transaction weight value is $twu(e)=93$ of the node "e", the real utility of the node "e" in the sliding window is $uti(e)=40$, because the set $minUti$ =80, satisfies twu(e)> $minUti$, so the generated data pattern I={e}, and because $uti(e)<minUti$, the data pattern I is not a high utility pattern. At this time, the current node "e" has a parent node that is not the root node. Therefore, it is necessary to create a SGRHT-Table and a SUT-Tree for data pattern I, and perform sub-mining.

For the creation of SGRHT-Table in sub-mining, the sum twu of all data items on the path root-b-c-e and path root-b-e is counted. If the sum of twu is no smaller than $minUti$, the items are saved to a SGRHT-Table in the order of the dictionary table. For the creation of the SUT-Tree, you need to read all the paths and the Utility_cache on the node again. For example, from the two paths obtained in figure 4, the items not in SGRHT-Table are deleted, and the data item "b" is obtained, and twu = 93. For each of the two paths, the utility of the item "b" is 32 and 16, so the total utility is 48; the utility of the item "e" is 10 and 30, so the total utility is 40. Add data item b to the SUT-Tree, and put 40 as the base data pattern utility on the new tail node's utility list of SUT-Tree, and put 48 on the corresponding to the tail node's utility list, the result shown in figure 5.

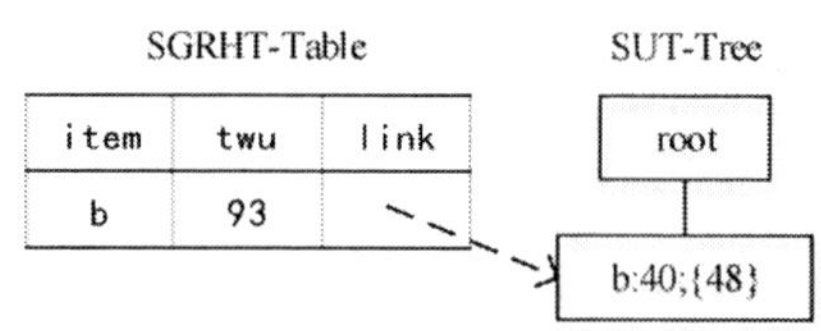

Figure 5. Pattern {e} of SGRHT-TableE and SUT-TreeE.

For the data item "b" in SUT-Tree in figure 5, after calculation, twu({eb})=93, uti({eb})=40+48=88, since the $minUti$ =80, satisfies twu({eb})> $minUti$, so the generated data pattern Is={eb}, and since uti({eb})> $minUti$, the data pattern Is is a high utility pattern. Since the node "b" has only one parent and the parent is the root node, the current sub-mining process ends.

2.3. Revision of GRHT-Table and update of UT-Tree

After the first high utility pattern mining, when a new batch of data arrives, the window slides, replacing a batch of data with the longest dwell time with a new batch of data. At this time, the GRHT-Table needs to be revised and the UT-Tree is updated. The specific process is shown in Procedure 3.

Procedure 3 revision of *GRHT-Table* and update of *UT-Tree*
Input a *GRHT-Table*,a *UT-Tree*, a *TN-Table*, a batch of data(*BD*)
Output the latest *GRHT-Table*, the latest *UT-Tree*, the latest *TN-Table*

```
0: if(GRHT-Table has included sw batches of data) //sw is the size of window
1:    find the longest dwell time batch of data(Ldwell- BD) from GRHT-Table
2:    for each(item Qi in Ldwell- BD)
```

```
    set Qi's estimated-utility(twu) to 0
  end for
  delete link list
  for each(item Qi in BD)
    calculate the estimated-utility(twu) of Qi,twu(Qi)
    if(Qi doesn't exist in GRTH-Table)
      create a new item for Qi into  the GRHT-Table
    end if
    set twu(Qi) as the estimated-utility of Qi into the GRHT-Table
  end for
  for each(item Qj in GRHT-Table)
    if(all estimated-utilities of Qj are 0)
      delete Qj from GRHT-Table
    end if
  end for
  find tail nodes(TD)of the longest dwell time batch of data from TN-Table
  for each(node Ni in TD)
    find the path, P, from Ni to root
      for each(node Nj on the P)
        if(Nj isn't a shared node)
          delete Nj from P
        end if
      end for
    delete Ni from TN-Table
  end for
  create a TN-Tree for BD, then merge TN-Tree to UT-Tree
  for each(tail nodes TN on TN-Tree)
    put TN into the TN-Table
  end for
end if
```

In the process of constructing GRTH-Table and UT-Tree in section 2.1, GRHT-Table has read three batches of data, and in section 2.2, the first mining process ends, and then read in the fourth batch of data. When reading the fourth batch of data, find the first batch of data in the GRTH-Table with the longest dwell time, set the twu corresponding to all the data items to 0, and then clear the all pointers in the list in the GRTH-Table. The data items in the fourth batch of data are arranged according to the dictionary table, and then the data items that don't exist in the table are added to the GRTH-Table, and the twu corresponding to each data item is filled in the original position of the first batch of data, and the GRTH-Table is shown in table 5. As can be seen from Table 5, there is a redundancy item "f" that is unrelated to the current data field to be processed. Revise the GRTH-Table and delete the redundant item so that the algorithm does not traverse the item redundant again during the mining process.The revised GRTH-Table is shown in table 6.

Table 5. GRTH-Table after reading the fourth batch of data.

Item	T[1]	T[2]	T[3]
a	24	51	63
b	0	123	109
c	10	123	0
d	24	72	63
e	0	0	46
f	0	0	0

Table 6. Revised GRTH-Table.

Item	T[1]	T[2]	T[3]
a	24	51	63
b	0	123	109
c	10	123	0
d	24	72	63
e	0	0	46

Through the above process GRTH-Table has been revised, the update process of UT-Tree will be detailed below. Find the tail node "b" and the tail node "f" of the first batch of data from the TN-Table. First, clear the batch utility of tail node "b" corresponding to the UT-Tree. Since the node b has the child node "c", it indicates that the node on the node "b" to the root path is shared by other batch data, so cannot be deleted. For the tail node "f", since all batch utility on f are empty and "f" has no child nodes, the node "e" and node "f" that is not shared with other paths is deleted. Next, create a new structure TN-Tree for the fourth batch of data and merge it into UT-Tree. Clear all the tail nodes of the first batch of data in the TN-Table, and fill the tail nodes of the fourth batch of data to the corresponding position. The updated UT-Tree and TN-Table are shown in figure 6.

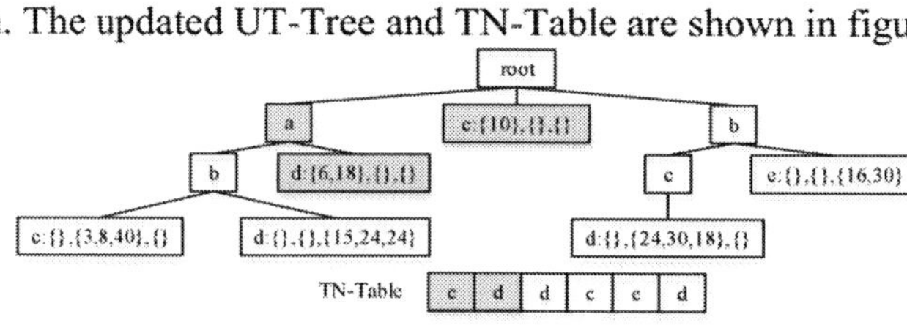

Figure 6. Updated UT-Tree and TN-Table.

3. Comparative experiment and result analysis

In order to verify the performance improvement of the HUIGRT algorithm proposed in this paper in space-time. In this paper, compares the algorithm with the existing high efficiency algorithm HUPMS and HUM-UT on the three datasets with different sparse: Mushroom、T10.I4.D100K and Retail.

Three different sparse datasets are derived from the raw datasets of the FIMI website (http://www.cs.rpi.edu/~zaki/Workshops/FIMI/data/) to simulate real data streams, its characterized are shown in table 7, where I represents the dataset contains the number of different data items, T represents the number of transactions in the dataset, AS represents the average length of the transaction, and DS represents the density of the data.

Table 7. Characteristics of three datasets.

Dataset	I	T	AS	DS
Mushroom	119	8124	23	19.33%
T10.I4.D100K	1000	100000	10	1%
Retail	16470	88162	10.3	0.06%

Since the data items in the raw dataset lack internal utility and external utility, the generation method in [17] is used, and the internal utility of the items in the transaction item set are randomly generated from 0 to 10; The external utility is randomly generated from a logarithmic positive distribution between 0.01 and 10. In this paper, different minUT are used as the only change conditions, and the runtime and memory consumption are used as evaluation criteria. The experiment is implemented in C# language, the compiler is Visual Studio 2013, the hardware platform is Intel dual-core 2.60GHz, 12G memory, and the operating system is Windows7. Specific experiments are described below.

3.1. Runtime comparison experiment on different datasets

In order to mining high utility patterns for the all three algorithms on a given dataset, first you need to set a suitable minimum utility threshold (minUT) for the given dataset. If the minUT setting is too large, the dataset under the threshold does not contains the high utility patterns, and if the minUT setting is too small, it will cause the mining time to be too long. For the data streams simulated by the three datasets, set the size of window sw= 3, the size of batch data b = 200, use three algorithms respectively, and set different minUT for experiments. The runtime comparison results are shown in (A), (B) and (C) in figure 7.

(A) Mushroom (B) T10.14.D100K (C) Retail

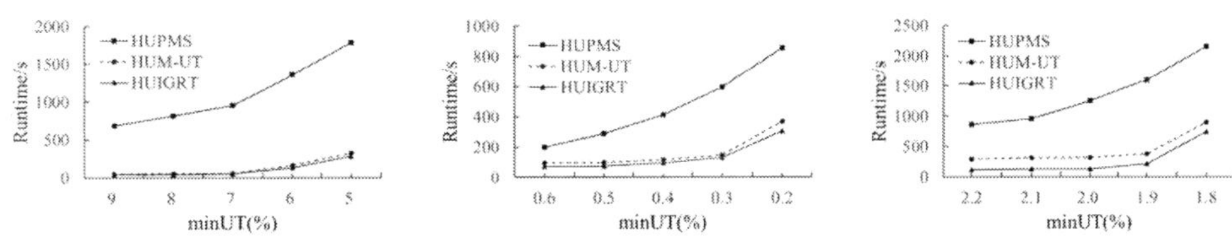

Figure 7. Runtime on different datasets.

The specific runtime are shown in table 8, table 9, and table 10.

Table 8. Specific runtime on mushroom.

minUT/%	HUPMS	HUM-UT	HUIGRT
9	681	48	40
8	815	52	42
7	956	64	51
6	1357	164	132
5	1787	321	281

Table 9. Specific runtime on T10.14.D100K.

minUT/%	HUPMS	HUM-UT	HUIGRT
0.6	198	93	72
0.5	289	98	74
0.4	412	117	93
0.3	596	146	129
0.2	856	367	305

Table 10. Specific runtime on retail.

minUT/%	HUPMS	HUM-UT	HUIGRT
2.2	861	291	117
2.1	955	306	125
2.0	1251	323	127
1.9	1596	380	213
1.8	2150	898	743

It can be seen from figure 7 that the HUIGRT and HUM-UT algorithm has the least runtime compared with HUPMS, this is because HUPMS needs to generate a large number of candidates at the beginning and then mine high utility patterns from the candidates, but HUIGRT and HUM-UT directly mine high utility patterns from UT-Tree to avoid generating candidates, thereby reducing the runtime of the algorithm. From the runtime curves of HUIGRT and HUM-UT, the proposed of HUIGRT algorithm has less runtime compared with HUM-UT. This is because the algorithm HUM-UT contains extraneous redundancy items in the mining process. Traversing these redundant items will generate additional time consumption, while HUIGRT removes irrelevant redundancy items, thus reducing the runtime of the algorithm.

Combined with the experimental data in table 8, table 9, and table 10, as the density of the dataset decreases, HUIGRT is more efficient in improving the runtime than the HUM-UT. On the densest Mushroom, HUIGRT has a significant improvement in runtime compared to HUM-UT, and when $minUT$ =6%, the runtime is the most efficient, improved by 19%; On the T10.14.D100K with the second highest density, when $minUT$ =0.5%, HUIGRT has the highest efficiency in improving the runtime of HUM-UT, improved by 25%. On the lowest density Retail, the best contrast is achieved when $minUT$=2.0%, improved by 61%. This is because the distribution of data items in a data streams transaction with a lower density is more dispersed. For the algorithm HUM-UT, the more redundant items in the data field to be processed, and the HUIGRT algorithm on the GRHT-Table, the more redundant items are deleted after the revision, reduce the more time it takes to traverse these redundant items. During the experiment, the data streams simulated by the Mushroom, and the redundancy items deleted by the algorithm HUIGRT after the GRHT-Table revision is 56; the T10.14.D100K deleted by the algorithm HUIGRT after the GRHT-Table revision reached 11817; for the Retail, the algorithm HUIGRT deleted the most redundant items, which was 213483, so the time efficiency improvement was the most obvious.

In addition, as can be seen from figure 7, as the minUT is continuously decreasing, the runtime of the three algorithms is increasing, because the number of patterns whose utility is more than the minUT will increase. You need to create more SUT-Tree and perform more mining operations, so the runtime will increase as the minUT decreases.

3.2. Memory consumption comparison on different datasets

In order to verify that the HUIGRT algorithm is superior in memory efficiency to the HUPMS algorithm and the HUM-UT algorithm. Next, the experimental parameters in section 3.1 are used to ensure that sw = 3 and b = 200 are unchanged, and the memory consumption of the three algorithms on different datasets is recorded. The experimental results are shown in (A), (B) and (C) in figure 8.

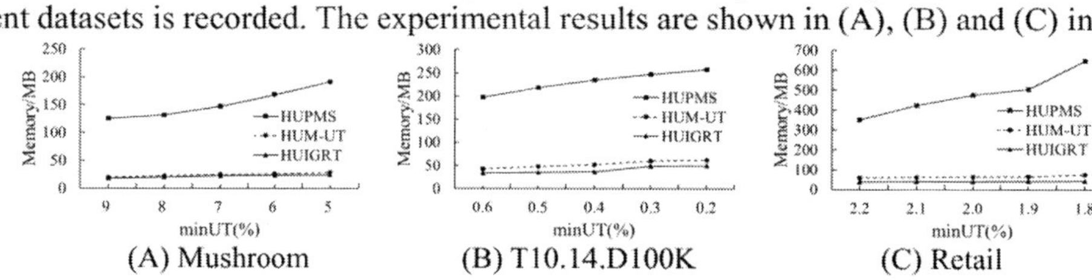

Figure 8. Memory consumption on different datasets.

The specific memory consumption are shown in table 11, table 12 and table 13.

Table 11. Specific memory on retail.

$minUT$/%	HUPMS	HUM-UT	HUIGRT
9	125	20	18
8	131	23	20
7	147	25	22
6	168	26	23
5	191	28	24

Table 12. Specific memory on retail.

$minUT$/%	HUPMS	HUM-UT	HUIGRT
0.6	198	43	34
0.5	218	48	36
0.4	235	52	37
0.3	247	61	49
0.2	258	63	50

Table 13. Specific memory on retail.

minUT/%	HUPMS	HUM-UT	HUIGRT
2.2	351	61	39
2.1	424	62	40
2.0	478	65	42
1.9	507	67	43
1.8	650	77	46

It can be seen from figure 8 that the HUIGRT and HUM-UT algorithm consumes the least memory compared to HUPMS, this is because the algorithm HUIGRT uses the established UT-Tree to get the real twu. When the minimum utility is more than the real twu, it is not need to create a SUT-Tree. The two-phase algorithm HUPMS uses overestimated twu, which are often more than the real twu. HUPMS creates more trees than HUIGRT, resulting in HUPMS consuming more memory than HUIGRT. From the memory consumption curves of HUIGRT and HUM-UT, the proposed of HUIGRT algorithm has less memory consumption compared with HUM-UT.*This is because* during the mining process, the global header table of the HUM-UT contains extraneous redundancy items. The storage of these redundant items requires additional memory, while the HUIGRT removes extraneous redundancy items, thereby reducing the memory consumption of the algorithm.

Combined with the experimental data in table 11, table 12 and table 13, it can be seen that HUIGRT is more efficient in memory efficiency than HUM-UT as the dataset density decreases. On the densest Mushroom, HUIGRT has a significant improvement in memory efficiency compared to HUM-UT, and the memory efficiency is maximized when $minUT$ =5%, the memory efficiency is improved by 14%. On the T10.14.D100K with the second highest density, when $minUT$ =0.4%, HUIGRT has a significant improvement in memory efficiency compared to HUM-UT, the memory efficiency is improved by 29%. On the lowest density Retail, when $minUT$ = 1.8%, the best contrast is achieved, the memory efficiency is improved by 40%. This is because for the dataset with lower density, the more redundant items are included in the algorithm HUM-UT, and the more redundant items are deleted after the algorithm HUIGRT modifies the GRHT-Table, reducing The redundancy memory will also occupy more memory, so the memory efficiency is more obvious.

In addition, as can be seen from figure 8, as the minUT is continuously decreasing, the memory consumption of the HUPMS algorithm will increase rapidly. This is mainly because the HUPMS algorithm uses the overestimation of the twu to determine whether to create a SUT-Tree. Therefore, in the process of decreasing the minUT, the total number of trees that the HUPMS algorithm far more than one-phase algorithm HUIGRT and HUM-UT, so the memory consumption is far more than the algorithms HUIGRT and HUM-UT.

4. Conclusion

Aiming at the problem that the existing high utility pattern mining algorithm has a large number of unrelated redundancy items, resulting in low space-time performance, this paper proposes HUIGRT algorithm based on GRHT-Table. HUIGRT adopts a new GRHT-Table stores the data items and transaction utility in the current pending data domain, avoiding the extra time consuming due to window sliding. The comparison experiments on three different sparse datasets of Mushroom, T10.I4.D100K and Retail show that the HUIGRT algorithm proposed in this paper is better than the existing high efficiency algorithms HUPMS and HUM-UT.

Acknowledgments

This work was supported by the National Social Science Foundation of China (15BGL048).

References

[1] Z. N. Feng, Y. N. Zhu. A Survey on Trajectory Data Mining: Techniques and Applications. IEEE Access, 4, 2056-67(2017)

[2] G. Pyun, U. Yun, K. H. Ryu. Efficient frequent pattern mining based on Linear Prefix tree. Knowledge-Based Systems, 55, 125-39(2014)
[3] P. Claude, S. Jérémy, F. Frédéric, et al. Frequent pattern mining in attributed trees: algorithms and applications. Knowledge & Information Systems, 46(3), 491-514(2016)
[4] A. Khubaib, A. Rodina, K. Kwangman. Enabling far-edge analytics: performance profiling of frequent pattern mining algorithms. IEEE Access, 5, 8236-49(2017)
[5] Zida, Souleymane, FournierViger, et al. EFIM: a fast and memory efficient algorithm for high-utility itemset mining, Knowledge & Information Systems, 51(2), 1-31(2017)
[6] H. Yao, H. Hamilton, C. Butz. A foundational approach to mining itemset utilities from databases. Proceedings of the 2004 SIAM International Conference on Data Mining. Society for Industrial and Applied Mathematics, 482-86(2004)
[7] L. Ying, W. K. Liao, A. Choudhary. A two-phase algorithm for fast discovery of high utility itemsets. Pacific-Asia Conference on Advances in Knowledge Discovery and Data Mining. Springer-Verlag, 689-95(2005)
[8] L. Wang, S. Q. Xiong,Y. F. Chang, et al. An Algorithm for Mining High Utility Patterns Based on Pattern-growth, ACTA AUTOMATICA SINICA, 41(9), 1616-26(2015)
[9] W. W. Wu, B. E. Shie, P. S. Yu, et al. Mining top-K high utility itemsets. ACM SIGKDD International Conference on Knowledge Discovery and Data Mining, 78-86(2012)
[10] H. Li, H. Q. Chen, Q. Y. Yang, et al. Real-time Visualization Method for Log Data streams. Journal of System Simulation, 28(9), 2133-37(2016)
[11] H. F. Li, H. Y. Hung, and S. Y. Lee. Fast and memory efficient mining of high-utility itemsets from Data streams: with and without negative item profits. Knowledge & Information Systems, 28(3), 495-522(2011)
[12] C. Q. Jin, W. N Qian, A. Y Zhou. Analysis and Management of Streaming Data:A Survey. Journal of Software, 15(8), 1172-81(2004)
[13] C. F. Ahmed, S. K. Tanbeer, B. S. Jevong. Efficient Mining of High Utility Patterns over Data streams with a Sliding Window Method. Studies in Computational Intelligence, 295, 99-113(2010).
[14] H Ryang, and U. Yun. High utility pattern mining over Data streams with sliding window technique", Expert Systems with Applications, 57(C), 214-31(2016)
[15] C. F. Ahmed, S. K. Tanbeer, B. S. Jeong, et al. Choi. Interactive mining of high utility patterns over Data streams. Expert Systems with Applications,39(15),11979-91(2012)
[16] L. Wang. Pattern Mining Algorithms over Data streams. Dalian: Dalian University of Technology(2013)
[17] C. F. Ahmed, S. K. Tanbeer, B. S. Jeong, et al. Efficient Tree Structures for High Utility Pattern Mining in Incremental Databases. IEEE Transactions on Knowledge & Data Engineering, 21(12), 1708-21(2009)

IOP Conf. Series: Materials Science and Engineering **715** (2020) 012023 doi:10.1088/1757-899X/715/1/012023

DBSCAN-KNN-GA: a multi Density-Level Parameter-Free clustering algorithm

Bin Mu*, Meng Dai and Shijin Yuan

School of Software Engineering, Tongji University, 4800 Caoan Road, Shanghai, China

*E-mail: binmu99@163.com

Abstract. DBSCAN is a popular tool to analyse datasets which can effectively discover clusters with arbitrary shapes. However, it requires two input parameters which are difficult to be determined, according to the fact that the performance of clustering result depends heavily on user-specified parameters. In addition, it uses global parameters which are not appropriate to those multi-density datasets. Aiming at these problems, we propose a parameter-free algorithm to perform DBSCAN with different density-level parameters. We select some classical datasets and a TLC taxi trip record used for experiments to compared our proposed algorithm with the original DBSCAN to evaluate the performance of our improved DBSCAN. The results show that the proposed algorithm is capable for efficiently and effectively detecting clusters automatically with variable density-levels. Compared with original DBSCAN, the proposed algorithm can discover more noise points and its execution accuracy is higher.

1. Introduction

Clustering is a powerful unsupervised learning technique and one of the fundamental tasks in data mining which is used in applications such as machine learning, pattern recognition, image analysis, information retrieval, bioinformatics, data compression, and computer graphics. Clustering is a method of dividing data into different groups such that objects in the same group have the similar characteristics while are dissimilar to each other when they are not in different groups. There are many kinds of clustering algorithms which can be classified into partitioning, hierarchical, density, model based and grid-based methods [1]. K-means and k-medoids are famous partitioning clustering algorithms. The two algorithms are sensitive to k, the number of partitions. It is difficult to get the best value of k, and it is easier to get locally optimal result by using iteration. Balanced Iterative Reducing and Clustering using Hierarchies (BIRCH) [2] are hierarchical clustering algorithm, which clusters dataset by CF tree and is suitable to process large dataset. Model-based clustering algorithms attempt to optimize the fit between the given data and some mathematical models, e.g. decision trees and neural networks [3]. The grid-based clustering approach using a multiresolution gird data structure, e.g. STatistical INformation Grid-based method (STING) [4] and CLustering In QUEst (CLIQUE) [5], processes data very fast and is typically independent of the number of data objects, yet dependent on only the number of cells in each dimension in the quantized space [6]. DBSCAN [7], one of density-based clustering algorithm, does not need the number of clusters like k-means and convex sample set like BIRCH and it can discover clusters of arbitrary shape and distinguishing noises. However, it needs two parameters Eps and minPts, and the speed of processing data is not effective enough.

The parameters Eps and minPts play the significant role in DBSCAN on which the clustering results heavily depend, so it is significant to determine the exact values of them. Many scholars are devoted to discovering a general approach to determine Eps and minPts. Martin Ester and Hans-Perter Kriegel [7] apply k-dist plot to determine the parameter Eps and minPts. The main idea is to gain k-th nearest neighbor matrix. Their experiments indicate that there is no difference when k is larger than 4, so they draw 4-dist graph and discover the "valley" in a graphical representation to get the threshold point. They suggest users estimate the percentage of noise and use the threshold point to calculate Eps [7]. Hou and Gao presented a parameter-free algorithm based on the DSets (Dominant sets) and DBSCAN algorithm. They use histogram equalization to the pairwise similarity matrix of input data and make DSets clustering results, expand the clusters from DSet with DBSCAN to generate the clusters of arbitrary shapes without any parameter input. Actually, they don't need to set the value of Eps, but minPts is set by default [8]. Gaonkar and Sawant, at first, get the k-dist values of every point in the dataset and then, unlike other authors, they sort it in ascending order and find a sharp change between two smooth curves from graph to determine whether the slope can be Eps. They cluster the datasets with different densities [9]. Wang et al. introduce adaptive DBSCAN in accordance with data. Their idea is based on Gaonkar and they make an improvement on calculating minPts [3]. Instead of calculating the number of data points in the eps-neighborhood of every point in the dataset, they select points with the same density level, ignoring the points in denser region.

Similar to the above improvements, we are also dedicated to solving the problem of acquiring Eps and minPts. Our proposed algorithm DBSCAN-KNN-GA combining KNN algorithm, genetic algorithm and DBSCAN, can acquire Eps and minPts automatically without any human factors. KNN algorithm [10] is proposed by Cover is one of a simple but effective classification methods in data mining. The core idea of KNN algorithm is that provided that most of the k nearest samples in the feature space of cluster belongs to a certain category, the cluster also belongs to this category, so we can apply KNN to classify dataset as several small clusters to discover different densities in DBSCAN rather than the global density to discover more noises in the dataset. By drawing k-dist plot, we can gain the rough range of Eps may exist by observation or calculate slopes. To acquire the exact value of Eps, we can use genetic algorithm [11] which find the optimal solution by imitating the natural selection and genetic mechanism. Genetic algorithm has good global search ability and can search all solution space quickly without failing into fast descent trap of local optimal solution. We can set the fitness function to find the optimal solution of Eps by iteration. Thus, DBSCAN-KNN-GA can automatically obtain Eps and minPts according to datasets and then pass them to DBSCAN to get the clustering results. We conduct several experiments to validate our proposed algorithm. We select some public well-known datasets to verify the feasibility of our proposed algorithm, that is whether it can automatically acquire Eps and minPts according to the different datasets and effectively cluster data. Then we select taxi trip record datasets to compare our proposed algorithm with original DBSCAN and GA-DBSCAN [12]. On one hand, the dataset is large enough to compare the efficiency of the algorithm; one the other hand, it is an uneven density distribution dataset so that we are able to verify whether the proposed algorithm can acquire multiple set of Eps and minPts and compare the accuracy of clustering results.

The rest paper is organized as follows. Section 2 gives a detailed description of traditional DBSCAN. In Section 3, we present our proposed algorithm called DBSCAN-KNN-GA in detail. Section 4 shows experimental results of proposed algorithm to verify its superiority. Finally, Section 5 concludes the paper and some directions for future research.

2. DBSCAN

Density Based Spatial Clustering of Application with Noise is a density-based clustering method. It fits in processing spatial dataset on account of its capability of detecting clusters of arbitrary shape at different scales. The main idea is: For each class of data objects in the cluster, for a given value minPts, this object Eps field must be greater than or equal to this number [13]. That is, the neighbourhood within Eps has to contain at least the number of minPts points. A brief description of DBSCAN is given below. Database is D, the radius of each point is Eps, and the minimum number of points in a field is minPts.

Definition 1. Eps-neighborhood

For each point in the database, the neighborhood of a point P within a given radius Eps is called eps-neighbourhood of P, denoted as:

$$N_{eps}(P) = \{Q \in D \mid dist(P,Q) \leq Eps\} \quad (1)$$

where *dist* (*P*, *Q*) is the distance between point P and point Q.

Definition 2. Core point

If P could be core point, its eps-neighborhood should have contained at least minPts points.

$$|N_{eps}(P)| \geq minPts \quad (2)$$

Definition 3. Boundary point

Suppose P is a core point, and Q is in the eps-neighborhood of P, but Q is not a core point. In this case, Q is a boundary point.

Definition 4. Noise

If P is neither a c ore point, nor a boundary point, P will be noise, or outlier point. Figure 1 show the concept map of core point, border point and noise.

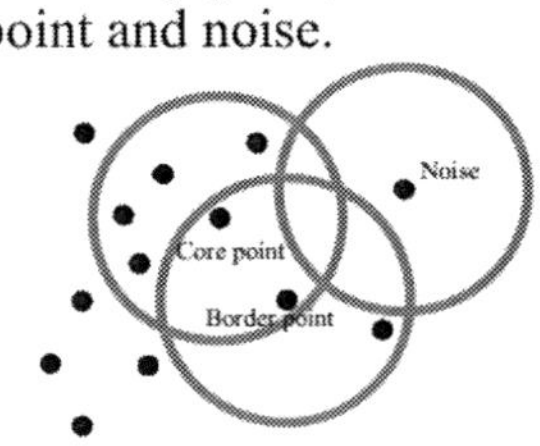

Figure 1. Core point, border point and noise.

Definition 5. Directly density-reachable

If P is a neighbour of Q and Q is a core point, then the relationship from P to Q is called directly density-reachable.

Definition 6. Density-reachable

Given a chain P_1, P_2, …, P_n, where $P = P_1$, $Q = P_n$, if any relationship from P_i to P_{i+1} is directly density-reachable, then the relationship from P to Q is called density-reachable. Figure 2 shows the concept map of directly density-reachable and density-reachable.

Definition 7. Density-connected

If P is density-reachable to Q and Q is density-reachable to P, then the relationship between P and Q is density-connected.

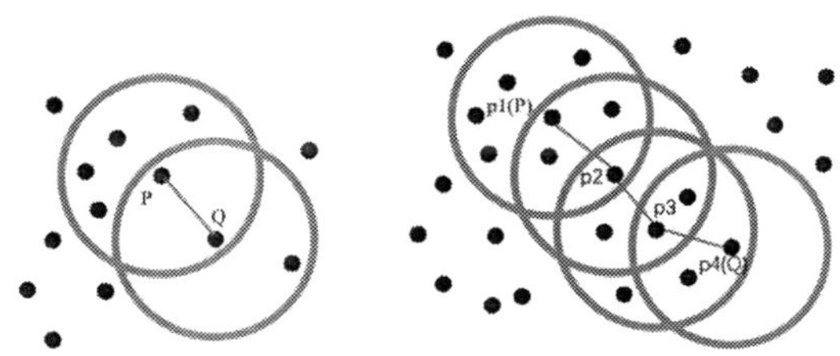

Figure 2. Directly density-reachable (a) Density-reachable (b).

The steps of DBSCAN is given below:

Step 1: Find an unvisited point P, mark P as visited.

Step 2: Get $N_{eps}(P)$, if $|N_{eps}(P)| < minPts$, mark P as noise and go to Step 1.

Step 3: Create a new collection C, put P into C. Traverse each point Q in $N_{eps}(P)$, and remove Q from $N_{eps}(P)$. If Q is not visited, mark Q as visited. Otherwise, go to Step 5.

Step 4: Get $N_{eps}(Q)$, if $|N_{eps}(P)| \geq minPts$, put $N_{eps}(Q)$ into $N_{eps}(P)$.

Step 5: If Q is not yet a member of any collection, put Q into C. If is not empty, go to Step 3. If exist an unvisited point P, go to Step 1.

The traditional methods are based on experiences to determine the values of Eps and minPts. Some improved methods are based on k-dist graph to find the valley in the graph sorting points in the descending order or the knees sorting points in the ascending order. The latter which can discover multiple Eps is better than the former because an important property of many datasets is that their intrinsic cluster structures are unable to be characterized by global density parameters [3]. Figure 3 shows that the cluster could not be detected by global density parameter. Under this circumstance may global density parameters are big enough to cluster several small groups into a large one. This paper use KNN to observe multi density-level of eps, utilize genetic algorithm to adjust parameter to acquire the accurate eps and minPts.

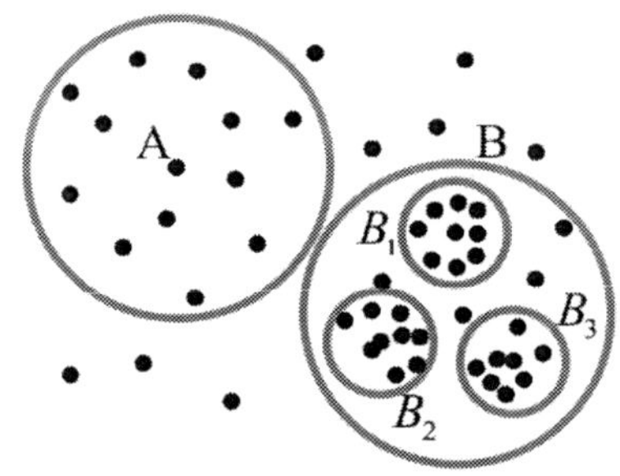

Figure 3. Cluster with different level of density.

3. DBSCAN-KNN-GA

DBSCAN-KNN-GA is aimed to determine the value of Eps and minPts to improve DBSCAN. The main strategy of DBSCAN-KNN-GA is to narrow the range of density radius by means of KNN (k-nearest neighbor), utilize fitness function to approach the value of density radius and terminate in some constraints. The concrete method is as follows: initialize the distance matrix D, where presents the distance between point i to point j. The matrix D is symmetric diagonally.

$$D=\begin{bmatrix} 0 & d_{12} & \cdots & d_{1n} \\ d_{21} & \ddots & & d_{2n} \\ \vdots & & \ddots & \vdots \\ d_{n1} & d_{n2} & \cdots & 0 \end{bmatrix}$$

Then we will draw k-dist graph in ascending order. In k-dist graph, it only considers the distance of a point from its k-th nearest neighbor point, instead, we calculate the average of the distance from a point to all its k-nearest neighbor (KNN). Considering all of the KNN neighbors of a point and the average leads to a smooth curve with the removal of noise and it is easier to detect the threshold of the density levels. For density levels [a, b], we can transform D to D' which contains only zero and one.

$$d_{ij}^{'}=\begin{cases}1, a\le d_{ij}\le b\\0, others\end{cases} \tag{3}$$

In matrix D', we can discover the candidate core points. We can suppose all the points in D as a population, and each row in represent the gene of point in this population, and each value can be thought of as a gene fragment. We can calculate the frequency of each gene fragment, and donate it as its fitness.

$$f_{ij}=\begin{cases}\dfrac{d_{ij}}{\sum_{i=1}^{n} d_{ij}}, d_{ij}^{'}=1\\0, d_{ij}^{'}=0\end{cases} \quad (i=1,2,\ldots,n) \tag{4}$$

Due to the particularity of genes that each gene segment is equivalent, we can simply consider the sum of each gene fragment's fitness as each point's fitness.

$$f_i = \sum_{j=1}^{n} f_{ij} (i = 1, 2, \ldots, n) \quad (5)$$

The fitness of whole population is the maximum of each point's fitness.

$$f = \max_{i=1}^{n}(f_i) \quad (6)$$

Our strategy is to find core points and remove them from the population. The probability of the selected core point is P_i.

$$P_i = \frac{f_i}{\sum_{j=1}^{n} f_j} (i = 1, 2, \ldots, n) \quad (7)$$

We will set a value to determine the max times of iteration K. In every iteration, will calculate the maximum of the population fitness, marked as f'_k,

$$f'_k = \max_{i=1}^{n}(f_i) \quad (8)$$

where i is the value of times of iteration. Record the max population fitness f in all iterations. If $f'_k - f > \beta$, it means the iteration is moving on, $f = f'_k$. Otherwise, it means we have found the optimal core point. Table 1 The procedure of DBSCAN-KNN-GA shows the procedure of DBSCAN-KNN-GA.

Table 1. The procedure of DBSCAN-KNN-GA.

Step1: Draw KNN graph and determine the range of Eps may exist.
Input: raw dataset
Output: range of *Eps* (a, b)
Process:
(1) Read raw dataset and save as point matrix D, remove the repeated points.
(2) Calculate distance matrix *Dist* for every two different point.
(3) Sort distances for every point to the other points in *Dist* as new matrix *SortedDist*.
(4) Normalize the matrix *SortedDist*, discover k-nearest distances for every point and draw KNN graph.
(5) Find valley or knee from KNN graph according to slope between adjacent points in the graph. For each valley or knee, calculated corresponding range (a, b) in which Eps exists.
Step 2: Determine two parameter Eps and minPts in DBSCAN.
Input: *Dist*, Iteration K, threshold β, range of *Eps* (a, b)
Output: Eps, minPts
Process:
(1) Read distance matrix *Dist* and convert it to special matrix which only contains zero or one according to range of *Eps* (a, b). Set the iteration number k as 0.
(2) Calculate fitness for every point as matrix *Fitness* and then acquire probability matrix *Prob* which represents the probability of each point to become candidate core point. Find the point with the highest probability and remove it from the group.
(3) Acquire the maximum value from *Fitness* as the fitness f'_k of whole group. If $f'_k - f > \beta \wedge k < K$, the value of k increments and return to process 2.

(4) Determine the threshold Eps by calculating average of d_{ij} of candidate core points. MinPts is assigned by average of Eps-neighborhood of every point.
Step 3: DBSCAN algorithm
Input: *Dist*, a list of (minPts, Eps)
Output: (point, cluster, visited, isCorePoint, isNoise)
Process:
1) Select a point *p* from raw *Dist* to start its neighborhood searching and mark it visited. If *p* is not a core point, mark it as noise point, else mark it as core point.
2) If *p* a core point, then create a cluster *c*, adding all of neighbors of *p* in *c* for recursive calls.
3) Repeated process 2 and 3 until all the point is marked visited.
4) Select other Eps and minPts while Eps is bigger than previous. Repeat process 1, 2 and 3 for those point marked as noise.

4. Experiments

In this section, we verify our proposed algorithm and evaluate the performance comparing it with DBSCAN. We perform the experiments in two steps: the first step is to utilize several classical datasets to support and verify DBSCAN-KNN-GA algorithm, the second step is to apply it to taxi trip dataset to analyse and evaluate the clustering performance.

In first step, we choose three datasets in this paper, one is classical iris dataset, one is from Compound and Aggregation of Computing School of the East Finland University [14], the other is the experimental data of Chameleon algorithm- to verify the feasibility of DBSCAN-KNN-GA. To estimate the range of the Eps value change, the k-dist graphs were constructed for the studied data. The k-dist graphs of Iris and Synthetic are shown in figure 4.

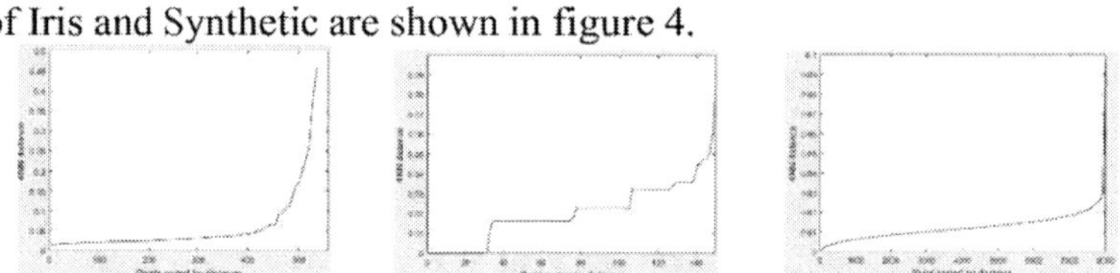

Figure 4. K-dist graph of Iris, Aggregation and t5.8k.

Table 2. DBSCAN-KNN-GA on different inputs on Iris, Aggregation and t5.8k.

Eps	MinPts	Total Points	Cluster Points	Noise Points	No. of Clusters
0.25	4	150	145	5	3
0.085	3	788	788	0	5
3.6	4	8000	7153	847	6

In short to find all possible Eps values we have to calculate the slopes at regular interval and then find the difference between slopes values at the same regular interval. By setting certain threshold value we can get different Eps values automatically based on this threshold value while discarding those with higher thresholds [3]. In figure 4, some knees could be mistaken for as their sharp slope. for points of the same density level, the range of variation will not be huge while a sharp change is expected to see between two density levels. Thus, there will be several smooth curves connected by greatly variation ones [6]. After obtaining the ranges of Eps, we can acquire Eps and minPts using fitness function shown in table 2 DBSCAN-KNN-GA on different inputs on Iris, Aggregation and t5.8k. Figure 5 shows the results of DBSCAN-KNN-GA algorithm.

- http://glaros.dtc.umn.edu/gkhome/cluto/cluto/download

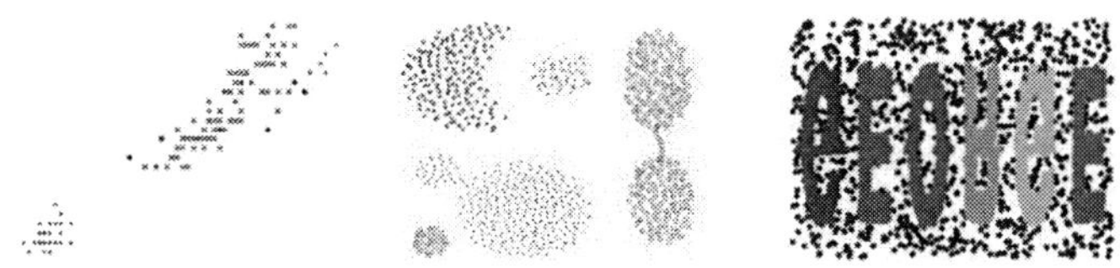

Figure 5. Result of Iris, Aggregation and t5.8k.

In second experiment, we adopt TLC trip record data[-] as our data source. We select six datasets with increasing number of data points, which are 4162, 8481, 13761, 25667, 41682, 86473, to carry out the clustering simulation experiments. We perform several experiments to observe the relationship between the accuracy and the parameter. The relationship between the accuracy and the parameter is shown in figure 6.

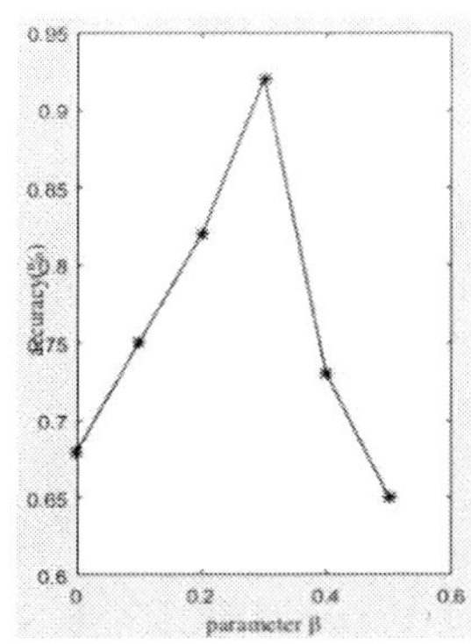

Figure 6. The relationship between the parameter β and the correct rate.

Table 3. Parameter adaption value.

Parameter / Data	β	Eps	minPts
4162	0.3	0.246	4
8481	0.3	0.187	5
	0.3	0.266	4
13761	0.35	0.167	4
	0.35	0.274	4
25667	0.3	0.143	6
	0.3	0.185	4
	0.35	0.347	5
41682	0.3	0.175	5
	0.35	0.315	5
86473	0.3	0.148	4
	0.35	0.266	6
	0.3	0.382	4

Table 4. Accuracy of the algorithm execution (%).

Data / Accuracy	4162	8481	13761	25667	41682	86473
DBSCAN	95.32	92.16	89.62	86.45	83.16	81.43
GA-DBSCAN	97.63	95.45	93.31	90.41	88.42	87.43
DBSCAN-KNN-GA	98.52	96.47	95.46	93.94	91.86	90.78

[-] https://www1.nyc.gov/site/tlc/about/tlc-trip-record-data.page

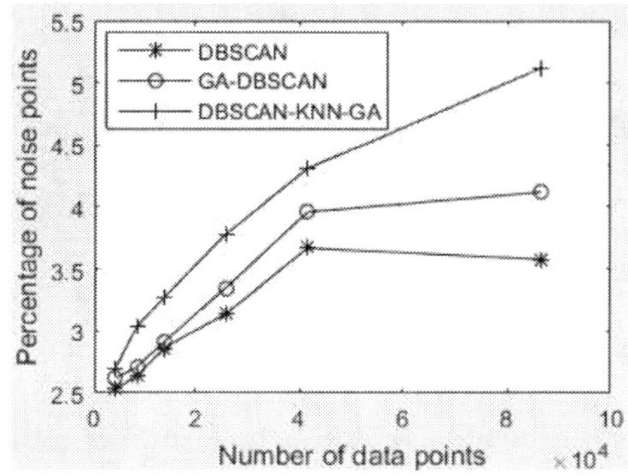

Figure 7. Contrast map with percentage of noise points.

We select DBSCAN algorithm and GA-DBSCAN as comparison. These clustering methods are experimented in the selected six datasets. The values of β, Eps and minPts of DBSCAN-KNN-GAs in experiments are recorded in table 3. We can observe that datasets are clustered with different densities. The accuracy rate and percentage of noise points are shown in table 4 and figure 7. We can see that the accuracy of the K-GA-DBSCAN algorithm execution is obviously higher than the other algorithms. The percentage of noise points is higher than the other algorithms because the minimum of local Eps is usually smaller than the global Eps, so many points may become core point or border point for those DBSCAN algorithm using global Eps, but are recognized as noise point in our proposed algorithm. In this way, we can distinguish the areas with concentrated density from those with sparse density. The execution time of each method on each dataset in the above experiments is recorded and their average execution time is compared. We can find that DBSCAN-KNN-GA saves manual calculating time and eliminate the interference of human factors and it can be further optimized through parallel and distributed computing.

5. Conclusion

This paper focuses on how to automatically determine multiple set of parameters Eps and minPts. The original DBSCAN requires Eps and minPts that are difficult to be determined and it simply uses the global Eps and minPts so that the clustering result of multi-density database is inaccurate. For these problems, we proposed a new algorithm using KNN algorithm and genetic algorithm. Based on k nearest neighbor matrix, we draw k-dist plot and reduce the value range of Eps, then with the iteration of genetic algorithm, we can select the optimal population and determine accurate Eps and minPts and finally pass the parameters to DBSCAN algorithm. We select classic datasets and real taxi trip record dataset for experiments, we compare our proposed algorithm with original DBSCAN algorithm and GA-DBSCAN in terms of clustering result accuracy, percentage of noises points and algorithm running speed. The experiments result shows that our proposed algorithm acquires more accurate parameters automatically has better clustering effect especially for those clusters of very different densities.

The proposed algorithm is lack of efficiency when encountering large datasets as it will cost a lot to establish distance matrix and perform DBSCAN algorithm. In the future research, we are looking forward to using paralyzed calculating or distribute calculating to reduce the algorithm execution time and improve processing efficiency.

References

[1] Tan, Pang-Ning, Michael Steinbach, Vipin Kumar. Introduction to Data Mining. (2005)
[2] *T.* Zhang, R. Ramakrishnan, M. Livny. BIRCH: An Efficient Data Clustering Method for Very Large Databases. ACM SIGMOD Record, 25, 2 (1996)
[3] Gaonkar, M. Naik, K. Sawant. AutoEpsDBSCAN: DBSCAN with Eps Automatic for Large Dataset. ICACTE, 2(2), 11-16 (2013).

[4] *W.* Wang, J. Yang, R. Muntz. STING: A Statistical Information Grid Approach to Spatial Data Mining. VLDB, 41 (1997)
[5] *A.* Rakesh, J. Gehrke, D. Gunopulos, P. Raghavan. Automatic subspace clustering of high dimensional data for data mining applications. ACM, 27 (1998)
[6] *S.* Suman, P. Rani. A Survey on STING and CLIQUE Grid Based Clustering Methods. IJARCS, 8(5) (2017)
[7] E. Martin, H. Kriegel, J. Sander, X. Xu. A density-based algorithm for discovering clusters in large spatial databases with noise. KDD, 96(34), 226-231 (1996)
[8] H. Jian, H. Gao, X. Li. DSets-DBSCAN: a parameter-free clustering algorithm. IEEE T IMAGE PROCESS, 25(7), 3182-93 (2016)
[9] *W.* Wang, Y. Wu, C. Tang, M. Hor. Adaptive density-based spatial clustering of applications with noise (DBSCAN) according to data. ICMLC, 1, 445-451 (2015)
[10] G. G. Enas, S. C. Choi. Choice of the smoothing parameter and efficiency of k-nearest neighbor classification. Statistical Methods of Discrimination and Classification. Statistical Methods of Discrimination and Classification, 235-244 (1986)
[11] *D.* Jong, K. Alan. Analysis of the behavior of a class of genetic adaptive systems (1975)
[12] *X.* Hu, L. Liu, D. Yang, M. Li. A MapReduce-based improvement algorithm for DBSCAN. J Algorithm Comput Tech, 12(1), 53-61 (2018)
[13] *Y.* He, H. Tan, W. Luo, H. Mao, D. Ma, S. Feng, J. Fan. Mr-dbscan: an efficient parallel density-based clustering algorithm using mapreduce. ICPADS, 473-480 (2011)
[14] *A.* Gionis, H. Mannila, P. Tsaparas. Clustering aggregation. TKDD, 1(1), 4 (2007)

IOP Conf. Series: Materials Science and Engineering **715** (2020) 012024 doi:10.1088/1757-899X/715/1/012024

A new method for detecting communities in network based on the affinity between nodes

Jin Zeng, Chenxi Shao*, Xingfu Wang and Fuyou Miao

University of Science and Technology of China, China

*E-mail: cxshao@ustc.edu.cn

Abstract. It has been found that many networks display community structure, analysis of the com-munity structure in the network is meaningful for understanding the real-world network. Recently, many community discovery methods have been proposed, most of which are based on the assumption that nodes in the network can only belong to a community at most. The proposed method in the paper is based on the affinity between nodes and it does not agree with the assumption, i.e., the method is able to discover overlapping community structures in networks. We apply the method to various real-world networks and artificial networks, the experimental results confirm that it has better performance on networks with overlapping community structures and non-overlapping community structures.

1. Introduction

Complex networks are abstracted from complex systems, the vertice in the network represents the individual in the system and the edge represents the relationship between individuals. Analysis of network community structure is helpful to understand the system. For example, communities may be related to functional modules in biochemical networks or individuals with common interests in social networks [1]. Previous researches demonstrate that there are community structures in most complex networks, that are group nodes densely connected within and loosely connected with the rest of the network [2,3]. The existed community discovery methods are roughly divided into the following categories: Modularity optimization method [4,5,6], because of the flaw of the definition of Modularity, such methods are exposed to resolution limits [7]; Methods based on propagation dynamics [8,9], this type of method holds the idea that the information remains in the same community with high probability after several times dissemination, they are fast but poor robustness; Methods based on graph bisection which requires prior-information. The methods mentioned above are effective for detecting non-overlapping community but rarely able to detect network overlapping community which is ubiquitous in real-world networks. For example, individuals could belong to different social circles in social networks. Another, a large fraction of proteins belong to several protein complexes simultaneously in protein networks [10].

In this paper, we propose a localized community detection method based on the affinity between nodes. In order to measure the quality of community structures, the stability of community structure has been defined. When node joined or was remove from a community, the stability of the community would change. Group of nodes with the largest local stability is considered as a suggested community.

The proposed method does not limit the number of communities to which the node belongs, i.e., it is effective for detecting overlapping community. The method has been applied to various real-world networks and artificial networks. The community structures identified by the method are in agreement with the actual communities present in real-world networks. Moreover, it achieves greater score in artificial networks and is effective for networks with overlapping and non-overlapping community structures.

2. Definition

The relationship between nodes is close or distant in complex networks, i.e., it is easy for information propagates from the node to some nodes but difficult to the rest. For example, it is easy to exchange information with friends but difficult with strangers in social networks. We proposed a hypothesis that there is affinity between individuals in systems. We proposed a hypothesis that there is affinity between individuals in systems. Under the affinity, the relationship between individuals becomes close or alienated that explains the existence of the community structures in the network. Intuitively, the greater the affinity, the closer the relationship between the nodes. For two nodes in the network, the greater the ratio of the number of paths shorter than k to the product of their degrees, the closer they are. For simplicity, we set k=2. Therefore, the affinity factor f(i,j) between two nodes is been defined as

$$f(i,j) = \frac{\sum_{k=1}^{n} A_{ik} A_{jk}}{\left(\left(\sum_{k=1}^{n} A_{ik}^2\right)\left(\sum_{k=1}^{n} A_{jk}^2\right)\right)^{\frac{\theta}{2}}} \quad (1)$$

where A is adjacency matrix of the network, θ controls the strength of influence of nodes degree. The affinity factor f (i , j) is positively related to the number of common neighbors (paths with length 2) and negatively correlated with the product of degrees of the two nodes. Since most nodes are not sharing any neighbor node, we construct a virtual node connected with each node in the network. Therefore, the Eq. (1) is changed to

$$f(i,j) = \frac{\sum_{k=1}^{n} A_{ik} A_{jk} + 1}{\left(\left(\sum_{k=1}^{n} A_{ik}^2 + 1\right)\left(\sum_{k=1}^{n} A_{jk}^2 + 1\right)\right)^{\frac{\theta}{2}}} \quad (2)$$

The affinity A (i , j) is positively related to the affinity factor and negatively related to the shortest distance d between the two nodes. It is defined as

$$A(i,j) = \frac{f(i,j)}{d^2} \quad (3)$$

As mentioned above, the definition of community is that community is a group of nodes which connected densely within and loosely connected with the rest, but it is fuzzy not a quantitative measure. However, a large room has been given to researchers. Based on the definition, the stability of community is proposed. The community is considered as stable structure, when the community expands to a certain scale, the stability would be degraded if it continues to expand. In the process of community expansion, the new node that joins the community will be the frontier, which causes a change of the affinity within and outside the community. Therefore, the stability is positively related to the sum of affinity of the inside for the frontiers of the community and is negatively related to the sum of affinity of the outside. For simplicity, we just take account of the affinity between neighbor-nodes. Consequently, the affinity of the inside for the frontiers of the community g Fin(g) is formulated as

$$Fin(g) = \sum_{i \in f} \sum_{j \in (n \cap g)} A(i,j) \quad (4)$$

The affinity of the outside for the frontiers is formulated as

$$Fout(g) = \sum_{i \in f} \sum_{j \in (n-g)} A(i,j) \quad (5)$$

where f is the set of frontiers of the community, n is the set of neighbor nodes of node i. Consequently, the stability S(g) of community g is formulated as

IOP Conf. Series: Materials Science and Engineering **715** (2020) 012024 doi:10.1088/1757-899X/715/1/012024

$$S(g)=\frac{F_{in}(g)}{F_{in}(g)+F_{out}(g)} \tag{6}$$

3. Method

Initially, each node in the network is considered as an initial community. Firstly, select node as seed community according a certain rule. In the process of seed community expansion, select the best node as a candidate of the member of the community at each step. In order to determine whether the candidate should join the community, the fitness of node i is been defined as

$$\text{fitness}(g,i) = s(g \cup i) - s(g) \tag{7}$$

Join the node that is connected to the community and whose fitness is the largest and greater than zero into the community at each step until the fitness of all nodes connected to the community is less than 0 As a consequence the stability of the community obtained by the expansion of the seed is the local maximum. It is not necessary to pursue global optimal value of the stability, because the whole network as a community, the stability S(g) will be the largest, but such a solution is obviously meaningless. Then re-select the seed and repeat the process until all nodes in the network have their own community. As for the strategy of selection of the seed, it is a solution to choose randomly a node that have not yet belonged to any community as the seed, but it is low-efficiency. Since there are often leader nodes in communities, the nodes are ranked by the sum of affinity for their neighbor nodes (The greater the affinity for other node, the greater the probability that the node is leader), then select node that is not affiliated with any community as the seed in turn. In this way, it is high-efficiency to select as few nodes as possible to achieve the goal of dividing the whole network into communities. In the process of the expansion of the seed, some nodes may be divided into multiple communities. When the expansion of the seed is over, if there is a community obtained by other seed is similar to the community (we agree that two communities are similar when the jaccard coefficient of them exceeds 0.5), the two communities should be merged. We can describe our proposed method in the following steps.

(1) Rank the nodes in descending order according to the sum of affinity for their neighbors.

(2) Select the top ranked nodes that does not belong to any community as the seed.

(3) Join the neighbor node whose fitness is the greatest in the seed community and mark the node as belonging, then update the seed community. Repeat the process until the fitness of all nodes connected to the seed community less than 0

(4) Compare the community with each of existed community, merge them into a single community if the jaccard coefficient of them exceeds 0.5

(5) Repeat step (2) to step (4) until all nodes in the network have their own community.

The algorithm is an agglomerative algorithm, i.e., the community is gradually expended from the seed. In the process of expansion, it costs $o(\mu s < k >^2)$ time to compute the fitness of a node, the candidate should be selected from μs neighbor nodes. Therefore, each expansion costs $o(u^2 s^2 < k >^2)$, where μ is the ratio of the number of frontiers to the size of the community, $< k >$ is the average degree of the nodes in the network and s represents the average scale of the communities. Because the seed community need to be updated after joining new node, the time complexity is $o(u^2 s^3 < k >^2)$. Although the time complexity of the algorithm is proportional to the cube the average scale of the communities, compare with the whole network, the average scale of the communities is much smaller.

4. Experiment

We apply our proposed method and other representative methods to the real-world networks (section 4.1) and artificial networks with different scales (section 4.2). The normalized mutual information(NMI) and modularity Q are used as quantitative measure to evaluate the performances of various methods.

4.1. Real-world network

The first is the Zachary karate network which captures 34 members of a karate club, documenting 78 pairwise links between members who interacted outside the club. Because of the conflict between the president and the coach, the club is divided into two communities. The division is shown in Fig.1. The performances of the proposed method and other methods are shown in table.1. As Table.1 shows, none of these method divides the network into two communities.

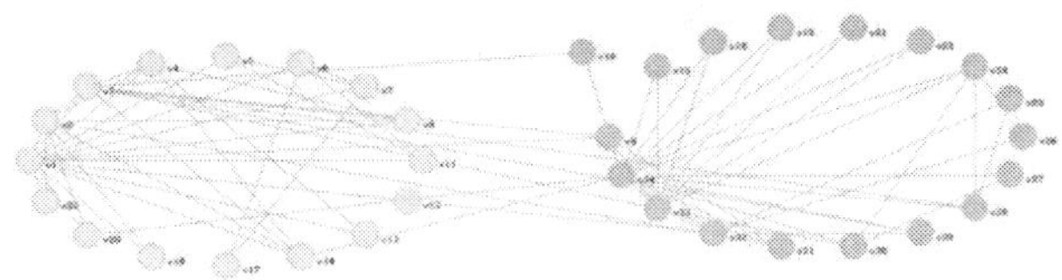

Figure 1. The natural division of the Zachary karate network that includes two community structures.

It does not indicate that these methods are inaccurate, contrarily, most mainstream methods divide it into four communities. The reason that the communities identified by the methods are different from the natural communities is that community structure is hierarchical. Most methods, in pursuit of accuracy, tend to divide the natural communities further. The community structures identified by our method is shown in figure 2. It is worth mentioning that the node 10 is wavering, i.e., the node 10 belong to two communities. Because the classification of node 10 differs from the standard, the NMI between identified communities and natural communities is less than LPA and Fast-unfolding.

Table 1. Some indexes of communities identified by the methods.

	NMI	Modularity	The number of communities
Our method	0.633	0.418	4
Fast-unfolding	0.677	0.393	4
LPA	0.699	0.385	3
CNM	0.524	0.277	5

It is not a mistake that the proposed method put the node 10 into two communities, many methods that can detect overlapping community, also consider the node 10 is a wavering node. As shown in figure 2, the number of connections between node 10 and two communities is equal.

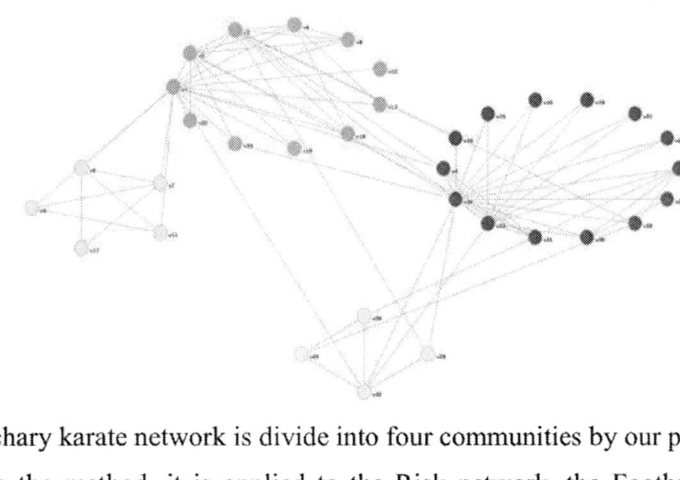

Figure 2. The Zachary karate network is divide into four communities by our proposed method.

To further validate the method, it is applied to the Risk network, the Football network and the Dolphins network. The experimental results can been seen in table.2, table.3 , table.4.

Table 2. The normalized mutual information(NMI) between natural communities and communities identified by the methods.

Network	Our method	LPA	Fast-unfolding	CNM
Risk	0.908	0.778	0.869	0.839
Dolphins	0.467	0.888	0.474	0.403
Football	0.911	0.830	0.892	0.861

Table 3. The modularity Q of communities identified by the methods.

Network	Our method	LPA	Fast-unfolding	CNM
Risk	0.667	0.572	0.631	0.486
Dolphins	0.521	0.378	0.500	0.439
Football	0.581	0.553	0.604	0.538

Table 4. The number of communities identified by the methods, the figure in parentheses represents the number of natural communities.

Network	Our method	LPA	Fast-unfolding	CNM
Risk(6)	7	5	7	6
Dolphins(2)	13	2	7	7
Football(12)	13	9	9	9

As tables show, In Risk network and Football network, the results using our method are comparable with the standard, but the performance on Dolphins network is a disappointment. It can be found that Fast-unfolding and CNM also perform poorly, while LPA is excellent which is related to the principle of the LPA. The LPA assigns a label to every node in networks, and adopts the label that a maximum number of its neighbors have at every iteration. Therefore, LPA is disadvantageous to the small community in the network, i.e., small community is easy to be merged with large community adjacent to it. The standard division of Dolphins network is two communities with comparable size, although there are many connections between the two communities. From table.2, the modularity of the communities identified by LPA is obviously less than the other three. The other methods divide the network into more fine-grained communities, so the communities identified with lower NMI and greater modularity.

Through the above experiments, we found that what we proposed in this paper is a method of pursuing fine-grained community. In order to achieve the greatest local maximum of the stability of community, as far as possible to divide the whole network into small communities with higher stability. As a consequence, it would reduce the time complexity of the method.

4.2. Artificial network

We analyzed the proposed method on a class synthetic benchmark networks with planted partition [11]. The significance of community structure is controlled by a mixing parameter $\mu \epsilon$ [0,1], where smaller values give clearer community structure. Compare with GN-Network, the size of community and the degree of the node are no longer fixed, but obey the power law distribution in the benchmark. Using the method proposed in [11], the synthetic networks with different μ, different scales and overlapping communities are generated. The experimental results are shown in figure 3. Figure 3 shows that the mixing parameter μ=0.5 is a critical point. The community structure is no longer defined in the strong sense when μ is greater than 0.5. The communities identified by our method are comparable with the standard division, while the other methods are also 90% similar to the standard partition when μ <0.5. When μ >0.5, the NMI between the standard partition and communities

IOP Conf. Series: Materials Science and Engineering **715** (2020) 012024 doi:10.1088/1757-899X/715/1/012024

identified by the various methods drops rapidly, where the LPA is most sensitive, it is already not working when μ=0.6(dividing the whole network into a community). Compared with the other method, the performance of our method decreases slowly.

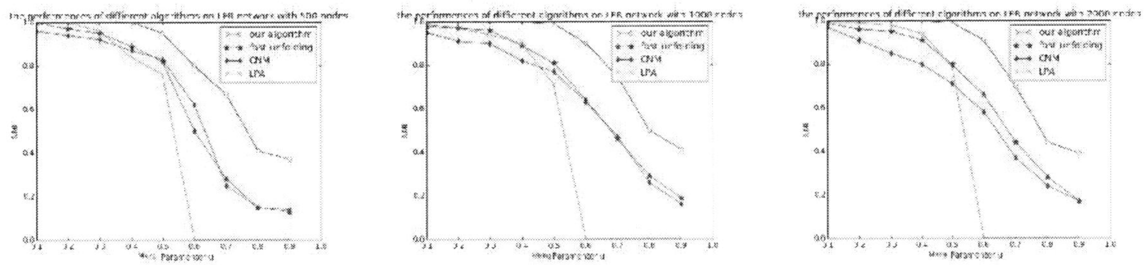

(a) Networks with 500 nodes (b) Networks with 1000 nodes (c) Networks with 2000 nodes

Figure 3. The normalized mutual information between planted communities and communities identified by the various methods in the benchmark with different size.

It is still about 75% similar to the standard when μ=0.7, which indicates that the proposed method is more robust. Furthermore, as the scale of network grows, the running time of the other methods significantly increased, while the proposed method does not. It confirms that the time complexity of the method is related to the average scale of communities, not the network's. We test the proposed method on the benchmark networks with overlapping com-munity, and the OSLOM [12], COPRA [13] are used as the contrast. The performance of the three methods is shown in Fig.4. As the figure shows, as the fraction of overlapping nodes increases, there is no significant fluctuation in the performance curve of our method and the OSLOM, while the COPRA suddenly fails when the fraction of overlapping nodes reaches about 0.7. The reason is that The COPRA is the same as the LPA, based on label propagation, so it has lower running time than the proposed method and OSLOM, but is unable to deal with networks with a high proportion of overlapping nodes. The performance of the three methods is comparable when the mixing parameter μ<=0.5, all of which can divide the network currently. When μ>0.5, the COPRA fails suddenly, the performance of the proposed method and the OSLOM also decreased, but the decline trend of the method proposed is slower. Comparing with the OSLOM, our method is slightly backward in accuracy, but has lower running time.

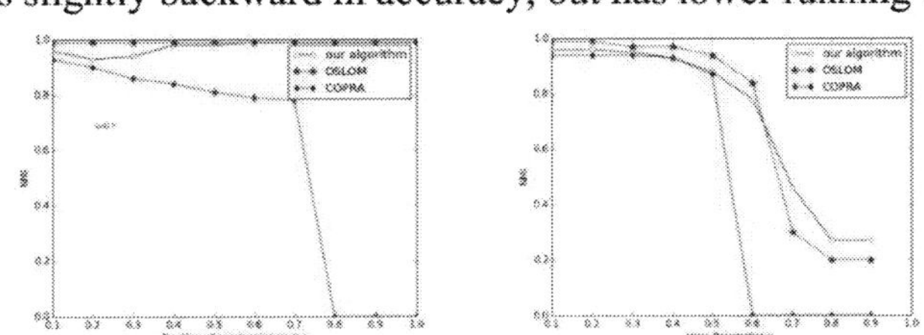

(a) (b)

Figure 4. The number of node is 1000. The left is the change curve of performance corresponding to the fraction of the number of overlapping nodes and the right corresponds to the mixing parameter.

5. Conclusion

We have introduced a community detection method, which based on the affinity between nodes and the stability of community. The proposed method is effective for networks with overlapping community structures and non-overlapping community structures, and the time complexity is related to the average scale of community structures. Since the scale of communities and the size of whole network are not in an order of magnitude, the method could be applied to large-scale networks.

Acknowledgement

The work is supported by the NNSFC No.61472381, No.61174144, No.61472382, No.61772490, No.61572454

References

[1] G. Palla, I. Der enyi, I. Farkas, T. Vicsek, Nature 435, 814 (2005)
[2] Fortunato, Santo. Community detection in graphs[J]. Physics Reports, 2009, 486(3):75-174.
[3] Newman M E J. The Structure and Function of Complex Networks[J]. Siam Review, 2003, 45(2):167-256.
[4] Newman M E. Fast algorithm for detecting community structure in networks.[J]. Phys Rev E Stat Nonlin Soft Matter Phys, 2004, 69(6 Pt 2):066133.
[5] Blondel V D, Guillaume J L, Lambiotte R, et al. Fast unfolding of communities in large networks[J]. Journal of Statistical Mechanics, 2008, 2008(10):155-168.
[6] Clauset A, Newman M E, Moore C. Finding community structure in very large net-works.[J]. Physical Review E, 2004, 70(2):066111.
[7] Fortunato S, Barthlemy M. Resolution limit in community detection.[J]. Proceedings of the National Academy of Sciences of the United States of America, 2007, 104(1):36-41.
[8] Raghavan U N, Albert R, Kumara S. Near linear time algorithm to detect community structures in large-scale networks.[J]. Physical Review E Statistical Nonlinear & Soft Matter Physics, 2007, 76(3 Pt 2):036106.
[9] Pons P., Latapy M. (2005) Computing Communities in Large Networks Using Random Walks. In: Yolum ., Gngr T., Grgen F., zturan C. (eds) Computer and Information Sci-ences - ISCIS 2005. ISCIS 2005. Lecture Notes in Computer Science, vol 3733. Springer, Berlin, Heidelberg.
[10] Palla G, Dernyi I, Farkas I, et al. Uncovering the overlapping community structure of complex networks in nature and society.[J]. Nature, 2005, 435(7043):814.
[11] Lancichinetti A, Fortunato S, Radicchi F. Benchmark graphs for testing community detection algorithms[J]. Physical Review E Statistical Nonlinear & Soft Matter Physics, 2008, 78(4 Pt 2):046110.
[12] Lancichinetti A, Radicchi F, Ramasco J J, et al. Finding statistically significant communities in networks[J]. Plos One, 2011, 6(4):e18961.
[13] Gregory S. Finding overlapping communities in networks by label propagation[J]. New Journal of Physics, 2009, 12(10):2011-2024.

Optimal mission planning of active space debris removal based on genetic algorithm

Yang Chen[1], Yuzhu Bai[1], Yong Zhao[1,*], Yi Wang[1] and Xiaoqian Chen[2]

[1]College of Aerospace Science and Engineering, National University of Defense Technology, Changsha 410073, China
[2]National Innovation of Defense Technology, Chinese Academy of Military Science, Beijing, 100081, China

*E-mail: zhaoyong@nudt.edu.cn

Abstract. With the increasing space activities, the space debris has seriously deteriorated the safe performance of the on-orbit spacecraft and has attracted much attentions. To reduce the rising influence of the space debris and improve the safe performance of the space mission, the three-stage removal strategy for space debris is proposed in this paper. Firstly, the multiple spacecrafts including one main spacecraft and some following spacecraft for space debris removal mission is developed. Then, the fuel, time and the quantity of the following spacecraft are defined as the constraints. Moreover, using the minimum fuel consumption as the optimal object, the mathematical model of the debris removal problem is established. Finally, the genetic algorithm is applied to solve this problem. Compared with the prior space debris removal strategy, the proposed three-stage space debris removal can effectively reduce the fuel consumption. Numerical simulation verifies the effectiveness of the proposed space debris removal scheme.

1. Introduction

Due to the human space activities, space debris becomes the major sources of the space pollutions. The space debris contains the scrapped rocket arrow and satellite body, the jet of the rocket, the discards during the execution of the space mission, the fragments generated by the collision between the space objects and so on. Furthermore, because of the launch of the satellite, the amount of space debris has considerable increase. According to the reports from Space Surveillance Net (SSN), more than 18,000 space targets have been catalogued. Moreover, 90% of the space targets are space debris [1] and the large amount of space debris threats the safety of the spacecraft in orbit.

For example, Iridium 33 satellite collided with the Russian scrap satellite Cosmos 2251 on February 10, 2009, which seriously influenced the American Iridium constellation. In 1991, Kessler proposed the Kessler Syndrome: Collision of space targets caused a large amount of space debris. Subsequently, the resulting space debris may collide with other spacecraft to lead to the mission failure and produce more debris. Therefore, the number of space debris will continue to increase. Furthermore, when space debris grows to the alert values, even that the humans cannot implement any space activities, space collisions will occur and the amount of space debris will still grow [2]. Based on the related research by NASA and ESA, the fragments should be actively removed every year since

2020. Moreover, the number of space debris need to be controlled and the normal space activities should be guaranteed in the future [3].

At present, the most feasible methods for space debris removal are as follows: the space robot arm capture method [4,5], the space rope net capture method [6], the increasing resistance method [7], and the laser cleaning method [8]. Furthermore, some articles focus on mission planning for debris removal. Braun et al. proposed four debris removal mode and analyzed different types of propulsive systems carried by mission spacecraft. Finally, they obtained the optimal task sequence by exhaustive method [9]. Considering the fuel and time constraints, Madakat optimized the mission planning for LEO debris removal [10]. However, these methods only use the two-stage system of spacecraft: mission satellite and space depot. The mission satellite needs to release the debris to the target orbit and transfer to next debris to continue the mission, resulting in greater fuel consumption. To solve this problem, this article proposed a three-stage debris removing method. The main satellite carries several small propulsive satellites. The small satellites push space debris into lower orbit and the main satellite only need to maneuver among the target debris. Through this way, the space mission could reduce the fuel cost. Finally, considering the limitation of the number of propulsive satellites, time and fuel, the genetic algorithm is applied to design the optimal trajectory. Compared with the prior space debris removal sequence, the optimal three-stage space debris removal sequence can effectively reduce the fuel consumption.

2. The three-stage debris removal mission

The three-stage removal strategy for space debris contains multiple spacecrafts including one main spacecraft and some following spacecrafts. The Main spacecraft is mission satellite and the following spacecrafts is propulsive satellites. Space depot can provide replenishment for mission satellite. The mission spacecraft carries several small propulsive satellites, maneuvering between the target debris, approaching the debris, and releasing the propulsive satellite. The propulsive satellite uses electric propulsion, pushing space debris into lower orbit and burning together with space debris. In the three-stage system, small propulsive satellite burns up after removing one debris and do not need to return. In this way, some fuels will be saved.

Considering that orbital maneuvers between different orbital heights and different orbital planes consume a large impulse, this article select the debris on Sun-synchronous orbit (SSO) with similar altitude and inclination. The space debris and mission satellite run on SSO, and the space depot circles on the orbit 100km below the Sun-synchronous orbit. When the propellant or the spacecraft carried is insufficient, it is necessary for mission satellite to return to the space depot for replenishment.

The entire task process is divided into the following steps

1. Mission satellite transfer to target debris.
2. Mission satellite release propulsive satellite.
3. Propulsive satellite push debris to deorbit.
4. Mission satellite transfer to the next target debris.

It is essential to notice that the space depot is heavy and its maneuver will consume huge propellant. Therefore, the space depot does not execute maneuver and the mission satellite will transfer to approach space debris.

3. Maneuver strategy between debris

3.1. Two-impulse optimal rendezvous model

Before building maneuver model, several simplifying hypotheses are made:

1. The orbital elements and mass of debris are known.
2. Ignoring the propellant for attitude control during the entire process.
3. Number of propulsive satellites needed for removing any debris is known.
4. Ignoring the effect of perturbations.

This article adopts two-impulse rendezvous maneuvering model under J2000 coordinate frame. Given the location of the two targets of space and the orbital maneuver time, the transfer orbit between the two targets can be obtained by solving the loopy Lambert problem.

In this paper, the method of solving the loopy Lambert problem proposed by Han Chao is used [11].

Define $t(z)$ as shown in Eq. (1):

$$t(z)=\frac{1}{\sqrt{\mu}}\left(x^3(z)S(z)+A\sqrt{y(z)}\right) \tag{1}$$

where $x(z),S(z),y(z)$ is universal variable z, and μ is gravitational constant of earth.

Figure 1 shows the relationship between $t(z)$ and the variation of the universal variable z.

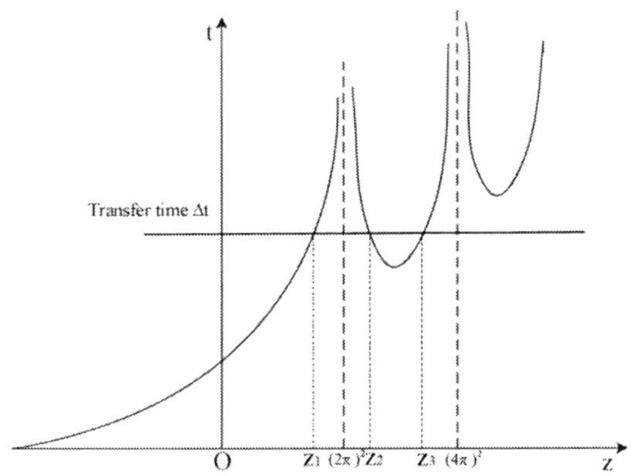

Figure 1. Function curve of transfer time and universal variable.

From figure 1, transfer time t and the universal variable z has the relationship of approximate periodicity. When given the transfer time Δt, there may be one or more corresponding universal variable z。 Calculate the $\Delta v_1, \Delta v_2$ corresponding to the universal variable, then the impulse required for the orbital transfer can be calculated by Eq. (2)

$$\Delta v=\Delta v_1+\Delta v_2 \tag{2}$$

Take the orbit with the smallest speed impulse as the maneuvering transfer orbit, that is, the loopy Lambert transfer orbit with minimum fuel consumption.

However, this method can only calculate the impulse for a given transfer time. When the given transfer time varies, the impulse will change with it. Therefore, there is a certain time to minimize the velocity impulse, and the maneuver time need to be designed. We optimize this problem with particle swarm optimization algorithm (PSO) and obtain the optimal transfer orbit with the smallest impulse, meeting the time constraint. Through priori calculation, the PSO will converge to better results when iterating 6-7 times. In order to speed up the calculation, in this paper, the number of particle groups is set to 20, and the maximum number of iterations is 10. Each Particle swarm optimization calculation takes 0.025s. So far, the required impulse Δv_i^j and time t_i^j from the i-th debris to the j-th debris can be obtained.

3.2. Mission costs

Let I_{sp} be specific impulse of the mission satellite. The initial mass is M_s. The mass of the carried propellant and propulsion are M_f and M_p respectively. num represents the number of propulsive satellites. The total number of debris to be cleaned is n, and the number of propulsion spacecraft for

IOP Conf. Series: Materials Science and Engineering **715** (2020) 012025 doi:10.1088/1757-899X/715/1/012025

each piece of debris is $\{num_1, num_2 ... num_n\}$. t_i^j, t_j and t_f are maneuver time from the i-th debris to the j-th debris, clearing time for the j-th debris and refueling time respectively. The maneuver from the i-th debris to the j-th debris can be divided into two cases:

a. maneuver to the next target directly

In this situation, fuel and propulsive satellite are enough. Once maneuver between two targets need twice impulse. M_l represents the left mass of mission satellite after last task. Let $\Delta v1_i^j$, $\Delta v2_i^j$ be the first and second impulse respectively, By the Tsiolkovsky formula, we can obtain the fuel consumption by Eqs. (3), (4) and (5).

$$dM1_i^j = M_l(1-\exp(-\frac{\Delta v1_i^j}{I_{sp}g})) \tag{3}$$

$$dM2_i^j = \left(M_l - dM1_i^j\right)(1-\exp(-\frac{\Delta v2_i^j}{I_{sp}g})) \tag{4}$$

$$dM_i^j = dM1_i^j + dM2_i^j \tag{5}$$

where $dM1_i^j, dM2_i^j$ are fuel consumption of twice impulse respectively. dM_i^j is the fuel consumption from i-th to j-th debris.

Propulsive satellite is needed for one-time removing mission. The mass change caused by the consumption of propulsive satellite is $M_p \times num_j$.

The whole mass decrement of mission satellite is shown in Eq. (6):

$$\Delta M_i^j = dM_i^j + M_p \times num_j \tag{6}$$

The time for one debris is got by Eq. (7)

$$T_j = t_i^j + t_j \tag{7}$$

where t_i^j and t_j are maneuver time from the i-th debris to the j-th debris and clearing time for the j-th debris respectively.

b. maneuver to the space depot for replenishment

In this situation, fuel or propulsive satellite is not enough. According to the orbit model, the optimal transfer orbit can be calculated through PSO, that is, the maneuver impulse to space depot and impulse from depot to the next debris. Record the four maneuver impulses as $\Delta v1_i^P, \Delta v2_i^P, \Delta v1_P^j, \Delta v2_P^j$ in accordance with the time. When maneuvering to space depot, the mass of mission satellite returns to the original state. P represents space depot, the fuel consumption can be calculated by Eqs (8) ~ (12).

$$dM1 = M_l(1-\exp(\frac{\Delta v1_i^P}{I_{sp}g})) \tag{8}$$

$$dM2 = (M_l - dM1)(1-\exp(\frac{\Delta v2_i^P}{I_{sp}g})) \tag{9}$$

$$dM3 = M_s(1-\exp(\frac{\Delta v1_P^j}{I_{sp}g})) \tag{10}$$

$$dM4 = (M_s - dM3)(1-\exp(\frac{\Delta v2_P^j}{I_{sp}g})) \tag{11}$$

$$dM_i^j = dM1_i^j + dM2_i^j + dM3_i^j + dM4_i^j \tag{12}$$

where $dM1_i^j, dM2_i^j, dM3_i^j, dM4_i^j$ are fuel consumption of four impulse respectively. dM_i^j is the fuel consumption from i-th to j-th debris

The whole mass decrement of mission satellite is shown in Eq. (13):

$$\Delta M_i^j = dM_i^j + M_p \times num_j \tag{13}$$

Fuel to be filled at space depot is shown in Eq. (14):

$$M_b = M_f - M_r \tag{14}$$

where M_r is the remaining fuel of mission satellite.

The time for one debris is got by Eq. (15)

$$T_j = t_i^P + t_f + t_P^j + t_j \tag{15}$$

where t_f and t_j are refueling time and clearing time for the j-th debris respectively. t_i^P and t_P^j are transfer time from i-th debris to space depot and from space depot to j-th debris respectively.

4. Optimization model

The number of space debris to be cleaned is n, and they are numbered 1-n in order. The target of the task is all the n debris are completely removed. The order of removing the depot can be represented by an array of the n numbers. For example, N = [3,4,1,2] means that the 3rd, 4th, 1st, and 2nd debris are cleaned in order. All the possible paths are the full arrangement of the n numbers. The optimal array or the optimal path can be found after calculating the fuel consumption in each arrangement. Set the initial position of the mission satellite the same as the first debris.

4.1. Calculation process

When a sequence of N, that is, the order of debris removal is given, from the first debris, the mission spacecraft execute the orbital maneuver in sequence to remove debris. After removing the first debris, the fuel consumption and propulsive spacecraft consumed by orbital maneuvers and debris removal can be obtained according to the maneuver strategy. When the next target is to be removed, a judgment is needed: if the mission satellite carries insufficient propulsive satellites or the remaining fuel is insufficient to transfer to the next target, it will maneuver to the space depot for replenishment. The maneuver impulse and time consumption for the mission satellite transfer from debris to debris or space depot can be determined by the orbital model. Through such process, the fuel and time consumption for the given mission sequence can be obtained.

The calculation process can be summarized as the following steps:

Step1: Remove the first debris.

Step2: Judge whether fuel or propulsive satellite is enough.

Step3: If the condition of step 2 is met, remove the next space debris.

Step4: If the condition of step 2 is not met, maneuver to space depot for replenishment and remove the next space debris.

In all the steps, the particle swarm optimization algorithm is needed to find the optimal transfer orbit.

4.2. Constrains

In this problem of mission planning, some constrains need to be satisfied.

Let $t_{\max}$ be the maximum transfer time for each piece of debris. Therefore, the transfer time need to satisfy $t_i^j \le t_{\max}$. Total mission time need not to be limited as each process of the task is constrained. Let M_r be the remaining fuel in mission satellite. The constraint of propulsion in fuel consumption is less than surplus fuel, that is $dM_i^j \le M_r$. The third constraint is the limit of propulsive satellite. Mission satellite must carry enough propulsive satellites to remove debris, that is $num_j \le num$. num represents the remaining amount of propulsive satellite.

Take fitness function $\phi(N)$ as the whole fuel consumption and the target of optimization is to find N with minimum fitness function. The optimization model of space debris removal is shown as Eq.(16):

IOP Conf. Series: Materials Science and Engineering **715** (2020) 012025 doi:10.1088/1757-899X/715/1/012025

$$\begin{aligned} &find \quad N=[n_1,n_2..n_n] \\ &min \quad \phi(N)=\sum_{i=1}^{n-1} dM_i^{i+1} \\ &s.t \quad \begin{cases} n_i \le n \\ n_i \ne n_j \\ \sum_{i=1}^{n} n_i = \dfrac{n(n-1)}{2} \quad i,j=(1,2..n), i\ne j \\ t_i^j \le t_{max} \\ dM_i^j \le M_r \\ num_j \le num_r \end{cases} \end{aligned} \tag{16}$$

5. Genetic algorithm design

It can be analyzed that the problem to be solved in this paper is a TSP problem in fact. When there are more debris to be removed, the problem of "combination explosion" will occur, which greatly increases the amount of calculation. In the method of solving the TSP problem, the genetic algorithm has great advantages in solving this problem. This paper uses genetic algorithm to solve this problem.

Step1. Code

In this paper, positive integer coding is used. The chromosomal gene represents the order of debris removal. For example, there are 7 debris to be removed, then the chromosomes [6, 5, 7, 2, 3, 1, 4] indicate that these targets are removed in order.

Step2. Select

The random sampling method is selected for selection. The fitness function is $\phi(N)$, which represents the fuel consumption.

Step3. Crossover

The crossover method is selecting a position in chromosomal gene randomly, and crossing over the parental gene after the selected position. If the chromosome after the crossover has a repeated gene, the repeated gene is replaced with other gene. E.g.

select a position in chromosomal gene randomly

FatherN1 = [3, 6, 8, 7, 4, ||5, 1, 2]   FatherN2 = [6, 8, 7, 1, 2, ||4, 3 ,5]

Crossover and generate offspring

SonN1 = [3, 6, 8, 7, 4, ||4, 3, 5]   SonN2 = [6, 8, 7, 1, 2, ||5, 1, 2]

Two repeat genes appear in the two offspring, replace them with other genes

SonN1 = [3, 6, 8, 7, 4, ||1, 3, 2]   SonN2 = [6, 8, 7, 1, 2, ||5, 3, 4]

Step4. Mutation

The mutation operation is selecting two points on the chromosome randomly and exchanging the genes represented by the two points.

6. Simulation and analysis

All the spacecrafts run on near-circular orbit. Select 9 pieces of space debris on SSO at the height of 850 km. Orbital inclination is 98.8212 and RAAN is 70. Space depot runs on the same orbital plane 100km under the debris orbit. The true anomaly of debris is shown in table1.

Table 1. Debris orbital parameter.

Serial number	Space debris1	Space debris2	Space debris3	Space debris4	Space debris5	Space debris6	Space debris7	Space debris8	Space debris9
True Anomaly	15	46	94	138	162	217	239	277	334

The mass of the mission satellite is 1000kg, the propellant carried is 300kg, the specific impulse is $I_{sp} = 300s$, and gravity constant is $g = 9.8$. The number of propulsive satellites is 5, and each mass is 100kg. The number of propulsive satellites required for removing each piece of debris is $num = [2,2,3,1,1,3,2,1,2]$. Time needed for debris cleaning is $t_j = 1000s$. The refueling time is $t_f = 8000s$. The maneuver time constraint is less than four orbital periods.

Before optimization, the fuel and time consumption are calculated in the condition that the mission schedule is not planned. Set the debris removal path N = [1,2,3,4,5,6,7,8,9]. The results are as follows, mission satellite need replenishment for 3 times and the fuel and time consumption of the entire task is 624.28kg and 352830s respectively.

$$N = [1,2,3,4,5,6,7,8,9]$$
$$dM = 624.28kg$$
$$dt = 352830s$$
$$n = 3$$

Optimize the debris removal path using genetic algorithm. The GA population size is 20, and number of generations is 100. The optimal result is obtained by calculation.

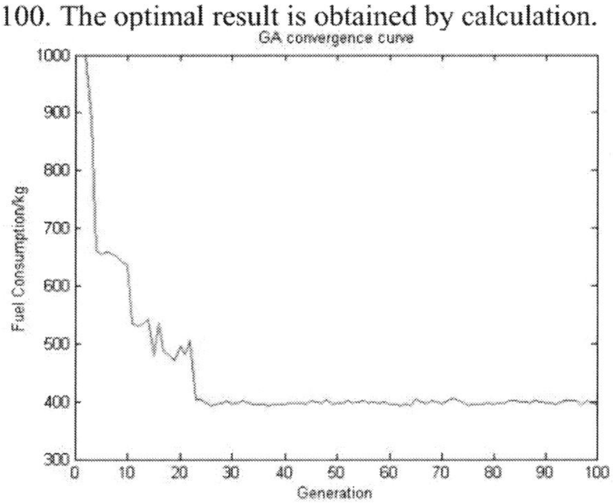

Figure 2. GA convergence curve.

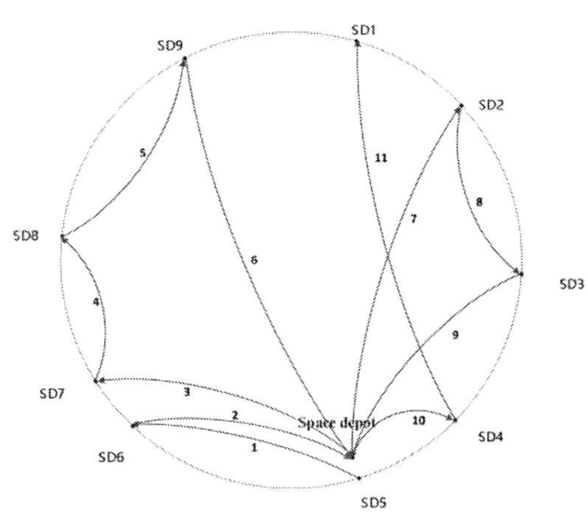

Figure 3. Optimal mission satellite removing path.

Figure 2 shows iterative process of genetic algorithm. The result is convergent after 25 generations. As shown in figure 3, the optimal debris removal path $N = [5,6,7,8,9,2,3,4,1]$. The minimum fuel consumption during the refueling process is 402.14 kg. 280.02kg of fuel is supplied by space depot. Mission satellite needs to be replenished three times. Total task time is 316330s. It can be seen from the experimental results that the optimal fuel consumption will reduce by 35.6%(222.14kg) compared to the initial removal path, and the total mission time reduces by 10.3%.

Table 2. optimal results for debris removal.

Removing sequence	Impulse(m/s)	Fuel consumption(kg)	Transfer time(s)
SD5	0	0	1000
SD6	192.9	57.16	24467
SD7	102.7+164.4	18.64+54.38	38549+30132
SD8	103.1	23.16	30882
SD9	158.1	28.76	30586
SD2	250.2+53.81	26.15+18.14	38644+3910
SD3	167.7	43.84	24710
SD4	65.22+80.37	9.808+26.97	37500+19900
SD1	333.8	95.15	33350
Total		402.14	316330

Table 2 shows the detailed consumption of the whole mission. The plus sign indicates mission satellite maneuver to space depot for replenishment. From the figure, fuel consumption is large when satellite need replenishment. The optimal removing sequence of the mission satellite is in the order of sequentially adding a true anomaly angular position. When the service spacecraft returns to the space depot, the order of removing is no longer in full compliance with this law. Under the constraint of propulsive satellite, mission satellite needs to maneuver to space depot more often, increasing fuel consumption of the mission. Transfer time is close to the critical value as longer maneuvering time bring smaller impulse.

7. Conclusion

In this paper, the active space debris removal mission planning is studied and the three-stage system is proposed to solve this problem. Firstly, the multiple spacecraft including one main spacecraft and some following spacecraft for space debris removal mission is developed. Then, the fuel, time and the quantity of the following spacecraft are defined as the constraints. Moreover, using the minimum fuel consumption as the optimal object, the mathematical model of the debris removal problem is established. Finally, the genetic algorithm is applied to solve this problem. Compared with the prior space debris removal sequence, the optimal three-stage space debris removal sequence can effectively reduce the fuel consumption. Numerical simulation verifies the effectiveness of the proposed space debris removal scheme. Furthermore, the debris removal mission planning strategy can also be applied to refueling problem. However, this developed strategy ignores the influence of perturbation. In the future, the influence of the perturbation and the safe constraints should be studied.

Acknowledgments

This work is supported by the Major Program of National Natural Science Foundation of China under Grant Numbers 61690210 and 61690213, the National Natural Science Foundation of China (No.11723211), and the National Science Foundation of China (No. 51675525).

References

[1] The NASA Orbital Debris Office. Orbital Debris Quarterly News [J]. *Orbital Debris Quarterly News*, 2018, **22(1)**.
[2] Klinkrad H. Space Debris – Models and Risk Analysis[M]. *Springer*, 2006
[3] Liou J C, Johnson N L, Hill N M. Controlling the growth of future LEO debris populations with active debris removal[J]. *Acta Astronautica*, 2010, **66(5)**:648-653.

[4] Bosse A B, Henshaw C G, Pipitone F, et al. SUMO: spacecraft for the universal modification of orbits[J]. Proceedings of SPIE - The International Society for Optical Engineering, 2004, 5419.
[5] Obermark J, Henshaw C G. SUMO/FREND: vision system for autonomous satellite grapple[J]. *Proceedings of SPIE - The International Society for Optical Engineering*, 2007, 6555:65550Y-65550Y-11.
[6] Kawamoto S, Makida T, Sasaki F, et al. Precise numerical simulations of electrodynamic tethers for an active debris removal system[J]. *Acta Astronautica*, 2006, **59(1)**:139-148.
[7] Toyoda K, Reiso S, Cho M. Experimental investigation of space debris removal method using electrostatic force in space plasma[C]// *Aiaa Aerospace Sciences Meeting Including the New Horizons Forum and Aerospace Exposition*. 2013.
[8] Rubenchik A M, Barty C P, Beach R J, et al. Laser Systems for Orbital Debris Removal[J]. 2010, 1278(1278):347-353.
[9] Braun V, Lüpken A, Flegel S, et al. Active debris removal of multiple priority targets[J]. *Advances in Space Research*, 2013, **51(9)**:1638-1648.
[10] Madakat D, Morio JRO, Vanderpooten D. Biobjective planning of an active debris removal mission[J]. *Acta Astronautica*, 2013,**84**:182-188
[11] Chao H. Research on Algorithm of Loopy Lambert Transfer in Space Rendezvous[J]. *Chinese Space Science & Technology*, 2004.

IOP Conf. Series: Materials Science and Engineering **715** (2020) 012026 doi:10.1088/1757-899X/715/1/012026

Statistical testing data generation for UAS

Jianwei Sui*, Yunzhan Gong, Dahai Jin and Yawen Wang

State Key Laboratory Of Networking And Switching Technology, Beijing University of Posts and Telecommunications, Beijing, China

*E-mail: jwsui@bupt.edu.cn

Abstract. The reliability of UAS(Unmanned Aircraft Systems) is of vital importance in practical applications. The most effective method to guarantee the reliability of UAS is to test UAS with multiple types of testing data. In this paper, we propose a method for the testing data generation of UAS in consideration of the combination and weight of the input data, which improves the availability of the method in UAS. Then we make four experiments to verify the effectiveness of the proposed method: (1) the random testing data generation experiment; (2) testing data generation considering the combination of input data experiment; (3) testing data generation considering the combination and weight of input data experiment; (4) the history data from the actual usage situation. Experiments results show that the method proposed in this paper is closer to the reality. At last, we analyse the code coverage by using Code Test System(CTS). And the analysis reflects that the proposed method has a better coverage than other methods.

1. Introduction

UAS reliability test is to test software on purpose of verifying whether UAS achieves reliability requirements and evaluating UAS reliability level. [1] The most commonly used method of reliability test is statistical testing method based on usage model, building software usage model and generating test cases based on the usage model. [2] Markov model is a famous model which is usually used to generate operation sequences in the traditional test method. [3] However, the test data generated by the method is redundant due to the disregard of interaction between operations; And the importance and using frequencies of the data are also the factors that cannot be ignored. On account of the two problems above, this paper proposes a method which considers the interaction among different operations and the weight of operation data. At last, we perform experiments which show that the method proposed in this paper could reduce the redundancy of test data and improve the testing efficiency while guaranteeing the coverage ratio of test data. [4][5]

2. Usage model and related definition

2.1. Usage model

In this paper, the directed graph is used to represent the usage model. Figure 1 shows all the states during the running process of a UAS software with an initial state as start and an ending as end.

2.2. The interaction between operations and the input data types

In the actual software, as shown in figure 2, the input data of each step may have a lot of types. Also, as is mentioned before, some interactions usually exist between different operations or steps.

IOP Conf. Series: Materials Science and Engineering **715** (2020) 012026 doi:10.1088/1757-899X/715/1/012026

Considering these two factors(the diversity of input data types, the interaction between operations or steps), generating test data for each operation sequence is required. [6][7] And while making experiments in actual situation of software, we found that strong interactions usually only exist in adjacent operations, while weak or even none interactions exist in non-adjacent operations. As a result, it makes a more efficiency way to generate test data with only interactions between adjacent operations counted. [8][9]

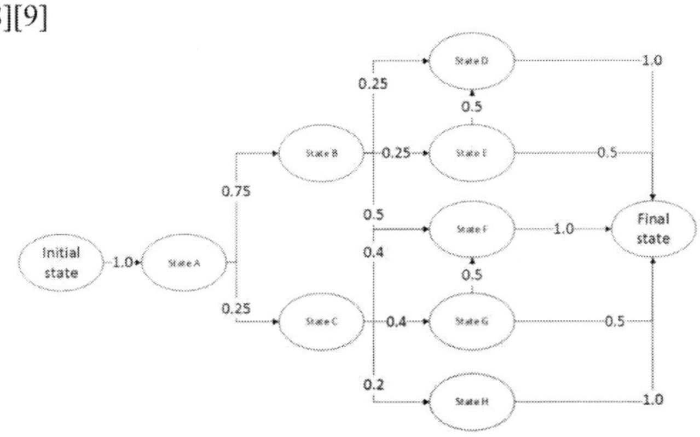

Figure 1. Markov usage model for a UAS software.

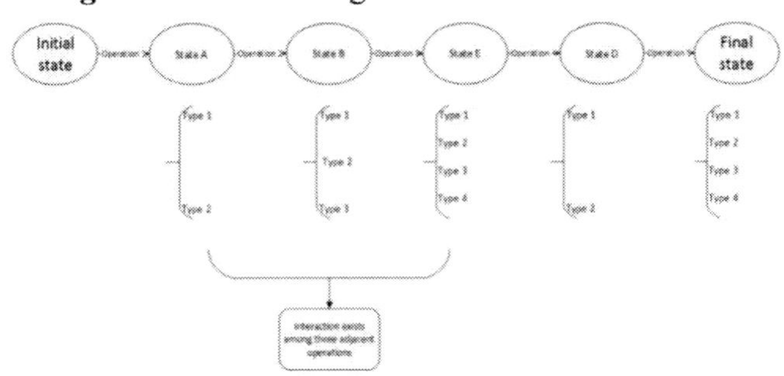

Figure 2. One operation sequence from Markov usage model.

2.3. Mathematical definition

Definition 1. We define n-gram $(v_1, v_2, ..., v_n), v_1 \in V_1, v_2 \in V_2, ..., v_n \in V_n$, as a testing sequence.

Definition 2. $A = (a_{i,j})_{m \times n}$ is a $m \times n$ matrix, of which the jth column represents operation f_j of operation sequences. All of the elements in the jth column are taken from the collection $V_j (j = 1, 2, ..., n)$.

Definition 3. Adjacent Matrix: In order to facilitate the handling of Markov model, the adjacent matrix G is usually used according to the Markov model, and $G[i][j] = p_{ij}$. p_{ij} is the transition probability of the way from point i to point j. $G[i][j] = 0$ shows there is no access from point i to point j.

Definition 4. Operation Sequence Weight: An operation sequence as $k_i, i \in [1, p]$, the Weight of Operation Sequence $k_i, i \in [1, p]$ is defined as $Weight(k_i) = \prod_i p_i$, with p_j representing the transition probability of jth path in operation sequence.

3. Analysis for N-Dimensional adjacent algorithm

3.1. Outline of method

Based on the Markov model, this algorithm can generate test data and the weight of test data for each operation sequence in the usage model. In algorithm 1, the adjacent matrix G of Markov model is got to traverse all paths of the usage model and calculates the transition probability of each operation sequence.

3.2. Algorithm 1

Input: initial state: start; the final state: end; adjacent matrix G ={the adjacent matrix of Markov model }; initial path; initial weight
Output: K={all paths from start to end }, Weight = { weight of all operation sequences }

```
getPaths(start, end, G, path, weight) {
hasFlag[start]=true;
for(each node 'i' in adjacent matrix G) {
if(G[start][i]==0 || hasFlag[i]) {continue;}
if(i==end) {
record the result;
continue;}
getPaths(i, end, G, path+"->"+i, weight*G[start][i]);
hasFlag[i]=false;}
}
```

3.3. Algorithm 2

Input: operation sequence k, number of types in each operation number={a[1],…,a[n]}, weight of every data type, dimension N.
Output: N-dimension coverage array A.

```
for(each element in array number){
m[i]=number[i]*number[i+1]*...number[i+N-1];
}
m=max{m[i]};
initial array A[m][n];
calculate the element of the first N columns of A;
for(every line of A[m][n]){
calculate the weight of each test data;
}
calculate the total weight of test data;
for(every line of A[m][n]){
calculate the proportion for each test data;
}
```

3.4. Test data evaluation

In order to evaluate whether the test data generated by the proposed method can reflect the actual operation situation accurately, GFI(Goodness of Fit Index) is imported as to quantify the accuracy.[10][11]

Definition 7: Defining

$$GFI = \frac{\chi^2_{independent} - \chi^2_{proposed}}{\chi^2_{independent} - \chi^2_{ideal}}, \chi^2 = \sum_{i=1}^{n} \frac{(b_i - a_i)^2}{a_i}$$

a_n represents the times of test data n actually occurs, and b_n represents the times that test data n is assigned. Assuming that the $\chi^2_{independent}$ meaning there is no correlation is the maximum value of χ^2. χ^2_{ideal} meaning the ideal condition with the best correlation with reality is the minimum value. The formula is modified as :

$$GFI = \frac{\chi^2_{independent} - \chi^2_{proposed}}{\chi^2_{independent}}, \chi^2_{proposed} < \chi^2_{independent}$$

From the formula, we can see that the value of GFI is between 0 and 1.

4. Experiment

Experiments are made to compare the proposed method with the other methods, and the GFI between method and actual situation shows that the method proposed in this paper performs better than other methods when applied in operation sequence with strong interaction. Also, the coverage ratio shows that the proposed method could achieve good performance with less test data.

4.1. Analysis of GFI

Table 1. GFI Table.

Sequence number	Method 1	Method2	Method3	Reality
1	93	287	153	152
2	24	287	153	160
3	146	287	247	225
4	48	287	243	159
5	50	287	150	149
6	4	287	158	160
7	0	287	247	251
8	90	287	241	243
9	150	287	120	109
10	0	287	122	121
11	44	287	170	192
12	140	287	167	187
13	40	287	373	410
14	53	287	372	400
15	0	287	610	595
16	0	287	549	612
17	0	287	333	313
18	140	287	337	301
19	48	287	561	625
20	97	287	564	540
21	48	287	287	340
22	0	287	241	221
23	51	287	421	354
24	54	287	272	272
25	52	287	120	123
26	54	287	131	131
27	51	287	175	183
28	0	287	120	185
29	0	287	291	273
30	2	287	256	217
31	97	287	434	434
32	48	287	449	449
total	1624	9184	9067	9086
GFI	0	0.6328	0.96	1

In order to verify that the method proposed in this paper have better GFI, four groups of experiments are designed. In method 1, only use algorithm 1 is used to traverse the Markov model on purpose of getting all operation sequences, and test data for each path is generated randomly without input data diversity and the interaction between steps; In method 2: algorithm 1 and algorithm 2 are both used to generate test data for operation sequence with input data diversity and the interaction among operations considered, ignoring the operation sequence weight. In method 3: we take the operation sequence weight into consideration compared to method 2, and test data is assigned according to the weight; And group 4 is the actual usage of a software as the control group.

As shown in figure 2, taking path 'Initial—A—B—E—D—Final' as an example, the weight of the path is 0.093. Assuming that the total number of test data is 200000, and we can see that the actual number of test data generated by the method is 18600 according to the weight 0.093. And from the path, we can get the number of test data combination is 192. While according to the analysis before, only 32 kinds of combination of test data is actually needed when take the interaction between operations into consideration. The result is shown as table 1, in method 1, a lot of useless test data is wasted on the 160 invalid operation combination with the lowest efficiency. Method 2 performs better and wastes less test data when compared to method 1, with considering the interaction between adjacent operations. Obviously, method 2 is more efficient than method 1. And in method 3, by considering both the interaction among sequences and the weight of operation sequence, it assigns more test data to the operation sequence with higher weight and wastes less test data in invalid operation sequences, making it closest to reality.

As is mentioned before, GFI is imported to quantify the accuracy with reality. As shown in table 1, for method 1, 'GFI = 0' shows that method 1 divorces from reality seriously without considering the interaction between operations; for method 2, 'GFI = 0.6328' shows that method 2 can be partly consistent with the reality; for method 3, 'GFI = 0.96' shows that method 3 is basically in line with the reality.

4.2. Code coverage analysis

Coverage ratio is an important indicator for the quality of test data in software testing. In this section, we use the tool CTS to analyse the method coverage ratio and branch coverage ratio. The CTS is a code testing system developed by author's laboratory supporting executing and analysing programs. CTS can feeds back a report that shows the method coverage ratio and branch coverage ratio of current program. At first, test data is generated by method 3, then execute the tested program with the test data in CTS and get the method coverage ratio and the branch coverage ratio. [12]

As shown in table 2, the test data generated by the proposed method could cover over ninety percent of total methods and eighty-eight percent of branches. This shows that the method proposed in this paper could achieve high method coverage ratio and branch coverage ratio with less test data and higher efficiency.

Table 2. The coverage ratio of code.

Total methods	**Covered methods**	**Total branches**	**Covered branches**
25	23	67	59
Cover Ratio	92%	**Cover Ratio**	88.1%

5. Related work

Model-based testing is widely used in software testing. Conformiq DesignerModels supports creating models as UML and QML state machines. [13] And results can be exported to other test management tools. BPM-Xchange can generate test cases from models based on different criteria, and it also supports importing models from other tools and exporting test cases to Excel. [14] The DTM (Dialogues Testing Method) tool could choose test data based on structural coverage and execute based on custom models. [15] fMBT (free Model-Based Testing) can generate test data from model which is written in Python, while also supporting several heuristic strategy. [16]

6. Conclusion and future work

In this paper, in order to solve the problem of test data redundancy in model-based testing, a method which takes the interaction between operations and the weight of operation into account is proposed from a practical point of view. And GFI and coverage ratio are imported to quantify the quality of the test data generated by the proposed method. At last, experiments are made on CTS to show the GFI and coverage ratio, of which the result shows that the method proposed in this paper performs better.

While applied to the software of which interaction only exists in adjacent operations, the proposed method performs well. However, on the other hand, while applied to the software of which interaction doesn't only exists in adjacent operations, it may cause the degradation in performance. And we'll improve this situation in future work.

References

[1] IEEE Reliability Society, "IEEE Recommended Practice on Software Reliability", IEEE Std 1633-2008, New York, 27 June 2008.
[2] R.C.Cheung, "A user-oriented software reliability model", IEEE transactions on software Engineering, SE-12(1):118–125, Mar. 1980.
[3] J.A.Whittaker and M.G.Thomason, "A Markov Chain Model for Statistical Software Testing," IEEE Trans. Software Eng., Vol. 20, No. 10, pp.812–824. 1994
[4] Liang Shi, Baowen Xu, Changhai Nie. Combinatorial Design Approaches for Automatic Test Generation. Journal of Electronics(China), 2005, 22(2): 205-208
[5] Anderson J A, Lorenz C D, Travesset A.General Purpose Molecular Dynamics Simulations Fully Implemented on Graphics Processing Units[J].Journal of Computational Physics,2008 227(10):5342—5359.
[6] Zhang D , Sui J , Gong Y . Large scale software test data generation based on collective constraint and weighted combination method[J]. Tehnicki vjesnik - Technical Gazette, 2017, 24(4).
[7] Kuhn R, Lei Y, Kacker R. Practical combinatorial testing: Beyond pairwise. IT Professional, 2008,10(3):19−23.
[8] Grindal M. Handling combinatorial explosion in software testing [Ph.D. Thesis]. Linköpings Universitet, 2007.
[9] Wang ZY, Nie CH, Xu BW, Shi L. Optimal test suite generation methods for neighbor factors combinatorial testing. Chinese Journal of Computers, 2007,30(2):200−211 (in Chinese with English abstract).
[10] Wang ZY, Nie CH, Xu BW. Generating combinatorial test suite for interaction relationship. In: Pezzè M, ed. Proc. of the 4th Int'l Workshop on Software Quality Assurance (SOQUA 2007). New York: ACM Press, 2007. 55−61.
[11] Yan J, Zhang J. A backtracking search tool for constructing combinatorial test suites. The Journal of Systems and Software, 2008, 81(10):1681−1693.
[12] Ma F, Zhang J. Finding orthogonal arrays using satisfiability checkers and symmetry breaking constraints. In: Ho TB, Zhou ZH, eds. Proc. of the 10th Pacific Rim Int'l Conf. on Artificial Intelligence (PRICAI 2008). LNAI 5351, Berlin/Heidelberg: Springer-Verlag, 2008. 247−259.
[13] P. Joshi, M. Naik, K. Sen, and D. Gay, "An effective dynamic analysis for detecting generalized deadlocks," in Proceedings of the eighteenth ACM SIGSOFT international symposium on Foundations of software engineering, Santa Fe, New Mexico, USA, 2010, pp.327-336.
[14] Bryce RC, Colbourn CJ. The density algorithm for pairwise interaction testing. Software Testing, Verification and Reliability, 2007. 17(3):159−182.
[15] Meagher K, Moura L, Zekaoui L. Mixed covering arrays on graphs. Journal of Combinatorial Designs, 2007,15(5):393−404.
[16] Zekaoui L. Mixed covering arrays on graphs and tabu search algorithms [MS. Thesis]. Ottawa-Carleton Institute for Computer Science, University of Ottawa, 2006.

Research on UAV flight safety interval under wind influence

Zekun Wang, Xiangxi Wen*, Mingong Wu, Shidai Qiu and Ang Li

Air Traffic Control and Navigation College , Air Force Engineering University, Xi'an Shanxi 710051, China

*E-mail: wxxajy@163.com

Abstract. With the progress of aerospace technology, unmanned aerial vehicles (UAV) are gradually showing incomparable advantages over manned aircraft. However, the safety interval of formation flying has always puzzled us. In order to solve this problem, this paper mainly studies the collision risks of three UAVs in flight considering the two factors of wind and positioning error. It provides a reference for the research on the of flight safety interval UAV formation.

1. Introduction

At present, unmanned aerial vehicles (UAV) have attracted extensive attention in both civil and military fields due to their small size, low cost, high survival rate, simple structure, flexible use, strong adaptability and no risk to life. The formation flying of UAVs refers to a certain formation arrangement or task allocation organization mode for multiple UAVs to meet the task requirements. During the formation flying process, each UAVs must maintain a specified distance and height. However, how to control the safety interval inside the UAV formation is an urgent problem to be solved at present.

Domestic research on UAV formation flight is still in its infancy. Ref.[1]suggests that for larger UAVs that can accommodate the load of airborne radars and generate the required electrical energy, equipped with airborne radars is a good choice to resolve conflicts. Ref.[2] proposes a mathematical framework of UAV perception&avoidance system, which is defined and analyzed by differential game and survivability theory. Ref.[3] also mentions the concept of perception and collision avoidance system, which is divided into two categories: cooperative and non-cooperative. Ref.[4] quantifies the collision probability between two UAVs when flying in formation. Ref.[5] makes a general analysis and discussion on the safety interval of aircrafts flying in low-altitude airspace.

Some scholars abroad have suggested that UAVs cannot avoid conflicts according to the principle of "seeing and avoiding"[6]. Ref.[7] transforms the conflict resolution problem into a nonlinear optimization problem. Jenie proposed a conflict detection and resolution method for unmanned aerial vehicles[8].Based on the uncertainty of the future UAV position[9], Xiang J proposed a method to evaluate the collision probability between the space point and the UAV airbag at the mid-term horizon [10]. After that, Jenie used Monte Carlo simulation to evaluate the safety of UAVs in different airspace densities[11].

There are still many difficult problems to be solved in the research of formation flying of UAVs.This paper starts with the influence of wind on UAV flight, calculates the collision probability between UAVs by changing the size and direction of the wind, thus determining the UAV flight safety interval in this case.

2. Modeling of UAV formation flight

2.1. Dynamic model

In this paper, Euler-Newton method[12] is used for specific analysis. If the speed of the UAV along the positive direction of the three coordinate axes of the ground coordinate system is u, v, w, then:

$$\dot{x} = u \tag{1}$$

$$\dot{y} = v \tag{2}$$

$$\dot{z} = w \tag{3}$$

$$\dot{u} = m^{-1}\left(\cos\phi\sin\theta\cos\psi + \sin\phi\sin\psi\right)\sum_{i=1}^{4} T_i \tag{4}$$

$$\dot{v} = m^{-1}\left(\cos\phi\sin\theta\sin\psi - \sin\phi\cos\psi\right)\sum_{i=1}^{4} T_i \tag{5}$$

$$\dot{w} = \text{-}gm^{-1}\left(\cos\phi\cos\theta\right)\sum_{i=1}^{4} T_i \tag{6}$$

The state variables in this model are: x, y, z, u, v, w. The control variables are: ϕ, Ψ, θ, T (Ψ is the yaw angle; θ is pitch angle; ϕ is the roll angle). And random interferences include: wind, navigation errors, and reaction delay.

2.2. Risk assessment model

The formation of unmanned aerial vehicles will be affected by various weather conditions when flying, of which wind is the most important. Sometimes the sudden change of wind speed will change the interval between UAVs in the formation, especially the UAVs on the windward side will be greatly affected. At this time, ground personnel cannot control the position of UAVs in time. Therefore, leaving a certain margin of safety can better ensure the flying safety of the formation.

Because of the irregular configuration of the UAV, the method in Reich collision model is adopted to abstract the UAV into a cuboid, and the length, width and height of the cuboid are l_x, l_y, l_z, so the collision critical values in all directions are respectively:

$$d_i = \left(l_{i1} + l_{i2}\right)/2, \quad i = x, y, z \tag{7}$$

This paper mainly studies the influence of wind speed and navigation positioning error on formation, and takes the formation of three unmanned aerial vehicles arranged in a horizontal row as the research formation. The formation formation is shown in the figure 1.

Figure 1. UAV formation.

UAV's positioning error δ follows normal distribution[13], $\delta \sim N\left(\mu, \sigma^2\right)$. The distance between the two agents can be expressed as: $d_{ij} = d + \left(\delta_i + \delta_j\right)$, then there is:

$$d_{12} \sim N\left(\mu_1 + \mu_2 + x_0, \sigma_1^2 + \sigma_2^2\right) \tag{8}$$

Assuming that the probability density function of the distance between the two machines is $f_d(x)$, the probability distribution function of collision is $P(x)=\int_{x_1}^{x_2} f(t)dt$. Where x1=0, x2 is the minimum safe distance between the two agents[14].

Several assumptions and settings are made:

• The force exerted on an aircraft decreases by 1/4 for each passing aircraft.

• The moment when the wind starts to act on the UAV is 1 second later along the wind direction.

• The response delay of each UAV is 3 s, that is, each UAV will move for 3 s under the action of wind force within one change.

Take No.1 as the origin and the horizontal right direction as the positive direction to establish the axis, then:

$$F_x = WS = \frac{\gamma}{2g} v^2 S \tag{9}$$

$$a = \frac{\gamma}{2mg} v^2 S \tag{10}$$

Each variable is defined as follows: γ is the gravity per unit volume of air, g is the gravity acceleration, W is the wind pressure per unit area, S is the force area of unmanned aerial vehicle subjected to wind, F_x is the wind force along axis x, m is the mass of unmanned aerial vehicle, and v is the wind speed.

If the initial distance between the two UAVs is d_{ij}, then:

$$d'_{ij} = \begin{cases} d_{ij} - \frac{1}{2}at^2 + (\delta_i + \delta_j), t \in [0,1) \\ d_{ij} + \frac{1}{2}\cdot\frac{3}{4}a(t-1)^2 - \frac{1}{2}at^2 + (\delta_i + \delta_j), t \in (1,3] \end{cases} \tag{11}$$

Since formation flying is unsafe as long as there is collision between two UAVs, the probability of collision between UAVs *i* and *j* is defined as P_{ij}, and the probability of safe flying of the whole formation is:

$$P_{safe} = (1-P_{12}) \cdot (1-P_{23}) \tag{12}$$

3. Experimental analysis

When UAV formation flies at low altitude, the wind direction and wind speed are always changing. The influence of wind in different directions and different sizes on UAV formation is different. For formation flying side by side, as long as the projection of wind speed in the lateral direction is not zero, it will affect the formation. In this paper, two typical wind directions are selected to study, one is the left side wind, and the other is the left rear wind that forms an angle of 45° with the track. The collision risk is simulated by Matlab 2016a. Each parameter is set to: $m = 5\,\text{kg}$, $S = 0.1\,\text{m}^2$, $\gamma = 12\,\text{N/m}^2$, $g = 9.8\,\text{m/s}^2$, $l_x = l_y = 1.35\,\text{m}, l_z = 0.6\,\text{m}$, $x_1 = 0$, $x_2 = 1.5$.

When the initial distance d0 is 4-7 m, the collision probability of UAVs 1 and 2 (P_{12}), 2 and 3 (P_{23}) and the probability of safe flight of the whole formation (P_{safe}) are calculated respectively under the influence of two kinds of wind as the wind speed increases from 0 to $9\,\text{m/s}$.

When the wind is on the left, the probability change trend at different initial distances is shown in figure 2:

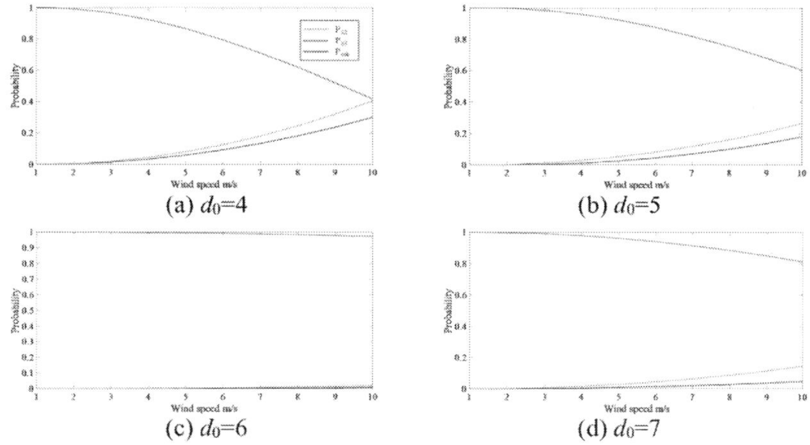

Figure 2. Left wind influence.

Obviously, when the initial distance and wind direction are fixed, P_{12} and P_{23} shows an upward trend with the increase of wind speed, while the probability of safe flight of the whole formation shows a downward trend, and P_{12} rises faster than P_{23} .

A comprehensive comparison of figure 2(a)-(d) shows that when the wind direction is constant, the larger the initial interval, the smaller the slope of the curve, that is to say, the smaller the impact of the wind on flight safety. As can be seen from figure 2(d), when the initial distance is more than $7\ \mathrm{m}$, the left wind from 0 to $9\,\mathrm{m/s}$ has little effect on the formation safety.

When the wind direction is from the left rear and the wind speed and track form an included angle of 45°, the probability change trend at different initial distances is shown in figure 3:

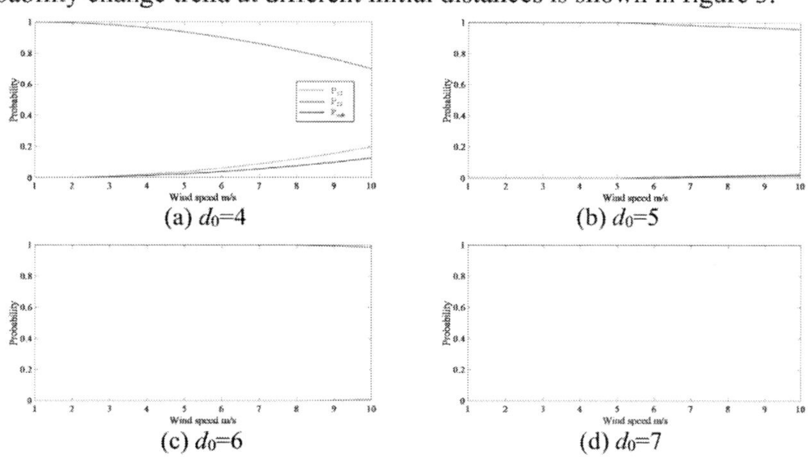

Figure 3. Left rear side wind (track angle 45°).

As shown in the figure, the same conclusion obtained from experience is that when the wind direction is left rear, the influence of wind on flight safety is obviously lower than that of left wind, i.e. the larger the angle between the wind direction and the lateral direction of formation, the smaller the influence, and the safer the flight.

4. Conclusion

After selecting a specific formation, this paper studies the change rule of the probability of UAV formation flying safely with the initial interval under the premise of only considering the influence of wind and positioning error. The research finds that:

(1)As the wind speed increases, the collision probability between two adjacent aircrafts increases, and the formation flying safety probability decreases.

(2)For horizontal formation, under the condition of certain wind speed and initial interval, assuming the angle between wind speed direction and lateral direction is $\theta \in \left[0, \frac{\pi}{2}\right]$, the larger θ, the higher the probability of formation flying safely.

(3)Under the environment of low-altitude flight, using the 10L type of four-rotor UAV, when the internal interval of the formation is $7\,\mathrm{m}$, the safety probability of the formation flight has basically met the requirements and can ensure the normal communication between the main engines and wingmen in the formation.

Acknowledgment
This work is supported by National Natural Science Foundation of China (Grant No. 71801221) and Nature Science Basic Research Plan in Shaanxi Province of China (Grant No. 2018JQ7004).

References

[1] Cai Z H, Yang L M, Wang Y X, et al. Analysis for whole airspace flight key factors of unmanned aerial vehicles[J]. Journal of Beijing University of Aeronautics and Astronautics, 2011, 37(2): 175-179.

[2] CUI J H, WEI R X, ZHANG X Q. Dynamic decision-making method for safety region of sense and avoid system for unmanned aerial vehicle [J]. Control and Decision, 2014,18(12):219~2200.

[3] ZUO Qing-hai, PAN Wei-jun, ZHUO Xing-yu, TAN Jing-wei. Research on Sense and Avoiding Method of Unmanned Aircraft System[J]. Equipment manufacturing technology, 2016,35(11):65~69.

[4] ZHANG Y B. The UAV formations research based on probability calculations and state predicted[D]. Dalian University of Technology, 2013,3:18.

[5] CENG G Z. Safety Interval analysis of aircraft flying in low altitude airspace [J]. Technology Innovation and Application, 2015,6(3):18.

[6] Udovic A, Jong H D, Vielhauer J. Validation of Unmanned Aircraft Systems' Integration into the Airspace (VUSIL I and II)[J]. Sae International Journal of Aerospace, 2011, 4(2):1216~1227.

[7] Yang J, Yin D, Niu Y, et al. Unmanned aerial vehicles conflict detection and resolution in city airspace[C].USA: IEEE, 2015.

[8] Jenie Y I, Kampen E J V, Ellerbroek J, et al. Taxonomy of Conflict Detection and Resolution Approaches for Unmanned Aerial Vehicle in an Integrated Airspace[C]. IEEE Transactions on Intelligent Transportation Systems, 2017, 18(3):558~567.

[9] Qu Y, Zhang Y. Cooperative localization against GPS signal loss in multiple UAVs flight[J]. Journal of Systems Engineering and Electronics, 2011, 22(1):103~112.

[10] Xiang J, Yang L, Luo Z. Flight safety measurements of UAVs in congested airspace[J]. CJA, 2016, 29(5):1355~1366.

[11] Jenie Y I, Kampen E J V, Ellerbroek J, et al. Safety Assessment of Unmanned Aerial Vehicle Operations in an Integrated Airspace[C] USA:AIAA Infotech & Aerospace, 2016.

[12] LIU C L. Research on modeling and control for quadrotor UAV[D]. Hubei University of Technology, 2016.

[13] QU Y L. Research on Collision Risk Modeling of Air Traffic[D]. Nanjing University of Aeronautics and Astronautics, 2011.

[14] ZHANG Z N, LIANG Y W. Bayesian network-based study on collision risk in free flight [J]. China Safety Science Journal, 2014, 24(9):40~45.

Mobility-assisted adaptive routing for intermittently connected FANETs

Xianfeng Li*, Fan Deng and Jiaojiao Yan

Shenzhen Key Lab of Information Theory & Future Network Architecture, School of EECS, Peking University Shenzhen Graduate School, 518000 Shenzhen, China

*E-mail: lixianfeng.sz@pku.edu.cn

Abstract. Flying Ad Hoc Network (FANET), a self-configuring network consisting of Unmanned Aerial Vehicles (UAVs), can undertake different kinds of civilian and military tasks. As a particular instance of FANETs, intermittently connected FANETs (IC-FANETs) are characterized by a highly dynamic network topology and intermittent connectivity. Existing routing algorithms have been proposed for FANETs given that the underlying network is well connected, which are not suitable for such network environments. In this paper, we present a geographic routing algorithm called mobility-assisted adaptive routing (MAAR) targeting IC-FANETs. Different from conventional geographic routing protocols exploiting other location services to obtain the location information, MAAR combines routing strategy with location service in order to reduce the packet latency and routing overhead. MAAR updates the information of neighbour nodes dynamically according to nodal mobility and introduces a new metric called IDM to update the positions of other nodes in the network adaptively. In addition, we use a store-carry-and-forward paradigm to handle the technical issues of networks suffering from communication interruptions. When the forwarder node does not have any solution to find an appropriate relay node, mobility is exploited. Simulation results demonstrate that the MAAR scheme performs remarkably in terms of packet delivery ratio, throughput and routing overhead.

1. Introduction

The last decade has witnessed the widely use of Unmanned Aerial Vehicles (UAVs) in various civilian and military applications[1]. In most cases, such as reconnaissance operation, mapping, search and rescue, several UAVs form an ad hoc connected group to complete difficult tasks in a large area. Due to the high degree of UAVs' mobility and the low density, the network is not fully connected, and a contemporaneous path can never be assured to exist between any two nodes. We call this class of network intermittently connected Flying Ad Hoc Network (IC-FANET), which is a type of Delay/Disruptive Tolerant Network (DTN)[2].

In IC-FANETs, the sparse and highly dynamic topology results in frequent link disconnections, which makes the design of routing mechanism more challenging. Existing routing algorithms in FANETs[3-5] assume that there usually exists at least one complete path between source and destination, which may be well suited for fully connected environment. However, they are not able to forward packets efficiently in intermittent, partially connected network environments. Consequently, a routing layer with high efficiency and resilience is of critical importance for IC-FANETs.

Compared to topology-based routing scheme, position-based routing scheme is more suitable for this kind of networks. The scheme utilizes the geographic position of mobile nodes to forward data packets, and eliminates the need to consider whether there exists a contemporaneous source-destination path. Previous works usually use routing and location services separately. The location information is maintained by extra location-based services in a distributed manner. When routing packets, to obtain the destination position, the source node launches a request to the location services. This approach increases end-to-end delay. What is more, the overhead produced by inquiring the location service is not considered when evaluating the performance of a protocol.

To this end, in this paper, we propose a geographic routing protocol called mobility-assisted adaptive routing (MAAR) specific to IC-FANETs, which combines routing strategy with location service effectively. In order to achieve high accuracy with low overhead, MAAR updates the information of neighbour nodes and the positions of all nodes in the network using different adaptive strategies respectively. Specifically, based on nodal mobility, it adjusts the frequency of beacons dynamically, which is used to update the positions and velocities of neighbours. At the same time, in order to maintain the positions of all nodes in the network, the Position Update messages recording the updated location information of other nodes are broadcast. We define a new metric IMP to measure the importance degree of the Position Update message. When IMP reaches a certain threshold, the message is propagated to its neighbours. In this way, not only the source node is able to obtain the destination's position immediately, but also forwarder nodes can update the destination's position in a packet if they own fresher location about the destination.

To route packets, MAAR adopts the store-carry-forward paradigm to overcome intermittent connections. When failing to find any node to forward data packet according to movement direction and location, the forwarder nodes are allowed to carry the packet for a certain distance until the suitable node or destination emerges within its communication range. In other words, nodal mobility is exploited to help transmitting data packets.

The remainder of this paper is organized as follows. Section 2 reviews the related work. Section 3 gives details on the proposed routing protocol specific to IC-FANETs. Simulation setup and results are discussed in Section 4. Finally, the paper is concluded in Section 5.

2. Related works

A massive body of studies have been dedicated in ad hoc networks. In this section, we briefly give an overview of relevant research in two categories of routing protocols, geographic routing and delay-tolerant network routing.

2.1. Geographic routing

Brad Karp et al.[6] propose Greedy Perimeter Stateless Routing (GPSR) that makes greedy forwarding decisions using only information about forwarder node's immediate neighbours in the network topology. Once the greedy forwarding is impossible, the algorithm switches to the perimeter mode. Seung Hyeon el at.[7] propose a new geographic routing protocol called Geographic Routing Protocol for Aircraft Ad Hoc Network (GRAA), which update the whole topology of network by the information sent from base station. In order to choose the best intermediate nodes during routing, GRAA predicts the neighbours' location before forwarding data packets. In[8], the Location-Aided Routing (LAR) is presented which limits the search for a new route to a small request zone of the network. However, GPSR, LAR, and the routing protocols in[9-11] need the help of some kinds of location service to retrieve the destination position before routing packets. Both[12] and[13] propose adaptive position update schemes for geographic routing protocol. However, the schemes are insufficient for IC-FANETs to handle the disruptive connections effectively.

2.2. Delay-tolerant networking (DTN)

In most basic routing algorithms of DTN Networks like Epidemic algorithm[14], PRoPHET algorithm[15], the overhead is one of the most considerable issues. Spyropoulos et al.[16] introduce a

IOP Conf. Series: Materials Science and Engineering **715** (2020) 012028 doi:10.1088/1757-899X/715/1/012028

limited flooding-based routing scheme called Spray and Wait, including two phases: spray phase and wait phase. In spray phase, a source node generates L packet copies, then spreads these copies to L different relaying nodes. In wait phase, these L relaying nodes carry the packet copy and forward it only to its destination. The mechanism is popular owing to its simplicity, but at the same time a large number of copies generated in spray phase do not make the overhead performance improved obviously. Furthermore, spreading packet copies without accurate direction increases the end-to-end delay. In[17], Plankton considers to dynamically reduce the number of copies that a message can be replicated.[18] combines multi-hop and mobility-assisted routing protocols from a resource allocation point of view.

The mechanisms in[19-21] take routing decisions based on geographical location data designed for Vehicular delay-tolerant network (VDTN), but they are evaluated using vehicles that move along streets with static destinations, which are inadequate for IC-FANETs. Kuiper et al.[22] present a location-aware routing for delay-tolerant networks (LAROD). A beacon-less strategy is used to achieve low overhead. However, the setting of delay timer for selecting new custodian introduces substantial delay, the discussion about which is missing.

3. Mobility-assisted adaptive routing protocol

The aim of Mobility-assisted Adaptive Routing Protocol (MAAR) is to adapt the intermittent networks, enhancing the packet delivery ratio while not generating too high overhead. MAAR integrates location service into routing mechanism, the advantage of which is to reduce the delay at the packet source. At the same time, the destination's position in a packet is capable to be updated during the transmission, which makes the information more accurate and thus increases the delivery rate. In this section, we detail the two main components of MAAR: *dynamic location update strategy* and *mobility-assisted forwarding strategy*.

Like other geographic routing protocols, MAAR makes routing decisions based on location information. Due to the high mobility of the nodes in IC-FANETs, the topology changes frequently. The key issue is how to maintain the accurate information about other nodes. Period broadcasting of beacons is not satisfied for this kind of networks. Because beaconing in a fixed short interval wastes a great deal of communication overhead for those nodes that do not exhibit significant dynamism. On the other hand, if beacons are propagated in a fixed long interval, it is very likely that outdated locations are utilized, which is adverse to routing packets effectively. Consequently, MAAR employs a dynamic location update strategy, which consists of *adaptive beacon update* and *global position update*. The former provides position and velocity information to one-hop neighbours while the latter is destined to make nodes known the positions of all possible destination.

3.1. Adaptive beacon update

Assume that all nodes are aware of their own positions and velocities. Each node broadcasts beacons to notify neighbours of its updated position and velocity information. Similar to[11], in order to achieve both position accuracy and low overhead simultaneously, the transmission frequency of beacons is adapted to the nodes' motion. That is to say, highly mobile nodes are required to sending updates frequently since their positions are changing rapidly. On the contrary, for those nodes that move slowly, the beacon generation rate does not need to be frequent due to the relative stable topology.

Denote $(x^N_{t'},y^N_{t'})$ and $(vx^N_{t'},vy^N_{t'})$ as the position and velocity of node N at last beaconing time t'. In a real world, UAVs cannot move in a random trajectory because of the kinematic and dynamic constraints. Thus, the estimated location at current time t can be predicted based on the existing information, by using the following equations:

$$\begin{cases} x^N_p = x^N_{t'} + (t - t') * vx^N_{t'} \\ y^N_p = y^N_{t'} + (t - t') * vy^N_{t'} \end{cases} \quad (1)$$

Then node N is able to compute the location estimation error, that is the distance between the actual and predicted positions:

$$d = \sqrt{(x_t^N - x_p^N)^2 + (y_t^N - y_p^N)^2} \quad (2)$$

If the deviation d is greater than an acceptable threshold *DTH*, it can be concluded that the nodal motion (speed or direction) has changed to a great extent. A new beacon needs to be sent to broadcast its current actual position and velocity. In the same way, when a node intends to acquire the current positions of neighbours, it can be predicted according to equation (1).

However, if node N moves at a constant speed for a long time, there is no difference between the actual position and the estimated value. According to above rule, a new beacon update will not be sent from node N. This gives rise to the problem that the new neighbour which enters into the transmission range of node N after last broadcast, is unaware of the existence of node N. Therefore, MAAR specifies that if mobile nodes have not broadcast its information in a certain period of time *TTH*, a new beacon also needs to be sent.

3.2. Global position update

In an IC-FANET environment, the topology is prone to be partitioned. This condition means that inquiring destination's position by any extra location service will bring lots of delay and offer outdated information. Thus, MAAR integrates the location service with routing mechanism. In MAAR, all nodes are regarded as location servers, convenient for gaining the destination's position when needed. Each node owns a global position table, which maintains positions of other nodes in the network it knows. Since the nodes in IC-FANETs move very quickly, the position data often needs to be updated. If every node exchanges all known position information with others, the size of the message would be quite large while valid information may be small. The reason is that most positions are very likely to be duplicate with last delivered message.

Hence, in MAAR, each UAV uses a waiting queue to store the information of those nodes that positions have been updated since last broadcast, including the id, position and updated time. Then the Position Update message containing such information is propagated to neighbours when certain conditions are satisfied. The amount of information in the waiting queue not only depends on the time, but equally on the nodal mobility and other factors. Periodical broadcast cannot guarantee low overhead and high accuracy at the same time. Consequently, MAAR transmits the Position Update messages dynamically according to the quantity of information.

We present a metric called *Importance Degree of Message (IDM)* to measure the quantity of message, synthetically considering the timeliness, effectiveness and amount of the information, which can be expressed as follows:

$$IDM = \sum_{k=1}^{n} N_k * T_k \quad (3)$$

where k and n are the k_{th} node and the number of nodes in the waiting queue, N_k is the number of the neighboring nodes fulfilling the condition that the distance between which and node k is greater than the transmission range, T_k shows the time difference between current time and the update time of node k.

Suppose that the current node is node i, node k is one of the nodes in its waiting queue. Among node i's neighbors, the nodes beyond the transmission scope of node k probably hold outdated information about node k. If broadcasting the update message lets these nodes have access to update the position of node k, then this broadcast is deemed as useful. Consider the simple example in figure 1. Current node I has the updated position of node P in the waiting queue. Among its neighbours, node A and B are without the transmission range of node P while node M can communicate with P directly. Thus, the value of N_p is 2. The larger the value of N_k is, the more useful the message becomes. Thus, N_k reflects the effectiveness of the information. In equation (3), the time difference T_k embodies the timeliness of information, which shows the freshness of the information about node k. A great value of T_k indicates that the information is getting older, the position update message is more prone to be sent. If there are more nodes in the waiting queue, the message is more informative. That is to say, the value of n represents the amount of information.

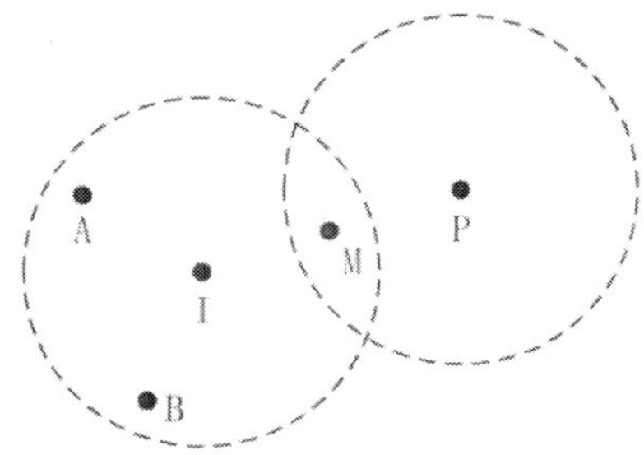

Figure 1. An example of N_k.

When the metric *IDM* exceeds an acceptable threshold *ITH*, a Position Update message is sent in order to inform neighbors of the updated positions of nodes in the waiting queue. Upon receiving a Position Update message, the node compares the information in the message with its own global position table. If the position in the message is fresher, update the position table and the waiting queue.

The main goal of the global position update mechanism is to provide a rough position about destination at source node, which may be more or less inaccurate. Furthermore, during the packet transmission, the destination also keeps moving. Actually, as long as the location provided by source node points to the right direction, we consider the mechanism to play a significant role. In addition, the mechanism also makes it possible to update the destination's position in the packet when forwarded. Obviously, the nodes nearer to destination may have more precise information of the destination with high probability. In MAAR, the destination in the data packet would be replaced when the forwarding nodes have more recent information. In this way, the accuracy of the destination's location will increase as the distance from destination decreases.

3.3. Mobility-assisted forwarding strategy

Due to the frequent disruption and sparse network density in IC-FANETs, communication interruptions often emerge when the data packets are in transit. The node holding the packet sometimes fails to find an appropriate forwarding node. Under such a situation, adopting the store-forward routing paradigm will make massive data packets discarded, which is inappropriate for IC-FANETs. To this end, MAAR designs a forwarding strategy, which combines the improved greedy forwarding scheme with the store-carry-forward paradigm, to increase the efficiency of data transmission with the help of mobility.

MAAR makes improved greedy forwarding decisions and selects the forwarding nodes among the current node and its neighbours, taking distance and moving direction into account simultaneously. We give preference to those neighbours that are close to destination and move towards destination. The reason for this is that there exist frequent disconnections in IC-FANETs. If we consider the distance between nodes as the only metric and forward the packets to the neighbour which is closet to destination but moves away from destination, once a suitable forwarding node is impossible for it, the data packets will be carried to farther region by the neighbour.

We prioritize current node holding the packet and its neighbours according to distance and direction when choosing next hop. Those nodes close to destination and move roughly towards destination own the highest priority to be chosen as next hop, and the second is current node if its moving orientation directs the region of destination. If there is no nodes that satisfy the above conditions, nodes at closer range from destination will be chosen. Otherwise, current node will carry the data packet itself.

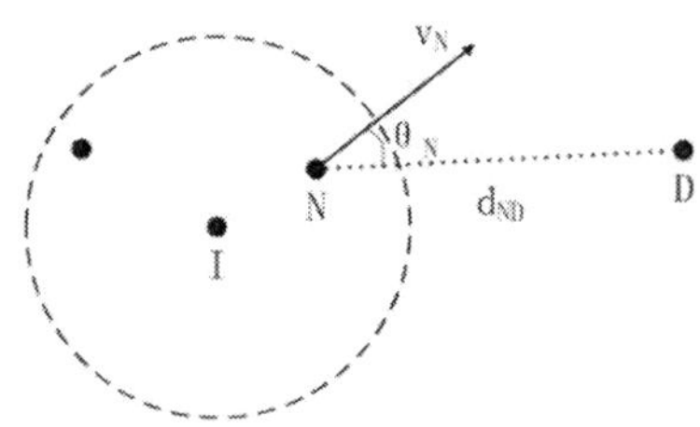

Figure 2. Forwarding example.

In order to better expound the mobility-assisted forwarding strategy, we use figure 2 to explain some notations used in the rest of the discussion. Assume that node I holds the data packet, node N is one of its neighbors. Denote d_{ND} as the distance between node N and destination node D. v_n refers to the velocity of node N and Θ_N is the angle deviation, i.e., the angle between the moving direction of node N and the line segment of *ND*. All the above data can be obtained or computed using the information in node I's neighbor table easily. When a node has data to forward, we use Algorithm 1 to make forwarding decision.

When current node I has data to transmit, it checks whether there are neighbours close to and move towards destination or not firstly. Θ_{th} is a certain threshold, which is the maximum angle that allows the motion to deviate. If exists, the data packet will be forwarded to the node having the shortest estimated time among those nodes. Otherwise, node I computes its own angle deviation. If it is smaller than Θ_{th}, node I will carry the packet by itself and the forwarding operation will not be triggered. Making the most of node I's mobility, packet gets close to destination. If not, it indicates nodes meeting the requirements of distance and moving direction simultaneously are unavailable. Next, we only take the distance into consideration to settle for second best. The node closest to destination is chosen as next hop. If fails to find any node close to destination, node I carry the packet until an appropriate forwarder appears.

Algorithm 1 Forwarding Rule

1: Find the neighbor subset A that satisfies both conditions $d_{ND} < d_{ID}$ and $\Theta_N < \Theta_{th}$;
2: **if** A is not empty **then**
3: $N^* = \arg\min_{N \in A} \left\{ \frac{d_{ND}}{v_n * \cos \Theta_N} \right\}$;
4: forward the packet to node N^*;
5: **else if** $\Theta_{ID} < \Theta_{th}$ **then**
6: current node I carry the packet;
7: **else if** neighbor subset B satisfying the condition $d_{ND} < d_{ID}$ is not empty **then**
8: $N^* = \arg\min_{N \in B} d_{ND}$;
9: forward the packet to node N^*;
10: **else**
11: current node I carry the packet.
12: **end if**

4. Performance evaluation

4.1. Simulation environments

Simulation study is conducted on the NS-3 simulator. We evaluate the performance results of MAAR, in comparison to GRAA. UAVs are deployed in a region of size 3000m*3000m. The total number of mobile nodes is in the range of[20, 50]. The Random Walk Model without pause time is used to model nodes' mobility. We use two-ray ground model as the radio model and select IEEE 802.11 DCF as the medium access control (MAC) layer protocol. Radio communication range is set to 250m. The MAC layer data rate is set to 2Mbps. We randomly select 10 source-destination pairs in each scenario. The generated constant bit rate (CBR) is 4 packets/s. The simulation time is 500s. The parameters *DTH* and *TTH* are set as 20m and 4s respectively.

To evaluate the performance of MAAR, the following evaluation metrics are used:

• Packet Delivery Ratio: the ratio of the number of data packets delivered to the total number of data packets generated.

• Overhead: the ratio of protocol packets and data packets.

• Average Throughput: the average bytes of data packets of network received in one second.

4.2. Simulation results and discussions

To evaluate the impact of parameter *IDM*, we evaluate MAAR with different values of *ITH* firstly. Table 1 shows the delivery ratio and overhead under different settings. As the value of *IDM* increases, the transmission interval of Position Update message increases, and the information in the global position table become older. This incurs less accurate positions provided when forwarded, which results in the performance degradation in terms of packet delivery ratio. On the other hand, the overhead ratio decreases owing to the less-frequent transmission interval. It is a trade-off between delivery ratio and overhead. In the following-up experiments, we set the value of *IDM* to 200.

Table 1. Performance for different values of *ITH*.

Value of *ITH*	100	150	200	250
Delivery Ratio	90.7%	89.5%	88.4%	86.1%
Overhead	14.57	13.13	12.31	11.75

Figure 3 studies the performance of MAAR compared with GRAA while varying the mobility of nodes in the network. The maximum moving speed is set in the range of[5, 35] m/s and the number of nodes is 30. Figure 3(a) shows that, MAAR outperforms GRAA in delivery ratio under same mobility, increasing by 15%-51%. Note that for both protocols, packet delivery ratio accordingly decreases with the increase of moving speed in the definite velocity range. This is because when the network is sparse and intermittent, the high nodal mobility leads to less opportunities that both forwarder and receiver in their communication range while they have each other's location information. Therefore, when the maximum speed exceeds some extent, which is 20m/s for both MAAR and GRAA in this experiment, the performance degrades faster. In figure 3(b), we observe that the overhead of both MAAR and GRAA decrease with the increase of nodal velocity. The reason is the same as why packet delivery ratio decreases in high nodal velocity. It is notable that the overhead of MAAR is always much lower than GRAA. Because MAAR adopt the adaptive beacon update strategy, which decrease the number of beacon packets significantly when the location of nodes changes slowly. Figure 3(c) shows the average throughput of whole networks in one second. The simulation result is similar to packet delivery ratio.

In figure 4, we vary the number of nodes from 20-50 with maximum moving speed fixed at 20m/s. Owing to the use of store-carry-forward mechanism, MAAR holds the packet delivery ratio at high level, regardless of the number of nodes. While the performance of GRAA increases slowly and reach top in 40 nodes, and rapidly degrade with larger number of nodes. The reason is that, as the number of nodes increase, the protocol packets for neighbour information update increase simultaneously, which occupy the wireless channel and influent the transmission of data packets. However, MAAR adopt the adaptive beacon update strategy, which dramatically degrade overhead and consequently improve the performance of packet delivery ratio.

IOP Conf. Series: Materials Science and Engineering **715** (2020) 012028 doi:10.1088/1757-899X/715/1/012028

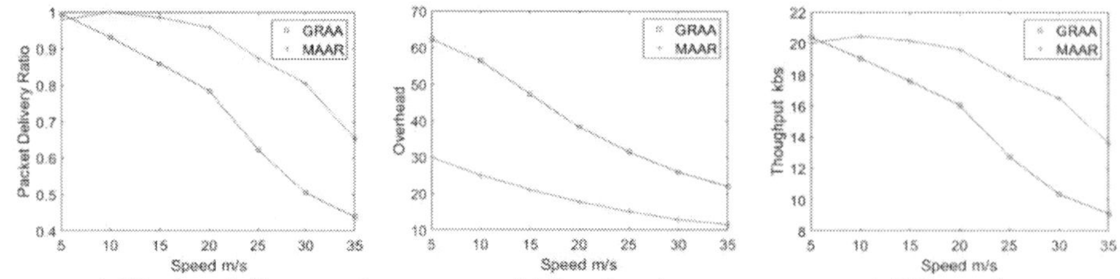

(a)Packet delivery ratio. (b)Overhead. (c)Throughput.

Figure 3. Performance with varying mobility.

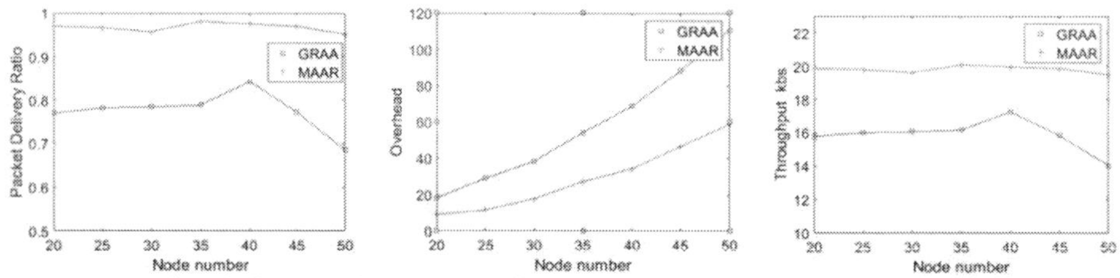

(a)Packet delivery ratio (b)Overhead. (c)Throughput.

Figure 4. Performance with varying density.

5. Conclusion

IC-FANETs suffer from recurrent disconnections due to the high degree of nodes' mobility and low node density. In this paper, we present the mobility-assisted adaptive routing(MAAR) for IC-FANETs, which combines routing strategy and location service. Different self-adaptive strategies are used to update the information of neighbor nodes and other nodes of the network respectively. Besides, we use the combination of the improved greedy forwarding method and store-carry-forward paradigm to make effective forwarding decisions. The simulation results show that MAAR can efficiently adapt to the networks with disruptive connectivity and perform better than GRAA.

Acknowledgments

The work is sponsored by Shenzhen Science and Technology Innovation Committee (SZSTI) under the grant for basic research (No. JCYJ20170818090006804).

References

[1] Bekmezci, Ilker, Ozgur Koray Sahingoz, and Şamil Temel. "Flying ad-hoc networks (FANETs): A survey." Ad Hoc Networks 11.3 (2013): 1254-1270.

[2] Fall, Kevin. "A delay-tolerant network architecture for challenged internets." Proceedings of the 2003 conference on Applications, technologies, architectures, and protocols for computer communications. ACM, 2003.

[3] Rosati, Stefano, et al. "Dynamic routing for flying ad hoc networks." IEEE Transactions on Vehicular Technology 65.3 (2016): 1690-1700.

[4] Lei, Lei, et al. "Link availability estimation based reliable routing for aeronautical ad hoc networks." Ad Hoc Networks20 (2014): 53-63.

[5] De Rango, Floriano, Francesca Guerriero, and Peppino Fazio. "Link-stability and energy aware routing protocol in distributed wireless networks." IEEE Transactions on Parallel and Distributed systems 23.4 (2012): 713-726.

[6] Karp, Brad, and Hsiang-Tsung Kung. "GPSR: Greedy perimeter stateless routing for wireless networks." Proceedings of the 6th annual international conference on Mobile computing and networking. ACM, 2000.

[7] Hyeon, SeUng, Ki-Il Kim, and SangWoo Yang. "A new geographic routing protocol for aircraft ad hoc networks." 29th Digital Avionics Systems Conference. IEEE, 2010.

IOP Conf. Series: Materials Science and Engineering **715** (2020) 012028 doi:10.1088/1757-899X/715/1/012028

[8] Ko, Young-Bae, and Nitin H. Vaidya. "Location-Aided Routing (LAR) in mobile ad hoc networks." Wireless networks 6.4 (2000): 307-321.
[9] Noguchi, Taku, and Takahiko Kobayashi. "Adaptive location-aware routing with directional antennas in mobile adhoc networks." 2017 international conference on computing, networking and communications (ICNC). IEEE, 2017.
[10] Ghafoor, Kayhan Zrar, et al. "Improved geographical routing in vehicular ad hoc networks." Wireless Personal Communications 80.2 (2015): 785-804.
[11] Peters, Kevin, et al. "A geographical routing protocol for highly-dynamic aeronautical networks." 2011 IEEE Wireless Communications and Networking Conference. IEEE, 2011.
[12] Chen, Quanjun, Salil S. Kanhere, and Mahbub Hassan. "Adaptive position update for geographic routing in mobile ad hoc networks." IEEE Transactions on Mobile Computing 12.3 (2013): 489-501.
[13] Zhang, Qiang, et al. "Contention-based adaptive position update for intermittently connected VANETs." 2014 IEEE Global Communications Conference. IEEE, 2014.
[14] Vahdat, Amin, and David Becker. "Epidemic routing for partially connected ad hoc networks." Duke Technical Report, 2000.
[15] Lindgren, Anders, Avri Doria, and Olov Schelen. "Probabilistic routing in intermittently connected networks." International Workshop on Service Assurance with Partial and Intermittent Resources. Springer, Berlin, Heidelberg, 2004.
[16] Spyropoulos, Thrasyvoulos, Konstantinos Psounis, and Cauligi S. Raghavendra. "Spray and wait: an efficient routing scheme for intermittently connected mobile networks." Proceedings of the 2005 ACM SIGCOMM workshop on Delay-tolerant networking. ACM, 2005.
[17] Guo, Xiang Fa, and Mun Choon Chan. "Plankton: An efficient dtn routing algorithm." 2013 IEEE International Conference on Sensing, Communications and Networking (SECON). IEEE, 2013.
[18] Liu, Cong, and Jie Wu. "Efficient adaptive routing in delay tolerant networks." 2009 IEEE International Conference on Communications. IEEE, 2009.
[19] Cheng, Pei-Chun, et al. "GeoDTN+ Nav: geographic DTN routing with navigator prediction for urban vehicular environments." Mobile Networks and Applications 15.1 (2010): 61-82.
[20] Soares, Vasco NGJ, Joel JPC Rodrigues, and Farid Farahmand. "GeoSpray: A geographic routing protocol for vehicular delay-tolerant networks." Information Fusion 15 (2014): 102-113.
[21] Leontiadis, Ilias, and Cecilia Mascolo. "GeOpps: Geographical opportunistic routing for vehicular networks." 2007 IEEE International Symposium on a World of Wireless, Mobile and Multimedia Networks. Ieee, 2007.
[22] Kuiper, Erik, and Simin Nadjm-Tehrani. "Geographical routing with location service in intermittently connected MANETs." IEEE Transactions on Vehicular Technology 60.2 (2011): 592-604.

IOP Conf. Series: Materials Science and Engineering **715** (2020) 012029 doi:10.1088/1757-899X/715/1/012029

LEO IoT based big data management and analysis platform design for intermodal containers

Jieyin LYU[1,2,*], Shouqin Zhou[2], Hao Liu[2] and Xiangmo Zhao[1]

[1]Chang'an University, School of Information Engineering, Xi'an, P.R.C.
[2]Shenzhen CIMC Intelligent Technology Co.,Ltd., Shenzhen, P.R.C.

*Email: jieyinlv@chd.edu.cn

Abstract. In view of the global information management requirements of intermodal containers, combined with the development status, this paper proposes an IoT model based on LEO satellites, and a big data management and analysis platform based on the model. This paper designs and analyses the scheme of the platform, and proposes a middleware approach to solve the problem of multi-source heterogeneous data access of containers intelligent terminals. Then,this paper designs business processes such as user and device registration, middleware registration, historical data access, and timely data access for the platform. The big data analysis and management platform has been prototyped and verified; the platform has the advantages of compatibility, openness and convenience. This study has certain value.

1. Introduction

Intermodal transport is usually a continuous and comprehensive form of cargo transport organization, using containers as the transport unit and organically combines different modes of transport. It is suitable for a variety of transportation modes such as waterways, roads, railways and aviation. In international trade, around 90% of goods are completed through intermodal transport. As the main carrier of intermodal transport, the location, status, security and other data in the container transportation process are often the content of the intermodal transport participants, usually using the means of the Internet of Things for collection, transmission and management, so called intelligent container [1-3].

At present, there are at least three problems in the intermodal transport container Internet of Things. This paper designs a big data[4] management and analysis platform based LEO satellites IoT [5] for intermodal containers to solve the below problems.

1) Platform data access method: Container transportation has the characteristics of global circulation and operation. It requires the transmission of IoT communication data with global coverage, low power consumption and low cost. However, the usual RFID and cellular communication cannot meet the above requirements, and the combination of LEO satellites and LPWA can fill the shortcomings of the technology. In recent years, the global LEO satellite industry has developed rapidly, such as Orbcomm Satellite constellation, Tianqi Satellite constellation, etc. At the same time, NB-IoT, LORA and other LPWA technologies have also appeared.
2) Platform data access protocol: The custom communication protocol used for data transmission between the container intelligent terminal and the container data operation management

platform. Due to the lack of a unified communication protocol and data format, the generated data is multi-sourced. The data of the structure makes it impossible for the container operation and management platform to collect the data generated by the intelligent terminals produced by various manufacturers quickly and conveniently. How to solve the problem of global container multi-source heterogeneous data access is a problem in the current container management platform.

3) Data interaction between platforms: In the entire business chain of container intermodal transport, including cargo owners, freight forwarding companies, car companies, carriers, ports, customs, factories, highways and management departments, etc., participate in division of labor and cooperation. However, there are many platforms in the industry chain at present, there is no unified communication protocol between the platforms, and the data formats are different, leading to information barriers, poor communication, and etc.

2. LEO satellites based IoT

The LEO satellites based IoT model is shown in figure 1. It consists of intermodal container intelligent terminal [6], transmission channel, big data management and analysis platform and third-party service platform.

In this model, the intelligent terminal is manufactured by various terminal manufacturers. Data transmission channels can be mobile cellular networks or LEO satellites. The big data management and analysis platform can be used to access, manage and publish data. The third party service platform can obtain data from the big data management and analysis platform, and carry out corresponding business interaction and trade. Considering the commercial requirements of low power, low cost and full coverage, Orbcomm Satellite Constellation is adopted for LEO satellite data transmission channel and NB-IoT for LPWA low power WAN for cellular mobile network data transmission channel.

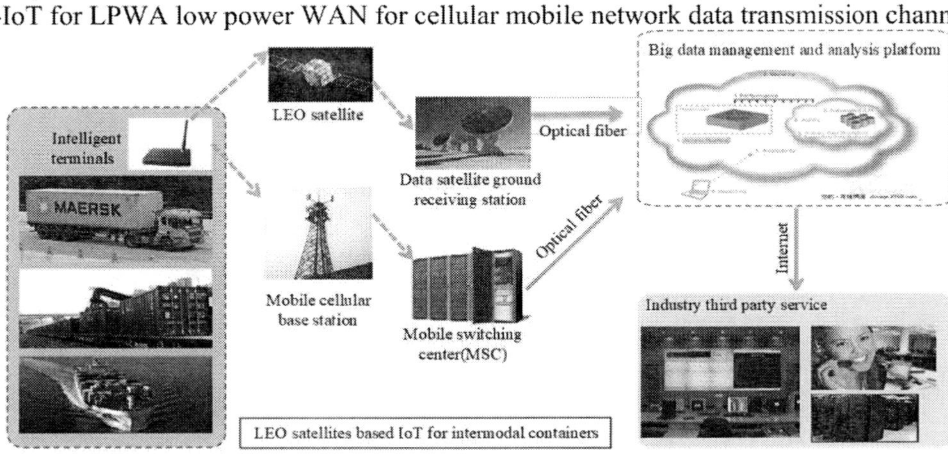

Figure 1. LEO satellites based IoT model.

3. The platform design

The big data platform is a public service platform and a data storage center. It does not care about the specific data content transmitted. Therefore, the commercial value of data transmitted and the specific data content are not visible to the big data management and analysis platform. The big data platform gives unified data and interface standards. Terminal manufacturers develop their own middleware to encrypt, decrypt and access big data management and analysis platform.

The big data management and analysis platform is an open data platform that includes four main modules: data center, data collection, data release, data distribution, and operation management. The design of the big data platform needs to include these four modules as well as middleware communication protocols and typical business processes.The scheme of the platform is shown in figure 2.

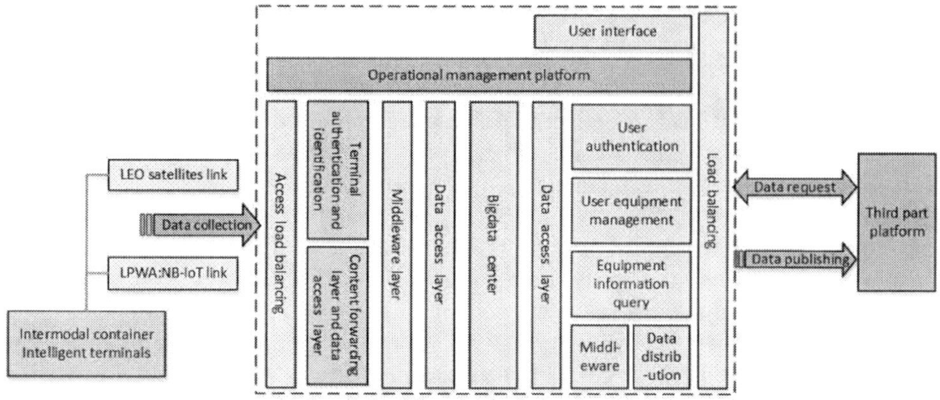

Figure 2. The scheme of the platform.

The main technical difficulties of the big data platform are twofold. On the one hand, it is the demand for the expansion of heterogeneous services, that is, facing different link providers need to establish corresponding data transmission protocols, facing different devices need to accommodate the corresponding middleware. On the other hand, it is multi-user demand for scalability in data distribution.

Therefore, the big data platform adopts the architecture of the cloud platform. On the one hand, a vertical extension mechanism is adopted in the access module for data collection to support different application layer transmission protocols of multiple link providers in the future. On the other hand, in the data distribution module, the middleware layer adopts a horizontal expansion mechanism to improve the system's concurrency and load capacity, and to support unlimited expansion of middleware.

3.1. Data center module design

Data center module includes data access layer and big data center. Data access layer mainly defines the interface of writing and reading information to big data center, which can be used by data collection module, data publishing module and management platform, such as connecting database interface, closing database interface, inserting intelligent terminal information interface, inserting user information interface, querying user information interface, querying user equipment interface, etc. The scheme of the data center module is shown in figure 3.

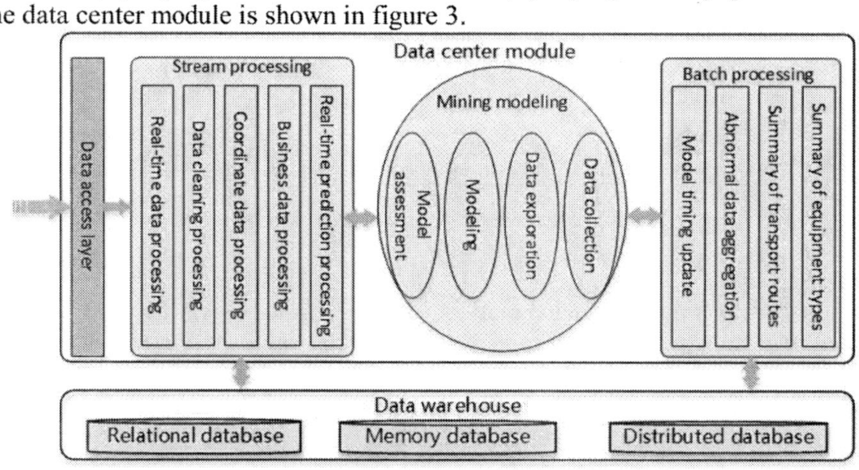

Figure 3. The scheme of the data center module.

Data stored and analyzed in big data centers come from data access layers, including data sent by intelligent terminals, user data, user-device binding relationship data and so on. Big data center database uses distributed storage architecture. Redis is used for real-time data, and MongoDB is used for business and other data with low real-time requirements. The two databases are selected according to different business needs. At the same time, Hadoop extracts the data from both of them regularly for big data analysis.

Real-time data of big data platform,which accesses through distributed message middleware, is real-time processed ,saved to data warehouse, and pushed to management platform or third-party platform for display.

Through regular batch processing, data are extracted from distributed database for statistical analysis, mining and modeling processing (for real-time processing, real-time prediction based on the model established by batch processing), and the statistical results are stored in the database for platform query display.

3.2. Data collection module design

Because the data generated by different intermodal containers and their intelligent terminals are multi-source and heterogeneous, so the data collection module adopts the open design idea, in order to collect the data stream transmitted by different transmission protocols between different link providers, and convert it into a unified data format and store it in the data center module.

Data collection module mainly includes four levels: terminal authentication and identification, content forwarding layer, middleware layer and data access layer. The specific design of each layer is as follows.

- Terminal authentication and identification: When the data of the intelligent terminal is forwarded from the terminal access layer to the authentication layer, the terminal ID is verified, that is to say, its validity is verified; invalid ID information is eliminated directly, and a valid ID terminal further identifies the middleware adapted by the terminal manufacturer.
- Content forwarding layer: After terminal authentication and identification, the access load layer will deal with data, then it is sent to the relevant middleware for parsing and standardization, and responds to the communication link public server.
- Middleware layer: After receiving the data package from the upper layer, the middleware encapsulates the data in the specific format required by the data through the analysis function module defined by the device manufacturer and the data re-encapsulation module, and transmits the data to the next layer through TCP/IP protocol.
- Data access layer: After receiving the data from the middleware, the data access layer calls the related interface functions of the data access layer and stores the data in the big data center.

3.3. Data release module design

Data distribution module is designed and developed for the third party. It consists of different business modules. Each business module provides a logical way to realize business.Business module receives requests from access layer, calls corresponding business logic methods, updates database content through database access layer interface, and generates business response required by users.In big data platform, an application server host only carries one business module, and each module runs independently. When new business needs arise, only the corresponding business module and data access layer interface need to be added, in order to have high scalability.

This paper provides four business modules, including user authentication, user equipment management, device information query, data distribution. The data in the big data center is sent to the third party by HTTP protocol through PUSH or PULL.

- User authentication: The big data center accepts and processes HTTP requests distributed by the access layer, extracts and encapsulates relevant parameters, and verifies the user's identity. When the user's authentication is successful, the system will allocate a unique Session Id, and store the Session Id in the database user table to generate a successful login response package. After that, every user's request needs to carry Session Id, and the system will authenticate user requests according to Session Id; when user authentication fails, the login failure package will be generated directly.

- User equipment management: Users need to manage their own devices, such as add, change, delete, they can login to a web page we provide third-party developers for device management operations.
- Device information query: After identification, the user establishes a connection with the platform. At this time, if the user needs to query the device owned by him, he sends a device information request to the background through HTTP. After business analysis of data access layer, the request is sent to the device information query service cluster, which retrieves the big data center according to the content of the request, and then returns the data to the third party platform.
- Data distribution: While the data access layer writes the data from the intelligent terminals into the big data center. According to the specific business, it can transmit a signal to the data distribution module. The data distribution module directly pushes the intelligent terminals data to the third party platform.

3.4. Operation management module design

The operation management module is mainly responsible for the functions at the operation level of the big data platform. On the one hand, it provides system management such as state control between the system and the server hosts; on the other hand, it provides business-level management functions such as middleware deployment and management, traffic statistics, user management and registration.

4. Data access middleware design

Middleware can shield complex operating systems and network protocols in heterogeneous systems, and provide standard program interfaces and protocols. In order to be compatible with different data protocols of terminal manufacturers, big data platforms need to customize uniform data writing specifications. The terminal manufacturer needs to develop the corresponding middleware to access to the big data platform. The data protocol of the terminal manufacturer needs to be converted into the standard data writing protocol of the open platform through the middleware. Data reading of big data platform requires terminal manufacturers to develop corresponding reading middleware. Reading middleware converts the standard data format of big data platform into the data format customized by terminal manufacturers for customers to use. The middleware controller assigns an independent port to each terminal access parsing middleware, and packages the middleware into dynamic link library (DLL) files, Jar or independent programs for platform programs to read and parse. If it is a stand-alone program, the data can be transmitted to the platform through TCP/UDP, HTTP, MQTT, CoAP and other communication protocols after the data analysis is completed. Middleware data flow for dynamic link library (DLL) and independent programs are shown in figure 4.

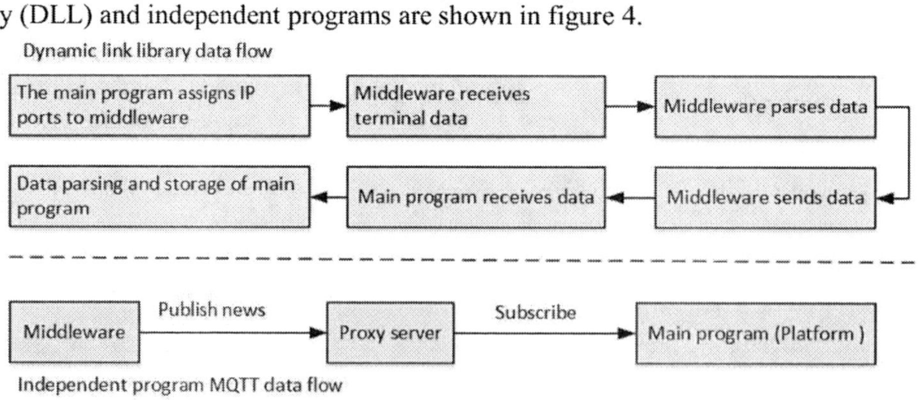

Figure 4. Middleware data flow.

By formulating the standard of intermodal container terminals accessing data middleware, the intelligent terminal can access big data management and analysis platform according to the unified standard protocol, realize the fast communication and data sharing between the middleware and the platform, which is conducive to the unified centralized management of intermodal container terminal data. The big data platform provides unified interface to the outside world through API, and carry out timely and effective data analysis, and it can improve the efficiency and security of intermodal transport logistics.

5. Typical business processes design

The main business of big data management and analysis platform includes user and terminal device registration, middleware registration, historical data access and instant data access. Because the big data platform designed in this paper is a standardized public open platform for the whole intermodal container industry, it is necessary to standardize the design of its key business processes.

5.1. User and terminal device registration process

User and device registration business is carried out between operation management platform and user interaction platform. User interface shown in figure 2 is their information exchange interface. User interaction platform carries out new user registration, user information filing, platform device number application, platform device number and hardware binding application, etc. Operations management platform audits user information filing and platform equipment number filing applications. User and terminal device registration process is shown in figure 5.

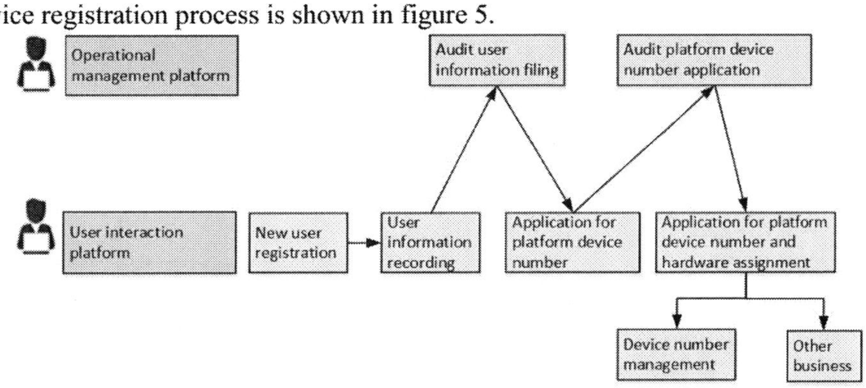

Figure 5. User and terminal device registration process.

5.2. Middleware registration process

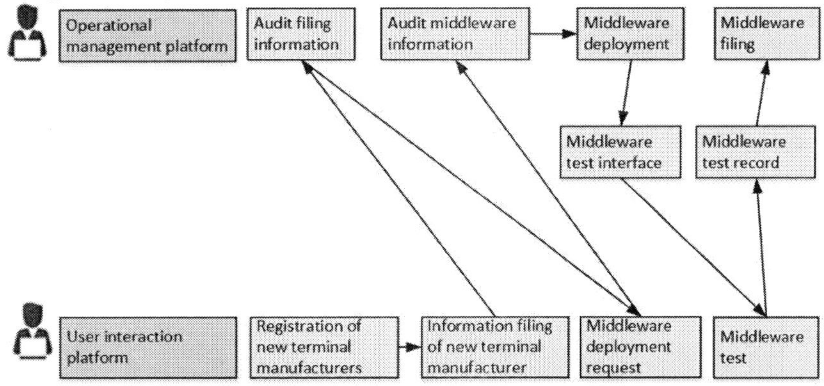

Figure 6. Middleware registration process.

The Middleware is developed by terminal equipment manufacturer according to the standard of big data management and analysis platform. After the development of middleware, middleware registration is needed in big data management and analysis platform. Then, middleware can be used in big data platform.The registration of middleware is also accomplished by information exchange between operation management platform and user interface platform through user interface.The middleware registration can only be carried out after the its terminal manufacturer registers. Middleware registration process is shown in figure 6.

5.3. Data access process

Data access includes real-time data access and historical data access, it refers to data access of big data management and analysis platform by operation management platform or third party platform. This part mainly shows the call and analysis process of middleware. It is shown in the figure 7.

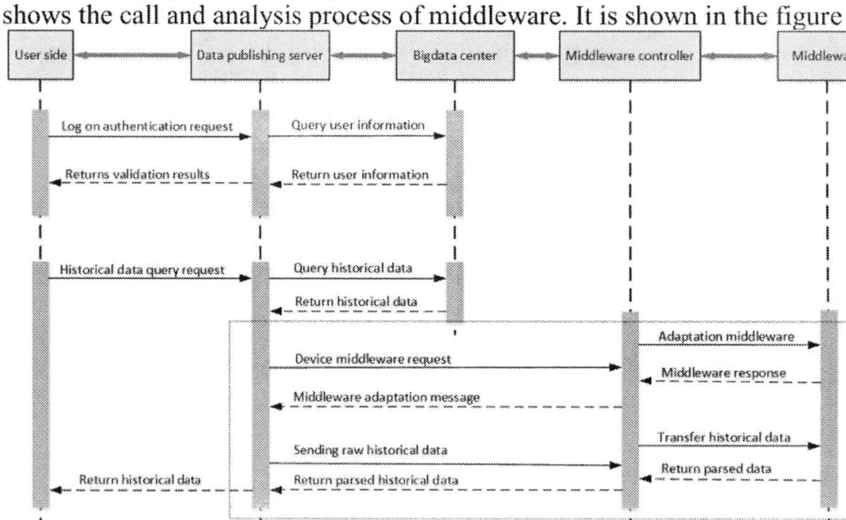

Figure 7. Historical data access process (Calling middleware).

6. Platform prototype development and verification

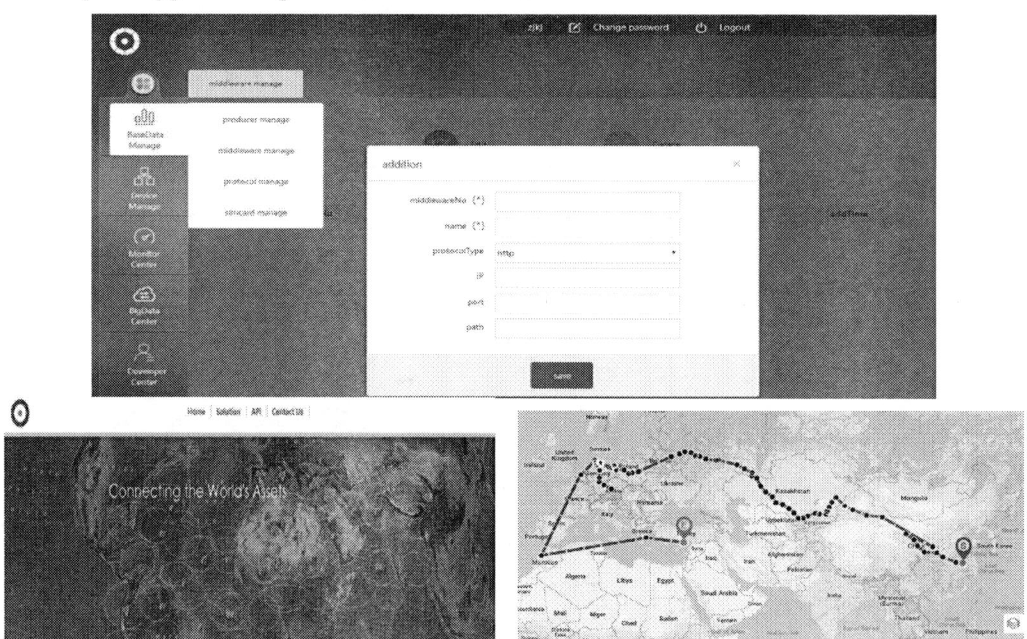

Figure 8. Prototype and verification.

According to the above design, this paper develops the prototype of big data management and analysis platform. The platform prototype application is carried out in combination with "The Belt and Road Initiative" container intermodal transport. The platform prototype and its validation applications are shown in figure 8.

7. Conclusion

As the main means of intermodal transport, Containers connect intermodal transport in series all over the world. With the development of globalization, international intermodal freight logistics trade is more frequent, and the iterative progress of science and technology is gradually meeting the increasingly strong demand of human beings for information acquisition such as location, status and safety of containers worldwide. However, there are still many areas to be studied in the unified data access, processing and distribution platform for data operation management and analysis.

The big data management and analysis platform based on LEO satellite Internet of Things proposed in this paper has the characteristics of openness, simplicity and low cost.No matter who is the terminal manufacturer and what technology the terminal adopts, as long as the terminal manufacturer provides the middleware of the middleware protocol standard, it can access the big data management and analysis platform proposed in this paper easily.As a public service platform for data acquisition, storage and distribution, the platform serves the customers of the whole supply chain of interimodal container transportation. The solution proposed in this paper adopts the integration of LEO satellite and LPWA, which has the characteristics of low cost and low power consumption.

This study provides a better design idea for the global operation and management of container intermodal transport and the cost reduction and efficiency enhancement of the global freight logistics supply chain.

Acknowledgments

Supported by the National Key Research and Development Program of China (2017YFB0503702,2018YFB0505201 and 2017YFC0804809), Shenzhen Municipal Technology Program(JSGG20170414114406355),Introducing Innovative Scientific Research Team Plan of Dongguan City (201636000200039), and Shenzhen Satellites Based Equipment IoT Technology Engineering Laboratory ([2016]724). Thanks to my research team.

References

[1] Lang W , Jedermann R , Mrugala D , et al. The "Intelligent Container"—A Cognitive Sensor Network for Transport Management[J]. IEEE Sensors Journal, 2012, 11(3):688-698.
[2] Dittmer P , Veigt M , Scholz-Reiter B , et al. The Intelligent Container as a Part of the Internet of Things[C]. IEEE CYBER 2012. IEEE, 2012.
[3] Hidalgo Fort E , Garcia Oya J R , Munoz Chavero F , et al. Intelligent Containers Based on a Low-Power Sensor Network and a Non-Invasive Acquisition System for Management and Tracking of Goods[J]. IEEE Transactions on Intelligent Transportation Systems, 2017:1-6.
[4] Xuejiao Chen, Xiaofeng Hu, Yang Liu. Summary of Research on Big Data [J]. Journal of System Simulation, 2013(S1):142-146.
[5] Qu Z , Zhang G , Xie J . LEO Satellite Constellation for Internet of Things[J]. IEEE Access, 2017:1-1.
[6] Colombaroni C , Fusco G , Isaenko N , et al. Optimization of container operations at inland intermodal terminals[C].IEEE International Conference on Models & Technologies for Intelligent Transportation Systems. IEEE, 2017.

A scheme of D2D-based delay analysis and vehicle relay algorithm in VANET

Yinze Zhao* and Lingyun Jiang

College of Telecommunications & Information Engineering, Nanjing University of Posts and Telecommunications, Nanjing, China

*Email: zhaoyinzemvp@163.com

Abstract. The D2D-based VANET (Vehicular Ad-hoc Network) makes up for the defects of the traditional IEEE 802.11p network, and has the advantages of low latency and high throughput. In order to reduce the delay of message transmission between vehicles, a D2D-based hybrid V2I/V2V (Vehicle-to-Infrastructure/Vehicle-to-Vehicle) IoV (Internet of Vehicle) architecture was considered, which firstly clusters vehicles on the road according to motion consistency. The message is transmitted between the clusters, and in addition to being transmitted through the RSU (Road Side Unit), it can also be relayed by establishing a D2D link between the vehicles. Based on the SINR (Signal to Interference plus Noise Ratio) and the minimum capacity constraint, the optimal transmission path is selected by predicting the delay of each cellular link and D2D link to reduce the end-to-end delay. In addition, for accurate modeling and analysis, a three-step method is used to predict and compare end-to-end delays. The simulation proves that this relay algorithm significantly reduces the delay compared with other methods.

1. Introduction

Recently, the rapid increase in the number of vehicles has intensified the problems of traffic accidents and congestion., which led to the increasing attention of ITS (Intelligent Transportation Systems). Various ITS-based applications, including traffic monitoring, autopilot, etc., will lead to smarter, more convenient and more efficient modes of travel[1][2][3]. 3GPP (3rd Generation Partnership Project), METIS (Mobile and Wireless Communications Enablers for the Twenty-Twenty (2020) Information Society) and other alliances are working hard to coordinate efficient and reliable in-vehicle communication in next-generation networks[4][5]. During the driving process, some safety-related information such as CAM (Cooperative Awareness Messages) and DEMN (Decentralized Environmental Notification Messages) usually need to be transmitted in the surrounding vehicles in a cyclical or event-triggered manner. In this case, strict delay and reliability requirements are required. For example, in the METIS project, it is necessary to control the end-to-end delay to within 5 milliseconds and the reliability to 99.999% for 1600-byte packet transmission[6].

In VANET (Vehicular Ad-hoc Network), the RSU (Road Side Unit) can be used as a wireless access point to transmit vehicle information to the cloud or destination vehicle. The existing standardized protocol is the IEEE 802.11p protocol for the V2I (Vehicle-to-Infrastructure) and the V2V (Vehicle-to-Vehicle) communication links[7]. However, some recent studies[8][9][10] show that some problems exist in the IEEE 802.11p protocol due to the nature of its physical layer, such as unpredictable channel access delay, unguaranteed QoS (Quality of Service), and poor scalability. And the IEEE 802.11p networks can only provide intermittent and short-term V2I connections due to its limited transmission range and lack of roadside units. On the contrary, cellular networks cover a wide range and can flexibly control network resources flexibly. D2D communication, as one of the key technologies of 5G, allows users to directly communicate with cellular spectrum resources under the control of cellular systems[11]. Compared with IEEE 802.11p protocol transmission, D2D has wider communication range, higher transmission rate and better high-speed mobile performance[12]. Therefore, a cellular network supporting D2D will become a kind of Promising solutions in order to achieve efficient and reliable vehicle communication.

At present, there have been some studies on the introduction of D2D in VANET. In [13], Sergio and Tralli control the cellular receiver causing interference when the D2D transmitter re-uses cellular resources by limiting the transmit power of the D2D user. In [14], an interference-limited area control scheme is proposed to protect the D2D receiver from cellular interference, where D2D users are not allowed to share spectrum with cellular users located in interference-limited areas and the SINR (Signal to Interference plus Noise Ratio) is above a predetermined threshold at the D2D receiver. In addition, a three-step approach was proposed to design power control and spectrum allocation to maximize system throughput while minimizing the SINR of cellular and D2D links in [15]. Although these documents all consider interference and their optimization goals are mostly throughput, reliability, delay is a very important factor that has to be considered in many applications of IoV (Internet of Vehicle).

In [16][17], the delay performance of D2D communication is studied. And in [18], a new joint mode selection algorithm is proposed according to the vertex coloring method, aiming at achieving higher spectral efficiency and reducing communication delay. However, high mobility in the vehicle environment can cause wireless channels to change rapidly over time [19]. The traditional method for D2D communication under the assumption of full CSI (channel state information) is no longer applicable because it is difficult to track channel changes on such a short time scale.

Therefore, this paper considers a D2D-based IoV architecture, which first clusters vehicles according to motion consistency to increase the continuous connection time of the vehicle to reduce the probability of disconnection. And then, a D2D-based vehicle relay algorithm is proposed based on the SINR and the minimum capacity constraints to reduce the end-to-end delay. Finally, it is verified by simulation to compare the performance of the relay algorithm with the other.

The rest of the paper is arranged as follows. The second section introduces the system model and proposes a clustering algorithm based on motion consistency. The third section analyzes the delay and proposes a D2D-based vehicle relay algorithm. The simulation and analysis are carried out in four sections, and the fifth section is summarized in this paper.

2. System model

2.1. Problem description

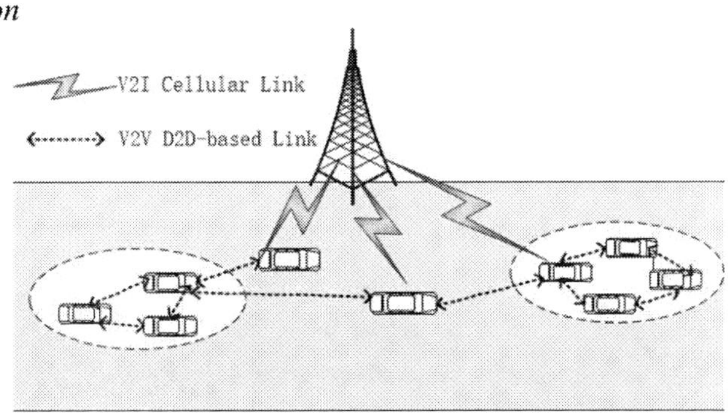

Figure 1. A D2D-based hybrid V2I/V2V VANET architecture.

We consider a VANET that supports D2D, as shown in figure 1, where is a user that communicates with the roadside base station via the V2I link, called a cellular user, and a user that performs V2V communication through the D2D link, called D2D user. We first clusters vehicles on the road to maintain a stable and long-lasting connection. The cluster head vehicles can be connected via V2I and V2V links. Due to the low use of uplink resources, D2D users use orthogonally allocated cellular users. The uplink spectrum not only improves spectrum utilization, but also makes interference at the base station easier to manage.

The data transmission between vehicles can be divided into cellular links and D2D links. Our goal is to select the optimal transmission path of the message to minimize the delay. It is assumed that the entire data transmission delay is D for a period of time, that is, the objective function is:

$$\min \sum_{v\in\Omega} D = \min \sum_{v\in\Omega} (D_{D2D} + D_{V2I}) \tag{1}$$

Since the D2D link is based on the cellular spectrum, it is necessary to consider interference between the cellular V2I link and the D2D pair. Let $h_{p,B}$ be the channel power gain between the pth cellular user and the base station, which can be expressed as:

$$h_{p,B} = g_{p,B}\beta_{p,B}AL_{p,B}^{-\gamma} \cong g_{p,B}\alpha_{p,B} \tag{2}$$

where $g_{p,B}$ represents the small-scale fast fading power component, and assumes that the exponential distribution is a unit mean, $L_{p,B}$ represents the distance from the pth cellular user to the RSU, A is the path loss constant, and γ is the decay exponent. $\beta_{p,B}$ is a log-normal shadow fading random variable with a standard deviation ξ.

We assume that large-scale fading components for the channel, such as the path loss of the link, are known to the RSU because they usually depend on the location of the user and vary less [20]. We can measure the parameters α_q and $\alpha_{p,q}$ at the receiver of the D2D user and send it to the RSU periodically, so that the RSU can calculate the link parameter $\alpha_{p,B}$ between the cellular user and the RSU and the link parameter $\alpha_{q,B}$ between the D2D user and the RSU.

Thereby, the SINR value γ_p^c of the pth cellular user and the SINR value γ_q^d of the qth D2D user received at the RSU can be obtained:

$$\gamma_p^c = \frac{P_p^c h_{p,B}}{\sigma^2+\sum_{q\in Q}\rho_{p,q}P_q^d h_{q,B}} \tag{3}$$

$$\gamma_q^d = \frac{P_k^d h_k}{\sigma^2+\sum_{p\in P}\rho_{p,q}P_p^c h_{p,q}} \tag{4}$$

where P_p^c and P_q^d represent the transmit power of the pth cellular user and the qth D2D user respectively, σ^2 represents the noise power, and $\rho_{p,q}$ represents the spectrum allocation indicator, that is, when the D2D user reuses the pth cellular user, $\rho_{p,q} = 1$, and other times $\rho_{p,q} = 0$.

The traversal capacity of the pth cellular user is:

$$C_p = \mathrm{E}\left[\log_2\left(1+\gamma_p^c\right)\right] \tag{5}$$

In addition to considering delay optimization, we also consider the QoS difference between the V2I link and the D2D link. Because the V2I link can connect the vehicle to the Internet, there may be transmission of large-capacity data, so the minimum capacity threshold is set for each cellular user.

In summary, the constraint is:

$$\mathrm{E}\left[\log_2\left(1+\gamma_p^c\right)\right] \geq \gamma_0^c, \forall p \in P \tag{6}$$

$$\Pr\left\{\gamma_q^d \leq \gamma_0^d\right\} \leq P_0, \forall q \in Q \tag{7}$$

$$0 \le P_p^c \le P_{max}^c, \forall p \in P \tag{8}$$

$$0 \le P_q^d \le P_{max}^d, \forall q \in Q \tag{9}$$

$$\sum_{p \in P} \rho_{p,q} \le 1, \rho_{p,q} \in \{0,1\}, \forall q \in Q \tag{10}$$

$$\sum_{q \in Q} \rho_{p,q} \le 1, \rho_{p,q} \in \{0,1\}, \forall p \in P \tag{11}$$

where γ_0^c represents the minimum capacity threshold of the cellular user, and γ_0^d represents the minimum SINR threshold required for the D2D user to establish a reliable D2D link. $\Pr\{\gamma_q^d \le \gamma_0^d\}$ gives the probability of interruption due to the inability to reach the SINR requirement, P_0 is the tolerable outage probability of the D2D link, and P_{max}^c and P_{max}^d are the maximum transmit power threshold of the cellular user and the D2D user respectively.

2.2. Vehicle clustering algorithm

In order to maintain the performance of long connections, we divide the vehicles in the RSU into multiple clusters. The clustering principle is as follows: 1) Ensure the distance between the member vehicles in a single cluster to be similar, that is, within a certain threshold range. 2) Guarantee the vehicles classified into the same cluster remain for a period of time. Firstly, in order to satisfy the first principle, the initial division can be initially based on the proximity range. Secondly, in order to satisfy the second principle, the vehicles that meet the requirements can be divided into new clusters according to the speed of the member vehicles, and the vehicles that do not meet the requirements are deleted from the cluster.

Suppose there are N cars on the road, we divide them into M clusters, and ensure that the driving directions of the vehicles in the same cluster are the same, assuming that the number of cars contained in the kth cluster isN_k, where $k = 1,2,...,M$, the speed of the ith car in the kth cluster is $v_{k,i}$, where $i = 1,2,...,N_k$. So the average speed of the kth cluster is:

$$\overline{v_k} = \sum_{i=1}^{N_k} v_{k,i}\, , i = 1,2,...,N_k \tag{12}$$

Algorithm 1: Vehicle clustering algorithm

Input: Number of clusters M, Vehicle distance $d_{k,i}$
Output: Clusters Ω_k, Cluster head $\Omega_{k,head}$
Initialize: Divide N vehicles into M clusters, each cluster is Ω_k, according to $d_{k,i} \le L_k, k = 1,2,\cdots,M, i = 1,2,\cdots,N_k$

Calculate the average speed of the vehicle in the kth cluster $\overline{v_k}$, $k = 1,2,\cdots,M$
for k=*1* to M do
    for i=*1* to N_k do
        if $\overline{v_k} \ge v_{k,i}t + L_k + d_{k,i}$ and $\overline{v_k} \le v_{k,i}t + L_k - d_{k,i}$ then
            if $d_{k,i} > d_{k,max}$ then $\Omega_{k,head} \leftarrow \Omega_{k,i}$, $d_{k,max} \leftarrow d_{k,i}$
            else if $d_{k,i} < d_{k,min}$ then $d_{k,min} \leftarrow d_{k,i}$
            end if
        end if
        else $\Omega_k \leftarrow \Omega_k - \Omega_{k,i}$
        end if
    end for
end for

As shown in algorithm 1, the vehicles with similar geographical locations are first classified into the same initial cluster, and then the average speed of the vehicles in each initial cluster is calculated. We set a certain time threshold to determine whether the vehicles in the initial cluster will run at a certain time. Outside the range of the cluster, the vehicles still in the cluster classify them into new clusters and judge whether they are cluster heads, thereby obtaining new clusters and cluster heads of each cluster.

3. Delay analysis and relay algorithm

3.1. End-to-end delay analysis

In this system, the source vehicle can use a D2D-based V2I/V2V link to transmit messages to the forward or backward target vehicle and we can select the best transmission mode by expecting the delay of each path. For accurate analysis, we consider the different data transfer phases and their associated delays. Since our clustering algorithm is based on motion consistency, we can assume that the data information in a single cluster is shared, and the data transmission delay in the cluster. The control is within a certain range, and the data is transmitted between the clusters through the cluster head.

If the source cluster and the target cluster are in the same RSU range, only the D2D connection needs to be made between the cluster and the cluster. Since the traveling directions of the vehicles in the same cluster are the same, we can obtain the probability density function of cluster size[21]:

$$f_c(x) = \frac{x^{\lambda-1}e^{-\frac{x}{\theta}}}{\theta^{\lambda}\Gamma(\lambda)}, x > 0 \tag{13}$$

where $\lambda = \frac{\mathrm{E}\{C_k\}^2}{\mathrm{E}\{{C_k}^2\}-\mathrm{E}\{C_k\}^2}$, $\theta = \frac{\mathrm{E}\{{C_k}^2\}-\mathrm{E}\{C_k\}^2}{\mathrm{E}\{C_k\}}$, and C_k indicates the range of the vehicle cluster.

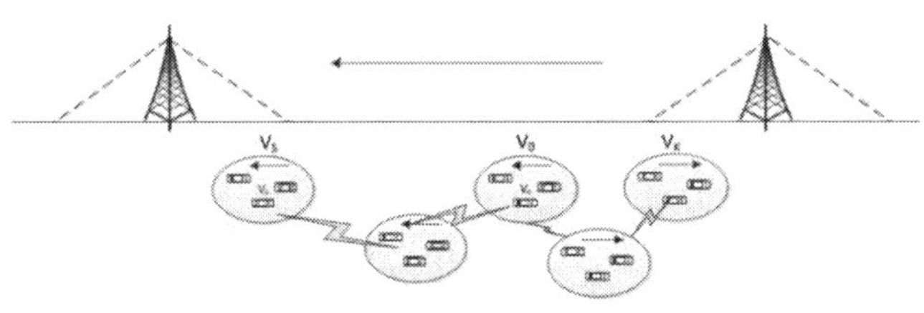

Figure 2. Source vehicle sending messages to the nearest forward RSU.

For accurate modeling, the message transmission can be divided into three phases: the source vehicle to the nearest forward RSU phase, the intermediate vehicle or RSU relay phase, and the transmission phase within the RSU of the message to the destination vehicle. In the first phase, the data packet is from the source vehicle to the nearest forward RSU stage, as shown in figure 2. This phase can transmit messages in two ways, one is transmitted from the source vehicle to the RSU by way of forward vehicle relay. The vehicle in the range is relayed to the backward RSU through the backward vehicle and then transmitted through the RSU. Assuming that the source vehicle position is V_0, the vehicle cluster set to the nearest forward RSU is $\{1,2,\cdots,S\}$, and the time from the *s-1* cluster to the *s* cluster is t_s. There is SAC delay and wireless communication delay and we can discuss it in two cases. When the distance between communication vehicles is less than D2D communication range, the D2D link can be directly established to communicate between vehicles. In this case, there is no SAC delay, so $t_s = d_s/v_d$, where d_s represents the *s* cluster to the *s+1* cluster and v_d represents the D2D communication rate. In addition, when there is no other vehicle in the communication range of the vehicle carrying the message, it needs to store and forward the message data, and the SAC delay generated at this time is $(d_s - L)/v_{s,head}$, where L represents the communication distance of D2D and $v_{s,head}$ represents the speed of the cluster head vehicle of the *s*th vehicle cluster. In summary, t_s can be expressed as:

$$t_s = \begin{cases} \frac{d_s}{v_d}, d_s \le L, s \in S \\ \frac{d_s - L}{v_{s,head}} + \frac{L}{v_d}, d_s > L,\ s \in S \end{cases} \quad (14)$$

Let V_0 be the location of the source vehicle, C_0 be the location of the cluster head of the source cluster, and thus the transmission delay in this cluster can be obtained as $(V_0 - C_0)/v_d$. Similarly, Let V_S be the position of the vehicle connected to the forward RSU, C_S be the cluster head position of the cluster connected with the RSU, and the transmission delay in this cluster is $(C_S - V_S)/v_d$, so the total delay in this mode can be expressed as:

$$T_{d,beg} = \sum_{s \epsilon S} t_s + \frac{V_0 - C_0}{v_d} + \sum_{s \epsilon S \setminus \{0,S\}} \frac{C_S}{v_d} + \frac{C_S - V_S}{v_d} \quad (15)$$

Next, we discuss the delay in the backward direction transmission, assuming that the vehicle cluster set of the source vehicle to its nearest backward RSU is $\{1,2,\cdots,K\}$, and V_K is the vehicle position connected to the backward RSU, C_S is the cluster head position of the cluster where the vehicle is located, d/v_c is the transmission delay between RSUs, where d represents the distance between RSUs, and v_c represents the speed of cellular transmission, so the transmission delay between the reverse vehicle and the RSU can be obtained for:

$$T_{c,beg} = \sum_{k \epsilon K} t_k + \sum_{k \epsilon K} \frac{C_k}{v_d} + \frac{C_K - V_K}{v_d} + \frac{d}{v_c} \quad (16)$$

In summary, the expected delay for this phase is

$$\mathrm{E}\{D_{h,beg}\} = \int_0^a \left(T_s + \frac{V_0 - x}{v_d}\right) f_c(x)dx + \int_a^{+\infty} T_{c,beg} f_c(x)dx \quad (17)$$

where $T_s + \frac{V_0 - x}{v_d}$ represents the delay experienced by the vehicle at the initial position at x through the D2D vehicle relay, and a represents the distance threshold at which the relay transmission can be used, which is obtained by the above analysis: $T_s = \sum_{s \epsilon S} t_s + \sum_{s \epsilon S \setminus \{0,S\}} \frac{C_S}{v_d} + \frac{C_S - V_S}{v_d}$, $a = V_0 + \sum_{s \epsilon S \setminus \{0\}} C_S - V_S + \sum_{s \epsilon S} t_s v_d$.

The second stage is analyzed below. In the message transmission process, it is necessary to select a suitable relay vehicle or choose to transmit through the base station. We select the delay path with the smallest delay by predicting the delay of each path.

Assuming that the V_x position is a certain vehicle position before the first RSU and has not yet connected with the RSU, the V_S position is the position of the carrying message vehicle connected with the last RSU before reaching the destination vehicle. When the message is transmitted from the V_x position to the V_S position through the vehicle relay, the delay of the message transmitted in the carrying vehicle cluster is $(V_x - C_x)/v_d$, and the delay of the last carried message in the cluster is $(C_S - V_S)/v_d$. Whereby the total delay at this stage can be expressed as:

$$T_{d,rel} = \sum_{s \epsilon S} t_s + \frac{V_x - C_x}{v_d} + \sum_{s \epsilon S \setminus \{0,S\}} \frac{C_s}{v_d} + \frac{C_S - V_S}{v_d} \quad (18)$$

The delay of the message transmission between the RSU is:

$$T_{c,rel} = d/v_c \quad (19)$$

where d represents the distance between RSUs, and v_c represents the transmission speed of the cell. Similarly, the latency expectation experienced at this stage is:

$$\mathrm{E}\{D_{h,rel}\} = \int_0^b \left(T_s + \frac{V_x - x}{v_d}\right) f_c(x)dx + \int_b^{+\infty} T_{c,rel} f_c(x)dx \quad (20)$$

where $b = V_x + \sum_{s\epsilon S\setminus\{0\}} C_s - V_S + \sum_{s\epsilon S} t_s v_d$.

In the final stage, when the message is transmitted to the RSU where the target vehicle is located, the relay vehicle needs to choose whether to continue to select the vehicle relay forward, or to transmit to the RSU, and then transmit it to the destination vehicle by the RSU.

The transmission situation in the destination vehicle base station is exactly the opposite of the source vehicle transmission process, and it is easy to get:

$$\mathrm{E}\{D_{h,end}\} = \int_0^c \left(T_s + \frac{V_y - x}{v_d}\right) f_c(x)dx + \int_c^{+\infty} T_{c,end} f_c(x)dx \tag{21}$$

where $c = V_y + \sum_{s\epsilon S\setminus\{0\}} C_s - V_S + \sum_{s\epsilon S} t_s v_d$, $T_{c,end} = \sum_{k\epsilon K} t_k + \sum_{k\epsilon K} \frac{C_k}{v_d} + \frac{C_K - V_K}{v_d} + \frac{d}{v_c}$.

As summarized in the review, the total end-to-end delay can be obtained as follows:

$$\mathrm{E}\{D_h\} = \mathrm{E}\{D_{h,beg}\} + \mathrm{E}\{D_{h,rel}\} + \mathrm{E}\{D_{h,end}\} \tag{22}$$

3.2. D2D-based vehicle relay algorithm

Through the above delay analysis, we propose algorithm 2 to get the optimal transmission path of the message according to the source vehicle, the target vehicle and the information of the intermediate vehicle. According to the previous clustering algorithm, the vehicles on the road are divided into corresponding clusters, and then the message transmission direction and the vehicle traveling direction are considered to be consistent. If they are consistent, the D2D link and the cellular link are respectively calculated according to the formula (15) (16). Considering that the message transmission distance may be long, it is necessary to pass through a plurality of RSUs, and the above method is used to select the best relay vehicles and RSUs.

Algorithm 2: Vehicle relay algorithm

Input: Source vehicle position V_0, target vehicle position V_e
Output: Message transmission path, including relay vehicle collection, RSU collection

Cluster vehicles
if $V_0 < V_e$ then
  Calculate $T_{d,beg}$ and $T_{c,beg}$ according to (15)(16)
  if $P_p^c \le P_{max}^c$, $P_p^d \le P_{max}^d$, $T_{d,beg} < T_{c,beg}$ then
    Add relay vehicle collection, Calculate $T_{d,rel}$, $T_{c,rel}$
    while $T_{d,rel} < T_{c,rel}$ do
      Add relay vehicle collection, Calculate new $T_{d,rel}$, $T_{c,rel}$
    end while
  end if
  while $V_e > V_y + d$ do
    Add RSU collection
  end while
  Calculate $T_{d,end}$, $T_{c,end}$
  if $T_{d,end} < T_{c,end}$ then
    Add relay vehicle collection
  end if
end if

4. Simulation results and analysis

This section simulates the performance of the clustering algorithm and the vehicle relay algorithm proposed in the previous two sections. Based on the Veins vehicle networking simulation platform, the simulation environment is set in the traffic simulator SUMO, and then we develop C++ program to simulate the clustering algorithm and the relay algorithm respectively in the network simulator OMNeT++. We assume that the simulation scene is a 20km highway. This section of the highway starts with a total of 200 vehicles. The initial position of each vehicle on this road is obtained by a random function. In order to match the actual situation, it is assumed that each vehicle is in the process of driving. The speed range is [10~35] m/s, and the initial velocity is randomly generated between [10~35]. During the running of the vehicle, the vehicle is changed every 10 seconds to shift at a certain acceleration within the above speed range, assuming an acceleration of $2m/s^2$. Each time we randomly select 20% of the vehicles in these vehicles to generate data packets, and at the same time randomly select the destination vehicles that the data packets need to transmit, these data packets will be transmitted to the destination vehicle through RSU or vehicle relay. In order to compare performance fairly, keep the algorithm 100 times and take the average value each time keeping other scenes consistent. table 1 is the specific simulation parameter settings[22].

Table 1. Simulation parameters.

Parameter	Value
Minimum capacity of D2D users γ_0^c	0.5bps/Hz
SINR threshold of D2D users γ_0^d	5dB
Maximum transmit power of Cellular users P_{max}^c	17dBm
Maximum transmit power of D2D users P_{max}^d	23dBm
Noise power σ^2	-114dBm
RSU distance L	5km
Cellular range d_c	500m
RSU antenna height	25m
RSU antenna gain	8dBi
RSU receiver noise figure	5dB
Vehicle receiver noise figure	9dB

Firstly, in order to analyze the performance of the dynamic clustering algorithm in this paper, we compare it with the location-based clustering algorithm from the average continuous communication time and delay of the vehicles in the cluster. As shown in figure 3, the abscissa is the vehicle maximum speed, the ordinate of figure 3 (a) is the average continuous communication time of the vehicles in the cluster, and the ordinate of figure 3 (b) is the average delay. It can be seen that there is no significant difference between the performance of the geolocation clustering algorithm and the dynamic clustering algorithm when the vehicle is slow. However, as the speed increases, the advantages of the dynamic clustering algorithm become more and more significant, mainly because the geolocation clustering algorithm does not consider the moving speed of the vehicle, and can only ensure that the vehicles in the cluster are in a relatively constant position in a short time. The dynamic clustering algorithm based on motion consistency considers the moving speed of the vehicle and predicts the trajectory of the vehicle during a period of time, thus ensuring the connection of the vehicles in the cluster for a long time. However, when the maximum speed of the vehicle is large, the speed of the vehicle changes significantly, which causes the performance of our algorithm to degrade. Finally, it is no different from the location-based clustering algorithm.

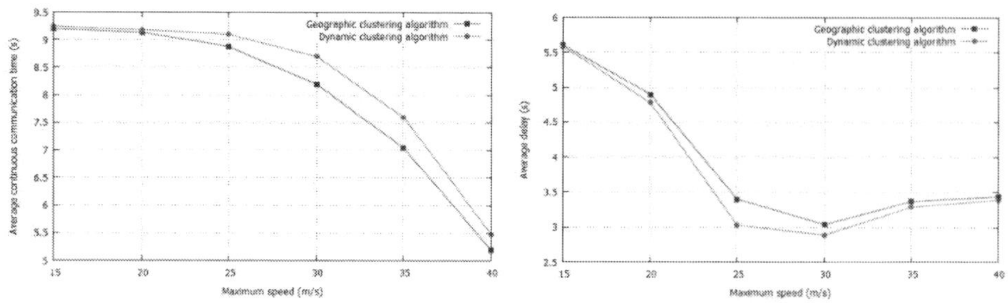

Figure 3. (a) Average continuous communication time. (b) Average delay of successful packet transmission.

Next, we compare the D2D relay algorithm of this paper with the RSU transmission and random relay algorithm. As shown in figure 4, we use the delay as an indicator to compare the maximum speed of different vehicles and the number of different vehicles. As shown in Fig. 4(a), when the vehicle speed changes little, the SAC delay is large, so the delay of using the RSU transmission is lower than other algorithms. However, as the vehicle speed increases, the number of relay vehicles that can be selected for message transmission increases. The D2D relay algorithm in this paper will choose the optimal relay vehicle, so the performance will be better than the random relay selection and RSU transmission. As shown in figure 4(b), the performance of the RSU transmission is the worst, the random relay performance is slightly better than that of the base station transmission, and the D2D relay algorithm in this paper is optimal. And as the number of vehicles increases, the performance of the RSU transmission gradually deteriorates. The performance of several relay algorithms gradually becomes better, and then tends to be stable, mainly because of the increasing number of vehicles lead to resource shortage and performance degradation. In contrast, the relay strategy of this paper is better because of the SAC delay to be reduced or even disappeared due to the increase of vehicles. However, when the number of vehicles increases to a certain amount, due to the SINR interference constraint, it is necessary to select a relay vehicle with small interference, so the performance will not increase all the time, and will reach a relatively stable state in the later stage.

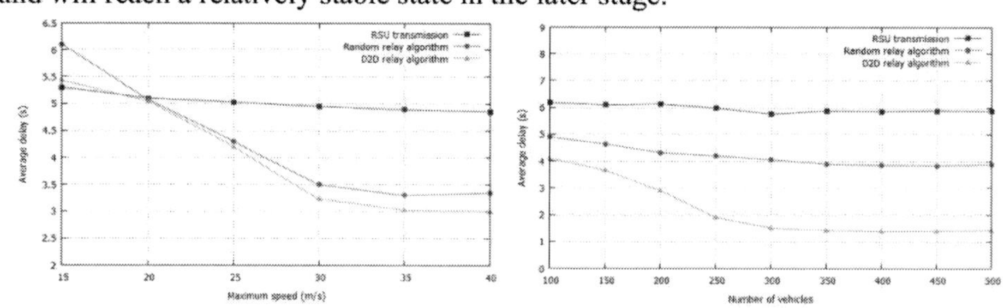

Figure 4. (a) Average delay at different maximum speeds. (b) Average delay for different vehicle numbers.

5. Conclusion

In this paper, a hybrid V2I/V2V vehicle networking system architecture based on D2D is considered. Firstly, the vehicles on the road are clustered according to the motion consistency and based on this, a D2D-based vehicle optimal relay algorithm is proposed. The simulation proves that the relay algorithm in this paper significantly reduces the end-to-end delay in the message transmission process, especially in a certain range, the more the number of vehicles, the better the performance of this relay algorithm, the lower the delay. And when the number of vehicle reaches a certain amount, the delay will tend to be stable.

References

[1] D. Jiang, L. Huo, Z. Lv, H. Song and W. Qin, "A Joint Multi-Criteria Utility-Based Network Selection Approach for Vehicle-to-Infrastructure Networking," in IEEE Transactions on Intelligent Transportation Systems, vol. **19**, no. 10, pp. 3305-319 (2018).

[2] M. M. Rana, "Attack Resilient Wireless Sensor Networks for Smart Electric Vehicles," in IEEE Sensors Letters, vol. **1**, no. 2, pp. 1-4 (2017).

[3] Y. Xia, X. Qin, B. Liu and P. Zhang, "A greedy traffic light and queue aware routing protocol for urban VANETs," in China Communications, vol. **15**, no. 7, pp. 77-87 (2018).

[4] A. Soua and S. Tohme, "Multi-level SDN with vehicles as fog computing infrastructures: A new integrated architecture for 5G-VANETs," 2018 21st Conference on Innovation in Clouds, Internet and Networks and Workshops (ICIN), Paris, pp. 1-8 (2018).

[5] G. Luo et al., "Cooperative vehicular content distribution in edge computing assisted 5G-VANET," in China Communications, vol. **15**, no. 7, pp. 1-17 (2018).

[6] L. Liang, G. Y. Li and W. Xu, "Resource Allocation for D2D-Enabled Vehicular Communications," in IEEE Transactions on Communications, vol. **65**, no. 7, pp. 3186-197 (2017).

[7] Y. Yang, D. Fei and S. Dang, "Inter-vehicle cooperation channel estimation for IEEE 802.11p V2I communications," in Journal of Communications and Networks, vol. **19**, no. 3, pp. 227-38 (2017).

[8] A. A. Almohammedi, N. K. Noordin, A. Sali, F. Hashim and M. Balfaqih, "An Adaptive Multi-Channel Assignment and Coordination Scheme for IEEE 802.11P/1609.4 in Vehicular Ad-Hoc Networks," in IEEE Access, vol. **6**, pp. 2781-802 (2018).

[9] R. Tomar, M. Prateek and H. G. Sastry, "Analysis of beaconing performance in IEEE 802.11p on vehicular ad-hoc environment," 2017 4th IEEE Uttar Pradesh Section International Conference on Electrical, Computer and Electronics (UPCON), Mathura, pp. 692-96 (2017).

[10] B. Bloessl, M. Segata, C. Sommer and F. Dressler, "Performance Assessment of IEEE 802.11p with an Open Source SDR-Based Prototype," in IEEE Transactions on Mobile Computing, vol. **17**, no. 5, pp. 1162-175 (2018).

[11] R. Wang, Jia Liu, G. Zhang, Shuanghong Huang and Ming Yuan, "Energy efficient power allocation for relay-aided D2D communications in 5G networks," in China Communications, vol. **14**, no. 6, pp. 54-64 (2017).

[12] L. Shi, L. Zhao, K. Liang and H. Chen, "Wireless Energy Transfer Enabled D2D in Underlaying Cellular Networks," in IEEE Transactions on Vehicular Technology, vol. **67**, no. 2, pp. 1845-849 (2018).

[13] S. Cicalò and V. Tralli, "QoS-Aware Admission Control and Resource Allocation for D2D Communications Underlaying Cellular Networks," in IEEE Transactions on Wireless Communications, vol. **17**, no. 8, pp. 5256-269 (2018).

[14] X. Cao, L. Liu, Y. Cheng, L. X. Cai and C. Sun, "On Optimal Device-to-Device Resource Allocation for Minimizing End-to-End Delay in VANETs," in IEEE Transactions on Vehicular Technology, vol. **65**, no. 10, pp. 7905-916 (2016).

[15] D. Feng, L. Lu, Y. Yuan-Wu, G. Y. Li, G. Feng and S. Li, "Device-to-Device Communications Underlaying Cellular Networks," in IEEE Transactions on Communications, vol. **61**, no. 8, pp. 3541-551 (2013).

[16] M. Salehi, A. Mohammadi and M. Haenggi, "Analysis of D2D Underlaid Cellular Networks: SIR Meta Distribution and Mean Local Delay," in IEEE Transactions on Communications, vol. **65**, no. 7, pp. 2904-916 (2017).

[17] W. Xu and Y. Wang, "Heterogeneous Statistical-Delay QoS and Security Provisioning for D2D Underlay Cellular Networks," 2018 IEEE 87th Vehicular Technology Conference (VTC Spring), Porto, pp. 1-5 (2018).

[18] Y. Li, M. C. Gursoy, S. Velipasalar and J. Tang, "Joint Mode Selection and Resource Allocation for D2D Communications via Vertex Coloring," GLOBECOM 2017 - 2017 IEEE Global Communications Conference, Singapore, pp. 1-6 (2017).

[19] H. Chour, Y. Nasser, H. Artail, A. Kachouh and A. Al-Dubai, "VANET Aided D2D Discovery: Delay Analysis and Performance," in IEEE Transactions on Vehicular Technology, vol. **66**, no. 9, pp. 8059-071 (2017).
[20] W. Sun, E. G. Ström, F. Brännström, K. C. Sou and Y. Sui, "Radio Resource Management for D2D-Based V2V Communication," in IEEE Transactions on Vehicular Technology, vol. **65**, no. 8, pp. 6636-650 (2016).
[21] Y. Zhuang, J. Pan, Y. Luo and L. Cai, "Time and Location-Critical Emergency Message Dissemination for Vehicular Ad-Hoc Networks," in IEEE Journal on Selected Areas in Communications, vol. **29**, no. 1, pp. 187-96 (2011).
[22] L. Liang, G. Y. Li and W. Xu, "Resource Allocation for D2D-Enabled Vehicular Communications," IEEE Transactions on Communications, vol. **65**, no. 7, pp. 3186-197 (2017).

IOP Conf. Series: Materials Science and Engineering **715** (2020) 012031 doi:10.1088/1757-899X/715/1/012031

An improved authentication scheme for Internet of things

Zhongliang Xie[1,2] **and Lingyun Jiang**[1,2,*]

[1]Institute of Internet of Things, Nanjing University of Posts and Telecommunications, China

[2]Institute of Communication Technology, Nanjing University of Posts and Telecommunications, China

*Email: jiangly@njupt.edu.cn

Abstract. Designing a secure authentication scheme is of great importance to the practical applications of the Internet of Things (IoT). In 2017, Wang et al. proposed an improved mutual authentication scheme for the embedded devices and a server over Kalra et al.'s and Chang et al.'s. And they provided a formal proof for their scheme. However, their scheme is still insecure and cannot resist against key compromise impersonation (KCI) attack. In the paper, we first demonstrate the shortcoming of Wang et al.'s scheme, and then we carry out an improvement over Wang et al.'s scheme to address the revealed issue by using an auxiliary server. Keywords. Authentication, Cryptanalysis, Elliptic curve cryptography, Internet of Things

1. Introduction

The Internet of Things (IoT) is considered to be the technological and economic wave of the global information industry after the Internet [1]. The use of IoT applications [2-5] brings many conveniences to people. In these IoT applications, the things can collect sensitive information from the environment, which is then used to analyze and make decisions [6]. However, the security issues of the IoT are more challenging than that of the traditional Internet [7]. Most of the IoT devices are resource-constrained, and it is difficult to implement complex encryption algorithms on them. If the security issues cannot be addressed, it will hinder the massive use of IoT applications [8]. Therefore, it is important to design a secure and lightweight authentication scheme for the IoT.

Traditional Public Key Cryptosystems (PKC) have to deal with the distribution and management of certificates which are too complex to be utilized for the IoT devices [9]. In 1984, Shamir [10] first introduced the idea of ID-based cryptosystems (IBC). IBC avoid time-consuming and complex certificate mechanisms by using the entity's ID or email address as its public key, which are suitable for IoT scenarios. An ID-based cryptosystem often uses elliptic curve cryptography (ECC [11]) for the reason that ECC can obtain the same security strength with a smaller key size [12]. Truong et al. [13] proposed an ID-based authentication for mobiles on ECC. Liao et al. [14] proposed a secure ECC-based RFID authentication scheme. The scheme is integrated with ID-verifier transfer protocol. Shah et al. [15] used a variant of the three-way authentication mechanism to authenticate the IoT device and the server. They used the secure vault to store keys and change the secure vault according to the data exchanged between the server and the device after every session. Yao et al. [16] proposed a lightweight attribute-based encryption scheme for the IoT on ECC. Aman et al. [17] proposed a mutual authentication scheme for the IoT systems using physical unclonable functions. Hao et al. [18]

proposed a collaborative PHY(physical-layer)-Aided technique for end-to-end IoT devices. They used the devices' unique PHY features (carrier frequency offset (CFO), in-phase/quadrature-phase imbalance (IQI)) and the collaboration of several intermediate nodes to achieve an elevated resistance to computation-based attacks.

In 2015, Kalra et al. [19] proposed a lightweight authentication scheme by using elliptic curve cryptography. In their scheme, they implement mutual authentication between the embedded devices and the cloud server. However, Chang et al. [20] found that Kalra et al.'s scheme cannot achieve mutual authentication and the server cannot calculate the session key, so Chang et al. proposed an improved scheme to address these problems. In 2017, Wang et al. [21] found that Chang et al.'s improved scheme is still insecure. The adversary can impersonate the server to communicate with the device and can further obtain session keys which may bring huge damage (Kalra et al.'s scheme cannot withstand this attack as well). Therefore, Wang et al. proposed a modified scheme to solve the problem and gave a formal proof for their improved scheme. However, Wang et al.'s scheme is still not safe. In this paper, we demonstrate that Wang et al.'s scheme cannot resist against key compromise impersonation (KCI) attack [22]. The adversary can impersonate the legitimate device to communicate with the server. And we offer an improved scheme to address the revealed issue without losing the security features and advantages of the Wang et al.'s scheme.

The paper is organized as follows. Section II reviews Wang et al.'s scheme. Section III shows the insecurity of Wang et al.'s scheme. Section IV is our improved scheme and the analysis of it. Section V compares our scheme with relevant schemes. Section VI presents the conclusion.

2. Review Wang et al.'s scheme

In this section, we review the scheme proposed by Wang et al. Some intuitive notations are listed in table 1.

Table 1. Notations and descriptions.

D_i	i th device
S	the server
$\mathcal{A}$	the adversary
ID_i	identity of device D_i
$H(.)$	A one-way hash function maps any string to a l_H -bit long string. (i.e., $H:\{1\}^* \to \{1\}^{l_H}$)
$h(.)$	A one-way hash function maps any string to a l_h -bit long string. (i.e., $h:\{1\}^* \to \{1\}^{l_h}$)
$\mathcal{O}$	point at infinity
G	generator point of order n . (i.e., $n \cdot G = \mathcal{O}$)
EXP_Time	expiration time for a specific device
X	secret key of the server
SK	session key

2.1. System initialization phase

Before the system begins, the server S selects a secret key X with l_h -bit long, a one-way hash function $H(.)$, an elliptic curve E and a generator G on E . And then S publishes $(H(.), E, G)$ as system parameters and keeps the key X secret.

2.2. Registration phase

In this phase, the embedded device D_i is registered with the server S . D_i sends its ID_i to S . Then S calculates

$$CK = h\left(R_i \parallel X \parallel EXP\_Time \parallel ID_i\right) \tag{1}$$

$$CK' = CK \times G \tag{2}$$

$$T_i = R_i \oplus H\left(X\right) \tag{3}$$

$$A_i = h\left(R_i \oplus H\left(X\right) \parallel CK'\right) \tag{4}$$

$$A_i' = A_i \times G \tag{5}$$

where R_i is a random l_H -bit string generated by S . S stores $\{A_i', T_i, ID_i, EXP\_Time\}$ in its database and sends CK' back to the device via a secure channel. Upon receiving CK' from S , the device D_i securely stores ID_i and CK' .

2.3. Login and authentication phase

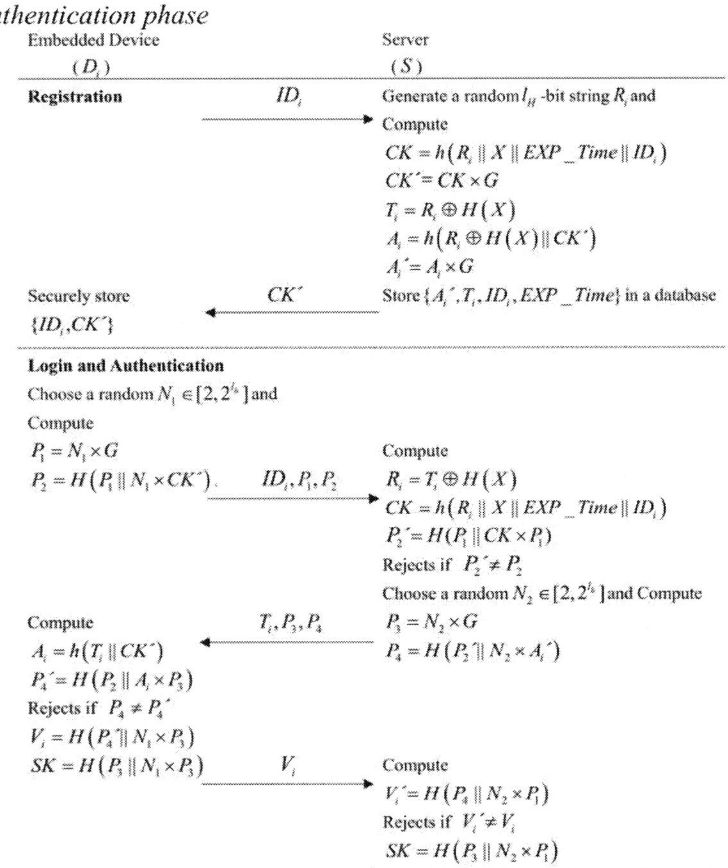

Figure 1. The scheme of Wang et al.

When D_i wants to login to S , it chooses a random number N_1 with l_H -bit long. It then computes

$$P_1 = N_1 \times G \tag{6}$$

$$P_2 = H\left(P_1 \parallel N_1 \times CK'\right). \tag{7}$$

D_i sends P_1 and P_2 to S along with its ID ID_i . S searches its database to get the record of D_i by exploiting ID_i . S reconstructs CK with the help of T_i , EXP_Time , X and calculates $P_2' = H(P_1 \parallel CK \times P_1)$, and rejects if P_2' is unequal to the received P_2 . Then, S chooses a random number N_2 with l_H -bit long and calculates

$$P_3 = N_2 \times G \tag{8}$$

$$P_4 = H\left(P_2' \| N_2 \times A_i'\right) \tag{9}$$

$\{T_i, P_3, P_4\}$ are delivered to D_i. D_i computes $A_i = h(T_i \| CK')$ and $P_4' = H(P_2 \| A_i \times P_3)$ and rejects if $P_4 \neq P_4'$. Then D_i computes

$$V_i = H\left(P_4' \| N_1 \times P_3\right) \tag{10}$$

$$SK = H\left(P_3 \| N_1 \times P_3\right) \tag{11}$$

V_i is transmitted to S. S computes $V_i' = H(P_4 \| N_2 \times P_1)$ and compares V_i' with the received V_i. If $V_i' = V_i$, S accepts the device D_i's login request and keeps the session key $SK = H(P_3 \| N_2 \times P_1)$, otherwise rejects. The specific scheme of Wang et al. is shown in figure 1.

3. Analysis of the Wang et al.'s scheme

In this section, we demonstrate Wang et al.'s scheme cannot withstand a realistic attack, namely key compromise impersonation (KCI) attack. In the case of KCI attack [22], if the adversary $\mathcal{A}$ gets the key of a communicating party, it can impersonate the compromised party to communicate with the other parties, or impersonates the other parties to communicate with the compromised party. A scheme that free from this form of "reverse impersonation" is said to resist against KCI attack, that is, in the scheme, the adversary $\mathcal{A}$ can only use the key to impersonate the compromised party.

Here, we show the damages of KCI attack. If the adversary $\mathcal{A}$ impersonates the device D to communicate with the server, it may send some false information to the server, causing the server to fail to make the right decisions (e.g., temperature alarms, etc.). At the same time, according to the access control policy, the data (such as some sensitive medical data) stored on the server by the device D is only accessible by the device D itself, and the server has no access rights, which is very common in practical applications. If the server is compromised, the adversary $\mathcal{A}$ may not be able to get the device's data, but if the adversary $\mathcal{A}$ impersonates the device D to communicate with the server as well as gets the session key, the adversary $\mathcal{A}$ will get the data that the device D stored on the server.

Under the scheme of Wang et al., if the adversary $\mathcal{A}$ breaks into the server (e.g., using zero-day attacks like Heartbleed [23, 24]), and gets the server's secret key X and stored database information $\{A_i', T_i, ID_i, EXP\_Time\}$. The adversary $\mathcal{A}$ can impersonate D_i as follows:

Step 1: The adversary $\mathcal{A}$ chooses a random $N_1 \in [2, 2^{l_h}]$ and computes $P_1 = N_1 \times G$, $P_2 = H(P_1 \| N_1 \times CK')$, where $CK' = h(T_i \oplus H(X) \| X \| EXP\_Time \| ID_i) \times G$. Then $\mathcal{A}$ sends $\{ID_i, P_1, P_2\}$ to the server.

Step 2: Upon receiving $\{ID_i, P_1, P_2\}$ from $\mathcal{A}$, the server will find no abnormalities and sends $\{T_i, P_3, P_4\}$ back to $\mathcal{A}$.

Step 3: $\mathcal{A}$ then computes $A_i = h(T_i \| CK')$, $P_4' = H(P_2 \| A_i \times P_3)$, $V_i = H(P_4' \| N_1 \times P_3)$, $SK = H(P_3 \| N_1 \times P_3)$ and sends V_i to the server.

Step 4: After receiving V_i, the server computes V_i' and will find V_i' equal to the received V_i. So the server will accept the adversary $\mathcal{A}$'s login request and computes the session key $SK = H(P_3 \| N_2 \times P_1)$, which is equal to the session key that $\mathcal{A}$ holds.

Therefore, Wang et al.'s scheme is insecure. In the KCI attack scenario, it assumes that the server was compromised, but it was discovered a long time later. This exists in real-world. For example, JPMorgan Chase's server was found to be attacked a few months later after it was compromised [25].

4. The proposed scheme

As stated in [26], when an authentication server also acts as a registry, it is impossible for the authentication server to withstand KCI attack. Inspired by the study of Camenisch et al. [27] and Wang et al. [12], we use one more server (i.e., an auxiliary server) to solve the insecure problem of Wang et al.'s scheme.

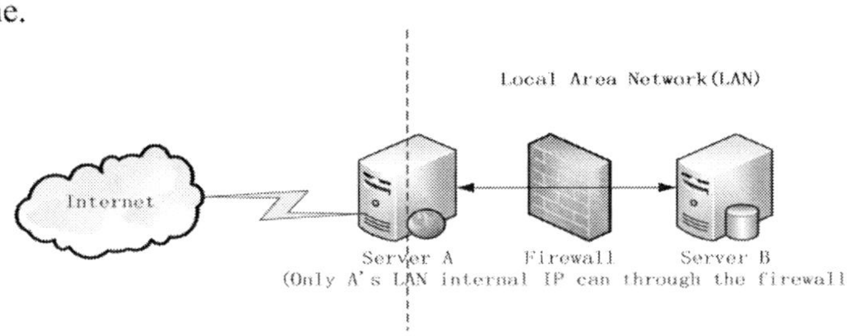

Figure 2. A secure way to deploy servers.

For security reasons, we can deploy two servers (e.g., A and B) on the same Local Area Network (LAN). Therefore, both A and B have an IP address on the LAN. In addition, one (assumed A) is also deployed on the Internet, so A also has a public IP address, and others can access A through the Internet. A firewall inside the LAN can be used to implement an access policy that only A's LAN internal IP is allowed to access B. In this way, it is impossible to access B through the Internet. As a result, it is a good choice to store important or sensitive data on B. Details can be seen in figure 2.

4.1. Details of the improvement

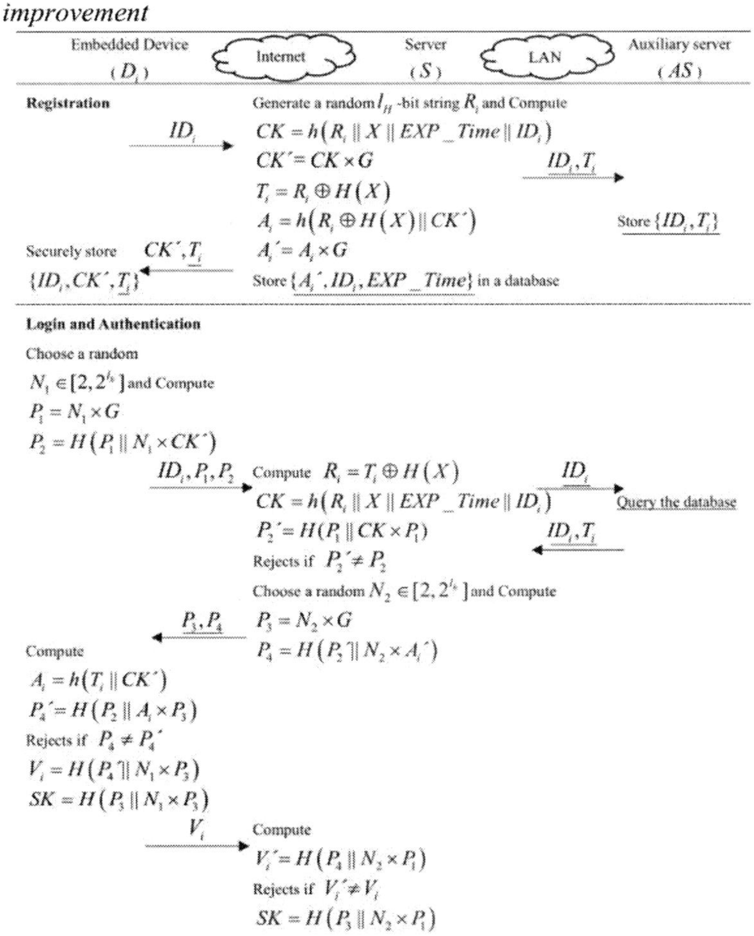

Figure 3. Our improved scheme. The bottom of the place we modified was marked with a solid line.

IOP Conf. Series: Materials Science and Engineering **715** (2020) 012031 doi:10.1088/1757-899X/715/1/012031

Figure 3 shows our improved scheme, the bottom of the place we modified was marked with a solid line. Based on the above analysis, we decided to deploy the auxiliary server AS and server S on the same LAN. The server S has both a public IP address and a LAN IP address. We set the policy to only allow the server S 's LAN IP address to access the auxiliary server AS , and the server S is restricted to only send the ID to the auxiliary server AS .

In the device registration phase, the server S does not store T_i anymore, instead, it sends T_i through a secure channel to the auxiliary server AS for storage together with the device's ID ID_i . Besides, the server will not only send CK' to the device D_i over the secure channel but also send T_i to the device D_i .

In the login and authentication phase, the server S extracts the device's ID ID_i from the login request and sends it to the auxiliary server AS by using its LAN IP address. The auxiliary server AS . queries the database to get the corresponding T_i based on the ID ID_i and sends ID_i, T_i back to the server S . The message that the server S responds to the device D_i is only P_3, P_4 , and there will be no more T_i .

4.2. Analysis of our improved scheme

Theorem 1: Our improved scheme retains the security features and advantages of Wang et al.'s scheme.

Proof: In our proposed scheme, only the storage location of T_i has been changed, that is, T_i is stored in the device and the auxiliary server separately. Other than that, no change has been made elsewhere. Therefore, our scheme inherits the security features and advantages of Wang et al.'s scheme. Details of the security features and advantages of the Wang et al.'s scheme can be seen in [21].

Theorem 2: Our improved scheme can resist against KCI attack.

Proof: Now if the adversary $\mathcal{A}$ breaks into the server S and gets the server's secret key X as well as database information, but because of lacking of T_i , and the adversary $\mathcal{A}$ cannot intercept T_i (only P_3, P_4 are sent by the server S , no more T_i) from the channel, so the adversary $\mathcal{A}$ will not be able to reconstruct CK , which means $\mathcal{A}$ cannot calculate CK' , let alone P_2 and V_i . As a result, the adversary $\mathcal{A}$ cannot impersonate the device D_i to communicate with the server S , which means that the adversary $\mathcal{A}$ will not be able to conduct KCI attack unless $\mathcal{A}$ breaks into the auxiliary server AS as well. In fact, it is almost impossible to break into two servers with different protection measures at the same time. Further, when we adopt the above deployment or other methods to ensure the security of the auxiliary server AS , it is impossible for the adversary $\mathcal{A}$ to carry out KCI attack, which means that our scheme can resist against KCI attack.

5. Efficiency and comparison

Table 2. The comparisons of relevant schemes.

Schemes	Computation overhead		Communication cost		Storage cost	
	Device	Server	Device	Server	Device	Server
Chang et al. [20]	$5T_P+4T_H$	$5T_P+5T_H$	$4L$	$3L$	$3L$	$4L$
Wang et al. [21]	$5T_P+5T_H$	$5T_P+6T_H$	$4L$	$3L$	$2L$	$4L$
Ours	$5T_P+5T_H$	$5T_P+6T_H$	$4L$	$2L$	$3L$	$5L$

In this section, we compare our scheme with relevant [20] and [21]. Without loss of generality, for the sake of simplicity, here we assume that all values (device ID, the output of hash functions, points on an ECC curve, etc.) have the same length L. Let T_P and T_H denote the time required for the ECC point multiplication and the hash operation, respectively. Other lightweight operations are ignored. Table 2 shows the results of the comparison.

As can be seen from the table, the efficiency of our scheme is slightly lower than that of Chang et al. [20] and Wang et al. [21]. But considering that our scheme solves the security flaws of their schemes, the loss of efficiency is worthwhile.

6. Conclusion

In this paper, we prove that Wang et al.'s scheme is insecure and unable to resist against KCI attack. The adversary can impersonate a legitimate device to communicate with the server, which may bring many security risks. Using an auxiliary server successfully solves the defects of Wang et al.'s scheme while retaining the security features and advantages of the Wang et al.'s scheme. In general, our improved scheme can provide greater security for the embedded devices and the server.

Acknowledgments

This work was supported by the Natural Science Foundation of China under Grant 61871446.

References

[1] S. Chen, H. Xu, D. Liu, B. Hu, and H. Wang, "A vision of iot: Applications, challenges, and opportunities with china perspective," IEEE Internet of Things journal, vol. 1, no. 4, pp. 349–359, 2014.

[2] L. Da Xu, W. He, and S. Li, "Internet of things in industries: A survey," IEEE Transactions on industrial informatics, vol. 10, no. 4, pp. 2233–43, 2014.

[3] S. Li, T. Tryfonas, and H. Li, "The internet of things: a security point of view," Internet Research, vol. 26, no. 2, pp. 337–359, 2016.

[4] A. GhaffarianHoseini, N. D. Dahlan, U. Berardi, A. GhaffarianHoseini, and N. Makaremi, "The essence of future smart houses: From embedding ict to adapting to sustainability principles," Renewable and Sustainable Energy Reviews, vol. 24, pp. 593–607, 2013.

[5] Y. Yuehong, Y. Zeng, X. Chen, and Y. Fan, "The internet of things in healthcare: An overview," Journal of Industrial Information Integration, vol. 1, pp. 3–13, 2016.

[6] N. Li, D. Liu, and S. Nepal, "Lightweight mutual authentication for iot and its applications," IEEE Transactions on Sustainable Computing, vol. 2, no. 4, pp. 359–370, 2017.

[7] D. E. Kouicem, A. Bouabdallah, and H. Lakhlef, "Internet of things security: a top-down survey," Computer Networks, 2018.

[8] K. Sha, W. Wei, T. A. Yang, Z. Wang, and W. Shi, "On security challenges and open issues in internet of things," Future Generation Computer Systems, vol. 83, pp. 326–337, 2018.

[9] D. Wang, B. Da, J. Li, and R. Li, "Ibs enabled authentication for iot in ion framework," in Global Internet of Things Summit (GIoTS), 2017. IEEE, 2017, pp. 1–6.

[10] A. Shamir, "Identity-based cryptosystems and signature schemes," in Workshop on the theory and application of cryptographic techniques. Springer, 1984, pp. 47–53.

[11] N. Koblitz, "Elliptic curve cryptosystems," Mathematics of computation, vol. 48, no. 177, pp. 203–209, 1987.

[12] D. Wang, H. Cheng, D. He, and P. Wang, "On the challenges in designing identity-based privacy-preserving authentication schemes for mobile devices," IEEE Systems Journal, vol. 12, no. 1, pp. 916–925, 2018.

[13] T.-T. Truong, M.-T. Tran, and A.-D. Duong, "Improvement of the more efficient and secure id-based remote mutual authentication with key agreement scheme for mobile devices on ecc," in 2012 26th International Conference on Advanced Information Networking and Applications Workshops. IEEE, 2012, pp. 698–703.

[14] Y.-P. Liao and C.-M. Hsiao, "A secure ecc-based rfid authentication scheme integrated with id-verifier transfer protocol," Ad Hoc Networks, vol. 18, pp. 133–146, 2014.

[15] T. Shah and S. Venkatesan, "Authentication of iot device and iot server using secure vaults," in 2018 17th IEEE International Conference On Trust, Security And Privacy In Computing And Communications/12th IEEE International Conference On Big Data Science And Engineering (TrustCom/BigDataSE). IEEE, 2018, pp. 819–824.

[16] X. Yao, Z. Chen, and Y. Tian, "A lightweight attribute-based encryption scheme for the internet of things," Future Generation Computer Systems, vol. 49, pp. 104–112, 2015.

[17] M. N. Aman, K. C. Chua, and B. Sikdar, "Mutual authentication in iot systems using physical unclonable functions," IEEE Internet of Things Journal, vol. 4, no. 5, pp. 1327–1340, 2017.

[18] P. Hao, X. Wang, and W. Shen, "A collaborative phy-aided technique for end-to-end iot device authentication," IEEE Access, vol. 6, pp. 42279–93, 2018.

[19] S. Kalra and S. K. Sood, "Secure authentication scheme for iot and cloud servers," Pervasive and Mobile Computing, vol. 24, pp. 210–223, 2015.

[20] C.-C. Chang, H.-L. Wu, and C.-Y. Sun, "Notes on secure authentication scheme for iot and cloud servers," Pervasive and Mobile Computing, vol. 38, pp. 275–278, 2017.

[21] K.-H. Wang, C.-M. Chen, W. Fang, and T.-Y. Wu, "A secure authentication scheme for internet of things," Pervasive and Mobile Computing, vol. 42, pp. 15–26, 2017.

[22] H. Krawczyk, "Hmqv: A high-performance secure diffie-hellman protocol," in Annual International Cryptology Conference. Springer, 2005, pp. 546–566.

[23] C. Lee, L. Yi, L.-H. Tan, W. Goh, B.-S. Lee, and C.-K. Yeo, "A wavelet entropy-based change point detection on network traffic: a case study of heartbleed vulnerability," in Cloud Computing Technology and Science (CloudCom), 2014 IEEE 6th International Conference on. IEEE, 2014, pp. 995–1000.

[24] Z. Durumeric, F. Li, J. Kasten, J. Amann, J. Beekman, M. Payer, N. Weaver, D. Adrian, V. Paxson, M. Bailey et al., "The matter of heartbleed," in Proceedings of the 2014 conference on internet measurement conference. ACM, 2014, pp. 475–488.

[25] M. Goldstein, N. Perlroth, and D. Sanger, "Hackers' attach cracked 10 financial firms in major assault," The New York Times, 2014.

[26] D. Wang, N. Wang, P. Wang, and S. Qing, "Preserving privacy for free: efficient and provably secure two-factor authentication scheme with user anonymity," Information Sciences, vol. 321, pp. 162–178, 2015.

[27] J. Camenisch, R. R. Enderlein, and G. Neven, "Two-server password-authenticated secret sharing uc-secure against transient corruptions," in IACR International Workshop on Public Key Cryptography. Springer, 2015, pp. 283–307.

Fault diagnosis of power transformer based on tree ensemble model

Yunfei Liu[1], Jing Li[1,*], Lin Qiao[2], Shuo Chen[2], Sai Liu[3] and Jiahua Liu[3]

[1]College of Computer Science and Technology, Nanjing University of Aeronautics and Astronautics, Nanjing 210016 Jiangsu, China
[2]State Grid Liaoning Electric Power Supply Co, LTD, Shenyang 110004 Liaoning, China
[3]Nari Group Corporation/State Grid Electric Power Research Institute, Nanjing 211000 Jiangsu, China

*E-mail: lijing@nuaa.edu.cn

Abstract. During the running operation of the oil-immersed transformer, some gases may be dissolved in the insulating oil which can be used to diagnose the incipient failure of the power transformer. This is the Dissolved Gas Analysis (DGA). This paper proposes a power transformer fault diagnosis method based on tree ensemble model (Extreme Gradient Boosting, XGBoost): constructing a large number of classification and regression trees (CART) to fit the residuals obtained by each learning. Compared with the commonly used SVM and BPNN methods, our method has a significant improvement in accuracy, F1-score, precision and recall.

1. Instruction

Power transformers are used to convert and transfer power generated by power plants to customers worldwide. Therefore, it is important to ensure that the power transformers are in good running condition so that they can provide reliable and continuous power, which is very necessary in the modern world [1]. Now, many power companies have implemented various running state assessment and maintenance measures on the running state of power transformers. Dissolved Gas Analysis (DGA) is one of them [2].

Transformers in normal state will decompose a very small amount of gases due to aging and cracking of insulating oil or solid insulation. These gases are mainly H_2, CH_4, C_2H_6, C_2H_4, C_2H_2, CO, CO_2 [3]. When the power transformer fails, the concentration of some gases will increase rapidly. However, gas concentration fluctuation range is large, which has a certain impact on the fault diagnosis of the power transformer. Therefore, in addition to the ordinary gas concentration, the ratios between different gases are also used to diagnose the fault of the power transformer. Commonly used ratios are IEC ratio (CH_4/H_2, C_2H_2/C_2H_4, C_2H_4/C_2H_6), Rogers ratio (CH_4/H_2, C_2H_2/C_2H_4, C_2H_4/C_2H_6, C_2H_6/CH_4), Dornenburg ratio (CH_4/H_2, C_2H_2/C_2H_4, C_2H_2/CH_4, C_2H_6/C_2H_2), which are all DGA methods [4][5]. DGA methods mentioned above can make estimation of the running state of power transformer, unfortunately, although they are simple to implement, they have problems such as incomplete coding and excessive boundaries. Besides, these methods often give different diagnosis results, indicating that these methods are very inaccurate [6]. These problems have driven many researchers to study methods based on machine learning to diagnose transformer faults. Support vector machines (SVM) [1][7], neural networks(NN)[8][9], classification regression trees(CART) [10], principal component analysis(PCA) [7], fuzzy logic inference system(FLIS) [11] and other technologies [12] have been gradually applied to power transformer fault diagnosis and have achieved certain results. However, since transformer data is often very rare and difficult to collect, these intelligent algorithms, although able to achieve fairly good diagnosis results, tend to face problems such as overfitting.

XGBoost is an emsemble learning method based on regression trees that improves the traditional Gradient Boosting Decison Tree (GBDT). The excellent features of XGBoost drive us to explore the potential of XGBoost in power transformer fault diagnosis. The main contributions of this paper are as follows:

(1) Our DGA data is collected from multiple sources, and the ratios between different gases are combined to form a relatively large DGA data set.

(2) The use of ensemble learning (XGBoost) for transformer fault diagnosis and classification can allow the existence of a small number of missing values in the DGA data of transformer (this is very common due to the harsh running environment), Through the ensemble learning ability of XGBoost, a large number of classification and regression trees which can process highly oblique and polymorphic continuous data are constructed to diagnose the fault type of power transformer.

2. Extreme gradient boosting (XGBoost)

Extreme Gradient Boosting (XGBoost), a scalable machine learning system for tree boosting, is a novel classifier based on ensemble of classification and regression trees (CART) [13]. Before training, the data will be sorted firstly and then saved as a block structure. The subsequent iterations can use this structure repeatedly to greatly reduce the number of calculation and improve the classification speed of power transformers.

The data set with m-dimensional features can be represented as $D=\{(x_i,y_i)\}(|D|=n, x_i\in\mathbb{R}^m, y_i\in\mathbb{R})$, x_i means the features vector, y_i means the label of the fault type of power transformer. XGBoost is a tree ensemble model using K additive function to predict the output, and it can be defined as eq. (1).

$$\hat{y}_i=\sum_{k=1}^{K} f_k(x_i), f_k\in F \tag{1}$$

$F=\{f(x)=w_{q(x)}\}(q:\mathbb{R}^m\to T, w\in\mathbb{R}^T)$ is a collection of classification and regression trees (CART), q is a mapping function from sample data x_i to the leaf nodes of the CART, which is used to represent the structure of a tree, T indicates the number of leaf nodes in a tree. Each f_k is equivalent to a q mapping and the weight of its leaf nodes (score). This weight is a continuous value and contributes to an efficient optimization algorithm. w_i indicates the weight of the i-th leaf node.

In order to get the final XGBoost model, we need to train a set of regression trees. The objective function of this ensemble learning model is defined in eq. (2). l represents differentiable convex loss function, which is used to measure the difference between the real label y and the predicted label $\hat{y}$. Ω is a regular term that is used to smooth the learned weight and punish the complexity of the model to prevent overfitting.

$$L(\phi)=\sum_i l(\hat{y}_i,y_i)+\sum_k \Omega(f_k) \tag{2}$$

$$\Omega(f)=\gamma T+\frac{1}{2}\lambda\|w\|^2$$

This model is trained in an iterative manner. Let $\hat{y}_i^{(t)}$ be the prediction result of the i-th sample of the t-th iteration, the objective function is shown in eq. (3), f_t represents the new tree created by the t-th iteration. f_t is selected by eq. (3) to enhance the model. The residual between the result of previous iteration and real label is fitted by the new tree f_t.

$$L^{(t)} = \sum_{i=1}^{n} l(y_i,\hat{y}_i^{(t-1)} + f_t(x_i)) + \Omega(f_t) \tag{3}$$

In the gradient boosting process, XGBoost uses a second-order Taylor expansion to optimize the objective function. The simplest form is shown in eq. (4), where $g_i = \partial_{\hat{y}^{(t-1)}} l(y_i,\hat{y}^{(t-1)})$ and $h_i = \partial^2_{\hat{y}^{(t-1)}} l(y_i,\hat{y}^{(t-1)})$, they are the first and second derivatives of the loss function respectively.

$$\begin{aligned} L^{(t)} &\simeq \sum_{i=1}^{n}\left[l(y_i,\hat{y}^{(t-1)}) + g_i f_t(x_i) + \frac{1}{2}h_i f_t^2(x_i)\right] + \Omega(f_t) \\ &\simeq \sum_{j=1}^{T}\left[\left(\sum_{i\in I_j} g_i\right) w_j + \frac{1}{2}\left(\sum_{i\in I_j} h_i + \lambda\right) w_j^2\right] + \gamma T \end{aligned} \tag{4}$$

When the structure of a tree $q(x)$ is given, the optimal weight of its leaf nodes can be calculated by eq. (5), the quality of the tree structure can be calculated by eq. (6).

$$w_j^* = -\frac{\sum_{i\in I_j} g_i}{\sum_{i\in I_j} h_i + \lambda} \tag{5}$$

$$\tilde{L}^{(t)} = -\frac{1}{2}\sum_{j=1}^{T}\frac{(\sum_{i\in I_j} g_i)^2}{\sum_{i\in I_j} h_i + \lambda} + \gamma T \tag{6}$$

3. XGBoost-based power transformer fault diagnosis model

The method proposed in this paper is to use XGBoost to diagnosis the fault type of the power transformer based on the dissolved gas in the transformer insulating oil. For training, we use the data provided by NARI and combine some existing data from the network. Features and fault types of DGA data sets are shown in table 1.

Table 1. Features and fault types of DGA data sets.

DGA gas content (input)	Hydrogen(H_2),Methane(CH_4),Ethylene(C_2H_4), Ethane(C_2H_6),Acetylene(C_2H_2)
Ratios between different gases (input)	C_2H_2/C_2H_4,CH_4/H_2,C_2H_4/C_2H_6,$C_2H_2/(C_1+C_2)$,$H_2/(H_2+C_1+C_2)$, $C_2H_4/(C_1+C_2)$,$CH_4/(C_1+C_2)$,$C_2H_6/(C_1+C_2)$,$(CH_4+C_2H_4)/(C_1+C_2)$
Transformer fault category (output)	0- Partial Discharge (PD) 1- Low energy discharge (D1) 2- High energy discharge (D2) 3- Thermal fault $t < 300°C$ (T1) 4- Thermal fault $300°C<t<700°C$ (T2) 5- Thermal fault $t>700°C$ (T3) 6- No Fault (NF) 7- Undefined (UD)

We believe that transformer faults are related to the DGA gases and ratios between them. These 9 ratios are shown in table 1, in which C_1 and C_2 represent first-order and second-order hydrocarbons. C_1->CH_4, C_2->$C_2H_4+C_2H_6+C_2H_2$.

The DGA data of power transformer is input to XGBoost, and XGBoost builds the first CART according to the pre-processed data through the approximate algorithm in the split finding algorithm, and the predicted value is compared with the real value, then the residual is used as the new label information together with the sample data to construct the next CART tree to fit the residual.

The details of our fault diagnosis model for power transformer are as follows:

1. Input: The input data includes gas concentration, ratios and fault type.
2. Output: Set XGBoost to select softmax objective and return the diagnosed category by probability matrix.
3. Loss function: XGBoost only requires the loss function to be a differentiable convex loss function. We use merror (multiclass classification error rate) and mlogloss (defined as the negative log-likelihood of the true labels given a probabilistic classifier's predictions) as our loss function.

4. Experiments and results

4.1. Data set and experiment setup

Due to the lack of existing DGA data sets, we collected and organized the data from multiple network resources. It may also brought data quality problems, but this is not our main consideration. Each sample in the DGA data consists of a series of concentration of dissolved gases, which can be seen in the table 2.

Table 2. Content of DGA data set.

H2	CH_4	C_2H_4	C_2H_6	C_2H_2	C_2H_2/C_2H_4	......	Fault
117	17	3	1	1	0.333	......	0
595	32	18	4	65	3.611	......	1
10	63	35	176	0.001	0	......	3

In order to eliminate the dimensional impact between features, data standardization is required to resolve the comparability between features. Through eq. (8), data in the DGA dataset is normalized to [0,1]. For entire experiment, we used a computer, Intel(R) Xeon(R) W-2102 CPU, 32G memory, python 3.6 and the operating system is Ubuntu 16.04.

4.2. Results and analysis

In this experiment, we tested the XGBoost classifier on a relatively large data set, and conduct in-depth analysis for them. We compare the XGBoost model with commonly used SVM and BPNN algorithms. 5-fold cross-validation is used to judge the pros and cons of the current model, and finally select the optimal XGBoost transformer fault diagnosis model.

Before training, we divided all the data into training set and testing set according to 7:3. The optimal parameters are max_depth=10, eta (learning rate)=0.22, gamma=0.18, subsample=0.69, colsample_bytree=0.56, min_child_weight =2, max_delta_step=10, the best number of regression trees required is 48. As the number of CART increases, merror and mlogloss rapidly decrease and stop at 48, then merror and mlogloss never again decrease significantly, and tend to be stable. Because the samples in the data set are not balanced, we not only evaluate the model from the accuracy rate, but also from precision, recall, and f1-score. The performance of optimal XGBoost model for each category is shown in the following table 3.

Table 3. Performance of optimal XGBoost diagnosis model.

Catagory	precision	recall	f1-score	samples
0	0.75	0.41	0.53	29
1	0.60	0.69	0.64	52
2	0.89	0.85	0.87	93
3	0.81	0.73	0.77	70
4	0.88	0.91	0.90	261
5	0.87	0.56	0.68	93
6	0.79	0.81	0.80	137
7	0.90	0.97	0.93	450
avg/total	0.86	0.86	0.85	1185

It can be seen that in the categories with a large number of samples, the precision, recall, and f1-score scores are higher than those with fewer samples. For example, the UD category has the most samples, and the scores of the corresponding evaluation indicators are also the highest. Scores for categories with a less samples are less than expected. It means that imbalance and quality of the data set do have a certain impact on the results of the diagnosis model.

In the experiments of SVM and BPNN, the optimal weight parameters of BPNN are cross-validated to obtain. Grid search is also used for SVM. The parameters of the optimal SVM model obtained by grid search are {C=1000, gamma=0.3, kernel='rbf'}. In the experiment of BPNN, we also obtain the optimal model parameters by setting the grid search to compare the learning rate and the number of layers and cells. The SVM model is not expected on our testing set, and several scores are quite low, BPNN performs slightly better than SVM on testing data set, but several evaluation indicators are still much lower than XGBoost.

In addition to comparing precision, recall, and f1-score, accuracy is also a very important indicator. To some extent, it is the most important indicator. The average accuracy and top accuracy of the three models are shown in the following table 4.

Table 4. Accuracy of e model.

Model	avg	top
XGBoost	0.88	0.91
BPNN	0.71	0.74
SVM	0.56	0.62

As can be seen from table 4, XGBoost is significantly better than SVM and BPNN in the four evaluation indicators of accuracy, precision, recall, and f1-score. The average accuracy is 0.17 higher than BPNN and 0.32 higher than SVM.

5. Conclusion

This paper proposes a fault diagnosis method for power transformer based on XGBoost (Tree Ensemble Model). The XGBoost-based transformer diagnosis model is compared with the commonly used SVM and BPNN methods and the experimental results show that the XGBoost-based transformer fault diagnosis model has a significant improvement in accuracy, F1-score and recall.

Acknowledgement

This research is supported by the State Grid Liaoning Electric Power Supply CO., LTD, and we are grateful for the financial support for the "Key Technology and Application Research of the Self-Service Grid Big Data Governance (SGLNXT00YJJS1800110)".

Reference

[1] Z. Sahri, R. Yusof, Journal of Computer and Communications, 2, 22-31 (2014)
[2] A. Peimankar, S.J. Weddell, T. Jalal, et al. Swarm & Evolutionary Computation, 36, 62-75 (2017)
[3] Z. Sahri, R. Yusof, J. Watada, Industrial Informatics IEEE Transactions, 10(4):2093-2102(2014)
[4] A. Febriyanto, T.K. Saha. Power Engineering Conference. IEEE (2008)
[5] D.E.A. Mansour, Electrical Insulation & Dielectric Phenomena. IEEE (2012)
[6] R. Naresh, V. Sharma, M, Vashisth, IEEE Transactions on Power Delivery, 23(4):2017-24(2008)
[7] W. Shi, Y. Zhu, T. Huang, et al. Journal of Signal Processing Systems, 86(2-3):221-36 (2017)
[8] V. Miranda, A.R.G. Castro, IEEE Transactions on Power Delivery, 20(4):2509-16(2005)
[9] J.L. Guardado, J.L. Naredo, P. Moreno, et al. IEEE Power Engineering Review, 21(7):71-71(2007)
[10] X. Huang, L.I. Wenjunzi, S. Tong, et al. High Voltage Engineering (2016)
[11] S. Mofizul Islam, T.Wu, G. Ledwich. IEEE Transactions on Dielectrics and Electrical Insulation, 7(2):177-86(2000)
[12] S.S.M. Ghoneim, I.B.M. Taha, N.I. Elkalashy. IEEE Transactions on Dielectrics & Electrical Insulation, 23(3):1838-45(2016)
[13] T. Chen, C. Guestrin, Proceedings of the 22nd ACM SIGKDD International Conference on Knowledge Discovery and Data Mining. ACM (2016)

IOP Conf. Series: Materials Science and Engineering **715** (2020) 012033 doi:10.1088/1757-899X/715/1/012033

Real time outlier monitoring for power transformer fault diagnosis based on isolated forest

Li Shen[1] , Hongjun Du[1], Shuji Liu[1], Shuo Chen[1], Lin Qiao[1], Sai Liu[2], Jiahua Liu[2], Kexin Li[3] and Jing Li [3,*]

[1]State Grid Liaoning Electric Power Supply CO, LTD. Shenyang, 110004 Liaoning, China

[2]NARI Group Corporation/State Grid Electric Power Research Institute, Nanjing, 211000 Jiangsu, China

[3]College of Computer Science and Technology, Nanjing University of Aeronautics and Astronautics, Nanjing, 211106 Jiangsu, China

*E-mail: lijing@nuaa.edu.cn

Abstract. In order to improve the accuracy and efficiency of transformer fault detection, this paper uses the isolated forest algorithm combined with the historical transformer characteristic gas data to establish the characteristic gas outlier recognition model, and then uses the uncoded ratio method to establish the abnormal event strategy and abnormal event library for the transformer historical fault information. Finally, the state of the transformer is diagnosed based on the established outlier detection model and the exception event library. The experimental results show that the proposed method has a great improvement in the detection efficiency and stability of outliers, and is more accurate in transformer fault diagnosis combine with abnormal event database.

1. Introduction

With the continuous development of artificial intelligence, the scale of the power grid in the power industry is also constantly increasing, and the requirements for power quality of users are also constantly improving. The safe and stable operation of equipment in the power grid is a prerequisite for meeting the needs of users [1]. However, in the actual operation process of the power equipment, certain faults will always occur due to the equipment itself or external factors. If the equipment fails, it will bring irreparable damage to the grid operation.

Transformers are the key equipment for power system energy conversion and transmission. They often work in various stress environments, such as electricity, heat, high voltage, high current, moisture, and oxygen. The internal main insulation properties will suffer irreversible aging and eventually fail. Failure to find and take appropriate measures in time will result in failure of the transformer and unnecessary economic losses. Therefore, timely grasping the operating state of the transformer equipment has special significance for the safe operation of the power supply system [2].

As a method to identify the internal discharge and overheat abnormality of oil-immersed transformers, the dissolved gas analysis [3] has become a means of transformer anomaly detection. Since there are many hydrocarbons with different molecular weights in the transformer oil, the molecule contains chemical groups such as CH_3, CH_2, CH, and C-C bonds and C-H bonds. Under the action of electricity or heat, some C-C bonds and C-H bonds are broken to form unstable hydrogen atoms and hydrocarbon radicals, which are re-synthesized into hydrogen and low molecular hydrocarbon gases. Due to the different degrees of oil temperature and discharge, the gases produced are also different. Therefore, the dissolved gas method in the oil can detect and diagnose the abnormality of the transformer according to the dissolved gas composition and the content of the transformer oil [4].

2. Related work

As a common method for detecting the abnormal state of the transformer, the dissolved gas method in the oil is used as the statistical threshold by the dissolved gas content in the oil. The threshold value is derived from the statistics of a large number of transformer historical characteristic gas detection data, and is judged to be abnormal when the monitored value exceeds the specified threshold data. The key to this method is that for the threshold setting, when the historical data of the transformer is large, the method can be well recognized, but if the historical data is lacking, the detection effect of the method will be greatly reduced. In response to the problems of the statistical threshold method, many scholars began to study other methods to detect the abnormal state of the transformer. Such as based on characteristic gases [5-7], based on IEC three ratios [8-10], based on the non-coding ratio method [11].

Based on the characteristic gas method, fault diagnosis is performed based on the difference in gas production rate, gas type, and gas concentration [6] of the transformer. Although this method can intuitively and conveniently detect the abnormal state of the transformer, there is no quantitative concept, and information such as temperature change law is needed to accurately locate the fault type. The three-ratio method selects two gases with similar solubility coefficient and diffusion coefficient from C_2H_2, C_2H_4, CH_4, H_2 and C_2H_6 gases according to the dependence of the relative characteristics of different characteristic gases on the transformer temperature during transformer failure, forming three contrast values [8]. Transformer fault diagnosis is performed according to different coding combinations by different coding representations. Although this method has a quantitative concept, the calculation is relatively simple. However, certain conditions are required for use, and there are cases where there is no corresponding failure in the ratio. On this basis, a codeless rate method [11] is proposed. This method solves the problem of fault coding of three-ratio method to some extent, and the problem that some three-valued method fault cannot be diagnosed. However, this method also diagnoses transformer faults based on the ratio between the corresponding characteristic gases, and does not fundamentally solve the problem of lack of ratio coding between the characteristic gases. Later, based on support vector machine [12], extreme learning machine [13], deep learning [14] and other machine learning methods are also used for transformer fault diagnosis. Although the machine learning-based approach improves the efficiency and accuracy of transformer fault diagnosis to a certain extent, these methods rely on transformer fault data. When a new fault occurs, only the existing empirical data does not identify the fault well.

3. The proposed solution

In this paper, firstly, based on the historical transformer characteristic gas data, the isolation forest detection algorithm is used to establish the outlier detection model, and then the current transformer characteristic gas real-time data is detected, and the detected abnormal value is combined with the transformer abnormal event library to judge the current transformer state. The overall steps for transformer outlier monitoring and abnormal state diagnosis based on isolated forests are shown as figure 1:

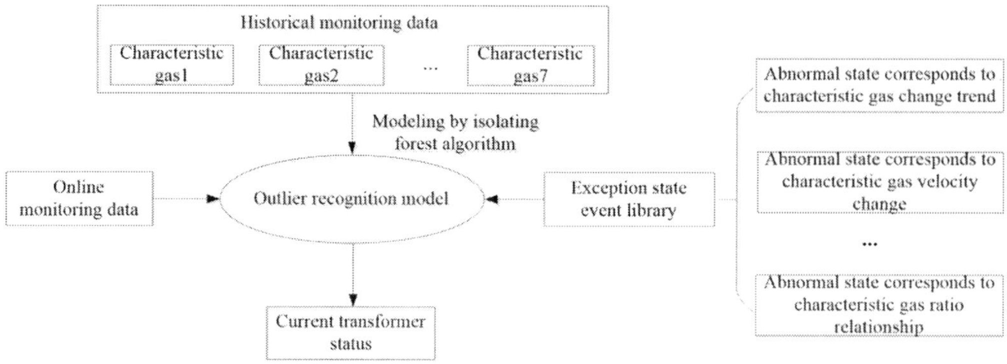

Figure 1. Real-time monitoring of abnormal values of transformers based on isolated forests and abnormal state diagnosis process.

3.1. Outlier detection model based on isolated forest

The key to the Isolation Forest algorithm is the construction of the isolation tree. First, the original data is used to establish the isolation tree, and then the isolation tree is combined into an isolated forest for the identification and detection of outliers. In the isolated forest, in order to detect outliers, it is necessary to define the calculation method for the isolation tree and path length.

Isolation Tree: An isolation tree can be thought of as a binary search tree. If T is a node of the isolation tree, then T is either a leaf node or a child node (T_l,T_r) with a test instance. The test at the node T consists of the attribute p and the segmentation value q. The segmentation attribute and the segmentation value are used to segment different data, the data record smaller than the segmentation value is divided into the left child node, and the data record larger than the segmentation value is divided into the right child node. Repeat the above process until there is only one data in the child node or the maximum height of the tree has been reached. It is generally considered that the data corresponding to the shorter path is recorded as an abnormal value, and the record corresponding to the longer path is a normal value.

Path length: The path length h(x) of the data record x refers to the number of edges encountered during the traversal process from the root node to the traversal of the isolation tree until the leaf node is encountered.

Outlier score: Since the constructed isolation tree and the binary search tree have the same structure, here we use the average length that was not successfully searched in the corresponding binary search tree as the normalization factor. From the binary search tree, for a given sample size of sample size φ, the average length of the corresponding binary search tree is:

$$c(\varphi)=\begin{cases}2H(\varphi-1)-2(\varphi-1)/\varphi & \text{if } \varphi>2,\\ 1 & \varphi=2,\\ 0 & \text{otherwise.}\end{cases} \tag{1}$$

where, h(φ) is the harmonic function, which can be calculated by $\ln(\varphi)+0.5772156649$ (Euler constant). After the normalization factor is obtained, we normalize the path length, and then calculate the anomaly of data x is worthy of classification. The anomaly is worthy of classification as follows:

$$s(x,\varphi)=2^{-\frac{E(h(x))}{c(\varphi)}} \tag{2}$$

$$E(h(x))=\Sigma_{i=1}^{n}h_i(x)/n \tag{3}$$

where h(φ) represents the path length of the data x on the i-th tree, and E(h(x)) represents the average of the path lengths of the data x in different isolation trees.

The Isolation Forest algorithm can be divided into two steps when performing outlier detection: the first step, the training process. The original data set is randomly sampled to obtain multiple sub-data sets, the sub-data set is used to construct the isolation tree, and then the constructed isolation tree is used to form the isolation forest. The second step, the prediction process, can also be called the outlier scoring process. The data of the detected isolated forest is used to calculate the outlier score, and the outlier score is compared with the set threshold to judge the abnormal data. For the calculation method of the abnormal value score established above, it can be known that when all the abnormal values of the data to be detected are scored. When both are close to 0.5, it means that the data set has no obvious outliers; when $E(h(x))$ is closer to 0, the outlier value s is closer to 1, which means that the data is basically regarded as abnormal data; when $E(h(x))$ is closer to the sample size φ , the outlier score is closer to 0. If the outlier score is much smaller than 0.5, the data is considered to be normal data.

3.2. Fault diagnosis model

The outlier identification model based on the isolated forest can detect the abnormal value in the characteristic gas data of the transformer, but cannot determine whether the state of the transformer corresponding to the abnormal value is normal. Therefore, this paper proposes a real-time diagnosis of transformer faults based on the model of anomaly strategy and anomalous event library.

(1) Abnormal state strategy

An abnormal state identification library is established according to the content of the characteristic gas and the proportion of the characteristic gas, and the characteristic gas abnormal data and the abnormal state of the transformer are matched to achieve the purpose of identifying the abnormal state of the transformer. According to the change of the characteristic gas value, an abnormal state strategy is formulated, and the judgment criteria are as follows:

(a) When the values of all the characteristic gases are simultaneously decreased or increased, it is judged that there is no abnormality;

(b) When the value of some characteristic gases increases and the proportion increases or decreases significantly, it is judged to be abnormal;

(c) When the content of one or several gases is increased and the combined characteristic gas combination can be matched with the abnormal state in the abnormal state library, it is determined to be an abnormal state;

(d) When one or several characteristic gases increase and do not match all events in the abnormal state library, it is judged to be an abnormal state and relevant experts are required to perform auxiliary judgment.

(2) Abnormal state event library

The abnormal state event library is a combination of typical characteristic gases corresponding to various abnormalities of the transformer. The abnormal event library can be used to accurately analyze the abnormal state of the transformer. The main body of the abnormal state event library is the characteristic gas, and the abnormal event library can be continuously enriched according to the actual situation. The exception status event library is shown in table 1:

Table 1. Abnormal state event library.

Abnormal state event	**Characteristic gas**							**C_2H_2/C_2H_4**	**C_2H_4/C_2H_6**	**CH_4/H_2**	**Abnormal state event**
	H_2	CH_4	C_2H_2	C_2H_4	C_2H_6	CO	CO_2	<0.1	<1	Irrelevant	T1
O	2	1	—	1	2	—	—	<0.1	>1&<3	Irrelevant	T2
OP	2	1	—	1	2	1	1	<0.1	>3	Irrelevant	T3
PD&OP	1	1	2	—	2	1	2	>0.1<&<3	Irrelevant	<1	D1
SDO	1	—	1	—	—	—	—	>0.1<&<3	Irrelevant	>1	D1&O
OA	1	2	1	2	2	—	—	>3	Irrelevant	<1	D2
AOP	1	2	1	2	2	1	1	>3	Irrelevant	>1	D2&T

Note: 1 means the main gas, 2 means the secondary gas, - means irrelevant, O means oil overheating, OP means oil and insulation paper overheated, PD&OP means partial discharge in oil and paper insulation, SDO means spark discharge in oil, OA means oil arc, AOP means arc in oil insulating paper, T1 means thermal fault of T<300 °C, T2 means thermal fault of 300°C<T<700°C, T3 means thermal fault of T >700°C, D1 means high energy discharge, D2 means low energy discharge, T means thermal fault.

4. Experiment

This paper takes the characteristic gas history test data of No. 3 oil-immersed transformer of 220KV substation in a certain area of Nanjing as an example. Firstly, the time domain analysis of the characteristic gas data of the above transformer is carried out to obtain whether the characteristic gas is abnormal or not.

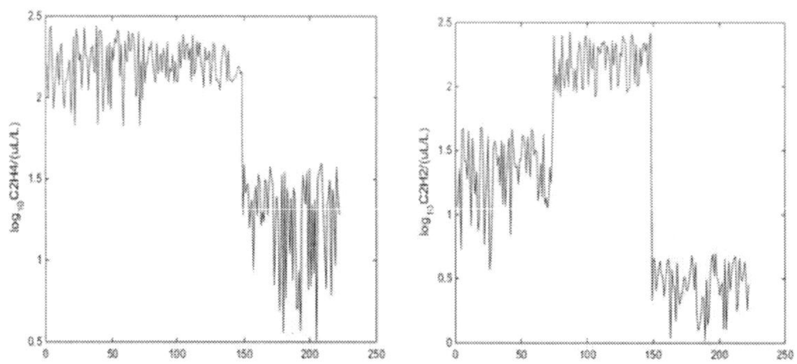

Figure 2. Partial characteristic gas time domain analysis.

As shown in figure 2, it can be seen from the results of the above time domain diagram that there is a significant step in the characteristic gas data of the transformer, indicating that there is an abnormal value in the characteristic gas, and the state of the transformer may be abnormal. The transformer characteristic gas data is modeled by an outlier detection algorithm based on the isolated forest to identify outliers in the characteristic gas. According to the historical characteristic gas data of the transformer, the isolation forest is established. The detection effect of the algorithm on the historical data is shown in the figure 3.

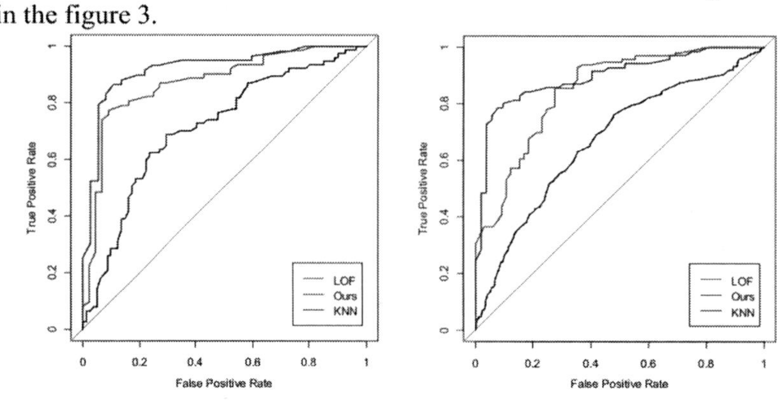

Figure 3. Comparison of outlier detection accuracy of different algorithms on different datasets.

According to the outlier detection model established by the isolated forest algorithm on the characteristic gas of the transformer, the characteristic gas data of other transformers are detected, and the abnormal data of the characteristic gas of the detected transformer is used to diagnose the fault of the transformer. Using the same transformer characteristic gas data, the method proposed in this paper is compared with other transformer abnormality diagnosis methods. The results are shown in table 2:

Table 2. Detection results and comparison with other methods.

Time Point	Guide method	Method of this paper	Transformer real state
1	abnormal	Discharges of High Energy	Discharges of High Energy
2	abnormal	Discharges of Low Energy	Discharges of Low Energy
3	abnormal	Thermal fault	Thermal fault
4	normal	normal	normal
5	abnormal	Thermal fault T>700°C	Thermal fault T>700°C
6	normal	Partial Discharge	Discharges of Low Energy
7	normal	normal	normal
8	abnormal	Thermal fault 300°C<T<700°C	Thermal fault 300°C<T<700°C
9	abnormal	Discharges of High Energy	Discharges of High Energy
10	abnormal	Thermal fault T<300°C	Thermal fault T<300°C

Through the above test results, it can be found that compared with other transformer abnormal state diagnosis, the isolated forest-based transformer outlier detection method and fault diagnosis model proposed in this paper can not only detect the transformer with or without fault, but also accurately detect the transformer fault type.

5. Conclusion

Based on the isolated forest algorithm, a transformer characteristic gas outlier recognition model and transformer fault detection method based on anomaly strategy and abnormal event library are proposed. Firstly, the characteristic gas data of the transformer is modeled by isolating the forest, and the abnormal value identification of the characteristic gas of the transformer is completed. Then, the abnormal value identified is classified according to the abnormality strategy. Finally, the classification result is performed according to the abnormal event library to diagnose the type of fault in the transformer. When the fault of the transformer is diagnosed, the method proposed in this paper only diagnoses the data with abnormality. This method can reduce the computational cost in the diagnosis process. Combined with the exception event library, the specific fault type of the transformer can be detected more accurately.

Acknowledgments

The research was supported by the State Grid Liaoning Electric Power Supply CO, LTD. and we are grateful for the financial support for the "Key Technology and Application Research of the Self-Service Grid Big Data Governance (SGLNXT00YJJS1800110)".

References

[1] F. Wan, W. Chen, X. Peng, et al. Study on the gas pressure characteristics of photoacoustic spectroscopy detection for dissolved gases in transformer oil. International Conference on High Voltage Engineering and Application, 286-89(2012)

[2] X. Zhao, Y. Li. An On-Line Monitoring System for Gases Dissolved in Transformer Oil Using Wireless Data Acquisition. Power and Energy Engineering Conference, 1-4(2009)

[3] X. Liu. Application of Dissolved Gas Online Monitoring System in Transformer Oil in Gezhouba Power Station. Hydropower & Pumped Storage (2017)

[4] J. Dai, H. Song, G. Sheng, et al. Dissolved gas analysis of insulating oil for power transformer fault diagnosis with deep belief network. IEEE Transactions on Dielectrics & Electrical Insulation, 24(5), 2828-35(2017)

[5] P. Wang, W. Chen, F. Wan, et al. Simultaneously anayze fault characteristic gases extracted from transformer oil by Raman spectroscopy. IEEE International Conference on High Voltage Engineering and Application, 1-4(2016)

[6] W. Chen, L. Zhao, S. Peng, et al. Analysis of Dissolved Gas in Transformer Oil Based on Laser Raman Spectroscopy. Proceedings of the Csee, 34(15), 2485-92(2014)

[7] D. Yang, X. Chen, Z. Ma. Fault detection of transformers based on Raman spectra of the dissolved gas in transformer oil. Power Electronics and Motion Control Conference, 2431-36(2016)
[8] X. T. Zhao, Q. Shen, X. U. Jia-Xiang, et al. Fault Diagnosis for Power Transformer Based on IEC Three-ratio and Extension Association with Combined Weights. Power System & Clean Energy, (2013).
[9] P. U. Lijuan, N. Liu, H. Liu, et al. Statistical Analysis of IEC Three-ratio Method Based on Fault Data of Oil-immersed Transformers. Shanxi Electric Power(2015)
[10] L. F. Lin, Z. Q. Wen, S. C. Deng, et al. Research on Fault Diagnosis for Power Transformer Based on Fuzzy Three-Ratio Method and Case-Based Reasoning. Transformer (2015)
[11] H. Xue, H. P. Zhang. Fuzzy No Coding Ratio in Transformer Oil Gas Detection in the Application. Science Technology & Engineering (2012).
[12] Y. Zhang, B. Zhang, Y. Yuan , et al. Transformer fault diagnosis based on support vector machine. Telecom Power Technology, 6, 405-08(2016)
[13] L. L. Wang, F. Pei, Y. L. Zhu. Transformer Fault Diagnosis Based on Online Sequential Extreme Learning Machine. Applied Mechanics & Materials, 721, 360-65(2015)
[14] T. Yang, J. Huang, X. U. Kui, et al. Diagnosis method of power transformer fault based on deep learning. Power Systems & Big Data, (2018)

Design and implementation of fault injection based on abstract syntax tree of C Program

Hang Yu[1], Huiquan Gong[2] and Yawen Wang[1,*]

[1]State Key Laboratory of Networking and Switching Technology, Beijing University of Posts and Telecommunications, 100876, Beijing, China
[2]Faculty of Information Technology, Beijing University of Technology, 100022, Beijing, China

*Email: wangyawen@bupt.edu.cn

Abstract. Mutation testing is a fault-based technique to evaluate test quality by injecting bugs in programs under test. Traditional manual fault injection has many disadvantages such as inefficiency, large workload, low compilation success rate and so on. To improve the efficiency of fault injection, this paper aims to develop an automatic tool that supports fault injection for programs written in C. This paper first introduces the types of mutation operators and injection rules. Then, it proposes a framework, which mainly includes four modules: parser, fault location extraction, fault generation and data model. Finally, experiments show that our tool not only aids to produce a large number of program faults but also provides syntax-correct mutant programs, which would be further used in mutation testing.

1. Introduction

To improve safety and reliability of the system, it is necessary to pass software test to detect potential faults in the program before putting it into production. Mutation testing is a widely used test method. As a white-box unit testing technology, it has designed corresponding fault injection rules for a variety of programming languages. For example, Fortran[1], C[2] and Java[3]. The core idea of mutation testing is to inject manually generated faults into the program, and then test these injected faults using test suite. If the test can find all injected faults, then the test suite is sufficient.

Although the mutation testing can effectively evaluate the fault detection capability of the test suite, it also faces performance problems. One problem that prevents mutation testing from becoming a practical testing technique is that even for a small program, it is still possible to generate a large number of fault programs. At present, many people have been researching cost reduction techniques, it is divided into three types by Offutt and Untch[4].

Another issue is related to labor costs such as human prejudge problems and equivalent mutants[5]. Human prejudge problem refers to verifying the output process of the source program in each case. The equivalent mutants mean that the two pieces to code have different syntax, but the semantics are the same. It is an undecidable problem and generally needs to be done manually, which is very time consuming and inefficient. Offutt found through empirical research that among a large number of variants generated, equivalent mutants generally account for 10% to 40%[6].

This paper implements an automatic fault injection tool to design a rapid and effective fault generation method by accelerating the fault injection method, the fault injection process and the mutation testing process.

2. Design of automatic fault injection

2.1. Mutation operators and injection rules

In the mutation test, the process of fault injection is essentially the process of grammatical conversion of the program code. The grammar rule used to describe the conversion process is called the mutation operator. There are 18 mutation operators[7] used in this paper. They are *AROR*, *ASOR*, *BIOR, CORP*, *LOOR*, *REOR*, *VARP*, *TIAS*, *TDBS*, *TDGS*, *TDCS*, *LOCN*, *ARER, STER*, *RERP*, *LORP*, *BSRP* and *CSRP*.

When injecting a fault, it is generally required to abide by the following rules: 1) The injected fault should be similar to the error that may occur when the programmer writes the code; 2) The injected fault must be grammatically and semantically correct, that is, the generated fault code can be compiled and passed; 3) The injected fault can be tested, that is, some algorithms or techniques can detect these faults, and the difficulty of detecting the injected faults cannot be too simple. Otherwise, it is easy to bring a certain difficulty to the accurate measurement of the positioning effects.

2.2. Business process design of fault injection

After analyzing the relevant mutation operator[7] and according to the structure of the abstract syntax tree, the automated fault injection framework can be divided into four modules.

1) Parser module: This module calls the C compiler and extracts information such as the abstract syntax tree and code text. There are already many mature tools that can do this.

2) Fault location extraction module: The input of the module is an abstract syntax tree of the source code and a mutation operator, and output is a set of syntax tree nodes and fault can be injected at the location of a node.

3) Fault generation module: The input of the module is the output of the previous module and the corresponding mutation operator. And the output is a fault set that can be injected at the position. A fault is represented by a text replacement instruction that represents a string operation, including insertion, deletion, and replacement.

4) Data model module: In the process of fault injection, we have designed many data structures and defined several operations on them. They each serve a specific module, and some run through several modules. This module will be described separately in the following subsections.

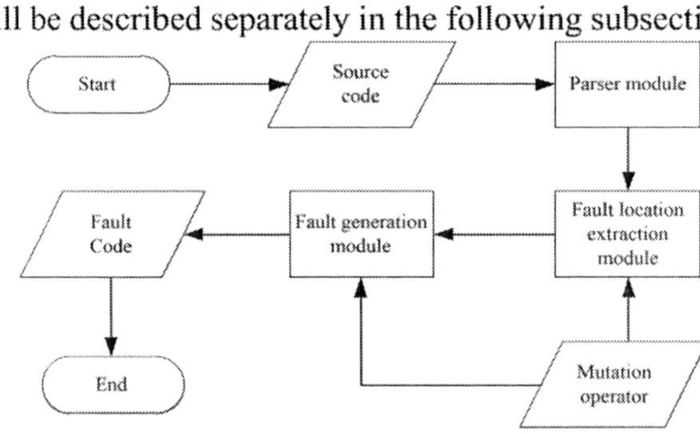

Figure 1. Fault injection process.

Figure 1 shows the specific business process. First, the parser module can get the abstract syntax tree of the program to be injected. Then, based on the given mutation operator, fault location extraction module analyzes the syntax tree structure and extracts the syntax tree node corresponding to the fault location. And it outputs a fault source object set indicating a fault injection position. And then, the fault generation module analyzes the mutation operator to be injected and the extracted fault location, generates a corresponding replacement code and outputs the variant object in the form of a text replacement instruction. Finally, the generated fault text replacement instruction is applied to the source file code portion to generate the corresponding fault program.

2.3. Design of fault location extraction

Fault injection needs to extract the abstract syntax tree first, and then extracts the fault location based on this. So the next step is to design a function to identify and extract the set of fault location nodes from the abstract syntax tree. For example, when the *AROR* mutation operator is specified, the abstract syntax tree will be traversed in breadth, and each *SimpleNode* will be judged to determine whether it conforms to the rules of *AROR* injection.

Figure 2 shows the specific process. In this process, range option requires users to specify the scope of the program to be injected when using the module interface, and the type option requires users to specify mutation operator to be injected into the fault. The last return option provides two types of fault location output mode, first is to output directly through the return value, and the other is that user provides the fault location manager object, and the module adds extracted fault location to the manager.

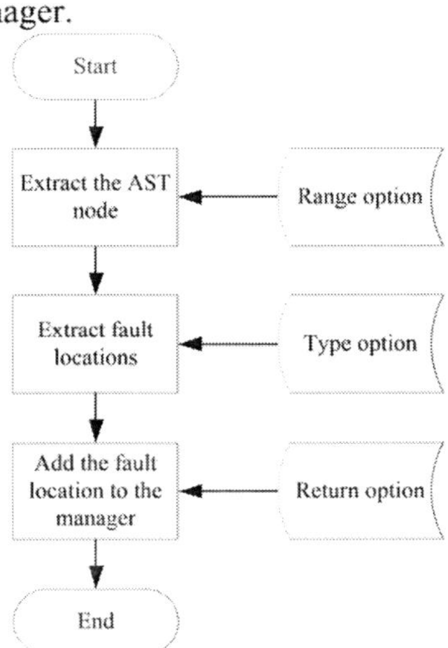

Figure 2. Fault location extraction process.

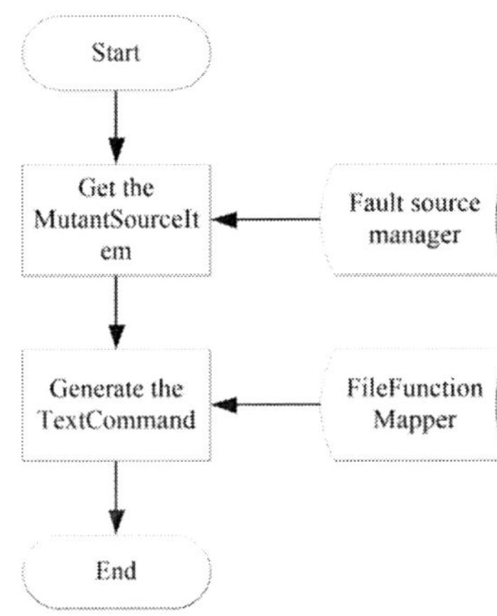

Figure 3. Fault generation process.

2.4. Design of fault generation

After extracting the fault injection location, the next step is to generate the fault code. The specific process is to calculate the corresponding fault replacement code according to the code conversion rules of mutation operator on the given fault source, and output the corresponding text conversion instruction. So the end result is a text conversion instruction. Essentially, the process of program mutation is the process of program text replacement.

The specific process is shown in figure 3. First, the fault output module receives the fault source manager as an input, traverses and extracts the fault source object in the manager space; and then, analyzes the source code text and the location information provided by the fault source. After that, generates the corresponding replacement code. Finally, outputs the text replacement instruction set corresponding to the fault source.

2.5. Data model

The automatic fault injection tool involves the following main data structures:

1) Abstract syntax tree node *SimpleNode*: it's an abstract data type representing the structure of the program code.

2) Text index library *TextIndexLib*: this class provides a mapping from the code index of the *SimpleNode* to the string order of the program code. It ensures the correspondence between the syntax tree node and the program code.

3) File function mapper *FileFunctionMapper*: it provides the access interface for the fault injection process. For example, it can provide information such as code text, an abstract syntax tree and so on.

4) Fault type *MutantType*: the enumeration type, that is, the 18 mutation operators described above, used to describe the type of mutation operator.

5) Fault source *MutantSourceItem*:it indicates the location of the fault to be injected, and it includes two basic attributes, namely the mutation operator type and the fault location node.

6) Code change instruction *TextCommand*: this instruction regards the fault generation process as a string replacement operation on the original code, and it has three attributes: *command*, *location* and *text*. The *command* is a string operation instruction, which may be replacement, delete or insert. The *location* is the position of the string to be replaced and *text* is a replacement string.

3. Experimental test and result analysis

3.1. Case study

As shown in figure 4, this is a very simple code that defines three variables and has a nested loop. Then, we performed several tests on this code and analyzed the results.

```
1 int main() {
2   int i, j;
3   int count = 0 * 2;
4   for(i = 0; i < 5; i++)
5     for(j = 0; j < 5; j++)
6       count++;
7   return 0;
8 }
```

Figure 4. Sample code.

As shown in table 1, it gives the mutation operator and the number and location of faults corresponding to it. Six mutation operators are specified for this code, namely *VARP*, *AROR*, *TIAS*, *LOCN*, *REOR* and *CORP*. The fault location is represented by a binary coordinate (*line, column*), where *line* represents the number of rows in which the fault is located, and *column* represents the number of columns of the starting character at the location of the fault.

Table 1. Test results.

Mutation operator	Number of fault locations	Fault location
VARP	7	(7,4) (5,10) (5,6) (6,11) (6,15) (6,14) (6,7)
AROR	4	(3,14) (5,14) (6,15) (7,14)
TIAS	2	(6,3) (5,2)
LOCN	2	(5,10) (6,11)
REOR	2	(5,10) (6,11)
CORP	7	(5,8) (6,13) (6,9) (3,16) (3,14) (5,12) (9,9)

Table 2 shows partial mutation operators and its corresponding fault location and replacement instructions. The format of the replacement code is *command (start, end, text)*, and the *command* is a text replacement command. There are three types of *REPLACE*, *INSERT*, and *DELETE*. *start* indicates the starting position of the fault to be injected into the source code, and *end* indicates the end position. And *text* is the fault code to be replaced.

Table 2. Partial test results.

Mutation operator	Fault location	Replacement code
AROP	5,14	[REPLACE(55, 57, "--")]
	6,15	[REPLACE(74, 76, "--")]
	7,4	[REPLACE(86, 88, "--")]
	3,14	[REPLACE(36, 37, "+"), REPLACE(36, 37, "%"), REPLACE(36, 37, "-"), REPLACE(36, 37, "/")]
TIAS	6,3	[INSERT(81, ";")]
	5,2	[INSERT(61, ";")]

At the same time, it generated 41 fault files. Comparing the results of the above two tables, we can find that the generated fault codes and their numbers are correct.

3.2. Project experiment

We tested three projects, qlib, deco and aa200c. Table 3 shows partial test results of aa200c.

Table 3. Partial test results of aa200c.

File	Func	Killed	Survivor	Invalid	Valid	Score
rotate	epmat	1,932	0	12	1,932	1
moonjpl	domoon	7,632	0	49	7,632	1
deltat	update	1,144	0	7	1,144	1
rplanet	doplanet	85	0	0	85	1
nutate	nutate	3,431	1	0	3,432	0.999,708,6
dms	dms	859	89	91	948	0.906,118,1

File indicates the source file name in aa200c, *Func* is the function name in the file, *Killed* and *Survivor* respectively indicate the number of faults that have not been detected and detected. *Invalid* indicates the number of fault programs that have compiled errors, and *Valid* indicates the number of fault programs compiled successfully. *Score* indicates the fault detection rate.

As can be seen from table 4, the tool generated a large number of fault programs, and the compilation success rate is also high. Except for a few generated fault programs that cannot be compiled, most of them are valid and can be used for mutation testing.

Table 4. Fault compilation accuracy rate of the tested project.

Project name	Total number of generated faults	Compilable fault number	Invalid number of faults	Compilation success rate
aa200c	127,386	124,083	3,303	97.407%
deco	12,708	12,538	170	98.662%
qlib	7,320	7,000	320	95.628%

According to the above two tables, we can see that the automatic fault injection tool realizes a predetermined function. It can generate a large number of fault programs, which have a high compilation success rate.

4. Conclusion

This paper implements an automatic fault injection tool. It has been proved by experiments that automatic fault injection greatly improves efficiency and has a high compilation success rate. However, this tool still has some shortcomings. It cannot eliminate equivalent faults, and it currently does not support statement substitution faults, because the 18 mutation operators[7] implemented by the tool only cover 59 of the 77 mutation operators of the standard C89. These things need to be further researched.

This paper is supported in part by the National Natural Science Foundation of China under Grant No.U1736110 and No.61702044.

References

[1] A.T. Acree, T.A. Budd, R.A. DeMillo, R.J. Lipton, and F.G. Sayward, Mutation Analysis (Technical Report GIT-ICS-79/08, Georgia Inst. of Technology, 1979)

[2] H. Agrawal, R.A. DeMillo, B. Hathaway, W. Hsu, W. Hsu, E.W. Krauser, R.J. Martin, A.P. Mathur, and E. Spafford, Design of Mutant Operators for the C Programming Language (Technical Report SERC-TR-41-P, Purdue Univ., 1989)

[3] P. Chevalley, Applying Mutation Analysis for Object-Oriented Programs Using a Reflective Approach , Proc. Eighth Asia-Pacific Software Eng. Conf., 267 (2001)

[4] A. J. Offutt, R. H. Untch, Mutation 2000 : Uniting the Orthogonal, Proceedings of the 1st Workshop on Mutation Analysis, 34-44,(2001)

[5] Y. Jia, M. Harman, An Analysis and Survey of the Development of Mutation Testing, IEEE Trans. Software Eng., **37**:649-78,(2011)

[6] Offutt J, Pan Jie. Automatically detecting equivalent mutants and infeasible paths[J]. Software Testing, Verification and Reliability, 165 (1997)

[7] H. Song, Y. Wang, Y. Gong, A Mutation-based Fault Injection Method for C Program, (2014). http://www.paper.edu.cn/releasepaper/content/201412-725

REDT: Remote exploitation detection technology for network infrastructure

Xiaokang Yin, Shengli Liu*, Fangfang Jia and Da Xiao

State Key Laboratory of Mathematical Engineering and Advanced Computing, Zhengzhou, CO 450000, China

*Email: mr.shengliliu@gmail.com

Abstract. An attacker exploits a remote vulnerability to execute arbitrary code on network infrastructure. It seriously threatens the security of network infrastructure. To detect the exploitation attack in a timely and effective way, in this paper, we present REDT, a remote exploitation detection technology based on the effect of exploitation attack. This method determines whether there is an authentication bypass exploitation by detecting the remote authentication process and saves attack traffic to the anomalous traffic database. With the database, we can replay the attack and analyze the exploitation. In the experiment, the detection accuracy rate reached 97.67%. In addition, this method does not rely on the signatures of the vulnerability and shellcode, so it can detect undisclosed vulnerabilities exploitation and provide abnormal traffic for the study of unknown vulnerabilities.

1. Introduction

With the increasing exposure of vulnerabilities in network infrastructure such as routers, switches, firewalls, and Internet of Things (IoT), the security of network infrastructure has attracted widespread attention from security researchers [1-5]. Unlike the conventional software, the firmware running on the network infrastructure can't directly be patched to repair the vulnerability. The administrator has to download the patched firmware from the vendor support site and replace the original one to repair the vulnerabilities, which is tedious and complicated. Therefore, most administrators rarely upgrade the firmware [6]. So, there are a lot of vulnerabilities in the running firmware. How to efficiently and accurately detect exploitation for embedded systems has grown up to be an important research topic [7-11].

Currently, a lot of studies have been made to detect remote exploitation. Packet vaccine mechanism was introduced into exploitation detection by [12]. This method randomizes address-like strings in packet payloads to carry out fast exploitation detection, vulnerability diagnosis and signature generation. But it also relies on the signatures of shellcode or vulnerabilities. Michalis Polychronakis et al. [13] had studied more than 1.2 million polymorphic code injection attacks targeting production systems, captured using network-level emulation. They focused on the analysis of the structure and operation of the attack code, as well as the overall attack activity in relation to the targeted services. They provide a great reference for extracting exploit signature. Alex Shenfield et al. [14] put forward a novel approach to detection of malicious network traffic using artificial neural networks. This method greatly enhances the accuracy and usability of the intrusion detection system. However, this method is hard to learn the signatures of exploitation from the encrypted network traffic. At the DefCamp 2017 security conference in Bucharest, Romania, Kirill Shipulin [15] had explained how to turn available signatures against the IDS to paralyze its operation. Attackers can construct malformed packets with public exploit signatures and send them to IDS. So, this indicates that remote exploitation detection based on signature is not perfect and can be abused by attackers.

To the best of our knowledge, most of the already known methods are based on the signatures of shellcode or vulnerabilities in essence. To address the problem of lack of samples and hysteresis problem of signature-based exploitation detection on network infrastructure, we proposed REDT, a remote vulnerability exploitation detection system based on remote login authentication bypass. To evaluate the efficacy and performance of REDT, we have implemented it as a proof-of-concept in python. We evaluated REDT in terms of accuracy in identifying the exploitation with the exploits published on the exploit-db. In the experiment, REDT obtained an average accuracy 97.67%, and an average false alarm rate 2.75%. What's more, the novel approach can detect the zero-day exploitation attack and discovery undisclosed vulnerability on network infrastructure. It also saves the exploitation network traffic into abnormal traffic database. The database can be used to analyze the attack and extract vulnerability information.

The contributions of this paper are as follows:

1. We proposed a remote exploitation detection technology for network infrastructure.
2. We analyzed the remote login process and the difference between normal login and remote exploitation.
3. We designed and implemented REDT in python as a proof-of-concept. We verified it in LAN environments with the public exploits on the internet.

The remainder of this paper is structured as follows. Section 2 reviews the challenges of exploitation detection. Section 3 describes the proposed approach to authentication bypass exploitation attack. Section 4 presents the experiment and evaluates REDT with public exploits. Finally, section 5 provides the conclusion.

2. Challenge

In this section, we describe the challenges faced in exploitation detection.

2.1. Difficult to extract the vulnerability signature

The triggering conditions of different vulnerabilities are various. And different vulnerabilities have different triggering conditions. In addition, after the new vulnerabilities are discovered, it still takes some time to study and analyze the vulnerabilities to obtain much more accurate characteristics.

2.2. Exploit diversity

Even for the same purpose of the attack, the exploits are also different. The triggering conditions of the same vulnerability are also diverse. For example, Cisco IOS IP Options vulnerability, CVE-2007-0480, can be triggered with up to seven types of packets for this vulnerability. Therefore, attackers could develop a great number of exploits to exploit this vulnerability in different routers. For network infrastructure, such as Cisco routers, switches and IoT, due to differences in different versions, resulting in different addresses, functions, etc., exploits cannot be directly generated by using tools like Metasploit. Attackers have to develop shellcode for each version manually. So, it is difficult to extract the signature of exploit developed manually.

2.3. Obfuscation exploit signature with encryption

Figure 1 presents the structure of common shellcode. The attacker can use a simple encryption method (XOR, NOR operation, etc.) to confuse the features. Figure 2 shows the structure of the encrypted shellcode. The attacker encapsulates the decoder at the front end of the shellcode and shellcode decodes the encrypted payload through the decoder. After the encryption, the original characteristics of the shellcode no longer exist. Qinghua Zhang et al. [16] proposed an approach to detect self-decrypting exploit code. This proposed method combines static analysis and emulated instruction execution techniques to detect such highly obfuscated polymorphic exploit code. But non-exploit code that uses code obfuscation, code encryption, and self-modification for legitimate purpose will interfere the detection process resulting in false alarm.

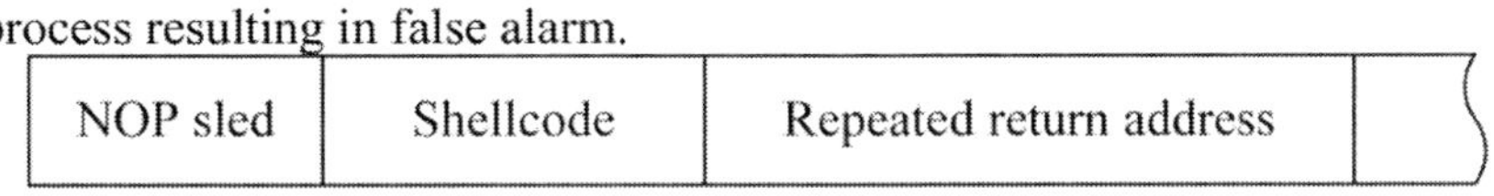

Figure 1. Structure of shellcode [16].

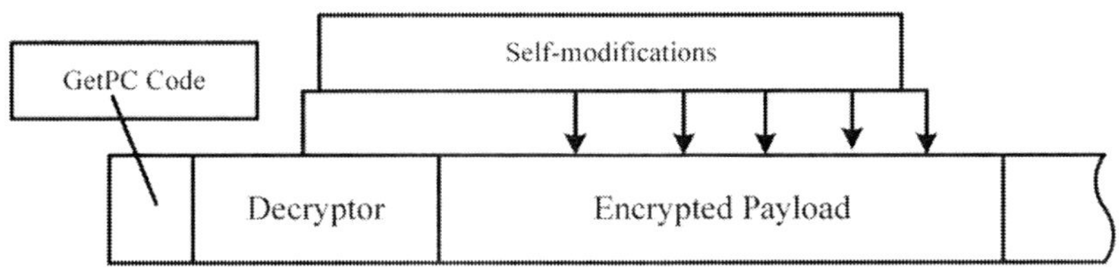

Figure 2. Structure of encrypted shellcode [13].

2.4. Bypasses IDS using signatures

As mentioned in the previous section, available IDS signatures can be turned against the IDS to paralyze its defenses. Attackers construct a large number of malformed packets containing public vulnerabilities and exploits signatures and send them to IDS, which leads to IDS paralysis. So, the real attack can bypass the defense of IDS.

3. Detection process

In general, the entire process of exploitation attack on network infrastructure could be divided into five steps [17]: a) scan device and collect fingerprint, b) bypass remote login authentication and gain the control, c) check and modify the configuration, implant backdoor or update firmware and so on, d) clear attack traces, e) restore remote login authentication. The following heuristic experiences exist in the network environment: A network infrastructure configured correctly with remote login service on must have the login authentication mechanism. Attackers often use exploit to attack the network infrastructure for the purpose of gaining the control. After the network infrastructure under attack, attacker can login without password or with arbitrary password. In order to achieve the purpose of the attack and conceal their own aggressive attack, attackers will complete the attack within a short time.

3.1. Normal and abnormal authentication

This paper takes telnet as an example to illustrate the remote login authentication process and analyze the difference between normal and abnormal login. Telnet establishes connection and transfers data with the TCP protocol. During the authentication phase, the server requests login authentication password from the client and the client must answer it with correct username, password. After authentication, the client sends commands to the server, and the server executes the commands and returns the result to the client.

After attackers attacks the network infrastructure with login authentication bypass exploitation, the telnet login authentication process is destroyed. We analysed the public exploits for network infrastructure and found that the login authentication bypass attacks can be divided into two types: login authentication with an arbitrary password and login authentication without password. Arbitrary password login still contains the authentication request process, but the attacker can input any password or directly enter the 'enter' to login the network infrastructure. Login authentication without password returns a command shell once the telnet connection is established without asking for password. And the entire process misses the authentication phase. After login to the network infrastructure, the attacker can totally control it, view and modify the configuration, implant backdoor, replace firmware and so on.

3.2. Detection algorithm and process

We divide the entire remote exploitation detection system into five modules: traffic processing, authentication anomaly detection, anomalous traffic database, attack alert and restore attack process. figure 3 show the detection processes.

Traffic processing module is placed at the middle of internet and infrastructure device. It captures all the traffic passing through them, stores the network traffic into the traffic buffer queue and authentication traffic queue. Traffic collection and filtering module extracts the authentication-related packets with certain rules and forwards them to the authentication anomaly detection module. The filter criteria are based on the IP address and TCP port. Since our goal is to detect attacks against the infrastructure device, we need the IP address of infrastructure device to be protected, which can be one or more. So, the traffic buffer queue stores all the data traffic that communicates with the protected device and the authentication traffic queue stores the remote authentication login traffic. As we all known, currently, the remote login protocol mainly is Telnet, SSH and HTTP/HTTPS for infrastructure device not containing the private protocols. Those protocols have specific TCP port. The port, for Telnet, is 23, for SSH, is 22, and for HTTP/HTTPS is 80 or 443.

The Authentication anomaly detection module analyses the authentication traffic and judges whether the login authentication process is broken or not. The authentication process starts from the three-way handshake of TCP. As we mentioned above, when there is a remote authentication bypass exploitation, the authentication login process will be broken, and attacker can login the device without password or any password. To accurately detect the any password login attack, we'd better put the correct remote login password into REDT, so, it will know whether people login the device with correct password or not. Then we can know whether there is a remote authentication bypass exploitation or not.

When an attack is detected, the attack alert module will send an alert message. The message contains the attack IP address, time and the IP address of attacked device. If REDT is integrated into IDS, it can be more intelligent, for example, send an alert message to administrator with SMS or Email.

The abnormal traffic database stores the suspected attack traffic. When the authentication anomaly detection detects the authentication process is broken, it will notify the traffic buffer queue to stores all the traffic in the queue into the abnormal traffic database. In order to capture all attack traffic, there is delay storing the traffic to collect more attack packets. Then we can know what the attacker does after login the device and provide traffic to analyze the attack means. That traffic may contain the attack traffic with exploit. From that traffic, we can restore the exploit and vulnerability that attacker used to bypass the authentication login. If the vulnerability is not disclosed, we will discover the zero-day vulnerability.

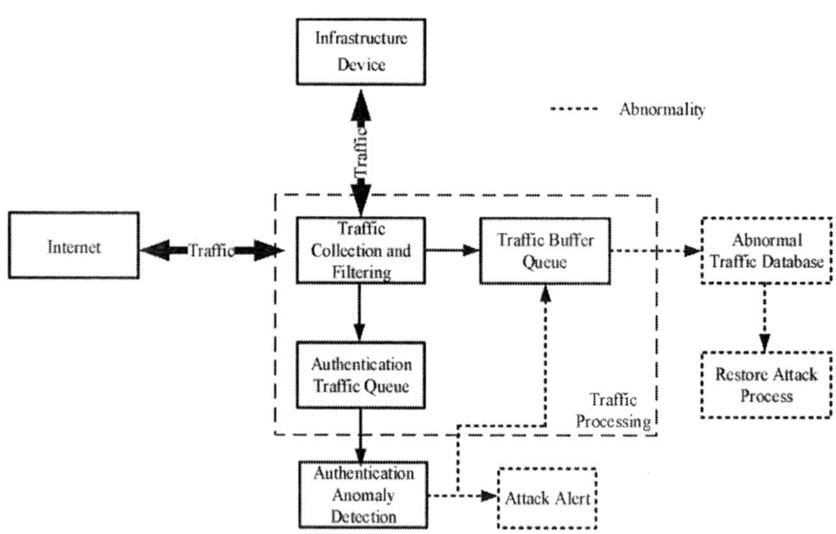

Figure 3. Exploitation detection process based on authentication bypass.

The restore attack process is an offline program and it is to restore the attack process with the traffic stored in the abnormal traffic database. It is used to support the researcher to analyse the attack traffic and verify whether the attack used a published vulnerability or zero-day vulnerability.

4. Experiments and analysis

In this section, we first present the environment used in the experiments in detail, and then analyze the experiment result and evaluate the proposed approach.

4.1. Environment

To evaluate the proposed approach, we use an automation test program to randomly perform simulated exploitation attack, normal login, and error login. Within a period of time, the automation test program randomly executes one operation and records the operation and operation time, then writes them into the file named attack_log. The attack_log contains the operation time and the event, for example {2018.08.08.15:20:15 attack 192.168.1.1 router}. REDT will saves the detection result into file named de_attack_log with detection time and IP of attackers. The de_attack_log contains the detection time, the IP addresses of attack and the IP addresses of attacked devices, for example {2018.08.08.15:20:16 192.168.1.168 attack 192.168.1.1}. By comparing the time between attack_log and de_attack_log, we can judge the detection whether is correct or not. But in the experiment, in order to capture the communication traffic among all devices, one of the ports of switch is configured as a listening port. So, all traffic forwards to the listening port. Table 1 shows the vulnerabilities used in the experiments. The protocol in the table shows the device can't correctly deal with this protocol leading to the vulnerability.

Table 1. Vulnerabilities used in this experiment.

System	CVE	Vulnerability Type(s)	Protocol	CVSS Score
Cisco IOS	CVE-2007-0480	Denial of Service, Overflow, Execute Code	IP Option	10.0
Cisco IOS	CVE-2007-2586	Execute Code, Overflow	FTP	9.3
Cisco IOS	CVE-2017-6736 (36-44)	Execute Code	SNMP	9.0
Cisco IOS Cisco IOSXE	CVE-2017-3881	Execute Code	CMP	10.0
Cisco ASA	CVE-2016-6366	Execute Code, Overflow	SNMP	8.5

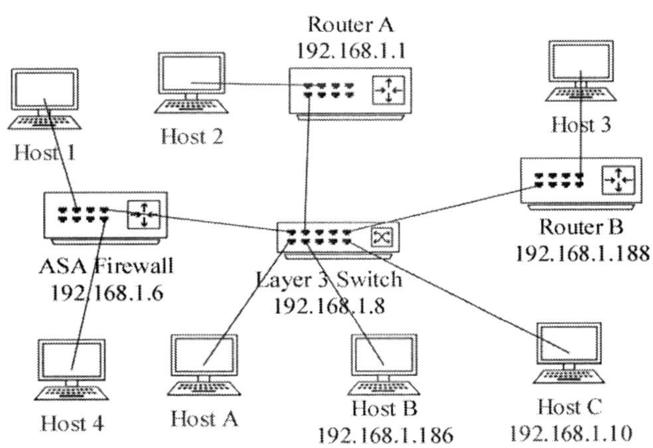

Figure 4. Network topology in the experiment.

Figure 4 presents the experiment network topology. In this topology, Host 1, 2, 3 and 4 communicate with each other with the local network software to simulate the real network. The specific operation is as follows: 1. Host A runs the exploitation detection system and records the time and attack traffic when detects an attack happened. 2.Host B attacks router, switch and ASA firewall in turn with public exploits and records the attack time. 3. Host C logins the network infrastructure, modifies the configuration, and simulates unauthorized users' login with error password. After the experiment, compare file attack_log with file de_attack_log.

4.2. Analysis of results

There are 473 times automated operation with program during the experiment, including exploitation attack, normal login and error login. Table 2 shows the experiment data. Among 473 times operations, there are 218 times exploitation attacks, 215 times normal login and 40 times error login. Comparing the attack_log with de_attack_log, we found that the exploitation detection system correctly detected 214 times attack, regarded normal login as attack with 6 times and considered error login as attack with 1 time by mistake.

Table 2. Experimental data.

Times Operation	Actual Operation	Attack detected
Exploitation attack	218	214
Normal login	215	6
Error login	40	1
Total	473	221

Table 3. Result of exploitation detection.

Accuracy	97.67%
Precision	96.83%
Recall	98.17%
False Alarm Rate	2.75%

Table 3 presents the result of remote exploitation detection system. The experimental result shows that this method can effectively detect the authentication bypass attack with accuracy 97.67% on network infrastructure, and in the experimental, the attack alarm can be issued as long as the attacker logins to the device. And the detection latency is very small about second level. Since we are just concerned about the authentication process packets and have filter the packets with TCP port, actually the packets that we need to analyze is very small.

This method is different from the signature-based remote exploitation detection method. It does not consider the specific vulnerability characteristics. It is not necessary to spend a lot of time analyzing the vulnerabilities and exploits. And it does not care about how the shellcode is constructed and whether the shellcode is encrypted. As long as the exploitation attempts to bypass the login authentication, it can accurately detect the exploitation attack. What's more, it can discover that an attacker uses an undisclosed vulnerability to launch zero-day attacks on network infrastructure.

4.3. Inadequacy

As REDT is based on the abnormality of remote authentication login process after being attack, it can detect the remote authentication bypass exploitation. Due to the integrity of the login authentication process, the loss of authentication packets will lead to false alarm. For the DoS attack or other type exploitation attack, it is unable to detect. For login authentication bypass with an arbitrary password, the detection effect of this method is limited, because the authentication process is not destroyed, that is, the login is no different from the administrator's login under normal circumstances. So, we need to preload the correct password into the program and judge the password that has logged in the device is the correct password or not.

5. Conclusion

In this paper, we proposed a remote exploitation detection method based on the exploitation effect which bypasses the remote login authentication on network infrastructure, such as, routers, switches and firewalls. We implemented this method in python and set up a network environment to evaluate it. This approach accurately detected login authentication bypass attack on network infrastructure in time and reached accuracy 97.67%. What's more, this method has the ability to detect the undisclosed vulnerability exploitation on network infrastructure. This approach also has the characteristics of a small load, low detection delay and does not depend on specific network infrastructure. When a new vulnerability or exploit is published, it is no need to analyze the vulnerability to extract the signature and update the exploitation detection system. The exploitation detection system also stores the exploitation traffic into abnormal traffic database to help security researcher replay the attack behavior and analyze the triggering conditions of new vulnerabilities and exploits.

Acknowledgments

The authors would like to thank the funding of the National Key R&D Plan under Grant 2016YFB0801505 and 2016YFB0801601.

References

[1] Cui, Ang, and Salvatore J. Stolfo. 2011. Defending embedded systems with software symbiotes. International Workshop on Recent Advances in Intrusion Detection. (Berlin, Heidelberg, 2011), RAID'11, Springer-Verlag, pp. 358–377.

[2] Fournaris, A. P., Pocero Fraile, L. and Koufopavlou, O. 2017. Exploiting Hardware Vulnerabilities to Attack Embedded System Devices: a Survey of Potent Microarchitectural Attacks. Electronics, 6(3), 52, 2017, doi: 10.3390/electronics6030052.

[3] Cui, Ang, Michael Costello and Salvatore J. Stolfo. 2013. When Firmware Modifications Attack: A Case Study of Embedded Exploitation. NDSS. 2013.

[4] D'Orazio, Christian J., Kim-Kwang Raymond Choo and Laurence T. Yang. 2017. Data exfiltration from Internet of Things devices: iOS devices as case studies. IEEE Internet of Things Journal 4.2, 2017, pp. 524-535, doi: 10.1109/JIOT.2016.2569094.

[5] Granjal, Jorge, Edmundo Monteiro and Jorge Sá Silva. 2015. Security for the internet of things: a survey of existing protocols and open research issues. IEEE Communications Surveys & Tutorials 17.3, 2015, pp. 1294-1312, doi: 10.1109/COMST.2015.2388550.

[6] Lindner, Felix. Cisco IOS router exploitation. Black Hat USA, 2009. Accessed: April 15, 2018. [Online] Available: https://www.blackhat.com/presentations/bh-usa-09/LINDNER/BHUSA09-Lindner-RouterExploit-SLIDES.pdf

[7] Davi, Lucas, Patrick Koeberl and Ahmad-Reza Sadeghi. 2014. Hardware-assisted fine-grained control-flow integrity: Towards efficient protection of embedded systems against software exploitation. Proceedings of the 51st Annual Design Automation Conference. ACM, 2014, doi: 10.1145/2593069.2596656.

[8] Francillon, Aurélien. 2009. Attacking and protecting constrained embedded systems from control flow attacks. Diss. Institute National Polytechnique de Grenoble-INPG, 2009.

[9] Papp, Dorottya, Zhendong Ma and Levente Buttyan. 2015. Embedded systems security: Threats, vulnerabilities, and attack taxonomy. Privacy, Security and Trust (PST), 2015 13th Annual Conference on. IEEE, 2015, doi: 10.1109/PST.2015.7232966.

[10] Shao, Zili, et al. 2006. Security protection and checking for embedded system integration against buffer overflow attacks via hardware/software. IEEE Transactions on Computers 55.4, 2006, pp. 443-453, doi: 10.1109/TC.2006.59.

[11] Parameswaran, Sri and Tilman Wolf. 2008. Embedded systems security—an overview. Design Automation for Embedded Systems 12.3, 2008, pp. 173-183.
[12] Wang, X., Li, Z., Xu, J., Reiter, M. K., Kil, C. and Choi, J. Y. 2006. Packet vaccine: Black-box exploit detection and signature generation. In Proceedings of the 13th ACM conference on Computer and communications security, ACM, 2006, pp. 37-46 , doi:10.1145/1455518.1455523.
[13] Anagnostakis, K. G., and E. P. Markatos. 2009. An empirical study of real-world polymorphic code injection attacks. In Proceedings of the 2nd USENIX conference on Large-scale exploits and emergent threats, 2009.
[14] Shenfield, A., Day, D. and Ayesh, A. 2018. Intelligent intrusion detection systems using artificial neural networks. ICT Express, 2018, doi: 10.1016/j.icte.2018.04.003.
[15] Kirill Shipulin. 2018. Turning IDS signatures against an IDS itself: a new evasion technique. DefCamp9, 2018. Accessed: May 23, 2018. [Online] Available: https://def.camp/speaker/kirill-shipulin/
[16] Zhang, Q., Reeves, D. S., Ning, P. and Iyer, S. P. 2007. Analyzing network traffic to detect self-decrypting exploit code. In Proceedings of the 2nd ACM symposium on Information, computer and communications security. ACM, 2007, pp. 4-12, doi: 10.1145/1229285.1229291.
[17] Erickson, Jon. 2008. Hacking: the art of exploitation. No starch press, 2008. p.140

IOP Conf. Series: Materials Science and Engineering **715** (2020) 012036 doi:10.1088/1757-899X/715/1/012036

Video stalling detection method

Shushi Yu* and Na Wang
College of Information Engineering, Communication Engineering, Shenzhen University, China

*Email: 739536243@qq.com

Abstract. ISP providing network access services has a huge user group; therefore the competition between the ISP providers is extremely fierce. Under the situation that video occupies the most of network flow, providing high satisfaction network services for video users can effectively improve the competitiveness of ISP, in which video Stalling plays an important index. For protecting the privacy of users, most video websites use HTTPS as the application layer protocol for video transmission, so ISP could not get the playback of video users by parsing HTTP packets, and could not directly detect the video users. In this assay, a new method, protocol transport layer Video Stalling Detection Method based on HTTPS is proposed. This method can effectively match the YouTube, Youku, Letv, Tencent video and other major video sources of the stalling event.

1. Introduction

The main medium of internet communication in 3G era was text, and 4G era was pictures and small videos. With the growing of 5G technology which proposed faster speed of the internet and proposed the release of the internet flow, video will become the mainstream media in the future. It can be predicted that video flow will account for a larger share in near future, with data reporting that the proportion of video flow will rise from 60% in 2016 to 78% in 2021 [1].

As a provider of Internet services, ISP has always been highly competitive [2].The arrival of 5G will be an opportunity for them to re-divide the market, the one who provide a better user experience who will be able to have more users, so it is particularly important to improve the competitiveness of ISP. As video is the main body of network traffic, the user group behind it is huge. For these users, faster network speed means less tolerance of video stalling, therefore improving the viewing experience of video users is an effective means to improve the competitiveness of the ISP industry. Video stalling detection becomes extremely important at the meanwhile. In reference [3], the author estimate the time and frequency of the stalling time (QOS metric) by analysis the packets captured in the transport layer, and modelling the download time, the throughput of network packets and the video buffer. The modelling accuracy of video buffer is the highest; its main method is to compare the playback times of video frames with the time stamps of received packets by parsing data packets. On this basis, reference [4] adds some additional parameters, including video playback time, playback delay and so on, to improve the estimation accuracy and improve the model. In reference [5], the author makes statistical analysis on average packet loss, loss period, loss distance, RTT and other user satisfaction (QOS) metrics by using machine learning algorithms such as support vector machine (SVM) (SVM), neural network (KNN) and Logistic regression algorithm, and eventually realized the classification of user satisfaction, but this method needs to add the observation flow actively and consume extra bandwidth.

All of the above methods need to parse the HTTP package. With the user privacy has been concerned, more and more video providers abandon the former HTTP as the application layer protocol, and use HTTPS as the application layer protocol. This makes it infeasible that obtain the user buffer state by parsing the HTTP to evaluate the video Stalling probability of the user. In Reference [6-8], scholars use the relevant models and algorithms of machine learning to apply to video stalling detection. However, machine learning methods often require a large number of training samples in order to achieve a higher prediction accuracy of the model, which means that a large number of tagged video is required, that is undoubtedly a huge labor and material resources of the project.

In this paper, we propose a new method for detecting and evaluating video stalling based on HTTPS video of ISP. By capturing the TCP/IP packet information of user's video in real time and using the Stalling matching algorithm, the outlier points of video stalling can be matched. The decision analysis is added to the matching segment to realize the matching of the stalling fragment. By collecting and experimenting on various video sources of various video websites, the final results show that this method has a good effect on the estimation of stalling. The main features of this method are:

- Able to adapt to HTTPS protocol.
- Use simple and efficient models.

2. Video stalling detection method

Video Stalling refers to the slow buffering of video in the process of watching video, such as network fluctuations, which makes the downloaded video segment consume faster than the cache speed, and leads to the end of the client buffer depletion and then video pause. The video data message is downloaded from the video source server to the client buffer, and it will go through a dynamic interactive process from transmission to play. Next, we will analyse the phenomenon of video stalling in the bottom layer of the client and the transport layer.

The performance of video client-side is that the client-side buffer segment is depleted and the client-side buffer fragment is less than the maximum playback threshold, which results in the video not being able to play normally [3]. As shown in figure 1, the horizontal axis represents the timestamp and the vertical axis indicates the client video buffer size.

•Video starts playing from Click to download when the playback threshold θ is reached. Before that, the video was in a stall state with time Δt_1.

•Video stalling during the period from video buffer exhaustion to download to playback threshold θ with time Δt_2.

In these two stages, the video is in a stalling state. As shown by the red dotted line in the figure 1.

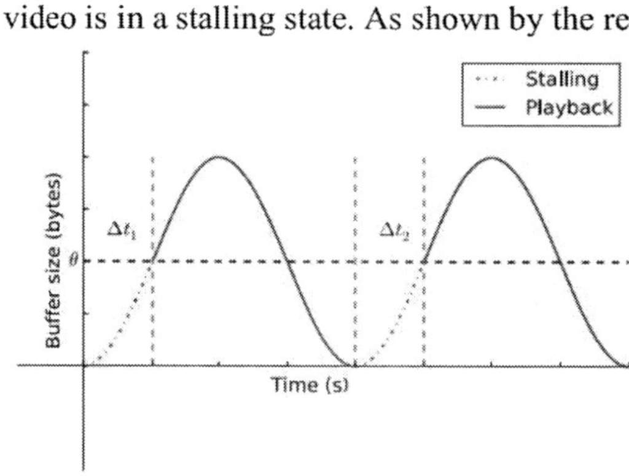

Figure 1. Video buffer changes.

Combining buffer changes, download speeds, and message accumulations. In figure 2, the horizontal axis represents the timestamp, and the three zigzag line segments represent the TCP_Segment_Sum, Buffer_Size, and Download_Speed. We can see that after 140 seconds of speed limits, the client buffer changes within a minimum range and tends to 0 at some point, according to the description of figure 1, when the video appears to be a stalling phenomenon. The description of the video stalling is:

1)The buffer is depleted;

2)The download speed is too low for a certain period of time;

3)The slope of the curve in the Cumulative Variation of the Video Packet diagram begins to decrease.

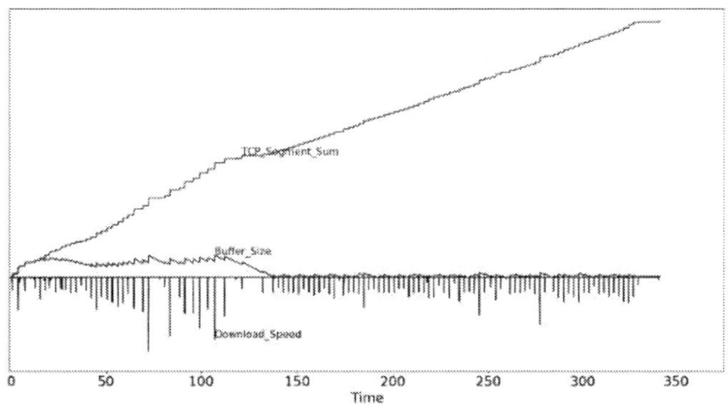

Figure 2. Buffer size, video download speed, the cumulative variation of the video packet.

The description in 3) means that the download of TCP values Q_d in accumulative graph will decrease in a certain period of time T. It's different from the front area, we call this region an abnormal segment. Next we prove that video Stalling is related to unit time T and the download values Q_d.

In the case of speed limit, we assume that the video buffer from the beginning to the time of occurrence of stalling is T and the average download speed of video is V_d, the average video playback speed is V_p, the client buffer threshold is Q. Then, T shows below:

$$\mathrm{T} = \frac{Q}{V_p - V_d} \tag{1}$$

During video playback, for a given Q and V values, when the network fluctuates, causing V_d Smaller, then:

$$\mathrm{T} = \begin{cases} \infty, while\ V_p - V_d < \ni \text{ ,} where\ \ni is\ Infinite\ small, and\ \ni\ > 0 \\ \partial, while\ V_p \gg V_d,\ \ where\ \partial\ is\ Constant \end{cases} \tag{2}$$

T shows the time of stalling happening. When the network fluctuates, the average download speed is less than the average playback speed, and the video will occur after the time T, T is a constant.

Besides, the formula can be written in another form. We assume that in duration time T, the video download values is Q_d, the amount of video playback is Q_v ,the buffer threshold value is Q, and the Stalling event is S which is a Boolean value. When S equalling 1 means a Stalling occurs, then:

$$\mathrm{S} = \begin{cases} 1, while\ Q_v - Q_d > Q \\ 0, while\ \ Q_v - Q_d \le Q \end{cases} \tag{3}$$

From the estimation of T in the above formula and the Boolean judgement of the Stalling event, it can be seen that whether the video stalling occurs or not, relates to the download quantity Q_d within T in this time period. So we can do this by T and Q_d to carry out parameter estimation to fit the stalling event and realize the stalling matching. Of which L_{tcp} it represents the cumulative length of TCP fragments in a certain period of time, and lambda denotes the threshold value. When T value is greater than the threshold T_0, it is considered as an abnormal fragment, that is, the fragment of Stalling may occur. Here, why do we achieve the best match, we construct a matching function F. F to achieve the best segment matching for the input fragment.

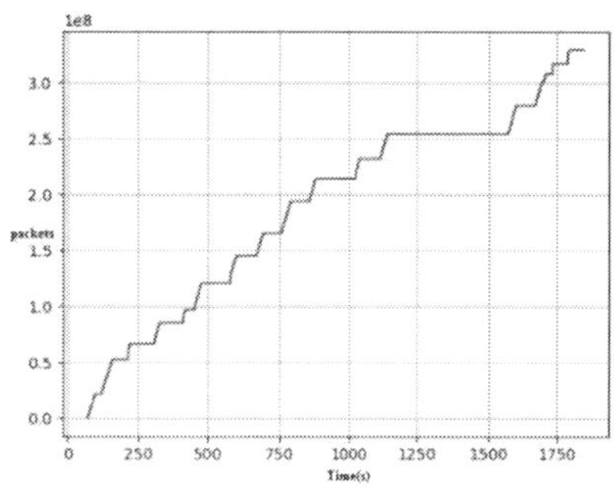

a. TCP Message Accumulation Graph.

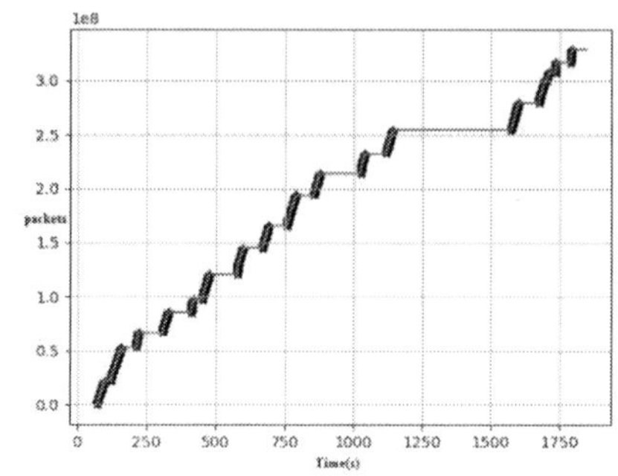

b. Matching TCP Message Accumulation Graph.

Figure 3. Matching analysis.

Through the matching algorithm, the optimal matching is achieved, as shown in figure 3. The horizontal axis represents the timestamp, the vertical axis represents the cumulative length of the TCP/IP package. Figure 3a means when the client plays, the accumulated information of video message length captured in the experiment. Figure 3b is the result of the Stalling matching algorithm for the accumulated message shown in figure 3a. After estimating the parameters of the time length T in the stalling matching algorithm and the download quantity Q, the best match is realized, and leads to the matching of stalling by getting the length of the download interval in the playback process. The experimental results show that the accuracy of this method is over 97% for stalling event matching.

3. Experiment and the result

3.1. Video source and collection distribution

In order to verify the effectiveness of the method, we sampled videos of different duration of major mainstream websites with equal probability, as shown in figure 4. In view of the characteristics of the method, we only need to detect the short-time Stalling. From figure 5, we can see that 50% of the Stalling time distribution is about 1 second; 90% of the Stalling length distribution is less than 15 seconds; less than 2% of video stalling are longer than 50 seconds, mainly because some of them are too long due to some uncontrollable factors during the experiment:

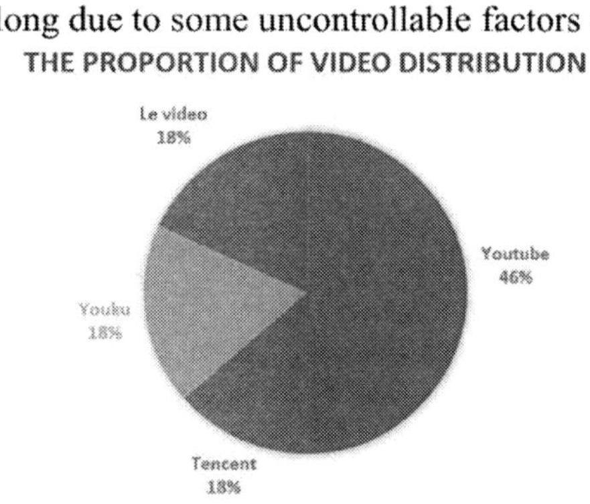

Figure 4. Stalling times distribution.

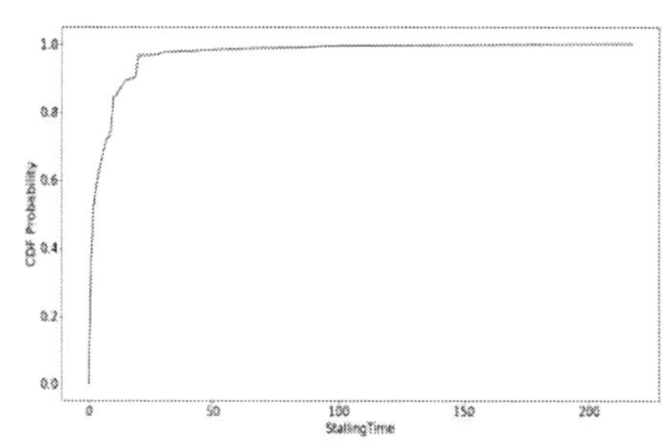

Figure 5. Stalling times distribution.

3.2. Stalling detection accuracy

Next, we will show the results of the whole experiment. The main purpose of this experiment is to detect different Stalling time, and the experimental objects include all the main video sources. For ISP, the key point of this paper is whether or not we can effectively identify the Stalling events of video users. We illustrate this by analyzing the detection accuracy of different video sources (domestic and foreign) with different effective playback time distribution.

The following is a histogram of the accuracy rate of different video websites and different video Stallings in the course of the experiment:

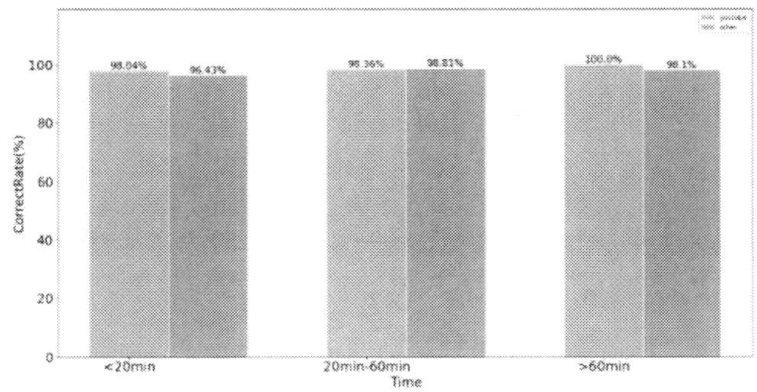

Figure 6. The horizontal axis shows the video duration, the vertical axis represents the accuracy of the Stalling decision, and the left and right histograms represent the YouTube and other video sources, respectively.

We draw columnar diagrams for different video transmission technologies at home and abroad, and for different playback time long video sources. As can be seen from figure 6, this method has a very high accuracy for different video sources and different video types.

4. Conclusion

The experiment is to have the stalling detection of the video transmitted by HTTPS protocol based on ISP. In the process of the experiment, through the Wireshark capture video message to simulate the ISP side, in the case of non-parsing video messages we only use the message basic information to realize video stalling detection. After that, through the video capture and experiment of 4 mainstream video sources, a total of 66 videos were collected including YouTube, and various user behaviours were added several times. The result of abnormal analysis shows that the total matching rate of stalling is 98.42%, and the matching rate of domestic video is 97.99% and that of foreign video is 99.04%.

References

[1] L. Ioannis, et al, Artificial Neural Networks for Traffic Prediction in 4G Networks(International Wireless Internet Conference Springer, Cham,2014).
[2] D. Jamie, COMPETITIVE EFFECTS ON QUALITY COMPETITION: A Study of the U.S. Internet Service Provider Market (Economics, 2018).
[3] S. Raimund, T. Hobfeld, and P. Casas, Passive YouTube QoE Monitoring for ISPs. (IEEE, Sixth International Conference on Innovative Mobile & Internet Services in Ubiquitous Computing,2012).
[4] C. Pedro, R. Schatz, and T. Hoßfeld, Monitoring YouTube QoE: Is Your Mobile Network Delivering the Right Experience to your Customers? (Wireless Communications & Networking Conference IEEE, 2013).
[5] M. Umar Nasir, Monitoring Network Congestion in Wi-Fi, based on QoE in HTTP Video Steaming Services(Lunds universitet,elektro- och informationsteknik,2016).
[6] D. Giorgos, et al, Measuring Video QoE from Encrypted Traffic(The 2016 ACM ACM, 2016).
[7] D. Tsilimantos, T. Karagkioules, S. Valentin, Classifying flows and buffer state for YouTube's HTTP adaptive streaming service in mobile networks, arXiv:1803,00303(2018).
[8] A. Vaneet, et al, Prometheus: toward quality-of-experience estimation for mobile apps from passive network measurements (Workshop on Mobile Computing Systems , 2014).

Discovering malware based on co-clustering host-domain graphs

Wenqian Zhang[1,*], WeiNa Niu[2], Jinhong Zhang[3] and Jingqi Li[4]

[1]The University of Electronic Science and Technology of China ,Chengdu,Sichuan, China
[2]The Sichuan University ,Chengdu,Sichuan, China
[3]The University of Electronic Science and Technology of China ,Chengdu,Sichuan, China
[4]The University of Electronic Science and Technology of China ,Chengdu,Sichuan, China

*Email: a306810397@gmail.com

Abstract. Malware domain discovery based on passive DNS graph analysis supplements existing methods through DNS request and response data analysis. However, the method does not consider the IP relevance and the high complexity of the suspicious computing process. This paper proposes a malware domain discovery method based on passive DNS and IP relevance. The method calculates reputation score, which combines the shortest path from malicious domains to unknown domain with the malicious IP ratio to determine whether the domain malicious.

1. Introduction

Attackers often use malware to control infected hosts to steal sensitive information. At present, behavior-based malware detection method has high accuracy by running malware in sandbox environment to analyze its dynamic behavior characteristics. However, this method is not only time-consuming but also has huge resource overhead.

In order to improve the recognition rate of Issa's mothed, a malware domain name discovery method using global domain name and IP correlation is proposed in this paper. Firstly, the request domain name record is filtered according to the domain name whitelist, and the suspicious domain name is retained. Secondly, the passive DNS database is queried to generate the domain name-IP two-part graph. Then, according to domain name and IP blacklist, the selection of seed domain node and the calculation of suspicious IP ratio of unknown domain name are carried out respectively. Finally, the unknown domain name prestige value is inferred by combining the suspicious score of domain name and the suspicious IP ratio of connection, and the domain name larger than the threshold value is identified as malware domain name.

2. Related work

At present, there are many methods for malware domain name identification for DNS traffic at home and abroad, but in general, it mainly focuses on primary and passive DNS data analysis, host-domain name and domain name-IP graph reasoning.

IOP Conf. Series: Materials Science and Engineering **715** (2020) 012037 doi:10.1088/1757-899X/715/1/012037

In the enterprise network, the host-domain map reasoning and active DNS data analysis are combined to detect the malware domain name. According to the belief propagation algorithm, the suspicious domain name is searched in the host-domain name map [1]. Xu L proposes an APT malware detection method based on malicious DNS and traffic analysis [2], which extracts features based on active DNS records and combines decision tree algorithms to identify malicious domain names.

Compared with previous studies using DNS analysis to detect malicious domain names, the method in this paper is quite different. Passive DNS analysis method is used, and compared with reference [3], [1], and [2], it is not necessary to obtain large-scale data for characteristics. Compared with the reference [1], the analysis based on the requested domain name record requires a relatively small amount of data and is highly targeted. However, when calculating the suspiciousness of the unlabeled domain name, the method needs to calculate the path of all the seed nodes to the current domain name, and only considers the influence of the domain name on it, and the time complexity and the false positive rate are high [4].

3. Malware domain name discovery based on passive DNS domain name and IP affinity

This section describes the proposed method for discovering potential malware domains based on passive DNS domain names and IP affinity[5]. The flow chart of the proposed method is shown in figure 1. It mainly consists of two parts: 1. The pre-processing stage is responsible for filtering the normal domain name according to the Alexa value and the popular domain list in the monitored network; 2. Using graph reasoning to identify the potential malicious domain name. wherein the domain name suspicious score is determined by the global domain name relevance and the proportion of connected malicious IP.

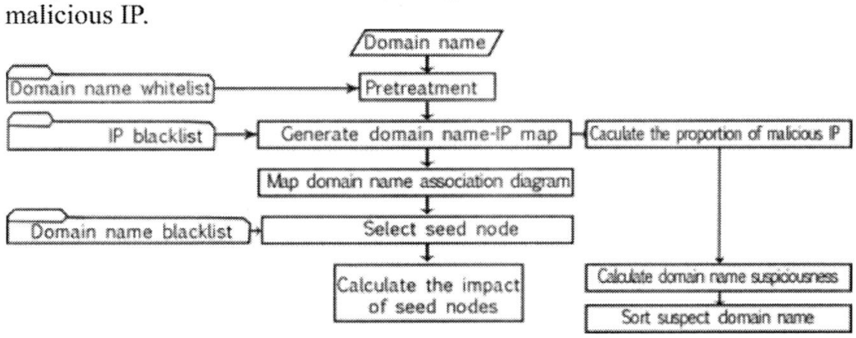

Figure 1. Method flow chart.

3.1. Generating a domain name - IP diagram

First, according to the list of suspicious domain names accessed by the host in the enterprise network, the IP list corresponding to the domain name is generated by the A record in the passive DNS database. Secondly, the undirected bipartite graph is used to represent the domain name-IP map, which represents the set of suspicious domain names accessed, indicating the corresponding IP set of the domain name in the passive DNS database, and the matrixrepresents the connection relationship between the domain name and the IP. For example, the domain name-IP relationship is as shown in figure 2, and its adjacency matrix E is as shown in equation (1).

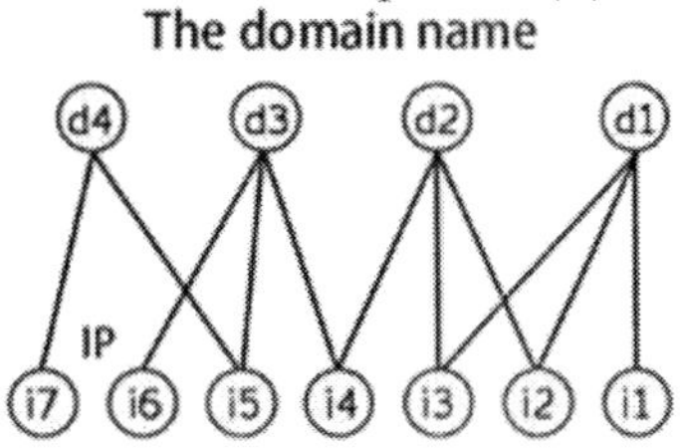

Figure 2. Domain name - IP Diagram.

IOP Conf. Series: Materials Science and Engineering **715** (2020) 012037 doi:10.1088/1757-899X/715/1/012037

$$E=\begin{bmatrix}1&1&1&0&0&0&0\\0&1&1&1&0&0&0\\0&0&0&1&1&1&0\\0&0&0&0&1&0&1\end{bmatrix} \tag{1}$$

3.2. Mapping domain name association diagram

Because some domain names have the same IP, there is a correlation between the domain names in the domain name-IP map. Building a domain name association graph by single mode mapping $G(D\ W)$' $D=\{d_1,d_2,...,d_n\}$ Indicates the set of suspicious domain names accessed, and the edge weight matrix to indicate the strength of the association between domain names, as shown in equation (2). The domain name association diagram of the example figure 2 is shown in figure 3.

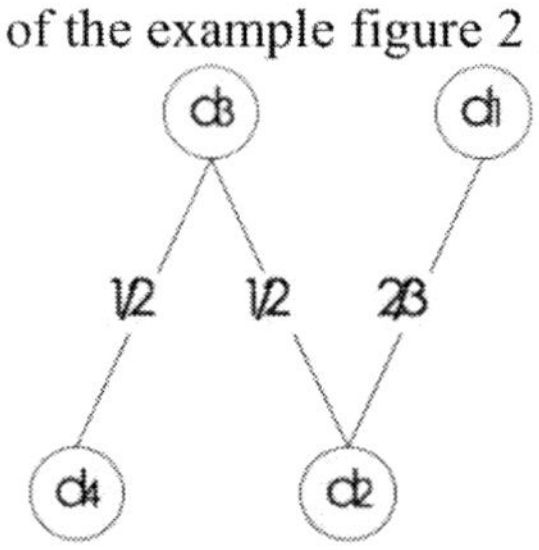

Figure 3. Illustrates the domain name association diagram.

3.3. Calculating suspiciousness

The suspicious value of each unlabeled domain name is calculated based on the correlation between the domain names and the proportion of malicious IPs corresponding to the domain name. Intuitively, the longer the path to an unmarked domain name to a known malicious domain name, the smaller the association between the domain name and the malicious domain name, or even negligible. Therefore, for each seed node, only the weight of the shortest path is selected and substituted. Assume $S=\{S_1,S_2,...,S_n\}$ that it is a seed node, that is, a known malicious domain name set; the impact value of the seed node on the unlabeled domain name is obtained by formula (3):

$$P(d,S)=\frac{\sum_{i=1}^{n}\frac{1}{2^{len-1}}P(d,S_i)}{2} \tag{2}$$

$P(d,s_i)$ Indicates the value of the seed node's impact on the domain name, *len* indicates the shortest path length from the seed node *Si* to the domain name d. The hypothesis $Ld_1,d=\{d_1,d_2,...,d\}$ is the shortest path from the seed node *Si* to the domain name d , which $P(d,s_i)$ is the product of the shortest path weight from the seed node *Si* to the domain name d .

The suspicious value of an unlabeled domain name d is obtained by equation (4):

$$M(d)=\alpha P(d,S)+\beta R_{IP\_M} \tag{3}$$

α,β is the parameter, satisfy $\alpha+\beta=1$; $P(d,S)$ indicates the impact value of the seed node on the domain name d, R_{IP_M} indicating the proportion of malicious IP in the IP list corresponding to the domain name d. Assume $I=\{i_1,i_2,...,i_n\}$ is the domain name corresponding IP list, $IM=\{i_1,i_2,...,i_m\}\subset I$ is the domain name corresponding IP list, so $R_{IP\_M}=\frac{m}{n}$.

4. Experimental results and analysis

4.1. Experimental environment and assessment methods

Randomly select several hosts to randomly access some blacklisted domain names in the lab environment, replay the traffic generated by the public APT malware running in the lab network environment, collect the network traffic generated every day, and take it as an example. verification. In order to quantitatively compare the recognition effects, the value is selected as the evaluation standard as shown in the formula (4). Indicates the number of malicious domain names that are correctly identified. The non-malicious domain name is determined as the number of malware domain names, and the malware domain name is identified as the number of non-malicious domain names. The machine running the passive DNS-based malware domain name recognition algorithm runs the 64-bit Windows 10 operating system, the programming language is python 3.5, and the development environment is PyCharm.

$$F1 = 1 - \frac{FP + FN}{2TP + FP + FN} \tag{4}$$

4.2. Experimental setup

On a daily basis, the statistical laboratory's domain name access for half a year keeps the records in the top 1/3 of the visit in the list of popular domain names. There are 59 suspicious domain names and 22 malicious domain names, accounting for 37.288%. Then, according to the passive DNS database, obtain a historical IP list corresponding to the suspicious domain name (a total of 833 different IP addresses), and query the malicious IP list to obtain the proportion of malicious IP corresponding to the domain name, compared with the method based on passive DNS graph reasoning. There are four seed nodes in this paper. Next, the shortest path from each seed node to the unlabeled domain name is calculated. The product of the weights of all reachable paths does not need to be calculated, which greatly saves time overhead. Finally, the suspicious calculation is based on the proportion of malicious IP and the degree of association between the seed node and the untagged domain name to identify a malicious domain name that is associated with a known malware domain name. The machine running the passive DNS-based malware domain name recognition algorithm runs the 64-bit Windows 10 operating system, the programming language is python 3.5, and the development environment is PyCharm.

The experimental parameters of the method and Issa method are set as shown in table 1 below:

Table 1. Parameter setting.

Method	threshold	α	β
Issa Method	1.8	-	-
Method of this paper	0.4	0.4	0.6

4.2.1. Comparison of recognition accuracy

In table 2, the two methods judge the non-malicious domain name error as the number of malicious domain names has not changed, and the number of correctly identified malware domain names has changed slightly, but a large number of malware domain names are not recognized, mainly due to This experiment uses the public APT malware dataset.

Table 2. Comparison of recognition accuracy.

method	TP	FP	FN	F1
Issa Method	7	1	11	53.85%
Method of this paper	8	1	10	59.26%

Compared with Issa's passive DNS map-based malware domain name discovery method, the malware domain name detection method in this paper reduces the number of malicious domain names that are misidentified, mainly because there is a domain name that is related to known malicious domain names. Not large, but the domain name returned a large number of malicious IPs in the historical request, and Issa's method based on passive DNS graph reasoning did not consider the IP address of the domain name.

4.3. Comparison of detection time

In the table 3, when verifying the test data, the overall time consumption of Issa's method is nearly twice that of this method, mainly because the method first filters some normal domain names, which reduces the workload; secondly, when calculating the impact of seed nodes on unlabeled domain names, this method uses only the shortest path, further reducing the path traversal time. Combined with the data in table 2, the recognition accuracy of the method is further improved on the basis of reducing the recognition time.

Table 3. Comparison of detection time.

Method	Detection time(seconds)
Issa Method	17.2
Method of this paper	5.8

5. Conclusion

The continued outbreak of malware poses a serious threat to network security, and most of the software uses DNS to establish a connection between the attacked host and C & C. The main recognition method based on DNS traffic is to extract features by analyzing DNS request and response packets, and the domain name recognition method based on passive DNS graph reasoning can effectively supplement it. In order to alleviate the recognition method of Issa, this paper proposes a malicious domain name discovery method based on passive DNS and IP correlation, which combines the proportion of malicious IP when calculating the domain name prestige value. At the same time, the calculation process of known malicious domain name influence value is improved. The experimental results show that the recognition accuracy of the proposed method is improved by nearly 6% and the detection time is reduced by nearly two times compared with Issa's method.

Acknowledgments

This work was supported by National key research and development plan of China under grants No.2016QY04W0800 and Basic research business fees of central colleges of China under grants No.20826041B4252.

References

[1] Zhao G, Xu K, Xu L, et al. Detecting APT malware infections based on malicious DNS and traffic analysis[J]. IEEE Access, 2015, 3: 1132-42.
[2] Manadhata P K, Yadav S, Rao P, et al. Detecting malicious domains via graph inference[C]//European Symposium on Research in Computer Security. Springer, Cham, 2014: 1-18.
[3] Soltani S, Seno S A H, Nezhadkamali M, et al. A survey on real world botnets and detection mechanisms[J]. International Journal of Information and Network Security, 2014, 3(2): 116.
[4] Bilge L, Kirda E, Kruegel C, et al. EXPOSURE: Finding Malicious Domains Using Passive DNS Analysis[C]//Ndss. 2011.
[5] Antonakakis M, Perdisci R, Dagon D, et al. Building a Dynamic Reputation System for DNS[C]//USENIX security symposium. 2010: 273-290.

IOP Conf. Series: Materials Science and Engineering **715** (2020) 012038 doi:10.1088/1757-899X/715/1/012038

Egonet based anomaly detection in E-bank transaction networks

Yuan Wang[1,2], Liming Wang[1,*] and Jing Yang[1]

[1]Institute of Information Engineering, Chinese Academy of Sciences, Beijing, China
[2]School of Cyber Security, University of Chinese Academy of Sciences, Beijing, China

*Email: wangliming@iie.ac.cn

Abstract. While most of the risks to online banking come from outside, some abnormal accounts can also cause losses to banks or their customers. In this work, we used a data set of 3,411,486 customers collected from a top Chinese bank to mine the abnormal accounts of online banking customers. In order to better express the transaction activities between user accounts, the concept of egonet is used to represent the account relationship in the transaction network. Then, we found that the characteristics of the egonet-based account model conform to the power law distribution, and based on this, anomaly detection is performed. When this method is applied to actual transaction data, we find abnormal account relationships from the data.

1. Introduction

Today, online banking (E-banking) has become one of the financial services that most people visit frequently. At the same time, various security issues are increasing. Although people have done a lot of work on identity authentication and communication security in e-banking [1], there are still some security issues that these solutions cannot solve.

Authentication is at least in the following two situations: an account that is stolen by phishing software or created by a malicious user. Most solutions detect these anomalous behaviours by detecting deviations from previously established baselines [2], [3], [4], [5], [6]. These jobs really help solve the first problem, that is, some users have lost certificates, or their accounts have been stolen [7]. However, it does not recognize abnormal accounts for special purposes. Therefore, based on the assumption that abnormal behaviour is less than normal behaviours, the anomaly detection work in this paper analyzes whether the behaviour of the account is like that of other accounts.

Due to cooperation with a large Chinese bank, we could analyze online banking customer behaviour based on anonymous ground truth data sets. In the analysis, it is found that these transactions between accounts naturally constitute a network; one time of transaction is two nodes and an edge between them, because the transaction is only meaningful to both parties. In this paper, we propose a method to detect abnormal transaction behaviour by checking the characteristics of the egonets of user accounts in the transaction network, and then use the Mahalanobis distance to measure the degree of deviation of the user account from normal behaviour.

Main contributions of this work are:

1)By constructing a transaction network between transaction accounts, the behaviour of an account

IOP Conf. Series: Materials Science and Engineering **715** (2020) 012038 doi:10.1088/1757-899X/715/1/012038

is represented as an egonet model.

2)It is found that the characteristics of the user's egonet model obey the power-law distribution, which gives us the confidence to declare outliers that are abnormally deviated.

3)We apply the method to real transaction data and show that it indeed spots the abnormal account relationships.

The reminder of this paper is organized as follows: In Section 2, we describe the data set and the reference methods. Section 3 examines the account behaviour and summarizes the findings we derived and implications. Section 4 presents the anomaly detection results. Section 5 discusses the related work, while Section 6 concludes this paper.

2. Data description

Online banking customers can generally access online banking through a browser to perform various financial services for individual customers or corporate customers, like various inquiries, money transfers, fee payments, investments. During the login process, each request from the customer is stored as a record in the data.

Our data comes from a very large online banking system in China that provides online services to millions of customers every day. The data contains various information about the customer's behaviour, such as time stamp, account ID, payer ID, payee ID, operation (e.g. login, money transfer), amount login IP, login area, operation status (success or failure). Our data includes 3,411,486 customers and 23,852,308 sessions in total. A session is a period during which a client logs in to log out, using a session ID assigned by the system. We call a session as a transaction session if there exists at least one transaction (money-moving operation) in a session; otherwise call it as a non-transaction session.

3. Analysis of transaction accounts

In this section, we analyze the user accounts in the trading network and extract the egonet attributes of the user accounts from the transaction network.

3.1. Transaction network

In this section, we use the concept of interaction graph in social networks to model transaction relationships as transaction graphs. The transaction relationship is usually used to describe the transaction activity between the payer and the payee. Payer and payee are envisioned as nodes and each node can act as either a payer or a payee in one transaction activity. For two nodes x and y, the activity of node x paying money to node y can be denoted as a directed edge $e = (x, y)$. Then, we represent the transaction network as a directed graph $G = (V, E)$, where V is the set of all nodes and E is the set of all edges. The outdegree and indegree of node x are the number of edges going out of x and the number of edges coming into x in G, respectively. Consider that our data only includes the transfer-out record without the transfer-in record for interbank transactions, we only build the transaction graph for inner-bank transactions.

The distribution of outdegree and indegree is shown in figure 1. In this transaction graph, the outdegree follows a power-law distribution fitted with parameter $a = 1.646$, and the indegree with parameters $a = 1.724$, meaning also a heavy-tailed distribution. This power-law distribution of indegree and outdegree indicate that the transaction graph is a scale-free network in complex network theory [8]. We find that 64.83% accounts have higher outdegree than indegree. The maximum outdegree is 1,525, which means that an account has transferred to thousands of payees in 12 days. The maximum indegree is 440, which indicates that many accounts transfer funds to an account.

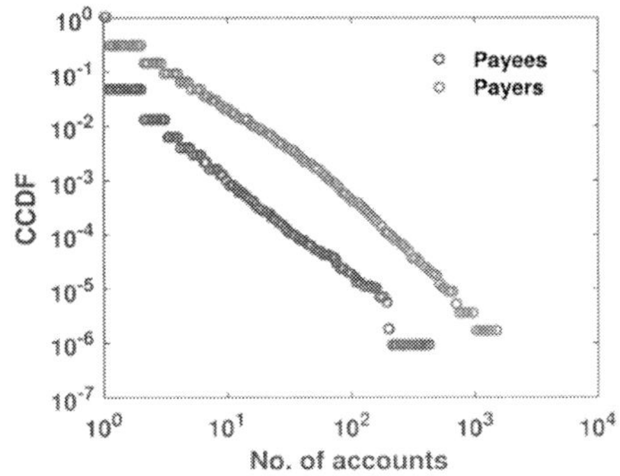

Figure 1. Outdegree and indegree per account.

3.2. Feature extraction based on egonet

What statistical information/characteristics extracted from the transaction network have a significant impact on the detection results. Intuitively, features which help spot the type of anomalies we are interested in should be selected. As mentioned above, a transaction combines two accounts in the transaction network, which is applicable to the egonet model, namely the 1-step neighbourhood of a node. In this paper, we want to find user accounts with unusual operations in transaction activities. To accommodate the detection of all these anomalies, we extract 4 features based on egonet from account-level information as followings.

1)N_i: number of neighbours (degree) of $egonet_i$.

2)E_i: number of edges in $egonet_i$.

3)W_i: total weight of $egonet_i$.

4)$\lambda w,i$: principal eigenvalue of the weighted adjacency matrix of $egonet_i$.

3.3. Laws and observations

Due to the lack of abnormal samples, the second important issue in anomaly analysis is the accurate description of normal behaviour. Therefore, it is important to find patterns for egonet-based transaction neighbours and then use them to calculate deviations.

As shown in figure 2, all features we extract follow a power law or have a power law tail at least. The power-law-type distribution is called heavy-tailed or fat-tailed if $a < 2$. In this case, the variance of the random variable is infinite. Furthermore, when $a < 1$, the mean of the random variable is also infinite [9].

In our experiments, the a in the number of nodes of every egonet is about 1.92 (Figure 2(a)); the a in the number of edges of every egonet is about 1.76(Figure 2(b)); the a in total weight of every egonet is about 1.53(Figure 2(c)); the a in principal eigenvalue of the weighted adjacency matrix of every egonet is about 1.12(Figure 2(d)).Therefore, it can be concluded that all of these features have theoretical averages, and the existence of these mean values is important because we can use them as criteria for judging user behavior anomalies.

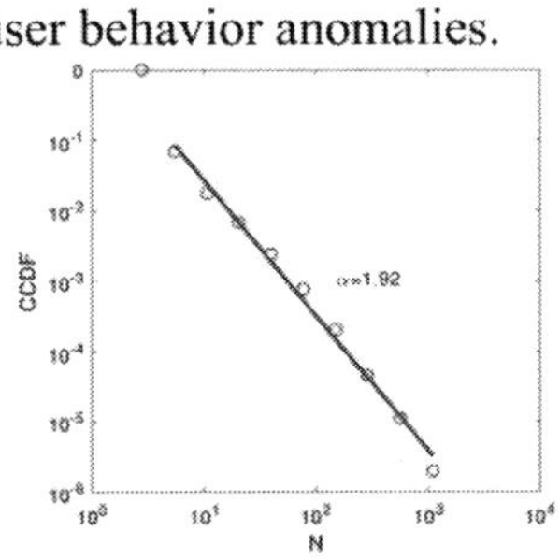

(a)

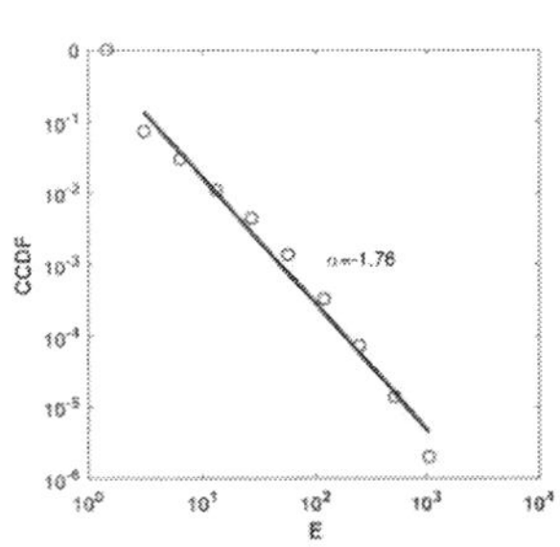

(b)

IOP Conf. Series: Materials Science and Engineering **715** (2020) 012038 doi:10.1088/1757-899X/715/1/012038

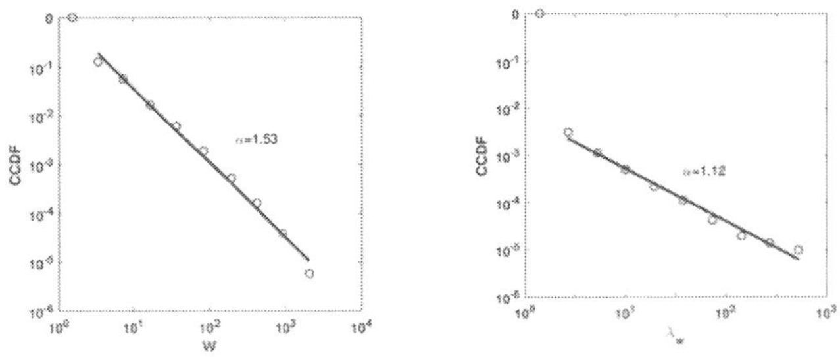

(c) (d)

Figure 2. Description of Transaction Network: (a)No. of Nodes of every egonet;(b)No. of Edges of every egonet;(c)Total weight of every egonet; (d)Principal eigenvalue of the weighted adjacency matrix of every egonet.

4. Abnormal transaction account detection

4.1. Anomaly detection

Usually, it is a good idea to use LOF-based detection algorithm to detect outliers far away from communities. However, as mentioned in Section 3, we find that the characteristics of accounts behaviours follow power law distributions and it also tells that we can get mean values of these characteristic variables in theory. Intuitively, the Mahalanobis distance-based algorithm has more power to detect and rank all these anomalies.

After extracting the accounts, Mahalanobis-based abnormal scores are used to estimate the abnormal of online banking accounts. Intuitively speaking, it is believed that the farther away an account's characteristics values are from the mean, the more abnormal it will be.

4.2. Result analysis

To evaluate the detection result, we manually analyze the top 1000 results according to Mahalanobis-based abnormal scores and use the LOF-based abnormal scores in comparison.

The AUC of the ROC plot of Mahalanobis-based detection in figure 3 is 0.998, while the AUC for LOF-based is 0.821. It is shown that Mahalanobis-based detection method has a better performance than LOF-based. After analysis, it tells us an unexpected result: some accounts are tightly connected to form a community where users have abnormal money moving. These behaviours are abnormal for common customers but will not be detected by LOF-based algorithms.

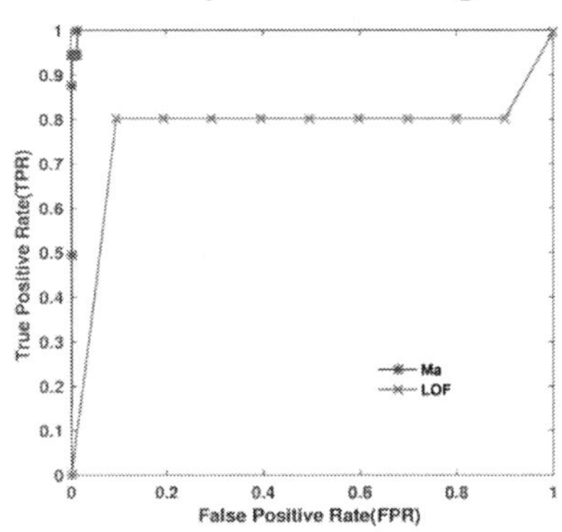

Figure 3. ROC plot of Mahalanobis-based and LOF-based.

5. Related work

Because of privacy, confidentiality and business interests, there is little research on customer behaviour analysis of online banking.

Most research work is based on customer behaviour analysis to detect online banking fraud [10], [11], [12], [13], [6]. Their research usually models the behaviour of each customer and monitors whether it deviates from normal behaviour [14]. However, only a small percentage of these efforts systematically analyze customer behaviour based on actual data and perform anomaly detection. Wei et al. [6] introduced a systematic online banking fraud detection method using transaction data from a large Australian bank. His detection method is mainly through the analysis of web logs of users accessing online banking. Carminati et al. [11] developed a semi-supervised and unsupervised fraud and anomaly detection method based on a real-world dataset of a large Italian national bank.

Compared to these studies, our research relies on the transaction dataset that includes more details about the transaction. Our method focuses on building transaction relationships in transaction networks, but their approaches are based primarily on the user's own historical behaviour. In addition, most of these works use fraudulent data generated by simulations due to the lack of publicly available real-world fraud samples. Our work based on ground truth data reveals some of the anomalous behaviour that is occurring in the online banking system.

6. Conclusion

In this work, we use the data sets collected from a top Chinese bank to mine the abnormal accounts of online banking customers. The online banking accounts were analyzed under transaction network, which focuses on account transaction activities. Then, we propose a method that based on egonet model to extract the egonet characteristics of each account. It shows that egonets of accounts obey some patterns, which gives us the confidence to declare as outliers the ones that deviate. We apply the method to real transaction data and show that it indeed spots the abnormal behaviours like abnormal community connected by transactions.

References

[1] S. Kiljan, K. Simoens, D. D. Cock, M. V. Eekelen, and H. Vranken. A survey of authentication and communications security in online banking. ACM Computing Surveys (CSUR), 49(4):61, 2017.

[2] A. Abdallah, M. A. Maarof, and A. Zainal. Fraud detection system: A survey. Journal of Network and Computer Applications, 68:90–113, 2016.

[3] N. A. Azeem Ush Shan Khan and M. N. Qureshi. Real-time creditcard fraud detection using artificial neural network tuned by simulated annealing algorithm. Institute of Doctors Engineers and Scientists, ACEEE (A Computer division of IDES), 2014.

[4] S. Pulakkazhy and R. Balan. Data mining in banking and its applications-a review. Journal of Computer Science, 9(10):1252–59, 2013.

[5] J. T. S. Quah and M. Sriganesh. Real-time credit card fraud detection using computational intelligence. Expert Syst. Appl., 35(4):1721–32, 2008.

[6] W. Wei, J. Li, L. Cao, Y. Ou, and J. Chen. Effective detection of sophisticated online banking fraud on extremely imbalanced data. World Wide Web, 16(4):449–75, 2013.

[7] Li S, Romdhani I, Buchanan W. Password pattern and vulnerability analysis for web and mobile applications[J]. ZTE Communications, 2016, 14(S0): 32-36.

[8] A. Mislove, M. Marcon, K. P. Gummadi, P. Druschel, and B. Bhattacharjee. Measurement and analysis of online social networks. 2007.

[9] J. Gao, Y. Cao, W.-w. Tung, and J. Hu. Multiscale analysis of complex time series: integration of chaos and random fractal theory, and beyond. John Wiley & Sons, 2007.

[10] G. Cabanes, Y. Bennani, and N. Grozavu. Unsupervised learning for analyzing the dynamic behavior of online banking fraud. In 2013 IEEE 13th International Conference on Data Mining Workshops (ICDMW), pages 513–20. IEEE, 2013.

[11] M. Carminati, R. Caron, F. Maggi, I. Epifani, and S. Zanero. Banksealer: A decision support system for online banking fraud analysis and investigation. Computers & Security, 53:175–86, 2015.

IOP Conf. Series: Materials Science and Engineering **715** (2020) 012038 doi:10.1088/1757-899X/715/1/012038

[12] K. N. Karlsen and T. Killingberg. Profile based intrusion detection for internet banking systems. 2008.
[13] S. Kovach and W. V. Ruggiero. Online banking fraud detection based on local and global behavior. In Proc. of the Fifth International Conference on Digital Society, Guadeloupe, France, pages 166–71, 2011.
[14] V. Jyothsna, V. R. Prasad, and K. M. Prasad. A review of anomaly based intrusion detection systems. International Journal of Computer Applications, 28(7):26–35, 2011.

IOP Conf. Series: Materials Science and Engineering **715** (2020) 012039 doi:10.1088/1757-899X/715/1/012039

Effects of residual stresses and shape deformities on failure of rollover protective structures

J. Abou-Hanna*, S. Tahamtam and V. Kumareshan

Mechanical Engineering Dept., Bradley University, Peoria, Illinois, USA

*E-mail: jannah@bradley.edu

Abstract. Rollover Protective Structures (ROPS) are manufactured typically from components that require forming such as pipe bending process. The manufacturing process creates residual stresses and shape deformities in the pipe structure. These parameters in turn may affect the strength of the pipe to bear the load during rollover. Conventional modeling does not capture the effects of residual stresses and shape irregularities in the pipe resulting from forming. A detailed analysis is therefore required to determine the significance of the manufacturing process on the strength of ROPS. Finite Element Method (FEM) is employed to simulate the bending process. Residual stress and deformation data from bending process analysis are used to further investigate failure of the structure under different loading conditions. The maximum load the structure can withstand, before it loses its stiffness, under a specific displacement is referred to here as failure load. Results revealed that there is a significant difference in the failure loads due to presence of bending process induced residual stresses and due to shape deformities. The model with the residual stresses and shape deformities has higher failure loads as compared to the conventional model. It also has been noted that residual stress is the main contributing factor to this increase in failure load. Parametric study was performed to investigate the effect of pipe wall thickness and radius of the formed bend on the failure loads.

1. Introduction

In order to improve operator safety, farm and mining vehicles have to be equipped with roll over protective structures (ROPS). In general design and testing of the ROPS, the structure is analyzed using finite element methods for sufficient stiffness and energy absorption before the loading causes the deflection to cross the limits specified by the standards [1]. However these results could be unrealistic considering the fact that the residual stresses created in the structure during the manufacturing process could produce significant impact [2]. So a simple model of a bent pipe 'L' shape is considered to study in detail the effect of these residual stresses on the failure loads under loading conditions similar to lateral and longitudinal loading of ROPS testing [1]. Finite element methods can satisfactorily capture the residual stresses created in a bending process [3]. Since the main objective was to determine the impact of the residual stresses concentrated in the bend, the standard loading conditions are slightly modified to meet the needs. The finite element package Abaqus/CAE 6.9-1 was used for the entire analysis of this study.

2. Analysis model

The analysis considered three cases as listed in table 1.

IOP Conf. Series: Materials Science and Engineering **715** (2020) 012039 doi:10.1088/1757-899X/715/1/012039

Table 1. Model cases used in the analysis.

Case	Residual Stresses	Shape Imperfections
Case – I	No	No
Case – II	No	Yes
Case – III	Yes	Yes

The model used for Case-I is shown in figure 1. The bending process, shown in figure 2, is an un-stretched bending process which produces moderate strains in the final model [4]. While Case-II model uses the shape at the end of bending process, Case-III includes that shape and the residual stress created in the bending process as well, and considers the spring back that occurs once the dies are removed.

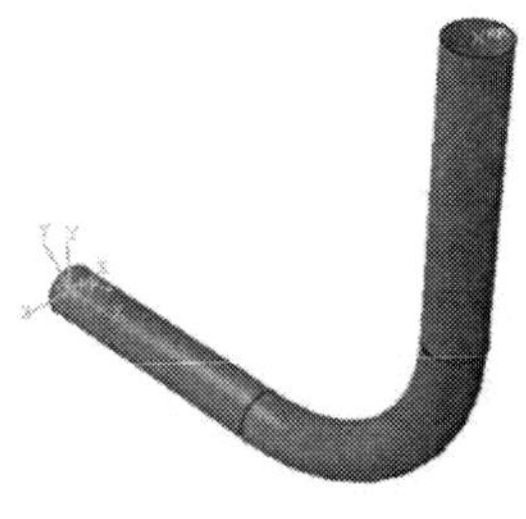

Figure 1. Conventional model design used for the analysis.

Figure 2. Model for pipe bending process.

Material: The material model used for the analysis is A500 Grade C, cold formed, carbon steel structural tubing. The true stress - true strain curve is constructed out of the engineering stress - engineering strain curve by assuming that the necking occurs at 21% elongation and the reduction in cross section at failure is 45%. The curve beyond the 21% strain is extrapolated using the above assumptions. Also a ductile damage model is constructed and included in the material properties with the damage occurring at 26% true strain. Ductile damage initiation and propagation functions of Abaqus are used to reproduce the material damage [5]. The assumed material property is as shown in figure 3.

Assumptions in Model: A few assumptions in the model are made to simplify the analysis. During the bending process the friction between the rollers and the pipe is assumed to be zero. The ends of the pipe are constrained as shown in figure 4. The nodes on the edge are coupled to a reference point at the center to ensure they do not lose their circularity during bending process. This simplifies the transfer of the model data from Abaqus/Explicit model for bending process to ROPS failure static load analysis model using Abaqus Standard solver. Abaqus/Explicit is a dynamic analysis procedure best suited for simulating forming operations. The spring back and failure load analysis however utilized Abaqus/Standard static equilibrium procedure.

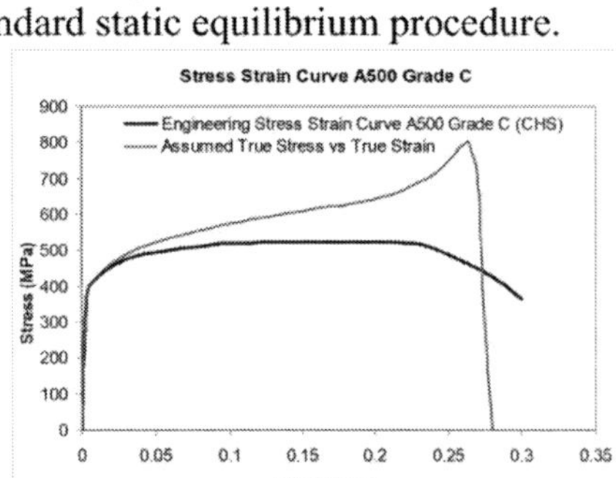

Figure 3. Actual engineering stress-strain curve and assumed true stress-strain curve for A500 Grade C.

Figure 4. Coupling constraints at the ends of the pipe.

IOP Conf. Series: Materials Science and Engineering **715** (2020) 012039 doi:10.1088/1757-899X/715/1/012039

Loading Conditions: A depiction of the lateral loading of ROPS is as shown in figure 5. Only half of the entire structure (an 'L' shaped pipe) is considered for this analysis. Also the chief concentration was on the bend of the pipe. The lateral loading on ROPS forces the bend near the direction of application of force to open up and the other to close. In Load-1condition used in the analysis as shown in figure 6a, the ends are pushed together closing the bend and in Load-3 condition, as shown in figure 6b, the ends are pulled apart and the bend opens up.

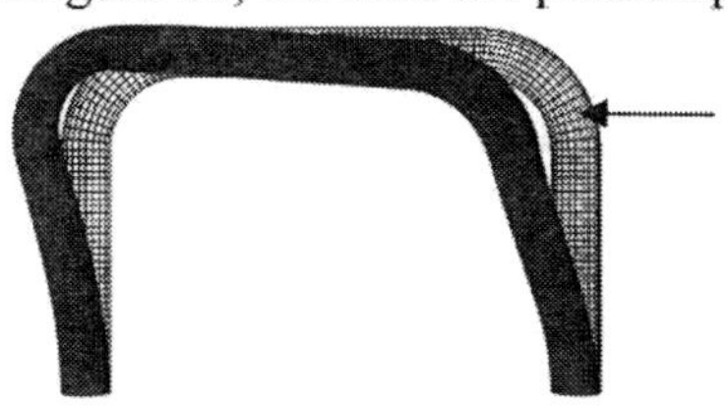

Figure 5. Lateral Loading on ROPS structure.

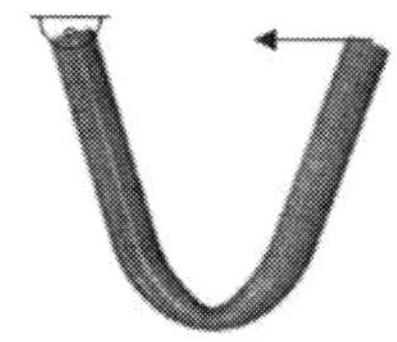

Figure 6a. Load-1 case-displacement causes the bend to close.
Figure 6b. Load-3 case-displacement; bend to open.

ROPS longitudinal loading is shown in figure 7. The equivalent scenario in this analysis is shown in figure 8 (Load-2), which is a modification of the standard longitudinal loading on ROPS used for estimating the failure load of the bend. Here the tube is mainly subjected to twist mode.

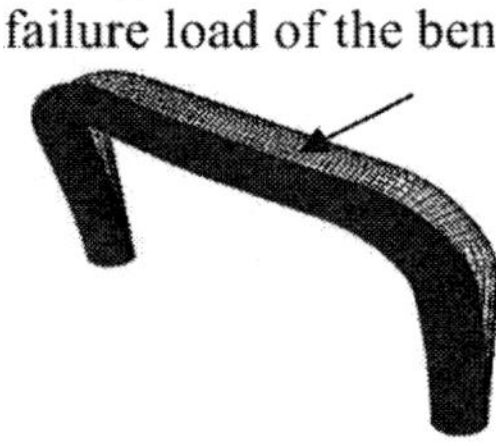

Figure 7. Longitudinal loading on ROPS structure.

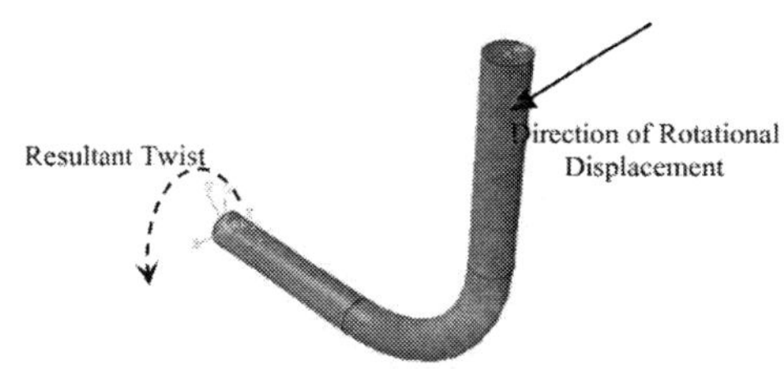

Figure 8. Loading condition (Load-2).

The point of application of the load is the center of the pipe end. The center point is coupled to the edge restricting any local deformities of the edge. The other end of the pipe is similarly coupled to a center point where the pipe is pivoted or fixed based upon the loading condition. The load cases are analyzed in detail for the model with the dimensions shown in figure 9.

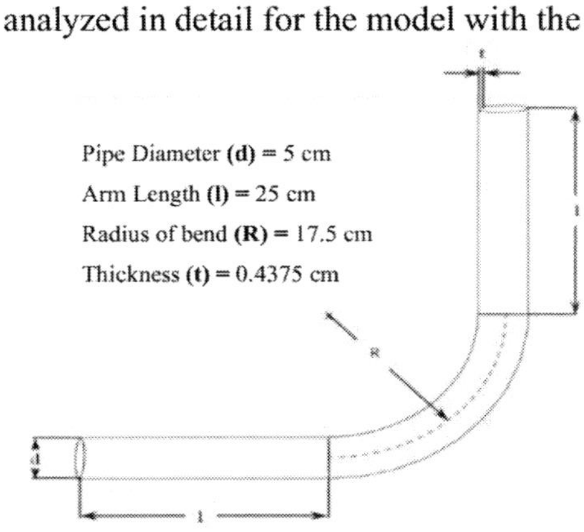

Figure 9. Model dimensions.

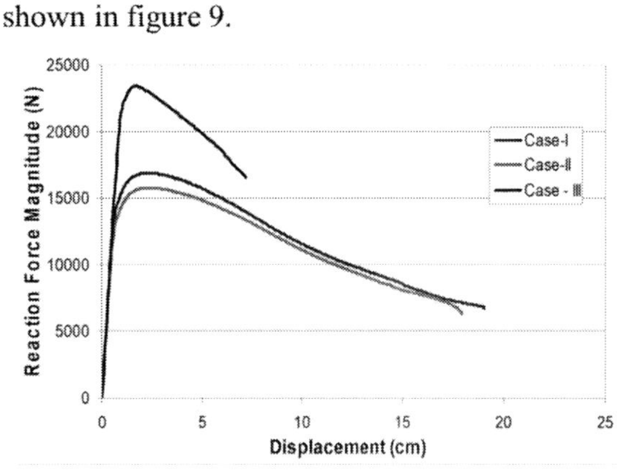

Figure 10. FR vs. displacement for Load-1.

3. Results

Load-1 Case: The Load-1 condition is shown in figure 6a. Here the point of load application is pulled with a constant displacement until instability is realized by drop in reaction load or instability of the structure as it exhibits its inability to support the loading condition. Load-1 is applied to all the three cases (Case-I, II & III) and the reaction force (F_R) at the displaced point is shown in figure 10. As seen from the graph in figure 10, Case-III is much stiffer than Case-I and Case-II. Case-III showed a 39% higher failure load than the other two cases. Moreover the failure load was reached at a shorter displacement compared to Case-I and Case-II. Case-II showed a little lower failure load compared to Case-I, but both had similar load - displacement profile. This clearly indicates that residual stress is the major contributing factor for the higher failure load in Case-III and not the shape deformities.

In Case-I the failure occurred mainly at the inner surface of the bend, figure 11 where the stress concentration was very high. In Case-II, the shape change, from perfect circular cross section to slight flattening at the sides during bending process, caused stress concentration at the sides and subsequently failure at the sides as well, and resulted in lower failure load, figure 12. Failure is depicted by strain exceeding the 26% limit, with elements losing their stiffness as was shown in the material model.

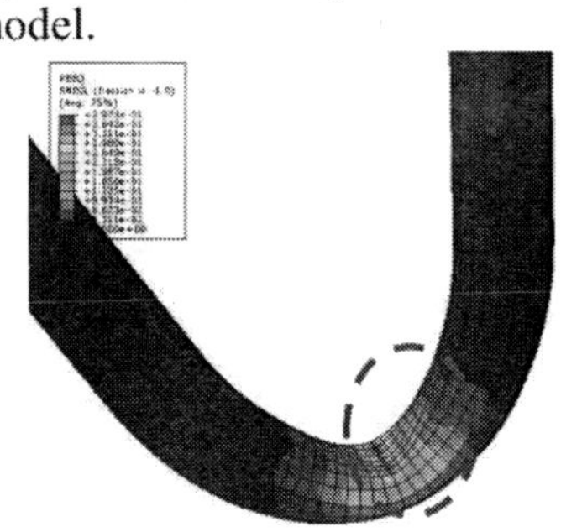

Figure 11. PEEQ of Case-I under Load-1.

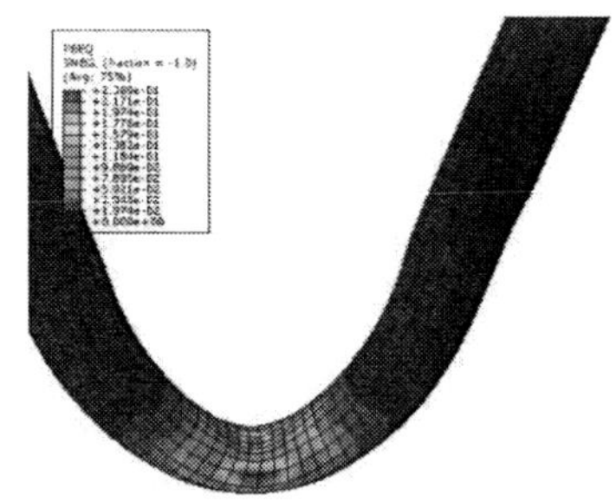

Figure 12. PEEQ of Case-II under Load-1.

In Case-III, when the failure load occurred the strains in the model never exceeded the failure strain of 26%. So the structure first buckled under loading, and it then progressed to damage point at the outer surface of the bend. Failure occurred at the outer surface of the bend as shown in figure 13. As seen from figure 14, end of the bending process, the outer surface had a positive strain and hence compressive residual stresses. So stretching along the bending direction required a higher load to overcome the residual stresses. Once the material starts yielding under load the strain further increases and causes the damage to progress faster thus failing at a lower displacement than the other cases, figure 10. It was noted that Case-III reached the failure load after 1.64 cm of displacement while Case-I reached the failure load after 2.34cm, 30% lower displacement than case of stress-free perfect cross section.

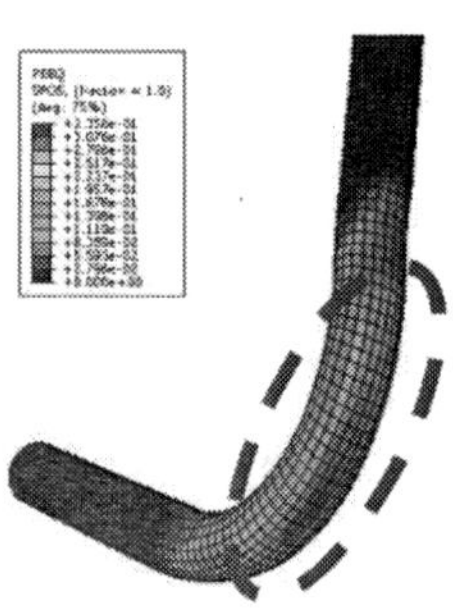

Figure 13. PEEQ of Case III under Load-1.

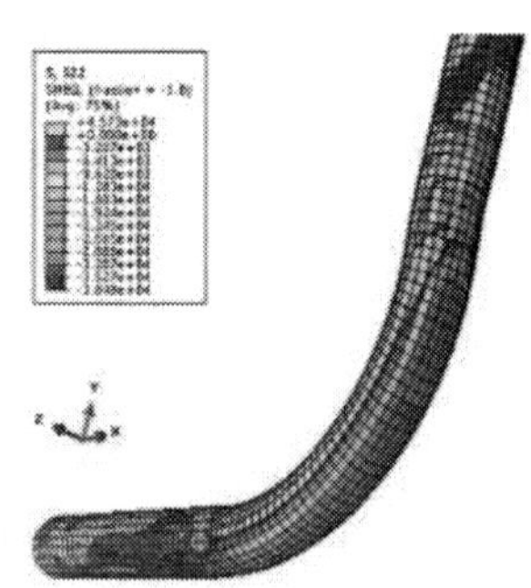

Figure 14. Case III under Load-1.

It has been reported [2, 6, 7] that the FEA results showed a little lower failure loads and a little higher displacement than actual ROPS testing, at failure. Thus Case-III can be assumed to be a more realistic model, relative to the other two cases.

Load-2 Case: The Load-2 condition is shown in figure 8. Here the point of load application is displaced with a specified rotational displacement. This Load-2 scenario is applied to all the three cases (Case-I, II & III), and the reaction moment (M_R) at the point is monitored as in figure 15.

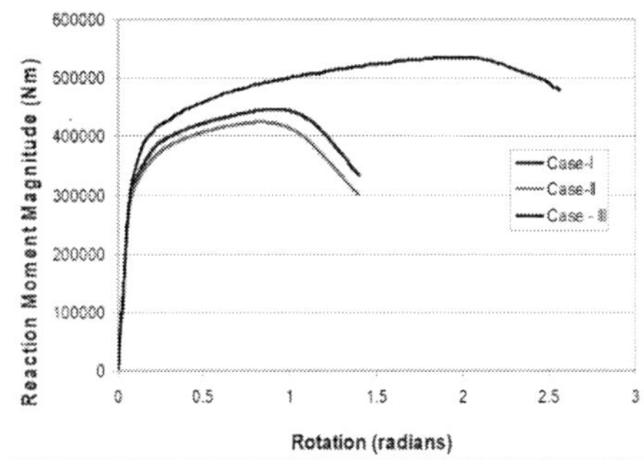

Figure 15. Reaction moment (M_R) Load-2.

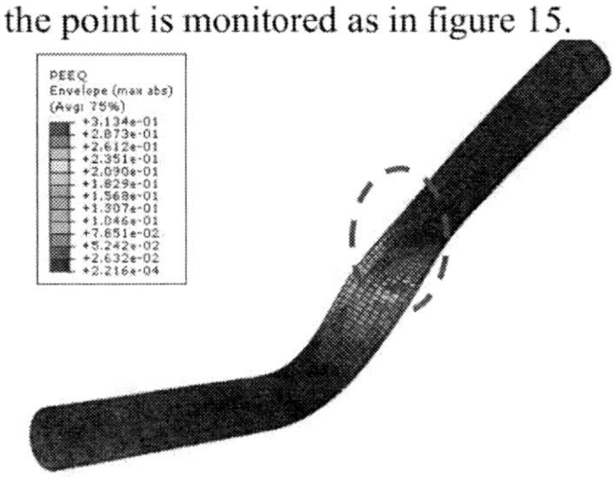

Figure 16. PEEQ of Case-I under Load-2.

Again, as seen in the previous Load-1 condition, Case-III provides a higher resistance to the twist loading as compared to the stress-free models, Case-I and Case-II. This again proves that residual stresses play a key role in causing a 20% difference in failure load. Case-I and Case-II had inelastic buckling at about 7% strain, and then they progress to ductile failure. Both cases had failure at the same location, shown in figure 16. Case-I and Case-II had similar failure pattern but had small variation in the failure moment due to the shape deformities in Case-II.

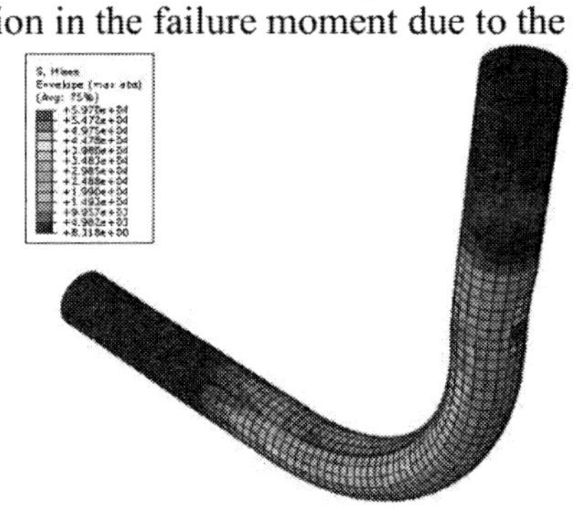

Figure 17. PEEQ of Case-III under Load-2.

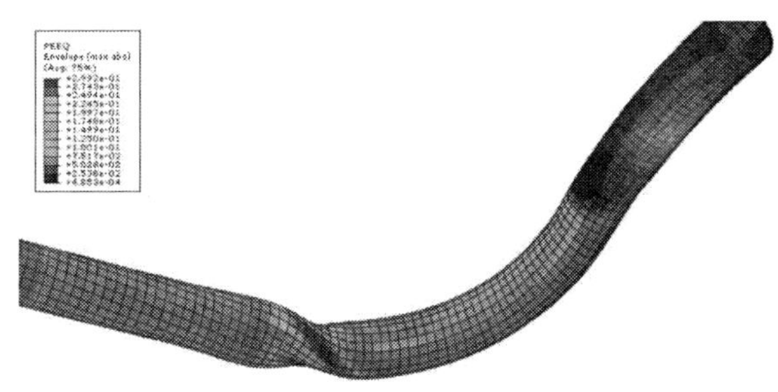

Figure 18. Mises stress distribution of Case-III.

However, Case-III did not fail at the bend (as shown in figure 17) but instead buckled near the bend, and it resulted in higher reactive moment resistance; huge rotational displacement ensued before failure was reached. Since the 26% plastic strain damage limit was never reached, the failure mode here was buckling instead of plastic failure. As seen from figure 18 the residual stresses are concentrated at the bend, Case III. The failure load at the bend region was higher because of the residual stresses and consequently the higher yield strength. Buckling occurred near the bend instead of the bend where the material strength is lower. Since Case-III showed a failure mode different from that of Case-I and Case-II, and since the failure did not occur at the bend, a comparison of load vs. rotational displacement among the cases, as shown in figure 15, is not justified. But it can be said with confidence that Case-III has higher failure loads than the other two cases, if one were not to identify type or mode of failure.

Load-3 Case: Load-3 condition is shown in figure 6b. Load-3 acts to open the bend and the reaction force (F_R) indicating the material resistance is monitored and reported in figure 19.

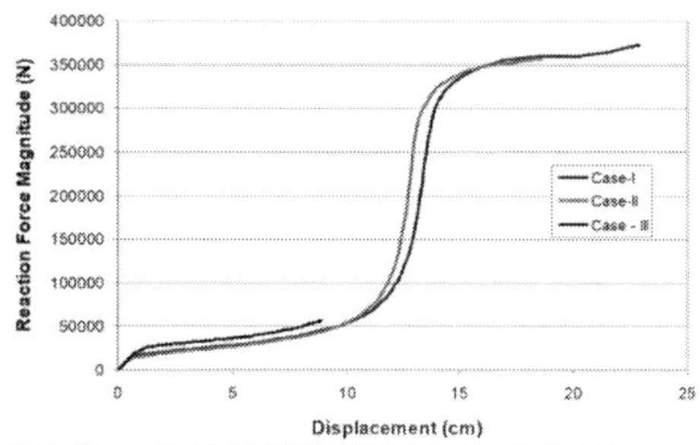

Figure 19. Reaction Force (F_R) vs. Displacement.

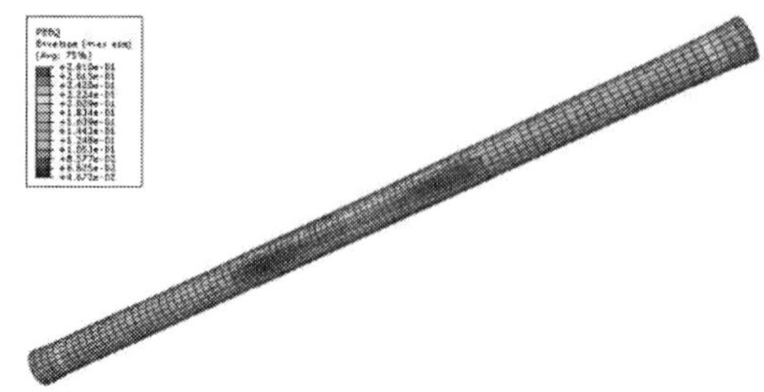

Figure 20. PEEQ, Load-3 in Case-I.

Analyzing this load case in detail shows that Case-III exhibits a higher reactive force but fails at a very low displacement compared to the other two cases. In the other two cases, the bend continues to open to the fullest extent as shown in figure 20 of Case-I. Thus the failure load of Cases I and II is equivalent of a straight pipe subjected to a uniaxial tensile pull under axial limit load. Case-II shows a similar behavior but as seen from figure 19 Case-II undergoes a lower displacement even though the loads are almost the same. The midsection of the pipe as highlighted in figure 21 was the region subjected to the bending process. This region showed a reduction in cross section area during loading in comparison to the rest of the pipe. Case-I had a uniform reduction in cross area throughout and hence Case-II reached failure load at lower displacement than Case-I.

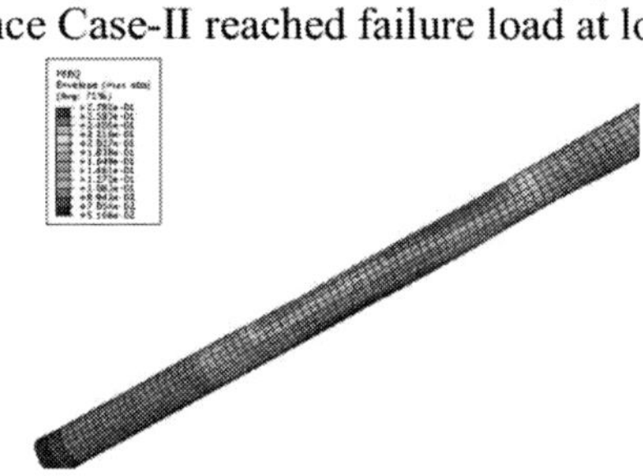

Figure 21. PEEQ, Load-3 in Case-II.

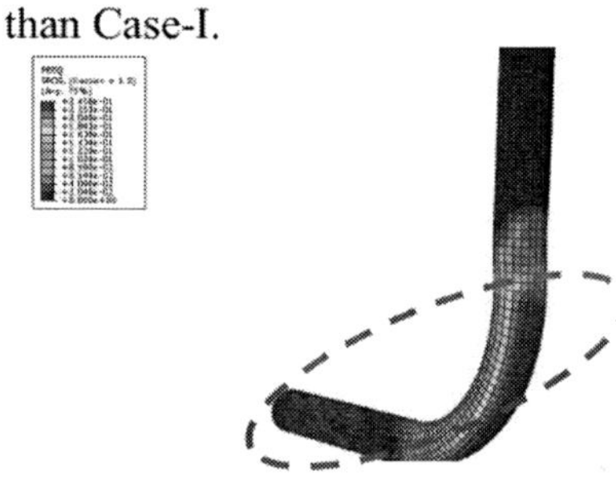

Figure 22. PEEQ of the bent pipe after bending process.

Case-III, as shown in figure 20, failed at a lower displacement. The outer surface of the bend in Case-III had positive strain due to stretching during the bending process as shown in figure 22. When Load-3 is applied these elements are stretched further in the same direction and damage strain was reached at a lower displacement than the other two cases. Though the failure occurred at a lower displacement the reactive force was slightly higher than other two cases for the same displacement (around 8.5 cm).The reason for the high reactive force is the compressive residual stresses in the bend region. Figure 23 shows the PEEQ at instability for Case-III. ROPS design and analysis takes into account the deflection limit which is the maximum displacement the structure can undergo before it interferes with operator safety. Conventional design and analysis do not consider residual stresses in the structure. It should be noted that if the deflection limit considered in conventional design and analysis is not properly chosen there is a chance that a false peak load value would be achieved as a result. For example if from the graph on figure 19 a deflection limit of 12 cm is chosen for a ROPS then one would be overestimating the failure load and failure displacement if residual stresses and deformities were not included.

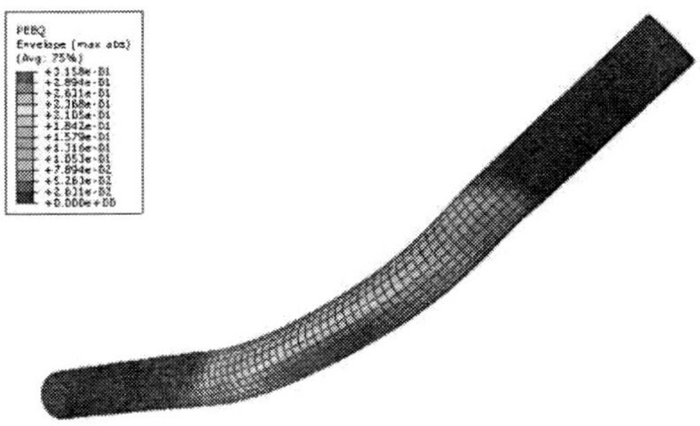

Figure 23. PEEQ- Load-3 in Case-III.

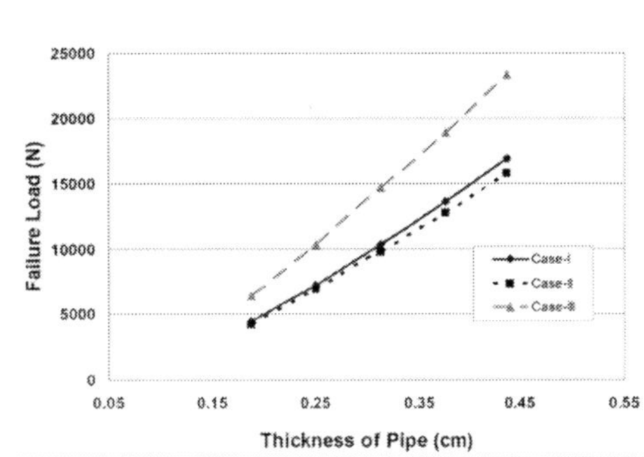

Figure 24. Failure load vs. thickness under Load-1.

4. Parametric study

Thickness of Pipe: Thickness of the pipe is varied, and the failure load under each loading condition and under each case was recorded. The trend with varying thickness is observed to be linear under each loading condition and is shown in figures 24, 25 & 26.

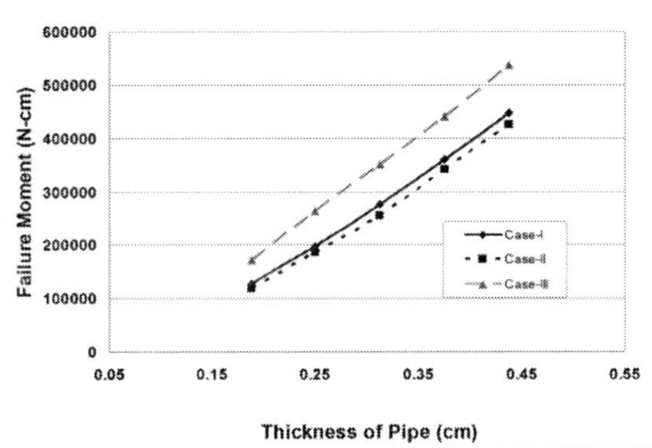

Figure 25. Failure load vs. thickness- Load -2.

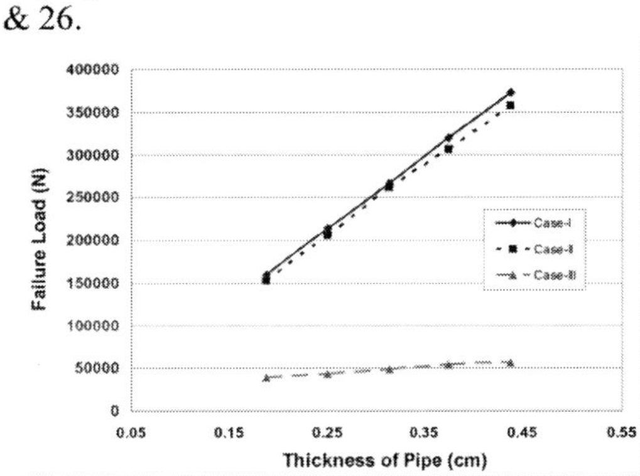

Figure 26. Failure load vs. thickness- Load -3.

Radius of Bend: The radius of bend considered for the thickness analyses and for the general study was 3.5x radius of the pipe. For bend radius smaller than that, the model had strains exceeding the failure limits in bending process as shown in figure 27. Therefore, lower bend radius was not entertained.

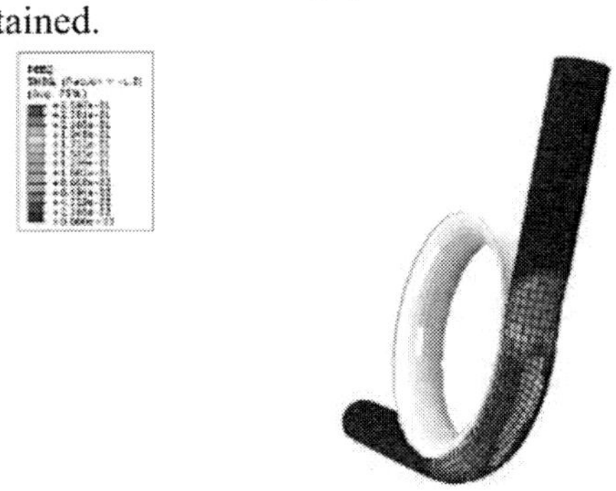

Figure 27. PEEQ showing failure at the bend.

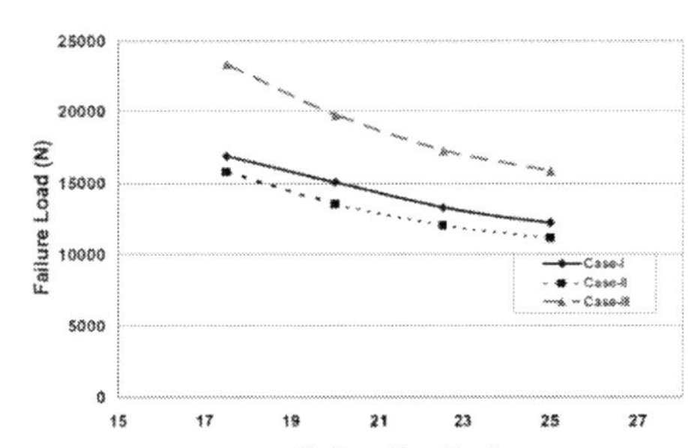

Figure 28. Failure load vs. bend radius, Load-1.

Figure 28 show that the Load-1 case has an almost linear response to the radius of bend. As seen from figure 29 the load for Case-III remains almost a constant for all different bend radii. This is due to the fact that Case-III has buckling behavior for failure and since the cross section area of the pipe remains a constant, the buckling load remains the same.

Figure 30 shows that for Case-I and Case-II the failure load remains almost a constant throughout the different bend radii for Load-3 situation. As seen from figure 20 and figure 21 failure in these cases is equivalent to subjecting a straight pipe to tension and hence there is no variation when the bend radius is changed. Case-III, due to the residual stresses, showed significantly less loads. But as the radius of the bend increased, the residual stresses and strains dropped causing the failure load to approach that of Case I and Case II, figure 30.

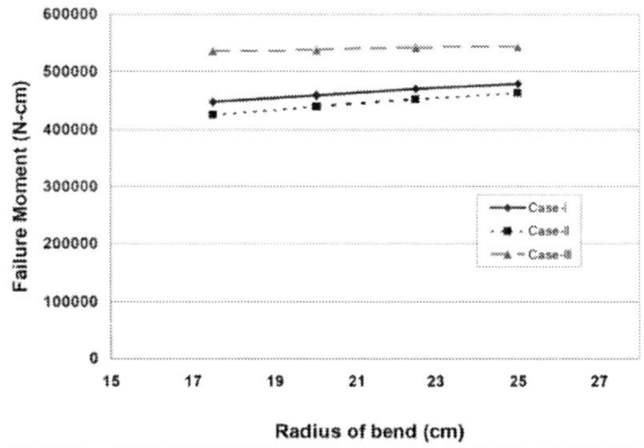

Figure 29. Failure load vs. bend radius- Load-2.

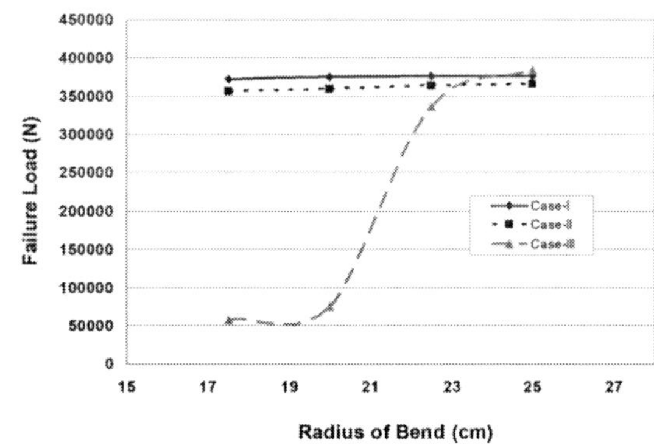

Figure 30. Failure load vs. bend radius- Load-3.

5. Summary/Conclusions

As seen from the above results, it is clearly evident that the residual stresses are significant in determining the failure load and deflection limits for the ROPS structure. For the same lateral loads the deflection limit for the model with residual stresses could be less by as much as 40% as seen from Figures10 &20. The above analysis is limited to a particular material model, and the behavior might be different under different materials having larger or smaller ductility. If a different bending process is used then the residual stresses and the results may vary. This study did not attempt to include residual stresses present due to the manufacturing of the straight pipe.

References

[1] Earth-moving machinery -- Roll-over protective structures, "Laboratory tests and performance requirements," ISO Standard 3471, 2008

[2] Clark B, Thambiratnam, DP & Perera N, "Analytical and Experimental Investigation of the Behaviour of Rollover Protective Structures," The Structural Engineer, 84(1): 29-34, 2006

[3] Kang MS, et al., "A Study on Residual Stress Measurement of Aluminum Pipe Using Raman Spectroscopy," Materials Science & Technology Conference and Exhibition, Pittsburgh, Pennsylvania Oct 2008, pp 771-779

[4] Lindgren M, "Residual Stresses in Longitudinal Direction in a Bending Operation," 4th ICFG Workshop on Process Simulation in Metal Forming Industry, Shanghai, 2004

[5] Abaqus CAE User's Manual, V **6.11**

[6] Yeh RE, "Elements of ROPS Engineering", Volume 76, Issue 46 of Technical report system, American Society for Metals ASM, 1976

[7] Thambiratnam, DP, Clark B, and Perera N, "Dynamic Response of a Rollover Protective Structure," Computer-Aided Civil and Infrastructure Engineering, 23(6): 448-464, 2008. doi: 10.1111/j.1467-8667.2008.0055

IOP Conf. Series: Materials Science and Engineering **715** (2020) 012040 doi:10.1088/1757-899X/715/1/012040

Study on the impact of LNG carrier entering and leaving port on the passing capacity of narrow channel

Tai-kun Huang*, Ran Pang, Yuan-zhan Wang, Xing-yan Yang and Han Jia

[1]Civil Engineering School, Tianjin University, China
[2]CCCC First Harbor Consultants Co., Ltd, Tianjin, China

*Email: huangtaikun@fdine.net

Abstract. As a clean energy source, LNG has been widely used. LNG ships need to enter and leave the port area through the channel. With the development of the port, the ship traffic will gradually increase. As a manual excavation of narrow and long channel, its passing capacity will become a constraint factor for LNG receiving stations and port development. In this paper, the system simulation method is used to study the impact of LNG ship entry and exit on the passage capacity of the channel, which can provide basis and reference for the planning, site selection and channel construction of the LNG receiving station.

1. Introduction

As a clean energy source, LNG has been widely used. LNG ships need to enter and leave the port area through the channel. With the development of the port, the ship traffic will gradually increase. As a manual excavation of narrow and long channel, its passing capacity will become a constraint factor for LNG receiving stations and port development. The impact of LNG ships on port channel throughput always becomes a bottleneck restricting the development of large ports. An objective and reasonable estimation of the channel's ability to pass can provide important theoretical and technical support for port planning, design and decision-making. Because the factors affecting the passage ability of the channel are more complicated, the estimation formulas are mostly empirical formulas, and mainly for the river channel, the most famous formula is the West German formula proposed at the 1969 International Shipping Conference [1]. The port and shipping system, which consists of ships, waterways, berths and loading and unloading operations, is a stochastic service system with many random factors and complex dynamic connections.

Zhang Baohua [2] based on the theory of FujiiMiping, combined with the theory of maritime traffic engineering, analyzed the ability of the Laotieshan waterway in Dalian; Liu Jingxian[3] redefines the passage ability of the channel and proposes to enter through a section of the port channel per unit time. The maximum number of ships in the port and the outbound vessel is used as the channel passing capacity, and the channel saturation is used as an evaluation index of the channel service level to study the degree of patency of the inbound channel. Li Yunbin[4] studied the model of the passing ability of the channel under static and dynamic conditions based on the ship domain model and the queuing theory model. Huang Taikun et al. [5] established a simulation model of channel passing capacity. The model considers the ship random arrival, tide conditions, hydrometeorology and other factors, and calculates the ship density, service level, loading and unloading efficiency, waterway waiting,

IOP Conf. Series: Materials Science and Engineering **715** (2020) 012040 doi:10.1088/1757-899X/715/1/012040

anchorage waiting. The influence of various factors on the ability to pass the two-way channel and the complex channel.

The port navigation operation system is a complex discrete event system with high internal correlation. It is difficult to reflect the random dynamic changes of various elements in the port operation process by traditional statistical methods. It is a new method to calculate the port channel passing ability by computer system simulation[6-8]. In this paper, using computer system simulation technology, considering the factors such as ship density, service level, loading and unloading efficiency, and channel waiting, LNG ships have influence on the passage capacity of large ports, and provide basis for channel construction and site selection of LNG receiving stations.

2. Project overview

Dagang Port Area is selected as an analysis object, shown in figure 1. According to forecasts, its development is divided into three stages: near, medium and far. Recently, some berths in the port area were completed and put into operation, with a cargo throughput of 136 million tons. In the medium term, the cargo throughput was 216 million tons. In the long run, all berths in the port area were fully developed and the cargo throughput was 316 million. At present, a 100,000-ton waterway has been opened, with a length of about 46km and a width of 300m. The design has a bottom elevation of -14.6m, which is a manual excavation type long and narrow channel.

3. Simulation model establishment

In this section, the model assumptions, model constraints were explained, and model building was performed.

3.1. Model assumptions and simplifications

To simplify the problem and facilitate modeling, the following assumptions are made.

(1)The storage capacity of the terminal yard is not limited to the berthing of the ship. The ship is loaded with cargo and unloaded.

(2)Regardless of the two-way navigation of the ship in the channel.

(3)The impact of non-cargo ships such as social ships and working ships and passenger ships is not considered.

3.2. Model constraints

To simplify the problem and facilitate modeling, the following assumptions are made.

3.2.1. Model boundary

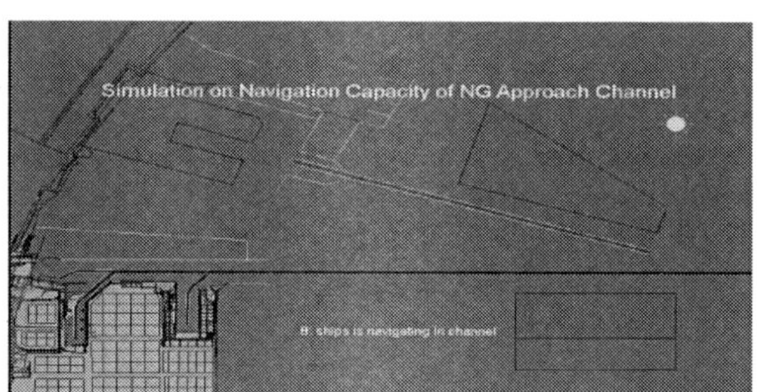

Figure 1. Model construction.

(1)LNG ships have the first priority, followed by large ships with tidal requirements and other ships.

(2)After the ship arrives at the port, it will be parked at the corresponding anchorage. When there is a free berth, it will enter the channel from the anchorage location.

(3)From the anchorage center into the waterway, ships above 100,000 tons enter the navigation channel from the starting point of the channel 46+000. Ships under 50,000 tons can be driven into the channel from 38+000 by anchorage.

(4)LNG does not sail at night, other ships can sail at night.

(5)The LNG ship mobile safety zone is 1nmile before and after.

3.2.2. Ship

Count the number of cargo ships (excluding social ships and work ships) arriving at the port and obtain a histogram of frequency distribution (Figure 2).

In probability theory and statistics, the Poisson distribution, named after French mathematician Siméon Denis Poisson, is a discrete probability distribution that expresses the probability of a given number of events occurring in a fixed interval of time and/or space if these events occur with a known average rate and independently of the time since the last event.

The number of ships arriving in port is subject to the Poisson distribution. For the statistics of ships arriving in port, the number of ships arriving in port per day is tested as a random variable.

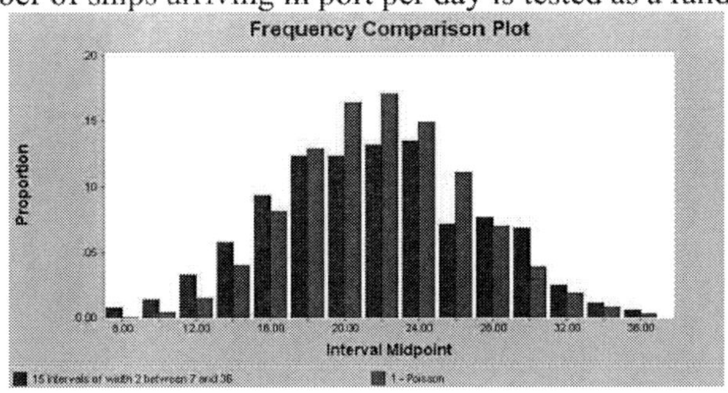

Figure 2. Distribution of ships arriving at the port.

4. Simulation experiment design and results analysis

In this section, three different simulation experiments were designed and the experimental results were analyzed.

4.1. Experimental design

Three sets of simulation experiments were performed (Table 1). In the first experiment, the port has recently developed 136 million tons, three LNG berths, and unloaded 5 million tons of LNG annually. In the second experiment, the LNG ship mobile safety zone is 1 nautical mile. The port has recently developed 216 million tons, three LNG berths, and 15 million tons of LNG. In the third experiment, the LNG ship mobile safety zone is 1 nautical mile. The port has recently developed 316 million tons, 3 LNG berths, and 15 million tons of LNG.

4.2. Simulation experiment results and analysis

In the first experiment, the number of ships entering and leaving the channel was relatively small, so the impact of the LNG ship entering and leaving the port was relatively light. The average channel time of the ship was only 2.13 hours/ship.

In the second experiment, with the development of the port area, the density of the ship entering and leaving the port increased by 3.94 hours per boat. The impact (including direct and indirect effects) of LNG ships entering and leaving the port has become larger.

In the third experiment, due to the high density of the ship, the average waiting time of the ship was as high as 7.31 hours per boat, and the impact became more significant(Figure 3).

Since many of the systems are random factors, people tend to pay special attention to the random distribution of some statistical indicators. In the second experiment, the average daily voyage of channel was 83.7 ships. Under weather conditions permitting navigation, The ratio of free time to navigable time of channel is only 5.19(Figure 4), that is an average of 8.93 vessels sailing in the fairway, and reached up to 27 . The average waiting time for the vessel to enter and exit the port is 3.94 h/ship, which is acceptable.

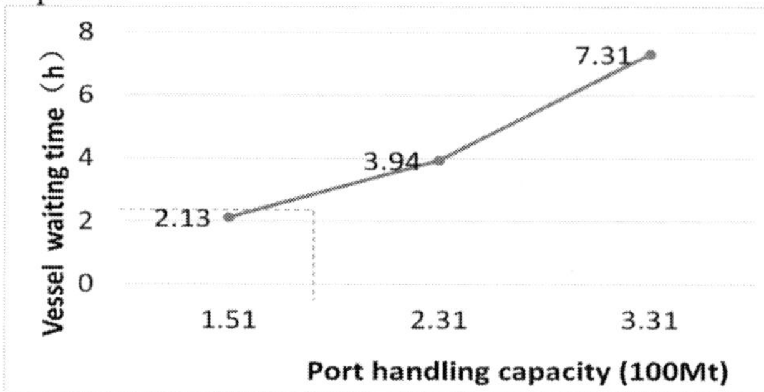

Figure 3. Ship waiting time with throughput curve.

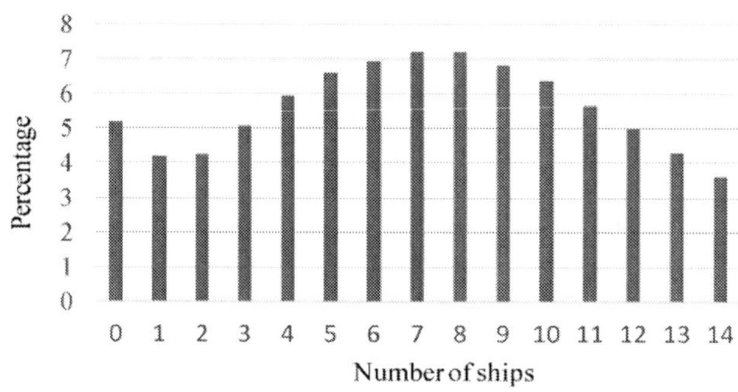

Figure 4. The ratio of free time to navigable time with ships curve.

Table 1. Simulation test results.

Conditions		Experiment 1	Experiment 2	Experiment 3
Other cargo types	Transportation volume (100 million tons)	1.36	2.16	3.16
	Arrival ship (ship)	8512	14068	19731
	Average waiting time for the ship (h/ship)	2.13	3.94	7.31
Directly affected by LNG ships	Proportion of ship directly affected by LNG ship (%)	33.4	47.6	54.3
	The average time to influence each ship (h / ship)	3.77	5.12	7.53

5. Conclusions and recommendations

(1)Actual statistics show that the applied port-to-port ships meet the Poisson distribution.

(2)When setting up the mobile safety zone and opening other ships at night, it is feasible to reach 200 million tons of port area and handle 15 million tons of LNG.

(3)As the port throughput increases, the ship waiting for the channel time also increases. When it reaches about 300 million tons, because the ship waits for a channel time of nearly 7.31 hours per boat, the impact is greater, and it is considered to open a special channel for LNG.

(4)In order to improve the passage capacity of the channel, it is recommended to open the night navigation function of the LNG ship under the condition of meeting the navigation rules.

References

[1] D.Y Yu, J.L Zhao, Y Deng, et al(2010). Qingdao Port main channel passage ability research. China Water Transport, 10 (3), 35-37.
[2] B.H Zhang (2009). Research on fairway capacity of Lao-Tie Hill based on ship field, Dalian University of Technology, Dalian, China.
[3] Y.Q Wen, J.X Liu (2010). Computing Model of Traffic Capacity of Harbor Public Channel. Navigation of China, 33(2), 35-39.
[4] Y.B Li (2008). Evaluation and Prediction of the Passing Capacity (Saturation) of the Main Channel of Tianjin Port, Wuhan University of Technology,Wuhan, China.
[5] T.K Huang, Y.Z Wang, S.W Li, et al (2015). Dynamic system simulation on throughput capacity of compound channel of seaport. Journal of Dalian Maritime University, 41(1), 20-26.
[6] D Jagerman, T Altiok (2003). Vessel Arrival Process and Queuing in Marine Ports Handling Bulk Materials. Queuing Systems, 45(3), 223-243.
[7] M Kia, E Shayan, F Ghotb (2002). Investigation of Port Capacity under a new approach by computer simulation. Computer & Industrial Engineering, (42), 533-540.
[8] P Legato, R.M Mazza (2001). Berth planning and resources optimization at a container terminal via discrete event simulation. European Journal of Operational Research, 133(3), 537-547.

IOP Conf. Series: Materials Science and Engineering **715** (2020) 012041 doi:10.1088/1757-899X/715/1/012041

Analysis on regional difference of meteorological S&T innovation capability in China

Danna Shen[1,*] and Yan Li[2]

[1]China Meteorological Administration, 10081, China
[2]Zhongnan University of Economics and Law, 430073, China

*Email: s.dn@163.com

Abstract. This paper selects the relevant meteorological data in 2014-2017 of the seven regions of Chinese mainland divided in accordance with the administrative area to construct the index system of meteorological S&T innovation capability, and then by means of Entropy TOPSIS method to get the values of C_i of the seven regions, and finally calculates the coefficient of variation of those C_i values based on the analysis of the meteorological S&T innovation capability, the innovation environment, the innovation input and output to analyze the developmental differences across regions. The results of this research prove that from 2014 to 2017, the meteorological S&T innovation ability of South China, North China and East China was relatively strong, while that of other regions was relatively weak. Besides, the meteorological S&T innovation ability of each region fluctuated in different years, but by 2017, the differences in innovation ability of meteorological S&T among regions were narrowing.

1. Introduction

Technological innovation, an evolving concept, is a process to create new values by applying scientific discoveries and technological inventions to production systems [1]. And meteorological S&T is a very important part of the scientific and technological innovation system. For example, the accuracy of meteorological forecasting directly affects the national ability to prevent and mitigate natural disasters, to deal with climate changes and to develop and utilize climate resources, greatly concerning national economic and social security as well as people's health and standard of living. Meanwhile, every major advancement in meteorological undertakings is inseparable from the great discoveries of science and technology. Therefore, the forecast accuracy of meteorology is closely related to the progress and development of meteorological S&T. Research on meteorological S&T innovation is of great importance.

In recent years, numerous researches on regional innovation ability have been done by scholars at home and abroad with two main research approaches. The first is to establish an index system for evaluating regional innovation ability, and then adopt suitable methods to measure innovation ability and finally analyze the regional differences in innovation ability. For instance, Zhao S L, et al (2013) used AHP to analyze the innovation activities of governments, colleges, research institutions and companies [2]. Sleuwaegen L, et al(2014) took advantage of data of 2007, 2009 and 2011 and the recursive structural equation model to analyze the regional innovation capabilities of 83 EU countries [3]. Chen J, et al (2017) collected data of various provinces in China from 2000 to 2014 to calculate the index of regional ecological innovation by constructing the index system and adopting projection, and analyzed the ecological innovation abilities of 31 provinces [4]. Zhang Aili, et al(2016) established a panel clustering model specific to the unbalanced development of regional innovation capability in China, and deeply analyzed the development differences in innovation ability of three categorical regions [5]. Wu Dan, et al(2018) constructed the DEA-Malmquist index model to analyze the time and space differences in national S&T innovation ability between China and other 10 countries [6].

In the field of meteorological S&T innovation, quite a few Chinese scholars have also conducted relevant researches from the perspectives of its policies and countermeasures. On the whole, many researches on innovation ability and S&T innovation ability done by Chinese domestic and overseas scholars are involved in numerous fields except the meteorological field. The scanty papers themed on meteorological S&T are mostly based on policies and recommendations. There haven't enough or none papers analyzing regional innovation ability on the basis of data. Hence, a series of in-depth studies should be done on the levels and differences of regional meteorological S&T innovation ability in China. What's more, most of the previous studies on regional differences are from the view of 31 Chinese provinces, autonomous regions, and municipalities directly under the central government because the results of analyzing that of individual province, autonomous region and municipality are scattered and insufficient. As we all known, China can be divided into seven major regions according to the geographic administrative areas. As a consequence, this paper combines the connotation of meteorological S&T with the statistical data to construct an index system for meteorological S&T innovation capability, and then analyze the meteorological S&T innovation capability of these seven regions by means of Entropy TOPSIS method, and finally put forward some related suggestions for the results.

2. Index system and data resource

With reference to the global innovation index, national innovation ability, urban innovation capability and other evaluation indicators, combined with the connotation of meteorological S&T innovation ability, considering the availability of data, this paper constructs the index system of meteorological S&T innovation ability, and finally forms three secondary indexes and nine indexes of the third level, which comprehensively covers various kinds of indexes, such as, the research funding, scientific and technological talents, dissertations, achievements, awards, the data volume of meteorological data sharing services, and the number of projects undertaken by the per capita. The specific index system is shown in table 1.

Table 1. Index system of meteorological S&T innovation ability.

Grade I Index	Grade II Indexes	Grade III Indexes
Meteorological S&T Innovation Ability	S&T Innovation Environment	The proportion of the number of meteorological professionals in total employment personnel
		The proportion of the number of talents in or above master's degree in total employment personnel
		The data volume of meteorological data sharing services
	S&T Innovation Input	Ratio of meteorological S&T funds in GDP
		Annual growth rate of meteorological S&T funds
		The amount of scientific research funds owing by meteorological professionals
	S&T Innovation Output	The number of SCI papers published by professionals per capital
		The number of projects taken by professionals per capital
		The number of awards won by professionals per capital

IOP Conf. Series: Materials Science and Engineering **715** (2020) 012041 doi:10.1088/1757-899X/715/1/012041

On account of the above index system, this paper selects the related data of 31 provinces, autonomous regions and municipalities directly under the central government in China for relevant analysis. And the data used in this study are derived from the China Meteorological Statistics Yearbook 2014-2017, China Statistical Yearbook and the 2017 Statistical Communiques on National Economic and Social Development of 31 provinces, autonomous regions and municipalities.

3. Theoretical model

TOPSIS (Technique for Order Preference by Similarity to an Ideal Solution) was originally developed by C.L. Hwang and K. Yoon in 1981. Its basic principle is to sort by measuring the distance between the evaluation object and the optimal solution and the worst solution. If the evaluation object is closest to the optimal solution and farthest from the worst solution, it is the best; otherwise it is not optimal. In addition, adopting the TOPSIS method can make sure the chosen objects for research are obtained from statistical measurement data, avoiding the interference of subjective factors, and can objectively and comprehensively evaluate the chosen objects in the case of multiple indexes [7]. The basic steps are as follows:

① Indicator evaluation: the comprehensive evaluation of m evaluation units and n indicators to construct decision matrix

$$A = \begin{bmatrix} a_{11} & \cdots & a_{1n} \\ \vdots & \ddots & \vdots \\ a_{m1} & \cdots & a_{mn} \end{bmatrix} m \times n \tag{1}$$

② Normalization: namely the dimensionless process of value A. Use the formula $b_{ij} = \frac{x_i - \min x_i}{\max x_i - \min x_i} * 0.9 + 0.1$ for normalization to get

$$B = \begin{bmatrix} b_{11} & \cdots & b_{1n} \\ \vdots & \ddots & \vdots \\ b_{m1} & \cdots & b_{mn} \end{bmatrix} m \times n \tag{2}$$

③ Giving weight with entropy weight method to each indicator, we can get the weight matrix W_j, then the weighted normalization matrix T = WB is as follows

$$T = \begin{bmatrix} t_{11} & \cdots & t_{1n} \\ \vdots & \ddots & \vdots \\ t_{m1} & \cdots & t_{mn} \end{bmatrix} m \times n, \text{ where } t_{ij} = W_j b_{ij}, \ (i = 1,\ldots,m; j = 1,\ldots,n) \tag{3}$$

④ Determining positive ideal solution T^+ and negative ideal solution T^-

$$T^+ = \left\{t_j^+ | j = 1,\ldots,n\right\} = \left\{\max t_{ij} | j = 1,\ldots,n\right\}$$
$$T^- = \left\{t_j^- | j = 1,\ldots,n\right\} = \left\{\min t_{ij} | j = 1,\ldots,n\right\} \tag{4}$$

⑤ Calculate the distance between the evaluation units and positive ideal solution and negative ideal solution

$$D_i^+ = \sqrt{\sum_{j=1}^{n} (t_{ij} - t_j^+)^2}, D_i^- = \sqrt{\sum_{j=1}^{n} (t_{ij} - t_j^-)^2}, (i = 1,\ldots,m) \tag{5}$$

⑥ Calculate the relative closeness degree between the evaluation units and the ideal solution

$$C_i = \frac{D_i^-}{D_i^+ + D_i^-} \quad (i = 1,\ldots,m) \tag{6}$$

The value of C_i is between [0,1]. The closer the value C_i of evaluation object is to 1, the closer it is to the optimal solution, which is at a higher level; the closer to 0, the farther it is to the optimal solution, which is at a lower level [6].

4. Empirical analysis

The mainland of China can be divided into seven major regions according to the geographical administrative areas, namely, North China (Beijing, Tianjin, Hebei, Shanxi, Inner Mongolia), Northeast China (Liaoning, Jilin, Heilongjiang), East China (Shanghai, Jiangsu, Zhejiang, Anhui, Fujian, Jiangxi, Shandong), Central China (Henan, Hubei, Hunan), South China (Guangdong, Guangxi, Hainan), Southwest China (Chongqing, Sichuan, Guizhou, Yunnan, Tibet), Northwest China (Shaanxi, Gansu, Qinghai, Ningxia, Xinjiang). And we can calculate the values of C_i of provinces, autonomous regions and municipalities by the way of Entropy TOPSIS method and average them to get the values of C_i of the seven regions. With that, the values of C_i of regional S&T innovation abilities, environment, input and output can also be calculated, and finally we can calculate the variation coefficient of C_i for relevant analysis.

4.1. Regional meteorological S&T innovation ability

The measurement results of the meteorological S&T innovation ability are shown in table 2.

Table 2. The values of C_i of meteorological S&T innovation ability of the seven regions in 2014-2017 and their rankings.

Regions	2014	Ranking	2015	Ranking	2016	Ranking	2017	Ranking
North China	0.2788	1	0.2287	5	0.2632	2	0.2761	1
Northeast	0.2278	4	0.1414	7	0.2072	5	0.1940	6
East China	0.2385	3	0.2620	2	0.2260	4	0.2166	3
Central China	0.2179	5	0.1502	6	0.1211	7	0.1966	5
South China	0.2489	2	0.2986	1	0.2837	1	0.2063	4
Southwest	0.1727	6	0.2305	3	0.1622	6	0.2540	2
Northwest	0.1649	7	0.2302	4	0.2402	3	0.1829	7

On account of the measurement results, the value of C_i is used to describe the meteorological S&T capability of each region, which reflects their abilities on meteorological S&T. For example, in 2017, the C_i value of meteorological S&T innovation ability in North China was 0.2761, ranking first; while the C_i value in Northwest was 0.1829, ranking the last, which indicates the highest level of the meteorological S&T innovation ability in North China and lowest one in Northwest in 2017.

Based on the above results, it is easier to see that in terms of the time dimension, in 2014, meteorological S&T innovation ability was the strongest in North China and the weakest in Northwest. The strong ability of meteorological S&T innovation in Beijing and Tianjin was the main reason for North China ranking the first, which boosted the innovation capability of the entire region. However, except Shaanxi, the other provinces in the northwest region were all with weaker meteorological S&T innovation ability, especially Gansu in a countdown position all over the country. In 2015, the rankings of these regions changed slightly. Compared with that of 2014, the abilities in North China, Northeast China and Central China all decreased, while the capabilities in East China, South China, Southwest China and Northwest China all improved. Among which, the northwest region enjoyed the biggest increase in ability owing to the certain improvement of the abilities of meteorological S&T innovation in Gansu, Qinghai, Ningxia and Xinjiang, particularly the first two. In the meantime, the capabilities of various regions in North China have declined greatly, in particular, Tianjin. In 2016, the regional meteorological S&T innovation ability ranked in the following order: South China, North China, Northwest China, East China, Northeast China, Southwest China and Central China. South China was still continuing their leading position. But for Central China, the scores were continuously declining from 2014 to 2016. It was obvious to see the decreasing meteorological S&T innovation capabilities of Henan, Hubei and Hunan from the table. With the comparison of 2016, the C_i value of the seven regions in 2017 were 0.0129, -0.0132, -0.0094, 0.0775, -0.0774, 0.0918, -0.0537. It could be easily seen that the largest increase in meteorological S&T innovation abilities was in Southwest, secondly in Central China; correspondingly, the largest declining in that was in Southwest China and then the Northwest China.

4.2. Regional meteorological S&T innovation environment

The measurement results of the meteorological S&T innovation environment are shown in table 3.

Table 3. The values of C_i of meteorological S&T innovation environment of the seven regions in 2014-2017 and their rankings.

Regions	2014	Ranking	2015	Ranking	2016	Ranking	2017	Ranking
North China	0.2483	3	0.2179	2	0.2174	3	0.2281	1
Northeast	0.2051	4	0.1164	6	0.2408	2	0.1362	6
East China	0.2641	2	0.2609	1	0.2468	1	0.2124	2
Central China	0.1518	6	0.1128	7	0.1161	5	0.1204	7
South China	0.2906	1	0.2139	3	0.2171	4	0.1771	4
Southwest	0.1045	7	0.1975	4	0.0812	7	0.2095	3
Northwest	0.1725	5	0.1563	5	0.1109	6	0.1501	5
Average Nationwide	0.2070	—	0.1940	—	0.1773	—	0.1847	—

Further, from 2014 to 2017, the C_i values of S&T innovation environment in North China and East China were above the Chinese national average value, while that in Central China and Northwest China were below the average. North China and East China both ranked on the top list, indicating that relatively speaking, they were equipped with more meteorological professionals and talents in or above master's degree, much data volume of meteorological data sharing services as well. Obviously, Central China and Northwest China performed rather worse on these aspects, especially in the latter, where the C_i value of meteorological S&T innovation environment was much lower than the national average.

4.3. Regional meteorological S&T innovation input and output

The measurement results of the meteorological S&T innovation input and output are respectively shown in table 4 & 5.

Table 4. The values of C_i of meteorological S&T innovation input of the seven regions in 2014-2017 and their rankings.

Regions	2014	Ranking	2015	Ranking	2016	Ranking	2017	Ranking
North China	0.2153	4	0.1959	5	0.2601	2	0.1932	2
Northeast	0.1722	7	0.1081	7	0.0989	6	0.0918	6
East China	0.2418	3	0.2023	4	0.1589	4	0.1453	5
Central China	0.2077	5	0.1660	6	0.0784	7	0.0886	7
South China	0.2628	2	0.3001	2	0.2331	3	0.1581	3
Southwest	0.2813	1	0.2411	3	0.1519	5	0.2514	1
Northwest	0.1864	6	0.3704	1	0.3007	1	0.1497	4
Average Nationwide	0.2270	—	0.2315	—	0.1905	—	0.1614	—

Table 5. The values of C_i of meteorological S&T innovation output of the seven regions in 2014-2017 and their rankings.

Regions	2014	Ranking	2015	Ranking	2016	Ranking	2017	Ranking
North China	0.3079	1	0.2422	3	0.2856	2	0.3587	1
Northeast	0.2192	3	0.1670	5	0.1993	5	0.2487	3
East China	0.1863	5	0.2756	2	0.2171	4	0.2439	4
Central China	0.2346	2	0.1555	6	0.1359	7	0.2858	2
South China	0.1948	4	0.3548	1	0.3668	1	0.2304	5
Southwest	0.1082	7	0.1674	4	0.1963	6	0.2001	7
Northwest	0.1330	6	0.1279	7	0.2181	3	0.2073	6
Average Nationwide	0.1934	—	0.2145	—	0.2299	—	0.2526	—

When it comes to Southwest China, its investment level of meteorological S&T innovation had been declining until 2016, however, its output level of that had been rising, which showed its meteorological S&T innovation efficiency was continuously increasing before 2016. But in 2017, more investment but no increase in output indicated that its efficiency of meteorological S&T innovation declined compared to the previous years. As far as East China is concerned, the continuous reduction of investment in 2014-2017 as well as the obvious fluctuation in output proved its efficiency instability of meteorological S&T innovation. In terms of South China, there had been taking a great downward trend of investment on meteorological S&T innovation since 2015, while its corresponding output had not changed much but fell sharply in 2017. The decline might be caused by the decline of the investment. The exact reason for that might be the significant reduction of funds in meteorological S&T invested by Shanghai, Jiangsu and Zhejiang in 2017 in the comparison of the previous years, leading to the much decrease of investment of the whole region, which in turn had affected the output of meteorological S&T innovation. With regard to the northwest region, its investment on meteorological S&T innovation had fallen the most since 2015 but with its output increasing year by year, which proved its efficiency of that was constantly improving. For North China, it experienced the increase in 2016 and then the decrease in investment as well as the continuous increase in output, reflecting that its meteorological S&T innovation efficiency has also increased to some extent.

4.4. Differences analysis of the regional meteorological S&T

The coefficient of variation of C_i values in each region from 2014 to 2017 was calculated so as to measure the change trend of differences in meteorological S&T innovation capability, environment, input and output. The calculation results are shown in table 6.

Table 6. Regional differences analysis on meteorological S&T innovation: the coefficient of variation of c_i values in each region from 2014 to 2017.

	2014	2015	2016	2017
Meteorological S&T Innovation Ability	0.1841	0.2575	0.2652	0.1575
Meteorological S&T Innovation Environment	0.3249	0.3050	0.3985	0.2375
Meteorological S&T Innovation Input	0.1782	0.3851	0.4562	0.3681
Meteorological S&T Innovation Output	0.3348	0.3818	0.3207	0.2146

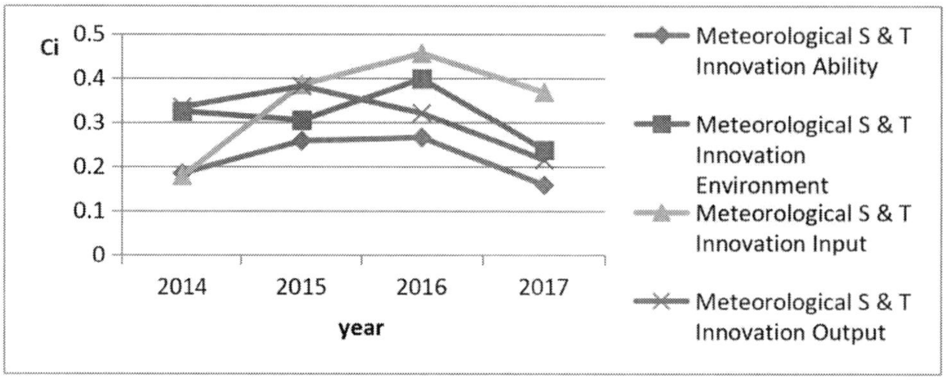

Figure 1. Regional Differences Analysis on Meteorological S&T Innovation: The Coefficient of Variation of C_i Values in Each Region from 2014 to 2017.

It can be seen from table 6 that there existed an increase in the coefficient of variation of meteorological S&T capacity from 2014 to 2015, showing the regional differences in meteorological S&T innovation capabilities in various regions increased in 2015. By 2016, the coefficient of variation further increased but with small increasing range, which indicated the further expansion of regional differences in meteorological S&T innovation abilities among these seven regions. However, in 2017, the coefficient of variation reached the minimum of the research range to prove the differences of the meteorological S&T innovation abilities of the seven regions were narrowing down to the smallest one in recent years.

It is observed from figure 1. that the meteorological S&T innovation environment differences of the seven regions had shrunk in 2015 when compared with that in 2014, but suddenly increased in 2016, and reached the minimum in 2017. The reason is that C_i values of meteorological S&T innovation environment of each region were relatively concentrated in 2017 with fewer changes of that in North China and Central China than previous years, some decrease in East China, South China and Northeast China and certain increase in Northwest and Southwest China, which leads to the small degree of dispersion of C_i values in various regions, showing that the differences of meteorological S&T innovation environments in various regions are gradually decreasing, further explaining the tendency to balance of the environment differences. Meanwhile, it can be found that the differences in input and output of meteorological S&T innovation in various regions in 2017 were also decreasing, too.

5. Conclusions and suggestions

We can clearly see via the above analysis that among the seven regions divided by the China's geographic administrative areas, from 2014 to 2017, the meteorological S&T innovation ability of North China are all higher than the Chinese national average, while that of Central China were lower than the national average. Intuitively, in 2014-2017, changes of the innovation ability are relatively small for East China but highly fluctuated for Central China. We can infer from the C_i coefficient of variation that the continuous narrowing of the differences in meteorological S&T innovation capabilities among the seven regions means their tending to be in a balanced development for meteorological S&T innovation ability. In the meantime, East China and North China ranks in the forefront and Central China and Northwest China are logging behind in S&T innovation environment. With regard to the meteorological S&T innovation efficiency, Northeast China takes the last but Central China takes the top. There has certain volatility in meteorological S&T innovation efficiency in various regions.

From the above empirical analysis of the meteorological S&T innovation capabilities of the seven regions, we can know that they are distinctly different from each other in this aspect with configuration to be further improved in S&T innovation resources. However, it should be noted that the resource endowments vary from place to place, and whether S&T innovation capabilities can be fully developed requires further discussion. But the regional meteorological S&T innovation capabilities can be coordinately developed. The specific strategies are as followed: First, establish inter-regional financing mechanisms to narrow regional differences. Areas with strong meteorological S&T innovation abilities, such as South China and East China, should try to maintain their advantages of resource endowments to reduce fluctuations. At the same time, they should also build good interaction mechanisms with areas with lower resource endowments to realize sharing and interaction of resources and technologies so as to narrow regional gaps and achieve the comprehensive improvement of meteorological S&T innovation capabilities, such as, firstly, constructing a number of cross-regional, large-scale, multidisciplinary research projects, or large-scale key research infrastructures to effectively reduce costs, mitigate risks, accelerate technology diffusion and expand external resource input to improve overall energy efficiency; and secondly, strengthening the input on meteorological innovation and technology, especially on basic research and development funds. The breakthroughs and innovations of core technology are derived from the continuous investment in basic research [8]. The development of meteorological S&T is more inseparable from the research of basic science. Hence, we should attach greater emphasis to the investment in applied basic research, and strive to narrow the gap between major core technology and international advanced levels, and provide strong support for the construction of meteorological powers. And the third is to pay more attention to the opening and cooperation of scientific and technological innovation. Technological innovations today present new characteristics-more international and collaborative. All people are likely to benefit from the comprehensive innovation. In conclusion, China should learn the experiences from home and abroad, inside and outside the industry to increase the sharing level of meteorological data in a further way, enhance the construction of the infrastructure platform of meteorological S&T and promote the opening and sharing of meteorological S&T resources.

References

[1] Zhang Laiwu. Scientific and Technical Drives the Transition of Economic Development Mode. J.China Soft Science. 12(2011)1-5.
[2] Zhao S L, Song W, Zhu D Y, et al. Evaluating China's regional collaboration innovation capability from the innovation actors perspective—An AHP and cluster analytical approach. J.Technology in Society. 35(2013)182-190.
[3] Sleuwaegen L, Boiardi P. Creativity and regional innovation: Evidence from EU regions. J.Research Policy. 43(2014)1508-1522.
[4] Chen J, Cheng J, Dai S. Regional eco-innovation in China: An analysis of eco-innovation levels and influencing factors. J.Journal of Cleaner Production. 153(2017)1-14.
[5] Zhang Aili, Li Yueming. Analysis the Differences of China's Regional Innovation Capabilities: Based on Grade Division. J.Science & Technology Progress and Policy. 33(2016)28-33.
[6] Wu Dan, Hu Jing. Research on Temporal and Spatial Differences Evaluation of National Science & Technology Innovation Ability-Based on the Comparative Analysis between China and other Ten Countries in the World. J.Science & Technology Progress and Policy. 35(2018)128-136.
[7] Yuan Xiaoling, Zhong Yunyun, Guo Yiqun. Analysis of the Measurement and Evolution of China's Regional Economic Development Differences——Empirical Study Based on TOPSIS Method. J.Inquiry into Economic Issues. (2010)33-39.
[8] Liu Xielin, He Yubing. Basic Research is the Source of Industrial Core Technological Innovation in China. J.China Soft Science. (2011)104-1

IOP Conf. Series: Materials Science and Engineering **715** (2020) 012042 doi:10.1088/1757-899X/715/1/012042

Information cascade prediction based on T-DeepHawkes model

Shijie Wang, Lihua Zhou* and Bing Kong

School of Information, Yunnan University, 650500, Kunming, China

*E-mail: lhzhou@ynu.edu.cn

Abstract. It is an important research point of social network analysis to predict future propagation range of information based on its early propagation characteristics. DeepHawkes model combines the Hawkes model in the traditional methods with deep learning, which not only inherits the high interpretability of Hawkes model, but also contains the high prediction ability of deep learning. However, the DeepHawkes model ignores the effect of the text content of the information on the propagation. Therefore, on the basis of DeepHawkes model, this paper further considers the influence of the text content on the diffusion, and proposes a T-DeepHawkes model which merges a topic classification model into DeepHawkes model. The experimental result shows that T-DeepHawkes model has a better prediction accuracy.

1. Introduction

The emergence of Twitter, Facebook, WeChat and, Sina Weibo and other online social platforms has greatly promoted the generation and transmission of information, which profoundly changed the way of information transmission. Such a huge amount of information contains a wealth of knowledge, such as users' preferences, retweeting behavior and the relationships between the users. It has extremely wide application value for mining of these knowledge in viral marketing, online advertising, information recommendation, rumor control and other aspects. So, it has very important research significance [1]. However, as social platforms are usually large-scale open systems, they will be affected by external factors [2], such as network topology, concern relationship, user interest, release time, privacy protection, etc. Due to the large scale of network users, the dynamic diffusion of information, the greater randomness of the transmission path, and the uneven "popularity" of different information, the accuracy of information cascade prediction is full of challenging [3].

In the prediction problem of information cascade, the current methods are mainly divided into feature-based methods [1-7] and generation methods [8-10]. The feature-based methods firstly extract various features of information based on existing experience, and then predict the popularity of future information through training regression/classification model. The generation approach dedicates to describe and model the process of an information notice, and make people easy to understand the underlying mechanisms governing the popularity dynamics of information [2]. However, the predict power performance of generation methods are not ideal, because they are not optimized by popularity prediction. So Cao et al. proposed DeepHawkes model by combining the high interpretability of Hawkes model with the high predictability of deep learning [2]. But DeepHawkes model aims to model the process of information transmission, ignoring the influence of the text content on message transmission. In fact, the text content of the message is also an important factor in the prediction problem of message transmission.

So in this paper, we propose T-DeepHawkes (Topic-DeepHawkes) model which considers the effect of text content by a topic classification model. T-DeepHawkes model contains the topic of the messages and merges the influence of cascade and text content. The main work of this paper including: (1) The T-DeepHawkes model considering the topic is proposed, and the classification model is used to extract the topic of text content of every message, and the deep learning model is used to learn the representation vector of the topic. T-DeepHawkes model incorporates the topic of messages into DeepHawkes model, which models the information diffusion process more comprehensively and improves the accuracy of popularity prediction. (2) The sufficient experiments are carried out on the real data set, and comparing the prediction results of multiple algorithms, which verified the effectiveness of the proposed method.

2. Related work

This section mainly introduces the methods of information popularity prediction, including feature-based methods and generation methods.

Feature-based methods manually extract various features related to information popularity, such as time features [1], structural features [3] and content features [6], and regard popularity prediction task as a regression or classification problem. Zhu et al. [3] proposed the concept of propagation acceleration on information propagation. Pinto et al. [5] predicted the future popularity of news based on the historical information of popularity. Cheng et al. [6] found that in the study of information cascade prediction, temporal and structural characteristics are the key predictors of cascade size.

The generation method regards the information popularity accumulation as an incentive process of user retweeting behaviour, and then models the process. Based on the self-excitation point theory, Zhao et al. [7] designed a statistical model to predict the information popularity. Shen et al. [9] used the reinforcement Poisson process (RPP) to model three factors in social networks. DeepHawkes model [3] combines the high interpretability of feature-based methods with the high predictability of generation methods, bridging the gap between prediction and understanding of information cascade. However, this model only models the information cascade propagation and ignores the effect of information content on information propagation.

3. T-DeepHawkes model

The T-DeepHawkes model includes model component for cascade, model component for topics and the fusion of the above two components. The T-DeepHawkes model takes the information cascade and the text content of the message as input, and the output of the model as the predicted value of the cascade retweeting increment.

3.1. DeepHawkes component for cascading

The DeepHawkes component for cascade adopts the model proposed by Cao et al. [3]. The component takes the information cascade as input and converts the input cascade into a set of diffusion paths, and each diffusion path depicts the information retweeting process during the observation time. The component makes an analogy between the explicable factors of Hawkes process through the three components: user embedding, retweeting path encoding and time decay. For the details of user embedding, path encoding and time decay in the DeepHawkes components of cascade, please refer to literature [3].

3.2. Topic component for topics

The DeepHawkes component for topics takes the text content of the information as input, extracts the topic of the message through the LDA (Latent Dirichlet Allocation) topic model, and constructs the topic embedding matrix. The component builds a topic path for all the topics included by a message, and passes to the cyclic neural network GRU for encoding, so as to simulate the impact of topics on message propagation.

(1) Topic embedding

Different messages involve different topics. In this paper, topics in messages are extracted by LDA topic classification model, and topic information embedding matrix is constructed. The construction process is as follows:

LDA Topic Modelling. Input the document-word frequency matrix into the LDA topic classification model, and construct the topic-word frequency matrix and document-topic matrix. Topic-word frequency matrix stores the probability of each topic generating different words, and document-topic matrix stores the probability of each message corresponding to different topics.

Build Topic Embedded Matrix. Word2Vec method [11] is used to generate the word vector of each word, and then the average word vector of all keywords under each topic is taken as the representation vector of this topic. Each column of the topic embedding matrix $B \in \mathbb{R}^{L' \times K}$ represents the representation vector of the topic, L' is the dimension of the topic representation vector, and K is the number of topics.

(2) Topic path coding

We suppose that a message contains n topics $t_1, t_2, \ldots, t_n$, and these topics can build up a topic path $\{r_1, r_2, \ldots, r_n\}$. This topic path is represented as a one-hot vector $q' \in$ RK, where K is the total number of topics. All topics share an embedded matrix $B \in R^{L' \times K}$ where L' represents the dimensions of the topic vector. Topic embedding matrix B converts each topic to its representation vector:

$$y = Bq' \tag{1}$$

Similar to user embedding matrix A, topic embedding matrix B is also studied under the supervision of future popularity in the training process. Therefore, the topic embedding matrix B learned is optimal for popularity prediction. The expression vector of the topic path is represented by the output $h'_{j'}$ of the last GRU unit on the topic path, and j' represents the number of topics that this message contains.

3.3. *Fusion components*

The pooling layer integrates the output of the last GRU of each path (cascade path and topic path) in the DeepHawkes components by a sum pooling, comprehensively reflecting the influence of cascade and topic on information diffusion. The cascading-topic c_y^i representation of the message m^i is defined as:

$$c_y^i = \sum_{j=1}^{R_T^i} \lambda_{f(T-t_j^i)} h_j^i + h'_{j'} \tag{2}$$

where K^i is the number of topics covered by message m^i, and $h'_{j'}$ is the output of the topic path. The output of the sum pooling layer is passed to the full connection layer as its input. The output layer has only one output unit, and its output value is: $\Delta\hat{R}_T^i = MLP(c_y^i)$. The minimization objective function is defined as: $obj = \frac{1}{M}\sum_{i=1}^{M}\log(\Delta\hat{R}_T^i - \Delta R_T^i)^2$, where $\Delta\hat{R}_T^i$ is the predicted incremental popularity of the message m^i, ΔR_T^i is the true incremental popularity and M is the total number of messages.

4. Experiment and results

4.1. *Experiment preparation*

Dataset. This experiment used a real dataset crawled from Sina Weibo. In this system, users' posting time has obvious daily and weekly distribution patterns, and the number of blog posts is shown as Weibull distribution. The retweeting and evaluation behaviors of blog posts have a strong correlation, and the retweeting probability of blog posts is higher than the evaluation probability [12]. The crawl time of dataset is from 8:00 on June 20, 2018 to 16:00 on June 21, 2018. The crawled data contains the message ID, the message publisher ID, the retweeter ID, the time interval between the message retweeting time, the message publishing time (in seconds), and the content of the message. The dataset has 131888 messages, and the average R_T^i is 70.2, average path length is 1.29.

Baseline model. Three models such as feature-linear, DeepCas and DeepHawkes are used as Baseline models in this paper. Feature-Linear [6] is a cascade prediction model based on time features, structural features and time decay, etc. DeepCas [10] is a cascade prediction model based on representation learning, and DeepHawkes [3] is a cascade prediction model combining deep learning with Hawkes model.

Evaluation Indicators. Similar to literature [3], this paper uses the Mean Square log-transformed Error [3] and the median Square log-transformed Error [3] as evaluation indexes, defined as: $MSLE = \frac{1}{M}\sum_{i=1}^{M} SLE^i$, M is the number of messages, SLE^i is the logarithmic error of message mi, $SLE^i = (\log \widehat{\Delta R_T^i} - \log \Delta R_T^i)^2$, $\widehat{\Delta R_T^i}$ and ΔR_T^i is the predicted and true increment of message popularity, respectively. $mSLE$ is the median of SLE^i which can effectively reduce the impact of outliers, $mSLE = median(SLE^i, ..., SLE^M)$.

4.2. *Parameter settings*

Obvious Time. The task of this paper is to predict the retweeting increment at a certain point in the future according to the retweeting situation observed during the observation time. This paper sets the observation time as 1 hour, 2 hours and 3 hours respectively.

Training Set, Verification Set and Test Set. Observation time of dataset starts from 13:00 on June 20, 2018. The first 70% of the data set is set as the training set, the middle 15% as the verification set, and the last 15% as the test set.

Topics. In this experiment, the topic number is set as $K = 480$ through the analysis of the dataset.

Other Parameter. LDA super-parameter is set to the default parameter $\alpha = 50/K$、$\beta = 0.01$. The dimension of user embedded matrix vector and topic embedded matrix vector is set as $L = L' = 50$. The related parameters of the cascade component are consistent with those in DeepHawkes model [2].

4.3. *Experiment results*

Table 1 shows the performance and T-DeepHawkes and other algorithm on test dataset. T-DeepHawkes obtained the best MSLE and mSLE in all observation windows of the dataset, indicating that it is reasonable to introduce topic content features on the basis of the original model.

Table 1. The performance of baseline model and T-DeepHawkes on dataset.

	Dataset					
T	1h		2h		3h	
Evaluation Metric	*MSLE*	*mSLE*	*MSLE*	*mSLE*	*MSLE*	*mSLE*
Feature-linear	4.387	0.971	4.014	0.817	3.955	0.797
DeepCas	3.744	1.066	3.351	0.876	3.258	0.907
DeepHawkes	2.596	0.801	2.387	0.699	2.359	0.723
T-DeepHawkes	**2.582**	**0.776**	**2.353**	**0.654**	**2.297**	**0.712**

5. Conclusion

On the basis of DeepHawkes model, this paper considers the influence of the text content of information on the diffusion, integrates LDA topic mode into DeepHawkes model, and proposes T-DeepHawkes model. T-DeepHawkes model not only considers the cascade factor, but also considers the text content of the information, so as to model the information diffusion process more comprehensively. T-DeepHawkes further improves the accuracy of popularity prediction while inheriting the high interpretability power of DeepHawkes. The validity of T-DeepHawkes model is verified by experiments on the real data set. In this paper, neural networks are used to autonomously learn the embedding matrices of users and topics, which contain user behaviors, preferences, correlations between topics and other information. However, the accuracy of the model proposed in this paper is less, and how to further improve the prediction accuracy remains to be studied.

Acknowledgments
The authors thank the reviewers for their helpful comments and suggestions. This research was supported by the National Natural Science Foundation of China (61762090, 61262069, 61472346, 61662086), The Natural Science Foundation of Yunnan Province (2016FA026, 2015FB114), the Project of Innovative Research Team of Yunnan Province, Program for Innovation Research Team (in Science and Technology) in University of Yunnan Province (IRTSTYN).

References

[1] Ding Zhaoyun, Jia Yan, and Zhou Bin. Survey of Data Mining for Microblogs [J]. Journal of Computer Research and Development, 2014, 51(4): 691-706 (in Chinese)
[2] Cao Qi, Shen Huawei, Cen Keting, et al. DeepHawkes: Bridging the Gap between Prediction and Understanding of Information Cascades [C] //Proceedings of the 26th ACM International Conference on Information and Knowledge Management. Singapore: ACM, 2017: 1149-58
[3] Zhu Hailong, Yun Xiaochun, Han Zhishuai. Weibo Popularity Prediction Method Based on Propagation Acceleration [J]. Journal of Computer Research and Development, 2018, 55(6): 1282-93 (in Chinese)
[4] Li Wei, Wang lihong, Li Ruiguang. Topic-oriented measurement of microblogging network[J]. Journal on Communications, 2013, (11)： 171-78(in Chinese)
[5] Pinto H, Almeida J M, and Gonçalves M A. Using Early View Patterns to Predict the Popularity of Youtube Videos [C] //Proceedings of the Sixth ACM International Conference on Web Search and Data Mining (WSDM '13). Rome: ACM, 2013:365–74.
[6] Cheng Justion, Adamic Lada A., and Kleinberg J. Can Cascades be Predicted? [C] //Proceedings of the 23rd International Conference on World Wide Web (WWW '14). Seoul: ACM, 2014: 925–36.
[7] Zhao Qingyuan, Erdogdu, Murar A.,et al. SEISMIC: A Self-Exciting Point Process Model for Predicting Tweet Popularity [C] //Proceedings of the 21th ACM SIGKDD International Conference on Knowledge Discovery and Data Mining (KDD '15). Sydney: ACM, 2015: 1513–22
[8] Yang Shuanghong , Zha Hongyuan. Mixture of mutually exciting processes for viral diffusion [J]. Proceedings of the 30th International Conference on International Conference on Machine Learning. JMLR, 2013, 28(2): 1-9
[9] Shen Huawei, Wang Dashun, Song Chaoming, et al. Modeling and Predicting Popularity Dynamics via Reinforced Poisson Processes [C] //Proceedings of the Twenty-Eighth AAAI Conference on Artificial Intelligence (AAAI'14). Canada: 2014: 291–97.
[10] Li Cheng, Ma Jiaqi, Guo Xiaoxiao. DeepCas: an End-to-end Predictor of Information Cascades [C] //Proceedings of the 26th International Conference on World Wide Web Companion (WWW '17). Australia: ACM, 577–86.
[11] Zhang Zhifei, Miao Duoqian, Gao Can. Short text classification using latent Dirichlet allocation [J]. Journal of Computer Applications, 2013: 33(6): 1587-1590(in Chinese)
[12] David M. Blei, Andrew Y. Ng, Michael I. Jordan. Latent Dirichlet Allocation [J]. Journal of Machine Learning Research, 2003, 3: 993-1022

IOP Conf. Series: Materials Science and Engineering **715** (2020) 012043 doi:10.1088/1757-899X/715/1/012043

Fine-grained analysis method for Android volatile memory

Peijun Feng, Qingbao Li and Zhifeng Chen*

State Key Laboratory of Mathematical Engineering and Advanced Computing, Zhengzhou, China

*E-mail: xiaohouzi06@163.com

Abstract. Android dominates the mobile operating system market. Volatile memory analysis of Android devices has been the focus of research on mobile forensics technology. However, due to the semantic gap between the kernel and the volatile memory allocator, existing Android volatile memory analysis methods are coarse-grained. With the volatile memory capacity of Android devices increasing, these methods cannot satisfy the need of Android volatile memory analysis accuracy. In this paper, we first discuss the address space layout of Android processes and the management mechanism of Jemalloc, the default Android volatile memory allocator. Then, we bridge the semantic gap by utilizing the boundary auto alignment feature of the data structure of Jemalloc. Finally, we propose a Fine-grained Analysis Method for Android volatile Memory, called FAMAM. Experimental results shows that FAMAM has an accurate data analysis capability as well as a good robustness. In addition, we successfully use FAMAM to discover new storage patterns for username and password of Wechat.

1. Introduction

Android has dominated the mobile market in recent years and been the focus of research in the field of mobile device forensics. As more and more Android applications using a variety of data encryption to protect data, for example WeChat[1], the most popular social network application in China, uses SQLCipher[2] to encrypt the entire database, even if the investigators could acquire the application database by some acquisition methods based on non-volatile memory forensics, they cannot direct access to application data, which has brought a huge challenge for Android mobile forensics.

Volatile memory is a dynamic reflection of the current state of the OS. It not only records the system kernel data such as the state of processes and networks, but also retains the application data. In addition, application data is stored in plain text in volatile memory, resulting in the ability to bypass database encryption for application data acquisition. So the research on volatile memory analysis is of great value for mobile forensics[3].

1.1. Motivation

The existing Android volatile memory analysis methods are coarser-grained, which simply take entire process physical address area even the entire volatile memory image as a large chunk of data for analysis, resulting in the inaccuracy of Android memory forensics analysis.

The dynamic data stored in the heap of process, such as username and password of applications, histories of browser and records of instant messaging software, are the most interesting information to investigators. However, existing forensics researches on Android heap data concentrated primarily on serialized Java objects contained in the heap and not on the way heap objects are managed, which may lead to a fragmentary forensics result.

1.2. Contribution

In this paper, we make the following contributions:

•We discuss the kernel structure of Android process and the management mechanism of Jemalloc(Jason Evans malloc)[4], and explain how the Android heap structures are organized and where the data is located in volatile memory.

• Based on the above analysis results, we bridge the semantic gap between the kernel and the volatile memory allocator by utilizing the boundary auto alignment feature of the data structure of Jemalloc.

• We propose a Fine-grained Analysis method based on volatile Memory Forensics for Android heap data, called FAMAM.

• We successfully use FAMAM to analyze the heap data of Wechat, especially discover a new pattern for username and password of Wechat.

1.3. Outline

The rest of this paper is structured as follows: Section 2 shows the related works; Section 3 presents the principle of FAMAM. Section 4 presents the implementation of FAMAM. Section 5 tests and evaluates FAMAM in terms of functionality and robustness. Section 6 presents the new pattern of username and password of Wechat discovered by FAMAM. Section 7 concludes the paper.

2. Related works

Lime[5] implemented by Sylve is a widely used volatile memory acquisition tool for Linux/Android. Lime extracts device volatile memory through a LKM (Loadable Kernel Module) injected in system kernel and supports dumping memory image by SDcard or TCP connections with 99.46% accuracy. Stüttgen[6] proposed a robust volatile memory acquisition method with minimal target impact for Linux/Android, which enhances the versatility of Lime in different kernel versions. In this paper, we use Lime to acquire the volatility memory images of Android device.

Volatility[7] is a open source memory forensics framework, which obtains detailed system state by analyzing the kernel structures. All functions of Volatility are implemented by plugins written in Python, so it is highly extensible. However, these functions mainly focus on the kernel level, especially the analysis of kernel structures, processes and networks, while there are only a few functions on user applications.

Zhou et al.[8] introduces some methods to dump the Android volatility memory and used some software tools to analyze the memory to extract the encrypted chats and deleted messages of WeChat. However, they simply took the entire volatile memory image as a large chunk of data for analysis, resulting in the inaccuracy of forensics result.

Macht[9] and Leppert[10] both perform Android volatility memory forensics by a comprehensive bottom-up methods and analyze the heap data of a set of applications. However, their analysis concentrated primarily on serialized Java objects contained in the heap and not on the way heap objects are managed, leading to a fragmentary forensics result.

Ntantogian et al.[11] discover common patterns for usernames and passwords of mobile apps. However they still use the string matching method to search the entire memory image, which is coarse-grained, and have not found the new pattern that we found.

Cohen[12], Ligh[13] and Block[14] et al have conducted detailed researches on the fine-grained volatile memory analysis technology, especially the process heap data structure and its management methods, which make great contributions to volatile memory forensics research work. However, it is a pity that these researches are aimed at Windows, MacOS or Linux. Since the memory allocator used by Android is different from the memory allocator used by other systems, the forensic analysis method is completely different in the detailed implementation.

IOP Conf. Series: Materials Science and Engineering **715** (2020) 012043 doi:10.1088/1757-899X/715/1/012043

3. Principle analysis

In this section, we first introduce the result of our analysis on the Android process address space layout and system memory allocation mechanism from a volatile memory forensic perspective, and then we present the bridge principle of semantic gap between the kernel and Jemalloc.

3.1. The Android process address space layout

Android adopts Linux kernel and inherits the process management mechanism of Linux. Each Linux/Android process corresponds to a process controller structure, called task_struct. The mm pointer of the task_structure points to the memory descriptor structure of the process, managing the virtual address space of the process, called mm_struct. The pgd pointer of the mm_struct points to global page table of the process, and the mapping from virtual address to physical address is implemented through the multi-level page table. The address space layout of Linux/Android process is shown in figure 1.

The Linux/Android process address space contains the following data regions.

• Code segment stores the program code and some statically allocated read-only constant variables, such as character constants.

• Bss segment stores statically allocated uninitialized global variables.

• Data segment stores statically allocated initialized global variables.

• Brk segment stores dynamically allocated data smaller than 128KB.

• Mmap segment stores dynamically allocated data larger than 128KB.

• Stack segment stores dynamically temporary created local variables.

Without considering the shared memory, there are two ways to dynamically allocate memory for Linux processes: brk and mmap. Brk allocates small contiguous chunks of memory and increases the value of brk pointer of mm_struct. Mmap allocates discontinuous large chunks of memory and concatenates all mmap regions in a process into a double linked list. However, for Android processes, we find that the value of brk pointer of mm_struct is equal to the value of start_brk pointer, that means the Android processes only use mmap to allocate memory. In other words, the Android heap is the mmap area.

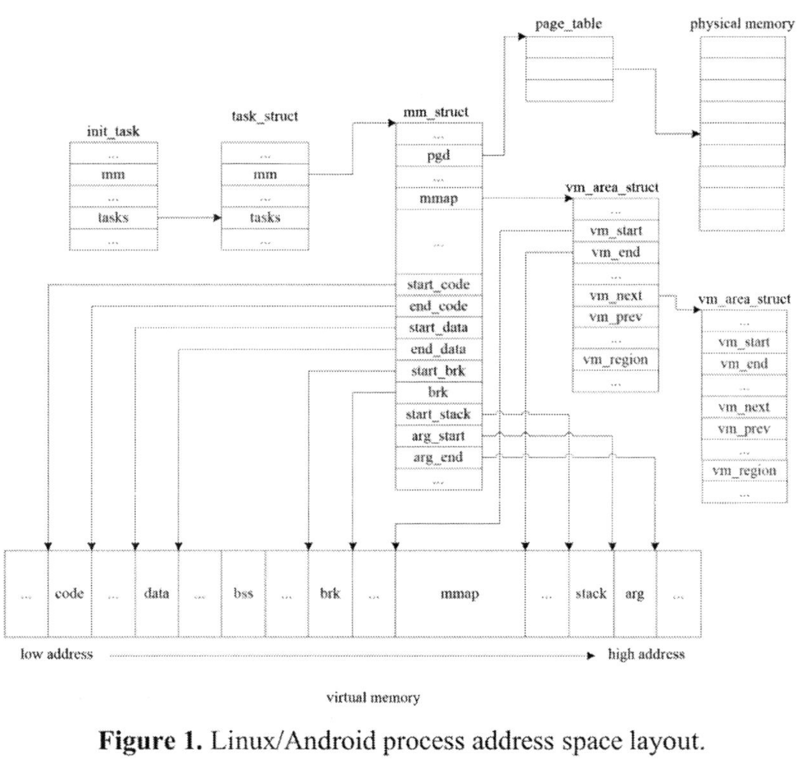

Figure 1. Linux/Android process address space layout.

Android forensics investigators can locate the task_struct of process according to the process name or PID by traversing the tasks list. The mm_struct and the virtual address of various data areas are further obtained according to the offset among various data structures as shown in figure 1. Finally, the physical address of each data area are obtained through the page table transformation.

The mobile device user behaviors data, such as instant messages, browser history, and GPS geographic location, which are concerned by investigators, are almost all dynamically generated and stored in the heap of processes. However, the inner structure of the heap is still a black box for investigators. Heap contain not only various kinds of data, but also management and control information of data structures. Indiscriminately regarding heap as data sources of evidence, is not enough to support the accuracy of digital evidence. Therefore, the research of heap data analysis method is particularly important, and further analysis of the Jemalloc, the default Android heap manager, is required.

3.2. The Android process address space layout

Since version 5.0, Android has adopted Jemalloc as the default memory allocator. Jemalloc is originally a new generation memory allocator developed by Jason Evans for FreeBSD with powerful multi-core and multi-threaded allocation capability. Jemalloc manages the memory data through the data structures of Arena, Chunk, Run and Region. The relationships of these structures are presented in figure 2.

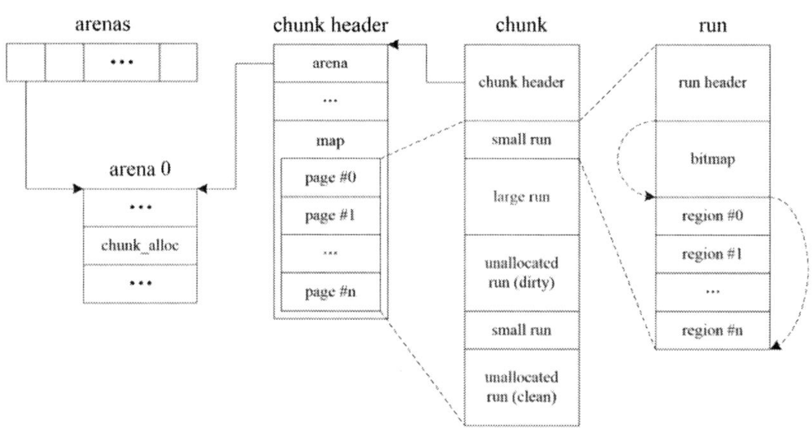

Figure 2. The relationships of data structures in Jemalloc.

Arena is the top-level structure in Jemalloc, and the data structure used to describe the state of Arena is arena_s. Jemalloc divides the heap memory into several separate areas, each of which is managed and maintained through Arena structure. In Android platform, the number of Arena is limited to no more than two by the makefile of Jemalloc. Arena divides the heap memory area managed by itself into several Chunks, and manages Runs and different types of Regions through these Chunks.

Chunk is the secondary level structure in Jemalloc, size of which defaults to 4MB and must be aligned to the 4MB boundary. The data structure for the Jemalloc to record the state of chunk is arena_chunk_s, but this structure is just a header for the whole chunk, containing only some statistics information and a dynamic array map. Chunk and its internal Runs are made up of many pages, and the allocation state, properties, offsets, and other tag information of these pages are recorded in the map array of Chunk. Therefore, in addition to recording page state, the map array also determines the base address retrieval of Runs.

Run is the heap memory allocation main body of Jemalloc, size of which must be aligned to the page size (default to 4KB). The data structure for the Jemalloc to record the state of chunk is arena_run_s, and similar to arena_chunk_s, this structure is also just a header for the whole Run. Run is internally divided into Regions of the same size and records the state of these Regions by the bitmap of its header.

Region is the basic heap memory data storage unit in Jemalloc. As shown in table 1, according to the storage space size, Jemalloc divides the Region into three types: small, large and huge. Small Region and large Region are managed by Run. However, huge Region is directly managed by Arena, which is equivalent to Chunk.

Table 1. Spacings and sizes of different types regions.

Types	Spacing	Sizes
small	N/A	[8B]
	16B	[16B, 32B, 48B, ..., 128B]
	64B	[192B, 256B, 320B, ..., 512B]
	256B	[768B, 1024B, 1280B, ..., 3840B]
large	4KB	[4KB, 8KB, 12KB, ..., 4072KB]
huge	4MB	[4MB, 8MB, 12MB, ...]

3.3. The bridge principle of the semantic gap between kernel and Jemalloc

From mobile forensics investigators' perspectives, we have to locate the physical address of every data structure both in the kernel and Jemalloc, so that we can extract the accurate data from the volatile memory. However, due to the address space of Android process is layout by the kernel and Jemalloc resides in the system software level which is above the kernel level, there is a semantic gap between the kernel and Jemalloc. In addition to that, although the physical addresses of kernel level data structures can be resolved step by step through a system.map file, the physical addresses of Jemalloc data structures cannot be located independently because the address of Arenas, a global vector group which contains the top-level data structure in Jemalloc, is not exported. Hence, the key to solve this problem is to accurately locate the physical address of Arena structure in Jemalloc.

As shown in figure 2, there is a pointer named arena in Chunk points to the Arena that the Chunk belongs to, and this provides a new idea for Arena structure positioning. Because the size of Chunk defaults to an integer multiple of 4MB and the bound of that is aligned with 4MB. So each 4MB starting address in the Android process address space is likely to be a Chunk header structure.

Therefore, as shown in figure 3, taking advantage of the boundary auto alignment feature of Chunk structure, we can bridge the semantic gap between the kernel and Jemalloc, and accurately locate the physical address of each data structure in Jemalloc.

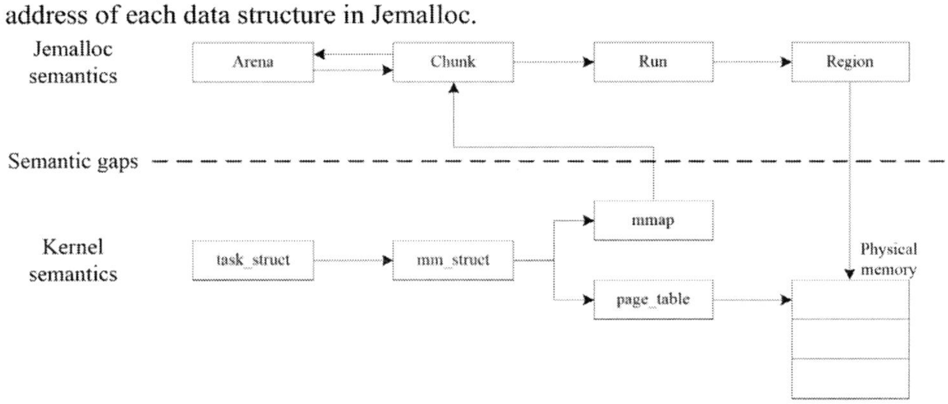

Figure 3. The bridge of the semantic gaps between the kernel and Jemalloc.

4. Implementation

Through the analysis on the bridge principle of the semantic gap between the kernel and Jemalloc in previous section, based on the memory acquisition tool Lime and analysis framework Volatility, we purpose a Fine-grained Analysis method for Android volatile Memory, called FAMAM.

The overall framework of FAMAM is shown in figure 4, including three modules: memory acquisition module, data locating module and data analysis module.

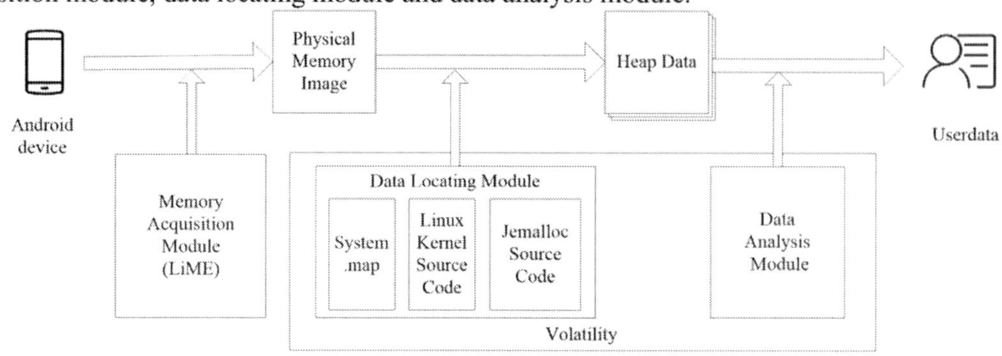

Figure 4. The overall framework of FAMAM.

4.1. Memory acquisition module

FAMAM uses Lime to acquire the physical memory image of Android device, this process can be divided into the following steps:

• Download the device kernel source from the official website according to the device model, modify the kernel source (hijacking system calls, etc.), and set the Root permission back door.

• Turn on the LKM loadable option in the kernel configuration, recompile the kernel, and generate a new boot.img.

• Unlock the device (if needed), enter the fastboot mode, and brush in the new boot.img to enable the device to load the LKM • Brk segment stores dynamically allocated data smaller than 128KB.

• Install Lime on the device, configure the device kernel source path, compile and generate lime.ko kernel module.

• Get the ADB shell with a Root permission using the preset back door.

• Load the lime.ko into the device with a optional transmission mode (SDcard or TCP) and export the device physical memory image which can be analyzed by Volatility.

4.2. Data locating module

The locating of Android process heap data requires not only the kernel data structures, but also data structures of Jemalloc memory allocator. Therefore, FAMAM needs to combine several methods to locate the process heap data structure, including System.map, Linux kernel source and Jemalloc source.

• System.map is a symbol table of Linux kernel, recording the address of symbols such as functions and global variables of kernel. With the help of system.map, the task_struct address of init_task, the first process of Linux kernel, can be found.

• Linux kernel source code defines kernel data structures such as process control block task_struct, memory control block mm_struct and virtual memory management unit vm_area_struct, which can be used to accurately calculate the offsets among various kernel data structures. By tracking tasks process list of init_task, we can not only get the page table address of any process and implement the transformation from virtual address to physical address, but also locate all data regions of any process.

• Jemalloc source code defines memory heap data structures such as Arena, Chunk, Run and Region, which we need to calculate the offset between various heap data structures. Particularly noteworthy is that for each mmap area of the target process, doing an alignment operation of the 4MB boundary can get an address of the region belongs to the Chunk, further confirm the address point to a chunk_header structure, the locating of the chunk structure can be realized.

Based on the above analysis, the pseudo-code of Android process heap data locating algorithm is shown in table 2.

Table 2. The pseudo-code of data locating algorithm.

```
/* Gets the start address of init_task from system.map */
struct task_struct *p = read_system_map(init_task);
/* Gets the start address of target process according to PID */
struct task_struct *entry = NULL;
list_for_each_entry (entry, &(p->tasks), tasks) {
   if (PID == entry->pid)
      entry = p;
}
/* Locate and export the target process mmap areas*/
if (entry) {
   struct mm_struct *mm = entry->mm;
   struct vm_area_struct *vma = mm->mmap;
   while (vma) {
      export(vma->vm_start, vma->vm_end);
      vma = vma->next;
   }
}
/* Locate the chunk structure in the exported mmap areas */
for_all_exported_areas {
   struct arena_chunk_s *chunk_ptr = chunk_addr2base(start_addr);
   if (chunk_ptr is a chunk_header)
      export this area as a chunk;
}
/* Locate the data area in the exported chunk structure */
for_all_exported_chunks {
   if (chunk contains a huge region)    //huge region
      acquire the data in the huge region;
   else {                               //large or small region
      struct arena_run_s * runs[] = get_runs_in_chunk(maps);
      for_all_runs_in_chunk {
         regions[] = get_regions_in_run(bitmap_info);
         acquire the data in the regions;
      }
   }
}
```

4.3. Data analysis module

Application data in memory is usually stored in plain text, so that regular expression matching can be used to mark and analyze application data in memory with specific format or special meaning. As the heap data locating algorithm purposed in this paper can directly locate the application data within the Region structure, the data used to record the data structure relations outside the Region structure will not be taken as the application data. In this case, the regular expression matching method can be used to obtain a higher accuracy.

Zhou[8], Zhang[15] and Wu[16] et al. have studied the extraction and analysis techniques of WeChat application data from different perspectives. Based on their analysis results of WeChat, through further experimental verification, we confirm that the process of WeChat is com.tencent.mm and the chat text of that stored in memory is in this format:

<Unix timestamp><Username><Content><Type>

• Unix timestamp field represents the timestamp of text sent or received, which is a hexadecimal number of 6 bytes in length.

• Username field represents the username of the text sent or received and can be obtained via WeChat unencrypted database IndexMicroMsg.db.

• Content field represents the specific content of the text.

• Type field represents the received and sent types, 0x7B00 for received and 0x1BFF for sent.

5. Test and evaluations

This section tests and evaluates FAMAM in terms of functionality and robustness, and then presents a new storage pattern of application authentication information such as username and passwords. Functional testing is used to test whether FAMAM can acquire user data for the Android application processes. Robustness tests are used to test the ability of FAMAM to maintain access to user data for Android application processes under special situations.

The test environment is as follows: the test device is Samsung Galaxy Note3, device model is SM-N9008V, baseband version is N9008VZMUDOH2, memory size is 3G, kernel version is 3.4.0, Jemalloc version is 3.6.0, Lime version is 1.8, Volatility version is 2.6, WeChat version is 6.6.5, the forensic analysis platform is A Lenovo PC (Lenovo-A6810f) with Ubuntu 14.04 OS.

5.1. Functional test

The chat text of WeChat is selected as the test object to evaluate the functional validity of FAMAM. The test steps are as follows:

• Initialize the test device (restore factory settings).

• Perform steps in Section 4.1 to prepare for the acquisition of the physical memory image of text device.

• Install WeChat on the test device, login the test account, randomly send 20 chat text messages, and receive 20 chat text messages, and then randomly delete 5 messages sent and 5 messages received.

• Perform the acquisition step to export the acquired physical memory image to the forensic analysis platform.

• Use the plugins implemented in this paper to analyze the physical memory image.

Based on the analysis result of the regular expression format of WeChat chat text, FAMAM is used to locate and analysis the WeChat chat text in the physical memory. The result is shown in figure 5. It can be seen that the plugins have successfully acquired WeChat process heap data structure information and chat texts.

Comparing the acquired text chat records with the 40 texts set in the text steps, it is found that they are all same. The result indicates that the function of the FAMAM method proposed in this paper is effective.

There are some researchers have studied the forensics and analysis methods on WeChat. Zhou[8] performed a dump and coarse-grained analysis of Android volatile memory on WeChat. Zhang[15] and Wu[16] studied the logical acquisition and analysis methods on WeChat. In the test environment used in this paper, FAMAM is compared with their acquisition methods, and the results are shown in table 3.

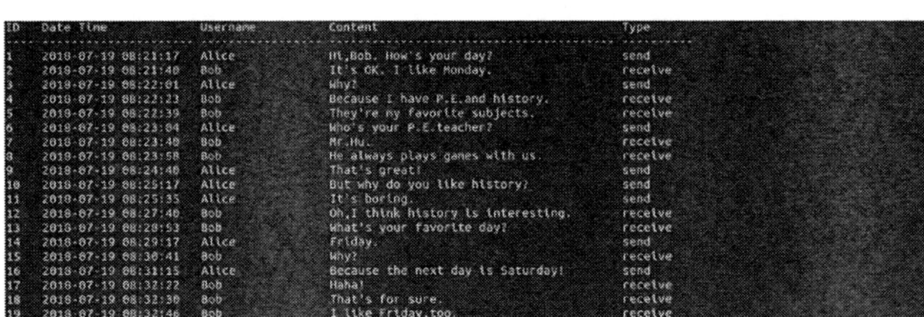

ID	Date Time	Username	Content	Type
1	2018-07-19 08:21:17	Alice	Hi,Bob. How's your day?	send
2	2018-07-19 08:21:40	Bob	It's OK. I like Monday.	receive
3	2018-07-19 08:22:01	Alice	Why?	send
4	2018-07-19 08:22:23	Bob	Because I have P.E.and history.	receive
5	2018-07-19 08:22:39	Bob	They're my favorite subjects.	receive
6	2018-07-19 08:23:04	Alice	Who's your P.E.teacher?	send
7	2018-07-19 08:23:40	Bob	Mr.Hu.	receive
8	2018-07-19 08:23:58	Bob	He always plays games with us.	receive
9	2018-07-19 08:24:40	Alice	That's great!	send
10	2018-07-19 08:25:17	Alice	But why do you like history?	send
11	2018-07-19 08:25:35	Alice	It's boring.	send
12	2018-07-19 08:27:40	Bob	Oh,I think history is interesting.	receive
13	2018-07-19 08:28:53	Bob	What's your favorite day?	receive
14	2018-07-19 08:29:17	Alice	Friday.	send
15	2018-07-19 08:30:41	Bob	Why?	receive
16	2018-07-19 08:31:15	Alice	Because the next day is Saturday!	send
17	2018-07-19 08:32:22	Bob	Haha!	receive
18	2018-07-19 08:32:30	Bob	That's for sure.	receive
19	2018-07-19 08:32:46	Bob	I like Friday,too.	receive
20	2018-07-19 08:33:23	Alice	What are you plan to do afternoon?	send

Figure 5. The analysis results of WeChat processes.

Table 3. Results of comparison test.

Method	Sent		Received		Total	
	undeleted	deleted	undeleted	deleted	undeleted	deleted
Zhou	15/15	5/5	0/15	0/5	15/30	5/10
Zhang	15/15	0/5	15/15	0/5	30/30	0/10
Wu	15/15	0/5	15/15	0/5	30/30	0/10
FAMAM	15/15	5/5	15/15	5/5	30/30	10/10

It can be seen from table 3 that compared with other WeChat data forensics and analysis methods, the FAMAM proposed in this paper can completely and effectively acquire the undeleted and deleted chat text records received and sent of WeChat, which performs better in term of data acquisition ability. The logical acquisition method of Zhang[15] and Wu[16] is implemented by analyzing WeChat database, but WeChat database need a root privilege to access while enables encryption security and deletion policies. Even if the key of WeChat database can be obtained, only undeleted data can be acquired. Similar to FAMAM, the acquisition method of Zhou[8] is also implemented based on memory forensics, but it simply takes the whole memory as a large block of data and does not analysis the data structure of memory, which is coarse-grained, so that it only can find out the text format of sent message.

5.2. Robustness test

In this paper, the ability of FAMAM to acquire user data of Android applications in the context of some special scenarios are tested, so as to evaluate the robustness of FAMAM. We still use WeChat as the test application, chat text as the test data. The test steps are consistent with those in the functional test except for perform the acquisition step after let the device enter special scenarios. These special scenarios include application background running, application account relogin, application restart, device reboot and device shutdown. The test results are shown in table 4.

It can be seen from the test results that FAMAM can acquire undeleted data and deleted data when the application process is background running. However, FAMAM cannot acquire deleted data when the account relogins, the application restarts, the device reboots, and the device shutdowns.

Table 4. Results of comparison test.

Scenario	Sent		Received		Total	
	undeleted	deleted	undeleted	deleted	undeleted	deleted
background running	15/15	5/5	15/15	5/5	30/30	10/10
Account relogin	15/15	0/5	15/15	0/5	30/30	0/10
App restart	15/15	0/5	15/15	0/5	30/30	0/10
Device reboot	15/15	0/5	15/15	0/5	30/30	0/10
Device shutdown	15/15	0/5	15/15	0/5	30/30	0/10

6. New discovered patterns

As discovered by Ntantogian[11], the common pattern for application username and password is shown as table 5.

Table 5. The common pattern for username and password discovered by Ntantogian.

Username	Password
j_username=	j_password=
username=	password=
userid>	password:
login i:type=	pass i:type=

Unfortunately, we are failed to use these patterns to find username and password of Wechat. As shown in figure 6, although we find a pattern for username (username=), we do not find the actual username behind the pattern. This indicates that the Wechat has taken security measures to protect the username and password stored in memory.

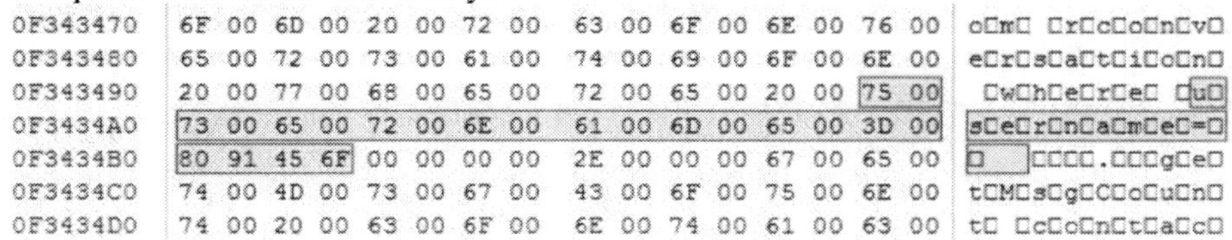

Figure 6. The search result of using a common pattern for Wechat username.

However, we notice that there is a 8 bytes magic symbol "(0x)80 91 45 6F" appears behind the common username pattern. We use FAMAM to quickly locate the heap data address space of Wechat processes, and find that this magic symbol appears at other addresses in the data space, as shown in figure 7 and figure 8. More importantly that, we find the actual username and password of our test account behind this magic symbol.

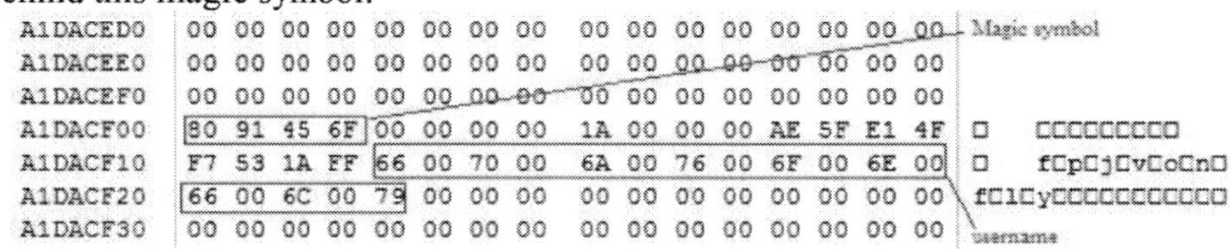

Figure 7. The search result of using a common pattern for Wechat username.

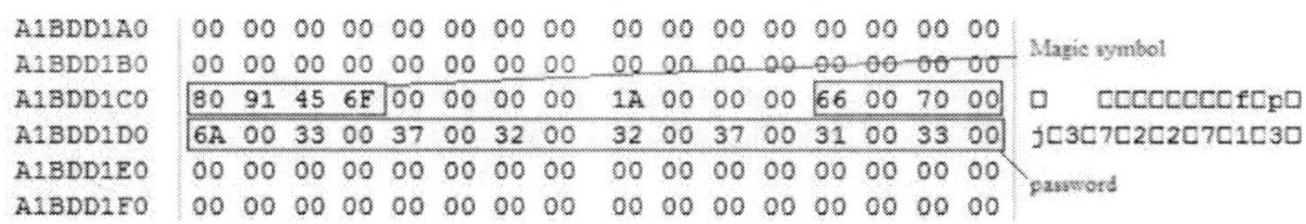

Figure 8: The new pattern for WeChat password discovered by FAMAM.

This interesting finding indicates that the patterns for username and password of Wechat are as follows:

- The pattern for magic symbol: <common pattern>=<magic symbol>(8 bytes).
- The pattern for username: <magic symbol> (8 bytes) <junk> (16 bytes) <username>.
- The pattern for password: <magic symbol> (8 bytes) <junk> (8 bytes) <password>.

7. Conclusion

In order to solve the inaccuracy of forensics analysis caused by coarser-grained analysis of Android volatile memory, this paper proposes a Fine-grained Analysis Method for Android volatile Memory, called FAMAM. We analyze the address space layout of Android processes and the heap data management mechanism of Jemalloc memory allocator. FAMAM takes advantage of Lime, the famous Linux memory extract tool, to extract the memory image of Android mobile devices. According to the analysis results of the Android process address space layout and Jemalloc heap data management mechanism, the localization of Android process heap data is realized. The heap data of Android process is parsed using the method of regular expression matching. The functional test shows that FAMAM can accurately acquire the undeleted and deleted chat texts of Wechat. Compared with several other analysis methods for Wechat, FAMAM has a more accurate analysis capability. The robustness test proves that FAMAM still has the data acquisition ability under some special scenarios. More importantly, we successfully use FAMAM to discover new storage patterns for username and password of Wechat.

At the same time, FAMAM has the same limitations as Lime because it uses Lime to extract memory images: it requires a Root privilege access to device and the enable of LKM loading. For the vast majority of Android devices, it need to recompile kernel to do so. Future work will focus on reducing the limitations of this approach.

Acknowledgement
This work is supported by project 61802432 National Natural Science Foundation of China.

References
[1] WeChat. Tencent company. http: //weixin.qq.com
[2] Full Database Encryption for SQLite. https://www.zetetic.net/sqlcipher/
[3] Ligh M H, Case A, Levy J, et al. The Art of Memory Forensics: Detecting Malware and Threats in Windows, Linux, and Mac Memory[C]// Wiley Publishing, 2014.
[4] Jason Evans malloc (Jemalloc) . https://wikipedia.org/wiki/C_dynamic_memory_allocation#jemalloc.
[5] Sylve J, Case A, Marziale L, et al. Acquisition and analysis of volatile memory from android devices[J]. Digital Investigation, 2012, 8(3): 175-84.
[6] Stüttgen J, Cohen M. Robust Linux memory acquisition with minimal target impact[J]. Digital Investigation, 2014, 11: S112-S119.
[7] The Volatility Foundation. http://www.volatilityfoundation.org.
[8] Zhou F, Yang Y, Ding Z, et al. Dump and analysis of Android volatile memory on Wechat[C]// IEEE International Conference on Communications. IEEE, 2015:7151-56.
[9] Macht H. Live memory forensics on android with volatility[J]. Friedrich-Alexander University Erlangen-Nuremberg, 2013.
[10] Leppert S. Android memory dump analysis[J]. Student Research Paper, Chair of Computer Science, 2012, 1.
[11] Ntantogian, C., Apostolopoulos, D., Marinakis, G. Evaluating the privacy of Android mobile applications under forensic analysis[J]. Computers & Security, 2014, 42(4):66-76.
[12] Cohen M. Forensic analysis of windows user space applications through heap allocations[C]// IEEE Symposium on Computers and Communication. IEEE, 2015: 237-44.
[13] Ligh M H, Case A, Levy J, et al. The Art of Memory Forensics: Detecting Malware and Threats in Windows, Linux, and Mac Memory[C]// Wiley Publishing, 2014.
[14] Block F, Dewald A. Linux memory forensics: Dissecting the user space process heap[J]. Digital Investigation, 2017, 22:S66-S75.
[15] Zhang L, Yu F, Ji Q. The Forensic Analysis of WeChat Message[C]// IEEE International Conference on Instrumentation & Measurement. IEEE, 2016:500-03.
[16] Wu S, Zhang Y, Wang X, et al. Forensic analysis of WeChat on Android smartphones[J]. Digital investigation, 2017, 21: 3-10.

Reservoir inflow prediction using a hybrid model based on deep learning

Bin Luo[1], Yang Fang[2], Handong Wang[1] and Di Zang[2,*]

[1]ChangJiang survey, planning, design and research co., LTD. Wuhan China
[2]Department of Computer Science and Technology Tongji University. Shanghai China

*E-mail: zangdi@tongji.edu.cn

Abstract. Reservoir inflow prediction plays a significant role in the field of hydrological prediction. Accurate and reliable prediction of reservoir inflow is the key to flood control decision. In this paper, we combine the deep belief network (DBN) with the Long Short-Term Memory (LSTM) to present a hybrid model based on deep learning (HDL) for reservoir inflow prediction. We take a full consideration of the basin flow and rainfall factors, which significantly affect the inflow flow. According to the rainfall data, we divide the corresponding flow data into two cases: rain and no rain. The proposed approach consists of three parts: we apply the DBN to learn the characteristics of the flow data and get predicted values of the reservoir inflow in case of rain and in case of no rain respectively. Then, we use the basin rainfall data and adopt the LSTM to fit the differential between the predicted inflow value generated by DBN in case of rain and the real inflow value, and get the predicted differential. Finally, the outputs of these three parts are added to obtain the final predicted result. Experiments are evaluated by the historical flow and rainfall data of a reservoir in China, and the results have proved that our method is effective and has higher prediction accuracy.

1. Introduction

Reservoir inflow prediction is one of the most important tasks and challenges in hydrology. Hydrological phenomenon is of high fuzziness, complexity and randomness since it is influenced by many uncertainty factors such as rainfall, the characteristics of topography, river channel and soil distribution of the catchment area. There are quite a few methods have been proposed for hydrological forecasting, which can be classified into 3 major categories: statistical methods based on hydrology, machine learning methods and deep learning methods.

Over the past decades, there have been plenty of traditional methods for the hydrologic inflow prediction such as process-based hydrology model [1], auto-regressive moving average (ARMA), and auto-regressive integrated moving average (ARIMA) [2]. ARIMA model is also applied to forecast the stream flow of the Hirakud reservoir in [3]. Moreover, a probabilistic method named SCHADEX was presented for flood volumes estimation in [4]. Generally, these conventional methods usually rely on a number of assumptions for natural environment. However, it is hard to reflect the stochastic and nonlinear nature of hydrological data due to various factors such as climate, natural geography and human activities that significantly affect them.

Afterwards, considering the nonlinear characteristics of the real data of the hydrology, a great deal of machine learning approaches are used for forecasting in the field of hydrology. For example, some classic models are Bayesian-based methods [5], [6], Support Vector Machines (SVM) [7], [8], fuzzy inference systems [9], and Artificial Neural Network [10]. To achieve better prediction accuracy and efficiency. Cheng et al. [11] developed an artificial neural network model based on quantum-behaved particle swarm optimization for daily reservoir runoff prediction. Hu et al. [12] applied principal component analysis (PCA) and improved the traditional BP algorithm to predict river runoff. Although these methods can improve the accuracy of hydrological prediction to a certain extent, they are belong to "shallow" learning category and have certain deficiencies such as large number of parameters, time-consuming, and easily falling into local optimum, which cannot be ignored in practical application.

Recently, several deep learning methods, which can achieve better prediction performance owing to its "deeper" representations, have been proposed as well. Bai et al. [13] employed deep belief networks (DBNs) to represent multiscale features and proposed a hybrid model to handle the daily reservoir inflow prediction. Liu et al. [14] presented a integrated deep learning model based on stacked auto encoders (SAE) and back propagation neural networks (BPNN) and adopt it to predict the stream flow. Wu et al. proposed a context-aware attention LSTM (CA-LSTM) network to extended the original LSTM network for flood prediction [15].

Deep learning has a powerful ability to automatically extract complex features of data, thus it is a very great way to apply it to analyse a large number of historical hydrological data. However, it has not been widely applied in reservoir inflow prediction so far. The existing deep learning network models are generally weak in the analysis for specific problems, and have such shortcomings as low prediction accuracy, simple network structure and poor scalability.

To improve the accuracy of reservoir inflow prediction, in this paper, we propose a hybrid model based on deep learning (HDL) and apply it to predict the reservoir inflow. We first take the historical flow data as the input of a DBN model to obtain the predicted inflow value F1 in the case of no rain. Then, the DBN model is used again to generate the predicted inflow value F2 in case of rain according to the mapping relationship between the flow data and the reservoir inflow learned in case of no rain. Next, we use the rainfall data as the input of a LSTM to fit the differential between the real inflow value and F2, and obtain a predicted value of the differential ΔF in case of rain. Finally, we add up these results F1, F2 and ΔF generated by three modules to get the predicted reservoir inflow value. The experimental results have demonstrated that our proposed approach achieves higher accuracy than the comparative methods in reservoir inflow prediction.

2. Dataset description and processing

In this paper, the data sets used for experiments is the historical hydrological data of a reservoir in China from 7/7/2010 to 31/10/2017, which consists of three parts: the first part is the flow data collected by 3 control stations of 3 tributaries within the catchment of the reservoir; the second part is the rainfall data collected by 123 rainfall control stations located between three tributaries and the reservoir; and the third part is the inflow data collected by a control station of the reservoir. Data in flood periods from May to October every year except 2010 are available for experiments. Here the sampling interval of data is 6 hours so that there are four recorded time points in a day: 2 am, 8 am, 14 pm and 20 pm. Therefore, the flow indicates the instantaneous flow measured every 6 hours and rainfall indicates the depth of water after a 6-hour rainfall record. To reflect the changes of flow and rainfall, the previous 8 days of flow or rainfall data are used as input of the model to predict the reservoir inflow for next 24 hours. Hence, the length of the input time series is 32. Namely, to predict the reservoir inflow at 2 am on May 2, 2017, the historical data of 2 am on May 1, 2017 and its 31 previous time points are needed. Data from 2010 to 2016 are selected as training samples and the data of 2017 are used as a test samples in our work.

For the data processing, firstly, the original data need to be pre-processed to remove abnormal elements which are hard to fit and seriously affect the prediction accuracy. Then, we classify the data of tributary flow and the reservoir inflow (flow data) at each time point into two categories (rain and no rain) according to the sum of the rainfall values recorded by rainfall control stations at corresponding time points. To be specific, the flow data are classified into the rain category if the sum of rainfall values is greater than 10mm, otherwise the flow data are classified into the no rain category if the sum of rainfall values is less than or equal to 10mm. In addition, we adopt the method of coarse-grained clustering to divide the catchment area of the rainfall control stations into several regions with obvious boundaries according to the position information of all rainfall control stations. In this paper, we get four regions by dividing, and then sum up the rainfall values recorded by the rainfall control stations in each region at each time point to get four sum values, which are used as the input data of the LSTM module.

3. Presented approach

Figure 1 shows the system architecture of our approach, which mainly includes four major steps: first of all, we build a DBN model to extract and learn the sophisticated characteristics of historical tributary flow and the reservoir inflow data (flow data), and generate the predicted reservoir inflow value F1 in the case of no rain. Secondly, the historical flow data are mapped to generate the predicted reservoir inflow value F2 in case of rain by the same DBN model using the parameters obtained in the first step. After that, we calculate the difference between the value of real reservoir inflow and the value F2 to get the differential Δ. Thirdly, we develop a LSTM model to learn and capture the corresponding relation between the rainfall data after clustering and the Δ, so as to generate the predicted differential value ΔF. At last, the predicted result of reservoir inflow is the sum of the above three parts.

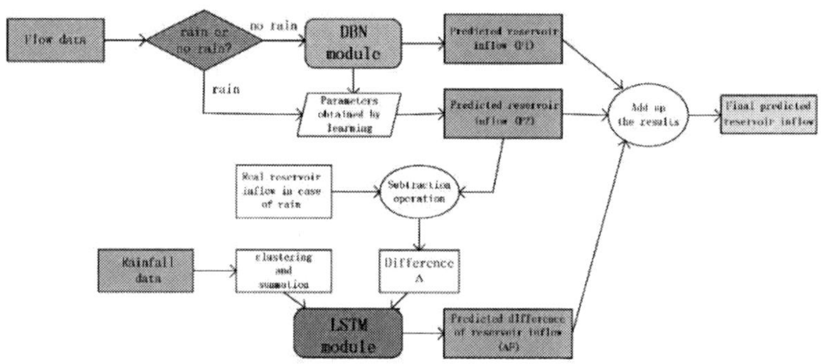

Figure 1. The system architecture of proposed approach.

3.1. DBN module for inflow prediction in case of no rain

The DBN module is utilized to predict the reservoir inflow in case of no rain in the first step. We selected the reservoir inflow data belonging to no rain category as the label for the DBN module, and the corresponding input data is a one-dimensional vector transformed from the historical flow data. As mentioned in section 2, the length of the input time series is 32, thus the length of the input data is 128 since the historical flow data come from 3 tributaries and a reservoir. If there exist data belonging to the rain category in the input data, we replace them with the average of their adjacent values.

The structure of the DBN module is shown in figure 2, DBN is a probability generation model, which is composed of restricted boltzmann machines (RBM). Our DBN model consists of a RBM and an artificial neural network (ANN), wherein RBM is composed of a visible layer and a hidden layer, and neurons are bidirectional and fully connected between the visible and hidden layers. Besides, ANN is composed of two hidden layers with hidden size of 1000 and a fully connected layer. The size of the input visible layer in RBM is the length of the input data, and the size of the RBM hidden layer is set to 1000. The output of each fully connected layer is defined as the following formula:

$$y^l = f(w^l x^{l-1} + b^l) \tag{1}$$

where, y^l, w^l and b^l are the output, the weight and the bias of layer l, respectively. x^{l-1} is the output of layer l-1. f is the activation function, Here, the activation functions of the hidden layers in RBM and ANN are all sigmoid activation functions, which can be defined as:

$$f(x) = \frac{1}{1+e^{-x}} \tag{2}$$

The last output fully connected layer has no activation function, which is a linear output.

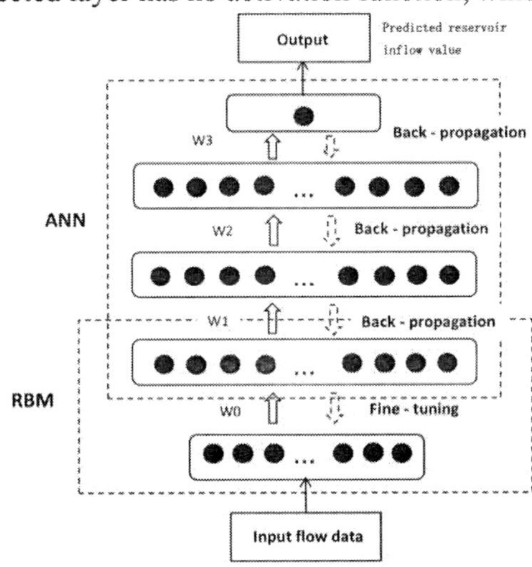

Figure 2. The structure of the DBN module.

The training process of DBN network consists of pre-training and fine-tuning. In the process of pre-training, the RBM is trained unsupervised to make feature vectors map to different feature spaces and ensure that feature information is retained as much as possible. After the RBM has been trained, the output feature vectors of RBM are input to ANN to initialize the weight parameters of ANN network, and then fine tune the entire DBN network through back propagation.

We adopt the mean square error (MSE) as the loss function to calculate the difference between the ground truth and the predicted result. Assuming there are N training samples, y' and y respectively represent the predicted value and the ground truth, then the loss function L is defined as:

$$L = \frac{1}{N}\sum_{i=1}^{N}(y' - y)^2 \tag{3}$$

In this paper, the loss function of both DBN and LSTM are optimized by Adam method. In this part, we employ the DBN to find the corresponding relationships between the training data and the training label by exploring and extracting their sophisticated natures, and finally get the reservoir inflow value F1 in case of no rain.

3.2. DBN module for inflow prediction in case of rain

Based on the same DBN model, to predict the reservoir inflow in case of rain, namely, the reservoir inflow to be predicted belongs to rain category. Similarly, we replace those input data belonging to the no rain category with the average of their adjacent values. Then, the corresponding relationships between the data and the label learned in section 3.1 are directly applied here for analysis in case of rain. In other word, the input data is mapped to generate the predicted reservoir inflow F2 in case of rain according to the training parameters obtained in section 3.1. Finally, we calculate the difference between the ground truth in case of rain and the corresponding predicted value F2 to get the differential Δ.

3.3. LSTM module for reservoir inflow differential prediction

Considering that the reservoir inflow should be equal to the sum of the tributary flows and the total rainfall of the reservoir basin, however, we find that it is difficult to establish a reasonable mapping relationship between the original rainfall data and the reservoir inflow. Since the differential Δ mentioned in 3.2 can be regarded as the result of rainfall, inspired by the excellent performance of

IOP Conf. Series: Materials Science and Engineering **715** (2020) 012044 doi:10.1088/1757-899X/715/1/012044

LSTM in modeling the dependencies of sequential data, we develop the LSTM module to fit the differential of reservoir inflow based historical rainfall data.

LSTM is a special kind of recurrent neural network, whose key construction unit is the cell. A LSTM cell is composed of three gate structures (input gate, forget gate and output gate) and a state vector transmission line, in which the state vector transmission line is responsible for long-term memory and the three gates are responsible for short-term memory selection. Therefore, the input vector can be deleted or added by gate setting, so that information can be selectively passed. What's more, the problem of gradient disappearance can be solved since the forgetting gate and the output gate can forget part of the information.

The structure of the LSTM network model is shown in figure 3, our LSTM module consists of one cell which is unrolled by time step. As described in section 2, we take the sum of rainfall of the four regions as the input data of the LSTM module. Owing to we use the data of the previous 8 days to make the prediction and there are 4 time points every day, the number of time step is 32 (32 cells unrolled by time), that is, for each prediction, the sum of the four rainfall data of 32 historical time points needs to be input, so the input is 32 one-dimensional vectors. The input data of LSTM is entered sequentially, one input in the one time step. At the same time, the differential Δ obtained in the second 3.2 section is used as the label for training. It is worth noting that here the data and labels used for network training are set to 0 if there is no rain. Considering that it is unreasonable to use rainfall values to fit negative values, thus set the differentials less than 0 to zero as well.

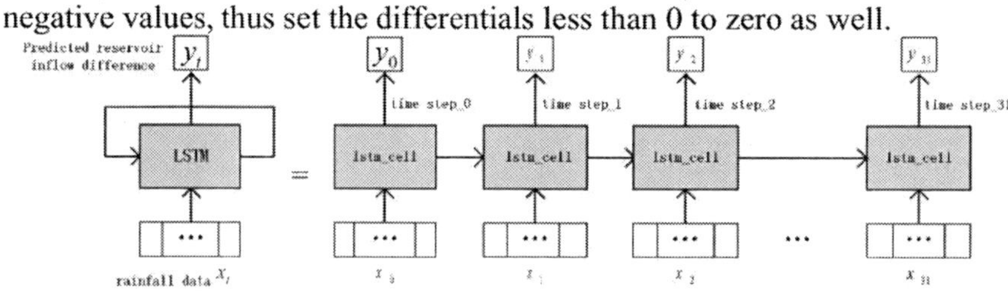

Figure 3. The structure of the LSTM module.

In the LSTM model, first of all, the LSTM has to decide what information should be forgotten from the cell state, which is determined by a sigmoid layer called the "forget gate". Assuming h_{t-1} represents the output of the previous cell, x_t represents the input of current cell, the output of the forget gate can be written as:

$$f_t = \sigma(W_f \cdot [h_{t-1}, x_t] + b_f) \tag{4}$$

where f_t, W_f, b_f respectively represent the output, the weight and the bias of forget gate of current cell, σ represents the sigmoid activation function. The output is a value in the range 0 to 1, where 1 means "completely retained" and 0 means 'completely forgotten'.Next, the LSTM decides what information to be saved in the cell state, which consists of two parts: first, a sigmoid layer called the "input gate" determines what information needs to be updated, second, a tanh layer generates a new candidate value vector $C_t^{'}$. Then, the two parts are combined to make an update to the cell's state. The input gate of current cell i_t and the new candidate value vector $C_t^{'}$ can be represented as formula (5) and formula (6) respectively:

$$i_t = \sigma(W_i \cdot [h_{t-1}, x_t] + b_i) \tag{5}$$

$$C_t^{'} = \tanh(W_c \cdot [h_{t-1}, x_t] + b_c) \tag{6}$$

where W_i, b_i respectively represent the weight and the bias of input gate of current cell. W_c, b_c respectively represent the weight and bias of $C_t^{'}$ of current cell.

The old state C_{t-1} is multiplied by f_t to forget the information we decided to forget before, and then added by $i_t * C_t^{'}$ to get the new state C_t which can be defined as follows:

$$C_t = f_t * C_{t-1} + i_t * C_t^{'} \tag{7}$$

Finally, the LSTM decides what to output. Firstly, the sigmoid layer called "output gate" is used to determine which part of the cell state will be output. The cell state is then passed through the tanh layer (to get a value in the -1 to 1 range) and multiplied by the output of the sigmoid layer. Therefore, the output gate o_t and the final output h_t of the current cell can be represented as formula (8) and formula (9) respectively:

$$o_t = \sigma(W_o \cdot [h_{t-1}, x_t] + b_o) \tag{8}$$

$$h_t = o_t * \tanh(C_t) \tag{9}$$

The expression of loss function to be optimized in LSTM model is the same as formula (3). Through the training and learning of LSTM network, predicted differential value ΔF is obtained in case of rain. Finally, the predicted value F of reservoir inflow obtained by our method is: F1+F2+ΔF.

4. Results and evaluation

Our experiments are conducted on the server with i7-5820K CPU, 48GB memory and NVIDIA GeForce GTX1080 GPU, and we employ the TensorFlow framework of deep learning to implement our approach. The learning rates of the DBN model and LSTM model are set to 0.0001and 0.001 respectively, and the total number of network iterations of both DBN and LSTM is 20000.

In order to evaluate the performance of the proposed approach, first, we calculate three error values: error >=10%, error >=15% and error >=20%, which respectively represent the percentage of data with an error of 10% or more in the total data, the percentage of data with an error of 15% or more in the total data, and the percentage of data with an error of 20% or more in the total data. Moreover, we calculate three performance evaluation criteria: Mean Relative Error (MRE), Mean Absolute Error (MAE) and Root Mean Square Error (RMSE) which are defined as formula (10), formula (11) and formula (12) respectively:

$$\text{MRE} = \frac{1}{N}\sum_{i=1}^{N}\frac{|y' - y|}{y} \tag{10}$$

$$\text{MAE} = \frac{1}{N}\sum_{i=1}^{N}|y' - y| \tag{11}$$

$$\text{RMSE} = \sqrt{\frac{1}{N}\sum_{i=1}^{N}(y' - y)^2} \tag{12}$$

Where y' represents the predicted value, y represents the real value, N represents the number of samples in test set.

We compare our approach with the methods of conventional ANN, the single DBN model and single LSTM model. In the comparison model, ANN has two hidden layers with 128 neuron nodes. All comparative models don't take into account the impact of rainfall data on reservoir inflow, only use historical flow data (flow data of 3 tributaries and inflow data of the reservoir) of the previous eight days to predict reservoir inflow. Besides, these models also don't analyze the flow data separately in terms of rain or no rain. Tables 1-2 show the comparison results of error and comparison results of MRE, MAE and RMSE.

Table 1. Comparison results of error.

Models	error >=10%	error >=15%	error >=20%
Our method	0.177143	0.075714	0.037143
ANN	0.311429	0.181429	0.105714
DBN	0.225714	0.104286	0.052857
LSTM	0.220000	0.098571	0.051724

Table 2. Comparison results of MRE、MAE、RMSE.

Models	MRE(%)	MAE	RMSE
Our method	6.29	1254.39	1963.68
ANN	8.95	1889.29	3006.75
DBN	6.99	1470.64	2376.26
LSTM	6.83	1466.65	2339.94

As shown in tables, our proposed method gets the lowest error compared with the other three methods. In terms of three error values, the results of error >=10%, error >=15% and error >=20% obtained by our model are respectively 0.177143, 0.075714 and 0.037143. It can be seen from the results that our model can achieve a high accuracy rate of 97% with an allowable error of more than 20%. For the results of three criteria, the values of MRE, MAE and RMSE yielded by our method are also the lowest of the listed methods, which are 6.29 %, 1254.39 and 1963.68, respectively. In the comparison model, the traditional machine learning model ANN has the highest prediction error, while DBN and LSTM perform much better which can be attributed to their powerful deep feature representation capability.

In our model, we simultaneously take advantage of the excellent performance of DBN and LSTM. On the one hand, we apply the deep structure of DBN to digging the latent features of flow data, on the other hand, we take full advantage of LSTM's ability to learn about long-term dependence and apply it to learn and explore sophisticated features of the rainfall time series data. In addition, the coarse-grained clustering of rainfall data and the classification of flow data according to rain or no rain are significantly contribute to improvement of prediction accuracy. Figure 4 shows the fitting curves of the predicted reservoir inflow generated by our model and its corresponding real reservoir inflow. In the figure, the x axis represents time series, the y axis represents the value of reservoir inflow, and the predicted reservoir inflow is represented by the red curve and the real reservoir inflow is represented by the blue curve. As can be seen from the figure, these two curves almost completely overlap, which not only indicates that our model has a high degree of fitting, but also reflects the high prediction accuracy of our method.

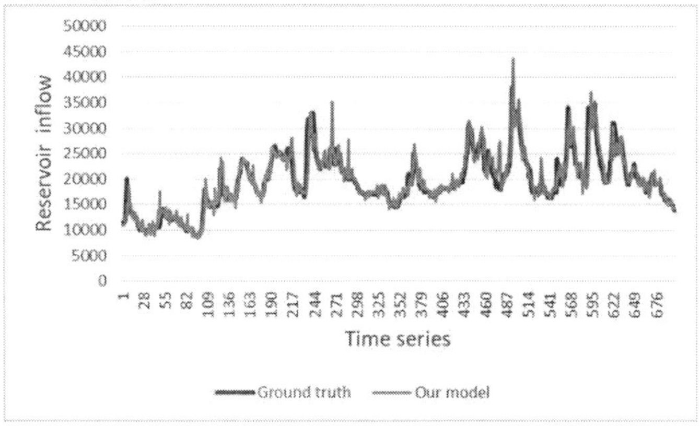

Figure 4. The fitting curve of the predicted value and the ground truth.

5. Conclusion

In this paper, a hybrid model based on DBN and LSTM has been proposed for reservoir inflow prediction. To improve prediction accuracy, in terms of data processing, we classify the flow data into two categories according to the situation of rain or no rain. Additionally, we adjust the granularity of rainfall data by clustering and dividing the rainfall control stations according to their positions. In terms of model design, we first employ DBN to discover effective characteristics from flow data, and obtain the predicted values of the reservoir inflow in the case of rain and no rain respectively.

Furthermore, in view of the impact of rainfall data on the reservoir inflow, we develop LSTM, which has excellent performance in fitting time series data, to deal with the problem of fitting between rainfall data and the differential of reservoir inflow in case of rain, so as to obtain the predicted differential value of reservoir inflow in case of rain. Finally, the final reservoir inflow value is obtained by organically combining the results of the above three parts. The experimental results have demonstrate that the proposed approach not only achieves higher prediction accuracy, but also has higher robustness and generalization.

Acknowledgement

This work is supported by the National Natural Science Foundation of China under grants 61876218, 61872271 and 61573259; and the National Key R&D Program of China, grant number [2018YFC1508006].

References

[1] R.J Zhao. The Xinanjiang Model Applied in China. Journal of Hydrology, 135(1–4), 371-81(1992).
[2] Samir. Parameters Estimate of Autoregressive Moving Average and Autoregressive Integrated Moving Average Models and Compare Their Ability for Inflow Forecasting. Journal of Mathematics and Statistics, 8(3), 330-38 (2012).
[3] A. Rath, K.S.Bhoi, S.Samantaray, P.C. Swain. Flow forecasting of hirakud reservoir with ARIMA model. International Conference on Energy, Communication, Data Analytics and Soft Computing (ICECDS). IEEE, Chennai, India, 9 pages (2017).
[4] Emmanue Paquet, Federico Garavaglia, Rémy Garçon and Joël Gailhard. The schadex method: A semi-continuous rainfall–runoff simulation for extreme flood estimation. Journal of Hydrology, 495(15), 23-37 (2013).
[5] S. Han, P. Coulibaly. Bayesian flood forecasting methods: A review. Journal of Hydrology, 551, 340-51(2017).
[6] Yirui Wu, Weigang Xu, Jun Feng, S. Palaiahnakote and Tong Lu. Local and Global Bayesian Network based Model for Flood Prediction. 2018 24th International Conference on Pattern Recognition (ICPR). IEEE, Beijing, China, 6 pages (2018).
[7] B. Bell, B. Wallace and D. Zhang. Forecasting river runoff through Support Vector Machines. 2012 IEEE 11th International Conference on Cognitive Informatics and Cognitive Computing (ICCI*CC). IEEE, Kyoto, Japan, 7 pages (2012).
[8] M. C. Wu, G. F. Lin and H. Y. Lin. Improving the forecasts of extreme streamflow by support vector regression with the data extracted by self-organizing map. Hydrological Processes, 28(2), 386-97 (2014).
[9] Y. H. Lin, C. C. Chiu, P. C. Lee and Y. J. Lin. Applying fuzzy grey modification model on inflow forecasting. Engineering Applications of Artificial Intelligence, 25(4), 734-43 (2012).
[10] M. T. Sattari, K. Yurekli and M. Pal. Performance evaluation of artificial neural network approaches in forecasting reservoir inflow. Applied Mathematical Modelling, 36(6), 2649-57 (2012).
[11] C. T. Cheng, W. J. Niu, Z. K. Feng, J. J. Shen and K. W. Chau. Daily reservoir runoff forecasting method using artificial neural network based on quantum-behaved particle swarm optimization. Water, 7(8), 4232-46 (2015).
[12] H. Daidi, S. Yuan and G. Mingtao. PCA-IBP model application in medium and long-term runoff forecasting. International Conference on Computer Science and Network Technology (ICCSNT). IEEE, Harbin, China, 4 pages (2016).
[13] Y. Bai, Z. Chen, J. Xie, and C. Li. Daily reservoir inflow forecasting using multiscale deep feature learning with hybrid models. Journal of Hydrology, 532, 193-206 (2016).
[14] F. Liu, F. Xu, and S. Yang. A Flood Forecasting Model Based on Deep Learning Algorithm via Integrating Stacked Autoencoders with BP Neural Network. 2017 IEEE Third International

IOP Conf. Series: Materials Science and Engineering **715** (2020) 012044 doi:10.1088/1757-899X/715/1/012044

Conference on Multimedia Big Data (BigMM). IEEE Computer Society, California, USA, 4 pages (2017).

[15] Y. Wu, Z. Liu, W. Xu, J. Feng, S. Palaiahnakote and T. Lu. Context-Aware Attention LSTM Network for Flood Prediction. 2018 24th International Conference on Pattern Recognition (ICPR). IEEE, Beijing, China, 6 pages (2018) .

IOP Conf. Series: Materials Science and Engineering **715** (2020) 012045 doi:10.1088/1757-899X/715/1/012045

Optimal mission planning for refuelling LEO satellites with peer-to-peer strategy

Jing Yu and Dong Hao*

China Aerodynamics Research and Development Center, Mianyang, 621000, China

*E-mail: haodongcardc@outlook.com

Abstract. The problem of optimal mission planning for refuelling LEO satellites with Peer-to-Peer strategy is studied in this paper. Specifically, satellites are distributed over different orbital slots of LEOs. Fuel-sufficient ones are required to exchange fuel with fuel-deficient ones. The purpose of this paper is to find the optimal pairs, as well as the rendezvous trajectories. A Hybrid Optimal Control Model is proposed and the existing rendezvous method is employed to address the model. Numerical simulations will demonstrate the effectiveness of the developed model.

1. Introduction

On-Orbit Refueling (OOR) plays a key role in extending the lifespan of space assets. As the enabling technologies for servicing single spacecraft have been increasing mature (e.g. Orbital Express [1]), the mission planning problem for OOR is issued, in which multiple Servicing Spacecrafts (SScs) are required to refuel multiple satellites. There are number of reported works devoted in On-Orbit Refueling (OOR) mission planning, i.e. developing optimal mission sequence and trajectories for SSc refueling multiple targets, aiming at improving the economic returns and fuel consumptions [2]-[5].

A system of multiple satellites can be served by a single servicing spacecraft in one by one pattern or by a distributed peer-to-peer (P2P) strategy, i.e. satellites exchange fuel amongst themselves in pairs, with the fuel-sufficient satellites providing fuel to the fuel-deficient satellites [6]. P2P strategy is a distributed and robust refueling pattern, which could offer protection against failures to some extent. Systematical studies have been performed on P2P OOR mission planning by Dutta and Tsiotras [7]-[10]. Based on Dutta's work, Yu et al. solved P2P mission planning problem with time window constraint in [11]. Recently, hybrid optimal control (HOC) theory has been applied to the solution of space mission planning [12].

In this paper, Optimal Mission Planning for Refuelling LEO Satellites with Peer-to-Peer Strategy is studied, a HOC model and the corresponding solution are proposed.

2. Problem description

Problem statement: $2N$ satellites distributed over different orbital slots of a LEO circular orbit are considered, as depicted in figure 1, where N of them are fuel-sufficient, active and called SSc, while the others are fuel-deficient, passive and called OSc. One SSc can exchange fuel with only one OSc. After a fuel exchange takes place between the active and the passive satellite, the active satellite returns to its original available orbital slots. The purpose of this paper is to find the optimal pairs, as well as the rendezvous trajectories.

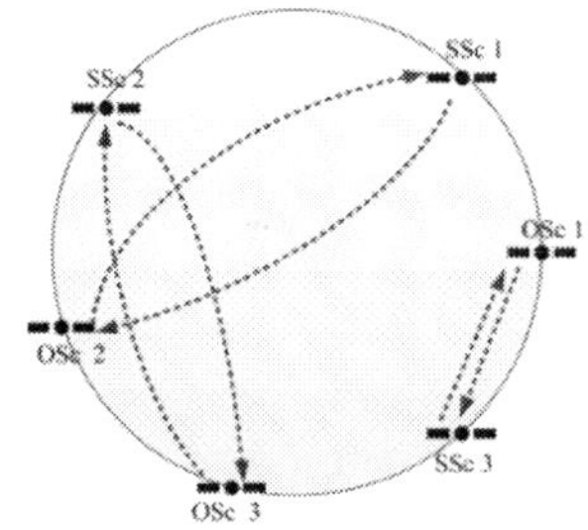

Figure 1. P2P refueling mission scenario.

3. Mathematical description of the HOC problem

Supposing there are N OScs and M SScs, a resource set $Q_s=\{s_1,s_2,...s_M\}$ (M SScs) and a task set $Q_t=\{t_1,t_2,..t_N\}$ (N OScs) can be defined. The categorical state space for the problem is $Q=Q_s\times Q_t$. Each element $q_{ij}=(s_i,t_j)\in Q$ represents a categorical/discrete state. It is clear that each element $q_{ij}=(s_i,t_j)\in Q$ has a definite physical meaning, i.e. the j th OScs would be serviced by the i th SSc. For this paper, we have a categorical state space as Eq. (1) (0 represents the initial location of the SSc):

$$Q=\begin{Bmatrix} (1,0) & (1,1) & \dots & (1,n) \\ (2,0) & (2,1) & \dots & (2,n) \\ \dots & \dots & \dots & \dots \\ (n,0) & (n,1) & \dots & (n,n) \end{Bmatrix} \tag{1}$$

The directed graph about this problem could be depicted as figure 2:

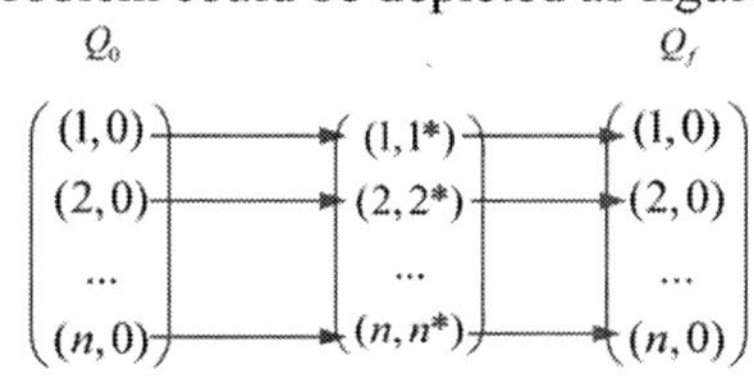

Figure 2. Directed graph for P2P on-orbit refueling mission.

A mission plan could be described as $\mathbf{q}=\{\mathbf{q}_1,\mathbf{q}_2,...\mathbf{q}_n\}$, where $\mathbf{q}_i=\{q_{i,0},q_{i,i^*},q_{i,0}\}$, and $(i=1,2,...n)$. Actually, the initial and final states of the system are definitely defined beforehand, so we just need to find $q_{i,i^*},(i^*=1,2,...n)$ only.

Associated with each state $q\in Q$ is a continuous-time controlled dynamical system: $\dot{\mathbf{x}}=f(\mathbf{x},\mathbf{u},t,q)$. In this paper, the dynamical system is the orbital rendezvous. A typical rendezvous process includes four major phases: phasing, far range rendezvous, close range, and mating. In this paper, each phase is completed by two-impulse maneuver, and the decision variable is given as: $\boldsymbol{u}=(\boldsymbol{t},\Delta\boldsymbol{v})$, where $\boldsymbol{t}$ describes the maneuver time and $\Delta\boldsymbol{v}$ denotes the velocity increment for each maneuver.

The constraints considered in this paper are:

- C_1: Each maneuver should be performed within certain communication time window.
- C_2: One SSc can exchange fuel with one and only one OS.

The total velocity increment is expected to be as small as possible. Let $\Delta v_{ito},\Delta v_{iback}$ represent the maneuver cost for servicing and return respectively, so we can get Eq. (2):

$$J_1=\min\sum_{i=1}^{n/2}(\Delta v_{ito}+\Delta v_{iback}) \tag{2}$$

Summing up, the P2P OOR mission planning problem could be modeled as Eq. (3):

$$(H)\begin{cases} find & \mathbf{q}, \mathbf{u} \\ \text{optimize} & J_1 \\ \text{subject to} & \dot{\mathbf{x}} = f(\mathbf{x}, \mathbf{u}, t, q) \\ & C_1 \cap C_2 \end{cases} \tag{3}$$

4. Solution

To address the HOC model above, two key problems should be considered and resolved:

(1) Task assignment: deciding which SSc should complete which task.

(2) Trajectory generation: deciding the transferring trajectory for each pair.

A two-level hybrid optimization strategy is proposed to solve the provided P2P refueling problem. The up-level is mission assignment, assigning fuel-deficient satellites to fuel-sufficient satellites in optimal pairs. It is expected to redistribute the fuel as equal as possible. All possible pairs are enumerated to find the best one. The low-level is to find the maneuver time and the corresponding velocity change for each rendezvous segment, minimizing the rendezvous cost. The "orbital rendezvous with complex constraints method" developed in [13] is employed for the lower-level.

5. Simulations

In this section, simulation cases are carried out to demonstrate the effectiveness of the proposed method. 6 satellites are taken into consideration, where 3 satellites are fuel-sufficient (numbered 1,2,3) and the other 3 are fuel-deficient (numbered 4,5,6). Perturbation are out of consideration. In orbital transfer, only phasing is considered. The orbit radius is r = 6878000 m. The orbital inclination is 60 degree. The initial RAAN is 60 degree. The time period of planning is 5 days. The initial phase angle and available fuel are given in table 1. 20 ground stations are selected for this case in table 2.

Table 1. Satellites parameters.

Satellite number	**1**	**2**	**3**	**4**	**5**	**6**
The initial phase angle (degree)	36	100	270	60	90	300
Available fuel (unit)	30	36	40	20	25	12

Table 2. Parameters of ground stations.

Station number	**1**	**2**	**3**	**4**	**5**	**6**	**7**	**8**	**9**	**10**
Longitude (degree)	38.54	40.132	41.727	43.223	34.250	36.842	25.476	36.443	23.482	38.537
Latitude (degree)	73.536	85.718	94.093	113.830	91.700	106.966	106.473	113.364	103.549	113.235
Station number	**11**	**12**	**13**	**14**	**15**	**16**	**17**	**18**	**19**	**20**
Longitude (degree)	28.3	24.3	50	33.7	39.6	26.5	25.6	36.6	23.9	31.6
Latitude (degree)	117.8	108.8	123.3	96.6	115.9	101.5	109.6	119.6	109.9	102.4

The solutions calculated by the methodology proposed in this paper are given in table 3. The total maneuver cost is 219.458 m/s.

Table 3. Maneuver parameters for case 1.

	SSc	**1**	**2**	**3**
	OSc	**5**	**4**	**6**
Forward trip	Time window for the first maneuver (s)	[69710, 70112]	[68722, 69106]	[66164, 66632]
	Station number	5	14	19
	Time window for the second maneuver (s)	[159606, 160015]	[160128, 160457]	[156434, 156741]

	Station number	1	1	5
	Total maneuver cost (m/s)	48.049	35.041	26.617
Return trip	Time window for the first maneuver (s)	[239397, 239724]	[239817, 240183]	[235904, 236389]
	Station number	19	7	11
	Time window for the second maneuver (s)	[330672, 331210]	[329604, 330214]	[327351, 327659]
	Station number	1	1	16
	Total maneuver cost (m/s)	47.568	35.730	26.453

Simulations show that the model proposed in this paper could address P2P OOR mission planning problem effectively.

6. Conclusions

In OOR mission, P2P is a strategy, in which satellites exchange fuel amongst themselves in pairs. It is a distributed and robust refuelling pattern, offering protection against failures to some extent. The problem of optimal mission planning for refuelling LEO satellites with Peer-to-Peer strategy is studied in this paper. The purpose of this paper is to find the optimal pairs, as well as the rendezvous trajectories. A Hybrid Optimal Control Model is proposed and the existing rendezvous method is employed to address the model. Simulation cases are carried out to demonstrate the effectiveness of the proposed model.

Acknowledgment

This work was supported by National Natural Science Foundation of China under Grant No. 71601179.

References

[1] R.B. Friend, Orbital express program summary and Mission overview, Proc. SPIE (2008). http://dx.doi.org/10.1117/12.783792 .

[2] X. Chen, J. Yu, Optimal mission planning of GEO on-orbit refueling in mixed strategy. Acta Astronaut. 133 (2017) 63–72.

[3] Y. Zhou et al., Mission planning optimization for multiple geosynchronous satellites refueling, Adv. Space Res. 56 (2015) 2612–2625.

[4] K.T. Alfriend, D. Lee, N.G. Creamer, Optimal servicing of geosynchronous satellites, J. Guid. Control Dyn. 29 (1) (2006) 203-206.

[5] J. Zhang, Y.Z. Luo, G.J. Tang, Hybrid planning for LEO long-duration multi-spacecraft rendezvous mission, Sci. China Tech. Sci. 55 (2012) 233–243.

[6] A. Dutta, N. Arora and R. P. Russell, Peer-to-Peer Refueling Strategy Using Low-Thrust Propulsion, ournal of spacecraft and rockets, 2012, 49(5): 944-954

[7] Dutta, A., and Tsiotras, P., "Asynchronous Optimal Mixed P2P Satellite Refueling Strategies," Journal of the Astronautical Sciences, Vol. 54, Nos. 3–4, 2006, pp. 543–565.

[8] Dutta A, Tsiotras P. "An Egalitarian Peer-to-Peer Satellite Refueling Strategy." Journal of Spactcraft and Rockets. 2008, 45(3): 608-618.

[9] Tsiotras, P., and Nailly, A., "Comparison between Peer-to-Peer and Single Spacecraft Refueling Strategies for Spacecraft in Circular Orbits," Info tech at Aerospace Conference, AIAA Paper 05-7115 Sept. 2005.

[10] Dutta A. Optimal cooperative and non-cooperative peer-to-peer maneuvers for refueling satellites in circular constellations. Ph.D. Thesis, Georgia Institute of Technology, 2009.

[11] Jing Yu, Qi Ouyang, Xiao-qian Chen and Li-hu Chen. Optimal peer-to-peer maneuvers for refueling satellites in circular constellations (C). 2012 International Conference on Spacecraft Structures, Materials and Mechanical Testing, Xiamen, China, Dec 2012

IOP Conf. Series: Materials Science and Engineering **715** (2020) 012045 doi:10.1088/1757-899X/715/1/012045

[12] Jing Yu, Xiao-qian Chen, Li-hu Chen, Dong Hao. Optimal scheduling of GEO debris removing based on hybrid optimal control theory[J]. Acta Astronautica, 2014, 93: 400–409.
[13] Jing Yu, Xiao-qian Chen, Li-hu Chen. Orbital rendezvous mission planning with complex constraints (J). Proc IMechE Part G: Journal of Aerospace Engineering, 2015(230).

IOP Conf. Series: Materials Science and Engineering **715** (2020) 012046 doi:10.1088/1757-899X/715/1/012046

Rapid entry trajectory planning without segmentation

Li Jun and Jiang Zheng-yu*

College of Aerospace Science and Engineering, National University of Defense Technology, Changsha 410073, China

*E-mail: jiang_nudt@sina.com

Abstract. A new rapid entry trajectory planning algorithm for hypersonic glide vehicles is presented in this paper. Firstly, build a new technique to calculate the boundary of the control variable that satisfy the path and control constraints without using quasi-equilibrium glide condition (QEGC). Secondly the optimal control variable is expressed as function of current state. Then combine two of them by control coordination. Finally, introduce one or two constant parameters to satisfy the final state condition. The approach is tested using the CAV model. This algorithm is efficient and will provide feasible trajectory without reversal of bank angle. It can apply for entire entry process, segmentation and the QEGC is unnecessary.

1. Introduction

Due to trajectory generation directly based on the optimization method[1-3] often requires a huge amount of computation, techniques such as segmentation strategy, quasi-equilibrium glide condition and lateral azimuth error deadband are widely used in the onboard trajectory generation of hypersonic glide vehicles.

The entry trajectory of the space shuttle[4] is divided into five phases that fitting with different formulas, and try to approximate constraint boundary. Zimmerman[5] divided the trajectory into two phases. The first phase adopted constant heat flow rate constraint to generate analytic solution, while the second phase did not consider the constraint and generated the trajectory by shooting method. Shen[6] employed the quasi-equilibrium glide condition to convert the path constraint into the constraint of bank angle, and single-point or two-point bank angle reversal was used for lateral planning. Li ZhaoYing[7] divides the entry trajectory into the initial descent phase, the constant heat flow rate phase, the constant overload phase and the interim phase, and each phase adopts different analytical formulas and adjusts some parameters online.

Although trajectory segmentation simplifies the planning in single phase, it increases the total number of parameters to be adjusted and the complexity of guidance, which is not conducive to the further development of the method, and the specific profile form will also limit the flight capability of the spacecraft itself.

Moreover, the application range of quasi-equilibrium glide condition is limited[8], which cannot be applied to the whole entry process. And reversal of bank angle usually brings about control tracking problems and has certain limitations on large lateral range[9].

In order to establish an efficient approach that avoid these problems, a new technique is invented in this paper to calculate the boundary of the control variable that satisfy the path and control constraints, which imply a function that restrict the arbitrary control variables to the available control variables. And an optimal control model is established, but instead of trying to find an explicit solution, the optimal control variable is regard as the function of the state variables. The composition of the two functions will lead to an initial value problem. The only thing left is to satisfy the final state condition while integrate the initial value problem. By introduce one or two parameters, and represent the entire problem as to solve the zero point of a function of those parameters, it can be done easily.

2. Entry dynamics

Ignoring the earth rotation, the hypersonic unpowered glide entry dynamics can be expressed as equation (1-6). The effect of the earth's rotation can be taken into account later.

$$\dot{r} = v\sin\theta \tag{1}$$

$$\dot{v} = -D - g\sin\theta \tag{2}$$

$$\dot{\theta} = \frac{1}{v}[L\cos\upsilon + (\frac{v^2}{r} - g)\cos\theta] \tag{3}$$

$$\dot{\Psi} = \frac{1}{v}[\frac{L\sin\upsilon}{\cos\theta} + \frac{v^2}{r}\cos\theta\sin\Psi\tan\phi] \tag{4}$$

$$\dot{\lambda} = \frac{v\cos\theta\sin\Psi}{r\cos\phi} \tag{5}$$

$$\dot{\phi} = \frac{v\cos\theta\cos\Psi}{r} \tag{6}$$

where λ and ϕ are the longitude and latitude, θ is flight path angle, Ψ is the heading angle, υ is the bank angle. The lift and drag accelerations, denoted as L and D, are given by equation (7)

$$\begin{cases} \mathrm{L}=\mathrm{C}_L(\alpha,\mathrm{Ma}(v,r))\rho(r)v^2\dfrac{S_r}{2m} \\ \mathrm{D}=\mathrm{C}_D(\alpha,\mathrm{Ma}(v,r))\rho(r)v^2\dfrac{S_r}{2m} \end{cases} \tag{7}$$

The density variation with altitude is given by equation (8)

$$\rho=\rho_0\exp[(r-R_0)/h_s] \tag{8}$$

3. Path and control constraints

Considering heat flow, dynamic pressure, and total overload as mainly constraints, they are given by equation (9-11)

$$\dot{Q} = k_Q\rho^{0.5}v^{3.15} \le \dot{Q}_{\max} \tag{9}$$

$$q = \frac{1}{2}\rho v^2 \le q_{\max} \tag{10}$$

$$n = \sqrt{L^2 + D^2}/g_0 = \frac{\sqrt{C_L{}^2 + C_D{}^2}\rho V^2 S_r}{2mg_0} \le n_{\max} \tag{11}$$

From (8) and (9), altitude constraint at a given speed are equation (12)

$$h \ge h_{\min}(v) = \max\left(2\mathrm{h}_s\log\left(\frac{\mathrm{K}_Q\sqrt{\rho_0}v^a}{\dot{Q}_{max}}\right), \mathrm{h}_s\log\left(\frac{\rho_0 v^2}{2\mathrm{q}_{\max}}\right)\right) \tag{12}$$

3.1. Control constraints

Using (3) divided by (1) and arrange to get equation (13)

$$\sin(\theta)d\theta = \left(\frac{C_L\cos(\upsilon)\rho S_r}{2m} + \left(\frac{1}{r} - \frac{g}{v^2}\right)\cos(\theta)\right)dr \tag{13}$$

Assuming that the vehicle drops at a constant value of $C_L\cos(\upsilon)$ and is just tangent to the boundary curve of the $h_{\min}(v)$ in (11), as shown in figure 1.

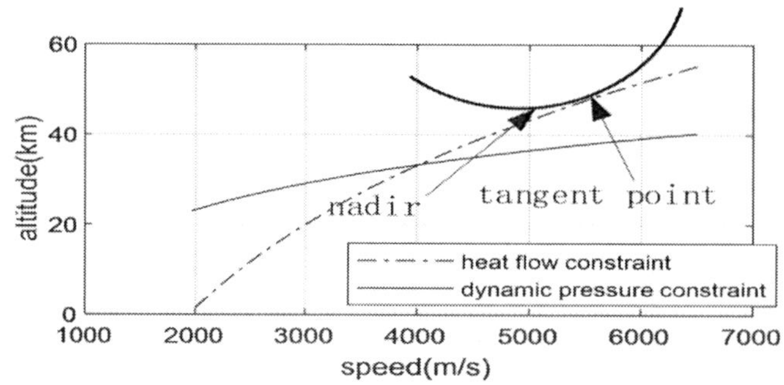

Figure 1. h-v constraints.

This tangent condition can be approximated to (14)

$$h_{low} \geq h_{\min}(v_{low}) \tag{14}$$

where h_{low} and v_{low} are the height and velocity of the local nadir. Integrate by assuming $k_h = (1/r - g/v^2)\cos(\theta)$ is constant will provide equation (15)

$$\cos(\theta_0)\text{-}\cos(\theta_1) = [-\frac{C_L\cos(\upsilon)e^{-\frac{h}{hs}}h_s\rho_0 S_r}{2m} + k_h h]_{h0}^{h_1} \tag{15}$$

At local nadir, $\theta_1 = \theta_{low} = 0$, and k_h , h_1 can be estimated based on the current state equation (16)

$$\begin{cases} k_h = k_{h0} = \left(\dfrac{1}{r_0} - \dfrac{g}{v_0^2}\right)\cos(\theta_0) \\ h_1 = h_{\min}(v_0) \end{cases} \tag{16}$$

Apply with will provide equation (17)

$$(C_L\cos(\upsilon))_{\min} = \frac{2m(k_h(\text{h}_0\text{-h}_1) + \cos(\theta_0) - 1)}{\text{h}_s\text{S}_r\rho_0(\exp(-h_0/h_s) - \exp(-h_1/h_s))} \tag{17}$$

The maximum value of C_L which constrained by (10),can denoted as equation (18)

$$(C_L)_{\max} = C_{L\max}(r, v) \tag{18}$$

Within (16) and, all path constraints are satisfied.

For testing accuracy of this constraint-handling technique, we force the $C_L\cos(\upsilon)$ to be the minimum that provided by (16) and (15) in simulation, it turns out that the h-v curve almost precisely track the $h_{\min}(v)$ without crossing it, see figure 2 which just indicated that our method is accurate and effective.

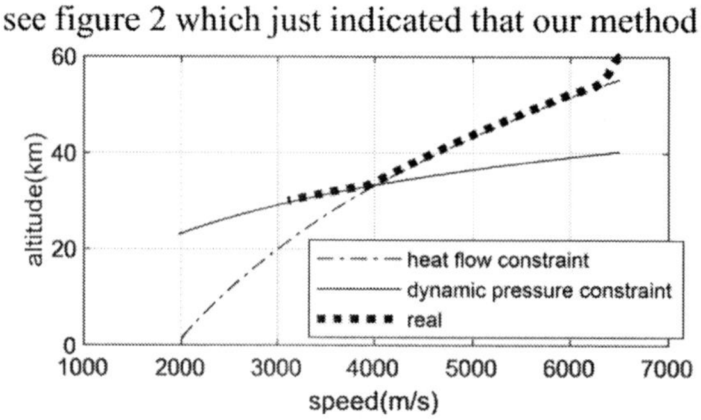

Figure 2. h-v curve that track the boundary.

4. Optimal control model

By dividing the planning problem into vertical and lateral, the corresponding control variable are $C_L\cos(\upsilon)$ and $C_L\sin(\upsilon)$.

4.1. Lateral planning

Define line-of-sight error $\Delta\psi = \psi_{\text{Los}} - \psi$, where ψ_{Los} is the angle of rotation from the current line-of-sight direction counterclockwise to the direct north direction.

Notice the range differential $\dot{R} \approx v\cos(\theta)$, take the derivative of $\Delta\psi$, and apply (4) equation (19)

$$\frac{d\Delta\psi}{dR} \approx \frac{\dot{\psi}_{Los}}{V\cos(\theta)} - \frac{C_L \sin(\upsilon)\rho S_r}{2m\cos^2(\theta)} - \frac{\tan(\phi)\sin(\psi)}{r} \overset{def}{=} u_z \tag{19}$$

Use equation (20) performance index

$$\min J = \frac{1}{2}\int_0^{R_{rest}} u_z^2 dR \tag{20}$$

such that equation (21)

$$\Delta\psi\big|_{R=R_{rest}} = 0 \tag{21}$$

then according to optimal control theory[10] equation (22)

$$u_z = -\frac{\Delta\psi}{R_{rest}} \tag{22}$$

where R_{rest} can be estimation by spherical trigonometry. According (21) and (18), we have equation (23)

$$C_L \sin(\upsilon) = -\frac{2m\left(u_z - \frac{\dot{\psi}_{\text{Los}}\sec(\theta)}{V} + \frac{\sin(\psi)\tan(\phi)}{r}\right)}{\rho\sec^2(\theta)S_r} \tag{23}$$

4.2. Vertical planning

Similar to lateral planning, considering θ is very small, vertical dynamic respect to range can denoted by equation (24)

$$\begin{cases} \frac{dr}{dR} = \frac{\dot{r}}{\dot{R}} = \tan(\theta) \approx \theta \\ \frac{d\theta}{dR} = \frac{\dot{\theta}}{\dot{R}} = \frac{C_L\cos(\upsilon)\rho S_r}{2m} - \frac{g}{v^2} + \frac{1}{r} \overset{def}{=} u_y \end{cases} \tag{24}$$

Use the equation (25) performance index

$$\min J = \frac{1}{2}\int_0^{R_{rest}} u_y^2 dR \tag{25}$$

such that equation (26)

$$\begin{cases} r\big|_{R=R_{rest}} = r_f \\ \theta\big|_{R=R_{rest}} = \theta_f \end{cases} \tag{26}$$

Solve (24) with (25) to get equation (27)

$$u_y = -\frac{2\left(3r - 3r_f + R_{rest}(2\theta_0 + \theta_f)\right)}{R_{rest}^2} \tag{27}$$

combine (23) and (26) will provide equation (28)

$$C_L\cos(\upsilon) = \frac{2m\left(u_y + \frac{g}{v^2} - \frac{1}{r}\right)}{\rho S_r} \tag{28}$$

Given any state, there are corresponding control variables lead the vehicle to the target. It can be represented as equation (29)

$$\mathbf{u}_{\text{opt}} = \mathbf{opt}(\mathbf{x}) \tag{29}$$

5. Construct an initial value problem

5.1. Control coordinator

The control variables calculated from (27) and (22) may be out of boundary that calculated from (16) and (17). The control coordinator force them into it, while maintain their continuity. New control variables denote as C_{LV} and C_{LL}, corresponding to $C_L\cos(\upsilon)$ and $C_L\sin(\upsilon)$.

The algorithm employed by the control coordinator can simply describe as follow, ordered by priority.

(1) keep safe, in other words $C_{LV} \geq (C_L\cos(\upsilon))_{\min}$

(2) try to keep lateral to be optimum, in other words minimize $|C_{LL}-(C_L\sin(\upsilon))|$

(3) try to keep vertical to be optimum, in other words minimize $|C_{LV}-(C_L\cos(\upsilon))|$

Given any state and any control variables, there are new corresponding control variables that guarantee safe. It can be represented as a function

$$\mathbf{u} = \mathbf{cc}(\mathbf{x}, \mathbf{u}_{org})$$

when there is enough lift force, the coordinator will be an identity function of uorg.

5.2. Initial value problem

Given any state, there are corresponding control variables C_{LV} and C_{LL} that can guarantee safe and lead the vehicle to the target. It can be represented as a function

$$\mathbf{u} = \mathbf{f}(\mathbf{x}) = \mathbf{cc}(\mathbf{x}, \mathbf{opt}(\mathbf{x}))$$

Dynamics (1)~(6) can denoted as

$$\dot{\mathbf{x}} = \mathbf{F}(\mathbf{x}, \mathbf{u})$$

Combine (30) and (31) will provide an initial value problem

$$\dot{\mathbf{x}} = \mathbf{F}(\mathbf{x}, \mathbf{f}(\mathbf{x}))$$

5.3. Reach the final state condition

Since the path constraints is solved, only the final state condition needs to be considered when integrating the equation (32). Due to the terminal time is uncertain, but the terminal energy $E = v^2/2 - \mu/r$ is known at the given altitude and speed, it is more convenient to integrate the equation with energy as the independent variable[11]. At the end point of integration, since the energy is known and there usually is enough lift force at the end, the altitude can be accurately satisfied, so that the speed can be accurately satisfied.

The remaining major problem is the horizontal position error. Note that the constraint control coordinator always gives priority to the lateral optimization when there is enough lift available, which enables the horizontal trajectory (or its extension line) to pass through the desired position (λ_f,ϕ_f) accurately. Therefore, the horizontal position error is actually a voyage problem. By adjusting the resistance, the terminal point can be just landed at point (λ_f,ϕ_f).

To do this, simply introduce two factor $[k_1, k_2]^T$ and multiply the control variables before passing it into the coordination function. That is, update (30) with

$$\mathbf{u} = \mathbf{f}(\mathbf{x}) = \mathbf{cc}(\mathbf{x}, [k_1, k_2]^T \otimes \mathbf{opt}(\mathbf{x}))$$

where ⊗ represent multiply by elements. Now, Define the integral end error function

$$g(k_1, k_2) = R^*_{rest}\cos(\Delta\psi^*)$$

where R^*_{rest} and $\Delta\psi^*$ represent range-to-go and line-of-sight error at integral end. If endpoint is in front of (λ_f,ϕ_f), $\Delta\psi^* < 90^0$, otherwise $\Delta\psi^* \geq 90^0$. In particular, if endpoint is exactly (λ_f,ϕ_f), then

$$g(k_1, k_2) = 0$$

So the entire trajectory planning problem will be solved by find roots of (35).

6. Testing and discussion

For test the performance of the algorithm, the general aircraft model[12] CAV-H were adopted. It has a reference area of 0.48 m^2, and a mass of about 907kg. And the maximum heat flow density constraint $\dot{Q}_{\max}=2MW/m^2$, maximum dynamic pressure constraint $q_{\max}=100kPa$, maximum total overload constraint $n_{\max}=3g_0$, attack range $\alpha\in[10^0,20^0]$, attack change rate $|\dot{\alpha}|\le 2^0/s$, bank angle $|\upsilon|\le 60^0$, bank angle rate $|\dot{\upsilon}|\le 20^0/s$. The initial conditions of the simulation are set as, the initial height $h_0=60km$, speed $v_0=6500m/s$, the flight path angle $\theta_0=0$, the initial longitude and latitude are both 0, the heading angle $\psi_0=90^0$, the target longitude and latitude(800,200), the target altitude is 30km, the target speed is 1965m/s, and target flight path angle is 0.

The zero point of $g(k_1,k_2)$ is not unique, in this example we set $k_1=1$, and random choose initial value of k_2 from [1, 5], then find zero point of $g(1,k_2)$ through Newton's method. After 7 iterations (8 times of integrals), it converges to $k_2=2.057$. The terminal position and height error are both 0, and the velocity error is 0.3m/s, which is a relatively accurate feasible solution. The corresponding trajectory parameter curve is shown in figure 3~ figure 10.

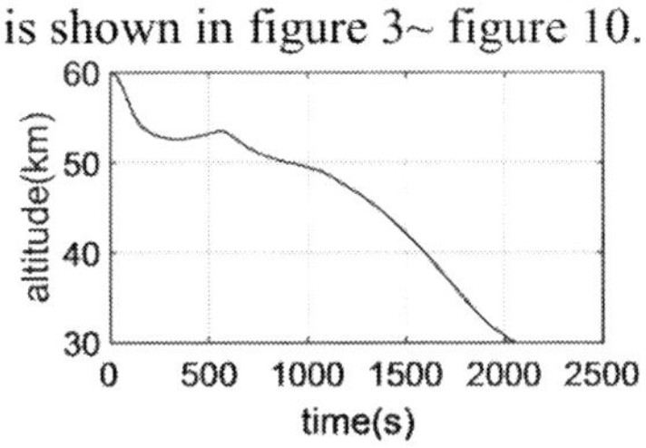

Figure 3. Altitude.

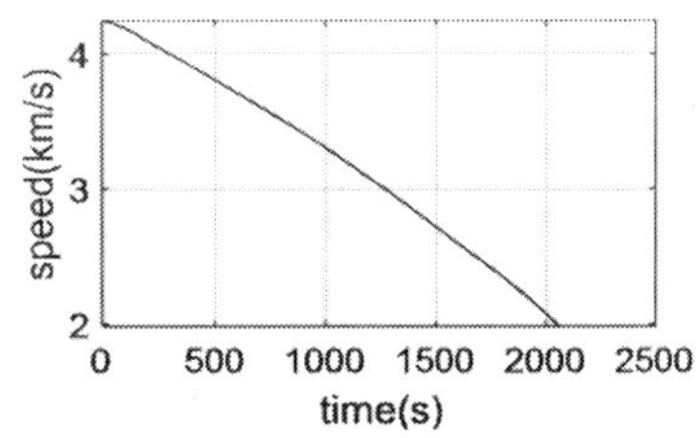

Figure 4. Speed.

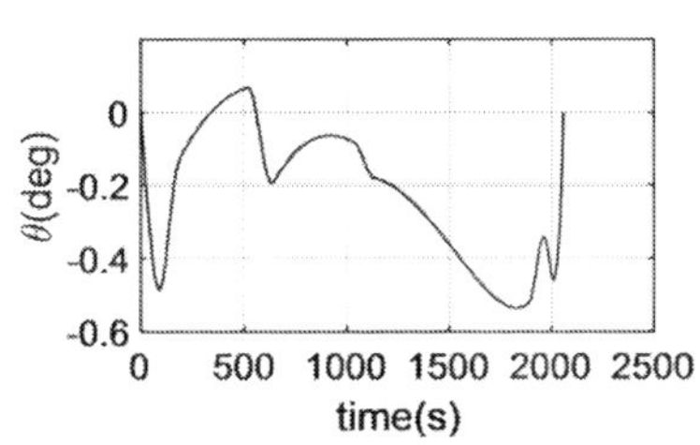

Figure 5. Flight path angle.

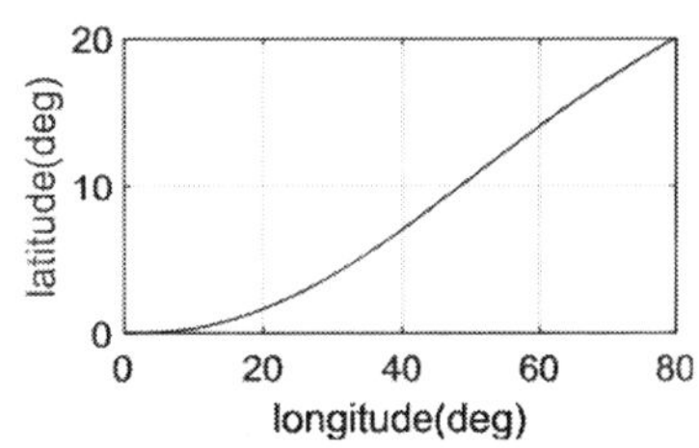

Figure 6. Horizontal trajectory.

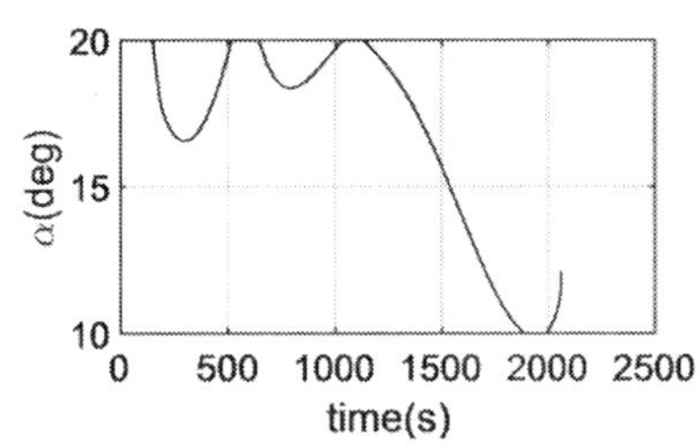

Figure 7. Angle of attack.

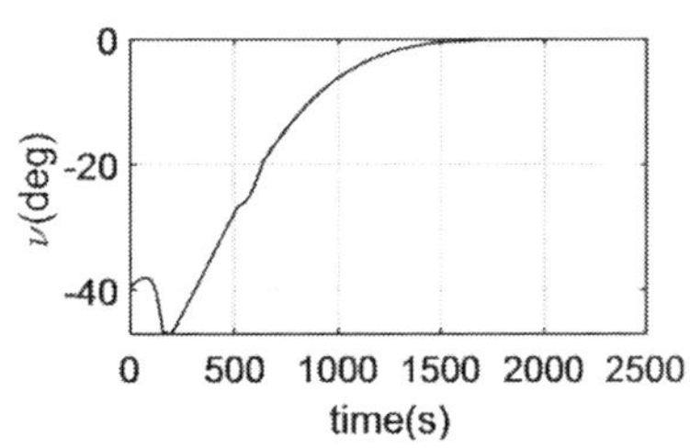

Figure 8. Angle of bank.

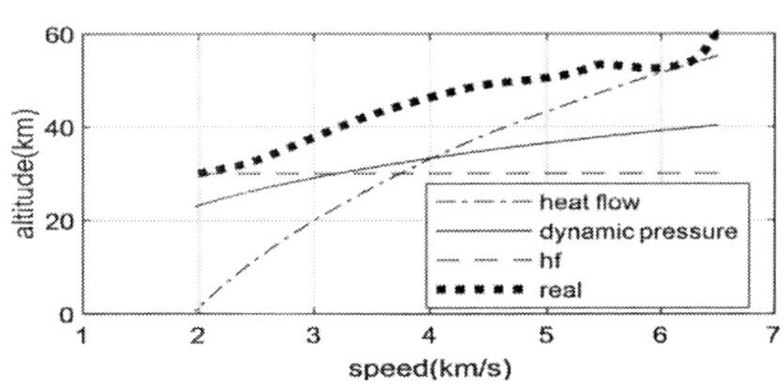

Figure 9. H-V profile.

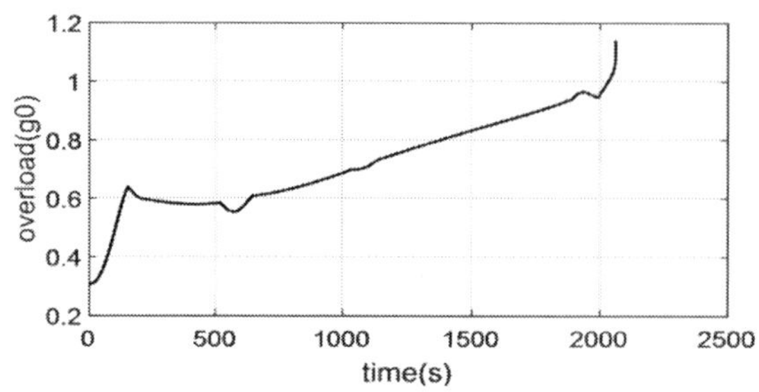

Figure 10. Total overload.

7. Conclusion

In this paper, a new trajectory planning method for hypersonic glide entry is proposed, and the relationship between path constraints and control variables is derived directly based on the dynamic equation. Compared with the previous method, it has the following advantages: (1) it is not limited by the QEGC, does not need segmentation, and has good applicability in the whole entry flight. (2) no need for bank angle, provide smooth trajectory and continuous control quantity. (3) reliable convergence, good calculation performance, suitable for onboard calculation.

References

[1] V. Istratie, *AIAA Atmospheric Flight Mechanics Conference and Exhibit*, 4069 (2001)
[2] K. Bollino, M. Oppenheimer, D. Doman, *AIAA Guidance, Navigation, and Control Conference and Exhibit*, 6691 (2006)
[3] J. Rea, *AIAA guidance, navigation, and control conference and exhibit*, 5640 (2003)
[4] J. C. Harpold, D. Gavert, J. Guid. Control Dyn, **6**, 442 (1983)
[5] C. Zimmerman, G. Dukeman, J. Hanson, J. Guid. Control Dyn, **26**, 523 (2002)
[6] Z. Shen, P. Lu, *AIAA Guidance, Navigation, and Control Conference and Exhibit*, 5736 (2003)
[7] Z. Y. Li, X. L. Huan, H. F. Li. J. Aeronaut, **34**, 755 (2013)
[8] J. Hanson,R. Jones,*AIAA Guidance, Navigation, and Control Conference and Exhibit*,4561(2002)
[9] E. M. Yong, G. J. Tang, L. Chen, J. Aeronaut,**29**,46 (2008)
[10] D. G. Hull, *Optimal control theory for applications* (2013)
[11] A. J. Roenneke,A. Markl, J. Guid. Control Dyn, **17**, 916 (1994)
[12] T. H. Phillips, *Schafer Corp. for AFRL and AFSPC*, **27** (2003)

Multi-echelon inventory allocation of multi-indenture spare parts considering maintenance ratio

Jun Xia*, Yunwen Feng, Xiaofeng Xue and Cheng Lu

School of Aeronautics, Northwestern Polytechnical University, 710072 Xi'an, China

*E-mail: tomxj375@163.com

Abstract. The existing multi-indenture spare parts maintenance has a certain maintenance proportion, a multi-echelon inventory allocation method with maintenance ratio for multi-indenture spare parts was proposed. Multi-echelon inventory allocation model for multi-indenture spare parts was established, which integrates maintenance rate into METRIC (Multi-echelon technique for recoverable item control) model, and then the marginal analysis method was applied to solve this model. First, the theory of multi-echelon inventory allocation model with maintenance ratio for multi-indenture spare parts was elaborated, and the mathematical model was established with the spare parts inventory cost as the optimization objective, the fleet availability as the constraints. Then, the procedure of solving the model by using marginal analysis method is discussed. Finally, taking an aircraft multi-indenture LRU (Line repairable units) as the research object, multi-echelon inventory allocation was investigated by considering maintenance ratio. The feasibility and effectiveness of the developed method are verified by an example analysis.

1. Introduction

Spare parts plays an important role in maintaining the reliability and Airworthiness of aircraft. The composition of spare parts of civil aircraft is complex. Aircraft is composed of many systems and the system is a combination of LRU, LRU also contains many SRU. Therefore, the failure of an aircraft system is often the result of the interrelated failure between LRU and SRU.

Many scholars have studied the allocation of maintenance resources considering the maintenance ratio. In order to realize the allocation of civil aircraft maintenance resources, Boeing and Airbus Company takes the maintenance ratio into account and evaluates the demand for repairable parts [1]. In METRIC theory, maintenance ration is also introduced into multi-echelon spare parts inventory allocation. For example, Patriarca and other researchers use METRIC theory to study the maintenance resource allocation of civil aircraft [2]; De Smidt-Destombes explore the application of METRIC theory in multi-echelon spare parts inventory allocation of redundant systems [3]; Basten realize multi-echelon spare parts inventory allocation under LORA combined with METRIC theory [4]; Xue Tao introduced scrap flow into spare parts inventory allocation model, established a multi-echelon and multi-indenture spare parts inventory allocation model [5]; Li Yongkai introduced scrap into VARI-METRIC model complete the multi-echelon inventory allocation of repairable parts [6]; Feng Yunwen studied the multi-echelon inventory allocation of civil aircraft spare parts under the premise of considering the maintenance ratio [7-9].

Although the above research introduces maintenance ration into spare parts inventory allocation model, but there is no further study on the allocation scheme under multi-echelon and multi-indenture repair mode. Therefore, this paper introduces the factor of maintenance ratio into spare parts inventory allocation model, and modeling the spare parts inventory allocation under different repair modes in initial state of civil aircraft. The model is solved by the marginal analysis method, and then the spare parts of multi-echelon allocation mode considering maintenance ratio is studied.

2. Model description and assumptions

In general, the maintenance of civil aircraft mainly adopts two-level repair mode, including the Local Warehouse (LW) and the Central Department (CD), in which the terminal is the LW maintenance site of spare parts. And the CD is located behind the terminal, which is a repair center with higher repair capability. The schematic diagram of aircraft maintenance location and spare parts composition is shown in figure 1.

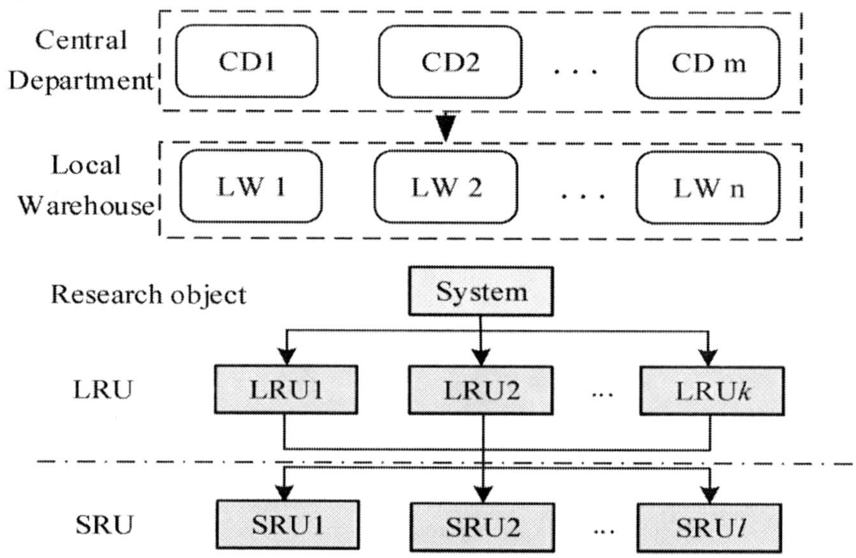

Figure 1. Spare parts inventory and composition chart.

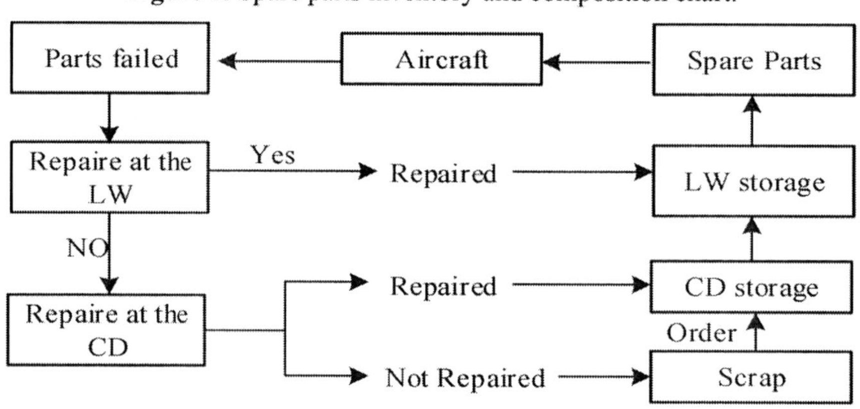

Figure 2. Flow chart of spare parts maintenance and supply.

When the aircraft's spare parts fail, the parts will be disassembled. If the parts are LRU, it will be sent to the LW for repair, and a spare part will be called from the LW for installation. Because of the limited maintenance capacity of the LW, if the LRU can't complete repair in the LW, it will be sent to the CD for repair. If the failure part is SRU, send it to the CD for repair, and call a spare part from the

CD. If it can't be repaired, choose scrap and purchase a new spare part. The repair process of spare parts is shown in figure 2.

In order to simplify the modeling process, there are some reasonable assumptions for multi-echelon inventory allocation method:

(1)The failure of LRU and SRU follows the Poisson Process, and different fault units are independent of each other in repair quantity.

(2)Each LRU does not belong to each other and SRU is independent of each other.

(3)The invalid parts can be repaired perfectly.

(4)When repaired at the LW and CD, the faulty parts are served according to the first-come-first-served principle.

(5)There is no lateral supply.

3. Establishment of model

3.1. Annual demand for spare parts

The average annual demand of LRU is:

$$m_i = \frac{\mathrm{FH \times N \times QPA_i}}{\mathrm{MTBUR_i}} \tag{1}$$

where: i=1, 2, 3...k, k is the number of LRU, FH is the Annual Flight Time of Aircraft, N is the Fleet size, QPA is the installation Number of LRU in Single Aircraft, MTBUR is the mean time between unscheduled removal.

(1) Average annual demand of LRU in LW

Calculating the annual average demand of LRU for each LW based on the distance between terminals:

$$m_{0j} = \frac{\mathrm{D_j}}{\sum_{\mathrm{j=1}}^{\mathrm{n}} \mathrm{D_j}} m \tag{2}$$

(2) Average annual demand of SRU in LW

The average annual demand of SRU is the average annual demand of LRU multiplied the maintenance ration of LRU in the LW, and then multiplied the SRU demand probability generated by LRU repair:

$$m_{ij} = m_{0j} \lambda_{0j}^{\mathrm{line}} \lambda_{ij}^{\mathrm{cause}} \tag{3}$$

where, m_{ij} is the annual average demand of *i-th* SRU at terminal j, $\lambda_{0j}^{\mathrm{line}}$ is the LRU maintenance ration of terminal j, $\lambda_{ij}^{\mathrm{cause}}$ is the probability of *i-th* SRU fault caused by terminal j's LRU fault.

(3) Average annual demand of LRU in CD

Equal to the sum of the LRU requirements for CD applications that occur at each terminal.

$$\mathrm{m_{00}} = \lambda_{00}^{\mathrm{shop}} \sum_{\mathrm{j=1}}^{\mathrm{l}} \mathrm{m_{0j}} (1 - \lambda_{0j}^{\mathrm{line}}) \tag{4}$$

(4) Average annual demand of SRU in CD

Equal to the sum of all LW requirements, and plus the number of SRU required for CD repair.

$$m_{i0} = \lambda_{i0}^{\mathrm{shop}} \left(\sum_{j=1}^{l} m_{ij} (1 - \lambda_{ij}^{\mathrm{line}}) + m_{00} \lambda_{i0}^{\mathrm{cause}} \right) \tag{5}$$

3.2. Expected demand for spare parts

The expected shortage of spare parts is as follows:

$$EBO(s) = \sum_{\mathrm{x=s+1}}^{\infty} (x - s) \frac{(u)^x e^{-u}}{x!} \tag{6}$$

Expected demand of SRU in CD is:

$$\begin{gathered} \mu_{i0} = m_{i0} \mathrm{MSPT_{i0}} \\ \mu_{i0}^{*} = m_{i0}^{*} \mathrm{PT_{i0}} \end{gathered} \tag{7}$$

where, $\mathrm{MSPT_{i0}}$ is the average repair time of *i-th* SRU CD, $\mathrm{PT_{i0}}$ is the purchase time of *i-th* SRU.

The expected demand of LRU in CD consists of two parts: the repairing spare parts and the spare parts that are delayed due to the shortage of the upper level.

$$\mu_{00}=m_{00}\mathrm{MSPT}_{00}+\sum_{i=1}^{l} f_{i0}EBO(s_{i0}|m_{i0}\mathrm{MSPT}_{i0}) \tag{8}$$

where

$$f_{i0}=\frac{m_{00}\lambda_{i0}^{\mathrm{cause}}}{m_{i0}} \tag{9}$$

The expected demand of LRU in LW consists of the repaired materials, the materials in transit, the materials that are delayed due to the shortage of the upper level and the material that are delayed in repairing due to the shortage of the next level.

$$\mu_{0j}=m_{0j}\left(1-\lambda_{0j}^{\mathrm{line}}\right)\tau_{0j}+m_{0j}\lambda_{0j}^{\mathrm{line}}MSPT_{0j}$$
$$+f_{0j}EBO(s_{00}|E[X_{00}],Var[X_{00}])+\sum_{i=1}^{l} EBO(s_{ij}|E[X_{ij}],Var[X_{ij}]) \tag{10}$$

where

$$f_{0j}=\frac{m_{0j}(1-\lambda_{0j}^{\mathrm{line}})}{m_{00}} \tag{11}$$

$$Var[X_{00}]=m_{00}\mathrm{MSPT}_{00}+\sum_{i=1}^{l} f_{i0}(1-f_{i0})EBO(s_{i0}|\, m_{i0}\mathrm{MSPT}_{i0})$$
$$+\sum_{i=1}^{l} f_{i0}^{2}VBO(s_{i0}|\, m_{i0}\mathrm{MSPT}_{i0}) \tag{12}$$

$$Var[X_{ij}]=m_{ij}[(1-\lambda_{0j}^{\mathrm{line}})\tau_{ij}+\lambda_{ij}^{line}\mathrm{MSPT}_{ij}]$$
$$+f_{ij}(1-f_{ij})EBO(s_{i0}|m_{i0}\mathrm{MSPT}_{i0})+f_{ij}^{2}VBO(s_{i0}|m_{i0}\mathrm{MSPT}_{i0}) \tag{13}$$

$Var[X_{00}]$ is the variance of the base repair LRU, $Var[X_{ij}]$ is the variance of *i-th* SRU that the terminal j is repairing or replenishing.

3.3. Optimization model

In this paper, spare parts inventory cost as the optimization objective, the fleet availability as the constraints, a multi-echelon inventory allocation method is put forward:

$$\begin{cases} \min C \\ \text{s.t. } A>A_{min} \end{cases} \tag{14}$$

where

$$A=\prod_{i=1}^{k}\left(1-\frac{EBO_i(s)}{\mathrm{N}\times\mathrm{QPA_i}}\right)^{QPA_i} \tag{15}$$

C is the spare parts inventory cost, A is the fleet availability, A_{min} is the minimum fleet availability requirements.

4. Solving process based on marginal analysis

The multi-echelon inventory allocation model solved by the method of marginal analysis. The solution steps are as follows.

Step. 1 Determine the spare parts included in the research object, calculate the average annual demand of the CD, the average annual demand of the LW and the corresponding scrap volume.

Step. 2 Calculate the expected demand, the expected shortage, and the fleet availability of CD and LW.

Step. 3 Calculate the marginal increment of spare parts, calculate the maximum value of $\Delta i=[EBO(S_i)-EBO(S_i+1)]/C_i$, Increase the unit number of spares parts according to the maximum marginal increment.

Step. 4 Calculate the total investment cost C, if $A<A_{min}$, go to step.3; otherwise, end of the calculation.

5. Numerical example

A typical numerical example is provided to illuminate the application of our model. The fleet operation information is Fleet size N=10, Annual flight hours FH=3300h. There are 3 terminals and one Central Department, and the distance between terminals is shown in figure 3. The maintenance ratio information and specific parameters of each spare part are shown in table 1 and table 2.

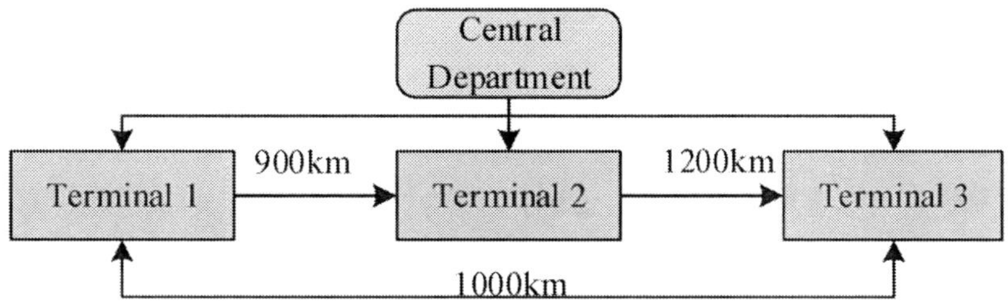

Figure 3. Spare parts maintenance structural diagram.

Table 1. Spare parts parameters.

Spare Parts	MTBUR/h	SR	QPA	C/$	λ_{0j}^{line}	λ_{00}^{shop}	λ_{ij}^{line}	λ_{i0}^{shop}	λ_{ij}^{cause}
LRU1	60000	0.3	2	16901	0.6	0.83	0.2	0.9	0.15
LRU2	25000	0.2	12	19400	0.55	0.75	0.2	0.9	0.15
LRU_SRU1	41500	0.1	6	15947	0.6	0.75	0.2	0.9	0.15
SRU1.1	50000	0.2	2	14622	0.4	0.8	0.15	0.8	0.1
SRU1.2	60000	1	2	1364	0.4	0.6	0.2	0.7	0.15
SRU1.3	50000	1	2	618	0.5	0.8	0.25	0.9	0.15
LRU_SRU2	35000	0.5	4	27062	0.6	0.8	0.2	0.9	0.15
SRU2.1	50000	1	2	1364	0.6	0.75	0.2	0.75	0.18
SRU2.2	50000	1	2	618	0.6	0.8	0.3	0.8	0.1
SRU2.3	50000	1	2	1264	0.6	0.8	0.2	0.85	0.15
SRU2.4	33333.3	1	6	1510	0.6	0.8	0.2	0.9	0.18
LRU_SRU3	40000	0.5	4	21337	0.6	0.65	0.2	0.9	0.15
SRU3.1	46666.67	1	1	3015	0.6	0.8	0.15	0.8	0.17
SRU3.2	13333.33	1	8	1090	0.6	0.7	0.2	0.85	0.2
SRU3.3	210000	1	6	2448	0.6	0.8	0.3	0.75	0.15
LRU3	15343.07	0.1	4	2470	0.65	0.8	0.2	0.9	0.15

Table 2. Spare parts parameters.

	$MSPT_{i0}$	PT_{i0}	$MSPT_{00}$	PT_{00}	$MSPT_{ij}$	τ_{ij}	τ_{0j}	$MSPT_{0j}$
LRU/SRU	6	10	6	11	8	7	7	5

Based on the above data, according to formulas (1) to (5), the annual average demand and scrap amount can be obtained, and then the expected demand, expected shortage, and fleet availability are calculated according to formulas (6) to (13). Then, by using this data to establish an inventory allocation model. The inventory allocation and costs at the CD and LW are shown in table 3.

As can be seen from table 2, when the spare parts inventory quantity is 52, the constraint is met, and the fleet availability is 0.9711, and the total cost is 375,989 US dollars.

In order to verify the feasibility and effectiveness of this model, based on the spare parts parameters in table 1 and fleet operation information, the traditional METRIC model is solved by marginal analysis method without considering the influence of maintenance ratio. The relationship between fleet availability and spare parts costs is shown in figure 4.

Table 3. Spare parts inventory allocation.

	CD	LW1	LW2	LW3
LRU1	0	0	0	0
LRU2	3	1	1	1
LRU_SRU1	3	1	0	1
SRU1.1	0	0	1	1
SRU1.2	2	0	1	1
SRU1.3	2	0	1	1
LRU_SRU2	2	0	0	0
SRU2.1	1	0	1	1
SRU2.2	2	0	0	1
SRU2.3	1	0	1	1
SRU2.4	2	0	1	1
LRU_SRU3	2	0	0	0
SRU3.1	0	0	1	1
SRU3.2	2	0	1	1
SRU3.3	1	0	0	1
LRU3	3	1	1	1
Costs/$	225725	37817	46717	65730

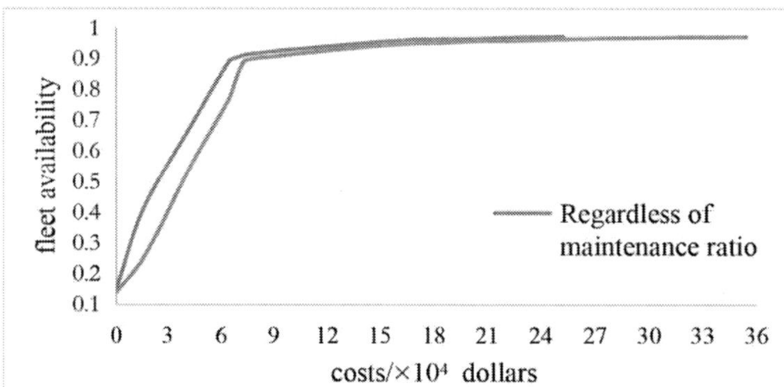

Figure 4. Costs and Fleet Availability Relationship Curve.

From figure 4, it can be seen that the fleet availability gradually reaches the minimum availability requirement with the increase of costs, but the growth rate of fleet availability gradually decreases. And, after the fleet availability reaching 0.90, the increase in fleet availability will cost more money. Therefore, the minimum fleet availability should adjust according to the budget to suit the airlines.

And it shows the spare parts inventory allocation considering maintenance ratio has a slightly lower fleet availability than that without considering maintenance ratio, but the spare parts cost is higher than the spare parts cost without considering maintenance ratio. The main reason for this result is that the multi-echelon inventory allocation without considering maintenance ratio is considered to have unlimited repair capability and no scrap. SO, compared with the multi-echelon inventory allocation considering maintenance ratio, the allocation quantity of multi-echelon inventory allocation is smaller and the spare parts costs are lower. And the consideration factor is too idealized, the allocation result did not conform to the actual engineering. Therefore, the multi-echelon inventory allocation model considering maintenance ratio is much more fit the engineering reality.

6. Summary

In this paper, the engineering practices factors of maintenance ratio are introduced into spare parts allocation calculation, and a multi-echelon inventory allocation model of spare parts considering maintenance ratio is constructed, which is solved by the marginal analysis. Also, the spare parts of a civil aircraft are used to calculate the multi-echelon inventory allocation, and the feasibility and validity of the proposed method are verified by comparing the multi-echelon inventory allocation without considering the maintenance ratio.

The model considers the maintenance proportion, the membership relationship of spare parts and multi-echelon inventory in actual engineering. It is more suitable for practical application, and it can optimize the configuration of complex spare parts for aircraft and reduce the operating cost of airlines.

Acknowledgments

The work is supported by the National Natural Science Foundation of China (No. 5187051969) and Special Scientific Research Project of Civil Aircraft.

References

[1] A. Sleptchenko, A. A. Hanbali , H. Zijm, Eur. J. Oper. Res 271, 97-108 (2018).
[2] P. Riccardo, C. Francesco, D. Gravio G, J. Air. Tra. Man 57, 206-216 (2016).
[3] Karin S. de Smidt-Destombes, Matthieu C. van der Heijden, Aart van Harten, Int. J. Pro. Eco 118, 260-268 (2009).
[4] Basten RJI, van der Heijden MC, Schutten JMJ. Eur. J. Oper. Res 222, 474-483, 2012.
[5] T. Xue. Northwestern Polytechnical University (2014).
[6] Y.K. Xue. Northwestern Polytechnical University (2018).
[7] Y.W. Feng, C. Lu, X.F. Xue, J. Northwestern. Polytech. Univ 36, 582-589 (2018).
[8] Y.W. Feng, Y.K. Li, X.F. Xue, C. Lu, J. Northwestern. Polytech. Univ 35,827-833 (2017).
[9] Y.W. Feng, C. Lu, X.F. Xue, Y.C. Liu, J. Northwestern. Polytech. Univ 36, 582-589 (2018).

IOP Conf. Series: Materials Science and Engineering **715** (2020) 012048 doi:10.1088/1757-899X/715/1/012048

Research on integrated logistical support systems of medium altitude long endurance UAS

Deng Haoran

CH UAV Technology Co. Ltd. China Academy of Aerospace Aerodynamics, China

*E-mail: 453302458@qq.com

Abstract. The Medium Altitude Long Endurance (MALE) UAS in modern warfare is becoming more and more widespread, which makes the UAS more and more frequently used in warfare. The equipment characteristics of UAS are embodied in the following aspects. The composition of UAS is complex, Ground Stations is distributed deployment, Mission payloads can flexibly change, and the whole system need highly cooperative ability between air and ground. That put forward higher requirements for integrated logistical support system of UAS. Aiming at the unique support characteristics of UAS, this paper constructs the integrated support system of UAS based on the integrated support system of traditional aircraft, which is mainly reflected at five aspects: support organization, support resources, support function, support process and support management. This systems can ensure the integration of all support links in the actual process of UAS support, and meet the operational and mission requirements of UAS to the maximum extent.

1. Introduction

Since the 21st century, the Medium Altitude Long Endurance (MALE) UAS has developed rapidly, its combat scope and application form are more and more extensive, and it has more and more influence on the war form and military revolution. The major military countries in the world have begun to research, develop, purchase and apply large UAS, which are mainly used in target indication, military mapping, battlefield reconnaissance, target surveillance, low speed or fixed target strike, damage assessment, communication relay, electronic warfare, information attacks, targeted killings and so on.[1]

MALE UAS is a kind of power-driven and reusable flight system. Due to UAS needs to be reused, it needs corresponding support system. UAS are widely used and frequently used in warfare, which requires UAS to ensure sufficient sorties, so we need to have a standardized and effective comprehensive security system.

Due to the different equipment characteristics and use modes between UAV and traditional aviation products, the traditional aviation integrated logistical support system can not be fully applicable to UAS. Therefore, this paper studies and constructs an integrated logistical support system for MALE UAS.

2. Composition of UAS

The UAS is composed of UAV, Measurement and Control and Information Transmission Subsystem, Mission Load Subsystem and Integrated Support Subsystem, as shown in figure 1.

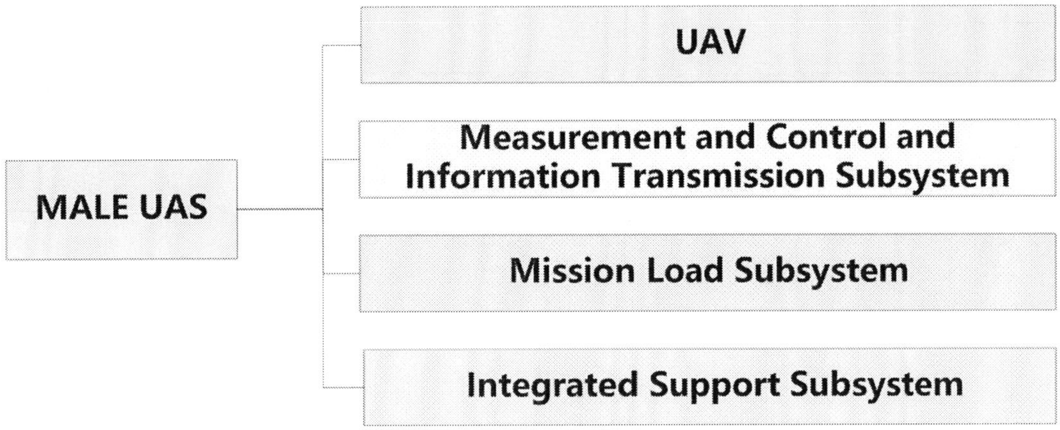

Figure 1. Composition of UAS.

2.1. UAV

UAV is an important part of the UAS. It is the basic platform for realizing the tactical and technical requirements of the system. It is used to carry the airborne equipment of Measurement and Control and Information Transmission Subsystem and various mission loads equipment.

2.2. Measurement and control and information transmission subsystem

Measurement and control and information transmission subsystem mainly completes the real-time transmission of UAS remote control, telemetry and reconnaissance information. It consists of Ground Control Station (GCS), Ground Data Terminal (GDT) and Airborne Data Terminal (ADT).

2.3. Mission load subsystem

UAV can mount and replace a variety of reconnaissance loads and weapon loads, including EO/IR, radar, communication reconnaissance load, aeromagnetic load, missile, bomb and other types of loads.

2.4. Integrated support subsystem

The Integrated Support Subsystem is a comprehensive support system to ensure the normal use and successful mission of the above three subsystems, and to provide support for the daily use, maintenance, spare parts supply, logistics management of UAS. This part is also the focus of this paper.

3. Support characteristics of UAS

According to the composition of UAS, we can see the difference between UAV and traditional aircraft, and its integrated support system cannot copy to the UAV completely. According to the system composition, operation characteristics and equipment characteristics of UAV, the support characteristics of UAS are analysed as follows:

3.1. The composition of UAS is complex.

According to the composition of the above-mentioned UAS, the UAS is more complex than that of the traditional aircraft, and the collaborative work between subsystems has high requirements. On the one hand, due to no pilots on the UAV, the information connection, information control and information collection between computers on aircraft platforms are more complicated. More functions of condition monitoring, fault diagnosis, automatic flight, autonomous judgment and autonomous decision-making are added to the UAV, which makes the UAV more automated, intelligent and autonomous[2]; on the other hand, the GCS belongs to the large-scale ground electronic system and remotely operates the UAV which has powerful function, advanced technology and complex structure. Therefore, UAS covers many technical fields such as control, communication, command and control, and technology integration is more complex.

3.2. Distributed deployment of ground stations
In the deployment of UAS, not all ground stations are placed in the same site, but distributed deployment because of the need of mission. For example, the US Air Force Predator completes the UAV take-off and landing control at the front base through the take-off and landing control station and line-of-sight link. After take-off, the UAV link is established by the remote control station in the U.S. mainland and SATCOM, and the UAV mission is completed by the local mission commander, pilot and payload operator. For UAVs, flexible deployment and transshipment are needed in different bases according to operational tasks. Therefore, UAS has strong characteristics of distributed deployment and flexible configuration in operation and deployment.

3.3. Flexible replacement of mission load
According to the diversity of UAV mission requirements, UAV mounted mission loads can be flexibly changed, such as EO/IR, SAR, communication load, mapping camera, geomagnetic probe and other loads. Therefore, in operation, the ground station will configure and use different ground software to operate the load; in support, all kinds of load support equipment, support tools, spare parts, consumables, human resources, technical manuals should be complete.

3.4. High requirement of synergy ability
On the one hand, any link of UAS has malfunction, which will destroy the system work between air and ground, and lead to mission failure. Therefore, when UAV carries out its mission, the UAV in the air and the station on the ground need to cooperate highly with each other. On the other hand, when maintaining and repair the UAS, it needs aircrew and ground crew. The two groups of members work together in fault detection, fault detection, system testing, maintenance and repair, so they need a high degree of collaboration between personnel.

4. Planning of integrated logistical support system for UAS
In order to ensure the effective use of UAS, the integrated support system of UAS is to implement planning, organizing, commanding, controlling and coordinating management activities in all aspects of UAS support by means of modern information technology and management technology.

4.1. Top-level design of integrated support system for UAS
Integrated support system is a complex system with clear operation and mission. Its fundamental mission is to ensure and maintain the capability of UAS to carry out operation and training tasks. Specifically, in order to meet the needs of different support tasks, the integrated support system of UAS is an organic whole with the intersection and integration of various support activities, which has certain functions and interrelated support elements, according to the UAS support rules and support principles.[3]

The Integrated support system consists of support organization, support activities, support resources, support process, support management. Support activities, support organizations and support resources are the basis of support process. Support organizations are the foundation of site, support resources are the basis of resources, and support activities are the basis of work. Support management is a management activity based on the support process, which aims to ensure the effective control and operation of the support system, as shown in figure 2. Therefore, the purpose of this paper is to describe comprehensively the static structure composition relationship and dynamic system operation behavior of the support system.[4]

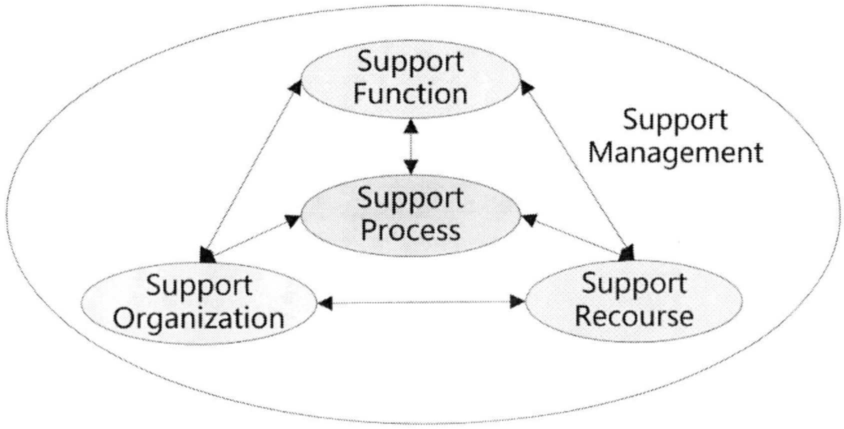

Figure 2. Top-level design of integrated support system for UAS.

4.2. Support organization for UAS

Support organization refers to the establishment of equipment support organizations. The support organization is composed of the support sites, which are the places to complete the support activities. The support organization of UAS can be divided into two levels according to the level of maintenance organization: depot level (D-level) and organization level (O-level), as shown in figure 3.

O-level support organizations consist of UAS operators and maintenance personnel of their affiliated units. They carry out operation and maintenance support, and only complete short-term and simple support work. They generally undertakes the operation of UAV support activities and equipment replacement and maintenance activities, which is the forefront of UAV mission and maintenance support. O-level support organizations mainly include runway, hangar, ground station, ammunition room, spare parts storehouse, maintenance room, tool room, fuel station, weather station, command hall and other places.

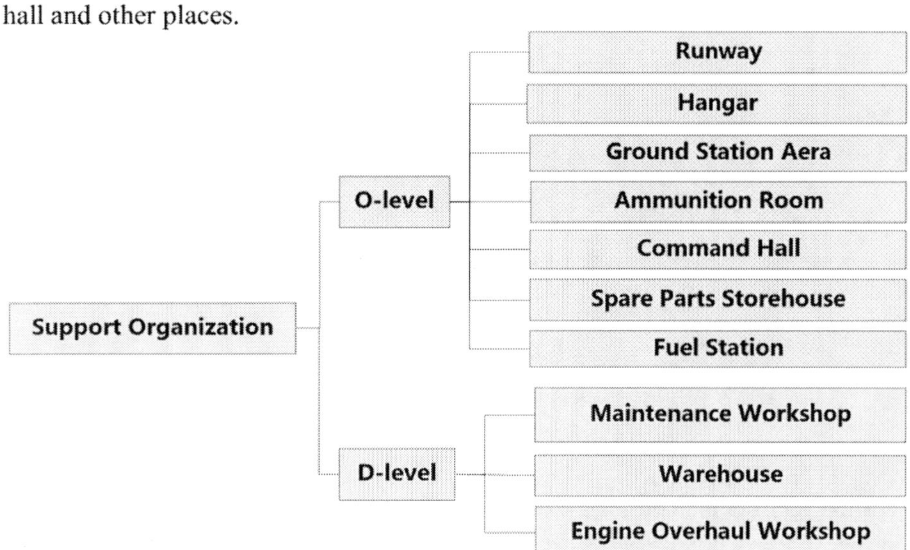

Figure 3. Composition of support organization.

D-level support organizations have higher repairing capacity, undertake system overhaul and repair of large parts, spare parts manufacturing and support work that basic level cannot complete, mainly including maintenance workshop, warehouse, engine overhaul workshop and other places, as shown in figure 4.

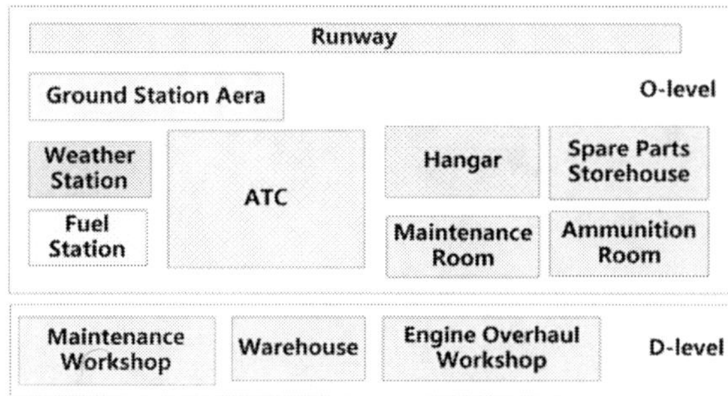

Figure 4. Schematic diagram of support organization.

Different from traditional aircraft, ground station is an important part of UAS, and its deployment location needs to be considered. Generally, it requires good visibility, stable power supply and so on. Ground stations can be deployed at fixed front-line bases or distributed, and multiple ground stations can be deployed at different front-line locations.

4.3. Support recourses

Support resources are the material basis for UAS operation and maintenance, including support facilities, support equipment, support tools, manpower, spare parts, consumables and technical manuals, as shown in figure 5.

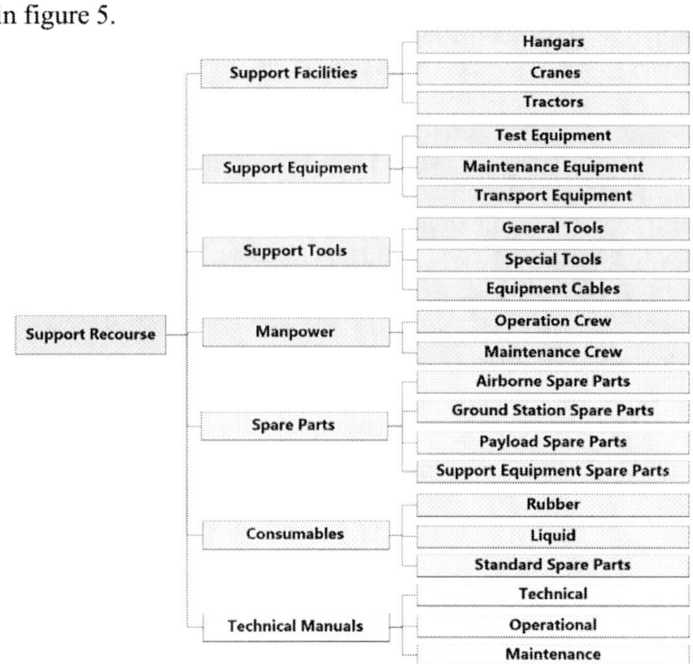

Figure 5. Composition of support resources.

Support facilities are permanent and semi-permanent buildings and supporting equipment needed for the operation and maintenance of UAS, such as hangars, factory buildings, cranes, tractors and refueling trucks. In order to ensure the uninterrupted power supply of the ground station in the course of the mission, besides the UPS configuration of the ground station itself, it is necessary to provide generators for the ground station.

Support equipment is all the equipment (mobile and fixed) needed for the operation and maintenance of the system, such as test equipment, maintenance equipment, transport equipment, calibration equipment, test equipment, etc.

IOP Conf. Series: Materials Science and Engineering **715** (2020) 012048 doi:10.1088/1757-899X/715/1/012048

Support tools are all the tools needed for system operation and maintenance, such as general tools, special tools and equipment cables.

Manpower is the number of personnel needed for the maintenance of the system and the technical level of professional aircraft. The manpower of UAS is mainly divided into operation crew and maintenance crew. Operation crew is also divided into ground crew and air crew. The ground crew mainly supports the traction and release of UAV, and the air crew mainly carries out tasks in the ground station. Maintenance crew can be divided into mechanical technicians, avionics technicians, weapons technicians and so on.

Spare parts and consumables are replaceable parts and consumables for system maintenance. Spare parts include airborne spare parts, ground station spare parts, payload spare parts and support equipment spare parts. Consumables include rubber, liquid, standard spare parts and so on.

Technical manuals are the manuals needed for the operation and maintenance of the system. It can be divided into technical, operational and maintenance manuals according to categories.

4.4. Support function

Support function refers to the function that the support system needs to complete, mainly including the operation support function and maintenance support function, as shown in figure 6.

The operation support functions means that when equipment is stored, transported and used, the support system can provide corresponding support functions to ensure that equipment can be used at any time, including pre-flight inspection, post-flight inspection and re-fly inspection.

Maintenance support function is that when equipment fails, the support system can provide maintenance function for the malfunctioning equipment to maintain and restore the equipment in good technical condition, mainly including corrective maintenance and preventive maintenance. Among them, corrective maintenance can be divided into replacement maintenance, direct maintenance and fault isolation. Preventive maintenance can be divided into engine hour maintenance, flight hour maintenance, sortie maintenance and calendar time maintenance according to maintenance type.

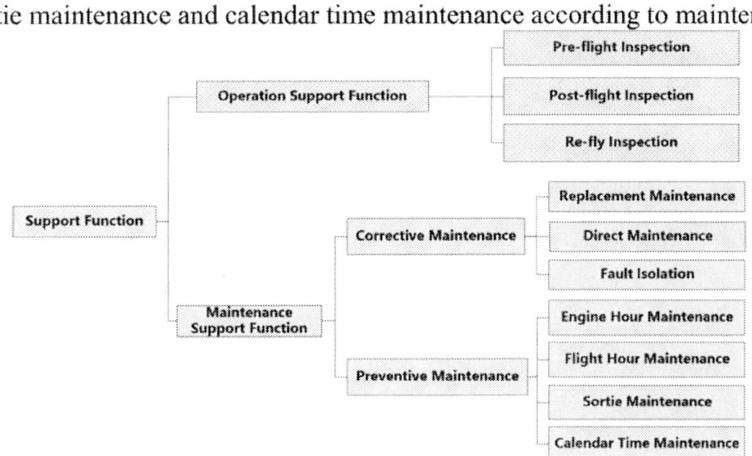

Figure 6. Composition of support function.

4.5. Support process

Support process refers to the process of ensuring the UAS can normally perform tasks, including the operation support process, maintenance support process and supply support process, as shown in figure 7.

Mission process is the core of operating UAS. Tasks mainly include mission type, mission area, mission time and other requirements. According to the mission type, it can be divided into two kinds: one is strike mission, which has fixed-point reconnaissance and clearance mission with clear information about the precise location or specific scope of the target before the mission is executed; the other is reconnaissance mission, which can be used for wide-area reconnaissance mission in areas of potential high-value target activity. The two tasks can be combined and converted in real time. After defining the mission, choose the UAS that can carry out the mission, including selecting the UAV and ground station, take off the UAV after completing the pre-flight support, carrying out reconnaissance or strike mission, and landing back to the hangar after the mission.

Operation support process refers to pre-flight support and post-flight support. Pre-flight support carries out flight path planning according to mission content, chooses payload or mounted weapons according to mission type, fills fuel according to mission time, and carries out pre-flight inspection and ground station preparation. Post-flight support mainly carries out post-flight inspection, unload weapons and withdraw the ground station.

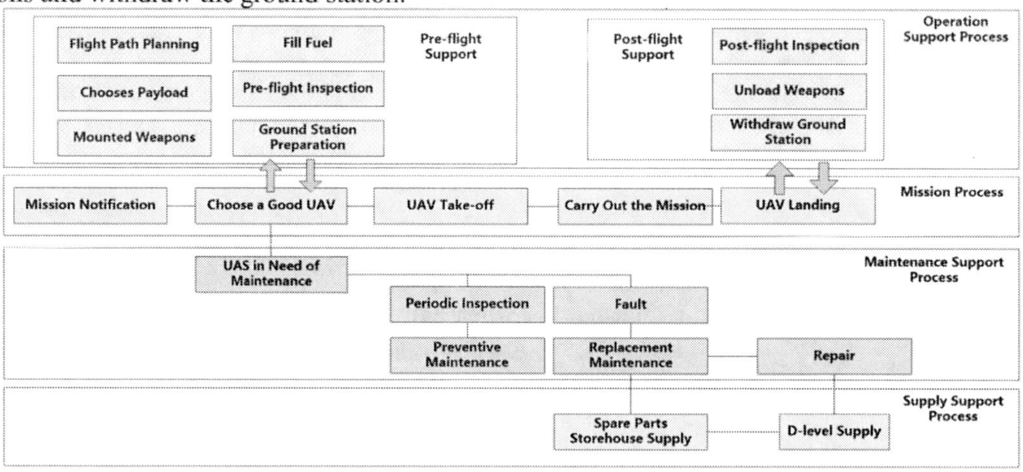

Figure 7. Schematic diagram of support process.

Maintenance support process refers to UAV cannot be used normally, there are two situations: first, UAS has a fault, replacing the LRU at the O-level, fault LRU is sent to the D-level for repair; the other is preventive maintenance, also known as periodic inspection, when the operation of UAV accumulates to a certain engine hours, flight hours, sorties or calendar times need to be maintained accordingly.[5]

Supply support process refers to the process of supplying spare parts and consumables for UAV maintenance in time. O-level warehouse supplies spare parts and consumables for direct repair and D-level warehouse. D-level warehouse provides spare parts and consumables for replacement and maintenance at the O-level.

4.6. Support management

Support management is a series of management processes on the basis of support organization, support resources, support activities and support process, which makes the whole support system can be effectively implemented, mainly including: support record, maintenance management, quality control, data analysis, inventory management, organizational structure, fund management, etc., as shown in figure 8.

Support records refer to the support data generated during the operation and maintenance of UAS, mainly including UAV matching tables, flight confirmation records, flight data, flight sorties statistics, maintenance records, power-on time records, engine hours records, transport records, etc. Support records can effectively monitor the status of every link of UAV's daily use, so that every work can be recorded and traceable.

Maintenance management is divided into failure maintenance management and preventive maintenance management. Failure maintenance management records the replacement or direct maintenance of UAS due to failure. Preventive maintenance management refers to reminding maintenance personnel to carry out corresponding maintenance operations when UAV accumulates to a specific fixed inspection period.

Quality control means that the quality responsible department should supervise and manage the maintenance of UAS, coordinate the relevant resources, provide corresponding technical support, and complete the relevant support activities on time and according to the requirements.

Figure 8. Composition of support management.

Data analysis refers to the status analysis of flight data according to the system or equipment, timely attention to the health status of UAS, through the analysis of data, we can locate faults and find potential faults in time.

The organizational structure refers to all the departments responsible for the operation, maintenance, management and supervision of UAS. Each department has a clear division of responsibilities and cooperates to complete the work of UAS.

Funds management refers to the cost management generated in the life cycle of UAS, which mainly includes manpower cost, procurement cost, maintenance cost, supply cost and so on.

5. Conclusion

Firstly, this paper describes the four major components of UAS, analyses and refines the four support characteristics of UAS, which are complex composition, distributed deployment of ground station, flexible change of mission payload and high requirement of collaborative ability. Based on the above, an integrated support system of UAS is established, and the several key elements of support system is described: support organization, support resources, support function, support process and support management. Through the above research, a comprehensive support system of UAS with static structure and dynamic system operation is formed, which can effectively and reliably support the operation and maintenance of UAS.

References

[1] H. R. Deng, Z. Y. Zhu, J. H. Wang, Research on Support Mode of Medium Altitude Long Endurance UAS Based on the Mode of Military Trade, MEAMT 2018, 189, 1-8 (2018)

[2] X. F. Shen, F. Chang, Analysis on Integrated Support Technology of UAV System, Aviation Safety and Equipment Maintenance Technology—Collected Papers of Seminar on Aviation Safety and Equipment Maintenance Technology, 726-731 (2014)

[3] B Xing, T. L. Song, J. H. Cao, Research on Key Techniques of Modelling and Simulation for the Equipment Support System of Systems, Computer Measurement & Control, 4, 108-111 (2016)

[4] F. Xue, Study and design of the architectures of the aviation equipment unified support system, Electronic Design Engineering, 19 20-22 (2011)

[5] G. Xu, K. Xue, Discussion on Flight Support Flow of Large UAV, Science and technology economic guide, 4, 245-246 (2017)

IOP Conf. Series: Materials Science and Engineering **715** (2020) 012049 doi:10.1088/1757-899X/715/1/012049

Design of a relief materials delivery system based on UAV

Zhu Wenjian*, Zhong Sidong, Chen RongJie, Xie Jingchang, Li Yeqian and Cao Huiru

School of Electrical and Computer Engineering, Nanfang College of Sun Yat-sen University, Guangzhou 510970, China

*Email: 15622626377@163.com

Abstract. The quality and efficiency of traditional rescue work are seriously affected by the destruction of roads and road traffic. In view of these existing problems, the current work investigates the delivery of relief materials using an unmanned aerial vehicle (UAV). At present, studies investigating the delivery of emergency relief materials are few, the technology is not mature, the disadvantage is bigger. In this paper, the UAV material delivery system can realize the function of automatic navigation and automatic delivery of materials. The experimental results show that, in a large number of delivery tests, 97% of deliveries contained a delivery error within 4 m, 64% of deliveries contained a delivery error within 2 m. The average delivery error was 1.86 m. The results of the experiment show that the accuracy of the UAV material delivery system is high, and the test results can provide reference for the further study of the robot.

1. Introduction

Naturally-occurring disasters, such as earthquakes, floods and mudslides, are unavoidable, causing great loss to people and threatening their livelihood and safety. In the face of these emergencies, those in danger often require emergency supplies (water, food, medicine, etc.). However, the emergencies often result in the destruction of roads and traffic jams, slowing the arrival time of rescue teams and seriously affecting the quality of any relief effort. Emergency supplies struggle to reach the scene quickly, such that they are unable to guarantee the safety of victims. In recent decades, UAV has developed rapidly, tending to be associated with military applications. In recent years however, the rapid rise of civilian unmanned aerial vehicles (UAVs) has had a great impact on a number of fields, including communications, agriculture, environmental protection and the film and television industry[1-4]. At present, UAV is developing into a technology for use in logistics transportation, by companies including UPS[5] and Amazon[6], some of whom have already registered the relevant patents. After two years of clandestine research and development of its UAV program, Google has conducted a public demonstration in Australia of the use of UAV in the delivery of relief materials[7]. The application of UAVs to rescue work can solve the problems associated with traditional methods. Moreover, the new approach can greatly improve the efficiency of rescue efforts and better ensure the safety of the lives of the victims.

There are few studies on the use of UAV for the delivery of emergency relief materials, and they have not been popularized in the fields. Recently, the large heavy-duty (up to 1.2 ton) unmanned aerial vehicle developed by SF-express was demonstrated. The UAV achieves unmanned autonomous control, can automatically plan routes, take off and land at the press of a button, and has the ability to provide emergency treatment autonomously. In the case of a major natural disaster, the UAV can be used to deliver medicine and emergency supplies[8]. In addition, Seguin C et al. proposed UAV for the safe and timely delivery of flotation devices to swimmers, improving the quality and speed of first aid and keeping lifeguards away from the dangerous maritime environment[9]. Zipline's autonomous fixed-wing UAV has now become part of the medical supply infrastructure in Rwanda, transporting blood products from central distribution centres to hospitals across the country, helping to address the challenging issue of rural healthcare[10].

The speed quality of traditional ground rescue work are inherently limited. Nonetheless, studies investigating the use of UAV for the delivery of emergency relief materials are few. Hence, the aim of this paper is to design the hardware and software systems of a UAV, to be used as a relief delivery system in the case of emergency events. The software system is used to obtain the real-time position and delivery request sent by the client, while the hardware system performs automatic navigation and delivery. In order to analyze the novel UAV delivery system, experimental tests are carried out to investigate whether the system can realize automatic fixed-point delivery with the desired accuracy. Hence, the content of this research paper will provide reference for the future application of UAV in rescue work.

2. Embedded hardware system

2.1. Design of mechanical delivery device

The mechanical delivery device, shown in figure 1, is a key component of the UAV delivery system. The frame of the delivery device is built from PMMA material and provides support for the battery, motor, motor drive, embedded system, global positioning system (GPS) module, ESP8266 Wi-Fi module and SIM900A communication module. The delivery device contains six storage compartments, which can carry 5 kg of emergency rescue materials. Six independent deliveries to multiple target locations can be carried out during each mission. A hardware module storage bin, used to store hardware modules such as the Micro Control Unit (MCU), is located at the center of the delivery device. In this way, ensures the safety of the UAV delivery system.

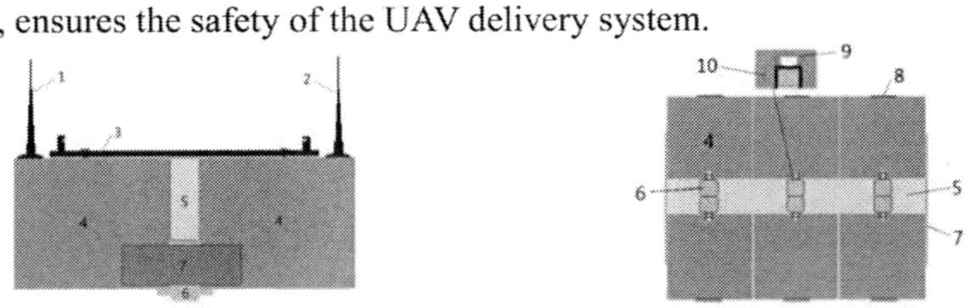

(a) Front view (b) Vertical view

1-2. Signal amplifier 3. Storage space of battery 4.The supplies warehouse
5. The hardware warehouse 6. Motor 7. Fixed plate 8. Hinges 9. Door slot 10. Shaft

Figure 1. Structure diagram of delivery device.

The delivery device is designed to contain six storage compartments for emergency rescue materials, each with a corresponding delivery control door slot. The door slot width is 5mm. The structure of the rotating shaft is a cuboid, as shown in figure 2, with a cross-sectional area of 3x8 mm^2. Before delivery, the axis of the rotating shaft in 0°Angle. That is to say, the 8mm part of the rotating shaft is relative to the door slot, and the conveyor cabin is closed. During delivery, the control motor rotates 90°. The 3mm part of the rotating shaft is relative to the door slot. At this point the hatch opens to deliver supplies to the customer. This is the operating principle of material delivery system.

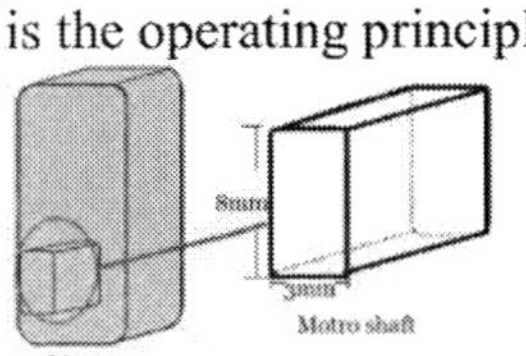

Figure 2. Structure of rotating shaft.

2.2. *Hardware circuit*

The hardware circuit includes the main control system, including a high-performance MCU based on the ARM cortex-m3 kernel by ST. This has the advantages of high performance, low cost, low power consumption and powerful software support. The GPS module contains a u-blox neo-7n chip - with high sensitivity, low power consumption, small footprint and extremely high tracking sensitivity - greatly expanding coverage and achieving high positioning precision in the narrow urban sky and dense jungle environment. The ESP8266 Wi-Fi communication module is equipped with an ultra-low-power Tensilica L106 32-bit RISC processor, and the CPU clock speed can reach up to 160 MHz. It has the advantages of low power consumption, high integration and stable performance. The SIM900A is a two-frequency GSM/GPRS module that is an SMT package. It has stable performance, small size and low cost advantage.

The MCU of the hardware circuit connects with the GPS, Wi-Fi and SIM900A modules through the USART interface and implements the data transmission and positioning functions. When the MCU receives a delivery instruction, it will deliver a PWM signal to the L298 drive. The drive motor will then rotate 90°, delivering the relief materials at the target location.

3. System framework and algorithm design

3.1. *System framework*

The delivery system is composed of the hardware and software systems. Through computer and Internet of Things (IoT) technology, data transmission between these systems is achieved, such that the software system can control the hardware system. The client, the administrator control center and the embedded hardware system communicate with each other through the public cloud platform. The client sends a delivery request and geographic location to the public cloud platform. The administrator control center obtains the user information from the public cloud platform, and issues a delivery instruction to the embedded hardware delivery system through the cloud platform.

A distressed person in need of relief supplies sends a delivery request and geographic location to the control center through the client application, as shown in figure 3. When the administrator control center receives the information, it automatically sends the UAV delivery robot, equipped with rescue supplies, to the target location. By delivering the necessary goods and medicine quickly, the safety of the distressed person is ensured. Subsequently, the robot moves on to next delivery.

Figure 3. System framework.

3.2. *Algorithm design based on GPS threshold*

The biggest distance for a delivery to occur is defined as l_1. When the UAV delivery robot flies to the center of the site, l_1 is the radius of the circle, within which, the robot will deliver the relief materials. The real-time displacement of the delivery robot from the delivery destination is l. When the following conditions are met, the delivery robot will deliver the rescue materials:

$$l < l_1 \tag{1}$$

As shown in figure 4, the initial position of the UAV delivery system is $A(J_A, W_A)$. The position of the materials requested by the person in danger is $B(J_B, W_B)$. The origin O (0, 0) is positioned at the center of the earth. By using the trigonometric theorem, the distance from the drone to the delivery locate is l.

$$l = 2R * \arcsin\left(\sqrt{\sin^2\left(\frac{W_A - W_B}{2}\right) + \cos(W_A) * \cos(W_B) * \sin^2\left(\frac{J_A - J_B}{2}\right)}\right) \quad (2)$$

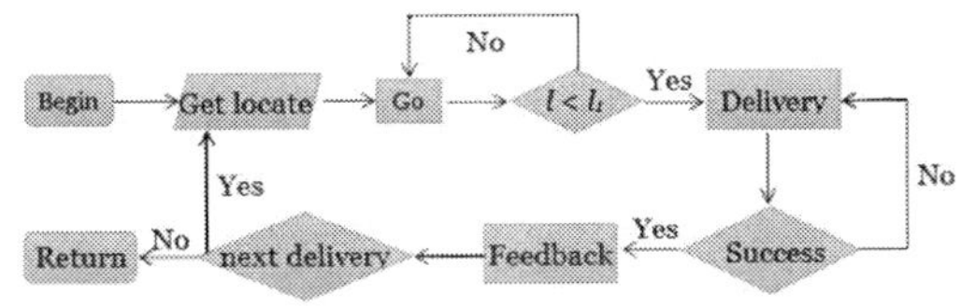

Figure 4. Automatic delivery implementation schematic.

4. Test results and analysis

4.1. Mechanical delivery tests

100 delivery tests were carried out on the 6 storage compartments of the mechanical delivery device. The test results show in figure 5, that the delivery success rate of each storage compartment is above 95%. Thus, the mechanical and structural design of the delivery device achieves the desired results successfully.

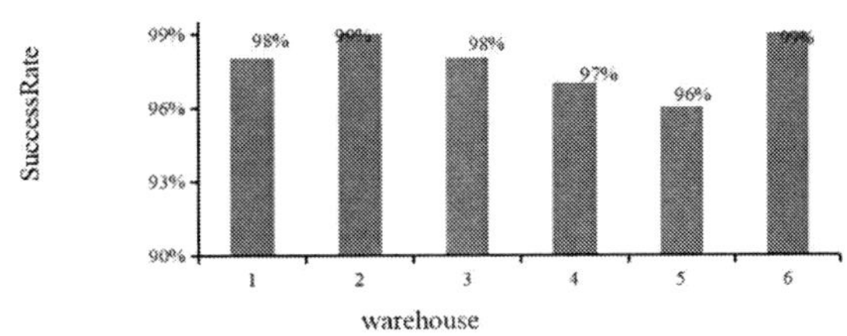

Figure 5. Validation test.

4.2. Delivery accuracy tests

An experiment is designed in order to verify the delivery accuracy of the UAV delivery robot. As shown in figure 6, the target delivery points are set at A, B and C (L_{AB}=L_{AC}=L_{BC}). R is the delivery tolerance error value. The starting point of the UAV is located in the center of the three target delivery points, to investigate its autonomous navigation and relief material delivery performance. The relevant parameters of the accuracy test are shown in table 1.

Table 1. Accuracy test parameters.

Parameters	R	L_{AB}	Number of groups	Number of targets
Value	2-8 m	30 m	10	3

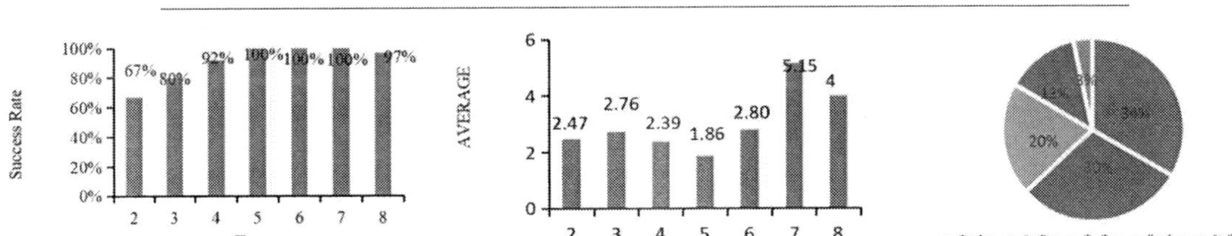

(a)success rate (b)average error (c) R=5 Error Proportion

Figure 6. Precision test analysis.

The results of the delivery accuracy test are shown in figure 6(a). When R=5-7 m, the delivery success rate of the UAV delivery robot reaches 100%. With R<5 m, the delivery success rate gradually decreases. The mean error of the delivery is R, at different values is shown in figure 6(b). When R=5 m, the mean error of the delivery is 1.86 m. Figure 6(c) shows the proportion of the delivery error range when R=5 m. The proportion with a delivery error less than 3 m is 94% and the proportion with a delivery error less than 4 m is 97%. When R=5 m, the delivery success rate of the UAV is the highest and the delivery error range is smallest. Therefore, the allowable delivery error of the delivery robot should be set at 5 m, where the deliveries are the most accurate.

4.3. Verification of UAV delivery robot

The UAV delivery robot is based on the Four-axis drone, as shown in figure 7, and delivers a payload up to 5 kg. During a mission, the robot can carry out six independent deliveries to multiple locations, carrying a diverse array of delivery materials and emergency relief supplies.

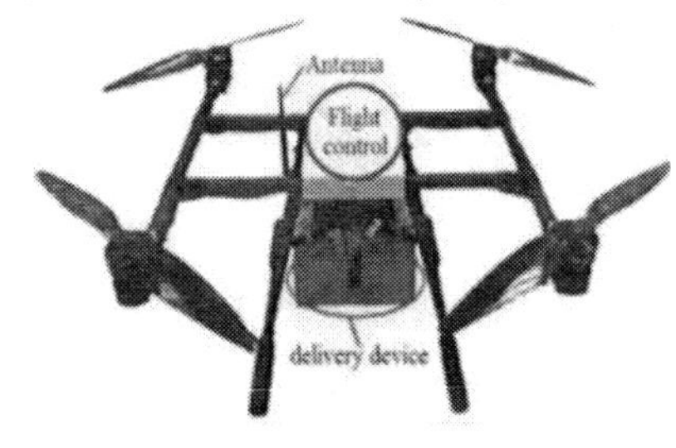

Figure 7. Delivery robot.

The software of the delivery system, based on Internet of Things technology, is divided into client and administrator control center applications. The client application has functions such as login, registration, localization and material request. When someone is in distress, the user can send a material delivery request command through the client to the control center with the real-time geographic location of the victim. The control center can display the client's request and position, and send the target position to the UAV delivery system, such that it can automatically move to the destination and provide emergency relief materials to the victims.

5. Conclusion

In this paper, an emergency relief material delivery robot is designed based on UAV technology. The mechanical structure is designed and establishes the relevant mathematic model and automatic delivery algorithm, and carries on the precision test of the delivery robot. The delivery success rate of the six storage compartments of the mechanical delivery device was found to reach more than 96%, with most of the storage compartments near 100%. When R<5 m, the allowable error R decreases, the success rate of automatic delivery of the delivery robot decreases, and the actual delivery error increases slightly. R increases the delivery success rate to 100%, but the error of delivery will be greater. The delivery accuracy tests investigated the most suitable parameters for the delivery robot. When the delivery tolerance error is R=5 m, the delivery success rate of the delivery robot is 100%. The actual average delivery error is 1.86 m, where the vast majority of the actual delivery errors are within 2 m.

Although the experimental results of the UAV delivery robot studied in this paper are satisfactory, further research is required to investigate the optimal delivery parameters and achieve more accurate deliveries.

Acknowledgments

This work is supported by Special Funds for the Cultivation of Guangdong College Students' Scientific and Technological Innovation.("Climbing Program" Special Funds.) (NO. pdjhb0649), Guangdong university students innovation and entrepreneurship training program (NO.201812619016), Guangzhou science and technology program (No. 201804010427) and the Water Resource Science and Technology Innovation Program of the Guangdong Province, China (No.2016-18).

References

[1] Floreano Dario, Robert J. Wood, science, technology and the future of small autonomous drones. Nature. 2015;521:460–466.
[2] Szantoi Z, Smith S E, Strona G, et al. Mapping orangutan habitat and agricultural areas using Landsat OLI imagery augmented with unmanned aircraft system aerial photography[J]. International Journal of Remote Sensing, 2017, 38(8-10): 2231-45.
[3] Yang F B, Xue X Y, Zhang L, et al. Numerical simulation and experimental verification on downwash air flow of six-rotor agricultural unmanned aerial vehicle in hover[J]. International Journal of Agricultural & Biological Engineering, 2017, 10(4):41-53.
[4] Coveney S, Roberts K. Lightweight UAV digital elevation models and orthoimagery for environmental applications: data accuracy evaluation and potential for river flood risk modelling[J]. International Journal of Remote Sensing, 2017, 38(8-10):3159-80.
[5] UPS. UPS tests residential delivery via drone launched from atop package car. United Parcel Service.https://pressroom.ups.com/pressroom/ContentDetailsViewer.page?ConceptType=PressReleases&id=1487687844847-162 (2017).
[6] CBS News. Amazon unveils futuristic plan: delivery by drone. 60 Minutes Overtime. https://www.cbsnews.com/news/amazon-unveils-futuristic-plan-delivery-by-drone/(2013).
[7] Google has secretly developed a drone program aimed at delivering relief supplies. http://tv.cntv.cn/video/C10319/894c628d62b2467388f5b2c5bd68279e(2015).
[8] SHEN ZHEN PRESS GROUP. ShunFen completed the country's first large-scale drone delivery demonstration.http://szsb.sznews.com/PC/layout/201712/27/colA07.html#c268011
[9] Seguin C, Blaquière G, Loundou A, et al. Unmanned aerial vehicles (drones) to prevent drowning ☆[J]. Resuscitation, 2018, 127.
[10] Ackerman E, Strickland E. Medical delivery drones take flight in east africa[J]. IEEE Spectrum, 2018, 55(1):34-35.

IOP Conf. Series: Materials Science and Engineering **715** (2020) 012050 doi:10.1088/1757-899X/715/1/012050

Experimental investigation of cutting temperature during drilling of float glass specimen

Ankit Sharma[1,*] and Vivek Jain[2]

[1]Chitkara College of Applied Engineering, Chitkara University, Punjab, India
[2]Department of Mechanical Engineering, Thapar Institute of Engineering and Technology, Patiala, India

*E-mail: ankit.sharma@chitkara.edu.in

Abstract. In today's industrialization, hard and brittle material such as float glass are widely used in various applications such as micro-fluidic devices, bio-medical parts, automotive glass, and optical lenses. Due to its difficult-to-machine characteristic, drilling of float glass leads to defects related to drilled hole quality. These defects exemplify as chipping, crack and surface roughness. All these defects are directly depend upon the temperature which could affect the drilling quality. Hence, authors tried to investigate the influence of cutting temperature of float glass specimen using rotary ultrasonic drilling. The experimentation is performed at two conditions with constant value of ultrasonic amplitude (20 µm), feed rate (6 mm/min) and spindle rotation speed (5000 rpm). These conditions are without coolant and with-coolant drilling. It is noticed that maximum cutting temperature reached with coolant drilling is 50.56 °c and without coolant is 62.11°c. Consequently, it is revealed that drilling with coolant (water) should be adopted to improve the hole quality. Since, it reduced the formation of stresses during drilling operation which are induced due to rise in temperature. Microstructure images are carried out to visualize the hole quality.

1. Introduction

Continuous growth in the technology motivates the researchers to make enhance the machining and drilling quality. The demand for the efficient machining is continuously increasing especially for brittle materials i.e. float glass. Owing to high chemical and optical stability and incomparable thermo-mechanical properties, glass is widely used in optical lenses, automotive glass, reflectors, solar panels, and metrological instruments[1]. To facilitate good quality with high surface integrity during drilling, float glass should not possess any defects. But difficult to machine materials like float glass suffered numerous types of defects during drilling for instance chipping, surface roughness, and cracks which deteriorates the final hole's quality[2-3].

At present, conventional machining process are not able to fulfill the demands of the current scenario due to generation of high temperature, low surface finish and high forces during drilling process. Therefore, hybrid machining processes has been becoming a potential candidate for hard, brittle, and tailor-made materials like float glass. The rotary ultrasonic machining (RUM) process is one of these non-conventional hybrid processes which combine the material removal mechanism of conventional drilling and vibration drilling.

Ding et al.[4] compared the results produced by conventional drilling (CD) and RUM in terms of hole quality with varying rotational speed and feed rate. The results revealed that RUM generates defect free holes with superior surface integrity. Huang et al. [5]highlighted that the use of coolant during drilling had a significant effect on the surface roughness of drilled hole on using high rotational speed and ultrasonic power during drilling with ultrasonic actuations. Zhang et al. [6]used compressed air coolant during RUM and CD of optical K9 glass and outlined that lesser cutting forces are produced on using ultrasonically actuated tool. Similarly, Cong et al.[7] investigated RUM for cutting force, surface roughness, and temperature using cutting fluid and cold air as coolant. It was found that coolant played a significant role in lowering the temperature generated during drilling. In another study [8], compared twist drilling and rotary ultrasonic drilling during machining of carbon fiber reinforced plastics. It was found that rotary ultrasonic process performed better than twist drilling in terms of better surface finish, low temperature, and high hole quality. On the same context, Kuruc et al.[9] investigated the rotary ultrasonic process during milling of polycrystalline CBN (cubic boron nitride) and outlined that RUM has the ability to machine hard material with less surface roughness. The rotational speed came out to be most influential parameter affecting the surface texture of workpiece.

In the light of above-mentioned literature, it can be stated that RUM has the ability to produce defect free high-quality surfaces. With this view, present study investigates the rotary ultrasonic machining process in terms of temperature produced during drilling of float glass 'with coolant and 'without coolant' conditions. The parameters like rotational speed, feed rate, vibration power was kept constant. Further, microstructure analysis has been carried out to determine the effect of coolant on hole's surface quality.

2. Material and methods

A rotary ultrasonic machining (RUM) set up are used to make holes in float glass material. A complete experimental set up includes ultrasonic generator, horn tool assembly, thermal camera, fixture, coolant nozzle, personal computer and float glass specimen as shown in figure 1. RUM has been working with 20 KHz of vibration frequency and 20 μm of vibration amplitude.

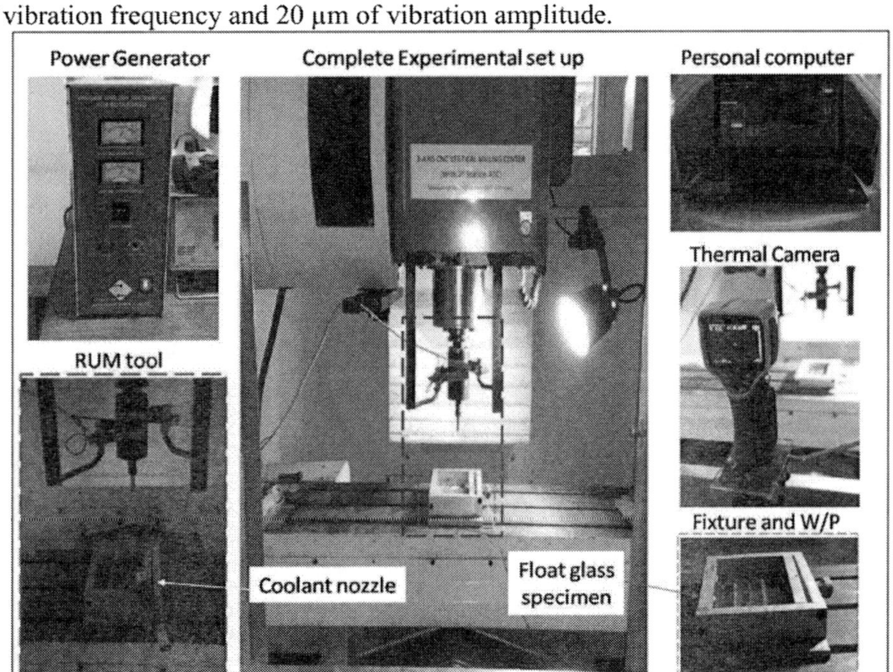

Figure 1. Complete RUM experimental set up.

In this study, diamond coated hollow abrasive tool is used to perform whole experimentation. The abrasive mesh size is in range of 100-150. Tool's dimension as tool length (50.75 mm), tool outer diameter and inner diameter are 7.86 mm and 5.89 mm, respectively and abrasive coated lateral face is around 3.15 mm. The float glass specimen is composite of silica sand, salt cake, soda ash, limestone, dolomite and small broken glass pieces. It has a compressive strength of 212 MPa, poison's ratio of 0.23, tensile strength of 50 MPa and Density (ρ) of 2500 Kg/m^3. Fixed designed experimental factors are illustrated as shown in table 1.

Table 1. Designed experimental factors.

S. no.	Parameters	Unit
1	Spindle rotation speed (N)	5000 rpm
2	Feed rate (fr)	6 mm/min
3	Coolant type	Water
4	Depth of cut or W/P thickness (T)	5 mm
5	Vibration amplitude	20 μm
6	Glass specimen size (L*B*T)	148*84*5 mm^3
7	Tool type	Diamond coated hollow abrasive tool

After completing all the drilling aspects while considering experimental setup capacity, the designed experimentation is performed at two conditions. These are (a) without coolant drilling, (b) With coolant drilling. In this study, overall 10 holes are created in which 5 holes are drilled at without coolant condition and remaining 5 holes are drilled with coolant condition. Water is acquired as a coolant which is widely used in glass drilling based industries. For the study, machining time is constant for both the drilling condition. Consequently, machining time (Tm) is estimated as 50 sec using Tm= T/fr. Where, Hole depth (T) is 5 mm, feed rate (fr) is 0.1 mm/sec.

Here, the cutting temperature at both the conditions is noticed using infrared camera (Keysight Technologies-U5855A) as shown in figure 1. 'Keysight true IR analysis software' is used further to collect thermo-graphic images. The temperature recording is performed by focusing the camera light on drilling zone (contact region between tool and workpiece). The value of emissivity and humidity during observation are 0.90 and 50 %, respectively. The range of temperature quantification is 0 to 350°c. In each second, 8 frames are taken place. As, the total machining time is 50 sec, in total 400 frames are has been carried out.

3. Results and discussion

Figure 2 illustrates the thermo-graphic images at various stages during rotary ultrasonic drilling. Figure 2 (a) depicts the thermo-graphic images showing the initial drilling temperature (29.75^{o}c) at without coolant condition. Figure 2 (b) shows the image of maximum drilling temperature (62.11^{o}c) at without coolant condition. Figure 2 (c) and figure 2(d) illustrate the images of initial drilling temperature (29.70^{o}c) at with-coolant condition and maximum drilling temperature (50.56^{o}c) at with-coolant condition.

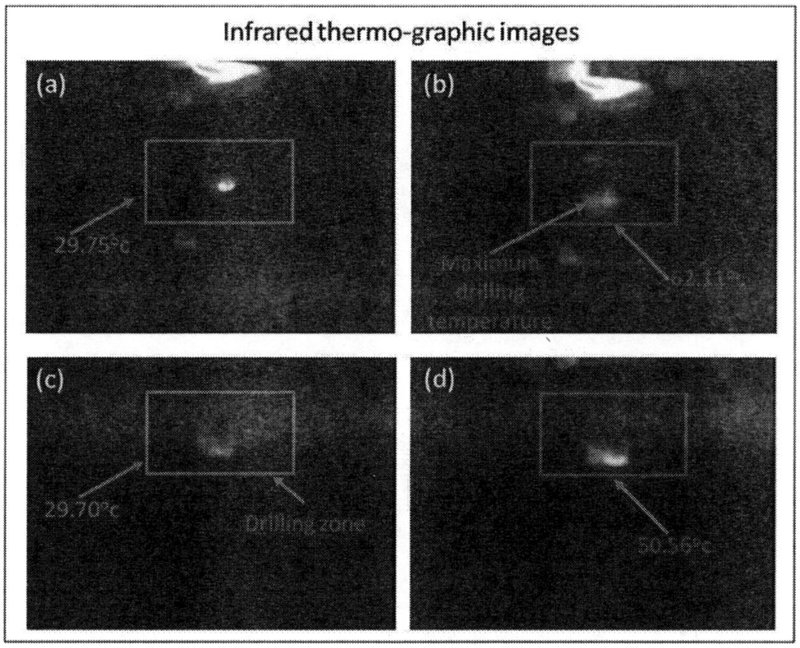

Figure 2. Illustration of thermo-graphic images at various stages during rotary ultrasonic drilling: (a) Initial drilling temperature at without coolant condition, (b) Maximum drilling temperature at without coolant condition, (c) Initial drilling temperature at with coolant condition, (d) Maximum drilling temperature at with coolant condition.

After visualizing the thermo-graphic images, it is noticed that as compare to 'without coolant condition', at 'with coolant condition' the temperature is reduced by 11.55 °c. Figure 3 shows the comparison of drilling/cutting temperature with respect to time for without coolant and with coolant conditions.

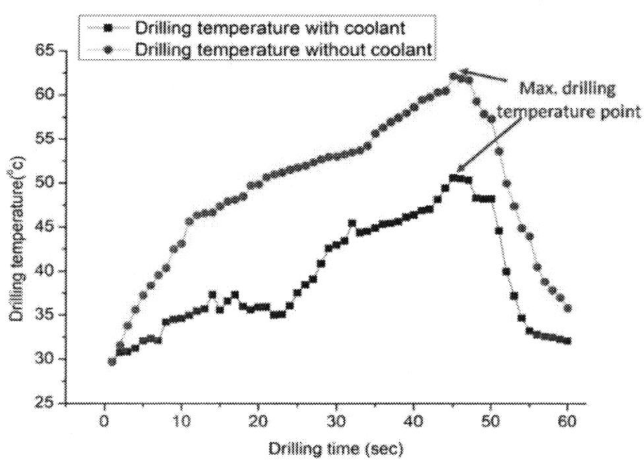

Figure 3. Comparison of drilling temperature with respect to time for without coolant and with coolant conditions.

This reduction in temperature is because of heat dissipation in case of 'with coolant condition' is far superior to other case. It is cleared that from the initiation of drilling till conclusion of drilling, there is a drastic effect of presence of coolant is visualized which leads to decrease in thermal energy caused by friction produced between the tool face and float glass periphery. This friction could also affect the quality of the hole specifically at the corners [10]. Figure 4 shows the microscopic image at two drilling conditions (without coolant and with-coolant).

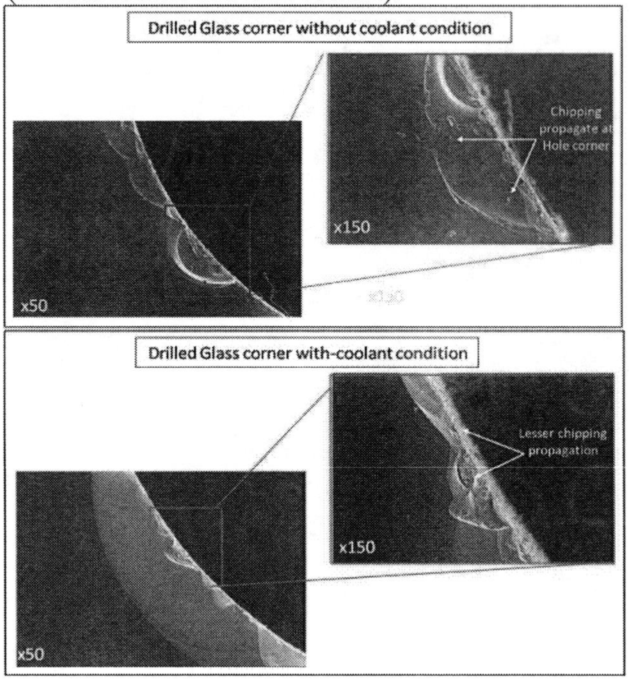

Figure 4. Microscopic image at two drilling conditions (without coolant and with-coolant) with 50x and 150x magnification.

As shown in microscopic images, it is noticed that use of coolant could also aid in reducing the hole quality. More chipping amount is occurred seen nearby float glass edges at 'without coolant condition'. It means the reduction in cutting temperature would be the prime concern that can also raise the float glass application. It can be visualize that the surface corner damages would ultimately effected by the cutting temperature. Thus, the implementation of cutting fluid, which acts as a coolant, is very crucial during drilling of brittle materials. Consequently, it is revealed that drilling with coolant (water) should be adopted to improve the hole quality. Since, it reduced the formation of stresses during drilling operation which are induced due to rise in temperature.

4. Conclusion

An experimental investigation has been performed to confirm the influence of cutting temperature of float glass specimen using rotary ultrasonic drilling. The key conclusion has been drawn as:

- Drastic effect of coolant is visualized which leads to decrease in thermal energy caused by friction produced between the tool face and float glass periphery.
- Maximum cutting temperature reached with coolant drilling is 50.56 °c and without coolant is 62.11°c.
- Cutting temperature is reduced by 11.55 °c by using 'with coolant condition' drilling.
- It is affirmed that the damages occurred at the drilled corner ultimately affected by the cutting temperature.

5. Future scope

A study could happen to quantify the optimum amount and flow rate of coolant required during drilling. A study on effect of 'without coolant and with-coolant' conditions on tool wear can also consider.

References

[1] P. D. Gaudio, H. Behrens, J. Deubener, Viscosity and glass transition temperature of hydrous float glass, J Non Cryst Solids **353**, 223–236 (2007). doi: 10.1016/j.jnoncrysol.2006.11.009
[2] A. Sharma, V. Jain, D. Gupta, Multi-shaped tool wear study during rotary ultrasonic drilling and conventional drilling for amorphous solid **0**,1–10 (2018). doi: 10.1177/0954408918776724
[3] A. Sharma, V. Jain, D. Gupta, Characterization of Chipping and Tool wear during drilling of Float glass using rotary ultrasonic machining, Measurement **128**, 254–263 (2018). doi: 10.1016/j.measurement.2018.06.040
[4] K. Ding, Y. Fu, H. Su, Experimental studies on drilling tool load and machining quality of C/SiC composites in rotary ultrasonic machining, J Material Process Technol **214**, 2900–2907, (2014). doi: 10.1016/j.jmatprotec.2014.06.015
[5] D. Lv, Y. Huang, Y. Tang, H. Wang, Relationship between subsurface damage and surface roughness of glass BK7 in rotary ultrasonic machining and conventional grinding processes, Int J Adv Manuf Technol **67**, 613–622, (2013). doi: 10.1007/s00170-012-4509-1
[6] C. Zhang, W. Cong, P. Feng, Z. Pei, Rotary ultrasonic machining of optical K9 glass using compressed air as coolant: A feasibility study, Proc Inst Mech Eng Part B J Eng Manuf **228**, 504–514, (2014). doi: 10.1177/0954405413506195
[7] Q. Feng, W. L. Cong, Z. J. Pei, C. Z. Ren, Rotary ultrasonic machining of carbon fiber-reinforced polymer: Feasibility study, Mach Sci Technol **16**, 380-398, (2012). doi: 10.1080/10910344.2012.698962
[8] F. D. Ning, W. L. Cong, Z. J. Pei, C. Treadwell, Rotary ultrasonic machining of CFRP: A comparison with grinding, Ultrasonics **66**, 125–132, (2016). doi: 10.1016/j.ultras.2015.11.002
[9] Kuruc M, Vopát T, Peterka J Surface roughness of poly-crystalline cubic boron nitride after rotary ultrasonic machining. In: Energy Procedia **100**, 877-884, (2015).
[10] G. Deng, Z. Lu, G. Yao, J. Liu, Z. Li, D. Zhang, Cutting temperature and resulting influence on machining performance in rotary ultrasonic elliptical machining of thick CFRP, International Journal of Machine Tools and Manufacture **123**, 160–170, (2017).

IOP Conf. Series: Materials Science and Engineering **715** (2020) 012051 doi:10.1088/1757-899X/715/1/012051

Dual stage control of a servo system consisting of a piezoelectric actuator and a linear motor for electrical discharge machining

Muyad Bani Melhem[1], Milan Simic[1], Yong Feng[2] and Songlin Ding[1,*]

[1]School of Engineering, RMIT University, Melbourne, Australia
[2]School of Electrical Engineering and Automation, Harbin Institute of Technology, Harbin, China

*E-mail: songlin.ding@rmit.edu.au

Abstract. Gap width is the most important factor that affects the stability of the electrical discharge machining (EDM) process, material removal rate and surface finish of the workpiece. This research is to develop a new hybrid positioning system which consists of a linear servo motor and a piezoelectric actuator (PZT) for high efficiency EDM processes. In this new system, the servo motor provides the macro feeding for the workpiece while the PZT feeds the workpiece in micro scale at high frequency. To reduce the delay caused by separate movements of the linear motor and PZT, a new control algorithm was developed to synchronize the movements of both the motor and PZT. Cutting experiment shows that an increase in MRR of 1.5 times was achieved by using the proposed algorithm.

1. Introduction

Due to the ultra-hardness of polycrystalline diamond (PCD), electrical discharge grinding (EDG), one variation of electrical discharge machining (EDM) process, is considered the most effective machining technique to manufacture PCD products used in biomedical and manufacturing industry [1, 2, 3]. EDG overcomes the most common problems in the traditional grinding process, such as tool wear, significant grinding force and low material removal rate [4, 5]. However, since EDM is a stochastic process which depends on multiple interactive independent parameters, it makes the maintaining of a favorable spark gap extremely difficult. The spark gap between the workpiece and the electrode is the main factor that affects the machining performance [6]. The unstable gap distance will result in abnormal discharge pulses and the accumulation of debris in the gap which, in turn, negatively affect the stability of the machining process [7]. Therefore, keeping a constant gap distance is critical in the control of a stable and efficient EDM process.

2. Dual stage feeding mechanism

Various actuators have been used to feed the workpiece in EDM process, for example, linear servo motor, DC motor and stepper motor. Due to the very small spark gap which is in the range of 20-50μm and the high frequency changes in the working condition in the gap, the need for more accurate and fast actuators is continuously increasing. The piezoelectric actuator (PZT) is one of the most suitable actuators because of its high precision and high response frequency. However, the working range of

PZT is usually very short. A feeding system which consists of PZT alone does not meet practical requirements. An EDG process should have both high response frequency and long working range.

A hybrid dual stage feeding mechanism which consists of a servo motor to satisfy the long-range requirement and one PZT to realize high frequency response was recently developed by Hu et al. [8], and different algorithms have been developed to distribute the movements between the linear motor and the piezoelectric actuator. For example, Tong et al. [9] used a novel algorithm by stopping one actuator (PZT) to move the other (a linear motor); this method did increase the machining speed, however, because the PZT and the linear motor could not move simultaneously, the time was still wasted when the PZT actuator paused for the linear motor to move to compensate the distance after the PZT reached its full range. Kunieda et al. [10] built a simulator in order to investigate the effect of using PZT on the machining speed and stability. The PZT was used to respond to the machining process and the linear motor to keep a dynamic range of the PZT to around 16μm, but it is unclear how both actuators will move to perform the required tasks. Zhu et al. [11] used PZT as a capacitor, based on inverse piezoelectric effect to adjust the discharge gap; the feeding process was dependent on the charging and discharging of two capacitors (PZT and external adjustable capacitor). Once the discharge happened, the PZT would retract and it would move forward again after the charging of the capacitors. This process could undoubtedly increase the stability of the machining by preventing the arc and short circuit from happening, however, it was unable to ensure enough power delivered to the gap in order to complete the machining process, therefore it is unable to significantly increase the speed and efficiency.

In order to satisfy the need for large stroke and high precision, a piezoelectric actuator had been combined with step motor to form a drive with high positioning accuracy and repeatability [12]. Piezoelectric actuator was used as a micro driven system with micro steps which was mounted on a macro driven table for macro steps. Based on inverse piezoelectric effect the piezoelectric actuator was used to adjust the discharge gap. Due to the short range of PZT, the retract distance, in case arcing, was not sufficient to eliminate that spark status. We propose a new algorithm to drive the dual stage mechanism consisting of a linear servo motor and a piezoelectric actuator. In this system, the linear motor is responsible for long travel distance while the PZT is responsible for fast feeding. The new control strategy of macro-micro feeding obtains the best machining efficiency. Experiments were conducted to validate the effectiveness of new system by measuring the machining speed.

3. New dual stage control method

The proposed dual stage feeding mechanism consists of a linear motor as macro drive and a piezoelectric actuator as micro drive. The workpiece was hold by the PZT. The PZT was mounted on the top of the linear motor which moved in the same direction. The main goal of this structure was to maintain a stable machining process by controlling the gap width at desired level. The linear motor fed the piezo actuator to guarantee the PZT has a buffer distance from the electrode which was in the in the range from 25μm to 100μm. The buffer range ensured that at any machining status the piezo actuator has a minimum distance of 25μm to respond to any short circuit spark, and the piezo has enough backward and forward range to move the workpiece and respond to the changes in the spark gap. Once the PZT reached its forward limit (100μm), it was required to return to the left limit (25μm) and start another cycle. At the same time when the PZT was returning to the "25μm" position, the linear motor would compensate the difference in workpiece position by 75μm to ensure that the workpiece is always at a distance 20μm away from the tool, as shown in figure 1.

4. Experimental setup

The EDG experimental system consists of five sub-systems: power source system, the rotating electrode, fluid system, the dual stage feeding mechanism and control system, as shown in figure 2. The main power supply is a transistor type power supply with 90V as Voc and 10A the maximum current. Fluid system contains the machining tank, which would be filled up with the dielectric before

IOP Conf. Series: Materials Science and Engineering **715** (2020) 012051 doi:10.1088/1757-899X/715/1/012051

the machining starts. This fluid is flushed from a storage to the tank by a pump. The rotating electrode disk is mounted on a hydraulic motor.

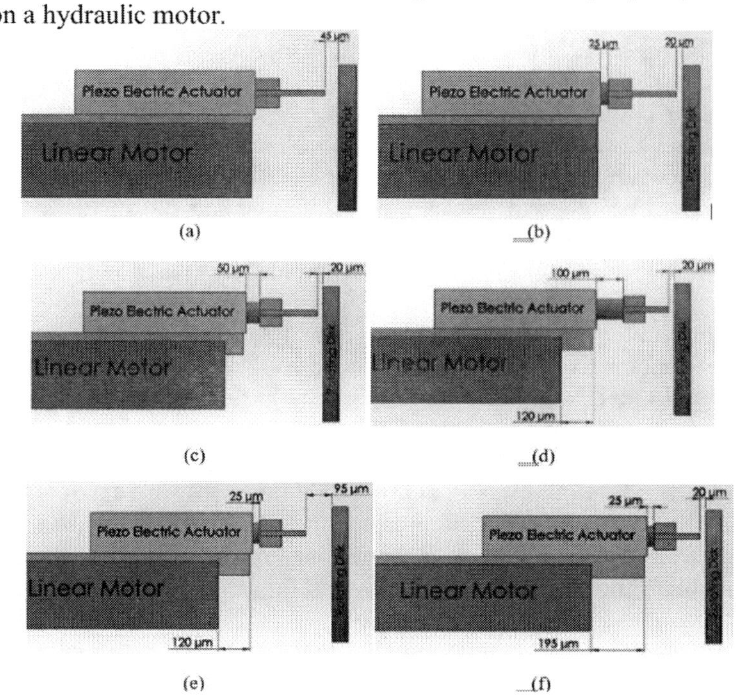

Figure 1. Schematic drawing for the feeding process of the dual stage feeding mechanism. (a) initial position where the PZT is at zero position and linear motor at 45μm from the rotating disk. (b) the PZT moved by 25μm to make the gap being 20μm. (c) the machining started by moving both actuators PZT and linear motor. (d) the PZT reached the full range position. (e) the PZT moved backward to the 25μm position. (f) the linear motor moved forward to reduce the gap to 20μm and the machining started again.

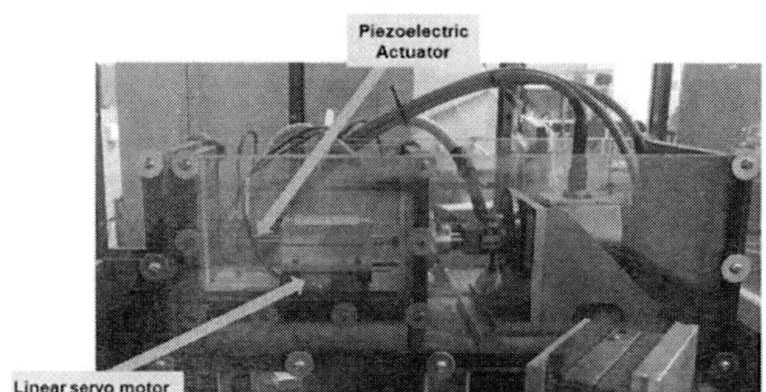

Figure 2. Set up of the EDG experiment system.

Dual stage feeding mechanism consists of a linear motor and a piezo electric actuator, positioned on the top of the linear motor. Workpiece is mounted on the tip of the feeding mechanism. The main controller was built by using LabView software. Linear motor was controlled through AKD factory-built power controller. Position feedback signal was read by a linear encoder through an analog circuit. Piezoelectric actuator was controlled through THORLAB controller which provided the actuator with the required power and read the position using strain gauges. This controller is connected to the main controller device using USB port.

Frequency is an essential parameter to determine the effectiveness of using the PZT in high-speed applications. The bandwidth of a piezo controller can be estimated based on the following parameters:

• The maximum current the controllers can produce (0.5A)
• The load capacitance of the piezo. (20μF)
• The desired signal amplitude (V).
• The absolute maximum bandwidth of the driver, which is independent of the load being driven.

The band width of the PZT is calculated using Slew rate, the change in the charge with respect to time is:

$$slew\,rate = \frac{dv}{dt} = \frac{I_{max}}{C} \tag{1}$$

$$slew\,rate = \frac{0.5\,A}{20\,\mu F} = 25V/ms \tag{2}$$

The bandwidth of the system usually refers to the system's response to a sinusoidal signal of a given amplitude A.

$$V(t) = A\sin 2\pi f t + A \tag{3}$$

$$\frac{dv}{dt}_{t=2n\pi} = 2\pi A f_{max} \tag{4}$$

Using Equations 1 and 4, the maximum frequency can be calculated as:

$$f_{max} = \frac{I_{max}}{2\pi AC} = \frac{I_{max}}{\pi V_{pp}C} \tag{5}$$

Substituting the values of I_{max}, V_{p-p} and C

$$f_{max} = \frac{I_{max}}{\pi V_{pp}C} = \frac{0.5\,A}{\pi\,(20\mu F)(75V)} \approx 106Hz \tag{6}$$

5. Control and test performance of EDG

The spark gap is critical in controlling the EDM process, therefore, it is important to control the feeding mechanism to keep the gap at the desired value. Proposed control algorithm was based on the simultaneous moving of the linear motor and the piezoelectric actuator to respond to the changes in spark status. As shown in figure 3, the control process starts by putting both actuators (PZT and linear motor) at the zero position. Then the linear motor moves to the position of 45 μm from the workpiece, while the piezo moves to the position of 25 μm (this gives the PZT a backward buffer of 25μm to use if short circuit, or aching status occurs). This makes 20μm gap between the workpiece and the tool.

During the machining process the piezoelectric actuator will respond to the changes in gap caused by the volume of removed material, from the workpiece, and the changes in debris concentration in the gap. Since PZT will move forward, the position of PZT will increase from 25μm to 100μm eventually. During this movement the linear motor will move slowly to compensate the enlarged distance caused by removed material. The position of PZT will always be more than 25μm, which gives the PZT enough range to move backward with high speed, responding to the short circuit status.

In the case of arc spark, the PZT will move backward by about 10μm to increase the spark gap allowing the current to decrease and the spark back to normal status.

In order to test the performance of the new algorithm, experiments were conducted and machining results of using the current algorithms [7] were compared with the results of our algorithm with the same machining parameters. The current control algorithm is based on moving each actuator individually. The PZT will respond to the changes in the spark status. When PZT reaches its limit, it will stop and return to its zero-position waiting for the other actuator, linear motor, to move and compensate the distance of materials removed by the PZT (figure 4).

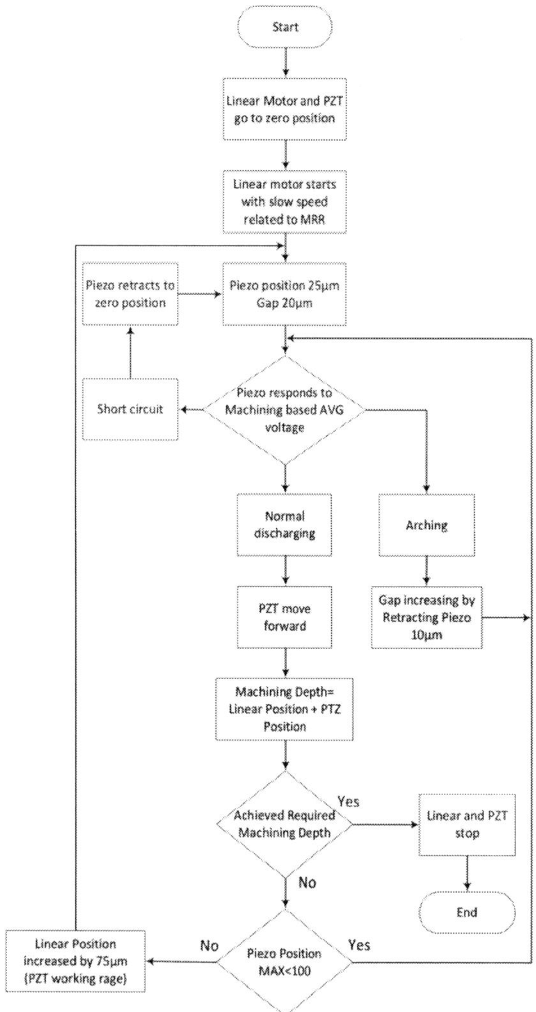

Figure 3. Block diagram of the proposed control of dual stage feeding mechanism.

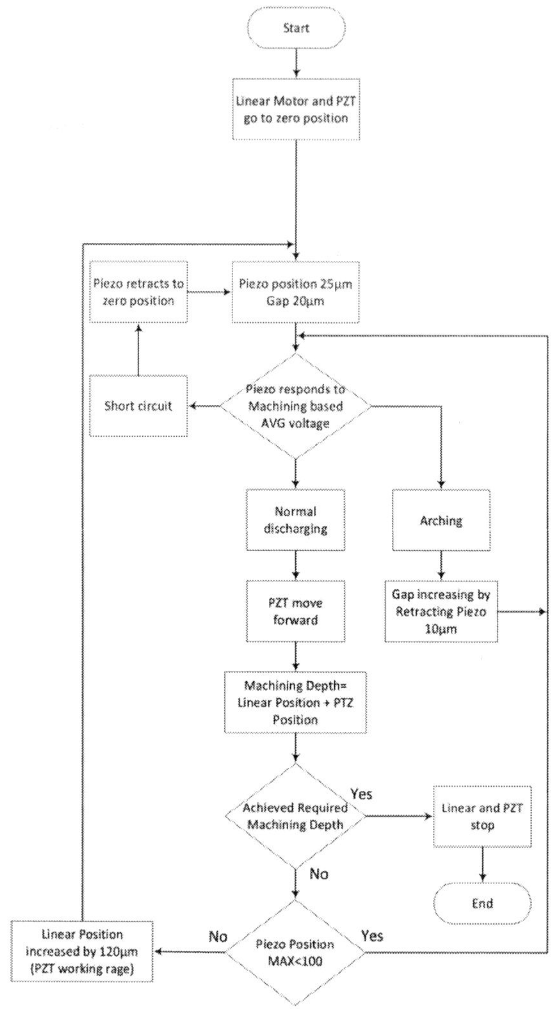

Figure 4. Conventional control of dual stage feeding system.

6. Results and discussions

In this experiment, workpiece was a copper strip with the dimension of 3mm x 1mm (figure 5). The machining parameters are shown in table 1:

Table 1. EDM machining parameters.

Discharge Voltage	90V
Pulse on time (T_{on})	40us
Pulse off time (T_{off})	80us
Discharge current	4Amp

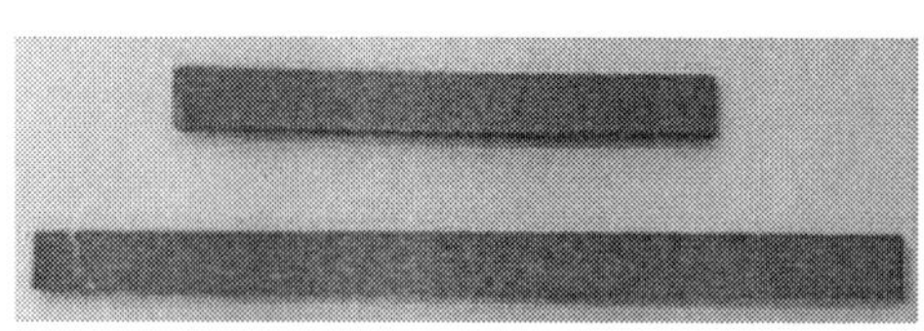

Figure 5. Machined copper strip used in the experiment.

Novel algorithms were implemented using LABVIEW software. We are using average gap voltage as feedback variable for the machining process. Positions of both actuators were measured during the discharging process. Machining time was also measured in order to investigate performances of both algorithms.

Figure 6 shows the position of the actuators based on the proposed algorithm. The blue line represents the position of the linear motor during the machining process whereas the orange one represents the PZT position. The linear motor was moving forward with a constant speed while the PZT was responding to spark status. As shown in the figure, when the PZT reached its maximum range the PZT would backward to zero and the linear position increase by the PZT range (100μm). During the entire machining distance of 1300μm, the PZT reached its maximum range 4 times, and the overall machining time was 440 sec.

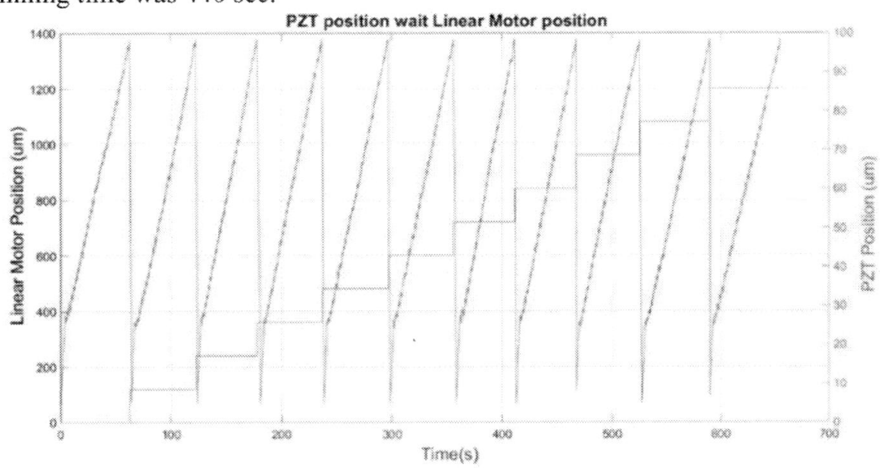

Figure 6. PZT and linear motor position while feeding 1300μm (the proposed algorithm).

The results of using conventional algorithm was shown in figure 7. As shown in the figure, the PZT reached its maximum limit around 10 times, which required the PZT to stop and the linear motor to move. It consumed 660 sec to machine the same specimen length (1300μm).

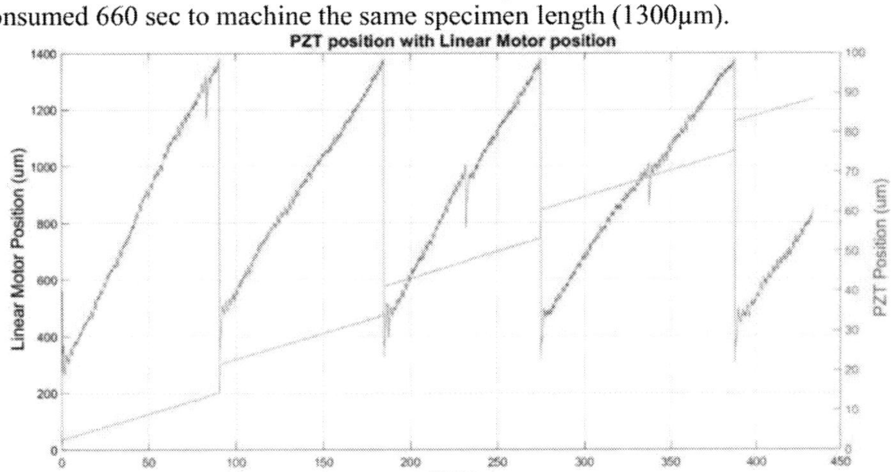

Figure 7. PZT and linear motor position while feeding 1300μm (existing algorithm).

The material removal rate (MRR) of the new algorithm was 2.95μm/s, while MRR of the conventional algorithm was 1.96μm/s. This difference in the MRR was due to the excess time caused by the stop of the PZT waiting for the linear to move. There was no discharging during this waiting time. Therefore, the more times the PZT stopped, the more efficient the new algorithm would be.

7. Conclusion

In this paper a new dual stage control algorithm was proposed to drive the linear motor and PZT. Two machining processes were conducted using the conventional and the new algorithms. The new algorithm was found to be 1.5 times faster than the conventional one, and MRR was one and half bigger compared with the conventional one for machining the same copper strip under the same machining conditions.

Acknowledgments

The research is supported by the Australian Research Council (Discovery Project DP180100762).

References

[1] M.Z. Rahim, S. Ding, J. Mo, "Electrical discharge grinding of polycrystalline diamond - effect of machining parameters and finishing in-feed", Journal of Manufacturing Science and Engineering, Transactions of the ASME, **137(2)**, 021017 (2015)

[2] K. Inthavong, J. Tu, Y. Ye, S. Ding, A. Subic, F. Thien, "Effects of airway obstruction induced by asthma attack on particle deposition", Journal of Aerosol Science, **41 (6)**, 587-601(2010)

[3] S. Ding, D.C.H. Yang, Z. Han, "Boundary-conformed machining of turbine blades", Proceedings of the Institution of Mechanical Engineers, Part B: Journal of Engineering Manufacture, **219 (3)**, 255-263 (2005)

[4] G. Li, S. Yi, S. Sun, S. Ding, "Wear mechanisms and performance of abrasively ground polycrystalline diamond tools of different diamond grains in machining titanium alloy", Journal of Manufacturing Processes, **29**, 320-331 (2017)

[5] G. Li, M.Z. Rahim, S. Ding, S. Sun, Performance and wear analysis of polycrystalline diamond (PCD) tools manufactured with different methods in turning titanium alloy Ti-6Al-4V, International Journal of Advanced Manufacturing Technology, **85 (1-4)**, 825-841 (2016)

[6] S. Ding, R. Yuan, Z. Li, K. Wang, "CNC electrical discharge rough machining of turbine blades", Proceedings of the Institution of Mechanical Engineers, Part B: Journal of Engineering Manufacture, **220(7)**, 1027-1034 (2006)

[7] M.Z. Rahim, S. Ding, J. Mo, "Electrical discharge grinding of polycrystalline diamond - effect of wheel rotation" Machining Science and Technology, **20 (1-2)**, 62-78 (2016)

[8] B. Hu, J. PT Mo, S. Ding, M. Brandt, "Model-driven design of a fast material removal electrical discharge machine", Cogent Engineering, **3(1),** 123380 (2016)

[9] H. Tong, Y. Li, Y. Wang, D. Yu. "Servo Scanning 3D Micro-EDM Based on Macro/micro-dual-feed Spindle." International Journal of Machine Tools and Manufacture, **48(7)**, 858-69 (2008)

[10] M. Kunieda, T. Takaya, S. Nakano. "Improvement of Dry EDM Characteristics Using Piezoelectric Actuator." CIRP Annals - Manufacturing Technology, **53(1)**, 183-86, (2004)

[11] Y. Zhu, Q. Zhang, X. Fu, G Bao. "Study on Discharge Characteristics of Piezoelectric Self-Adaptive Micro-EDM System" Applied Mechanics and Materials, **34-35**, 492-496 (2010)

[12] M. Zhou, F. Ha, I. Soichiro, "A time-varied predictive model for EDM process," International Journal of Machine Tools & Manufacture, **48**, 1668–1677 (2008)

IOP Conf. Series: Materials Science and Engineering **715** (2020) 012052 doi:10.1088/1757-899X/715/1/012052

High precision material removal of copper surface by jet electrochemical machining

Yiming Feng, Ying Yan, Ping Zhou* and Dongming Guo

Key Laboratory for Precision and Non-traditional Machining Technology of Ministry of Education, Dalian University of Technology, Dalian 116024, China

*Email: pzhou@dlut.edu.cn

Abstract. High-precision machining technology has attracted increasing attention with the rapid development of modern industry. Jet electrochemical machining is a promising method for forming complex 3D shape surface by controlling the path and dwelling time of electrode nozzle moving over workpiece surface. The machining accuracy mainly relies on the stability of the material removal rate and the smoothness of the machined surface. In this paper, a new jet electrochemical machining method, which is based on a newly developed acid electrolyte, is proposed to improve its machining performance. The effects of voltage, interelectrode gap, electrolyte temperature and flow rate are discussed in this work. Stable machining can be achieved by selecting the optimum process parameters, and the surface roughness (Ra) can be reduced to 14.8 nm. Moreover, the obtained material removal rate distribution agrees well with the Gaussian function, which is recognized to have high capability to form complex 3D surface through controlling the movement of the tools.

1. Introduction

Copper is widely used in precision devices in the fields of aerospace, electronics, military products, instrumentation due to its good electrical and thermal properties [1]. Since the surface quality and structural accuracy of precision devices directly affect the performance of the device, the research on precision machining for high stability and high surface quality of copper is of great significance.

Electrochemical machining (ECM) is a controlled anodic electrochemical dissolution process, and the major advantages of ECM are its high material removal rate, almost no tool wear and good surface quality without the occurrence of re-solidified layers, heat-affected zone, or strain hardening [2, 3]. Liu et al. studied the ECM of γ-TiAl alloy through experiments, the surface roughness (Ra) of the micro-pores is about1 μm by changing the voltage, the interelectrode gap (IEG), the flow rate and temperature of electrolyte [4]. Klocke et al. changed the current density of the machining region to obtain micro-pores with a surface roughness (Ra) of 0.4 μm for Ti-6Al-4V alloy [5]. Another investigation on technological capabilities of ECM by Klocke et al. revealed that the small frontal gap was an important factor in achieving high material removal rate and high machining precision [6].

Jet electrochemical machining (Jet-ECM) is an important variation of traditional ECM and is effective to manufacture complex structure with the help of multi-dimension mechanical motion [7]. Liu et al. studied the effects of process parameters of Jet-ECM of TB6 titanium alloy (Ti-10V-2Fe-3Al). The surface roughness (Ra) of the micro-pores and long grooves are 0.231 μm and 0.273 μm respectively [8]. Mitchell-Smith et al. studied the Jet-ECM of Titanium grade 5 (Ti-6Al-4V). In their work, with the help of ultrasonic method the surface roughness Ra was reduced from 0.245 μm to 0.168 μm [9]. However, the existing Jet-ECM results cannot meet the processing requirements of high-precision devices with surface roughness of ten nanometers or several nanometers.

In this paper, a new type of electrolyte is used to investigate the effect of process parameters on the Jet-ECM machining performance of copper, such as machining voltage, interelectrode gap, electrolyte temperature and flow rate. And the optimum processing parameters are selected to achieve stable precision machining.

2. Experimental

2.1. Principle of Jet-ECM

As shown in figure 1, Jet-ECM removes material by jetting electrolyte from the nozzle toward the workpiece while applying voltage on cathode (nozzle) and anode (workpiece). Unlike traditional ECM, a jet of electrolyte is applied as a tool to selectively remove material on the workpiece [10]. With the help of multi-dimensional motion, it is possible for Jet-ECM to manufacture any 3D structure in theory [7].

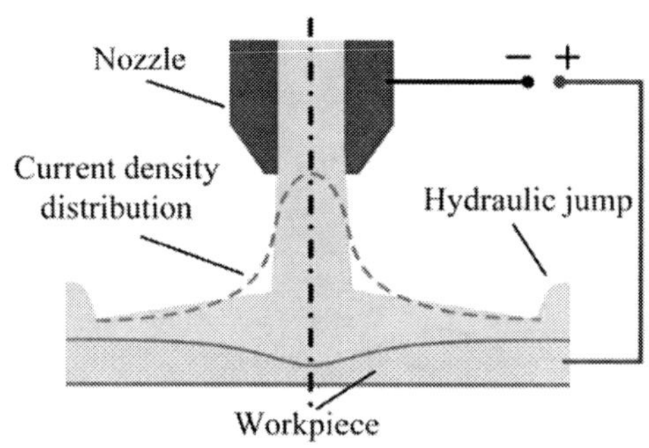

Figure 1. Schematic of Jet-ECM process.

In Jet-ECM, the dissolution of anode is confined at the exposed surface directly below the electrolyte jet due to the localization of the machining current density [11]. This is achieved by the so-called hydraulic jump phenomenon [12]. When the rapid flowing electrolyte hits the workpiece, a fast thin layer of electrolyte is formed on the workpiece surface near the nozzle, and then, the thickness of electrolyte at remote area suddenly increases and becomes subcritical. With the help of this fast thin electrolyte layer, the current density is concentrated and good machining accuracy can be obtained in Jet-ECM [13].

2.2. Experimental equipment

The schematic diagram of the Jet-ECM experimental system used in this paper is shown in figure 2. The machining groove and the nozzle are fixed on the CNC XY two-axis and Z single-axis displacement platform by the fixture with the displacement accuracy of 100 nm respectively. The voltage is supplied by the high frequency pulse power supply. The electrolyte circulation system supplies the electrolyte and the flow rate and temperature can be controlled by it.

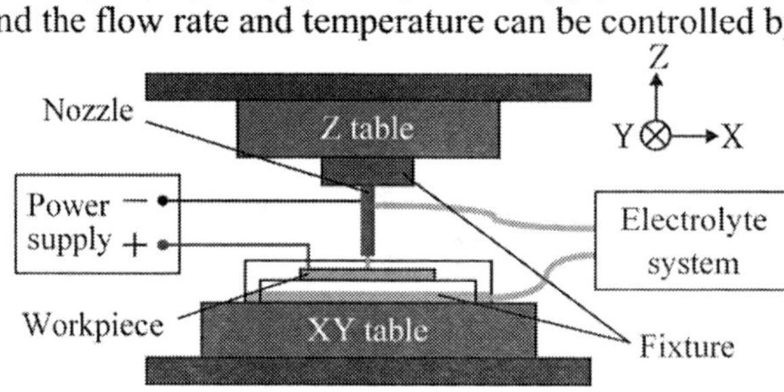

Figure 2. Experimental equipment.

In addition to excellent electrical conductivity, the nozzle material also needs to have good acid and alkali resistance, corrosion resistance and tensile strength. The metallic nickel meets these requirements and a cylindrical nozzle of nickel with an inner diameter of 1 mm is selected for this study. The fixture is made of polytetrafluoroethylene (PTFE) because of its excellent acid and alkali resistance and electrical insulation.

2.3. Electrolyte

In the Jet-ECM, the electrolyte not only transmits electrons as a conductive medium, but also performs a chemical reaction with the help of electric field, so that the anode dissolution can be performed smoothly and efficiently. At the same time, the electrolyte will also take away the electrochemical reaction products and heat in the interelectrode gap in time to ensure material renewal rate and cooling efficiency.

The neutral electrolyte is commonly used in modern industries produce hydroxide precipitates during processing, which can deteriorate the surface roughness and local short circuits [14, 15]. Using acidic electrolyte can cause the removed metal to dissolve into solution instead of forming precipitates [16], which will result in better surface roughness [17]. The acidic electrolyte require lower voltages than the neutral electrolyte [14] due to its good electrical conductivity. Moreover, the acidic electrolyte can achieve higher removal rate [18], so it is more suitable for precision machining [19].

In this paper, a new type of acidic electrolyte is used in the Jet-ECM, with phosphoric acid as the main electrolyte, assisted with a small amount of additives. The specific component and proportion of the new electrolyte are shown in table 1.

Table 1. Composition and proportion of electrolyte.

Phosphoric acid	Ethanol	Lactic acid	BTA	Ammonium acetate
850 ml/L	90 ml/L	60 ml/L	6 g/L	3 g/L

When copper is removed in phosphoric acid, a soluble and viscous phosphate film layer is formed on the surface. The mucosa layer can act as a concentration polarization, which can slow down the dissolution rate of the depressions and accelerate the dissolution rate at the protrusions. And lower surface roughness can be achieved in this way. The removal process is the combination of an anodic electrochemical process and a phosphate film formed on the metal surface [20].

Ethanol as a hydroxyl additive is a good organic solvent, which can increase the fluidity of the electrolyte and promote the release of bubbles. As a carboxylic acid additive, lactic acid is a good acid-base buffer and reaction stabilizer. Benzotriazole (BTA) and ammonium acetate are good corrosion inhibitors [21], and if the anode is removed too quickly, it may cause erosion. However, the effects of different additives are not completely separated, and they can be matched to each other for multiple functional effects [22].

3. Results and discussion

The Jet-ECM experimental system is used with the new electrolyte for initial processing. The initial surface of the machined workpiece is very flat. In this paper, the depth of the pit is used to indicate the amount of material removed. The measurement of the pit profile is performed by the Keyence VK-X200 series 3D laser scanning confocal microscope. The surface roughness is measured by Zygo Nexview5022S 3D optical surface profiler.

A single pit processing experiment is performed. The two-dimensional height distribution of a single pit is shown in figure 3(a) and the three-dimensional topography distribution of the pit is shown in figure 3(b). As shown in figure 4, the longitudinal section of a single pit conforms to a Gaussian distribution by shape fitting, and the removal rate of single pit is about 10 μm/min.

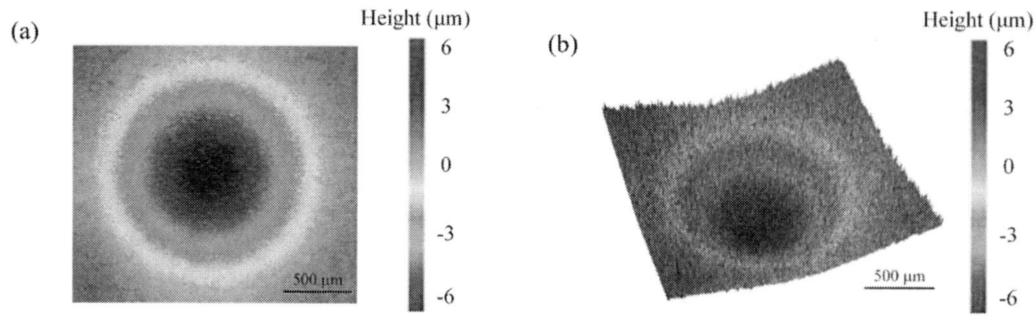

Figure 3. Machining result of single pit. **a.** 2D height distribution. **b.** 3D topography distribution.

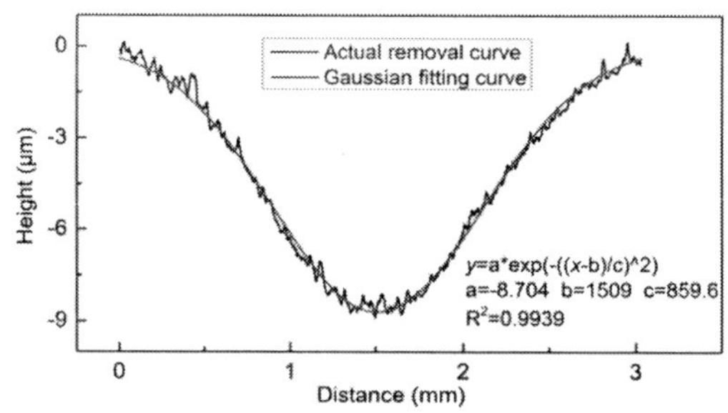

Figure 4. The longitudinal section in a Gaussian distribution.

3.1. Voltage

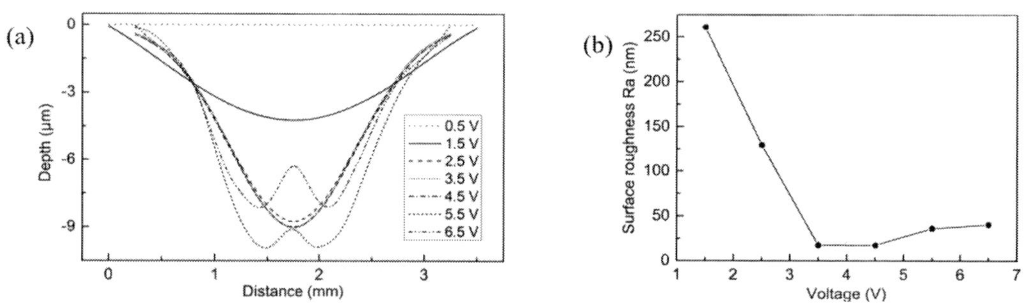

Figure 5. Influence of voltage on machining result. **a.** On removal profile. **b.** On surface roughness.

As shown in figure 5, when the voltage is 0.5 V, the current density in the interelectrode gap is too small to occur significant material removal. When the voltage is increased to 1.5 V, the material removal occurs in all low current density regions. The removal range is enlarged and the surface roughness is deteriorated from the initial surface. As the voltage increases, the current density increases gradually and the removal only occurs in the high current density regions. The longitudinal section of the pit agrees well with the Gaussian distribution. The removal depth is gradually increased and the surface roughness is gradually reduced.

When the voltage is increased to 3.5-4.5V, the removal is in a saturated state. The removal depth is no longer increased and the surface roughness Ra is stabilized at about 17 nm. However, when the voltage is increased to 5.5 V or more, the influence of the cathode on the current density distribution is remarkably increased which causes the change of removal profile and the deterioration of surface roughness. Moreover, cavitation appear on the machined surface, which can cause some energy to be lost, so that the depth of removal is reduced.

3.2. Array pits processing

With comprehensive consideration of the effect of process parameters on the machining performance, the optimal parameters such as voltage 3.5 V, IEG 500 μm, electrolyte temperature 45 °C and flow rate 50 mL/min are selected. The processing of 3×3 array pits are performed with the optimal processing parameters, and the processing time of each pit is 60 s.

The Corning Tropel FlatMaster 200 is used to measure the machining results of the array pits. The macro profiles of the pits with top-view and reverse-view are shown in the figure 6. The removal profile and removal depth of the 3×3 array pits are very stable, and the error of the removal depth is within the range of 0.3 μm. The average value of the surface roughness is 14.19 nm.

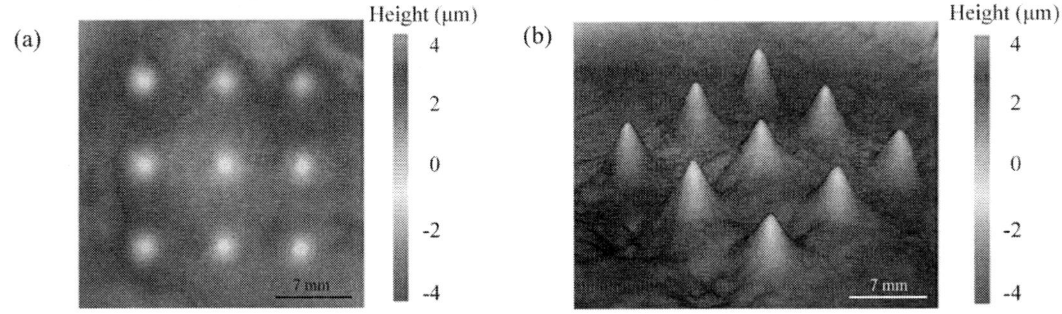

Figure 6. Macro profiles of the pits. **a**. 2D depth of array pits (top-view) **b**. 3D profile array pits (reverse-view).

4. Results and discussion

In this paper, the Jet-ECM with a new type of acidic electrolyte is studied through experiments. The influence of process parameters such as voltage, IEG, electrolyte temperature and flow rate on the machining performance of copper are analyzed. According to the experimental results, the following conclusions are as the following:

1. Jet-ECM using the new acidic electrolyte proposed in this paper can further reduce the surface roughness, thereby enable the machining of high precision and ultra-smooth surface of copper. Changing the voltage causes the current density to change, which affects the removal rate, removal profile, and surface roughness.

2. With comprehensive consideration of the effect of process parameters on the machining performance, optimal parameters of voltage 3.5 V, IEG 500 μm, electrolyte temperature 45 °C and flow rate 50 mL/min are selected. The machining results show high accuracy (Ra=14 nm) combined with relatively high removal stability (depth error within 0.3 μm), which is acceptable from the viewpoint of high-precision machining industry.

Acknowledgement

National Natural Science Foundation of China (51875078, 51605079), Science Fund for Creative Research Groups of NSFC (51621064).

References

[1] K. Shan. Analysis of the Polishing Ability of Electrogenerated Chemical Polishing. Precision Engineering, **47** (2017), 122–30

[2] J. McGeough. Principles of Electrochemical Machining. *Chapman and Hall Ltd/John Wiley & Sons*, London, (1974)

[3] K. Rajurkar. New Developments in Electro-chemical Machining. *CIRP Annals 1999: Manufacturing Technology*, **48**, 2 (1999), 567–79

[4] J. Liu. Experimental Investigation on Electrochemical Machining of γ-TiAl Intermetallic. Procedia CIRP, **35** (2015), 20–24

[5] F. Klocke. Technological and Economical Capabilities of Manufacturing Titanium- and Nickel-based Alloys Via Electrochemical Machining (ECM). *Key Engineering Materials*, **504–506**

(2012), 1237–242

[6] F. Klocke. Experimental Research on the Electrochemical Machining of Modern Titanium- and Nickel-based Alloys for Aero Engine Components. Procedia CIRP, **6** (2013), 369–73

[7] A. Schudert. Research on Generation of Three-dimensional Surface with Micro-electrolyte Jet Machining. Procedia CIRP, **42** (2016), 384–89

[8] W. Liu. Jet Electrochemical Machining of TB6 Titanium Alloy. International Journal of Advanced Manufacturing Technology, **90**, 5-8 (2017), 2397-409

[9] J. Mitchell-Smith. ElectroChemical Jet Machining of Titanium: Overcoming Passivation Layers with Ultrasonic Assistance. Procedia CIRP, **42** (2016), 379–83

[10] W. Natsu. Research on generation of three dimensional surface with micro-electrolyte jet machining. CIRP Journal of Manufacturing Science and Technology, **1** (2008), 27–34

[11] K. Yoneda. Numerical analysis of cross section shape of micro-indents formed by the electrochemical jet machining. Journal of the Japan Society of Electrical Machining Engineers, **29,** 63 (1996), 1–8 (in Japanese).

[12] H. Chanson. Numerical analysis of cross section shape of micro-indents formed by the electrochemical jet machining. European Journal of Mechanics B/Fluids, **28**, 2 (2009), 191–210

[13] T. Kawanaka. Mirror-like finishing by electrolyte jet machining. CIRP Annals - Manufacturing Technology, **64** (2015), 237–40

[14] M. Sen. A review of electrochemical macro- to micro-hole drilling processes. International Journal of Machine Tools & Manufacture, **45** (2005), 137–52

[15] S. Sharma. Electrochemical Drilling of Inconel Superalloy with Acidified Sodium Chloride Electrolyte. International Journal of Advanced Manufacturing Technology, **19**, 7 (2002), 492–500

[16] J. Asm. *Metals Handbook, Machining*. ASM International. (1989)

[17] J. Kozak. Study of Electrochemical Jet Machining Process. Journal of Manufacturing science and Engineering-Transactions of the ASME, **118**, 4 (1996), 490–98

[18] V. Subburam. Effect of Acidified Electrolyte on the performance of Electrochemical Micromachining. *Proceedings of the Second International Conference on Advances in Materials Processing and Characterisation (AMPC 2013)*, **vol. II** (2013), 799–806

[19] C. Jackson. Shaped Tube Electrolytic Machining (STEM Drilling). SME Technical Paper, MR 69–109. Product ID: TP69PUB243. (1969)

[20] D. Li. Research on the Copper Electrochemical Polishing in Phosphoric Acid. Surface Technology, **42**, 3 (2013), 1–4, 37

[21] K. Shan. Chemical Components and Its Functions of Polishing Solution Used in Electrogenerated Chemical Polishing of Copper. Journal of Mechanical Engineering, **52**, 5 (2016), 88–94

[22] B. Du. Development of Electrochemical Polishing Technology. Surface Technology, **36**, 2 (2007), 56–58.

IOP Conf. Series: Materials Science and Engineering **715** (2020) 012053 doi:10.1088/1757-899X/715/1/012053

Research on airline operational risk management and control method

Shidai Qiu*, Dengkai Yao, Zekun Wang and Guhao Zhao

Air Traffic Control and Navigation College, Air Force Engineering University, 710051 Xi'an Shanxi, China

*E-mail: 1250901147@qq.com

Abstract. Starting from the reality of airlines, this paper studies the operational risk and control methods of airlines, establishes relevant risk index system, and formulates corresponding control methods. Based on the four aspects of human, aircraft, environment and management, the risk index system is established. The weight of the index and the score of each system are determined by using the analytic hierarchy process and the fuzzy comprehensive evaluation respectively, and the airline risk evaluation model is established. In the formulation of airline management and control methods, genetic algorithm is used to compare different flight task combinations, and the advantages of this method are obtained.

1. Introduction

In recent years, with the vigorous development of civil aviation, the demand of customers for civil aviation transportation is growing, and the demand is getting higher and higher. The daily operation of airlines involves many factors and is a complex large-scale system. Therefore, it is difficult to evaluate the operational risk of airlines, and it is even more difficult to formulate corresponding control methods for the operational risk of airlines. These two problems are becoming increasingly prominent and need to be solved urgently.

Y.T. Wang et al.[1] established a flight risk assessment system from the operational point of view based on flights, and then Y.T. Wang[2] revised the degree of mutual influence of various indicators in airline safety risk assessment. In order to analyze the influence of operational risk factors, Y.T. Wang[3] systematically analyzed and calculated the coupling situation of flight operational risk. Literature[4] analyzed the fatigue degree of crew members and the risk value of experience. D.L. Li[5] pioneered the airline safety risk assessment based on dispatch release. T. Liu[6] combines the theory of organizational behavior and risk management to explore effective measures of risk management and control from the perspective of organization, and elaborates the existing operational system of airlines in an all-round way. Z.W. Ji[7] evaluates airline safety risk based on Bayesian network.

Beyond seas, Experts such as M. Bineid[8] focus on the development of reliability methods for civil aircraft dispatching, starting from the feasibility of airline flight dispatching. M. Janic[9] based on the airline database for many years, programmed by himself, and studied the risk and safety assessment of civil aviation. R.E. Braithwaite[10] carries out risk assessment and control for local airlines based on Australian airline database and related information.

The above research lacks unified evaluation criteria. This paper evaluates the operational risk of airlines by combining qualitative and quantitative analysis around the four systems of human, aircraft, ring and management, and puts forward corresponding control strategies.

2. Airline operational risk assessment

2.1. Analytic hierarchy process

(1) Constructing judgment matrix

Taking the subsystem of human factors as an example, the judgment matrix of the two comparisons of the subsystem of human factors is established.

$$F = (f_{ij})_{n \times n}(i, j = 1, 2, \cdots, n) \tag{1}$$

The judgment matrix of human factor subsystem is as follows:

$$C = \begin{pmatrix} 1 & 2 & 3 & 4 & 4 \\ 1/2 & 1 & 2 & 3 & 3 \\ 1/3 & 1/2 & 1 & 2 & 2 \\ 1/4 & 1/3 & 1/2 & 1 & 1 \\ 1/5 & 1/4 & 1/3 & 1/2 & 1 \end{pmatrix} \tag{2}$$

(2) The weight of each index can be obtained by AHP method, As shown in table 1:

Table 1. The classification rule of unsafe area.

First level index	weigh	Second level index	weight
Human factors C	0.394	Fatigue degree(C1)	0.414
		Skill level(C2)	0.257
		Tacit understanding degree(C3)	0.153
		English ability (C4)	0.088
		Psychological quality (C5)	0.088
Machine factors M	0.258	Aircraft properties (M1)	0.311
		Maintenance (M2)	0.019
		Communication navigation	0.111
		Aircraft MEL term (M4)	0.298
		Insecurity incidents (M5)	0.261
environmental factors E	0.213	Airspace condition of route (E1)	0.034
		Weather condition(E2)	0.299
		Airport condition (E3)	0.317
		Communication coordination(E4)	0.195
		Response status (E5)	0.155
Management factors P	0.135	Relevent system(P1)	0.332
		Training status(P2)	0.233
		Organization structure(P3)	0.241
		Departmental cooperation(P4)	0.098
		Emergency plan(P5)	0.096

2.2. Fuzzy comprehensive evaluation

Taking the subsystem of human factors as an example, the safety level of each risk index is defined as:

$$H=\{1\text{ (Extremely poor)},2\text{ (Poor)},3\text{ (Medium)},4\text{ (Good)},5\text{ (Excellent)}\}$$

Applying expert scoring method to obtain evaluation matrix:

$$A=\begin{bmatrix}0.3&0.4&0.4&0.6&0.3\\0.7&0.2&0.5&0.1&0.5\\0.6&0.4&0.3&0.2&0.5\\0.1&0.8&0.3&0.3&0.5\\0.6&0.6&0.6&0.1&0.1\end{bmatrix}$$

Therefore, the fuzzy comprehensive evaluation matrix is:

$$C=W\cdot A=[0.3671\quad 0.4696\quad 0.4552\quad 0.3578\quad 0.5285]$$

The risk assessment grade is calculated.

$$C_i=\frac{\sum_{j=1}^{m} b_j^k\cdot j}{\sum_{j=1}^{m} b_j^k}=3.09 \tag{3}$$

Therefore, the anthropogenic factors in the operational risk of airlines belong to level 3, that is, the degree of risk is moderate. Similarly, $M_i=2.42, E_i=2.92, P_i=3.44$,Then it can form a general secondary evaluation matrix: $[0.3002\quad 0.2843\quad 0.2673\quad 0.1492]$ Finally, the overall airline operational risk score can be obtained:

$$C=\frac{\sum_{j=1}^{m} b_j^k\cdot j}{\sum_{j=1}^{m} b_j^k}=2.906 \tag{4}$$

That is to say, the operational risk of the whole airline is close to the medium level.

3. Genetic algorithms

3.1. Mathematical model

In the calculation of airline operational risk evaluation model, the lowest risk index is used to calculate. The first-level fuzzy comprehensive evaluation formula is as follows:

$$B_i=W_i^T\cdot R_i \tag{5}$$

For the weight of each risk index, the relationship between each risk index and the score set.

Similarly, two-level fuzzy comprehensive evaluation:

$$B_i=W^T\cdot R_i=(b_1,b_2,\cdots,b_n) \tag{6}$$

For each subsystem corresponding weight, for each subsystem corresponding to the relationship between scoring set. So the objective function is:

$$MAX(F(C))=\sum_{i=1}^{n}\frac{\sum_{j=1}^{m} b_j^k\cdot j}{\sum_{j=1}^{m} b_j^k} \tag{7}$$

The total score representing a group of flight tasks depends on the score of the person, the machine, the loop and the management in the task. Firstly, the risk scores of human, aircraft, loop and tube are obtained by using the fuzzy comprehensive evaluation. Then, according to the task combination, the risk scores of a group of tasks are obtained by using the fuzzy comprehensive evaluation again. Finally, the total risk scores of N groups of missions are obtained. The fitness function is chosen as (penalty function):

$$Fit(F(C))=\frac{10000}{(1-c+F(C))\cdot\partial} \tag{8}$$

3.2. Selection operator

Roulette betting is used for selection, and the generation gap is 0.8, which means that 80% of the excellent groups will be selected to replicate. Next, the operation steps of roulette are introduced.

Firstly, the individual fitness of each individual is calculated, and then the overall fitness of the population is calculated. Then, the probability of each individual being selected is calculated.

$$P_i = \frac{per(i)}{\sum_{i=1}^{m} per(i)} \tag{9}$$

In this paper, the single-point crossover method is adopted, and the crossover probability is set to 0.7. According to the principle of single-point crossover, if an individual's coding length is V, there are V-1 points to be selected as crossover points, and then V-1 new individuals are generated by single-point crossover.

3.3. Variant operator

The mutation operation is an important operation in the later stage of genetic algorithm. Genes of organisms will mutate in the genetic evolution of successive generations, and the new generation of chromosomes that are likely to mutate has a better advantage than the previous generation of chromosomes. The mutation operation minimizes the possibility of destroying the better patterns that have been formed in the population, and generates the more excellent patterns as much as possible. Therefore, the setting of mutation probability is particularly important, assuming that the mutation probability is constant, 0.001.

Mutation operation: firstly, the numbering codes of the four subsystems of human, machine, loop and tube are generated, and the gene numbers at this position are mutated. The gene values of the mutated individuals are replaced by other alleles, and a new generation of individuals is generated.

The whole algorithm flow is shown in figure 1:

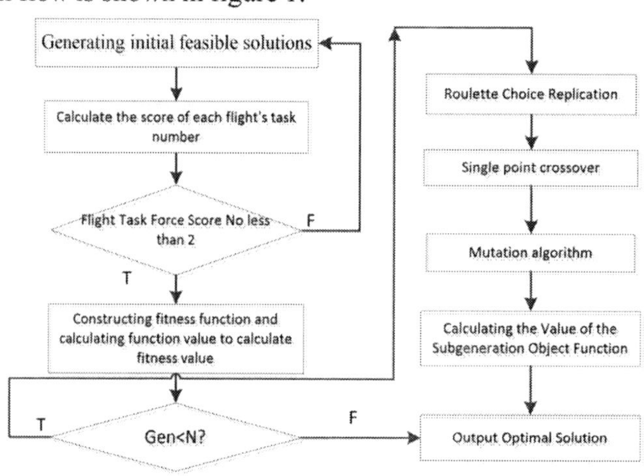

Figure 1. Algorithm procedure.

4. Examples verification

In this paper, the data of auspicious airlines to develop flight plans and corresponding flight tasks, the existing personnel, aircraft, flight environment, management departments are 10. This paper tries to search for a better control scheme by genetic algorithm.

The default chromosome 1 is the initial feasible solution, which is calculated from the initial feasible solution.

$$F(C) = \sum_{n=1}^{10} f(C_n) = 23.323 \tag{10}$$

The concrete calculation results are shown in table 2:

Table 2. Primary chromosome scoring.

Serial number	1	2	3	4	5
Systematic	2.901	2.683	3.983	1.981	2.947
Serial number	6	7	8	9	10
Systematic	3.401	1.713	1.819	3.926	3.482

According to the data in the table, the operational risk score of the combination of serial number 4, 7 and 8 is less than 2. If these three combinations want to fly, they will not only cause great potential safety hazards, but also cause considerable losses to the interests of airlines.

In the algorithm, the number of individuals is set to 100, the maximum genetic algebra is 200, the chromosome length is 4, the generation gap is 0.8, the crossover probability is 0.7, and the mutation probability is 0.001.

By calculating, the model is solved as a new flight task group, i. e. the combination of human, aircraft, loop and management after adjustment. The trend of optimal individual fitness is as follows in: figure 2 and figure 3.

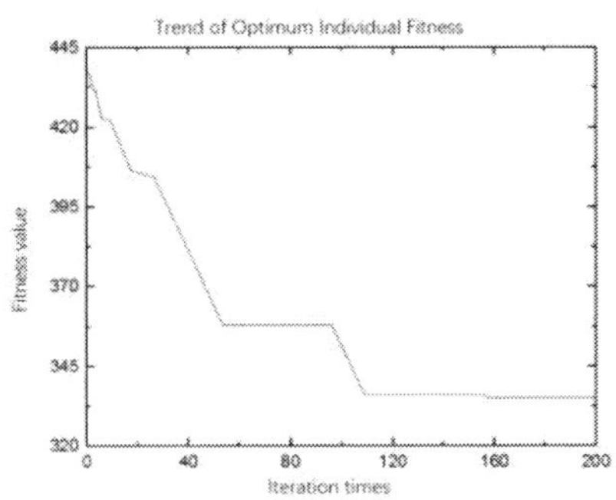

Figure 2. The solution is related to the population.

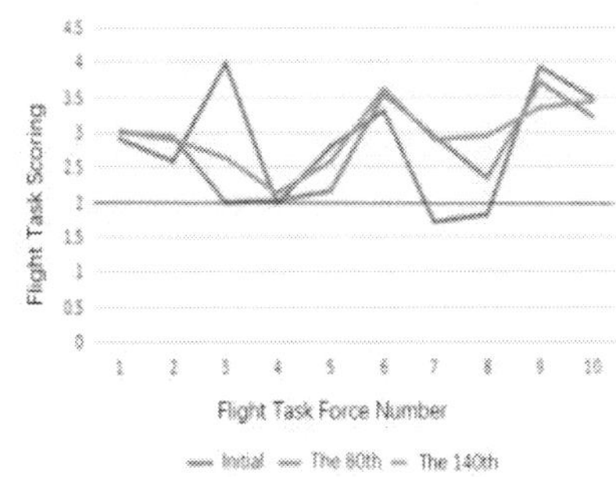

Figure 3. Compare flight mission ratings.

As shown in figure 2, about the 50th generation of solutions began to converge, and now the 80th and 140th generation solutions are used to calculate the new flight task scores by using the fuzzy comprehensive evaluation in the previous section. The new flight task scores are drawn together with the initial solutions, as shown in figure 3.

By comparison, only task 3 did not reach score 2 in the 80th Generation Solution, and the rest exceeded score 2. The overall trend tended to be stable, but there was no obvious advantage. It only increased the number of flights that exceeded score 2, and the effect was not satisfactory.

In the 140th generation solution, all tasks have been fulfilled with more than 2 points, and the airline's benefits have been maximized. According to the trend of the image, the flight task score of the 140th generation solution is not only higher than that of the 80th generation solution and the initial solution, but also more stable and concentrated, and the fluctuation is significantly reduced, which means that the flight task score has been improved and maintained at a relatively safe level. This also proves that the risk of airline operation control is further reduced, and the effect of control strategy is more intuitive. The total score of airline operational risk management and control method can be obtained by further calculation, The comparison of results is given in table 3.

Table 3. Comparison of results.

	Initial solution	The 80th generation	The 140th generation
Fitness value	435.144	365.296	339.536
Risk score	23.323	27.875	29.951

Through the above examples, it can be seen that three flight task groups in the initial feasible solution can not reach 2 points, and one flight task group in the eightieth generation solution can not reach 2 points. These flight tasks pose a great threat to the operation safety of the entire airline. In the optimized results, all flight tasks in the 140th generation have reached 2 points, which achieves the desired results.

5. Conclusion

This paper actively searches for the risk indicators under the subsystem, and then establishes the airline operational risk index system and the airline operational risk evaluation model. The genetic algorithm is used to optimize the airline risk management and control strategy and search for the optimal solution, which provides support for the safe and efficient operation of airlines. However, genetic algorithm has a general problem, the stability of its operation results is poor, and it is difficult to get the optimal solution, which is also the direction of later research to provide more stable and accurate algorithm and conclusions.

References

[1] Y.T. Wang, R. Li, H.H. Wang, et al. Journal of Tianjin University of Technology,**33**(2014)
[2] Y.T. Wang, F. Lu, J. Zhang, et al. Chinese Journal of Safety Sciences, **24**(2014)
[3] Y.T. Wang, J.X. Tang and Y.F. Zhao, Chinese Journal of Safety Sciences, (2017)
[4] N.M. Liu, K. Liu and J. Yang, Journal of Civil Aviation, (2017)
[5] D. Li, Risk Assessment of Airline Delivery (2015)
[6] T. Liu, P. Xiang, Airline Operation Control Risk Management Research(2013)
[7] Z.W. Ji, Y.B. Yang and Y.X. Sun, Journal of China Civil Aviation Flight Academy, **27**(2016)
[8] M. Bineid, J. P. Fielding, Aircraft Engineering and Aerospace Technology, **75**(2003)
[9] M. Janic, Journal of Air Transport Management, **6**(2000)
[10] R.E. Braithwaite, Caves, J.P.E. Faulkner, Journal of Air Transport Management, **4**(1998)

Why a run-to-fail strategy is not good practice for SIS?

Keyur G. Vora

Engineering Design Head (Control & Instrumentation) - RPMG Reliance Industries Limited, INDIA

*E-mail:keyurvora2363@gmail.com

Abstract. The Maintenance & Reliability engineers understand pain area that the portion before the "flat" part of the "Bathtub" is where a high number of premature failures can occur: commonly referred to as "infant mortality". One of the fundamentals of the Probability of Failure on Demand (PFD_{avg}) calculations in SIL verification analysis, is the concept of a constant failure-rate during useful life of the equipment. Probabilistic calculations assume that the failure rate of the devices used in a Safety Instrumented System (SIS) remains constant during the "flat" (constant failure rate) portion of the "bathtub" curve. Most manufacturers will perform stringent testing to weed-out, weaker units that could fail prematurely, leading to unwanted warranty claims. When equipment reaches to flat portion of it's lifetime called as "wear-out" phase then the failures start to rise dramatically & then the integrity of the Safety Instrumented Functions (SIF) is compromised.During the Operation and Maintenance phase of the SIS, periodic performance assessments need to be made in order to ensure the as-designed, installed and commission SIS and its SIFs, still meet the performance requirements of the Safety Requirement Specification (SRS) [1]. Using a run-to-fail strategy will not enable the SIFs to meet the performance requirements once wear-out occurs. This paper will explain why a run-to-fail strategy is not good practice for a SIS, which will lead to the de-graded performance of the SIFs. This paper addresses both the design and operational impacts of a run-to-fail strategy.

1. Introduction

Over the past decade or so, automation has been one of the dominant factors in enabling end users in the process, chemical and petro chemical industries to be able to streamline their costs and improve efficiency; often at the expense of personnel. Most modern plants today have less man power than plants of 1980 and even the 1990s. This means that the burden of running and maintaining a modern plant has fallen on fewer and fewer plant personnel. Coupled with the shortage of skilled employees, this places a significant burden on plant personnel to maintain and improve their skill set in order to maintain the technologically more complex instrumentation and automation systems. Aside from the Basic Process Control System (BPCS), there's the plant safety system (referred to as the Safety Instrumented System (SIS).

The advent of the IEC 61511 [2] & ANSI/ISA 84.91.01 [3] standards for the process industries has provided a path to improvements in safety by introducing the concept of a Safety Lifecycle (SLC) and moving away from a strictly prescriptive methodology to a more performance based methodology, with the emphasis being placed on reducing risk and mitigating the potential for hazards that could lead to the loss of life, destruction of property and plant assets. The purpose of this paper is not to define the application of the standard but to examine one of the most common errors made by operations and maintenance: having a run-to-fail policy for the SIS equipment. The tools such as ProSET, as exSILentia [4] streamlines the Process Safety Management work process and the Safety Instrumented System Functional Safety Lifecycle.

2. The nature of field equipment

Reliability Engineers know that most devices, either commercial or consumer-based, follow what is commonly known as the "Bathtub Curve". The figure 1 below illustrates an example of the bathtub curve that demonstrates the typical behaviour of control system operating time interval (in hrs) over life time.

System	Op.Hours	λ
12	33	0.001045
1	96	0.000567
14	196	0.00037
13	240	0.000874
30	409	0.000237
6	614	0.000203
11	831	0.0002
15	1045	0.000212
16	1282	0.000201
10	1540	0.000194
7	1815	0.000191
19	2106	0.000191
25	2414	0.000191
21	2740	0.000192
2	3091	0.00019
24	3471	0.000188
27	3886	0.000185
26	4348	0.00018
3	4862	0.000177
20	5431	0.000176
23	6056	0.000178
9	7499	8.66E-05
5	8339	0.00017
28	9270	0.000179
8	10305	0.000193
22	11460	0.000216
18	12751	0.000258
29	13351	0.000833
4	13853	0.001992
17	13990	Inf

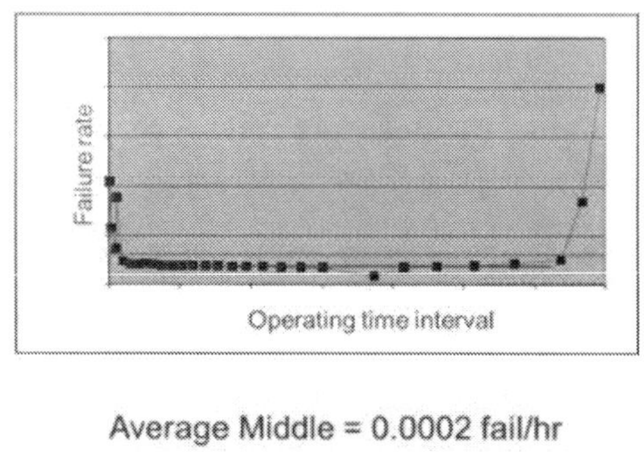

Figure 1. Example of Failures Over Time of 30 Units.

The blue line indicates the average middle value of failures and it can be seen that this "flat" portion of the curve has a constant failure rate over the average. The blue line indicates the average middle value of failures and it can be seen that this "flat" portion of the curve has a constant failure rate over the average.

2.1. *The importance of the bathtub curve*

Most manufacturers recognise the problems with premature failures and so will devise stress tests to expose the weaker units, causing them to fail. Examples of such tests are: Temperature cycling, Shock and vibration, Humidity, EMI/RFI etc. This will avoid weaker units getting out into circulation and presenting warranty failures for the manufacturers. The Bathtub Curve is important because it identifies two key areas:

- Infant Mortality
- Wear Out

Infant Mortality refers to the first part of the curve where a high level of failures can occur prematurely. Conversely, at the back-end of the Bathtub we can see a sharp rise in failures. This portion of the curve is known as "Wear Out". At some point there is a sudden increase in failures as the normal operational stresses overcome the strength of the device, which then leads to failure. This portion is known as wear out and is very important for a number of reasons. After "useful life", equipment/component must be checked, repaired or replaced or refurbished.

2.2. *Probabilistic calculations: come basic details*

For SIS, the primary concern is a failure on demand of any of its associated SIFs. A SIF has a defined Safety Integrity Level (SIL), which defines how much Risk Reduction the SIF will provide for a given hazard. For the application of layers of protection analysis (LOPA) technique [5], the semi quantitative method is applied for the events involving significant risk where analysis of barrier effectiveness & gaps is critical for risk management. The Safety Lifecycle is specifically geared towards ensuring that the SIS

IOP Conf. Series: Materials Science and Engineering **715** (2020) 012054 doi:10.1088/1757-899X/715/1/012054

and its SIFs will be available to perform its intended function and to provide the given level of Risk Reduction required. If a SIF fails, then this could lead to an undesirable outcome (an Accident).

2.2.1. Probability of an accident

If a hazard occurs and the automatic protection function operates correctly, no accident occurs. However, if a hazard occurs and the automatic protection function fails to operate, an accident occur. So, obviously, it is important that we know the probability of failure for an automatic protection function.

$$P(\text{Hazard}) * \text{PFD of SIF} = P(\text{Accident}) \quad (1)$$

2.2.2. How a SIF achieves its SIL

From the detailed design of the SIS and its SIFs, a SIL verification calculation is performed to ensure the SIF will meet its target SIL. For process industry applications, SIFs normally operate in Low Demand Mode and there are three design barriers that need to be met before a SIF will achieve its target SIL:

(a)PFD_{avg} – a measure of the defence against random hardware failures.

(b)Architecture Constraints – imposes minimum hardware redundancy requirements

(c)Systematic Capability - to show sufficient defense against systematic failures within the equipment chosen for the safety application.

If any one of the three barriers are not met, then the SIF will fail to meet its target SIL. IEC 61511 [2] requires that equipment selection be justified. Two alternative means are allowed, IEC 61508 compliance and Prior Use. The first requirement is to therefore meet the target PFD_{avg} for the SIF, which includes all the equipment: Therefore, in order, to calculate the overall PFDavg, we must first calculate the individual PFDavg for each piece of equipment for the SIF and sum them together:

$$PFD_{avg}\,(\text{SIF}) = PFD_{avg}\text{ Sensors} + PFD_{avg}\text{ of Logic Solver} + PFD_{avg}\text{ Final Element} \quad (2)$$

We must also consider any interfaces and/or signal conditioning as well, since the failure rates of these components will contribute to the overall PFD_{avg} result.Taking the simplified formula,

$$PFD_{avg} = \underline{\lambda d TI} \quad (3)$$

where: λd = Dangerous Failure Rate, TI = Proof Test Interval.

This formula assumes perfect Proof Testing (i.e. where any dangerous undetected faults are revealed and fixed before the SIF Demand occurs). The PFDavg of the SIF will deteriorate over time due to imperfect proof testing. If the equipment is maintained on a run-to-fail basis, then once the Useful Life has been exceeded the PFDavg is no longer valid because the premise of Constant Failure rate is not applicable. Beyond Useful Life the SIF will move into the "Wear Out" phase of the Bathtub. Understanding this and building in the Useful Life into a plant's mechanical integrity program is key to ensuring the ongoing integrity of the SIS and its SIFs.

3. SIL validation/operation & maintenance/modification/de-commissioning

After SIS installation, SIS validation is done to validate that the installed and mechanically completed, SIS and its associated safety instrumented functions, meets the requirements as stated in the Safety Requirements Specification. The objective of SIS Operation & Maintenance is to describe how the SIS shall be operated and maintained to ensure that it functions in accordance with the Safety Requirements Specification, throughout the SIS operational lifetime. All activities concerning operation of the SIS shall be performed by competent personnel. Operators shall have the proper competence and training on the function and operation of the SIS and should undergo regular assessments, as part of the ongoing competency requirements.

A maintenance program shall be available, which includes descriptions for maintaining and testing the SIS, to maintain the required integrity level. The maintenance routines should also describe compensatory measures required (if any) for the testing to reveal faults that are not automatically detected by the SIS diagnostics. In general, it is not allowed to operate with impaired barriers. Necessary actions to correct or compensate the impaired barrier(s) shall be taken upon overrides & failures and the repair and restore times should be in accordance with the SRS [1]. It is important to record the repair and restore times since this should be a key metric for measuring performance. In order to ensure that the SIS is performing in accordance with the design intent and hence the required integrity level, it is necessary to record non-conformance between the expected behaviour and actual behaviour of the SIS. This will also include ensuring replacement of devices that have reached the end of useful life. This shall be analysed and where necessary, a safety analysis performed and corrective action taken, via management of change procedures and updates to the SRS [1], if required.

It is important that the person responsible for the SIS is able to easily extract functional test documentation and installation trip and shutdown reports. The carrying out of audits and statistical analysis on these data are essential to ensure that the SIS is performing and being maintained as intended, and to ensure that the installation is being operated at an acceptable risk level. The assurance that planned testing is carried out on time and as specified, and that any backlogs are investigated and corrective actions taken, is vital for ensuring the performance of the SIS.

While doing SIS modification, it shall be ensured that modifications to any safety instrumented system are properly planned; reviewed and approved prior to making the change & to ensure that the required safety integrity of the SIS is maintained due to any changes made to the SIS. Personnel affected by the change shall be informed and trained prior to implementation of the change or start-up of the process, as appropriate.

SIS de-commissioning is to ensure that prior to decommissioning any SIS from active service, a proper review is conducted and required authorisation is obtained. It is to ensure that the safety instrumented functions remain appropriate during decommissioning activities. Management of change procedures shall be implemented for all decommissioning activities. The impact of SIS decommissioning on adjacent operating units and facilities or other field services shall be evaluated prior to decommissioning.

4. Challenges & issues after SIF installation in the field

The Final Commissioning and Start-up Phase starts when systems or parts of systems are mechanically completed (pre-commissioning), and is concluded when all systems are handed over to operations and finally accepted by the customer. The Operational Phase starts when the installation is handed over and accepted by operations. The De-Commissioning Phase starts with the decision to shut down the field and remove the installation. Even though, the IEC standards & guidelines clearly specify how to address work after mechanical installation of SIS, the following are the major issues, which can cause reliability concerns, safety issues & non-compliance to standards & codes after mechanical completion of SIF and handing over to operations.

a) Incomplete/Non-availability of Safety Requirement Specification (SRS) [1], resulting in non-availability of critical parameters including process safety response time.
b) Time availability after mechanical completion & before plant start-up/commissioning.
c) Non-availability of certified personnel to carryout SIL validation and subsequent phases
d) Run-to-fail of SIS equipment
e) Risk & fear of operating staff to carryout live SIF loop testing
f) Shutdown availability to carryout SIF test.
g) Implementation of Modification in SIS after mechanical completion & SIL validation
h) Handling of Software upgrades

Above issues are to be addressed during planning stage of SIL study. The SRS [1] document is very important and shall contain three main types of requirements:

(1)Functional requirements and response times of SIF
(2)Integrity requirements like PFDavg and SIL
(3)Operating prerequisites and constraints

Time availability of SIF mechanical installation shall be kept to carryout SIL validation activities by a competent, independent, authority/auditor team who was not involved in this activity. It is a good idea to carryout SIL validation activities immediately after mechanical completion and before plant start-up, during the commissioning stage. Proper format for SIF proof testing shall be developed that

IOP Conf. Series: Materials Science and Engineering **715** (2020) 012054 doi:10.1088/1757-899X/715/1/012054

captures the SIL validation requirement. Later, it will be challenging to carryout SIL validation, as the plant is already started. The operation team will perceive this activity as a risk after plant start-up as any operational upset or trip of the plant/unit can result in production / profit loss & safety / reliability issues. Good planning & adherence to standards/codes, documentation/procedures will eliminate other issues as described above from (c) to (h).

5. Summary

The paper has highlighted the importance of understanding Useful Life, as demonstrated by the bath tub curve for SIF components, from which it can be clearly seen that the PFDavg of the SIF will deteriorate over time due to imperfect proof testing. If the equipment is maintained on a "run-to-fail" basis, then once the Useful Life has been exceeded the PFDavg is no longer valid.

Beyond Useful Life the SIF will move into the "Wear Out" phase of the Bathtub. Understanding this and building in the Useful Life into a plant's mechanical integrity program is key to ensuring the ongoing integrity of the SIS and its SIFs. Replacement of devices within the SIF that have reached the end of its useful life needs to be built in to any mechanical integrity program. As inconvenient as this may seem, this paper has illustrated that SIL integrity is compromised once the end of useful life of the SIF equipment has been reached.

Therefore, it is important to carryout good planning & execution of SIS maintenance at every phase of the project. The non-compliance to any phase or Run to Fail strategy may result in unplanned failures, Safety & Reliability issues, company's downtime and profit loss, safety and/or statutory/regulatory issues.

References

[1] Safety Requirement Specification (SRS) guidelines of IEC 61511 Standard for a Safety Instrumented System (SIS) that incorporates all the analysis done during the Risk Assessment, HAZOP/PHA and LOPA reviews. Visit https://www.sp.se/sv/index/services/functionalsafety/Documents/Safety%20requirements%20specification%20guideline.pdf

[2] [IEC 61511-1] International Electro technical Commission (IEC) 61511-1 Functional Safety – Safety instrumented systems for the process industry sector – Part 1: Framework, definitions, system, hardware and software requirements:2016 edition

[3] [ANSI/ISA 84.91.01-2012] American National Standards Institute ANSI/ISA 84.91.01-2012 – Identification and Mechanical Integrity of Safety Controls, Alarms and Interlocks in the Process Industry

[4] ProSET & exSILentia software tools

[5] "Layer of Protection Analysis (LOPA)– Simplified Process Risk Assessment", CCPS 2001

Design of non-contact fuel consumption real-time monitoring system for BDS positioning special vehicle

Pengwei Zhang[1,]* Chengliang Guo[1], Yongjie Wu[1], Wenjun Wang[1], Bingyang He[1,2] and Yuanfa Zhang[3]

[1]No.32139 Troops of PLA, Beijing, China
[2]Army Academy of Armored Forces, Beijing, China
[3]Naval Research Academy, Beijing, China

*E-mail: zhpw1119@gmail.com

Abstract. The fuel consumption measure is inaccurate and no real-time with the three-scale mete in the special vehicles, so the real-time fuel measurement system is developed with the technologies of ultrasonic sensing, single-chip microcontroller, BDS positioning and wireless communication. An ultrasonic oil level sensor is stuck on the bottom of each tank to detect the remaining oil of tanks accurately. It can avoid destroying the existing equipment structure. BDS positioning module is applied to monitor the location of vehicle. This technology is China complete autonomy. The special wireless transmission module and special network node are set up for transmitting the data of fuel consumption and position. It can guarantee the transmission secure and confidential. And a fuel consumption detection algorithm is studied with BDS measuring distance. The prototype is tested in four different road conditions. The error rate of oil detection on a highway with good conditions is 3.92%. And the measured fuel consumption value is close to the manufacturer's nominal value. The analysis of experimental data shows that the design of the detection system is reasonable and the fuel consumption detection algorithm is correct and reliable, and real-time fuel detection and positioning for the special vehicles can be realized.

1. Introduction

In order to carry the fuel as much as possible and avoid excessive fuel concentration in the special vehicles, the multiple fuel tanks are arranged in a corresponding space of the vehicle chassis. The adjacent fuel tanks are connected to each other by oil pipes, and all the tanks are divided into left groups and right group. Since the left and right group of fuel tanks are on the same level, a capacitance oil level sensor is fixed in the rear fuel tank of each group to detect the remaining total oil volume of the tanks. The result of detection is displayed in the meter dial for the driver[1,2]. But this meter dial can only display the proportion of remaining oil in the total tank volume, so there is a lack of precision. In the other way, the component of capacitance oil level sensor is easily jammed in the tank, which may cause the driver wrong judgment and become a common hidden danger of safe driving. At the same time, when the vehicle passes through the precipitous road, the float of capacitance oil level sensor will vibrate with the vehicle vibrating. This must lead to an inaccurate oil detection result [3]. Especially for military special vehicles with many irregular fuel tanks, the original oil detection device can lead to greater errors with the inclination of road and the bump of running. This will affect the work efficiency gravely. In addition, the accuracy of the action plan will be seriously restrict because the headquarter can not grasp accurately the position, remaining fuel capacity and endurance of each special vehicle in real time.

In this paper, the fuel consumption real-time monitoring system is designed with the technologies of ultrasonic sensing, single-chip microcontroller, BDS positioning and wireless communication without damaging the existing equipment[4-6]. It can help the driver to sense the vehicle oil situation, and help the headquarter to remotely monitor the vehicle running state in real time, accurately and comprehensively.

2. General designing plan of the system

The non-contact real-time fuel consumption detection and location system for special vehicles is mainly composed of two parts. As shown in figure 1 and 2, One is the vehicle terminal for measuring fuel and positioning vehicle, the other is the remote monitoring center. The ultrasonic oil level sensor [7] is stuck on the bottom of the fuel tank to detect the remaining oil level of tanks accurately. The detection data of multiple fuel tanks and the positioning data of Beidou are calculated and controlled by the single-chip microcomputer. The results are displayed on the LCD panel installed in the cab in real time. At the same time, the data transmission is stored in the cloud server with wireless communication technology (such as GPRS) [8]. The number of command or dispatch center can monitor the running status of vehicles in real time, comprehensively and accurately with the PC, tablet and mobile phone client.

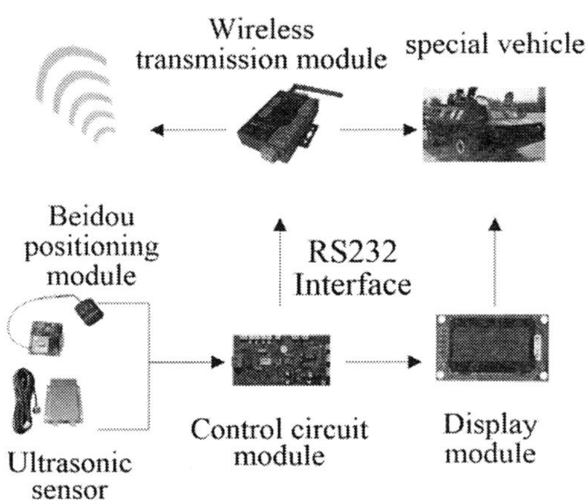

Figure 1. Schematic diagram of vehicle terminal oil-measuring and positioning device.

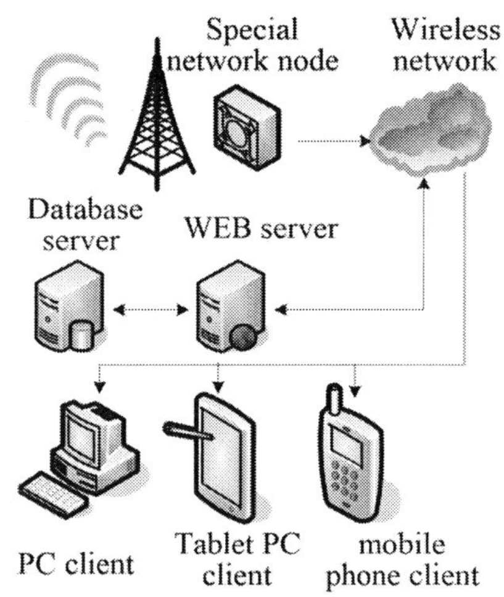

Figure 2. Schematic diagram of Long-distance monitoring center.

3. Design of vehicle terminal oil-measuring and positioning device

The working process of the oil consumption detection and location device is as follows. Several ultrasonic oil level sensors detect the remaining oil data of each tank separately. At the same time, Beidou positioning module collects the vehicle position information data. The data is processed by the single-chip microcomputer in the control circuit module. The result are displayed on the LCD display in real time and sent to the network server by the wireless transmission module and the special network node.

3.1. Hardware design

The hardware includes ultrasonic oil level sensor, Beidou positioning module, display module, wireless transmission and control module. The ultrasonic oil level sensor is stuck on the bottom of the outer surface of the vehicle fuel tank. The other modules are integrated in a small box and installed in the workbench area of the vehicle cab. The control circuit design is shown in figure 3.

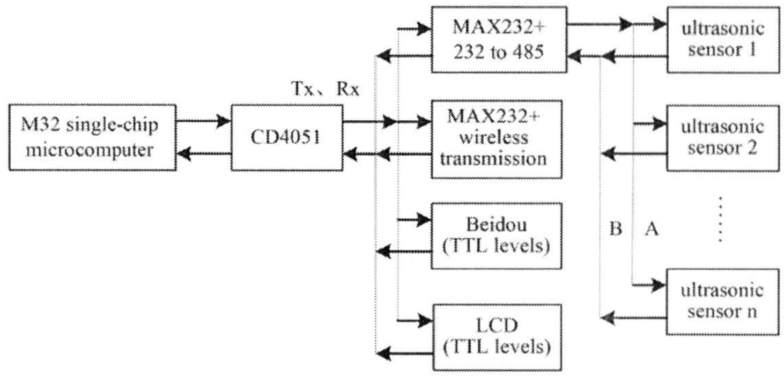

Figure 3. Structure diagram of the hardware design.

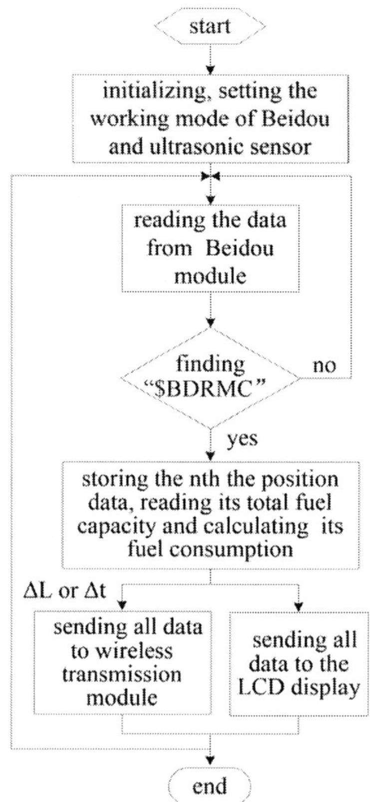

Figure 4. Flow chart of fuel consumption detection algorithm.

ATmega32 MCU controls each device or module such as ultrasonic oil level sensor, wireless transmission, Beidou positioning and LCD display by the CD4052 chip which is a double 4-channel analog multiplexer/demultiplexer. And it communicates with one of the devices or modules separately. First, the TXD of ATmega32 MCU is extended to TXD1, TXD2, TXD3 and TXD4 by CD4052. Then, two groups of them, such as, TXD1, RXD1 and TXD2, RXD2 are converted TTL/CMOS input levels into RS232 levels by the MAX232, and can accept ±12V inputs. The conversion circuit convert RS232 into RS485 on the port of TXD1 and RXD1, and ATmega32 MCU can communicate with many of ultrasonic sensors. At the same time, TXD2 and RXD2 communicate directly with wireless transmission module. On the other hand, TXD3 and RXD3, TXD4 and RXD4 communicate with Beidou positioning module and LCD display respectively.

3.2. Software part design

The flow of software design is shown in figure 4. Firstly, the MCU and other modules are initialized, and the working mode of Beidou and ultrasonic sensor is set. Then it enters the circular work program. The data from the Beidou module are read and judged. If the value of the string "$BDRMC" from the received signal is true, that is to say, if $BDRMC = 1, the position data are temporarily stored, and the program continues to execute after the string is cleared. Otherwise the program returns to read the data from Beidou module again. Thirdly, the data from the ultrasonic sensor are read and processed, and the data of the total fuel and fuel consumption are gotten. Fourthly, the position information, fuel quantity and fuel consumption data are sent to the LCD display in real time. At the same time, the data is sent to wireless transmission module. The frequency of sending data is set to the vehicle driving distance ΔL (eg 2 kilometers, as set) or the time interval Δt (eg, 10 minutes, as appropriate). The data are saved in the remote server and used for the client at any time. Finally, the return loop continues to receive and process data or exit the program.

4. Design of remote monitoring center

The remote monitoring center is used to receive information transmitted by the wireless data transmission network, and the information is processed and transmitted to the user, including the database server, the WEB server, and the user client. The database server and the WEB server access the wireless data transmission network through a dedicated line, and receive the transmitted data of fuel capacity and the vehicle positioning. And these data are processed and outputted to the user client. User clients include tablet computers, mobile phones and desktop computers. The self-developed monitoring software is installed to realize remotely management in the user client.

5. Experiments and results analysis

Based on the above design idea, a prototype of fuel consumption real-time monitoring system for BDS positioning special vehicle is developed, which is shown in figure 5. It is installed on a common special vehicles, and the experimental data are collected in executing the task. The nominal fuel consumption of special vehicles is 18 liters per 100 kilometers from the manufacturer in the experiment. It runs for 2 hours under the different road conditions around Pinggu District of Beijing. The remaining fuel of the tank is collected every 10 minutes. The number of actual fuel consumption is showed in table 1.

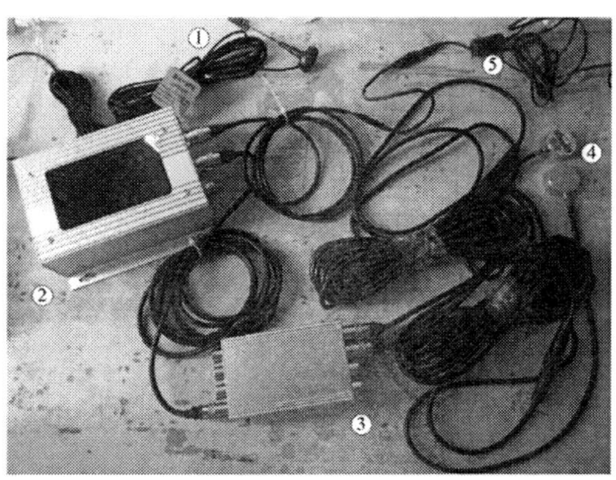

1 Antenna of Beidou, 2 control circuit and LCD display, 3 adapter of ultrasonic sensor, 4 ultrasonic sensor, 5 input wiring of power

Figure 5. Real-time fuel measurement prototype with BDS positioning.

In order to verify the reliability of the prototype detection, the measured average fuel consumption is multiplied by the actual driving mileage in four different road conditions. The calculated fuel consumption is compared with the actual fuel consumption, and the error rate is obtained. These data are shown in table 2.

Table 1. Fuel consumption measured value in different road conditions(Litres / 100 kilometres).

time(minute)	highway	common road	urban road	country road
10	19.2	21.2	22.1	23.7
20	18.6	20.3	23.4	24.5
30	18.8	19.8	24.1	25.0
40	17.6	20.5	22.5	24.8
50	19.2	22.1	23.3	23.9
60	18.3	20.7	24.7	25.3
70	18.2	19.9	22.3	24.9
80	17.7	19.5	23.6	25.2
90	17.8	20.2	24.2	23.8
100	18.4	20.6	23.7	24.6
110	19.0	21.5	24.3	26.1
120	19.5	19.5	23.6	25.3
average fuel consumption	18.53	20.48	23.48	24.76

Table 2. Error rate of measuring fuel consumption relative to actual value.

road conditions	mileage (km)	calculated fuel consumption $(10^{-2} \cdot km^{-1})$	actual fuel consumption $(10^{-2} \cdot km^{-1})$	error rate (%)
highway	136.4	93.7	74.3	87.6
common road	25.30	19.19	17.45	21.69
urban road	26.3	21.2	20.2	24.9
country road	-3.8	-9.48	-13.61	-12.89

Analysis of the experimental data shows that:

(1) On the highway, the measured fuel consumption data is relatively stable due to better road conditions, and the fuel consumption value is closest to the manufacturer's standard value. On the other hand, the measured fuel consumption is slightly smaller than the actual fuel consumption, and the error rate is the smallest. Analysis reasons: According to the literature [9], the nominal fuel consumption of the manufacturer is usually measured at the same speed on a good road surface. On the highway, running speed is closest to the constant in the whole distance. But it inevitably reduces the speed for the toll stations or a few congested roads. While multiple brakes and speeds increase the fuel consumption. Therefore, it is consistent with the theoretical basis that the measured fuel consumption is close to and greater than the nominal value of the manufacturer, and the calculated fuel consumption is less than the actual consumption of oil.

(2) On the common road and urban road, although the pavement texture is better, the vehicles run slowly for the narrow pavement and large crowded vehicles. Especially on urban roads, the frequent brake-starting operation increases fuel consumption for signal lights. Therefore, the measured average fuel consumption is greater than the manufacturer's nominal value, and the measured fuel consumption is also less than the actual fuel consumption, and the error is relatively greater.

(3) On country road, the texture of surface is poor, and the resistance from pavement is large. That results a great difference between the measured average fuel consumption data and the nominal value of the manufacturer. However, the people and cars in rural roads are few, which avoid the loss of fuel from the frequent brake-starting operation. Therefore, although the measured fuel consumption is also less than the actual fuel consumption, the error rate is lower than on the common road and urban roads.

6. Conclusion

In this paper, a prototype of real-time fuel consumption monitoring and positioning system for special vehicles is designed with the technology of ultrasonic sensor, single-chip microcontroller, Beidou positioning and wireless communication. For the structural characteristics and confidentiality requirements of special vehicles, the fuel monitoring system is studied with ultrasonic sensor to detect the remaining oil of tanks, Beidou module to locate vehicles and tactical or GPRS internet to transmit data. The design principle of hardware and software is discussed. And the experimental result of the prototype is analysed. All of these works underpinned the rationality and reliability of the system design. It can provide a novel idea for the scientific management and precision guarantees of vehicle.

References

[1] C. F. Yan, L. T. Lin, P.J. Wu, C.C. Huang. VTCA 2018, 128, 343-351 (2018).
[2] B. Mashadi, Y. Amiri-rad, A. Afkar, M. Mahmoodi-kaleybar. Journal of Central South University. 21, 1058–1066(2014).
[3] L. Y. Fan, Q. Dong, C. Chen, et al. Journal of Mechanical Science & Technology, 28, 4319-4330(2014).
[4] F. W. Meng, D.K. Yang, Y. Q. Wang, Y. X. Zhang. IISA 2017. 686(268-273).
[5] J. F. Zeng, J. Su. Computer Applications and Software, 34, 324-328(2017).
[6] G. L. Ma, P.W. Zhang, Z. H. Huang, W. J. Wang. Automation & Instrumentation. 33, 68-71(2018).
[7] Jenny Terzic, Edin Terzic, Romesh Nagarajah, et al. Ultrasonic Fluid Quantity Measurement in Dynamic Vehicular Applications. 11-35(2013).
[8] Chhabi Rani PanigrahiArun, K. Pujari, Sudip Misra, et al. Proceedings of ICACIE 2017, 2, 433-439 (2018).
[9] N. Lan, J. Long, G. P. Deng. Highways & Automotive Applications. 9-12 (2018).

Calibration method of curvature distortion in step height measurement by atomic force microscopy

Wei Li*, Xixi Yu, Sitian Gao, Yushu Shi, Shi Li, Lu Huang and Shu Zhang

National Institute of Metrology, Beijing, China

*E-mail: liwei@n im.ac.cn

Abstract. Atomic force microscopy (AFM) is widely used to characterize the surface topography in nanomaterials and biology research since the high resolution in 3 dimensions. Height measurement is important in the investigation of graphene thickness. Due to some AFMs use tube scanner, the curved scanning distortion will cause height measurement error. Step height standards calibrated by metrological AFM are used to calibrate the Z axis of conventional AFMs. However, for step structure with large width, the method specified in ISO 5436 is not suitable. The distortion of the curvature scanning is discussed for different samples, and for large step width the influence is significant. So the substrate correction method is proposed to eliminate the scanning curvature and the comparison with several other fitting correction methods is also discussed.

1. Introduction

Due to the 3-dimension characterization capability with high resolution, atomic force microscopes (AFMs) are commonly used to image the material and biological surface topography, and also some mechanical and electrical properties with special tips and modules[1,2]. The height measurement is an importance parameter for AFM in the measurement of the thickness of nano material like graphene [3,4]. The piezoelectric scanner in AFMs needs to be calibrated for the hysteresis for qualitative measurement of the height of nano dimensional structures. The calibration artefacts standards are traced to SI unit by metrological AFMs which can directly trace the measurement value to the laser wavelength by three interferometers in X, Y and Z axis of the scanner. Some AFM utilizing tube scanners has bow scanning distortion. This effect can be eliminated by AFMs with special scan heads or flexture stage scanning AFM [5,6]. However, the bow distortion in tube scanning AFM has to be corrected by fitting methods. The scanner calibration methods for height measurement have been studied, however the methods to correct the bow distortion has not been studied thoroughly. In this paper the influence of the bow distortion of tube scanner on step height is studied and some step height analysis methods are discussed. A substrate correction method is introduced to accurately measure the step height by correction the bow error.

2. Methods

2.1. Experimental materials

Step height standards artefacts with different height are measured with AFM (Dimension Icon, Veeco). The AFM scans the sample with a tube scanning head. The Z axis of AFM scanner is calibrated with standards. The artefacts are Cr coated on quartz substrate. The maximum scanning range of the AFM is 90 μm in lateral direction. A serial of step height standards are measured to investigate the influence of the bow error on the height measurement.

2.2. Step height evaluation methods

The step height is usually defined according to ISO 5436 as shown in figure 1. The lateral length of measured step profile is at least 3 times the width of the step W. The height of the step is expressed as h=(A+B)/2-C, where A, B and C are the average height of the w/3 region of the profile. According to the ISO 5436, the width of the step is limited by the scanning range of AFM when using a standard to calibrate the step height value. For some samples the width is 100μm, larger than the scanning range of common AFMs, then only one sidewall of the step can be scanned in one scanning profile. The definition is not applicable for these samples. For the Si atomic layer height standards with single edge artefacts like terrace structure, different methods are proposed to evaluate the step height [7].

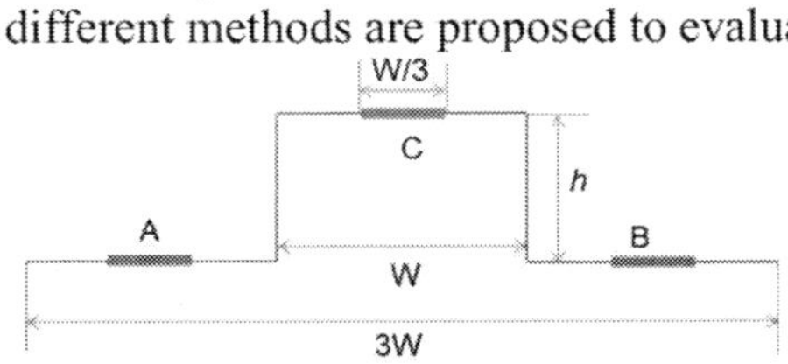

Figure 1. Step height definition according to ISO 5436.

2.2.1. Polynomial fitting method

For tube scanning AFM, the measured height is influenced by the bow error. The fitting methods proposed by Yang et al can be used to corrected bow error [8]. The image was processed line by line. First, a profile curve is obtained by averaging several consecutive lines, which significantly reduces the burr of the curves. Second, the Sobel edge detection algorithm was used to extract the edge of step of the curve, and then the curve is bisected by the edge (or as the complete step divided into three parts). The step part was shifted down successively according to a unit shifting distance to obtain the shifted curve (reasonable number of moving steps). For each move, 6 points around the edge were removed for the edge effect by the tip and the higher-order polynomial was fitted by the least square method. Then the residual of the shifted curve and the fitting curve was calculated, and the minimum value of the residual was taken to obtain the optimal shifted curve and the number of moving steps. Finally, the original curve minus the optimal shifted curve gives the step curve, which removes the bow error. The product of the number of steps and unit shifting distance is the height of the step.

2.2.2. Substrate subtraction method

For the bow error is caused by the instrument and not related to the sample, a substrate subtraction method is introduced here to eliminate the influence of the scanner bow error. The sample is scanned with AFM successively with the step is shifted a certain range laterally to scanning the left and right sidewall respectively. Then the two images of the sample are subtracted to obtain the final step height profile as shown in figure 2. Then the bow distortion can be eliminated by subtracting the two images. This method is direct and does not require complex programs. Here we used the SPIP software. Then the residual profile is the step difference as shown in figure 2. The height can be evaluated with standard methods from the corrected curve.

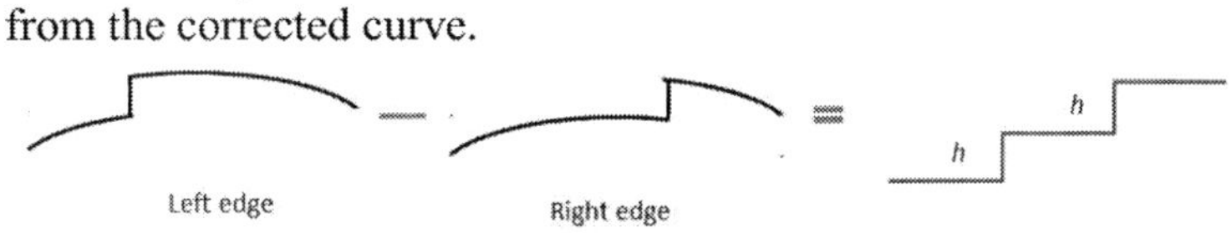

Figure 2. Step height evaluation method for bow distorted image.

3. Results and discussion

The bow error due to the tube scanning in AFM is demonstrated firstly. A 50 nm step height is measured and the 3D topography is shown in figure 3. The distortion can be obviously seen and a section profile of the topography cross the step is also shown in figure 3.

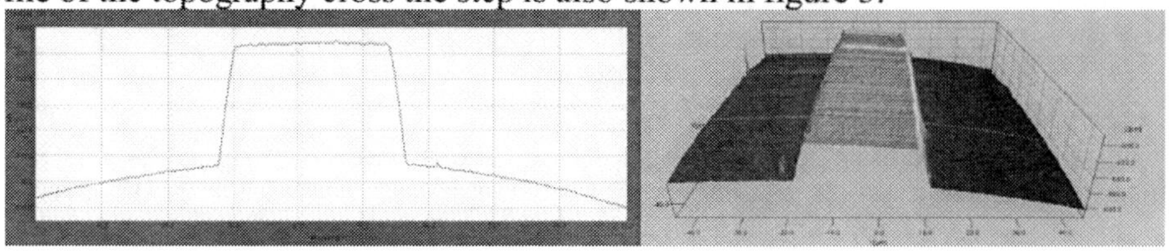

Figure 3. Step height image of a 50 nm height sample.

The standard w/3 evaluation method is applied to the results and the height is 51.8 nm. However, due to the bow error of the scanner, the height is not accurate. The substrate is quadratic and with the polynomial fitting method, the substrate is removed from the raw data. The height is 45.7 nm. By shifting the sample laterally and get another image, the substrate can be eliminated and the height is 46.0 nm, consistent with the polynomial fitting method. 100 nm and 20 nm step height standards are also studied and the results demonstrated that the step heights calculated with the two methods are equivalent as shown in table 1.

Table 1. Comparison of step height measured with different methods.

W/3	Polynomial fitting	Subtracting substrate
23.1 nm	17.7 nm	18.3 nm
51.8 nm	45.7 nm	46.0 nm
91.2 nm	84.5 nm	85.6 nm

The methods are also used for a 44 nm height sample with width larger than the scanning range of the AFM. The scanned result and the fitting process are shown in figure 4 and figure 5.

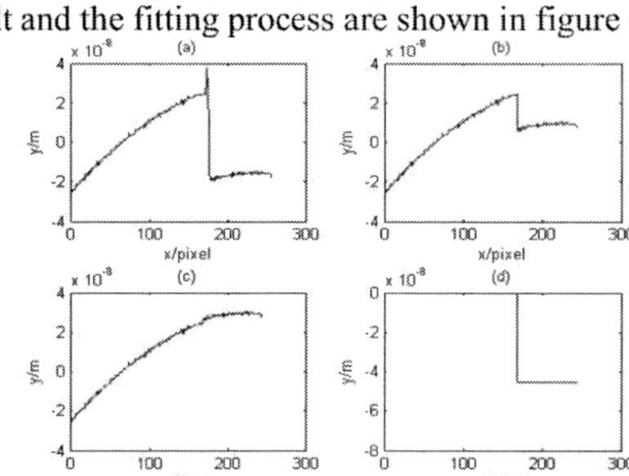

Figure 4. Scheme of shifting fitting correction method. (a) The original scanning profile. (b) The curve shifted by a certain distance. (c) The red curve fitted and the curve shifted by an optimal distance. (d) The adjusted step profile.

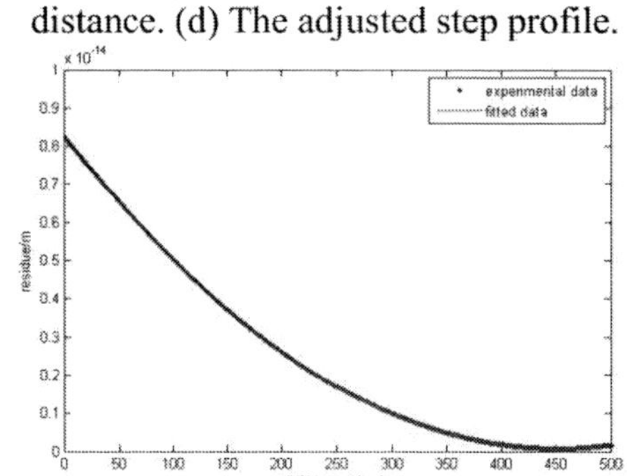

Figure 5. The relationship between the shifting distance and residual error.

IOP Conf. Series: Materials Science and Engineering **715** (2020) 012056 doi:10.1088/1757-899X/715/1/012056

As can be seen from figure 6, as the unit shifting distance increases, the step height fluctuates around the result measured by the minimum unit shifting distance, and the error is getting larger and larger. The error of unit shifting distance less than 0.9 nm will not exceed 0.2 nm, and the error will not exceed 0.025nm when the unit shifting distance is less than 0.1nm. Considering the error and operation time, 0.1 nm is selected as the unit shifting distance. The result is 43.4 nm.

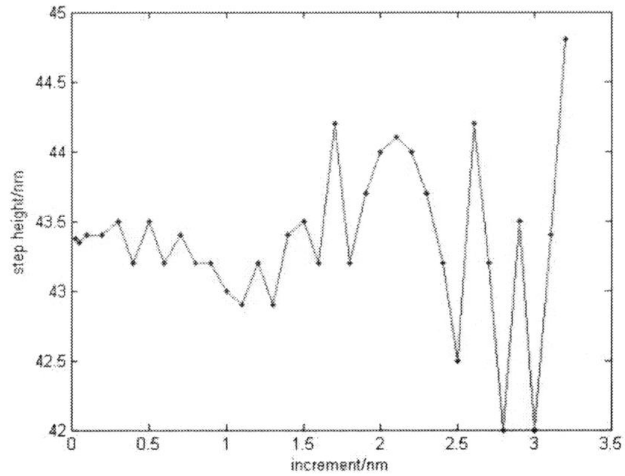

Figure 6. The influence of the unit shifting distance during the polynomial fitting on the step height.

For the step sample wider than the AFM lateral scanning range, the two edges of the step are scanned respectively with AFM. Then the two images are subtracted with each other. The residual curve is shown in figure 7 and then the step is obtained by removing the bow error. Just like the shifting fitting method, the continuous averaging multiple lines, the middle section was shifted down or up according to certain unit shifting distance, and the polynomial of order 1 was fitted to calculate the residual of the shifted curve and the fitted curve. Finally the step profile curve minus the optimal curve gets the fine step. The result is 44.4 nm. The height can also be obtained by fitting the 3 step with linear fitting and then the difference is the height between steps.

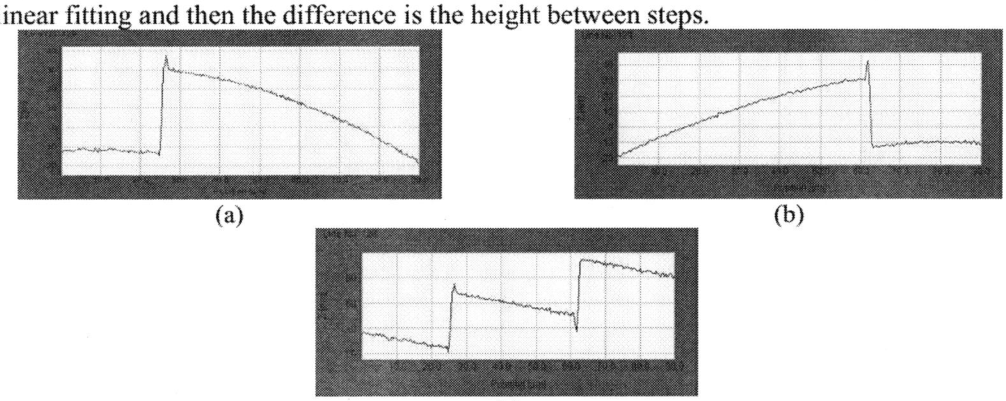

(a) (b)

(c)

Figure 7. Scheme of substrate removal correction method. (a) The original scanning profile of the left edge. (b) The curve of the right edge. (c) The profile of step with the bow error removed.

4. Conclusion

A method to eliminate the bow error of tube scanning AFM is introduced. The method is compared with the common standard W/3 methods and the polynomial fitting method by measuring some step height standards. The step height results of the method are consistent with the fitting method within 1 nm. The method is also applied to sample with step structure wider than the scanning range of AFM. This method provided a convenient way to correct the bow error and calibration method for AFM.

Acknowledgments
Support of the project by National Key R&D Program of China (No.2016YFF0200404) and National Key R&D Program of China (No 2017YFB1104702) are gratefully acknowledged. We thank Dr. Chenying Wang in Xi'an Jiaotong University providing the step samples for calibration.

References
[1] X. Li, J. Sun, X. Zhou, G. Li, P. He, Y. Fang, M. Li, J. Hu, J. Vac. Sci. Technol. B **21**. **1070**, (2003)
[2] Y. Yao, L. Ren, S. Gao, S. Li, J. Mater. Sci. Technol. **33**, 815–820, (2017)
[3] A.K. Geim, K.S. Novoselov, Nat. Mater. 6, 183, (2007)
[4] C. Shearer, A. Slattery, A. Stapleton, J. Shapter and C. Gibson, Nanotechnology, 27, 125704, (2016)
[5] Y. Shi ,W.i Li , S. Gao, M. Lu , X. Hu, Ultramicroscopy 190,77–80, (2018)
[6] J. Kwon, J. Hong, Y. Kim, D. Lee, K. Lee, S. Lee, S Park, Rev. Sci. **74 (10)**, 4378–4383, (2003)
[7] F. Joseph, T. Vincent, K., Rainer, D. Ronald, V. Theodore, Nanotechnology **10** ,428-433, (1999)
[8] S. Yang, C. Li, C. Yang, Z. Jiang, Meas. Sci. Technol. **25**, (2014)

IOP Conf. Series: Materials Science and Engineering **715** (2020) 012057 doi:10.1088/1757-899X/715/1/012057

Complete coverage of perovskite materials over ZnO Nanorods with Multiwalled Carbon Nanotubes (MWCNTs) as Hole Transport Material (HTM)

Albertus Bramantyo[1,*], Nji Raden Poespawati[1], Arief Udhiarto[1], Kenji Murakami[2] and Masayuki Okuya[2]

[1]Department of Electrical engineering, University of Indonesia, Kampus Baru UI Depok, Jawa Barat, Indonesia

[2]Graduate School of Integrated Science and Technology, Shizuoka University 3-5-1 Johoku, Naka-ku Hamamatsu, Japan

*E-mail: bramantyoalbertus@gmail.com

Abstract. In this research, ZnO nanorods (NRs) have been fabricated for perovskite solar cell (PSC) application. An issue about the use of ZnO material for PSC application is lower photovoltaic outputs than TiO_2 material. Such outcome is due to unstable ZnO/perovskite interface. Other factor is due to partial coverage of the perovskite over the ZnO material. By using 1-step method and mixture of DMF and DMSO as the solvent for perovskite solution, complete coverage of the ZnO NR was achieved. Multiwalled carbon nanotubes (MWCNTs) and copper iodide (CuI) were used as hole transport materials (HTMs) for comparison purpose. The experiment and characterization were done in Shizuoka University, Hamamatsu campus, Japan. The ZnO NRs were grown by implementing 2-steps method of seed layer deposition by spin coating and NRs growth through chemical bath deposition. Methylammonium lead iodide ($CH_3NH_3PbI_3$ or $MAPbI_3$) perovskite was grown through 1-step method by mixing equimolar lead iodide (PbI_2) and methylammonium iodide (CH_3NH_3I or MAI) in DMF and DMSO solvents (volume ratio of 7 to 1 for DMF and DMSO, respectively). The current-voltage (I-V) curve for both materials showed higher J_{SC} and V_{OC} for MWCNTs at 5.34 mA/cm^2 and 0.21 V, respectively compared to CuI materials (J_{SC} = 3.40 mA/cm^2 and V_{OC} = 0.14 V). Lower V_{OC} is the result of higher series resistance due to non-passivated ZnO/$MAPbI_3$ interface. Another factor to consider is recombination of electron and hole at the interface. MWCNT material is an alternative material to act as HTM because of high conductivity and lower electrical resistance compared to CuI. Complete coverage of the perovskite over ZnO material helped improving the production of photogenerated electrons.

1. Introduction

Perovskite solar cell (PCS) technology has developed from its first application of $MAPbBr_3$ nanocrystalline in DSSC (dye sensitized solar cell) device by Kojima et al. [1] in 2009 with PCE (power conversion efficiency) of 3.8 %. Recent development has shown that TiO_2 material is the ETM (electron transfer material) that deliver the highest PCE at 22.1 % by Yang et al. [2]. ZnO material is an alternative to TiO_2 material because of its similar Fermi level bandgap with TiO_2 material while possessing higher electron mobility. Despite such advantages over TiO_2 material, the highest PCE of ZnO based PSC is found at 17.3 % by Li et al. [3]. One explanation about the lower PCE for ZnO based PSCs is due to presence of defects in the ZnO/perovskite interface which increases the number of electron-hole recombination.

In order to reduce the electron-hole recombination in ZnO based PSCs, surface passivation can be done by applying a dopant on top of the ZnO ETM. Deposition of TiO_2 material was reported by many groups of researchers to form ZnO-TiO_2 core-shell structure as the ETM for PSC application [4-6]. The advantages of ZnO-TiO_2 structure compared to pure ZnO is larger surface area, higher impedance of charge-transfer resistance R_{CT}, and increased chemical stability which gives increased photovoltaic parameters. A research conducted by Wang et al. explored the possibility to coat ZnO material with SnO_2 to reduce hysteresis effect and stabilize the decomposition of $MAPbI_3$ perovskite[7]. Another material capable of passivating the surface of ZnO for PSC application is aluminium (Al)[8].

In this research, ZnO nanorods (NRs) was developed to provide direct electron pathway for photogenerated electrons. MWCNTs (Multiwalled carbon nanotubes) was utilized as HTM (hole transport material) because of the high conductivity and low resistivity. MWCNTs also provided protection for the perovskite material from water moisture. Although the use of MWCNTs have been reported by many researchers, most of them were applied for TiO_2 based PSCs[9-13]. This research fabricated MWCNTs as an HTM alternative other than Spiro-OMeTAD. Complete coverage of the perovskite over the grown ZnO NRs was achieved.

2. Experimental

2.1. Seed layer deposition

FTO substrates were cleaned ultrasonically through separate use of distilled water, acetone, and ethanol at 5 minutes each. The seed layer solution was made from 10 mL of 0.30 M zinc acetate dihydrate with 2-methanol as solvent. Ethanolamine was dropped 10 times into the seed layer solution. The seed layer solution was heated at 60 °C for 30 minutes. The seed layer was spin coated from the low speed of 500 rpm to high speed of 3000 rpm for 1 minute and then heated at 100 °C for 10 minutes to remove left over solution on top of the substrate. This cycle was repeated until 3 times and finally annealed at 350 °C for 1 hour.

2.2. ZnO NRs growth

The growth solution for the ZnO NRs was formed from mixture of equimolar 0.050 M zinc acetate dihydrate and HMT (hexamethylenetetramine) at 50 mL of distilled water solvent for each solution. The substrate containing spin coated seed were kept at vertical position. Chemical bath deposition method was used at 85 °C from room temperature. When the desired temperature was reached, the growth condition was kept for 3 hours. After the growth process was over, the samples were annealed at 350 °C for 1 hour.

2.3. Perovskite 1-step deposition

The perovskite solution was made from mixture of equimolar 1.2 M of PbI_2 and MAI in 1 mL DMF and DMSO solvents (volume ratio of DMF and DMSO were 7:1). The solution was heated at 100 °C for about 30 minutes. The perovskite solution was spin coated into the ZnO NRs grown substrates at 1000 rpm for 1 minute. During the spin coating, 200 – 300 μL of diethyl ether was dropped in order to remove leftover solution and smoothened the perovskite surface. The perovskite was heated at 100 °C for 30 minutes in low humidity environment inside a glovebox (less than 20 % humidity).

2.4. HTM deposition and PSC fabrication

Doctor blading method was applied to deposit the HTMs. CuI and MWCNT powders were grinded separately to form pastes with chlorobenzene as the solvent. Each HTM pastes were applied onto different substrates with the deposited perovskite. After the paste was dried, platinum plate was put on top of the HTM. The HTM was sandwiched between the perovskite and the Pt plate. Solar simulator was used to characterize the photovoltaic parameters.

2.5. Characterizations
Scanning electron microscopy (SEM) images of the samples were captured using SEM JOEL 7001F. The X-ray diffraction (XRD) chart was measured using type D5000 machine. Finally, The I-V characteristics was carried out using a calibrated solar simulator (JASCO CEP-25 BX solar cell evaluation system) under incident light intensity 1.5 AM, 100 mW cm^{-2}.

3. Results and discussions
Complete coverage of the perovskite over ZnO NRs had been attempted before. However, usage of pure DMF as solvent for the perovskite solution resulted in wires or fibers of $MAPbI_3$ perovskite. Such morphology left some parts of ZnO NRs' surface not covered by the perovskite which contributed to short circuit device. Fast crystallization of the perovskite during the processing of the 1-step perovskite in pure DMF solution gave rise to formation of perovskite crystallites. When the perovskite solution was spin coated, the perovskite crystallites could not cover the whole surface of the ZnO NRs because the mixture of PbI_2 and MAI have already formed the $MAPbI_3$ crystallites before the solution was spin coated. One way to slowed down the crystallization of perovskite during the solution processing was done by mixing DMF and DMSO solvents. In this manner, the perovskite crystallites have not been converted completely and when the spin coating was finished, most if not all of the ZnO NRs' surface was covered by the perovskite crystallites. SEM images for the perovskite using mixed DMF and DMSO solvents is shown in figure 1.

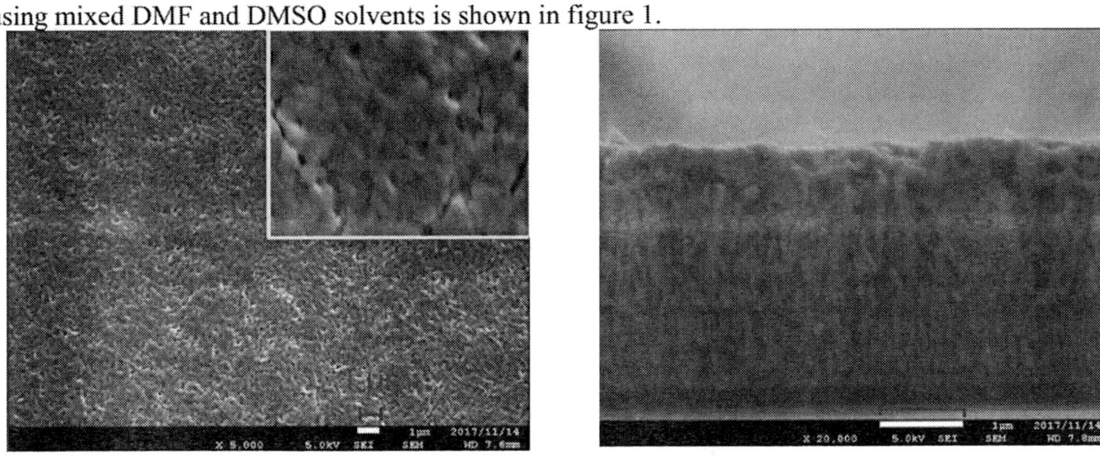

(a) (b)

Figure 1. SEM images (a) upper view, and (b) cross section view of the ZnO/Perovskite interface with mixed DMF and DMSO solvents.

The XRD profile for the perovskite spin coated onto the ZnO NRs is shown in figure 2. There were two dominant $MAPbI_3$ peaks in (110) and (220) directions which corresponded to growth in two-dimensional X and Y axis. Meanwhile, there were two peaks for ZnO NRs in (100) and (002) directions. Growth along the horizontal and vertical directions were represented in the (100) and (002) peaks, respectively. Stronger intensity of the (002) ZnO peak meant that the growth alongside the vertical direction was dominant or the ZnO NRs were vertically aligned.

HTM is an important part in a PSC device. Spiro-OMeTAD is one of the most commonly used HTM due to its ability to generate high efficiency PSCs. However, Spiro-OMeTAD is quite expensive and the need for other alternative HTMs that are more cost friendly while being able to deliver high efficiency. CuI is an alternative HTM that is often used for ZnO based PSCs. MWCNTs was compared with CuI in this research for PSCs application. A Pt plate was pressed on top of the HTM to form the metal contact. An insulator mask was used to prevent the Pt plate from touching the areas that were not covered with the HTM while a hole at about 0.25 cm^2 was provided for calculation of photovoltaic process. Figure 3 shows the PSC configuration that was calculated.

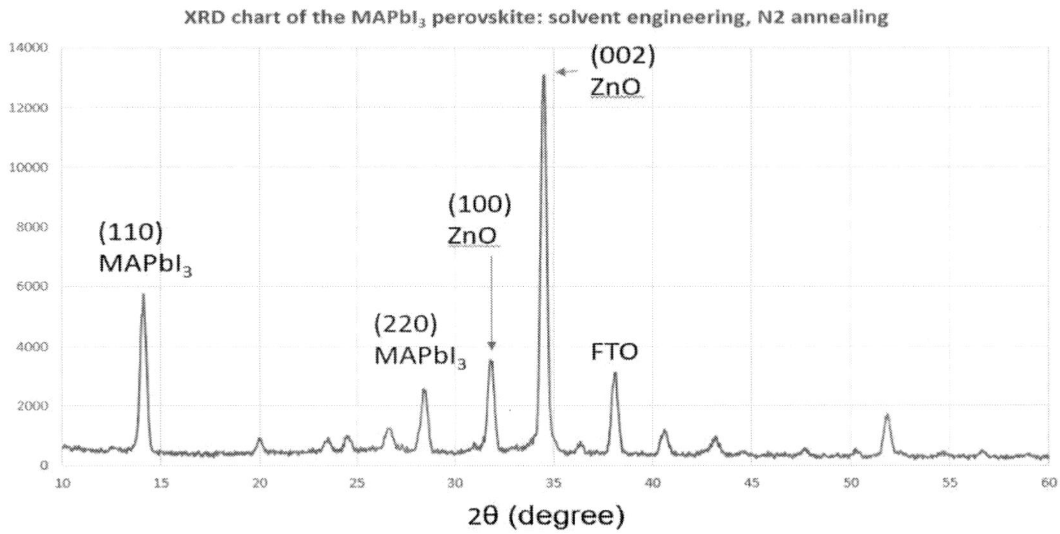

Figure 2. XRD chart of the $MAPbI_3$ perovskite deposited on ZnO NRs.

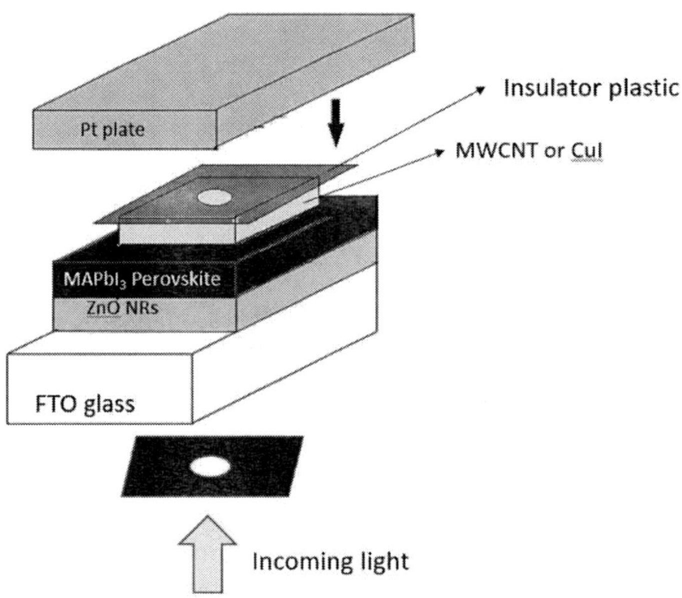

Figure 3. Configuration of the fabricated PSC device.

Photovoltaic measurement of the PSCs can be seen in table 1. The PSC with MWCNTs have larger outputs compared to CuI based PSCs. Such outcome was expected because of higher conductivity of the MWCNTs compared to CuI. The generated holes could move faster from the perovskite to MWCNTs compared to the perovskite to CuI. Additional observations about the I-V curve can be seen in figure 4. The I-V curve shows that the shunt resistance (R_{SH}) for the PSC was low that the I-V curve exhibited nearly flat lines for both MWCNTs and CuI. Low R_{SH} meant that there was a defect in the interfaces. It was possible that unstable ZnO/$MAPbI_3$ interface became the source for electron-hole recombination which gave less than desired results. Another source for interface defects was faulty deposition of the HTM by doctor blading where the HTM was spread unevenly in the active area. The application of Pt plate by pressing the plate onto the HTM could leave empty spaces along the HTM/Pt interface that lead to short circuit device.

Table 1. Photovoltaic parameters of the PSCs with CuI and MWCNTs.

HTM	J_{SC} (mA/cm^2)	V_{OC} (V)	FF	PCE (%)
CuI	3.3953	0.1417	0.2835	0.1364
MWCNTs	5.3409	0.2055	0.3026	0.3322

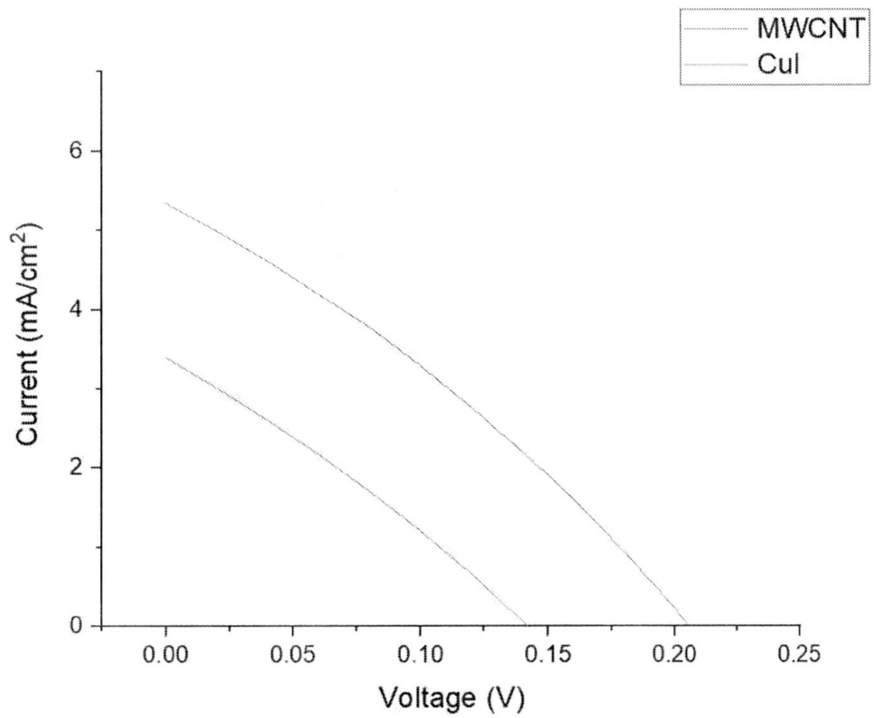

Figure 4. I-V curve of the PSCs with MWCNTs and CuI.

4. Conclusions

Complete coverage of $MAPbI_3$ perovskite over the ZnO NRs was done through preventing the finalization of perovskite crystallization before the spin coating was conducted. In this way, the perovskite was able to cover the ZnO NRs completely while still in the crystallization process. The slowing down of the perovskite crystallization was done through mixing of the perovskite solvents between DMF and DMSO. The photovoltaic outputs of MWCNTs and CuI were compared and calculated. MWCNTs delivered higher outputs due to higher conductivity and lower resistivity. Less than ideal photovoltaic parameters occurred due to defects between the interfaces.

References

[1] Kojima, A., et al., *Organometal Halide Perovskites as Visible-Light Sensitizers for Photovoltaic Cells.* Journal of the American Chemical Society, 2009. **131**(17): p. 6050-51.

[2] Yang, W.S., et al., *Iodide management in formamidinium-lead-halide–based perovskite layers for efficient solar cells.* Science, 2017. **356**(6345): p. 1376-1379.

[3] Li, S., et al., *Mesoporous PbI2 assisted growth of large perovskite grains for efficient perovskite solar cells based on ZnO nanorods.* Journal of Power Sources, 2017. **342**: p. 990-997.

[4] Garcia, V.J., C.M. Pelicano, and H. Yanagi, *Low temperature-processed ZnO nanorods-TiO2 nanoparticles composite as electron transporting layer for perovskite solar cells.* Thin Solid Films, 2018. **662**: p. 70-75.

[5] Zhong, M., L. Chai, and Y. Wang, *Core-shell structure of ZnO@TiO2 nanorod arrays as*

electron transport layer for perovskite solar cell with enhanced efficiency and stability. Applied Surface Science, 2019. **464**: p. 301-310.

[6] Wang, Y., M. Zhong, and L. Chai, *Effects of the concentration of PbI2 and CH3NH3I on the perovskite films and the performance of perovskite solar cells based on ZnO-TiO2 nanorod arrays.* Superlattices and Microstructures, 2018. **123**: p. 189-200.

[7] Wang, P., et al., *Stabilization of organometal halide perovskite films by SnO2 coating with inactive surface hydroxyl groups on ZnO nanorods.* Journal of Power Sources, 2017. **339**: p. 51-60.

[8] Tseng, Z.-L., et al., *Surface Engineering of ZnO Electron Transporting Layer via Al Doping for High Efficiency Planar Perovskite Solar Cells.* Nano Energy, 2016. **28**: p. 311-318.

[9] Bag, M., et al., *Evidence for reduced charge recombination in carbon nanotube/perovskite-based active layers.* Chemical Physics Letters, 2016. **662**: p. 35-41.

[10] Cheng, N., et al., *Multi-walled carbon nanotubes act as charge transport channel to boost the efficiency of hole transport material free perovskite solar cells.* Journal of Power Sources, 2016. **332**: p. 24-29.

[11] Wang, J., et al., *Enhanced Photovoltaic Performance with Carbon Nanotubes Incorporating into Hole Transport Materials for Perovskite Solar Cells.* Journal of Electronic Materials, 2016. **45**(10): p. 5127-5132.

[12] Wei, Z., et al., *Hysteresis-free multi-walled carbon nanotube-based perovskite solar cells with a high fill factor.* Journal of Materials Chemistry A, 2015. **3**(48): p. 24226-24231.

[13] Lee, J., et al., *Hierarchically Structured Hole Transport Layers of Spiro-OMeTAD and Multiwalled Carbon Nanotubes for Perovskite Solar Cells.* ChemSusChem, 2015. **8**(14): p. 2358-2362.

IOP Conf. Series: Materials Science and Engineering **715** (2020) 012058 doi:10.1088/1757-899X/715/1/012058

Synthesis and experimental investigation of PZT nanocomposite films

Tomas Janusas, Giedrius Janusas*, Arvydas Palevicius and Justas Ciganas

Kaunas University of Technology, Faculty of Mechanical Engineering and Design, Studentu 56, 51424 Kaunas, Lithuania

*Email: giedrius.janusas@ktu.lt

Abstract. In this paper, synthesis of nanocomposite and experimental investigation of three samples of nanocomposite films are presented. The aim of this research is to determine influence of the PZT particle concentration in nanocomposite on the surface morphology, qualitative and quantitative chemical composition ant direct piezoelectric effect. The specimens were prepared by mixing the PZT nanopowder with PMMA in three different proportions: 85% of PZT and 15% of binder; 90% of PZT and 10% of binder; 95% of PZT and 5% of binder. Surface morphology and qualitative and quantitative chemical composition were evaluated using scanning electron microscope, which is equipped with the Energy Dispersive X-Ray Spectrometer. Electrical response of the pulse-excited specimens is also presented.

1. Introduction

Creation of the effective micromechanical fluidic systems increases efficiency of microdosing or fluid microflow control, resulting reduced analysis time of the biological fluids and increased applicability in biosensors. Active microfluidic components [1, 2] and platforms [3, 4] have been developed. However, some disadvantages are still unsolved. Ability to manipulate microvolume fluids or increase the flow speed is the key in the development of single particle biosensors [5]. Techniques of microfluid flow control using standing or traveling surface acoustic waves [6] or bulk acoustic waves [7] are very well known. These techniques are based on generation of acoustic waves in micromechanical fluidic devices using piezoelectric, electromagnetic and other active materials.

Thin layer PZT films have been used for a long time in MEMS systems [8], but only in 2018 P. Reichert et. al. [9] proposed to apply them in microfluidic devices to generate bulk acoustic waves, which enable controlling the bio particles effectively using low voltage and making minimal effect on temperature of analysed analyte. The quality factor of these micromechanical fluidic elements depends on the piezoelectric materials, that are used for formation of microfluidic devices. Researchers S. Siddiqui [10], K. Choi [11], H. Parangusan [12], S. Osho [13], J. Belovickis [14], and others, from all over the world are working on the development of piezoelectric composites. Efficiency, quality and application of the piezoelectric nanocomposites are the main areas of the research in this topic.

This research paper covers the investigation of piezoelectric nanocomposite material employed as microresonators, which could ensure more effective flow of microfluids when the system is excited by vibrations. The impact of different PZT nanoparticles concentration on the PZT nanocomoposite film properties was analysed. Thus, three unique specimens made of PZT nanocomposite material were fabricated and tested.

2. Materials and experimental setup

2.1. PZT nanocomposite synthesis

The fabrication of the PZT nanocomposite material starts with powder preparation. This composite material was designed by mixing PZT nanopowder with poly methyl methacrylate (PMMA) in three different proportions: 85% of PZT and 15% of binder (PZT85); 90% of PZT and 10% of binder (PZT90); 95% of PZT and 5% of binder (PZT95).

An oxalate/hydroxide co-precipitation method was used for synthesis of lead zirconate titanate [Pb $(Zr_{0.52}Ti_{0.48})O_3$] nanopowder. The materials used were lead (II) acetate $Pb(CH_3COO)_2$, titanium butoxide $Ti(C_4H_9O)_4$, zirconium butoxide $Zr(C_4H_9O)_4$ (80 % solution in n-butanol), oxalic acid dihydrate $C_2H_2O_4{\bullet}2H_2O$, 25% ammonia solution and deionized water. The detailed description of synthesis is presented in paper [15].

Layer (60 μm thickness) of PZT nanocomposite materials (PZT85, PZT90, PZT95) was formed on the copper foil using screen printing technique with the polyester monofilament screen mesh (48/70) and dried in electrical oven at 100 °C for 30 min (figure 1).

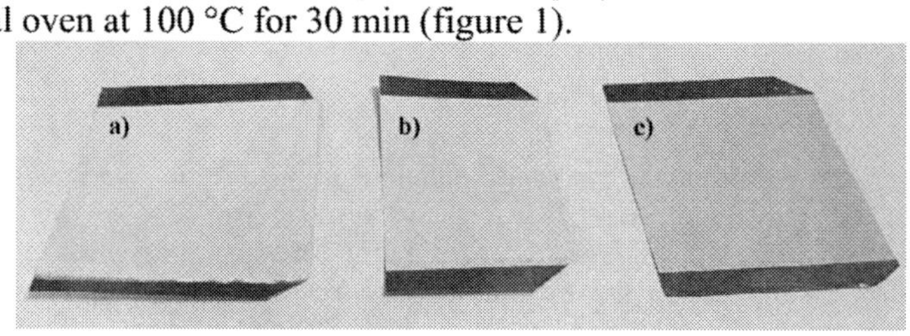

Figure 1. Prepared samples: a) PZT85, b) PZT90, c) PZT95.

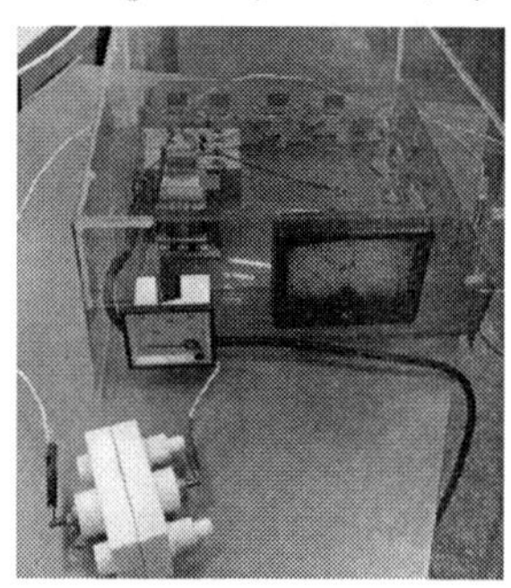

Figure 2. High voltage set for the pole alignment of PZT coatings.

In order to improve piezoelectric characteristics of the formed PZT nanocomposite with PMMA binder electrical pole alignment was performed using a high voltage generator and a custom–made holder, shown in figure 2. An element with a PZT coating was placed in the special holder between positive and negative poles. The high voltage generator was set at 5 kV current and held for 30 min. The poling technique aligns a positive pole on one side of the PZT coating and a negative pole on the other side. Prepared specimens were tested and compared with the results presented in papers [15, 16].

2.2. Surface morphology and chemical composition

Surface morphology and chemical composition of the formed PZT nanocomposite materials were evaluated using scanning electron microscope (SEM) FEI Quanta 200 FEG, equipped with the Energy Dispersive X-Ray Spectrometer (EDS) detector XFlash 4030 from Bruker. Samples were investigated under controlled pressure water steam atmosphere. Maximal achievable resolution for high-vacuum (< 6e-4 Pa) is 1.2 nm, for low-vacuum (10 to 130 Pa) – 2.5 nm and for extended vacuum mode (10 to 4000 Pa) – 3 nm. The EDS detector allows detecting elements ranging from boron (5) to americium

(95). The chemical analysis can be determined at the chosen point, along the line or distribution on the surface – mapping also is a possibility. A modern 30 mm^2 area solid state drift detector is cooled with Peltier element and provides 133 eV (at Mn K) energy resolution at 100,000 cps. Moreover, X-Ray spectroscopy method allows analysing energy distributions, i.e. the energy differences between the various quantum states of a system and the probabilities that the system jumps between these states.

2.3. Dynamic and electric characterisation

The dynamic and electric characteristics of the multilayer specimens were investigated using the laser triangular displacement sensor: sensitive head LK-G82 and control unit LK–GD500. The measured vibration amplitude was collected with a USB oscilloscope PicoScope 3424. The experimental setup is presented in figure 3. It was used to investigate the electrical response of the multilayer specimens upon calibrated mechanical pulse of 10 N.

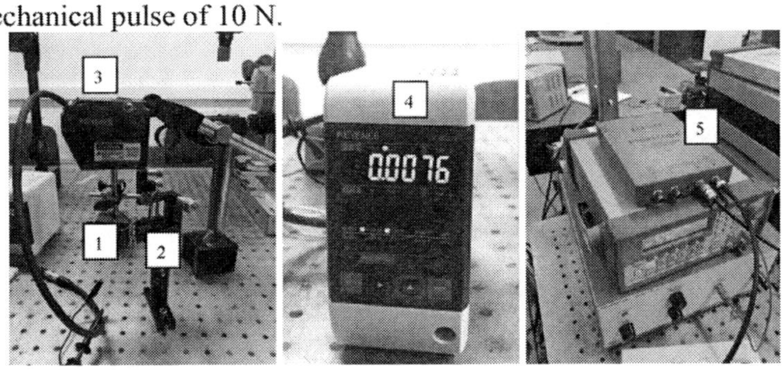

Figure 3. Experimental setup for dynamic and electric characterization of specimens: 1 – specimen; 2 – holder; 3 – sensor head LK-G82; 4 – control block LK-G500; 5 – USB oscilloscope PicoScope 3424.

3. Results

In this research, three samples of nanocomposite films were fabricated and investigated. The SEM views and EDS spectrums of PZT85, PZT90 and PZT95 nanocomposite films on Cu substrate are presented in figure 4 – figure 6.

The surface morphology of the first sample PZT85 (figure 3a) shows non even distribution of PZT particles of the size up to 1 μm. Also small pileups of 5 – 10 μm in diameter were observed on top of the surface. The SEM view of the nanocomposite PZT90 showed better distribution of PZT particles (figure 5a). Whereas, the surface view of the nanocomposite PZT95 showed even distribution of PZT particles. Finally, the results showed that a PMMA binder loaded with bigger amount of the PZT particles ensures the higher surface quality, thus forming a much more promising sensing and actuating. Simultaneously, formation of active microfluidic devices using thermal formation process [17] could be too complicated.

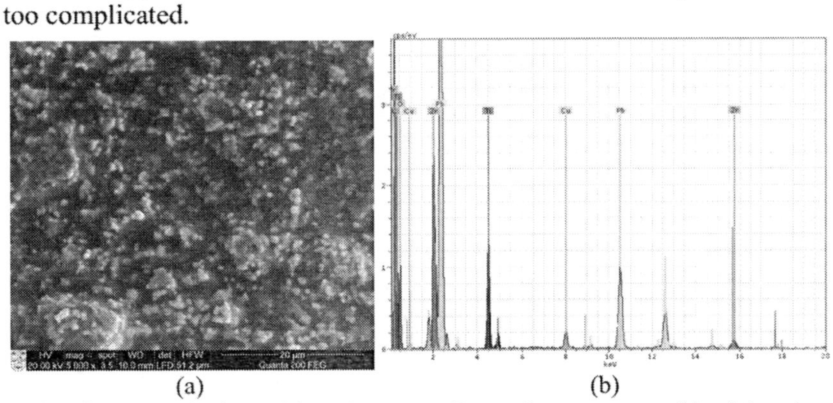

(a) (b)

Figure 4. Surface morphology (a) and energy dispersive spectrum (b) of the piezoelectric nanocomposite PZT85.

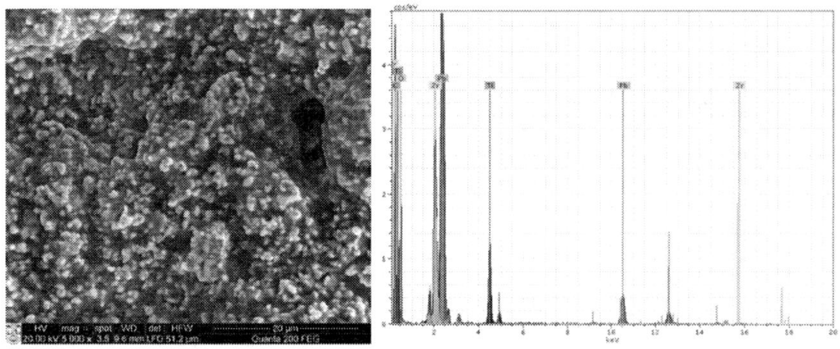

(a) (b)

Figure 5. Surface morphology (a) and energy dispersive spectrum (b) of the piezoelectric nanocomposite PZT90.

EDS analysis gives qualitative and quantitative determinations of the elemental composition of the piezoelectric nanocomposites. These microanalyses show the dominance of lead, with a Lα peak at 10.6 keV, Lβ peak at 12.6 keV and Mα peak at 2.3 keV, and oxygen with a Kα peak at 0.5 keV (Figure 4.b – Figure 6.b). Other peaks show the smaller quantities of carbon (Lα peak at 0.3 keV), zirconium (Kα peak at 15.8 keV, Kβ peak at 17.7 keV, Lα peak at 2 keV and Lβ peak at 2.1 keV) and titanium (Kα peak at 4.5 keV, Kβ peak at 4.9 keV, Lα peak at 0.4 keV and Lβ peak at 0.5 keV). In addition, minor amount of copper with a Kα peak at 8 keV, Kβ peak at 8.9 keV, Lα peak at 0.93 keV and Lβ peak at 0.95 keV was detected in the specimen of PZT85. The detailed information of the normalized concentration in weight percent of the chemical composition of the piezoelectric nanocomposites PZT85, PZT90 and PZT95 is presented in table 1.

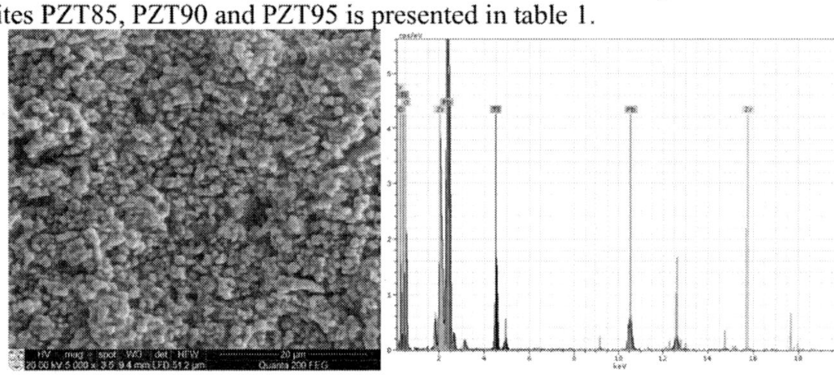

(a) (b)

Figure 6. Surface morphology (a) and energy dispersive spectrum (b) of the piezoelectric nanocomposite PZT95.

Table 1. Chemical composition of the piezoelectric nanocomposites PZT85, PZT90 and PZT95.

Chemical elements	The normalized concentration in weight percent (norm. wt., %)		
	PZT85	PZT90	PZT95
Carbon	17.6	24.9	11.8
Oxygen	55.3	31.2	28.5
Titanium	3.6	5.2	7.1
Copper	0.9	0	0
Zirconium	5.4	12.2	13.4
Lead	17.2	26.5	39.2

Direct piezoelectric effect of the three fabricated PZT nanocomposites was investigated using experimental setup described in Section 2.3. All specimens were pulse-excited using calibrated mechanical pulse of 10N. Open circuit voltage generated by all three specimens was measured and collected using USB oscilloscope PicoScope 3424 (figure 7). The specimen coated by PZT95 nanocomposite film showed the best results. It generated electrical pulse of 0.36 mV. Specimen with PZT90 film generated voltage of 0.08 V and specimen with PZT90 film generated 0.2 V electrical pulse. In comparison with previously achieved results [11] generated voltage increases. However, application area of this piezoelectric nanocomposite became more complicated due to its bigger microhardness.

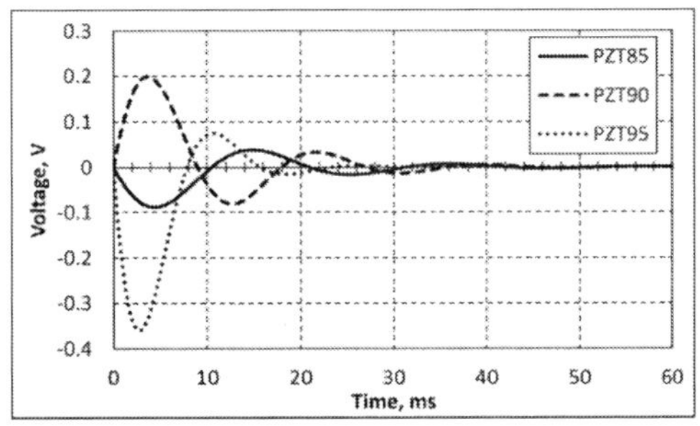

Figure 7. Generated voltage by pulse-excited PZT nanocomposite films.

4. Conclusion

In this research, three samples of nanocomposite films were fabricated and investigated. PZT nanopowder was mixed with PMMA in three different proportions: 85% of PZT and 15% of binder; 90% of PZT and 10% of binder; 95% of PZT and 5% of binder. The SEM views showed that higher PZT concentration leads to better surface quality and even distribution of PZT nanoparticles. The microanalyses using EDS spectrums showed the dominance of lead and oxygen in the PZT nanocomposite. Other peaks showed smaller quantities of carbon, zirconium and titanium. Quantities of all chemical components vary and correlate with the concentration of PZT nanoparticles. The better surface quality of specimen PZT95 was confirmed by the measurement of the direct piezoelectric effect of PZT nanocomposites using pulse-excitation. It showed that PZT nanocomposite of the higher PZT particle concentration generates up to 4 times higher electrical potential. The change in chemical composition of the PZT nanocomposite helps to manipulate the mechanical and piezoelectric characteristics for different applications.

Acknowledgments

This research was funded by a grant S-MIP-17-102 from the Research Council of Lithuania.

References

[1] C.I. Rogers, J.B. Oxborrow, R.R. Anderson, L.F. Tsai, A.T. Woolley, Sens. Actuators B Chem. **191** (2014).
[2] J. Thomas, Z. Tehrani, B. Redfearn, Addit. Manuf. **9** (2016).
[3] R. Burger, L. Amato, A. Boisen, Biosens. Bioelectron. **76** (2016).
[4] X. Wang, D. Nilsson, P. Norberg, Biochim. Biophy. Acta (BBA) Gen. Subj. **1830** (2013).
[5] J.J. Gooding, K. Gaus, Angewandte Chemie International Edition **55**, 38 (2016).
[6] A. Mujahid, F.L. Dickert, Sensors **17**, 12 (2017).
[7] Y. Chang, Z. Hui, X. Wang, H. Qu, W. Pang, X. Duan, Sensors **18**, 2 (2018).
[8] S.T. Mc Kinstry, P. Muralt, Journal of Electroceramics **12**, 1–2 (2004).
[9] P. Reichert, D. Deshmukh, L. Lebovitz, J. Dual, Lab on Chip (2018).
[10] S. Siddiqui, D.I. Kim, L.T. Duy, M.T. Nguyen, S. Muhammad, W.S. Yoon, N.E. Lee, Nano Energy **15** (2015).
[11] K. Choi, W. Choi, C. Yu, Y.T. Park, Journal of Nanomaterials (2017).
[12] H. Parangusana, D. Ponnamma, M. Al Ali Al Maadeed, RSC Adv. **7** (2017).
[13] S. Osho, N. Wu, M. Aramfard, C. Deng, O. Ojo, Smart Mater. Struct. **27** (2018).

IOP Conf. Series: Materials Science and Engineering **715** (2020) 012058 doi:10.1088/1757-899X/715/1/012058

[14] J. Belovickis, M. Ivanov, V. Samulionis, J. Banys, A. Solnyshkin, S.A. Gavrilov, K.N. Nekludov, V.V. Shvartsman, M.V. Silibin, Physica status solidi (b) **255**, 3 (2018).
[15] E. Cekas, G. Janusas, A. Guobiene, A. Palevicius, A. Vilkauskas, S. Ponelyte-Urbaite, Sensors **18**, 11 (2018).
[16] G. Janusas, S. Ponelyte, A. Brunius, A. Guobiene, A. Vilkauskas, A. Palevicius, Sensors **16**, 11 (2016).
[17] A. Sodah, R. Gaidys, B. Narijauskaite, R. Sakalys, G. Janusas, A. Palevicius, P. Palevicius. Microsystem technologies (2018).

IOP Conf. Series: Materials Science and Engineering **715** (2020) 012059 doi:10.1088/1757-899X/715/1/012059

An automatic grading system for electronic medical records with neural network

Yi Liu

Department of Computer Science and Engineering, Shanghai Jiao Tong University, 800 Dongchuan Road, Shanghai, China

*E-mail: liuyi61ly@hotmail.com

Abstract. Automatically grading electronic medical records from clinical researchers is an important task on healthcare research. In this paper, we present a convolutional neural network framework to grade medical records by grading note. We regard the scoring process as a text pattern classification task with mapped sentences from grading concept rule to medical records. This framework involves two stages. The first stage is key medical concept matching between grading note and medical records. The second stage is text pattern classification which can predict whether the key concept in grading note is correct, missing or incorrect. The result shows that our neural network model performs better than other traditional machine learning grading methods. Our system makes a great progress for text pattern classification accuracy and performs much better baselines on grading process.

1. Introduction

Automatically grading electronic medical records (EMR) for clinical researchers has attracted much research attention in healthcare community. EMRs help doctors further understandings of patients' health conditions, identifying their best-matching patient category and determining patients' customized treatment plan [1-2]. Unfortunately, before writing an excellent EMR, it takes a lot of time to train a recorder. The most important part in this progress is grading EMRs written by recorders.

Each EMR is collected by the conversion between doctor and patient. It usually contains various types of information, including patient personal information, past medical history, present illness and symptom etc. A good EMR should include all the important information known as key medical concepts. The grading task should judge if a recorder captured the key medical concepts. As shown in figure 1, medical concept is called grading rubric. If a grading rubric can be found correctly in the corresponding EMR, it will be labelled as C. Similarly, I will represent incorrect and M will represent missing.

A quick approach called word-count based model first scans all the words in grading rubric and EMR and scoring according to the proportion of words common to both texts. However, without considering semantic information will reduce the accuracy of scoring. Thus, the semantic distance representation between two texts has been developed. The two most common ways for representing texts is via a bag of words (BOW) [3] or by their term frequency inverse document frequency (TF-IDF) [4]. Furthermore, the Word Mover's Distance [5] (WMD) is developed to improve the above approaches. These methods can reflect the EMRs quality in a certain degree but is interfered with a large amount of irrelevant information in EMRs. Also, they cannot map the grading rubric exactly to the corresponding sentence in EMRs. In this paper, we build an automatically grading system for EMR. It first matches grading rubric to sentence in EMR with closest sematic relatedness. Then, a convolutional neural network (CNN) is established for classifying new concatenated text pattern. We score the EMR by evaluating number of C labels as a percentage of the total number of labels in grading note.

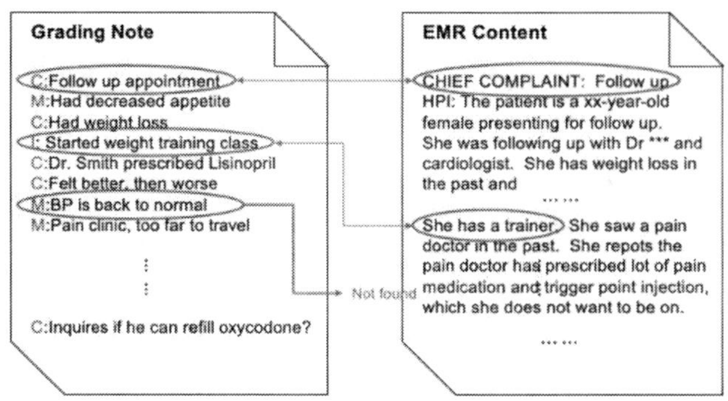

Figure 1. Grading rubric and medical record.

2. The proposed methodology of grading system

2.1. General framework of grading system

As shown in figure 2, automatically grading system (auto-grader) considers the grading process as a two-stage pipeline framework. The first stage is called grading rubric matching. We split each grading note into a list of grading rubrics and sentence level vectors. Meanwhile, we get sentence vectors for EMR by same method. Then, we map the sentence embedding in grading rubric to EMR by calculating relatedness score of two pairs of sentences. The second stage is regarded as text pattern classification. We present a slight variant of the CNN architecture to predict whether the key concept is correct, missing or incorrect. The score is calculated based on the prediction results.

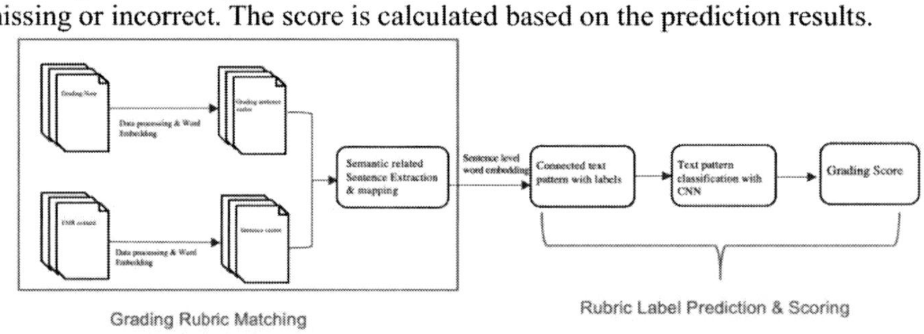

Figure 2. Workflow of auto-grading system with CNN model.

2.2. Grading rubric matching

2.2.1. Sentence embedding

The most important part of grading rubric matching is calculating sentence embedding through word vector. The supervised learning methods usually train a neural work and use the last hidden layer as sentence embedding [6]. However, these methods require additional annotated data. Here, we get sentence embedding v_s by weighted averaging word embedding v_w which is similar to Sanjeev's work [7].

2.2.2. Semantic relatedness sentences extraction and mapping

In order to judge whether grading rubric is correct, incorrect or missing, we need to find out if there's a sentence in EMR which contains exact same meaning of grading rubric. Sentence similarity is a basic way to compare the semantic relatedness of a pair of sentences. Here, we use the Cosine Similarity to measure the relatedness score of the grading rubric in grading rubric to corresponding sentence in EMR. Here $v_g = [x_1,x_2,x_3,\cdots,x_n]$ and $v_r = [y_1,y_2,y_3,\cdots,y_n]$. The cosine similarity is

$$CosSim(\boldsymbol{v_g},\boldsymbol{v_r}) = \frac{x_1 \cdot y_1 + x_2 \cdot y_2 + \cdots + x_n \cdot y_n}{\sqrt{x_1^2+x_2^2+\cdots x_n^2} \cdot \sqrt{y_1^2+y_2^2+\cdots y_n^2}} \quad (1)$$

where v_g stands for the embedding of grading rubric and v_r stands for sentence embeddings in EMR. For each v_g, we select sentence embedding v_r in EMR with the highest score and map them together.

2.3. Grading rubric labelling and scoring

2.3.1. Convolutional neural network for label prediction

The model architecture is similar to the CNN architecture of Yoon Kim's work [8]. Let $x_i \in R_k$ be the k-dimensional word vector corresponding to the ith word in grading rubric and $y_i \in R_k$ be the k-dimensional word vector corresponding to the ith word in the most related sentence in EMR. Let the max of sentence length be n (padded when sentence length is less than n) and the concatenated text pattern can be represented as equation (2),

$$T_{1:2n} = x_1 \oplus x_2 \oplus \cdots \oplus x_n \oplus y_1 \oplus y_2 \oplus \cdots \oplus y_n = t_1 \oplus t_2 \oplus \cdots \oplus t_{2n} \quad (2)$$

where $\oplus$ is the concatenation operator. In this paper, let $T_{i:j}$ refer to the concatenation of words $t_i,t_{i+1},\cdots,t_{i+j}$. The input layer becomes a matrix of word vectors corresponding to the words in the text pattern. There are 2n words in the text pattern, so the matrix is $2n \times k$.

This type of matrix can be static or non-static according to [9-10]. We use pretrained embedding to make the matrix static. For the convolutional layer, a filter $w \in R^{hk}$ is applied to a window of h to produce a new feature during convolution operation. For example, a feature α_i is generated from a window of words $t_{i:i+h-1}$ by

$$\alpha_i = f(w \cdot x_{i:i+h-1} + b) \quad (3)$$

Here $b \in R$ is a bias variable and $f(\cdot)$ is a rectified linear unit (ReLU). This filter is applied to each possible window of words in the text pattern $\{t_{1:h},t_{2:h+1},\cdots,t_{2n-h+1:2n}\}$ to produce a feature map $\alpha = [\alpha_1, \alpha_2,\cdots,\alpha_i,\cdots,\alpha_{2n-h+1}]$ and $\alpha \in R^{2n-h+1}$.

We then apply a pooling layer. A max-over-time pooling method [8] is used to extract the maximum value $\widehat{\alpha} = \max\{\alpha\}$ from the previous one-dimensional feature map α as the feature. Just like extracting one feature from one filter, this model can also use multiple filters with different window sizes to get multiple features. The pooling layer outputs the one-dimensional features and fully connects softmax layer. Similar to traditional neural networks, the output layer is defined as

$$y = \boldsymbol{w} \cdot (\boldsymbol{z} * \boldsymbol{r}) + b \quad (4)$$

Here $z = [\widehat{\alpha}_1,\widehat{\alpha}_2,\cdots,\widehat{\alpha}_m]$ and m represent that we have m filters. The softmax layer reflects the probability distribution on the final classification with n labels and can be set with the need of the task. We employ dropout to prevent neural networks from overfitting and $*$ is the element-wise multiplication operator and r is a vector of m Bernoulli random variables which have certain probability to be 0 or 1.

2.3.2. Grading electronic medical records

The percentage of concepts need to be captured by doctor and patient determines the score of each EMR. For each generated text pattern, we calculate the number of each label and get the percentage of label C as the final score of EMRs. As equation (8) shows,

$$Score = \frac{num_c}{num_c + num_I + num_M} \cdot 100 \quad (5)$$

where Score is the final score of EMRs and num_i ($i = C,I,M$) represent the number of each label of grading rubric.

3. Experiments and results

3.1. Dataset details

The EMRs are collected from our partner hospital and the data only permits noncommercial or academical use. The grading rubric and its label was done by several doctors. The data contain two parts. The first is EMRs which was written by recorders. These EMRs are based on conversation between clinicians and patients. They contain many abbreviated messages and incomplete expressions from oral information. The second part is grading notes which are written by doctor. A grading note contains grading rubrics which a recorder must record when he is listening conversation between clinicians and patients. We manually go through every grading rubric and label them by { C, I, M } for correct, incorrect or missing.

There are 1429 grading note and EMR pairs. Each grading note contains dozens to hundreds of rubrics. Firstly, we match every rubric with their most related sentence in EMRs. This process generates concatenated text pattern of two sentences with labels. We then use machine learning approaches to predict the labels of concatenated text pattern. Here, about 80000 concatenated text patterns will be the train data in the next stage.

3.2. Baseline model

To establish a baseline performance, we use weighted averaging the word vectors to get sentence embedding. We then concatenate both the sentence embedding pairs in grading rubric and EMR by calculating their cosine similarity. The label prediction problem can be regard as a vector classification problem. We train a nearest neighbor classifier (KNN), a logistic regression classifier (LR), a decision tree (DT) and a multi-layer neural network.

3.3. Hyperparameters and training

We initialize word embedding by using publicly available data which were trained on 100 billion words from Google News. The word embeddings are 300-dimensional word vectors trained by continuous bag-of-words architecture [11]. Words which are not present in the set of pre-trained embeddings will be initialized randomly. The class labels can be divided into two category that is 0 for C and 1 for M and I.For parameter setting, we refer to Yoon Kim's work [10] which use filter windows (h) of 3, 4, 5 with 128 feature maps each. We set the dropout rate of 0.5 and default batch size of 64. We select 80% of the text pattern as the train set and select test data set from the rest of 20%. Training is done through Adam. Cross entropy loss measures the performance of our CNN model. We calculate a separate loss for each class label per text pattern and sum the result.

3.4. Result and discussion

We compare our method with widely used text pattern classification methods. The experimental results in table 1. show that our auto-grader with CNN outperform the traditional methods on predicting three labels in the EMRs grading task. It proves that CNN-based approach can effective compose the semantic representation of texts.

Table 1. Results for text pattern classification.

Model	Accuracy
Sentence Embedding + KNN	64.27%
Sentence Embedding + LR	59.62%
Sentence Embedding + DT	69.57%
Sentence Embedding +Multi-Layer Neural Network	67.46%
Auto-grader with CNN	**70.46%**

For grading result comparison, we introduce mean absolute error (MAE) rate and root-mean-square error (RMSE). Mean absolute error is used to measure the difference between the scores we grade and the real scores. The mean absolute error rate MAE is defined as

$$MAE = \sum_{k=1}^{N} \frac{|real_{score} - pre_{score}|}{N} \tag{6}$$

where $\text{real}_{\text{score}}$ represent real score and $\text{pre}_{\text{score}}$ represent predicted score by our auto-grader. The root-mean-square error is used to measure the bias of grading result.

$$RMSE = \sqrt{\frac{1}{N}\sum_{k=1}^{N} (real_{score} - pre_{score})^2} \tag{7}$$

Here N is the number of EMRs need to be graded.

Table 2. Results for grading EMRs.

Model	MAE	RMSE
Word-count Based Model	13.92	16.89
Sentence Embedding +KNN	11.34	13.88
Sentence Embedding +LR	13.47	16.63
Sentence Embedding +DT	10.61	13.38
Sentence Embedding +Multi-Layer Neural Network	11.23	14.55
Auto-grader with CNN	**10.09**	**12.96**

From table 2, we can see auto-grader achieves better performance both in MAE and RMSE measurements.

4. Conclusion

In the present work we have described an automatic grading system for EMRs with convolutional neural networks. Our method can not only grade the EMRs through key medical concept but can also match the key concept with the corresponding sentence if it exists in EMR. We evaluated our work with other models such as word-count based model, linear regression, k-neighbors classifier and so on. The result shows that our automatically grading system performs better than machine learning baselines. We believe our CNN-based auto-grader will bring more inspiration to real word machine learning applications.

Acknowledgments

The author would like to thank Sabbir, Nadim, Kevin and Sarah's suggestions. Augmedix Inc. provides the environment and conditions for this study.

References

[1] Z. Huang, M.-L. Shyu, J. M. Tien, M. Vigoda, D. J. Birnbach. Knowledge-assisted sequential pattern analysis with heuristic parameter tuning for labor contraction prediction. IEEE J. Biomed. Health Informat., 18, 492 (2014)

[2] F. Liu, L. Zhou, C. Shen, J. Yin. Multiple kernel learning in the primal for multimodal Alzheimers disease classification. IEEE J. Biomed. Health Informat., 18, 984 (2014)

[3] Tomas Mikolov, Kai Chen, Greg Corrado, Jeffrey Dean. Efficient estimation of word representations in vector space. arXiv preprint arXiv:1301.3781. (2013a)

[4] S. Robertson. Understanding inverse document frequency: on theoretical arguments for IDF. Journal of documentation, 60(5), 503 (2004)

[5] M Kusner, et al. From word embeddings to document distances. International Conference on Machine Learning. (2015)

[6] J. mMueller, and Aditya Thyagarajan. Siamese Recurrent Architectures for Learning Sentence Similarity. In AAAI, 16 (2016)

[7] A. Sanjeev, Y. Liang, T. Ma. A simple but tough-to-beat baseline for sentence embeddings. (2016)

[8] Kim, Yoon. Convolutional neural networks for sentence classification. arXiv preprint

arXiv:1408.5882 (2014).
[9] R. Collobert, J. Weston, L. Bottou, M. Karlen, K. Kavukcuglu, P. Kuksa. Natural Language Processing (Almost) from Scratch. Journal of Machine Learning Research 12 (2011)
[10] R. Socher, J. Pennington, E. Huang, A. Ng, C. Manning. Semi-Supervised Recursive Autoencoders for Predicting Sentiment Distributions. In Proceedings of EMNLP (2011)
[11] T. Mikolov, I. Sutskever, K. Chen, G. Corrado, J. Dean. Distributed Representations of Words and Phrases and their Compositionality. In Proceedings of NIPS (2013)

IOP Conf. Series: Materials Science and Engineering **715** (2020) 012060 doi:10.1088/1757-899X/715/1/012060

Heart rate prediction model based on neural network

Minmin Luo[1,*] and Kaibin Wu[2]

[1]School of Biomedical Engineering, Southern Medical University, Guangzhou 510515, China

[2]Class of 2016th Biomedical Engineering (Imaging Engineering), School of Biomedical Engineering, Southern Medical University, Guangzhou 510515, China

*E-mail: 22809017@qq.com

Abstract. Heart is an essential organ of human body and heart rate (HR) is the most obvious heart activity in daily life. In order to predict heart rate, a heart rate prediction model based on LSTM (Long Short-Term Memory) neural network is proposed in this paper. This model combines five physiological parameters as input to ensure the validity of heart rate prediction. The results show that Adam-LSTM is a good method for heart rate prediction and reflects the tendency of heart rate change in daily life. At the same time, the experimental results show that the root mean square error of prediction value is small and the validity is high.

1. Introduction

In daily life, the most common change of physiological parameters perceived by us is the heart rate. Heart rate (HR) refers to the number of times the heart beats in one minute, which can be affected by many factors [1], mainly including physical activities, mental state, gender, age, physical health, heart diseases and so on. Therefore, the study of the heart rate and its influencing factors is of great importance.

So far, there have been a lot of researches on heart rate and the influencing factors[2], such as the heart rate prediction based on physical activities [3], the assessment of autonomic nervous system, and the measurement of energy consumption. Although these researches have been applied in many fields, they are performed on the basis of the relationship between a single factor and the heart rate and focus on exploring the direct relationship between the two, but few studies related to the heart rate and multiple influencing factors have been made. Internationally, Pawer[4]et. al, established a human activity monitoring model based on ECG signals, rather than HR; Meijer[5] et al. set up the relationship between heart rate and human activities in some special experimental scenarios. However, there were limited activities and singular variables in the experiment, which cannot fully reflect the actual heart rate changes. Heart rate prediction is to estimate the heart rate at the next moment based on now and other influencing factors. Reliable and effective heart rate prediction has great significance for the prevention and control of some cardiovascular diseases. Yet most of the current studies only considers estimating the heart rate at the next moment by the previous heart rate, without considering other physiological factors such as gender, age, physical activities, mental state and so on.

In this paper, LSTM[6] neural network is employed to establish a heart rate prediction model based on multiple influencing factors, so as to obtain effective heart rate prediction values and provide ideas for the subsequent heart rate anomaly detection system.

2. Method

2.1. Method overview

This paper uses Long Short-Term Memory (LSTM) to avoid gradient explosions in the circulating neural network. Three "gate" structural designs are used by LSTM to control the transmission of information in the neural network, to train the input compression depth data and compression force data, and to predict the change of CPP. The model can generally be divided into the following three steps: data preprocessing, model training, model prediction, as shown in figure 1.

2.2. Data preprocessing

The collected heart rate data is often unequal in length, and improper handling may increase the difficulty of training. To this end, this paper intercepts the fixed length of each set of data and performs five-point secondary smoothing on the data.

2.3. Model

2.3.1. Model structure

In the traditional feedforward neural network, each step of each parameter is independent and is in lack of correlation with the time. But in the heart rate predication, changes in physiological signals are closely related to time. The heart rate at the present moment depends on the changes of various physiological parameters before the moment t, so the time related cyclic recurrent neural network (RNN) is introduced in this paper(RNN[7]). Unlike feedforward neural network, RNN is a time series-based network model. The parameters in the network are not only related to the current input but also linked to the past, which can establish the time correlation between the previous information and the current environment. However, due to the role of activation function, the ordinary RNN will have the problem that the parameters in the neural network are not updated, which is known as gradient disappearance. Like all RNNs, LSTM can make any calculation that can be done by the traditional computer as long as there are enough network elements. Compared with the traditional RNN network, the LSTM neural network has an effective mechanism, that is, the forget gate, which avoids the problem of gradient explosion and gradient disappearance in the standard RNN network. In LSTM, a memory unit dedicated to saving historical information is added.. The historical information in the network will pass through three gates: the input gate, the forget gate, and the control of the output gate, solving the problem of the gradient explosion that occurs in the RNN. Nowadays, LSTM is mainly used in natural language translation, physiological signal prediction, image and subtitle and other fields that are in linear relationship with the time, and can perfectly simulate the problem of multiple input variables. Moreover, it has unique advantages to changes of heart rate signal prediction caused by multiple physiological parameter inputs.

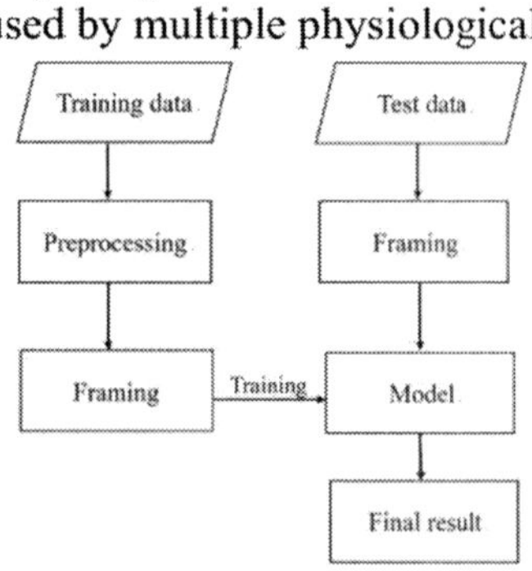

Figure 1. Algorithm flowchart.

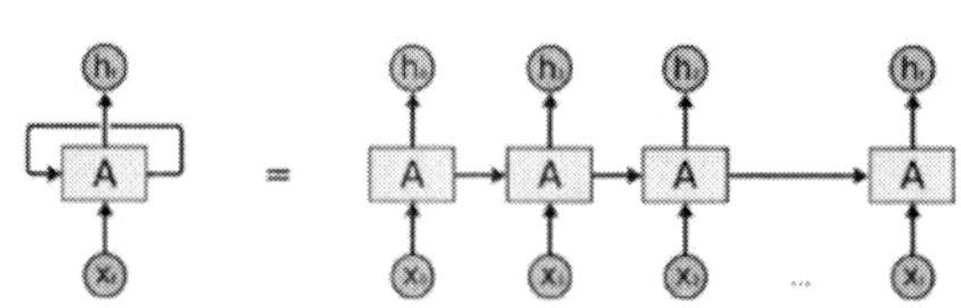

Figure 2. LSTM topology diagram.

The LSTM-based heart rate predication model in this paper consists of the input layer, hidden layer and output layer, as shown in figure 2. Relying on the long-term memory characteristic of LSTM and combined with the multivariate input, it is transformed into the supervised learning, which completes the inputs according to five physiological characteristics of the testees, namely, heart rate signal, gender, age, physical activities and mental state. At the same time, the model retains seven blank inputs, where more targeted physiological parameters can be designed for different testees. For example, hypertension patients may add the parameter of blood pressure, whereas the hyperglycemia suffers may choose blood glucose as the variable. In addition to the mental state, the input signals used must be recorded synchronously to further improve the accuracy. A portable active signal monitor is applied for the physical activity state to record the current activity signal of the testee (acceleration in three directions, x, y, z).

The preprocessed data is used as the input layer and the input data has 5 characterisistics. The first layer of the neutral network is the LSTM layer, which contains 50 neurons. Dropout[8] regularization process is performed by inputting the pooling layer to prevent overfitting. Besides, the dropout is connected to a fully connected layer, and the output is the heart rate predicted value. Through the operation of forward transmission, the neural network continues to transmit the information of the previous moment in the form of the memory stream, affecting the processing of each newly input data as well as the output of each phase. Adaptive moment estimation (ADAM)[9] is employed by the LSTM network in this paper to update the LSTM network parameters, and the results of heart rate prediction is obtained synchronously.

2.3.2. Model building

In this paper, the sequential model in the in-depth learning framework keras is introduced to linearly stack multiple network layers with the add function. The specific process is as follows:

(1) 1 layer LSTM network model is added, input data dimension is 50, with the activation function of tanh;

(2) A fully connected Dense layer is added, with the activation function of linear.

(3) Each layer is configured with the dropout rejection rate of 0.2 to prevent overfitting.

(4) A fully connected layer is added, the activation function of ReLU to connect the hidden layer and output.

The network structure of the model is shown in figure 3 and figure 4.

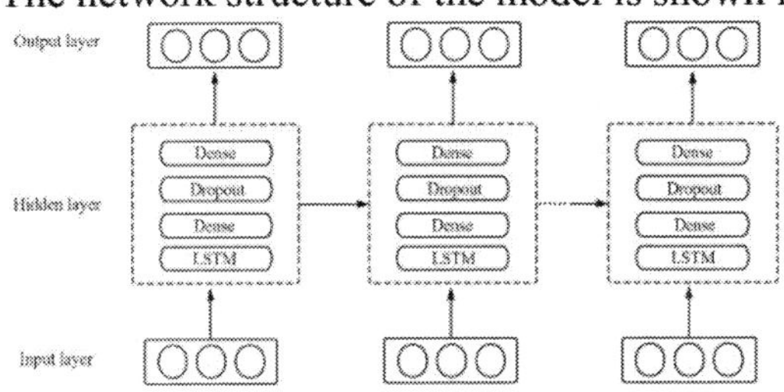

Figure 3. LSTM-based CPP prediction model structure.

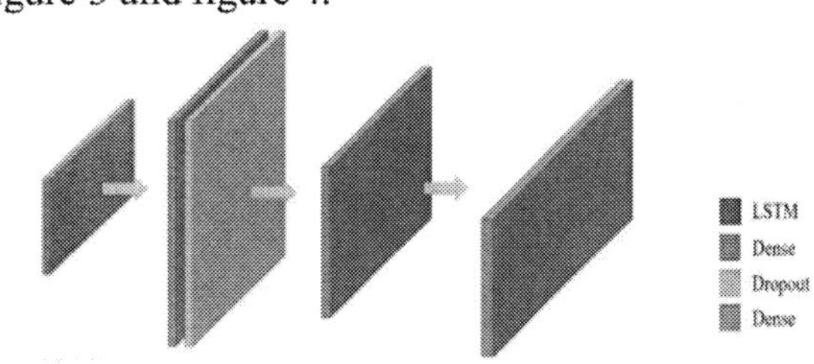

Figure 4. Heart rate prediction network.

2.3.3. Model training and testing

1. Model Parameter Setting

Before training, it is necessary to configure the model optimizer, loss function and indicator list. The mean absolute difference MAE is chosen in this paper as the loss function for the acquisition of the model output error.

As SGD[10] algorithm has a slower convergence speed, it is prone to oscillate at the same time of result prediction, so Adam is selected as the optimizer in this paper. The optimized Adam algorithm has a faster convergence speed, which also improves the accuracy.

2. Model Training

The number of samples batch size in each training is set as 128, with the training rounds epoch of 100; also, the test data set is specified for the model to output the record of each training.

3. Model Predication

The trained model needs to be verified on the test set, where the root mean square error (RMSE) is used to verify the accuracy of model prediction.

3. Result

The data of this paper was collected by the laboratory, a total of 48 samples, with the sampling rate of 50Hz, and the data length ranging from 90s to 580s. The testee is a healthy young man without any history of heart disease, who maintains a good mental state and the physical activity is slow walking. The recorded data is divided into two parts. The signal data in the first part is used as the data set for neural network training; and the remaining data (20%) is used as the test data set. There are 5 labels, including heart rate signal, gender, age, physical activity, mental state. Each label is standardized, and the label subordinates are free of mutual exclusion.

The predictions obtained by LSTM neutral network heart rate predicator from the training dataset and test dataset are shown in the figure 5. The predicted heart rate is represented by a solid blue line and the actual heart rate is indicated as a yellow dotted curve. The results show that after the completion of predictor training, the predicted heart rate value can be well in line with the actual heart rate value.

In order to compare the superiority of the ADAM algorithm in the prediction of heart rate in the ADAM-LSTM model, the ADAM and SGD gradient descent algorithms is predicted respectively under the same condition. The comparison of the LSTM predication evaluation optimized by the two algorithms is shown in table 1, where Root Mean Square Error (RMSD) and (Mean Absolute Percentage Error MAPE) are used to compare the validity of the algorithms. The result shows that ADAM is significantly better than SGD.

In the meantime, the LSTM heart rate predictor based on multivariate input works quite well, with a root mean square error less than 0.5, indicating high validity.

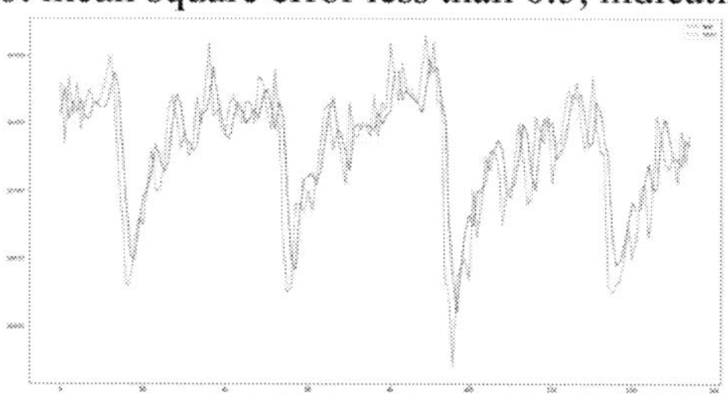

Figure 5. Heart rate prediction.

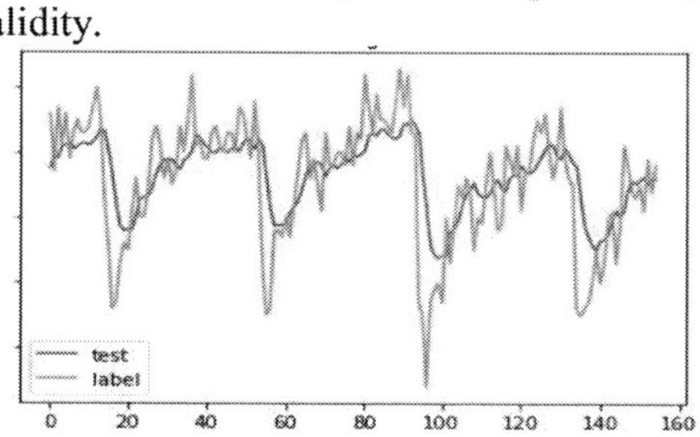

Figure 6. General heart rate prediction and training convergence curve.

4. Discussion

Table 1. Comparison of the effects of Adam and SGD optimization algorithms 3

Model	RMSE	MAPE/%
Sgd-LSTM	0.458	11.93
Adam-LSTM	0.208	6.592

Note: $RMSE = \sqrt{\frac{1}{n}\sum_{i=1}^{n}(yi - y_i^*)^2}$, $MAPE = \frac{1}{n}\sum_{i=1}^{n}\frac{|y^* - y| \times 100}{y}$

n is the size of the sample, y is the actual value of HR, y* is the predicted value of HR

When other parameters are the same, the effect of prediction by solely relying on the heart rate before the t moment is shown in the figure 6. Under the same training conditions, not only the prediction result is undesirable, but also the training convergence curve declines slower.

From the comparison between the actual data and the predicted data, we can see that although the mean absolute error of the prediction is in a small range, no complicated environment is introduced in the process of data collection. In more complicated cases, such as the testee's more intense physical

IOP Conf. Series: Materials Science and Engineering **715** (2020) 012060 doi:10.1088/1757-899X/715/1/012060

activities, more unstable mental state, and even the emergence of other diseases, the neural network prediction proposed in this paper has a great application prospect and potential. It can be seen that the stability of neural network prediction is still in need of improvement in different situations, which is also one of the priorities in the future work.

Acknowledgments
In preparation of this paper, I had to take the help and guidance of my tutor MinMin LUO, who deserves my deepest gratitude. I do appreciate her patience, encouragement and her professional guidance.

References

[1] Malik, M., Heart rate variability. Standards of measurement, physiological interpretation, and clinical use. Task Force of the European Society of Cardiology and the North American Society of Pacing and Electrophysiology. Circulation, 1996. 93(5): p. 1043-65
[2] La Rovere, M.T., et al., Baroreflex sensitivity and heart-rate variability in prediction of total cardiac mortality after myocardial infarction. ATRAMI (Autonomic Tone and Reflexes After Myocardial Infarction) Investigators. Lancet, 1998. 351(9101): p. 478-84
[3] Zhu Bing, Xiao Feng and Yan Chimin. Neural Network Heart Rate Predictor. Computer and Digital Engineering,2009. 37(7): 169-71
[4] Pawar, T., S. Chaudhuri and S.P. Duttagupta, Body Movement Activity Recognition for Ambulatory Cardiac Monitoring. IEEE Transactions on Biomedical Engineering, 2007. 54(5): p. 874-82
[5] Jéquier, E., K. Acheson and Y. Schutz, Assessment of energy expenditure and fuel utilization in man. Annual Review of Nutrition, 1987. 7(7): p. 187-208.
[6] Graves, A., Long Short-Term Memory. Neural Computation, 1997. 9(8): p. 1735-80
[7] Williams, R.J. and D. Zipser, A learning algorithm for continually running fully recurrent neural networks. 1989
[8] Srivastava, N., et al., Dropout: A Simple Way to Prevent Neural Networks from Overfitting. Journal of Machine Learning Research, 2014. 15(1): p. 1929-58
[9] Kingma, D. and J. Ba, Adam: A Method for Stochastic Optimization. Computer Science, 2014.
[10] Bottou, L., Large-Scale Machine Learning with Stochastic Gradient Descent. 2010.

IOP Conf. Series: Materials Science and Engineering **715** (2020) 012061 doi:10.1088/1757-899X/715/1/012061

The disc expansion model to determine the MIC of a planar bounded connected domain

Min Zhou

College of Engineer, China Agricultural University, 100083 Beijing, China

*E-mail: zhoumin2016@cau.edu.cn

Abstract. To determine the maximum inscribe circle (MIC) of a planar bounded connected domain which can be represented as a generalized polygon (with arcs edges), a disco expansion model is proposed. The model simulates a disc continues expanding process in a polygon without friction. When it cannot expand, the three contact points which are both on the circle and the sides of polygons are just the points to determine the MIC. The implementation algorithm of the model for convex polygon is also given. For generalized polygons with arcs or concave points, bottleneck lines as soft edge are adopted to partitioned the polygons first. Then, the proposed model can be applicable for generalized polygons with little modification.

1. Introduction

The solution of the maximum inscribed circle (MIC) in a planar bounded connected domain (PBCD) has wide application value. For example, in the measurement technique, the evaluation of roundness error is required to calculate the maximum inscribed circle of the contour [1]; and the MIC of the machining field is used as the constraint reference for the optional maximum diameter tool when the tool is selected for pocket milling [2], and so on. For the problem of the MIC determination, there are mainly the exploration method for the single facility maximin location problem (MAX G(Ω)) [3], the map algebra method [4], Voronoi diagram method [5], the method of central axis transformation [6], least squares method [7] and other methods [8]. The mainly problems of the above methods are described as follows: (1) MAX G (Ω) only requires the center of the MIC in the bounded domain; (2) The method of the central axis transformation is less efficient in the case of complex curves or too many points being on the contour; Voronoi diagrams, least squares, and other methods cannot be used directly in the presence of holes in the domain. Therefore, this paper proposes the simulation model of the disk expansion to determine the MIC of a planar bounded connected domain.

The MIC calculation of a PBCD can be transferred to the MIC determination of a generalized polygon which represents its boundary. It is easy to know that three non-collinear points can uniquely determine a circle. Thus, the MIC of a polygon can be determined by three points. These three points may be points on the edge of the polygon or concave vertices, and the edge or concave vertices determining the MIC are defined as touch elements. Obviously, the shortest distance from the center of the MIC to the three touch elements is equal, which is the MIC radius. Specially, when there are mutually parallel line segments in the polygon, if the MIC exists between the two parallel line segments, there are an infinite number of MIC, and the center line is a line segment parallel to the two parallel line segments. At this time, any MIC is determined by two touch elements, that is, the two parallel line segments. Based on the above principle, a simulation model of disc expansion is proposed and its implementation algorithm is given.

2. The simulation model of expansion disk

Let Ω be a PBCD, Γ be the boundary of Ω, P be a polygon representing Γ, and *P* is a convex polygon. C is a disc in Ω, and the disc is continuously free to expand outward under the unobstructed condition by the action of internal force *F*, which is called a free expansion disc, referred to as an expansion disc. The simulation model of expansion disk is as following:

(1) When C does not touch the boundary of Ω, the resultant external force on the disk is 0 (assuming all friction coefficients are 0).

(2) When the disc expands to touch a certain edge of *P* at point p (point p is called a contact point), the disc is subjected to a reaction force -F at point p in direction of $\overrightarrow{po}$ (as shown in figure 1), and the condition is not limited by the force state of the disc.

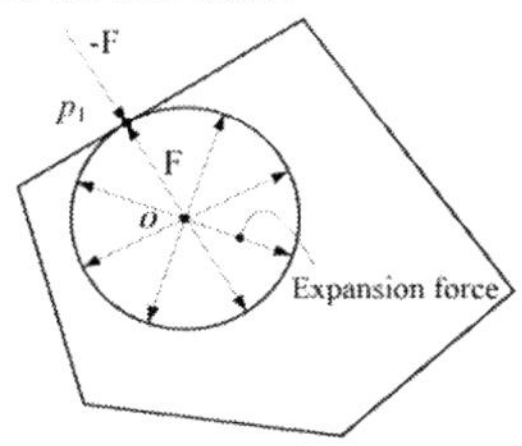

Figure 1. Force diagram of the expansion disk touching the point p_1.

(3) Under the action of external force, the center of the disc moves in the same direction as the direction of reaction force -F at contact point p, while the disc continues to expand in the direction of unobstructed.

(4) The speed at which the disc expands causes the disc to be always tangential to the edge of the obstacle edge until it hits the next obstacle edge.

(5) When the disc expands to the state that the radius cannot continue to increase when the center moves towards the direction of the resultant force of the reaction forces of contact points, and the disc stop expanding, then, the disc at this time is the MIC of *P*.

Based on the simulation model of the disk expansion, let the moving direction of center be $\boldsymbol{n}_o$. The figure 2 illustrates the expansion process of disc in convex polygon *P*. A expansion disk *C* (figure 2a) is expanding in a polygon P. When C expands to touch a certain edge of P at point p_1 (figure 2b), center of C moves towards to the direction of $\overrightarrow{p_1o}$ ($\boldsymbol{n}_o$), while the disc continues expanding. When C expands to touch another edge of P at point p_2 (figure 2c), the disc is affected by two reaction forces at directions $\overrightarrow{p_1o}$ and $\overrightarrow{p_2o}$, thus, the center *o* moves along $\boldsymbol{n}_o$, the direction of the resultant force, while the disc continues expanding. When C expand to touch the edge of P at p_3 as shown in figure 2d, the disc is affected by three reaction forces at directions $\overrightarrow{p_1'o}$, $\overrightarrow{p_2'o}$ and $\overrightarrow{p_3o}$, p_1' and p_2' are new contact points for disc and *P*. Since only three non-collinear points can uniquely determine a circle, it is necessary to judge whether the disc can continue to expand. If the disc continues to expand, and any contact point will becomes a line while distance between the center and the contact point/line no longer increases, then, the disc should stop expanding and the disc at this time is the MIC of P; otherwise, the circle continues to expand and the center moves towards the direction of the new resultant force till the disk cannot to continue expanding to obtain the MIC of *P* (figure 2e).

If *P* is not a convex polygon, when the contact point *p* is a concave vertex, the direction of the reaction is $\overrightarrow{po}$. Or, the polygon is a generalized polygon which has arcs, the simulation model of disc expansion is still adoptable. However, in above both cases, the simulation model of disc expansion may result in the final solution being a local optimal solution rather than a global optimal solution, as shown in figure 3. For this situation, the polygon can be first partitioned according to the bottleneck line, and the bottleneck line of the partition is regarded as a soft boundary which can intersect with the MIC. Then, the disk expansion model with little modification can be used to determine the MICs of each sub-region. And compare them. The MIC of the entire *P* can be obtained by comparing MICs of regional areas.

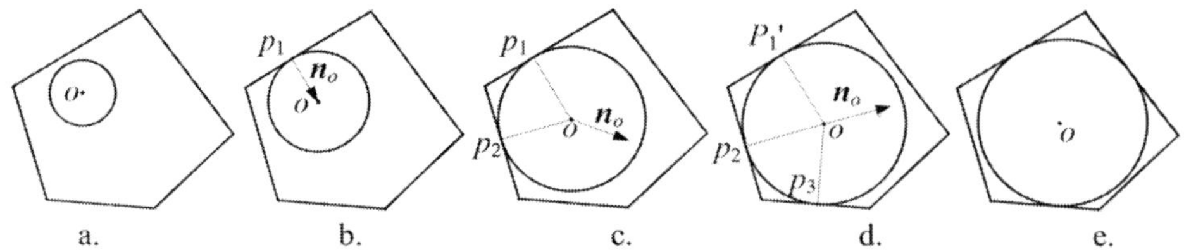

Figure 2. Schematic diagram of the solution for MIC of P by expansion disk method. (a. Initial disk; b. o moving in the direction of $\overrightarrow{p_1 o}$; c. o moving in the $\boldsymbol{n}_o$ direction; d. o moving in the $\boldsymbol{n}_o$ direction; e. MIC of *P*).

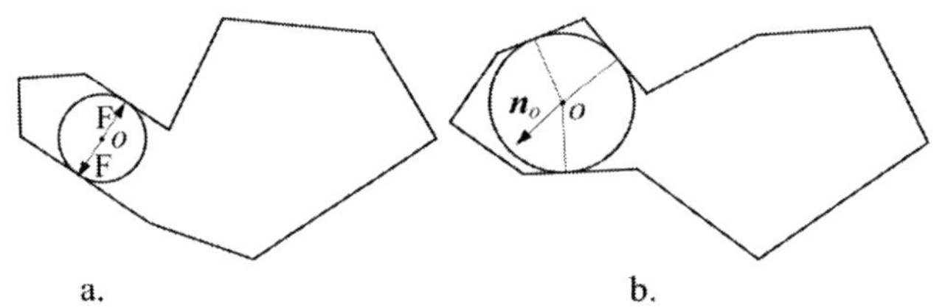

Figure 3. Schematic diagram of local optimal solution. (a. The resultant force of the reaction forces is 0; b. Having reached the condition of stopping expansion).

3. Algorithm for the disc expansion model

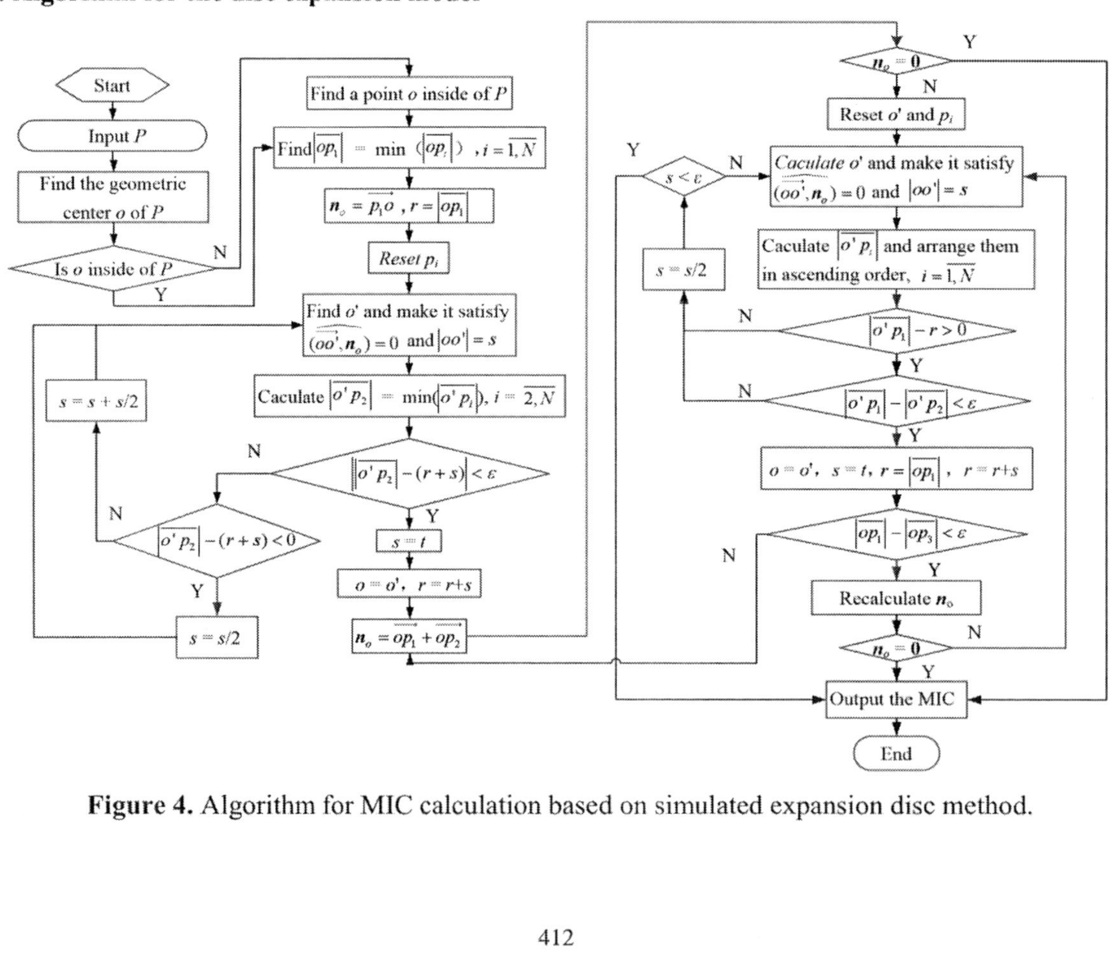

Figure 4. Algorithm for MIC calculation based on simulated expansion disc method.

Since the geometric center of non-porous polygons is very close to the center of its MIC in most cases, and the solution for geometric center determination is simple, this paper first finds the geometric center of the polygon (convex -hull CH(*P*) for generalized polygon) as a starting point for searching the center of the MIC to speed up the search. For polygon or CH(*P*) whose geometric center is not inside of *P*, find a point in *P* as a starting point. The specific algorithm is shown in figure 4.
where:

Step1. Enter *P* and find the geometric center o of *P* or CH(*P*). If *o* is not inside *P*, find a point in *P* as the initial center *o* of the expansion disk;

Step2. Calculate the distance from o to each edge of P, and let $|\overline{op_1}| = \min\{|\overline{op_i}|\}$,
where, $i = 1,2,\ldots,N$, *N* is the number of edges of *P*, p1 is the point on P and closest to *o*;

Step3. Let the vector of expansion disc's center o's moving direction $\boldsymbol{n}_o = \overrightarrow{p_1 o}$, and the radius of the expansion disc $r = |\overline{op_1}|$;

Step4. Find a point *o*' to make it on the extension line of *o*', that is $(\widehat{\overrightarrow{oo'}, \boldsymbol{n}_o}) = 0$, and make $|\overline{oo'}| = s$,
where, *s* is the moving step of the center.

Step5. Calculate the shortest distance from *o*' to each edge of *P* except the edge with p_1, and let $|\overline{op_2}| = \min\{|\overline{op_i}|\}$, $i = 1,2,\ldots,N$;

Step6. If the distance from *o*' to p_1 is equal to the distance from *o*' to p_2, that is $\left||\overline{o'p_2}| - (r+s)\right| < \varepsilon$,
where ε is an any small positive number, then go to Step10; otherwise, go to Step7.

Step7. If the distance from *o*' to p_2 is smaller than the distance from *o*' to p_1, that is, $\left||\overline{o'p_2}| - (r+s)\right| < 0$, then, reduce the step size to make $s = s/2$, and go back to Step4.

Step8. If the distance from *o*' to p_2 is bigger than the distance from *o*' to p_1, that is, $\left||\overline{o'p_2}| - (r+s)\right| > 0$, then, increase the step size to make $s = s + s/2$, and go back to Step4.

Step9. Update the values of step, center and radius to make $s = t$, $o = o'$, $r = r + s$.
where, t is the given value of step size and equal to the previously given step *s* in Step4.

Step10. Recalculate the moving direction of the center, which is the sum of the directions of the reaction forces of p1 and p2, that is, $\boldsymbol{n}_o = \overrightarrow{op_1} + \overrightarrow{op_2}$

Step11. Determine whether vector $\boldsymbol{n}_o$ is equal to **0**, if it is equal to **0**, it means that the edge with p_1 are parallel to the edge with p_2, and the MIC of *P* has been obtained, then, go to Step21; otherwise, go to Step12.

Step12. Find a point *o*' and make $\overrightarrow{oo'}$ and $\boldsymbol{n}_o$ in the same direction, that is $(\widehat{\overrightarrow{oo'}, \boldsymbol{n}_o}) = 0$, and let $|\overline{oo'}| = s$.

Step13. Calculate the distance from *o*' to each side of *P* and arrange them in ascending order;

Step14. If $|\overline{op_3}|$ is bigger than the radius obtained by Step10, namely, $|\overline{o'p_1}| - r > 0$, then go to Step16; otherwise, go to step15.

Step15. Reduce the value of step size to make $s = s/2$, and determine whether inequality $s < \varepsilon$ is satisfied. If it is satisfied, go to step21; Otherwise, go to step12.

Step16. Determine whether the value of $|\overline{o'p_1}|$ is equal to the value of $|\overline{o'p_2}|$, that is, whether $|\overline{o'p_1}| - |\overline{o'p_2}| < \varepsilon$ is satisfied, if yes, go to Step17; otherwise, go to Step15;

Step17. Update the values of step, center and radius to make $r = |\overline{o'p_1}|$, $s = t$, $o = o'$, $r = r+s$.

Step18. Determine whether $|\overline{op_1}|$ and $|\overline{op_3}|$ are equal, that is, whether $|\overline{op_1}|-|\overline{op_3}|<\varepsilon$ is satisfied, if yes, go to Step19; otherwise, go to Step10;

Step19. Recalculate the moving direction of the center: find the two sides with the two smallest angles among the three edges with the three contact points, then the moving direction of the center is just the direction of the resultant force of the reaction forces generated by the two sides.

Step20. Determine whether the vector $\boldsymbol{n}_o$ is equal to **0**, if yes, go to Step21; Otherwise, go to Step12.

Step21. Let $r_{MIC} = r$, $o_{MIC} = o$ ', output the value of the radius and the position information of the center.

4. Conclusion

A disc expansion model is proposed to determine the MIC of generalized polygon with arc lines. The polygon with concave point is portioned into convex polygons by bottleneck lines which are dealt as soft edges first. Then, the model is applicable. The experiment results also show the correctness of the proposed model and the corresponding algorithm.

References

[1] X.Q. Lei, C. Zhang, Y.J. Xue, J.S. Li, Measurement., 44: 345-350(2011)
[2] M. Zhou, G.L. Zheng, Z.C. Chen, Int J Adv Manue Tech, 83(1-4): 407-420(2016)
[3] F.G. Zhang, Melachrinoudis E, Comput Optim Appl, 19(2): 209-234(2001)
[4] Hu Hai, Liu Jingnan, Geo Inf Sci of Wuhan University, 31(008): 700-703(2006)
[5] Toussaint G.T., Int J Com Info Sci, 12(5): 347-358(1983)
[6] J.T. Chen, Dalian: Dalian University of Technology(2009)
[7] K. Kim, S. Lee, H.B. Jung, Int J Adv Manue Tech, 16(8): 559-563(2000)
[8] Y. Sun, R. Che, Opt.Precision Eng., 11(2): 181-187(2003)

Dynamic searchable symmetric encryption schemes with forward and backward security

Ke Huang, Xiaolei Dong, Zhenfu Cao and Jiachen Shen*

Shanghai Key Laboratory of Trustworthy Computer, East China Normal University, Shanghai 200062, China

*Email: jcshen@sei.dcnu.edu.cn

Abstract. Dynamic searchable symmetric encryption (DSSE) is helpful to users with limited storage. In DSSE, the users are able to perform search and update queries on the ciphertext which stored on the cloud server. However, DSSE may suffer from file-injection attacks and leak information after deletion. In order to solve these problems, DSSE schemes with forward security and/or backward security have been proposed. In this paper, we propose two efficient DSSE schemes. The first one supports forward security using binary tree as the basic data structure. The second one uses puncturable encryption to achieve both forward and backward privacy.

1. Introduction

With the development of cloud technology, when local storage resources are limited, user can consider storing files on the cloud server. However, since the external cloud server is untrusted, the file needed to be encrypted and uploaded to the cloud server. In order to guarantee the searchable of these encrypted files, searchable encryption is proposed. In short, searchable encryption ensure that user can search on encrypted files which stored on a cloud server. Dynamic searchable encryption allows the server to dynamically update the stored data without changing the confidentiality of the data and without losing the searchable feature. But most of the current dynamic searchable encryption schemes have the risk of being attacked by file injection. The adversary injects new files obtained some keywords into the encrypted database and then do some query operations. By that way, the adversary may learn the user's private information. Similarly, when deleting files, it will also cause some leakage of private information. Therefore, for the addition and deletion, forward security and backward security are proposed respectively to protect the confidentiality of files.

1.1. Contribution

We will propose two DSSE constructions in this paper. The first one is forward security with basic data structure binary tree. The second one achieves both forward privacy and backward privacy. In more details, our main contributions are as follows:

Binary tree. We use binary tree as a basic data structure to store encrypted database to improve query efficiency, include search queries and update queries; change the structure of leaf node on thr tree to satisfy our DSSE scheme.

Forward privacy. By improving the composition of the search token base on Bost and combine the data structure of binary tree, we construct an effienct DSSE scheme with forward security.

Backward privacy. Based on our forward security DSSE scheme, we use puncturable encryption to achieve backward security by adding a tag.

1.2. Related work

Searchable Encryption Symmetric searchable encryption (SSE) was first introduced by Song et al. [1]. Curtmola et al. [2] gave the modern definitions of SSE and corresponding security model. They designed the first reversed-index-based SSE construction. Chase and Kamara [3] focused on multi-maps, which was a base structure of their scheme. After that, SSE is a particular case of structured encryption.

However, the above schemes cannot support data update. To solve this problem, some Dynamic symmetric searchable encryption (DSSE) schemes have been proposed. Kamara and Papamanthou [4] designed the first sublinear dynamic scheme. Cash et al. [5] constructed a dynamic scheme optimized for large datasets.

Many DSSEs could suffer file-injection attack proposed by Zhang et al. [6]. In file-injection attack, the adversary can infer some information through client queries by inject some new documents into encrypted database. So, the need for guaranteeing forward security which protects security of new added parts raised. Similarly, we have backward security which protects security of new added parts and later deleted. The notions of forward and backward security were first introduced by Stefanov et al. [7]. The authors present an ORAM-inspired SSE construction. However, this was only designed to improve the performance of the scheme, rather than its security and the overhead of ORAM is too high for a practical SSE scheme [8]. Forward and backward security formal definitions for DSSE were given by Bost [9] and Bost et al. [10]. In [9], Bost proposed a DSSE scheme only support forward security. Later on, Bost et al. [10] proposed a DSSE scheme support both forward and backward security.

Puncturable encryption In [11], Green and Miers introduced a new form of encryption named puncturable encryption. Puncturable encryption is a way to provide forward secrecy in asynchronous messaging systems. Their scheme modifies the secret key every time a message is received, and modified key has decryption capability only for special messages. That means old messages remain safe. They also mentioned secure deletion as another application of their work.

In this paper, we will use puncturable encryption to securely delete entries in an encrypted database.

2. Preliminaries

2.1. Dynamic symmetric searchable encryption

We follow the database model given in the paper [10]. A database DB is a collection of keyword/index pairs. W is a set of keywords and *ind* stand for index of document. $DB = (ind_i, W_i)$, $ind_i \in (0,1)^l$ and $W_i \subseteq (0,1)^*$.

D is the number of documents in DB and W is the number of keywords and N is the number of index/keyword pairs. DB(w) represents the set of indices of documents which contain a keyword w.

A DSSE scheme consists of an algorithm Setup and two protocols Search and Update as described below.

EDB $\leftarrow$ Setup(DB,1^λ): Input a security parameter 1^λ and a database DB. The algorithm outputs an encrypted database EDB

(index, $\perp$) $\leftarrow$ Search(q, σ, EDB): The protocol is executed between a client and a server. The client asks for documents with given keyword. The server returns a collection of file indices of requested documents.

(σ, EDB) $\leftarrow$ Update(σ, op, in, EDB):The protocol runs between a client and a server. According to the op = (add, del), client send different tokens to server and the server do update on EDB.

2.2. Forward and backward privacy

Forward security (or forward privacy) is that an SSE scheme won't leak what keywords are associated whit the keyword/document pairs while doing update queries.

Backward security (or backward privacy) is that an SSE scheme can't get that document index when doing a search query after a keyword/document pair is deleted.

Forward privacy and Backward privacy were informally defined in [7]. In this paper, we use the formal definition of [10].

3. Our schemes

3.1. Data structure binary tree

3.1.1. Binary tree structure

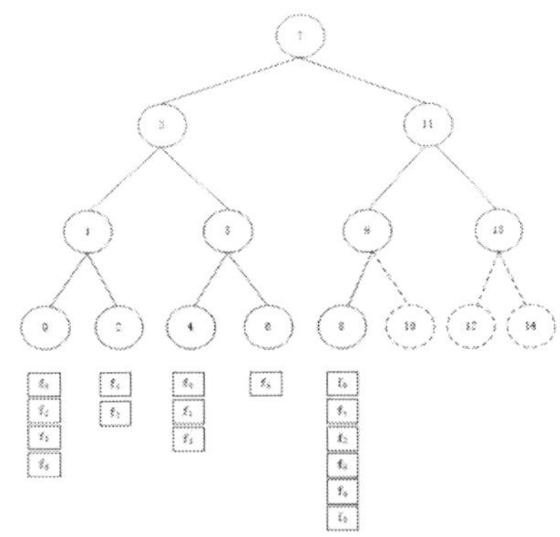

Figure 1. Basic data structure binary tree.

In our construction, a perfect binary tree is used as the basic data structure. As shows in figure 1. The value of a leaf node represents a keyword. But not every leaf node is real, as indicated by the dotted line in figure 1. In addition, a list is stored in the leaf node, which stores the file indices associated with that keyword.

3.1.2. Binary tree assignment and update

We define the following construction, search and update operations to manipulate the base data structure of our DSSE. As shown in the Algorithm 1.

Algorithm 1. Our Binary Tree

```
TreeCon (m)
Input number of keywords m
Output complete binary tree PBT

1: Construct a PBT with ⌈log(m)⌉ +1 levels.
2: Set the top m leaf nodes to be real from left to right.
3: return PBT

TreeAssign (cn, PBT)
Input complete binary tree PBT, an initial counter cn
Output assigned binary tree ABT

1: TreeAssignSub (cn, PBT.root)
2: return ABT

TreeAssignSub (cn, n)
Input a node n of PBT, an integer c
Output Assigned binary tree ABT

1: if n.left != ⊥ then
2:     TreeAssignSub (cn, n.left)
3: end if
4: n.value = cn
5: cn = cn + 1
6: if n.right != ⊥ then
6:     Create a new root node root_n.
7:     Create a VBT = ABT
8:     ABT.parent=VBT.parent =root_n
9:     TreeAssign (v, VBT)
10:    ABT.root = root_n
11:    ABT.root.value = v-1
7:     TreeAssignSub (cn, n.right)
8: end if
9: return ABT

TreeSearch (ABT.root, v)
Input Assigned binary tree ABT, a value v
Output a node n of ABT

1: if v = root.value then
2:     return n = root
3: else if v < root.value then
4:     TreeSearch (root.left, v)
5: else
6:     TreeSearch (root.right, v)
7: end if
8: return n

TreeUpdate (add, v, ABT)
Input op= add, number of keywords m, value v, ABT
Output updated ABT

1: if ABT = ⊥ then
2:     Create a node.
3:     Associate value v = 0 to this node.
4:     Set ABT to this node.
5: else if ABT.root.value = m-1 then

12:    m=m+1
13: else
14:    Execute line 12.
15: end if
16: return ABT
```

3.2. DSSE with forward privacy

In this section, we apply our binary tree to the Bost[9] scheme to improve the efficiency of search and update queries.

Now we are ready to present the DSSE scheme with forward security. The scheme is described in Algorithm 2.

Setup(1^λ): For a security parameter 1^λ, the algorithm outputs (PK, SK, K, **W**, m).

Search(w, σ, EDB): The protocol is executed between a client and a server. The client asks for documents with keyword w. The server returns a collection of file indices of requested documents.

Update(add, w, *ind*, σ, m, EDB): The protocol is performed jointly by a client and server. The client wishes to add a keyword w together with a file index *ind* to EDB.

Algorithm 2. DSSE with forward security

<u>**Setup**(1^λ)</u>

Client:
Input security parameter 1^λ
Output (PK, SK, K, **W**, m)

$K \leftarrow \{0, 1\}^\lambda$
(SK, PK) ← **KeyGen**(1^λ)
W ← empty map
return (PK, SK, K, **W**)

<u>**Search** (w, σ, **EDB**)</u>

Client:
Input keyword w, state of EDB σ
Output (K_w, ST_c, c, v)

$K_w \leftarrow F_K(w)$
(ST_c, c, order) ← **W**[w]
if (ST_c, c, order) = ⊥ **then**
  return ∅
end if
v = 2*order
Send (K_w, ST_c, c, v) to the server.

Server:
Input (K_w, ST_c, c, v), EDB
Output {*ind*}

Upon receiving (K_w, ST_c, c, v)
node **n** =**TreeSearch** (v)
{e} ← **n**.*list*
for i = c to 0 **do**
  $ST_{i-1} \leftarrow \pi_{PK}(ST_i)$
end for
for j = 0 to c **do**
  ind ← $e_j \oplus$ **H** (K_w, ST_j)
  Output the *ind*
end for

<u>**Update** (add, w, *ind*, σ, **EDB**)</u>

Client:
Input op = add, keyword/index pair w and *ind*, State of EDB σ
Output e

$K_w \leftarrow F_K(w)$
(ST_c, c, order) ← **W**[w]
if (ST_c, c, order) = ⊥ **then**
  ST_0 ← ***M***, c ← -1
end if
W[w] ← (ST_{c+1}, c+1)
e ← *ind* ⊕ **H** (K_w, ST_{c+1})
v = 2*order
Send (e, v, m_c) to the server

Server:
Input (e, v, m_c)
Output updated **EDB**

if $m_c > m_s$ **then**
TreeUpdate (add, v, **ABT**)
end if
node **n** ← **TreeSearch** (**ABT**.root, v)
n.*list*.add (e)

3.3. DSSE with forward and backward privacy

In this section, we apply PPKE [11] to scheme in section 4.2. So now we have a DSSE scheme with both forward and backward security. For details, we add a tag to the file index. When deleting a pair of keyword/index, the server modifies the tag and the client puncture the SK of PPKE. After that, that keyword/index pair couldn't be decrypted.

The scheme is described in Algorithm 3.

Setup(1λ): For a security parameter 1λ, the algorithm outputs (PK, SK, K, Ktag, W).

Search(w, σ, EDB): The protocol is executed between a client and a server. The client asks for documents with keyword w. The server returns a collection of file indices of requested documents. But only certain indices (not been deleted) will be returned.

Update(op, w, *ind*, σ, EDB): The protocol is performed jointly by a client and server. Op = (add, delete). If op = add, the client wishes to add a keyword w together with a file index *ind* to EDB. There are three cases same as the TreeUpdate. If op = delete, the client wishes to delete a keyword/index pair in the EDB.

Algorithm 3. DSSE with forward and backward security

```
Setup(1^λ)

Client:
Input security parameter 1^λ
Output (PK, SK, K, W, m)

K, K_tag ← {0, 1}^λ
(SK, PK) ← KeyGen(1^λ)
W ← empty map
return (PK, SK, K, K_tag, W)

Search (w, σ, EDB)

Client:
Input w, σ
Output (K_w, ST_c, c, v)

K_w ← F_K(w)
(ST_c, c, order, sk_0, PPK) ← W[w]
if (ST_c, c, order, sk_0, PPK) = ⊥ then
    return ∅
end if
v = 2*order
Send (K_w, ST_c, c, v) to the server.

Server:
Input (K_w, ST_c, c, v), EDB
Output {ind}

Upon receiving (K_w, ST_c, c, v)
node n =TreeSearch(2v)
e ← n.list
for i = c to 0 do
    ST_{i-1} ← π_PK(ST_i)
end for
for j = 0 to c do
    e_j' ←PPKE.Decrypt(e_j)
    ind ← e_j' ⊕ H(K_w, ST_i)
    Output the ind
end for

Update (op, w, ind, σ, EDB)

Client:
Input op = (add, del), keyword/index pair w and ind,
      State of EDB σ
Output e

if op =add then
    K_w ← F_K(w)
    (ST_c, c, order, sk_0, PPK) ← W[w]
    if (ST_c, c, order, sk_0, PPK) = ⊥ then
        ST_0 ← M, c ← -1
        (PSK_0, PPK) ← PPKE.KeyGen(1^λ)
        sk_0 ←PSK_0
    end if
    t ← F_Ktag (w, ind)
    W[w] ← (ST_{c+1}, c+1, order, sk_0, PPK)
    e' ← ind ⊕ H(K_w, ST_{c+1})
    e ← PPKE.Encrypt (e', t)
    v = 2*order
    Send (e, v, m_c, sk_0) to the server
else op = del
    t ← F_Ktag (w, ind)
    (ST_c, c, order, sk_0, PSK) ← W[w]
    (sk_0', sk') ← PPKE.IncPuncture(sk_0, t)
    W[w] ← (ST_c, c, order, sk_0', PSK)
    e' ← ind ⊕ H(K_w,ST_{c+1})
    e ← PPKE.Encrypt (e', t)
    v = 2*order
    Send (e, v, sk_0, sk') to the server
end if

Server:
Input (op, e, v, sk_0, m_c/sk')
Output EDB

if op =add then
    if m_c > m_s then
        TreeUpdate (add, v, PBT)
    end if
    node n ← TreeSearch (ABT.root, v)
    n.list.add(e)
else op =del
    node n ← TreeSearch (ABT.root, v)
    n.SK_n← [sk_0, sk_1, ... sk_n, sk_{n+1}=sk']
end if
```

4. Conclusion

In this paper, we give two DSSE schemes. The first one is forward security with basic data structure binary tree. This basic scheme is improved by [9] and the perfect binary tree is used to improve the efficiency of search and update queries. However, this scheme only meets the requirements of forward security. To address these problems, we propose the second DSSE construction that uses puncturable encryption [11]. It achieves both the forward and backward security. But the above schemes just support the queries on single-keyword. In the future, we would like to construct a DSSE scheme support multi-keywords.

Acknowledgments

This work was supported in part by National Nature Science Foundation of China (Grant No.61632012, 61672239, and U1509219) and in part by "the Fundamental Research Funds for the Central Universities". Zhenfu Cao and Jiachen Shen shared the corresponding authors.

IOP Conf. Series: Materials Science and Engineering **715** (2020) 012062 doi:10.1088/1757-899X/715/1/012062

References

[1] D. X. Song, D. Wagner, A.Perrig, Practical techniques for searches on encrypted data. IEEE S&P 2000, pp.44–55(2000)

[2] R. Curtmola, J. Garay, S. Kamara, R. Ostrovsky, Searchable symmetric encryption: improved definitions and efficient constructions. ACM ASIACCS 2006, pp.79–88 (2006)

[3] M. Chase, S. Kamara, Structured encryption and controlled disclosure. ASIACRYPT 2010, **6477**, pp.577–594 (2010)

[4] S. Kamara, C. Papamanthou, Parallel and dynamic searchable symmetric encryption. A.R. Sadeghi (ed.) FC 2013, **7859**, pp.258–274 (2013)

[5] D. Cash, J. Jaeger, S. Jarecki, C. S. Jutla, H. Krawczyk, M. C. Rosu, M. Steiner, Dynamic searchable encryption in very-large databases: Data structures and implementation. NDSS 2014 (2014)

[6] Y. Zhang, J. Katz, C. Papamanthou, The fallacy of composition of oblivious RAM and searchable encryption. USENIX Security Symposium, pp.707–720 (2016)

[7] E. Stefanov, C. Papamanthou, E. Shi, Practical dynamic searchable encryption with small leakage. NDSS, **71**, pp.72–75 (2014)

[8] M. Naveed, The fallacy of composition of oblivious RAM and searchable encryption. Cryptology ePrint Archive Report 2015/668 (2015)

[9] R. Bost, Σοφος: forward secure searchable encryption. ACM SIGSAC CCS2016, pp.1143–1154 (2016)

[10] R. Bost, B. Minaud, O. Ohrimenko, Forward and backward private searchable encryption from constrained cryptographic primitives. ACM SIGSAC CCS2017, pp.1465– 1482(2017)

[11] M. D. Green, I. Miers, Forward secure asynchronous messaging from puncturable encryption. IEEE S&P 2015, pp.305–320(2015).

Dynamic programming based energy management of HEV with three driving modes

Shota Inuzuka*, Tielong Shen and Takafumi Kojima

Sophia University, 7-1, Kioi-cho, Chiyoda-ku, Tokyo, Japan

*Email: shota526@eagle.sophia.ac.jp

Abstract. We solve a optimization problem for HEV energy management. In this study, HEV system has three driving modes(EV, parallel, series parallel) with two motors, one engine and two clutches. We change this problem into a combination optimization problem by discretizing each torque as a variable. It is difficult to solve this problem because the combinations of variable are so many. To this hard problem, we solved it efficiently by using dynamic programming and applying steady-state approximation. It does not require much computation time, and a global approximate optimal solution can be obtained.

1. Introduction

In recent years, the demand for vehicles powered by electricity such as electric vehicles(EV), hybrid electric vehicles(HEV), and fuel cell vehicles (FCV) is increasing. Especially HEV can be fully utilized even in conventional infrastructure equipment, so it is expected that the proportion in the market will further increase. Various methods for HEV energy management are reported such as Dynamic Programming(DP) [1], Model Predictive Control [2], Genetic Algorithm [3], etc.

DP for HEV energy management can obtain global optimal solution according to Bellman's optimal equation when a demand vehicle speed is given in advance. Many of these studies are optimization methods for parallel HEV([4],[5]). They show that optimization calculation can be performed by using SoC as a state variable. In this study, HEV model has three driving modes. When the driving mode is series parallel mode, it cannot be solved exactly only with SoC. You need to consider not only SoC but also engine rotation speed as state variables. However, it would take much time to solve it. So, we find approximate optimal solution only with SoC as a state variable by applying steady-state approximation to engine rotation speed.

2. HEV system modelling

In this study, the HEV model which has one engine and two motor/generators connected by planetary gear and two clutches is used for simulation. This model has three driving modes (which are EV, parallel, and series parallel modes) and they are decided by clutch signals. The whole HEV system is shown in figure 1.

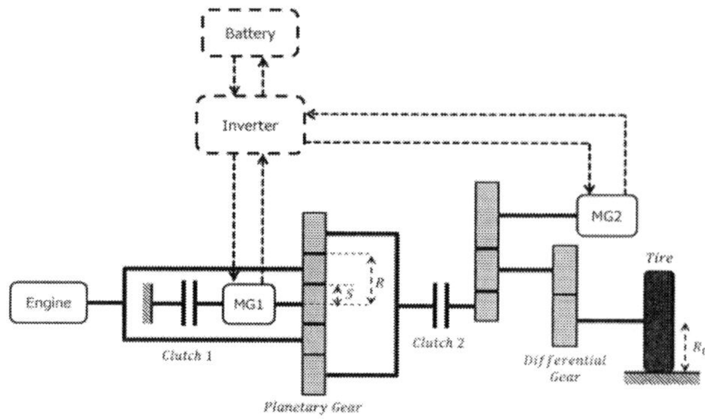

Figure 1. HEV system with two motors, one engine and two clutches.

You can understand to look at figure 1, the clutch 1 separates the system into part of engine and MG1 and part of MG2, the clutch 2 fixes MG1. First, if the clutch signals are [cl1,cl2] = [off,off], the drive shaft connect only with MG2, so this mode is EV mode. Second, if they are [cl1,cl2] = [on,on], the MG1 is fixed and the engine and the MG2 apply force to the drive shaft, so this mode is parallel mode. Finally, if they are [cl1,cl2] = [on,off], the MG1 and the MG2 and the engine apply force to the drive shaft, so this mode is series parallel mode. Therefore, the clutch signals(cl1,cl2), the engine torque(τ_e), the MG1 torque(τ_{MG1}), and the MG2 torque(τ_{MG2}) as inputs generate the vehicle speed(v), the engine rotation speed(ω_e), the MG1 rotation speed(ω_{MG1}), and the MG2 rotation speed(ω_{MG2}) as outputs.

$$[v, \omega_e, \omega_{MG1}, \omega_{MG2}] = f_{powertrain}(cl1, cl2, \tau_e, \tau_{MG1}, \tau_{MG2}) \tag{1}$$

Derivative of fuel consumption($\dot{mf}$) and SoC can be calculated using these torques, rotation speeds, the fuel consumption map which is function of τ_e and ω_e, and the motor/generator energy loss maps which are function of $\tau_{MG1/MG2}$ and $\omega_{MG1/MG2}$.

$$\dot{mf} = f_{FCmap}(\tau_e, \omega_e) \tag{2}$$

$$S\dot{o}C = f_{battery}(\tau_{MG1}, \tau_{MG2}, \omega_{MG1}, \omega_{MG2}) \tag{3}$$

3. The optimization problem

We solve the optimization problem as follows:

$$\min_{u_k} \sum_{k=1}^{T} (\dot{mf}_k + \lambda S\dot{o}C_k)\Delta t \tag{4}$$

$$\text{subject to} \quad v_{d_k} = v_k \tag{5}$$

$$|SoC_0 - SoC_T| \le 0.0005[\text{-}] \tag{6}$$

$$SoC_{min} \le SoC_k \le SoC_{max} \tag{7}$$

$$P_{b_{min}} \le P_{b_k} \le P_{b_{max}} \tag{8}$$

$$0 \le \tau_{e_k} \le \tau_{e_{max}}(\omega_{e_k}) \tag{9}$$

$$\tau_{MG1_{min}}(\omega_{MG1_k}) \le \tau_{MG1_k} \le \tau_{MG1_{max}}(\omega_{MG1_k}) \tag{10}$$

$$\tau_{MG2_{min}}(\omega_{MG2_k}) \le \tau_{MG2_k} \le \tau_{MG2_{max}}(\omega_{MG2_k}) \tag{11}$$

where $u_k = [cl1_k, cl2_k, \tau_{e_k}, \tau_{MG1_k}, \tau_{MG2_k}]$. v_{d_k} is the demand vehicle speed which is given in advance. SoC_0 and SoC_T are initial and terminal value of SoC and SoC_0 is equal to 0.163[-]. Equation (7) and (8) are constraint equations on the maximum and minimum value of SoC and electric power of battery. Equation (9) to (11) are constraint equations on the maximum and minimum value of each torque which is function of each rotation speed. We solve the energy management problem to minimize the energy equivalent of equation (4).

4. DP algorithm

4.1. Dynamic programming

Although each torque as a variable is a continuous value, we change it into a combination optimization problem by discretizing part of them. DP is known as a solution to it. It is effective and can find a global optimal solution by repeatedly storing and using partial optimal solutions according to Bellman's optimal equation. It can be described as:

$$V^{\#}_{x_k} = \min g_{x_k} + V^{\#}_{x_{k+1}} \tag{12}$$

where x_k is a state variable at the k step. g_{x_k} is a partial cost from x_k to x_{k+1}. $V^{\#}_{x_k}$ is a partial optimal cost from x_k to x_T. In this algorithm, you need to calculate it from T step back to 0 step because you must refer to $V^{\#}_{x_{k+1}}$ to calculate $V^{\#}_{x_k}$.

4.2. Steady-state approximation

In this algorithm, we apply a steady-state approximation which is ignoring the transient state of engine rotation speed. When u_k is redefined as $u_k = [cl1_k, cl2_k, \tau_{e_k}, \omega_{e_k}]$, each torque and rotation speed is function of redefined u_k and v_d under the condition of Equation (5), by applying this approximation.

$$[\tau_{e_k}, \tau_{MG1_k}, \tau_{MG2_k}, \omega_{e_k}, \omega_{MG1_k}, \omega_{MG2_k}] = f_{approx}(u_k, v_d) \tag{13}$$

When $[cl1_k, cl2_k] = [off, off]$, which means EV mode, ω_{MG2_k} can be calculated from v_d and the gear ratio between the drive shaft to MG2, and $\omega_{e_k} = \omega_{MG1_k} = 0$. τ_{MG2_k} can be calculated from $\dot{v}_d$, and $\tau_{e_k} = \tau_{MG1_k} = 0$. When $[cl1_k, cl2_k] = [on, on]$, which means parallel mode, ω_{MG2_k} can be calculated as with EV mode, and $\omega_{MG1_k} = 0$. And then, ω_{e_k} can be calculated from planetary gear constraint which is $(s + r)\omega_{e_k} = s\omega_{MG1_k} + r\eta\omega_{MG2_k}$. τ_{e_k} is determined from u_k, and τ_{MG2_k} can be calculated from $\dot{v}_d$ and τ_{e_k}, and $\tau_{MG1_k} = 0$. Finally, when $[cl1_k, cl2_k] = [on, off]$, which means series parallel mode, ω_{MG2_k} can be calculated as with the other modes. ω_{e_k} is determined from u_k, and ω_{MG1_k} can be calculated from planetary gear constraint. τ_{e_k} is determined from u_k, and τ_{MG1_k} and τ_{MG2_k} can be calculated from equations of motion in this model when $\dot{\omega}_{e_k} = 0$ and $\dot{v} = \dot{v}_d$. $\dot{\omega}_{e_k} = 0$ indicates steady state of engine rotation speed because we apply a steady-state approximation.

You can execute the calculation of DP with SoC as the state variable in one dimension by applying the approximation.

4.3. Algorithm

In this section, we show the algorithm in detail. In this algorithm, you leave partial optimal solutions for discrete points of each SoC at each time, and they are used next calculation. Therefore, you have to discretize $SoC(= x)$. In addition, as mentioned earlier, we discretize inputs to make combinatorial optimization problems. So, τ_e and ω_e are discretized. For every combination of u_k, each torque and rotation speed is calculated by equation (13) at x_k, and x_{k+1} can be calculated by equation (3). Among these combinations, those satisfying constraint equations from (6) to (11) and $V^{\#}_{x_{k+1}} \neq \emptyset$ become candidates for the partial optimum solution. Of course, x_{k+1} does not reach exactly the discrete point of x because SoC is continuous value. Therefore, you need to approximate the closest point. Let U_{x_k} be the set of candidates, it can be described as:

$$U_{x_k} = \{u_k \mid eq.(6) \sim (11), V^{\#}_{x_{k+1}} \neq \emptyset\} \tag{14}$$

For every element of U_{x_k}, the partial cost which is regarded as g_{u_k} can be calculated by equation (2) and (3). Let G_{x_k} be the set of g_{u_k}, it can be described as:

$$G_{x_k} = \{g_{u_k} | g_{u_k} = (\dot{mf}_k + \lambda \dot{SoC}_k)\Delta t, u_k \in U_{x_k}\} \tag{15}$$

You need to evaluate not only g_{u_k} but also $V^{\#}_{x_{k+1}}$ to select a partial optimal solution at x_k. You can calculate and store $V^{\#}_{x_k}$ and $u^{\#}_{x_k}$ which is the partial optimal solution according to Bellman's optimal equation by using G_{x_k} and $V^{\#}_{x_{k+1}}$ at x_k as below:

$$V^{\#}_{x_k} = \min\ V^{set}_{x_k} \tag{16}$$

$$u^{\#}_{x_k} = \operatorname{argmin}\ V^{set}_{x_k} \tag{17}$$

$$\text{where}\ \ V^{set}_{x_k} = \{V_{x_k} \mid V_{x_k} = g_{u_k} + V^{\#}_{x_{k+1}}, g_{u_k} \in G_{x_k}\} \tag{18}$$

You have to execute this calculation to all x_k discretized, and from $k = T$ to $k = 0$. You can execute the calculation efficiently by using $V^{\#}_{x_{k+1}}$, and that is a good point of this algorithm. When $u^{\#}_{x_0}$ is selected according to this algorithm, you can call the optimal solution by referring $u^{\#}_{x_k}$ at x_k which is calculated by $u^{\#}_{x_{k-1}}$ and equation (3) and (13), and approximated to the closest x_k discretized.

5. Simulation results

In this study, we use a vehicle speed data($= v_d$) as below:

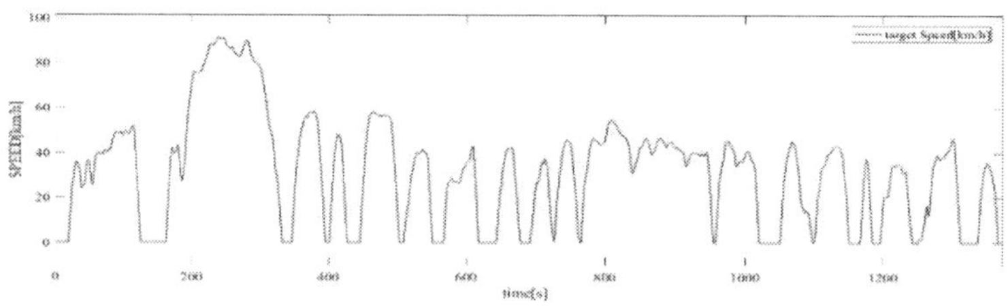

Figure 2. A vehicle speed data($= v_d$).

We solve the optimization problem for this vehicle speed data with hyperparameters as below:

Table 1. Setting of hyperparameters.

Hyperparameter	Range of value
Step size of x: Δx [%]	0.1,0.01
Step size of τ_e: $\Delta\tau_e$ [Nm]	1,5
Step size of ω_e: $\Delta\omega_e$ [rpm]	50,100

In addition, the step time (Δt) is equal to 1[s], and equivalent coefficient (λ) is equal to 0.0129[-] which means that the electric energy is converted to the weight of fuel consumption through price of them. We show the simulation result as below:

Table 2. Simulation results for each combination of hyperparameters.

$\Delta x, \Delta\tau_e, \Delta\omega_e$	Calculation time [min]	Sum of cost [g]	Distance per liter [km/ℓ]
0.1,1,50	10.1	147.0	59.53
0.01,5,50	22.4	145.4	60.19
0.01,1,100	47.4	145.3	60.25
0.01,1,50	70.0	145.2	60.25

When $[\Delta x, \Delta\tau_e, \Delta\omega_e] = [0.01,1,50]$, each power plant's operating point is as follows:

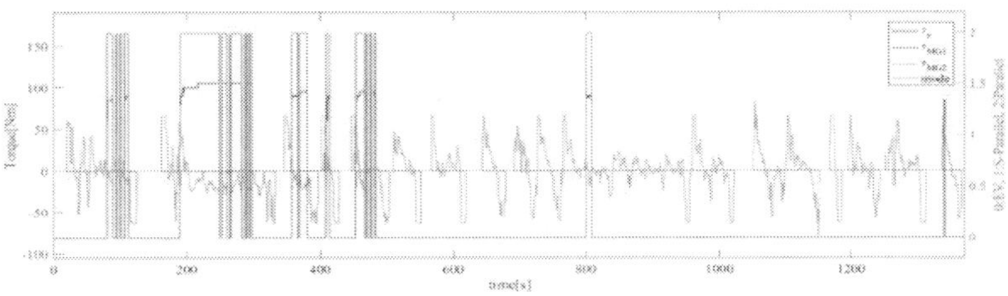

Figure 3. Torques of τ_e, τ_{MG1}, and τ_{MG2}, and driving mode.

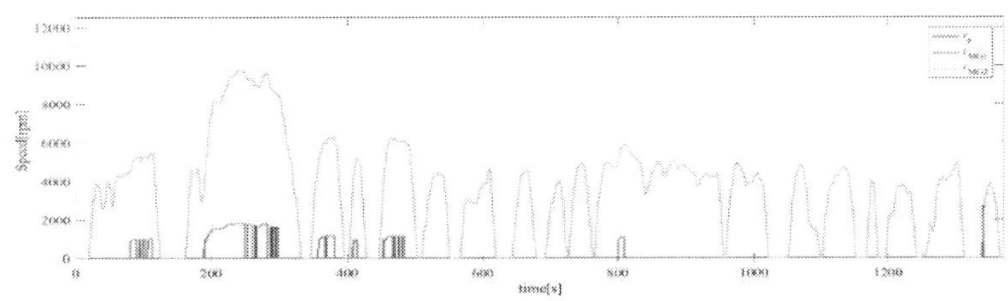

Figure 4. Rotation speeds of ω_e,ω_{MG1}, and ω_{MG2}.

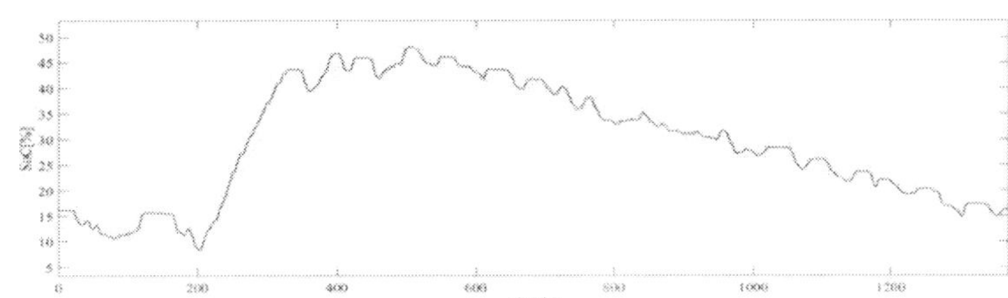

Figure 5. Optimized SoC route.

6. Discussion

Looking at the result, you can see that the calculations have been completed quickly and global approximate optimal solutions can be obtained because the engine runs when the vehicle speed is relatively high as shown in figure 2, 3, and 4. It is a clever use to run the engine at that timing because thermal efficiency of the engines is efficient when the rotation speed is high. In HEV energy management optimization problems, it is important when engines run because the electric energy you can use is limited. In figure 5, you can see that equation (6) is satisfied. Looking at table 2, the distance per liter is too efficient because we apply the steady-state approximation to engine rotation speed, so we don't calculate the cost in transient state.

7. Conclusions

We show that you can solve the HEV energy management optimization problems efficiently by using DP and applying steady-state approximation to the complex problem that the HEV model has three modes. It is strict to solve with SoC and ω_e as the state variables, but that is tough calculations. In this study, the state variable is only SoC, and ω_e is approximated. Although you can obtain just approximate optimal solution, it is global optimal solution according to Bellman's optimal equation and can be calculated quickly.

References

[1] T. Leroy, F.Vidal-Naquet and P.Tona 2014 Stochastic Dynamic Programming based Energy Management of HEV's: an Experimental Validation *IFAC* vol 47 p 4813-18

[2] Borhan H, Vahidi A, Phillips A, Kuang M, Kolmanovsky I and Cairano S 2012 MPC-Based Energy Management of a Power-Split Hybrid Electric Vehicle *IEEE* vol 20 p 593-603

[3] Fang L, Qin S, Xu G, Li T and Zhu K 2011 Simultaneous Optimization for Hybrid Electric Vehicle Parameters Based on Multi-Objective Genetic Algorithms *energies* vol 4 p 532-544
[4] Lee H, Kang C, Park Y and Cha S 2016 Study on Power Management Strategy of HEV using Dynamic Programming *World Electric Vehicle Journal* vol 8 p 274-80
[5] Wang X, He H, Sun F and Zhang J 2015 Application Study on the Dynamic Programming Algorithm for Energy Management of Plug-in Hybrid Electric Vehicles *energies* vol 8 p1-20

IOP Conf. Series: Materials Science and Engineering **715** (2020) 012064 doi:10.1088/1757-899X/715/1/012064

Improved non-local mean denoising algorithm based reliability of reference pixel

Junghyun Lee[1] and Jechang Jeong[2,*]

[1]Department of Electronics and Computer Engineering, Hanyang University, Republic of Korea
[2]Department of Electronic Engineering, Hanyang University, Republic of Korea

*Email: jjeong@hanyang.ac.kr

Abstract. In this paper, we propose a method for improved the non-local mean denoising algorithm by using the similarity in comparison with reference pixels. The noise in image is eliminated by the calculation to follow values as the patterns are similar in non-local mean denoising alogirthm. In the conventional algorithm, all the comparison results of the pixels are included in the denoising process, and even the low similarity result is included. The method causes the denoised effect to incorrect. In this paper, we propose an algorithm that reduces the influence of low-similarity reference pixels or improves the effect of high-similarity reference pixels. The proposed method improves both the object quality and the subject quality.

1. Introduction

The noise on image is added from the acquiring process in nature to the image of the digital, and also when the image is on the processes to communicate and compress. There are various method for image de-noising. Image denoising process has a goal that suppress the noise and preserve the shape and features of the original image.

There are 2 kinds of methods for image denoising process. One is a method on spatial domain and the other is a method on transform domain.

Typically, the denoising method on spatial domain are Gaussian filter[1], median filter[2], and bilateral filter[3]. The methods are categorized local denoising algorithm because they conjugate the neighboring pixels as reference pixels for denoising calculation. Budes' non-local mean(NLM) denoising algorithm is representative of non-local denoising method. This method conjugates the whole pixels in the image for denoising calculation.

On the other hand, the denoising method on transform domain are wavelet transform[4-6] as orthogonal wavelet transform and non-orthogonal wavelet transform. The method are efficient tool for compression and denoising with thresholding process.

The conventional NLM denoising algorithm[7] calculates denoised pixel value through comparing the pixel and the other whole pixels. The comparison results of the pixels are included in the calculation. The low similarity reference pixel causes inaccuracy of the calculation.

In this paper, we propose the improved NLM denoising algorithm. Our proposed algorithm

IOP Conf. Series: Materials Science and Engineering **715** (2020) 012064 doi:10.1088/1757-899X/715/1/012064

supresses the effect of low-similarity reference pixel on the denoising calculation. Also, we apply the method on fast NLM denoising algorithm. In section 2, conventional denoising NLM algorithm and fast NLM denoising algorithm[8-9] will be discussed in details. Next, in section 3, we propose the method to improve NLM denoising algorithm. Finally, experimental results and the conclusion are shown in Section 4 and 5.

2. Non-local mean denoising algorithm

2.1. Conventional non-local mean denoising algorithm

The non-local mean denoising algorithm by Buades et al [7] belongs to the non-locally method. Unlike local denoising algorithm[1-3], the whole pixels participate in denoising calculation in non-local denoising algorithm.

The patch unit(P_i) that is constructed by current pixel(u_i) and the 8 neighboring pixels is used for denoising calculations in NLM denoising method. The weight is defined according to the similarity between current patch and reference patch. The more similar the current patch is to the reference patch, the higher the weight. The weight is calculated by the similarity between patches as follow :

$$w_{ij} = \frac{1}{z_i}\exp(-\frac{1}{h^2}\| P_i - P_j \|^2) . \quad (1)$$

where the $\| \ \|^2$ operation is Euclidean distance operator between two patches, z_i is the normalized coefficient as eq. (2), and h is the smoothing coefficient.

$$z_i = \sum_j \exp(-\frac{1}{h^2}\| P_i - P_j \|^2) \quad (2)$$

The denoised pixel value $\widehat{u_i}$ is calculated by the weight for current pixel u_i as follow :

$$\widehat{u_i} = \sum_{j \in I} w_{ij} u_j \ . \quad (3)$$

The whole pixels on the image affect the denoising calculation regardless of the weight.

2.2. Fast non-local mean denoising algorithm

The fast non-local mean denoising algorithm by Darbon et al [8] is faster on the calculation than conventional non-local mean denoising algorithm, because the method skips a process that is square operations of pixel difference.

$$w_{ij} = \frac{1}{z_i}\exp(-\frac{1}{h^2}\sum_{\delta \in \Delta}(v(i+\delta) - v(j+\delta))^2) \quad (4)$$

The difference of patches on eq. (2) can be replaced the difference of pixels on patch Δ as like eq. (4). A symbol δ, means the order of pixel on the patch Δ. A new vector d_x, is defined by the difference of each starting point as like eq. (5).

$$d_x = j - i \quad (5)$$

The square operations are eliminated by a cumulative summation of squared difference pixel value. A new image S_{d_x}, is generated by the cumulative summation of squared difference pixel value as like eq. (6).

$$S_{d_x}(p) = \sum_{k=0}^{p}(v(k) - v(k+d_x))^2 \quad (6)$$

The value p has the domain $p \in [[0, n-1]]$, and the value n is number of pixels on the image. The patch Δ is 1-dimension array as like $\Delta = [[-P, P]]$, and a new location vector $\hat{p}$ is defined as :

$$\hat{p} = i + \delta_x \quad (7)$$

Then the eq. (4) can be rearranged with the vector $\hat{p}$ as :

$$w_{ij} = \frac{1}{z_i}\exp(-\frac{1}{h^2}\sum_{\hat{p}=s-P}^{s+P}(v(\hat{p}+d_x))^2) \quad (8)$$

The eq. (8) can be rearranged with the image S_{d_x} as :

$$w_{ij} = \frac{1}{z_i}\exp(-\frac{1}{h^2}(S_{d_x}(s+P) - S_{d_x}(s-P))) \quad (9)$$

When 512x512 size image is calculated by NLM algorithm, the fast NLM algorithm is about 80 times faster than conventional algorithm.

3. Methods

3.1. Reliability of reference

The weight is calculated by the comparison between the patch including the current pixel and the other patches in the conventional NLM denoising algorithm. If the reference patch is not similar the current patch as including much noise or different pattern, it causes inaccuracy of the denoising calculation. Therefore, we call it is a low-reliability reference patch. We can calculate accurately the denoised pixel value, when we eliminate the effect of the low-reliability reference patch on the calculation. Not only elimination of low-reliability reference pixel, but also emphasis on high-reliability reference pixel is efficient for denoising.

3.2. Elimination of low-reliability reference

For elimination of low-reliability reference, we add the sorting and truncating process on the conventional NLM denoising algorithm, as figure 1.

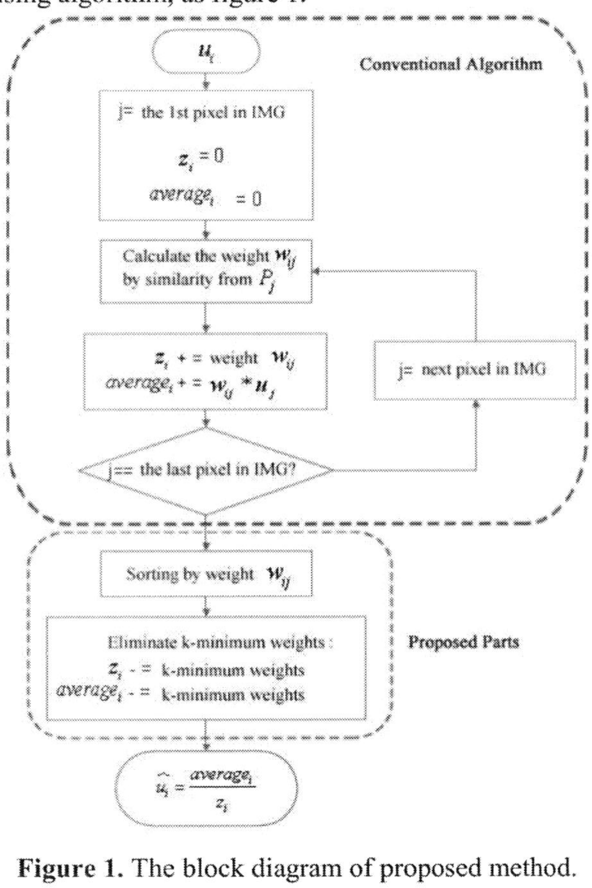

Figure 1. The block diagram of proposed method.

We calculate the weight w_{ij}, that is the comparison result between the current patch and the reference patch, and the normalized coefficient z_i,, that is the result of multiplying the weight by the pixel value. Sort by the weight w_{ij}, and we give the zero weight to the k-minimum weights reference. After elimination of low-reliability reference, we calculate the denoised pixel value by eq. (3). In this paper, we use k from 2 to 7 for finding optimized k value.

3.3. Emphasis on high-reliability reference

In the fast NLM denoising algorithm, we apply the method to emphasize on high-reliabitly referene. The conventional fast NLM denoising algorithm makes an image by the cumulated summation of squared difference pixel value. Calculate a weight from difference vector value from new coordinate on the image.

In the algorithm, we calculate the weight by inverse proportion with normalized coefficient z_i. It has smaller weight, the more similar it means. Therefore, by thresholding method[5] to the squared difference pixel value, we make the weight value is zero.

4. Experimental results

In this paper, we experiment on MATLAB with 4 images (Lena, Barbara, Boat, and House, 512x512 size). Each image uses white Gaussian noise variance from 10 to 60 in 10 steps.

For the experiment of section 3.2, we use the coefficient k from 2 to 7. For the experiment of section 3.3, we use hard thresholding method and soft thresholding method, when the squared difference pixel value is less than 5 reference values (5e-5, 1e-4, 5e-4, 1e-3, and the average value).

4.1. Elimination of low-reliability reference

Table 1 tabulates the object quality, PSNR, against the conventional NLM denoising algorithm in dB. The experimental results of elimination of low-reliability reference by section 3.2. From Exp 1 to Exp 6, the number of eliminate reference is from 2 to 7.

Table 1. PSNR [dB] result of comparison conventional and proposed algorithm.

Lena	σ=10	σ=20	σ=30	σ=40	σ=50	σ=60	Barbara	σ=10	σ=20	σ=30	σ=40	σ=50	σ=60
NLM	**31.7503**	**31.5685**	**29.5833**	27.5571	26.1599	25.1718	NLM	**30.9347**	**30.0076**	27.1362	24.3918	23.0086	22.2417
Exp 1	31.7502	31.5599	29.5758	27.5833	26.2181	25.2466	Exp 1	30.9347	30.0042	27.1451	24.4432	23.0882	22.3305
Exp 2	31.7502	31.5495	29.5674	27.6093	26.2755	25.3162	Exp 2	30.9347	30.0005	27.1560	24.4968	23.1680	22.4170
Exp 3	31.7502	31.5378	29.5593	27.6333	26.3243	25.3765	Exp 3	30.9347	29.9961	27.1671	24.5489	23.2430	22.4958
Exp 4	31.7501	31.5247	29.5491	27.6555	26.3700	25.4282	Exp 4	30.9346	29.9917	27.1793	24.6024	23.3154	22.5695
Exp 5	31.7501	31.5117	29.5378	27.6742	26.4089	25.4724	Exp 5	30.9346	29.9866	27.1910	24.6535	23.3857	22.6390
Exp 6	31.7500	31.4972	29.5273	**27.6899**	**26.4445**	**25.5101**	Exp 6	30.9346	29.9810	**27.2037**	**24.7033**	**23.4515**	**22.7030**
Boat	σ=10	σ=20	σ=30	σ=40	σ=50	σ=60	House	σ=10	σ=20	σ=30	σ=40	σ=50	σ=60
NLM	**30.8755**	**29.8155**	27.2676	25.0202	23.7105	22.9366	NLM	**30.9193**	**29.9986**	28.0072	26.8232	24.2501	23.0025
Exp 1	30.8755	29.8127	27.2729	25.0547	23.7713	23.0099	Exp 1	30.9193	29.9955	28.0168	26.8711	24.3112	23.0739
Exp 2	30.8755	29.8091	27.2796	25.0911	23.8320	23.0788	Exp 2	30.9193	29.9920	28.0255	26.9232	24.3755	23.1422
Exp 3	30.8755	29.8056	27.2851	25.1258	23.8894	23.1412	Exp 3	30.9193	29.9883	28.0363	26.9745	24.4316	23.2198
Exp 4	30.8754	29.8010	27.0912	25.1611	23.9435	23.1999	Exp 4	30.9132	29.9841	28.0485	27.0189	24.4965	23.2903
Exp 5	30.8754	29.7967	27.2968	25.9420	23.9947	23.2532	Exp 5	30.9132	29.9799	28.0601	27.0642	24.5570	23.3662
Exp 6	30.8754	29.7916	**27.3034**	**25.2253**	**24.0424**	**23.3023**	Exp 6	30.9132	29.9745	**28.0711**	**27.1210**	**24.6102**	**23.4222**

IOP Conf. Series: Materials Science and Engineering **715** (2020) 012064 doi:10.1088/1757-899X/715/1/012064

From the table, the proposed algorithm shows better noise reduction effect to the much noisy image against the conventional algorithm. When the noise variance is 20, the PSNR results are less only 0.0028~0.07 dB by conventional algorithm. Especially, when the noise variance is 60, the PSNR results are more from minimum 0.3383 dB to maximum 0.4613 dB.

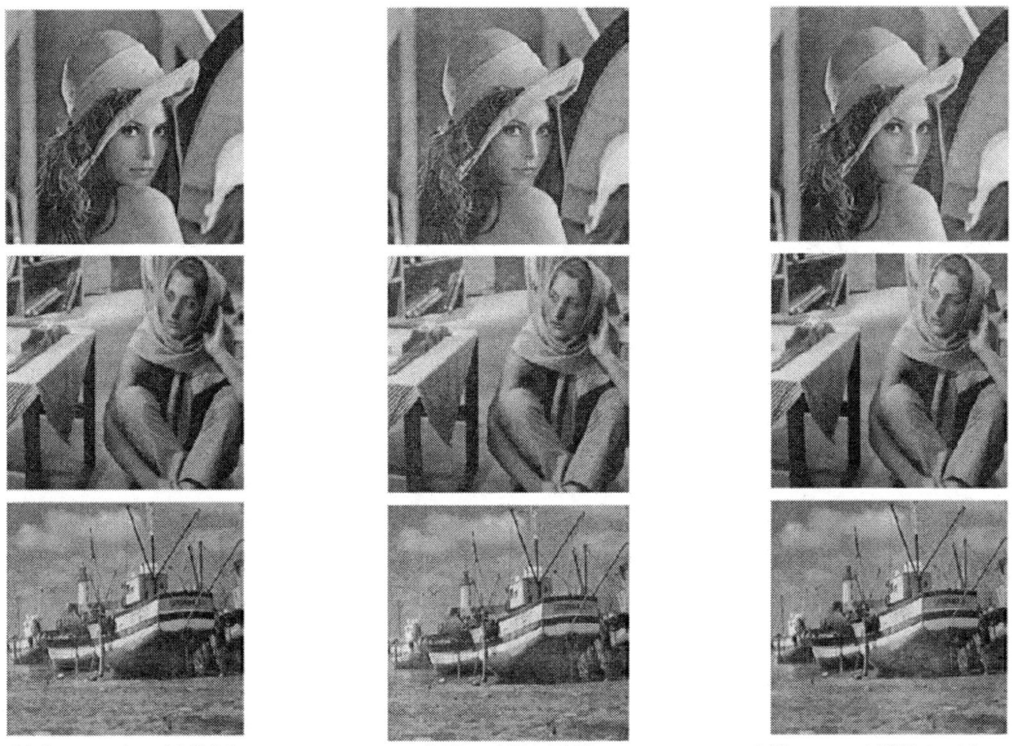

(a)Conventional NLM. (b)Fast NLM [9]. (c)Proposed NLM with soft thresholding.

Figure 2. The comparison of conventional NLM denoising algorithms and proposed algorithm.

The proposed method when it uses soft thresholding effects efficient and almost same as conventional fast NLM algorithm on subjective quality, and the comparison of the result images are identified by figure 2. From figure 3, the part of Barbara image, we can verify the improvement of subject quality on proposed algorithm. There are vertical, diagonal stripe patterns and flat patterns on the part. The left side diagonal stripe patterns and the right-down side vertical stripe patterns are more restored from proposed algorithm by conventional algorithm.

The proposed method is robust to much noisy image, and is especially efficient when there are repeated patterns.

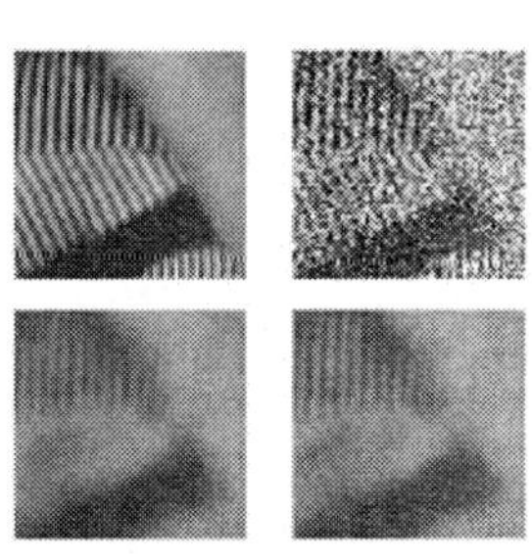

(a)	(b)
(c)	(d)

(a)Original Image.

(b)Noisy Image.

(c)The Result of Conventional algorithm.

(d)The Result of Proposed algorithm.

Figure 3. The part of Barbara image (σ=60, k=7).

4.2. Emphasis on high-reliability reference

Table 2 tabulates the PSNR result of using hard thresholding method against the conventional fast NLM denoising algorithm in dB. Table 3 tabulates the PSNR result of using soft thresholding method. Each table, from Experiment 1 to Experiment 5, the threshold coefficients are set-up the values(5e-5, 1e-4, 5e-4, 1e-3, and average value).

Table 2. PSNR [dB] result of comparison fast NLM algorithm and proposed algorithm (with hard thresholding).

Lena	σ=10	σ=20	σ=30	σ=40	σ=50	σ=60	Barbara	σ=10	σ=20	σ=30	σ=40	σ=50	σ=60
NLM	30.5100	**29.5547**	**28.4873**	**27.0489**	25.9776	25.1240	NLM	29.8891	**28.2418**	26.5242	24.3667	23.0797	22.3047
Exp 1	30.5094	29.5545	28.4873	27.0489	25.9776	25.1240	Exp 1	29.8888	28.2418	26.5242	24.3667	23.0797	22.3047
Exp 2	30.5079	29.5545	28.4873	28.4873	25.9776	25.1240	Exp 2	29.8883	28.2415	26.5243	24.3667	23.0797	22.3047
Exp 3	30.4890	29.5506	28.4871	28.4871	25.9777	25.1240	Exp 3	29.8762	28.2388	26.5243	24.3668	23.0797	22.3047
Exp 4	30.4636	29.5430	28.4867	28.4867	25.9777	25.1241	Exp 4	29.8597	28.2338	**26.5248**	24.3670	23.0797	22.3047
Exp 5	**31.9822**	29.2888	28.3873	28.3873	**25.9818**	**25.1300**	Exp 5	**30.8882**	28.0792	26.4852	**24.3989**	**23.1036**	**22.3182**
Boat	σ=10	σ=20	σ=30	σ=40	σ=50	σ=60	House	σ=10	σ=20	σ=30	σ=40	σ=50	σ=60
NLM	29.8614	**28.2591**	26.6070	24.8786	23.7313	22.9784	NLM	29.8055	**28.7113**	27.8966	26.7636	25.6637	24.6411
Exp 1	29.8608	28.2591	26.6070	24.8786	23.7313	22.9784	Exp 1	29.8041	28.7113	27.8966	26.7636	25.6637	24.6411
Exp 2	29.8593	28.2590	26.6069	24.8786	23.7313	22.9784	Exp 2	29.8019	28.7112	27.8966	26.7636	25.6637	24.6411
Exp 3	29.8467	28.2567	26.6070	24.8787	23.7313	22.9783	Exp 3	29.7999	28.7089	27.8965	26.7637	25.6637	24.6411
Exp 4	29.8319	28.2522	**26.6071**	**24.8788**	23.7313	22.9783	Exp 4	29.7957	28.7048	**27.8967**	26.7638	25.6637	24.6412
Exp 5	**30.3549**	27.9575	26.5100	24.8761	**23.7422**	**22.9868**	Exp 5	**30.3966**	28.4853	27.8159	**26.7911**	**25.6997**	**24.6533**

Table 3. PSNR [dB] result of comparison fast NLM algorithm and proposed algorithm (with soft thresholding).

Lena	σ=10	σ=20	σ=30	σ=40	σ=50	σ=60	Barbara	σ=10	σ=20	σ=30	σ=40	σ=50	σ=60
NLM	30.5100	29.5547	28.4873	**27.0489**	**25.9776**	**25.1240**	NLM	29.8891	28.2418	**26.5242**	**24.3667**	**23.0797**	**22.3047**
Exp 1	30.5117	29.5551	28.4873	27.0489	25.9776	25.1240	Exp 1	29.8907	28.2420	26.5242	24.3667	23.0797	22.3047
Exp 2	30.5146	29.5557	28.4874	27.0489	25.9776	25.1240	Exp 2	29.8928	28.2424	26.5241	24.3666	23.0797	22.3046
Exp 3	30.5595	29.5640	28.4878	27.0489	25.9776	25.1239	Exp 3	29.9203	28.2488	26.5239	24.3663	23.0795	22.3046
Exp 4	30.6419	29.5788	**28.4883**	27.0488	25.9774	25.1239	Exp 4	29.9719	24.2599	26.5232	24.3659	23.0794	22.3045
Exp 5	**32.4275**	**30.1172**	28.3648	26.9079	25.8763	25.0529	Exp 5	**31.1186**	**28.5785**	26.1038	24.0607	22.9116	22.2105
Boat	σ=10	σ=20	σ=30	σ=40	σ=50	σ=60	House	σ=10	σ=20	σ=30	σ=40	σ=50	σ=60
NLM	29.8614	28.2591	**26.6070**	**24.8786**	**23.7313**	**22.9784**	NLM	29.8055	28.7113	27.8966	**26.7636**	**25.6637**	**24.6411**
Exp 1	29.8623	28.2594	26.6070	24.8786	23.7313	22.9784	Exp 1	29.8073	28.7117	27.8966	26.7636	25.6637	24.6411
Exp 2	29.8644	28.2595	26.6069	24.8785	23.7312	22.9783	Exp 2	29.8099	28.7118	27.8967	26.7636	25.6636	24.6410
Exp 3	29.8919	28.2650	26.6070	24.8784	23.7311	22.9783	Exp 3	29.8255	28.7129	27.8971	26.7635	25.6635	24.6410
Exp 4	29.9440	**28.2745**	26.6067	24.8780	23.7310	22.9782	Exp 4	29.8811	**28.7211**	**27.8977**	26.7631	25.6633	24.6409
Exp 5	**30.3997**	28.2491	26.2595	24.6363	23.5818	22.8888	Exp 5	**30.5124**	29.1315	27.8536	26.5844	25.4842	24.5581

From table 2 and table 3, when the white Gaussian noise variance is 10 and using threshold coefficient is average value, the proposed method improves the reduction effect on PSNR, from minimum 0.4935 dB to maximum 1.9175 dB.

When the noise variance over 30, using hard thresholding method is better reduction noise effect and using average value for coefficient is the best efficient. But, when the noise variance over 30, using soft threshodling method is worse reduction noise effect against conventional algorithm.

In terms of subject visual quality, the irregular patterns as the hat shape of Lena image, are smoothed by the proposed algorithm. But the reduction noise effect is efficient when there are repeated patterns as Barbara image.

5.Conclusion

This paper presents improved NLM denoising algorithm. The proposed methods induce more reduction noise effect by the consideration of reliability from referen7ce pixel / patch. The conventional NLM denoising algorithm use the whole reference pixel / patch on the image for denoising calculation. The low-reliability reference bring about inaccuracy of the calculation.

The elimination of low-reliability reference method, when the noise variance over 30, is better PSNR result from minimum 0.0158 dB(σ=30) in average to maximum 0.3884 dB(σ=60) in average. The emphasis on high-reliability reference method on fast NLM algorithm, when the noise variance are 10~20, using soft thresholding method is better 0.7624 dB PSNR result in average. And when the noise variance are 50~60, using hard threshoding method is better 0.0111 dB PSNR result in average.

Acknowledgments

This work was supported by the ICT R&D program of MSIP/IITP. [2014-0-00670, Software Platform for ICT Equipment].

References

[1] M. Lindenbaum, M. Fischer, A. Bruckstein, Pattern Recognition, **27(1)**, 1-8 (1994)
[2] L. Alvarez, P.L. Lions, J.M. Morel, SIAM J. on num. analysis, **29(3)**, 845-866 (1992)
[3] L. Yin, R. Yang, M. Gabbouj, Y. Neuvo, IEEE Trans. Cir. and Sys., **43(3)**, 157-192 (1996)
[4] A. Chambolle, RA. De Vore, N.Y. Lee, B.J. Lucier, IEEE Trans. Image Process. **7(3)**, 319-335 (1998)
[5] J. Portilla, V. Strela, M.J. Wainwright, E.P. Simoncelli, IEEE Trans. Image Process. 12(11), 1338-1351 (2003)
[6] J.K. Romberg, H. Choi, R.G. Baraniuk, IEEE Trans. Image Process. **10(7)**, 1056-68 (2001)
[7] A. Buades, B. Coll, M. Morel, SIAM Journal on Multi. Modeling and Simul. **4(2)**, 490-530 (2005)
[8] J. Darbon, A. Cunha, T. Chan, S. Osher, G. Jensen, *Biomedical Imaging, IEEE International Symposium on*, 1331-34 (2008)
[9] D. Donoho, IEEE Trans. Info. Theory **41(3)**, 613-627 (1995)

Study of low and high temperature oxidation of n-pentane at supercritical conditions in a jet stirred reactor

Yifu Yang

Veritas Collegiate Academy, Chesapeake, VA 23320-3026, USA

*E-mail: cyang826@gmail.com

Abstract. The intermediate/low and high temperature oxidation of n-pentane from atmospheric pressure to supercritical pressures (1-100 atm) was studied by using Chemkin, a commercial software. A recently developed n-pentane model was used to predict the temperature evolutions of major and intermediate species, like n-pentane, O_2, CO, CO_2, CH_2O, C_2H_4, CH_3CHO at varied pressure and temperature conditions. It is found that as the environmental pressure increases, the difference between high temperature oxidation and low temperature oxidation becomes less obvious because the negative temperature coefficient (NTC) is decreasing (at 10 atm and 100 atm, high temperature oxidation occurs right after low temperature oxidation). At supercritical pressures, the high temperature oxidation is governed by the branching reactions of RO_2, and the addition reaction of QOOH with oxygen dominates the reaction paths at both low and high temperature oxidations because the addition reaction between QOOH and oxygen is still more important than the decomposition reaction of QOOH at high temperature and 100 atm. According to the pathway analysis and sensitivity analysis, the branching reactions of RO_2 are dominating at both low and intermediate to high temperatures.

1. Introduction

Recently, research field on chemical kinetics modeling of hydrocarbon oxidation has been moved toward the fuel with larger molecules such as n-pentane, butane, n-octane and isooctane [1-22]. Internal combustion engines have been widely in service as both industrial and infrastructural purposes. However, the efficiency of this kind of engines has been a problem troubling the engineers for a long time. The complicated internal combustion engines which are using complex fuels, such as gas turbines and pistons, could run into a problem named engine knock. The main cause of this problem is the low temperature property of the fuel. During the reaction, the activity will drop back down when it reaches a certain temperature [21].

Kinetics reaction mechanisms are complex and require a large amount of calculations. They also do not occur within only one reaction path, instead, a lot of branches are going through in the reactions. Calculations are supposed to be extremely complicated. As a solution for further understanding of the oxidation properties of these fuels mentioned above, computer programs can simulate the process of the reaction with an unapproachable speed by humans. And, the results would be purely theoretical without any unpredictable factors influencing the results.

As one of the smallest alkanes which have rich low temperature reactivity, n-pentane has been investigated in experiments and simulations at both low and high temperatures [21-26]. A recent n-pentane model [23] shows a good agreement with experimental results. However, the low and high temperature chemistry of n-pentane has not been studied at high pressure to supercritical pressure conditions in both experiments and simulations. It is noted that supercritical combustion has a tremendous potential in gas turbine and air-breathing engines due to its high fuel reactivity, high combustion efficiency, and ultra-lean flammability.

Therefore, the goal of the present study is to understand the low and high temperature oxidation of n-pentane from atmospheric pressure to supercritical pressures (1-100 atm) by using Chemkin software. A recently developed n-pentane model [21] was used to predict the temperature evolutions of major and intermediate species, like n-pentane, O_2, CO, CO_2, CH_2O, C_2H_4, CH_3CHO at varied pressure and temperature conditions.

2. Modelling

A recent literature model recently proposed by Bugler et al. [21] was used to carry on the present simulations and kinetic analysis. All simulations were performed in the perfectly stirred reactor model within CHEMKIN-PRO software employing the transient solver [27], and the numerical results were compared at the following simulation conditions in table 1.

Table 1. Simulation conditions.

Case	Equivalence ratio	n-C_5H_{12} (%)	O_2 (%)	N_2 (%)	Residence time (s)	Temperature range (K)	Pressure(atm)
1	0.5	1	16	83	1	350-1050	1
2	0.5	1	16	83	1	350-1050	10
3	0.5	1	16	83	1	350-1050	100

3. Results

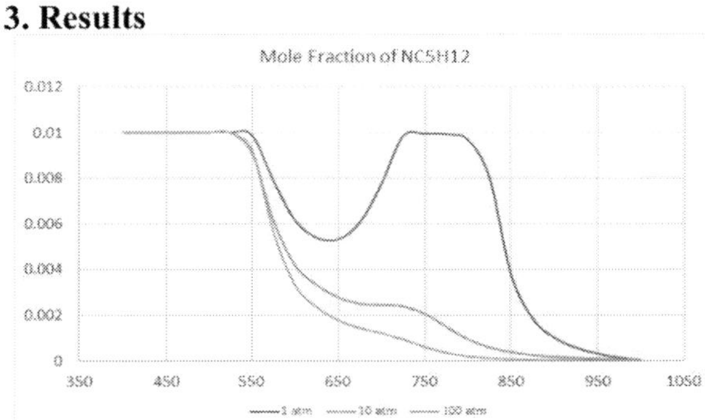

(a)

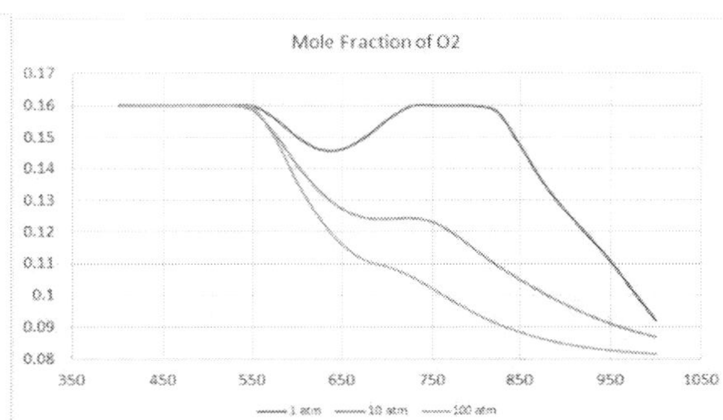

(b)

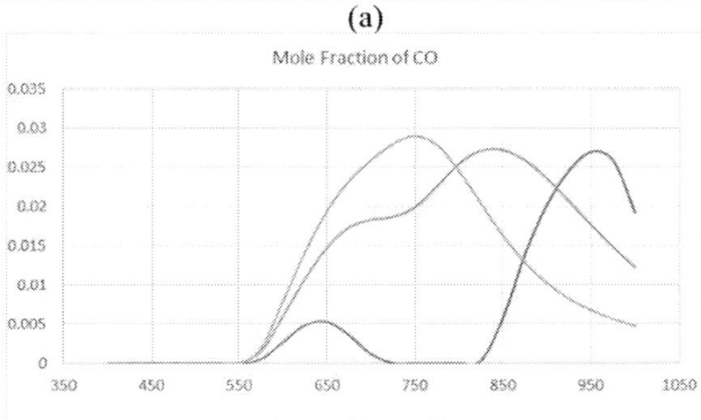

(b)

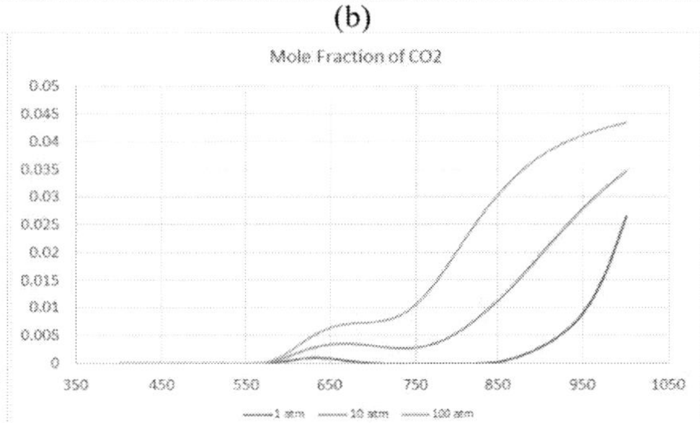

(d)

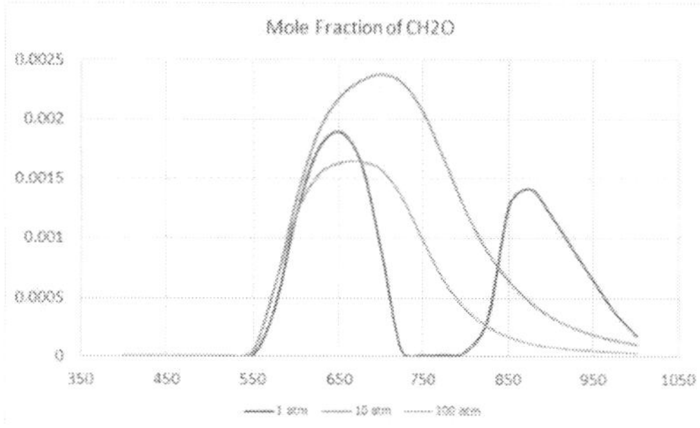

(e)

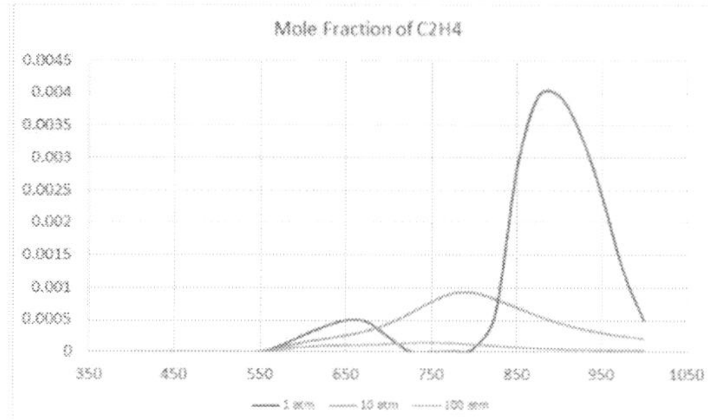

(f)

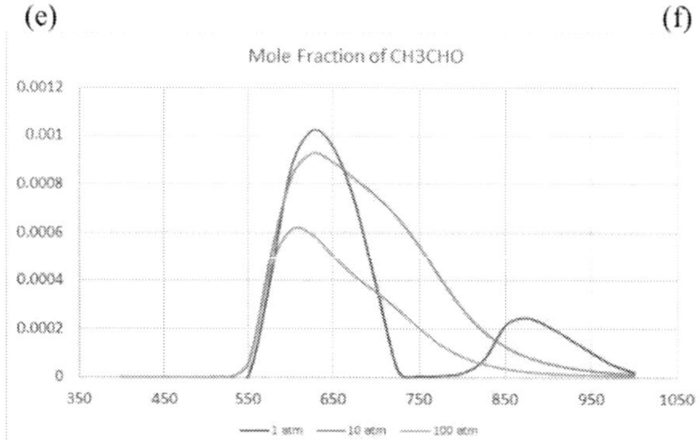

(g)

Figure 1. Variations of the mole fractions of important species (n-C_5H_{12}, O_2, CO, CO_2, CH_2O, C_2H_4 and CH_3CHO) under different environmental pressures.

In the reaction, n-pentane (n-C_5H_{12}) is the reactant and it has a very active low temperature oxidation property. As shown in figure 1, when the environmental pressure increases, the influence of low temperature chemical property of n-C_5H_{12} increases. Under the pressure of 1 atm, the oxidation reaction starts to occur at 550K, and the reactivity keeps increasing until the temperature reaches somewhere around 640K. From there the reactivity starts to drop back and finally almost stops oxidation at about 720-730K and remains not reacting due to its low temperature oxidation property until the temperature reaches about 800K. Above 800K, high temperature oxidation starts to occur, and the n-pentane continues to be obviously consumed until it is all gone. But the difference between low and high temperature oxidation weakens because the negative temperature coefficient (NTC) is decreasing (at 10 atm and 100 atm, high temperature oxidation occurs right after low temperature oxidation). Also, as the environmental pressure increases, the initial temperature of high temperature oxidation is lowered.

Formaldehyde (CH_2O) and ethylene (C_2H_4) are two of the main products of low temperature oxidation of n-C_5H_{12} and they have very similar temperature evolution profile. It was interesting to notice that the typical CH_2O and C_2H_4 low and high temperature oxidation peaks merged to a single peak at supercritical pressures. According to the graph, in high pressure environments, as the environmental pressure increases, the peak area widens but the maximum value of mole fraction decreased. However, we can see that at the normal pressure (1 atm), when the temperature reached 1000K, the mole fraction of CH_2O drops down to somewhere near zero. It is because of the decomposition of formaldehyde into aldehyde (HCO) in high temperature. Furthermore, the decomposition continues, producing carbon monoxide (CO) and then carbon dioxide (CO_2). Also, acetaldehyde (CH_3CHO) has similar evolution profile with CH_2O.

As the environmental pressure increases, the difference between high temperature oxidation and low temperature oxidation becomes less obvious. Corresponding to the graph of formaldehyde, under normal pressure (1 atm), the amount of carbon monoxide starts to increase at the time of decomposition of formaldehyde. CO starts to decrease after 850 - 900K due to its combining to

Oxygen (O_2) generating CO_2. CO_2 has very similar properties to CO. Corresponding to the decreasing of CO at the moment, CO_2 starts to increase in amount after 850-900K.

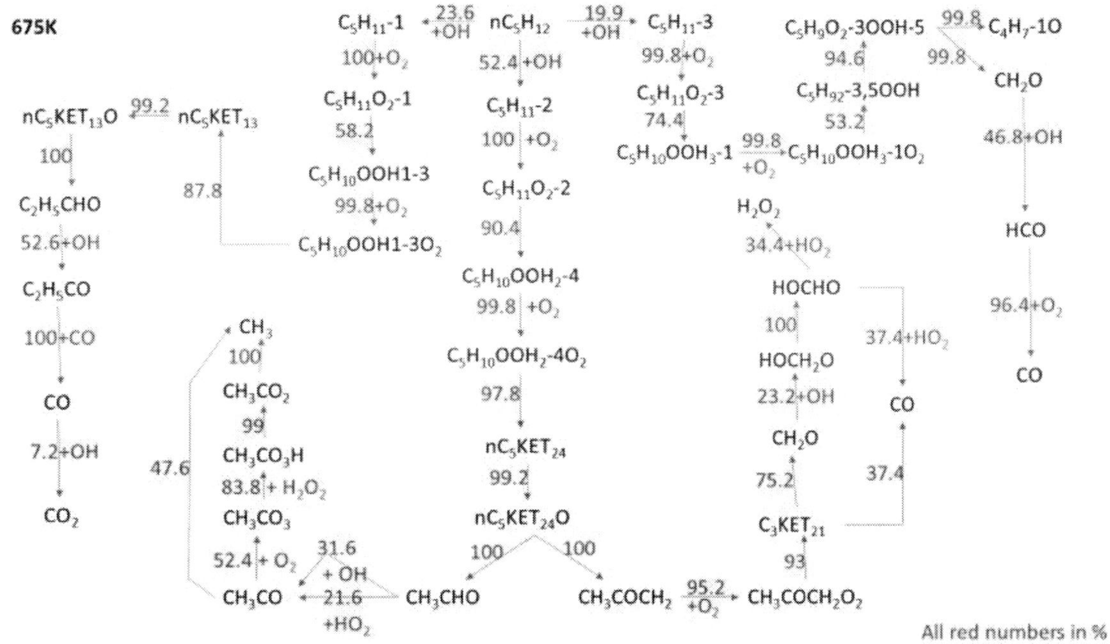

(a)

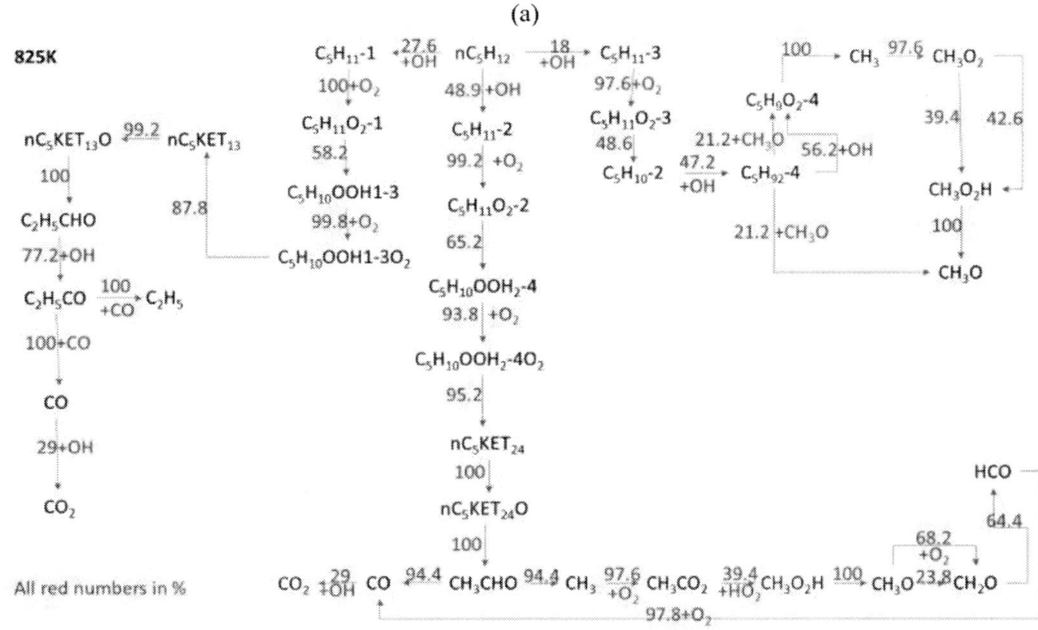

(b)

Figure 2. Reaction paths analyze for n-pentane decomposition to three isomers (C_5H_{11}-1, C_5H_{11}-2 and C_5H_{11}-3) and the following reactions after them at both low/intermediate temperature (675K) and high temperature (825K) at 100 atm.

According to the pathway analysis at 675K in figure 2 (a), n-pentane reacts with hydroxy and produces three different isomers of n-pentane radical (C_5H_{11}). 23.6% of n-pentane becomes C_5H_{11}-1, 52.4% becomes C_5H_{11}-2 and 19.9% goes to C_5H_{11}-3. And these isomers combine with oxygen, producing $C_5H_{11}O_2$ (known as RO_2 in the general reaction trend). In a relatively low temperature environment (675K, for example), the main consumption reaction of $C_5H_{11}O_2$ is the isomerization reaction to form $C_5H_{10}OOH$ by 90.4% (known as QOOH in the general reaction trend). Under low temperature and high pressure, QOOH mainly combines with oxygen, producing O_2QOOH. And the other decomposition path of QOOH is negligible. As a result, we can say that pressure can obviously promote low temperature oxidation.

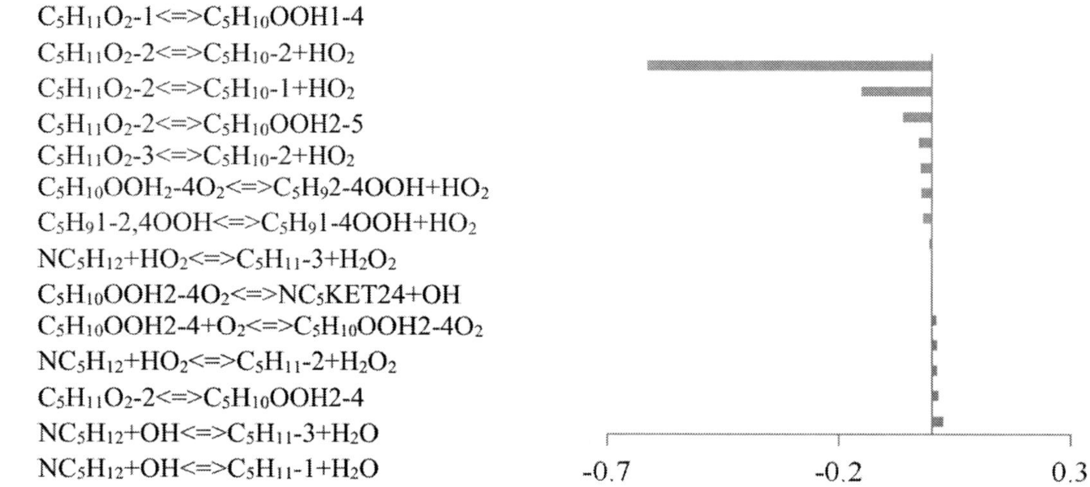

(a)Sensitivity at 675K

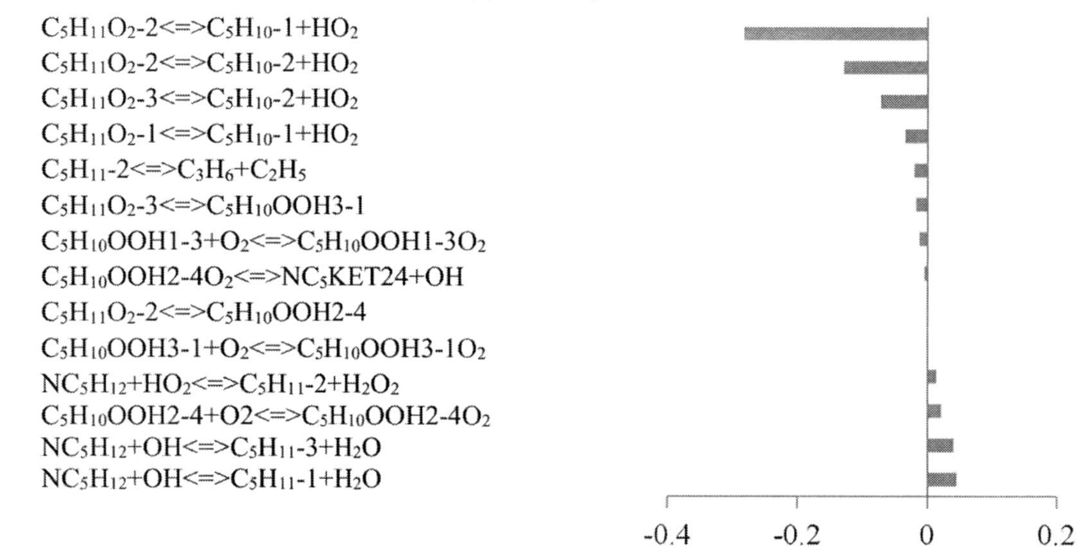

(b) Sensitivity at 825K

Figure 3. (a) (b) Sensitivity analysis of n-pentane at 675 and 825K, respectively, at 100 atm.

Moving onto the reaction paths under higher temperature (825K) in figure 2 (b), same as 625K environment, n-pentane reacts to produce three isomers of its radical, 27.6% becomes C_5H_{11}-1, 48.9% becomes C_5H_{11}-2 and 18% goes to C_5H_{11}-3. It is known that the typical pathway under high temperature is $C_5H_{11}O_2 = HO_2 + C_5H_{10}$, however, at 100 atm, the major consumption reaction of $C_5H_{11}O_2$ becomes the isomerization reaction to form $C_5H_{10}OOH$ (by 65.2%), which is different from the typical high temperature oxidation at lower pressures. In addition, the addition reaction between QOOH and oxygen is still more important than the decomposition reaction of QOOH at high temperature and 100 atm. Therefore, at supercritical pressures, the high temperature oxidation is governed by the branching reactions of RO_2, and the addition reaction of QOOH with oxygen dominates the reaction paths at both low and high temperature oxidations. It explains why the peaks of intermediate species (such as CH_2O, C_2H_4, CH_3CHO, etc.) at low and high temperatures merge to one single peak at supercritical pressures.

In figure 3 (a) at 675K, according to the sensitivity analysis of the mole fraction of n-pentane and the results of the net reaction rate, the reaction $C_5H_{11}O_2$-1<=>$C_5H_{10}OOH$1-4 is the most sensitive reaction. Other isomerization reactions, such as $C_5H_{11}O_2$-2<=>$C_5H_{10}OOH$2-5 and $C_5H_{11}O_2$-2<=>$C_5H_{10}OOH$2-4, are also very significant. Meanwhile, the reactions $C_5H_{11}O_2$-2<=>C_5H_{10}-2+HO_2, $C_5H_{11}O_2$-2<=>C_5H_{10}-1+HO_2 and $C_5H_{11}O_2$-3<=>C_5H_{10}-2+HO_2 are decomposition reactions that are the significant promoting ones that will yield HO_2, and thus enhance the overall reactivity. Therefore, at low temperature, the branching reactions of RO_2 dominate reaction paths. In addition, the H abstraction reaction of n-pentane with OH and HO_2 are also impotent. In figure 3(b) at 825K, similarly according to the sensitivity analysis of the mole fraction of n-pentane and the results of the net reaction rate, decomposition reactions, such as $C_5H_{11}O_2$-2<=>C_5H_{10}-1+HO_2, $C_5H_{11}O_2$-2<=>C_5H_{10}-2+HO_2, $C_5H_{11}O_2$-3<=>C_5H_{10}-2+HO_2 and $C_5H_{11}O_2$-1<=>C_5H_{10}-1+HO_2, take the most significant role. The isomerization reactions of RO_2, which are generally important at low temperatures, are still sensitive at intermediate temperatures at 100 atm. Therefore, at 825K, the branching reaction of RO_2 will dominate the reaction paths. The importance of RO_2 branching reactions in both low and intermediate temperatures in the sensitivity analysis are corresponding with the pathway analysis at the same temperatures.

4. Conclusion

The present study is to understand the low and high temperature oxidation of n-pentane from atmospheric pressure to supercritical pressures (1-100 atm) by using Chemkin software. A recently developed n-pentane model [23] was used to predict the temperature evolutions of major and intermediate species, like n-pentane, O_2, CO, CO_2, CH_2O, C_2H_4, CH_3CHO at varied pressure and temperature conditions. As the environmental pressure increases, the difference between high temperature oxidation and low temperature oxidation becomes less obvious because the negative temperature coefficient (NTC) is decreasing (at 10 atm and 100 atm, high temperature oxidation occurs right after low temperature oxidation). At supercritical pressures, the high temperature oxidation is governed by the branching reactions of RO_2, and the addition reaction of QOOH with oxygen dominates the reaction paths at both low and high temperature oxidations because the addition reaction between QOOH and oxygen is still more important than the decomposition reaction of QOOH at high temperature and 100 atm. According to the pathway analysis and sensitivity analysis, the branching reactions of RO_2 are dominating at both low and intermediate to high temperatures.

References

[1] S. Sharma, S. Raman, W.H. Green, J. Phys. Chem. A 114 (2010) 5689–5701.
[2] H. Zhao, J. Fu, F.M. Haas, Y. Ju, Effect of prompt dissociation of formyl radical on 1,3,5-trioxane and CH2O laminar flame speeds with CO2 dilution at elevated pressure, Combust. Flame. 183 (2017) 253-260.
[3] F. Zhang, T.S. Dibble, J. Phys. Chem. A 115 (2011) 655–663.
[4] Miyoshi A, 2011 J. Phys. Chem. A 115 3301–3325.
[5] Zhao H, Yang X, Ju Y, 2016 Combust. Flame. 173 187-194.
[6] Zhang Z, Zhao H, Cao L, Li G, Ju Y, 2018Energy. Fuels.
[7] Zádor J, Klippenstein SJ, Miller JA, 2011 J. Phys. Chem. A 115 10218–10225.
[8] Villano SM, Huynh LK, Carstensen HH, Dean AM, 2011 J. Phys. Chem. A 115 13425–13442.
[9] Felsmann D, Zhao H, Wang Q, Graf I, Tan T, Yang X, Carter EA, Ju Y, Kohse-Höinghaus K,

2017 Proc. Combust. Inst. 36 543-551.
[10] Tran LS, Pieper J, Carstensen HH, Zhao H, Graf I, Ju Y, Qi F, Kohse-Höinghaus K, 2017 Proc. Combust. Inst. 36 (2017).
[11] Miyoshi A, 2012 Int. J. Chem. Kinet. 44 59–74.
[12] Goldsmith CF, Green WH, Klippenstein SJ, 2012 J. Phys. Chem. A 116 3325–3346.
[13] Villano SM, Huynh LK, Carstensen HH, Dean AM, 2012J. Phys. Chem. A 116 5068–5089.
[14] Karwat DMA, Wagnon SW, Wooldridge MS, Westbrook CK, 2013 Combust. Flame 160 2693–2706.
[15] Jalan A, Alecu IM, Meana-Pañeda R, et al., 2013 J. Am. Chem. Soc. 135 11100–11114.
[16] Villano SM, Carstensen HH, Dean AM, 2013 J. Phys. Chem. A 117 6458–6473.
[17] Zádor J, Huang H, Welz O, Zetterberg J, et al., 2013 Phys. Chem. Chem. Phys. 15 10753–10760.
[18] Cai L, Pitsch H, 2014 Combust. Flame 161 405–415.
[19] Pelucchi M, Bissoli M, Cavallotti C, et al., 2014 Energy Fuels 28 7178–7193.
[20] Ranzi E, Cavallotti C, Cuoci A, Frassoldati A, Pelucchi M, Faravelli T, 2015 Combust. Flame 162 1679–1691.
[21] Bugler J, Somers KP, Silke EJ, Curran HJ, 2015 J. Phys. Chem. A 119 7510-7527.
[22] Zhang K, Banyon C, Togbé C, Dagaut P, Bugler J, Curran HJ, 2015 Combust. Flame 162 4194–4207.
[23] Bugler J, Marks B, Mathieu B, et al., 2016 Combust. Flame 163 138–156.
[24] Bugler J, Somers KP, Silke EJ, Curran HJ, 2015 J. Phys. Chem. A 119 7510–7527.
[25] Zhao H, Wu L, Patrick C, Zhang Z, Rezgui Y, Yang X, Wysocki G, Ju Y, 2018Combust. Flame. 197 78-87.
[26] Zhao H, Dana A, Zhang Z, Green W, Ju Y, 2018 Energy.165 727-738.
[27] <http://www.reactiondesign.com/products/chemkin/chemkin-pro/>

IOP Conf. Series: Materials Science and Engineering **715** (2020) 012066 doi:10.1088/1757-899X/715/1/012066

Research on flutter reliability of nonlinear binary-wing

Jing Kui Li*, Yi Rui Qi and Yun Kai Cheng

Shenyang Aerospace University, Shenyang 110000, China

*E-mail: lijingkui@sau.edu.cn

Abstract. In this paper, the aeroelastic analysis of the nonlinear binary wing model under unsteady aerodynamic conditions, the critical flutter speed was calculated by using the v-g method. Based on the critical flutter speed, the implicit limit state function of the nonlinear flutter reliability of the wing was obtained, and the response surface method was used for function fitting. The flutter reliability index was calculated by the checking point method, the design checking point and failure probability were obtained, and the flutter reliability sensitivity was analyzed. Finally, through the analysis of an example, it was proved that this method is feasible.

1. Introduction

With the development of China's aviation industry, the aeroelastic problem of aircraft has received more and more attention. The research focus on wing flutter analysis has also extended from classical linear flutter analysis to binary wing flutter analysis with nonlinear links.

The research on the reliability of wing flutter is mainly based on the wing parameters.The traditional flutter reliability analysis is performed on the determined wing parameters, but in practical applications, the wing parameters are uncertain due to various factors.Therefore, the uncertainty of the parameters should be considered in the reliability analysis of the nonlinear flutter of wings. Pitt [1] studied the influence of structural parameter uncertainty on wing flutter with Monte Carlo Simulation (MCS) method, and proposed that the randomness of parameters should be considered in the analysis of wing flutter. Scarth [2] used the Polynomial Chaos Expansion (PCE) method to deal with the uncertainty of the layup direction of composites, and gave the influence of different layup directions on the flutter probability. Song [3-5] analyzed the probability characteristics of random parameters and flutter reliability sensitivity of wings in transonic flow by using the improved linear sampling technique combined with Monte Carlo Simulation method.

MCS method is suitable for the reliability analysis of implicit limit state function, but has the disadvantages of large sample size and large computational workload. The response surface method has the characteristics of small calculation, high calculation accuracy and easy implementation. Therefore, the response surface method was adopted in this paper to carry out reliability analysis and obtain the nonlinear flutter limit state function of the wing. It provided an efficient and feasible calculation method for the flutter analysis and reliability design of the wing structure.

2. Nonlinear flutter analysis of binary wing

Figure 1 shows the physical model of a binary airfoil, which oscillates in the pitch and plunge directions.The symbol α denotes the pitch angle; the plunge deflection is denoted by h; the elastic axis is located at a distance *ab* from the mid-chord, while the mass center is located at a distance $x_a b$ from the elastic axis, $b=c/2$ is the half chord of the wing.

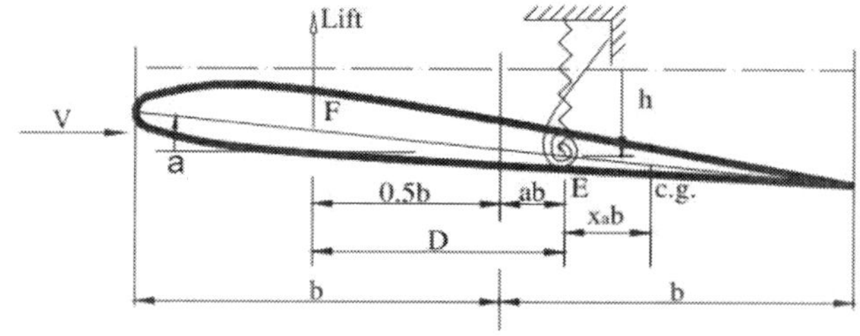

Figure 1. Airfoil model with pitch and plunge degrees-of-freedom.

The motion differential equation of binary wing can be obtained according to the Lagrange equation:

$$\left.\begin{aligned} m\ddot{h} + S_\alpha\ddot{\alpha} + c_h\dot{h} + K_h h = Q_h \\ S_\alpha\ddot{h} + I_\alpha\ddot{\alpha} + c_\alpha\dot{\alpha} + K_\alpha\alpha + eK_\alpha\alpha^3 = Q_\alpha \end{aligned}\right\} \tag{1}$$

where: m is the wing mass of the unit length; S_α is the mass static moment of the unit length wing to the elastic axis; I_α is the mass inertia moment of the unit length wing to the rotating shaft; c_h and c_α are the structural damping coefficients; K_h and K_α are the structural stiffness coefficients; *e* is the nonlinear stiffness factor; Q_h and Q_α are the aerodynamic and aerodynamic moment acting on the wing.

According to Theodorsen's unsteady aerodynamic force theory [6], the aerodynamic force *L* and the aerodynamic moment T_α of unit span with the binary wing can be obtained:

$$\left.\begin{aligned} L = -\pi\rho b^2(v\dot{\alpha} + \ddot{h} - ab\ddot{\alpha}) - \\ 2\pi v b C(k)[v\alpha + \dot{h} + (\frac{1}{2} - a)b\dot{\alpha}] \\ T_\alpha = \pi\rho b^2[ab(v\dot{\alpha} + \ddot{h} - ab\ddot{\alpha}) - \frac{1}{2}vb\dot{\alpha} - \frac{1}{8}b^2\ddot{\alpha}] + \\ 2\pi\rho v^2 b^2(\frac{1}{2} + a)C(k)[\alpha + \frac{\dot{h}}{v} + (\frac{1}{2} - a)\frac{b\dot{\alpha}}{v}] \end{aligned}\right\} \tag{2}$$

where: $k = b\omega/v$ is the reduction frequency, the dimension is 1, and $C(k)$ is the Theodorsen function. The basic equation for the binary wing flutter analysis using the *v-g* method is:

$$\left[-\omega^2\overline{M} + i\omega\overline{D} + (1+ig)\overline{K} - q_d Q(k, M_\infty)\right]q = 0 \tag{3}$$

According to the calculation formula of the equivalent frequency, equation (3) can also be rewritten into:

$$\left[\left(\overline{M}\left(\frac{k}{b}\right)^2 + \frac{\rho_a}{2}Q(k, M_a)\right)\frac{-V^2}{1+ig} + \left(\frac{k}{b}\right)\frac{iV}{\sqrt{1+ig}}\overline{D} + \overline{K}\right]q = 0 \tag{4}$$

Solved the equation (4) and obtained the system eigenvalue:

$$p^2 = \left(-\frac{V^2}{1+ig}\right) = -V^2\frac{1-ig}{1+g^2} = c + id \tag{5}$$

Further the damping *g*, the velocity *V* and the frequency ω were respectively obtained:

$$g = -\frac{d}{c}, V = \sqrt{-\frac{c^2+d^2}{c}}, \omega = \frac{kV}{\pi b} \tag{6}$$

3. Reliability analysis

The implicit limit state function of binary wing flutter reliability can be expressed as follows:

$$Z = g(x) = 0.85V_f^* - V_f \tag{7}$$

where: V_f^* is the deterministic flutter critical speed obtained by the v-g method, V_f is the actual flight speed and $V_f = 0.85V_f^*(\mu)$, where $V_f^*(\mu)$ is the flutter critical speed when the random parameter takes the mean, so Z=g(x) is the implicit limit State function.

Basic random parameters: $x=(\mu,a,b,x_a,r_a,\omega_a,\rho_a)^T$, where: μ is the mass ratio; a_h is the dimensionless distance from the stiffness center of the wing to the midpoint of the string; b is the half chord Length; x_a is the dimensionless distance from the wing's gravity center to the stiffness center ; r_a is the dimensionless radius of gyration of the wing to the stiffness center; ω_α is the torsional mode frequency of the wing; ρ_α is the bending mode frequency of the wing.

The steps to study the flutter reliability of binary wing using the quadratic response surface method without crossing terms and the deterministic flutter analysis are as follows:

- A quadratic analytical expression (limit state function) was fitted by regression to replace the real response surface:

$$\hat{g}(x) = \alpha_0 + \sum_{j=1}^{7} \alpha_i x_i + \sum_{j=1}^{7} \alpha_{ii} x_i^2 \tag{8}$$

- Bucher test design method was used to extract the sample points which were needed to determine the pending coefficients in the response surface function. In the k-th iteration, $x_m^{(k)} = (x_{m1}^{(k)}, x_{m2}^{(k)}, \cdots x_{m7}^{(k)})$ was used as the test center, and the 2n+1 test points were extracted as follows:

$$x_m^{(k)} = (x_{m1}^{(k)}, x_{m2}^{(k)}, \cdots x_{m7}^{(k)}) \tag{9}$$

$$x_m^{(k)} = (x_{m1}^{(k)}, x_{m2}^{(k)}, \cdots, x_{mi}^{(k)} \pm f\sigma_i, \cdots, x_{m7}^{(k)}), i = 1,2,\cdots,7 \tag{10}$$

The mean value point $\mu_i=(\mu_1,\mu_2,\ldots,\mu_7)$ (deterministic flutter analysis) was taken as the first iteration test center point x_m^1 . Where σ_i is the standard deviation of the basic random variable, f is the interpolation coefficient, and this paper is taken as 3.

- The sample point $x_m^{(k)} = (x_{m1}^{(k)}, x_{m2}^{(k)}, \cdots, x_{mi}^{(k)} \pm f\sigma_i, \cdots, x_{m7}^{(k)}), i = 1,2,\cdots,7$ which was extracted from Bucher test design was called by $x=(\mu,a,b,x_a,r_a,\omega_a,\rho_a)^T$ (uncertainty flutter analysis) to iteratively solve the flutter critical velocity V_f^* and obtain the implicit limit state function $Z = g(x) = 0.85V_f^* - V_f$ of wing flutter reliability.
- The least square method was used to minimize the square sum of the response surface function $\hat{g}(x_j)$ and the implicit limit state function $g(x_j)$ at each sample point, and obtained the response surface function of the *k-th* iteration.
- JC checking point method was used to calculate the flutter reliability index of wings:

$$\beta^{(k)} = \frac{\mu_{g_k}}{\sigma_{g_k}} = \frac{g(x_1^*, x_2^*, \cdots, x_{15}^*) + \sum_{i=1}^{k} (\mu_i - x_i^*) \frac{\partial g}{\partial x_i}\Big|_{x_i^*}}{\sqrt{\sum_{i=1}^{k} (\sigma_i \frac{\partial g}{\partial x_i}\Big|_{x_i^*})^2}} \tag{11}$$

- It was judged whether the reliability index β satisfied the convergence condition $|\beta^{(k)}-\beta^{(k-1)}<\varepsilon|$, ε is a given accuracy requirement, and if the condition was satisfied, the response surface method converged, and the iteration ended, otherwise the execution was performed (2) Steps.
- At the end of the cycle, the failure probability P_f can be obtained from the reliability index $\beta^{(k)}$.

4. Sensitivity analysis of flutter reliability

The sensitivity is the partial derivative of the failure probability P_f to the distribution parameter of the basic random variable $x=(\mu,a,b,x_a,r_a,\omega_a,\rho_a)^T$. The purpose of reliability sensitivity analysis is to study the influence of the variation of the basic random parameters on the flutter reliability index or failure probability in the wing mechanics model. The reliability sensitivity can be expressed as [7-9]:

$$\varepsilon_{\mu_i} = \frac{\partial_\beta}{\partial_{\mu_i}} = \frac{(\mu_i - x_i^*)\frac{\partial^2 g}{\partial x_i^2}\Big|_{x_i^*}}{\sqrt{\sum_{i=1}^{k}\left(\sigma_i \frac{\partial g}{\partial x_i}\Big|_{x_i^*}\right)^2}} - \frac{\left[g(x_1^*, x_2^*, \cdots, x_{15}^*) + \sum_{i=1}^{k}(\mu_i - x_i^*)\frac{\partial g}{\partial x_i}\Big|_{x_i^*}\right]\sigma_i \frac{\partial g}{\partial x_i}\Big|_{x_i^*}\sigma_i \frac{\partial^2 g}{\partial x_i^2}\Big|_{x_i^*}}{\left[\sqrt{\sum_{i=1}^{k}\left(\sigma_i \frac{\partial g}{\partial x_i}\Big|_{x_i^*}\right)^2}\right]^3} \quad (12)$$

The mean sensitivity of the above-mentioned reflects the degree to which the mean value of the variable affects reliability.

$$\varepsilon_{\sigma_i} = \frac{\partial_\beta}{\partial_{\sigma_i}} = \frac{\left[g(x_1^*, x_2^*, \cdots, x_{15}^*) + \sum_{i=1}^{k}(\mu_i - x_i^*)\frac{\partial g}{\partial x_i}\Big|_{x_i^*}\right]^2 \sigma_i \left[\frac{\partial g}{\partial x_i}\Big|_{x_i^*}\right]^2}{\left[\sqrt{\sum_{i=1}^{k}\left(\sigma_i \frac{\partial g}{\partial x_i}\Big|_{x_i^*}\right)^2}\right]^3} \quad (13)$$

The standard deviation sensitivity of the above formula reflects the degree of influence of random parameter volatility on reliability.

5. Example analysis

The binary wing model shown in figure 1. was used and determined the random parameters $x=(\mu,a,b,x_a,r_a,\omega_a,\rho_a)^T$, the mean value and standard deviation are shown in table 1.

Table 1. Mean and standard deviation of basic random variables.

Random Variable	Mean Value	Standard Deviation
μ	20.000	0.4036
a	-0.1000	-0.0060
b	1.0000	0.0025
x_a	0.2500	0.0086
r_a	0.5000	0.0146
ω_a	62.800	0.1842
ρ_a	28.100	0.0734

Figure 2 shows the relationship between velocity V and structural damping coefficient g:

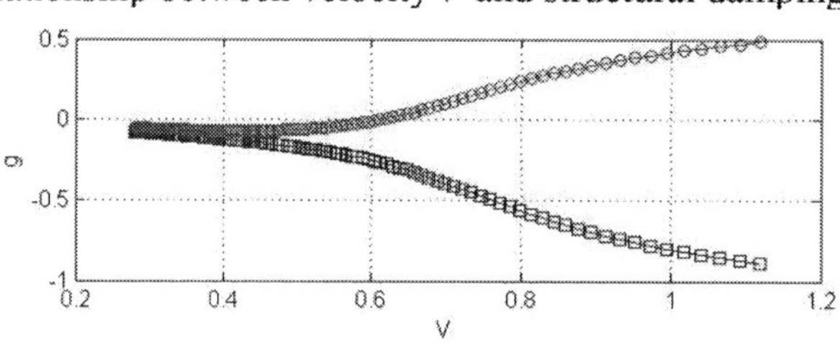

Figure 2. *v-g* figure.

It can be seen from figure 3. that the critical speed of the deterministic flutter of this binary wing is $V_f^* = 0.6126$, and the implicit limit state function describing the wing flutter reliability was established as $g(x) = 0.85V_f^* - 0.52071$ considering the parameter uncertainty. The implicit limit state function $g(x)$ was fitted to the explicit formula $\hat{g}(x)$ by using the quadratic response surface method:

$$\hat{g}(x) = -33605 + 19.1684\mu - 433.1843a + 25145b + 535.4303x_\alpha + 368.6466r_\alpha + 290.4442\omega_\alpha + 823.0207R_\omega - 0.4836\mu^2 - 2212.5a^2 - 12582b^2 - 1079.7x_\alpha^2 - 370.8255r_\alpha^2 - 2.3136\omega_\alpha^2 - 14.6502\rho_\alpha^2 \quad (14)$$

The flutter reliability index of the binary wing was calculated by the checking point method, and obtained the flutter reliability design checking point and failure probability. The mean value point and design checking point are shown in table 2.

Table 2. The mean point and the design verification point.

Random Parameters	Mean Value	Design Checking Point
μ	20.000	20.9319
a	-0.1000	-0.1113
b	1.0000	1.0037
x_a	0.2500	0.2601
r_a	0.5000	0.5145
ω_a	62.8000	62.9595
ρ_a	28.1000	28.1563

The flutter reliability calculation results at the mean point and design checking point of the binary wing are shown in table 3.

Table 3. Reliability index.

	Reliability Index β	Reliability R	Failure Probability P_f
Mean Point	1.7289	0.9581	0.0419
Checking Point	3.8447	0.9999	0.0011

The reliability at the design checking point of the binary wing is 4 % higher than that at the mean point. Therefore, through the research of flutter reliability, the design checking point of each parameter of the wing's aeroelasticity can be calculated, which can improve the flutter reliability of the wing and reduce the flutter amplitude.

The mean sensitivity and standard deviation sensitivity of failure probability to basic random parameters $x=(\mu,a,b,x_a,r_a,\omega_a,\rho_a)^T$ obtained from formulas (12)-(13) are shown in table 4, figure 3 and figure 4 are corresponding histograms.

Table 4. Basic random variable sensitivity.

xi	$\partial P_f/\partial \mu x_i$	$\partial P_f/\partial \sigma x_i$
μ	2.4651	-9.6675
a	-136.5169	430.8393
b	252.7235	-631.1765
x_a	59.7008	-121.6933
r_a	29.4352	-50.5179
ω_a	2.0187	-2.9871
ρ_a	4.5083	-5.8533

The sign of $\partial P_f/\partial \mu_{xi}$ and $\partial P_f/\partial \sigma_{xi}$ indicates the tendency of the influence of the change of the mean and standard deviation of the random parameters on the flutter failure probability of the wing.If the sign is positive, the probability of wing flutter failure tends to be safe with the increase of basic random parameters, and on the contrary, tends to fail. The absolute values of $\partial P_f/\partial \mu_{xi}$ and $\partial P_f/\partial \sigma_{xi}$ reflect the speed at which the change of random parameters affects the change of wing flutter failure probability. Therefore, the sign and absolute value of $\partial P_f/\partial \mu_{xi}$ and $\partial P_f/\partial \sigma_{xi}$ analysis results are the key to judge the wing flutter stability trend.

It can be seen from the sensitivity $\partial P_f/\partial \mu_{xi}$ of the failure probability to the mean of the random parameters that the increase of the values of μ, b, x_a, r_a, ω_a, ρ_a will cause the wing flutter stability to deteriorate and tend to fail; the increase in the value of a causes the wing flutter stability to be better.

From the sensitivity $\partial P_f/\partial \sigma_{xi}$ of failure probability to standard deviation of random parameters, it can be seen that the increase of values of μ, b, x_a, r_a, ω_a, ρ_a will lead to better flutter stability of wings. The increase of a value leads to poor flutter stability of the wing and tends to fail.

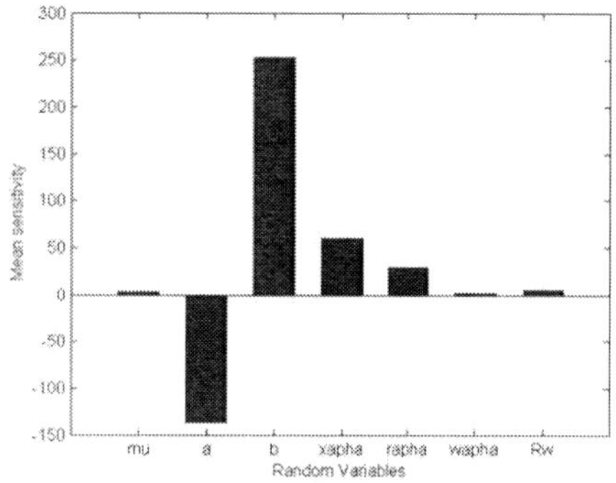

Figure 3. Reliability sensitivity of mean.

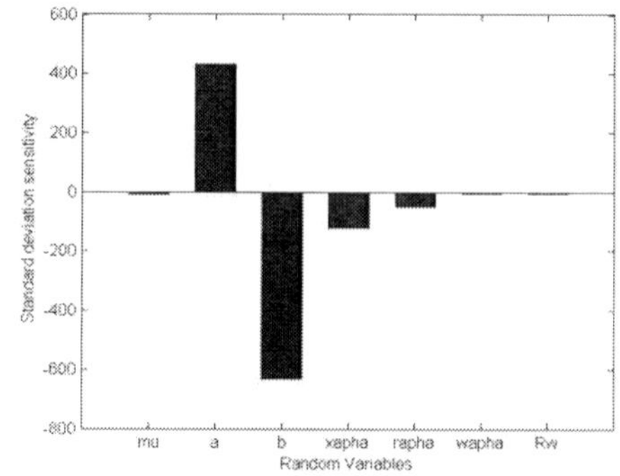

Figure 4. Reliability sensitivity of standard deviation.

6. Conclusions

The response surface method combined with deterministic flutter analysis was used to study the reliability of binary wing flutter, which solved the problem that the flutter limit state function of the wing aeroelastic system is difficult to explicitly calculated. The reliability sensitivity design was combined to obtain the influence law of random parameter changes on wing flutter reliability, the influence degree of random parameters on flutter critical speed was quantitatively determined, more comprehensive information was provided for the uncertainty flutter analysis, an important theoretical basis was provided for improving flight stability and reliability of the aircraft, and an efficient and feasible calculation method was provided for flutter analysis and reliability design of wing structures.

Acknowledgement

The authors wish to acknowledge the support of Liaoning Provincial Natural Science Foundation (201602560), Liaoning Provincial Education Department Fund (L201606), and Shenyang Science and Technology Project (18-400-2-07).

References

[1] Pitt D M, Haudrich D P, Thomas M J, et al. Probabilistic Aeroelastic analysis and its implications on flutter margin requirements, AIAA-2008-2198 [R]. Reston: AIAA,2008.

[2] Scarth C, Coper J E, Weaver P M, et al. Uncertainty quantification of aeroelastic stability of composite plate wings using lamination parameters [J]. Composite Structures, 2014, 116: 84-93

[3] Song S F, Lü Z Z, Zhang W W, et al. Random uncertainty of aeroelastic system [J]. Journal of Vibration Engineering, 2009, 22(3): 227-231 (in Chinese).

[4] Song S F, Lu Z Z, Zhang W W, et al. The uncertainty importance measures of the structural system in view of mixed uncertain variables [J]. Fuzzy Sets and Systems, 2014, 243: 25-35.

[5] Song S F, Lu Z Z, Zhang W W, et al. Reliability and sensitivity analysis of transonic flutter using improved line sampling technique [J]. Chinese Journal of Aeronautics, 2009, 22(5): 513-519.

[6] THEODORSEN T.General theory of aerodynamic instability and the mechanism of flutter: REPORT NO. 496[R]. US Government Printing Office,Washington， D.C.:Report Nation Advisory Committee for Aerodynamics,1949:412-433.

[7] Zhang Yi-min. Connotation and development of mechanical reliability-based design [J]. Journal of Mechanical Engineering, 2010, 46(14): 167-188.

[8] Gao Zongzhan, Liu Zhiqun, Jiang Zhifeng, et al. Strength Reliability and Parameter Sensitivity Analysis for Airfoil Spar Structure [J]. Journal of Mechanical Engineering, 2010, 46(14): 194-198 (in Chinese).

[9] WANG Changke, YANG Guangwu, ZHU Tao. Sensitivity Analysis of Car-Body Torsional Vibration [J]. Mechanical Design and Manufacturing, 2017(10): 81-84.

IOP Conf. Series: Materials Science and Engineering **715** (2020) 012067 doi:10.1088/1757-899X/715/1/012067

Investigation of flow characteristics of coronary slot stents using computational fluid dynamics

Xuan Zhi Wang, Syed H. Masood*, Syed H. Riza and Harsh Modi

Faculty of Science Engineering and Technology, Swinburne University of Technology, Hawthorn, Victoria, Australia 3122

*E-mail: smasood@swin.edu.au

Abstract. Coronary stents are metallic tubes inserted in coronary arteries to treat cardiovascular diseases. Palmaz-Schatz® Stent represents the original design from which many other slot-stent designs have been derived. In this paper, the impact of the geometry of the stent on its flow characteristics was investigated using the computational fluid dynamics (CFD) analysis. Seven design cases were selected for our research investigation, which were divided into three groups: (a) varying the slot length; (b) varying the slot width; and (c) varying the number of slots. Flow characteristics of stents with various geometries were studied by the CFD analysis using ANSYS CFX software. Results of the analysis revealed significant potential relationships between the geometry of the stent and its flow characteristics.

1. Introduction

Cardiovascular diseases constitute one of the major human health problems of modern times. Recently, the implantation of metallic coronary stents in the coronary artery has become an integral part of most interventional procedures for percutaneous revascularization [1]. The Palmaz-Schatz® stent (Johnson and Johnson Company, Warren, NJ, USA) is the most representative member of the family of slot stents. Much work [2, 3] has been done to investigate different flow behaviours of stents with various geometries. Tseng et al [4] emphasized the importance of computational fluid dynamics to obtain an improved endovascular aneurysm repair treatment outcomes and more accurate diagnosis, prognosis and prediction. Using CFD modelling, Morris et al [5] demonstrated the possibility of acquiring additional data for generating new patient specific insights that would help in better risk prediction and treatment planning. Gao et al [6] discussed the comparison between bioresorbable vascular scaffolds and metallic stents and concluded that bioresorbable vascular scaffolds are non-inferior to certain metallic components. Flow characteristics near stent strut configurations on femoropopliteal artery were studied by Paisal et al [7]. Their investigations show that by varying the design of stent model, different flow characteristics can be obtained and the stent with hexagon cross-section provides the best performance during surgical operations.

Currently, however, an appreciable research gap exists in investigating the flow characteristics of specific coronary stents with various geometries using CFD software. In this paper, a methodology for studying flow characteristics of Palmaz-Schatz type stents with various geometries by the CFD analysis using ANSYS CFX is presented in detail. Analytical results are discussed to reveal significant potential relationships between the geometry of the stent and associated flow characteristics.

IOP Conf. Series: Materials Science and Engineering **715** (2020) 012067 doi:10.1088/1757-899X/715/1/012067

2. Computational fluid dynamics model

2.1. Geometry and material properties

To study blood flow characteristics of different geometries, seven cases of varying designs of stents were modelled. It was assumed that all stents were equal in whole length L, thickness T and mean diameter D (See figure 1). Stent length, stent width, strut length, strut width, stent thickness, sector angle and numbers of slots vary in seven cases (See table 1).

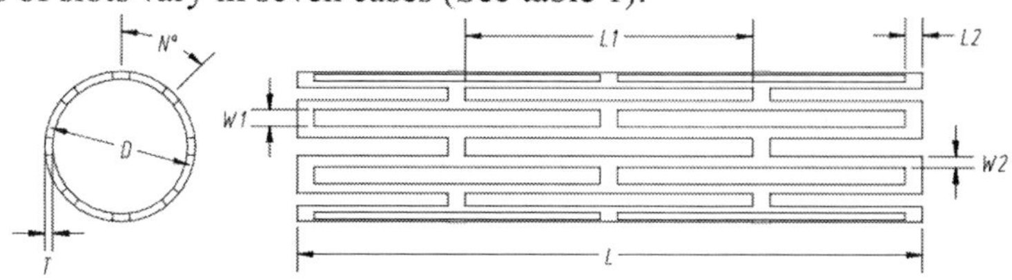

Figure 1. Geometric parameters of Palmaz-Schatz type stent.

Table 1. Values of geometric parameters in seven cases.

Stent design case	Slot length L_1 (mm)	Slot width W_1 (mm)	Strut length L_2 (mm)	Strut width W_2 (mm)	Stent thickness D (mm)	Sector angle N (°)	Number of slots
1	3.6200	0.2200	0.2200	0.1400	0.1000	45	24
2	3.4390	0.2200	0.3407	0.1400	0.1000	45	24
3	3.8010	0.2200	0.0993	0.1400	0.1000	45	24
4	3.6200	0.1980	0.2200	0.1620	0.1000	45	24
5	3.6200	0.2420	0.2200	0.1180	0.1000	45	24
6	3.6200	0.2933	0.2200	0.1867	0.1000	60	18
7	3.6200	0.1760	0.2200	0.1120	0.1000	36	30

To study the blood flow behaviour of the stent, a three-dimensional model of the blood flow domain was created. It was assumed that the stent had expanded against the stenosed artery to 4 mm, which was the diameter of the assumed healthy artery. Additional length was added to the two ends of the artery, respectively, to ensure the fully developed blood flow when it passes the domain of the stent. The minimum addition length (L) was determined by the following formulas [2]:

$$Re = \frac{\rho u D}{\mu} \tag{1}$$

$$L = 0.06 \times Re \times D \tag{2}$$

Blood behaves as a Newtonian fluid for shear rates great than 100 s^{-1} [8]. Shear rates in large arteries are in general considerably larger than this value [9, 10]. In this model, blood was assumed to behave like a Newtonian fluid due to the relatively large size of the artery and high speed of the blood flow. According to the previous assumption, the diameter of the flow domain D was 4 mm. The mass density ρ , speed of the blood flow u and dynamic viscosity of the blood μ were defined as 1060 kg/m^3, 0.105 m/s, and 3.7× 10^{-3} N s/m^2, respectively [2]. Thus, the Reynolds number Re was 120.32, and an additional length L should be larger than 28.88 mm. Due to different longitudinal recoils of stents with various geometry, lengths of various deployed stents were about 7.9 – 8.0 mm. Thus, the whole length of the model for each case was selected as 67.9 mm, which means the value of an additional length L for each case was approximately 30 mm.

The model is axisymmetric with respect to its central axis in all aspects, including the geometry, materials, loadings and boundary conditions. To take advantage of these symmetric characteristics and to reduce the size, scope and processing time of the model, a basic section of it was used in the analysis using CFX (see figure 2). Three-dimensional solid triangular elements were generated by

CFX-Mesh depending on the geometry of the model. Minimum and maximum edge lengths were determined as 0.05 mm and 1 mm, respectively. The inlet boundary condition and loading specifies inlet flow parameters. A steady-state velocity of 0.105m/s corresponding to the average blood flow velocity during one cardiac cycle was imposed on the inlet region for which the direction was along the positive direction of the *z* axis. The outlet boundary condition and loading specifies outlet flow parameters. An average static pressure 0 Pa was applied on the outlet region, which ensured the outlet pressure would not influence the blood flow pattern. It was assumed that there is no slip between the blood and luminal surface. To ensure accuracy, the advection scheme was set as High Resolution, the maximum number of iterations was defined as 100, and the residual target was chosen as 0.00001. Other parameters remained default values.

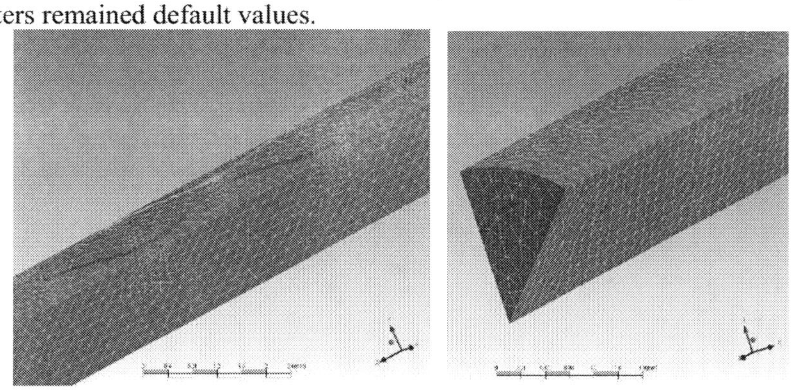

(a) Enlarged view of central part (b) Enlarged view of one end

Figure 2. Discretization of model.

3. Results and discussion

3.1. Velocity distribution in stents

CFD results indicated that the velocity of blood flow changed suddenly when the blood flow made contact with struts of a stent, which resulted in large pressures imposed on the lumen of the artery. In the analysis, velocity patterns of the blood flow with different geometries of stents were investigated. The line which was located at the centre of the slot was selected as the sample (see figure 3).

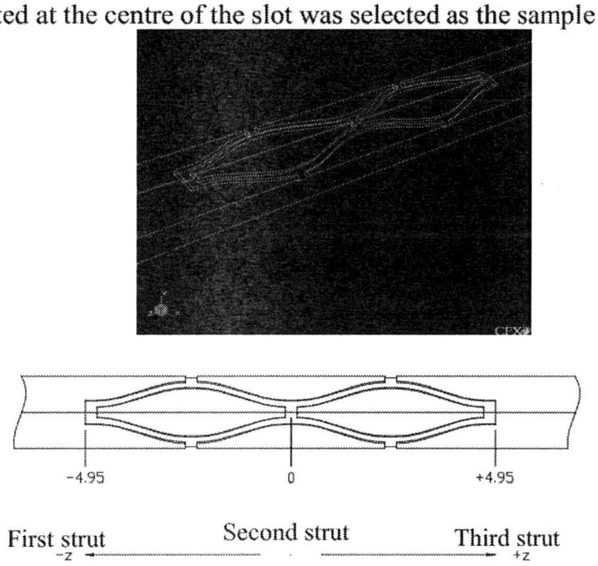

Figure 3. Line selected in CFD.

The effect of varying the slot length upon the velocity distribution in stents was investigated in Cases 1, 2 and 3. Figure 4 illustrates the velocity along the flow direction at the centre of the slot of Cases 1, 2 and 3. In Case 2, the velocity of blood flow was the lowest in the three cases when it passed through the first strut. But its velocity became highest of all when it made contact with the second strut. When the blood flow passed through the last strut, the velocity was larger than that in Case 1. In Case 3, the velocity of blood flow was the highest of all when it was at the first and last strut, while the velocity was least when it made contact with the second strut. This revealed that shorter slot length, i.e. larger strut length, results in the reduction and increase of velocity in first and second struts, respectively. The velocity of blood flow in the third strut seems to have no direct relation to the slot length.

The effect of varying the slot width upon the velocity distribution in stents was investigated in Cases 1, 4 and 5. Figure 5 illustrates the velocity along the flow direction at the centre of the slot of Cases 1, 4 and 5. The three velocity curves nearly overlap at the positions of three struts. The slot width was a circumferential dimension. The slot length, i.e. the distance between struts, of these stents after full expansion was similar, so the impact of the slot width on the velocity of axial flow was weak.

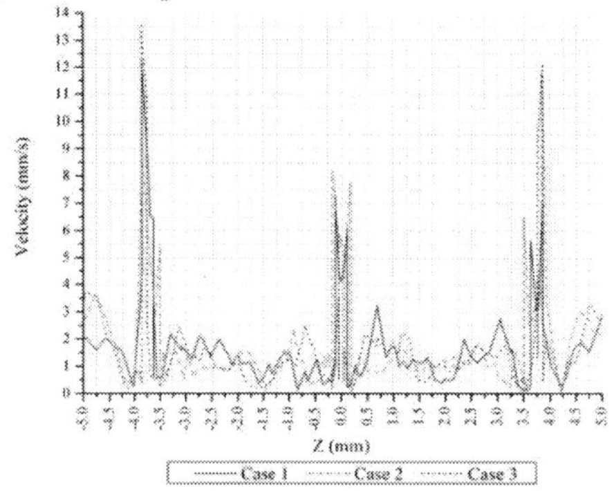

Figure 4. Velocity distributions in Cases 1, 2 and 3 (Varying slot length).

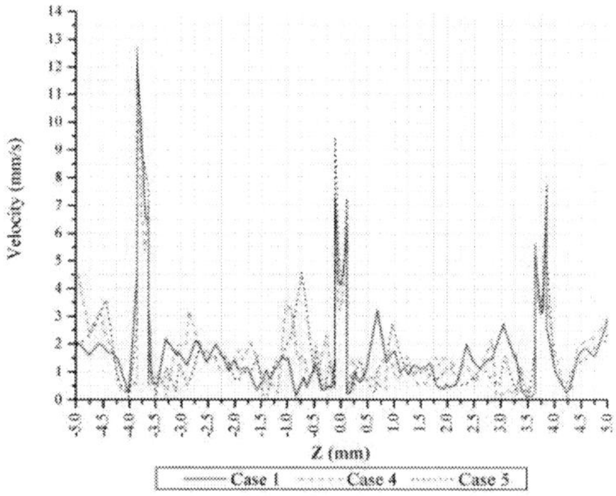

Figure 5. Velocity distributions in Case 1, 4 and 5 (Varying slot width).

The effect of varying the number of slots upon the velocity distribution in stents was investigated in Cases 1, 6 and 7. Figure 6 illustrates the velocity along the flow direction at the centre of the slot of Cases 1, 6 and 7. The number of slots has a significant influence on the velocity of the blood flow at the first strut, while the impact of the number of slots on the velocity at the last strut was weak. In Case 6, as the number of slots decreased, the velocity reduced at the first strut and increased at the second strut. To expand to the same diameters, Case 6 required more axial foreshortening, which reduced the actual slot length after full expansion. So the velocity pattern of Case 6 was similar to that of Case 2 to some extent. On the contrary, as the number of slots increased, the velocity increased at the first strut and decreased at the second strut. To expand to the same diameters, Case 7 required the least axial foreshortening, which enlarged the actual slot length after full expansion. So the velocity pattern of Case 7 was similar to that of Case 3 to some extent.

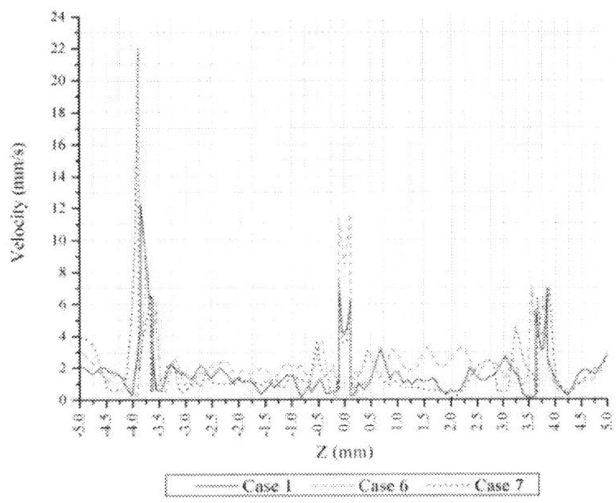

Figure 6. Velocity distributions in Cases 1, 6 and 7 (Varying number of slots).

3.2. Wall Shear Stress (WSS) distribution in stents

Stent geometry and expansion during deployment produce alterations in vascular anatomy that may adversely affect wall shear stress (WSS) [2]. The luminal surface of an artery with lower WSS is prone to form atherosclerosis [11]. Atherosclerosis is a disease characterised by the degeneration of the arteries because of the build-up of fatty deposits. Thus, a well-designed stent should have higher WSS distribution of flow in the stented artery. In the analysis, WSS patterns of the blood flow with different geometries were studied. The line which was located at the centre of the slot was selected as the sample, i.e. the line selected in the previous analysis. WSS was determined as the product of dynamic viscosity μ and shear rate $\dot{\gamma}$. Shear rate was calculated using the second invariant of the strain rate tensor, where u, v and w are the x, y and z components of velocity vector **u**, respectively. [2]

$$\dot{\gamma}=\left\{2\left[\left(\frac{\partial u}{\partial x}\right)^2+\left(\frac{\partial v}{\partial y}\right)^2+\left(\frac{\partial w}{\partial z}\right)^2\right]+\left(\frac{\partial u}{\partial y}+\frac{\partial v}{\partial x}\right)^2+\left(\frac{\partial u}{\partial z}+\frac{\partial w}{\partial x}\right)^2+\left(\frac{\partial v}{\partial z}+\frac{\partial w}{\partial y}\right)^2\right\}^{0.5} \quad (3)$$

$$WSS=\mu\dot{\gamma} \quad (4)$$

The effect of varying the slot length upon the WSS distribution in stents was investigated in Cases 1, 2 and 3. Figure7 illustrates WSS distributions along the flow direction at the centre of slot of Cases 1, 2 and 3. The WSS of Case 2 was the highest of all three cases at the first strut. The WSS of Case 2 was slightly less at the second strut and more at the third strut than that in Case 1, respectively. The WSS of Case 3 was the lowest of all three cases at all positions of strut. Thus, this revealed that increasing the slot length enabled a reduction in the WSS.

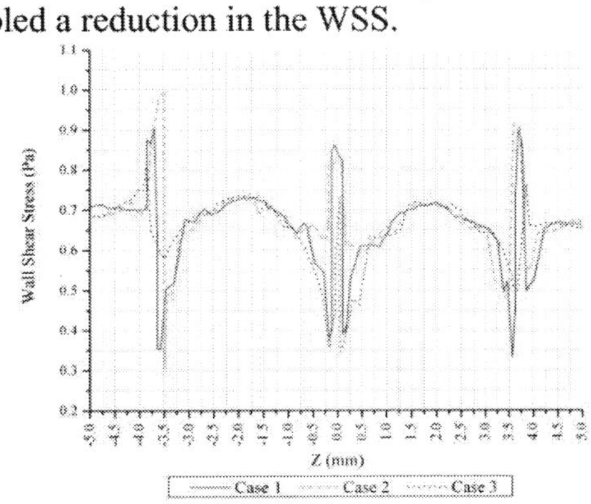

Figure 7. WSS distributions in Cases 1, 2 and 3 (Varying slot length).

4. Summary

In this paper, a parametric model of a certain Palmaz-Schatz® Stent has been developed by varying the slot length, slot width and number of slots. The CFD simulation for studying flow behaviours of stents with various geometries has been conducted by CFX software. A good design stent requires lower velocity and higher WSS of the blood flow in the stented artery. To achieve these goals, the following guidelines can be considered. Decreasing the slot length and number of slots can decrease the velocity at the first strut, while increasing the slot length and number of slots can decrease the velocity at the second strut. Decreasing the slot length can increase WSSs at the first and third strut. However, it revealed that the slot width has no significant effect on flow behaviours, because it is a circumferential dimension.

References

[1] A. Colombo, G. Stankovic, J.W. Moses, Selection of coronary stents. J Am Coll Cardiol **40(6)**:1021-1033. (2002)

[2] J.F. Jr LaDisa, L.E. Olson, I. Guler, D.A. Hettrick, S.H. Audi, J.R. Kersten, D.C. Warltier, P.S. Pagel, Stent design properties and deployment ratio influence indexes of wall shear stress: a three-dimensional computational fluid dynamics investigation within a normal artery. J Appl Physiol **97(1)**:424-430. (2004)

[3] T. Seo, L.G. Schachter, A.I. Barakat, Computational study of fluid mechanical disturbance induced by endovascular stents. Ann Biomed Eng **33(4)**:444-456. (2005)

[4] Fan Shuen Tseng, Tse Kiat Soong, Nicholas Syn, Chi Wei Ong, Leo Hwa Liang, Andrew Choong, Computational fluid dynamics in complex aortic surgery: applications, prospects and challenges. Journal of Surgical Simulation. 4. **1-4**. 10.1102/2051-7726.2017.0001. (2017)

[5] P.D. Morris, A. Narracott, H. von Tengg-Kobligk, et al Computational fluid dynamics modelling in cardiovascular medicine: Heart; 102:18-28. (2016)

[6] Runlin Gao, Yuejin Yang, Yaling Han, Yong Huo, Jiyan Chen, Bo Yu, Xi Su, Lang Li, Hai-Chien Kuo, Shih-Wa Ying, Wai-Fung Cheong, Yunlong Zhang, Xiaolu Su, Bo Xu, Jeffery Popma, Gregg Stone, Bioresorbable Vascular Scaffolds Versus Metallic Stents in Patients with Coronary Artery Disease: ABSORB China Trial. Journal of the American College of Cardiology. 66. 10.1016/j.jacc.2015.09.054. (2015)

[7] Muhammad Sufyan Amir Paisal, Syed Fadhil Syed Adnan, Ishkrizat Taib, Al Emran Ismail, Mohammad Kamil Abdullah, Normayati Nordin, Suzairin Md Seri and Nofrizalidris Darlis, Flow Characteristics Near to Stent Strut Configurations on Femoropopliteal Artery. IOP Conference Series: Materials Science and Engineering. 226. 012147. 10.1088/1757-899X/226/1/012147. (2017)

[8] S. Chien, S. Usami, H.M. Taylor, J.L. Lundberg, M.I. Gregersen, Effects of hematocrit and plasma proteins on human blood rheology at low shear rates. J Appl Physiol **21(1)**:81-87. (1996)

[9] D.N. Ku, D.P. Giddens, C.K. Zarins, S. Glagov, Pulsatile flow and atherosclerosis in the human carotid bifurcation. Positive correlation between plaque location and low oscillating shear stress. Arteriosclerosis **5(3)**:293-302. (1985)

[10] J.E. Jr Moore, C. Xu, S. Glagov, C.K. Zarins, D.N. Ku, Fluid wall shear stress measurements in a model of the human abdominal aorta: oscillatory behavior and relationship to atherosclerosis. Arteriosclerosis **110(2)**:225-240. (1994)

[11] A.M. Malek, S.L. Alper, S. Izumo, Hemodynamic shear stress and its role in atherosclerosis. JAMA **282(21)**:2035-2042. (1999)

IOP Conf. Series: Materials Science and Engineering **715** (2020) 012068 doi:10.1088/1757-899X/715/1/012068

Evaluating the impact of using various biodiesel blends on the performance of diesel engine at variable load conditions

Mahmoud Abu-Zaid

Faculty of Engineering, Mutah University, Jordan

*E-mail: abuzaid@mutah.edu.jo

Abstract. Recently, researchers use various types of alternative fuels, for diesel engines. Previous studies, reveals that biodiesel with renewable origin, is the most promising alternative fuel, with less impact on the environment. The objective of this paper, is to determine experimentally, the impact of using different biodiesel blends, extracted from Sun flower, Palm, and Corn oils, on the performance on diesel engine. The engine was operated at variable load and fixed speed. The following main parameters were determined: Torque, Brake power, Brake Specific Fuel Consumption (BSFC), Brake thermal efficiency, Fuel consumption, and Exhaust temperature. Results show that when all the biodiesels were used as a fuel to the engine, the engine has lower BSFC and exhaust temperature, and higher brake thermal efficiency, than when using pure diesel as fuel for the engine. This implies that using the biodiesel blends is more economically than using pure diesel as a fuel for the engine. In addition, it was concluded that using all these blends produces less NO_x emissions.

1. Introduction

The transportation sectors usually used a large quantity of crude oil products. The fuel usage in Jordan increases dramatically, with the increase in population. In addition, the prices of crude oil keep increasing. The major resource of fuel for many vehicles is diesel. One way to decrease the diesel fuel consumption is using renewable fuel. For CI engines bio diesel is the most suitable alternative fuel. It produces less harmful emissions, with no sulfur content.

Bio diesel is One of the main sources of renewable energy. Many researchers studied the use of bio diesel derived from animal fats and vegetable oils, in CI engines. Recently, Numerous studies has been performed to use bio diesel blends instead of diesel, to evaluate its effect on the performance of CI engines [1-4]. Puravi and Meiyanathan [5] performed tests on diesel engines using 100% bio diesel alone, he concluded that brake power and thermal efficiency increased significantly. Bhatt et, al. [6] studied the suitability of mahua oil as alternative fuel for CI engines. They reported that muhau could be easily substituted up to 20% in the diesel without any significant differences in the performance of CI engines.

Puravi and Meiyanathan [7] studied the performance of diesel engines using bio diesel blends, by varying the load and speed. The results shows that with 100% biodiesel blend alone, the brake power, and brake thermal efficiency were significantly increased. Kasundra and Gohi [8] compared different performance parameters of CI engine with different vegetable oils as a fuel.

Nadir et. al. [9] studied the effect of using propanol, n-butanol, and 1-pentanol in waste oil methyl ester (B100) on engine performance and exhaust emissions on a diesel engine running at different loads (0, 3, 6, and 9 KW) with a fixed engine speed (1800 rpm). They concluded that the addition of propane n-butanol, and 1-pentanolto B100 had the effect of increasing brake specific fuel consumption and exhaust gases.

Anbumani and Singh [10] investigated the use of vegetable oils as bio fuel for CI engines. Their study have revealed that among the different vegetable oils, sunflower blend at 15% by volume with diesel fuel exhibited best combustion and performance in terms of total fuel consumption, specific fuel consumption, brake thermal efficiency and cylinder peak pressure.

The objective of this study is to evaluate the impact of using different blends of bio diesel derived from: corn, sun flower, and palm oils, on the performance of diesel engine at variable load and constant speed.

2. Experimental apparatus and procedure

A single cylinder diesel engine, four stroke, direct injection, with compression ratio of 17:1, was used in the experiments. The dynamometer maximum speed is 67 rev./sec., with five different engine loads (0, 0.25, 0.5, 0.75, 1.0).

The production of bio diesel blends vegetable oil (Palm, or corn, or Sunflower) was prepared, using sodium hydroxide as alkaline catalyst. The following chemical properties of the blends: calorific value, cetane number, and density were determined.

The engine was started with diesel fuel, and until the engine reached the operating temperature. Then the engine operated using blends of bio diesel with diesel in proportion (20% bio diesel from vegetable oil, and 80% pure diesel).

The following parameters were recorded: instantaneous speed (rpm), fuel consumption, Torque, Exhaust temperature, at variable engine load and fixed engine speed conditions.

3. Analysis

The following formulas were used to determine the value of various parameter in this study:

$$\text{Brake torque } (T_b) = \text{Force x radius (N.m)} \quad (1)$$

$$\text{Brake power } (B_p) = (2 \times \pi \times N \times T_b)/(6000) \quad (2)$$

where N is r.p.m.

$$\text{Brake thermal efficiency } (\eta_{bth}) = B_p/(\text{mass flow rate of fuel x calorific value of the fuel}) \quad (3)$$

$$\text{Bake mean effective pressure (B.M.E.P.)} = (6000 \times B_p) / (L \times A \times n) \quad (4)$$

where L is the stroke, A is the cross sectional area of the piston, and n = N/2.

$$\text{Brake specific fuel consumption (B.S.F.C)} = \text{mass flow rate of fuel / Brake power} \quad (5)$$

4. Results and discussion

From the collected data during each experiment, torque, mean effective pressure, brake power, fuel consumption, brake specific fuel consumption, brake thermal efficiency, and exhaust temperature, were analyzed, and were plotted versus engine load, to evaluate the performance of the engine as a function of the engine load. The following results were obtained:

The results for the performance of the engine as a function of engine load keeping engine speed constant, the collected data during each set of experiments. The following parameters were determined: torque, mean effective pressure, brake power, fuel consumption, brake specific fuel consumption, brake thermal efficiency. and were plotted versus engine load (min. load, 1/4 load, 1/2 load, 3/4 load, and full load). The results are as follows:

Figure 1 shows, the torque as a function of load. In general the torque increase as the engine load increases. This is because as the load increases, more fresh charge enter the cylinder, producing larger force exerted on top of the piston, thus more torque produced. Pure diesel has the highest values. This is because diesel has the highest calorific value, when combustion occurs, produce large amount of thermal energy, which result in producing very high torque.

IOP Conf. Series: Materials Science and Engineering **715** (2020) 012068 doi:10.1088/1757-899X/715/1/012068

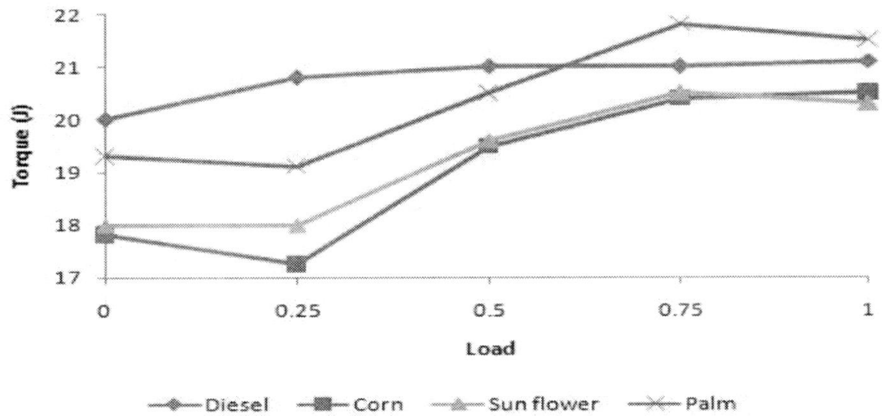

Figure 1. Torque as a function of engine load.

Figure 2 shows, the brake mean effective pressure as a function of engine load. As explained before it is a function of engine torque. As expected the BMEP increase as the engine load increases, similar to the variation of torque. It is clear that pure diesel has the highest values. Because it has the highest calorific value, as explained before for the torque parameter.

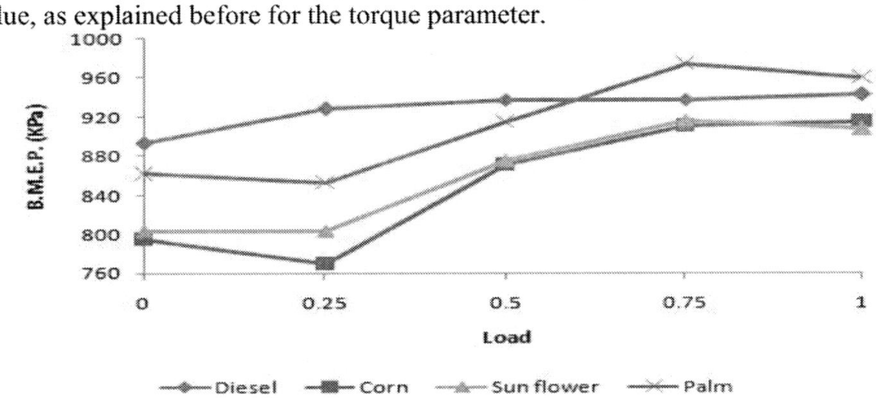

Figure 2. Brake mean effective pressure as a function of engine load.

Figure 3 shows, the fuel consumption rate as a function of engine load. Generally the fuel consumption increase as the engine load. This is because the amount of fuel required is higher. The values of all bio diesels are lower than pure diesel. This is Because of their lower calorific values compared to pure diesel.

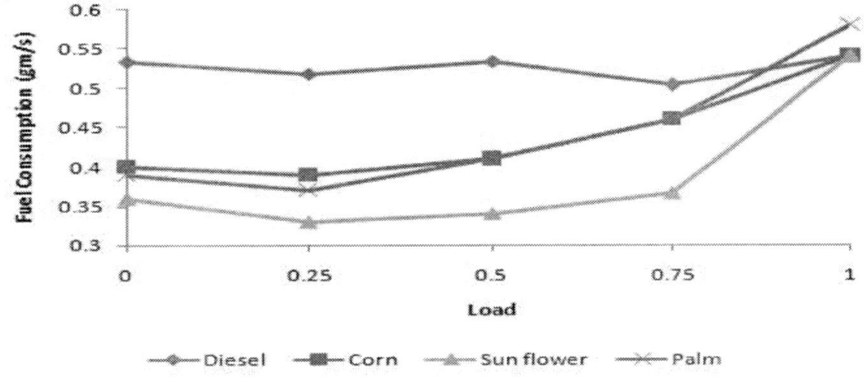

Figure 3. Fuel consumption as a function of engine load.

Figure 4 shows, the brake power as a function of engine load. It is clear that when load increases brake power also increases. The brake power is a function of torque and engine speed. As shown before the torque increases with the engine load, and the speed remains constant. Hence as the engine load increases the brake power increases. The values for pure diesel is the highest. This is because the fuel consumption rate, and is its calorific value is the highest.

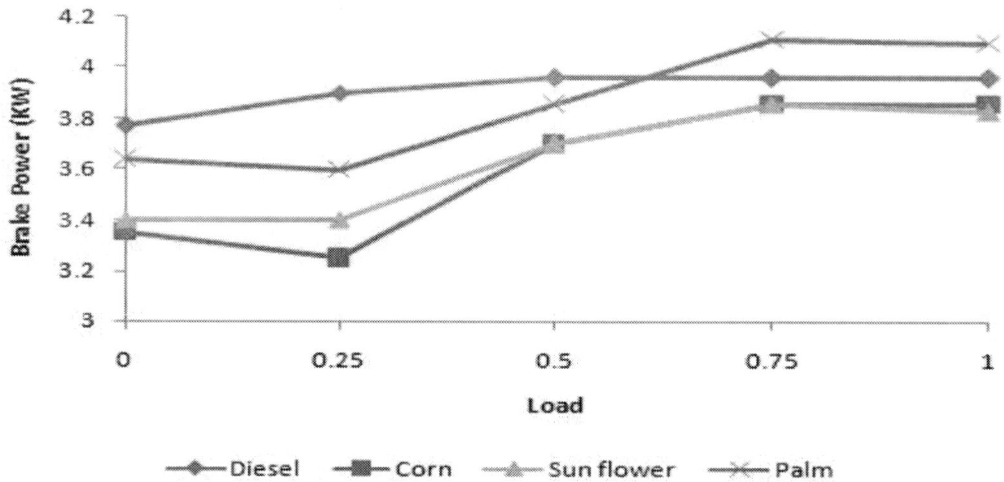

Figure 4. Brake power as a function of engine load.

Figure 5 shows, the brake specific fuel consumption as a function of engine load. This parameter is very important parameter, it shows the capability of the engine to convert the fuel to brake power, which reflects how good the engine performance is. It is clear that with increase in engine load, the BSFC decreases. One possible explanation for this could be due to an increase in brake power with load, as compared with fuel consumption. In addition, once the load reached at full load, the time taken for combustion was decreased, caused incomplete combustion. The values of BSFC of all bio diesels are lower than pure diesel. This shows that using the bio diesel blends is more economically than using pure diesel.

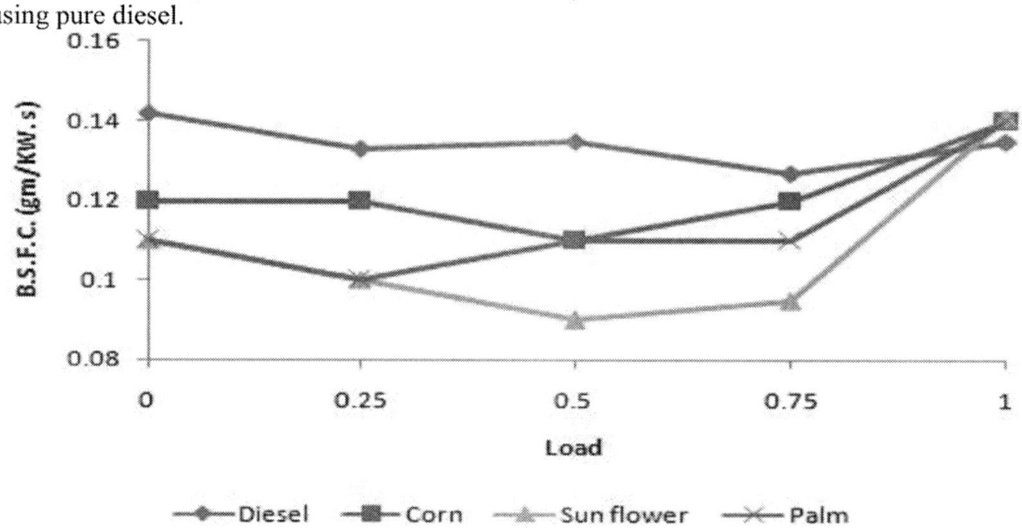

Figure 5. Brake specific fuel consumption as a function of engine load.

Figure 6 shows, brake thermal efficiency as a function of engine load. The brake thermal efficiency is defined as the ratio of brake power to the thermal energy produced from the combustion of the fuel supplied during the same interval of time. It is clear, that the brake thermal efficiency increases to about half load, and then decreases. One possible explanation for this could be due to an increase in brake power with load, as compared with fuel consumption. In addition, once the load reached at full load, the time taken for complete combustion was decreased. The values of brake thermal efficiency of all bio diesels are higher than pure diesel. This shows that using the bio diesel blends is more economically than using pure diesel.

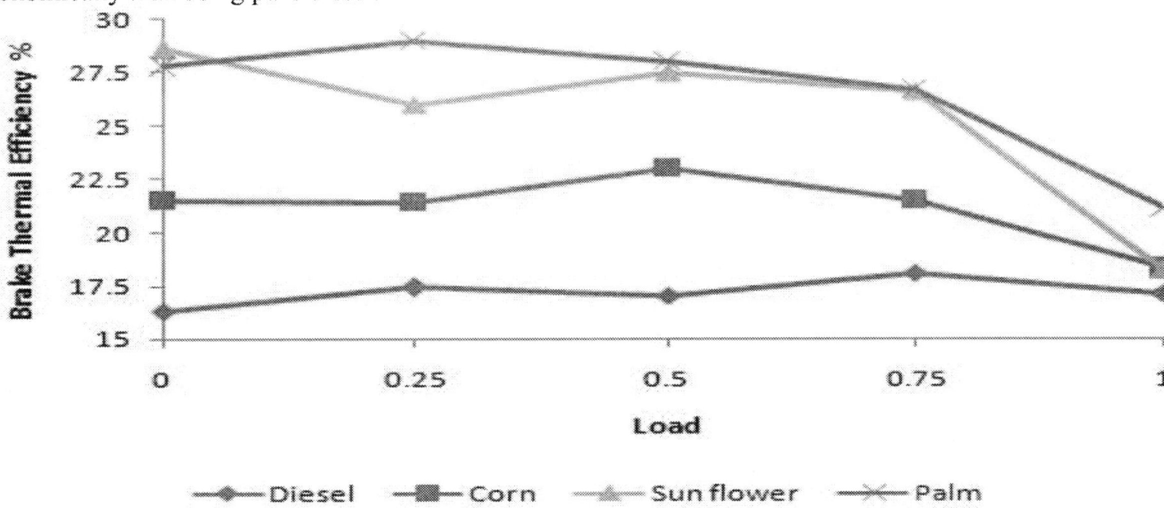

Figure 6. Brake thermal efficiency as a function of engine load.

Figure 7 shows, the exhaust temperature as a function of engine load. In general the exhaust temperature increases along with the increase in engine load for all fuels. The increase in exhaust temperature with load is obvious from the fact that more fuel is required to take additional load, which is resulting in producing more thermal energy. Hence increasing the temperature of the products of combustion inside the cylinder.

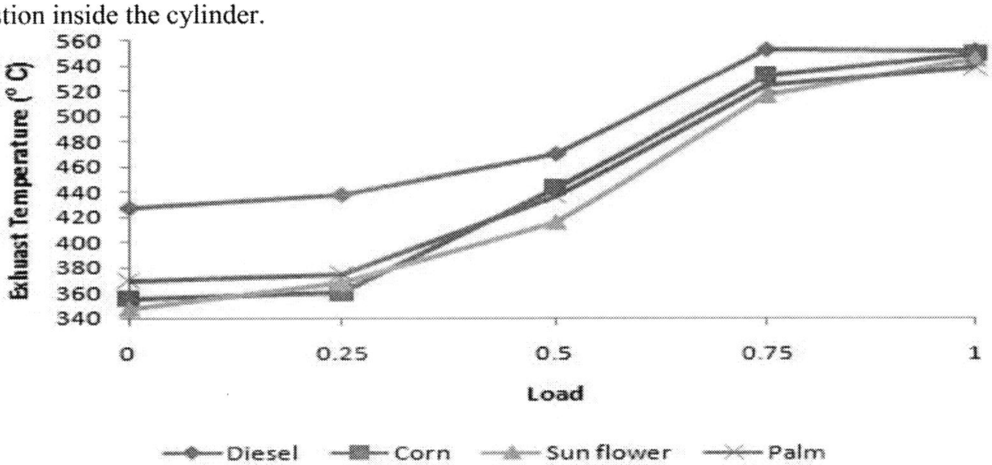

Figure 7. Exhaust temperature as a function of engine load.

5. Conclusions

In this study, a comparison of the performance for a compression ignition engine, using pure diesel and bio diesels fuel blends. Results shows that the use of bio diesel blends can be used in compression ignition engine with no modifications, and can replace pure diesel fuel. The engine consumes less fuel for corn and sun flower bio diesels than pure diesel. It was concluded that all bio diesel blends have lower BSFC and higher brake thermal efficiency than pure diesel. This implies that using the bio diesel blends is more economically than using pure diesel. In addition, All bio diesels blends produces less NO_x emissions than pure diesel.

References

[1] D. Jain, and A. Sharma, J. Mater. Environ. Sci., **4** (4) pp. 434-447, (2013).
[2] Z. Khan, A. Saeed, O. Gregory, and A. Ghafoor, Tribology in Industry, Vol. **38**, No. 2, pp. 197-213, (2016).
[3] D. Harreh, A. Saleh, A. Reddy, and S. Hamdan, J. of Eng., Vol. **2018**, Article ID 417205, (2018).
[4] H. Saikiran, and V. Sakthimurugan, J. of Chem. And Pharmaceutical Science, Special issue **5**, (2016).
[5] K. Puravi, and M. Meiyanathan, Int. J. of Scientific Eng. And applied Sciences, **(1)**, (9), pp. 308-312, (2015).
[6] G. Bhatt, A. Murthy, and K. Dutta, J. of Institute of Engineers, **Vol. 85**, pp. 10-14, (2004).
[7] K. Puravi, and M. Meiyanathan, Int. J. of Science Eng. and Applied Sciences, **Vol. 1**, No. 9, (2015).
[8] L. Kasundra, and Gohil, Int. J. of Eng. Trends and Tech., **Vol. 2**. No. 2, (2011).
[9] M. Nadir, J. Erolleri, and Atmanli, Int. J. of Energy Research. **Vol**. 40, No. 8, pp. 1134-1143, (2016).
[10] D. Anbumani, and F. Singh, J. of ARISER, **Vol. 5**, No.2, pp. 87-97, (2009).

Simultaneous optimization of saturation equations for two hydrocarbons and four hydrofluoroolefins refrigerants

Koemleng KAN[1,*], I Made Astina[2] and Prihadi Setyo Darmanto[2]

[1]Mechanical Engineering Department, Institute of Technology of Cambodia, Pochentong blvd., P.O. Box 86, Phnom Penh, CAMBODIA
[2]Energy Conversion Research Group, Faculty of Mechanical and Aerospace Engineering, Institut Teknologi Bandung, Jalan Ganesha 10, Bandung 40132, INDONESIA

*E-mail: komelng001@yahoo.com

Abstract. This research exhibits a different simultaneous optimization for saturation equations of six pure refrigerants from Hydrocarbon (HCs) and Hydrofluoroolefins (HFOs). The study objective is to model two different group of saturation equations in accordance with the same structural forms, one for n-butane (R600) and isobutane (R600a), and other for 2,3,3,3-tetrafluoropropan (R1234yf), trans-1,3,3,3-tetrafuoropropene (R1234ze(E)), cis-1,3,3,3-tetrafuoropropene (R1234ze(Z)), and trans-1-chloro-3,3,3-trifluoropropene (R1233zd(E)) refrigerants. Genetic algorithm with weighted least square regression was assigned to carry out the research objectives. The modeling equations for both groups are valid from triple points up to temperature nearby the critical points. The uncertainties respecting to each equation of R600, R600a, R1234yf, R1234ze(E), R1234ze(Z), and R1233zd(E) refrigerants are estimated to be 0.05%, 0.75%, 0.10%, 0.35%, 0.57%, and 0.22% in saturated liquid density; 0.30%, 2.68%, 0.86%, 1.48%, 4.13%, and 1.23% in saturated vapor density; and 0.73%, 2.15%, 0.08%, 0.47%, 0.97%, and 2.85% in vapor-pressures, respectively. The new ancillary equations can be used to calculate the saturation properties for engineering interests.

1. Introduction

Hydrocarbons (HCs) and hydrofluoroolefins (HFOs) have notable interest in scientific and industrial sectors. Regarding to the global warming issues, HCs and HFOs refrigerants are considered as an environment substance due to their zero-ozone depleting potential (ODP) and very small global warming potential (GWP). Saturation properties, as an important thermodynamic property of refrigerants are required for the precision of calculation of the cycle simulation and optimal design of the thermal systems such refrigerator, heat pump, and power systems. Saturation properties have also been widely used in calculating the non-dimensional parameters of heat and mass transfer analysis such as Bond number and Weber number in boiling and convection heat transfer calculation, Reynolds number in flow characteristic calculation, friction factor in pressures drop calculation and so on. On the other hand, saturation properties can be used for evaluating performance of equation of state (EOS) such as Helmholtz energy EOS and cubic EOS. The most reliable and accurate thermodynamic properties play important role in finding prospective candidate working fluids for engineering interests. Accurate analysis result for system design contributes in improvement of system efficiency and energy saving. High performance systems with ozone-friendly and negligible GWP refrigerant are expected to improve the global warming environmental issue. Therefore, it is very important to develop reliable saturation equations by applying different simultaneous optimization to get the same structural form equations for the four HFOs refrigerants consisting of R1234yf, R1234ze(E), R1234ze(Z), and R1233zd(E), and other two HCs refrigerants consisting of R600, R600a.

2. Data preparation and model development

2.1. Data selection for modelling

Experimental data are very important for property modeling of saturation equations. Assessment of all experimental input data of the HCs and HFOs were carefully observed through the existing experimental data to find the reliable data set for simultaneous optimizations. All data selections were prepared in the same standard unit system and scale. Data with older temperatures standard than ITS-90 were converted into ITS-90. All experimental input data of the HCs and HFOs refrigerants are summarized in table 1.

Table 1. Selected experimental data.

Fluids	Sources	Selected Points			Fluids	Sources	Selected Points		
		P_s	ρ'	ρ''			Ps	ρ'	ρ''
R1234yf	Tanaka[1]	11	-	-	R600a	Lim[27]	8	-	-
	Zhao[2]	-	10	10		Lim[28]	4	-	-
R1234Ze(E)	Gong[3]	10	10	-		Martinez[29]	9	-	-
	Mark[4]	26	-	-		Morris[30]	2	3	-
	Nicolas[5]	78	-	-		Sage[31]	8	4	7
	Tanaka[6]	18	18	-		Waxman[32]	16	-	7
	Yin[7]	15	-	-		Weber[33]	13	-	-
	Gao[8]	12	-	-		Kayukawa[34]	8	8	-
	Higashi[9]	-	7	14		Liu[35]	86	-	-
R1234Ze(Z)	Tanaka[1]	22	22	-		Glos[36]	20	23	9
	Fedele[10]	64	-	-	R600	Sliwinski[37]	-	10	10
	Higashi[11]	19	9	10		McClune[38]	-	11	-
R1233zd(E)	Hulse[12]	16	13	-		Haynes[39]	-	12	-
	Raabe[13]	-	8	-		Orrit[40]	-	36	-
	Cui[14]	-	12	12		Benoliel[41]	-	9	-
R1233zd(E)	Kondou[15]	-	10	9		Flebbe[42]	12	-	-
	Kondou[16]	-	-	7		Kratzke[43]	12	-	-
R600a	Beattie[17]	5	5	-		Magee[44]	-	100	-
	Connolly[18]	6	-	-		Sako[45]	22	-	-
	Miyamoto[19]	25	-	-		Holldorff[46]	20	-	-
	Higashi[20]	5	9	11		Glos[36]	20	23	9
	Gilliland[21]	5	-	-		Machin[47]	224	-	-
	Hipkin[22]	5	-	-		McClune[38]	-	7	-
	Hirata[23]	5	-	-		Orrit[39]	-	50	-
	Kahre[24]	-	5	-		Sliwinski[37]	-	9	7
	Lee[25]	8	-	-		Olds[48]	-	4	4
	Leu[26]	5	-	-		Haynes[39]	-	12	-
	-	-	-	-		Kay[49]	19	19	-

Ps: Vapor Pressure [MPa]; ρ': Saturated Liquid Density [kg.m^{-3}]; ρ'': Saturated Liquid Density [kg.m^{-3}]

IOP Conf. Series: Materials Science and Engineering **715** (2020) 012069 doi:10.1088/1757-899X/715/1/012069

2.2. Modelling frameworks

Different simultaneous optimization for two of HCs and other four of HFOs refrigerants are conducted for the same structural form of saturation equations by using genetic algorithm with weighted least square regression method. The optimization criterion of this method is called as fitness, which is sum of weighted least squares for the mathematical model was fitted. The main paradigm of simultaneous optimization is shown in figure 1. Normally, the modeling process start with trial structural forms of the equation to find better structure for each of studied fluids. Not only functional form that has the influence on model quality, but weighting factor has significant effects also. In the case, advancement by iteration of weighting factors may be assigned to accomplish desired deviations of the models from the experimental input data.

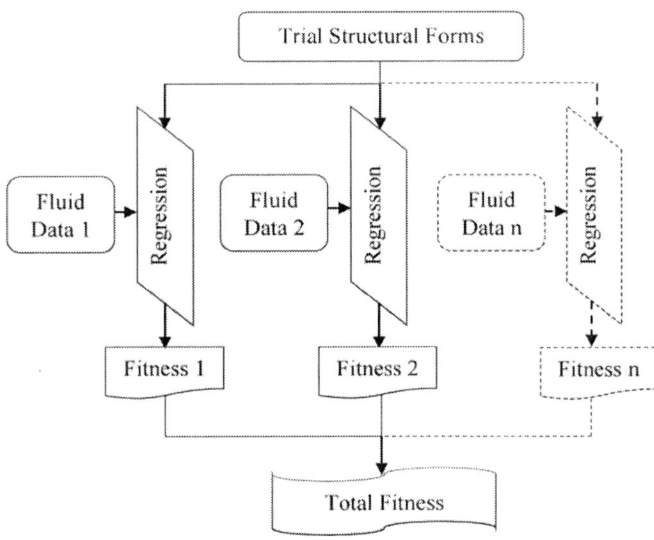

Figure 1. Main paradigm optimization [50].

3. Saturation equations

Saturation equations of two HCs and four HFOs refrigerants were differently simultaneous established. The equations of the HFOs consist of 3 terms and have the same functional form. The equations of the HCs consist of 6, 5, and 3 terms for saturated liquid density, saturated vapor density, and vapor pressure equations, respectively. Each equation group has the same functional form with the general form of ancillary equations as given in Eqs.(1)-(3). The numerical coefficients and functional constants for each refrigerant are listed in table 2.

$$\ln\left(\frac{P_s}{P_c}\right) = \sum_i A_i \frac{T_c}{T}\left(1 - \frac{T}{T_c}\right)^{t_i} \tag{1}$$

$$\left(\frac{\rho'}{\rho_c} - 1\right) = \sum_i B_i\left(1 - \frac{T}{T_c}\right)^{t_i} \tag{2}$$

$$\ln\left(\frac{\rho''}{\rho_c}\right) = \sum_i C_i\left(1 - \frac{T}{T_c}\right)^{t_i} \tag{3}$$

IOP Conf. Series: Materials Science and Engineering **715** (2020) 012069 doi:10.1088/1757-899X/715/1/012069

Table 2. Numerical coefficients and structural forms of equations.

Fluids	HC	R600	R600a	HFO	R1234yf	R1234ze(E)	R1234ze (Z)	R1233zd (E)
				Eq.(1)				
i	t_i	A_i	A_i	t_i	A_i	A_i	A_i	A_i
1	1.11	-18.104	-17.849	0.766	-0.6315	-0.6132	-0.6995	-0.4697
2	1.20	12.4259	12.257	1.016	-6.0585	-6.2635	-6.2247	-6.3603
3	3.50	-3.3681	-3.366	4.146	-3.6479	-4.6392	-4.2351	-5.0158
				Eq.(2)				
i	t_i	B_i	B_i	t_i	B_i	B_i	B_i	B_i
1	0.15	4.3635	-0.4270	0.15	0.1110	0.1507	0.0504	0.0200
2	0.25	-15.855	1.8797	0.4	2.2920	2.2616	2.4273	2.4555
3	0.41	38.155	-0.0164	1.641	0.4620	0.4374	0.4759	0.3373
				Eq.(2)				
i	t_i	B_i	B_i	t_i	B_i	B_i	B_i	B_i
4	0.516	-38.000	-0.8429	-	-	-	-	-
5	0.641	13.993	1.9459	-	-	-	-	-
6	3.891	0.2039	0.2081	-	-	-	-	-
				Eq.(3)				
i	t_i	C_i	C_i	t_i	C_i	C_i	C_i	C_i
1	0.11	-0.13369	-0.1323	0.37	-2.8147	-2.8108	-3.0410	-2.7704
2	0.537	-4.55516	-4.5684	1.37	-9.6775	-10.1414	-9.8409	-10.073
3	2.025	-11.8311	-11.546	4.73	-51.7094	-52.4962	-58.979	-53.1901
4	5	-31.6358	-31.172	-	-	-	-	-
5	9.375	-61.0603	-62.210	-	-	-	-	-
				Main Parameters				
	P_c	3.796	3.628	P_c	3.3822	3.632	3.533	3.573
	T_c	425.3	407.79	T_c	367.85	382.51	423.27	439.6
	ρ_c	225.3	229	ρ_c	475.55	486	470.62	480.23
	T_r	134.9	113.73	T_r	220	168.62	273	195.15

P_c : Critical Pressure [MPa]; T_c: Critical Temperatures[K]; ρ_c:Critical Density [kg.m^{-3}]; T_r: triple point temperature[K].

where A_i , B_i, C_i and t_i are numerical coefficients and predicted variable, respectively.

4. Assessment and discussion

The deviations of experimental data of vapor-pressures from the vapor pressure equation of HFOs and HCs refrigerants are shown in figure 2. The experimental vapor-pressure data of R1234yf are represented within ±0.2%, which are corresponding to the temperatures nearby triple point up to the critical point. Most of experimental data of R1234ze(E) can be reproduced with deviations less than 0.5%, except some data points at lower temperatures. The experimental data of R1234ze(Z) can be accurately represented with deviation less than 1% in the temperature range from 280 K up to near critical points, expect some data at lower temperatures (±1.5%). The deviation of data used for R1233zd(E) can be represented within ±2.5%, except 4 data points at temperatures around 300 K. The data for R600 can be exhibited with deviations less than 0.5% from lower temperatures up to higher temperatures, while the data from Magee et al. [44] and Machin et al. [47] can be represented within ±1.6%. Most of experimental data of R600a can be reproduced with deviations less than 0.5%, except some data points of Higashi et al. [20] and Weber et al. [33]. All data calculated from existing EOS are also plotted for the HCs and HFOs refrigerants.

For saturated liquid density equations, the deviations of measurement data from the equations of HFOs and HCs refrigerants are illustrated in figure 3. The experimental data of R1234yf can be accurately represented within ±0.05% in the temperature range between 280 K up to 350 K, while region nearby critical point has a maximum deviation of 0.27%. The data of existing models of R1234yf are also shown in the figure 3. Most of the deviations of data used for R1234ze(E) can be reproduced with good agreement within ±0.06%, except some data points nearby the critical temperatures have deviations higher than 0.4%. Some data points of R1234ze(Z) have deviations within ±0.06% and other data points have (±0.1%). Three data points at higher temperatures have deviations higher than 1%. The data calculated from EOS model of Akasaka et al. [51] have good deviations less than 0.5%. All the experimental data used in modeling of R1233zd(E) can be represented with deviations less than ±0.5%. The saturated liquid density of R600 refrigerants can be accurately reproduced with deviations of 0.0001% up to 0.11%, while the data of R600a can be represented with deviations less than 0.1%.

For the saturated vapor density equations, the experimental data of saturated vapor density from the equations of HFOs and HCs refrigerants are plotted in figure 4. The experimental data of R1234yf can be reproduced within ±0.3% in the temperature range between 293 K up to temperatures 358 K, while region nearby critical point has a maximum deviation of 1.79%. The existing models of R1234yf are also plotted. Most of the experimental data of R1234ze(E) can be represented with deviations within ±0.9%, except some data points have deviations higher than ±2% up to ±3.4%. The experimental data of R1234ze(Z) have deviations within ±0.6% and some data points at higher temperatures are off-scale from the criteria. The data calculated from model of Akasaka et al. [51] have good deviations. Most of experimental data of Cui et al. [14] and Kondou et al. [15-16] used in model of R1233zd(E) can be represented with deviations within ±0.3%, except the data of Raabe et al. [13]. The saturated liquid density of R600a refrigerants can be accurately reproduced with deviations less than ±0.5%, except the data of Sage et al. [31] and some data point nearby the critical temperatures. On the other hand, some experimental data of R600 can be represented with deviations less than 0.5%, except data of Kay et al. [49].

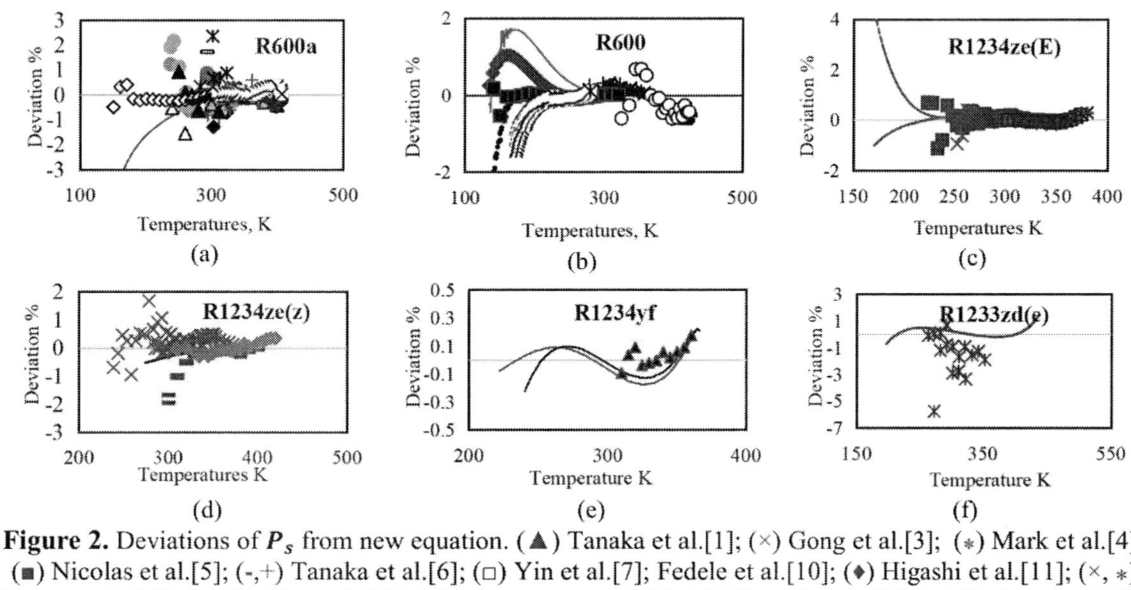

Figure 2. Deviations of P_s from new equation. (▲) Tanaka et al.[1]; (×) Gong et al.[3]; (∗) Mark et al.[4]; (■) Nicolas et al.[5]; (-,+) Tanaka et al.[6]; (□) Yin et al.[7]; Fedele et al.[10]; (♦) Higashi et al.[11]; (×, ∗) Hulse et al. [12]; (♦) Beattie et al.[17]; (×) Miyamoto et al.[19]; ( ) Higashi et al.[20]; (●) Hipkin et al.[22]; (●) Hirata et al.[23]; (●) Lee et al.[25]; (∗) Lim et al.[27]; (●) Lim et al.[28]; (●) Martinez et al.[29]; (●) Morris et al.[30]; (+) Sage et al.[31]; (●) Waxman et al.[32]; (▲) Weber et al.[33]; (◊) Kuyukawa et al.[34]; (-) Liu et al.[35]; (■) Glos et al.[36]; (×) Flebbe et al.[42]; (—) Kratze et al.[43]; (♦) Magee et al.[44]; (+) Holldorff et al.[46]; (∗) Machin et al.[47]; (○) Kay et al.[49] ; (—) Akasaka et al . [51]; (—) Miyamoto et al.[52]; (….) Sarin and Astina [53]; (—) Richter et al.[54]; (----) Akasaka et al.[55]; (—) Mondejar et al.[56]; (—) McLinden et al.[57]; (----) Thol et al.[58]; (- - -) Miyamoto et al. [59].

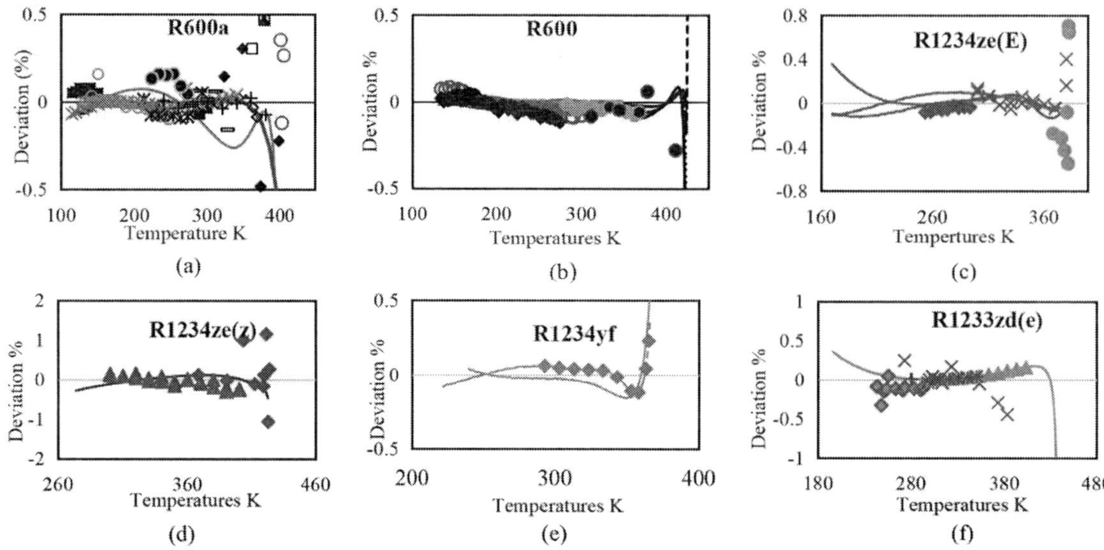

Figure 3. Deviations of ρ' from the new equations. (♦) Zhao et al.[2]; (♦)Gong et al.[3]; (×,▲) Tanaka et al.[6]; (●) Higashi et al.[9]; (♦) Higashi et al.[11]; (♦) Hulse et al.[12]; (×) Rabbe et al.[13]; (▲) Cui et al.[14]; (∗) Kondou et al.[15]; (+) Kondou et al.[16]; (○) Higashi et al.[20];(—) Kahre[24]; (▲) Morris et al.[30]; (□) Sage et al.[31]; (+) Kayukawa et al. [34]; (-,×) Glos et al.[36]; (●,◊) Sliwinski et al.[37]; (○,■) Haynes et al.[39]; (♦,○) Orrit et al.[40]; (∗) Benoleil et al.[41]; (×) Magee et al.[44]; (●,-) McClune et al.[38]; (●) Olds et al[48]; (—) Akasaka et al. [51] ;(—) Miyamoto et al. [52]; (….) Sarin and Astina [53]; (- - -) Richter et al. [54]; (—) Akasaka et al. [55]; (—) Mondejar et al. [56]; (—) McLinden et al. [57] ;(—) Thol et al. [58]; (—) Miyamoto et al. [59]; (—)Younglove and McLinden [60].

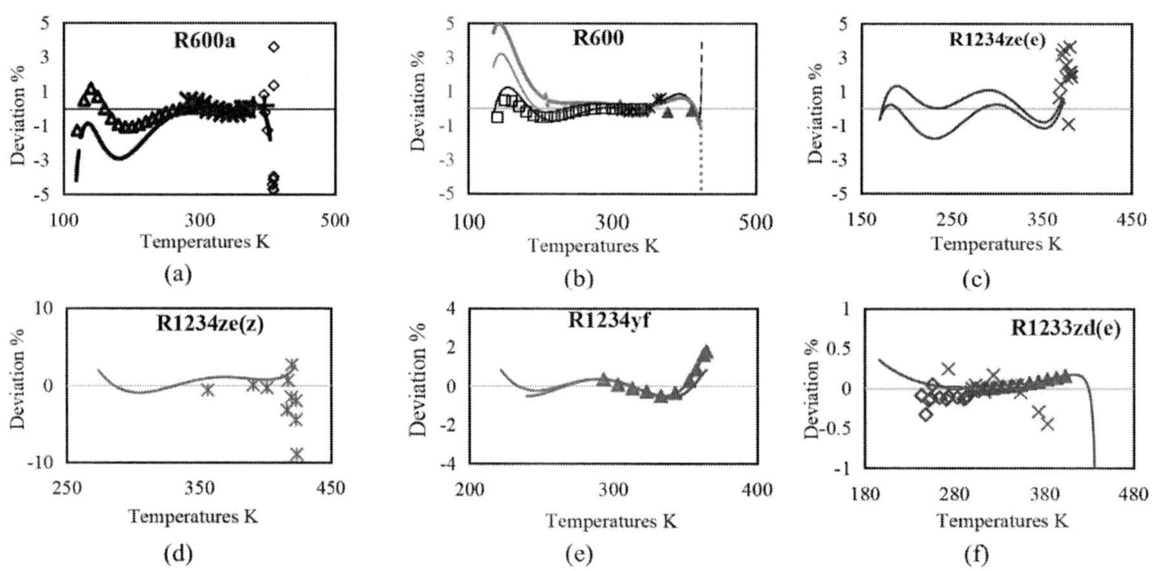

Figure 4. Deviations of ρ'' from the new equations. (▲) Zhao et al.[2]; (×) Higashi et al.[9]; (∗) Higashi et al.[11]; (◊) Hulse et al.[12]; (×) Rabbe et al.[13]; (▲) Cui et al.[14]; (∗) Kondou et al.[15]; (-) Kondou et al.[16];(◊) Higashi et al.[20]; (●) Waxman et al.[32]; (▲,□) Glos et al[36]; (■,∗) Sliwinski et al.[37]; (♦) Olds et al.[48]; (—) Akasaka et al. [51]; (- - -) Miyamoto et al. [52]; (----) Sarin and Astina [53]; (—) Richter et al. [54]; (….) Akasaka et al. [55]; (—) Mondejar et al. [56]; (----) McLinden et al. [57]; (—) Thol et al. [58]; (…..) Miyamoto et al. [59]; (—) Younglove and McLinden [60].

5. Conclusion

New saturation equations for saturation thermodynamic properties of two HCs (R600 and R600a) and four HFOs (R1234yf, R1234ze(E), R1234ze(Z), and R1233zd(E)) refrigerants that are valid for triple point temperatures up to temperatures nearby critical points have been simultaneously developed. The uncertainties respecting to different equations of R600, R600a, R1234yf, R1234ze(E), R1234ze(Z), and R1233zd(E) refrigerants are estimated to be 0.05%, 0.75%, 0.10%, 0.35%, 0.57%, and 0.22% in saturated liquid density; 0.30%, 2.68%, 0.86%, 1.48%, 4.13%, and 1.23% in saturated vapor density; and 0.73%, 2.15%, 0.08%, 0.47%, 0.97%, and 2.85% in vapor-pressures, respectively. The reliable and accurate experimental saturation property data of the HFOs and HCs are satisfactorily reproduced, except saturated vapor density data of R1233ze(E) and vapor-pressures data of R1233zd(E) that have lower accuracy.

Acknowledgement

The authors are grateful to AUN/SEED-net program (ASEAN University Network/ Southeast Asia Engineering Education Development Network) for financial support on the research.

References

[1] K. Tanaka, Y. Higashi, Int. J. Refrigeration. 33, 474-479 (2010)
[2] G. Zhao, S. Bi, A. P. Froba, J. Wu, J. Chem. Eng. Data. 59, 1366-1371 (2014)
[3] M. Gong, H. Zhang, H. Li, Q. Zhong, Int. J. Refrigeration 64, (2016)
[4] O. M. Mark, M. Thol, E. W. Lemmon, Int. C.R. Refri. Air. 1041, 2189-2197 (2010)
[5] G. D.Nicolas, J. S. Brown, L. Fedele, J. Chem. Eng. Data 57, 2197-2202 (2012)
[6] K. Tanaka, Int. J. Chem. Eng. Data 61, 1645-1648 (2016)
[7] J. Yin, Y.Zhou, G. Zhoa, Z. Ma, Fluid Phase Equilibria 460, 69-74 (2018)
[8] N. Gao, G. Chen, R. L. Y. Wang, Y. Hi, J. Therm Anal Calorim 122, 1469-1476 (2015)
[9] Y. Higashi, K. Tanaka, J. Chem. Eng. Data 55, 1594-1597 (2010)
[10] L. Fedele, G. N. Nicolas, J. S. Brown, Int. J. Thermophys. 35, 1-12 (2014)
[11] Y. Higashi, S. Hayasaka, Int. J. of Refrigeration 52, 100-108 (2015)
[12] R. J. Hulse, R. S. Basu, J. Chem. Eng. Data 57, 3581-3586 (2012)
[13] G. Raabe, J. Chem. Eng. Data 60, 2412–2419 (2015)
[14] J. Cui, S. Yan, S. Bi, J. Wu, J. Chem. Eng. Data 63, 751–756 (2018)
[15] C. Kondou, R. Nagata, N. Nii, S. Koyama, Int. J. Refrigeration 53, 80-89 (2015)
[16] C. Kondou, R. Nagata, N. Nii, S. Koyama, Int. Congress Refr., (2016)
[17] J. A. Beattie, S. Marple, D. G. Edwards, J. Chem. Phys.18, 127 (1949)
[18] J. F. Connolly, J. Phys. Chem. Ref. Data. 26, 1239 (1997)
[19] H. Miyamoto, K. Watanabe, J. Chem.Thermodyn. 36, 919-923 (2004)
[20] Y. Higashi, J.Chem. Eng. Data 51, 406-408, (2006)
[21] E. R Gilliland, H. W Scheeline, Ind. Eng. Chem. 32, 48-54 (1940)
[22] H. Hipkin, AIChE J.12, 484 (1966)
[23] M. Hirata, S. Suda, J. Jpn. Petrol. Inst. 9, 885 (1966)
[24] L. C. Kahre, J. Chem. Eng. Data 18, 267 (1973)
[25] B. G. Lee, J. Y. Park, J. S. Lim, Y. W. Lee, J. Chem. Eng. Data 45, 760 (2000)
[26] A. D. Leu, D. B. Robinson, J. Chem. Eng. Data 34, 315 (1989)
[27] J. S. Lim, J. Y. Park, B. G. Lee, Y.W. Lee, J. Chem. Eng. Data 44, 1226-1230 (1999)
[28] J. S. Lim, J. Y. Park, B. G. Lee, J. D. Kim, J. Chem. Eng. Data 45, 734-737 (2000)
[29] J. A. Martinez-Ortiz, D. B. Manley, J. Chem. Eng. Data 23,165 (1978)
[30] W. M. Morris, B. H. Sage, W. N. Lacey, Tech. Publ.,1128 (1939)
[31] B. H. Sage, D. C. Webster, W. N. Lacey, Ind. Eng. Chem. 29, 1309 (1937)
[32] M. Waxman, J. S. Gallagher, J. Chem. Eng. Data 28, 224 (1983)
[33] L. A. Weber, J. Chem. Eng. Data 34, 171 (1989)
[34] Y. Kayukawa, M. Hasumoto, Y. Kano, Chem. Eng. Data 50, 556-564 (2005)

[35] G. Liu, X. Feng, K. Zhang, B. An, Y. Duan, Fluid Phase Equilibra 382, 260-269 (2014)
[36] S. Glos, R. Kleinrahm, W. Wanger, J. Chem. Thermodyn. 36, 1037-1059 (2004)
[37] P. Sliwinski, Z. Phys. Chem. Neue Folge 63, 263 (1969)
[38] C. R. McClune, Cryogenics 16, 289 (1976)
[39] W. M. Haynes, M. J. Hiza, J. Chem. Thermodyn 9, 179 (1977)
[40] J. E. Orrit, J. M. Laupretre, Adv. Cryog. Eng. 23, 573 (1978)
[41] R. W. Benoliel, Some Physical Constant of Hydrocarbons, (Univ. Park 1941)
[42] J. L. Flebbe, D. A. Barclay, D. B. Manley, J. Chem. Eng. Data 27, 405 (1982)
[43] H. Kratzke, E. Spillner, S. Muller, J. Chem. Thermodyn. 14, 1175 (1982)
[44] J. W. Magee,T. O. D. Luddecke, Int. J. Thermophys. 19, 129 (1998)
[45] T. Sako, S. Horiguchi, H. Ichimaru, S. Nakagawa, J. Chem. Eng. Data 42, 169 (1997)
[46] H. Holldorff, H. Knapp, Fluid Phase Equil. 40, 113 (1988)
[47] W. D. Machin, P. D. Golding, J. Chem. Soc. Faraday. Trans. 85, 2229 (1989)
[48] R. H. Olds, H. H. Reamer, B. H. Sage, W. N. Lacey, Ind. Eng. Chem. 36, 282 (1944)
[49] W. B. Kay, Ind. Eng. Chem 32, 358–360, (1940)
[50] I M. Astina, H. Sato, Jurnal Teknik Mesin Indonesia 1(2), 51-60, (2006)
[51] R. Akasaka, Y. Higashi, A. Miyara, S. Koyama, Int. J. Refrig. 44, 168-176 (2014)
[52] H. Miyamoto, K. Watanabe, Int. J. Thermodyn. 22, 459 (2001)
[53] C. Sarin, I M. Astina, Journal. Teknik Mesin. 22, 44-89 (2006)
[54] M. Richter, E. W. J. Lemmon, Chem. Eng. Data 56, 3254-3264 (2011)
[55] R. Akasaka, Int. J. Thermopys. Data.32,1125-1147 (2011)
[56] M. E. Mondejar, M. O. McLinden, E.W. Lemmon, J. Chem. Eng. Data 60, 2477-2489, (2013)
[57] M.O. McLinden, M. Thol, E. W. Lemmon, Int. Refr. Air Cond. Conf, 1041 (2010)
[58] M. Thol, E.W. Lemmon, Int. J. Thermophys. 28, 2-16 (2013)
[59] H. Miyamoto, K. Watanabe, Int. J. Thermodyn. 23, 477 (2002)
[60] B. Younglove, A. McLinden, J. Phys. Chem. Ref. Data 23, 731-779 (1994)

The effects of flexible vortex generator on the wake structures for improving turbulence

Sharul Sham Dol[1,*], Hiang Bin Chan[2], Siaw Khur Wee[2] and Kumar Perumal[2]

[1]Department of Mechanical Engineering, Abu Dhabi University, Abu Dhabi, United Arab Emirates

[2]Faculty of Engineering and Science, Curtin University Malaysia, Miri, Sarawak, Malaysia

*E-mail:sharulshambin.dol@adu.ac.ae

Abstract. Turbulence is a very useful flow characteristic that is used in many engineering devices to facilitate mixing, heat transfer in heat exchanger and improving aerodynamics in propeller. This paper has numerically studied the spatial characteristics of the wake behind a free-oscillating flexible vortex generator (FVG) in order to evaluate its turbulence generation ability. As the wake is playing as the effective turbulence region, the spatial characteristics can reflect the ability of the FVG in turbulence generation when a larger wake size indicates a greater turbulence region. Two VGs are considered in this study; circular and flat plate cantilever. Each VG was submerged individually in a subcritical flow ($10^2 < Re_D < 10^5$) and the generated wake was studied. The wake behind the FVG was visualized via the lambda2 vortex identification criterion and it was compared with its respective rigid counterpart. From the comparison, it is found that the FVG has a larger wake compared to its rigid counterpart. The result demonstrates that the FVG can generate a larger turbulence region. The cause of the increasing wake size was found to be the weakening of downwash behind the FVG when it bends. The downwash becomes weaker as the deflection of the FVG increases.

1. Introduction

Turbulence plays an important role in those devices in facilitating mixing, heat transfer process and improving aerodynamics. In engineering applications, such as a heat exchanger and lift generation devices, the implementation of vortex generators (VGs) guarantees a better turbulence generation that helps to improve the overall performance of the exchanger and improving aerodynamics in airfoils. The implementation of vortex generators in engineering devices is a common yet promising strategy to generate turbulence. VG is normally a static structure that can generate turbulence by disturbing the flow; it can be in any shape, including winglets, cylinders, blades, blocks and others. As long as the structure can interrupt the flow, it can be generalized as a VG.

This study proposes to manipulate the structural properties of a VG, in particular, the stiffness of the VG to achieve flexibility; hence creating a flexible vortex generator (FVG) that can oscillate depending on the surrounding flow dynamics. It is expected that the turbulence and the wake characteristics behind the oscillating FVG will be different compared to a rigid vortex generator (RVG). In this study, circular cantilevers and flat plate cantilevers, with different aspect ratio (AR), are used as the VG.

The study of flow dynamics behind an oscillating object, such as a cylinder or cantilever, have received relatively less attention than its structural dynamics. In the current understanding of flow past an oscillating object, the vortex pattern behind the object, either in 2-dimensional or 3-dimensional cases, has established a respectively comprehensive information compared its turbulence characteristics aspect. The vortex pattern has been studied by a number of researchers [1-4], and the results agree well with the vortex pattern map founded by [5].

However, there are limited studies on the turbulence characteristics behind an oscillating object. Griffin [6] conducted an experimental study on a forced-oscillating cylinder. This work resembles considerably to the current work except that the cylinder was forced-excited instead of oscillating passively. They measured the velocity field using hot-wire anemometry and the velocity fluctuation that represents the turbulent regime was computed. It was found that the velocity fluctuation behind the forced-oscillating cylinder was greater than a stationary one.

Yong et al. [7] experimentally studied the flow dynamics behind a flexible finite cylinder in a single-phase flow using a water tunnel via ultrasonic velocity profiler measurements. They found based on the turbulence production analysis, the flexible cylinder can generate turbulence at maximum 2.27 times of the amount generated by the rigid cylinder under the same flow condition.

The present study focuses on the turbulence characteristic behind an oscillating FVG via analyzing the spatial aspects of the wake. Since the wake behind a VG can be regarded as the region where turbulence activities take place, the size of the wake should be considered as one of the aspects of the VG's turbulence generation ability. The work was done via the Fluid-Structural Interaction (FSI) simulation, which integrates both CFD and FEA. The model was validated with published data (e.g. [8]).

2. Numerical model

After reviewing some related studies, it is found that the RANS model is one of the popular fluid flow models that is used in simulating the FSI (e.g. [9]). In all of the mentioned papers, the classical equation of motion is used as the structural model.

Unsteady Reynolds-Averaged Navier-Stokes (URANS) equation is used to compute the fluid flow. (SST) k-ω turbulence model is used to model the turbulence. Details of the URANS equation, energy equation and (SST) k-ω turbulence model can be found and described extensively in [10, 11]. The typical (3 degree-of-freedom) equation of motion, Eq. (1) is used to model the structural motion:

$$m\ddot{x} + q\dot{x} + kx = \sum F_x \tag{1}$$

where F_x is the force in i-direction, m is the mass, q is the damping coefficient and k is the spring constant.

An important aspect of the numerical model is the method that couples the structural model with the fluid flow model. In the present work, a strong two-way FSI coupling method is used. Besides, an implicit coupling scheme is used during data transfer. This implies that the data is transferred iteratively within the two models until the pre-determined termination criteria are satisfied [10].

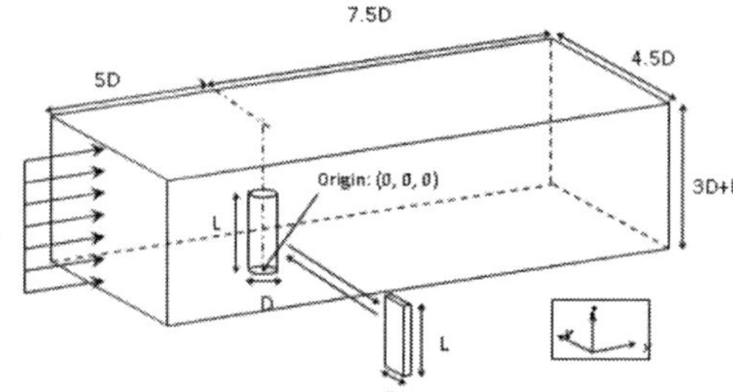

Figure 1. Schematic of the fluid domain (not to scale).

Figure 1 represents the simulation domain used in this study. Uniform flow was applied in the inlet of the domain. The boundary conditions of the flow domain are provided in [10]. In this study, the circular and flat plate cantilever are used as the VG. The fluid, flow and structural properties of the simulated cases are given in [10].

A mesh independency study was carried out to obtain the best mesh for the study. The $y+$ value was constantly monitored to ensure that it falls below 5 – a threshold or criteria from the Law of the Wall that allows the near wall mesh to model the boundary layer correctly. Unstructured polyhedral meshes are used in this study. The meshes that were used in this independency study is tabulated in [10]. Note that the mesh size is defined based on the diameter of the cantilever – the characteristic length of the present study. This allows the mesh to be universally applicable in any cases with the same configuration but with a different cantilever size.

The drag coefficient, pressure profile and velocity profile were monitored to get the optimal mesh size, as shown in figure 2. From the study, Mesh 4 is selected as the optimum mesh based on the analyses. The selection criteria are based on the convergences of each data monitored as the mesh size changes. The monitored results were converged at Mesh 4 and further refinement to Mesh 5 does not cause significant differences in the results (e.g. C_D). A structured hexahedral mesh is used on the VGs.

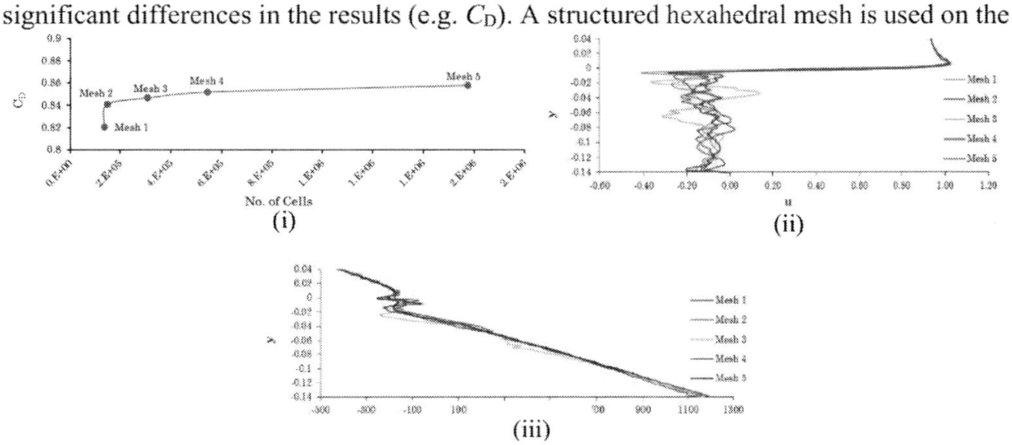

Figure 2. The mesh independency study: (i) drag coefficient, (ii) pressure profile at x/D=0.75 downstream and (iii) velocity profile x/D=0.75 downstream.

3. Results and discussion

The wake is visualized using Lambda2 criterion, a mathematical scheme developed by [12] to define vortex boundary. The Lambda2 criterion defines the boundary of a vortex only when the second eigenvalue of $S^2+\Omega^2$ is negative. This method has been used in other studies defining vortices (for example, in [13, 14]) and good results are obtained. In the following section, the Lambda2 iso-surfaces are presented. In other words, it is a 3-dimensional visualization. An abbreviation is introduced to facilitate the following discussion. With this abbreviation, the lengthy "AR=8 circular FVG" can be written as 8CF; where the "8" denotes the aspect ratio, the "C" denotes "circular" (it will be "P" for flat plate), and the "F" denotes "FVG" (it will be "R" for RVG). Using this concept, "AR=5 flat plate RVG" can be written as 5PR.

Firstly, the height of the wake is examined. In this discussion, the presented figures focus on the Lambda2 iso-surface in the x-z plane (side view). The wake behind the rigid and flexible circular RVG and FVG are presented. Figure 3 clearly shows that the wake of the circular FVG has greater height compared to the wake of circular RVG, regardless of AR. The wake of the FVG has increased 62.2%, 25% and 25% for AR=6, 8 and 10 cases respectively. Next, the wake of flat plate RVG and FVG are presented, in figure 4. Unlike the circular FVGs, the wake of flat plate FVG does not show significant height difference at lower AR. The height difference only becomes observable at greater AR; where the wake of the 6PF is greater than that of the 6PR by ~21.3%. As mentioned before, the wake represents the effective region where turbulence activities take place. Hence, a larger wake generated by the FVGs denotes a spatially greater turbulence compared to the RVGs.

The cause of the wake height increment will be discussed in the following discussion. It is known that the transportation of vortex depends on the velocity field of the flow, therefore, in order to examine the cause of the wake height increment, the mean streamlines behind the VGs are computed. The mean streamlines plot of the circular cases are shown in figure 5. Based on the mean streamline plots of the RVG case, a downwash (a downward fluid flowing motion) is present behind the circular RVG. The downwash is a common flow feature behind a cantilever, which is reported by all works that have studied the flow past a cantilever. By comparing the streamlines of the RVG case with its respective Lambda2 iso-surface, it is clear that the vortex structure is downwashed towards the ground plate by this unique flow feature and eventually it causes the wake height to be shorter than the total length of the RVG. On the other hand, the downwash is not present behind the circular FVG. As a result, the vortex structure is transported to the downstream without experiencing any downward motion. Thus, the wake of circular FVG has an equivalent height irrespective of its aspect ratio.

Figure 6 shows the streamline plots of the flat plate cases. In contrast to the circular case, the downwash behind the flat plate FVG does not vanish completely. It is weakened only at higher AR, which has explained the insignificant wake height difference between FVG and RVG at low AR. Now, it is clear that the weakening or vanishing of the downwash plays an important role in the wake height increment, but it raises another question: what governs the weakening or vanishing downwash? To answer this question, the evolution (time series) of the streamlines plot is examined. Note that the streamlines of the circular case do not possess significant changes with time. Therefore, the streamlines plot of the flat plate case is used to identify the relationship, as shown in figure 7. From the figure, it can be observed that the weakest downwash is obtained when the flat plate FVG is attaining a larger deflection. Conversely, the strongest downwash is attained when the flat plate FVG is at its smallest deflection. This observation suggests that the strength or scale of downwash is scaled with the degree of deflection of the FVG; the greater the deflection, the weaker the downwash.

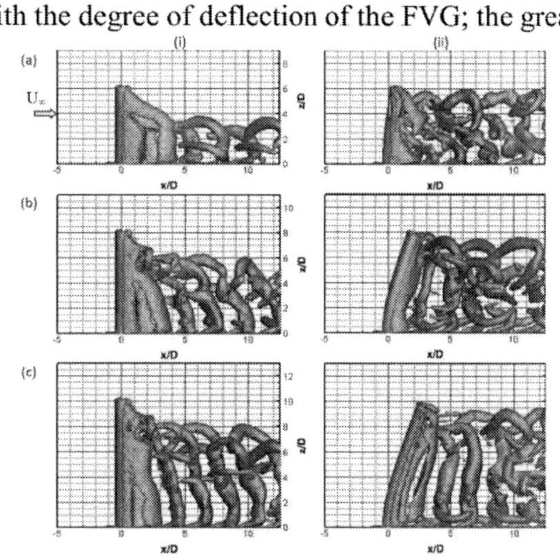

Figure 3. The lambda2 iso-surface at the *z-x* plane-of-symmetry (y/D=0), of the circular (i) RVG and (ii) FVG, with (a) AR=6, (b) AR=8 and (c) AR=10.

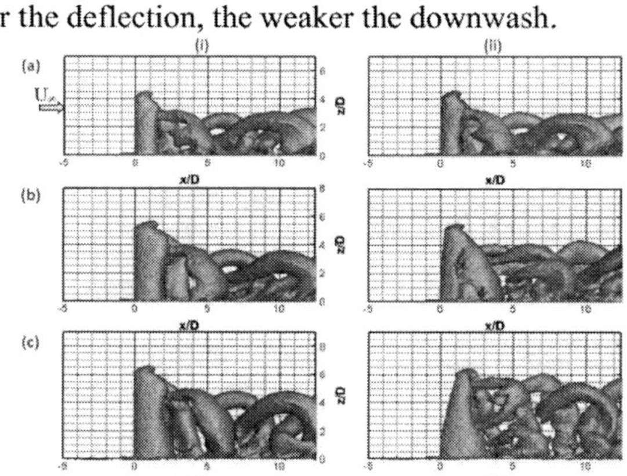

Figure 4. The lambda2 iso-surface at the *z-x* plane-of-symmetry (y/D=0), of the flat plate (i) RVG and (ii) FVG, with (a) AR=4, (b) AR=5 and (c) AR=6.

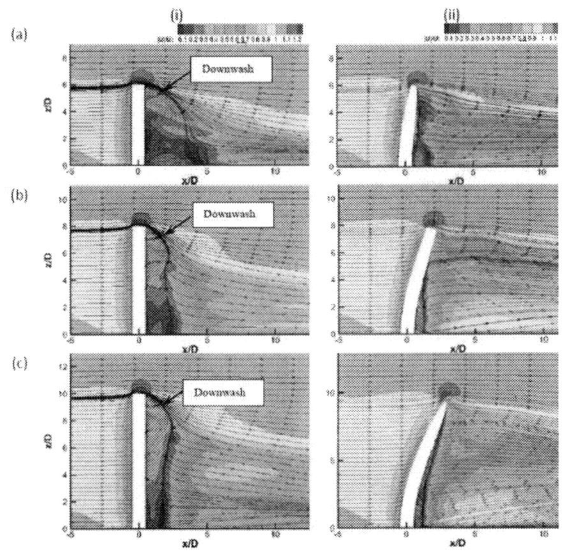

Figure 5. The streamlines of the normalized mean velocity magnitude at the plane-of-symmetry (y/D=0) of the circular (i) RVG and (ii) FVG, with (a) AR=6, (b) AR=8 and (c) AR=10.

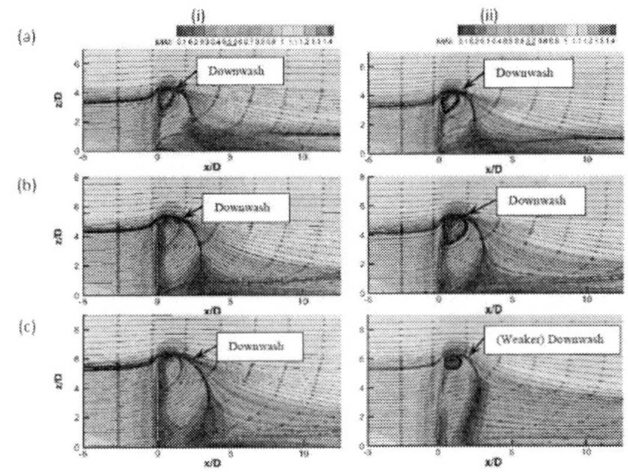

Figure 6. The streamlines of the normalized mean velocity magnitude at the plane-of-symmetry (y/D=0) of the flat plate (i) RVG and (ii) FVG, with (a) AR=4, (b) AR=5 and (c) AR=6.

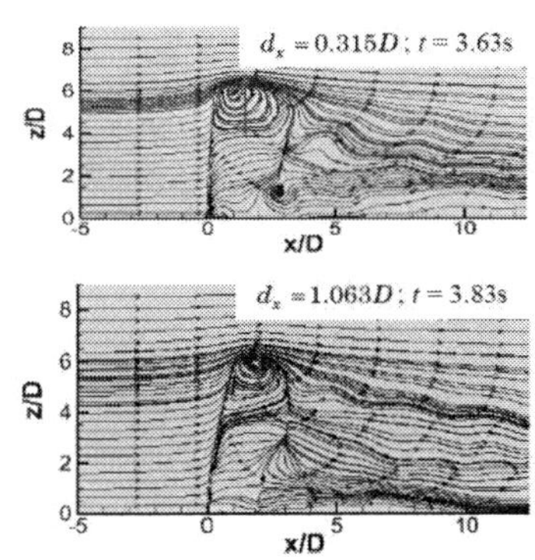

Figure 7. The streamlines plot at the plane-of-symmetry (y/D=0) of 6_{PF} at (top) t=3.63s and (bottom) t=3.83s.

4. Conclusion

In general, the FVG can generate wake with greater height compared to the RVG. In the current study, the circular FVGs have generated wake with 25.0~62.2% taller wake than the wake of the RVG; and the wake of the flat plate FVGs is at most ~21.3% taller than the wake of the RVG. The wake height increment is caused by the weakening or vanishing of downwash behind the FVG. The weakening or vanishing of downwash is scaled with the degree of deflection; the greater the deflection, the weaker the downwash. The width of the wake does not differ significantly despite a large transverse motion of the FVG. Since the wake represents the effective region where turbulence activities can take place, a larger wake generated by the FVGs denotes a spatially greater turbulence compared to the RVGs. Besides, this present study has demonstrated a different viewpoint to study turbulence generation ability other than the classical statistical study. The spatial analysis shed light on the possibility to actively control the spatial scale of turbulence, without compromising the turbulence strength, by simply altering the deflection of a VG in the system.

References

[1] Techet, A. H., F. S. Hover, and M. S. Triantafyllou. Vortical patterns behind a tapered cylinder oscillating transversely to a uniform flow. Journal of Fluid Mechanics 363, 79-96 (1998).

[2] Fujarra, A. L. C., C. P. Pesce, F. Flemming, and C. H. K. Williamson. Vortex-induced vibration of a flexible cantilever. Journal of Fluids and Structures 15, no. 3, 651-658 (2001).

[3] Yamamoto, C. T., J. R. Meneghini, F. Saltara, R. A. Fregonesi, and J. A. Ferrari. Numerical simulations of vortex-induced vibration on flexible cylinders. Journal of fluids and structures 19, no. 4, 467-489 (2004).

[4] Lam, K. M., P. Liu, and J. C. Hu.. Combined action of transverse oscillations and uniform cross-flow on vortex formation and pattern of a circular cylinder. Journal of Fluids and Structures 26, no. 5, 703-721 (2010).

[5] Williamson, C. H. K., and A. Roshko. Vortex formation in the wake of an oscillating cylinder. Journal of fluids and structures 2, no. 4, 355-381 (1988).

[6] Griffin, Owen M. The unsteady wake of an oscillating cylinder at low Reynolds number. Journal of Applied Mechanics 38, no. 4, 729-738 (1971).

[7] Yong, T. H., H. B. Chan, S. S. Dol, S. K. Wee, and P. Kumar. The flow dynamics behind a flexible finite cylinder as a flexible agitator. In IOP Conference Series: Materials Science and Engineering, 206, no. 1, p. 012033. IOP Publishing, (2017).

[8] Park, C. W., and S. J. Lee. Free end effects on the near wake flow structure behind a finite circular cylinder. Journal of Wind Engineering and Industrial Aerodynamics 88, no. 2, 231-246 (2000).

[9] Pan, Z. Y., W. C. Cui, and Q. M. Miao. Numerical simulation of vortex-induced vibration of a circular cylinder at low mass-damping using RANS code. Journal of Fluids and Structures 23, no. 1, 23-37 (2007).

[10] Dol, S.S., Wee, S.K., Chan, H.B., and Perumal K. Turbulence Characteristics behind a Flexible Vortex Generator. WSEAS TRANSACTIONS on FLUID MECHANICS 14, 1-7 (2019).

[11] Chan, Hiang Bin, Tshun Howe Yong, Perumal Kumar, Siaw Khur Wee, and Sharul Sham Dol. "The Numerical Investigation on the Effects of Aspect Ratio and Cross-Sectional Shape on the Wake Structure Behind a Cantilever (2006).

[12] Jeong, Jinhee, and Fazle Hussain. On the identification of a vortex. Journal of fluid mechanics 285, 69-94 (1995).

[13] Dol, S. S., Gregory A. Kopp, and Robert J. Martinuzzi. The suppression of periodic vortex shedding from a rotating circular cylinder. Journal of Wind Engineering and Industrial Aerodynamics 96, no. 6, 1164-1184 (2008).

[14] Dol, S. S., Salek, M., and Robert J. Martinuzzi. Energy Redistribution Between the Mean and Pulsating Flow Field in a Separated Flow Region. Journal of Fluids Engineering 136, 11, 111105-1 – 111105-9 (2014).

Effects of impact direction on energy absorption capability of aluminum honeycomb for road construction zone

Attisook Nokthong[1] and Sathaporn Chuepeng[2,*]

[1]Master of Engineering Program in Safety Engineering and Environmental Management, Faculty of Engineering at Sriracha, Kasetsart University, Chonburi, Thailand
[2]ATAE Research Group, Department of Mechanical Engineering, Faculty of Engineering at Sriracha, Kasetsart University, Chonburi, Thailand

*E-mail: schuepeng@eng.src.ku.ac.th

Abstract. This work studies the energy absorption capability of an aluminum honeycomb for road construction zone specifically designed as equipment for reducing severity of impact force. The design is considered when it is hit by a 1.5-ton car while running at the speed of 50 kilometers per hour. The research processes begin with the designing of honeycomb model using computer program, then simulating the force of impact acceptance with finite element method in order to search for the most suitable size and material. Afterwards, the prototype was created and tested on force acceptance using the methods of quasi-static test. The results illustrate that the aluminum honeycomb is capable of absorbing energy within designated condition of receiving force in the specified direction. The result will definitely be beneficial for the development of impact force severity reduction equipment that utilizes honeycomb aluminum as its main material.

1. Introduction

In road construction and maintenance, chances that cause workers to be injured or die can be accidentally occurred. One reason is caused by vehicles that unintentionally intrude into the work zone. Such accidents were continuously recoded for a decade [1]. The statistical data illustrates the declined numbers of accident, resulting from a development of more safety work conditions. This includes an improvement of work instruction in standardized work zone, utilization traffic control devices, protective equipment and imposing of stricter law enforcement on drivers.

To prevent vehicle to access the work zone, it is compulsory to conform to working standard called "Work Zone Safety" [2-4]. Further than that it also required for the accessibility protective equipment to be used as it can obstruct accident for workers and reduce severity of accident for drivers.

In the design of workpiece for impact force energy absorbing, according to related research, the structure of honeycomb has a characteristics of high-strength that can absorb energy well [5-11]. Therefore, the workpiece is designed into the shape of honeycomb and specified its design condition to be able to absorb kinetic energy from 1.5-ton car while running at 50 kilometers per hour. Therefore, the energy must be absorbed by 144 kJ following Eq. (1). Furthermore, honeycomb workpieces are also light-weight, commonly found in local market. The kinetic energy (E_k) can be expressed as:

$$E_k = \frac{1}{2} mv^2 \tag{1}$$

where m is mass and v is speed of a car [12].

In term of equipment for work zone accessibility protection and accident severity reduction, this research purpose is to design a workpiece that is able to absorb energy from impact force. It is a part of fundamental component of impact force severity reduction equipment that can be installed in many places on both fixed and movable objects. The design and construction of complete workpiece will be conducted further in this research work.

2. Method of experiment

The predesigned honeycomb with the computer program, SolidWorks[13], assists in creating the workpieces and simulating the impact force acceptance, that act against two sides of honeycomb including L-direction and W-direction as shown in figures 1 and 2, respectively.

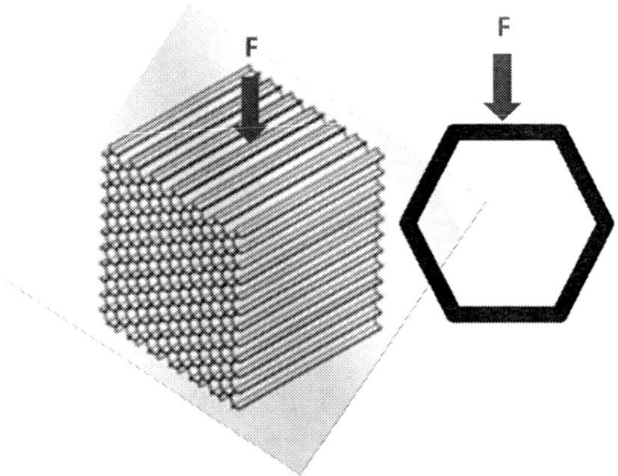

Figure 1. The L-direction of impact force.

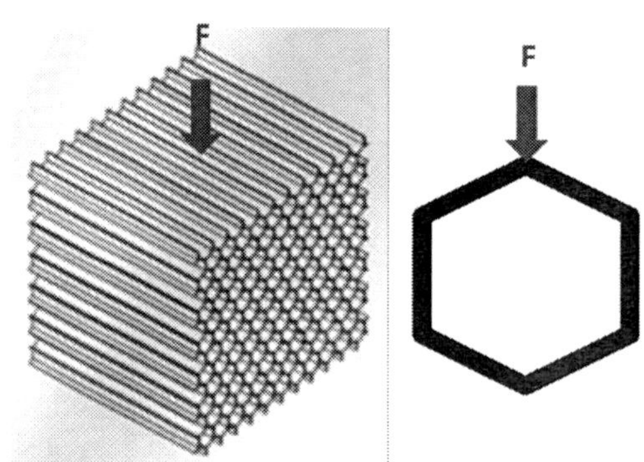

Figure 2. The W-direction of impact force.

Consequently, the analysis was performed to measure its capability to absorb energy and deform with finite element method by specifying the conditions that honeycomb must be able to absorb energy at least 144 kJ and have a maximum 80% deformation in the force direction. Once finding the appropriate size and material, the workpieces then produced in order to test the impact force acceptance using a method of quasi-static test by utilizing hydraulic puncher.

The designed honeycomb is 47 mm × 50 mm in cell size (Figure 3), constructed from 1 mm sheet aluminum grade A1100 (H14) with tensile strength of 140 MPa. It is then folded and welded into the shape of honeycomb with a size of 500 mm × 500 mm × 500 mm as shown in figure 4.

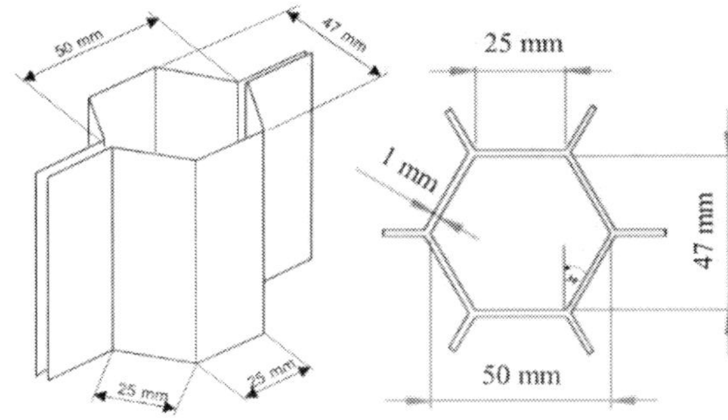

Figure 3. Dimension of the constructed aluminum honeycomb.

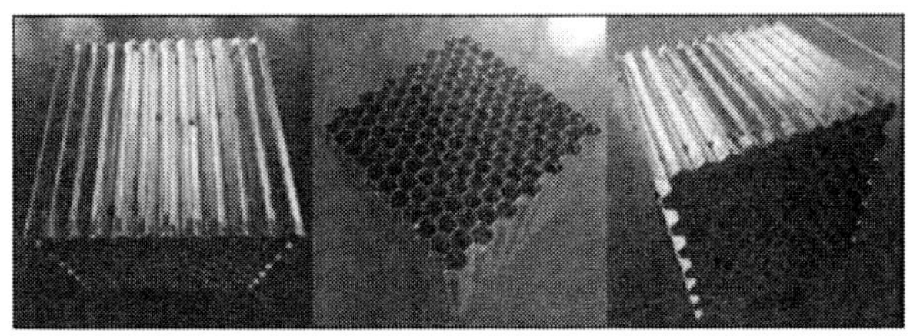

Figure 4. Complete constructed aluminum honeycomb.

Quasi-static test using hydraulic puncher as shown in figure 5 was gradually pressed on the aluminum honeycomb. The measurements of force and displacement were then used to calculate for absorption of the absorbed energy.

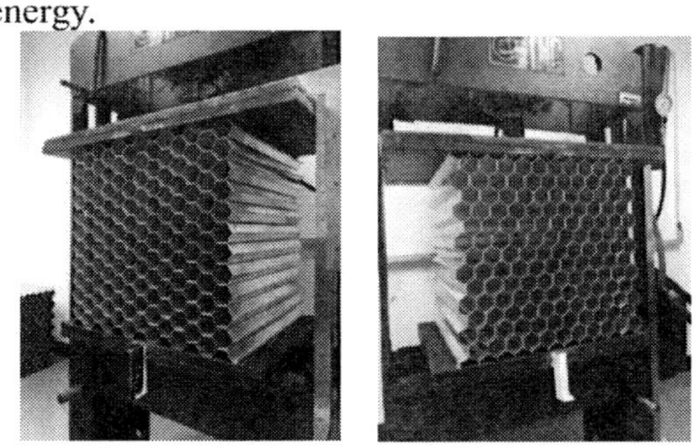

L-direction W-direction

Figure 5. Hydraulic puncher.

3. Results and discussion

3.1. Physical appearance after Quasi-static test

Test results of aluminum honeycomb force acceptance using quasi-static test to analyze for energy absorption capability of honeycomb are illustrated in figures 6(a) and 6(b), respectively for L-direction and W-direction.

Figure 6. The honeycomb appearance after quasi-static test (a) L-direction and (b) W-direction.

IOP Conf. Series: Materials Science and Engineering **715** (2020) 012071 doi:10.1088/1757-899X/715/1/012071

3.2. Force-displacement characteristics

Figure 7 shows the force-displacement characteristics of the honeycomb for both L-direction and W-direction in comparison. The forces acting on the two directions are likely characterized but differently in the deformation. The force acting in the L-direction results in greater deformation compared to the other direction. Meanwhile, the L-direction can resist greater force compared to the other direction. In the initial puncher pressing, the materials of the two workpieces are markedly displaced up to approx. 210 mm. Afterwards, the acting forces are mild increasing for displacement in the range of approx. 210 mm to 380 mm. In the last stage, a tiny displacement of the material are observed with a large amount of force acting on both honeycombs.

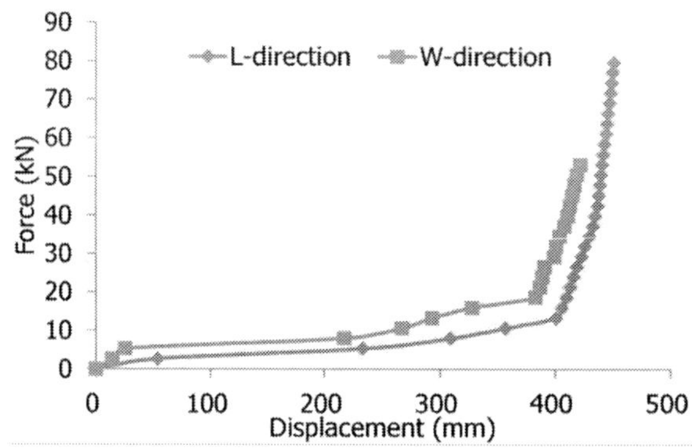

Figure 7. Force-displacement characteristics of aluminum honeycomb.

3.3. Displacement and energy absorption characteristics

In figure 8, when hydraulic puncher pressing against honeycomb in L-direction, honeycomb is gradually crushed until completion at the displacement of 451 mm (90% deformation in the direction of force), which is higher than designated criteria of 80%. While energy is absorbed by 4.6 kJ which is lower than the designated criteria.

For the case of W-Direction, its result resembles to those of the L-direction, as honeycomb is gradually crushed until completion with the displacement of 422 mm (84% deformation in the direction of force), which is greater than designated criteria. While energy is absorbed by 5.4 kJ lower than the designated criteria.

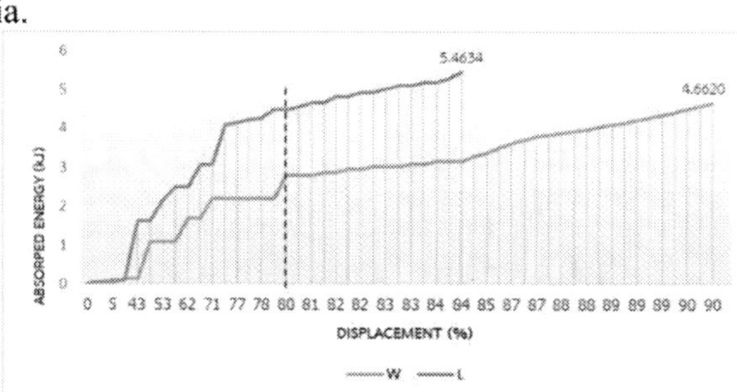

Figure 8. Energy absorbing result of aluminum honeycomb in quasi-static test.

3.4. Energy absorption according to force acceptant direction

Figure 9 indicates that L-direction absorbs less energy than that of the W-direction. With the quasi-static test performed, the L-direction demonstrates 15% less capability compared to the W-direction.

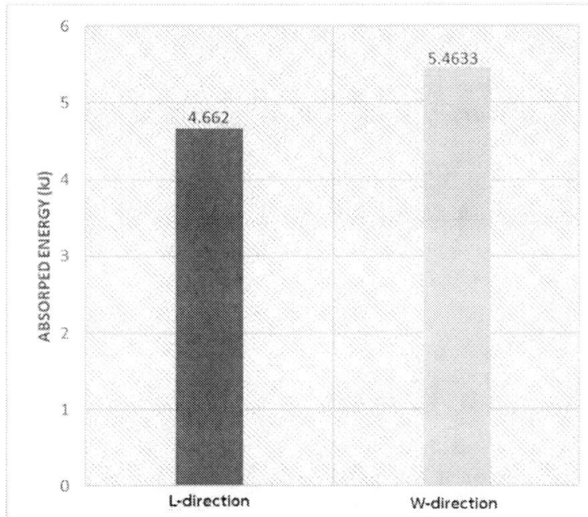

Figure 9. Aluminum honeycomb absorbed energy according to force acceptant direction.

4. Conclusion

The aluminum honeycomb is designed to mitigate severity in road construction zone, by first using computer program and building an appropriate model by utilizing materials that are locally available. The comparison of aluminum honeycomb energy absorbed capability using a quasi-static test demonstrates that aluminum honeycomb is capable of receiving impact force. Under the designated model when arranged to accept the force toward the W-direction, the results show greater absorbed energy but lesser deformation compared to that the force acting in the other direction.

As for the results of this research, aluminum honeycomb will be utilized as main component of impact force severity reduction equipment that will definitely be redesigned and developed further in the future, which is extremely beneficial for construction work and road maintenance to reduce the severity when accident occurs.

Acknowledgments

Kasetsart University Faculty of Engineering at Sriracha is acknowledged for support to this project. The authors wish to thank Bangkok Expressway and Metro Public Company Limited for the test apparatus.

References

[1] X. Wang, R. Katz and X.S. Dong, *"Fatal Injuries at Road Construction Sites among Construction Workers"*, The Center for Construction Research and Training, http://www.cpwr.com, (2018)

[2] Construction Safety Council, *"Work Zone Hazards Workbook"*, http://www.buildsafe.org, (2008)

[3] Land Transport Authority of Singapore, *"Code of Practice for Traffic Control at Work Zone"* 23rd Edition, (2010)

[4] S.G. Pratt, D.E. Fosbroke, *"Building safer highway work zones"*, DHHS (NIOSH) Publication No.2001-128, (2001)

[5] A. Baroutaji, M. Sajjiab and A. Olabic, *"On the crashworthiness performance of thin-walled energy absorbers: Recent advances and future developments"*, Thin-Walled Structures, **Vol.118**, 137-163 (2017)

[6] A. Baroutaji, D. Gilchrist, D. Smyth and A. Olabi, *"Crush analysis and multi-objective optimization design for circular tube under quasi-static lateral loading"*, Thin-Walled Structures, **Vol.86**, 212-131(2015)

[7] A. Dhiware, "*Design and crash analysis of frontal impact energy absorbing structures for supra SEAINDIA competition*", International Journal of Mechanical and Industrial Technology, **Vol. 2**,144-148 (2014)

[8] B. Hou, H.Zhao, S. Pattofatto, J. Liu and Y. Li, *"Inertia effects on the progressive crushing of aluminum honeycombs under impact loading"*, International Journal of Solids and Structures, **Vol.49**, 2254-2762 (2012)

[9] I. Ivañez, L.F. Cañadas and S.S. Saez, "*Compressive deformation and energy-absorption capability of aluminum honeycomb core*", Composite Structures, **Vol.174**, 123-133 (2017)

[10] Q. Liu, J. Fu, J. Wang, J. Ma, H. Chen, Q. Li and D. Hui, *"Axial and Lateral crushing responses of aluminum honeycombs filled with EPP foam"*, Composites Part B, **Vol.130**, 236-247 (2017)

[11] M. Asadia and B. Walker, *"Application of shell elements to create advanced finite element model for offset deformable barrier"*, International Journal of Vehicle Structures & Systems, **Vol.2**, 139-143 (2011)

[12] P. Booranasiri, et al., *"1-Physics for Scientists and Engineering with Modern Physics"* 1st ed., Pongwarinkarnpim, Bangkok,191-195, 257-261, 471-473 (2005)

[13] P. Techaunpai, V. Limtrakarn, S. Sucharitpakawatsakun and Y. Prathumwan, "*Apply Finite element method by Solidworks Simulation*" 1st ed., TPA Publishing, Bangkok, 1-8 (2009

Performance-based analysis for life safety from fire, Case study; large structural steel manufacturing building

Jutamas Laungta[1,]*, Prateep Chaisermtawan[2] and Supat Patvichaichod[2]

[1]Safety Engineering and Environmental Management program Kasetsart University Sriracha Campus Chonburi, Thailand.
[2]Department of Mechanical Engineering Kasetsart University Sriracha Campus Chonburi, Thailand.

*Email: Jutamas.lt@msn.com

Abstract. This study simulate the evacuation from structural steel manufacturing building and verified with Thai's regulation. According to the regulation, all occupants shall be in safe place within 5 minutes in case of fire incident. Two simulation scenarios were selected for this simulation. First, all exits can be used for evacuation; evacuation time is 4.23 minute after announcement. Second, fire caused by explosion of compressed air tank near Exit A, which completely block this exit A. Simulation result shown that evacuation time for this particular case is 7.06 minute, which is exceed maximum evacuation time defined in the regulation. Therefore, the following countermeasures are propose: Expand existing stair from 1.0 m to 1.5 m., Install an additional stair with the same width on the opposite side and modify firefighting procedure.

1. Introduction

Since fire incident in large facility building could cause huge property damage or even loss of employee's life. Allocating evacuation path to assembly point that use minimum time is important [1]. Focused their attention on analyzing the impacts of different evacuation behaviors and measuring their effificiency in terms of the evacuation rate [2]. Although the station covers large area and accommodates a large number of people, a peak passengers flow is relatively small. Therefore, the safety egress time is shorter [3]. A research of evacuation plan [4] for single floor administration office. As fire occurred with smoke and people nearby realized the situation. Therefore, they start to move away using an exit they familiar with. Test by varying number of occupants in each area to evaluate time of each scenarios. As stated in item 8 of government regulation in safety, health and environment management regarding fire prevention and counter measure [5]. Factory must provide evacuation routes which capable of handling all employees to safe place within 5 minute. The rule applied equally to any kind of factory, regardless of size nor type of industry. All industrial plant must comply with this standard. Pathfinder is a tool for simulate evacuation time and behavior of workers in predefined scenarios. In-depth analytical data will show if there is a bottleneck. These information is useful for evacuation paths improvement. Correction of bottleneck to comply with government regulation will finally result in a safer place for all employees.

2. Model building and calculation

2.1. Model building

Pathfinder by Thunderhead Engineering Consultants, is an emergency egress simulator that includes an integrated user interface and animated 3D results [6]. The software supports two movement simulation modes, "Steering" mode and "SFPE" mode. In "Steering" mode, occupants use a steering system to move and interact with others. This mode tries to emulate human behavior and movement as much as possible. SFPE mode uses a set of assumptions and hand-calculations as defined in the Engineering Guide to Human Behavior in Fire [7]. In a large structural steel manufacturing building, spark and hot particle always present in nearly every production phases. There are a lot of used gloves and rags contaminated with oil and grease. Furthermore, high power gas burners are also used with oxygen mixed and maintenance worker use LPG tank for cutting in the plant. Facility have total area of 23,526 m2, 4 floors and one gateway for overhead crane (named A1 and A2) at 24 m height. Production of steel involve 7 process as shown in figure 1, Electric arc furnace (EAF), Ladle furnace (LF), Continuous casting machine (CCM), Mold preparation, Refractories and Discharge area. The building has 6 exit as shown in table 1.

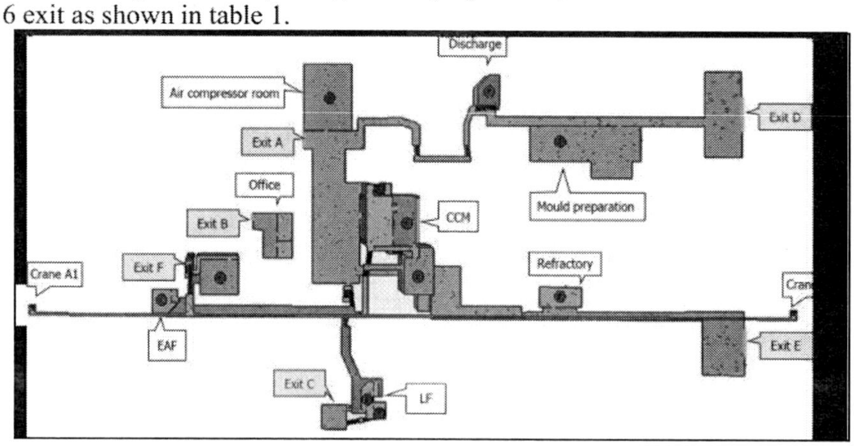

Figure 1. Layout of structural steel manufacturing.

Table 1. Detail of exit of structural steel manufacturing.

Area	Sector	Width(cm.)
CCM	Exit A	430
Office	Exit B	150
LF process	Exit C	120
	Staircase	100
Mould preparation	Exit D	950
Refractory	Exit E	760
Cabin Crane	Exit F	200

2.2. Simulation settings

1. In this simulation, 478 Employees (including outsource worker) are involve. This number derived from occupant load factor (m 2 /person) as defined in NFPA 101 [8]. Employees divided into 2 groups, production workers and maintenance workers by ratio of 70% and 30% respectively.

2. All employees and outsource workers are gentle with shoulder width of 45.58 cm., 182 cm. height, and walking speed is 1.19 m/s [7].

3.Behavior model has been set to "Steering" mode, Jam velocity is 0.25 m/s; controls the speed threshold at which occupants are recorded as being jammed at exit [7].

4. Maintenance workers must first go to control rooms they responsible for power source isolation. This should take approximately 1 minute before evacuate from the building. In contrast, production workers will evacuate instantly with nearest path to assembly point (if available).

5. There are two simulation scenarios, performance based analysis of the building and fire caused by explosion of high-pressure gas tank near exit A. In later scenario, exit A become unusable.

3. Simulation result and analysis

3.1. Simulation result

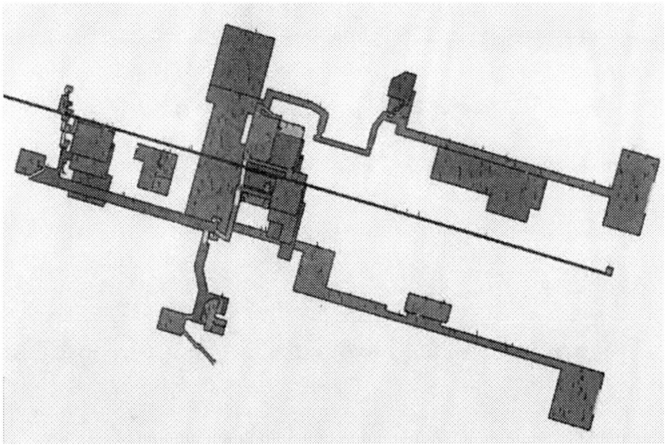

Figure 2. Simulate the evacuation from structural steel manufacturing building.

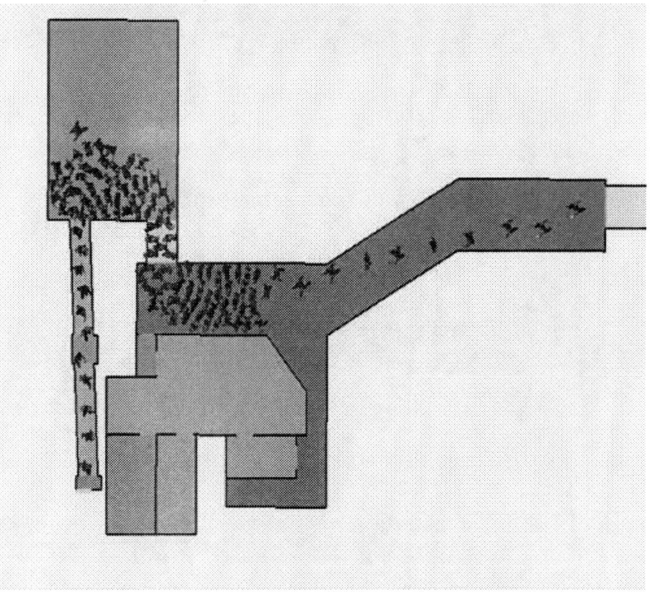

Figure 3. Bottleneck evacuation at Exit C.

Table 2. Number of occupant of each Exit.

	Exit B	Exit C	Exit D	Exit E	Exit F
Person	32	240	97	59	39

There are two simulation scenarios, performance based analysis of the building as illustrated in figure 2 with evacuation time of 4.23 minute. In another scenario, fire caused by explosion of high-pressure gas tank near exit A. In later scenario, exit A become unusable. As shown in table 2, vast majority of workers (240 persons) appears to use exit C instead as can be seen in figure 3. In this case, bottleneck at exit C occurred with total evacuation time of 7.06 minute. Production worker who wait for longest period is from CCM 1st floor. He wait for 275.8 s. for evacuation at exit C. Total evacuation time of

421.15 s. and distance of evacuation 118 m. Maintenance worker who responsible for electrical system isolation at CCM 2nd floor (same level with exit C) use exit C with waiting time of 194 s. and distance of 135 m. (which is longest waiting time among maintenance workers).

3.2. Analysis of simulation result

Figure 4. (1) Extend width of stair from 1 m. to 1.5 m.

Figure 5. (2) Construct additional stair.

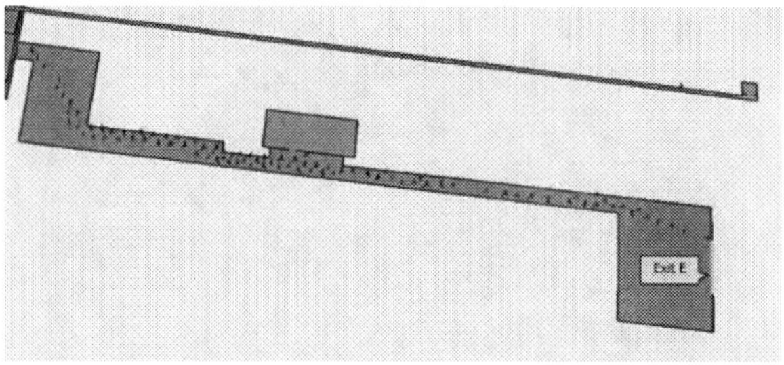

Figure 6. (3) Explicitly define evacuation path of workers who working at discharge and air compressor room to used exit E.

Evacuation in case of blocked exit A is fail to comply with regulation because bottleneck happens at exit C. Improvement of exit C can be made with 3 options, (1) Extend width of stair from 1 m to 1.5 m as in figure 4. (2) Construct additional stair shown in figure 5. (3) Explicitly define evacuation path of workers who working at discharge and air compressor room to used exit E which locate in opposite side of factory. Therefore, this exit will be a bit more crowded as show in figure 6. Simulation result for each improvement is 4.32 minute, 4.46 minute and 4.55 minute respectively, thus comply with this regulation.

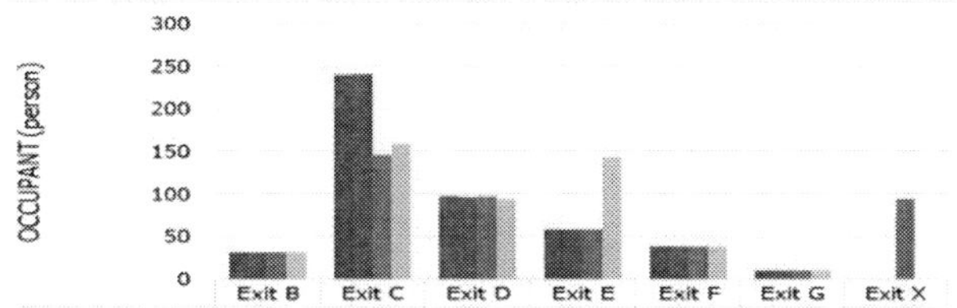

	Exit B	Exit C	Exit D	Exit E	Exit F	Exit G	Exit X
Exit A blocked	32	240	97	59	39	11	0
Option 1	32	241	96	59	39	11	0
Option 2	32	146	97	59	39	11	94
Option 3	32	159	94	143	39	11	0

Figure 7. Average waiting time, exit time and distance at exit A and after improvement.

From figure 7, evacuation time after exit C was improved indicate that option 1, 2 and 3 reduce average waiting time from 30.72 s/person to 17.17, 17.49 and 14.70 s/person (44.15%, 43.05% and 52.13%) respectively. Total evacuation also reduce from 108.26 s/person (base case) to 106.94 and 104 s/person or 1.2% and 3.93%.

4. Results and discussion

Further consideration for option 1 shown in figure 8. With the same amount of occupants, all exit handle nearly the same amount of persons but average flow rate (person/s) of exit C (bottleneck) is increase from 0.58 person/s to 0.94 person/s.

When consider option 2, the building has additional stair named exit X shown in figure 8. Occupants sharing between the 2 stairs can be realized. As a result, exit C become less chaos with amount of occupants reduced from 240 persons to 146 person (39% less) and flow rate at exit C and exit X is 0.54 and 0.39 person/s as shown in figure 9. Total evacuation time for this case is 4.46 min.

In option 3, there is no modification on physical structure of the building. Workers at discharge refractory and air compressor somehow inform to use exit E as shown in figure 8. This option reduce workload of exit C from 240 persons to 159 person. Even though workload of exit E increase from 0.41 person/s to 0.93 person/s as shown in figure 9, but overall evacuation time is, reduce to 4.55 min.

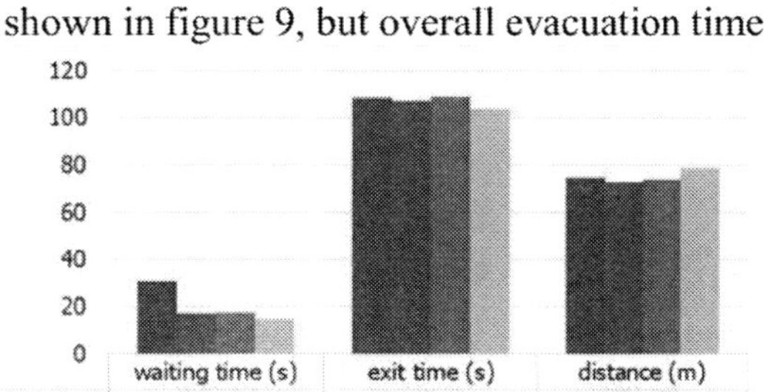

	waiting time (s)	exit time (s)	distance (m)
Exit A blocked	30.7105	108.2606	74.8278
Option 1	17.1559	106.9483	72.9011
Option 2	17.4927	108.9164	73.6152
Option 3	14.7043	104.0018	79.0329

Figure 8. Number of occupants at each exit.

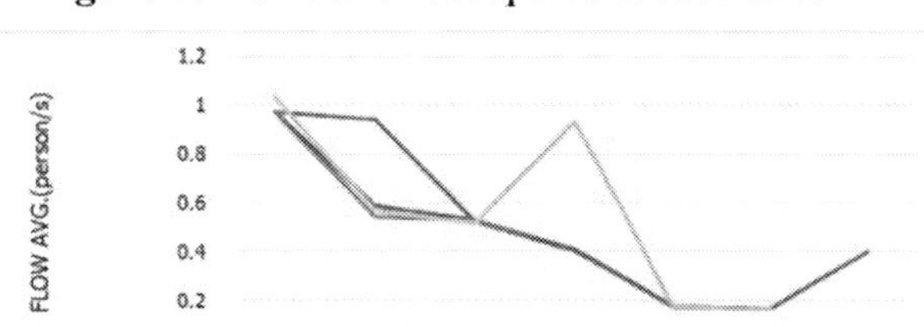

	Exit B	Exit C	Exit D	Exit E	Exit F	Exit G	Exit X
Exit A blocked	0.9697	0.5882	0.5301	0.4097	0.1781	0.1692	
Option 1	0.9697	0.9414	0.5246	0.4041	0.1781	0.1692	
Option 2	0.9697	0.5407	0.5301	0.4097	0.1781	0.1692	0.3983
Option 3	1.0323	0.5679	0.5137	0.9286	0.1781	0.1692	

Figure 9. Average flow rate for each improvement.

5. Conclusion

Simulation result indicate that present fire evacuation paths of the building might not meet maximum time as state in Thai' regulation.Width of stair for exit C should be expand from 1 m to 1.5 m. or provide additional stair for exit C Firefighting procedures and evacuation plan should be review and practice for real workers and real site should be setup properly for each scenarios. Fire exit sign should be clearly visible add proper lighting for evacuation paths should be consider and maintained.

Acknowledgments

This is project accomplished by Pathfinder with education license from Thunder Engineering. Layout supported by structural steel manufacturing building.

References

[1] Withayawut Jirasing. 2010. "A Numberical study of fire evacuation out of the library building of Kasetsart University, Si racha campus". Master Engineering thesis, Kasetsart University, Thailand.

[2] Vani and Sumam. 2013. "Agent based evacuation simulation using leader-follower model", Int. J. Sci. Eng. Res., 4 (8) (2013).

[3] Zhen-yu Li et al. 2016. "Numerical simulation of evacuation in a subway station", Procedia Engineering 135 (2016) 616 – 621, China.

[4] Somkait Vetpreukpitak. 2006. "Study of the evacuation plan using the fire dynamics simulation program and simulation of migration". Master's thesis, Kasetsart University, Thailand.

[5] Ministry of Labor "Thai government regulation in safety, health and environment management regarding Fire prevention and counter measure", December 2012.

[6] Thunderhead Engineering Consultants. "Flexible Movement Simulation". https://www.thunderheadeng.com/pathfinder,USA.

[7] Thunderhead Engineering Consultants. "Pathfinder 2018 Results User Manual". https://www.thunderheadeng.com/pathfinder/, USA.

[8] National Fire Protection Association. 2018. "NFPA 101 Life Safety Code 2018 edition", July 2018.

Study on the performance of inflatable decelerator under hypersonic aerodynamic load

Jingjing Wang, Li Yu*, Xue Yang and Xiaoshun Zhao

College of Aerospace Engineering, Nanjing University of Aeronautics and Astronautics, Nanjing, China

*E-mail: yuli_happy@163.com

Abstract. In order to understand the influence of hypersonic aerodynamic force and aerodynamic heat on the performance of flexible structure of inflatable decelerator, a unidirectional thermal-fluid-structure three field coupling model under the extreme conditions of reentry trajectory is established in this paper. The temperature distribution of thermal protection structure (TPS) and structure response are obtained. The numerical results show that the TPS has a good insulation effect, and the cooling contribution of insulation layer in TPS is up to 80.90%. The stress is concentrated at the contact position between the flexible structure and the blunt head. The stress of TPS layer fluctuates greatly.

1. Introduction

The inflatable membrane has a large deformation due to the influence of hypersonic aerodynamic force and aerodynamic thermal load in the reentry process of inflatable decelerator, which leads to the shape change of inflatable decelerator, and then changes the aerodynamic loads. It is a typical thermal-fluid-structure three field coupling problem.

Since the 1980s, a lot of researches on hypersonic thermal-fluid-structure coupling have been carried out. A hypersonic thermal-structure coupling simulation of stainless steel plates was conducted by Thornton and Dechaumphai [1]. Besides, Allan et al. [2, 3] carried out a thermal-fluid-structure coupling study on a cylindrical shell, and compared the numerical results with wind tunnel test results [4].In 2010, Culler and Adam [5] developed a prediction model to determine the strong and weak coupling relationship in thermal-fluid-structure coupling. Wu et al. [6] proposed a hierarchy solution for the hypersonic airfoil in thermal-fluid-structure coupling. [7].

Traditional thermal-fluid-structure coupling problems are mainly about rigid structures such as airfoils, control surfaces and turbine blades [8-10]. Andrew et al. [11] obtained the surface deformation of hypersonic IAD by fluid-structure coupling method. The thermal-fluid-structure coupling problems about flexible structures are more complicated. They involve nonlinear structural dynamics and fully coupled interaction between the compressible fluid flow, with heat flux, and the membrane structure undergoing large deformations. Therefore, some simplification and hypothesis are necessary in the simulation of the thermal-fluid-structure coupling problems about flexible structures. Xia et al. [12] compared the loose coupling and tight coupling results of the heating problem of a hypersonic tube. It showed that the loose coupling method was more efficient and had similar accuracy as the tight coupling method when the characteristic time of the flow field was much smaller than the heat transfer.

In this paper, the characteristic time of the flow field is much smaller than the heat transfer, so a loose coupling method is used in the simulation. The thermal-fluid-structure coupling of the inflatable decelerator is calculated. The temperature distribution of the TPS layers and structural results are studied.

2. Coupling method

Coupling computation is realized based on Workbench software in this paper. The process of coupling calculation is shown in figure 1.

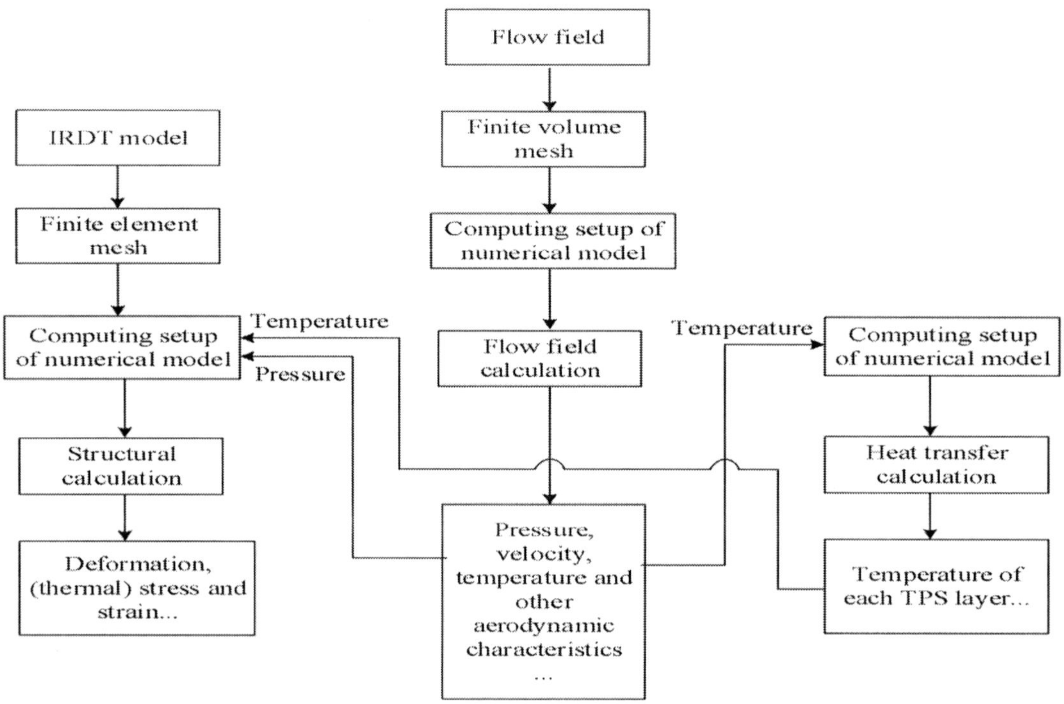

Figure 1. Coupling calculation process.

In this paper, S-A model is used as turbulence model to solve the hypersonic viscous flow. The wall of inflatable re-entry and descent technology (IRDT) is simulated as no-slip wall. The far field boundary is simulated as pressure far field. In heat transfer calculation, the radiation and convective boundary conditions are adopted at the outer and inner walls, respectively. The finite element method is used for structural calculation. Its boundary is inertia relief in static analysis. The physical quantities such as stress, displacement, heat flow and temperature on the coupling surface satisfy the following equation:

$$n \cdot \tau_f = n \cdot \tau_s ; u_f = u_s ; q_f = q_s ; T_f = T_s \tag{1}$$

The fluid mesh is inconsistent with structure mesh on coupled surface because the solvers have different calculation velocities. The data transfer between different physical fields on the coupled surface is mapped by the mapping algorithm of grid interface [13].

3. Method verification

The accuracy of coupling method is verified by a round tube in the hypersonic flow [3]. The comparison of numerical results in this paper and literature [3] are shown in figure 2 and figure 3. The distribution of surface pressure and tube temperature are consistent with the literature results. The average errors of pressure and stagnation temperature are 1.3% and 15%, respectively. The maximum stress appears at the tube stagnation point. Its error is 5.1%. The above results show that the coupling method used in this paper can accurately predict aerodynamic forces, aero-thermal loads and their effects on the structure.

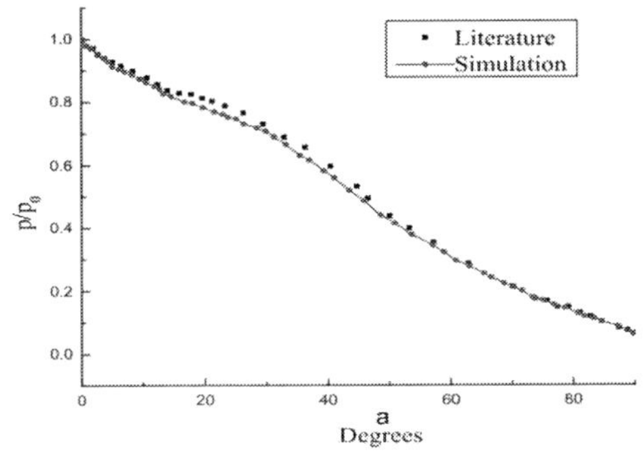

Figure 2. Surface pressure of the tube.

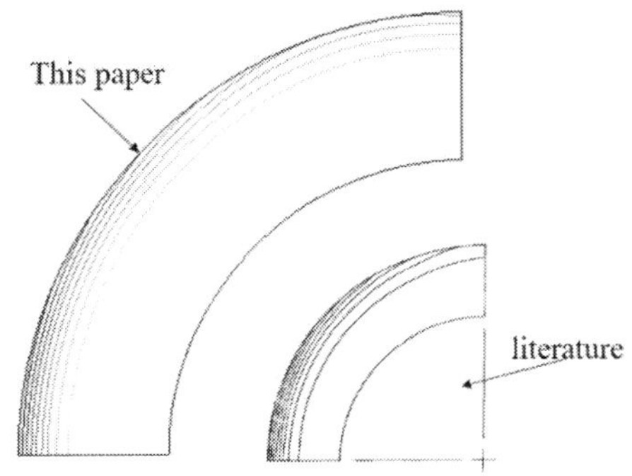

Figure 3. Temperature distribution in the tube.

4. Model

The IRDT simulated in this paper is a cone with six rings stacked inside (figure 4). The blunt head part is rigid and the cone part is flexible. The elastic modules of the blunt head part and cone part are 2e+11pa and 9e+10pa, respectively. The Poisson's ratios are 0.3 and 0.35, respectively. The thermal expansion coefficients are 1.2e-5/k and -2.5e-6/k, respectively. The TPS layers consist of a heat-shield layer, an insulation layer and a load layer. The calculation parameters are shown in table 1, the rings and load layer have same material parameters. This paper calculates two cases, one altitude, Mach number and attach angle are 82km, 24.6Ma and 0 degree, respectively. This case has the largest thermal load at the deceleration trajectory. Another altitude, Mach number and attach angle are 71km, 14.3Ma and 0 degree, respectively. This case has the largest force load at the deceleration trajectory.

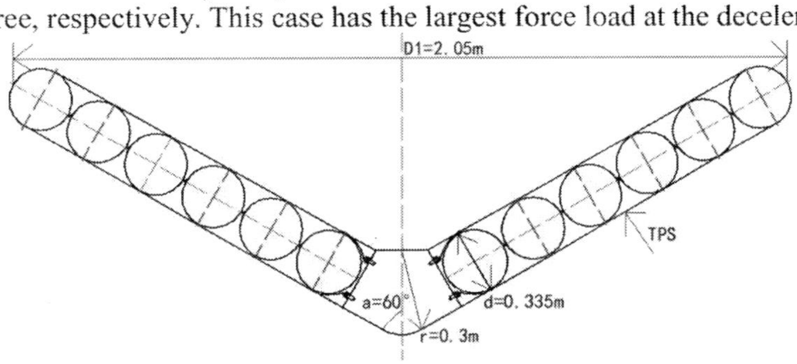

Figure 4. IRDT structure.

Table 1. Material parameters of TPS layers.

TPS layers	Thickness (mm)	Density (kg/m^3)	Thermal conductivity (w/(m.k))
Heat-shield layer	0.508	1362	0.146
Insulation layer	6	128	0.12
Load layer	0.3	1468	0.04

5. Results

5.1. Temperature distribution of TPS layers

Since IRDT is centrosymmetric, the temperature distribution of TPS layers is also centrosymmetric. The distribution curves along the meridian are shown in figure 5. It can be seen that the temperature of TPS layers are higher under the condition of 24.6Ma. The heat-shield layer can resist high temperature. The heat insulation effect of the insulation layer is obvious because its thickness is large. The load layer is used for absorbing load with low heat insulation effect. The contribution of each TPS layer for heat insulation is 6.87%, 80.90%, 12.23%, respectively. The heat insulation effect can be improved by changing the thickness or thermal conductivity of the TPS layers.

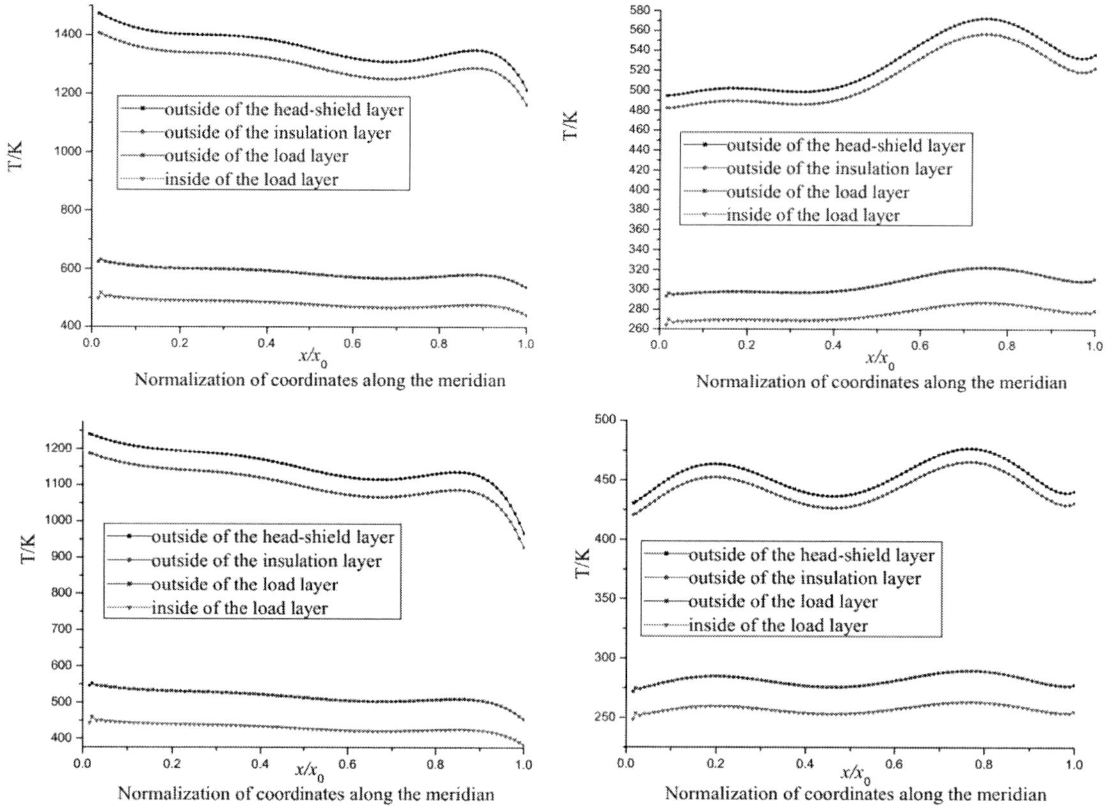

(a)Windward side. (b) Leeward side.

Figure 5. Temperature distribution curves of TPS layers along the meridian (up:24.6Ma down: 14.3Ma).

5.2 Structure response

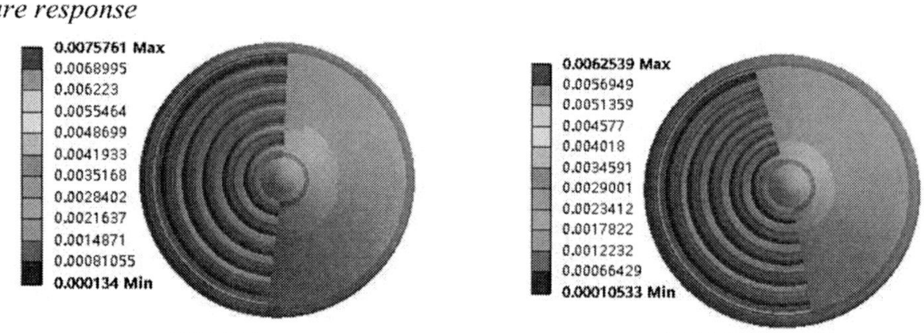

(a)24.6Ma. (b)14.3Ma.

Figure 6. IRDT deformation contours.

Figure 6 are IRDT deformation contours. It can be seen that the deformation of IRDT is small under an internal pressure of 150kPa. The deformation of IRDT is higher under the condition of 24.6Ma. The blunt head part has larger deformation because it has large thermal expansion coefficient and high temperature. The maximum deformation of rings is located at the outermost ring. The maximum deformation of TPS layer is located at gap which is connected with the rings.

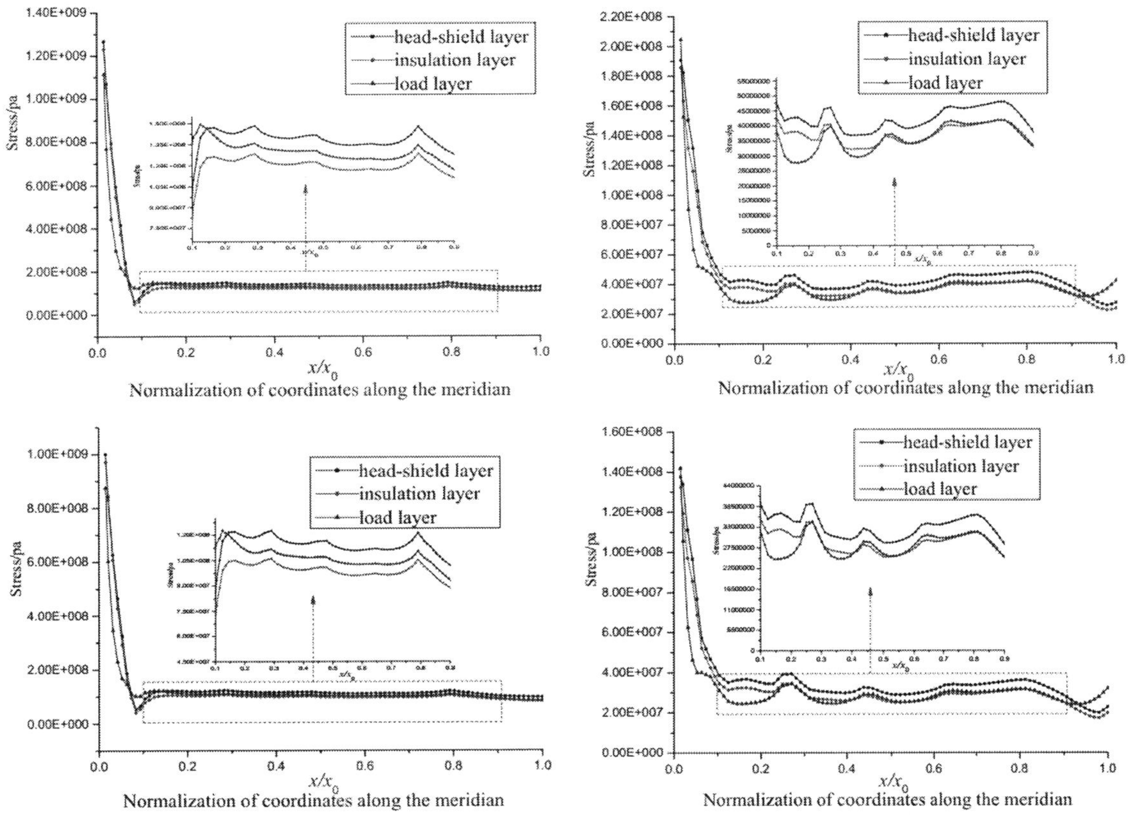

(a)Windward side. (b)Leeward side.

Figure 7. Stress distribution curves of TPS layers along the meridian(up:24.6Ma down: 14.3Ma).

Figure 7 shows the stress distribution curves of TPS layers along the meridian. The stress distribution law is consistent in two cases, while the values are larger under the condition of 24.6Ma. Stress concentrations are occurred at the contact position between the flexible structure and the blunt head. The change of the stress distribution curves of TPS layers are basically the same. The flexible structure between two rings would collapse when subjected to external aerodynamic loads (figure 8), leading to a wavy distribution of the stress. Under an internal pressure of 150kPa, the upwind pressure and internal pressure are almost equal. The outer layer and the inner layer both exert a force to the middle layer (figure 9 (a)), so the middle layer has the minimum stress at the windward side. At the leeward side, the internal pressure is greater than the external pressure (figure 9 (b)), so the stress increases from the inside to the outside.

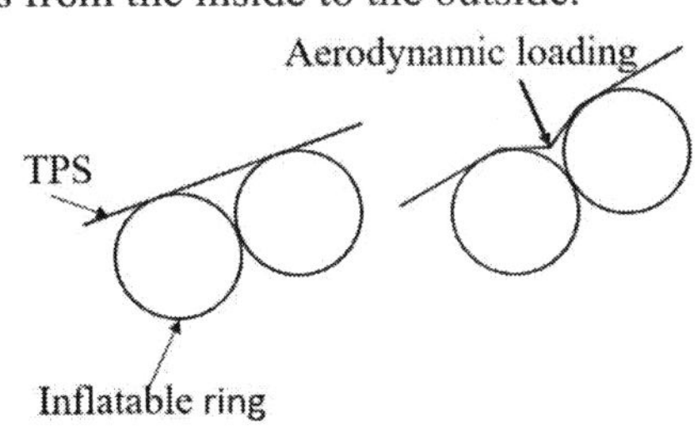

Figure 8. Force diagram of the TPS and ring.

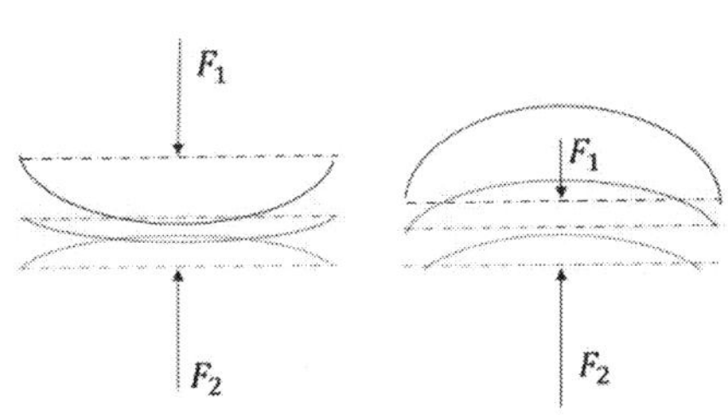

(a)Windward side (b)Leeward side

Figure 9. Force diagram of the TPS layers.

6. Conclusion

This paper establishes a unidirectional thermal-fluid-structure model to analysis the performance of IRDT under the maximum aerodynamic thermal load condition (Ma=24.6) and the maximum aerodynamic force load condition (Ma=14.3). The conclusions are as follows:

1) The TPS layers have a good insulation effect, and the cooling contribution of insulation layer is the most significant as its cooling contribution is up to 80.90%.

2) IRDT can still maintain a good aerodynamic shape under high inflated internal pressure.

3) The TPS layers have higher temperature and stress under the condition of 24.6Ma.

4) The maximum stress appears at the contact position between the flexible structure and the blunt head. Due to the interaction between TPS layer and rings, the TPS layer has a wavy distribution of stress. The insulation layer has the minimum stress at the windward side, while the load layer has the minimum stress at the leeward side.

Acknowledgments

Supported by the National Natural Science Foundation of China (11602018). A Project Funded by the Priority Academic Program Development of Jiangsu Higher Education Institutions.

References

[1] E. A. Thornton, P. Dechaumphai. Finite element prediction of aerothermal-structural interaction of aerodynamically heated panels. Presented at the AIAA 22nd Thermophysics Conference. Honolulu, Hawaii, AIAA-87-1610(1987)

[2] R. W. Allan, D. Promote, S. B. Kim. Application of integrated fluid-thermal-structural analysis methods. Thin-Walled Structures, **11**, 1 (1991)

[3] D. Pramote, A. T. Earl, R. W. Allan. Flow-thermal-structural study of aerodynamically heated leading edges. Spacecraft and Rockets, **26**, 4 (1989)

[4] R. W. Allan, S. H. Michael. Experimental study of shack wave intererence heating on a cylindrical leading edge at Mach 6 and 8. AIAA, **27**, 11 (1989)

[5] Culler, J. Adam. Coupled Fluid-Thermal-Structural Modeling and Analysis of Hypersonic Flight Vehicle Structures. Ohio State University. (2010)

[6] Z. G. WU, J. P. Hui, C. Yang. Hypersonic aerothermoelastic analysis of wings. Beijing University of Aeronautics and Astronautics, **3**, 270 (2005)

[7] C. Yang, Y. Xu, C. C. Xie. Review of Studies on Aeroelasticity of Hypersonic Vehicles. Acta Aeronautica et Astronautica Sinica, **1**, 1 (2010)

[8] Y. X. David, M. Chuh, P. S. Charles. Finite element two-dimensional panel flutter at high supersonic speeds and elevated temperature. AIAA, AIAA-90-0982 (1990)

[9] Y. H. Xie, S. Deng, S. Zhang. Study on Life Prediction for HP Turbine Blade of Heavy-Duty Gas Turbine Based on Analysis of Multi-Factor Coupling Physical Fields. Thermal Turbine, **3**, 151 (2011)

[10] L. Liu, Y. W. Gui, X. R. Geng. Study on the temperature field of hypersonic vehicle structure with aerothermoelasticity deformation. Acta Aerodynamica Sinica, **1**, 31 (2015)

[11] B. J. Andrew, H. Serhat, T. E. Karl. Uncertainty Analysis of Fluid-Structure Interaction of a Deformable Hypersonic Inflatable Aerodynamic Decelerator. Spacecraft and Rockets, **53**, 4(2016)

[12] G. Xia, X. J. Liu, W. K. Cheng. Numerical Simulation of Coupled Aeroheating and Solid Heat Penetration for a Hypersonic Blunt Body. National University of Defense Technology, **1**, 35 (2003)

[13] G. Paul, B. Rob, B. R. Hutchinson. Three-Dimensional Navier Stokes Predictions of Steady State Rotor / Stator Interaction with Pitch Change. Presented at the CFD '95: Third Annual Conference of the CFD Society of Canada, Banff, Canada (1995)

IOP Conf. Series: Materials Science and Engineering **715** (2020) 012074 doi:10.1088/1757-899X/715/1/012074

Calculation and analysis of gyroplane trimming characteristics

Y Xiang[1,*], J Wang[2], Y Cao[1] and N Wei[1]

[1]School of Aerospace Science and Engineering, Beihang University, Beijing 100191, China
[2]System Engineering Research Institute, CSSC, Beijing 100094, China

*E-mail: xiangyangxuedan@sina.com

Abstract. In order to analyse the trimming characteristics of gyroplane, a complete nonlinear dynamic equation is established, which is a seven-degree of freedom model. By using constraints, the quantity of unknown variables is equal to the quantity of equations. In the process of solving the equation, Levenberg-Marquardt algorithm is applied. After comparing the results of mathematical model to the flight test, it can be concluded that the seven degree of freedom model is valid.

Nomenclature

C_T = thrust coefficient
C_H = drag coefficient
C_Y = lateral force coefficient
C_Q = torque coefficient
$C_{L,HT}$ = lift coefficient of horizontal tail
$C_{D,HT}$ = drag coefficient of horizontal tail
$C_{L,VT}$ = lift coefficient of vertical tail
$C_{D,VT}$ = drag coefficient of vertical tail
e = offset value
M_S = mass moment
Q_{PR} = propeller torque
S_{HT} = effective area of horizontal tail
S_{VT} = effective area of vertical tail
T_{PR} = propeller thrust
Q_{PR} = propeller torque
(X_{PR}, Y_{PR}, Z_{PR}) = coordinates of propeller hub centre
δ_t = throttle position
η_s = lateral directional control of main rotor
η_c = longitudinal control of the main rotor
δ_r = rudder deviation
θ = angle of pitch
Ω = rotation speed of main rotor

1. Introduction

The gyroplane, or autogiro, is a class of rotorcraft, which is different from the helicopter. In fact, it was the first powered, heavier-than-air aircraft to fly successfully other than a conventional airplane. [1] Besides, it paved way for the development of helicopter both in engineering field like introducing cyclic pitch control and blades attached to the rotor hub by means of a hinge and in theory field. Unfortunately, from WWII broken out, helicopter attracts more and more attentions, while gyroplane gradually lost its attractions from the public and researchers.

IOP Conf. Series: Materials Science and Engineering **715** (2020) 012074 doi:10.1088/1757-899X/715/1/012074

Recent years, gyroplane meets new development opportunity since people realize its potential value. Compared to the helicopter and the fixed-wing aircraft, it owns many advantages, for example, excellent take-off and landing performance, simple structure, low cost, good safety, small noise etc.

From a scientific perspective, compering to substantial researches of other types of aircrafts, there have been a few studies of gyroplane. Although there are some engineering researches before WWII when rotor research has not focus on helicopter, they just own historical significances nowadays. Until 1991, UK CAA organized a program of research on gyroplane because of numerous fatal accidents. Funded by UK CAA, a team from Glasgow University tried to find the factors influencing gyroplane stability, control and handling quality. [2-4] However, there is no specific description about the mathematical model on their papers. Considering the few researches on this issue, there is a need to do more research to help the gyroplane technique develop. In aerodynamic research, it is a fundamental but vital step to establish a suitable equilibrium equation to analyze the trimming characteristics of gyroplane. This paper illustrates a seven-degree of freedom mathematical model to simulate gyroplane dynamic problem. Through using this model to analyze a kind of gyroplane named VPM M16, comparing the results from the model and the flight test, it can be concluded that this model is reliable and this mathematical model can be used to typical gyroplane.

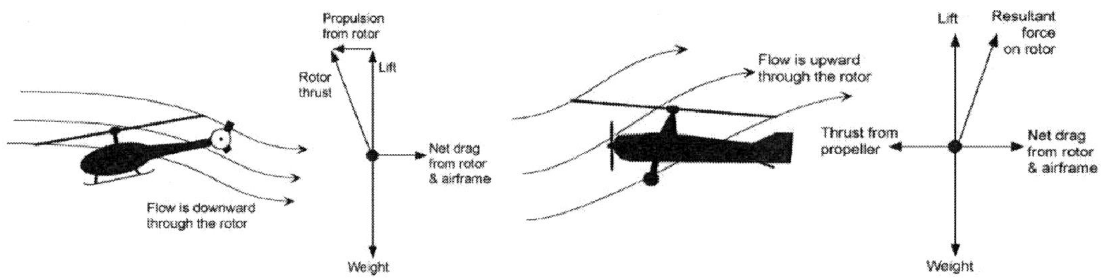

(a)Autogiro (b) Helicopter

Figure 1. Autogiro rotor (a) provides lift, with forward propulsion being provided by a conventional propeller, compared to the helicopter (b) where the rotor provides both lift and propulsion. [1]

2. Mathematical model

The most significant difference between gyroplane and helicopter is whether the main rotor connects to the engine. For gyroplane, its main rotor rotates only because of the inflow. So, for gyroplane, when it is in forward flight, the main rotor tilts backward, while for helicopter, its main rotor tilts forward. Gyroplane main rotor provides lift, with forward propulsion being provided by a conventional propeller, compared to the helicopter where the main rotor provides both lift and propulsion. The difference is shown on figure 1.

The gyroplane can be divided into five subsystems, main rotor, fuselage, horizontal tail, vertical tail and propeller. Through calculating the force and torque of every subsystem, summing them up together, we can get the final forces and torques.

2.1. Main rotor

Main rotor is the most complex part of the gyroplane. These assumptions are made when deriving the rotor aerodynamic force: the blade is rigid without considering the shimmy effect. In this way, the rotor aerodynamic model is composed of rotor force and torque model, rotor waving model and rotor induced velocity model. The blade twist angle is linearly distributed, and the incoming flow is evenly distributed.

Combined with the rotor blade element theory and the momentum theory, the aerodynamic model of the rotor is established by the formula derived from the blade radius and the azimuth integral along the blade. The integral format rotor aerodynamic model is based on the rotor lift line theory. The aerodynamic load of the section is calculated according to the flow velocity and elevation of the air

IOP Conf. Series: Materials Science and Engineering **715** (2020) 012074 doi:10.1088/1757-899X/715/1/012074

foil section. Then the aerodynamic model of the rotor aerodynamic force in the wind shaft system is derived. The expressions of the rotor tension, rear force, lateral force and rotor torque are as (1):

$$\begin{cases} T = C_T \rho \pi R^2 (R\Omega)^2 \\ H = C_H \rho \pi R^2 (R\Omega)^2 \\ Y = C_Y \rho \pi R^2 (R\Omega)^2 \\ M = \frac{k}{2} M_s \Omega^2 e \beta_{1c} \\ L = \frac{k}{2} M_s \Omega^2 e \beta_{1s} \\ Q = \frac{k}{2} M_s \Omega^2 e \beta_{1s} \end{cases} \tag{1}$$

These coefficients are derived from Johnson's book. [5]

2.2. Fuselage

The aerodynamic fuselage model is built based on wind tunnel test data. [6] The forces and the torques are as (2) (3):

$$\begin{bmatrix} F_x \\ F_y \\ F_z \end{bmatrix}_F = \begin{bmatrix} C_x \\ C_y \\ C_z \end{bmatrix}_F \frac{1}{2} \rho (u_F^2 + v_F^2 + w_F^2) \pi R^2 \tag{2}$$

$$\begin{bmatrix} M_x \\ M_y \\ M_z \end{bmatrix}_F = \begin{bmatrix} C_L \\ C_M \\ C_N \end{bmatrix}_F \frac{1}{2} \rho (u_F^2 + v_F^2 + w_F^2) \pi R^3 \tag{3}$$

All of these aerodynamic coefficients C_x, C_y, C_z, C_L, C_N, C_M are processed through the conversion and the non-dimension of the rotor radius, which are related to the angle of attack and the sideslip angle.

2.3. Horizontal tail

The aerodynamic horizontal tail model is built based on the combination of the wind tunnel test data and theory. In the wind axis, the lift and drag on the horizontal tail are as formula (4) (5). Then the conservation is conducted from the wind axis to the body axis. Lift and drag coefficient of the horizontal tail are related to the angle of attack and sideslip angle.

$$L_{HT} = \frac{1}{2} \rho (u_{HT}^2 + v_{HT}^2 + w_{HT}^2) S_{HT} C_{L,HT} \tag{4}$$

$$D_{HT} = \frac{1}{2} \rho (u_{HT}^2 + v_{HT}^2 + w_{HT}^2) S_{HT} C_{D,HT} \tag{5}$$

2.4. Vertical tail

The aerodynamic vertical tail model is built similarly to the horizontal tail. The lift and drag coefficient are as formula (6) (7).

$$L_{VT} = \frac{1}{2} \rho (u_{VT}^2 + v_{VT}^2 + w_{VT}^2) S_{VT} C_{L,VT} \tag{6}$$

$$D_{VT} = \frac{1}{2} \rho (u_{VT}^2 + v_{VT}^2 + w_{VT}^2) S_{VT} C_{D,VT} \tag{7}$$

Lift and drag coefficients of the vertical tail are related to the angle of attack and sideslip angle.

2.5. Propeller

The propeller model is built based on the parameters i.e. thrust, output power etc. provided by the engine manufacturer. In the body coordinate system, the aerodynamic forces and torques of the propeller are as (8) (9):

$$\begin{Bmatrix} F_{xPR} \\ F_{yPR} \\ F_{zPR} \end{Bmatrix} = \begin{Bmatrix} T_{PR} \\ 0 \\ 0 \end{Bmatrix} \tag{8}$$

$$\begin{Bmatrix} M_{xPR} \\ M_{yPR} \\ M_{zPR} \end{Bmatrix} = \begin{Bmatrix} Q_{PR} \\ 0 \\ 0 \end{Bmatrix} + \begin{bmatrix} 0 & -z_{PR} & y_{PR} \\ z_{PR} & 0 & -x_{PR} \\ -y_{PR} & x_{PR} & 0 \end{bmatrix} \begin{Bmatrix} F_{xPR} \\ F_{yPR} \\ F_{zPR} \end{Bmatrix} \tag{9}$$

2.6. Resultant forces and resultant torques

By adding up all of the forces and torques of these subsystems, the total force as (10) and moment as (11) of the gyroplane can be gotten.

$$\begin{Bmatrix} F_X \\ F_Y \\ F_Z \end{Bmatrix} = \begin{pmatrix} F_X \\ F_Y \\ F_Z \end{pmatrix}_R + \begin{pmatrix} F_X \\ F_Y \\ F_Z \end{pmatrix}_{PR} + \begin{pmatrix} F_X \\ F_Y \\ F_Z \end{pmatrix}_F + \begin{pmatrix} F_X \\ F_Y \\ F_Z \end{pmatrix}_{HT} + \begin{pmatrix} F_X \\ F_Y \\ F_Z \end{pmatrix}_{VT} \tag{10}$$

$$\begin{Bmatrix} M_X \\ M_Y \\ M_Z \end{Bmatrix} = \begin{pmatrix} M_X \\ M_Y \\ M_Z \end{pmatrix}_R + \begin{pmatrix} M_X \\ M_Y \\ M_Z \end{pmatrix}_{PR} + \begin{pmatrix} M_X \\ M_Y \\ M_Z \end{pmatrix}_F + \begin{pmatrix} M_X \\ M_Y \\ M_Z \end{pmatrix}_{HT} + \begin{pmatrix} M_X \\ M_Y \\ M_Z \end{pmatrix}_{VT} \tag{11}$$

Considering the main rotor is an independent freedom in this case, there is a group of seven-degree of freedom equilibrium equations as (12) (13) (14).

$$\begin{pmatrix} F_X \\ F_Y \\ F_Z \end{pmatrix} + \begin{pmatrix} 1 & 0 & 0 \\ 0 & \cos\phi & \sin\phi \\ 0 & -\sin\phi & \cos\phi \end{pmatrix} \begin{pmatrix} \cos\theta & 0 & -\sin\theta \\ 0 & 1 & 0 \\ \sin\theta & 0 & \cos\theta \end{pmatrix} \begin{pmatrix} 0 \\ 0 \\ m_G g \end{pmatrix} = 0 \tag{12}$$

$$\begin{pmatrix} M_X \\ M_Y \\ M_Z \end{pmatrix} = 0 \tag{13}$$

$$Q_R = 0 \tag{14}$$

3. Calculation of trim equations

Since there has been the seven-degree of freedom trim model, the next step is to solve this problem. This trim calculation is essentially to solve nonlinear equation. The nonlinear flight dynamic model is

$$x = f(x,u) \tag{15}$$

where x is the state vector, u is the control vector as (16) (17)

$$x = (u, v, w, \theta, \phi, \varphi, \Omega) \tag{16}$$

$$u = (\delta_t, \eta_s, \eta_c, \delta_r) \tag{17}$$

Using the constraints from Padfield [7], the ground velocity, the flight path angle, the velocity of flight deflecting path angle and sideslip angle are defined. Then the problem becomes solving seven-variable nonlinear equations as (18).

$$g_i(y_1, y_2, y_3, y_4, y_5, y_6, y_7) = 0, i = 1...7 \tag{18}$$

where y contains variables as (19)

$$y = (\delta_t, \eta_s, \eta_c, \delta_r, \theta, \phi, \Omega) \tag{19}$$

IOP Conf. Series: Materials Science and Engineering **715** (2020) 012074 doi:10.1088/1757-899X/715/1/012074

Considering the advantages of LMA (Levenberg-Marquardt algorithm) [8], the problem can be solved by it.

4. Mathematical model validation

In this paper, the example gyroplane is VPM M16, which is a normal configuration gyroplane. There are parameters of VPM M16 in table1.

Inputting these data into the model on Matlab and calculating the state variables and control variables in different ground speed situations. By comparing the variables from the model to the flight test [7], it can be concluded that the model is valid although there are some deviations. The comparison is shown on figure 2-6.

Rotor speed is consistent with the flight test counterpart both in tendency and number. The deficiencies of rotor speed between the model and the test may be due to the blade twist, which may manifest itself aeroelasticity in flight and is not incorporated in the rigid-blade model. Longitudinal and latitudinal tilt of main rotor compares favourably with the flight measurement especially at the high speed. Although there is a uniform error across the speed, this is less than five degrees. The pitch attitude shows the same trend which is downward as the speed increases. It is easy to understand this trend because as the speed increases, the inflow increases, to maintain the same lift, there is fewer need of angle of attack of main rotor. As for the roll attitude, the prediction of this model is not precious, but the results both in model and flight test are less than 2 degree and this number is almost not affected by velocity.

Table1. Parameters of VPM M16[9].

Item	Value
Mass(kg)	450
Ixx(kg.m2)	195
Iyy(kg.m2)	637
Izz(kg.m2)	442.5
Ixz(kg.m2)	-46
Airfoil	NACA 8H12
Main Rotor Blade Number	2
Radius(m)	4.267
Flapping Inertial(kg.m2)	51.6
Chord(m)	0.22
Radius(m)	0.86
Chord(m)	0.1
Fuselage Side Area(m2)	1.4
Plan Fuselage Area(m2)	1.6
Fuselage Frontal Area(m2)	0.59

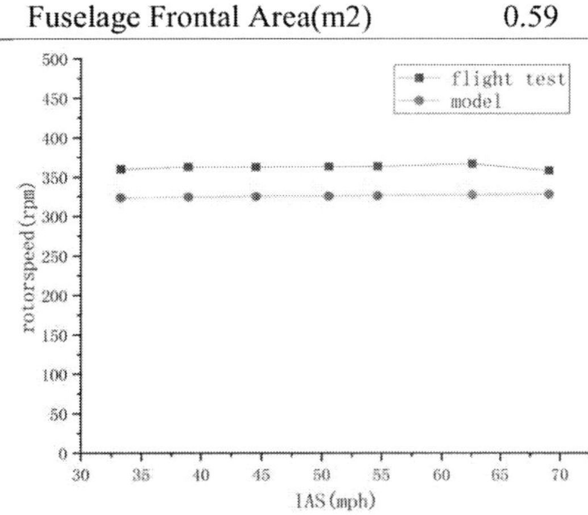

Figure 2. Rotor Speed Comparison.

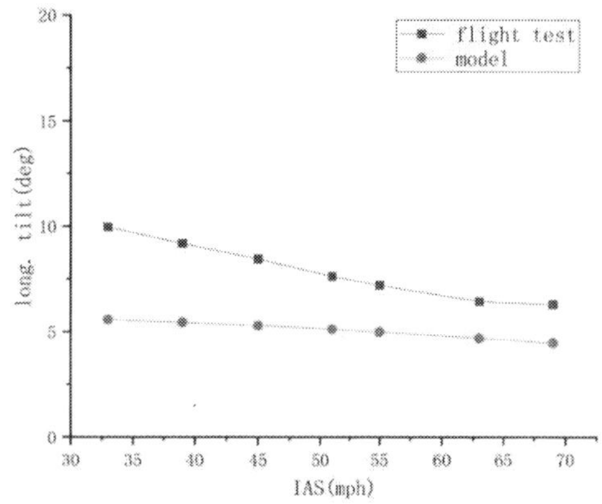

Figure 3. Longitudinal Tilt Angle of Main Rotor Comparison.

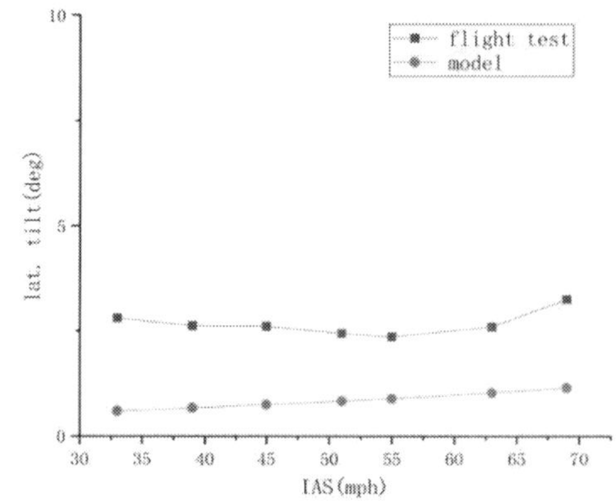

Figure 4. Lateral Tilt Angle of Main Rotor Comparison.

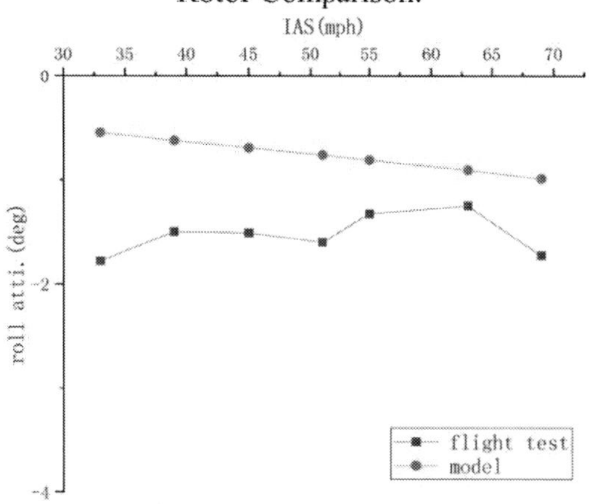

Figure 5. Pitch Attitude Comparison.

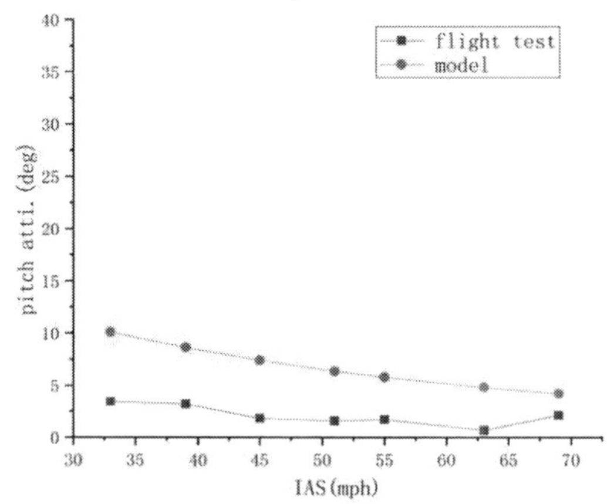

Figure 6. Roll Attitude Comparison.

5. Conclusion

By establishing five subsystems model separately and summing up the forces and torques together, there is a seven-degree-freedom model to simulate the trimming characteristics. Although there are some deviations between the model results and the flight test data, the differences are acceptable and explainable. Consequently, the seven-degree-freedom model is valid for simulating a typical gyroplane. The values about trimming showed a logical trend as the velocity changes. Consequently, the mathematical model is suitable to analyse a typical gyroplane aerodynamic problem.

References

[1] Leishman J G. Development of the autogyro:a technical perspective[J]. Journal of Aircraft, 2004, Vol.41.No.4:765-781.
[2] Houston, S. S., "Longitudinal Stability of Gyroplanes," [J] Aeronautical Journal, Vol. 100, No. 991, 1996, pp. 1–6.
[3] Houston, S. S., "Validation of a Non-Linear Individual Blade Rotorcraft Flight Dynamics Model Using a Perturbation Method," Aeronautical Journal, Vol. 98, No. 977, 1994, pp. 260–266.
[4] Houston, S. S., "Validation of a Blade-Element Helicopter Model for Large-Amplitude Manoeuvres," Aeronautical Journal, Vol. 101, No. 1001, 1997, pp. 1–7.
[5] Johnson,W., Helicopter Theory, Princeton Univ. Press, Princeton, NJ, 1980, pp. 237, 238.
[6] F. N. Coton and L. Smrcek, "Aerodynamic Characteristics of a Gyroplane Configuration",

Journal of Aircraft, Vol.35, No. 2, March-April 1998.
[7] Gareth D.Padfield. Helicopter Flight Dynamics: The Theory and Application of Flying Qualities and Simulation Modeling[M]. 2007: pp23-25.
[8] Moré J J. The Levenberg-Marquardt algorithm: Implementation and theory[J]. Lecture Notes in Mathematics, 1978, 630:105-116.
[9] UK CAA, Safety Regulation Group "The Aerodynamics of Gyroplane" CAA Paper 2009/02

IOP Conf. Series: Materials Science and Engineering **715** (2020) 012075 doi:10.1088/1757-899X/715/1/012075

Ideal performance analysis of the scramjet with Pre-jet fuel

Shengsheng Zhang[1,*], Yu Xue[2], Yang Yang[1], Haoliang Wang[1], Yubao He[1], Sihang Zheng[1], Ya Lv[1] and Yong Zhang[1]

[1]China Academy of Launch Vehicle Technology, Beijing 100076, China
[2]China United Gas Turbine Technology Co., LTD, Beijing 102209, China

*E-mail: zhangsshit@163.com

Abstract. In order to research the influence on the overall performance of scramjet with Pre-jet (jetting fuel or water vapor into the isolator) fuel, the governing equations of the different component in scramjet and the ideal performance analysis model with Pre-jet fuel are established. Moreover, Pre-jet fuel is considered based on the ideal Brayton cycle. The results of ideal performance analysis show that Pre-jet fuel doesn't burn at a given maximum temperature of the combustor exit. Due to the increasing of oil to air ratio with Pre-jet fuel, the outlet temperature of the isolator drops more. However, fuel potential of the combustion, the flux of working fluid and the engine thrust ratio are increased. Meanwhile, the influence of methane and water vapor on the engine performance is similar, because the Pre-jet fuel is small and there are some similarities between methane and water vapor.

1. Introduction

As the most important part of hypersonic vehicle, the scramjet can be applied in hypersonic aircraft, spaceflight aircraft and the future single-stage round-trip transportation system powered by combined cycle. However, there are still many problems and challenges in improving performance [1]. Such as, engine thermal protection and supersonic combustion.

The Pre-jet (jetting fuel or water vapor into the isolator) fuel is designed to realize the multipurpose and maximize the potential of fuel. It can be used for reference to the improvement of engine cooling and other performance, as well as the subsequent development of scramjet technology. Foreign research institutes have done some work in this field, including the central aero engine research institute in Russia and the Univesity of Florida in the United States. But there is very little research on this aspect in China. Only Peiyong Wang et al. Using the results of Hyshot scramjet test for reference and conducted numerical simulation of cold flow and combustion conditions respectively [2]. The initial goal of the Pre-jet fuel was to develop an effective aspirating propulsion cycle on a dual-mode scramjet using conventional hydrocarbon fuel when flight Ma>3.5 [3]. The main hypersonic aerodynamic problems including pressure loss caused by injection, mixing efficiency between fuel and main flow, interaction between intake system and combustion chamber, and flame stability [4-5].

Previous foreign researches mainly focus on the injection of liquid and gas fuel into the main flow of the inlet and the isolation section of supersonic or hypersonic, the location of the jetting tower bridge is either in the upstream of the combustion chamber or in the isolation section, Spraying fuel on the rear wall of the jetting tower bridge can improve mixing, flame stabilization and combustion efficiency. By fully considering fuel penetration depth, residence time, operating conditions, fuel physical and chemical properties, etc., the non-start phenomenon caused by combustion moving to the inlet can be avoided, as shown in figure 1 and figure 2 [6-15]. Therefore, a more flexible system is obtained with Pre-jet fuel, and the interaction between fuel and air occurs throughout the entire system of the inlet, isolator, and combustor. Despite the complexity of system is increased, the following advantages can be achieved:

1) Reduced the length of isolator and combustor, decreased the engine weight and cooling pressure;

2) Get a more flexible fuel control system, it is due to the possibility of distributing fuel in the pre-injection and direct injection into the combustor, moreover, achieving liquid and gaseous fuel injection through the placement of different injectors [16-17].

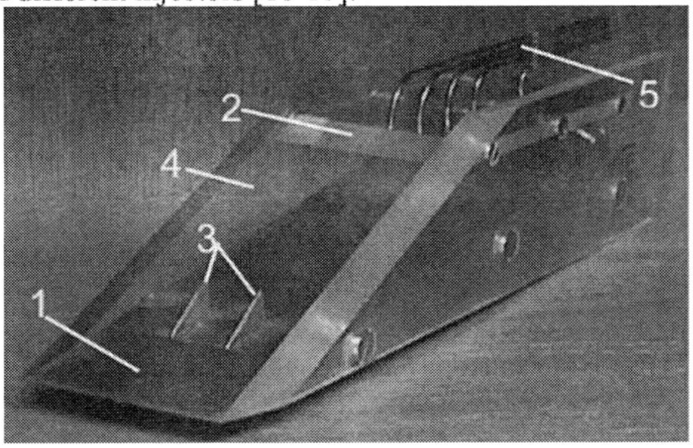

Figure 1. Tower-bridge installed inside the compression surface of inlet.

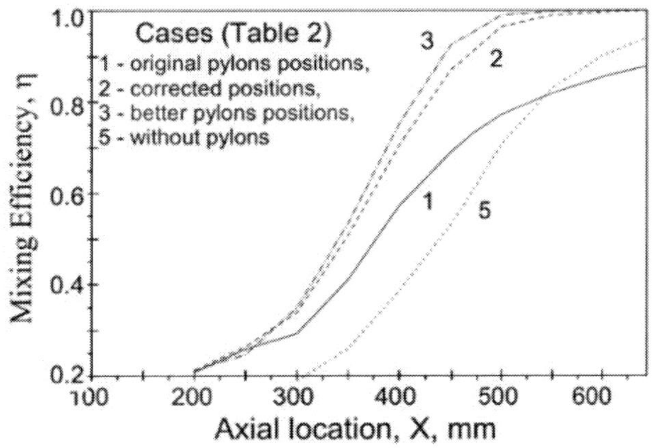

Figure 2. Mixing efficiency of ethylene injection at Ma=6.

A series of key problems need to be solved to realize the application with Pre-jet fuel in actual flight, for example, the variation of angle of attack and sideslip-offset cause the flow structural change and mechatronics.

Previous researchers have mostly focused on the study of improving combustion performance with Pre-jet fuel. However, few studies have been conducted to assess the impact with Pre-jet fuel on the inlet temperature of combustor, the ultimate heating capacity and the overall performance of scramjet. Based on the ideal Brayton cycle, a one-dimensional analysis method of scramjet with Pre-jet fuel was developed, in order to improve analysis efficiency, avoid a lot of simulation and test work, and realize the rapid prediction of the overall performance in the scramjet with Pre-jet fuel. It provides important reference and basis for the application in the scramjet with Pre-jet fuel technology.

2. Analysis of the thermodynamic process in the scramjet with Pre-jet fuel technology

2.1. Physical model

The effect on the overall performance of scramjet with Pre-jet fuel were researched, as shown as figure 3, the physical model of scramjet with Pre-jet fuel was established, containing equivalent isolator, expansion chamber and nozzle.

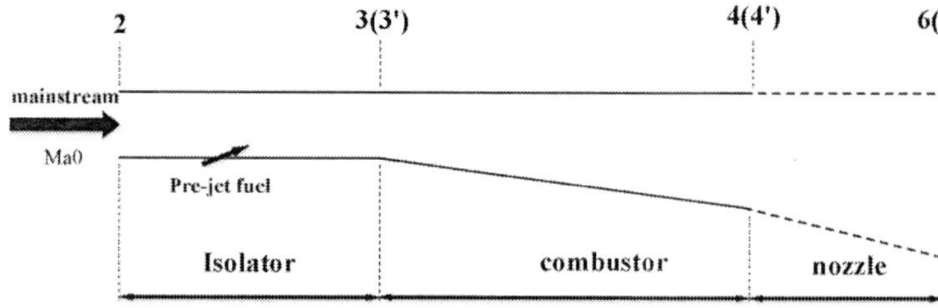

Figure 3. Schematic diagram of scramjet with Pre-jet fuel.

Cross section 2 — entrance of the isolator without or with Pre-jet fuel;
Cross section 3—exit of the isolator (or inlet of the combustor) without Pre-jet fuel;
Cross section 4—exit of the combustor without Pre-jet fuel;
Cross section 6—exit of the nozzle without Pre-jet fuel;
Cross section 3' — exit of isolator (or inlet of the combustor) with Pre-jet fuel;
Cross section 4' — exit of the combustor with Pre-jet fuel;
Cross section 6' — exit of the nozzle with Pre-jet fuel.

2.2. Analyses of thermodynamic cycle

According to mass, momentum, energy conservation and state equations, the governing equations of the working process of the scramjet with or without Pre-jet fuel are established respectively, as follow:

1) Entrance and exit of the isolator without Pre-jet fuel:

$$m_2 = m_3 \tag{1}$$

$$P_2A + m_2u_2 = P_3A + m_3u_3 \tag{2}$$

$$m_2\left(C_{p2}T_2 + u_2^2/2\right) = m_3\left(C_{p3}T_3 + u_3^2/2\right) \tag{3}$$

$$P = \rho R_g T \tag{4}$$

2) Entrance and exit of the isolator with Pre-jet fuel:

$$m_2 + m_{f1} = m_3^{'} = \rho_3^{'}u_3^{'}A \tag{5}$$

$$P_2A + m_2u_2 + +m_{f1}u_{f1}cos\theta = P_3^{'}A + m_3^{'}u_3^{'} \tag{6}$$

$$m_2\left(C_{p2}T_2 + u_2^2/2\right) + m_{f1}\left(C_{pf}T_{f1} + u_{f1}^2/2\right) = m_3^{'}\left(C_{p3}^{'}T_3^{'} + u_3^{'2}/2\right) \tag{7}$$

$$P = \rho R_g T \tag{8}$$

m — mass flow rate (kg/s), the subscript numbers represent different sections, the same below;
m_{f1}— mass flow rate of Pre-jet fuel (kg/s), $m_{f1} = m_2 * f_1$;
P — pressure (N/m^2);
A — cross-sectional area (m^2);
u — flow rate (m/s);
C_p— specific heat at constant pressure (J/kg.K);
T— specific heat at constant pressure (K);
ρ — density (kg/m^3);
R_g — gas constant (J/kg.K);
θ— the Angle between the spray direction and the axial direction (^{0}).

It is assumed that the compression process of air flow is completed within the inlet, the compression in the isolation section is negligible. It can be seen from the formula, there is no parameter change in the inlet and outlet of the isolation. Supposing that Pre-jet fuel is unburned, only reduced the inlet temperature of the combustor. The above formula(5)~(8), except for $\rho_3^{'}$、$u_3^{'}$、$P_3^{'}$ and $T_3^{'}$, The rest parameters are known that can be directly given or calculated by formulas (1) ~ (4). Four equations and four unknown parameters, the equations are closed and can be solved.

In general, some airflow properties kept constant in the scramjet, such as Mach number, flow area and static temperature, it is not difficult to modeling the combustion or heat addition process. But the main reasons for the preference of constant pressure processes are as follows: Firstly, from the perspective of aerodynamics, constant pressure process eliminated the possibility of boundary layer separation and the need for structure to withstand peak pressure; Secondly, the internal static pressure can be approximately regarded as a constant because the velocity of the combustor decreases to a small level for the scramjet; Thirdly, when chemical energy release an equal-area combustor operates at constant pressure processes due to separation of the boundary layer. This paper chooses constant pressure processes.

3) Entrance and exit of the combustor with Pre-jet fuel:

$$P_3^{'} = P_4^{'} \tag{9}$$

$$m_3^{'} + m_{f2} = m_4^{'} \tag{10}$$

$$m_3^{'}u_3^{'} - C_f A_w / (2A_3)\rho_3^{'}u_3^{'2}A_3 = m_4^{'}u_4^{'} \tag{11}$$

$$m_3^{'}\left(C_{p3}^{'}T_3^{'} + u_3^{'2}/2\right) + m_{f2} \times \left(\eta_b H_u + C_{pf}T_{f2} + u_{f2}^2/2\right) = m_4^{'}\left(C_{p4}^{'}T_4^{'} + u_4^{'2}/2\right) \tag{12}$$

$C_f A_w / A_3 = \text{drag} / \left(\rho_3^{'}u_3^{'2}A_3\right)$ — effective resistance coefficient of combustor, command $C_f A_w / A_3 = 0$;

H_u — heat value of methane(kJ/kg), command H_u =50010 kJ/kg [18];

η_b — combustion efficiency, command η_b =1.0;

m_{f2} — fuel mass flow of combustor (kg/s), command $m_{f2} = m_2 * f_2$.

The equations (5), (10) and (11) are obtained simultaneously:

$$u_4^{'} = u_3^{'}\left(1+f_1\right)/\left(1+f_1+f_2\right) \tag{13}$$

The maximum temperature T_{max} at outlet of the combustor is 2300K, meantime, outlet temperature of the combustor ($T_4^{'}$) is equal to T_{max}. Pre-jet fuel can reduce inlet temperature of the combustor without burning. The oil to air ratio f_2 of the combustor can be applied to explain the maximum energy injection of the combustor. Two equations (11) and (12) can be solved with two parameters.

4) Entrance and exit of the nozzle with Pre-jet fuel, the process is equivalent to the static pressure and frictionless. The specific equations are as follows:

$$T_{t4}^{'} = T_{t6}^{'} \tag{14}$$

$$P_{t4}^{'} = P_{t6}^{'} \tag{15}$$

$$P_0 = P_6 \tag{16}$$

$$m_4^{'}\left(C_{p4}^{'}T_4^{'} + u_4^{'2}/2\right) = m_6^{'}\left(C_{p6}^{'}T_6^{'} + u_6^{'2}/2\right) \tag{17}$$

3. Simulation of effect on the performance of the scramjet with Pre-jet fuel

3.1. Boundary condition

The hypersonic vehicle usually designs at the isodynamic pressure q_0. If the dynamic pressure (q_0) is too high, the structure stress and resistance on the vehicle is very heavy; on the other hand, if the dynamic pressure (q_0) is too low, the wing space required to maintain flight may be immense [19]. It is generally acknowledged the hypersonic vehicle must be designed in a relatively narrow dynamic pressure range, approximately 20000~90000N/m^2. Designing at the isodynamic pressure $q_0 = 50000\text{N}/m^2$ in this paper, the compression efficiency, combustion efficiency and expansion efficiency are given in accordance with the literature [20], relevant parameters are shown in table 1.

Table 1. Summary table of correlation parameters.

Numerical order	Parameters	Unit	Value
1	flight altitude	km	22.45
2	Mach number	-	6
3	compression efficiency	-	0.88
4	combustion efficiency	-	1.0
5	inlet Mach number of the isolator	-	2.8 (Ma0=6) 2.4 (Ma0=5)
6	heat ratio of free flow	-	1.36
7	heat ratio of compressive flow	-	1.34
8	heat ratio of the combustion	-	1.238
9	heat ratio of expansion	-	1.238
10	calorific value of gaseous methane	J/kg	50010
11	gas constant of mixed gas		calculate according to equation (19) [20]
12	inlet area of the isolator	m^2	0.15
13	inlet area of the combustor	m^2	0.15
14	outlet area of the combustor	m^2	0.225
15	total temperature of Pre-jet fuel	K	300
16	area of Pre-jet fuel	m^2	0.001

The gas is regarded as ideal gas, the heat capacity of common gases under constant pressure is calculated as follows:

$$C_P = C_0 + C_1\theta + C_2\theta^2 + C_3\theta^3 \quad \text{kJ}/(kg \bullet K) \quad \theta = \{T\}_K / 1000 \tag{18) [20}$$

Application range of equation (18): 250~1200K, the constant pressure specific heat capacities of methane and water vapor at constant pressure at 300K were calculated respectively, the results are shown in table 2.

Table 2. Constant pressure specific heat capacities of methane and water vapor.

Parameter	C_0	C_1	C_2	C_3	$C_p\{\text{kJ}/(kg \bullet K)\}$
Methane	1.2	3.25	0.75	-0.71	2.22333
Water vapor	1.79	0.107	0.586	-0.2	1.86944

The gas constants of mixed gas of methane and mainstream air in the isolator are calculated according to equation (19).

$$\text{R} = \frac{1.9857117\text{Btu}/\left(1\text{bm} \bullet R'\right)}{28.97 - f_1 \times 0.946186} \tag{19}$$

The unit conversion in equation (19) is as follows, when calculate outlet air flow parameters of the isolator with Pre-jet fuel, the oil to air ratio here refers f_1 pecifically.

$$1\text{Btu}/\left(1\text{bm} \bullet R\right) = 4.1868\text{kJ}/\left(\text{kg} \bullet \text{K}\right) \tag{20}$$

$$F_s = \left(1 + f_1 + f_2\right)V_6 - V_0 \tag{21}$$

$$I_{sp} = F_s / g_0 / (f_1 + f_2) \tag{22}$$

$$\text{W} = 0.5 \times \left[\left(1 + f_1 + f_2\right)V_6^2 - V_0^2\right] \tag{23}$$

$$\eta_{th} = \frac{W}{f_2(H_u\eta_b + u_{f2}^2/2)} \tag{24}$$

$$\eta_{tc} = \eta_{th} / \eta_b \tag{25}$$

$$\eta_P = F_s V_0 / W \tag{26}$$

$$\eta_0 = \eta_{th} \eta_P \tag{27}$$

W— mechanical power of scramjet (unit mass);

f_1、f_2 — oil to air ratio of isolator and combustor;

η_{th} — thermal efficiency of scramjet;

η_{tc} — thermodynamic cycle efficiency of scramjet;

η_p — propulsion efficiency of scramjet;

η_0 — total efficiency of scramjet.

3.2. Simulation study

Setting up the simulation platform based on the scramjet model at Matlab environment. Spray Angle is 300. The effects of methane and water vapor on the overall performance of the scramjet were investigated when there is Pre-jet fuel in isolator with inlet flow Mach number 5.0 and 6.0.

As shown in figure 4, because methane has more enthalpy than water vapor per unit mass ($C_{pf}T_{f1}$), when same Mach number and same oil to air ratio of isolator, outlet air static temperature of the isolator is reduced more by injecting methane gas. It can be seen that methane has more cooling capability. With the increasing of oil to gas air ratio, the difference of cooling effect between the methane and water vapor on the main stream becomes larger, this is mainly caused by the larger enthalpy difference between the methane and water vapor. Static temperature is shown in table 3.

Table 3. The influence of different oil to air ratio on the isolator outlet static temperature reduction with Pre-jet fuel.

f1	**0**	**0.01**	0.02	**0.03**	0.04	**0.05**	**0.06**
T_3 temperature reduction comparing without Pre-jet fuel (*K*) (Ma_0=5 methane)	0	5.66	13.21	22.22	32.34	43.21	54.53
T_3 temperature reduction comparing without Pre-jet fuel (*K*) (Ma_0=5 water vapor)	0	2.74	7.34	13.46	20.80	29.06	38.00
T_3 temperature reduction comparing without Pre-jet fuel (*K*) (Ma_0=6 methane)	0	3.40	10.41	20.15	31.79	44.64	58.02
f1	**0**	**0.01**	**0.02**	**0.03**	**0.04**	**0.05**	**0.06**
T_3 temperature reduction comparing without Pre-jet fuel (*K*) (Ma_0=6 water vapor)	0	0.1	3.66	9.96	18.33	28.18	38.94

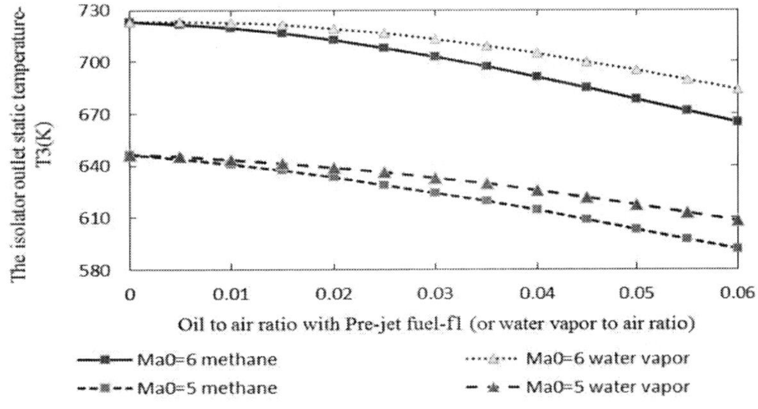

Figure 4. The relationship between the isolator outlet static temperature and oil to air ratio with Pre-jet fuel (or water vapor to air ratio).

As shown in figure 5, jetting methane or water vapor into isolator. It is assumed that the fuel jetted in the isolator didn't burn, due to its cold gas (T_f=300K), outlet static temperature of isolator (or inlet static temperature of combustor) reduced. Supposing that the combustor outlet temperature limit kept constant. The Pre-jet fuel can increase the temperature difference between the inlet and outlet of the combustor. Meanwhile, the combustion outlet is close to maximum mass flow with the temperature reached to maximum.

The results show that inlet static temperature of the combustor is low when the Mach number (Ma_0=5) is low. As a result, the maximum capacity of jetting oil to air ratio (*f2*) is relatively large. Under the same Mach number and the oil to air ratio of isolator, the combustion maximum oil to air ratio are close either jetting methane or water vapor. It can be seen from table 3, jetting methane only decreases about 20K more than water vapor when the maximum oil to air ratio *f1* is equal to 0.06. On the basis of $\nabla f \eta_b H_u = m C_p \nabla T$, Because the high heat value of methane and the similarity of constant pressure specific heat, the oil to air ratio change is very small. The oil to air ratio (or water vapor to air ratio) with Pre-jet fuel increases, while the difference between methane and water vapor temperature increases, the oil to air ratio (or water vapor to air ratio) difference between methane and water vapor in the combustor is also increased. As shown in table 4, compared without Pre-jet fuel, the case that oil to air ratio is 0.06 at Ma_0=6 (the oil to air ratio was 0.05825 when the equivalence ratio was 1.0), the maximum oil to air ratio of the combustor was increased by 6.76%. And as *f1* increases, *f2* increases almost linearly.

Table 4. Influence of different oil to air ratio with Pre-jet fuel on the ultimate oil to air ratio of combustor (taking methane as an example).

f1	**0**	**0.01**	**0.02**	**0.03**	**0.04**	**0.05**	**0.06**
f2 growth rate with Pre-jet fuel (Ma_0=5)	0	1.05%	2.16%	3.34%	4.56%	5.81%	7.09%
f2 growth rate with Pre-jet fuel (Ma_0=6)	0	0.921%	1.97%	3.10%	4.3%	5.53%	6.76%

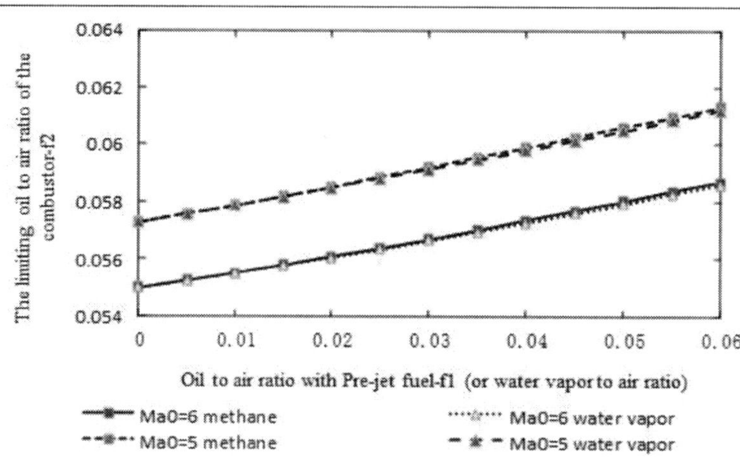

Figure 5. The relationship between the limiting oil to air ratio of the combustor and oil to air ratio with Pre-jet fuel (or water vapor to air ratio).

As shown in figure 6, the unmounted specific thrust increases with the increasing of oil to air ratio (or water vapor to air ratio). The main reason is increasing the flow rate of working medium, meanwhile, cooling can also play a role by increasing the oil to air ratio of the combustor, both functions increase propulsion. According to the analysis, the process of jetting methane or water vapor is similar, the detailed comparison between different oil to air ratio (or water vapor to air ratio) with or without Pre-jet fuel is shown in table 5.

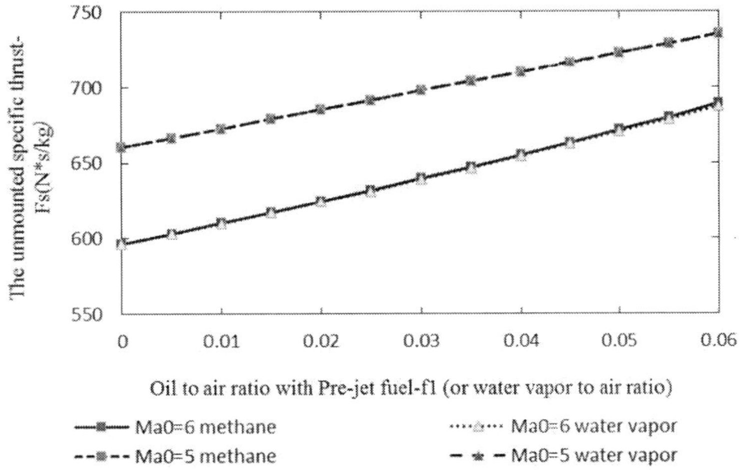

Figure 6. The relationship between the unmounted specific thrust and oil to air ratio with Pre-jet fuel (or water vapor to air ratio).

Table 5. Influence of different oil to air ratio with Pre-jet fuel on the unmounted specific thrust (taking methane as an example).

f1	**0**	**0.01**	**0.02**	**0.03**	**0.04**	**0.05**	**0.06**
Fs growth rate with Pre-jet fuel (Ma_0=5)	0	1.88%	3.77%	5.65%	7.54%	9.43%	11.33%
Fs growth rate with Pre-jet fuel (Ma_0=6)	0	2.29%	4.70%	7.23%	9.87%	12.63%	15.52%

As shown in figure 7, the specific impulse defined is the specific thrust of unit weight propellant, the propellant contains methane or water vapor with Pre-jet fuel. Although the Pre-jet fuel lowers the inlet temperature of the combustor, it can accommodate more fuel. However, the flow rate of the propellant is lower with Pre-jet fuel that this part of the propellant does not combust, but only plays a cooling role. With increasing of oil to air ratio (or water vapor to air ratio) with Pre-jet fuel, the specific impulse naturally decreases and fuel economy decreases.

As shown in figure 8, with increasing of oil to air ratio (or water vapor to air ratio) with Pre-jet fuel, the total engine efficiency is improved. In equation (25) the fuel only refers to the fuel injection to the combustor, and does not include the fuel with Pre-jet fuel. From the effect of input and output, the definition is flawed and the amount of Pre-jet fuel should be included.

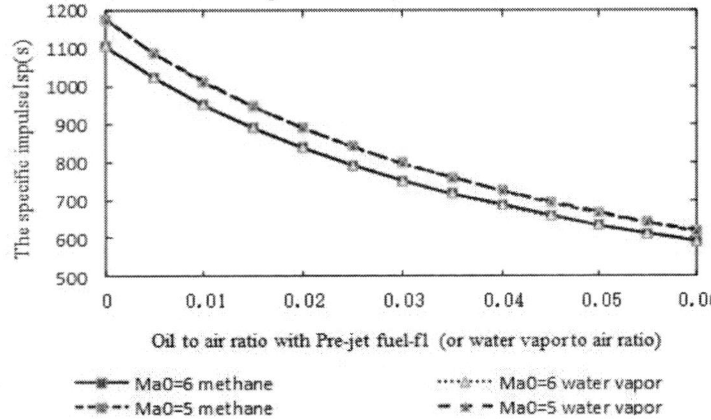

Figure 7. The relationship between the specific impulse and oil to air ratio with Pre-jet fuel (or water vapor to air ratio).

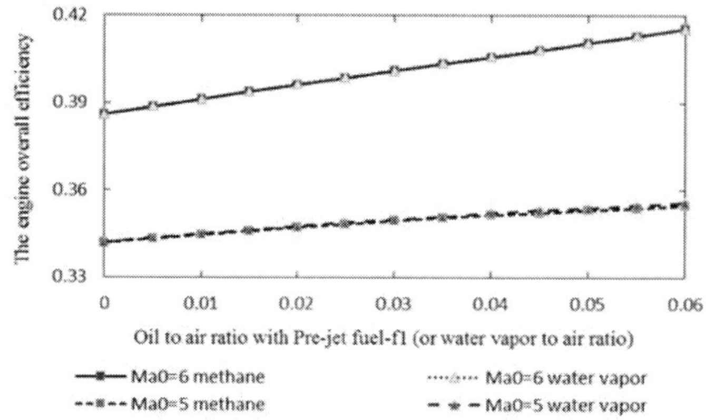

Figure 8. The relationship between the engine overall efficiency and oil to air ratio with Pre-jet fuel (or water vapor to air ratio).

4. Conclusion

The effect on the performance of the scramjet with Pre-jet fuel was analysed in this paper. Governing equations of the different component in scramjet and an ideal performance analysis simulation platform were established. Meantime, methane and water vapor with Pre-jet fuel was compared at different Mach number. There were combustion, shock wave, friction and fuel thermal dissociation losses in the actual scramjet with Pre-jet fuel. The following conclusions could be obtained based on ideal situation.

(1) Compared with the scramjet without Pre-jet fuel, with increasing of the oil to air ratio, the outlet static temperature of isolator was reduced more tempestuously, by contrast, the total engine efficiency was improved.

(2) Compared with the scramjet without Pre-jet fuel, when the oil to air ratio of isolator was 0.06 at Ma0=6 (the oil to air ratio was 0.05825 when the equivalence ratio was 1.0), the maximum oil to air ratio of the combustor was increased by 6.8%, and the unmounted specific thrust was increased by 15.5%.

(3) Compared with the scramjet without Pre-jet fuel, because the mass flow of working fluid increased with Pre-jet fuel, the specific thrust increased, meanwhile, the specific impulse and the fuel economy also reduced.

(4) Compared with the scramjet without Pre-jet fuel, the effect of methane on scramjet performance with Pre-jet fuel was similar to water vapor.

References

[1] Ning Li, Xuchang Li, Hongyu Xiao. A review of new technologies of scramjet [J]. Winged Missiles Journal. 2013, 7(7): 86-93.

[2] Peiyong Wang, Ming Chen, Fei Xin. CFD numerical simulation of Hyshot scramjet [J]. Journal of Aerospace Power, 2014, 29(5): 1020-1028.

[3] Vinsgradsv V A, Prudnikov A G. Injection of liquid into the strut shadow at supersonic velocities [R]. SAE Technical Paper, 1993.

[4] Seiner J M, Dash S M, Kenzakowski D C. Historical survey on enhanced mixing in scramjet engines [J]. Journal of Propulsion and Power, 2001, 17(6): 1273-1286.

[5] Sislian J P, Parent B. Hypervelocity fuel/air mixing in a shcramjet inlet [J]. Journal of propulsion and power, 2004, 20(2): 263-272.

[6] Schwartzentruber T E, Sislian J P, Parent B. Suppression of premature ignition in the premixed inlet flow of a shcramjet [J]. Journal of propulsion and power, 2005, 21(1): 87-94.

[7] Arai T, Kasahara J, Mukai K, et al. Experiments of Pre-Mixing Shock-Induced Combustion Scramjet with Forebody-Wall Fuel Injection [J]. AIAA Paper, 2002 (2002-5243).

[8] Star JB, Edwards, J R, Smart, MK, Baurle RA.Numerical Simulation ofScramjet Combustion in a Shock Tunnel. AIAA Paper No. 2005-0428, Jan. 2005.
[9] Shikhman Y M, Vinogradov V A, Yanovskiy L S, et al. The Demonstrator of Technologies—Endothermic Hydrocarbon Fueled Dual Mode Scramjet [J]. AIAA Paper, 1787- 2001, 2001.
[10] Buriko Y, Vinogradov V, Goltsev V,et al.Influence of Active RadicalConcentration on Self-Ignition Delay of Hydrocarbon Fuel/Air Mixture. Journal of Propulsion and Power, 2002, 18(5): 1049-1058.
[11] Livingston T, Segal C, Schindler M, et al. Penetration and spreading of liquid jets in an external-internal compression inlet [J]. AIAA Journal, 2000, 38(6): 989-994.
[12] Goldfeld M A, Starov A V, Vinogradov V V. Experimental study of scramjet module [J]. Journal of Propulsion and Power, 2001, 17(6): 1222-1226.
[13] GuoskovOV, KopchenovVI,et al.Numerical Researches of Gaseous Fuel Pre-Injection in Hypersonic 3-D Inlet. Journal of Propulsion and Power, 2001, 17(6): 1162-1169.
[14] Odam J. Scramjet experiments using radical farming [J]. 2004.
[15] Mudford N R, Mulreany P J, McGuire J R, et al. CFD calculations for intake-injection shock-induced-combustion scramjet flight experiments [J]. AIAA Paper, 2003, 7034.
[16] Owens M, Mullargili S, Segal C, et al. Effect of kerosene pre-injection on combustion flameholding in a Mach 1.6 airflow [J]. Journal of Propulsion and Power, 2001, 17(3): 605-611.
[17] Vinogradov V A, Shikhman Y M, Albegov R M, et al. Experimental research of pre-injected methane combustion in high speed subsonic airflow [J]. 12th AIAA International Space Planes and Hypersonic systems and technologies, Norfolk, Virginia, USA, 2003.
[18] Reynolds W C, PerkinsH C. Engineering thermodynamics [M]. McGraw-Hill, 1977.
[19] Bertin J J. Hypersonic aerothermodynamics [M]. AIAA, 1994.
[20] Weidao Shen, Jungeng Shen. Engineering thermodynamics [M]. Beijing: Higher Education Press 2011.

IOP Conf. Series: Materials Science and Engineering **715** (2020) 012076 doi:10.1088/1757-899X/715/1/012076

Supporting mechanism design of parachutes for 2.4m×2.4m transonic wind tunnel in CARDC

Dong Hao[1], Lin Zhang[1], Jing Yu[2,*], Daiyong Mao[1] and Xianwen Yang[1]

[1]High Speed Aerodynamics Institute, China Aerodynamics Research and Development Center, Mianyang, 621000, People's Republic of China
[2]Computational Aerodynamics Institute, China Aerodynamics Research and Development Center, Mianyang, 621000, People's Republic of China

*E-mail: yujinghd@hotmail.com

Abstract. In order to carry out the wind tunnel test of the parachutes in 2.4m×2.4m transonic wind tunnel of CARDC which is located at Mianyang, a supporting mechanism for the attack angle from 0 to 25 degrees at Mach 0.4 and 0.8 is designed and optimized in the present paper. Three section shapes (circle, ellipse, and airfoil) are utilized to analyze the aerodynamic characteristics and mechanical performances of the designed supporting mechanism. The aerodynamic and mechanical simulated results of the three different section shapes are compared and discussed. The optimized design is verified by wind tunnel experiments. The results show that the section shape of circle has the largest modal frequency of first order (53.497Hz), the smallest Mises stress (34.1MPa) and displacement (0.4735mm). It is the best section shape of the three for the supporting mechanism. The investigation in the present paper has a good significant meaning to the testing technologies of the wind tunnel experiments for parachutes.

1. Introduction

With the abilities to slow the motion of an object through the atmosphere by creating drag forces, the parachutes have been widely used and it is very necessary to accurately get the aerodynamic characteristics of the parachute [1-4].

Two approaches are usually utilized to obtain the aerodynamic performances of a parachute: numerical computation [5-10] and wind tunnel test [11, 12]. In the parachute's descent process of deployment, inflation and terminal, the aerodynamic environment is so complex that the numerical computational method usually owns large errors. For many years, the wind tunnel test is still an invaluable, accurate and cost-effective approach to investigate the aerodynamic characteristics of a parachute [13]. The construction of a series of asymmetrical wind tunnel model parachutes which was designed to a modular concept was described by Klimas et al. [14]. The effect of the arm ratio of the cross canopy on the transverse profiles of velocity, vorticity, and Reynolds stress was stated by Jin et al. [15]. The Stereoscopic Particle Image Velocimetry (SPIV) technique was adopted in his research and the flow structures of the cross parachute were measured and analyzed. The parachutes used in Mars exploration missions were also studied through the wind tunnel tests by NASA [16, 17] and Mars Science Laboratory [18-20]. Most previous literatures were on the topic of simulating the parachute aerodynamic behaviours. And only some of them were investigated in wind tunnels.

Especially, most of the wind tunnel tests were carried out under the condition of low speed flow and zero attack angle [21].

In the present paper, a supporting mechanism of parachutes for testing in 2.4m×2.4m transonic wind tunnel of CARDC is designed and optimized. The main configuration of the support mechanism is given according to the constraints of the testing model and facility. The section shapes (circle, ellipse, and airfoil) are adopted and the aerodynamic and mechanical analysis of them is also provided. The comparisons of the three section shapes are discussed and the section shape with best aerodynamic and mechanical performances is given. Summing up, the purposes of this paper are (i) to design three different section shapes (Section 2), (ii) to compare, discuss and verify the designation (Section 3), (iii) to give some conclusions in the last (Section 4).

2. Support mechanism design

In this section, the main configuration is designed according to the constraints of the testing model and facility (full model testing section of 2.4m×2.4m transonic wind tunnel). The circle, ellipse and airfoil shaped section are selected to design the support mechanism.

2.1. Main configuration

According to the constraints of the testing parachute model and facility, the main configuration of the supporting mechanism used in 2.4m×2.4m transonic wind tunnel is designed as figure 1 and the three dimensional coordinates of the main configuration is illustrated in table 1.

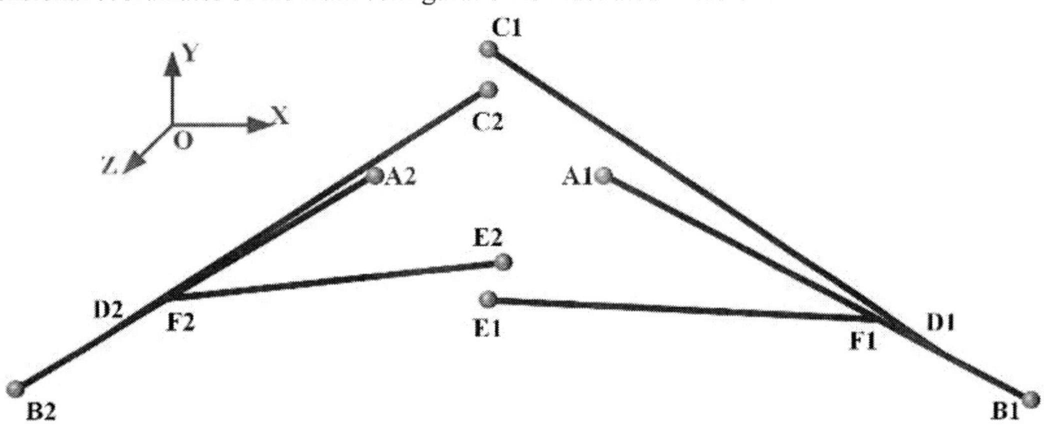

Figure 1. Sketch of parachute.

Table 1. Coordinates of the main configuration.

Point	Coordinate (mm)	Point	Coordinate (mm)
A1	(359, 0, 0)	A2	(-359, 0, 0)
B1	(1910, 0, 714)	B2	(-644, 0, 850)
C1	(0, 399, 0)	C2	(0, 240, 0)
D1	(1810, 0, 668)	D2	(-594, 0, 701)
E1	(0, -399, 0)	E2	(0, -240, 0)
F1	(1710, 0, 622)	F2	(-544, 0, 552)

2.2. Section design

For the purpose of getting a better design of the supporting mechanism, three different section shapes are adopted to design the supporting mechanism of the parachutes for the 2.4m×2.4m transonic wind tunnel of CARDC. The optimum section shape of the three is obtained according to the aerodynamic and mechanical simulating results. The three section shapes are circle, ellipse and airfoil, separately. The sizes of them are given in figure 2. In the part of the aerodynamic analysis, the reference length of the three section shapes is all 0.12m and the reference areas are respectively 0.011304m2, 0.005652m2, and 0.0026m2.

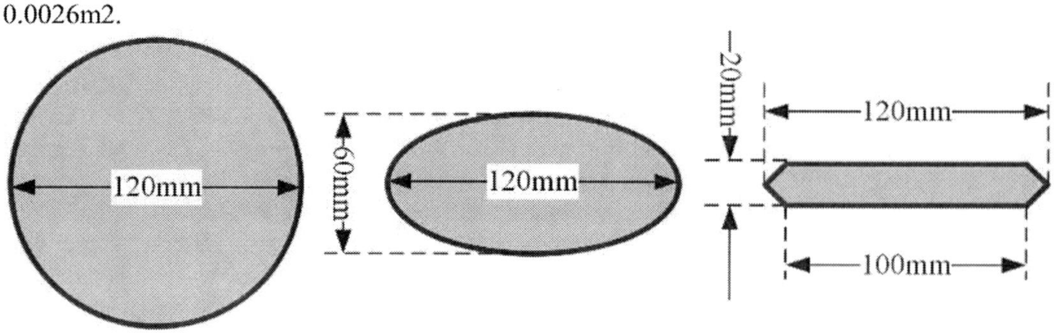

Figure 2. Sizes of three section shapes.

3. Results and discussions

3.1. Aerodynamic analysis

According to the real experimental condition, the worst set of Mach number 0.8 and angle of attack 25 degrees is adopted to carry out the aerodynamic analysis of the three different section shapes. In aerodynamic simulations, the total temperature and the total pressure are set as 298K and 160kPa. The mesh information of the three section shapes is illustrated in table 2 and the grid structures are plotted in figure 3. The elements of the section shapes are structurally modelled in ICEM software [22] and the aerodynamic results are computed in PHengLEIv2.1 which is a platform for hybrid engineering simulation of flows published by CARDC [23]. The maximum iteration step is set as 10000, the SA turbulence model [24] is utilized and the time-discrete format of LU-SGS is applied to solving Navier-Stokes equations in the aerodynamic simulation [25].

Table 2. Mesh information.

Items	Element number	Node number	Quality
Circle section	12948	12797	≥0.75
Ellipse section	6308	6232	≥0.75
Airfoil section	12782	12628	≥0.75

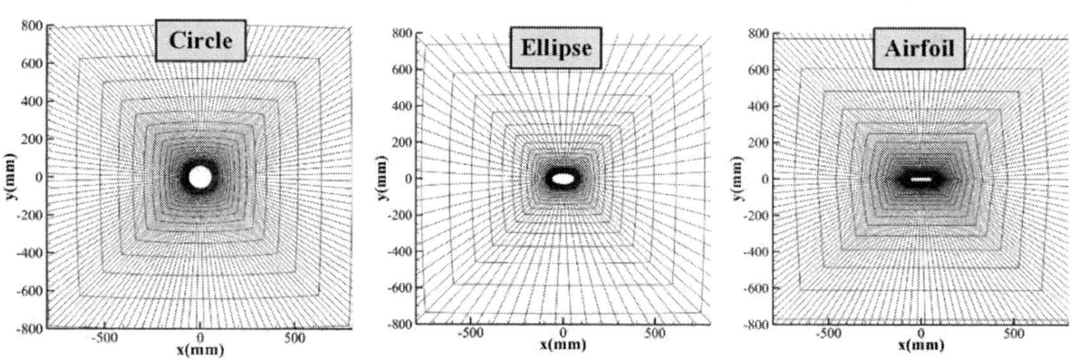

Figure 3. Grid structures.

The pressure fields of the three section shapes are shown in figure 4 and the Mach fields of the three section shapes are displayed in figure 5. As is seen from figure 4 and figure 5, the circle section shape has the largest influence areas of pressure and Mach number. The influence distance of the circle section shape is about 200mm and the installation position of the parachute in the wind tunnel is out of the influence scope. So both the three section shapes meet the aerodynamic requirements of the wind tunnel test for parachute. And the airfoil section shape has the smallest influence scope.

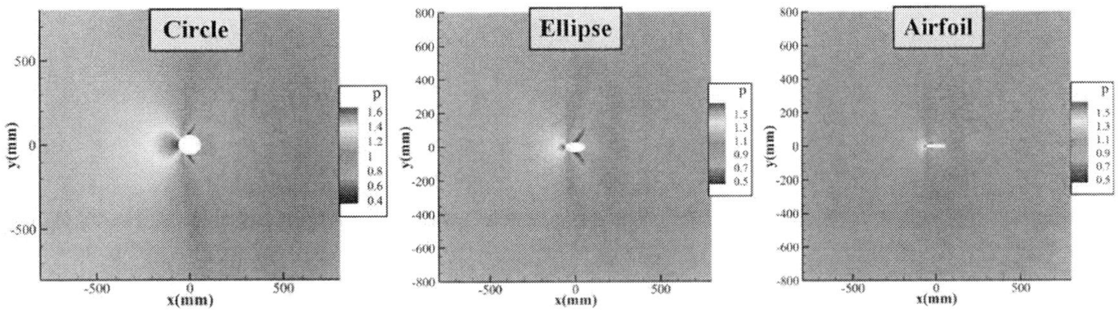

Figure 4. The results of the pressure fields.

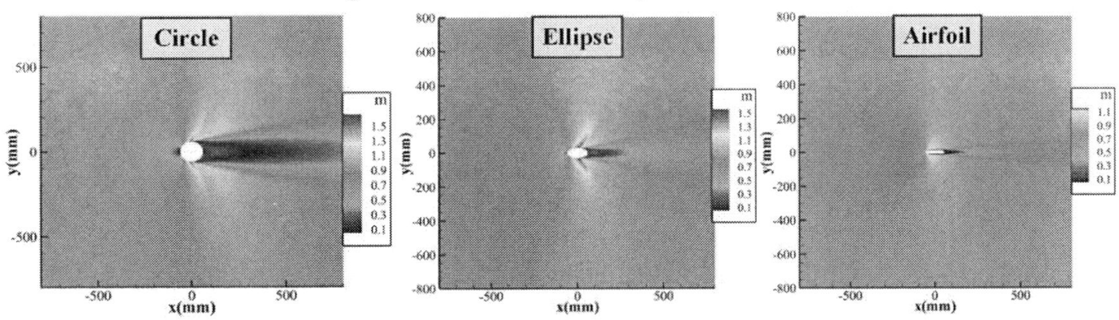

Figure 5. The results of the Mach fields.

3.2. Mechanical analysis

In this subsection, the mechanical properties of the main configuration with the three section shapes are discussed. The modal results and the static results are also described. In the mechanical analysis, the finite element model is carried out in ABAQUS and the tetrahedral element is used to obtain the grid. The material is 30CrMo and its material parameters are given in table 3. The mesh information of the mechanical finite element model is supplied in table 4.

Table 3. Parameters of 30CrMo.

Yong modulus (Pa)	Poisson ratio	Density (kg/m^3)	Yield strength (Pa)
2.11×10^{11}	0.279	7850	7.85×10^{8}

Table 4. Mesh information of mechanical finite element model.

Items	Element number	Node number	Element type
Circle section	66299	102054	Tetrahedral
Ellipse section	128775	197813	Tetrahedral
Airfoil section	138765	231468	Tetrahedral

In the mechanical analysis procedure, A1, A2, C1, C2, E1, and E2 are set as fixed condition. The load at B1 is (-20000N, 0, 0) and at B2 is (-8000N, 0, 0). The modal frequencies of the three section shapes are given in table 5 and the static results are described in table 6. The first modal shapes, the stress clouds, and the displacement clouds of the main configuration with the three section shapes are given in figure 6, figure 7 and figure 8, separately.

Table 5. Modal frequencies of three section shapes.

Modal order	Circle section	Ellipse section	Airfoil section
1st	53.497Hz	46.590Hz	21.935Hz
2nd	118.51Hz	59.674Hz	30.088Hz
3rd	164.17Hz	96.279Hz	42.565Hz
4th	181.67Hz	132.25Hz	46.652Hz
5th	237.70Hz	141.64Hz	71.475Hz

IOP Conf. Series: Materials Science and Engineering **715** (2020) 012076 doi:10.1088/1757-899X/715/1/012076

Table 6. Static results of three section shapes.

Items	Circle section	Ellipse section	Airfoil section
Maximum stress (MPa)	34.1	50.19	141.3
Maximum strain	2.53×10^{-4}	2.431×10^{-4}	6.365×10^{-4}
Maximum displacement (mm)	0.4735	1.047	2.759

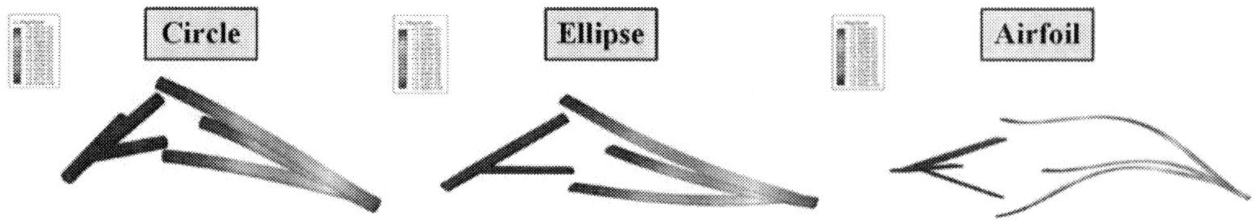

Figure 6. First modal results of three section shapes.

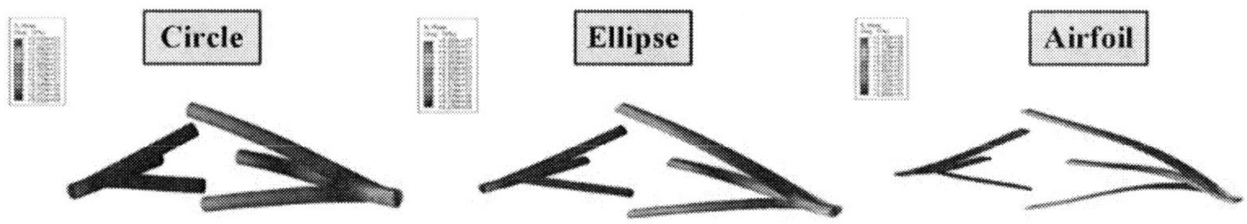

Figure 7. Stress results of three section shapes.

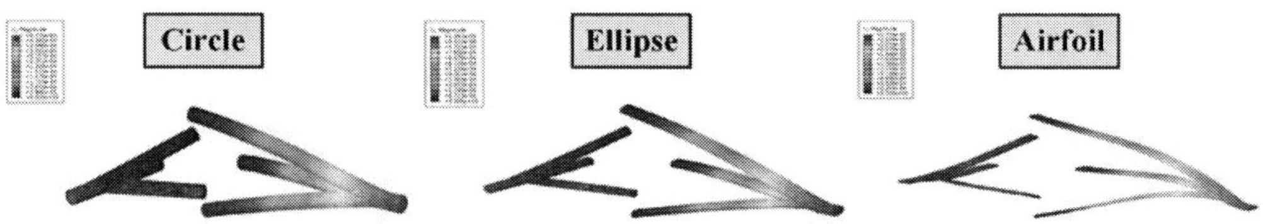

Figure 8. Displacement results of three section shapes.

From table 5, it can be concluded that the main configuration with circle section shape has the largest first modal frequency and the its second modal frequency is far larger than the other two. The circle section shape offers the largest stiffness of the main configuration. As is shown in table 6, the circle section shape has the smallest stress (34.1MPa) and the airfoil section shape has the largest stress (141.3MPa). The maximum strain of the circle section shape (2.53×10-4) is equivalent to the ellipse circle shape (2.431×10-4), less than the airfoil section shape (6.365×10-4). The maximum displacement of the circle section shape (0.4735mm) is the smallest, far less than the ellipse (1.047mm) and airfoil section shapes (2.759mm). As is seen in figure 6, figure 7 and figure 8, the maximum stress and displacement both occur at B1 part of the main configuration.

According to the results of the aerodynamic and mechanical analysis, in a summary, the circle section shape makes the main configuration own the largest stiffness and smallest stress and displacement. Its aerodynamic influence distance meets the requirements of the wind tunnel experiments.

3.3. Experimental verification

Considering the capacity of the manufacture and mechanical connection, the final supporting mechanism of the parachute for 2.4m×2.4m transonic wind tunnel is shown in figure 9. Two sensors are applied to obtaining the aerodynamic coefficients (4N6-70E and 4N6-64E). 4N6-70E is mounted at B1 part and 4N6-64E is positioned at B2 part. The state of the supporting mechanism in the experimental process is figured in figure 10. The measured data were sampled at 300 Hz and handled by 1 Hz lowpass filter. The directly obtained data include the influences of two parts: the contributions of the parachute and the effects of the supporting mechanism.

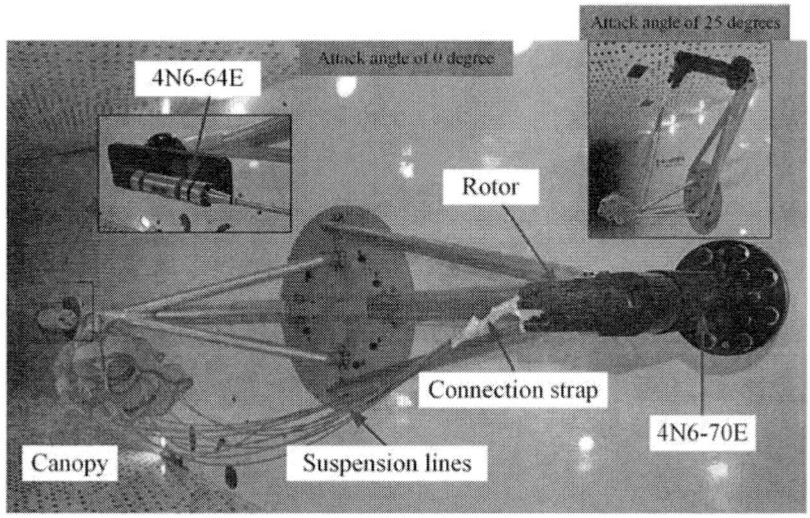

Figure 9. Final supporting mechanism.

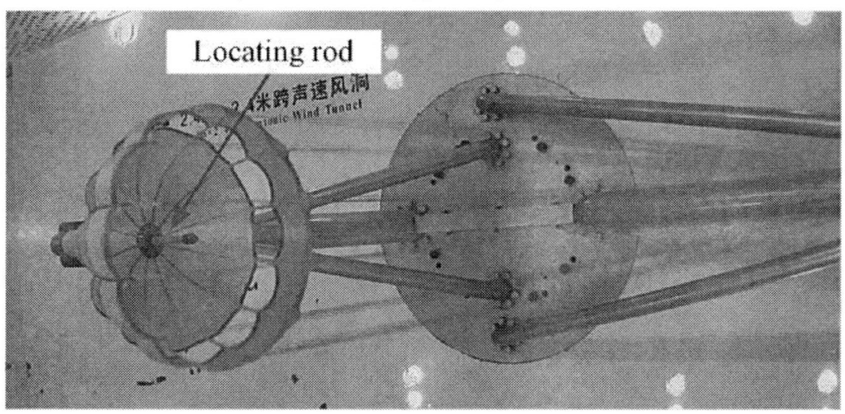

Figure 10. State of supporting mechanism in experimental process.

The results of the wind tunnel tests are given in figure 11. The measured data of the lift, drag and pitch moment coefficients are smooth and reasonable. The law of the aerodynamics versus the angle of attack is in accordance with physical reality. The designed supporting mechanism is effective on the wind tunnel tests of the parachutes in the 2.4m×2.4m transonic wind tunnel and achieves the requirements of the experiments.

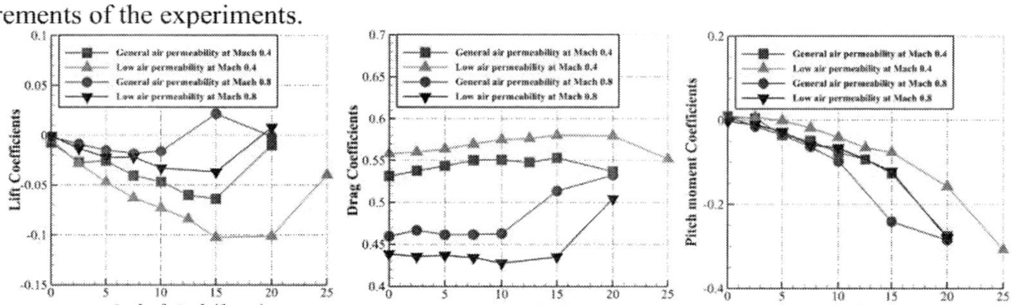

Figure 11. Results of wind tunnel tests.

4. Conclusions

In this paper, a supporting mechanism is designed for the wind tunnel tests of the parachutes in the

2.4m×2.4m transonic wind tunnel. In order to achieve an optimized design, circle, ellipse and airfoil section shapes are adopted and discussed. The aerodynamic influences of the three section shapes are given and the mechanical properties of the main configuration with the three section shapes are analyzed. As a result, the circle section shape is the most suitable one to design the supporting mechanism. The aerodynamic influence distance of the section shape is about 200mm and the circle section shape has the largest stiffness, smallest stress (34.1MPa), strain (2.53×10-4) and displacement (0.4735mm). The designed supporting mechanism is verified by the wind tunnel experiments of the parachutes and is able to assure the success of the wind tunnel tests. The investigation of this paper is significant to the development of the parachute applications.

Acknowledge
The author(s) disclosed receipt of the following financial support for the research, authorship, and/or publication of this article: This work was supported by National Natural Science Foundation of China (Grant No. 71601179).

References
[1] H.G. Heinrich, R.A. Noreen, Analysis of parachute opening dynamics with supporting wind tunnel experiments, AIAA 2nd Aerodynamic Deceleration Systems Conference, 1968. (DOI: 10.2514/6.1968-924)
[2] Z. Jin, S. Pasqualini, B. Qin, Experimental investigation of the effect of reynolds number on flow structures in the wake of a circular parachute canopy, Acta Mech. Sinica-Prc 30 (2014) 361-69. (DOI: 10.1007/s10409-014-0058-1)
[3] S. Steinberg, P.M. Siemers, R.G. Slayman, Development of the viking parachute configuration by wind-tunnel investigation, J. Aircraft 11 (1974) 101-07. (DOI: 10.2514/3.62016)
[4] R.H. Croll, C.W. Peterson, A computer-controlled video instrument for wind-tunnel testing of lifting parachutes, J. Aircraft 16 (1979) 269-74. (DOI: 10.2514/3.58516)
[5] K. Stein, R. Benney, V. Kalro, et al, Parachute fluid-structure interactions: 3-d computation, Comput. Methods Appl. Mech. Engrg. (2000) 373-86. (DOI: S0045-7825(00)00208-5)
[6] V.V. Lyalin, V.I. Morozov, A.T. Ponomarev, Parachute performance computer analysis, 16th AIAA Aerodynamic Decelerator Systems Technology Conference and Seminar, 2001. (DOI: 10.2514/6.2001-2064)
[7] G. Frost, M. Costello, Simulation of a mortar launched, parachute deployed battlefield imaging system, Journal of Dynamic Systems, Measurement and Control-Transactions of the ASME 126 (2004) 583-94. (DOI: 10.1115/1.1789974)
[8] K. Karagiozis, R. Kamakoti, F. Cirak, et al, A computational study of supersonic disk-gap-band parachutes using large-eddy simulation coupled to a structural membrane, J. Fluid. Struct. (2011) 175-92. (DOI: 10.1016/j.jfluidstructs.2010.11.007)
[9] J. Kim, Y. Li, X. Li, Simulation of parachute fsi using the front tracking method, J. Fluid. Struct. (2013) 100-19. (DOI: 10.1016/j.jfluidstructs.2012.08.011)
[10] X. Xue, H. Koyama, Y. Nakamura, et al, Effects of suspension line on flow field around a supersonic parachute, Aerosp. Sci. Technol. (2015) 63-70. (DOI: 10.1016/j.ast.2015.02.014)
[11] R.E. Rychnovsky, R.N. Everett, Wind tunnel tests. To evaluate a lifting transonic parachute, AIAA 5th Aerodynamic Decelearation Systems Conference, 1975. (DOI: 10.2514/6.1975-1365)
[12] W.B. Pepper, J.F. Reed, Parametric study of parachute pressure distribution by wind tunnel testing, J. Aircraft 13 (1976) 895-900. (DOI: 10.2514/3.58727)
[13] J. Barber, H. Johari, Experimental investigation of personnel parachute designs using scale model wind tunnel testing, 16th AIAA Aerodynamic Decelerator Systems Technology Conference, 2001. (DOI: 10.2514/6.2001-2074)
[14] P.C. Klimas, H.E. Widdows, R.H. Croll, Modular asymmetric parachute for wind tunnel testing, J. Aircraft 20 (1983) 89-90. (DOI: 10.2514/3.44833)

[15] Z. Jin, S. Pasqualini, Z. Yang, Experimental investigation of the flow structures in the wake of a cross parachute canopy, European Journal of Mechanics, B/Fluids (2016) 70-81. (DOI: 10.1016/j.euromechflu.2016.07.008)
[16] A. Witkowski, R. Bruno, Mars exploration rover parachute decelerator system program overview, 17th AIAA Aerodynamic Decelerator Systems Technology Conference and Seminar, 2003. (DOI: 10.2514/6.2003-2100)
[17] Y. Taeger, A. Witkowski, A summary of dynamic testing of the mars exploration rover parachute decelerator system, 17th AIAA Aerodynamic Decelerator Systems Technology Conference and Seminar, 2003. (DOI: 10.2514/6.2003-2127)
[18] D.S. Adams, T.P. Rivellini, Mars science laboratory's parachute qualification approach, 20th AIAA Aerodynamic Decelerator Systems Technology Conference and Seminar, 2009. (DOI: 10.2514/6.2009-2913)
[19] D.S. Adams, T.P. Rivellini, A.D. Steltzner, Mars science laboratory parachute inversion phenomenon and flight risk assessment, 20th AIAA Aerodynamic Decelerator Systems Technology Conference and Seminar, 2009. (DOI: 10.2514/6.2009-2956)
[20] D.S. Adams, N.P. Onufer, Mars science laboratory parachute development test program, 21st AIAA Aerodynamic Decelerator Systems Technology Conference and Seminar, 2011. (DOI: 10.2514/6.2011-2508)
[21] C.H. Zumwalt, J.R. Cruz, D.F. Keller, et al, Wind tunnel test of subscale ringsail and disk-gap-band parachutes, 34th AIAA Applied Aerodynamics Conference, 2016. (DOI: 10.2514/6.2016-3882)
[22] A. Inc., Ansys icem cfd 12.1 user manual, 2009.
[23] H. Xin, Z. Laiping, Z. Zhong, et al, Validation of the structured-unstructured hybrid cfd software hyperflow, Eighth International Conference on Computational Fluid Dynamics (ICCFD8), 2014.
[24] S. Benjanirat, L.N. Sankar, G. Xu, Evaluation of turbulence models for the prediction of wind turbine aerodynamics, 41st Aerospace Sciences Meeting and Exhibit, 2003. (DOI: AIAA-2003-0517)
[25] Y. Sun, Z.J. Wang, Y. Liu, Efficient implicit non-linear lu-sgs approach for viscous flow computation using high-order spectral difference method, 18th AIAA Computational Fluid Dynamics Conference, 2007. (DOI: AIAA 2007-4322)

Identification of real pitch difference in bolt-nut connection to improve anti-loosening performance and fatigue strength

Nao-Aki NODA*, Xi LIU, Shutaro KUBO, Yoshikazu SANO and Yasushi TAKASE

Mechanical Engineering Department, Kyushu Institute of Technology, 1-1 Sensui-Cho Tobata-Ku Kitakyushu-shi Fukuoka 804-8550, Japan

*E-mail: noda.naoaki844@mail.kyutech. jp

Abstract. In a wide industrial field, the bolt-nut joint is unitized as an important machine element and widely used. To ensure the safety of the structure and reduce cost, low-cost bolt-nut connections with good anti-loosening performance and high fatigue strength are always required. In our previous papers, the effect of pitch difference between a bolt and nut was studied. Under a small pitch difference, anti-loosening performance is not enough. Under a large pitch difference, fatigue strength is not good. Therefore, in order to improve both anti-fatigue and anti-loose performance, the choice of suitable pitch difference is very important although the manufacturing errors always appear in bolt-nut connection. Until now, the result of experiment and analysis without manufacturing error but is not match. Since even small manufacturing errors may affect anti-fatigue and anti-loose performance, in this paper how identification of real pitch difference is discussed after the pitch difference nut is produced. Also, the accuracy of the identified pitch difference is confirmed. By using the confirmed error, the machining accuracy may be improved to ensure the nut has good performance.

1. Introduction

The bolt-nut connections can be regarded as one of the most important materials joining techniques. They are widely used in various engineering fields, including aerospace, automotive and mechanical/civil engineering constructions. To ensure structures safety, high fatigue strength has been required, as well as anti-loosening performance. Many previous studies are focusing on the anti-loosening performance for newly developed bolt-nut connections [1-5], and several previous studies contribute toward improving fatigue strength [6-10]. This paper focuses on the effect of pitch difference between bolt-nut connection to improve the anti-loosing performance and fatigue strength. Compared with other special nuts, pitch difference bolt-nuts can be manufactured with lower cost to have good anti-loosening performance and high fatigue strength. Here, several pitch differences between the bolt and nut are designed as shown in figure 1 where the nut pitch is larger than the bolt pitch. When screwing a nut with pitch difference, torque is necessary before the nut contact with the clamped body. This torque in this screwing process is called prevailing torque [8,11] as shown in figure 2.

In this paper, a three-dimensional FEM simulation will be performed to calculate the maximum prevailing torque when the pitch difference nut is screwed. Then, the effect of pitch difference on the prevailing torque will be discussed. In this experimental study, after the pitch difference nut is manufactured, the accuracy of the pitch difference will be confirmed. By confirming correct pitch difference, the previous discussions [8] regarding anti-loosening and high fatigue strength can be more reliable. First, identify the nut rotating position where the prevailing torque appears during the nut screwing into bolt. This position can also be found theoretically by using axi-symmetric modelling of bolt-nut connection and also by using 3D modelling. In addition, the prevailing torque also can be focused to confirm the real pitch difference. Experimentally obtained T_p^{Real} using α and analytically obtained T_p^{FEM} can be compared to confirm the corrected pitch difference.

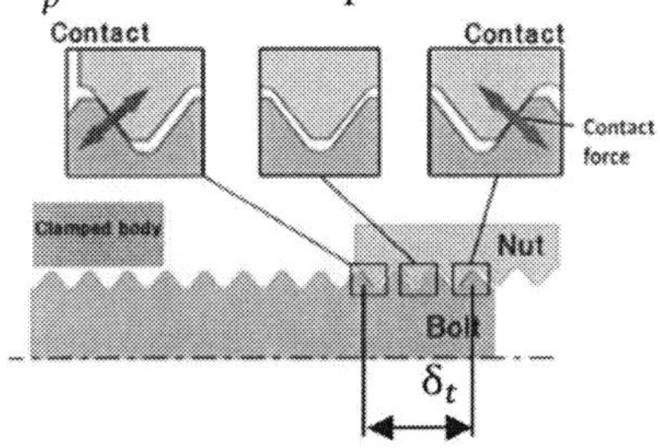

Figure 1. Contact status when the prevailing torque appears between bolt and nut.

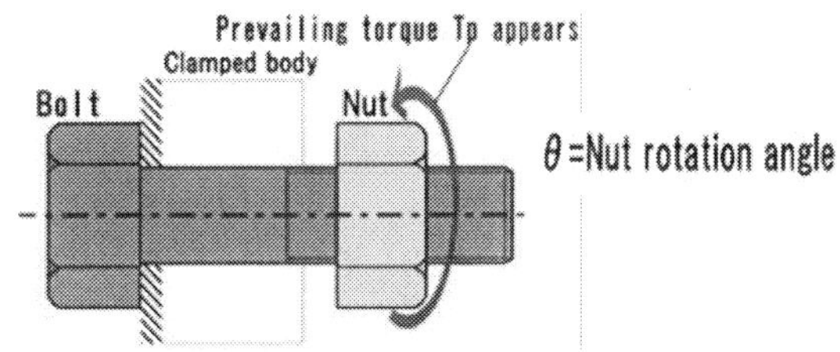

Figure 2. Schematic illustration for screwing process.

2. Identification of real pitch difference α from the nut contact position n_c

2.1. Specimen geometry

In this study, the Japanese Industrial Standard (JIS) M10 bolt-nut connections with strength grade 8.8 are employed. The bolt material is chromium molybdenum steel SCM435, and the nut material is medium carbon steel S45C quenched and tempered. Table 1 shows material properties. The standard M10 bolt-nut connection has the same pitch dimension of 1500 μm. Here, the nut pitch is assumed to be equal or slightly larger than the bolt pitch as shown in figure 3[7-9]. Several types of pitch differences, namely α=20μm, α=25μm, α=30μm, α=35μm, α=40μm and α=45μm are considered in this study. In addition, the horizontal clearance between bolt and nut Cx=60mm. Figure 4 is a cross-sectional photograph of the M16 nut illustrating chamfer, which is usually used for standard nuts larger than M16 (JIS).

Table 1. Material propertied of bolt and nut.

	Young's modulus E(GPa)	Poisson's ratio ν	Yield strengthσ_y(MPa)	Tensile strength σ_B(MPa)
SCM435 (Bolt)	206	0.3	800	1200
S45C(Nut)	206	0.3	530	980

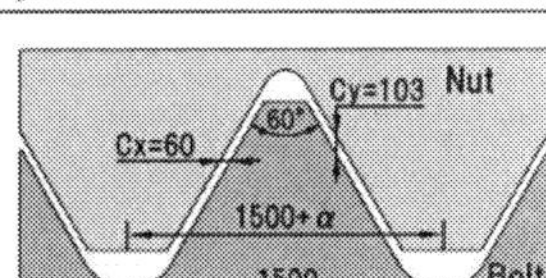

Figure 3. Pitch difference and clearance between bolt and nut threads.(unit:μm) in M10.

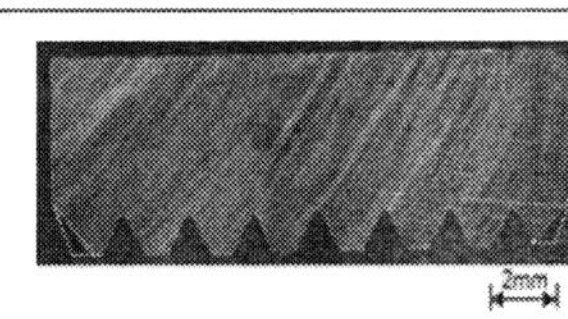

Figure 4. Chamfer usually used for standard nuts larger than M6(JIS).

2.2. How to obtain nut contact position n_c^{Theory}, n_c^{Exp} and prevailing torque T_p^{Exp}

Figure 5 is a schematic illustration of nut screwing process. In figure 5, position A is the one where the nut screwing starts and position B is the one where prevailing torque appears. Notations n_c^{Theory} and n_c^{Exp} can be defined as the number of nut rotation from A to B obtained theoretically and experimentally. Position C is in the one when prevailing torque is increasing, position D is the one where the nut is completely screwing into the bolt, and position E is the one where to continue to screw the nut for few turns after all the thread of nut is screwed into the bolt. Notation T_p^{Exp} denotes the prevailing torque during the nut screwing process.

In figure 5, the red arrow shows the contact force between bolt and nut threads during the screwing process. At position B in figure 5, the threads contact starts. As shown in figure 1, δ_t is defined as the distance where the contact takes place. The following Equations (1) and (2) can be obtained from the thread geometry.

$$n_c^{Theory}\alpha = 2C_x, C_x = \frac{C_y}{\tan\theta} \quad (1)$$

$$\delta_t = n_c^{Theory} p \quad (2)$$

To obtain n_c^{Exp} during screwing the nut in the experiment, a dial type torque wrench whose product name TOHNICHI DB 50N is used as shown in figure 6. Then, n_c^{Exp} is defined as the position when T_p^{Exp} =0.1Nm appears. To obtain the relation between T_p^{Exp} and the nut rotation angle θ experimentally,head of bolt fixed, T_p^{Exp} is measured to rotate the nut by using torque wernch at the interval of 45 degree. By considering both bolt and nut always have incomplete threads at the locations where the threads start, these thread starting positions are adjusted at position A as shown in figure 5 in this experiment. Then, this position is defined as θ = 0. This screwing experiment is conducted under lubrication surface state of Molybdenum disulfide grease spray PRO (manufactured by Azette Co., Ltd.) figure 7 shows T_p^{Exp} measured for M10 with α=35μm.

2.3. Relationship between n_c^{Theory} and α

Figure 8 shows the relationship between n_c^{Theory} and pitch difference α. The blue line shows the result of Equation (1) and the red line shows the result of 3D CAD modelling. Figure 8 shows the blue line and the red line coincide with each other within 5%. Therefore it may be concluded that Equation (1) can estimate n_c^{Theory} accurately without considering 3D chamfer geometry. Since the result of Equation (1) is almost consistent with 3D CAD model result, Equation (1) can be used as the theoretical result.

Figure 9 shows the result of n_c^{Exp} as green line. As shown in figure 9, n_c^{Exp} and n_c^{Theory} have some difference. This is because the bolt-nut connection experimentally used has some manufacturing error regarding the pitch difference α. As shown in figure 9, for example, the result of α = 20μm should be regarded as α = 30μm by considering n_c^{Exp}=4.25. Therefore, the black line can be drawn as the result of n_c^{Real} by correcting α.

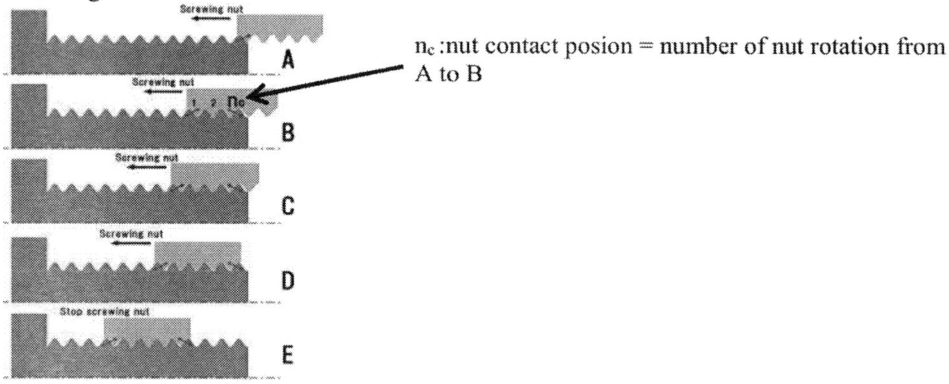

Figure 5. Screwing process of nut.

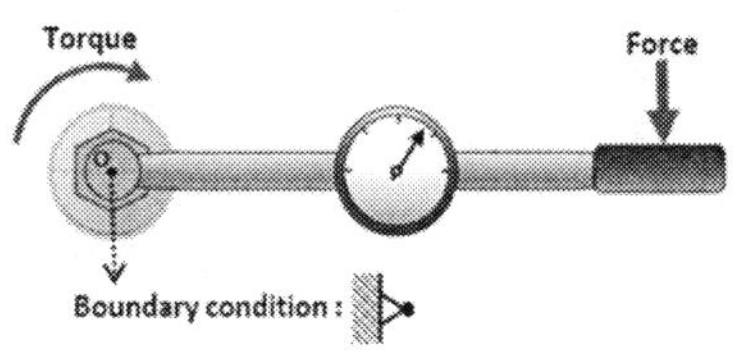

Figure 6. Torque measuring method and device (Torque wrench).

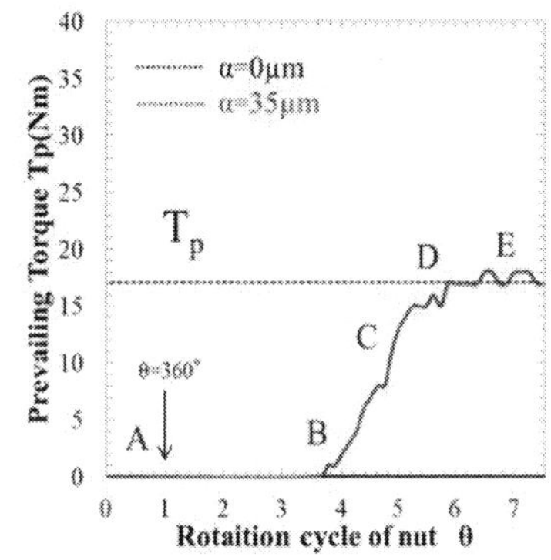

Figure 7. Prevailing torque Tp in the screwing process.

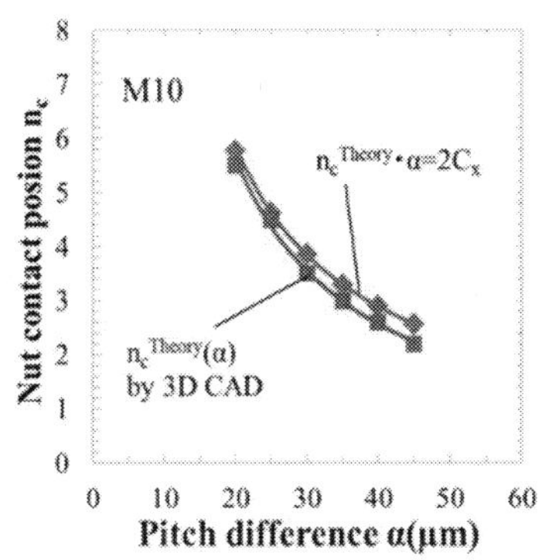

Figure 8. The starting position $\boldsymbol{n_c}$ of the prevailing torque in M10.

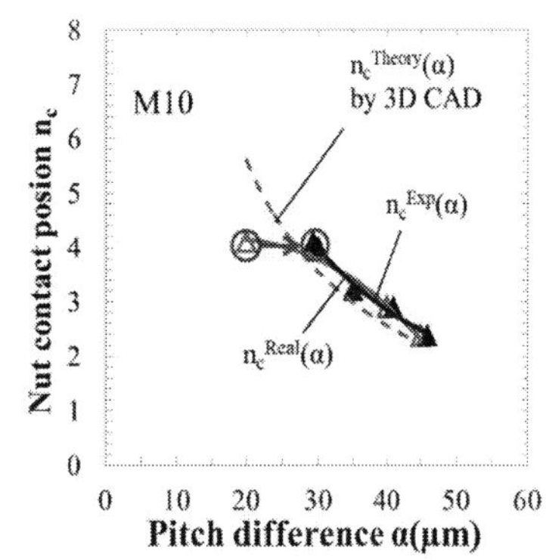

Figure 9. The starting position $\boldsymbol{n_c}$ of the prevailing torque in M10.

3. Confirmation of real pitch difference by calculating T_p^{FEM}

3.1. How to obtain T_p^{FEM}

From section 2, how to obtain real pitch difference α has been discussed in terms of n_c^{Theory}. To confirm the corrected value of α, in this section the prevailing torque T_p^{Exp} will be confirmed experimentally. A precise three-dimensional FEM model as shown in figure 10 is used for finite element analysis. To simplify the calculation, the hexagonal part of the bolt and the nut shapes are replaced by two cylinders. Especially, the helical thread of the bolt and nut were subdivided into smaller elements compared with other parts [4]. Contact types and material non-linearity are considered in the analysis. Boundary conditions are set as follows, one side of the bolt is fixed, a small screwing angel is applied on the side surface of the nut as shown in figure 11. The connecting method of the nut and the bolt is frictional contact. Then, $T_p^{FEM}$is calculated by FEM simulation.

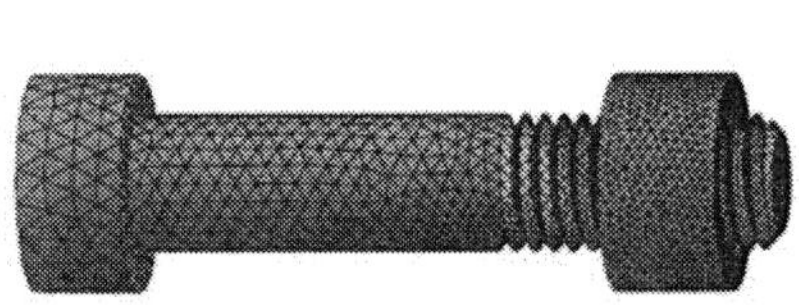

Figure 10. FEM model for M10 bolt and nut.

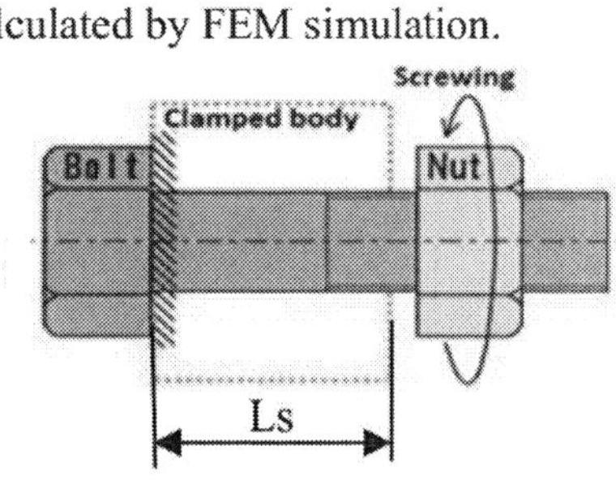

Figure 11. Boundary conditions.

3.2. Analytical result of T_p^{FEM}

Figure 12 shows that the relation between prevailing torque T_p^{FEM} and the nut rotation angle θ. When the nut screwing starts from position A in figure 5, the prevailing torque T_p^{FEM} increases from T_p^{FEM} =0. With increasing the prevailing torque T_p^{FEM}, the bolt is twisted with a small torsion angle. The prevailing torque $T_p(\alpha)$ can be defined as the saturated value in figure 12. The saturated point $T_p(\alpha)$ in figure 12 is proportional to the nut rotation angle θ as shown in the red dotted line.

Figure 13 shows the relationship between the prevailing torque and pitch difference α. The red line shows the result of $\mathrm{T_p^{FEM}}$. The green line shows the result of T_p^{Exp}, which does not coincide with result of $\mathrm{T_p^{FEM}}$. By collecting α as explained in figure 13, the black line show the result of T_p^{Real}. For example, the result of T_p^{Exp} when α = 20μm should be regarded as T_p^{Real} when α = 30μm by using n_c^{Real}. It is seen that T_p^{Real} coincide with T_p^{FEM} within a few percent error. It is confirmed that correct α can be obtained in terms of the prevailing torque.

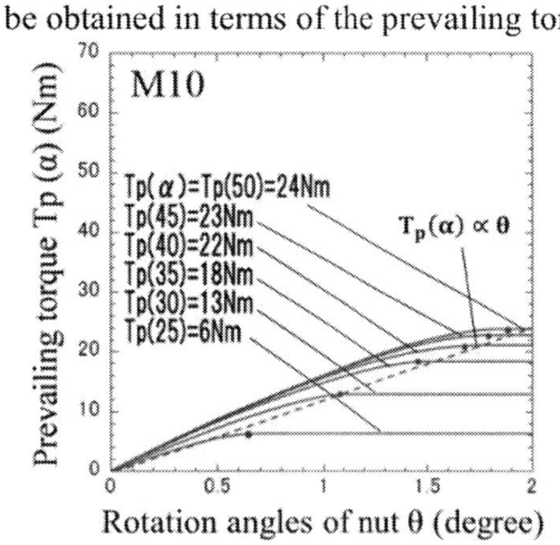

Figure 12. Prevailing torque T_p^{FEM} (α) in M10.

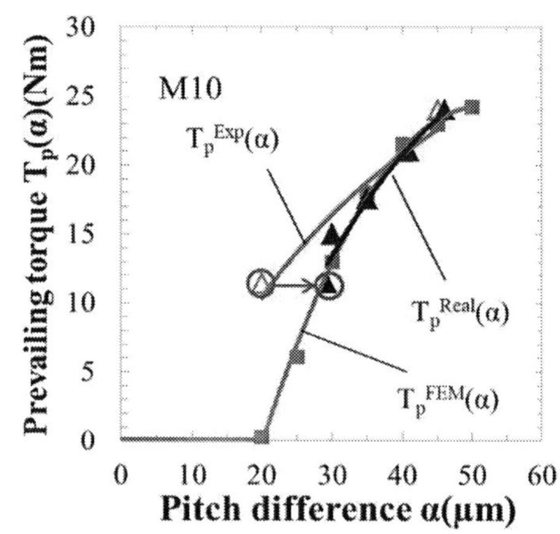

Figure 13. Comparison of prevailing torque $\mathrm{T_p^{EXP}}(\alpha)$, $\mathrm{T_p^{Real}}(\alpha)$ and $\mathrm{T_p^{FEM}}(\alpha)$ in M10.

4. Conclusion

1. The nut contact position n_c^{Theory} can be obtained by using Equation (1) assuming axi-symmetric geometry. Since the results of Equation (1) coinsides with the results of 3D CAD model considering nut chamfering, the nut chamfer geometry does not affect the contact position very largely.
2. Since bolt-nut connections have some manufacturing error, the real pitch difference can be obtained from the comparison between the experimentally obtained the contact position n_c^{Exp} and analytically obtained n_c^{Theory}. The difference can be regarded as the pitch difference error.
3. To confirm the corrected value of the pitch difference, the prevailing torque was confirmed in the experiment. By using the corrected pitch difference, the obtained T_p^{Real} relation coincides with T_p^{FEM} obtained by FEM simulation.
4. In the future, this method is used for calculating the real pitch difference of nuts and reflects the result of experiment considered manufacturing error.

References

[1] N-A. Noda, Y. Sano, Y. Takase, X. Chen, H. Maruyama, H. Wang and R. Fujisawa: Transactions of Society of Automotive Engineers of Japan, **46**, 121(2015)
[2] Hard Lock Industry KK, Looseness Preventing Nut, Patent Abstracts of Japan, Publication No.:

2002–195236, 2002-7-10

[3] M. Ishimura, H. Yamanaka, Y. Syoji, T. Kobayashi and T. Sawa: Jpn. Soc. Mech. Eng. A, **77**, 1444(2011)

[4] S. Izumi, T. Yokoyama, T. Teraoka, A. Iwasaki, S. Sakai, K. Saito, M. Nagawa and H. Noda: Trans. Jpn. Soc. Mech. Eng. A, **71**, 380(2005)

[5] S. Izumi, T. Yokoyama, A . Iwasaki, S. Sakai, Transaction of the Japan Society of Mechanical Engineers, Series A, **71**, 204(2005)

[6] Y. Akaishi, X. Chen, Y. Yu, H. Tamasaki, N-A. Noda, Y. Sano and Y. Takase: Transactions of Society of Automotive Engineers of Japan, **44**, 1111(2013)

[7] X. Chen, N-A. Noda, M.A. Wahab, Y. Akaishi, Y. Sano, Y. Takase and G. Fekete: Acta Polytechnica Hungarica, **12**, 61 (2015)

[8] N-A. Noda, X. Chen, Y. Sano, M.A. Wahab, H. Maruyama, R. Fujisawa and Y. Takase; Materials & Design, **96**, 476 (2016)

[9] X. Chen, N-A. Noda, M.A. Wahab, Y. Sano, H. Maruyama, H. Wang, R. Fujisawa and Y. Takase: Journal of the Chinese Society of Mechanical Engineers, **37**, 11 (2016)

[10] N-A. Noda, Y. Sano, Y. Takase, X. Chen, H. Maruyama, H. Wang and R. Fujisawa: Trans. Jpn. Soc. Mech. Eng., **81**, 1(2015)

[11] N-A. Noda, M. Kuhara, Y. Xiao, S. Noma, K. Saito, M. Nagawa, A. Yumoto and A. Ogasawara: Journal of Solid Mechanics and Materials Engineering, **2**, 801(2008)

IOP Conf. Series: Materials Science and Engineering **715** (2020) 012078 doi:10.1088/1757-899X/715/1/012078

The study of properties of metal rubber damper used in thruster

Fu-Wei Duan*, Yuan Tao, Cheng-Bin Ding, Ming-Fei Yu, Jin-Chuan Sun, Ming-Liang Wang and Lu Liu

Lanzhou Institute of Physics, 100# Feiyan Street, Chengguan District, Lanzhou City, Gansu Province China

*E-mail: sddfw@163.com

Abstract. Developed a new type of metal rubber damper to reduce the vibration of a certain type of thruster, researched the mechanical properties of this new damper and rubber damper with experiment. The results show that: the area of protection, the static load capacity and the energy consumption of metal rubber damper are better than rubber damper, it is useful to adjust the performance of the metal rubber damper by adjusting the density, height, pre-compression and other parameters. The study provides some reference for reducing vibration of thruster with metal rubber damper.

1. Introduction

The thruster in the project application will withstand the rigors of space environment and mechanical environment, the material physical properties of the rubber damper (narrow temperature range, high and low temperature resistance) limit, has been unable to meet the long life and high bearing capacity requirements [1]. To this end, a new type of metal rubber damper is used to replace the rubber damper. According to the application requirements of the damping effect of the thruster, this paper studied the metal rubber damper and rubber damper static and dynamic characteristics and the comparison of the theoretical which approved theoretical basis for the development of the use of the metal rubber damper.

2. Metal rubber damper installation

The metal rubber damper is composed of pressed metal rubber block with a circular cavity in the center which is arranged in the thrust ring on both sides. The screw makes some pre-compression of the metal rubber damper through the screw tightening. The installation of metal rubber damper is shown in figure 1.

From figure 1 we can see that when the thruster is subjected by axial vibration and shock, at least one damper play vibration damping function; when the thruster under radial vibration, part of the region of the two shock absorbers play a damping effect. By adjusting the bolt to control the pre-compression of the metal rubber damper and change its density and height to regulate the damping effect [2].

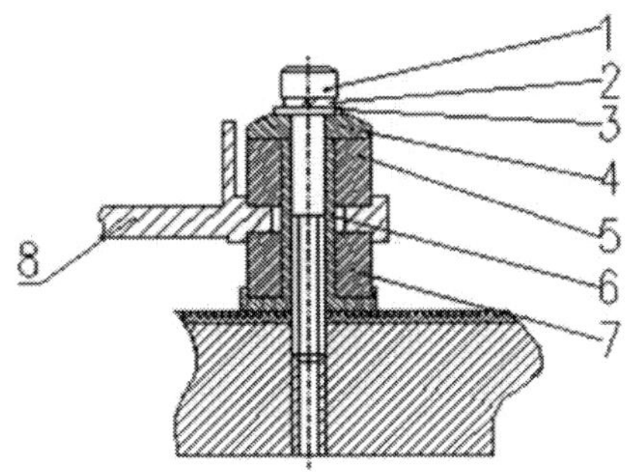

1- Screw; 2- Spring washer; 3- Flat pad; 4- Superstructure;
5,7- Metal rubber; 6- Sleeve; 8- Thrust ring.

Figure 1. Diagram of metal rubber damper installation.

3. Experimental research and comparison

3.1. Static experiment contrast

In order to analyse and compare the static characteristics of the metal rubber damper and rubber damper, the two kinds of damper are tested by axial and radial static loading. The experimental results show that the hysteresis loop of metal rubber damper is obviously better than that of rubber damper. In order to strengthen the static experiment effect, by selecting the wire material, wire diameter, density, height of metal rubber to make the limit points of metal rubber damper working range and the rubber damper working range limit point as close as possible [3, 4]. According to the practical engineering experience, the effective damping range of metal rubber damper is generally within 0.7 times of the maximum deformation. That is $A=A_{max}$(A-metal rubber damper vibration range; A_{max} -metal rubber damper maximum deformation) [5]. The vibration amplitude of the thruster vibration damper is about 1.1mm, determine the structure parameters of metal rubber damper closest to the rubber damper hysteresis loop according to the vibration amplitude: metal rubber damper density-$\rho=1.679g/cm^3$; wire diameter-$D=0.13mm$; pre compression-$Y=0.18mm$. The hysteresis loops of two types of dampers are shown in figure 2 and figure 3.

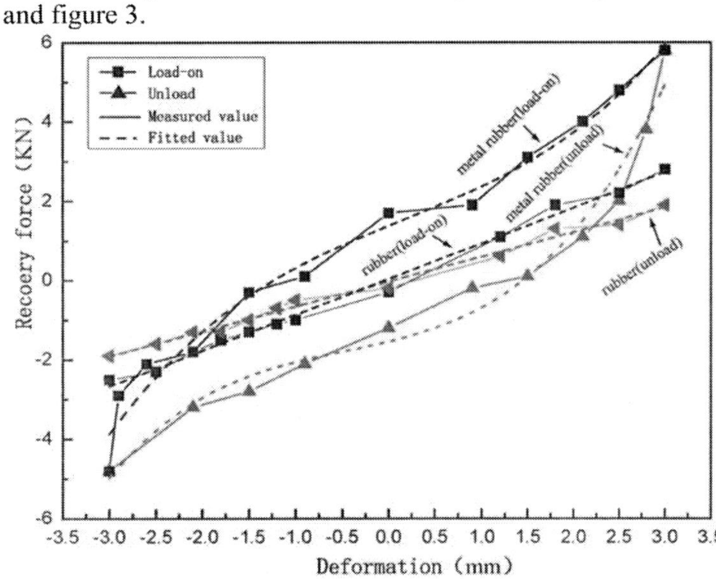

Figure 2. Axial static load hysteresis loop.

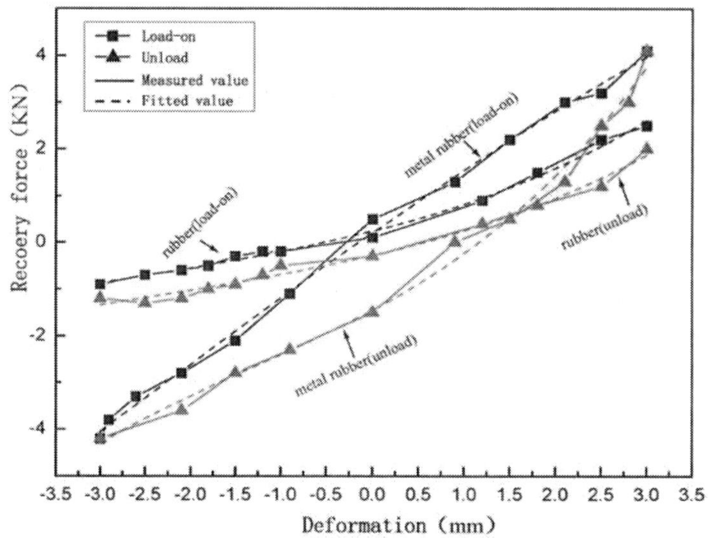

Figure 3. Radial load hysteresis loop.

According to figure 2 and figure 3: metal rubber damper static and dynamic hysteresis loops showed obvious nonlinear characteristics; and the static and dynamic hysteresis loop of rubber damper was approximately linear, and the surrounding area was less than that of metal rubber damper hysteresis loop area surrounded. In order to characterize the damper performance, the energy dissipation coefficient was introduced [1].

$$\Psi=\Delta S/W \quad (1)$$

$$W=F_{max}\cdot A_{max}/2 \quad (2)$$

Type: ΔS - hysteresis loop area (mm^2); W- the maximum deformation energy of the metal rubber damper; F_{max} - maximum recovery force (N); A_{max} - the maximum deformation of the metal rubber damper (mm).

The experimental results show that under the same static load conditions, the energy dissipation performance of the metal rubber damper is better than that of rubber damper. At the same time,the difference of damping of the metal rubber damper in axial and radial direction is less than the rubber damper. It can be found that the stiffness ratio of metal rubber damper in the axial and radial average is about 2.89 when the deformation value is 1mm; under the same conditions, the average stiffness ratio of the rubber damper in the axial and radial is about 9.766, which is in line with the performance for damping of the rubber absorber in the radial direction. When the vibration thruster is under different direction of impact, radial resonance frequency of the rubber damper is far less than the axial resonance frequency; while the resonance frequencies of the metal rubber damper in the two directions are similar. So the resonance interval of rubber damper is far greater than the resonance interval of the metal rubber damper.

Compared with figure 2 and figure 3, it can be found that the nonlinear characteristic of the hysteresis loop of metal rubber damper is very obvious in the axial static load test while the nearly linear characteristics in the radial static load test. The reasons are that the metal rubber damper volume is compressed, the density increases, the stiffness increased rapidly with axial loading, which performance is the nonlinear characteristics of hysteresis loop; with radial loading, the metal rubber damper deforms, but the size and density changes little makes the stiffness changes little, so the hysteresis loop is near linear.

3.2. Dynamic experiment contrast

3.2.1. Test apparatus

The experimental device is VS-300-2 electric drive vibration system. The system mainly consists of a vibration controller, a vibration generator, a power amplifier, a charge amplifier and a sensor. The experimental principle is shown in figure 4. During the test, the acceleration sensor 7 sends an acceleration signal to the vibration controller8, the acceleration sensor 2 and 5 collect vibration acceleration signal respectively that are on the shaker1 and object4 which is isolated by vibration damper 3.

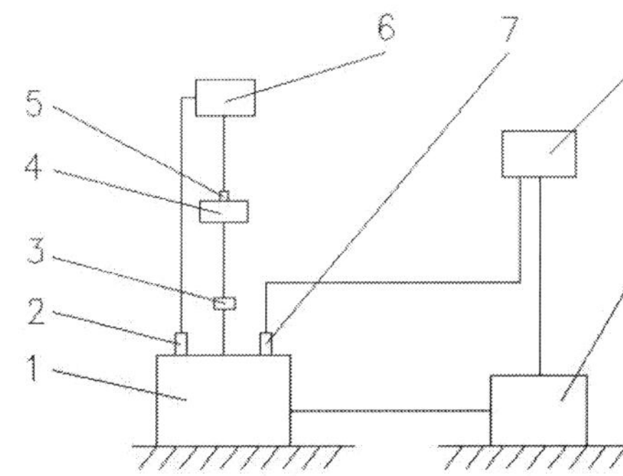

1- Shaker; 2,5,7- Acceleration sensor; 3-Damper; 4- Object isolated; 6- Charge Amplifier; 8- Vibration controller; 9- Power Amplifier.

Figure 4. Experimental schematic.

3.2.2. Dynamic experiment results analysis

In order to quantify the ability of transferring vibration of the damper, the transmission rate is introduced here. Smaller transmission rate means better damping effect. Figure 5 and figure 6 show the transmission rate curves of the two shock absorbers under dynamic experiment.

$$\eta=2\pi/\Psi \quad (3)$$

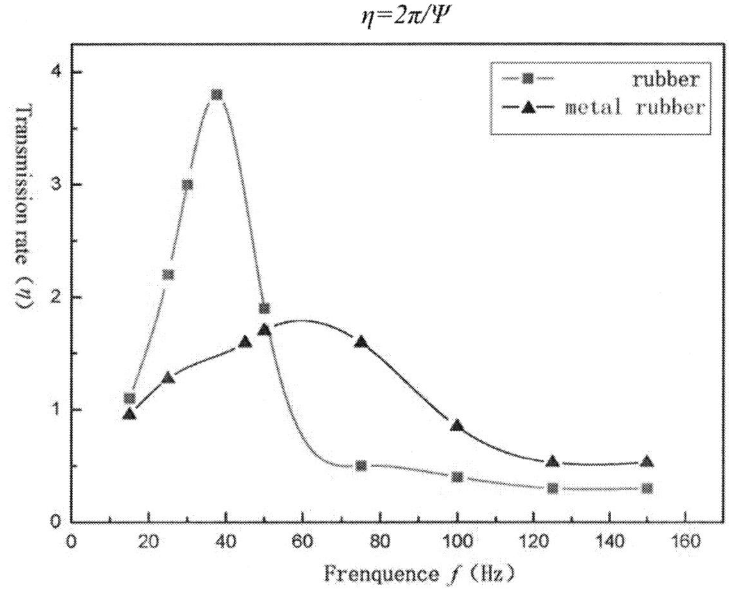

Figure 5. Transmission rate curve of axial vibration load test.

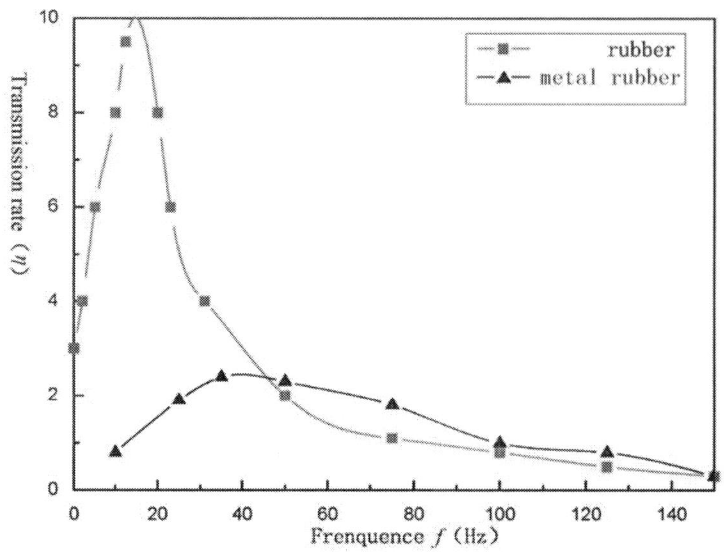

Figure 6. Transmission rate curve of radial vibration load test.

Figure 5 and figure 6 show that the transmission rate of rubber damper was significantly higher than that of metal rubber damper under the axial and radial load at the resonance frequency,the transmission rate was about 3:1 in axial vibration when the transmission rate was about 4.5:1 in radial vibration, this is because the rubber damper at almost has a damping effect on radial vibration while the metal rubber damper's damping effect is similar in the axial and radial vibration. Experimental results show that the damping effect of the metal rubber damper is better than that of the rubber shock absorber in the over resonance region, the transmission rate and resonance frequency in the vibration protection zone ($\eta < 1$) are larger than those of the rubber damper.

The literature [6] pointed out that good damping system must meet the three requirements of low frequency, small transmission rate when resonance, transmission rate decreases rapidly after passing the resonance region. Figure 5 and figure 6 shows: compared with the rubber damper, the metal rubber damper can't fully meet the three requirements. In order to know the effects on the damping performance of metal rubber damper structure parameters, vibration loading experiments were finished with two different parameters of metal rubber damper, damper transmission rate curves were shown in figure 7.

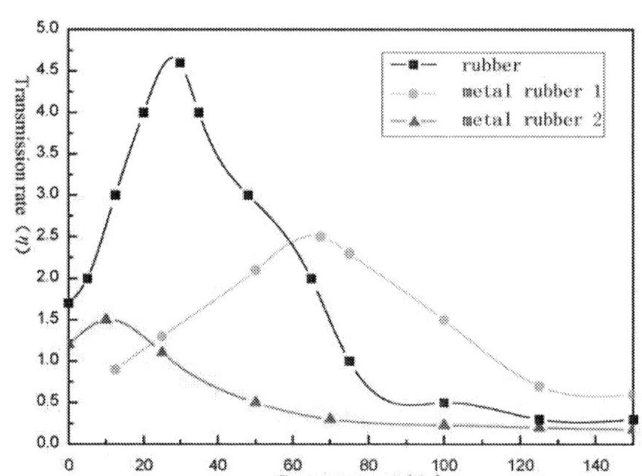

Metal rubber damper1: ρ=1.8g/cm^3,Y=0.17mm; Metal rubber damper2: ρ=1.5g/cm^3,Y=0.1mm.

Figure 7. Transmission rate curve of different structural parameters of metal rubber dampers and rubber damper load test.

It can be seen from figure 7 that the damping performance can be adjusted by adjusting the structural parameters of the metal rubber damper which can be used to achieve the ideal damping effect.

4. Conclusions

Through the experimental comparison of the metal rubber damper and rubber damper, the following 3 conclusions can be drawn.

(1)The bearing capacity, energy consumption, over critical and stability of metal rubber damper is superior to that of the rubber shock absorber.

(2)Metal rubber damper can meet the requirement of engineering application for vibration damping of thrusters.

(3)The structural parameters of metal rubber damper effect it's mechanical properties; different metal rubber dampers should be designed for different vibration isolation and vibration.

In addition, affected by the nature of itself, metal rubber damper cannot be fully restored to the original shape after impacted, so for the isolated object need precise positioning metal rubber damper should be carefully used.

References

[1] The design of metallic rubber component [M]. LI ZH Y, Translate. Beijing: National Defense Industry Press, (2000).

[2] Huang X Q, ZHang T SH,ZHang J CH. Research on vibration isolation characteristics of metal rubber [J]. Mechanical Science and Technology for Aerospace Engineering, 19(6): 977—980,(2000).

[3] Jiang H Y, Dong C F, Xia Y H, et al. Study on dynamic and static performance of metal rubber isolator of aero engine [J]. Journal of Aeronautics, 25 (6): 140 – 142, (2004).

[4] FU M G, Liu Y H, Cui M L, Cao Ming. Metal rubber damper for spacecraft [J. 21(5): 1174 – 1182, (2013).

[5] LI ZH Y, LU ZH R. Research on combined stiffness characteristic of metal rubber damper [J]. Journal of Harbin Institute of Technology, 37 (10): 1327—1329, (2005).

[6] Yan J K. Isolation techniques of mechanical vibration [M]. Shanghai: Shanghai Science Press, 282 – 289, (1986).

IOP Conf. Series: Materials Science and Engineering **715** (2020) 012079 doi:10.1088/1757-899X/715/1/012079

A recommendation of pension service based on user preferences and trust relationships

Kaiqi Zhang*, Chunshan Li and Dianhui Chu

Computer science and Technology, Harbin Institute of Technology, Weihai, China

*Email: 2931801774@qq.com

Abstract. Recommendation technology have been widely used in social networks and many e-commerce websites, but there are few recommendations of pension service and those recommendation results are not accurate enough. For more accurately recommending personalized pension services to users, this paper proposes a recommendation method of pension service based on the user preferences and the trust relationships between users. Firstly, by mining user socialization relationship information, the degree of trust relationships between users is calculated, and the Pearson correlation coefficient is used to obtain the user similarity, then their results are combined and used to find the neighbour user set of the target user. Secondly, a timeliness model of user ratings based on the user evaluation time are established, which is used to determine the user's preference for the service, and finally to predict the target user's rating value for the project. The experiment takes the Douban review data and the users social data as the case to verify the results. The experimental results show that compared with the traditional recommendation method, the recommendation method based on user preferences and trust relationships improves the accuracy of recommendation.

1. Introduction

With the rapid development of the Internet, more and more service information is available on the Internet. This makes it difficult for users to browse to the valid service information in a timely manner, and spends a lot of time in the invalid information[1]. Nowadays, recommendation systems are widely available in various e-commerce service websites, such as Amazon, CDNow, DangDang[2]. The purpose of these recommendation systems is to perceive the needs of users, provide service recommendations to users, and find the services that users really need from a large amount of information, thereby saving users' time and improving the efficiency of using information. However, the product recommendations of many websites that just recommend the current hot services directly to users, which not only ignores the user's individual needs, but also causes a large number of products to be forgotten by users. Not all the users have the same preferences as the mainstream users, but this type of product recommendation method cause that the product recommendation results obtained by all the users are roughly the same. This cannot satisfy the user's individual needs, and also loses the meaning of those recommendation results [3].

The recommendation system helps users to find projects they might be interested in from large amounts of data by mining binary relationships between users and projects, and generate personalized recommendations to meet individual needs. A good recommendation system can establish a close relationship with users. But how to accurately perceive user preferences,help users quickly find services that meet user needs in a large number of services is the focus and difficulty of current research. Technically, personalized recommendations can be based on The actual needs and interests of users meet the different needs of different users [4].

At present, the widely used personalized recommendation algorithm is user-based collaborative filtering algorithm[5]. In this algorithm, a set of users are selected by calculating the similarity between the target users and other users, and then according to the history information of the service that this set of users are interested in, the recommendation results are provided for the target users. However, the typical user-based collaborative filtering algorithm does not accurately perceive the user's true preferences. This is because it does not take into account the social relationships and trust relationships between users, the effect of the timeliness of user ratings on the accuracy of recommendation results is not considered. So the final personalized recommendation results will be biased. Therefore, in order to solve the above problems, this paper proposes to explore the trust relationship between users through social networks first, and then calculate the final user preferences based on users ratings and their time stamps, so as to recommend more accurate personalized service information for target users.

2. Calculate user similarity

2.1. Pearson correlation coefficient

There are many methods to calculate user similarity, common methods include cosine similarity, Pearson coefficient and adjusted cosine similarity. The cosine similarity is to calculate the angle between two vectors, but there is no decentralization, which requires a certain value on every dimension of the vector.However, not every two users will rate the same project, so some dimensions are empty and cannot be calculated, the cosine similarity does not take into account the factor of different rating standards among different users. The difference between the Pearson correlation coefficient and the adjusted cosine similarity is that the decentralization method is different. The Pearson correlation coefficient considers the average of each item's score by the user, while the adjusted cosine considers the average value of each user who rates the item. This paper completes the user similarity calculation by combining the Pearson correlation coefficient, social and trust relationships among users. Calculate the similarity between user u and user v using Pearson correlation coefficient, and calculate the formula as follows(1):

$$sim'(u,v)=\frac{\sum_{i\in I_u\cap I_v}(r_{u,i}-\overline{r_u})(r_{v,i}-\overline{r_v})}{\sqrt{\sum_{i\in I_u\cap I_v}(r_{u,i}-\overline{r_u})^2}\sqrt{\sum_{i\in I_u\cap I_v}(r_{v,i}-\overline{r_v})^2}} \tag{1}$$

where $r_{u,i}$ represents the rating of user u for item i, $\overline{r_u}$ represents the average rating of user u, $r_{v,i}$ represents the rating of user v for item i, and $\overline{r_v}$ represents the average rating of user v, $I_u \cap I_v$ represents a service that user u and user v jointly comment on.

2.2. Calculation of trust between users

In the recommendation system, the traditional recommendation method only considers the similarity of users, ignoring the influence of the association between the recommended objects on the recommendation results in many application scenarios [6]. The degree of trust between users is related to the interaction experience,when the experience between two users is positive, the degree of trust between users will be increase; on the contrary, the degree of trust declines [7, 8]. So under the same hobbies as the two users, the user prefers the service recommended by the user who has a social relationship with himself. The recommendation to integrate social information has become a research hotspot in the recommendation field [9]. In the pension system, it is necessary to fully consider the elderly with similar interests and social relationships, they would have a relatively fixed social circles, information interaction ability is strong, they trust each other, so add credibility in similarity calculation, the recommended results will be more accurate.

Trust between users is divided into direct trust and indirect trust. According to the user's social information, direct attention and interactive behavior is called direct trust. As shown in figure 1(a), A has a direct trust relationship with B. In social information, users have indirect trust with friends of people they following, as shown in figure 1(b), there is an indirect trust relationship between A and C.

Because the multi-layer indirect trust relationship is not reliable, so this paper only considers the two-layer indirect trust relationship. The direct trust of user u to user v is expressed as $t^1(u,v)$, the indirect trust of user u to user v is expressed as $t^2(u,v)$. The value of the trust is directly related to the number of interactions between users. The interaction behavior includes comments, likes and forwarding. Such data shows that users are more interested in a pension institution or a certain aged care service. By collecting and excavating such information in the pension service platform, it is more accurate to perceive the preferences of the elderly in order to recommend the pension institutions and service types that the elderly need. The formula for calculating the direct trust of user u to user v is as follows (2):

$$t^1(u,v)=\frac{\sum C_{u,v}\cup L_{u,v}\cup F_{u,v}}{N_u} \tag{2}$$

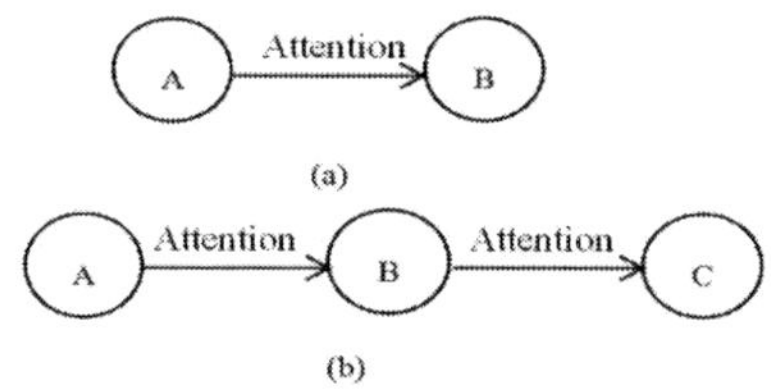

Figure 1. Direct trust and indirect trust relationship.

where $C_{u,v}$, $L_{u,v}$ and $F_{u,v}$ respectively indicate user u to comments, likes, and number of forwards to the user v, and N_u represents the total number of user u interaction information. The calculation formula for the indirect trust of user u to user v is as follows (3):

$$t^2(u,v)=\frac{\sum C_{u,v}\cup L_{u,v}\cup F_{u,v}}{N_u}\times\frac{\sum C_{h,v}\cup L_{h,v}\cup F_{h,v}}{N_h} \tag{3}$$

where h represents the intermediate user of user u and v, the user who is concerned by user u and who is pay attention to the user v. This paper considers that the similarity between user u and user v is expressed as follows (4):

$$sim(u,v)=\begin{cases} sim'(u,v)+t^1(u,v), \text{direct trust relationship} \\ sim'(u,v)+t^2(u,v), \text{indirect trust relationship} \\ sim'(u,v), \text{there is no trust relationship} \end{cases} \tag{4}$$

From this, the number of K users with the highest similarity to the user u are calculated as their adjacent user sets, which are denoted as S(u).

3. Forecast user rating

3.1. Establish the timeliness model of user rating

In the recommendation system, the user's preferences are constantly changing over time, and the similar types of recommendations over a period of time can reduce the user's satisfaction, so the recommendations should be made according to the dynamic user's preferences. In this section, it is believed that there is a non-linear diminishing trend of user preference for services over time. It is possible to recommend more accurate services for the elderly according to their dynamic user's preferences. The longer the time bucket from that user finished service evaluation to the current time, the lower the interest, and for the services that users have recently evaluated, they are more interested. So, the timeliness model of user rating should be established. However, the user's preference for a service is not directly proportional to the satisfaction after using the service. Some users are interested in a service, but after using it, the service experience is not good, so in this paper, the timeliness of user rating and the user rating are combined to calculate the user's preference value for the service as a formula(5):

$$P_{u,i} = \exp\{\frac{-time(u,i)}{time_i}\} \times r_{u,i} \tag{5}$$

In the pension service recommendation, time(u,i) indicates the timeliness of user rating of that the user u evaluate the service i, when the service evaluation time is in the most recent cycle, the value of time(u,i) is 0, when the service evaluation time is in the penultimate period, its value is 1 and so on. From the formula, it can be concluded that the closer the user evaluation time is to the current time, the smaller the time(u,i) value is, and the value of $\exp\{\frac{-time(u,i)}{time_i}\}$ approaches 1. $time_i$ indicates the user evaluation period. The shorter the user evaluation period, the faster the interest declines. Therefore, the newer evaluation, the higher the preference value. $r_{u,i}$ indicates the actual rating of user u for service i.

3.2. Predict user ratings for services

Through the steps described above, firstly we calculate K users that are most similar to the target user u, and calculate the degree of preference of these users to the target service, thereby the rating value of the target user u for the service is predicted, and finally the platform will recommend set of services with the top rating value for the user. The predicted scores of the target user u for the services can be calculated by the following (6):

$$S(u,i) = \frac{\sum_{v \in S(u)} P_{u,i} \times sim(u,v)}{\left|\sum_{v \in S(u)} sim(u,v)\right|} \tag{6}$$

4. Experimental results and analysis

This paper introduces the data used in the experiment, explains the test method and evaluation criteria, and gives the experimental results based on user trust and user evaluation time recommendation algorithm, and compares it with the user-based collaborative filtering recommendation algorithm.

4.1. Experimental data set

In this paper, the experimental data sources are the user data and movie evaluation data of Douban. The original data contains about 80,000 ratings of 986 users, and about 50,000 social behavior data of these users. All the scores of the movie are 1 to 5, and the higher the user evaluation score, the more the user likes it. In the movie data,the movie data attributes are mapped to the aged service data. The evaluation time of the movie indicates the user's evaluation time of the aged care service in the pension service platform, the movie user score indicates the old person's score on the aged care service, and the movie user social information indicates the social information between the elderly.

4.2. Experimental evaluation criteria

By collecting the user's evaluation data one week ago, by referring to the user's recommendation service, according to the user's next week's score record, it is judged whether the recommendation data is accurate, and the percentage of the recommended service in the service used by the user. In this paper, the average absolute error MAE is used to evaluate the prediction accuracy of the recommendation algorithm. The average absolute error of the user u is calculated as a formula(7):

$$MAE_u = \frac{\sum_{i=1}^{n} |P_{u,i} - R_{u,i}|}{n} \tag{7}$$

where n represents the number of recommended services that the user u has used and scored, and $P_{u,i}$ and $R_{u,i}$ represent the predicted score value and the actual score value of the user u for the service i.

IOP Conf. Series: Materials Science and Engineering **715** (2020) 012079 doi:10.1088/1757-899X/715/1/012079

4.3. Experiment process

The experimental code is implemented in Java and Python. The experiment process is as follows: First, the crawled data is processed, and each user evaluates the movie with no less than 20 data, and deletes illegal characters and sensitive information such as personal information. Secondly, the Pearson similarity between users is calculated based on the obtained score data, the social information of the user is found, and the degree of trust between the users is calculated. Third, the number of adjacent user sets is 10, 20, 30, 40, and 50, calculate the MAE changes when selecting different numbers of similar users. Fourth, the user evaluation period $time_i$ is selected as 10, indicating the length of the period in which the user evaluates the service in the pension service platform. Finally, the preference is calculated based on the evaluation time of the selected K similar users for the target service, predicting the predicted score value of the target user for the target service. The experimental results show that when K is 50, the experimental results are the best.

4.4. Experimental result

In this experiment, the user-based collaborative filtering algorithm is compared with the proposed algorithm. The user history evaluation data is used to predict the user's rating value for the service next week, and the recommendation accuracy of the two algorithms is judged according to the value of MAE. The experimental comparison data is shown in figure 2. Observing the two polylines in the figure 2 as shown in the figure 2 can be concluded that the recommended accuracy of the proposed algorithm is higher, and the larger the value of K, the smaller the value of MAE, the more accurate the recommendation.

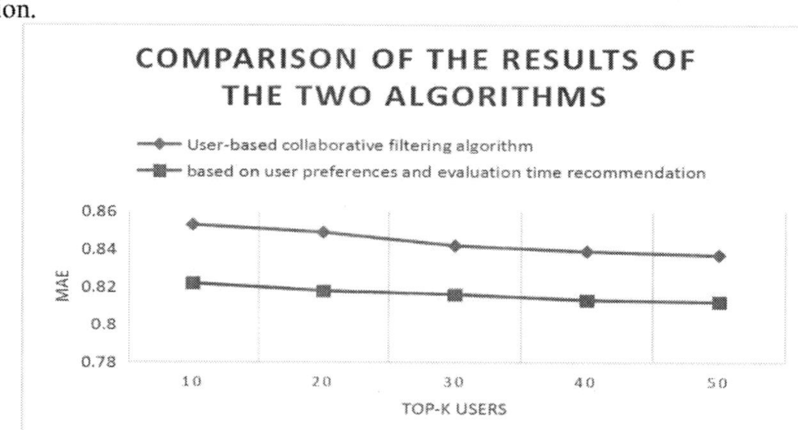

Figure 2. Comparison of the results of the two algorithms.

5. Conclusion

In this paper, the trust degree between users is calculated according to the social information of users, and the trust degree and the evaluation score are used as indicators for calculating the similarity of users. And we establish the timeliness model of user rating to calculate user preferences, and the scenario of that the user's preference for the service changes with time is fully considered, thereby a more accurate service is recommended for the user. It is proved by experiments that this method can obtain more accurate recommendation results compared with the traditional recommendation algorithm.

Acknowledgments

This work is supported in part by the National Natural Science Foundation of China(No.61772159), N atural Science Foundation of Shandong Province (No. ZR2017MF026), HIT Innovation Fund Proje ct(Grant No.HIT.NSRIF.201703), the Collaborative Innovation Projects for National Regional Center (Grant No.2017QYCX12).

References

[1] X.Y. Huang, R. Lu, Z.H. Wang, Z.F. Zhu, ICNC, 591-596 (2017)
[2] P. Wang, H.W. Ye, ICIIS, 152-154 (2009)
[3] R. Zhu, H.M. Wang, D.W. Feng, Journal of Software, 22, 852-864 (2011)
[4] G.J. Zou, J.S. Wang, H.L. Yuan, ICIEA, 1643-48 (2018)
[5] X. Wu, B. Cheng, J. Chen, IEEE Transactions on Services Computing, 1 (2017)
[6] L. Duan, H, Tian, Future Internet, 9, 63 (2017)
[7] W. Sherchan, S. Nepal, C,Paris, CSUR, 45, 47 (2013)
[8] X. Xia, J. Yu, S. Zhang , Security & Communication Networks, 9, 1-9 (2017)
[9] H.F. Liu, L.P. Jing, J. Yu, Journal of Software,29, 340-362 (2017)

IOP Conf. Series: Materials Science and Engineering **715** (2020) 012080 doi:10.1088/1757-899X/715/1/012080

Experimental study of effects of type of injectors on the performance of gas generator of liquid rocket engine

Taj Wali Khan* **and Ihtzaz Qamar**

Institute of Space Technology, Islamabad, Pakistan

*E-mail: tajwalikhan1@yahoo.com

Abstract. The role of the injector is atomization and mixing. It dictates the size and geometry of the combustion chamber and gas generator. Experiments are conducted with water-cooled adjustable length gas generator for a liquid rocket engine with two types of injectors at a lower mixing ratio. Tests are conducted at characteristic length of 1.75 and 4.27 m. Liquid oxygen and kerosene are used as propellants. Single pressure swirl coaxial injector and three like doublet impinging injectors are used. Combustion performance is studied and compared on the base of characteristic velocity. Characteristic velocity is found high at short length but is found low at larger length for pressure swirl injector as compare to impinging injectors.

1. Introduction

Atomization and mixing are two main paramétérs based on which type of injectors are selected and distributions of injectors are made in a liquid rocket engine. Various types of injectors and patterns have been used in past in different engines [1–3]. Popular injectors used in rocket engines are impinging injectors, pressure swirl injectors, and pintle injectors. Pressure swirl injectors are more widely used due to its performance over large range flow rates [4, 5]. Impinging injectors are simple in construction and can be used for higher oxidizer to fuel mixing ratio effectively [6]. Poor atomization with bigger droplet size requires more time in the chamber for complete evaporation and burning. The exit of unburned fuel droplet decreases the efficiency of combustion. The more characteristic length of the chamber is required for bigger droplets as compared to the smaller droplet.

In the gas generator cycle liquid rocket engine, the gas generator is used to produce gas to drive the turbine, which produces the power to run fuel and oxidizer pumps. The power of the turbine is dependent on the energy available in the gas and mass of the gas. The energy of the gas is a function of temperature and pressure of gas generator. More gas is required if the pressure and temperature of the gas are low. The temperature of the gas is limited by the material of the turbine [7]. If the performance of the gas generator is low then it uses more propellant to produce required pressure at the desired temperature. The use of extra propellant will decrease the overall specific impulse of the system, resulting in extra weight and volume. An increase in the weight is afforded at the cost of the payload of the vehicle. An increase of one kg of weight of system causes a reduction of one kg of payload if the system is the last stage of the vehicle. A comparative study is always required to select a proper type of injector and configuration for the desired project.

2. Experimental setup

The experimental setup used in the present experiments consists of portable test stand, high-pressure nitrogen bottles, manifold, pressure regulators, pressure sensors, k type thermocouple, PT100 thermocouple, adjustable length gas generator with replaceable head, two heads and data acquisition system.

Figure 1 shows the test stand which was used in these experiments for testing. The design of the test stand was such that it could test a small gas generator in horizontal position. The test stand had two tanks one for liquid oxygen and another for kerosene. Tank for liquid oxygen was a double wall with vacuum insulation. There were five switching valves, three for the on-off purpose of liquid oxygen, kerosene, purging supply to the gas generator and two for the liquid oxygen tank and kerosene tank for emergency evacuation. Two ball valves were used for the fuel tank, one at the filling point and the other in the supply line. One ball valve was used at the liquid oxygen filling point. Pressure sensors were used to measure pressure at injection and in the chamber.

Three high-pressure nitrogen bottles were used to pressurize the fuel tank. The flow rate of oxidizer was less in this case so the oxidizer tank was pressurized with one high-pressure bottle.

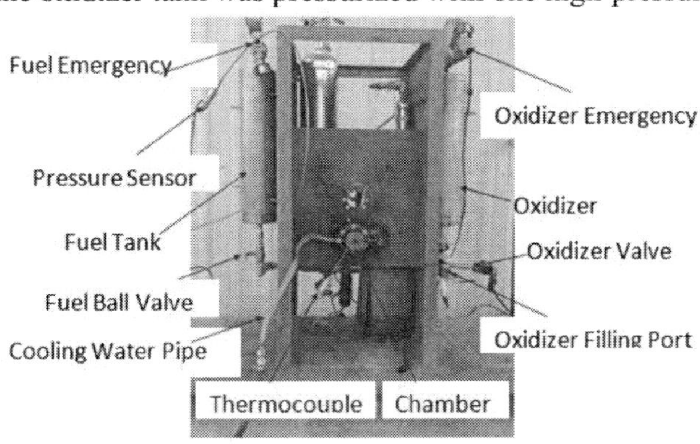

Figure 1. Test stand.

Figure 2 shows the bank of high-pressure bottles. The outlet of all three bottles combines in a manifold and a single outlet pipe from manifold passes through pressure regulator into the fuel tank. Figure 3 shows a single injector gas generator. This gas generator was water cool and the characteristic length was adjustable. Characteristic length was adjusted as 1.75 m and 4.27 m for different tests. Figure 3 shows, pressure ports, spark plug, K type thermocouple connection and water flow in and out connections. Every two parts of the gas generator were bolted together with the help of M10 bolts. Two heads were used in these tests one with pressure swirl injector and other with like doublet impinging injector. Spark plug was used for ignition in the gas generator.

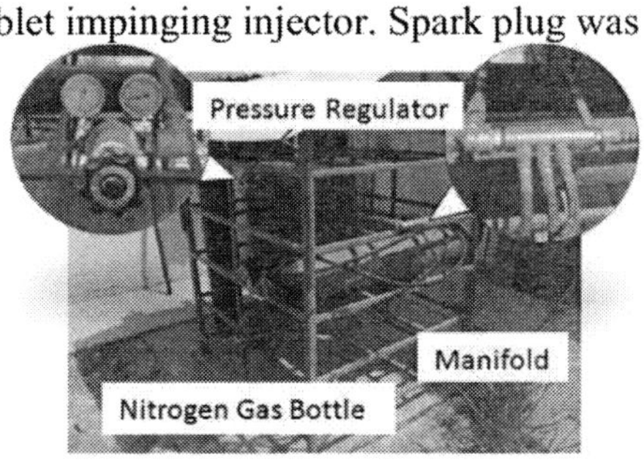

Figure 2. High pressure bottles bank.

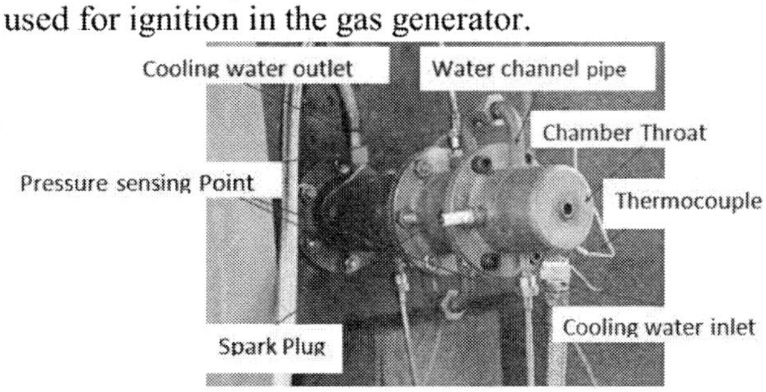

Figure 3. Adjustable length gas generator fitted for testing.

Figure 4 shows a pressure swirl injector fitted in replaceable head and detail of pressure swirl injector used in this gas generator respectively. Details of the impinging injector used in one head are shown in figure 5. Three doublet injectors one for oxidizer in middle and two for fuels at ends are prepared in a single piece which is fitted in place of pressure swirl injector.

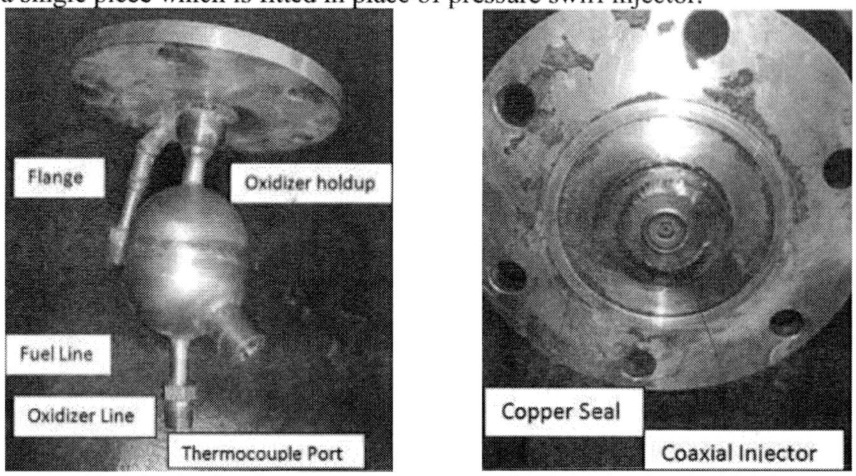

(a). Injection head ports (b). Injection head with injector

Figure 4. Replaceable injection head with pressure swirl injector.

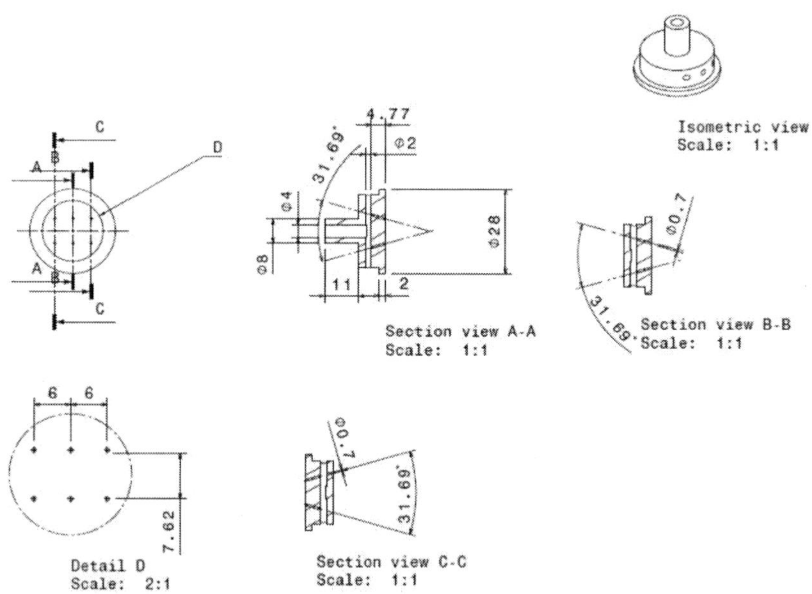

Figure 5. Detail of impinging injectors in a single piece.

3. Experimental procedures

The flow rates of the injectors were measured experimentally at the pressure of 6, 7 and 8 bar for both types of injectors and both paths with water. A plot shown in figure 6 and 7 is produced based on Eq.-1 to simulate flow rate of oxidizer and fuel at different pressures for pressure swirl and impinging injector respectively. Experimentally measured flow rates of water were used as a base to calculate the flow rate of kerosene and liquid oxygen. The discharge coefficient of pressure swirl atomizer has no or very less dependency on pressure drop and viscosity [8]. The discharge coefficient is independent of Reynolds number for Reynolds number of more than 4000, which was more in this case [9]. Pressure swirl injector used for fluid having low viscosity has a constant discharge coefficient [2]. Therefore results of Eq-1 were reliable where the discharge coefficient had been taken constant and the effect of viscosity on flow rate had been neglected. The outer passage was used for fuel and inner passage for oxidizer in the pressure swirl injector.

$$\dot{m} = water\ flow\ rate x \sqrt{\frac{fuel or\ LOX\ density}{density\ of\ water}} x \sqrt{\frac{new\ pressure}{pressure\ of\ water\ test}} \quad (1)$$

where $\dot{m}$ is mass flow rate in Eq-1.

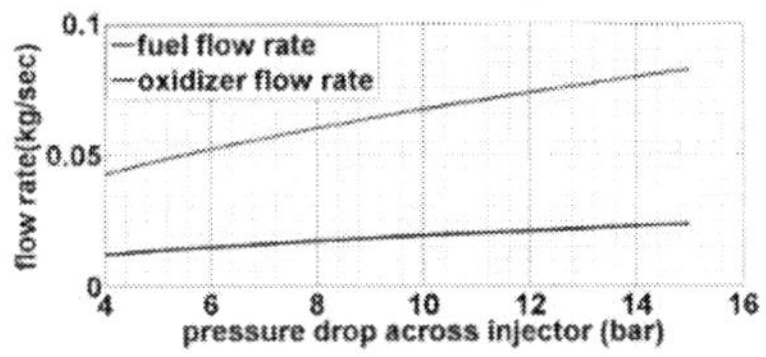

Figure 6. Flow rate variation with pressure drop for pressure swirl injector.

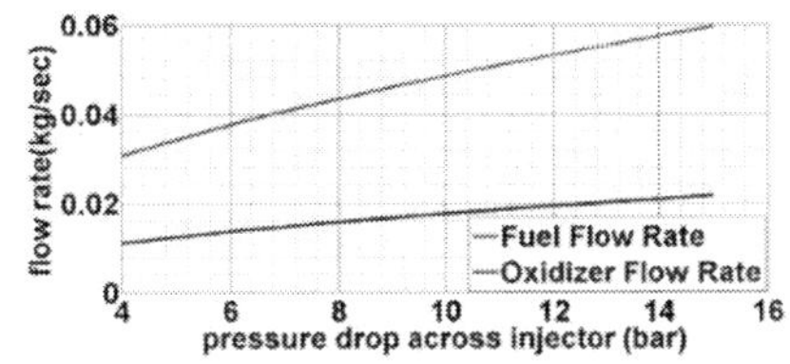

Figure 7. Flow rate variation with pressure drop for doublet impinging injectors.

The schematic diagram of the procedure is shown in figure 8. To save time liquid oxygen tank was time optimized chilled with opened vent and filled with liquid oxygen from the bottom side [10,11]. The vent was closed and the supply tank was disconnected. Liquid oxygen was passed from the oxygen tank through the injector into the chamber. The temperature of the liquid oxygen in the dome was noted and was kept under observation. The thermocouple used for this purpose was Pt100. The chilling process was continued till the temperature of the liquid oxygen reached -155 °C. Oxygen is in the liquid phase at -140 °C and 14 bar pressure [12]. Both liquid oxygen and kerosene valves were opened simultaneously. The spark plug was triggered to start ignition. At shutoff, both propellant valves were closed, the spark was off and the purging valve was open to make a smooth end.

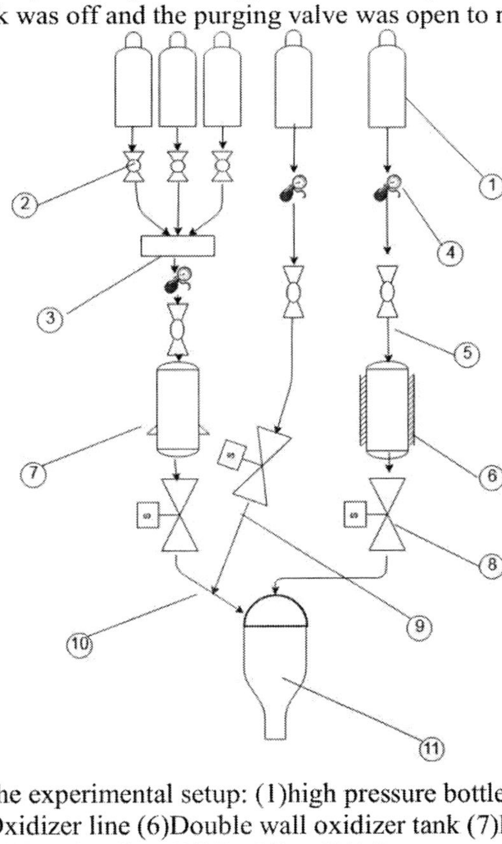

Figure 8. Flow chart of the experimental setup: (1)high pressure bottle (2)Ball valve (3)Manifold (4)Pressure regulator (5)Oxidizer line (6)Double wall oxidizer tank (7)Fuel tank (8)Solenoid valve (9)Purging line (10)Fuel line (11)Gas generator.

4. Results and discussion

Initial three important pressures are plotted in figure 9. Pressure drop for an initial fraction of second is low at the start of the test. As tanks pressures were regulated at a required pressure so pressure drops for both injectors were high due to low chamber pressure at the start. There was an increase in the mixing ratio from 0.31 to 0.37 as shown in figure 10. Drop in the mass flow rate was compensated by an increase in the mixing ratio to maintain a steady chamber pressure of 6.3 bars. There was drop in mass flow rate from 0.093 to 0.087 Kg/sec in the steady period of the test. Figure 9 and 10 are showing results of the test at a characteristic length of 1.75 m.

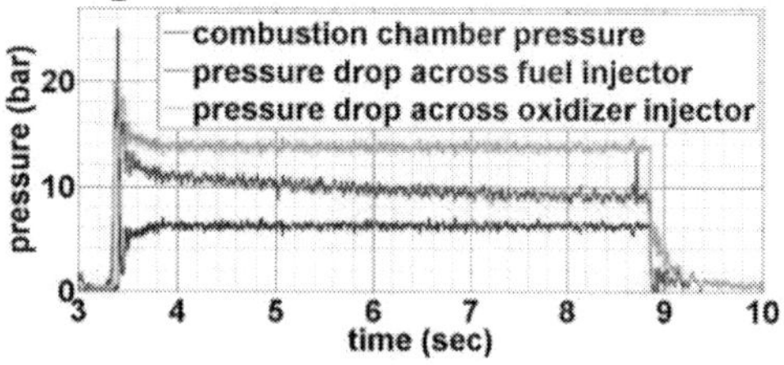

Figure 9. Chamber pressures with pressure swirl injector at characteristic length of 1.75 m.

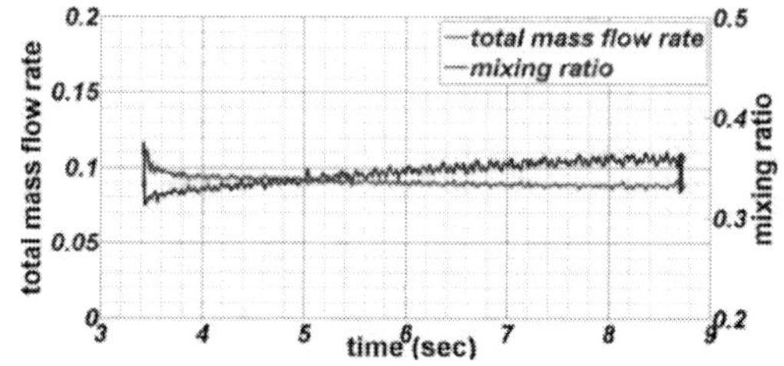

Figure 10. Mass flow rate and mixing ratio with pressure swirl injector at characteristic length of 1.75 m.

Figure 11 and 12 show results of the gas generator at the characteristic length of 4.27 m. Even at a lower mixing ratio as compared to the previous test with a characteristic length of 1.75 m, there was an increase in chamber pressure with an increase in the characteristic length of the chamber.

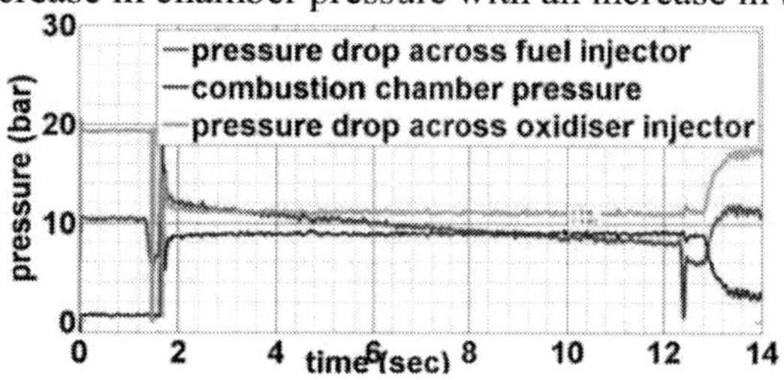

Figure 11. Chamber pressures with pressure swirl injector at characteristic length of 4.27 m.

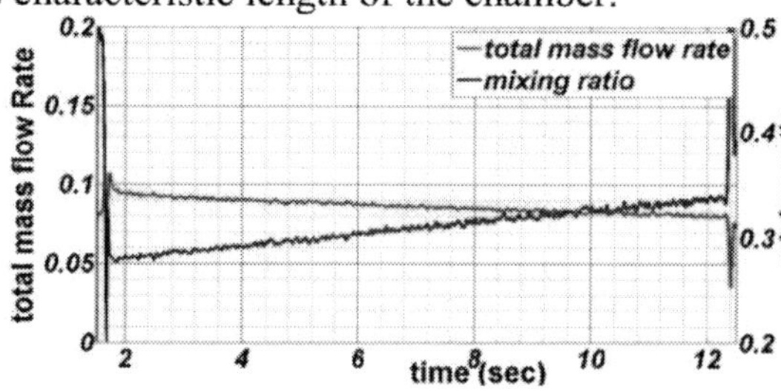

Figure 12. Mass flow rate and mixing ratio with pressure swirl injector at characteristic length of 4.27 m.

Combustion chamber pressure shown in figure 13 for the gas generator with the same characteristic length as shown in figure 9 is less. The mixing ratio as shown in figure 14 is less as compared to the gas generator with pressure swirl injector and same characteristic length.

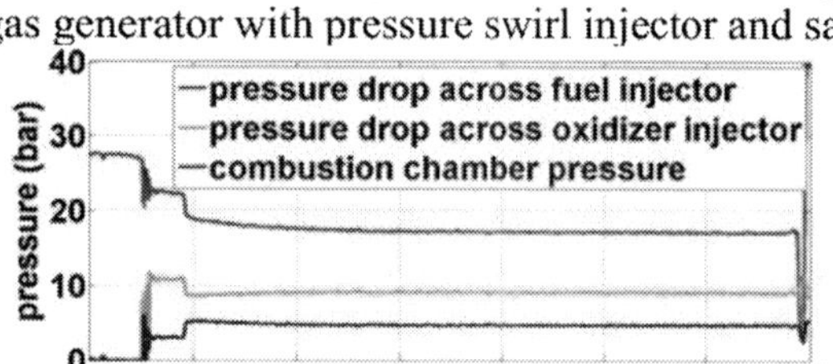

Figure 13. Chamber pressure with impinging injector at characteristic length of 1.75 m.

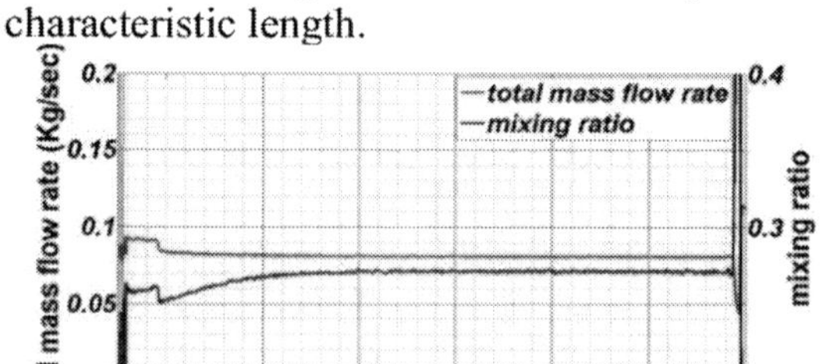

Figure 14. Mass flow rate and mixing Ratio with impinging injector at characteristic length of 1.75 m.

Figure 15 and 16 show the results of the gas generator with impinging injectors and a characteristic length of 4.27 m. Due to the pressure drop in the fuel tank, there was a decrease in fuel flow rate but an increase in the mixing ratio. Chamber pressure was stable between 7.4-7.2 bars.

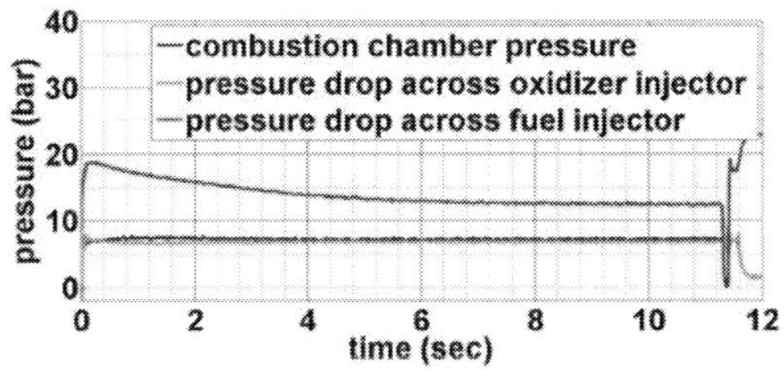

Figure 15. Chamber pressures with impinging injector at characteristic length of 4.27 m.

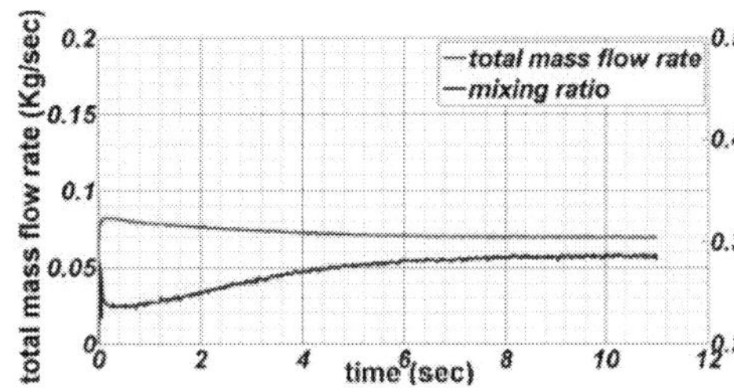

Figure 16. Mass flow rate and rixing ratio with impinging injectors at characteristic length of 4.27 m.

Figure 17 and 18 shows a comparison of characteristic velocity at characteristic lengths of 1.75 and 4.7 m respectively. Characteristic velocity is a measure of the performance of the gas generator and is given by the following equation as [3].

$$C^* = \frac{P_C A_T}{\dot{m}} \tag{2}$$

It is clear from these two figures that the performance of the gas generator with a pressure swirl injector was good at low characteristic length but was low in performance with a higher characteristic length of 4.7 m even mixing ratio was high as compared. The reason for the low performance of the impinging injector at lower characteristic length is the low mixing ratio. At low mixing ratio, the temperature of the chamber is low and fuel droplets require more time in the combustion chamber for complete burning. Moreover, droplet size may be more in the case of the impinging injector.

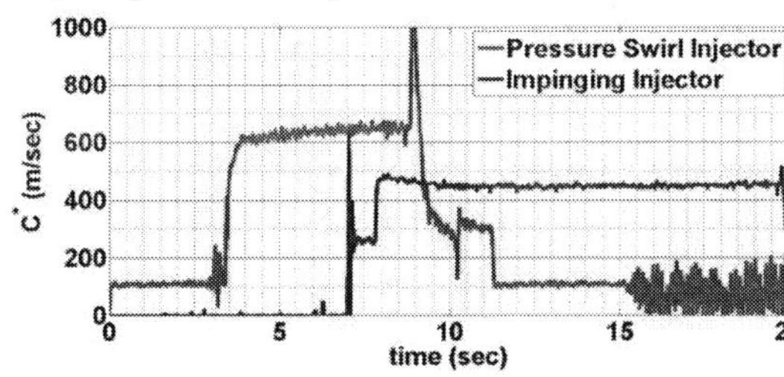

Figure 17. Characteristic velocity of gas generator with characteristic length of 1.75 m.

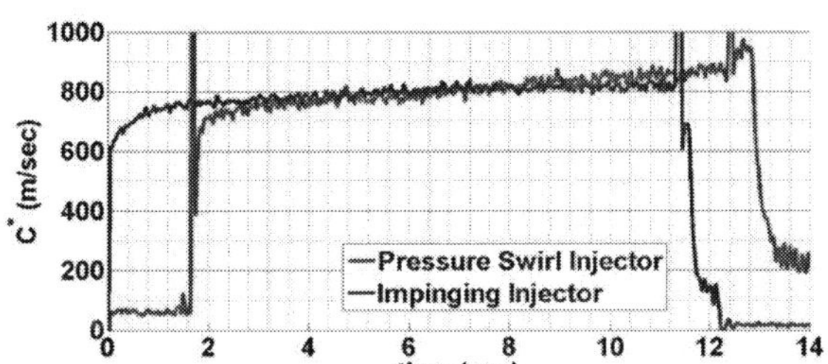

Figure 18. Characteristic velocity of gas generator with characteristic length of 4.27 m.

5. Conclusion

Tests were conducted to test the performance of the gas generator at two characteristic lengths of 1.74 and 4.27 m with two different injection head. Water cool gas generator with liquid oxygen and kerosene as fuel were used as propellants. Pressure swirl coaxial injector and impinging injectors were used. The performance of the pressure swirl injector at lower length was found better but the performance of the impinging injector was better with larger characteristic length.

Acknowledgement

We are thankful for the team of Propulsion and Engineering Research laboratory at Institute of Space Technology for their full support.

References

[1] D.H. Huang, D.K., Huzel, Design of Thrust Chambers and Other Combustion Devices, in: Modern Engineering for Design of Liquid-Propellant Rocket Engines, American Institute of Aeronautics and Astronautics,(1992)
[2] A.H. Lefebvre, Atomization and Sprays, Taylor and Francis(1989)
[3] G. Sutton, O. Biblarz, Rocket Propulsion Elements, 7th ed., John Wiley \& Sons, Canada(2001).
[4] X. Wei, G. Zhengyan, C. Pimin, Effect of Geometric Parameters on Mean Drop Sizes From Dual-Orifice Pressure Nozzles, Journal of Engineering for Gas Turbines and Power (2018)
[5] X. Wei, H. Yong, Improved Semiempirical Correlation to Predict Sauter, Journal of Propulsion and Power (2014)
[6] B. Huang, T. Yuan, Y.-T. Chen, Y.-H. Su, A Cold-flow Experimental Observation of the Two-stage Impinging Type Injector for Rocket Propulsion, in: 51st AIAA,Joint Propulsion Conference, Orlando (2015)
[7] M. Son, J. Koo, W.K. Cho, E.S. Lee, Conceptual design for a kerosene fuel-rich gas-generator of a turbopump-fed liquid rocket engine, Journal of Thermal Science. 21 (2012)
[8] J. Ballester, Discharge coefficient and spray angle measurement for small pressure swirl nozzles, (1994).
[9] K.U. Reddy, D.P. Mishra, Studies on Spray Behavior of a Pressure Swirl Atomizer in Transition Regime, Journal of Propulsion and Power. 24 (2008)
[10] Z. Kang, L. Yanzhong, X. Mengjian, W. Lei, Investigation of the chill-down behavior and thermal stress distribution of a cryogenic tank during the filling process, in: Physics Procedia, Elsevier B.V., (2015)
[11] K.A. Keefer, J. Hartwig, Development and validation of an analytical charge – hold – vent model for cryogenic tank chilldown, International Journal of Heat and Mass Transfer. 101 (2016)
[12] M.O.M. and D.G.F. E.W. Lemmon, "Thermophysical Properties of Fluid Systems" in NIST Chemistry WebBook, NIST Standard Reference Database Number 69, Eds. P.J. Linstrom and W.G. Mallard, National Institute of Standards and Technology, Gaithersbu, (n.d.).

IOP Conf. Series: Materials Science and Engineering **715** (2020) 012081 doi:10.1088/1757-899X/715/1/012081

Immediate measurement of fuel characteristics of bio-char using NIRS

Junjie Xue[1,2], Xiaoying Hu[1], Hui Liang[1], Xiaodong Wang[2], Haiyang LV[2], Zengling Yang[3] and Changqing Dong[1,*]

[1]National Engineering Laboratory for Biomass Power Generation Equipment, North China Electric Power University, Beijing 102206, China
[2]National energy biological power generation group corporation LTD, Beijing 100052, China
[3]College of Engineering, China Agriculture University, Beijing 100083, China

*Email: cqdong1@163.com

Abstract. Low temperature pyrolysis is effective to improve the fuel quality of wheat straw and corn stover. To determine the pyrolysis condition and fully utilizing the solid product (bio-char), the fuel characteristics of the bio-char should be measured. It costs a lot of time and money to finish all the measurement, which really retards the development of the biomass industry. This study is supposed to immediately predict the proximate analysis parameters, ultimate analysis parameters and heat value of sixty-five bio-char using NIRS. The results show the fuel characteristics of the bio-char vary in a very big range. The NIRS models can predict the VM, FC, ash, HHV, oxygen, carbon, hydrogen contents of the bio-char accurately. The Rc and Rcv^2 value are all bigger than 0.92 and the RPD are bigger than 3.59. The NIRS models can also tell the high or low content of moisture and nitrogen. Only for the sulphur, more work need to do to improve the accuracy of the NIRS models. Predicting the fuel characteristic of the bio-char only by scanning the NIR spectrum, which costs only 10 seconds, will really be helpful for the biomass industry.

1. Introduction

Biomass is with high moisture content, low energy density and high oxygen content, which makes it more difficult and expensive to use than fossil fuel [1-2]. Thermochemical treatment can improve the quality of biomass and make it a better solid fuel. Du et al reported the fuel properties of biomass after thermochemical treatment are even better than the coal with high volatile [3]. The solid product after thermochemical treatment is called bio-char, which is a kind of enrichment of carbon atoms. What's more, the fuel characteristics of bio-char is related to the thermochemical operating conditions [1]. To determine the proper operating conditions, the fuel characteristics, such as industrial analysis, element analysis, heat value of the bio-char are needed. It costs a lot time and money to get the data using traditional methods. Along with the thermochemical treatment, the bonds- C-H, O-H, C-O changed [2]. The information of C-H, O-H, C-O bonds have absorption in the near infrared (NIR) spectra. What's more, the NIR spectra contains more chemical and physical information of the sample [4], which makes is potential to realize the prediction of the fuel characteristics of bio-char using NIRS.

In this study, the fuel characteristics of sixty-five bio-char from different thermochemical treatment - thirteen different final temperature (from 200 to 500 °C, 25 °C temperature of a point), two different heating rate (10 K/min and 20 K/min) and two nitrogen purging rate (100 ml/min and 200 ml/min)- were predicted using NIRS.

2. Materials and methods

2.1. Biomass samples

Wheat straw (WS) and corn stover (CS) were used in this study, which were collected from China's Shandong province in 2013. After collection, the sample were milled to pass the 250μm sieve. The chemical composition, industrial composition, element composition and heat value of the sample were shown as table 1.

Table 1. Fundamental parameters of samples.

Sample/%	Cellulose	Semi-cellulose	Lignin	WSC	Moisture	VM	Ash
WS	41.49	20.54	19.14	0.98	3.35	70.75	7.49
CS	37.41	14.98	21.58	6.55	2.60	74.19	6.04
Sample/%	FC	C	H	N	S	O	Heat Value (MJ/Kg)
WS	18.41	43.50	5.81	0.63	0.48	42.10	17.81
CS	17.17	44.19	5.28	0.79	0.25	43.45	17.44

2.2. Bio-char preparation experiment and the calculation of the fuel characteristics

The bio-char was prepared using the tube furnace. According to the existing research [3, 5], a total of thirteen ending temperature (200-500 °C, per 25 °C), two heating rate (10 and 20 °C/min) and two nitrogen purging flow (100 and 200 ml/min) were chosen for the preparation, resulting in sixty-five bio-char samples.

The proximate analysis parameters are measured using thermal gravimetric analyser (TGA) from TA(USA) according to other researchers' method [6]. First, heat the furnace from room temperature to 105°C at heating rate of 10°C/min, hold for 40 min in nitrogen. The mass loss in this period is supposed to be the moisture. Second, heat the furnace from 105°C to 900°C at a heating rate of 10°C/min, hold for 20 min in nitrogen. The mass loss in this period is supposed to be the volatile matter (VM). Third, change the purge gas to be air, burn the furnace at 900°C and hold for 30 min. The mass loss in this period is supposed to be the fix carbon (FC). The remaining weight is ash.

The ultimate analysis parameters are measured using the elemental analyser from Elementar (Germany) according to ASTM E777 -08, AOAC Official Method 990.03 protein (crude) in Animal Feed and ISO351. The content of oxygen is calculated according to the formula: O%=100%-C%-H%-N%-S%.

The high heat value (HHV) is calculated using the formula [7]: HHV = 3.55C2 - 232C - 2230H+ 51.2C×H+ 131N+ 20 600.

All the measurements are repeated two times.

2.3. NIRS spectra collection and establishment of NIRS prediction model

Spectrum 400(PerkinElmer, USA)was used to collect the spectra (Figure 1). Sample cup (diameter = 1.5cm, height = 1cm) is put on the window of the integrating sphere. The cup will be filled up with sample and made flat with a blade. And it will spin when collect the spectra. Spectra of samples were collected in diffuse reflection mode in the range of 10 000-4 000 cm^{-1} at a resolution of 8 cm^{-1}. Internal reference was used for the spectrometer. Each spectrum is the average of 30 scans. Each sample was measured in triplicate and the results were averaged. The average spectrum would be used as final spectrum.

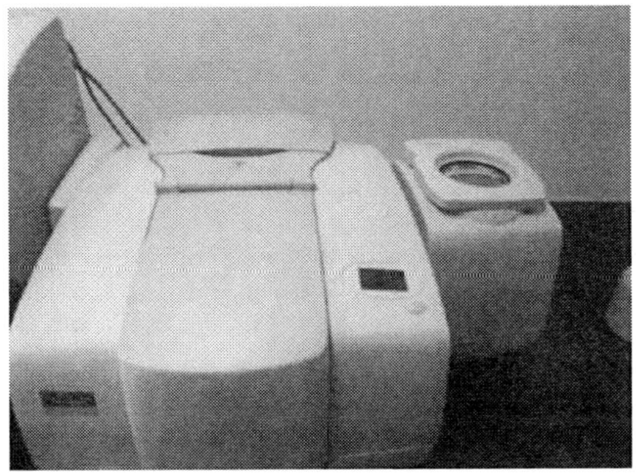

Figure 1. NIR spectrometer and sampling unit.

All NIR models were developed using PLS toolbox based on the MATLAB R2008a software (Math works Corporation, USA). Firstly, spectra were pretreated by autoscale, First derivative or Second derivative, smoothing (Sav-Gol), SNV or MSC respectively or some group of them. Then using original or pretreated spectra, a partial least squares (PLS) regression was calculated to determine the quantitative relation between the spectral variable and the straw analysis characteristics.

The best model was chosen according to Rc, RMSEC, Rcv and RMSECV. The accuracy of the model was evaluated by R, RMSEC, RMSECV, RSD and RPD. The RPD statistic compares the RMSEP with the range of the calibration parameter measured as, the standard deviation (SD) of the values obtained in the reference analysis [8]. The RSD is calculated as RMSEP*100%/Meanv. The RSD is the index of the accuracy of the prediction of the models.

3. Results and discussion

3.1. NIRS spectra of all bio-char samples

Figure 2a shows fourteen spectra of original and biochar samples of wheat straw, used to discuss the influence of different termination temperatures on the NIRS spectra of biochar. According to figure 2a, there are significant differences in NIRS spectra of bio-char at different termination temperatures, mainly including the following four significant changes:

1. The absorbance value increases sharply along with the increase of the final temperature. The main reason is the colour of the sample deepened along with the increase of final temperature, which results in the increase of the absorbance.

2. The peaks of 6600-7100, 5400-6200 and 4500-5000 cm^{-1} weaken gradually along with the increase of the final temperature and disappear at 300 °C, which are mainly comes from stretching vibration of N-H, O-H, C-H, S-H and C=O bonds in cellulose, water and protein in the straw.

3. The peak at 4600 cm^{-1} gradually forms after 300 °C, which is the information of the stretching vibration of C-H and C=O bonds from produced aldehyde.

4. The peaks of 5000-5400 and 4200-4400 cm^{-1} weaken gradually along with the increase of the final temperature and disappear at 425 °C, which are mainly comes from stretching vibration of C=O, O-H, N-H and C-H bonds from cellulose, starch and carboxylic acid compounds in the straw.

The PCA calculation results of the spectra were shown as figure 2b. Seen from the temperature, the distribution is a symmetrical u-shape. For the PC2 scores, it decreases and then increases with the increase of the temperature. The turning point is at 325 °C. While for the PC3, it increases and then decreases with the increase of the temperature. The turning point is at 275 °C. All these information reflect the different chemical composition and physical state of biochar after different terminal temperature treatment.

IOP Conf. Series: Materials Science and Engineering **715** (2020) 012081 doi:10.1088/1757-899X/715/1/012081

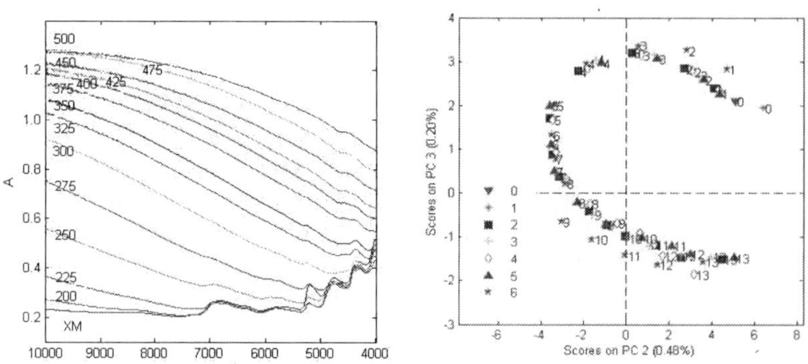

(a) (b)

Figure 2. NIRS spectra and the PCA distribution of bio-char.

3.2. Prediction results of the NIRS models

The proximate analysis parameters were shown as table 2. The moisture of bio-char is very low, with the average content of 2.76%. The contents of VM and FC have a very broad range, with a gap bigger than 60% and 50%, respectively. The maximum HHV of bio-char is 24.33MJ/Kg, which is even higher than some coal.

The statistical data of the best model are shown as table 2. Considering all statistical data, the NIRS models can predict the fixed carbon, ash content, HHV and volatile matter accurately. The Rc and Rcv^2 values are all bigger than 0.94 and the RPDs are bigger than 4.25. The prediction model are based on the information of the NIR spectra contains, which was discussed earlier. However, for the prediction of moisture, the model can only tell the high or low contents. More work is needed to improve the accuracy.

Table 2. Prediction results of the proximate analysis parameters and HHV using NIRS.

Proximate analysis parameters/%	Moisture	VM	FC	Ash	HHV(MJ/Kg)
Number of the sample	66	67	67	65	63
Min.	0.51	18.08	17.17	6.04	17.29
Max.	6.13	79.52	68.41	26.88	24.33
Mean	2.76	43.07	45.10	17.87	21.34
SD	1.19	19.69	16.41	6.07	1.90
CV	43.21	45.72	36.38	33.95	8.92
PLS Components	4	5	5	4	4
Pretreatment	Autoscale	Autoscale	Autoscale	Autoscale	Autoscale
Rc^2	0.72	0.98	0.97	0.96	0.96
Rcv^2	0.67	0.98	0.96	0.95	0.94
RMSEC	0.59	2.60	3.05	1.25	0.37
RMSECV	0.64	3.02	3.43	1.33	0.45
RSD	23.22	7.01	7.61	7.44	2.10
RPD	1.86	6.52	4.78	4.56	4.25

The ultimate analysis parameters were shown as table 3. The carbon content of bio-char increases from 43.50% to 68.60%, which is more than some coal [3].

The statistical data of the best model are shown as table 3. According to the data, the NIRS model can predict the oxygen content perfectly. The Rc and Rcv^2 value are all bigger than 0.99 and the RPD are bigger than 11. For the prediction of carbon and hydrogen content, the NIRS models can predict them accurately. The Rc and Rcv^2 values are all bigger than 0.92 and the RPDs are bigger than 3.59. The NIR spectra contain a lot of information of the C-H, O-H, C-O and C=O bonds, which is the basis of the prediction. For the prediction of the content of nitrogen, the NIRS model can only tell the high or low contents. Unfortunately, the NIRS model cannot predict the sulphur content. Because there is not much information of sulphur in the NIR spectra. And the content of sulphur is really low.

Table 3. Prediction results of the ultimate analysis parameters using NIRS.

Ultimate analysis parameters/%	C	O	H	N	S
Number of the sample	67	67	65	67	66
Min.	43.50	5.97	1.71	0.63	0.25
Max.	68.60	43.45	5.81	1.36	1.18
Mean	57.41	19.54	3.32	1.11	0.74
SD	6.78	11.41	1.05	0.18	0.22
CV	11.81	58.38	31.71	15.95	29.69
PLS Components	4	4	1	4	4
Pretreatment	Autoscale	Autoscale	Autoscale	SNV, Autoscale	Autoscale
Rc^2	0.94	0.99	0.93	0.84	0.42
Rcv^2	0.92	0.99	0.92	0.76	0.33
RMSEC	1.66	0.95	0.27	0.06	0.17
RMSECV	1.89	1.04	0.27	0.08	0.18
RSD	3.29	5.31	8.18	7.02	23.95
RPD	3.59	11.00	3.88	2.27	1.24

4. Conclusion

This study is supposed to immediately predict the fuel characteristics of sixty-five bio-char using NIRS. The results show the fuel characteristics of the bio-char vary in a very big range. The NIRS models can predict the VM, FC, ash, HHV, oxygen, carbon, hydrogen contents of the bio-char accurately. The Rc and Rcv^2 value are all bigger than 0.92 and the RPD are bigger than 3.59. The NIRS models can also tell the high or low content of moisture and nitrogen. Only for the sulphur, more work need to do to improve the accuracy of the NIRS models. Predicting the fuel characteristic of the bio-char only by scanning the NIR spectrum, which costs only 10 seconds, will really be helpful for the biomass industry.

Acknowledgments

The authors acknowledge the generous support from Fundamental Research Funds for the Central Universities (2019MS029, 2018ZD08).

References

[1] H Abdullah, D Mourant, CZ Li ; H Wu. Energy Fuels, 2010, 24: 5669-5676.
[2] T A. Lestander, M Rudolfsson. Green Chem., 2014, 16: 4906-4913.
[3] S Du, W Chen, J A Lucas. Bioresource Technol., 2014, 161: 333-339.
[4] B. Sander, J. J. Wagner, M. Jakob, L. Jane, E. S. Balling. Ind. Crop. Prod., 2010, 31(2): 321-326.
[5] S W Park, C H Jang, K R Baek, J K Yang. Energy, 2012, 45: 676-685.
[6] D. Medic, M. Darr, A. Shah, B. Potter, J. Zimmerman. Fuel, 2012, 91: 147-154.
[7] P McNamee, L I Darvell, J M Jones. Biomass Bioenergy, 2015, 5: 1-10.
[8] C Lomborg, J B Holm-Nielsen, P Oleskowicz-Popiel, K. H. Esbensen. Bioresource Technol., 2009, 100(5): 1711-1719.

Cross-modal retrieval by an end to end way

Bang Hu Yin* and Xiu Li

Tsinghua University, China

*Email: ybh16@mails.tsinghua.edu.cn

Abstract. Cross-modal retrieval has attracted most attention in the recent years. For the image and text, how to measure the semantic similarity among them is still a challenging problem in cross-modal retrieval task. In our paper, we propose a cross-modal retrieval method that uses a CNN to obtain the semantic similarity between image and text. Most methods to solve this problem employ two separate parts for each modality to obtain the semantic similarity between them. However, in our work we just use a CNN to get the semantic similarity without having to use separate networks. In addition, we are aiming to solve the problem between long text and image, so we use the topic model to process the text. We evaluate our approach on Wikepedia dataset.

1. Introduction

Convolutional Neural Networks has already made great success in image classification[1][2][3], object detection[4][5]. And now it has started to pivot from classical image classification and object detection problems to associating text descriptions to visual data. For example, Convolutional Neural Networks has made progress in image caption[6][7] which is aimed to produce the sentences describing the images. From these works, we can know that the deep Convolutional Neural Network can extract state-of-the-art image features and can obtain a better experiment result than other classical methods. It uses an image of the size $W \times H \times 3$ as input, and then put it into a convolutional layer which is consisting of C-channel filters. These filters will scan the input and then output a feature map with the size of $W1 \times H1 \times C$. Certainly, The result will be the input of the next convolutional layer.

The task of cross-modal retrieval is that according to queries in one modality, users would search for semantically relevant content in other modalities. In image-text cross modal retrieval, we need find images that best illustrate the topic of a textual query, or textual descriptions that best explain the content of a visual query. Therefore, it is a hard task in cross-modal retrieval that obtain the semantic similarity between queries and database entries from different modalities.

The standard solution to cross-modal retrieval is to explore a shared low dimensional latent feature space for these modalities using data pairs[8][9]. And then we can obtain the semantic similarity by processing feature vectors in the feature space. The efficient and effective distance measurements are Cosine distance[9][10] or Euclidean distance[11], both methods can be used to measure the semantic similarity. Besides, we can take other methods to get the semantic similarity. In[12], they train a network that can learn to judge video frame is matching with short audio clip. The positives are those frames and audios from one video at the same time, while those extracted from different videos are negatives. By this method, they get the best results at that time.

In our paper we propose an approach that obtain the semantic similarity of text and image by one network. We utilize the great power of Convolutional Neural Network to realize it. We aim to let the network output the similarity, not just get the feature vector of the image. In addition, we experiment on long texts not several sentences and transform them into vectors which are as inputs for network.

2. Related work

Many methods in the field of cross-modal retrieval have been proposed over recent years. Even if they are different from each other, they have a common feature that they try to search a shared subspace where the item of one modality can directly compare with others. And then these methods will transform each modality into vectors of common subspace. As a result, they can obtain the semantic similarity by computing the distance of two vectors.

One of the classical approaches towards is named Canonical Correlation Analysis(CCA)[10] which aims to search the common feature space that can maximizes the relevance between two types of data. Kernel canonical correlation analysis(KCCA)[13] uses the kernel trick based on CCA. And there are many other extensions of CCA. Just like CCA, Graph-based methods[14][15] also learn linear projections to discover a shared feature space by maximizing the cross-modal pairwise item correlation. Triplet ranking[16][17] is another effective approach. This method impose a constraint that make the distance between positive pairs be closer to the negative pairs. The cross-modal hashing[18][19] aims at searching a shared binary Hamming space where we can compare them directly. In TextTopicNet[20], they employ the topic level feature vectors of long texts to supervise the image feature learning of CaffeNet[28]. All these approaches could be divided into two parts: supervised and unsupervised approaches. The difference among them is if they utilize class or categorical information related to every image-document pair, of course, supervised methods use.

The method presented in our paper is the category of unsupervised methods. Our method is related to these cross-modal retrieval methods to some extent that we adopt LDA[21] to extract features of texts and use convolutional neural network to extract image feature vectors. However, our method differs from these methods in that we use the features of texts as one of inputs of the convolutional neural network with the image, besides, the output is the semantic similarity between the images and texts. Therefore, we needn't calculate the similarity with the features of images and texts, and we can get the similarity directly through the network.

3. Proposed method

Our goal is to learn the semantic similarity between the images and texts by the deep neural network. To achieve this goal, we will extract the texts features at the first through the LDA[21] topic model, and then we will accomplish the fusion of image features and text features in the neural network. Finally, we can get the semantic similarity between the image and text.

3.1. LDA model

For the text, we suppose that several hidden topics can generate the textual information by some ways. Therefore, we take advantage of the latent dirichlet allocation model to obtain these latent topics and represent the texts associated with images as a probability distribution over the set of discovered topics. LDA, a three-level hierarchical Bayesian model, is a generative probabilistic model of texts. The core of it is that texts can be seen as different mixtures over latent topics, and each topic is distinguished by a probability distribution over all words of texts. We can see the structure of the model in figure 1.

For a text corpus including M documents and a dictionary containing N words, we can form a new text consisting of several topics by these steps:

• Select parameter θ which is generated by Dirichlet distribution

• For word ω_n in the new text d:

- Select a topic z_n which is generated by multinomial distribution with the parameter θ

- Select a word ω_n from multinomial distribution $p(\omega_n \mid z_n, \beta)$

So the LDA model can be expressed as(K is the number of topics):

$$p(d \mid \alpha, \beta) = \int_{\theta} p(\theta \mid \alpha)(\prod_{n=1}^{N} \sum_{z_K} p(z_K \mid \theta) p(\omega_n \mid z_K, \beta)) d\theta \quad (1)$$

We can know that LDA obtains a more compact representation of text, so it reduces the dimensionality of data to a great extent. And we can easily interpret the meaning of the text vector. In our work, we need to extract the features of texts in advance which are inputs of the neural network. What we process are long texts, therefore, we need take some measures to present the texts in a compact representation.

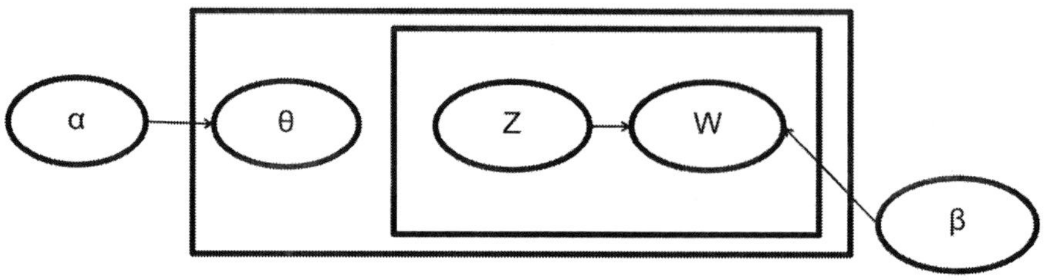

Figure 1. Graphical model of LDA

3.2. Network architecture

To achieve our goals, we design the network structure shown in figure 2. In our experiments we do with LDA topic model to obtain the compact representation of texts, and process the images by convolutional layers to get the feature vectors of image. Then the two 40-D image and text features are concatenated into a 80-D vector which is used to generate a 2-way classification result by two fully connected layers. We can judge whether the images correlate to the texts according to the results.

In experiment, we add some other layers behind each the convolutional layers for the purpose of improving the accuracy of retrieval, such as max-pooling, Relu, Local Response Normalization layers. After we get the trained model, we can calculate the semantic similarity of the new image and text through inputting them into the network. By this method, we achieve the goals that to learn cross-modality similarity by an end to end ways and we make no use of class and categorical information about each image-text pair.

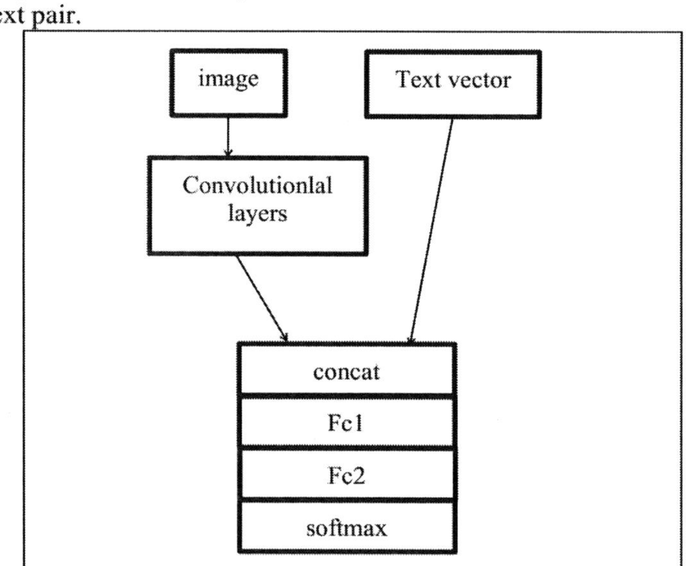

Figure 2. Network architecture.

4. Experiment and results

In this part we will describe our dataset, implementation details, and compare our method to other approaches so as to show the quality of the semantic similarity learned by our method. In our experiment, we utilize the texts of the dataset to train the LDA topic model so as to get the features of texts, then we input the images and text features into the network to obtain the semantic similarity, finally, we use the semantic similarity to finish the cross-modal retrieval task.

4.1. Dataset

We aim to solve the problem of text-image cross-modal retrieval by an end to end way. So we train our LDA and CNN models using Wikipedia articles contained in the Wikepedia ImageCLEF dataset[22]. We use the dataset which is the same as to TextTopicNet[20]. It consists of 35582 different articles which describe 100785 images. In order to test the quality of our models for cross-modal retrieval, we compare with other methods on a subset of Wikipedia dataset[10] which includes 693 pairs.

4.2. Implementation details

We do several experiments to choose the proper number of topics of LDA by varying it from 10 to 100, and finally we discover that 40 topics can obtain the best performance. One of the core ideas of this paper is that the fusion of image and text features, when we fuse the features of images and texts, we try to use the product of them, but it cannot reach the level of concating them. Finally , we use the softmax as our loss function. When we use it to calculate the semantic similarity, it will get a value between 0 and 1, which can be seen as the possibiltiy that the image and text are correlated to each other.

4.3. Comparison to other methods

We evaluate our models by two kinds of cross-modal retrieval tasks: (1)given image retrieve text, (2) given text retrieve image. For the process of retrieval, we input images and texts features into the network, and obtain the semantic similarity in the end of the network. We compare our method with other supervised and unsupervised approaches in the table 1.

Table 1. MAP comparison with other methods.

Method	Image query	Text query	Average
CCA[13]	19.70	17.84	18.77
PLS[23]	30.55	28.03	29.29
TextTopicNet[20]	39.58	38.16	38.87
SCM[10]	37.13	28.23	32.68
GMMFA[24]	38.74	31.09	34.91
CCA-3V[25]	40.49	36.51	38.50
GMLDA[24]	40.84	36.93	38.88
LCFS[26]	41.32	38.45	39.88
JFSSL[27]	42.79	39.57	41.18
ours	36.44	29.02	32.73

Our method outperforms some supervised methods and unsupervised approaches. Most importantly, our method achieves the goals of learning the semantic similarity by an end to end way. It simplifies the process of calculating the semantic similarity.

5. Conclusion

In our experiments, we train our models to learn cross-modality similarity by an end to end way without using two networks like other methods and our performance is better than some classical ways.

References

[1] Krizhevsky, Alex, Ilya Sutskever, and Geoffrey E. Hinton. "Imagenet classification with deep convolutional neural networks." Advances in neural information processing systems. 2012.

[2] Simonyan, Karen, and Andrew Zisserman. "Very deep convolutional networks for large-scale image recognition." arXiv preprint arXiv:1409.1556 (2014).

[3] Szegedy, Christian, et al. "Going deeper with convolutions." Proceedings of the IEEE conference on computer vision and pattern recognition. 2015.

[4] Ren, Shaoqing, et al. "Faster r-cnn: Towards real-time object detection with region proposal networks." Advances in neural information processing systems. 2015.

[5] Liu, Wei, et al. "Ssd: Single shot multibox detector." European conference on computer vision. Springer, Cham, 2016.

[6] Lu, Jiasen, et al. "Knowing when to look: Adaptive attention via a visual sentinel for image captioning." Proceedings of the IEEE Conference on Computer Vision and Pattern Recognition (CVPR). Vol. **6**. 2017.

[7] Chen, Long, et al. "Sca-cnn: Spatial and channel-wise attention in convolutional networks for image captioning." arXiv preprint arXiv:1611.05594 (2016).

[8] Kuang, Zhanghui, and Kwan-Yee K. Wong. "Reltively-Pired Spce Anlysis: Lerning Ltent Common Spce From Reltively-Pired Observtions." International Journal of Computer Vision113.3 (2015): 176-192.

[9] Sharma, Abhishek, et al. "Generalized multiview analysis: A discriminative latent space." Computer Vision and Pattern Recognition (CVPR), 2012 IEEE Conference on. IEEE, 2012.

[10] Rasiwasia, Nikhil, et al. "A new approach to cross-modal multimedia retrieval." Proceedings of the 18th ACM international conference on Multimedia. ACM, 2010.

[11] Hua, Yan, et al. "Cross-Modal Correlation Learning by Adaptive Hierarchical Semantic Aggregation." IEEE Trans. Multimedia 18.6 (2016): 1201-216.

[12] Arandjelovic, Relja, and Andrew Zisserman. "Look, listen and learn." 2017 IEEE International Conference on Computer Vision (ICCV). IEEE, 2017.

[13] Hardoon, David R., Sandor Szedmak, and John Shawe-Taylor. "Canonical correlation analysis: An overview with application to learning methods." Neural computation 16.12 (2004): 2639-664.

[14] Zhai, Xiaohua, Yuxin Peng, and Jianguo Xiao. "Learning cross-media joint representation with sparse and semisupervised regularization." IEEE Transactions on Circuits and Systems for Video Technology 24.6 (2014): 965-78.

[15] Zhuang, Yueting, et al. "Supervised coupled dictionary learning with group structures for multi-modal retrieval." AAAI. 2013.

[16] Frome, Andrea, et al. "Devise: A deep visual-semantic embedding model." Advances in neural information processing systems. 2013.

[17] Wang, Liwei, Yin Li, and Svetlana Lazebnik. "Learning deep structure-preserving image-text embeddings." Proceedings of the IEEE conference on computer vision and pattern recognition. 2016.

[18] Xu, Xing, et al. "Learning discriminative binary codes for large-scale cross-modal retrieval." IEEE Transactions on Image Processing 26.5 (2017): 2494-507.

IOP Conf. Series: Materials Science and Engineering **715** (2020) 012082 doi:10.1088/1757-899X/715/1/012082

[19] Zhu, Xiaofeng, et al. "Linear cross-modal hashing for efficient multimedia search." Proceedings of the 21st ACM international conference on Multimedia. ACM, 2013.
[20] Gomez, Lluis, et al. "Self-supervised learning of visual features through embedding images into text topic spaces." arXiv preprint arXiv:1705.08631 (2017).
[21] Blei, David M., Andrew Y. Ng, and Michael I. Jordan. "Latent dirichlet allocation." Journal of machine Learning research3.Jan (2003): 993-1022.
[22] Tsikrika, Theodora, Adrian Popescu, and Jana Kludas. "Overview of the Wikipedia Image Retrieval Task at ImageCLEF 2011." CLEF (Notebook Papers/Labs/Workshop). Vol. **4**. 2011.
[23] Rosipal, Roman, and Nicole Krämer. "Overview and recent advances in partial least squares." International Statistical and Optimization Perspectives Workshop" Subspace, Latent Structure and Feature Selection". Springer, Berlin, Heidelberg, 2005.
[24] Sharma, Abhishek, et al. "Generalized multiview analysis: A discriminative latent space." Computer Vision and Pattern Recognition (CVPR), 2012 IEEE Conference on. IEEE, 2012.
[25] Gong, Yunchao, et al. "A multi-view embedding space for modeling internet images, tags, and their semantics." International journal of computer vision 106.2 (2014): 210-33.
[26] Wang, Kaiye, et al. "Learning coupled feature spaces for cross-modal matching." Proceedings of the IEEE International Conference on Computer Vision. 2013.
[27] Wang, Kaiye, et al. "Joint feature selection and subspace learning for cross-modal retrieval." IEEE transactions on pattern analysis and machine intelligence 38.10 (2016): 2010-023.
[28] Jia, Yangqing, et al. "Caffe: Convolutional architecture for fast feature embedding." Proceedings of the 22nd ACM international conference on Multimedia. ACM, 2014.

Design and analysis of five probe flow analyser for subsonic and supersonic wind tunnel calibration

Akhila Rupesh[1], J V Muruga Lal Jeyan[2,*] and Sanas Uthaman[3]

[1] Research scholar , SME, Lovely Professional University, India
[2]Department of Aerospace Engineering, Lovely Professional University, India
[3]Bharat Electronics Limited, India

*E-mail: jvmlal@ymail.com

Abstract. The flow angularity in a wind tunnel plays a major role in test section when testing models. The models are being tested at a mixture of flow velocity. Usually a subsonic wind tunnel is calibrated using a normal Pitot tube and supersonic wind tunnels with high efficiency probes. The calibration of a wind tunnel includes determining the Mach number range ad pressure range throughout the test section of the flow throughout the pattern of operating velocity. In the analysis it is noted that the sensitivities of the calibrating instruments are maximum when it is either cone or wedge shaped. When the wedge or cone angles are maximum, more accurate results are also obtained, but it's not possible to have a maximum angle cone or wedge because of the detached wave and tunnel blockage factors. Hence in order to overcome this difficulties in measurement, a five probe flow analyzer has been designed. The five probe flow analyzer is used to determine the wind tunnel test section parameters like flow angularity, Pitot pressures, static pressures, wave angles, and the presence of test section noise. Over to these parameter, the stagnation states can also be determined using this analyzer. The proposed model is designed by getting optimum solutions from previous analysis and the existing data by considering all the aerodynamic and mechanical loading. This instrument designed is to be tested experimentally after carrying two and three dimensional computational analysis for Mach number ranging from 0.5 of 3 .Also it is designed to fit successfully for 0.3m and 0.6m test section wind tunnel to obtain reasonably merging solution with theoretical and experimental results.

1. Introduction

The change in kinetic pressure are accompanied by equal and opposite changes in static pressure. The static pressure defined for present purposes as the pressure sensed by a measuring device at rest relative to the fluid (i.e. moving with the fluid).[1] The total pressure defined as the pressure obtained when the fluid is brought to rest relative to the probe. It follows that the kinetic pressure can be determined directly by measuring high pressure. If velocity is required, then there is a need to determine the density as well. This is usually done by measuring the (absolute) pressure and temperature at some point in the flow field, and invoking the perfect gas equation: in practice the pressure changes arising from the motion in incompressible flow are negligible in comparison with the absolute pressures.[2][3]

2. Design of conical probes

Conical probes have been used determination of the Mach number flow direction and total pressure at supersonic and subsonic speeds. It is mainly because their shape will be offering less interference to flow the hemispherical types. The smaller the apex angle of the cone of the wider the range of Mach over which a smooth pressure response to obtained. At the same time, sensitivity of the side holes to change in the flow direction increases with angle of the cone, so that some compromise is necessary. [4]A cone angle of 60 degrees is found to be suitable for speeds above M = 1.5 although angle of 40 degrees have been used to give a slightly lower limit.

The multi-hole probe is devised by means of the AutoCAD with the specifications according to the needs and after taking into accounts various features such as tunnel blockage, model design and size and shock wave. It is planned in accordance with the conceptual evaluation and is endowed with the five-hole conical probe wedge type.

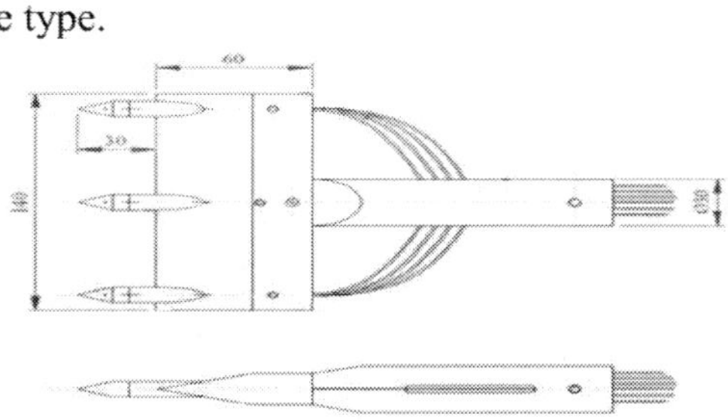

Figure 1. Schematic diagram of 5 probe flow analyzer.

The wedge type probe is endowed with the following dimensions: It is 140mm long and 60mm wide. The front facing edge of the wedge and tip of the multi-hole probe have a dimension of 30mm each .Each probe is effectively fitted with the probe adopter, multi-hole, and small adopter. The main adopter contains tube having ϕ30 along with length 130mm .It is coupled to the particular wedge from the yaw meter. The dimensions of the wedge are 140mm length, 60mm wide and also the thickness involving 20mm. It have 4 holes in front of the wedge with regard to inserting the particular conical probe and diameter is usually ϕ10mm. The actual conical probe contains the diameter involving ϕ10mm and length is usually 30mm. The actual probe adopter can be used to insert in the particular multi hole probe from the wedge possesses the particular diameter from the ϕ10mm along with 70mm length. The small adopter can be used to help conical pets from the probe together with along 30mm length and that is placed from the smaller adopter featuring a diameter involving ϕ10mm along with 55mm length.[5]

3. Methodology

The material selection was done by comparing the mechanical properties of stainless steel, aluminum and copper.It is clear that stainless steel is largely used because the corrosion resistance of its alloys depends on the preservation of the stainless passivation layer needing the least Chromium concentration of 11 % and exposure to an oxygenated atmosphere. Low temperature hot water systems are devised to restrict oxygen ingress which adversely affects the preservation of the passive layer. As it is a non-reactive base metal, Copper is also trivially prone to the creation of corrosion and assault by chlorides leached from insulating foam. Nevertheless, while scaling up the storage capacity is shored up, the yield strength of Copper makes it non-competitive economically vis-a-vis Stainless Steel substitutes. Copper and aluminum alloys are endowed with superior thermal conductivities and lesser electrical resistivity than stainless steel.[6] It is observed that the low thermal conductivity of stainless steel tends to enhance the scope of heating. The densities of the stainless steels are marginally greater than that of aluminum. Stainless steel emerges as the solitary alloy to possess higher properties at 200°C and above and hence is employed to originate the multi-hole probe. Computational fluid dynamic analysis and Nastron analysis were conducted for predicting the best material for manufacturing the multi-hole probe .[7]

IOP Conf. Series: Materials Science and Engineering **715** (2020) 012083 doi:10.1088/1757-899X/715/1/012083

3.1. CFD analysis: 2D analysis method

The following conditions are used to derive the real time conditions of the CFD computational analysis. The Mach number is varied with constant ambient temperature and ambient pressure. The output of the CFD is static pressure, static temperature, turbulence viscosity and velocity.

In MACH 1, the ambient pressure and ambient temperature is 1.01325 bar and 288K. figure (a), (b), (c) and (d) shows the static pressure, static temperature, turbulent viscosity and velocity magnitude of the probe.

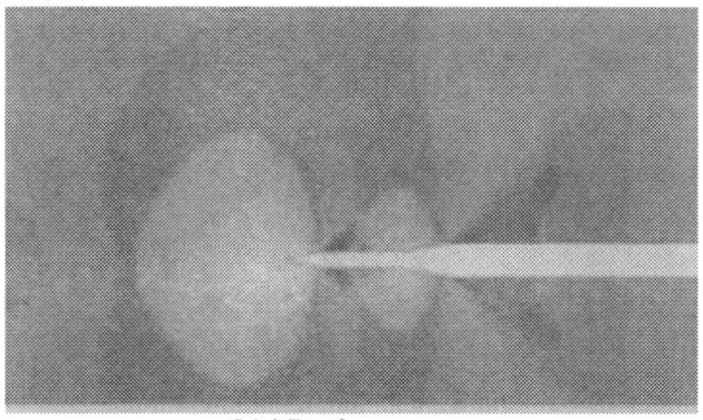

2(a)Static pressure

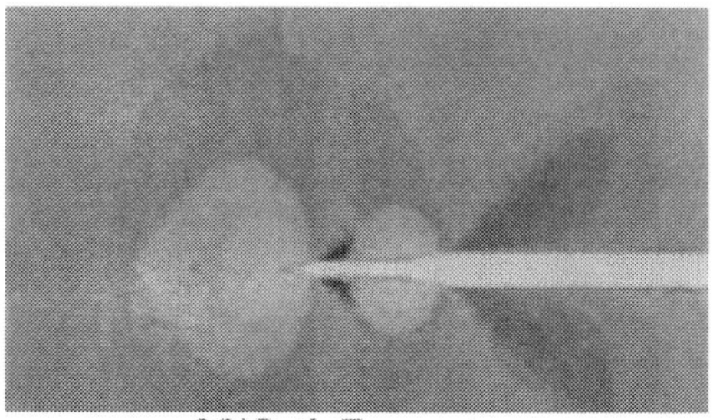

2(b)Static Temperature

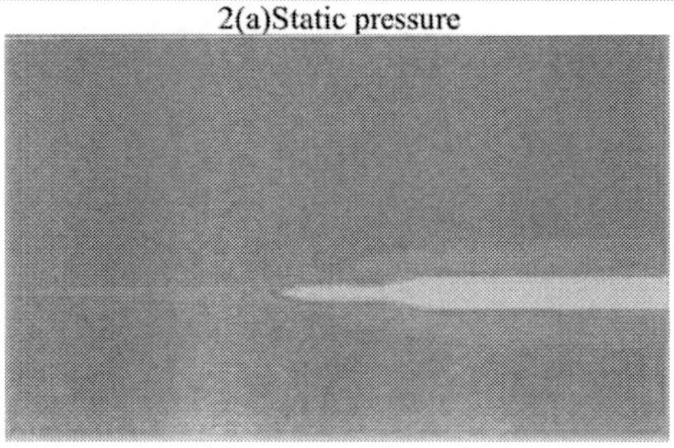

2(c)Turbulent Viscosity

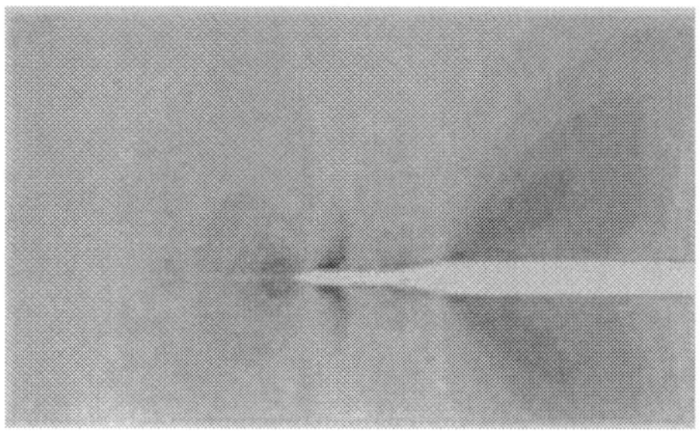

2(d)Velocity Magnitude

Figure 2. 2D Graphical output of the conical four-hole probe.

Figure 2(a) shows the static pressure output of the probe. It shows the detached wave is formed by the shock wave. It is not fitted with the pressure. So the pressure is accumulated in front of the probe tip in Mach1.it is clearly shows in figure 2(a) by the yellow color.Figure 2(b) shows the static temperature which also same as for the static pressure which is shown by the yellow color on tip of the probe, the Velocity of the air blocked by the detached wave. So the velocity is reduced at the tip of the probe which is shown by green color in figure 2(d).

In MACH 2, the detached wave is reduced, when Mach number increases. Figure 3(a), 3(b), 3(c) and 3(d) shows the static pressure, velocity magnitude, static pressure and turbulence viscosity of the probe.

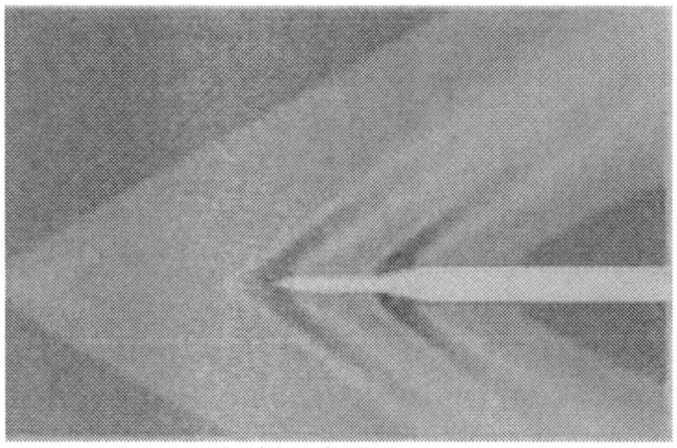

3(a)Static Pressure

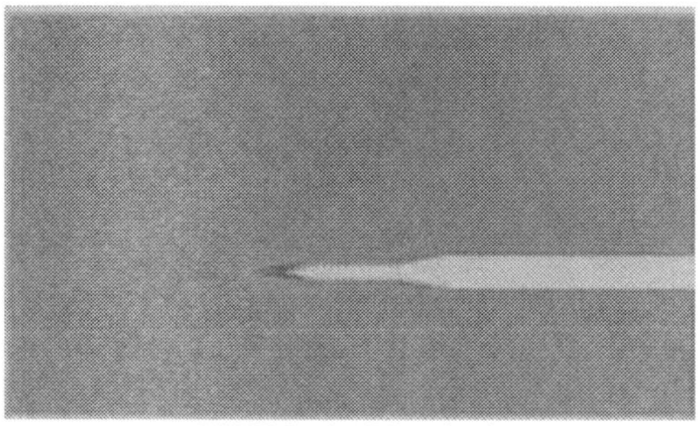

3(b)Velocity Magnitude

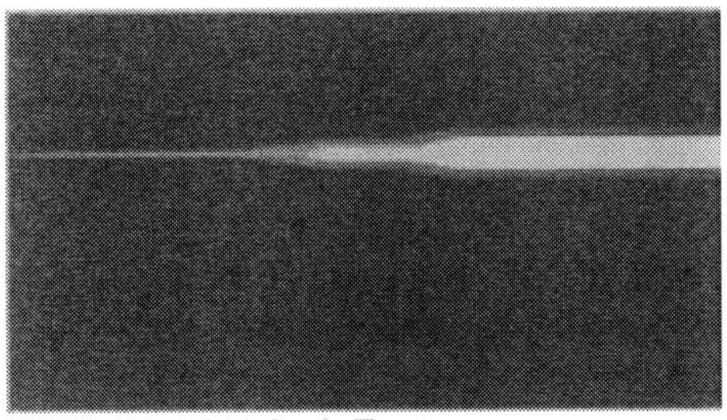

3(c) Static Temperature

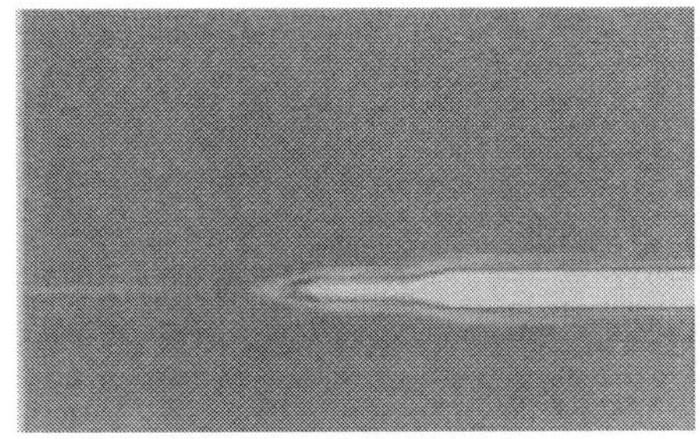

3(d) Turbulent Viscosity

Figure 3. MACH 2 output responses of the probe.

The Mach number increases, the shock wave coupled with the air pressure and create the shock wave angle with the probe. The figure 3(a) clearly shows the pressure fitted with probe tip and gives the angle of shock wave. The static pressure of probe with maximum value shows in red color. The static temperature varied in velocity varies which is shown by the blue color. Figure 3 (c) shows the variation of the static temperature with the minimum value which is shown figure 3 (c) as blue color. The turbulent viscosity shows in figure 3(d) with minimal range of variation shown by light green color.

From these almost all 2D graphical analysis shows, whether the Mach number increases, the own wind pressure clearly measured with the probe very accurately. Whenever the own air pressure increases the supersonic condition, the current velocity of an wind goes towards very high value which is to be also measured with the probe .

3.2. Nastron analysis

Table 1. Experiment results for Aluminium.

Mach	Velocity	Pressure	Displacement (max) mm			Displacement (min) mm			stress(max) kpa			stress(min) kpa		
	m/s	Pa	X	Y	Z	X	Y	Z	min prin	max pri	shear	min pri	max prin	Shear
0.2	68.6	3002.0	2.84E-02	3.51E-02	3.04E-02	3.02E-02	2.53E-02	3.49E-02	1.17E+05	4.46E+06	3.50E+06	-3.82E+06	-9.34E+04	1.34E-03
0.8	274.5	48032.8	2.60E-03	3.21E-03	2.78E-03	-2.76E-03	-2.31E-03	-3.19E-03	1.07E+04	4.08E+05	3.20E+05	-3.49E+05	-8.54E+03	1.22E-04
1	343.1	75051.2	3.24E-03	4.02E-03	3.47E-03	-3.45E-03	-2.89E-03	-3.99E-03	1.33E+04	5.10E+05	4.00E+05	-4.36E+05	-1.07E+04	1.53E-04
2	686.2	300204.9	2.84E+00	3.51E+00	3.04E+00	-3.02E+00	-2.53E+00	-3.49E+00	1.17E+07	4.46E+08	3.50E+08	-3.82E+08	-9.34E+06	1.34E-01
3	1029.3	675461.0	6.39E+00	7.91E+00	6.84E+00	-6.80E+00	-5.69E+00	-7.84E+00	2.62E+07	1.00E+09	7.87E+08	-8.58E+08	-2.10E+07	3.01E-01

In this analysis, the tabular columns are shows that the experimental investigation of the aluminium, copper and stainless steel material probes and which material is to be used for fabricating the multihole probe. From this experiment results, the high hardness and strength material is measured and that material is used for fabricating the multihole probe.[8][9] The inputs are pressure and velocity at different Mach number conditions and the outputs are displacement at minimum and displacement at maximum and stress at minimum and stress at maximum that used to select the material. The experimental results are shown in the below tables for different materials like aluminium, copper and stainless steel.

In table 1, the aluminum material probe is tested at the different mach number conditions and the outputs are displacement, stress and shear are measured. The displacement has three coordinates X, Y and Z with minimum and maximum. The stress at minimum and maximum are measured at different conditions of the Mach number by using the CFD analysis. Different mach numbers and pressures are given as input for probe and the outputs are displacement and stress with shear.

Table 2. Experiment results for Copper.

Mach	Velocity	Pressure	Displacement (max) mm			Displacement (min) mm			stress(max) kpa			stress(min) kpa		
	m/s	Pa	X	Y	Z	X	Y	Z	min prin	max pri	shear	min pri	max prin	shear
0.2	68.6	3002.0	1.75E-02	1.70E-02	2.05E-02	-2.04E-02	-2.02E-02	-2.14E-02	4.42E+04	4.54E+06	3.53E+06	-3.80E+06	-7.73E+04	2.31E+00
0.8	274.5	48032.8	2.12E+00	2.06E+00	2.48E+00	-2.46E+00	-2.45E+00	-2.59E+00	5.35E+06	5.49E+08	4.27E+08	-4.59E+08	-9.36E+06	2.80E+02
1	343.1	75051.2	4.38E-01	4.26E-01	5.13E-01	-5.09E-01	-5.06E-01	-5.36E-01	1.11E+06	1.14E+08	8.83E+07	-9.49E+07	-1.93E+06	5.78E+01
2	686.2	300204.9	1.75E+00	1.70E+00	2.05E+00	-2.04E+00	-2.02E+00	-2.14E+00	4.42E+06	4.54E+08	3.53E+08	-3.80E+08	-7.73E+06	2.31E+02
3	1029.3	675461.0	3.94E+00	3.83E+00	4.61E+00	-4.58E+00	-4.56E+00	-4.82E+00	9.95E+06	1.02E+09	7.94E+08	-8.54E+08	-1.74E+07	5.20E+02

In table 2, the copper material probe is used to analyze their strength and hardness at different Mach number conditions. The table 2 is shown that the copper material outputs at different Mach number and output are displacement and stress. From 0.2 to 3 Mach number conditions are checked and the velocity range is varied from 68 to 1030 m/s also the pressure is varied upto 68000Pa. The maximum displacement is occurred at maximum condition in Mach number 3.

In table 3, the stainless steel material is analyzed at different Mach number conditions from 0.2 to 3 and different velocity and pressure applied on the multihole probe. The experimental outputs of the probe are analyzed by using the Computational Fluid Dynamics (CFD).The shear is the component of stress coplanar with a material cross section. Shear stress arises from the force vector component parallel to the cross section.

The nastron analysis is conducted for different materials such as Aluminium, copper and stainless steel and its output are shown in the following tabular columns.

From the above all tabular colums of the materials,the stainless steel material has the best values in displacement and stress.From the figure, the displacement and stress both are depend on the material properties.Theoritically,the mach number 2.8 is given the best result and the performnace of the material is also very good at this condition.So,from the all conditions are satisfied for stainless steel matrial probe and its results are also good for all mach number conditions.Form the cbove results,the stainless steel material is used for fabricating the probe.

Table 3. Experiment results for stainless steel.

Mach	Velocity	Pressure	Displacement (minimum)			Displacement (minimum)			Stress(minimum)			Stress(maximum)		
	m/s	Pa	X	Y	Z	X	Y	Z	min prin	max pri	shear	min pri	max prin	Shear
0.2	68.6	3002.0	0.010	0.0092	0.011	-0.011	-0.011	-0.012	38590	4.55E+06	3.54E+06	-3.77E+06	-6.66E+04	2.345
0.8	274.5	48032.8	0.153	0.1469	0.180	-0.178	-0.175	-0.187	617500	7.28E+07	5.67E+07	-6.03E+07	-1.07E+06	37.52
1	343.1	75051.2	0.240	0.2296	0.281	-0.277	-0.274	-0.293	964800	1.14E+08	8.85E+07	-9.42E+07	-1.67E+06	58.62
2	686.2	300204.9	0.958	0.9182	1.123	-1.109	-1.095	-1.171	3859000	4.55E+08	3.54E+08	-3.77E+08	-6.66E+06	234.5
3	1029.3	675461.0	2.157	2.0660	2.526	-2.496	-2.464	-2.634	8683000	1.02E+09	7.97E+08	-8.48E+08	-1.50E+07	527.6

4. Summary and conclusions

The wedge type five hole conical probes flow analyzer is planned by their shock wave angle and designed according to indicated measurements. The planned wedge type probe (flow analyzer) setup is examined and effectively tried by employing the Computational Fluid Dynamics (CFD) investigation system. The composed wedge type probe is applied within the horizontal movement and measured the yaw axis angle moments. The distinctive sorts of materials, for example, aluminum, copper and stainless steel tests are applied to break down their execution in the continuous trials and their yields as far as displacement and stress are evaluated. Through this material execution analysis, the stainless steel material is found to yield the best come about contrasted with those of different materials.

References

[1] Sung-Hyun Kim, Young-Jin Kang, Rho-Shin Myong, Tae-Hwan Cho, Young-Min Park and In-Ho Choi (2009). '*Calibration of a Five-Hole Multi-Function Probe for Helicopter Air Data Sensors*', Journal of Aeronautical & Space Sciences, 10 (2), 43-52.

[2] Mehmet Yafsargun Dongdu, and Omer Turgay Goksel(2016). *'Calibration and Use of a Five-Tube Pressure Probe in the Swirling Flow Field'*, Journal of Engineering and Environmental Science, 22(5), 223-233.

[3] Armando Caldeira Pires, Rodrigo Queiroz Neiva, and Rafael de Mello Pereira (2007). '*Experimental Study Of A Five-Hole Pressure Probe for Measuring Three Mean Velocity Components*', Journal of Flow Measurements, 8(2), 1123-1128.

[4] P.Balakumar (2009). *'Stability of Supersonic Boundary Layers on a Cone at an Angle of Attack'*, International Journal of Computational Fluid Dynamics, 12 (3), 1-16.

[5] J. D. Hooper, and A. R. Musgrove (1997). *'Reynolds Stress, Mean Velocity, and Dynamic Static Pressure Measurement by a Four-Hole Pressure Probe'*, Elsevier Journal of Experimental Thermal and Fluid Science, 15 (3), 375-383.

[6] Phil Danielson (2005). *'Understanding Pressure and Pressure Measurement'*, Journal of Vacuum Technology, 1 (3), 127-132.

[7] Susheela V. Mallipudi, and Michael Selig (2004) , *'Use of a Four-Hole Cobra Pressure Probe*

to Determine the Unsteady Wake Characteristics of Rotating Objects', AIAA Journal, 3 (4), 89-93.

[8] N. Wildmann, S. Ravi, and J. Bange (2014). *'Towards higher accuracy and better frequency response with standard multi-hole probes in turbulence measurement with remotely piloted aircraft (RPA)'*, Journal of Atmospheric Measurement Technique, 7 (3) , 1027–1041.

[9] Sung-Hyun Kim, Young-Jin Kang,Rho-Shin Myong, Tae-Hwan Cho,Young-Min Park, and In-Ho Choi (2009). *'Calibration of a Five-Hole Multi-Function Probe for Helicopter Air Data Sensors'*, Journal of Aeronautical & Space Sciences, 10 (2), 43-51.

Current issues in uncertainty of dimensional tolerance metrology and the future development in the domain of tolerancing

Mussatayev Meirbek, Huang Meifa* and Tang Zhemin

School of Mechanical & Electrical Engineering, Guilin University of Electronic Technology, China

*E-mail: 1390385642@qq.com

Abstract. Uncertainty is an integral part of measurement result. Moreover, any problem in the specification of tolerance or in the measurement phase of tolerance is related to uncertainty, but few methods and lack of experimental data are available to integrate this knowledge. In this work we present the current situation of uncertainty in GD & T, analyzing at the same time a future prospective and including some specific areas to be taken into consideration.

1. Introduction

Production environment involving Smart Manufacturing, Intelligent Factory, or Industry 4.0 stage and competition among the producers tend to not only produce low cost products, but also maintain high performance level. Especially with development of space and robotics industry, a keenly felt need is that special and highly advanced applications are necessary. Prerequisite of these significant advances in technology can be traced to improved dimensional metrology.

Today Geometric Dimensioning and Tolerancing (GD & T) or dimensional metrology is one of the main parts of manufacturing and quality control processes. On the one hand every measurement is only an estimate of the true value, on the other hand, uncertainty, according to standardized rules, identifies where the true value of measurement result is within a certain range.

We have already witnessed that the ISO 14253 standard provides clear guidance about the necessity in allowing for the uncertainty of the measuring device by reducing the size of the acceptance band. Thus, there is significant interest in having access to instruments with smaller measuring uncertainties – unfortunately these are usually more expensive to purchase, and may involve additional expenses such as special air-conditioned rooms or a longer measuring period [1].

Professor Richard Leach says "Metrology is not just a process of measurement that is applied to an end product. It should also be one of the considerations taken into account at the design stage. According to the Geometrical Product Specification (GPS) model, tolerancing and uncertainty issues should be taken into account during all stages of design, manufacture and testing. The most compelling reason is that it is often considerably more expensive to re-engineer a product at a later stage when it is found that it is difficult to measure, compared to designing it at the start with the needs of metrology in mind" (p. 14) [1].

2. Current issues in uncertainty of dimensional tolerance metrology

In the past, the measuring uncertainty has been unattended when ascertaining compliance with a specification, as long as the uncertainty was one tenth of the specification width. This procedure is no longer admissible [1].

Figure 1 shows the impact of the measuring uncertainty on the acceptable process tolerance. Machine C which might be highly precise and with high accuracy, coordinates measuring machine with measuring uncertainty 2.5 μm, achieving the 10:1 inspection ratio, but it would allow a process tolerance of only 20 μm. It also may need stable temperature conditions. Another machine is Machine A that could be an articulated arm measuring machine and its cost is ten times cheaper than previous Machine C, but which can barely meet a 3:1 inspection ratio.

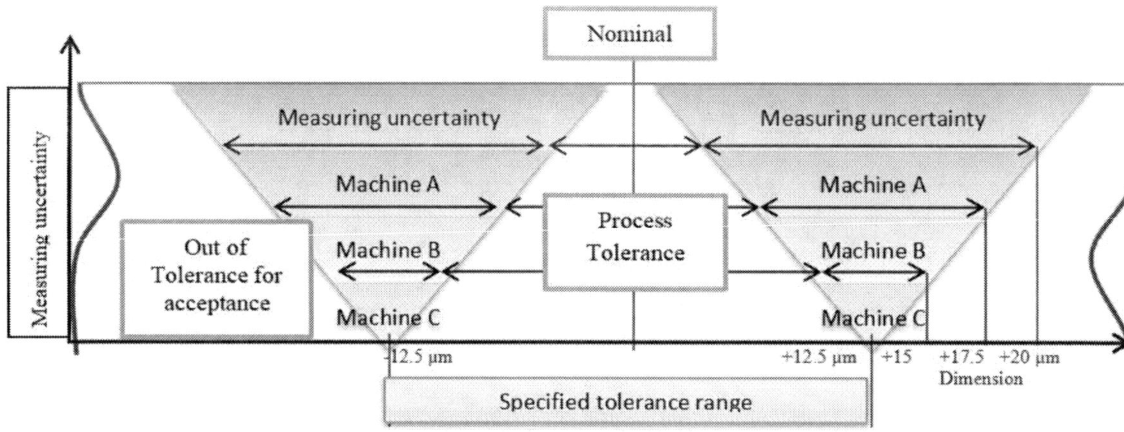

Figure 1. Impact of measurement uncertainty.

Table 1. Impact of measurement uncertainty.

	Measuring uncertainty	Specified tolerance range	For production		For customer
			Available tolerance	Costs	Available tolerance
Machine A	7.5 μm	25 μm	10 μm	High	40 μm
MachineB	5 μm	25 μm	15 μm	Average	35 μm
MachineC	2.5 μm	25 μm	20 μm	Low	30 μm

Figure 1 and table 1 are based on a diagram in VerlagModerneIndustrie 'Multisensor Coordinate Metrology ©www.sv-corporate-media.de.

The higher production costs, associated with producing components within this much reduced process tolerance of only 10 μm, may be acceptable if sufficient capital is available to purchase Machine C – it all depends on the actual circumstances of the job in hand, especially the value of the individual components, the number of items to be produced and the duration of the contract [1].

There are a lot of methods for evaluating and combining components of uncertainty. These standardized rules resulted in an ISO document [2-3]. Also, a lot of National Metrology Institutes (NMI) have their own guidelines for expression for uncertainty in measurement [4]. It is clear that uncertainty of measurement with fixed dimensional gages is not the same as with measuring instruments [5].

There are several contributors to the measurement uncertainty for an industrial length measurement [6]:

- the measurement equipment
- the application of the equipment
- the influence of the environment
- the influence of the device under test
- the repeatability of the measurement set-up

IOP Conf. Series: Materials Science and Engineering **715** (2020) 012084 doi:10.1088/1757-899X/715/1/012084

There are some main parameters that influence to workpiece measurement such as form error, roughness of the surface, shape (e.g. special uncertainties, if a radius of a partial arc should be measured), etc. Machine measurement has the repeatability of the machine, especially if positioning accuracy parameters are evaluated [6].

NMI has been working with estimating measurement uncertainty and has been developing this direction for a long time. Special emphasis has been placed on the uncertainties measurement involved in the dimensional calibrations gage blocks, gage wires, ring gages, gage balls, roundness standards, optical flats indexing tables, angle blocks, and sieves [7-8].

Other works are also available in dimensional metrology instrumentations' uncertainty. In paper [5] the author discussed the lack of instrumentation uncertainty available in GD & T. Since that time, and especially during the last decades, a lot of papers related to instrumentation uncertainty have been published [9-11].

The influences on the overall measurement uncertainty and the machining accuracy may be grouped as follows [6]:

- definition of the process,
- process parameters,
- environment,
- handling of the workpiece,
- workpiece characteristics,
- machine performance.

According to ISO/TS 17450-2 there are many main determination factors on how well part functionality can be ensured with dimensional metrology data:

1. The correlation uncertainty, which gives a quantity information about correspondence between dimensional measurand definition and the part function.
2. The specification uncertainty which quantifies how equivocal the measurand definition is.
3. The measuring uncertainty, which deals with quantity information about quality of measurement of the measurand.

The last factor among two others has been developed within the last decade. Careful consideration of all these factors in specification development process has/provides enormous potential for savings in manufacturing industry.

Until now correlation and specification uncertainty have largely been ignored. Only a few papers [6] proposed some methods to estimate the form error and the associated uncertainty that caused by the measuring process. However, today we are reaching the point when form errors and surface texture are important and both uncertainties significances might be the same as the measurement uncertainty. It is impossible for a designer using the current system – be it the ISO 1101 or the ANSI/ASME Y14.5 dialect – to express to which extent a surface texture should be considered or ignored in geometrical tolerances [12]. It means that control of correlations uncertainty is not possible with standardized specification language.

Taking dimensional metrology to the next level needs to enrich specification language. Consequently, it allows to unambiguously specify the wavelength domain at a design stage. In addition, it will require engineers at all points throughout the product development cycle to consider the ambiguities in measurement data definitions specifications and manage it to an appropriate level in order to ensure product functionality [12].

3. The future development in the domain of tolerancing

Today big manufacturing centers, especially among the NIM are situated in developed countries with well-equipped laboratories. They can carry out a lot of dimensional and geometric tolerance research and properly deal with uncertainty. However, with lack of experience and low financial support in developing countries' researchers cannot access and share the knowledge of specific fields, like GD & T. So, sharing knowledge among researchers is another activity to developing specific uncertainty measurement in dimensional metrology.

One of the goals of this paper is to initiate future discussions in the area of GD & T and to offer a shared vision of examples in common resulting from a regular exchange of views. There is a good example of realization of collaboration and joining researchers, thus leading to the creation of Tolerancing Research Groups across the whole France. It was initiated in France in the 1970s in the ENS de Cachan, by Professors Pierre Bourdet and André Clément, among others [13].

In order to initiate discussions about the future development GD & T that will unite research groups in the domain of tolerancing, this paper takes as a successful case creation of TRG in France and according to [13] it has some objectives. It consist of:

– comparing points of view on common scientific problems;
– exchanging solutions;
– bringing forward new research themes to respond to the needs of industry and others;
– promoting research into tolerancing and dimensional metrology;
– developing research in developing countries and defending a certain school;
– taking the responses proposed to the problems of tolerancing and metrology known to the industry;
– proposing solutions to the normalization organizations;
– producing collected written work in this domain, to respond to the diverse expectations of young researchers, industrialists, teachers and students.

As mentioned above, the lack of tools that can allow collaborating, share research ideas and discuss to related topics in dimensional metrology, lead to improve collaborative tools for developing GD & T in developing countries. The CIRP CAT, which refers to the Scientific Technical Committee Design (STC Dn) of the International Academy for Production Engineering is a good example of providing a technical forum for researchers, industrial practitioners, and policy makers to exchange ideas, share research findings, and discuss the various issues on the fields of dimensional and geometrical tolerancing, dimensional metrology, uncertainty management and standardization [14]. The Open Academy is a complete resource of free high quality video courses, lectures educational materials. IPPOP (Integration of Product – Process – Organization for engineering Performance improvement) is a successful project to share expertise in a collaborative design approach and to ensure the traceability of the geometric specifications [13]. Another platform to connect the world of science and make research open to all is ResearchGate.com. However, it is clear that these resources have still lack of information in GD & T domain. So, the best method for collaboration and exchanging knowledge is organizing trainings/workshops and practicing methods of Geometric Tolerancing of Products with implementation.

4. Conclusion

The important recent developments and interest among researchers in dimensional metrology is a solid basis to initiate future discussions among research groups GD & T of and uncertainty of the measurement. It is also an opportunity to develop this activity among developing countries. Taking into account that uncertainty evaluation which is one of the integral components of every research activities, concerned with measurement in GD & T, it is critical that the approach to estimating uncertainty is well defined and unilaterally applied.

Acknowledgements
This research is financially supported by the National Science Foundation of China (No. 51765012).

References

[1] David F and Hannaford J *Good 2012 Practice Guide No. 80*, pp 45-46. National Physical Laboratory.
[2] ISO/IEC 2008 ISO/IEC *GUIDE 98-3: Uncertainty of measurement - Part 3: Guide to the expression of uncertainty in measurement* JCGM (GUM:1995)
[3] *G103 - A2LA Guide for Estimation of Uncertainty of Dimensional Calibration and Testing*

Results 2008 ed Adams T M American Association for Laboratory Accreditation

[4] Taylor B Nand Kuyatt C E 1994 *Guidelines for Evaluating and Expressing the Uncertainty of NIST Measurement Results Technical Note 1297* National Institute of Standards and Technology

[5] Salsbury J 2007 Measurement Uncertainty when Calibrating Dimensional Measuring Instruments *NCSL International Workshop and Symposium*

[6] Jalid A, Hariri S, and Laghzale N E 2015 Influence of sample size on flatness estimation and uncertainty in three-dimensional measurement *International Journal of Metrology and Quality Engineering* 6, 102 EDP Sciences

[7] Doiron Tand Stoup J 1997 Uncertainty and Dimensional Calibrations *Journal of Research of the National Institute of Standards and Technology* vol **102**, №6.

[8] M3003 *The Expression of Uncertainty and Confidence in Measurement* UKAS publication sed 32012

[9] Amirkhanov A, Heinzl C, Kuhn C, Kastner Jand Groller M E 2013 Fuzzy CT Metrology: Dimensional Measurements on Uncertain Data *SCCG - 29th Proceedings Spring conference on Computer Graphics* pp 93-101

[10] Schwenke H, Knapp W, Haitjema H, Weckenmann A, Schmitt Rand Delbressine F 2008 Geometric error measurement and compensation of machines—An update *CIRP Annals - Manufacturing Technology* 57 pp 660–675

[11] Sanjid A M, Chaudhary K Pand Singhal R P 2006 Calibration of Gauge Blocks Using Automatic Phase Stepping Interferometer at NPLI with Nanometer Uncertainty *MAPAN – Journal of Metrology Society of India* vol **21** №2 pp103-109

[12] Nielsen H S 2005 Taking Dimensional Metrology to the Next Level *Proceedings of the 2005 International Dimensional Metrology Workshop*

[13] Villeneuve F and Mathieu L 2010 *Geometric Tolerancing of Products* (ISTE Ltd and John Wiley & Sons, Inc) pp 3-7

[14] Giovanni M and Stefano P 2018 Managing uncertainty in the new manufacturing era. *Procedia CIRP* 75 1-2

IOP Conf. Series: Materials Science and Engineering **715** (2020) 012085 doi:10.1088/1757-899X/715/1/012085

Design of the thrust chamber: dimensional analysis of the combustion chamber and the nozzle of rocket engine using LOX/LCH4 propellants

Hatem Houhou[1,*], Hemza Layachi[1] and Abd Elmajid Boudjemai[2]

[1]Algerian space agency, ASAL Algeria
[2]Satellite development centre, CDS Oran Algeria

*E-mail: hhouhou@asal.dz

Abstract. This paper presents a preliminary dimensional study of small thrust chamber using LOX/LCH4 as propellants. In this study, the mixture ratio of the propellants, the temperature and the pressure at the combustion chamber are used to determinate the geometric parameters of the nozzle and the combustion chamber such as the volume, surfaces, lengths, diameters, etc. our work is based on the most important required output of the propulsion engine which is the thrust force. The results of our work provides the flow areas at the exit of the nozzle and at the throat for different values of the thrust force (10-20-30 KN) and different mixture ratio of the propellants (2.7-2.8-2.9). Moreover, the volume of the combustion chamber is calculated for different proposed characteristic length (80-160-240 cm). Our results offer decisive guidelines for the design of these key parts of the thrust chamber which is required for the design of the whole rocket engine as future work.

1. Introduction

Space launch systems are composed of a great number of mechanisms assembled in order of subsystems. The performance of the vehicle depends on the specific performance of each subsystem [1, 2], which dependent on the design sets and parameters [3, 4]. Liquid-propellant rocket engines as subsystem, have been researched and studied since the late of 1930s, it can be varied depending to the details of the missions [5]. Its function is to provide thrust force to lift the vehicle which is resultant of chemical reaction of propellants. The liquid methane is classified as a no-toxic green propellant with high storage temperature (-161.6°C) comparing to the other fuels used in this field (example: liquid hydrogen -252.9°C) which mean that its storage tank need lower cost for the thermal insulation technologies [6]. Moreover, the tanks could be smaller since the density of the liquid methane is high [7], once more this is a saving of money and weight, that's explain our chose of liquid methane as a propellant for this study.

The combustion chamber is a key part of the thrust chamber where the chemical reaction take place to generate the gas with high pressure and high temperature used to produce the thrust force. To determinate the energy required for this thrust force, we need to take in consideration the mass flow rates of the propellants, their mixture ratio and the specific impulse are defined as decisive parameters for the design set.

Another main part of the thrust chamber is the nozzle, whose role is to create high exhaust velocity. In the design of this two parts, the parameters to determinate are the volume of the combustion chamber, the flow areas at the nozzle exit and at the throat. A useful parameter relative to chamber volume and the nozzle sonic throat area is the characteristic length, which is the length that a chamber of the same volume would have if it were a straight tube and had no converging nozzle section [8].

When designing higher rockets, it's better to compare data in deciding volume and shape instead of using characteristic chamber length, but for smaller rockets with fewer data available it's practical. Standards of characteristic chamber length lies between 0.8-3.0 m [9]. Again it's important to choose a value appropriate for one's propellant. A high value would in theory cause higher reaction and thus a better performance, however in reality this would cause loss of power.

In order to calculate the geometric parameters related to the design of the combustion chamber and the nozzle (volume of the combustion chamber, the flow area at the nozzle exit and at the throat), different desired engine specifications such as thrust force, specific impulse and characteristic length are combined with different pressure and temperature resultant of burns LOX/LCH4 with same mixture ratio.

2. Case of study

This study is based on the proposed engine specifications shown in the (table 1).

Table 1. Proposed engine specifications.

Parameters	Specification value
Thrust	10-20-30 (KN)
Specific impulse	250 (s)
Mixture ratio	2.7 - 2.8 - 2.9
Characteristic length	80 - 160 - 240 (cm)

Using the adiabatic flame temperature chart 'Figure 1', versus combustion chamber pressure for LOX/LCH4, we extract different combine of (Temperature (T_c)/Pressure (P_c)) to be used in next section.

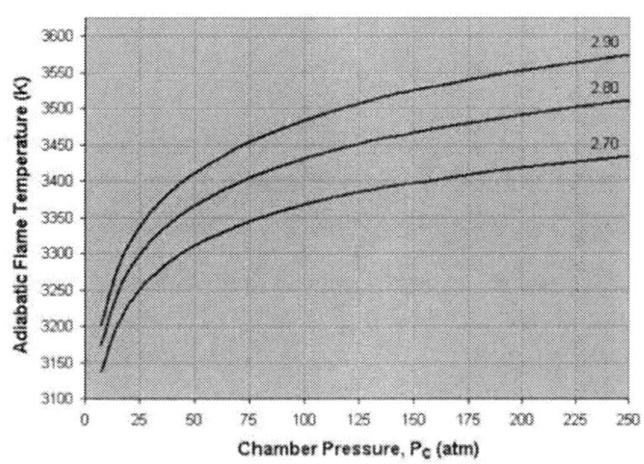

Figure 1. The adiabatic flame temperature chart [10].

3. Mathematical model

In this section, the mathematical equations used to calculate the desired parameters of our study are presented, including the volume of the combustion chamber, the nozzle and the throat exit areas.

3.1. Mass flow rate of propellants

The total mass flow of the propellants ($\dot{m}_{tot}$) is related to the thrust force (F), the specific impulse (Isp) and the gravity constant (g), described as follows:

$$\dot{m}_{tot} = \frac{F}{Isp \times g} \tag{1}$$

The mass flow of the oxidizer ($\dot{m}_{ox}$) and the fuel ($\dot{m}_{fu}$) is calculated using the mixture ratio (MR) and the total mass flow as shown in the following equations:

$$\dot{m}_{ox} = \frac{\dot{m}_{tot}}{\left(1+\dfrac{1}{MR}\right)} \tag{2}$$

$$\dot{m}_{fu} = \dot{m}_{tot} - \dot{m}_{ox} \tag{3}$$

3.2. Nozzle

At the nozzle throat, the pressure (P_t) and the temperature (T_t) are given as:

$$P_t = P_c\left(1+\frac{\gamma-1}{2}\right)^{-\gamma/(\gamma-1)} \tag{4}$$

$$T_t = \frac{T_c}{\left(1+\dfrac{\gamma-1}{2}\right)} \tag{5}$$

Mach number (M_n) is the ratio of the gas velocity to the local speed of sound. The Mach number at the nozzle exit is given by the perfect gas expansion expression:

$$M_n = \left(\left(\frac{2}{\gamma-1}\right)\left[\left(\frac{P_c}{P_a}\right)^{(\gamma-1)/\gamma} -1\right]\right)^{1/2} \tag{6}$$

where γ is the specific heat ratio and P_a is the atmospheric pressure.

The nozzle throat area (A_t) can be found if the total propellant flow rate is known and operating conditions have been selected. Assuming perfect gas law theory, we have:

$$A_t = \frac{\dot{m}_{tot}}{P_t}\left(\frac{RT_t}{\gamma M_n}\right)^{1/2} \tag{7}$$

assumed that the gas flow through the nozzle is an isentropic expansion, the expansion ratio (ε) is defined as follows:

$$\varepsilon = \frac{A_e}{A_t} = \frac{\left(\dfrac{P_c}{P_a}\right)\left(\dfrac{2}{\gamma+1}\right)^{1/(\gamma-1)}}{\left(\dfrac{\gamma+1}{\gamma-1}\left[1-\left(\dfrac{P_a}{P_c}\right)^{\gamma/(\gamma-1)}\right]\right)^{1/2}} \tag{8}$$

According to the equation (8), the nozzle exit area (A_e) is determinate as:

$$A_e = A_t\varepsilon = A_t \times \frac{\left(\dfrac{P_c}{P_a}\right)\left(\dfrac{2}{\gamma+1}\right)^{1/(\gamma-1)}}{\left(\dfrac{\gamma+1}{\gamma-1}\left[1-\left(\dfrac{P_a}{P_c}\right)^{\gamma/(\gamma-1)}\right]\right)^{1/2}} \tag{9}$$

3.3. *The combustion chamber*

The theoretically required volume of the combustion chamber (V_c) have relation with the mass flow rate of the propellants, the density of the produced combustion gases and the stay time, or combustion residence time needs for efficient combustion, expressed by the following equation:

$$V_c = \dot{m}_{tot} V_{av} t_s \tag{10}$$

where V_{av} is the average specific volume and t_s is the propellant stay-time.

Moreover, the volume of the combustion chamber can be calculated using the definition of the characteristic length (L_s):

$$V_c = A_t L_s \tag{11}$$

4. Result and discussion

Using the deferent extracted combine (T_c / P_c) from the adiabatic flame temperature chart 'Figure 1' and the proposed engine specification (table 1), the throat and the nozzle exit flow areas were calculated for deferent mixture ratio and deferent thrust force.

'Figure 2' and 'Figure 3' shows that as the thrust force is high as the flow areas are large which mean that the propulsion engine will be big. This result can be used as tools to decide the number of the used engines instead of using only one big to get the same required thrust force.

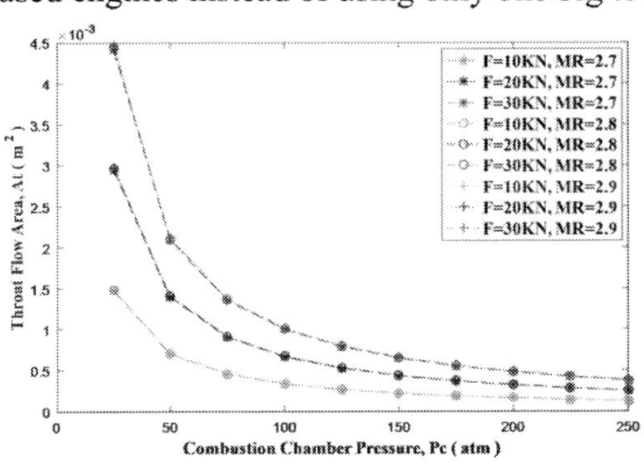

Figure 2. The throat flow area for deferent proposed thrust forces and mixture ratios.

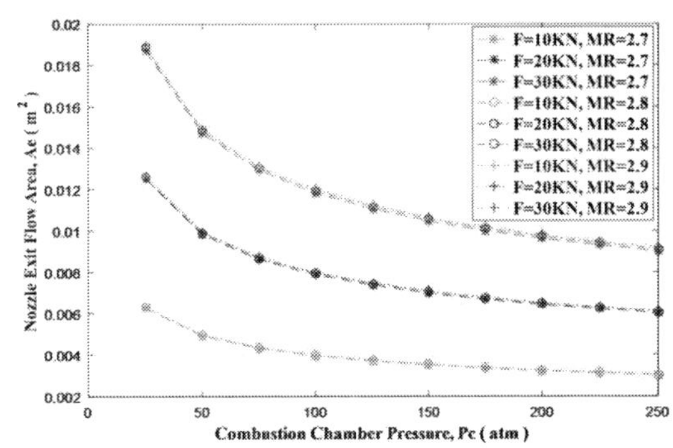

Figure 3. The nozzle exits flow area for deferent proposed thrust forces and mixture ratios.

The calculated throat flow area was combined with the proposed characteristic lengths of the combustion chamber to calculate its volume using the equation (11). 'Figure 4', 'Figure 5' and 'Figure 6' displays the volume of the combustion chamber for 80cm, 160cm and 240cm of characteristic length respectively. From this figures we note that the characteristic length is one of the decisive parameters to define the size of the combustion chamber, these lasts have proportional relationship (as the characteristic length is large, the volume the combustion chamber should be big enough) this is explained by the equation (10), since the throat area is proportion to the mass flow of the propellants and the specific volume, the characteristic length is function of stay-time needed for an efficient combustion.

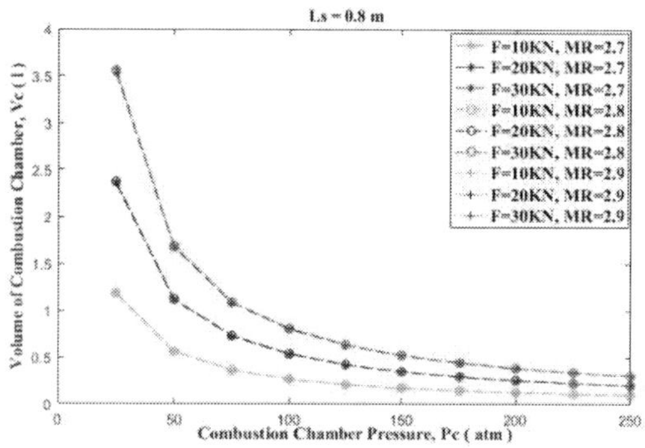

Figure 4. The volume of the combustion chamber for deferent proposed thrust forces and mixture ratios ($L_s = 0.8$ m).

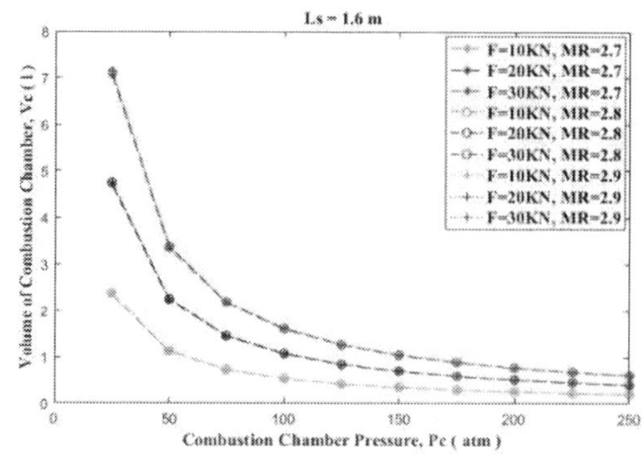

Figure 5. The volume of the combustion chamber for deferent proposed thrust forces and mixture ratios ($L_s = 1.6$ m).

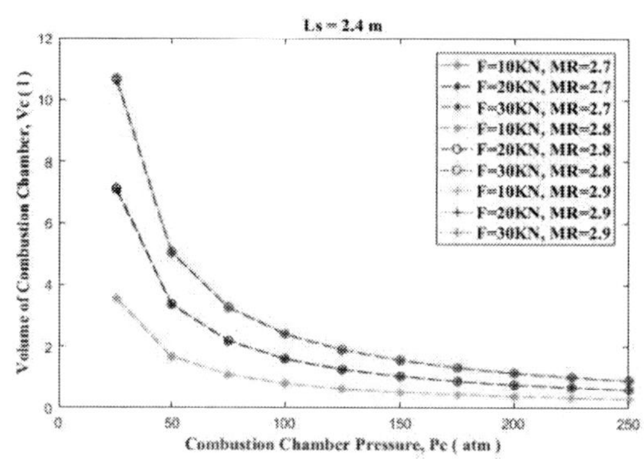

Figure 6. The volume of the combustion chamber for deferent proposed thrust forces and mixture ratios ($L_s = 2.4$ m).

5. Conclusion

This study was based on a deferent proposed engine specification combined with deferent thermodynamic parameter related to the combustion of LOX/LCH4 to determine the geometric parameters of the nozzle and the combustion chamber. The flow areas at the nozzle (throat and nozzle exit) designed to exhaust the gases resultant from the combustion and to satisfy the needed thrust, it is necessary to give the suitable dimensioning for this chamber, in this paper the different dimensions to respect in order to obtain that needed thrust were analyzed. Our results show that for high thrust force, the flow areas are large which mean that the propulsion engine to ensure that is big, this is such a problem for the manufacturing procedure but for that, using more than one small engine to get the desired thrust force is an alternative solution. Also the relation between the volume of the combustion chamber and the characteristic length is helpful for the efficiency of the combustion by deciding the ideal stay-time of the produced combustion gases.

References

[1] Krejci D and Lozano P 2018 Space Propulsion Technology for Small Spacecraft Proceedings of the IEEE 106 362-378
[2] Pamadi B, Tartabini P and Starr B 2004 Ascent, stage separation and glideback performance of a partially reusable small launch vehicle 42nd AIAA Aerospace Sciences Meeting and Exhibit 876
[3] Civek E and Özgören M K 2017 Space Launch Vehicle Design with Simultaneous Optimization

of Thrust Profile and Trajectory AIAA SPACE and Astronautics Forum and Exposition 5333
[4] Hutchinson V and Olds J 2004 Estimation of launch vehicle propellant tank structural weight using simplified beam approximation 40th AIAA/ASME/SAE/ASEE Joint Propulsion Conference and Exhibit 3661
[5] Casiano M J, Hulka J R and Yang V 2010 Liquid-propellant rocket engine throttling: a comprehensive review Journal of Propulsion and Power 26 897-923
[6] Clodic D and Toubassy J 2017 Method and device for liquefaction methane ed: Google Patents
[7] Sichler E, Gonzalez E, Montes J D, Picard R and Chandler F O 2018 Development of a Regeneratively Cooled Low Pressure Liquid Oxygen and Liquid Methane Rocket Engine 2018 Joint Propulsion Conference 4867
[8] Sutton G P and Biblarz O 2016 Rocket propulsion elements (John Wiley & Sons)
[9] Sørensen K A, Lupemba S K, Paludan J, Lippert E H and Fléron R 2017 Special Course
[10] Braeunig R A 2005 Rocket and space technology electronic resource

IOP Conf. Series: Materials Science and Engineering **715** (2020) 012086 doi:10.1088/1757-899X/715/1/012086

High precision displacement-measuring interferometer based on phase modulation technique and modulation index effect elimination

Vu Thanh Tung* and Vu Toan Thang

School of Mechanical Engineering, Hanoi University of Science and Technology, No 1 Dai Co Viet Road, Hanoi, Viet Nam

*Email: tung.vuthanh@hust.edu.vn

Abstract. A high precision displacement-measuring interferometer based on a phase modulation technique was developed. A PZT actuator was utilized to drive a mirror of a Michelson interferometer by applying a sinusoidal voltage to the PZT controller. The path difference between two arms of the interferometer was modulated leading to modulation in the phase of the interference signal with a frequency of 3 kHz. The first and second harmonics of the interference signal were detected at the modulation index of 2.63 rad, a special value when the values of the first and second orders of Bessel function are equal. The displacement was determined by the ratio of the first and second harmonic in which the effects of modulation index instability and intensity fluctuation were neglected. The phase modulation interferometer using a PZT actuator has no amplitude modulation effect compared with other modulation interferometers using EOM (electro-optics modulator) or modulation the injection current of a laser diode. Moreover, the direction of the displacement that was ambiguous of the traditional interferometers was clarified in a real time. A measurement precision of 60 nm was obtained using the phase modulation interferometer.

1. Introduction

Laser interferometers are widely utilized for displacement measurements with nanometer-order uncertainty because of their inherent accuracy and their traceability to the metric standard through the frequency of the laser source. Various signal processing techniques have been developed for displacement-measuring interferometers such as homodyne [1, 2], heterodyne [3, 4] and phase or frequency modulation techniques [5-7].

The homodyne interferometer technique is widely utilized in small-displacement measurements with very high measurement resolution. In particular, a measurement accuracy of 10 pm [8] and a resolution of sub-picometer [9] order have been reported. The interference signal of a homodyne interferometer is time independent, and therefore it enables an ultrafast response because interference converts instantaneously phase variations into intensity variations. The upper bandwidth limit is determined by the response time of the photodetector and the bandwidth of the signal-processing electronics. Therefore, homodyne interferometers have the potential to be used for high-speed applications. However, homodyne interferometers require highly stable laser intensity during each measurement. This means that the misalignment of the optics, disturbance of the environment or shifting of a measured point will strongly affect the measurement uncertainty [10]. A heterodyne interferometer is less sensitive to temperature and pressure variations but it is slower because of the delay introduced by electronic signal processing for phase acquisition. The maximum measurable speed of a heterodyne interferometer is limited by the heterodyne frequency [4]. A high-cost, volumnemous, voluminous, and complicated system are also disadvantages of heterodyne interferometers.

Among these techniques, the sinusoidal phase modulated (SPM) and sinusoidal frequency modulated (SFM) techniques have many advantages. The signal of SPM or SFM interference, which is a continuous function of time, is a series of harmonics of the modulation frequency. The phase shift, which is induced by the displacement of the target mirror in the interferometer, can be accurately extracted from the interference signal using an lock-in amplifier (LIA) [5-7]. Moreover, the measurement speed of an SPM or SFM interferometer is only limited by the modulation frequency, for which a very high frequency can be obtained by using an electro-optic modulator (EOM) or by modulating the injection of laser diodes. However, the disadvantaged feature of the SFM technique is the modulation index change when the unbalanced between two arms of the interferometer changes. The modulation index of the SPM interferometer is unchanged and the modulation index measurement is unnecessary during the operating time. However, the main problem remains of both the SPM and SFM is the residual amplitude modulation (RAM). The RAM reduces the accuracy of the measured modulation index value when the values is calculated from the amplitude of harmonics.

In this paper, a high precision displacement measuring interferometer using a phase modulation technique was proposed. The effect of the Bessel function values was neglected by using a suitable modulation index. The RAM effect is removed from the system because the phase of the interferometer is modulated using a PZT. Consequently, compared with other techniques, SPM is the most competitive for achieving high precision as well as a much wider measurement range.

2. Measurement principle

Figure 1 illustrates a sinusoidal phase modulation (SPM) Michelson interferometer. A laser beam goes through an isolator which protects the source from the reflection light. On beam splitter, the beam is divided into two paths, one goes to the reference mirror which is attached to a piezoelectric transducer (PZT). The movement of the reference mirror is modulated by sinusoidally modulating the applied voltage of PZT. Consequently, the phase of the interference signal is modulated. Another beam comes to the measurement mirror and returns the beam splitter. Two beams recombine and interfere on the beam splitter. The interference signal is detected using a photodetector.

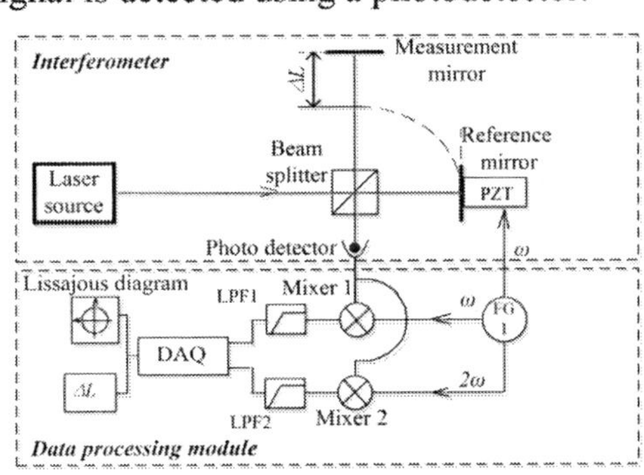

Figure 1. Phase modulation interferometer. FG: function generation; LPF: low-pass filter; DAQ: data acquisition.

The electric field in the reference arm is modulated sinusoidally and it can be expressed as:

$$E_r(r,t) = E_{0r} \times e^{i(\omega_0 t + m \sin \omega_m t)}, \tag{1}$$

where E_0 and ω_0 represent electric amplitude and carrier frequency of the laser source, ω_m and m are modulation angular frequency and a modulation index, respectively. The beam returning from the measurement mirror is represented by:

$$E_m(r,t) = E_{0m} \times e^{i\left(\omega_0 t + \frac{4\pi n}{\lambda_0}\Delta L\right)}, \tag{2}$$

where ΔL is measured displacement and λ_0 is the wavelength of the light source.

Since $I \propto E^2$, the interfering signal of two beams detected by the photodetector is written as

$$\begin{aligned} I &= \langle |E_r(r,t) + E_m(r,t)| \rangle^2 \\ &= \langle E_r(r,t) + E_m(r,t) \rangle \times \langle E_r^*(r,t) + E_m^*(r,t) \rangle \\ &= I_0\left[1 + \cos\left(\frac{4\pi n}{\lambda_0}\Delta L + m \sin \omega_m t\right)\right], \end{aligned} \tag{3}$$

where $I_0 = 2|E_{0r}|^2 = 2|E_{0m}|^2$ when the beam splitter divides the beam from the laser source into two beams propagating in the interferometer with the same intensity. From Eq. (3) m is constant and n can be determined then ΔL, the displacement of the measurement object, can be determined. However, to increase the measurement accuracy and determine the moving direction of the object, Lissajous diagram method is applied to this system. Using the Bessel function to expand Eq. (3) and it is give

$$I = I_0\left\{1 + \left\{\cos\left(\frac{4\pi n}{\lambda}\Delta L\right) \times \left[J_0(m) + 2\sum_{k=1}^{\infty} J_{2k}(m) \times \cos(2k\omega_m t)\right] - \sin\left(\frac{4\pi n}{\lambda}\Delta L\right) \times 2\sum_{k=1}^{\infty} J_{2k-1}(m) \times \sin\left[(2k-1) \times \omega_m t\right]\right\}\right\} \tag{4}$$

LIAs are used to obtain 1st and 2nd harmonic terms from Eq.(4)

$$I_1 = -I_0 J_1(m) \sin\left(\frac{4\pi n}{\lambda}\Delta L\right), \tag{5}$$

$$I_2 = I_0 J_2(m) \cos\left(\frac{4\pi n}{\lambda}\Delta L\right). \tag{6}$$

Equation (5) and (6) show that the 1st and 2nd harmonics of the interference signal are two quadrature phase signals. A Lissajous diagram obtained from the two signals can be used to clarify the direction of movement and to measure the phase shifting caused by displacement concurrently. The displacement ΔL is given by

$$\Delta L = \frac{\lambda}{4\pi n} \times \tan^{-1}\frac{I_1 \times J_2(m)}{I_2 \times J_1(m)}. \tag{7}$$

In Eq. (7), ΔL depends on the intensity of 1st and 2nd harmonics, and Bessel functions $J_1(m)$ and $J_2(m)$. Normally, the intensity fluctuation of laser source limits the measurement accuracy of homodyne interferometer. Using the ratio of 1st and 2nd harmonics (I_1/I_2) the effect of intensity fluctuation is neglected.

However, the Bessel functions, $J_1(m)$ and $J_2(m)$, which depend on the value of m can reduce the signal to noise ratio of the 1st and 2nd harmonics. In this research, a method to neglect the effect of the modulation index is proposed.

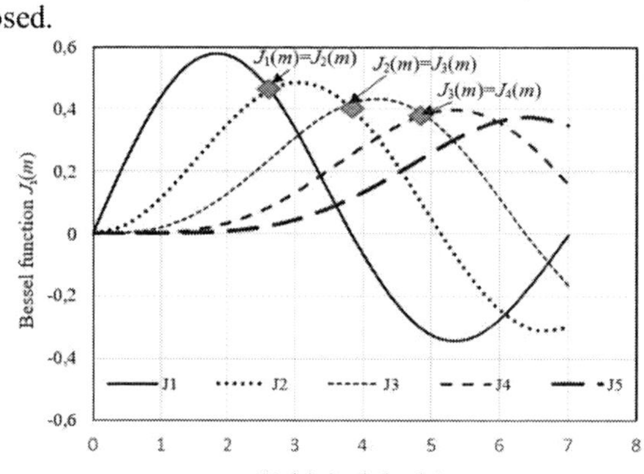

Figure 2. Bessel function.

Figure 2 shows the Bessel functions $J_1(m), J_2(m), J_3(m)$, and $J_4(m)$. There are some critical points where two consecutive Bessel functions are equal. $J_1(m) = J_2(m)$ when m=2,63 rad and $J_2(m) = J_3(m)$ when m=3,77 rad. In this research, the modulation index m=2,63 rad is used and Eq. (7) becomes

$$\Delta L = \frac{\lambda}{4\pi n} \times \tan^{-1}\left(\frac{I_1}{I_2}\right). \tag{8}$$

Equation (8) shows that the displacement ΔL is independent on the modulation index m. The Lissajous diagram is a circular and the normalized method for a nonstandard Lissajous diagram is unnecessary [5]. Therefore, the measurement uncertainties of modulation index measurement and approximation Bessel function value are removed from uncertainty sources of the proposed interferometer.

3. Measurement principle

The experimental system and the data processing module are shown in figure 3. A collimated laser diode (CPS532-C2, Thorlabs Inc.) was used as a light source for the interferometer. The movement of the reference mirror was sinusoidally modulated by a PZT actuator (PA4FKW, Thorlabs Inc.). The PZT actuator was driven by a voltage controller (PK4DMP1, Thorlabs Inc.) with the smallest increment of nanometer order. The interference signal was detected using a photodetector (PDA36A-EC, Thorlabs Inc.), figure (3a). A signal processing module was built by combining analog lock-in amplifiers and high-resolution data acquisition, figure (3b). The experimental condition is shown in table 1.

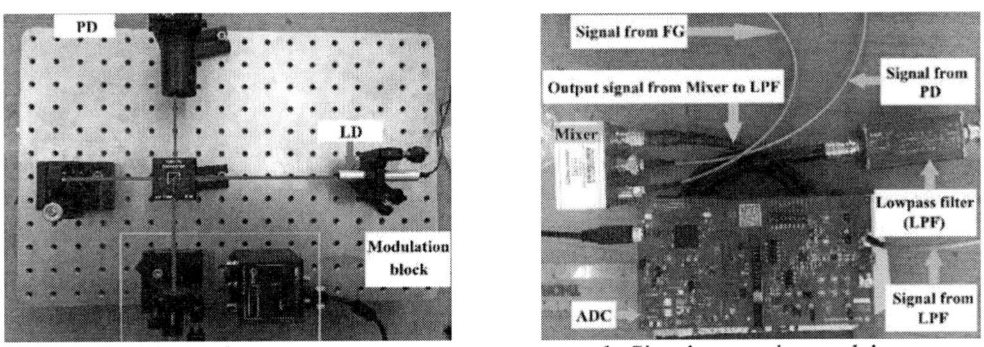

a. Experimental system b. Signal processing module.

Figure 3. Phase modulation interferometer system.

Table 1. Experimental condition.

Wavelength of laser source	532 nm
Maximum power	1 mW
Modulation frequency of PZT	3 kHz
Modulation index	2,63 rad
Resonant frequency of PZT	270 kHz
Spectral response range of detector	350-1000 nm
Frequency bandwidth of detector	DC-10 MHz
Resolution of ADC	24 Bit
Sample rates of ADC	512 kSPS

The proposed interferometer was used to measure a displacement which was generated by another PZT stage. The measuring result was compared with the reference displacement that was generated using another PZT (PK4DMP1, Thorlabs Inc.). The reference displacement can be determined directly by using a capacitive sensor (D100, Physik Instrumente). The resolution of the capacitive sensor is less than 0.5 nm, and the measurement accuracy was less than 2 nm with the measurement range of 10 μm. The interference signal and 1st and 2nd harmonics were shown in figure (4a) and figure (4b), respectively. The Lissajous diagram of 1st and 2nd harmonic was used to track the movement direction and to calculate the phase change due to the displacement of the object, figure (4c). The measured displacement obtained by the interferometer and reference displacement were depicted in figure (5a).

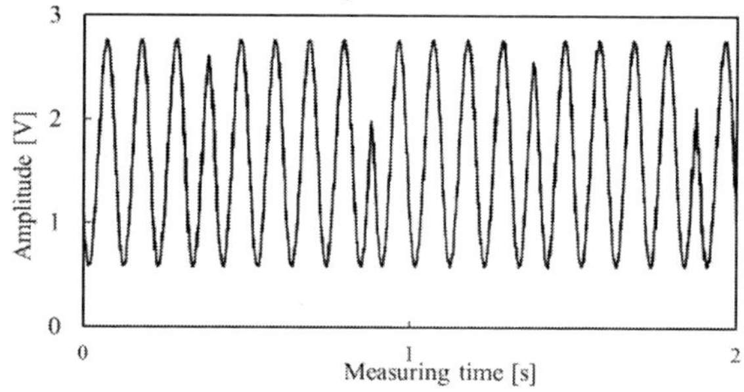

a. Interference signal.

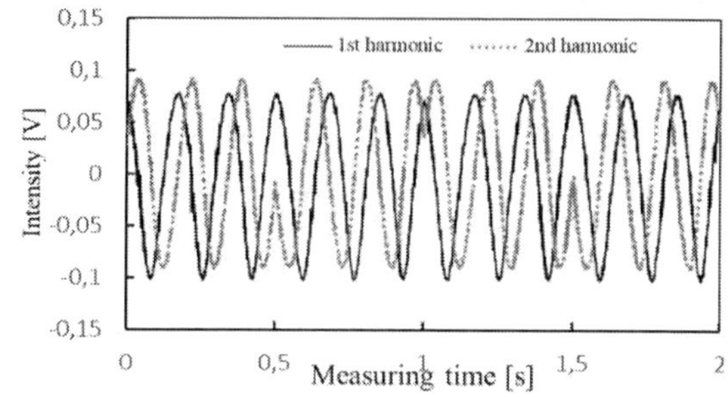

b. 1st and 2nd harmonics.

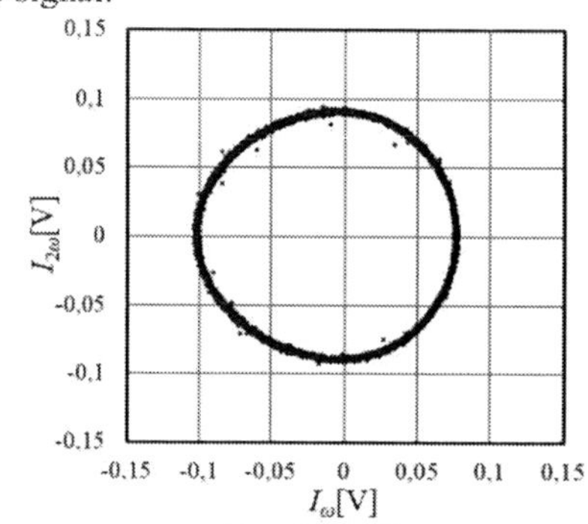

c. Lissajous diagram.

Figure 4. Demodulated signals of the phase modulation interferometer.

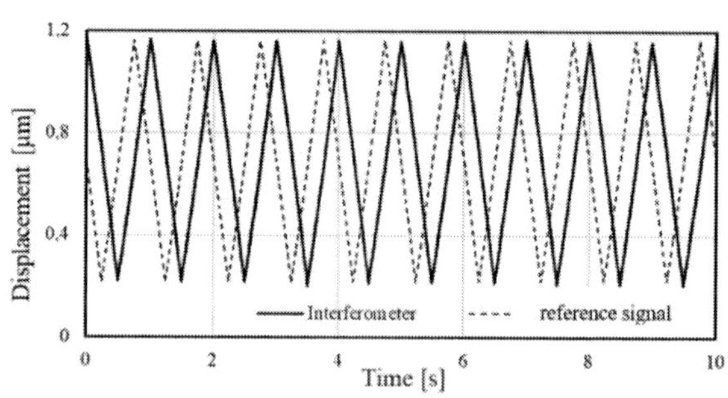

a. Displacement measurement results.

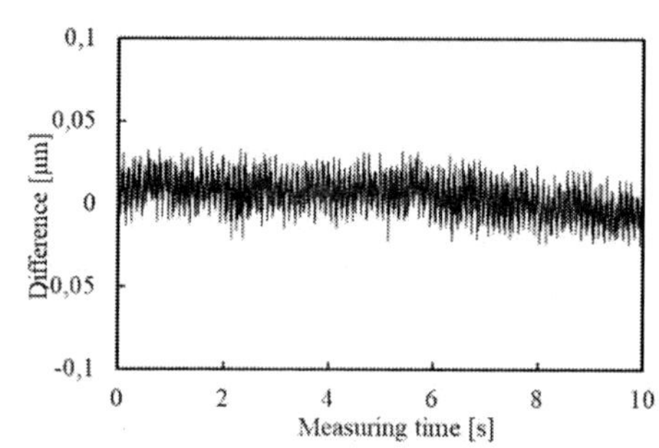

b. The difference between the measuring result.

Figure 5. The difference between the measuring result using the interferometer and the reference.

The experimental system was performed in an open space. However, 1st and 2nd harmonics were detected purely and then the displacement can be determined. It means that the phase modulation interferometer can work well even if there was the existence of the environment effect. To clarify the measurement accuracy, the difference of the displacement measurement results using the interferometer and the reference is shown in figure. (5b). The difference from peak to peak was about 60 nm. There were some uncertainty sources can be listed such as the refractive index fluctuation, vibration, and imperfectly optical polarization.

4. Conclusion

A phase modulation displacement measuring interferometer was successfully developed. The measuring system is compact, low-cost, and stable. The measurement accuracy was less than 60 nm. It can be used for industrial applications. For future work, the proposed interferometer should be compared concurrently with heterodyne interferometer to clarify clearly the measurement accuracy and measurement resolution.

Acknowledgments

This work was funded by Vietnam Ministry of Education and Training under project number B2018-BKA-07.

References

[1] P. Gregorči et al. 2009 Quadrature phase-shift error analysis using a homodyne laser interferometer Optics Express **17** 16322-31
[2] F. Petrů and O. Číp 1999 Problems regarding linearity of data of a laser interferometer with a single frequency laser Precision Engineering **23** 39-50
[3] C. C. Wu et al. 2008 Optical heterodyne laser encoder with sub-nanometer resolution Measurement Science and Technology **19** 045305
[4] F. C. Demarest 1998 High-resolution, high-speed, low data age uncertainty, heterodyne displacement measuring interferometer electronics Measurement Science and Technology **9** 1024-31
[5] Thanh-Tung Vu et. al 2016 Accurate displacement-measuring interferometer with wide range using an I2 frequency-stabilized laser diode based on sinusoidal frequency modulation, Measurement Science and Technology **27** 105201
[6] Thanh-Tung Vu et. al 2016 Sinusoidal frequency modulation on laser diode for frequency stabilization and displacement measurement Measurement **94** 927-933
[7] Duong Q. A. et al 2018 Iodine-frequency-stabilized laser diode and displacement-measuring interferometer based on sinusoidal phase modulation Measurement Science and Technology **29** 065204
[8] J. Lawall and E. Kessler 2000 Michelson interferometry with 10 pm accuracy Review of Scientific Instruments **71** 2669-2679
[8] M. Pisani 2008 Multiple reflection Michelson interferometer with picometer resolution Optics Express **26** 21558-21563.
[9] J. Ahn et al. 2009 A passive method to compensate nonlinearity in a homodyne interferometer Optics Express **17** 23299-23308
[10] K. N. Joo et al. 2010 High resolution heterodyne interferometer without detectable periodic nonlinearity Optics Express **18** 1159-1165

IOP Conf. Series: Materials Science and Engineering **715** (2020) 012087 doi:10.1088/1757-899X/715/1/012087

Metamorphic tensegrity structure for pipe inspection

Apicit Tantaworrasilp* and Robert C Richardson

School of Mechanical Engineering, University of Leeds, Woodhouse Lane, Leeds, LS2 9JT, United Kingdom

*Email: mnat@leeds.ac.uk

Abstract. This paper proposes a new concept of metamorphic tensegrity structure for pipe inspection. A new structural deformation technique is developed that can be employed to design a non-complex and lightweight robot or vehicle. This concept is to deform the cross-shaped tensegrity structure by using the cooperation of prismatic actuators (PAs) and passive joints with locking mechanisms called restrictor mechanisms (RMs). Two control levels are designed to control the structural deformation: A high-level controller for motion patterns and low-level controllers to locally control the PA lengths and restrict the cable lengths of four RMs. Experiments show that the proposed deformation technique can deform the shape and size of the tensegrity structures with the aim of creating a new generation of pipe inspection robots.

1. Introduction

Pipelines are the major equipment for transporting water, oils and gas from production sites to distribution sites. Most of them are installed in underwater or underground environments that are subjected to extreme conditions, for example, high temperature, dust, humidity and pressure. These lead to many problems in pipes such as cracks, corrosion, erosion, deposition, thermal cycling, pitting, shock loading and joint-failure [1]. Therefore, continuous inspection, maintenance and repair are strongly recommended. Often pipes are not easily accessed by humans, so effective equipment should be used to undertake the task instead. Several types of pipe inspection robots have been introduced to access the dangerous areas such as PIG type, wall-press type, wheel type, caterpillar type, Screw type, walking type, snake type, inchworm type and magnetic type [2]. The wall-press type of pipe inspection robots has high performance to negotiate vertical and curve pipelines because it can generate high traction forces between its track wheels and the inner wall pipes by using tension mechanisms that can extend and retract wheels. However, wall-press robots are generally heavy because of the rigid body and joints, leading to the need for high power actuators to generate high traction force for holding the robot during locomotion. High traction force is likely to damage the inner surface of the pipe. To alleviate these problems, we propose a new structure concept for a pipe inspection robot base on the tensegrity principle.

A tensegrity structure is one of the compliant structures. It consists of discontinuous compressive elements (struts) connected to a continuous network of tensile elements (cables) without any rigid joints to maintain a stable volume in space [3]. The tensegrity structure has unique and advantageous properties such as force distribution, lightweight mechanisms and compliant structure [4]. These properties are of interest in the field of robotics. The compliant capability to distribute external impact forces to their elements could be employed to design the compliant-structural robot to reduce or prevent the damage of robot when it faces high-external impact forces. Also, these structures would be safe for a human who works alongside or cooperates with the robot [5]. For these beneficial properties, if we can use the tensegrity structure to develop a lightweight pipe inspection robot, it would reduce the high traction force problem of existing wall-press pipe robots.

The basic locomotion of the tensegrity robot is to change the centre of gravity of the structure for rolling [6] and generate friction force between contact points for crawling [7]. There are several ways to deform the structure for locomotion. First, the use of actuator spooling the outer cables which are connected between struts generating high tension force to pull the strut close together resulting in rolling movement [8]. Another method is to make the robot rolling by adjusting the length of struts to unbalance the structure [9]. The adjusting strut length is also employed to design a duct inspection robot named DuCTTv2 [10] to move inside a pipeline. However, it can move only inside a straight pipe due to the limitation of the structural deformation. Therefore, this paper presents a new technique to deform the tensegrity structure by using the combination of the PAs and the RMs to find the new effective-deformable way of the tensegrity robot which can negotiate complex junctions in pipeline systems.

2. Conceptual design

2.1. Metamorphic tensegrity structure

The proposed concept of the metamorphic tensegrity structure is to deform the structure by using two linear motion types. The first type is the prismatic actuator (PA) that can exert forces in two directions ($F_{P1}$and F_{P2}) through actuation by a motor. The second type is the restrictor mechanism (RM) that can only exert restrictive force in one direction (F_R) by locking a cable as illustrated in figure 1. This basic tensegrity structure comprises of one PA and two RMs. They are connected by using inelastic cables. For the purpose of this investigation, one end side of the PA and RMs are fixed on a mounting plate. In this case, the PA will rotate to the left or right side as shown in figure 2 (depending upon the state of the RM).

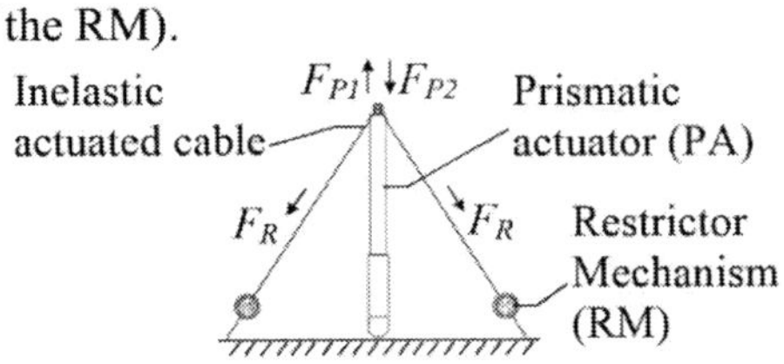

Figure 1. The basic tensegrity structure.

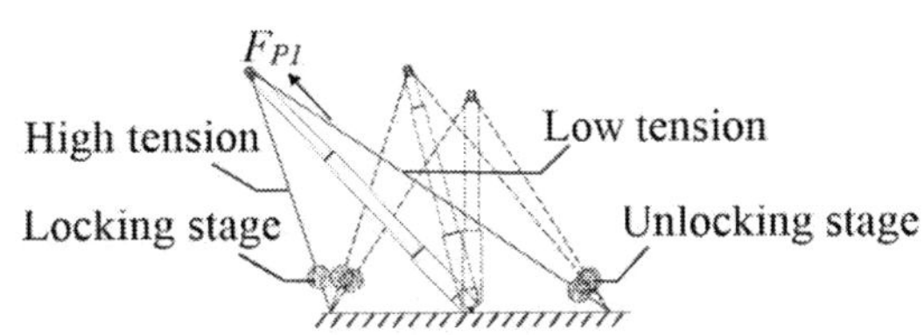

Figure 2. The structural deformation concept.

In order to change the PA position, the length of the PA and the cables will be controlled simultaneously. For example, to tend the PA on the left-hand side (see Figure 2) the RM on the left-hand side is locked, and the RM on the right-hand side is unlocked. Then, the PA length is extended. In this stage, the PA movement will rotate left side until either the cable length on the right-hand side reaches the target length and is locked or the actuator reaches its maximum length. To return to the starting position, the PA will retract its length to the initial length (minimum length), and then the cable on the right-hand side will be reeled inside the RM with the tension force (F_R).

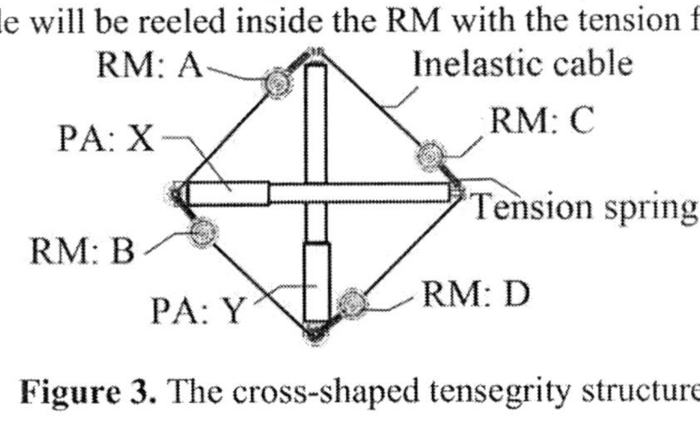

Figure 3. The cross-shaped tensegrity structure.

Table 1. The procedure design for shape changing.

PA:Y motion	RM				PA:Y action
	A	B	C	D	
UP	UL	L[a]	UL	L	E[c]
DOWN	L	UL[b]	L	UL	E
LEFT	L	L	UL	UL	R[d]
RIGHT	UL	UL	L	L	R
CW	UL	L	L	UL	E
CCW	L	UL	UL	L	E

[a] Lock, [b] Unlock, [c] Extend and [d] Retract

To thoroughly understand the movement behaviour of the proposed technique, we design a cross-shaped structure which is put on the backplane as illustrated in figure 3. It comprises of two PAs placed in X axis and Y axis, and four RMs (A to D) mounted between two PAs with inelastic cables to provide tension forces between them to form the symmetry structure. The one side of the RM is connected to the PA tip, and the opposite side of the RM is connected to an extension spring before connected to the other PA tip. The extension springs provide a structure compliant to prevent the RM and cable damage in case of over tension.

For the structure deformation analysis, we design control procedure of the locking stage of four RMs and the length of the PA in Y-axis (see in table 1) while the PA in the X-axis is fixed on the backplane because it is suitable for comparing with experimental results. The movement analysis of the proposed structure is shown in figure 4. For example, in figure 4 (a), the structural shape is changed to the top because the PA(Y) length is extended while the cables of RM (B) and RM (D) are restricted. It can be considered that the motion of the structures in figure 4 (a) and (b) could be used to make the structure moving inside a pipe like an inchworm robot. The structures in figure 4 (c) to (f) could be used to make the structure turning to the left and right directions which are useful to access complex junctions in the pipe.

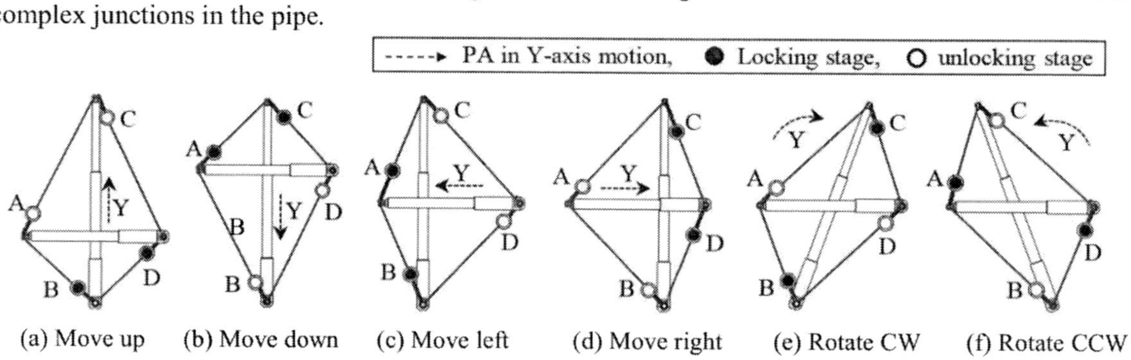

Figure 4. The shape-changing concept of the cross-shaped tensegrity structure.

2.2. Restrictor mechanism (RM)

One of the important parts of the proposed deformation technique is the RM because it is used to restrict the cable length which is spooled by the PA, and it can also passively roll the cable back to cable storage which is a useful feature to prevent the cable becoming stuck with objects surrounding them. The RM composes two main parts: a retractable part and a locking part. The spiral power spring is employed for the retractable part because it can generate a rotational force to reel the cable automatically without an actuator. A friction-based locking is used for the locking part because it can generate a high-frictional force to stop the rotational part, and it completely decouples between the frictional part and movable part leading to has a zero friction during rotational movements [11]. The conceptual design of the locking part is illustrated in figure 5.

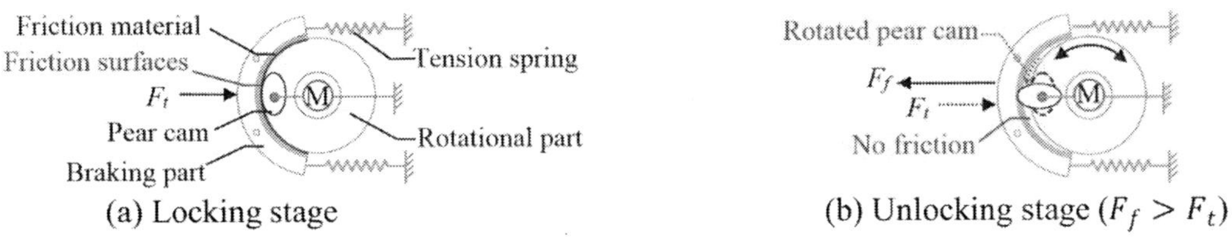

Figure 5. Locking part conceptual design; F_f is the forward force generated by a cam and F_t is the tension force generated by tension springs.

The locking part consists of a braking part, friction material, tension springs, pear cam and actuator. The braking part is designed to be curved to increase the friction force of the surface areas between the friction material and rotational parts as much as possible during locking stage as illustrated in figure 5 (a). The friction force values depend on two main factors: a tension force (F_t) relates to the spring constant of the tension springs and the material contact property of the friction material associated with the coefficient of friction (COF). For the unlocking stage, the friction surfaces between the friction material part and the rotational part are entirely decoupled (see in figure 5 (b)). The pear cam is rotated to push the breaking part out the rotational part. The pear cam is driven by the actuator to generate forward force (F_f) that must be higher than the spring tension force (F_t). Therefore, the selection of the tension springs and the actuator for the locking part should be proper.

3. Hardware design

The property requirements of the restrictor mechanism (RM) are to restrict cable lengths by using locking parts, to reel the cable keeping in the cable storage automatically and to measure the cable length accurately. Therefore, mechanical design of the RM consists of two main parts; locking parts and retractable parts including a cable length sensor.

3.1. Locking parts

The locking parts are designed by using the friction-based method. The design of the 3D model of the locking parts is illustrated in figure 6. A high-frictional force to stop rotational parts is generated by using two tension springs gripped between braking parts and a fixed base to create a pressing force. The moveable braking part is attached to a rubber sheet 1mm thick called a friction material that contacts to the rotational part surface in the locking stage. This braking part can be linearly moved in two directions (passively forward and actively backwards) via a slide guide attached to the fixed base. A miniature DC motor with 287:1 geared head is employed to generate the forward force against the pressing force through a hard coupling. This coupling is attached to a ball plunger mounted at the front surface. This ball plunger will push an unlocking guide during the unlocking stage to separate the friction material from the rotational part.

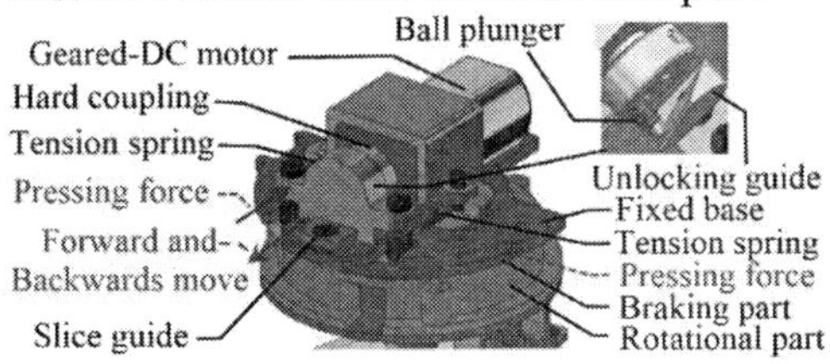

Figure 6. The 3D model of the locking parts.

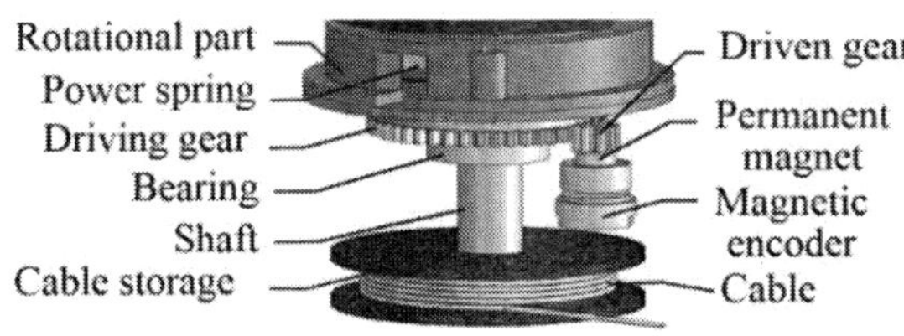

Figure 7. The 3D model of the retractable part and the magnetic encoder installation.

3.2. Retractable part

The retractable part is designed for retracting the inelastic cable (stainless steel 0.5 mm diameter, length 400 mm with 0.12 kN) into cable storage passively. Figure 7 shows the design of the retractable part. It composes of a spiral power spring which is installed inside the rotational part. This part is attached to the stainless-steel shaft (¼ inch diameter) impaled through an enclosure case. The miniature magnetic encoder (model: RM08ID0009B02L2G00, 512 counts per revolution, IP68) and a permanent magnet (4mm diameter) generating magnetic fields are employed. The magnetic encoder is placed on the bottom of the enclosure case. It senses the magnetic field variation generated from the permanent magnet mounted over the magnetic encoder. This magnet can be rotated by using a driven gear (8 teeth) which is driven by a driving gear (46 teeth) attached to the shaft. From these mechanisms, the magnetic encoder can sense the variation and direction of the magnetic field depending on the shaft rotation.

4. Control system design

In this section, the design of a simple control system for the cross-shape tensegrity structure will be described. It is divided into two levels; a high-level controller (HC) which is a master board and a low-level controller (LC) which is a slave board. The HC is used to send control procedures to six LC including two prismatic actuators and four restrictor mechanisms. To control the structure, the HC sends cable length target and prismatic actuator length target to the LC. The HC also receive data from the LC to process data and generate control signals. The PI controller is used to control the prismatic length. The PI parameters are found by using a trial and error method while the ON/OFF control is used to restrict the cable length of each RM. A serial communication I2C is used to communicate between the HC and the LCs.

5. Hardware integration and implementation

This section describes the assembly of the custom RM, the RM performance test, control board development and the implementation of a cross-shaped tensegrity structure. The RMs are made of two material: PLA material with 3D printing technology for non-frictional parts and metal material including aluminums, brass and stainless steel for frictional parts which must withstand the high-frictional forces between a rotational part and fixed part during rotating and braking process. The assembly process of the RM is illustrated in figure 8, and the physical specifications of the RM are summarised in table 2.

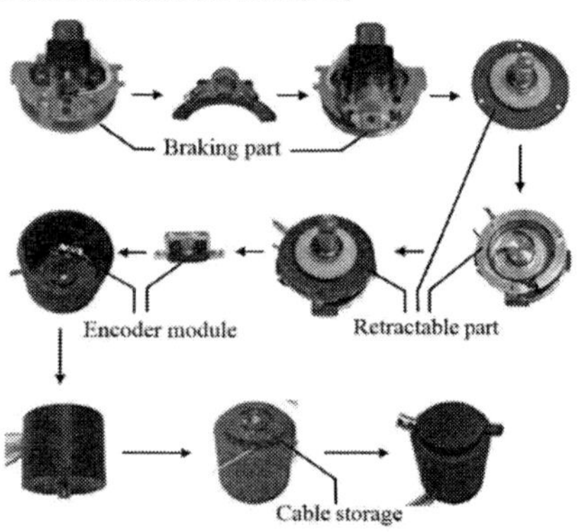

Figure 8. The RM assembly process.

Table 2. Physical specifications of the RM components.

Components	Values
Maximum cable length	400mm
Maximum power spring turn	Six revolutions
RM strain	8
Torsion spring	0.92N
Torsion spring constant	0.13 N/mm
Geared motor torque	0.23 Nm
Geared motor maximum speed	40rpm
Magnetic rotary encoder	512 counts per revolution

In order to evaluate the locking performance, a spring balance was employed to measure the rotational force of the RMs. The experimental result shows that RM can withstand the pulling force of 10N as illustrated in figure 9, and the relationship between cable lengths and the rotation force of four RMs are shown in figure 10. This experiment helps us to calibrate the rotational forces of each RM equally which improves the shape-changing performance.

One HC control board and six LC boards are developed to control the length of PAs and restrict the cable of four RMs. Each LC control board consists of a microcontroller (Arduino Nano V4), a motor driver (MAX14870 for the RM modules and VNH5019 for the PA modules) and an encoder counter (Counter click with LS7366R) to gather the encoder signal before sending to the microcontroller.

Figure 9. The locking performance test.

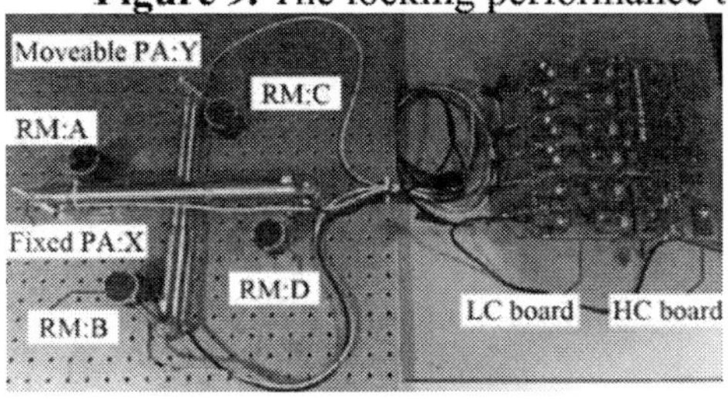

Figure 11. The experimental setup.

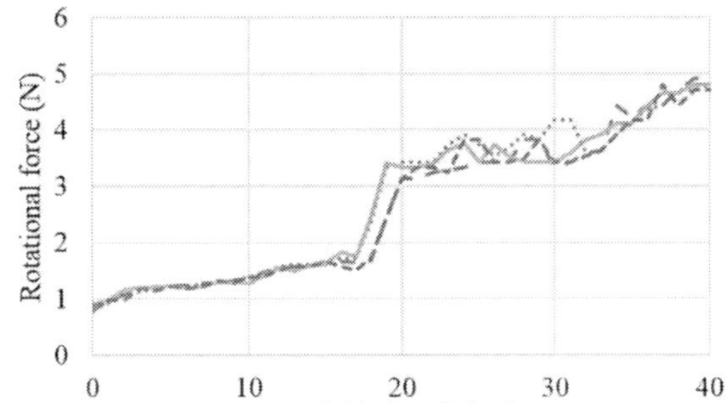

Figure 10. The relationship between the cable length and the rotational force of four RMs.

The cross-shaped tensegrity structure is constructed as illustrated in figure 11. Two PAs (LA35 series with 150 mm stroke, 350mm original length, 400N, 16mm/s and 0.1429mm/pulse) are placed on a backplane by fixing one PA in X-axis and another one in Y-axis which is moveable. The four RMs and tension spring are mounted between PAs to make the symmetry structure.

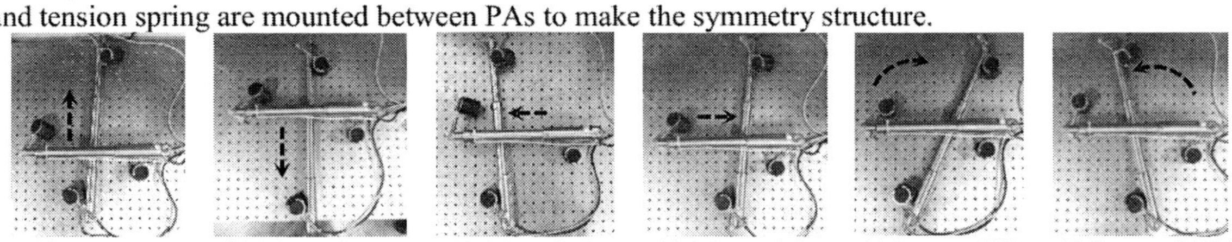

(a) Move up (b) Move down (c) Move left (d) Move right (e) Rotate CW (f) Rotate CCW

Figure 12. The experimental result of the cross-shaped tensegrity deformation.

In order to deform the cross-shape tensegrity structure, the control procedure in table 1 is utilised to generate control signals sending to each LC control boards. The experimental results in figure 12 (a) to (f) showed that the structure shape could be deformed similarly to the cross-shaped structure analysis. The custom RMs could restrict and reel the cables suitably, and the PA could be moved to the desired position resulting in the change of overall sizes. The PA movement behaviours could be used to design a robot which would be able to change its shape getting into different sizes of pipelines by extending and retracting its length against a pipe wall. Besides, the RM features could be employed to turn the structure in different angles to negotiate complex junctions in the pipelines.

6. Conclusions and future works

In this paper, we have presented the concept of a tensegrity structure deformation by using the cooperation between the RMs and the PAs. Unlike any other known tensegrity robot deformation technique, the proposed technique is to restrict the cable length by the RMs while the cable has been spooling by PAs leading to the design of a non-complex structure and control algorithm to deform the structure. The experimental results showed that the custom RM performed efficiently to restrict the cable lengths in order to deform the cross-shaped tensegrity structure, and it is sufficient to use the RMs and the proposed technique to design a pipe inspection robot based on tensegrity structure in the future.

References

[1] Shukla A and Karki H 2016 Application of robotics in onshore oil and gas industry—A review Part I, *Robotics and Autonomous Systems* vol. **75** pp. 490-507.
[2] Mills G, *et al.* 2017 Advances in the Inspection of Unpiggable Pipelines *Robotics* vol. **6**, p. 36
[3] Pugh A 1976 *an introduction to tensegrity* United States of America: University of California

[4] Skelton R E and de Oliveira M C 2009 *Tensegrity systems* vol. **1**. University of California USA: Springer.
[5] Kim, *et al.* 2014 Rapid prototyping design and control of tensegrity soft robot for locomotion *Robotics and Biomimetics (ROBIO) 2014 IEEE Int. Conf.* pp. 7-14.
[6] Sabelhaus A P, *et al.* 2015 System design and locomotion of SUPERball, an untethered tensegrity robot *IEEE Int. Conf. on Robotics and Automation (ICRA)* pp. 2867-73.
[7] Tietz B R, *et al.* 2013 Tetraspine: Robust terrain handling on a tensegrity robot using central pattern generators *IEEE/ASME Int. Conf. on Advanced Intelligent Mechatronics* pp. 261-67
[8] Chen L H, *et al.* 2016 Soft spherical tensegrity robot design using rod-centered actuation and control *ASME 2016 Int. Design Engineering Techical Conf. & Computers and Information in Engineering Conf. (IDETC/CIE 2016)* Charlotte North Carolina.
[9] Luo A, *et al.* 2016 Motion simulation of six-bar tensegrity robot based on Adams *IEEE Int. Conf. on Mechatronics and Automation* pp. 264-69.
[10] Friesen J M, *et al.* 2016 The second generation prototype of a Duct Climbing Tensegrity robot, DuCTTv2 *IEEE Int. Conf. on Robotics and Automation (ICRA)* pp. 2123-28.
[11] Plooij M, *et al.* 2015 Lock Your Robot: A Review of Locking Devices in Robotics *IEEE Robotics & Automation Magazine* vol. **22** pp. 106-17.

IOP Conf. Series: Materials Science and Engineering **715** (2020) 012088 doi:10.1088/1757-899X/715/1/012088

FFOCT imaging based on compressed sensing

Ning Mu*, Feng Liu, Hongxia Guo and Wanrong Gao

School of Electronic and Optical Engineering, Nanjing University of Science and Technology, Nanjing, Jiangsu, 210094, China

*Email: mu_ning@hotmail.com

Abstract. With the demand for high-resolution images, the storage space occupied by the images of the full-field optical coherence tomography (FFOCT) system is becoming more and more intensified. How to reduce the data volume of high-resolution image systems has become a research hotspot in the image processing industry. In particular, precision medicine requires high-resolution images of organs and tissues, and this data storage system requires a large storage space. For this reason, based on the high-resolution tomographic image obtained from our organ tissue, compressed sensing theory is used to perform compression reconstruction simulation. It is verified that the amount of data stored in the three-dimensional image of the diseased tissue is effectively reduced without changing the image resolution.

1. Introduction

Tomography technology can be used to study the complex internal structure of an object.Tomographic imaging could visualize the internal structure of an object, such as a floppy disk, air plastic, a casing, or a magnetic storage medium object. At present, optical coherence tomography is a research hotspot in the medical imaging world. Full-field optical coherence tomography (FFOCT) system can obtain optical tomographic images with micron resolution in three dimensions [1-2]. The key question in clinical application of FFOCT is whether it can obtain images with the same resolution as traditional pathological sections. Based on the FFOCT system developped in our lab, the high-resolution sectional images of human tissues including esophagus, intestine and uterus were obtained for the first time. However, as the resolution of the image obtained by this method is high, and this high-resolution image occupies a large storage space. How to reduce the amount of data stored in the high-resolution image system has become a research hotspot in the image processing industry. This is especially useful for precision medicine. In this paper, we have tried to solve the above problem based on compressed sensing (CS) theory [3-4].

2. The principle of compression and reconstruction

2.1. Signal sparse representation theory

Signals that are sparse or compressible under a certain transform domain are preconditions for compressed sensing theory and are the basis for many modern signal processing theories. How to design an effective transformation base ,which could make the number of projection coefficients of the signal in the dictionary less, is the premise of using CS theory to achieve signal acquisition [3-4].

Let $\mathbf{x}=[x_1,x_2,\cdots,x_N]^{\mathrm{T}}$ be a vector, and the matrix $\mathbf{A}=[a_1,a_2,\cdots,a_N]$ is a transformation base. The representation of the signal $\mathbf{x}$ on the transformation base $\mathbf{A}$ is:

$$\mathbf{x} = \sum_{n=1}^{N} s_n \mathbf{A}_n = \mathbf{A}\mathbf{s} \tag{1}$$

where $\mathbf{s} = [s_1, s_2, \cdots, s_N]^T$ is a $N \times 1$ weight coefficient vector. A vector $\mathbf{x}$ is said to be sparse or compressible if the number of non-zero elements in $\mathbf{s}$ is K or greater than K ($K \ll N$). At present, sparse dictionaries used to represent signals mainly include orthogonal base dictionaries and overcomplete redundant dictionaries. Commonly used orthogonal bases include Fourier orthogonal bases, short-time Fourier orthogonal bases, wavelet bases, etc. [5-6]. The sparsely decomposed signal representation theory under the overcomplete dictionary has become a hot topic in the field of signal sparse representation [7-12].

2.2. Non-correlated measurements of signals

The linear measurement of sparse signal $\mathbf{x}$ can be expressed as:

$$\mathbf{y} = \mathbf{\Phi}\mathbf{x} \tag{2}$$

where $\mathbf{\Phi} = [\varphi_1 \quad \varphi_2 \quad \cdots \quad \varphi_M]$ is measurement matrix. Substituting (1) into (2), we can get

$$\mathbf{y} = \mathbf{\Phi}\mathbf{x} = \mathbf{\Phi}\mathbf{A}\mathbf{s} \tag{3}$$

Equation (3) can be abbreviated as

$$\mathbf{y} = \mathbf{\Psi}\mathbf{s} \tag{4}$$

where $\mathbf{\Psi} = \mathbf{\Phi}\mathbf{A}$ is called perceptual matrix or recovery matrix. In the transformation process above, only by satisfying the restricted isometry property (RIP) [13], it is possible to recover the original signal $\mathbf{x}$ from the observation signal y through the reconstruction algorithm.

RIP guidelines are:If there is a constant $\delta_K \in [0,1)$. For any $\mathbf{s}$, equation (5) is established.

$$(1-\delta_K)\|\mathbf{s}\|_2^2 \le \|\mathbf{\Psi}\mathbf{s}\|_2^2 \le (1+\delta_K)\|\mathbf{s}\|_2^2 \tag{5}$$

We say the awareness matrix or recovery matrix $\mathbf{\Psi}$ meets the RIP criteria, and δ_K is the constant of $\mathbf{\Psi}$ in RIP.

Although the RIP criterion has a clear theoretical definition, it is difficult to determine whether a perceptual matrix satisfies the criterion, and it is difficult to design a measurement matrix from this criterion. In addition, the RIP criterion is a sufficient condition for signal recovery and is not a necessary condition. According to RIP guidelines, it has been found that many matrices satisfy the conditions, and an effective measurement matrix design method is also proposed. The common compression measurement matrices are: Gaussian random measurement matrix [14], very sparse projection matrix [15], sub-Gaussian random matrix [16], Bernoulli random matrix, and so on. We can use these measurement matrices to recover the signal.

2.3. Signal reconstruction algorithm

Signal reconstruction refers to the process of recovering the original signal from the compressed sampled observation signal. Signal reconstruction algorithm is the most important content of the three cores of compressed sensing theory, which directly affects the quality of the reconstructed signal and the complexity of the reconstruction. The premise of signal reconstruction is the sparse property or compressible nature of the signal. The essence is to solve the optimization problem of sparse constraints. Sparse constraints of the signal can be achieved with a minimum norm l_0. The optimization problem for directly solving the norm l_0 is an NP problem, it is difficult to solve the combination of all its signals, and even the reliability of the solution cannot be verified. Therefore, the norm l_0 is used instead of solving the norm l_0. The commonly used algorithms are: base tracking algorithm (BP), matching pursuit algorithm (MP) [17-18], smooth norm l_0 method (Smoothed, SL0) [19-20] and so on. The matching tracking algorithm has less complexity and fast calculation speed, which is a very effective algorithm.

Matching pursuit algorithm is a kind of greedy algorithm. After continuous improvement, the improved matching tracking algorithms mainly include OMP algorithm, Stagewise OMP (St OMP) algorithm, Regularized OMP (ROMP) algorithm and Compressive Sampling Matching Pursuit (Co Sa MP) algorithm, etc. Taking the OMP algorithm as an example, the idea of the matching pursuit algorithm is to obtain the sparse decomposition result of the signal when the convergence condition is satisfied by continuously searching for the largest correlation component between the residual vector and the measurement matrix. The idea of orthogonal matching pursuit is essential to solve K-sparse. For simplicity, assume $K=1$, and solve the position q of the unique non-zero element y_q in y and the column. Then iterate and find other K values. Match tracking uses "greedy" ideas to find these column vectors. The residuals are selected to be orthogonal during each iteration and then subtracted from the residuals to enter the next iteration until the iteration exit condition is satisfied. The specific flow of this algorithm is as follows[13]:

a: The initial margin is set to $r_0 = y$; the iteration number is set to $n=1$; the index value set is $\Lambda = \Phi$, $J = \Phi$;

b: Calculate the correlation coefficient u, and store the serial number of the maximum value in the correlation coefficient in J;

c: Update support set Φ_Λ , where $\Lambda = \Lambda \cup J$;

d: Get $\hat{S}$ from $\hat{S} = \arg\min \|Y - \Psi_\Lambda \hat{S}\|_2$,and use $r_{new} = Y - \Psi_\Lambda \hat{S}$ to calculate the margin;

e: If $\|r_{new} - r\| \ge \varepsilon$, let $r = r_{new}$, and $n = n+1$, then go to b; otherwise, the iteration is terminated.

3. Image compression reconstruction processing

In the experiment, based on the compressive sensing technology, we used the tomographic imaging res ults of human esophageal and intestinal cancer tissues. In the study, we used BP algorithm, OMP algor ithm and StOMP algorithm respectively, the sampling rate is 0.6. When using these algorithms, the PS NR and error of the BP algorithm are better than those of the OMP algorithm and the StOMP algorith m, which shows that the compressive sensing technology can be applied to medical image processing, and the BP recovery algorithm is more suitable, which is specifically given in table 1-5. The results of compression reconstruction experiments on FFOCT imaging were performed, as sho wn in figure 1-5.

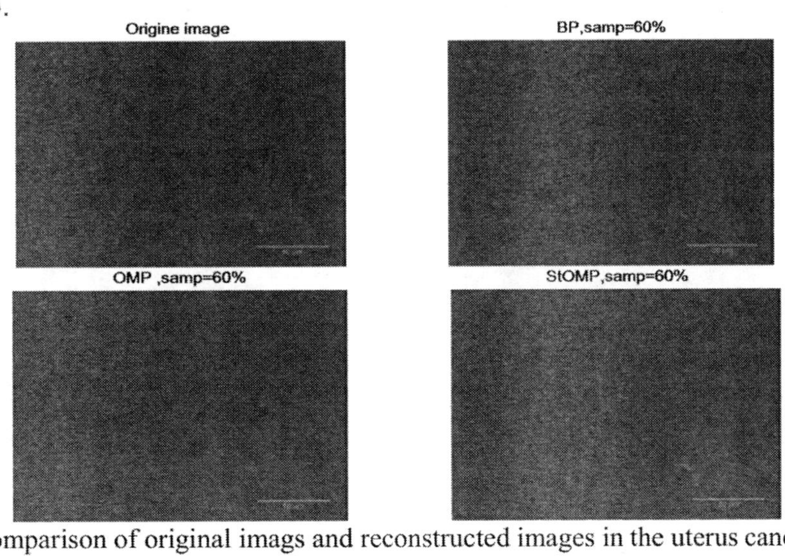

Figure 1. Comparison of original imags and reconstructed images in the uterus cancerous region.

Table 1. PSNR and error of each algorithm.

	BP	*OMP*	StOMP
PSNR(dB)	40.1828	36.2949	29.4387
ERROP	0.0388	0.0606	0.1335

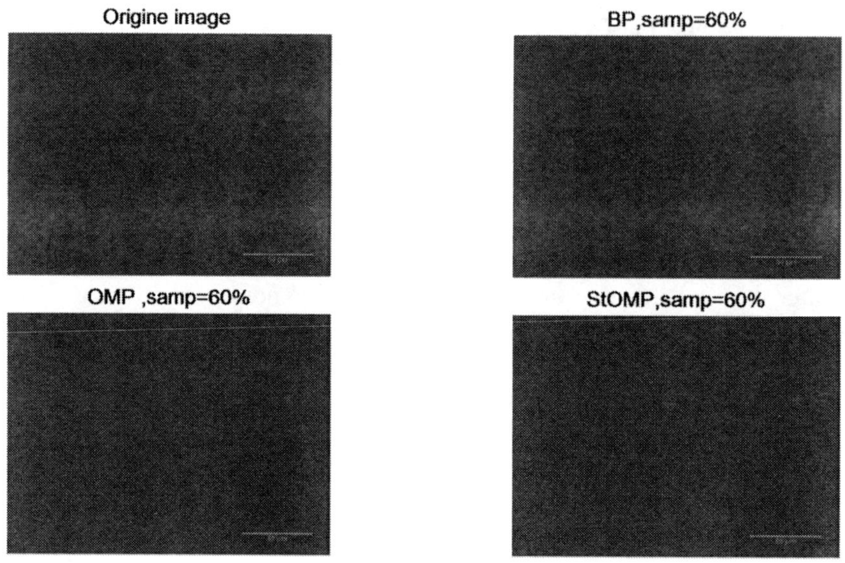

Figure 2. Comparison of original uterus area imags and reconstructed images.

Table 2. PSNR and error of each algorithm.

	BP	*OMP*	StOMP
PSNR(dB)	41.0862	37.1882	30.2238
ERROP	0.0370	0.0580	0.1292

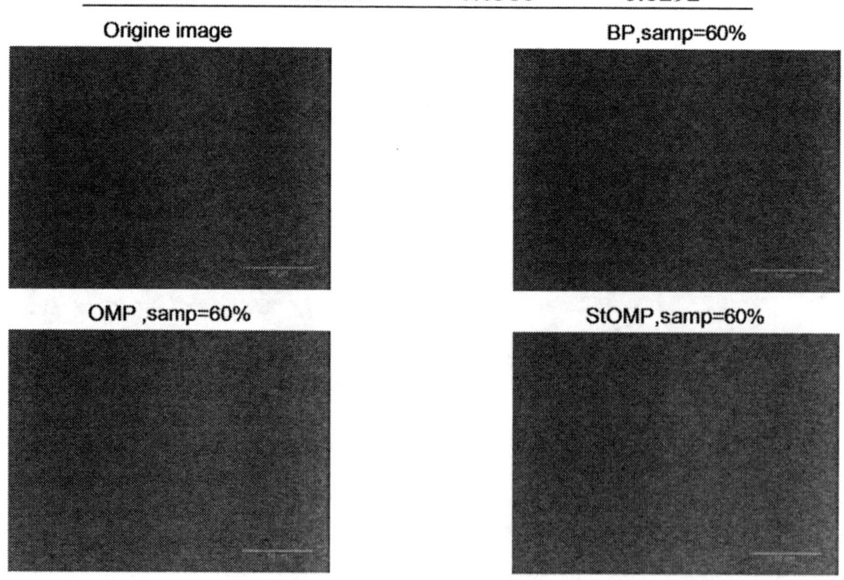

Figure 3. Comparison of original images of intesinal cancer area and reconstructed images.

Table 3. PSNR and error of each algorithm.

	BP	*OMP*	StOMP
PSNR(dB)	41.4594	37.5752	30.5356
ERROP	0.0364	0.0569	0.1280

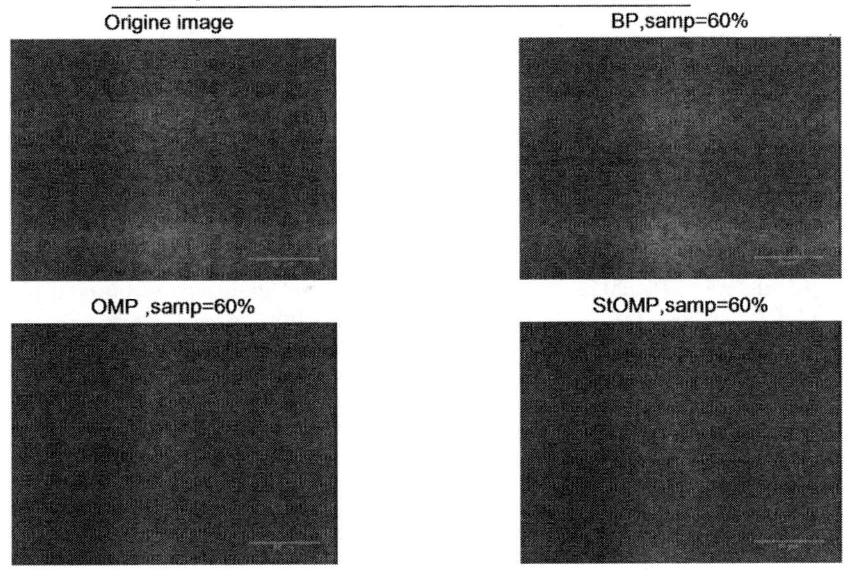

Figure 4. Comparison of original images of esophagus cancer area and reconstructed images.

Table 4. PSNR and error of each algorithm.

	BP	*OMP*	StOMP
PSNR(dB)	41.8340	37.9266	30.8289
ERROP	0.0356	0.0559	0.1265

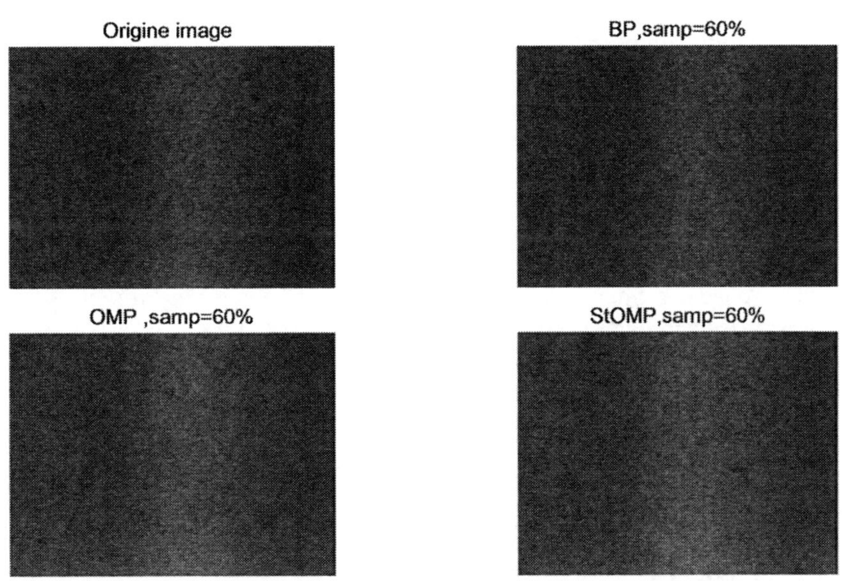

Figure 5. Comparison of original images of the intesinal area and reconstructed images.

Table 5. PSNR and error of each algorithm.

	BP	*OMP*	StOMP
PSNR(dB)	44.2399	40.3470	34.1429
ERROP	0.0410	0.0641	0.1309

4. Conclusion

The experimental results of compressive reconstruction studies of FFOCT imaging are illustrated in this paper. When adopting BP algorithm, OMP algorithm and StOMP algorithm respectively, the PSNR and error of BP algorithm are better than those of OMP algorithm and StOMP algorithm, which shows that compressive sensing technology can be applied to medical image processing, and BP recovery algorithm is more suitable.

References

[1] Bian Haiyi, Gao Wanrong, Zhang Xianling and Chen Chaoliang Reconstruction method based on the detected matrix for spectral-domain optical coherence tomography 2014 *Acta Optica Sinica* 34(2): 0211003

[2] Mandache D, Dalimier E and Durkin J R Basal cell carcinoma detection in full field OCT images using convolutional neural networks 2018 *IEEE Int. Symp. On Biomedical Imaging* p784-787

[3] Candés, Emmanuel J, Romberg, Justin K, Tao and Terence Robust uncertainty principles: Exact signal reconstruction from highly incomplete frequency information 2006 *IEEE T. Inform. Theory* 52(2) p 489-509

[4] Donoho and David L Compressed sensing 2006 *IEEE T. Inform. Theory* 52(4) p 1289-1306

[5] Lv Fangxu, Zhang Jingcheng and What Quan Adaptive compressed sensing reconstruction algorithm based on Fourier transform 2014 *J. Beijing Aviation University* 40(4)

[6] Rong Yanxia and Qiu Xiaohui Block compressed sensing alogrithm based on wavesform 2015 *Compu. tech. dev.* (5) p 29—32

[7] Knill C, Schweizer B and Sparrer S High range and doppler resolution by application of compressed sensing using low baseband bandwidth OFDM radar 2018 *IEEE Transactions on Microwave Theory & Techniques* 66(7) p 3535-46

[8] J C Ni, Q Zhang, Y Luo and L Sun, Compressed sensing SAR imaging based on centralized sparse representation 2018 *IEEE Sensors Journal* 18(12) p 4920-32

[9] Y Li and Y Liang, Compressed sensing in multi-hop large-scale wireless sensor networks based on routing topology tomography 2018 *IEEE Access* p 27637-50

[10] B Deka, S Datta and S Handique, Wavelet tree support detection for compressed sensing MRI reconstruction 2018 *IEEE Signal Proc. Let.* 25(5) p 730-734

[11] B Hamdaoui, B Khalfi and M Guizani, Compressed wideband spectrum sensing: concept, challenges, and enablers 2018 *IEEE Communications Magazine* 56(4) p 136-141

[12] R Mendoza-Smith, J W Tanner and F Wechsung A robust parallel algorithm for combinatorial compressed sensing 2018 *IEEE T. Signal Proces.* 66(8) p 2167-77

[13] Jian Wang, Wei-xing Sheng, Yu-bing Han and Xiao-feng Ma Adaptive beamforming with compressed sensing for sparse receiving array 2014 *IEEE Transactions on Aerospace & Electronic Systems* 50(2) p 823-833

[14] Emmanuel Candes, J Romberg and T Tao Stable signal recovery from incomplete and inaccurate measurements *2010 Communications on Pure & Applied Mathematics* 59(8) p1207-23

[15] Fang H and Zhang Q B Method of image reconstruction based on very sparse random projection 2007 *Computer Engineering & Applications*

[16] Fang H, Zhang Q and Sui A W A Method of image reconstruction based on sub-gaussian random projection 2008 *Journal of Computer Research & Development* 45(8) p 210-214

[17] Mallat S G and Zhang Z Matching pursuits with time-frequency dictionaries 1993 *IEEE T. Signal Proces*. 41(12) p 3397 - 3415

[18] Tropp J A and Gilbert A C. Signal recovery from random measurements via orthogonal matching pursuit 2008 *IEEE T. Inform. Theory* 53(12) p 4655-66

[19] Mohimani G H, Babaie-Zadeh M and Jutten C Complex-valued sparse representation based on smoothed l0 norm 2008 *IEEE Int. Conf. on Acoustics* p 3881-84

[20] Mohimani G H, Babaiezadeh M and Jutten C Fast sparse representation based on smoothed l0 Norm 2007 *Lecture Notes in Computer Science* 4666 p 389-396

IOP Conf. Series: Materials Science and Engineering **715** (2020) 012089 doi:10.1088/1757-899X/715/1/012089

A new automated drip irrigation and fertigation system for sugarcane crops

Diana Rose Rivera*, Sharaful-Ilmi Paduman, Leapheng Uon and Alvin Chua

Mechanical Engineering Department, De La Salle University, Manila Philippines

*Email: diana_rose_rivera@dlsu.edu.ph

Abstract. This paper presents a new automation system specifically design for sugarcane crops to optimize the irrigation and fertigation process against resource shortage and climate change. The system uses a programmable logic controller, sensors, actuators, and other mechanical equipment to control and manage the supply of water and nutrients based on the fertilizer requirement at sugarcane's development stages. The simulation was done using a commercially available tool Picosoft V6.22, a software compatible with Allen Bradley Programmable Logic Controller. The economic analysis was also provided to show the advantages of using an automated irrigation as compared to the manual irrigation. Result of the analysis shows a payback period of a month as compared to 4.3 months for the manual irrigation system.

1. Introduction

The needs of irrigation in sugarcane have been recognized for over 100 years. In the Philippines, sugarcane farming has Php 87 billion annual contributions to the economy, which comprises of 65,000 farmers, 700,000 workers with 5 million dependents [1]. Based on the report by Philippines Statistics Office in 2014, in terms of its volume of production, sugarcane is the highest among all major crops, with 25,029,880 metric tons per year followed by rice with 18,967,826 metric tons [2].

Currently, the challenges in the sugarcane farming involves the changing climatic conditions and weather patterns (wherein water is a major limiting factor for productivity), the use of water and fertilizer consumption, and sugar production at a lower cost [1].

1.1. Irrigation methodology

Surface irrigation methods are the earliest known method for agriculture. It is the easiest and most practice irrigation recorded in history. More than three-fourths of the irrigated agricultural land is surface irrigated. However, the soil infiltration, as well as the soil and crop flow resistance, are to be considered [3]. On the otherhand, unlike surface irrigation, sprinkler irrigation uses close pipelines for water distribution. The water is distributed over the area by using a mechanical device called sprinklers [2]. In sugarcane cropping, the use of sprinkler irrigation has 80% irrigation efficiency [4]. Moreover, another similar method to sprinkler irrigation is called center-pivot irrigation. Unlike the sprinkler method, this method is a well-engineered structure design to effectively deliver water to a large circular field. The water is fed from the pivot point at the center of the circle and each center has a main delivery pipe suspended over the field where sprinkles or spray nozzles can be spaced along the pipe [2]. This increases the irrigation efficiency of sugarcane to 85% [4]. Lastly, a well-known method to be used for the conservation of water is the drip irrigation method. This method reduces the amount of water consumption by allowing water to drip slowly to the roots of plants, either onto the soil surface or directly onto the root zone, through a network of valves, pipes, tubing, and emitters [3].

1.2. Nutrients for sugarcane crop fertigation

Since its introduction in the 1970's, drip irrigation for sugarcane cropping has increasingly gained popularity and as of today, this system is the most precise, efficient and practical method of delivering water and nutrients to crop with 95% irrigation efficiency [3]. According to [4], 50% was reduced in water consumption and an increase of 25 to 30% efficiency in terms of the use of fertilizer. As compared with the global average yield of 59.4 ton/ha, cane tonnage harvest under drip irrigation varies between 120 and 190 ton/ha. Another aspect that should also be taken for the improvement of sugarcane cropping is the fertigation system. Fertigation is the supply of the nutrients in the field, specifically nitrate (N), phosphate (P) and potassium (K). The proper amount of nutrients (NPK) yields to a good quality soil, which in the later stage, led to good quality crop and yield production. However, the improper use of these fertilizers led to poor quality and low quantity of crops [5].

Sugarcane is a giant crop that generally requires a higher number of nutrient elements. Its growth development is divided into four stages, namely, germination and establishment (1-80 days), tiltering or formative (81-130 days), grand growth period (131-300 days) and lastly, ripening and maturity (300-360 days) [4]. In sugarcane, nitrogen requirement is utmost during the tiltering (formative) phase. If the nitrogen is applied more than the optimum amount, sugarcane production may drop. At the beginning of vegetation growth, the application of nitrogen has no negative effect but if the application is behind, reduction in the sucrose content are seen which leads to a decrease in the sugar yield [4]. Phosphorus requirement for sugarcane is greater in the formative phase. Adequate phosphate must be made available in the soil during the formative phase for absorption, otherwise, this will result to a reduction in sugar concentration while excessive dosage above the optimal rates resulted to a decrease in sugar concentration, yield, pol percentage, and purity. On the contrary, if the application is too low, pol percentage and purity increases [4]. Application of potassium at around six months is best to improve sugar recovery. Nevertheless, if the application is too high, this also resulted in a negative effect on the percentage of sucrose in the cane [4].

1.3. Existing automation for irrigation

The advancement in agriculture in terms of reducing manual labor and water consumption is no longer a problem, however, there is also a necessity to look into modern methods in terms of the proper supply of nutrients in the field to maintain the soil quality. In the works of [6], the system was designed to analyze the soil nutrient content in real-time to make a crop prediction. The average percentage of the basic nutrients (NPK) was determined as well as the suitable crop for the particular soil type. In the paper of [7], an electrochemical sensor is developed to determine the NPK present in the soil. While in the paper of [8], automation is used to restore the standard levels of NPK and proposed an automatic fertigation dispensary system using a single nutrient fertilizer (urea). However, these studies in the fertigation systems are still few and different fertigation problems are still need to be addressed.

One current problem is the varying levels of nutrients to be supplied to the crop during its development stage. Therefore, this paper presents a novel automation of fertilizer and water deployment to sugarcane crops for optimal crop growth and utilization of resources. Currently, there is no automation system design to address the varying amount of nutrients requirements at every stage of sugarcane cropping. In this study, the automation system is consists of the sensors, controllers, and actuators integrated for performance. A ladder program was also developed to suit the automation methodology suggested.

2. New automation concept

2.1. Nutrient requirement

The dosage of fertilizer for the sugarcane crop depends on the current nutrients in the soil. In the Philippines, the optimum nutrient level for the soil is shown in table 1. The soil fertility rating should also be in standard level or higher to achieve the optimum harvest.

Table 1. Nutrient Requirement of Sugarcane after Planting.

Growth Stages of Sugarcane	Days after planting	Nutrients (kg/ha/day)			Reference
		N	P	K	
Germination and establishment	1-30 Days	1.2	0.10	0.20	[4]
	31-80 Days	1.5	0.40	0.24	[4]
Tiltering or formative	81-110 Days	2.00	1.00	0.40	[4]
	111-150 Days	0.75	0.30	0.75	[4]

2.2. Sensors and automation equipment

The new automated irrigation and fertigation system presented in figure 1 includes the water and fertilizer preparation area, dripping system, programmable logic controller, sensors, and the sugarcane field.

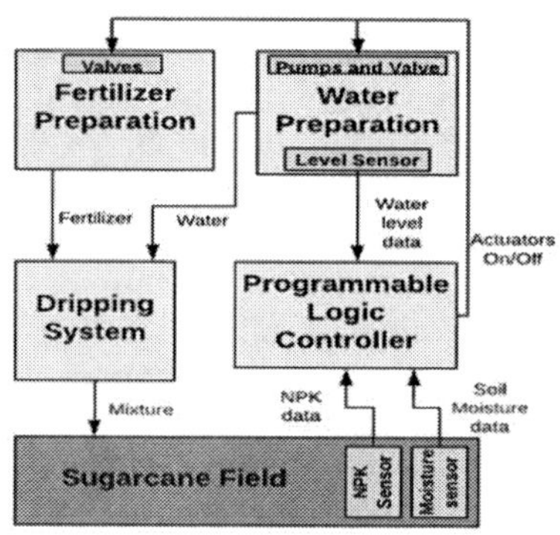

Figure 1. Automation system flow chart.

2.2.1. Water level sensor, NPK sensor, soil moisture sensor

Water level sensor is a mechanical device connected to a switch where the mechanical part is the actuator. This turns the connection on/off when the water level reaches a certain level. NPK sensor identifies the essential nutrient present in the soil. The electrochemical sensor, an NPK sensor type are connected to the analog input port of PLC controller board.

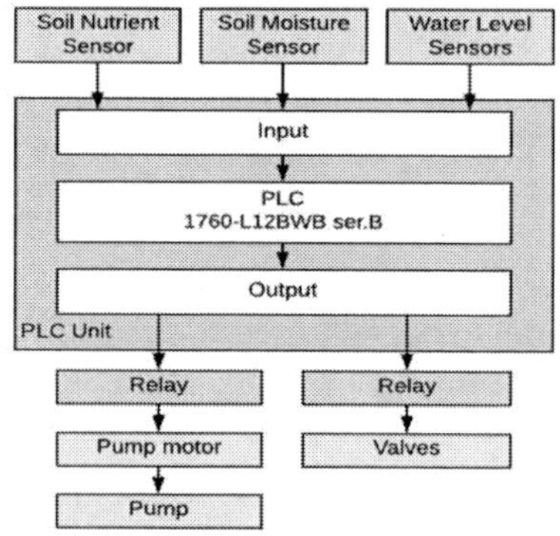

Figure 2. PLC block diagram.

2.2.2. Programmable logic controller

The flowchart shown in figure 3 describes the algorithm implemented for the irrigation and fertigation system. The type of PLC used in the proposed system is the Allen-Bradley (AB Type). The PLC unit includes the input, processing, and output unit as shown in figure 2. The signals from the sensors (moisture, nutrient, water) are transmitted to the input module of the PLC. The PLC are now capable of transfering the command to the processing unit to properly interpret the signals (if the NPK, moisture, and water level data is high or low). The processing unit sends signals to the output unit for the action to be done to the actuators (to open/close the valves and to turn on/off the water pump).

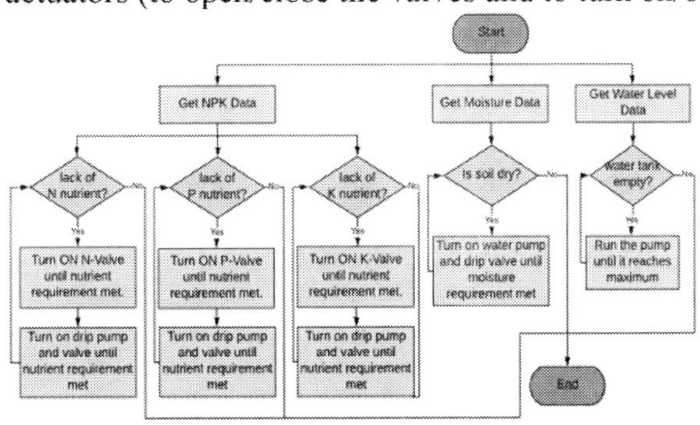

Figure 3. Algorithm of PLC ladder program.

3. Simulation

3.1. Program modules

Due to the limitation of the software used, the program is set to three modules: Module 1 is the supply of nutrients (NPK) to the field during the different sugarcane growth stages. Module 2 for monitoring of the nutrient content in the soil. Module 3 for the water level conditions in the tank and the automated drip irrigation system (water supply in the field).

3.2. Automation scenarios

The automation process is done using the Picosoft V6.22, a software compatible with Allen Bradley PLC. Four scenarios are made from the modules to do the simulation for the entire system. Scenarios 1 for the nutrient requirement per day during the germination and development period, Scenario 2 for checking of nutrient contents in the soil before scheduling the fertilizer supply, Scenario 3 for monitoring the water level in the tank, and Scenario 4 for the supply of water to the field. (*Simulation encoding upon request*).

4. Conclusion

Based on the proposed design simulated using the Picosoft Allen Bradley software, the simulation shows the automation process of supplying the exact amount of water and fertilizer requirements of the sugarcane crop can be obtained. The economic analysis (*economic analysis excel sheet upon request*) concluded that the proposed automation system can yield an optimistic profit as compared to the manual irrigation method.

Although the simulation and algorithm for the parameters in the growth of sugarcane have been designed and simulated, the design is only set for one hectare since the values obtained such as the fertilizer requirements are limited. If the system is to be scaled for use in larger farms, the appropriate size should be designed such as the size of water tanks and the dripping system.

Acknowledgments

The authors would like to acknowledge the financial support of the Department of Science and Technology - Engineering Research and Development for Technology (DOST-ERDT), Commission on Higher Education (CHED), and ASEAN University Networks (AUN).

IOP Conf. Series: Materials Science and Engineering **715** (2020) 012089 doi:10.1088/1757-899X/715/1/012089

Reference

[1] J. Armando N. Espino, "Recent Developments in Irrigation Practices for Sugarcane Industry," Laguna (2017).
[2] Crops, M., & Of, S. (2015).
[3] J. Holden and P. McGuire, *"Irrigation of Sugarcane Manual,"* (2014).
[4] Netafim, *"Sugracane Cultivation with drip irrigation,"* (2014).
[5] Energy, N., & Alliance, E. Agricultural Irrigation Initiative: Overview of Center Pivot Irrigation Systems (2015) https://doi.org/10.1109/SPAWDA.2008.4775854
[6] D. Vadalia, M. Vaity, K. Tawate, and D. Kapse, *"Real Time soil fertility analyzer and crop prediction,"* Int. Res. J. Eng. Technol., **4**, no. 3, pp. 3–5 (2017).
[7] R. Sindhuja and B. Krithiga, *"Soil Nutrient Identification Using Arduino,"* Asian J. Appl. Sci. Technol., **1**, no. 4, pp. 40–42 (2017).
[8] A. Amrutha, R. Lekha, and A. Sreedevi, *"Automatic soil nutrient detection and fertilizer dispensary system,"* Int. Conf. Robot. Curr. Trends Futur. Challenges, pp. 1–5 (2016). https://doi.org/10.1109/RCTFC.2016.7893418

IOP Conf. Series: Materials Science and Engineering **715** (2020) 012090 doi:10.1088/1757-899X/715/1/012090

Nonlinear control method of electro-hydraulic system driving Two-DOF robot arm with output constraint

Qing Guo[1, 2, 3,*], Yili Liu[1] and Dan Jiang[4]

[1]School of Aeronautics and Astronautics, University of Electronic Science and Technology of China, Chengdu, 611731, China
[2]State Key Laboratory of Fluid Power and Mechatronic Systems, Zhejiang University, Hangzhou, 310027, China
[3]Center for Intelligent Aircraft Systems Technology and Application, University of Electronic Science and Technology of China, 611731, China
[4]School of Mechanical and Electrical Engineering, University of Electronic Science and Technology of China, Chengdu, 611731, China

*E-mail: guoqinguestc@uestc.edu.cn

Abstract. This paper studies tracking control problem for a certain parameter electro-hydraulic system(EHS) to constrain the output cylinder position in prescribed performance. In the control design, a backstepping controller is adopted to guarantee the system stability and the dynamic tracking performance. The dynamic surface control (DSC) is used to eliminate the complexity explosion of virtual control repeatedly derivatives to make the controller and parameter design simple. A barrier Lyapunov function (BLF) is employed in the process of control design to avert the contravention of the time-varying output constraints. The comparative simulation results are given to demonstrate that the effectiveness of the proposed controller. Furthermore the smooth dynamic surface is obtained to guarantee the sufficient stable margin of EHS.

1. Introduction

Electro-hydraulic system is an interdisciplinary subject based on electrical control, hydraulic transmission and computer, which is widely used because of high control accuracy, fast response speed, high output power, flexible signal processing and easy to realize feedback of parameter[1]-[2]. EHS uses electro-hydraulic servo valve to convert small power signal into high power hydraulic power, and realizes the servo control of heavy machinery. With the continuous development of technology and the continuous improvement of the performance requirements, the stability, rapid and accurate response of the EHS become a basic requirement. Therefore, the research on the control of the electro-hydraulic system[3] becomes the inevitable trend.

EHS is a strict feedback model in nonlinear system, and the boundedness of output constraints is a important problem in the control of nonlinear systems. In the actual operation of the EHS, the violation of the constraint may reduce the performance, endanger the system, even damage the hardware. In order to cope with the constraints or limitations of the operating space and hardware capacity, and improve the performance of the actual system, the constraint control becomes one of the key areas of the control theory and engineering research. The existing research in the field of control has carried out a large number of studies on constraint problems[4]-[5]. A concept of invariant set is proposed to solve the constraint problem, but constraints can be satisfied only when the initial state is constrained to a invariant set[6]-[7]. Model predictive control mainly solves the state and control variables in a reasonable time interval with the optimal control algorithm[8]. Afterwards, the Lyapunov stability theory has been an important method of nonlinear control, then some scholars put forward the barrier Lyapunov function which is a kind of continuous function.

When barrier Lyapunov function[9] is used, as the system order number increasing, the steps in the process of virtual control iterative differential easily differential explosion phenomenon, resulting in a serious computational burden and complex controller design process of control system, it is more difficult to use in actual system. Therefore, the idea of the barrier Lyapunov function method based on the backsteping[10-12], dynamic surface control[13] to solve the time-varying output constraints of nonlinear systems[14] is proposed, this method adopts the first order filtering of the virtual control in the traditional backstepping method and introduces a new state, thus avoiding the emergence of the surge item[15].

Different from the existed work of EHS control method, the main contribution of this paper is given as follows:

1) The barrier Lyapunov function is proposed to handle the output constraint, which restrain the hydraulic cylinder position in desirable boundary.

2) The dynamic surface control is adopted to eliminate the differential expansion in the processing of backstepping controller design which guarantees the system stability and the dynamic tracking performance of EHS.

2. Plant description

As shown in figure 1, the control mechanism of EHS[16] is comprised by a electro-hydraulic servo valve, a hydraulic cylinder, a fixed displacement pump, a relief valve.

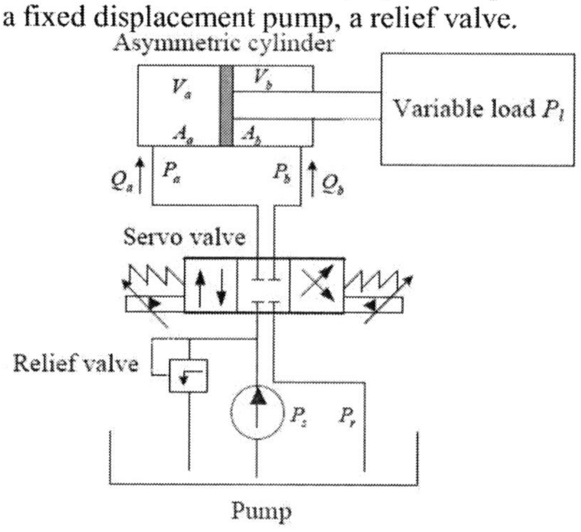

Figure 1. The control mechanism electro-hydraulic system.

By analyzing the structure and working principle of electro-hydraulic servo system, the load flow of electro-hydraulic servo valve can be obtained

$$Q_a = C_d w x_v \sqrt{\frac{1}{\rho}(P_s - \mathrm{sgn}(x_v)P_L)}\ , \tag{1}$$

where C_d is the discharge coefficient , w is the area gradient of the servo valve spool, ρ is density of hydraulic oil , P_s is the supply pressure of the pump , $P_L = P_a - P_b$ is the load pressure, x_v is the spool position of servo valve, $\mathrm{sgn}(\cdot)$ is the sign function.

As the functions appeared in the design of backstepping controller are required to be smooth, the sign function $\mathrm{sgn}(x_v)$ in (1) is replaced by the hyperbolic tangent function $\tanh(kx_v)$, where k is a positive constant.

$$Q_a = C_d w x_v \sqrt{\frac{1}{\rho}(p_s - \tanh(kx_v)P_L} \ . \tag{2}$$

Hydraulic cylinder flow-pressure continuity equation can be expressed as

$$Q_a = A_p \dot{y} + C_{tl} P_L + \frac{V_t \dot{P}_L}{4\beta_e} \ , \tag{3}$$

where $A_p = A_a = A_b$ is annulus area of symmetrical cylinder chamber, y is output displacement of the hydraulic cylinder, C_{tl} is the coefficient of the total leakage of the cylinder, V_t is the half-volume of cylinder, β_e is the effective bulk modulus.

The mechanical dynamic equation is given as

$$m\ddot{y} = P_L A_p - Ky - b\dot{y} - F_L , \tag{4}$$

where m is the load mass, F_L is the external load on the EHA, K is the spring constant, b is the viscous damping coefficient.

The state vector $x = [x_1, x_2, x_3] = [y, \dot{y}, p_L]$, u is input voltage of servo valve, then the dynamic model of EHS can be expressed as[17]

$$\begin{cases} \dot{x}_1 = x_2 \\ \dot{x}_2 = \dfrac{1}{m}(-Kx_1 - bx_2 + A_p x_3 - F_L) \\ \dot{x}_3 = -\dfrac{4\beta_e A_p}{V_t} x_2 - \dfrac{4\beta_e C_{tl}}{V_t} x_3 + \dfrac{4\beta_e C_d w K_{sv}}{V_t \sqrt{\rho}} \sqrt{p_s - \tanh(ku) x_3 u} \\ y = x_1 \end{cases} , \tag{5}$$

where Ksv is gain of servo valve. The output y = x1 is restricted in desirable boundary by feedback control to guarantee the tracking error of cylinder position Δx1 = x1 – yd in prescribed performance.

3. Control design

3.1. Useful technical lemmas and definitions

Remark 1[18]: The external load F_L is unknown plant dynamics, which is considered as a structural disturbance of EHS. Although the dynamic value of F_L depends on the variables $y, \dot{y}, \ddot{y}$, F_L is bounded by its upper boundary Δ_{FL} , i.e., $\|F_L(t)\| \le \Delta_{FL}$.

Remark 2[19][20]: Generally, the hydraulic parameters C_d, ρ , ω , b , β_e, C_{tl} are often perturbed by different hydraulic physical characteristic, but the other parameters are known.

Lemma 1[21]: Consider a positive constant $k_a \in R$, if $x \in R$ and $|x| < |k_a|$, the following inequality holds:

$$\ln \frac{k_a^2}{k_a^2 - x^2} \le \frac{k_a^2}{k_a^2 - x^2} \, . \tag{6}$$

Lemma 2[22]: For bounded initial conditions, if there exists a C^1 continuous and positive definite Lyapunov function $V(x)$, and satisfying $\kappa_1(\|x\|)\le V(x)\le \kappa_2(\|x\|)$, $\dot V(x)\le -\rho V(x)+c$, where $\kappa_1,\kappa_2 : R^n \to R$ are class K functions, and ρ and c are two positive constants, then the solution $x(t)$ is uniformly bounded.

3.2. Dynamic surface control with output constraint

Suppose all the states and parameters are measurable, let

$$d_L=-\frac{F_L}{m}, f_2=-\frac{K}{m}x_1-\frac{b}{m}x_2, g_2=\frac{A_p}{m}, f_3=-\frac{4\beta_e A_p}{V_t}x_2-\frac{4\beta_e C_{tl}}{V_t}x_3 .$$
$$g_3=\frac{4\beta_e C_d w}{V_t\sqrt{\rho}}\sqrt{P_s-\tanh(ku)x_3} \quad (7)$$

In this case, we can assure f2, f3, g2, g3 smooth, according to Remarks 1-2, simplify the dynamic model of EHS, the dynamic model of EHS is rewritten as follows

$$\begin{cases}\dot x_1=x_2\\ \dot x_2=f_2(x_1,x_2)+g_2x_3+d_L\\ \dot x_3=f_3(x_2,x_3)+g_3u\\ y=x_1\end{cases}. \quad (8)$$

In the process of designing the backstepping controller, the frequent derivation of the differential surge is inevitable, it is difficult to realize in practical situations. To solve this problem, we add dynamic surface control to eliminate the differential expansion to make the controller and parameter design simple. In order to ensure that the closed-loop signal in the closed loop is bounded and the output constraint is not violated, we use BLF to avert the contravention of the time-varying constraints.

The system state errors $z_i(t)$ (i = 1,2,3) are defined as follows

$$\begin{aligned} z_1&=x_1-x_{1d}\\ z_{i+1}&=x_{i+1}-\alpha_i, i=1,2\end{aligned}. \quad (9)$$

where α_i is stable virtual control variable.

According to the dynamic surface control method, a filter virtual function $\beta_i, i=1,2$ can be introduced. Let the stable virtual function α_i pass through a first order filter, it can be expressed as

$$\tau_i\dot\alpha_i+\alpha_i=\beta_i, i=1,2, \quad (10a)$$

where τ_i as the time constant , $\alpha_i(0)=\beta_i(0), i=1,2$.

So the filter error of the first order filter is $S_i=\alpha_i-\beta_i, i=1,2$. Then we have

$$\dot\alpha_i=\frac{\beta_i-\alpha_i}{\tau_i}=-\frac{S_i}{\tau_i}. \quad (10b)$$

Choose the barrier Lyapunov functions as

$$V=\frac{1}{2}\log\frac{k_a^2}{k_a^2-z_1^2}+\frac{1}{2}z_2^2+\frac{1}{2}z_3^2+\frac{1}{2}S_1^2+\frac{1}{2}S_2^2. \quad (11)$$

If the two filter virtual functions are designed as

$$\begin{aligned}\beta_1 &= -(k_a^2 - z_1^2)k_1 z_1 + \dot{x}_{1d} \\ \beta_2 &= \frac{1}{g_2}\left(-f_2 - k_2 z_2 - \frac{z_1}{k_a^2 - z_1^2} + \dot{\alpha}_1\right),\end{aligned} \tag{12}$$

and the controller is given by

$$u = \frac{1}{g_3}\left(-f_3 + \dot{\alpha}_2 - k_3 z_3 - g_2 z_2\right), \tag{13}$$

then

$$\dot{V} \le -\Gamma \ln\left(\frac{k_a^2}{k_a^2 - z_1^2}\right) - \Omega S_1^2 - \Gamma_2 z_2^2 - \Omega_2 S_2^2 - k_3 z_3^2 + \frac{\sigma_1}{2} + \frac{\sigma_2}{2} + \frac{d_{L\max}^2}{2}, \tag{14}$$

where

$$\Omega_1 = \frac{1}{\tau_1} - \frac{1}{2\left(k_a^2 - z_1^2\right)} - \frac{M_1^2}{2\sigma_1},\ \Gamma_1 = k_1\left(k_a^2 - z_1^2\right) - \frac{1}{2},\ \Gamma_2 = k_2 - \frac{\left|g_2\right|_{\max}^2}{2} - \frac{1}{2},\ \Omega_2 = \frac{1}{\tau_2} - \frac{\left|g_2\right|_{\max}^2}{2} - \frac{M_2^2}{2\sigma_2}$$

are positive constants, $k_i (i = 1,2,3)$ are the controller gains.

Let $c = \min\{2\Gamma_1, 2\Gamma_2, 2\Omega_1, 2\Omega_2, 2k_3\}$, then

$$\dot{V} = -cV + \sigma, \sigma = \frac{\sigma_1}{2} + \frac{\sigma_2}{2} + \frac{d_{L\max}^2}{2}. \tag{15}$$

Integrating two sides of (15), V yields

$$V(t) \le V(0)e^{-ct} + \sigma(1 - e^{-ct})/c. \tag{16}$$

Now according to (16), as let t → tf, the error convergence mainly depends on the item σ/c. Thus, the increased control gains ki(i = 1, 2, 4) and the reduced constant c can arbitrarily shrink this error convergence. This implies that the Lyapunov function V is ultimate bounded. Thus, the system state errors z_1, z_2, z_3 are stable.

4. Simulation

The block diagram of the proposed method is shown in figure 2. The simulations are carried out by Matalb / Simulink software. The tracking error boundary of output cylinder position is ka = ±2mm. Some known hydraulic parameters are xvmax = 5 mm, Lmax = 58 mm, ps = 40 bar, Ap = 4.91 cm2, Vt = 8.74×105 m3, Cd = 0.62, w = 0.024 m, Ctl = 2.5×10-11 m3/(sPa), βe = 7000 bar, ρ = 850 kg/m3, K = 10 N/m, b = 50 Ns/m, Ksv = 5 ×10-4 m/V, Tsv = 10 ms. The load mass is m = 3.347 kg. The control gains are set as k1 = 1000, k2 = 200, k3 = 1000, and k4 = 1. To illustrate the advantage of the proposed method, the traditional PI controller is given by $u = k_p(y_d - y) + k_i\int_0^t (y_d - y)dt$, where the control gain kp = 150, ki = 10.

The comparison results with PI control method are shown in figure 3-6. The position demands of hydraulic cylinder is selected as $y_d = 50\sin(2\pi t)$ mm. Figure 3 illustrates the dynamic tracking response of cylinder position. The dynamic tracking accuracy of the proposed BLF method is better than PI as shown in figure 4, whose output error is constrained in prescribed boundary $\left|\Delta x_1\right| < k_a = 2$ mm. The two dynamic surface errors S1 and S2 are shown in figure 5, which are smooth to avoid the complexity explosion caused by the virtual control repeatedly derivatives. Then the final control variable u is shown in figure 6.

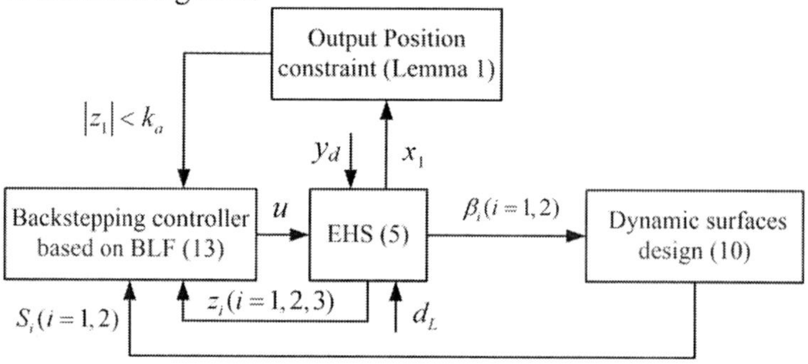

Figure 2. Block diagram of the proposed method.

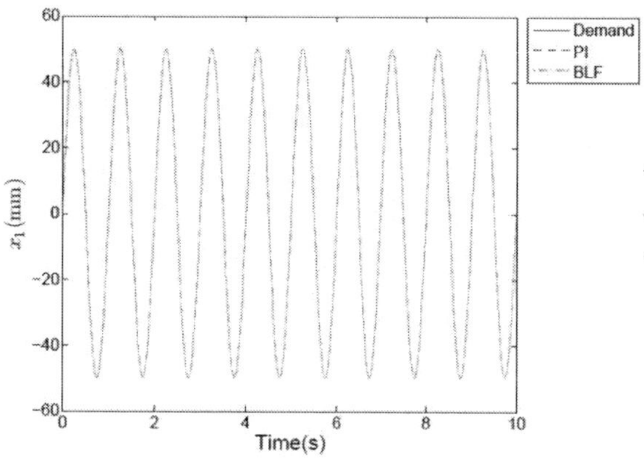

Figure 3. The dynamic response x_1.

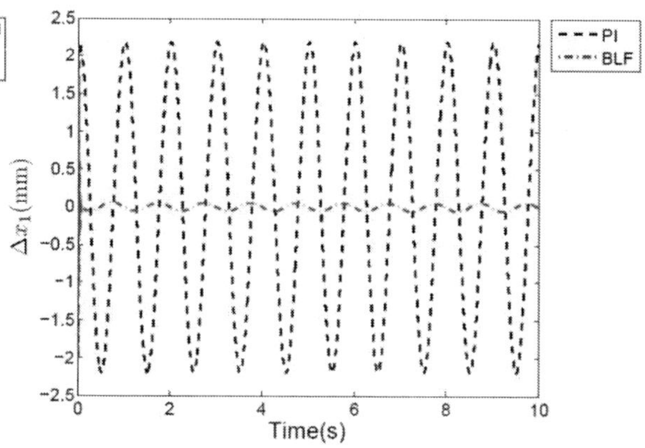

Figure 4. Tracking error Δx1.

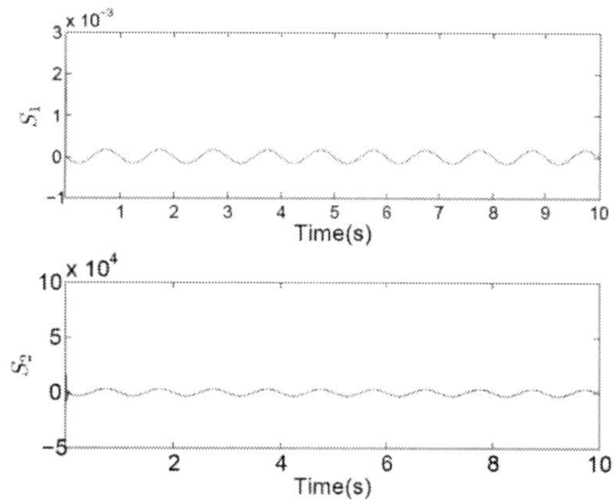

Figure 5. Two dynamic surface errors S1 and S2.

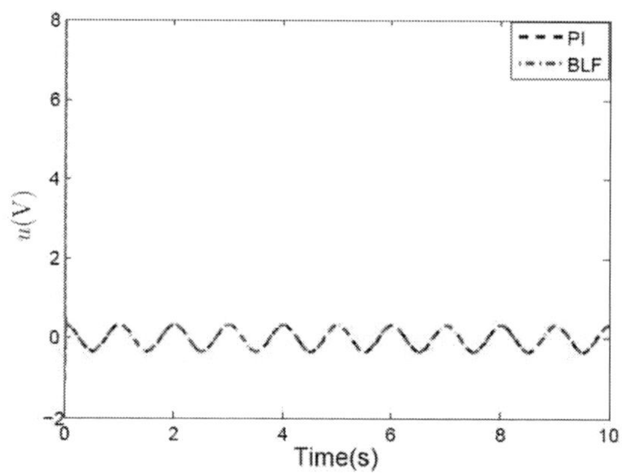

Figure 6. The control variable u.

5. Conclusion

In this paper, a backstepping controller is proposed for electro-hydraulic system to drive Two-DOF robotic arm. To eliminate the differential expansion, the dynamic surface control is employed. A barrier Lyapunov function is used to avert the contravention of the time-varying output constraints. The comparison simulation results with PI control have indicated that the presented control scheme can make the output well follow the target trajectory while ensuring the constraints satisfaction.

Acknowledgments
This work is supported by National Natural Science Foundation of China (51775089, 61305092 and 51205045), Sichuan Science and Technology Program (2018JY0565, 2018HH0101), Fundamental Research Funds for the Central Universities, China (ZYGX2016J160), Open Foundation of the State Key Laboratory of Fluid Power and Mechatronic Systems (GZKF-201515).

References
[1] Q. Guo, T. Yu, D. Jiang, "High-gain observer-based output feedback control of single-rod electro-hydraulic actuator," IET Control Theory & Appl., vol. 9, no. 16, pp. 2395-2404, 2015.
[2] Q. Guo, J. Yin, T. Yu, D. Jiang, "Coupled-disturbance-observer-based position tracking control for a cascade electro-hydraulic system," ISA Trans., vol. 68, pp. 367-380,2017.
[3] Q. Guo, Q. Wang Q, Y. Liu, "Anti-windup Control of Electro-hydraulic System with Load Disturbance and Modelling Uncertainty," IEEE Trans. Ind. Informat., vol. 14, no. 7, pp. 3097-3108, 2017
[4] W. He, Y. Chen, Z. Yin, "Adaptive Neural Network Control of an Uncertain Robot With Full-State Constraints," IEEE Trans. Cybern., vol. 46, no. 3, pp. 620-29, 2016
[5] Q. Guo, J. Yin, T. Yu, D. Jiang, "Saturated adaptive control of electrohydraulic actuator with parametric uncertainty and load disturbance," IEEE Trans. Ind. Electron., vol. 64, no. 10, pp. 7930–41, 2017
[6] T. Hu, Z. Lin, B. Chen, "An analysis and design method for linear systems subject to actuator saturation and disturbance," in 2002 Proc. American Control Conf., vol. 2, pp. 351-59, 2002
[7] T. Hu, Z. Lin, "Composite quadratic Lyapunov functions for constrained control systems," IEEE Trans. Autom. Control, vol. 48, no. 3, pp. 440-50, 2003
[8] A. Bemporad, F. Borrelli, M. Morari, "Model predictive control based on linear programming the explicit solution," IEEE Trans. Autom. Control, vol. 47, no. 12, pp. 1974-85, 2003
[9] K. P. Tee, S. S. Ge, E. H. Tay, " Barrier Lyapunov functions for the control of output-constrained nonlinear systems," Automatica, vol. 45, no. 4, pp. 918-27, 2009
[10] D. Won, W. Kim, D. Shin, "High-gain disturbance observer-based backstepping control with output tracking error constraint for electro-hydraulic systems," IEEE Trans. Control Syst. Technol., vol. 23, no. 2, pp. 787-95, 2015
[11] J. Zhou, C. Wen, Y. Zhang, "Adaptive backstepping control of a class of uncertain nonlinear systems with unknown backlash-like hysteresis," IEEE Trans. Autom. Control, vol. 49, no. 10, pp. 1751-59, 2004
[12] C. Kwan, F. L. Lewis, " Robust backstepping control of nonlinear systems using neural networks," IEEE Trans. Syst. Man Cybern.- Syst., vol. 30, no. 6, pp. 753-66. 2000
[13] D. Swaroop, J. K. Hedrick, P. P. Yip, " Dynamic surface control for a class of nonlinear systems," IEEE Trans. Autom. Control, vol. 45, no. 10, pp. 1893-99, 2002
[14] Q. Guo, Y. Zhang, B. G. Celler, and S. W. Su., " State-constrained Control of Single-rod Electrohydraulic Actuator with Parametric Uncertainty and Load Disturbance," IEEE Trans. Control Syst. Technol., DOI: 10.1109/TCST. 2017.2753167
[15] Q. Guo, Y. Liu, D. Jiang, "Prescribed Performance Constraint Regulation of Electrohydraulic Control Based on Backstepping with Dynamic Surface," Appl. Sci., vol. 8, no. 1, pp. 76, 2018
[16] Q. Guo, T. Yu, D. Jiang, "Robust H∞ positional control of 2-DOF robotic arm driven by electro-hydraulic servo system," ISA Trans., vol. 59, pp. 55-64, 2015
[17] Q. Guo, P. Sun, J. Yin, etal., "Parametric Adaptive Estimation and Back- stepping Control of Electro-hydraulic Actuator with Decayed Memory Filter," ISA Trans., vol. 62, pp. 202–14, 2016
[18] Q. Guo, Y. Zhang, B. G. Celler, "Backstepping Control of Electro-Hydraulic System Based on Extended-State-Observer With Plant Dynamics Largely Unknown, " IEEE Trans. Ind.

Electron., vol. 63, no. 11, pp. 6909-20, 2016

[19] J. Yao, Z. Jiao, D. Ma, " High-accuracy tracking control of hydraulic rotary actuators with modeling uncertainties," IEEE/ASME Trans. Mechatronics, vol. 19, no. 2, pp. 633-41, Apr. 2014

[20] Q. Guo, X. Li, D. Jiang, " Full-state Error Constraints Based Dynamic Surface Control of Electro-hydraulic System," IEEE Access, DOI: 10.1109/ACCESS.2018.2870956

[21] Z. Zhao, W. He, S. S. Ge, "Adaptive neural network control of a fully actuated marine surface vessel with multiple output constraints," IEEE Trans. Control Syst. Technol., vol. 22, no. 4, pp. 1536–43, Jul. 2014

[22] H. K. Khalil, Nonlinear Systems, 3rd ed. Upper Saddle River, NJ, USA: Prentice-Hall, 2002.

Data processing method in a context-aware system to provide intelligent robot services based on big-data

Yongseong Cho, Jonghyeok Mun, Yoosang Park, Jongsun Choi, Jaeyoung Choi*, and Yongyun Cho

School of Computer Science & Engineering, Soongsil University, 369 Sangdo-Ro, Dongjak-Gu, Seoul 156-743, Korea

*Email: choi@ssu.ac.kr

Abstract. In order to recognize the situation and to provide robot services based on big data, it is necessary to implement a machine learning model capable of analyzing big data and a method for providing this data to the analysis model for learning. Previous researches have focused on the implementation of various data analysis models to analyze the collected big data. However, in these researches, it is difficult to recognize the new data which is not learned by the analysis model, and further research is needed to solve it. In this paper, we propose a data processing method to provide context-aware information in context-aware system based on big data to provide intelligent robot services. The proposed data processing method can convert the data generated in the domain in which the robot is running to learn data of the machine learning model and provide it, thereby continuously improving the accuracy of the machine learning model. In the experiment, the context-aware information is provided through the prototype of the context-aware system with the proposed method.

1. Introduction

Since the start of the big data paradigm, various techniques for analyzing large amounts of data and obtaining valuable information have been continuously studied and developed [1,2,3,4,5]. These techniques enable to store and analyze large amount of unstructured data, which could not be processed in the past, and the meaningful information obtained through these techniques is utilized in various fields such as content industry, advertisement, and distribution [6,7,8]. In particular, artificial intelligence technologies based on machine learning have attracted a great deal of attention recently due to the evolution of hardware, the solution of over-sum problems, the implementation of in-depth neural networks, and the large amount of big data required for learning [9,10,11,12]. They can be applied to analyze the meaning of specific data by repeatedly learning large data. In addition, these technologies have been done in many fields such as speech recognition [13], image analysis [14,15], pattern analysis [16], and time series data analysis [17,18]. These artificial intelligence technologies are actively being developed by global IT companies, universities, and through many open source projects. IBM Watson [13], Tensor Flow [14] developed by google, Berkeley University's Caffe [15] and Deeplearning4J [16] are some of the representative enterprise projects.

On the other hand, the existing context-aware systems [19,20,21] based on artificial intelligence models are focused on the implementation of analytical models that can analyze specific data sets. In these cases, even if the implementation of the analysis model is completed, there is no way to continuously collect and provide the data sets for learning the analysis model. For this reason, it may be difficult to operate and manage the context-aware system. In addition, additional processing is required to convert the results of the artificial intelligence model into contextual information. Therefore, artificial intelligence-based context-aware systems need not only the implementation of analytical models, but also the collection and provision of data for analytical models, and concrete methods to convert the analyzed results into contextual information.

In this paper, we propose a data processing method to provide context - awareness information by using machine learning based analysis model in context-awareness system to provide intelligent robot service. The proposed data processing method is intended to improve the accuracy of the model for analyzing the big data, and allows to collect and provide the big data for learning the analysis model. The analysis model used in this paper is based on the RNN algorithm [22], which is effective for time series data analysis. In order to learn the analytical model based on the RNN algorithm, time series data of the service domain, in which the robot is operating, are needed. With the proposed method, data composed of time, environments, and index information related to execution of service can be collected at the time when context aware service is executed. The collected data are hereinafter referred to as 'snapshot data'. Every time the context aware service is executed, snapshot data is provided for learning the analytical model after a predefined period of time. In the experiment, we show that the analytic model is learned and contextual information is provided through the context-aware system with the proposed data processing method.

The composition of this paper is as follows. In Chapter 2, we will look at existing research on context-aware systems based on Big data. Chapter 3 introduces the outline and configuration modules of the context-aware system that utilize the proposed data processing method. And the method for collecting and providing big data. In Chapter 4, we show how the big data Analysis Model based on machine learning is learned through context-aware system that implements the proposed data processing method and context information is generated and provided. Chapter 5 includes conclusion remarks.

2. Related works

In this section, we introduce the existing research works on big-data based context-awareness systems and the characteristics of big data analysis techniques used in each study.

As a general method for analyzing big data, Hadoop based on distributed processing [23], data mining, and machine learning based artificial intelligence are used. Context-aware systems, that utilize Hadoop-based big data analysis, collect big data composed of environmental information around a user and biometric information provided by the wearable device. They recognize the user's surroundings with information obtained by processing data collected in units of day, week, and month. As a research related to this, Wireless Body Area Network (WBAN) [24] proposed a network structure for providing healthcare service by collecting and processing biometric information from sensors attached to human body.

The big data for context-aware monitoring (BDCaM) [25], which is a context-aware system for healthcare such as WBAN, uses sensor data, patient medical records and is able to provide customized medical services. BDCaM includes the Hadoop based system to analyze the collected big data, and big data analysis result is used to recognize the situation with the rule based context recognition method. Context-aware systems using these methods are mainly used when analysis results of big data are not needed in real time, and are useful when analyzing medical and life patterns.

On the other hand, context-aware systems that analyze big data based on artificial intelligence can be used when they want to process the collected real time data. These systems generally use artificial intelligence models learned from pre-collected data in a specific domain, and then analyze the current data with the model that has been terminated by the learning method. Context awareness component(CAC) [19] is the study which applied these methods. It applies the analysis result of sound data generated in the home to the existing context-aware system and focuses on improving the accuracy of context awareness in the system to provide intelligent robot services. In addition, it also utilizes artificial intelligence model that can analyze sound using CNN (Convolutional Neural Networks) [26] model, which are deep artificial neural networks that are used primarily to classify images (e.g. name what they see), cluster them by similarity (photo search), and perform object recognition within scenes. After collecting the voice data collected by the robot and learning the artificial intelligence model, each robot can recognize the situation using the artificial intelligence model.

3. Data processing method in Context-Aware system

In this chapter, we describe an outline of the intelligent robot's context-aware system that utilizes the proposed data processing method, the components that make up the system, and the flow that runs in conjunction with an external application to provide context aware services. The data processing method provided by the context-aware system of the proposed intelligent robot will be described in detail.

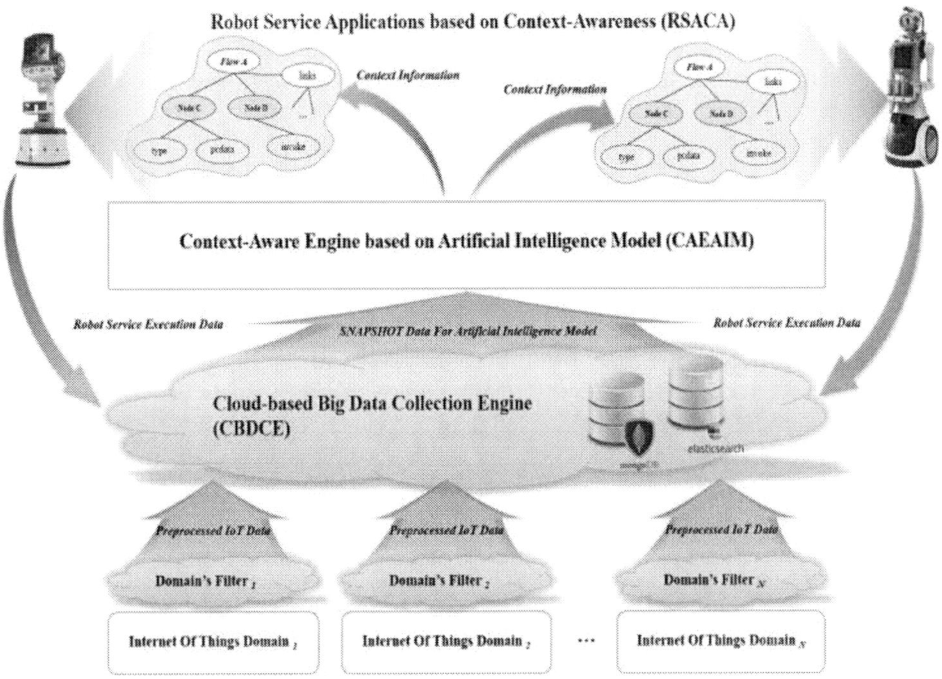

Figure 1. In this case simply justify the caption so that it is as the same width as the graphic.

3.1. Overview of Context-Aware system

Figure 1 shows an outline of the context-aware system that uses the proposed data processing method. At the top of figure 1 is Robot Service Applications based Context-Awareness (RSACA), which can provide intelligent robot services. RSACA is responsible for executing the robot service when a specific situation is recognized, and forwards the execution history of the robot service to the Cloud-based big data Collection Engine (CBDCE) at the bottom of figure 1. In the center of figure 1, there is Context-Aware Engine based Artificial Intelligence Model (CAEAIM) which can provide context recognition information. CAEAIM can provide context awareness information requested by RSACA and utilizes artificial intelligence model internally. At the bottom of figure 1, CBDCE is responsible for collecting environmental information and generating snapshot data. CBDCE has a cloud-based structure and contains a number of sub-modules. The environmental information and the snapshot data collected from the CBDCE are transferred to CAEAIM and used for learning and context awareness processing of the AI model. The following is a detailed description of each component.

3.1.1. Cloud-based big data collection engine

At the bottom of figure 1 is the Cloud-based Big Data Collection Engine (CBDCE). CBDCE is a big data collecting system that provides environment information collection and snapshot data generation for providing context aware information. It has a cloud-based structure that receives data from the subsystems for data collection of object Internet sensors located in each service domain, and can modify and process the data of each subsystem into the desired data form. Then, it receives the execution event of the robot service, and can generate and store the snapshot data by combining all environment information at the time when the robot service is executed and an index number indicating the executed service. The stored snapshot data is provided to the proposed context-aware system and used to learn the AI model.

3.1.2. Context-Aware engine based on artificial intelligence model

In the middle of figure 1, there is Context-Aware Engine based Artificial Intelligence Model (CAEAIM), which is the proposed context recognition system. CAEAIM receives snapshot data from CBDCE and controls learning of artificial intelligence model. The learned artificial intelligence model is used for context recognition processing and predicting information generation. After generation, these context recognition condition and prediction information are used as transition conditions for execution of robot service. And, as the system continues to operate, it controls the accumulated snapshot data to re-learn the artificial intelligence model according to a certain period of time, so that more accurate information can be returned. The components of CAEAIM and methods for generating prediction information are explained in detail in chapter 3.2 below.

3.1.3. Robot service applications based on context-awareness

At the top of figure 1, there is Robot Service Applications based Context-Awareness (RSACA) to provide context-aware robot services. The RSACA can receive intelligence from the CAEAIM with context-aware information or predictive information to control the intelligent robot. In this paper, we propose a workflow execution engine that can execute services described in a CAWL document. CAWL [27] is a context aware workflow language that can express the flow and control of robot services based on context awareness. When the robot service is executed, the RSACA transmits the contents and the execution time of the currently executed robot service to the CBDCE to generate snapshot data for providing a prediction-based service.

3.1.4. Proposed data processing methods

This chapter describes the data processing method of the proposed context-aware system. First, we explain the detailed process of providing snapshot data for learning AI model through big data collection module. The process of transforming the analyzed result of the learned artificial intelligence model into the intuitive situation information is explained in detail.

Figure 2 is a structure diagram to confirm the execution flow of the proposed data processing method. The lower part of figure 2 is a module that collects environmental data around the user, which is necessary for the situation recognition. The sub-collection modules in each area transmit the collected data to the Cloud-based Big Data Collection Engine(CBDCE). The transmitted data is delivered to the Environmental Data Processing Module (EDPM) and stored in the big data store. When the context-aware service is executed in advance, the context-aware service application transmits the label data to the CBDCE, and the label data is collected by the Labeled Data Processing Module(LDPM). Snapshot data accumulated during a certain period is transferred to Training Data Processing Module(TDPM) and Verification Data Processing Module(VDPM) of Context-Aware Engine based on the Artificial Intelligence Model(CAEAIM) is used as the data for learning and verification of the artificial intelligence model. The Context Information Transformation Module (CITM) transforms the analysis results of the AI model into contextual information according to predetermined rules and transmits the result of the context awareness to the context aware service application so that the service can be executed. Below is a description of the process in which the

IOP Conf. Series: Materials Science and Engineering **715** (2020) 012091 doi:10.1088/1757-899X/715/1/012091

snapshot data generated by combining environmental data and label data is used in the learning of the artificial intelligence model and the detailed process in which the analysis result of the artificial intelligence model is converted into the situation information.

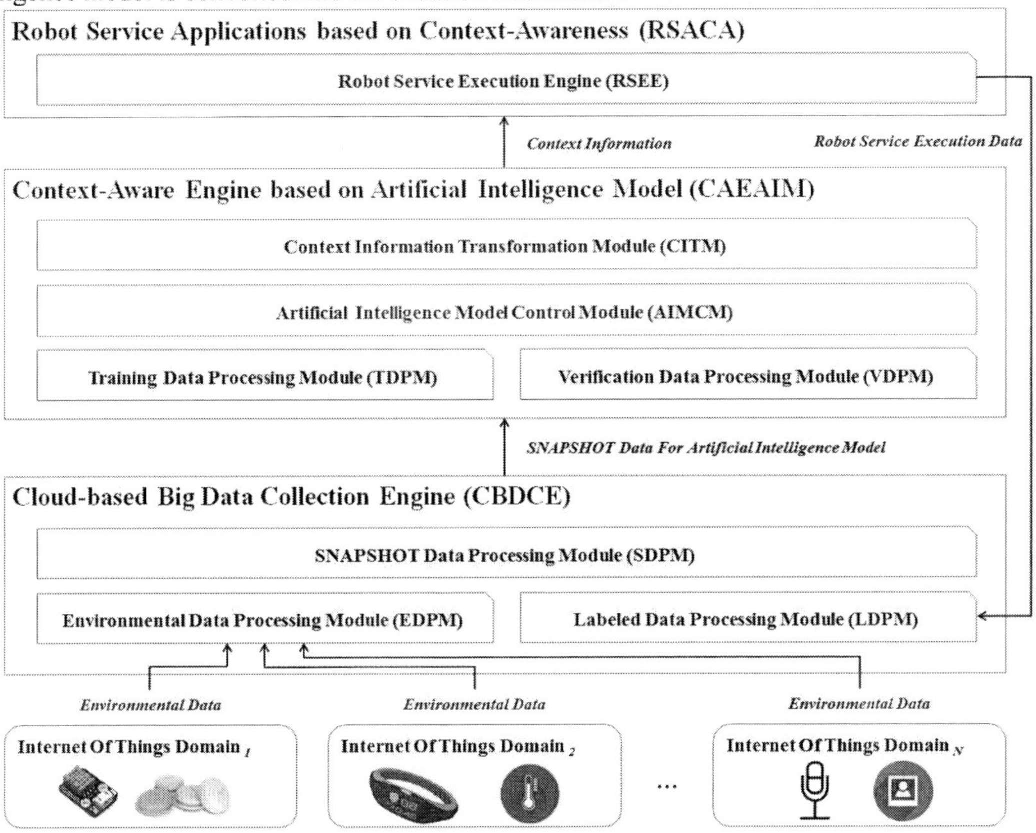

Figure 2. Structure of proposed data processing method.

Figure 3 shows the process of providing the data for the learning of the artificial intelligence model in the proposed data processing method. In figure 3, CBDCE is located at the bottom, and EDPM, which is a sub-module of CBDCE, stores current environmental data, and LDPM stores Label data where context aware service is executed. LDPM sends a signal to SDPM when new label data is input, and SDPM combines the latest EDPM data with LDPM data to generate new snapshot data. The snapshot data is generated by using the latest data when new label data is input, and accumulates for the period specified by the system. Cumulative snapshot data is transferred to CAEAIM's TDPM and VDPM. 80% of the accumulated snapshot data is transferred to TDPM and the remaining 20% of the data is transferred to VDPM. This is to make learning and verification of artificial intelligence model effective, and it can increase the reliability of artificial intelligence learning result. The data of TDPM and VDPM are used for learning of artificial intelligence model, data of TDPM is used for learning, and data of VDPM is used for verification of learning rate. The learning process of artificial intelligence model is conducted by AIMCM and it controls the learning by using criteria such as learning frequency, accuracy and error rate according to the setting of artificial intelligence model.

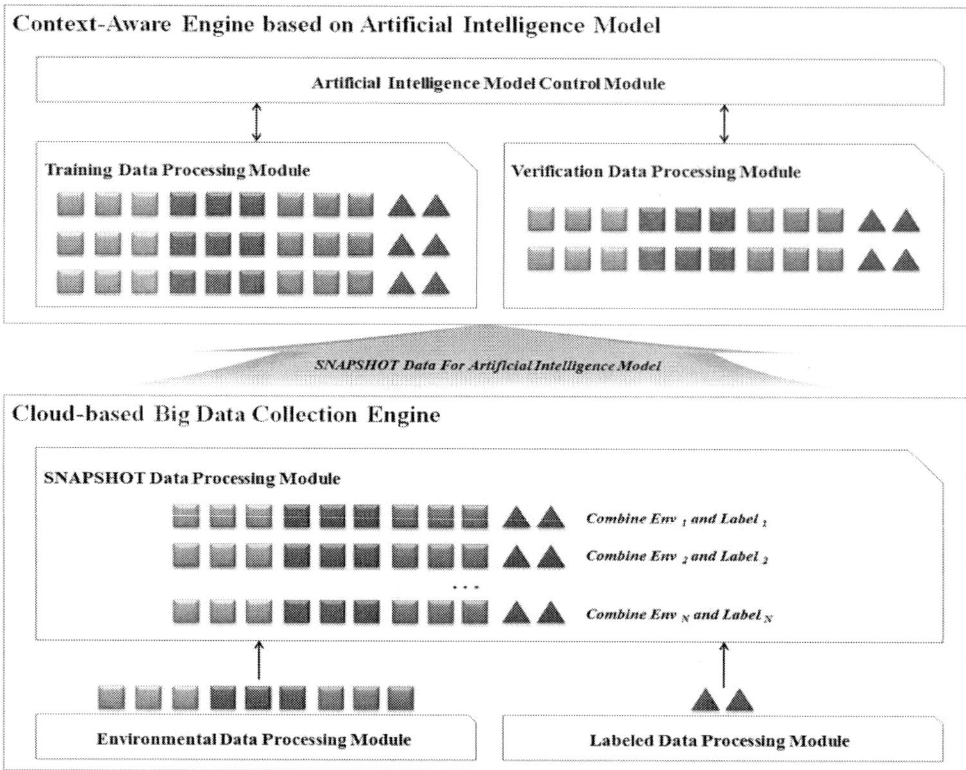

Figure 3. Generation of snapshot data and provision of learning data for artificial intelligence model.

4. Experiments and results

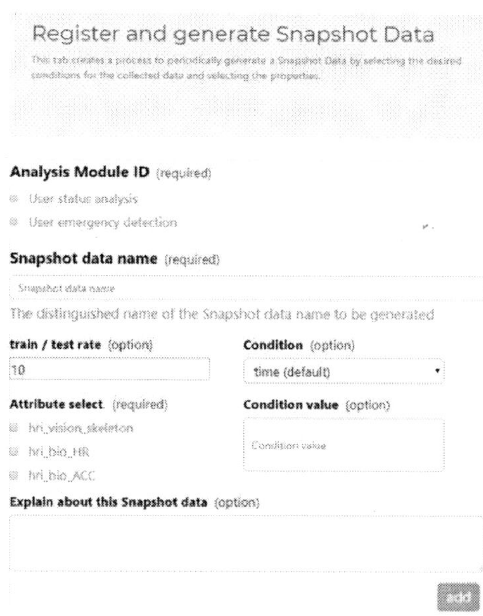

Figure 4. Configuration screen for collecting snapshot data.

In this chapter, through the prototype of the context-awareness system based on the artificial intelligence model that implements the proposed method of data processing of the context-awareness system, we show how the situation information is provided using the learning control process of the

IOP Conf. Series: Materials Science and Engineering **715** (2020) 012091 doi:10.1088/1757-899X/715/1/012091

artificial intelligence model and how the artificial intelligence model performs.

Figure 4 shows the screen for setting the context-aware system to generate snapshot data of the domain in which the robot is running. The settings allow you to select the data to be included in the snapshot data and control the preprocessing method of each object's Internet data.

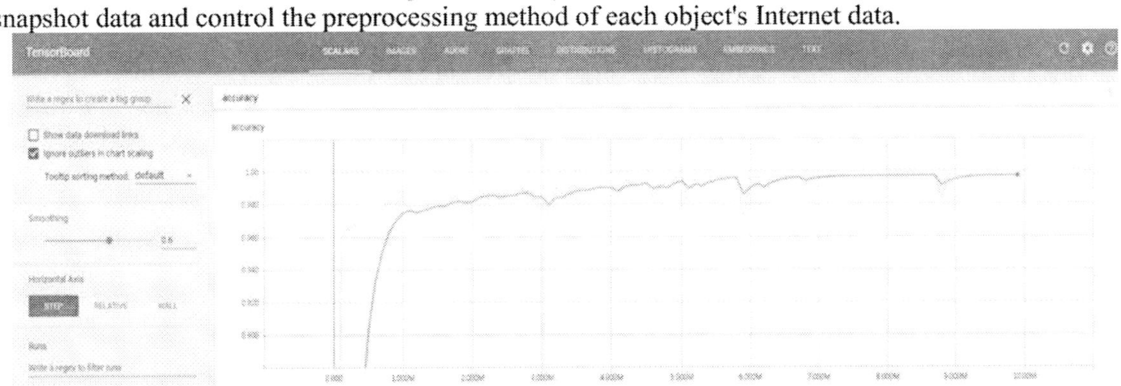

Figure 5. Learning result screen of artificial intelligence model.

Figure 5 shows the learning process of the AI model used in the context-awareness system of the intelligent robot. In the experiment, a TensorFlow-based artificial intelligence model is used. Figure 5 shows a TensorBoard that can visually express the learning result of artificial intelligence model. As shown in figure 5, the AI model is continuously increasing the accuracy, and the proposed data processing method can process data that has not been previously learned.

User Status Analysis

(A)

UDAM ID	UDAM Name	Using Sensor Data ID	Expression	Critical Value	TRUE Case Value	FALSE Case Value	Status
analysis_user_status	User Status Analysis	hri_vision_data	NEQ-N	5	NORMAL	EMERGENCY	DEACTIVATE

(B)

Log ID	Date	Sensor Data File	Processing Result	Expression Result	Conversion Result	Message
20181118_5	2018-11-18 16:51:19	EMPTY	null	EMPTY	null	Finish to job (job id : 2018_11_18_1)
20181118_4	2018-11-18 16:51:19	hri_vision_data_20181118_1.dat	5	false	EMERGENCY	Finish to send the conversion result
20181118_3	2018-11-18 16:51:19	hri_vision_data_20181118_1.dat	5	EMPTY	null	Receive result data (received data : 5)
20181118_2	2018-11-18 16:51:09	EMPTY	null	EMPTY	null	Start to job (job id : 2018_11_18_1)
20181118_1	2018-11-18 16:51:09	EMPTY	null	EMPTY	null	analysis_user_status Job(2018_11_18_1) registration completed.

Figure 6. Situational recognition result using artificial intelligence model

Figure 6 shows the result of providing context-awareness information through artificial intelligence model used in context-awareness system. Figure 6 (A), shows the information of the artificial intelligence model running in the context-aware system. The artificial intelligence model is a model that can recognize the risk situation by analyzing the attitude information of the user, and the situation recognition system returns the analysis result of the artificial intelligence model as "NORMAL" or "EMERGENCY". Figure 6 (B), shows the analysis results processed in the AI model. The data used in the experiment is data representing the state of the user's fall, and it can be confirmed that the "EMERGENCY" result is returned from the AI model.

5. Conclusion

In this paper, we have proposed a data processing method to provide context-aware information in context-aware system based on big data to provide intelligent robot services. The proposed data processing method, makes it possible to generate and accumulate the snapshot data, which combines the data obtained from various IoT devices in an intelligent robot service domain and log data related to service execution. This data is then learned by the RNN based analysis model. As a result, the analysis model of the situational awareness system can maintain consistently high accuracy for new data. The proposed data processing method is able to apply the result of the analysis model, which can be converted into the recognized situational information, for the engine to execute intelligent robot services. In the experiment, we showed how the analysis model learns to detect the user 's fall through the context-aware system with the proposed data processing method. We also verified the effectiveness of the proposed data processing method by using the learned analytical model to improve the accuracy of user fall detection. In the future, we will expand to a data providing framework for various artificial intelligence models.

Acknowledgments

This work was supported by the Industrial Convergence Core Technology Development Program (No. 10062501) funded by the Ministry of Trade, Industry & Energy (MOTIE), Korea and was supported by Basic Science Research Program (No. 2016R1C1B2009744) through the National Research Foundation of Korea(NRF) funded by the Ministry of Science and ICT(MSIT)

References

[1] Katal A, Wazid M and Goudar R H 2013 Big data: issues, challenges, tools and good practices Proc. Sixth Int. Conf. on Contemporary Computing (IC3).
[2] Liu B, Blasch E, Chen Y, Shen D and Chen G 2013 Scalable sentiment classification for big data analysis using naive bayes classifier Proc. IEEE Int. Conf. on Big Data.
[3] Sandryhaila A and Moura J M 2014 Big data analysis with signal processing on graphs: Representation and processing of massive data sets with irregular structure IEEE Signal Processing Magazine 31 (5) pp 80–90.
[4] Jha S, Qiu J, Luckow A, Mantha P and Fox G C 2014 A tale of two data-intensive paradigms: Applications, abstractions, and architectures Proc. IEEE Int. Conf. on Big Data.
[5] Li J and Tao F 2015 Big data in product lifecycle management Int. J. of Advanced Manufacturing Technology 81 (1-4) pp 667–684.
[6] Ramamoorthy S and S. Rajalakshmi 2013 Optimized data analysis in cloud using BigData analytics techniques." Proc. Fourth Int. Conf. on Computing, Communications and Networking Technologies (ICCCNT).
[7] Paul A, Ahmad A, Rathore M M and Jabbar S 2016 Smartbuddy: defining human behaviors using big data analytics in social internet of things IEEE Wireless communications 23 (5) pp 68–74.
[8] Chandramouli B, Goldstein J and Duan S 2012 Temporal analytics on big data for web advertising Proc IEEE 28th Int. Conf on Data Engineering.
[9] Kurnianingsih, Nugroho L E, Widyawan, Lazuardi L and Non-alinsavath K 2015 Ontology-based context aware for ubiquitous home care for elderly people Proc. 2nd IEEE Int. Conf. on Information Technology, Computer, and Electrical Engineering (ICITACEE).
[10] Barik R K, Priyadarshini R, Dubey H, Vinay Kumar V and Yadav S 2018 Leveraging Machine Learning in Mist Computing Telemonitoring System for Diabetes Prediction Advances in Data and Information Sciences Springer pp 95–104.
[11] Lemley J, Bazrafkan S and Corcoran P 2017 Deep Learning for Consumer Devices and Services: Pushing the limits for machine learning, artificial intelligence, and computer vision IEEE Consumer Electronics Magazine 6 (2) pp 48–56.
[12] Bazrafkan S and Corcoran P M 2018 Pushing the AI envelope: merging deep networks to accelerate edge artificial intelligence in consumer electronics devices and systems IEEE Consumer Electronics Magazine 7 (2) pp 55–61.
[13] Bakkar N, Kovalik T, Lorenzini I, Spangler S, Lacoste A, Sponaugle K, Ferrante P, Argentinis

E, Sattler R and Bowser R 2018 Artificial intelligence in neurodegenerative disease research: use of IBM Watson to identify additional RNA-binding proteins altered in amyotrophic lateral sclerosis Acta neuropathologica 135 (2) pp 227–47.
[14] Marburg A and Bigham K 2016 Deep learning for benthic fauna identification OCEANS 2016 MTS/IEEE Monterey. IEEE,.
[15] Shankar V and Chang S 2017 Performance of Caffe on QCT Deep Learning Reference Architecture — A Preliminary Case Study Proc. IEEE 4th Int. Conf. on. Cyber Security and Cloud Computing (CSCloud).
[16] Nicholson A C and Gibson A 2017 Deeplearning4j: Open-source, Distributed Deep Learning for the JVM deeplearning4j.org.
[17] Salazar O M, Ovalle D A and Duque N D 2015 Adaptive and personalized educational ubiquitous multi-agent system using context-awareness services and mobile devices Proc Int. Conf. on Learning and Collaboration Technologies Springer pp 301-12.
[18] Paola A D, Ferraro P, Gaglio S and Re G L 2016 Context-awareness for multi-sensor data fusion in smart environments Proc. Conf. of the Italian Association for Artificial Intelligence. LNCS 10037 pp 377-91.
[19] Dilsizian S E and Siegel E L 2014 Artificial intelligence in medicine and cardiac imaging: harnessing big data and advanced computing to provide personalized medical diagnosis and treatment Current Cardiology Reports 16 (1) 441.
[20] Liu R, Yabansu Y C, Yang Z, Choudhary A N, Kalidindi S R and Agrawa A 2017 Context Aware Machine Learning Approaches for Modeling Elastic Localization in Three-Dimensional Composite Microstructures Springer Integrating Materials & Manufacturing Innovation 6 (2) pp 160–71.
[21] Batarseh F A and Pithadia J 2017 Context-Aware User Interfaces for Intelligent Emergency Applications Proc. Int. Conf. on Modeling and Using Context pp 359–69.
[22] Cheng J, Sun Y and Meng M Q 2017 A dense semantic mapping system based on CRF-RNN network Proc. Int. Conf. on Advanced Robotics (ICAR).
[23] Khourdifi Y and Bahaj M 2017 Hadoop and MapReduce technology as a solution for Wireless Body Area Networks in e-Health Proc. Int. Conf. on Computing and Wireless Communication Systems (ICCWCS'17) Article No. 17.
[24] Wu T, Wu F; Redouté J-M and Yuce M R 2017 An autonomous wireless body area network implementation towards IoT connected healthcare applications IEEE Access 5 pp 11413–22.
[25] Forkan A R M, Khalil I, Ibaida A, Tari Z 2017 BDCaM: Big data for context-aware monitoring – a personalized knowledge discovery framework for assisted healthcare IEEE trans. on Cloud Computing 5 (4) pp 628–41.
[26] Maturana D and Scherer S 2015 Voxnet: A 3-D convolutional neural network for real-time object recognition Proc. IEEE/RSJ Int. Conf. on Intelligent Robots and Systems (IROS).
[27] Choi J, Cho Y and Choi J 2019 The Design of a Context-Aware Workflow Language for Supporting Multiple Workflows J. of Korea Society for Internet Information (in Korean) 10 (6) pp 145–57.

Partial soft body robots - a literature review

Yousef Amer[1], Simranjeet singh[1] and Linh Thi Truc Doan[1,2,*]

[1]School of Engineering, University of South Australia, Adelaide, Australia
[2]Deparment of Industrial Management, College of Engineering, Can Tho University, Viet Nam

*E-mail: linh.doan@mymail.unisa.edu.au

Abstract. One of the biggest concerns for every country is to look after their ageing population and provides them a safe life with safe medical equipment and technology. It is estimated that there will increase the ageing of the population in Japan, Western Europe, USA and India with 40%, 30%, 20% and 15%, respectively by the end of the year 2050. Hence, it will create a burden on aged care centres to hire more nursing staff members, which leads to increase the medication cost. One of the feasible solutions to solve this issue is to design the soft robots which will give care to patients at aged care centres. These robots can support patients by providing medicine to them without the assistance of the nursing staff during injury or illness. Therefore, this paper aims to investigate the need of the partial soft robotic arm which can look after patients like robots for interactive body assistance.

1. Introduction

With the advancement of technology, robots are used almost everywhere from the last century. Robots are a very powerful tool of present industry which can perform common functions without human interaction. They are used to increase safety by reducing the risk of injury during operations. According to Gupta and Arora [1], a robot is a computer-controlled device which moves objects and performs tasks while interacting with surrounding according to the program stored in it.

Robots can be classified based on different basis such as power source, control unit, use, mobility, coordinate system, a method of control, type of material etc. The structure of the robot consists of the mechanical unit, power source, tooling, drive system as seen in figure 1.

According to Hockstein et.al [2], Czech novelist Karel Capek introduced a word robot in a play named Rossum's Universal Robots (also called R.U.R) in 1920. Robot word derives from Czech word robota which is used for servant or labour. In 1954, a primitive arm was developed by American inventor George Devol Jr. This arm could perform tasks according to the programs stored in it. A first flexible multitask robot also known as Programmable Universal Manipulator Arm (PUMA) was introduced by American Mechanical engineer named Victor Scheinman in 1975. PUMA had a great capability of shifting and lifting the objects in any orientation. There were total 700,000 robots operating in the globe in 1995 [1]. Robots are currently used in every industry such as automobile, manufacturing, medical centres and etc.

Hirose and Ogawa [2] revealed that based on the previous results, Honda built the first humanoid robot with two legs which could walk easily in 1986. Honda's first aim was to create a bipedal robot which could walk easily in a straight line. Then the next challenge was to develop the new robot which can adapt itself in real surrounding conditions. ASIMO model was then developed by Honda which can walk and run on the even surface. It can climb stairs and remember face pour and other so many things.

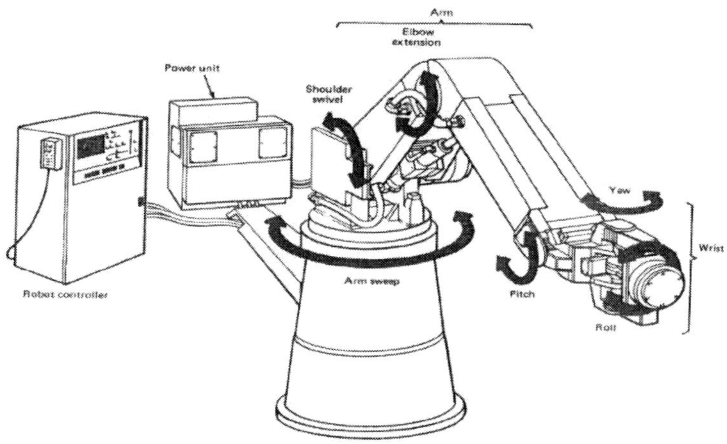

Figure 1. Industrial robot with key components [3].

When designing soft robotics arm, several issues need to be considered. First, current robots can lift only 60kg weight in one turn. Hence, this research aims to design the soft robotic arm which can lift more load. Second, the fabrication process is another key challenge while designing the soft robotic arm. Lastly, selecting an appropriate actuator, sensor, power source with great efficiency is also important. The main objective of the study is to develop or create partial soft robotic arm which can take care of patients without the support of nursing staff. Moreover, it can lift e patients like RIBA on its soft arms.

2. Soft robots

Soft robots are those kinds of robots which have a soft body and made up from soft materials. These are inspired by the soft body animals and so far, and several soft robots have been established. Different researchers define soft robots in different words. For instance, Rus and Tolley [4] stated soft robots as a system which has a capability of autonomous behavior, and they are primarily made up of soft materials with moduli in the range of soft biological materials. Kim, Laschi and Trimmer [5] defined soft robotics according to their original erection such as worm-like manipulators. Some of the characteristics of a soft robot are shown in table 1.

Table 1. Characteristics of soft robots [6].

Degree of freedom	Infinite
Actuators	Continuous
Materials strain	Large
Materials	Rubber, electroactive, polymer, silicon, elastomers
Accuracy	Low
Load capacity	Lowest
Safety	Safe
Working environment	Structured and unstructured
Controllability	Difficult
Path planning	Difficult
Inspiration	Muscular hydrostats

ASIMO and Nao robots (as seen in figure 2) are examples of soft robots.

Figure 2. ASIMO and Nao robots [7].

3. Literature review

Polygelines et.al [8] developed a soft robotic glove (as seen in figure 3) which utilizes a hydraulic soft actuator for assistance and at-home rehabilitation. These actuators were composed of elastomeric materials and it is mechanically programmed to perform twisting, bending and other functions. Combined fluidic pressure sensors are utilized to measure soft actuator's internal pressure and thus by mean of it controls the finger movement. In the end, they conclude that soft robotic glove gives more freedom to the user compared to existing equipment because of its open palm design.

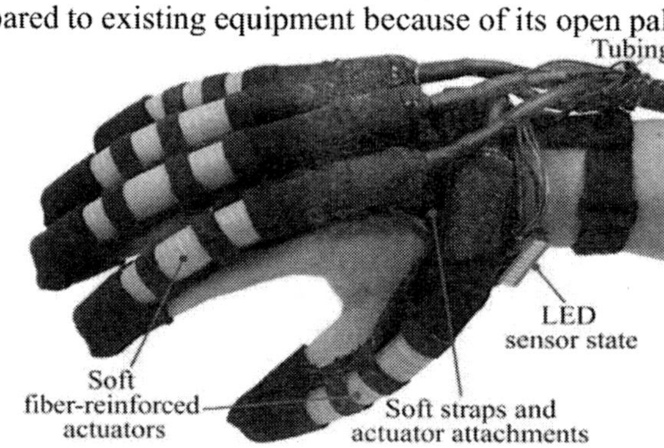

Figure 3. Soft robotic glove [8].

Ferris et.al [9] built a powered ankle-foot orthosis (as shown in figure 4) for people, which consists of carbon fiber and polypropylene shell, two artificial muscles. Dorsiflexion and plantarflexion torque are provided by these two artificial muscles about the ankle. The drawbacks of the previous prototype were covered by this new design as it provides more ease to don and doffs. This prototype can support patients during rehabilitation to recover after the neurological injury [10].

Mukai et al. [11] explained that there is a very high demand for soft robotics in medical and aged care centers to face the glitches which are caused due to an aging population. One of the most challenging parts of nursing care is to lift and transfer the patient from wheelchair to bed or vice-versa. To overcome this problem, they have developed a soft robot namely robot for interactive body assistance (RIBA). This robot can be used for heavy tasks needing human interaction. This robot is successful in transferring a patient from beds or wheelchairs (as seen in figure 5). RIBA can perform tasks according to the instructions given to it by the instructor or caregiver. The whole body of RIBA robot is made up of soft materials such as polyurethane foam and silicon elastomers for safety purposes. There are different types of sensors such as tactile or vision sensors are used in RIBA. Specifications of RIBA are shown in table 2.

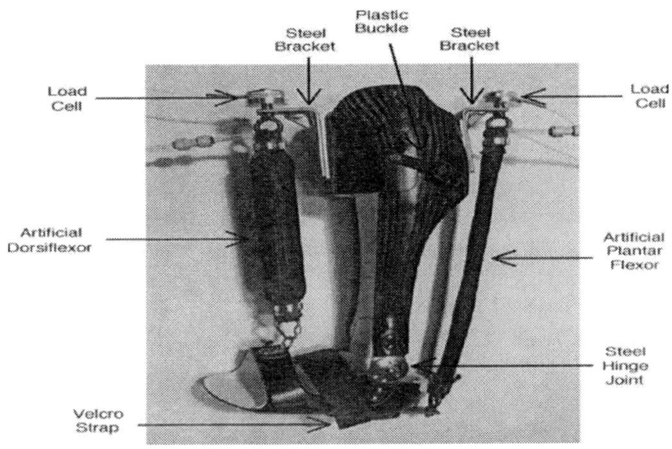

Figure 4. Pneumatically powered ankle-foot orthosis adopted from [9].

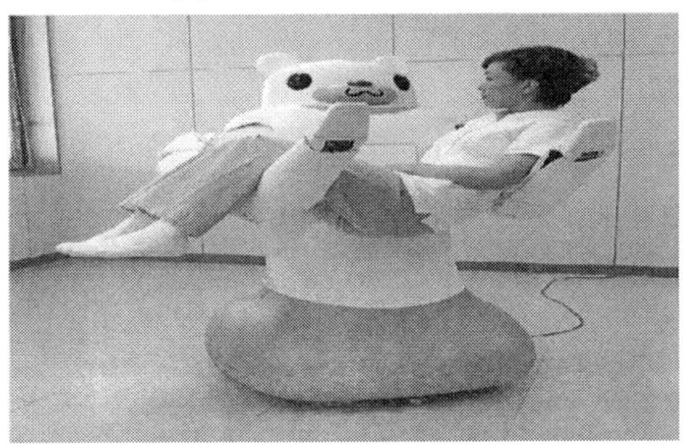

Figure 5. RIBA lifting the patient adopted from [11].

Specifications of RIBA:

Table 2. Specifications of RIBA [11].

DimensionsWidth	750 mm (when arms are folded)
Depth	840 mm
Height	1,400 mm
Weight inc. batteries	180 kg
D.O.F. Head	3 (only 1 in current use)
Arm	7 each
Waist	2
Cart	3 (with 4 motored wheels)
Base movement	Omnidirectional with omnidirectional wheels
Actuator type	DC motor
Payload	63 kg (tested value)
Operation time	1 hour in standard use
Power	NiMH batteries
Sensors Vision	2 cameras
Audio	2 microphones
Tactile	Upper arm (128 pts. each)
	Forearm (94 pts. each)
	Hand (4 pts. each)
	Shoulder pad (8 pts. each)

Dahl and Bulous [12] presented the various uses and applications of soft robotics in hospitals, medical care centers and social care sectors. They described that the traditional robot has a very low level of autonomy and mainly concentrates on the specialized platform for surgery or rehabilitation. These kinds of robots need qualified staff for ensuring the appropriate usage. Neuromata robot and da Vinci robot are the examples of these traditional robots. In hospitals, robots are using for transporting of equipment to supporting nursing staff and to help nurses for caring the patients. They further give examples of RI-MAN and cody robots which are very useful to helps nurses in lifting patients from bed with more safety (as seen in figure 6). Authors concluded that these kinds of soft robots play crucial roles in the medical centers and hospitals in upcoming years.

Figure 6. RI-MAN robot [13].

Causo et.al [7] explained that robots are used as a tutor for children. In their research, they review a different type of robots used in education. In the present world, robots are the source of attraction for smaller children and they like to play with these robots. Robots in the educational field can be classified according to the target users, robot role and venue of learning. The target users can be primary school students or secondary students. Likewise, robot role can be different depending upon the situation. They can be used as a companion or learning instructor. Robots can be categorized into three different form such as humanoid robot, semi-humanoid and pet-like robots. A humanoid robot is a robot with a body shape designed as a human body. Nao and Asimo robots are two examples of humanoid robots which look like a small child and attract the attention of other children to play with them. Nao robot is a humanoid robot which was launched in 2004 by Aldebaran robotics. It has 25 degree of freedom which helps Nao robot in smooth dancing and walking like a human being. It can be used to provide educational training for children. It has attached a camera, microphone and speaker. ASIMO (as seen in figure 7), a humanoid robot, was launched by Honda. It is used as a companion to small children. Asimo has specifications of 130 cm height, 50 kg weight and has 57 degrees of freedom.

Figure 7. ASIMO and Nao robots [7].

Tiro robot is a semi-humanoid robot (as seen in figure 8) which acts as a teaching assistant robot. Tiro robot provides teaching lessons to students in a class. Tiro has 130 cm height and 70 kg weight. It has in built camera, speakers and 16 pairs of ultrasonic sensors. The Robovie R3 robot has a height of 108 cm, weight 35 kg and has 17 degrees of freedom. It has 11 tactile sensors attached to its body. Apart from these sensors, I t consist of 2 USB cameras, 2 microphones, a speaker. It provides voice-based English conversation. Further, it performs other functions such as shaking hand, greeting, welcoming, kissing, hugging, singing, playing and entertaining etc. Maggie, a semi-humanoid robot, has 6 degreess of freedom. The examples of semi-humanoid robots are shown in figure 9. It is equipped with tactile sensors and 12 sets of bumpers, camera, speaker etc. The PaPeRo was launched by NEC and has height 38.5 cm, weight 5 kg. The degree of freedom for PaPeRo is 4. It has 2 facial recognition cameras and microphone for voice recognition. It can also be connected to the Internet to gain further information.

Figure 8. Examples of semi-humanoid robots[7].

Pleo the dinosaur, eMuu, probo are the examples of pet-like robot (as seen in figure 9). eMuu is one eye robot which can show its feelings according to the interaction with human being. The Huggable robot, which is companion robot, was launched by MIT Media Lab for various purposes such as healthcare, aged care, education etc. It has soft body sensitive skin with more than 150 sensors. During educational activities, it is controlled by a teacher on the internet.

Figure 9. Examples of pet-like robots [7].

A recent study carried out by [14] showed that children feel less pain and smile more when medical treatment is given to them with these robots. The main aim of this robot is to divert the minds of children from pain to other things. These robots also give companionship to elderly people in several tasks so that they cannot feel loneliness. These robots act as a fruitful source for the aging population.

According to Harwin, Rahman and Foulds [15], a few robot assistive applications have been developed for numerous applications since the 1960s. For example, upper-limb exoskeletons, gait exoskeletons and ankle rehabilitation robots are some examples of soft robots used in medical fields.

Moving further, soft robots are also used to reduce the service cost of nursing staff which is very high in the present world. It was reported by Guild [16] that the number of aged care workers had been increased from 262000 in 2007 to almost 352,000 in 2012 which represents the 34% increase in the size of the workforce of aged care in last five-year time interval. Almost 202,000 workers were employed in the residential care segment. Hence, it is also the one main reason why we need soft robots.

Gnanamanickam et.al [17] presented that more than half of the permanent citizen of Australia residing in residential aged care have dementia. People with dementia need great support from others at any time to divert their thinking. AUD 88,000 per person is the cost associated with the health and residential care for dementia patient in residential care segment. figure 10 shows robot supports dementia patients at aged care centers. Hence, soft robots are necessary in order to reduce this cost.

IOP Conf. Series: Materials Science and Engineering **715** (2020) 012092 doi:10.1088/1757-899X/715/1/012092

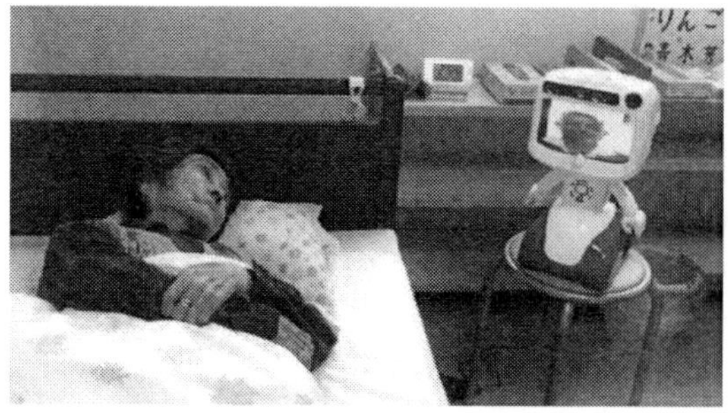

Figure 10. Robot supports dementia patients at aged care centers [18].

Soft robots are used for numerous applications in the medical field such as neurosurgery, surgery, serving patients, providing physiotherapy to patients, teaching children, orthopedics etc. apart from these, they support the patient during rehabilitation and helps during fracture or dementia problems. The percent applications of service robots in medical fields are depicted in figure 11. Karabegović and Dolecek [19] listed several other merits of robots in the medical field as follows:

a. These robots promote medical processes by delivering precise control of devices.
b. These robots reduce the risk factors and provide more safety to patients during surgery.
c. It increases the quality of surgery.
d. It can give better care to patients without the assistance of the nursing staff.

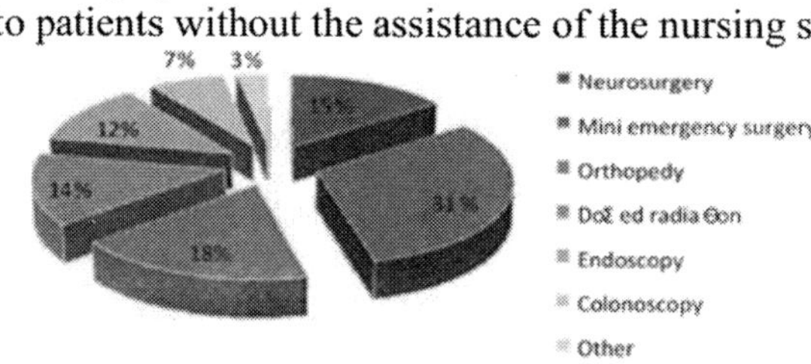

Figure 11. Percent applications of service robots in medical fields [19].

4. Conclusion and Further work

From the above discussion, it is concluded that soft robots are the major need of modern science and the medical field. Soft robots can support the patients without the assistance of the nursing staff. They can provide more safety in the workplace. Hence, the design of the partial soft body robots' part is necessary which can assist with the patients with more care and then it will lead to reducing strain on the nurses and doctors at various medical and aged care centres. In the next phase, 3D printing technique will be used to fabricate soft robots since it can provide high quality and simultaneously print several materials. This will lead to some significant advantages such as lifting the patients with more care and safety on its arms, handling heavy object without crushing it, more flexibility due to more degree of freedom.

References

[1] Gupta, A. and S. Arora, Industrial automation and robotics. 2009: Laxmi Publications.
[2] Hockstein, N.G., et al., A history of robots: from science fiction to surgical robotics. Journal of robotic surgery, 2007. 1(2): p. 113-18.
[3] Koren, Y. and Y. Koren, Robotics for engineers. Vol. 168. 1985: McGraw-Hill New York et al.
[4] Rus, D. and M.T. Tolley, Design, fabrication and control of soft robots. Nature, 2015. 521(7553): p. 467.
[5] Kim, S., C. Laschi, and B. Trimmer, Soft robotics: a bioinspired evolution in robotics. Trends in biotechnology, 2013. 31(5): p. 287-94.
[6] Trivedi, D., et al., Soft robotics: Biological inspiration, state of the art, and future research. Applied bionics and biomechanics, 2008. 5(3): p. 99-117.

[7] Causo, A., et al., Design of robots used as education companion and tutor, in Robotics and Mechatronics. 2016, Springer. p. 75-84.
[8] Polygerinos, P., et al., Soft robotic glove for combined assistance and at-home rehabilitation. Robotics and Autonomous Systems, 2015. 73: p. 135-43.
[9] Ferris, D.P., et al., An improved powered ankle–foot orthosis using proportional myoelectric control. Gait & posture, 2006. 23(4): p. 425-28.
[10] Xie, S., M. Zhang, and W. Meng, Soft Robots for Healthcare Applications: Design, Modelling, and Control. Vol. 14. 2017: Institution of Engineering and Technology.
[11] Mukai, T., et al. Development of a nursing-care assistant robot RIBA that can lift a human in its arms. in Intelligent Robots and Systems (IROS), 2010 IEEE/RSJ International Conference on. 2010. IEEE.
[12] Dahl, T. and M. Boulos, Robots in health and social care: A complementary technology to home care and telehealthcare? Robotics, 2013. 3(1): p. 1-21.
[13] Onishi, M., et al. Generation of human care behaviors by human-interactive robot RI-MAN. in Robotics and Automation, 2007 IEEE International Conference on. 2007. IEEE.
[14] Yasemin, M. and Y. Kasımo, Management of dental anxiety in children using robots Reduction of Dental Anxiety and Pain in Children using Robots,(May), 327–32.(2016). 2017.
[15] Harwin, W.S., T. Rahman, and R.A. Foulds, A review of design issues in rehabilitation robotics with reference to North American research. IEEE Transactions on Rehabilitation Engineering, 1995. 3(1): p. 3-13.
[16] Guild, A.C., Australia's aged care sector: economic contribution and future directions. 2016.
[17] Gnanamanickam, E.S., et al., Direct health and residential care costs of people living with dementia in Australian residential aged care. International journal of geriatric psychiatry, 2018.
[18] Üyesi, E.B.V.; Available from: http://www.vikilist.com/yaslilar-icin-bir-umut-olan-robot-dinsow-in-seri-uretimine-baslandi-4781.
[19] Karabegović, I. and V. Doleček, The role of service robots and robotic systems in the treatment of patients in medical institutions, in Advanced Technologies, Systems, and Applications. 2017, Springer. p. 9-25.

IOP Conf. Series: Materials Science and Engineering **715** (2020) 012093 doi:10.1088/1757-899X/715/1/012093

Modified extreme learning machine based motion control of robotic manipulators

Neha Kapoor* and Jyoti Ohri

Department of Electrical Engineering, National Institute of Technology, Kurukshetra, Haryana, India

*Email: ernehakapoor@rediffmail.com

Abstract. Precise trajectory tracking is a difficult task due to partially known and unknown dynamics and the disturbances present in the system. For improving the tracking performance of the robotic manipulator, this work proposes a novel PSO optimized kernel based Extreme Learning Machine (PSO-KELM) learning algorithm in which PSO is used to get the optimal values of the free kernel-parameters in KELM. The simulation results represent the good generalized performance and PSO-KELM outperforms the KELM and ELM based control techniques for the manipulator trajectory tracking. Comparative analysis of the proposed control schemes have been done with NN and SVM based controllers and various ELM based variants for the trajectory tracking problem in robotic manipulator.

1. Introduction

In modern times, the accuracy in position control in trajectory tracking of robotic manipulator is crucial to the commercial viability and widespread use in industries. Inherent characteristics of being highly non-linear and strongly coupled and the recent increasing demand for industrial robots has provided the researchers working in the area with a new challenging and time consuming problem of finding out the most accurate control input torque to robot actuators to take the robot to as close as to the desired trajectories. Hence, main motto of the work carried out in this paper is to propose the most recent and improved ELM based control techniques for the trajectory tracking problem of the manipulator. Joints are driven by actuators and hence, for practicality driving actuator dynamics have been taken into consideration. To validate the performance, proposed controllers are made to work with manipulators with planar elbow direct driven links and remotely driven links.

For improving the control performance for position control problem of robotic manipulator a tremendous research in [1-4 and references there in] has been done. However, most NN based forecasting methods cannot avoid a number of difficulties like choice of stopping criteria, learning rate and epochs, local minima etc.. Support vector machines (SVM) based control schemes having better learning, have shown an improved performance [5-8]. Another new most recent regression technique, proposed by Huang [9,10], having similar performance as SVM is Extreme Learning Machine (ELM) but needs lesser optimization constraints and has better generation with less operation time and hence, has been used in many research areas apart from motion control problem of robotic manipulators [11]. Although ELM is an efficient modern technique, but a values of non-optimal input weights and hidden biases may lead to over fitting. Some improvements in the ELM basic algorithm have been mentioned in [12]. However, how to get the optimal combination of the hidden layer parameters is still unresolved. As the feature mapping function of hidden neurons is unknown, like SVM, ELM can be extended to kernel ELM, which is called the kernel based extreme learning machine (KELM), which leads to its stability improvement too [11]. In KELM, no hidden layer parameter tuning is required, and however, the kernel parameters should be chosen properly for improving performance of the KELM algorithm. Kernel parameters in kernel functions of the proposed algorithm have been optimized using and meta-heuristic optimization technique, i.e. particle swarm optimization. Effectiveness of the proposed algorithm for the problem undertaken i.e. motion control of a robotic manipulator has been shown in the simulation study carried out in this work.

Other sections of the paper are organized as follows: Section 2 contains the basics of the ELM/KELM. Particle swarm optimization based KELM has been presented in Section 3. Preliminaries are contained in Section 4 and simulation study has been described in Section 5. Section 6 has performance analysis and the result compilation followed by the conclusions in the Section 7.

2. Kernel extreme learning machines

Recently developed ELM tends to give the best generalization performance at tremendously fast learning speed. Huang et al. [13] contains a detail study of ELM with hidden neurons, H and activation function, f(x) can approximate training data, K without any error. ELM can universally very well approximate a continuous function [14]. Also, further Huang et al. [10] proposed some modified forms of ELM. Huang et al. [10] suggested that (8) provide least training error, minimum weight norms having greatest generalization, and a unique solution, and thus involves no local minima as with the back propagation learning algorithm. With some of the modifications new proposed ELMs can be found in [15 & 16].

Radial basis, sine, cosine and exponential functions are some of the differential activation function [10] for which ELM works. Inspite of simplified learning approach of ELM, two major issues of ELM are: a) the number of hidden neurons (i.e. the size of the SLFNs). Usually users choose the optimal number of hidden nodes by trial-and-error, b) to reduce the computation complexity of ELM especially if given large number of training data and if large number of hidden nodes required. Also, ELM may suffer either from under-fitting or over-fitting. Out of these two problems, the over-fitting is more important because it can easily lead to wild estimate. An over-fitting model can give poor predictive performance by exaggerating minor fluctuations in the data. Hence, in this work emphasis has been given on these aspects and for this kernel ELM (KELM) has been used and implemented for motion control problem of robotic manipulator.

Recently, Huang et al. [11] propose to use orthogonal projection and kernel methods for the designing of ELM. This provided a more stable solution of ELM with improved generalization capabilities as comparison to least square solution based ELM. Thus, with this new kernel method output function of ELM is given by:

$$h(x)\alpha = h(x)A^T\left(\frac{I}{\rho} + AA^T\right)^{-1} \tag{1}$$

where all the symbols have their usual meanings and details of them can be found in [10, 11]. If hidden layer feature mapping h(x) is unknown, the kernel function is proposed. A kernel matrix for ELM can be represented as follows:

$$\chi_{ELM} = AA^T : \chi_{ELM_{ij}} = h(x_i).h(x_j) = B(x_i,x_j) \tag{2}$$

where $B(x_i,x_j)$ is a kernel function. Output function in (1) can be represented as:

$$\begin{bmatrix} B\,(x, x_1) \\ \vdots \\ B\,(x, x_K) \end{bmatrix} \left(\frac{I}{\rho} + \chi_{ELM}\right)^{-1} Y \tag{3}$$

Here, a kernel function corresponding to h(x) is used and hence, exact knowledge of hidden layer feature mapping and the number of hidden nodes is not required. List of kernel functions used in ELM is similar to that are used with support vector machines, such as linear, polynomial and radial basis

IOP Conf. Series: Materials Science and Engineering **715** (2020) 012093 doi:10.1088/1757-899X/715/1/012093

function. Compared with the ELM learning algorithm, the hidden layer feature mapping need not be known and the numbers of hidden neurons require not to be selected in the KELM. Moreover, the KELM learning algorithm achieves similar or better generalization performance and is more stable compared to traditional ELM.

3. Particle swarm optimization based KELM

PSO [17], a recently developed heuristic optimization technique, has been proposed as an alternative to genetic algorithm (GA) [18]. Animal behavior based PSO has many attractive features as compared to GA. ISimple concept, easy implementation, quick convergence and more accurate optimization made PSO a worldwide famous computational algorithm. PSO is based on changing the velocity and position of each particle toward its pbest (current position best) and gbest (global best) locations according to equations (4) and (5) respectively, at each time step

$$v_{id} = w * v_{id} + c_1 \in_1 (p_{id} - x_{id}) + c_2 \in_2 (p_{xd} - x_{id} \tag{4}$$

$$x_{id} = x_{id} + v_{id} \tag{5}$$

where, in a D-dimensional space, $\vec{x_i} = (x_{i1}, x_{i2}, ...x_{iD})$ is a present position vector, $\vec{p_i} = (p_{i1}, p_{i2}, ...p_{iD})$ is a best position vector, $\vec{v_i} = (v_{i1}, v_{i2}, ...v_{iD})$ is a velocity vector, c_1 and c_2 are the acceleration coefficients having constant value: 2, $\in_1$ and $\in_2$ are the random number generators between (0-1), w is the linearly changing inertia weight and d is the dimension ($1 \leq n \leq N$). Instead of choosing a constant of w in (4), it can be chosen as in (6)

$$w = (w_{max} - w_{min}) \left(\frac{maxiter - iter}{maxiter} \right) + w_{min} \tag{6}$$

Basic theory of particle swarm optimization (PSO) can be studied in author's paper [19].

PSO-KELM theory:

In KELM learning algorithm, the regulation coefficient C and kernel parameters should be chosen appropriately for improving the generalization performance of neural networks. In [11], these parameters are tried in a wide range and this proved to be a time consuming task. And in [20], a hybrid kernel function is proposed for improving the generalization performance of KELM. However, how to choose the optimal values of the parameters of the kernel function has not been resolved. In this paper, an optimization approach, naming PSO, has been introduced to the KELM for choosing the optimal parameters of kernel function. Steps involved in this particular technique are:

Step 1 Initiate the population (particle) based on the kernel function and the velocity and position of each particle.

Step 2 Evaluate the fitness function of each particle.

Step 3 The velocity and position of the particle are modified according to the basic PSO algorithm.

Step 4 Step 2 and Step 3 are iterated repetitively until the maximal iteration time is satisfied.

Step 5. The optimal parameters of kernel function can be determined. Then, based on the optimized parameters, the hidden layer kernel matrix is computed.

Step 6. Determine the final output weights β in terms of the following equation:

$\beta = H^T((1/C + HH\hat{}T)^{-1}T$.

4. Preliminaries

4.1. Manipulator dynamics

The general dynamics of n-link robotic manipulator is given by following nonlinear differential equation [21]

$$M(q)\ddot{q} + V(q,\dot{q}) + G(q) + F(q,\dot{q}) = \tau \tag{7}$$

with $q \in R^n$ is the joint position variables, τ is vector of input torques, M (q) is the inertia matrix which is symmetric and positive definite, $V(q,\dot{q})$ is the coriolis and centripetal matrix, G(q) includes the gravitational forces and $F(q,\dot{q})$ is the frictional forces.

Directly Driven Robotic Manipulator:
Dynamic equation of a 2 DOF directly driven robotic manipulator satisfying equation (7) chosen in this work have been given in below as [21]. Equations representing the dynamics of directly driven link are given in (8)

$$\begin{bmatrix} m_{11} & m_{12} \\ m_{21} & m_{22} \end{bmatrix}\begin{bmatrix} \ddot{q}_1 \\ \ddot{q}_2 \end{bmatrix} + \begin{bmatrix} v_{11}\ddot{q}_1 \\ v_{12}\ddot{q}_2 \end{bmatrix} + \begin{bmatrix} g_1 \\ g_2 \end{bmatrix} + \begin{bmatrix} F_1 \\ F_2 \end{bmatrix} = \begin{bmatrix} \tau_1 \\ \tau_2 \end{bmatrix} \tag{8}$$

In this equation,
$m11 = (m_1 + m_2)a_1^2 + m_1a_2^2 + 2m_2a_1a_2\cos q_2$; $m12 = m_2a_2^2 + m_2a_1a_2\cos q_2 = m21$; $m22 = m_2a_2^2$; $v11 =- m_1a_1a_2\left(2\dot{q}_1\dot{q}_2 + \dot{q}_2^2\right)\sin q_2$; $v21 = m_2a_1a_2\dot{q}_1^2\sin q_2$; $g11 = (m_1 + m_2)ga_1\cos q_1 + m_2ga_2\cos(q_1 + q_2)$;$g21 = m_2ga_2\cos(q_1 + q_2)$

where m_1 and m_2 are the masses and a_1 and a_2 are the lengths of the links 1 & 2 respectively and g is the gravity acceleration, these have been chosen as below
$m_1 = 1\text{kg};\ m_2 = 1\text{kg};\ a_1 = a_2 = 1\text{m};\ g = 9.8\text{m/s}^2$.

Remotely Driven Robotic Manipulator:
The robotic manipulator considered here is a planar elbow manipulator with remotely driven link, whose dynamic model has been derived by Spong et al [21]. Equations representing the dynamics of remotely driven link are given in (9).

$$\begin{bmatrix} m_{11} & m_{12} \\ m_{21} & m_{22} \end{bmatrix}\begin{bmatrix} \ddot{q}_1 \\ \ddot{q}_2 \end{bmatrix} + \begin{bmatrix} v_{11}\ddot{q}_1 \\ v_{12}\ddot{q}_2 \end{bmatrix} + \begin{bmatrix} g_1 \\ g_2 \end{bmatrix} + \begin{bmatrix} F_1 \\ F_2 \end{bmatrix} = \begin{bmatrix} \tau_1 \\ \tau_2 \end{bmatrix} \tag{9}$$

In this equation,
$m_{11} = m_1l_{c1}^2 + m_2l_1^2 + I_1$; $m_{12} = m_2l_1l_{c2}\cos(p_2 - p_1)$; $m_{21} = m_2l_1l_{c2}\cos(p_2 - p_1)$; $m_{22} = m_2l_{c2}^2 + I_2$;$v_{11} =- m_2l_1l_{c2}\sin(p_2 - p_1)$;$v_{12} = m_2l_1l_{c2}\sin(p_2 - p_1)$;$g_1 = (m_1l_{c1} + m_2l_1)g\cos(p_1)$; $g_2 = m_2l_{c2}g\cos(p_2)$
where the subscripts 1 & 2 indicate link 1 & link 2. In this case, q_i is the angle with respect to horizontal axis; m_i is the weight; I_i is the inertia; l_i is the length; l_{ci} is the distance from the joint to the centre of gravity; g is gravitational constant; τ_i is the torque input. The parameters chosen are as following:

$$m_1 = 10;\ m_2 = 5;\ I_1 = 0.2;\ I_2 = 0.2;\ l_{c1} = 0.25;\ l_{c2} = 0.5;\ g = 9.8.$$

4.2. Actuator dynamics
A robot may have several kinds of influencing actuators-electric, pneumatic, or hydraulic. In this case DC motor with permanent magnet has been considered as actuator motor as it is popular in robotic applications. Detailed study and the numeric values of the dynamics of the armature controlled dc motor that drives the joints of manipulator can be obtained from [22] and has the same form as the robotic manipulator dynamics given in equation (7) and are given by:

$$M_{RM}(q)\ddot{q} + V_{RM}(q,\dot{q}) + G_{RM}(q) + F_{RM}(q,\dot{q}) = \tau_{RM} \tag{10}$$

where,

$$M_{RM} = (J_M + R^2M);\ V_{RM} = \left(B + R^2V\right);$$
$$G_{RM} = R^2G \text{ and } F_{RM} = \left(RF_M + R^2F\right)$$

4.3. Uncertainties
- **Continuous Disturbance**: Disturbance changing continuously with time entering into the system is chosen as following:

$$T_d = 0.15 * \cos(5t) + 0.15 * \sin(5t) \tag{11}$$

- **Friction:** The LuGre model is a dynamic friction model presented in [25] is chosen here.
- **Payload changes:** An addition 30% extra load in mass of link 2 has been added as payload changes.

Predefined desired trajectories for joint 1 & 2 are given in (12) and (13) respectively below

IOP Conf. Series: Materials Science and Engineering **715** (2020) 012093 doi:10.1088/1757-899X/715/1/012093

$$q_1^d = 0.3\,sin(0.7t - \tfrac{\pi}{2}) + 0.3\,sin(0.1t - \tfrac{\pi}{2}) + 0.7; \tag{12}$$

$$q_2^d = 0.5\,sin(0.9t - \tfrac{\pi}{2}) + 0.5\,sin(0.1t - \tfrac{\pi}{2}) + 1.1; \tag{13}$$

Details of various performance evaluating indices opted here for performance evaluation can be found in [26].

5. Simulation example

In this ELM based control techniques error (e) and velocity error ($\dot{e}$) are given as input and the output is the control input torque (τ). In the simulation study, number of neurons in ELM is chosen by trial and error and is equal to 29. In KELM radial bias kernel is chosen with gamma (γ) as 0.57 whereas in the proposed PSO-KELM, PSO is used to optimize gamma (γ) and the regulation coefficient (C). In PSO-KELM control technique PSO parameters are chosen as: number of population-25, number of iterations-30, linearly varying inertia weights ranging from 0.9 to 0.4 for w_{max} to w_{min}, acceleration coefficients have been chosen constants =2. For this purpose range of γ is chosen from [0-4] and range of C is chosen from [200-400]. Cases constituted for the study under consideration are:

Case 1: under ideal condition with no uncertainties.

Case 2: with payload mass changes + LuGre friction + continuous disturbance + actuator dynamics.

Trajectory tracking for 2 dof directly driven robotic manipulator with the controllers naming, ELM, KELM and PSO-KELM of case 2 have been shown in figure 1(a) for joint 1 and (b) for joint 2 respectively. For clarity purpose, tracking results for these controllers have been zoomed in figure 2. figure 3 is representing the tracking error and control input torque, which clearly indicates the better tracking accuracy of PSO-ELM then KELM and ELM. Tracking results of case 1 are found to be similar to the results of case 2 and hence omitted. Also, tracking results for remotely driven manipulator have been found similar to directly driven manipulator and hence not shown here. Tracking results for NN and SVM based controllers for the motion control of a 2 dof robotic manipulator has been taken from the previous papers [23] (for NN) and [24] (for SVM) by the author.

Performance indices representing various performance evaluating factors for directly driven and remotely driven manipulators for case 1 & 2 have been given in table 1 and 2 respectively. Table 3 gives the execution time taken by the controllers.

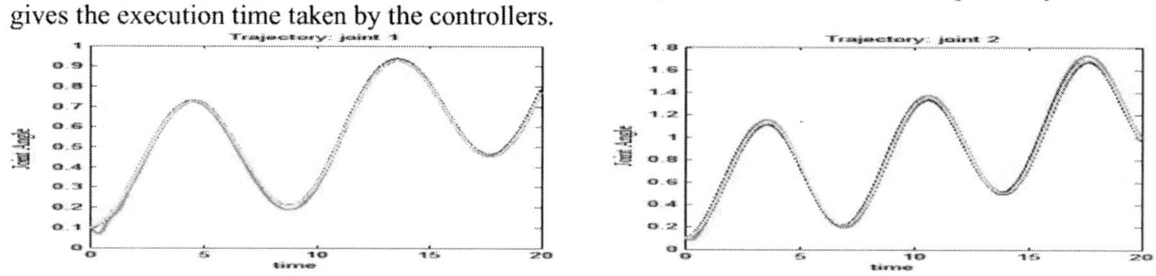

Figure 1. Trajectory to be Tracked with ELM, KELM and PSO-KELM.

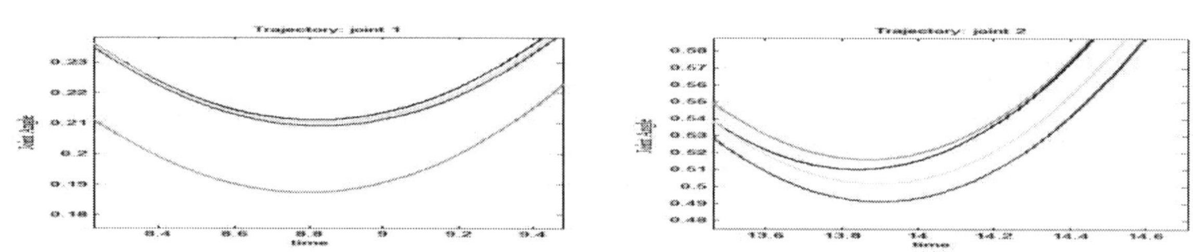

Figure 2. Trajectory Tracked with ELM, KELM and PSO-KELM.

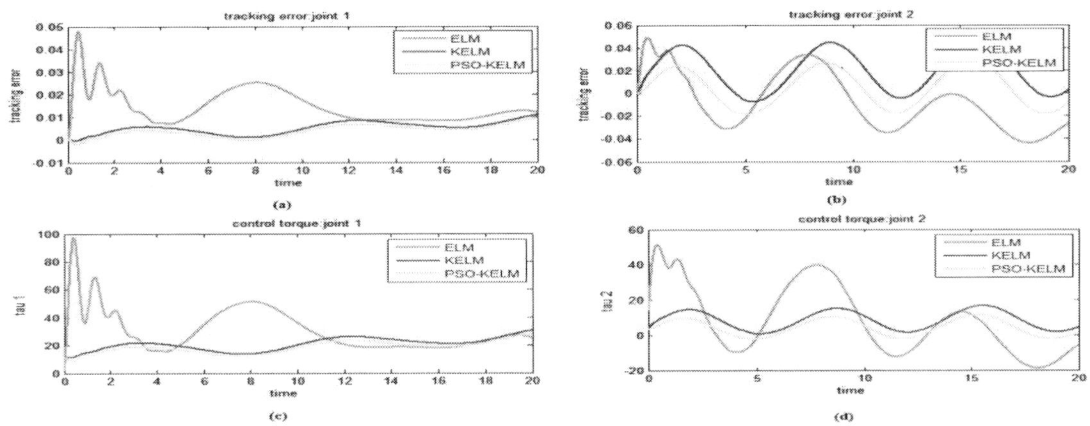

ELM and PSO-KELM (Zoomed).

Figure 3. (a) Tracking error: e1; **(b)** Tracking error: e2; **(c)** control input torque: τ1 and **(d)** control input torque: τ2.

Table 1. Quantitative performance indices for a 2 DOF robotic manipulator: Case 1.

	Controller	$\|e1\|_2$	$\|e2\|_2$	$\|\tau1\|_2$	$\|\tau2\|_2$	$c1_u$	$c2_u$
Directly Driven	NN	0.33	0.49	998.7	355.08	0.00055	0.00035
	SVM	0.2	0.401	1.90E+03	244.68	6.70E-04	4.90E-04
	ELM	0.17354	0.3880	1.5304e+03	946.0694	3.9242e-04	3.95e-04
	KELM	0.147	0.4301	1.348E+03	797.51	3.44E-04	8.60E-04
	PSO-KELM	0.1124	0.2322	1.022E+03	444.5	6.70E-04	4.90E-04
Remotely Driven	NN						
	SVM	1.16	1.57	3.30E+04	1.50E+03	3.20E-04	7.60E-04
	ELM	1.56	1.15	3.30E+04	1.50E+03	3.20E-04	7.60E-04
	KELM	1.07	1.10	1.6+3	9.27E+02	4.30E-04	4.70E-04
	PSO-KELM	0.983	1.02	3.20E+03	1.60E+03	3.30E-04	8.20E-04

Table 2. Quantitative performance indices for a 2 dof in the paper: Case 2.

	Controller	$\|e1\|_2$	$\|e2\|_2$	$\|\tau1\|_2$	$\|\tau2\|_2$	$c1_u$	$c2_u$
Directly Driven	NN	0.33	0.49	998.7	355.08	0.00055	0.00035
	SVM	0.1	0.45	845.4	275.8	0.00073	0.0000707
	ELM	0.19	0.45	1566.4	946.06	0.00073	0.0000707
	KELM	0.1498	0.3892	1460	810	0.00043	0.00022
	PSO-KELM	0.1006	0.347	1222	466.05	0.00067	0.00049
Remotely Driven	NN	2.01	2.22	1.74E+03	1032	5.50E-04	4.34E-04
	SVM	0.73	1.18	1.50E+03	946.06	3.90E-04	3.90E-04
	ELM	1.90	1.58	1.50E+03	946.06	3.90E-04	3.90E-04
	KELM	1.17	1.52	1.60E+03	987.73	4.40E-04	1.80E-04
	PSO-KELM	1.03	1.11	3.32E+03	1.07E+03	3.50E-04	7.50E-04

Table 3. Execution time, transient tracking error and the steady state error for directly driven manipulator: Case 2.

Controller	$time$(in seconds)	e_M	e_F
NN	40.17	0.0084	0.0343
SVM	4.66	0.0019	0.0074
ELM	0.567	0.0023	0.0077
KELM	0.055	0.0022	0.0056
PSO-KELM	16	0.0011	0.0007

By observing the results it can be observed that in joint 1, SVM is having the minimum tracking error index. It is having error lesser than KELM and even PSO-KELM. Also, the tracking error of SVM and ELM is almost same, but the tracking error reduces in KELM and it further reduces in PSO-KELM. Moreover, the tracking error for neural network based control technique is found to be highest. Control input torque indices tabulated in tables concludes that the control input torques reduce in PSO based KELM when compared to KELM and ELM. One of the major advantages of being reduced execution time in ELM based control technique can be observed. It can be observed in these results that the neural network based control technique is having the highest execution time, followed by PSO-KELM (table 3). In fact KELM based control technique has the minimum execution time. Moreover, transient error and steady state error (table 3) are found to be minimum in proposed controller i.e. PSO-KELM.

6. Conclusions

In this study, a novel PSO-KELM, having hybrid of KELM and PSO has been proposed. In PSO-KELM algorithm PSO approach has been used to get the optimal value of the kernel parameters in KELM. It has been observed from the outcomes of the comparisons of the proposed controller, that PSO-KELM is able to cope up with the unknown dynamics and the external disturbances in the robotic manipulator system. These proposed controllers not only overcome the drawbacks of some of the well known intelligent techniques but also enhances the control performance even while incorporating the actuator dynamics too. The proposed control schemes can also achieve better generalization performance with directly driven and remotely driven robotic manipulators.

References

[1] S. Purwar, I. N. Kar and A. N. Jha.: Neuro-sliding mode control of robotic manipulators, IEEE International Conference on Robotics, Automation and Mechatronics, pp. 595-600, Singapore, Dec. 2004.

[2] A. Ishiguro, T. Furuhashi, S. Okuma and Y. Uchikawa.: A neural network compensator for uncertainties of robotic manipulators, IEEE Transactions on Industrial Electronics, vol. **38**, pp. 565–70, Dec. 1992.

[3] Y. Fang, T. W. S. Chow and X. D. Li.: Use of a recurrent neural network in discrete sliding-mode control, IEE—Control Theory and Applications, vol. 146, no. 1, pp. 84–90, Jan. 1999.

[4] C.-M. Lin and C.-F. Hsu.: Neural-network hybrid control for antilock braking systems,IEEE Transactions on Neural Networks, vol. **14**, no. 2, pp. 351–59, Mar. 2003.

[5] G.B. Huang, L. Chen, C.K. Siew/.:Universal approximation using incremental constructive feedforward networks with random hidden nodes, IEEE Transactions on Neural Networks, 17 (2006) 879-92.

[6] D. Zu, T. Mei, M. Luo, K. Guan and D. Zhu.: Fuzzy SVM Controller for Robotic Manipulator Based on GA and LS Algorithm, Sixth International Conference on Fuzzy Systems and Knowledge Discovery, pp. 268-71, 2009.

[7] Z. Yibo and K. Liuting.: A Control Method for Flexible Manipulator Based on SVM, Sixth International Conference on Natural Computation (ICNC 2010), pp. 3462-64, 2010.

[8] D. Zu, T. Mei and L. Sun.: Fuzzy Support Vector Machines Control for 6-DOF Parallel Robot,

Journal of Computers, vol. **6**, no. 9, pp. 1926-34, 2011.
[9] G. B. Huang, Q.Y. Zhu, C.K. Siew.: Extreme Learning Machines, Pattern Recignition, 38(2006)1759-63.
[10] G.B. Huang, Q.Y. Zhu, C.K. Siew.: Extreme learning machine: theory and applications, Neurocomputing, vol. **70**, pp. 489-501, 2006.
[11] G-B Huang, H. Zhou, X. Ding and R. Zhang.: Extreme learning machine for regression and multiclass classification, IEEE Transactions on Systems, Man and Cybernetics-Part B: Cybernetics," Vol. **42**, No. 2, pp. 513-29, 2013.
[12] R. Rajesh and J. Siva Prakash.: Extreme learning machines - a review and state-of-the-art, International Journal of Wisdom Based Computing, vol. **1**, pp. 35–49, 2011.
[13] Huang G-B, Babri HA.: General approximation theorem on feed forward networks, IEEE proceedings of international conference on information, communications and signal processing, ICICS '97, Singapore, pp 698–702,(1997).
[14] Huang G-B, Chen L, Siew C-K.: Universal approximation using incremental feed forward networks with arbitrary input weights. Technical Report ICIS/46/2003, School of Electrical and Electronic Engineering, Nanyang Technological University, October, (2003) , Singapore
[15] Serre D.: Matrices: theory and applications. Springer,(2002) , New York.
[16] Rao CR, Mitra SK.: Generalized inverse of matrices and its applications. Wiley,(1971) ,New York.
[17] Kennedy, J. and Eberhart, R.: Particle Swarm Optimization, IEEE Int. Conf. Neural Networks,(1995) , pp. 942-1948.
[18] Goldberg, D. E.: Genetic algorithms in search, optimization, and machine learning, MA: Addison Wesley, (1989).
[19] Neha Kapoor and Jyoti Ohri.: Improved PSO Tuned Classical Controllers (PID and SMC) for Robotic Manipulators, International Journal of Modern Education and Computer Science, MECS Publications, vol. **1**, pp. 47-52, Jan 2015. DOI: 10.5815/ijmecs.2015.01.07.
[20] S.F. Ding, Y.A. Zhang, X.Z. Xu, and L.N. Bao.: A novel extreme learning machine based on hybrid kernel function, Journal of Computers, vol.**8**, no.8, pp. 2110–17, 2013.
[21] M. W. Spong and M. Vidyasagar, Robot Dynamics and Control, Wiley-India Edition, New York.
[22] Neha Kapoor and Jyoti Ohri.:Fuzzified PSO-SVM (fP-SVM) Controller for Motion Control of Robotic Manipulator, International Journal of Industrial and Systems Engineering, Inderscience Publications.(Accepted for publication, In Press).
[23] Neha Kapoor and Jyoti Ohri.: A Neural Network Based Novel Approach for Error Optimization in Path Tracking Control of a Robotic Manipulator, Proc. of National Conference Advancements in the Era of Multi Disciplinary Systems, AEMDS-2013, pp. 78-83, April 6-7, 2013. Published by Elsevier-2013. (ISBN: 978-93-5107-057-3).
[24] Neha Kapoor and Jyot iOhri.: PD-SVM Integrated Controller for Robotic Manipulator Tracking Control, Chinese Journal of Engineering, HindawiPublishingCorporation,vol.2014,ArticleID295210, pages 6, http://dx.doi.org/10.1155/2014/295210. (ISSN: 2314-8063 (Online)).
[25] B. A. Helouvry, P. Dupont and C. Canudas de wit.: A Survey of Models, analysis tools, and compensation methods for the control of machines with friction, Automatica, vol. **30**, pp.1083-183, 1994.
[26] L. Xu and B. Yao.: Adaptive Robust Precision Motion Control of Linear Motors with Negligible Electrical Dynamics, IEEE/ASME Transactions on Mechatronics, vol. **6**, no. 4, December 2001, pp. 444-52.

IOP Conf. Series: Materials Science and Engineering **715** (2020) 012094 doi:10.1088/1757-899X/715/1/012094

Photovoltaic intelligent gas sensors for the detection of acetone concentration over a wide range of measurement for biomedical applications and tasks of public safety

Yuriy Vashpanov*, Jung-Young Son, Gwanghee Heo and Kae-Dal Kwack

Department of Biomedical Engineering of Konyang University, 121 Daehak-ro, Nonsan, Chungnam, Korea 32992

*E-mail:vashpanov@hanyang.ac.kr

Abstract. This report presents experimental studies of electrical changes in photovoltaic sensitive elements based on a Por-Si: c-Si heterojunction during the adsorption of acetone molecules in a mixture of pure nitrogen. The porous silicon layer was formed by electrochemical etching on a p-type silicon wafer. The obtained samples were subjected to additional plasma-chemical treatment using hydrogen and fluoride ions after fabrication. Such surface treatment has a stabilizing effect on the electrical properties of the material and increases their adsorption sensitivity to polar molecules. The developed technology allows you to automatically change the sensitivity of photovoltaic sensors during the experiment, depending on the concentration of acetone. These photovoltaic sensors are able to expand the range of measured concentrations from 1 ppm to 10 vol%.

1. Introduction

Acetone is widely used in the chemical industry, cosmetics and biomedical researches. The most dangerous property of acetone is its high flammability. At temperatures above the flash point of acetone of -20 °C (-4 °F), air mixtures of 2.5 to 12.8 vol% acetone can explode or cause a flash. Static discharge can also ignite acetone fumes [1]. It is important to measure the acetone concentration over a wide range of measurements for the tasks of Occupations Safety and Health in industry [2]. Measuring the concentration of acetone in exhaled air by humans is important for the successful treatment of diabetes [3]. Acetone concentrations range from 300 to 900 ppb in healthy people and to more than 1,800 ppb in people with diabetes [4]. The acetone concentration reaches 1250 ppm in people with ketoacidosis [5]. Thus, measuring in wide range of concentrations of acetone is important for biomedical research too.

The creation of measuring systems for a wide range of gas concentration measurements is a complex technical task due to the limited range of sensor measurements. Currently, the attention of most researchers focused on nanotechnology semiconductor sensors [see, for example, 6-10]. The main advantages of such sensors are the possibility of their mass replication at their low cost and small size.

Most sensors for acetone operate at work temperatures above 175 °C [11]. Acetone gas sensors based on Al_2O_3-doped ZnO has an optical excitation-enhanced sensing properties at a temperature of 64 °C [12]. The nanorods bundles ZnO microspheres, consisting of closely packed nanorods with wideness of about 50 nm, have good response and recovery times (1s and 3s when the sensor was exposed to 100 ppm acetone at an temperature of 320°C) [13].

In recent publications [14-17], new methods for measuring acetone concentrations at room temperature are described, which greatly simplifies the design of the sensor. However, these modern gas sensors have a limited dynamic measuring range. For example, an IDE device with a nanogap has a dynamic range from 1000 to 10 ppm for acetone at room temperature. [18].

Previously, we published the results of research of photovoltaic sensors based on Por-Si-C-Si heterojunctions to detect polar molecules ammonia with dipole moment of 1.43 D in a wide range of concentrations [19]. The obtained samples underwent additional plasma-chemical treatment with fluorine and hydrogen ions after fabrication. Measurements by the method of secondary ion mass spectrometry showed that the concentration of fluoride ions on the surface of the material reaches 3–4 at%. Fluorine and hydrogen ions have a stabilizing effect on the electrical properties of the material surface. Fluorine is more active than oxygen, so contact with the air atmosphere is not accompanied by changes in the composition of the surface of the porous silicon. This operation eliminates the "aging" of porous silicon when interacting with air. Doping with hydrogen ions of the surface region of the material leads to saturation of dangling bonds of silicon and plays a significant positive role in luminescence in the visible region of porous silicon [20].

Acetone molecules have a more significant dipole moment of 2.93 D [21] compared to ammonia molecules. This report presents the results of using photovoltaic sensors to measure the concentration of acetone in an inert atmosphere of pure nitrogen at room temperature measurements. The technology of making samples and special electrical contacts to them and the methodology of the experiment are described in detail in the articles [19]. We used only acetone mixed with ultrapure nitrogen to supply the gas-dynamic generator to create an acetone gas mixture in the range from 1 ppm to 10 vol%.

This new gas detection method opens up a new way to create intelligent gas sensors in a wide range of measurements [22]. Such a developed method of measurement allows increase a measurement precision by automatically change the sensitivity in photovoltaic sensors during the experiment, depending on the gas concentration. Therefore, such sensors can extend the range of measured concentrations and the accuracy compared to known sensors.

2. Results and discussion

We used the light intensity of the samples from 2 to 200 lx when changing the concentration of acetone vapours in the mixture with high purity nitrogen from 1 ppm to 10 vol% (100,000 ppm) in a measuring chamber. Some experimental dependences the electromotive force on electrical contacts upon adsorption of acetone at different levels of illumination are shown in figure 1. Changes in light exposure lead to different ranges of acetone measurement (figure 1).

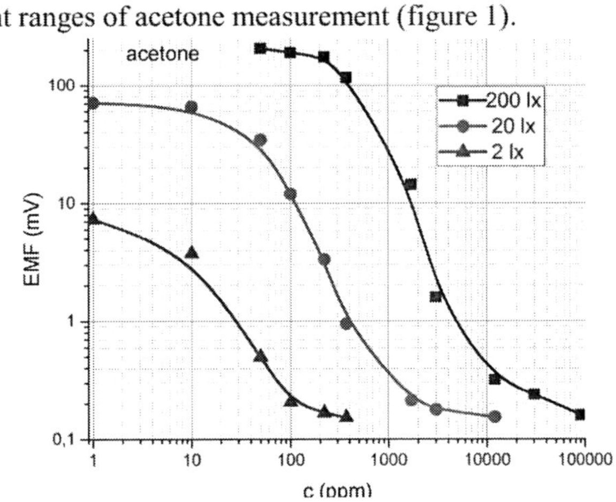

Figure 1. Dependences of the EMF values on acetone concentration at different illumination levels of 2, 20 and 200 lx.

So, the photovoltaic gas sensors can change sensitivities and ranges of measurement by changing illumination intensity (figure 1). Figure 2 shows a graph of light intensity and electromotive force, provided that maximum metrological accuracy can be achieved. These data can be the basis for the development of intellectual sensors for acetone with a wide dynamic measurement range and with the maximum possible accuracy of measurement.

Figure 2 includes a spline approximation of EMF-magnitude for sensors data from illumination levels based on the data of figure 1. We standardize parameter L based on these measurements. In this case, we have the maximal analysis accuracy for different ranges of acetone concentrations.

The approximation of the calibration curves can be described by the formula:

$$U(L)=-0.475+2.3739\ln(L+1.01) \tag{1}$$

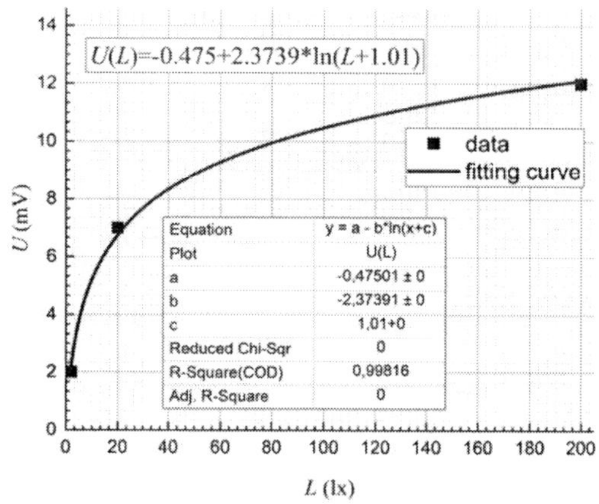

Figure 2. Dependence of the EMF magnitudes on different illumination levels with maximum sensitivity to acetone.

Previously for ammonia photovoltaic sensors, we developed software modules for control of the light intensity for maximum accuracy of gas measurements using the NI LabVIEW program and National Instruments equipment [22]. In our case, these modules need to be reprogrammed taking into account the data of formulas (1) and (2) for acetone sensors, which allow us to directly calculate and display the gas concentration in real time. We can also apply the developed algorithm for similar experiments with other polar gases.

The acetone concentration depends in our photovoltaic gas sensors on two parameters: the EMF magnitude (parameter U) and the light intensity (parameter L). The maximal precision can be achieved using Equation (2) by finding a two-parameter function $c(U,L)$ that describes the experimental data.

Fitting functions for the experimental data for acetone are been approximated using the following generalized two-parameter function:

$$c(U,L)=\alpha(L)\frac{\beta(L)^{\gamma(L)}}{[U-\delta(L)^{\gamma(L)}]} \tag{2}$$

We found fitting functions $\alpha(L)$, $\beta(L)$, $\gamma(L)$, and $\delta(L)$ for generalized two-parameter function from the experimental data and described them using the following relations:

$$\begin{aligned} \alpha(L) &\cong 11\cdot L \\ \beta(L)^{\gamma(L)} &\cong 1.2+0.22\ln(L-1.3) \\ \gamma(L) &\cong 0.2-\frac{0.1}{[1+(12\cdot L)^3]} \\ \delta(L) &\cong 0.11-\frac{0.06}{[1+(15\cdot L)^3]} \end{aligned} \tag{3}$$

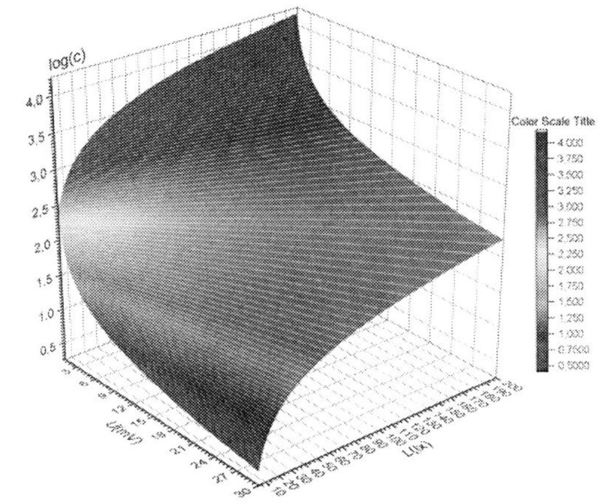

Figure 3. 3D-graph of the generalized two-parameter function for the experiment with acetone.

This function $c(U,L)$ (Figure 3) describes the acetone concentrations based on the experimental measurement of two signals.

Two experimental signals from our photovoltaic sensors are used as inputs to the engineered mathematical module of the NI LabView^{TM} program. The acetone concentrations calculate under this program and output on the front panel of the NI LabView^{TM} program in real time. The logic control module finds an optimal level of an illumination and minimizing error of the measurement. However, the maximum accuracy for analysis can be achieved in the event of slowly changing gas concentrations. In addition an iteration process increases the total measurement time.

3. Conclusions

We used the samples with a special plasma-chemical treatment with hydrogen and fluoride ions after fabrication. Such surface treatment has a stabilizing effect on the electrical properties of the material and increases their adsorption sensitivity to polar molecules of acetone. Experimental concentration dependences have been measurement for the electromotive force on contacts of photovoltaic sensors upon adsorption of acetone under different levels of illumination. We engineered special mathematical module of the NI LabView^{TM} program for automatic control and measurement of acetone concentration in the wide range from 1 ppm to 10 vol% (100000 ppm).

Acknowledgement

This work was supported by Priority Research Centers Program through the National Research Foundation of Korea (NRF) funded by the Ministry of Education (Grant Number: NRF-2018R1A6A1 A03025542).

References

[1] Acetone. Wikipedia. https://en.wikipedia.org/wiki/Acetone#cite_note-12#cite_note-12.

[2] ACETONE. The National Institute for Occupational Safety and Health (NIOSH) https://www.cdc.gov/niosh/ipcsneng/neng0087.html.

[3] V. Saasa, Th.Malwela, M. Beukes, M. Mokgotho, Chaun-Pu Liu and Bonex Mwakikunga Sensing Technologies for Detection of Acetone in Human Breath for Diabetes Diagnosis and Monitoring 2018 Diagnostics (MDPI) 8, n12, 17 p. doi:10.3390/diagnostics8020012.

[4] M. Righettoni, A. Tricoli, S. Gass, A. Schmid, A. Amann, S.E. Pratsinis. Breath acetone monitoring by portable Si:WO3 gas sensors. 2012 Anal. Chim. Acta 738 pp.69–75.

[5] J.C. Anderson Measuring breath acetone for monitoring fat loss. 2015 Obesity 23 pp.2327–2334.

[6] L. Ma, S.Y. Ma, X.F. Shen, T.T. Wang, X.H. Jiang, Q. Chen, Z. Qiang, H.M. Yang, H. Chen. PrFeO3 hollow nanofibers as a highly efficient gas sensor for acetone detection. 2018 Sens.

Actuators B Chem. 255 pp.2546–2554.
[7] S.T. Navale, Z.B. Yang, C. Liu, P.J. Cao, V.B. Patil, N.S. Ramgir, R.S. Mane, F.J. Stadler. Enhanced acetone sensing properties of titanium dioxide nanoparticles with a sub-ppm detection limit. 2018 Sens. Actuators B Chem. 255 pp.1701–1710.
[8] S. Kim, S. Park, S. Park, C. Lee. Acetone sensing of Au and Pd-decorated WO3 nanorod sensors. 2015 Sens. Actuators B Chem. 209 pp.180–185.
[9] S. Choi, B. Jang, S. Lee, B.K. Min, A. Rothschild, I. Kim. Selective detection of acetone and hydrogen sulfide for the diagnosis of diabetes and halitosis using SnO2 nanofibers functionalized with reduced graphene oxide nanosheets. 2014 ACS Appl. Mater. Interfaces 6 pp.2588–2597.
[10] A. Thamri, H. Baccar, FE. Annanouch, E. Llobet and A. Abdelghani. Methanol, Ethanol and Acetone Sensing Using AACVD-grown Tungsten Oxide Nanoneedles 2016 J Nanomed Nanotechnol 7 6 p. doi: 10.4172/2157-7439.1000380.
[11] Kun-Wei Kao, Ming-Che Hsu,Y uh-Hwa Chang, Shangjr Gwo and Andrew Yeh. A Sub-ppm Acetone Gas Sensor for Diabetes Detection Using 10 nm Thick Ultrathin InN FETs. 2012 Sensors 12 pp.7157-7168.
[12] Xuehai Guo, Guofeng Pan, Xin Ma, Xiangzhou Li, Ping He, Zhongqiu Hua, Haiqing Li Optical excitation-enhanced sensing properties of acetone gas sensors based on Al2O3-doped ZnO. 2017 Sensor Review 37 pp.364-370.
[13] Zheng LouYingliang Feng, Xiangwei Liu , Lili Wang and Tong Zhang. Acetone sensing properties of hierarchical ZnO urchinlike structures by hydrothermal process 2012 *Biomedical Engineering: Applications, Basis and Communications* **24**, No. 02, pp. 99-103.
[14] M.S. Lemraski, E. Nadimi Acetone gas sensing mechanism on zinc oxide surfaces: A first principles calculation. 2017 *Surface Science* 657 pp. 96-103.
[15] Ming-Yen Chuang, Yu-Ting, Ting-Wei Tung, Liang-Yu, Chang, Hsiao-Wen Zan, Hsin-Fei Meng, Yu-Tai Tao. Room-temperature-operated organic-based acetone gas sensor for breath analysis 2018 *Sensors and Actuators B: Chemical* **260** pp. 593-600.
[16] Qiulin Tan, Jiahua Fang, Wenyi Liu, Jijun Xiong, and Wendong Zhang. Acetone Sensing Properties of a Gas Sensor Composed of Carbon Nanotubes Doped With Iron Oxide Nanopowder. 2015 Sensors 15 pp.28502–28512.
[17] Y. Zhang, H. Xu, S. Dong. A fast response & recovery acetone gas sensor based on BiFeO3 nanomaterials with high sensitivity and low detection limit. 2018 J Mater Sci: Mater Electron 29 pp 2193–2200.
[18] Q.N. Minh, H.D. Tong, A. Kuijk, F. van de Bent, P. Beekman, and C.J.M. Van Rijn. Gas sensing performance at room temperature of nanogap interdigitated electrodes for detection of acetone at low concentration. 2017 RSC Advances, 7, pp. 50279-50286.
[19] Yu.A. Vashpanov, J.I. Jung, and K.D. Kwack Photo-EMF sensitivity of porous silicon thin layer – crystalline silicon heterojunction to ammonia adsorption. 2011 Sensors 11 pp.1321-1327.
[20] O. Bisi, S. Ossicini, and L. Pavesi. Porous silicon: A quantum sponge structure for silicon based optoelectronics. 2000 Surface Science Reports 38 pp.1–126.
[21] O. Dorosh and Z. Kisiel. Electric Dipole Moments of Acetone and of Acetic Acid Measured in Supersonic Expansion. 2007 ACTA PHYSICA POLONICA A 112 pp. S95-S104.
[22] Yuriy Vashpanov, Hyunseung Choo and Dongsoo Stephen Kim. Dynamic Control of Adsorption Sensitivity for Photo-EMF-Based Ammonia Gas Sensors Using a Wireless Network. 2011 Sensors 11 pp. 10930-10939.

An anti-peep passcode keypad based on hybird image and fuzzy adaptive visual distance adjustment

Hongchun Qu* and Linsheng He

Key Laboratory of Industrial Internet of Things & Networked Control (Ministry of Education), College of Automation, Chongqing University of Posts and Telecommunications, Chongqing 400065, China.

*E-mail: quhc@cqupt.edu.cn

Abstract. Because traditional Android mobile phone passcode keypad without anti-peeping function, so we propose a kind of anti-peep code passcode keypad based on hybrid image and fuzzy adaptive visual distance adjustment. First, replace the key bitmaps with hybrid images to solve the problem that password keyboard doesn't have anti-peeping function. Then, the fuzzy logic and adaptive principle are used to solve the user's use problem. Finally, it is found through experiments that the time to refresh 10 hybrid images is between 0.38s and 0.48s, and the blur effect of the image can be automatically adjusted according to the distance. The results show that the anti-peep code keyboard makes up for the problem that the traditional passcode keypad has no anti-peeping function.

1. Introduction

Market research firm Gartner pointed out in the 2017 mobile device system market share survey that Android system market share accounted for 85.9%, IOS system accounted for 14%. While most Android smartphones on the market have replaced traditional hardware buttons with full-touch screens, privacy leaks on mobile screens remain unresolved. Based on the huge smartphone market and mobile phone holding rate in the world, how to solve the privacy leakage caused by using the password keyboard in public places has attracted more and more people's attention.

Probably because most smartphones already have fingerprint recognition and other features, the cryptographic keyboard interface has not improved much. However, the passcode keypad is the basic input tool for the phone, so it is necessary to solve the anti-peeping problem of the passcode keypad. Wei Qin and Jing Yu [1] proposed a peep-proof display that prevents people from peeking from the side view while providing an image display for users looking from the front. However, the cost of hardware is relatively high, which is not conducive to the promotion of hardware. Nolan Jones and J.D. Sherry proposed a system and method involving the use of graphical password verification devices or computer system users [2]. The user sequentially selects the graphic image and generates a password based on the combination of the attributes of the selected image. The generated password is compared to the stored password to verify the identity of the user and grant access to the device. However, the software is essentially a graphic image corresponding to a number, so the effect of the user's anti-peeping effect in public places is not obvious.

In this paper, we present an anti-peep passcode keypad based on hybrid image [3] and fuzzy adaptive visual distance adjustment. First, replace the traditional key bitmap with hybrid images.

Second, using the fuzzy logic principle [4-5], the face detection algorithm [6-8] and the eye positioning detection algorithm [9], the distance from the eye to the screen is divided into fuzzy sets, and then use the fuzzy function to establish a function of the hybrid images values. Third, according to the adaptive principle, the initial parameters of the fuzzy function are dynamically adjusted by using the data parameters input by the user. It was found through experiments that the software refreshed ten hybrid images for about 0.36s to 0.43s. The advantage of the fuzzy adaptive adjustment anti-peeping passcode keypad is that the hybrid images will see different image effects at different distances. At the same time, the digital images are hybrid with nine different digital images, so the method is sufficient for the user to use, and the problem that the passcode keypad does not have the anti-peeping function can be solved. At the same time, the software is low in cost and easy to promote and improve.

This article will introduce the process and results of our project. In Section 2, we discussed the solution to the problem. The experiment will be presented in Section 3, followed by conclusions and future work.

2. Method

In view of the current analysis of cryptographic keyboard defects and the study of the dominant unlocking method, we choose a hybrid image to replace the traditional key bitmap, because the viewer will see different image blur effects when viewing the hybrid images at different distances. The whole process of the model is based on fuzzy logic and hybrid images, and finally combined with the adaptive principle [10] to achieve the use of most people.

2.1. Get the distance from the eyes to the screen

The face and eyes show a certain distance on the photo, the face is close to the camera, and the distance between the eyes and the eyes becomes larger; the face is far away from the camera, and the distance between the eyes and the eyes becomes smaller. Moreover, the distance from the eye to the camera is approximately the same as the distance from the eye to the screen, and the calculation method of the eye-to-screen distance is proposed [11]. Since the maximum viewing angle φ of the mobile phone camera is constant, the width W of the photograph taken by the camera is also constant. Therefore, the distance from the eyes to the screen can be further obtained by obtaining the distance L between the eyes and the distance E between the eyes projected on the photograph. The actual eye-to-screen distance can be obtained by the following formula (1).

$$D = \frac{L}{2\tan\frac{\varphi E}{2W}} \tag{1}$$

Li zhuqi et al. proposed that the maximum error of the distance was 2.4 cm when the distance from the eye to the screen was 20 cm to 50 cm. We found through experiments that the maximum error of the distance is 2.9cm when the distance from the eye to the screen is 20cm to 70cm, which verifies the feasibility of the (1) method.

2.2. Hybrid image

Hybrid images see different blur effects at different distances, so it has a certain deceptive on vision. The way of the Android platform implements image blurring are mainly through two algorithms, FastBlur[12] and RenderScript[13]. After the developer weakened the algorithm further, it was found that the image blurring time achieved by the Fastblur algorithm was faster than the RenderScript algorithm. For example, a picture with a pixel of 324*376, the time of Fastblur blurring is 2ms, but the time of Rendscript blurring is 13ms. The Fastblur algorithm mainly uses a one-dimensional Gaussian filtering algorithm to quickly blur the image. The image separation algorithm achieves high frequency and low frequency separation of the image. Then, the image fusion algorithm is used to realize the hybrid of high-frequency images and low-frequency images of different images, which replaces the traditional key bitmap.

2.3. Fuzzy logic

By obscuring the principle of collections, variables can belong to multiple collections at the same time, and each collection part occupies a variable. Blur specific discrete values to obtain a fuzzy set. Then we interpret the set by fuzzy rules of different angles to calculate the fuzzy value to obtain a certain value. This process is called defuzzification [14].

In this application, we enter the distance from the eye to the screen as input data. Because researchers have demonstrated that the human visual system's perception of images is non-uniform and non-linear [15], we use Gaussian functions as membership functions. Here, we set up six membership functions and convert the distance from the eye to the screen into six different memberships. Here, we use a single-valued output membership function, which is essentially a pre-fuzzy output function. Suppose u is the degree to which the output geometry is true, x is the exact single value of the output geometry, and the final aggregate and defuzzified [16] output is as follows:

$$f(x) = \frac{\sum_{i=1}^{n} u_i x_i}{\sum_{i=1}^{n} u_i} \tag{2}$$

2.4. Adaptive principle

Since the fuzzy logic is only based on the initial data, the blurring effect of the image is adjusted for different distances. Because everyone has a different understanding of the blurring effect of the image, so this type of adjustment may only apply to individuals or to certain types of people. Therefore, it is necessary for the software to adopt adaptive learning method to further achieve adaptive adjustment for most people.

Because the data that we adjust the initial parameters are directly from the user, so we adjust the learning method to achieve adaptive adjustment of fuzzy logic parameters [17]. First, we will give the initial parameters of the distance and image blur effect according to the principle of fuzzy logic. Then, by obtaining the setting of the user's fuzzy parameters and the distance from the eye to the screen, the model parameters are continuously optimized (the number of optimizations of the fuzzy parameters is determined according to the blurring effect of the user on the key bitmap). Achieve the most appropriate image blur effect that the user thinks is appropriate.

3. Experiments

In this section, we will test whether the software can implement image blending adjustment based on fuzzy logic in the first section. In the second subsection, the test software is able to implement adaptive adjustments to the initial parameters of different people. The experiment included an MI 5 smartphone and a Huawei mate 9 mobile phone using the test application algorithm described above to achieve a series of eye-to-screen distance measurements and image blend values (data from twelve people).

3.1. Hybrid images effect adjustment based on fuzzy logic

As shown in figure 1, The distance from the eye to the screen is different, and the effect of seeing the hybrid images is different. Explain as follows, you can look at a person from a distance, you can distinguish his or her gender, but you can't know if he or she has wrinkles or white hair, etc. When you are close enough to this person, you can have more clear understanding. It is verified that the software can realize the image blurring effect with the distance from the eye to the screen.

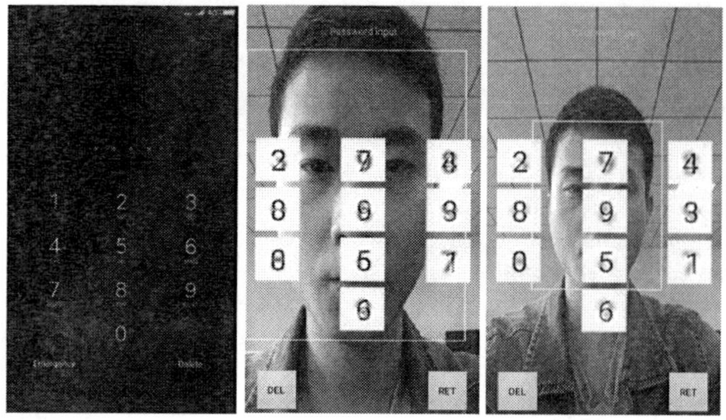

Figure 1. On the left is the traditional passcode keypad, the middle is the image blending effect obtained by the close-range software, and the right is the image blending effect obtained by the long-distance software.

3.2. Adjustment between distance and blur effect

We decided to measure the relationship between the distance and the image blend value σ over a screen distance of 18 cm to 80 cm, we start from the lower boundary and measure the distance by 1 cm after each step. The reason for the setup is discussed in Section 4. We start from the lower boundary and measure the distance by 1 cm after each step. The test application saves the data results in a table and performs the analysis on the computer while completing a series of measurements. As shown in figure 2, the relationship between the distance from the eyes to the screen and the hybrid images value σ are obtained by twelve people.

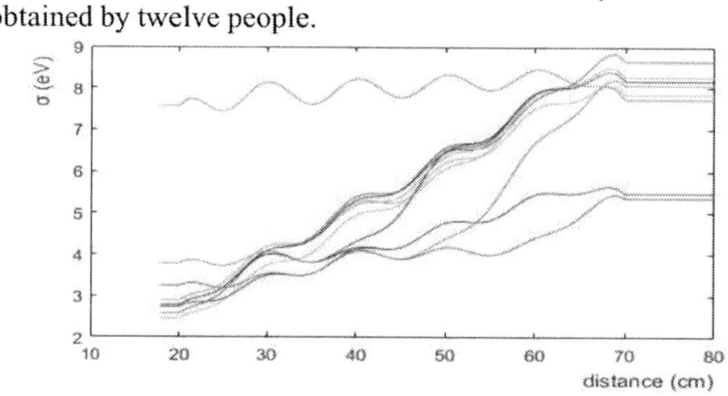

Figure 2. Through the analysis of data collection (data from twelve people), the relationship between distance and image blend values is plotted.

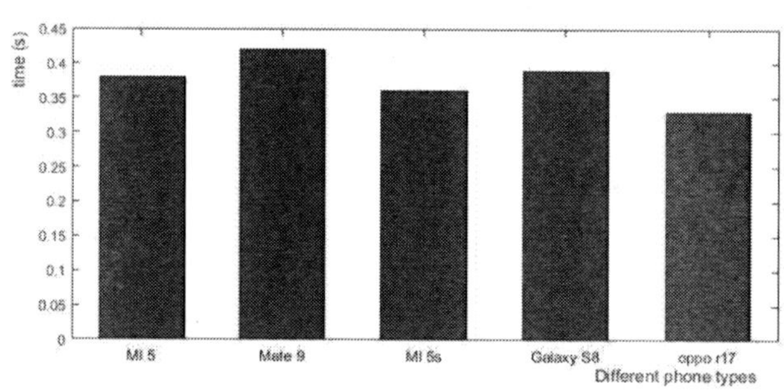

Figure 3. Time required for different types of mobile phones to refresh once.

Through the data analysis of figure 2, the adaptive anti-peep code cipher keyboard can adaptively adjust the fuzzy logic parameters according to the user's usage habits and the user settings of the fuzzy parameters. According to the data of figure 3, different types of mobile phones can also achieve the software to respond in less time.

4. Results

In our experiments, it was found that a single image refresh took approximately 33ms to 44ms. At the same time, the blurring effect of 10 images was changed, only the first time was about 0.5s, and the MI5 response time was about 0.38s to 0.44s. Since the mobile phone may be affected by other factors, such as hardware and Android version, the hybrid images refresh threshold are currently set to 0.65 seconds.

Our current adjustment distance is mainly between 18 cm and 80 cm. First, the distance from the eye to the screen is substantially above 18 cm, and the face detection has a minimum threshold. Secondly, 80 cm can be applied to the longest distance that most people stretch their arms. In figure 2, the feasibility of adaptively adjusting the image effect is verified. In experiments and studies, it was also found that in the case of crowding, the distance from the eyes to the screen was mainly between 25 cm and 49 cm.

5. Conclusion

In this paper, we propose an anti-peep passcode keypad based on hybird image and fuzzy adaptive visual distance adjustment. Fuzzy adaptive adjustment of the blended image is achieved by obtaining the distance from the eye to the screen. We also provide experimental proof that the software can be used in the range of 18 cm to 80 cm. Considering the other factors of different types of mobile phones, the refresh time of the current setting software is about 0.62 seconds. The utility model solves the defects that the traditional passcode keypad does not have the anti-peeping function, the application cost is low, the promotion is easy, and the user who needs to protect the privacy information has great attraction. But the project has some drawbacks. For example, a user may not be able to adjust the program adaptively with one or two adjustments. When the camera using the software is always on, the screen acquired by the camera will be used as the background for the password input, which may make the user feel uncomfortable. These are all areas for improvement.

References

[1] BOE Technology Group Co Ltd .(2016). Peep prevention display. Washington: Patent and Trademark Office. No. 9,507,188.

[2] NIC Inc .(2013). System and method for authenticating a user using a graphical password. Washington: Patent and Trademark Office. No. 8,347,103.

[3] Oliva, A.(2013). The art of hybrid images: Two for the view of one. Art & Perception, 1 (1-2), 65-74.

[4] Zadeh, L.A. (2015). Fuzzy logic—a personal perspective. Fuzzy Sets Syst, 281, 4-20.

[5] Yager, R.R., & Zadeh, L. A. (2012). An introduction to fuzzy logic applications in intelligent systems. In: R.R. Yager, & L. A. Zadeh, eds. An introduction to fuzzy logic applications in intelligent systems. Vol. 165. Germany: Springer Science & Business Media.

[6] Yang, S., et al.(2016). Wider face: A face detection benchmark. In: E.B. Terrance, & W. Alexander, eds. CVPR 2016, 2016 June 26 - July 1 in Seattle, WA. Washington: IEEE. 5525-33.

[7] Wang, Y.Q.(2014). An analysis of the Viola-Jones face detection algorithm. Image Processing On Line, 4, 128-48.

[8] Sun, X., Wu, P., & Hoi, S.C.(2018). Face detection using deep learning: An improved faster RCNN approach. Neurocomputing, 299, 42-50.

[9] Meng, S.H., et al.(2017). Research on Eye Detection and Fatigue Early Warning Technologies. In: K. Pavel, & A. Enrique, eds. ECC 2017, 9-11 October 2017 in Spain. Berlin: Springer, 3-9.

[10] Åström, K.J., & Wittenmark, B.(2013). Adaptive control. Chelmsford: Courier Corporation.

[11] Li, Z., et al.(2016). Look into My Eyes: Fine-grained Detection of Face-screen Distance on Smartphones. In: L. Xiang, & J. Taeho, eds. MSN 2016, HeFei 16-18 Dec 2016. Washington: IEEE, 258-65.
[12] Huang, H.Y., & Tsai, W.C.(2014). Blurred Image Restoration Using Fast Blur-Kernel Estimation. In: IH-MSP, Kitakyushu 27-29 August 2014. Washington: IEEE. 435-38
[13] Guihot, H.(2012). Pro Android Apps Performance Optimization. Berlin: Springer.
[14] Qu, H., et al.(2018). A Lightweight Intrusion Detection Method Based on Fuzzy Clustering Algorithm for Wireless Sensor Networks. Advances in Fuzzy Systems, 2018, 0-12
[15] Panetta, K., Gao, C., & Agaian, S. (2016). Human-visual-system-inspired underwater image quality measures. IEEE J. Ocean. Eng, 41 (3), 541-51.
[16] Bobyr, M.V., et al.(2017). A method of defuzzification based on the approach of areas' ratio. Appl. Soft. Comput, 59, 19-32.
[17] Golden, R.M.(2018). Adaptive Learning Algorithm Convergence in Passive and Reactive Environments. Neural computation, 30 (10), 2805-32.

Self-Parking management system based on geomagnetic detection and parking lock

Yingfang Zhang[1,*], Qi Zhang[2], Libin Zhang[1] and Jilong Shi[1]

[1]College of Computer Science, Jilin University, Jilin 130000, China
[2]College of Physics, Jilin University, Jilin 130000, China

*E-mail: zhangyf2116@mails.jlu.edu.cn

Abstract. With the increasing number of vehicles worldwide and the increasing demand for parking spaces, effective management methods are in urgent need to improve the utilization efficiency of parking spaces. In spite of many previous studies focusing on how to detect the status of parking spaces and guide vehicles, the schemes proposed by these studies can only provide certain parking information rather than effectively manage parking spaces. In this paper, a system based on intelligent parking lock is designed to realize functions of query, reservation and management, based on which, drivers can reserve the free parking space in advance. This system can ensure that the reserved parking space is not misoccupied, which can also automatically close the parking lock and perform functions such as billing after the vehicle leaves.

1. Introduction

With the rapid development of global economy and the acceleration of urbanization, the number of cars is also on the increase year by year.

As to large parking lots, traditional parking solutions, whether to use RFID technology or video recognition technology, are not only inseparable from the manual management at the entrance, but also unfavorable for drivers to quickly find suitable parking lots and idle parking spaces. Thus, It has become increasingly difficult to meet the existing demand.[1]

With the proposal and gradual development of the concept "smart city", how to make parking lots more efficient and realize intelligent management has become an urgent problem to be solved.

The intelligent parking lot management scheme based on geomagnetic detection and parking lock proposed in this paper is aimed at the large parking lots. It makes use of the parking lock and the geomagnetic wireless detection module to manage the usage rights of the parking space.

The way of putting up a rail at the entrance is abandoned, which may cause peak congestion, and instead, ZigBee technology is used, which has the characteristics of low power consumption and self-organizing network to form a local area network.

The unified management of the field coordinator and the exchange of information through GPRS technology and back-end server help to realize the grouping management of parking spaces, which, simultaneously, facilitates the later maintenance. Due to the fact that it is difficult to form a local area network for scattered parking spaces, a single parking lock is directly connected to the back-end server through GPRS. In addition, the driver can view the parking lot map in a visualized way through the webpage, make parking reservation and realize unlocking, which greatly improves drivers' efficiency of finding parking lots and idle parking spaces.

2. General description of the system

This self-service parking management system is mainly divided into three levels. As the picture shows, the top layer is the web page and web server for customer reservation, the second layer is the parking lot coordinator, and the third layer is the IoT parking lock.

The parking lot coordinator contains the GPRS module to communicate with the server. At the same time, there are ZigBee modules on the coordinator and parking locks to form a LAN to realize communication between them. Thereby, the coordinator can monitor the state of the parking spaces and control the opening and closing of the parking locks.[2]

There is a Mysql database on the server, which stores the basic information of each parking lot and the parking space status in each parking lot. The reservation client displays the current parking lot information and the status of the parking spaces to users through the querying database,so that users can view and reserve an idle parking space(as shown in figure 1).

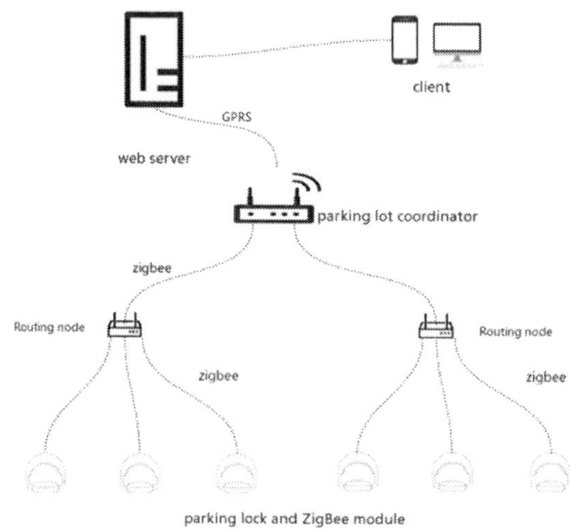

Figure 1. Parking lot and ZigBee module.

3. System software design

To simplify software design of the client, the whole software of the B/S structure is designed in the system. The user gets access to the system by firstly logining in the system on the brower.There are two types of servers supporting the system. One is the application server which provides not only interface between the server and users but also processing procedures, the other is the datebase server which saves information of users and all kinds of parking lots including the map of parking spaces.

To sum up, the website is mainly made up of 5 functions:

1) Querying a specific parking lot and displaying the state of all parking spaces, including occupied and idle ones;
2) Choosing and reserving an idle parking space;
3) Unlocking the parking lock on the program;
4) Payment;
5) Querying parking history.

The key is to send messages to the parking lot coordinator when unlocking the parking locks via this application and to receive messages from the coordinator to inform the application that the car has left. The following flow chart mainly describes the processing flow. First, users can login in the application, on which he or she can view the state of all parking spaces and reserve a proper one. Once arriving at the reserved parking space, he can unlock the parking lock and then drive into the parking space. At the same time, the application will start timing. Once the car leaves, the parking coordinator will inform the application to stop timing and calculate the cost. After this, the application will update the parking status for next usage(as shown in figure 2).[3]

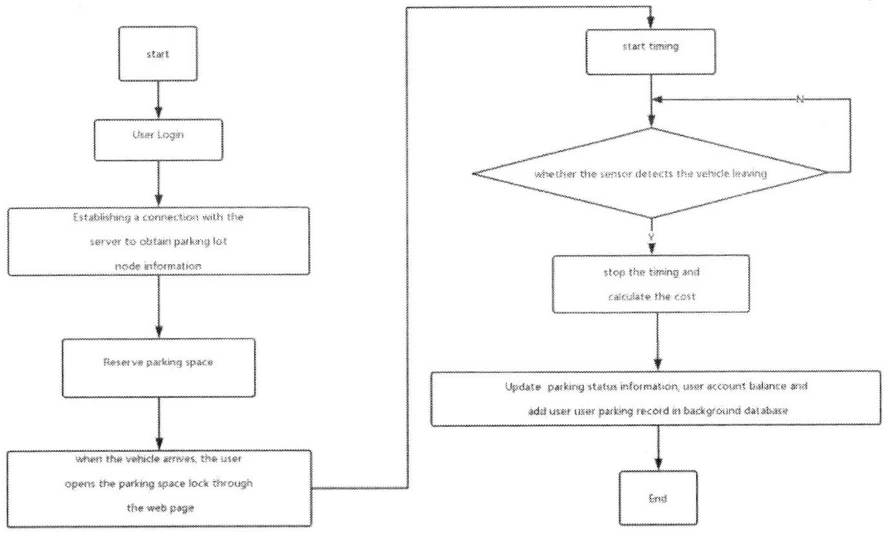

Figure 2. Normal operations of the system.

Specific details of some main functions are as follows.

3.1. Querying a parking space

Users can search for information about nearby parking lots, including the parking rates. The system provides a map of the city, by means of which a user can query information of all parking lots that are saved on the database and updated in time. If a user specifies a parking lot, he can view the state of all parking spaces including the distribution of all parking spaces. Therefore, users can choose a suitable parking space according to positions of free parking spaces. What's more, once users choose a free parking space, the system provides users a route to the parking space.

3.2. Booking a parking space

There is a button named special information of each parking space. Once you press the button named "free", the system will deal with your booking request. The specific procedures are as follows. Firstly, our system will identify your booking information. If you have booked a parking space, you are not permitted to book another one and the system will return with a failure. After checking your legal request, the system will actually deal with your request. Secondly, the system will access the database to modify the information of the booked parking space. At the same time, the system will start to count the seconds immediately. Finally, the system will return with a successful booking including a tip that "if you can't open the lock of the booked parking space in 15 minutes, the system will cancel your completed booking".If users who have reserved parking space can not arrive at his reserved parking space in time,this parking space will be released by the system.If his parking space has been released,the user needs to make a new reservation if necessary.

3.3. Unlocking the lock

After you press the button named "unlock the lock", the system will execute the final verification. If you unlock the lock after 15 minutes, the system will return with a failure. Under such a circumstance, you must book another free parking space again. Passing the checking of your request, the system will start to count the seconds immediately and return with a success. Then, the lock will be unlocked within a short time.

3.4. Exiting the parking space and payment

Once your car leaves the parking space, related sensor of the geomagnetic detection will send signal to the system server and the system will record the stop time and calculate the seconds of parking at the space. According to the length of seconds, the system will calculate accurate cost. The user can pay the cost through the third-party payment software such as wechat.Then the system modifies the parking space status to idle for next use。 Besides, these information including cost will be recorded on the database. The service ends until you receive a message about the cost.

3.5. Querying parking history

Considering the mistake of the system and transparent service of increasing users' cost, users can browse all histories of parking including the cost, the time of starting and ending as well as the location of the parking space. In addition, the system provides a interface to feedback to the administrator. Taking this measure, users can timely respond to the administrator. And the system can minimize the occurrence of accidents or losses.

4. System hardware design

The hardware of the system mainly includes intelligent parking lock and area coordinator. Intelligent parking lock is mainly responsible for vehicle detection and control of the car, and regularly sending messages to the regional coordinator. Regional coordinator is responsible for collecting management information within the scope of each parking lot, and sending it to the server. This design can facilitate the maintenance of the management system, which is easy to find out problems in time.[4]

4.1. Parking detection technology

At present, the commonly used technologies include RFID, photoelectric sensor, ultrasonic detection technology, geomagnetic detection and so on. RFID technology is currently widely used in parking lot management systems, but this technology requires each car to be equipped with radio frequency cards, which is of too much trouble. On the other hand, both photoelectric detection technology and ultrasonic detection technology are easily misjudged by environmental interference. Therefore, in this paper, geomagnetic detection technology is chosen, which can detect the uniform magnetic field generated by the earth within a certain range, and it is sensitive to any metal object. Moreover, it can detect the distortion or change of the geomagnetic field when the vehicle drives in and out. Through measurement in the experiment (as shown in figure 3), it is found that the most useful distance between magnetic field and metal object is -0.3m to 0.6m. Thus, we decide to install it in the center of the parking lot to detect the vehicle accurately (as shown in figure 4).[5]

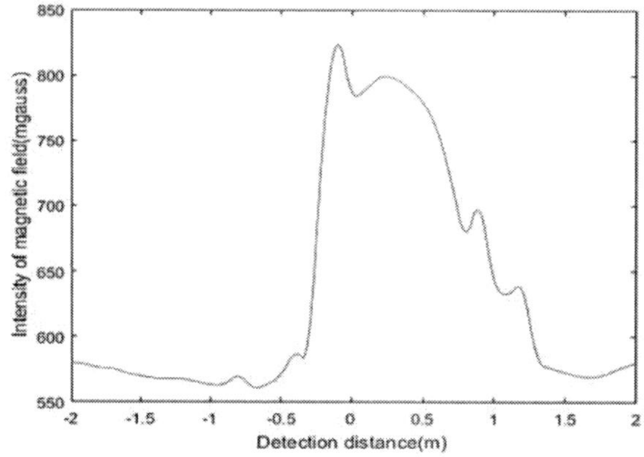

Figure 3. The relationship between intensity of magnetic field and distance (Z).

AMR module for detecting geomagnetic changes is installed in the center of t1he parking lock. After detecting the change of the geomagnetic field, parking locks will judge whether the disturbance is from the vehicle. If a vehicle is judged to have entered or left, the MCU will judge the working state of the parking lock at that time and execute corresponding commands, such as controlling the lifting or falling of the parking lock, and then send information to the regional coordinator through the local area network.[6][7]

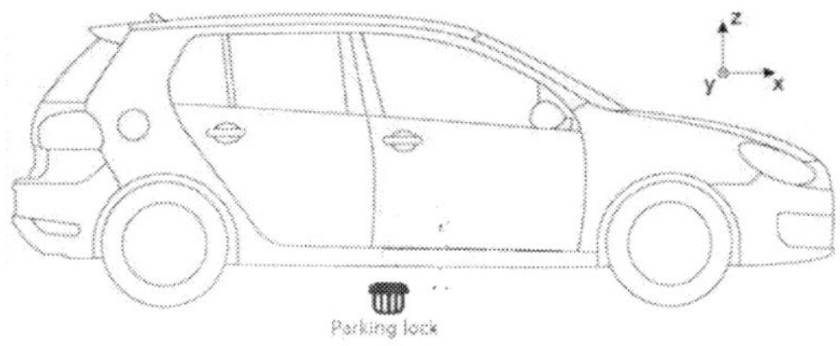

Figure 4. The geomagnetic sensor detects the parking condition.

4.2. Parking lock control scheme for parking lock.
Due to stepper motor has the advantages of simple drive control and no cumulative error, it is selected for the control of the mechanical arm, which has a high control accuracy, but also works stably and reliably. And because the lifting speed of the mechanical arm is required not to be high, it also avoids the problem that the torque of stepper motor reduces to too small with the acceleration of speed.[8]

4.3. Local area network construction technology of parking lock
Zigbee is used to build the parking lock network. Each parking lock will be equipped with a zigbee module to access the network. In areas where signal transmission is inconvenient, Zigbee modules can be placed separately in appropriate areas to ensure signal coverage.

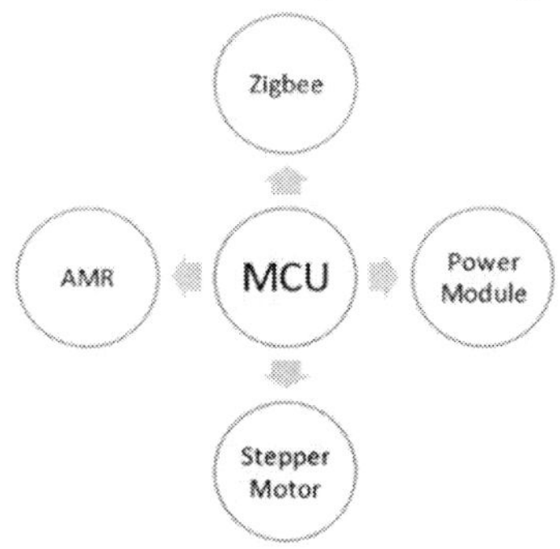

Figure 5. Design of parking lock.

In order to realize these functions on the parking lock, STM32 MCU is taken as the control body of the parking lock, and besides, AMR sensor, stepper motor, zigbee network module and power module are added to achieve the design functions mentioned abover(as shown in figure 5).

4.4. Design of the regional coordinators
The main components of the regional coordinators are STM32 MCU, Zigbee module, GPRS module, power module, and system management module (LED display). At regular intervals, the regional coordinator will collect information and working status of all parking locks from its zigbee network, and then send these messages to the server(as shown in figure 6). When the regional coordinator receives a message from the server, it will then send this message to the appropriate intelligent parking lock.[9] In order to facilitate the control of the parking lock, we add a LED screen on the parking lot coordinator, and managers can reset the status of parking locks or force the malfunctioning parking locks to open through simple interactions.

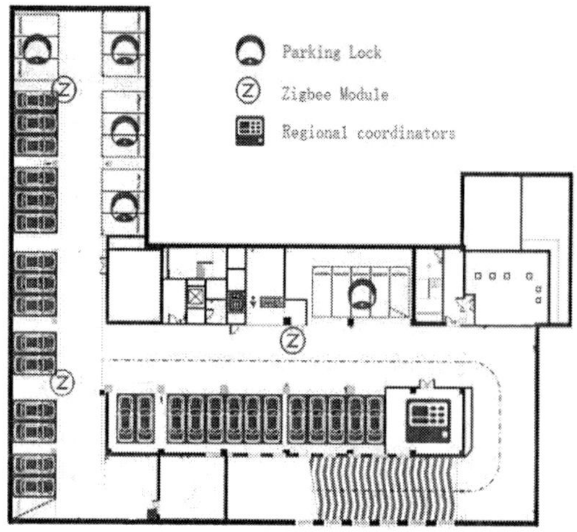

Figure 6. System installation diagram.

5. Conclusions

In this paper, an efficient parking management system design idea is put forward. The parking locks and coordinators are used to gather the required information, so that the drivers can find the parking information on the web page. Meanwhile, driver don't need to install extra application because of B/S architecture. When it comes to the management of parking, every parking lot can be a smallest unit to manage, which benefits from the design of the parking lot coordinator. Besides, parking locks controlled by the embedded system will show different behaviors according to the requests of the drivers. It is likely to provide the driver with a suitable parking lot with a route guiding there. Finally, the driver can park the car in a short time without looking for a free parking lot under the guidance of embedded systems and servers.

Acknowledgement

Here and now, I would like to express my sincere thanks to all those who have helped us complete this project and make this thesis possible and better. First of all, I am deeply grateful to my alma mater, Jilin University, for providing us with an opportunity and a platform to exercise ourselves.Then thanks to my mentor for his instructive suggestions and patient guidance during the completion of the project and the thesis.Finally,I am very grateful to my lovely team members for their hard work and determination to overcome difficulties.

References

[1] Xu, X. L. Yun, and L. Song 2017. Design of Intelligent Stereo Parking Lot Management System Based on RFID. Journal of Electronic Design Engineering, 25(7), 119-122.
[2] S. Deese, and J. Daum 2018. Application of Zigbee-Based Internet of Things Technology to Demand Response in Smart Grids. Journal of IFAC PapersOnline, 51(28), 43-48
[3] J.H. Zheng, and J. X. Guo 2018. Design and Implementation of Parking lot Management System Based on ASP. Net. Journal of Scientific and Technological Innovation, 22(35), 86-87.
[4] Y. L. Xing, X. H. Wang, and J. Y. Li 2018. Research and Design of Wireless Parking Detector Based on Geomagnetism. Journal of Langfang Normal University (Natural Science Edition), 18(2), 18-24.
[5] X. S. Li, G. H. Yu 2006. Application of AMR sensors to vehicle detection. Journal of University of Science and Technology Beijing, 28(6), 587-590.
[6] Z. Y. Zhang, H. Y. Wang, L. Y. Zhu, and S. J. Deng 2016. Design of Geomagnetic Vehicles Detecting System Based on ZigBee Wireless Sensor Network. Journal of Electronic

Measurement Technology, 39(7), 110-15.
[7] H. H. Dong, X. Z. Wang, C. Zhang, R. S. He, L. M. Jia, Y. Qin 2018. Improved Robust Vehicle Detection and Identification Based on Single Magnetic Sensor.Journal of IEEE Access, 6(1), 5247-55.
[8] Z. C. Chen. Control Technology of Stepper Motor Driver and its Design for Application[D]. Xiamen University, 2008.
[9] L. Wang 2015. Based on the Network Technology of Car Parking Management System Design and Application Field. Journal of Logistics Engineering and Management, 37(2), 52-53.

IOP Conf. Series: Materials Science and Engineering **715** (2020) 012097 doi:10.1088/1757-899X/715/1/012097

Solids velocity measurement using electric capacitance sensor assemblies

Usama Abrar[1,*], Liu Shi[1], Nasif Raza Jaffri[1], Qin Li[2], Hassan Raza Sindhu[3] and Mian Waqas Omar[4]

[1]School of Control and Computer Engineering North China Electric Power University Beijing 102206 China
[2]Institute of Mechanics Chinese Academy of Science Beijing 100190 China
[3]Department of Electrical Engineering Cowan University, Joondalup Perth Western Australia 6027
[4]Department of Computer Engineering Bahria University Islamabad Pakistan

*E-mail: uabrar@gmail.com

Abstract. This paper covers the application of the cross-correlation method for measuring the velocity of solid particles. Capacitive electrodes are used as primary sensors to measure the time required by the solid particles to cover a known distance between the electrodes. The capacitive variations of both the electrode sensors are stored in computer synchronously for offline estimation of the velocity of the solid particles. Single glass marble is used as a solid particle to conduct the performance in the lab for its good signal to noise ratio (SNR) and vivid graphs. Matlab R2018a is used as a programming tool to perform the cross-correlation algorithm. The distance between the sensors is adjusted optimally. The results were realistic to ensure the correctness of the system.

1. Introduction

Batch flow and continuous flow are two major techniques used to flow solid particles in industry. In the first one, the solid particles are divided into a number of bulks and are processed together, whereas the particles are processed without interruption in the later one. Their usage is varied from industry to industry and also depends upon the measuring and processing limitations [1]. In a variety of industrial processes, solid particles are dealt with the manner of batch flow. Feeding coal to the furnace is an important example of this technique, this can be done by using the conveyer belt but most of the times it is done pneumatically [2], [3]. Monitoring is the backbone of control part of the modern industry. In the absence of proper monitoring and control, it is impossible to achieve maximum efficiency from any industrial process and thus resource wastage takes place. Recently, the coal process industry has taken a few steps further by implying online monitoring, data logging and control of coal fed to the furnace by pneumatic means. Lately, there has been an emphasis on the precision and accuracy of these online measurements of solid particles flow. In these measurements, the parameters like flow rate, flow pattern, humidity, density and mass of the flowing solid are of main interests. There are other important notable parameters as well such as mass flow rate, pipeline erosion time that fundamentally depends upon the mass flow rate of the solid particles [4], [5]. A large number of techniques are available to measure the flow rate of solid particles depending upon sensing methods, sensitivity, and technology used. Digital imaging, optical sensing, acoustic techniques, radiometric methods, electrostatic sensors, and capacitive sensors are the notable techniques used for the measurement of Gas/Solid particle flow rate measurement. In case of dilute concentration of solid particles acoustic technique, optical sensors, and digital imaging work better. But there are also some techniques that work better even in high concentration like radiometric method, however, there is a chance of radiation leakage which can be avoided by proper shielding [6], [7]. But if an economical, trustworthy, harmless and robust solution is required, the electrostatic and capacitive sensors are the best choices. Capacitance measurement method is extensively employed in online and offline assessments of flow parameters. The capacitive measurement method is economical, accurate, and robust. Moreover, it has non-invasive nature among the rest measurement techniques [8]. Despite the fact of being simple robust and dynamic in its response several factors try to adversely affect the system response thus inculcating errors - the multiphase complex and variable flow, non-uniform sensitivity distribution of the electrode pair, the deposition of the solid particles on the internal walls of the channel's area where the electrodes are attached- moisture, solid composition, particles size, and contents [9].

This paper covers the preliminary investigations on the velocity of the solid particles flowing through the channel using electric capacitance measurement and applying the algorithm of cross-correlation to calculate the time required by the particles to flow through the channel. This paper focuses on the time required to create similar variation patterns in the capacitance values as the dielectric flows through the channel. The flow of the solid particles is based upon the assumption that the particles are moving parallel to the axis of the pipe and perpendicular to the plane of the sensor such that every passing particle has to flow through both the sensors after an interval of time. For a known distance between the two sensors, the ratio between the distance and time to cover that distance shows the mean velocity of the solid particles flowing through the channel.

1.1. Pneumatic conveying

The Pneumatic Conveying (derived from 'Pneumatikós' depending upon wind or air) is a most commonly used technique of material transportation in the industry [10]. It is generally an automatic system and is based upon the phenomenon of transportation of solid particles –powdered or granular within a piping system along with gas/air stream. The transportation of solid particles is based upon the high, low or negative pressure of air/gas. The solid particles are separated from the air on arrival at the destination [11]. The schematic of the pneumatic conveying system is shown in figure 1.

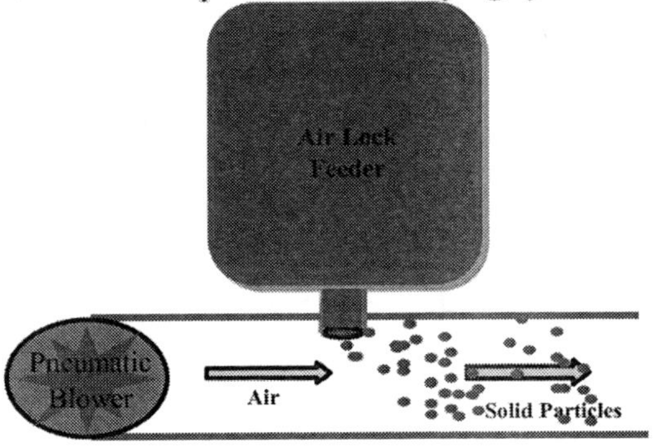

Figure 1. Pneumatic flow of solid particles.

The history of pneumatic conveying of the solids finds its traces in the era of ancient Greeks. Hero, a Greek Physicists, (10AD-85AD) gave the concept of Pneumatic Conveying [12]. The lead pipes for water transportation used by Romans and gas transportation bamboo tubes used by Chinese are from a prehistoric era. The granular solids transportation by the pneumatic system was introduced in 1920 [11]. This system was introduced in cement and coal industry in the1930s [13]. Since 1930-1960s remarkable efforts were done in the promotion and improvement of pneumatic conveying efficiency in

Japan, Germany, and USA [14]. Despite the fact that pneumatic conveying may cause damage to both pipes and particles and can transport to a limited distance while consuming high power [11], [15]. It is widely used in agricultural, chemical, mining food and mineral industry [16]. The reason is its ease of automation and control, dust-free transportation, low labor, and maintenance cost and last but not the least possibility of conveying of toxic and hazardous materials safely.

Solids are different from one another by their shapes, viscosity, density, stickiness, brittleness etc. Thus, various solids display different flow behaviors. Moisture content in the solid particles produces cohesive strength. The cohesiveness is also visible by temperature variation. Few solids are sensitive to rise in temperature. Others are reactant to a constant temperature. Cohesiveness is also dependent upon particle size. Finer powders are more cohesive also they have higher wall friction and thus difficult to flow smoothly [17]. If the solid particles are allowed to stay at rest for a longer time, the compaction loads due to head pressure can produce a strong cohesive bond. A chemical reaction, crystallization, or adhesive bonding can also cause this. Sometimes, after a cohesive arch is broken up, say by somehow initiating flow, the material can revert to its original flow condition and not exhibit a similar cohesion if left at rest again. On the other hand, some materials will time and again, bridge and rat-hole even after flow is re-initiated [18].

2. Mathematical description of velocity measurement

The velocity of the solid particles flowing through a channel can be computed using the cross-correlation technique [19]. Figure 2 elaborates the principle of calculating the velocity of solid particles using the cross-correlation technique. The cross-correlation of the two signals a(t) and b(t) is defined by equation 1

$$C(\tau) = \frac{1}{T}\int_0^T a(t)b(t+\tau)d\tau, \tag{1}$$

where $T \rightarrow \infty$.

Equation 1 shows the similarity between a(t) and b(t) while shifting b(t) along the time axis to $b(t+\tau)$. For a certain value of $\tau = \tau_{max}$ where $C(\tau)$ gives the maximum value, both a(t) and shifted b(t) are said to be most similar. If a(t) is considered as a signal from the first sensor while b(t) is considered as a signal from the second sensor. Since both the sensors are attached along the axis of the pipe in such a way that the solid particles flowing from the first sensor after a delay passes through the second sensor [20], the signals from both the sensors are the same with a difference of time delay which is the time required by the solid particles to flow from first sensor to the second sensor. If the distance between the sensors is obtained, the average velocity of the particles can be found by taking the fraction between the distance D and the time τ_{max}. In real time scenario the data is stored from both the sensors as discrete time values, so the cross-correlation of discrete time signals is defined by equation 2.

$$C[m] = \frac{1}{T}\sum_{n=0}^{T} a[n]b[n+m], \tag{2}$$

where $T \rightarrow \infty$.

However, it is impossible to find cross-correlation while extending $T \rightarrow \infty$. The modified version of cross-correlation is shown in equation 3.

$$C[m] = \frac{1}{L}\sum_{n=0}^{L} a[n]b[n+m], \tag{3}$$

where L is the length of the frame size, used to calculate the cross-correlation. It is the number of samples in a summation. m is the number of delayed samples-time delay [8]. In other words, L number of samples of a[n] is multiplied by L number of samples of shifted $b[n+m]$. If Δt is the sampling interval, The time to obtain peak of the cross-correlation function is given by $\tau_{max} = m * \Delta t$. The average velocity of the solid particles is given by equation 4.

$$V = D/\tau_{max}\ . \tag{4}$$

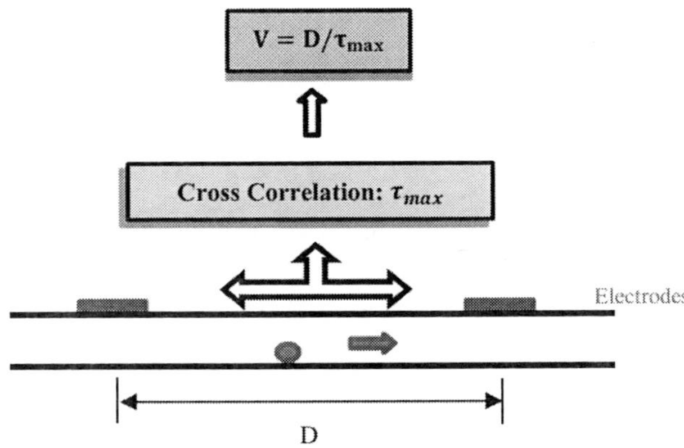

Figure 2. Velocity of solid particles using cross correlations.

The choice of the suitable distance between the two sensors is important. If the sensors are placed at a farther distance, the slow data acquisition rate may occur [21]. However, the better similarity between the signals is possible if the distance between the sensors is small. Although there is a chance of cross-interference of the signals of the two sensors if they are very close to each other. So the distance between the sensors is chosen depending upon the similarity of the signals and the dynamic behaviour of the system [4], [22].

3. Experimental setup

An Electric Capacitance Assembly system consists of three main components: the sensor, the data acquisition circuit, and the computer. The sensor is a single or more pair of metallic electrodes. The size, shape and the mutual distance between the pairs and among the electrodes of the individual pair are adjustable as per requirements. These electrodes are mounted on the insulated pipe which is extraneously covered with an earthed metallic sheet. The data acquisition circuit: typically excite the capacitive electrode pairs and converts the capacitance into the voltage which is conditioned and digitized for data acquisition. The computer stores and analyses the data by applying different mathematical techniques [22].

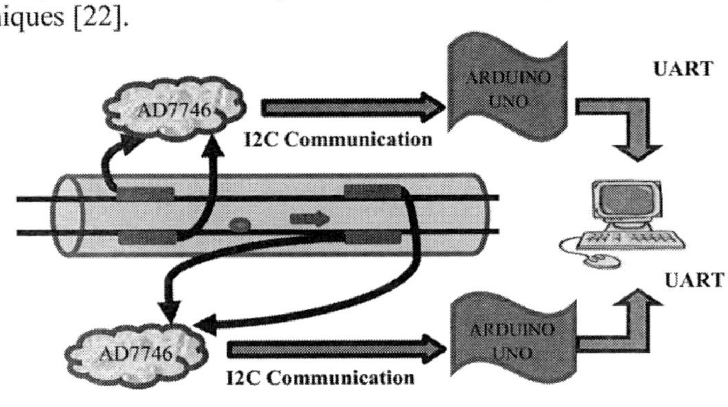

Figure 3. Block diagram of experimental setup.

The block diagram of the experimental setup is shown in figure 3. The glass marble flows through a rectangular, insulated and inclined channel- fixed on a stand and the angle of inclination can be varied as per requirements. The dimensions of the channel are $1\times2\times60$ cm^3. Two pairs of electrodes 'Sensor 1' and 'Sensor 2' of dimensions 2×1 cm^2 and 1×1 cm^2 respectively are fixed on the lower and upper surface of the channel 16.7 cm (distance between their centers) apart such that the glass marble flowing through 'sensor 1' must flow through 'sensor 2'. The channel is enclosed in a metallic shielding attached to a common ground with protective metallic shielding attached at the upper and lower ends of both the sensors to minimize the effects of the noise signals.

Eval-AD7746EBZ is a capacitance to digital converter evaluation board which is primarily used as data acquisition circuit for the measurement of capacitance between the electrodes. The main significance of AD7746 is its high resolution of 24-bits down up to 4aF and accuracy up to 4fF. Its full-scale capacitance rage is ±4pF. The maximum sampling rate of the device is 90Hz. The capacitance to be measured is directly connected to its terminals- Excitation terminal and Input terminal. It communicates through the I2C interface with microcontroller (Arduino-UNO). The registers were configured according to the Data Sheet for single mode capacitance measurement using I2C communication protocol. The raw data from AD7746 is received by Arduino and sent to the Computer through UART where the values were fed in the formula to compute the capacitance of both the sensors. The computer-program also gives the desired sampling interval to the Microcontroller that starts dropping off the data and thus stores at the required sampling rate [23]. The flow diagram of the data acquisition and storage is shown in figure 4.

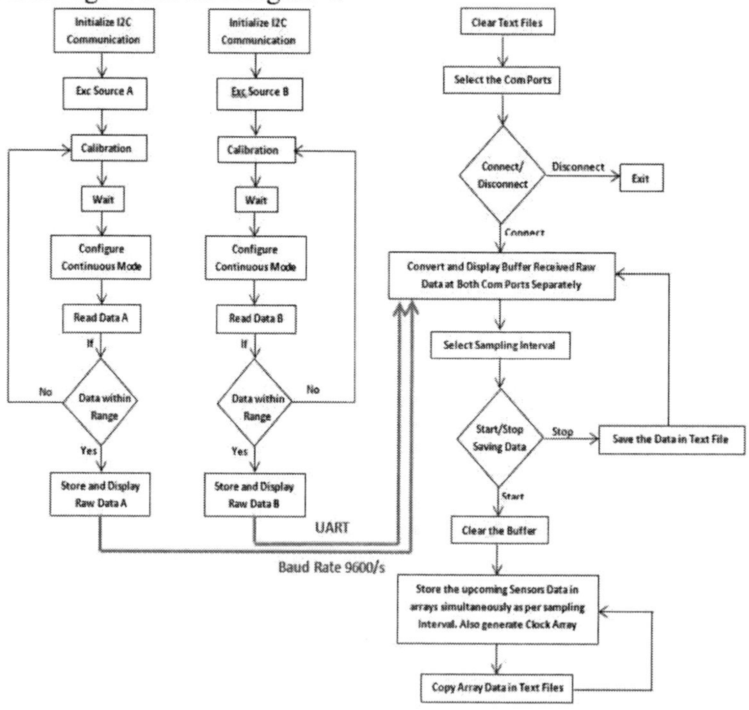

Figure 4. Flow diagram for data acquisition and data storage.

The data obtained from the microcontroller through UART at a baud rate of 9600/s is displayed on the GUI. The sampling interval is adjusted and as the command is given for data storage, it clears the buffer, generates a time clock variable and then starts storing the next upcoming data according to the set time interval. The intermediate samples are discarded and not stored in respective arrays thus ensuring time synchronized data storage from both the sensors. The arrays are then stored in respective files for further processing.

4. Results and discussions

Figure 5 shows the capacitance values of sensor 1 and Sensor 2 as a single spherical glass marble of size 7 mm is allowed to flow through the channel. The sampling interval of 10 ms is adjusted for the record of data. The spikes in both the signals indicate the instants as the glass marble flows through the sensors respectively. The separation between the pair of the electrodes of individual sensor is approximately 1 cm (outer diameter of the pipe). The marble behaves as a dynamically flowing dielectric medium between the electrodes of the sensor. The nature of the material (glass) and its size comparative to the diameter of the channel (space between the electrodes) is believed to give good signal to noise ratio. During the time glass marble has not reached the sensor 1, its capacitance value remains close to 0.4764 pF as the glass marble reaches the region of capacitance field, a sudden rise in the graph of sensor 1 is observed, the glass marble rolls down towards sensor 2 who until now has an average value maintained around 0.3128 pF the slight variations in both the signals are because of the noise that can be reduced by subtracting the average value of the signal from individual stored sample of the data. The difference in the average value of both the sensors is because of the size of the electrodes for both the sensors. Since sensor 1 is double in size as compared to sensor 2. However the variation in the average value of the capacitance for both the sensors is 0.1636 pF (52.3%). Also the spike of sensor 1 is wide in the initial then the graph sharpens suddenly. This is because in the design of sensor 1.

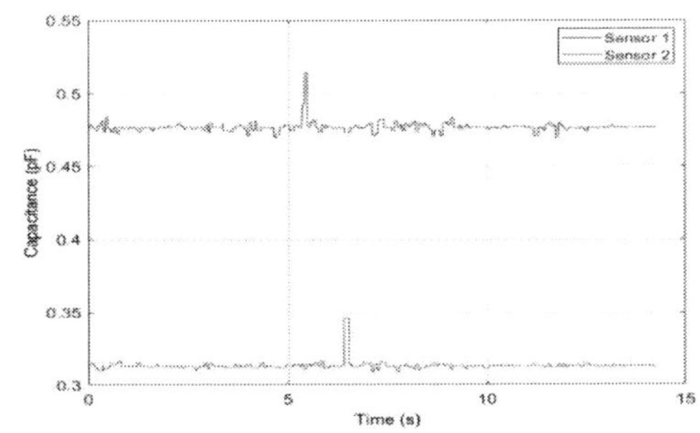

Figure 5. Output signals of capacitive sensors when the distance is 16.7 cm.

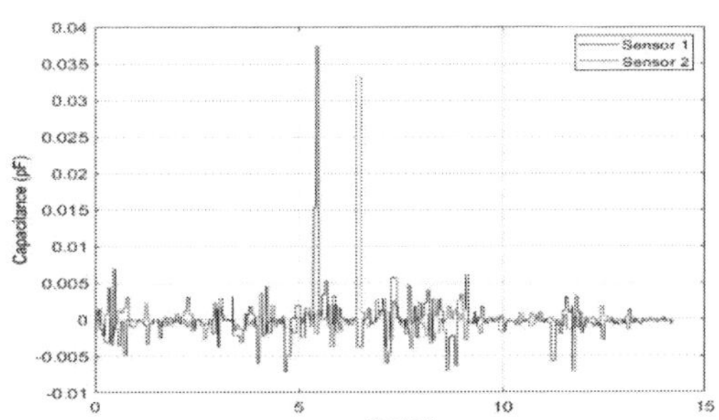

Figure 6. Output signals of both the sensors after noise reduction.

Two pieces of metal of approximate size 1x1cm^2 are placed over lapping each other which shows abnormality in the shape of the spike. However Sensor 2 electrode is a pair of single piece. The time is recorded for the flow of the glass marble through both the sensors. Figure 6 shows the normalized output of both the sensors after the noise reduction by subtracting the average value from individual signals.

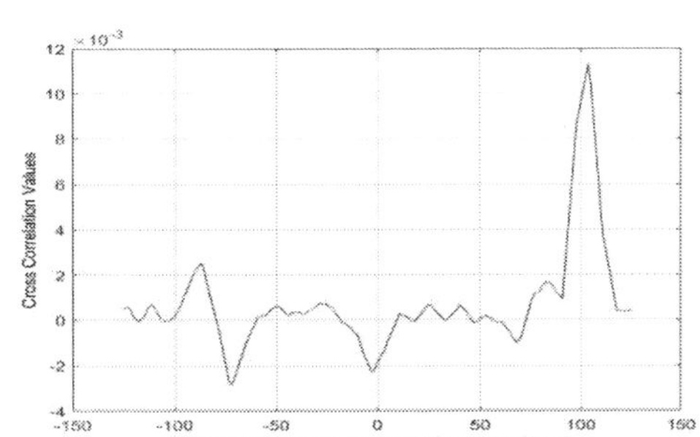

Figure 7. Typical cross correlation between the two sensors when the distance between them is 16.7 cm.

The cross-correlation is performed on the output signals of both sensors. Maltab R2018a is used to compute the cross correlation between the signals using equation 3. A window size of 125 discrete samples is used to compute cross correlation between the signals. Figure 7 shows the output of the cross correlation curve. Where maximum peak of the curve is obtained at 104th discrete time interval. This means that on shifting the window samples of Sensor 2 by discrete steps of 10 ms each it is 104th instant where both the curves match the most in their pattern. The pattern trend relies on the instant as glass marble flows through the sensor. So the peak of curve in figure 7 indicate the time required by the rolling glass marble to flow from sensor 1 to sensor 2 in discrete intervals. The separation between the two sensors is chosen as the distance between the centers of the electrodes of both the sensors which are 16.7 cm. The total time required by the glass marble to flow between Sensor 1 and Sensor 2 is 1.04 s. The velocity of the marble is found to be 0.16 m/s. Table 1 shows different iterations to calculate the velocity of the glass marble through the channel. The average velocity of the glass marble slowly rolling along a constant inclined surface is found to be 0.15 m/s.

Table 1. Velocity of glass marble using cross correlation.

Sr. No	Time Estimated (s)	Distance Between the Sensors (cm)	Velocity Estimated (m/s)
1	1.04	16.7	0.16
2	1.18	16.7	0.14
3	1.09	16.7	0.15

4.1. Limitations of the sensor

Table 2 shows the limitation of the sensor designed that is how fast moving solid particle can be detected by this sensor. Several iterations were performed on free rolling glass marble along the inclined surface.

Table 2. Effect of angle of inclination of channel on velocity of rolling solids.

Sr. No	Angle(θ°)	Distance Between the Sensors(cm)	Time (s)	Velocity (m/s)
1	6.87	16.7	1.64	0.101
2	8.84	16.7	0.350	0.477
3	10.58	16.7	0.330	0.506
4	11.83	16.7	0.320	0.521
5	14.85	16.7	0.250	0.668
6	18.69	16.7	0.160	1.043

At horizontal surface where angle of inclination was $\theta^{\circ}=0$ there was no rolling in the glass marble. The rolling in the glass marble started when the angle of inclination was increased and on further increase resulted in the increase in the velocity of the glass marble. The reason is in the absence of forced response-when no input force is exerted on the marble, the only driving force is the rectangular component of gravity force acting along the plane of inclined channel. The magnitude of this force linearly increases as Cosine function of the angle of inclination acting along the plane of channel.

Keeping sampling interval fixed at 10 ms for data collection from AD7746 whose maximum output sampling rate as per data sheet is 90 samples/sec, the maximum velocity that can be computed by this capacitive sensor is 1.043 m/s. In the absence of the external force applied such velocity can be achieved as the angle of inclination of the channel is increased up to $\theta^{\circ}=18.69^{\circ}$ As the angle is further increased. The value of m from equation 3 becomes gradually 0. i.e. not computable by the sensor for its limitations of sampling. Reason is solid particle is flowing through the distance D of equation 4 faster than the sampling frequency. Further higher velocity can be found either by using a higher frequency capacitance data acquisition circuit or by changing the dimensions of the sensors. Figure 8 shows the relationship between the velocity and angle of inclination of the channel.

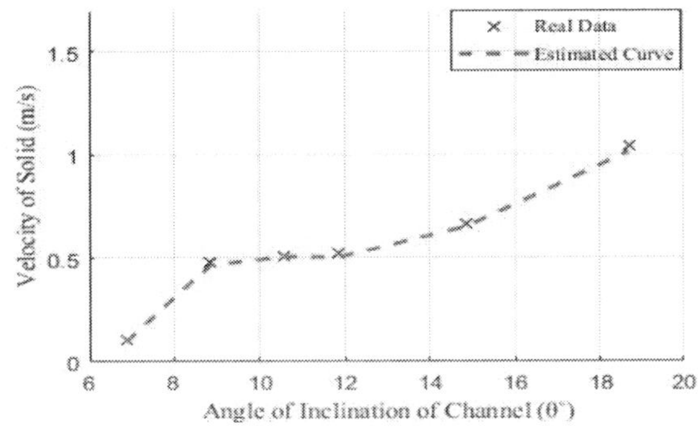

Figure 8. Effect of inclination of channel on the velocity of the solid.

$$V(\theta) =-0.0004354\theta^4 + 0.02364\theta^3 - 0.4634\theta^2 + 3.926\theta - 11.7 \quad (5)$$

Using curve fit tool and data from table 2, the relationship between the velocity of the glass marble and the angle of inclination is shown in equation 5.

5. Conclusion

An instrumentation system based upon the capacitance sensors has been developed in this paper. Experiments have been carried out on a rectangular channel. The results demonstrated that the average velocity of the glass marble was 0.15 m/s. This shows that the system is capable of data acquisition and offline measurement of the velocity of the particles. The instrumentation system is designed based upon AD7746 Evaluation Board and the microcontroller which can be installed in the industrial environment to provide particle flow parameters to system control units. In future, the system will be used to evaluate the performance of the instrumentation system on a dense phase pneumatic conveyor of pulverized fuel particles.

Acknowledgments

The authors wish to extend their gratitude to NSFC for sponsoring the research (No 61571189, 61871181) and to the State Administration of Foreign Experts Affairs for supporting the 111 project (ref: B13009).

References

[1] T. R. Dias, W. R. Melchert, M. Y. Kamogawa, F. R. P. Rocha, and E. A. G. Zagatto, "Fluidized particles in flow analysis: potentialities, limitations and applications," Talanta, vol. 184, no. December 2017, pp. 325–331, 2018.

[2] J. wei Zhou, C. long Du, and Z. ling Ma, "Influence of swirling intensity on lump coal particle pickup velocity in pneumatic conveying," Powder Technol., vol. 339, pp. 470–78, 2018.

[3] H. Lu, X. Guo, P. Li, K. Liu, and X. Gong, "Design optimization of a venturi tube geometry in dense-phase pneumatic conveying of pulverized coal for entrained-flow gasification," Chem. Eng. Res. Des., vol. 120, pp. 208–17, 2017.

[4] H. Seraj, M. F. Rahmat, and M. Khalid, "Measurement of velocity of solid/air two phase fluid using electrostatic sensors and cross correlation technique," Sci. Iran., vol. 20, no. 3, pp. 786–92, 2013.

[5] S. M. Rao, K. Zhu, C. H. Wang, and S. Sundaresan, "Electrical capacitance tomography measurements on the pneumatic conveying of solids," Ind. Eng. Chem. Res., vol. 40, no. 20, pp. 4216–26, 2001.

[6] C. Hartmann, J. Wang, D. Opristescu, and W. Volk, "Implementation and evaluation of optical flow methods for two-dimensional deformation measurement in comparison to digital image correlation," Opt. Lasers Eng., vol. 107, no. September 2017, pp. 127–41, 2018.

[7] M. B. Lima, S. I. E. Andrade, I. S. Barreto, L. F. Almeida, and M. C. U. Araújo, "A digital

image-based micro-flow-batch analyzer," Microchem. J., vol. 106, pp. 238–43, 2013.

[8] W. Zhang, C. Wang, W. Yang, and C. H. Wang, "Application of electrical capacitance tomography in particulate process measurement - A review," Adv. Powder Technol., vol. 25, no. 1, pp. 174–88, 2014.

[9] H. Q. Che, M. Wu, J. M. Ye, W. Q. Yang, and H. G. Wang, "Monitoring a lab-scale wurster type fluidized bed process by electrical capacitance tomography," Flow Meas. Instrum., vol. 62, no. September 2017, pp. 223–34, 2018.

[10] K. Malone and H. W. Fowler, A Dictionary of Modern English Usage., 2nd Editio., vol. 42, no. 3. Oxford University Press, 1983.

[11] G. E. Klinzing, "A review of pneumatic conveying status, advances and projections," Powder Technol., vol. 333, pp. 78–90, 2018.

[12] E. Papadopoulos, "Heron of Alexandria (c. 10–85 AD)," in Distinguished Figures in Mechanism and Machine Science, Dordrecht: Springer Netherlands, 2007, pp. 217–45.

[13] C. Ratnayake, "A Comprehensive Scaling Up Technique for Pneumatic Transport Systems," The Norwegian University of Sceince and Technology (NTNU), 2005.

[14] G. E. Klinzing, "Historical review of pneumatic conveying," KONA Powder Part. J., vol. 2018, no. 35, pp. 150–59, 2018.

[15] L. Cai, X. Pan, C. Xiaoping, and Z. Changsui, "Flow characteristics and stability of dense-phase pneumatic conveying of pulverized coal under high pressure," Exp. Therm. Fluid Sci., vol. 41, pp. 149–57, 2012.

[16] V. Mills, David. Jones, Mark. Agarwal, "Pneumatic Conveying of Coal and Ash," in Handbook of Pneumatic Conveying Engineering / Edition 1, 2004, pp. 170–208.

[17] W. Chen and A. W. Roberts, "A modified flowability classification model for moist and cohesive bulk solids," Powder Technol., vol. 325, pp. 639–650, 2018.

[18] H. Lu, X. Guo, Y. Jin, and X. Gong, "Effect of moisture on flowability of pulverized coal," Chem. Eng. Res. Des., vol. 133, pp. 326–34, 2018.

[19] M. O. Agolom, G. Lucas, and R. O. Webilor, "Measurement of velocity profiles in transient single and multiphase flows using inductive flow tomography," Flow Meas. Instrum., vol. 62, no. June 2017, pp. 246–54, 2018.

[20] X. Li, A. J. Jaworski, and X. Mao, "Comparative study of two non-intrusive measurement methods for bubbling gas-solids fluidized beds: Electrical capacitance tomography and pressure fluctuations," Flow Meas. Instrum., vol. 62, no. May 2017, pp. 255–68, 2018.

[21] K. Grudzień, Z. Chaniecki, and L. Babout, "Study of granular flow in silo based on electrical capacitance tomography and optical imaging," Flow Meas. Instrum., vol. 62, no. September 2017, pp. 186–95, 2018.

[22] M. Sun, S. Liu, J. Lei, and Z. Li, "Mass flow measurement of pneumatically conveyed solids using electrical capacitance tomography," Meas. Sci. Technol., vol. 19, no. 4, 2008.

[23] Z. Huang, J. Zhu, and L. Lu, "An AD7746-Based Data Acquisition System for Capacitive Pressure Sensor in Weather Detection Application," Key Eng. Mater., vol. 483, pp. 461–64, Jun. 2011.

IOP Conf. Series: Materials Science and Engineering **715** (2020) 012098 doi:10.1088/1757-899X/715/1/012098

Analysis and design of integrated intrusion detector performance testing system

Fei Sun[1,*], Xiaofeng Wang[1], Ming Tan[1] , Xianfeng Hu[2] and Rui Hu[3]

[1]Department of Computer Science and Technology, Hefei University, Hefei, Anhui 230601, China
[2]Science and Technology Informatization Department , Anhui Provincial Public Security Department, Hefei, Anhui 230000, China
[3]National University of Defense Technology, Hefei, Anhui 230037, China

*E-mail: sunfei75@163.com

Absract. The integrated intrusion detector performance testing system is mainly used to simulate the performance of active infrared,passive infrared, infrared and microwave composite detectors.The device has the advantages of complete function, simple operation, stable performance, high control precision,accurate collection and good security. In addition, the wireless transmission technology is adopted, and the remote operation can be carried out. The main goal of this paper is to develop the intrusion detector detection system according to the standard technical requirements and test methods,to meet the needs of product detection,and finally to improve the competitiveness of domestic related enterprises in the international market.The device can test whether the intrusion detector has the problem of false alarm and false alarm,test the anti-interference ability and the performance parameters of the intrusion detector,and then improve the product quality of the intrusion detector. It is very helpful to develop new products.

1. Introduction

With the rapid development of economy and the increasing improvement of people's living standard, people are demanding more and more safety,and the understanding of security and security prevention system is becoming more and more profound. In the intrusion alarm system,the main control unit has been relatively mature in technology, in the future will be integrated more ways,combined with network,cable,wireless integration and other directions.The detector technology is varied and the space for research and development of the technology is relatively large.Therefore,the innovation of intrusion alarm system is embodied in the technical change of the detector.

Intrusion detector as the most important detection device of anti-theft alarm system also has a rapid development. The domestic market mainly consists of active infrared detector passive infrared detector microwave and passive infrared composite detector magnetic switch detector[1]. As a key part of the whole alarm system, intrusion detection determines and affects the performance of the alarm system to a large extent[2], such as detection range, detection sensitivity, false alarm rate,etc. So it is necessary to detect the performance of the intrusion detector. In the course of many years' research and development, the problems of false alarm, false alarm and interference have always been the technical problems that need to be solved in the anti-theft alarm system industry[3]. For these problems the corresponding detection methods, detection standards, detection instruments have been launched and constantly improved, how to effectively design a set of intrusion detector simulation test equipment is particularly critical. The device can test whether the intrusion detector has the problem of false alarm and false alarm, test the anti-interference ability and the performance parameters of the intrusion detector, and then improve the product quality of the intrusion detector[4]. It is very helpful to develop new products.Therefore, the development of intrusion detector inspection equipment has a very practical significance.

At present, there is a lack of integrated detection equipment for intrusion detection performance in the market. As far as our province is concerned,there is no inspection organization that can do the testing of this kind of products.Therefore,our main goal should be to develop intrusion detector detection system according to the standard technical requirements and test methods, to meet the needs of product detection,and finally to improve the competitiveness of domestic related enterprises in the international market.

2. System architecture and design principle

2.1. Performance test device of active infrared intrusion detector

The performance test device of intrusion detector is mainly used to simulate the performance test of active infrared intrusion detector. It is composed of frequency conversion motor, deceleration mechanism, rotating chain, cylindrical object, special installation fixture, safety protection device, rotary table, optical power meter, etc. According to GB 10408. 1-2000 standard, the transmitter angle test (as shown in figure 1), receiver angle test, response time test and detection range test are carried out. In the process of related tests, the detector is observed to alarm, and the test sample is judged to be up to standard [5].

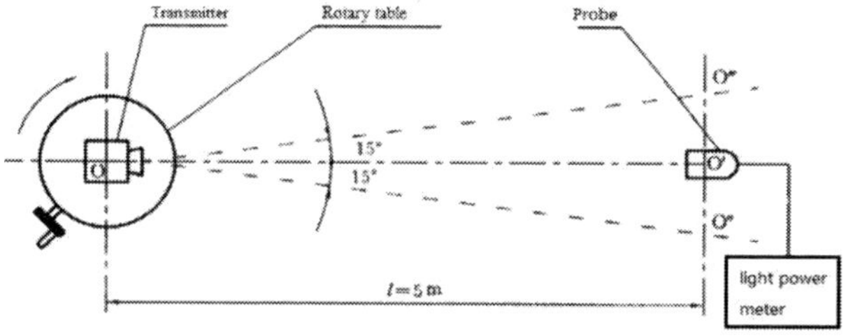

Figure 1. Principle diagram of transmitter beam angle.

The rotary table is mainly used for angle test of transmitter and angle test of receiver. The rotary table is mainly composed of stepping motor, stepping driver, converting gear and rotating disk. The stepper motor is an open-loop control unit which converts the electric pulse signal to angular displacement or linear displacement by changing the gear to drive the rotating disc (indicated by the scale on the disk). In the case of non-overload, the speed and stop position of the motor depend only on the frequency and the number of pulses of the pulse signal, and are not affected by the change of the load, when the stepping driver receives a pulse signal. It can drive the stepping motor to rotate a fixed angle according to the set direction, and can control the angular displacement by controlling the number of pulses, so as to achieve the purpose of accurate positioning. Positioning accuracy is determined by the step angle and the fine fraction on the drive[6].

The test should be carried out in the darkroom, with the transmitter giving priority to DC power supply with a rated voltage of 12 V (or AC power supply with a rated voltage of 220 V). The transmitter is fixed on the workbench so that its optical system node passes through the center of the workbench roundabout. At a distance of 5m from the transmitter, an optical power meter probe is placed so that photosensitive surface is perpendicular to the optical axis O-O′ , and the transmitter is fine-tuned to maximize the output of its optical power. Write down the reading P1, and then turn the worktable to the left to the right by 15 degrees, respectively.The radiative power P2 and P2′ of the

O-O"and O-O" transmitters are measured by the optical power meter, and the maximum values are taken between them.

According to the formula $N = 10\lg(P_1 / P_2)$, the value of N should be more than 20 dB.After that, the transmitter is rotated 90 degrees along the O-O′ axis, and the radiation power values on the plane perpendicular to the original test plane can be obtained by repeating the above tests. The test and calculation results should still meet the requirements of this standard.

The single-chip microcomputer is used to control the stepping motor to drive the rotation angle of the worktable. The angle of rotation is obtained by the encoder and transmitted to the single-chip microcomputer system and finally to the PC for display. The optical power meter is selected according to the range of beam power transmitted by the transmitter,and the value obtained by the optical power meter can also be transmitted directly to the PC for display. Finally, the results of two records are used to calculate whether the final results meet the requirements. The schematic diagram is shown in figure 2.

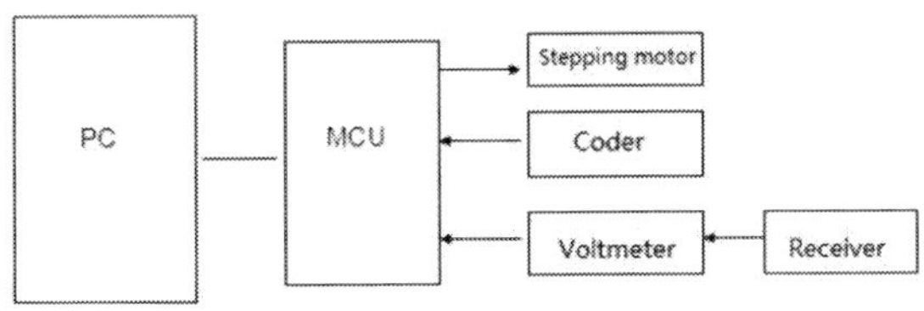

Figure 2. Control principle block diagram of angle test for receiver.

2.2. Response time measurement

According to GB 10408.1-2000 standard, the response time test is carried out at the detection distance specified by the manufacturer. The transmitter and receiver are fixed at both ends, and the distance between the supports can be adjusted arbitrarily. The cylindrical object is fixed with a special mounting fixture in the rotating chain, which drags the metal barrel through the beam at a predetermined speed. The frequency conversion motor is connected with the chain through the reducer,and the frequency conversion motor accelerates in a certain time,which reduces the impact of the cylindrical object and has the function of safety protection. The cylindrical object achieves the set speed and then supplies the power to the detector, avoiding the influence of the velocity not meeting the requirement on the test results. The whole control system can monitor the test results by remote operation and ensure the safety of the testers.

In adjusting the beam axis between transmitter and receiver, a cylindrical object with diameter of 200mm, its length should be able to fully block the beam and pass through the beam at a speed greater than 10m/s perpendicular to the beam axis. The detector should not produce an alarm state. When the object passes through the beam at a speed less than 5m/s, the detector should generate an alarm state and calculate the alarm duration. When the detector operates at the detection distance specified by the manufacturer, the duration of the radiation signal being completely or partially obscured by a given percentage of the detector shall be greater than 40 (1 ±10%) ms, detector shall produce an alarm state, The duration of the radiation signal being completely or partially obscured by a given percentage is less than 20 (1 ±10%) ms, detectors should not generate an alarm state. The duration of alarm should be more than 1 s when alarm state is generated.

2.3. Intrusion detector analog reference target device

The passive infrared detector works by detecting the infrared radiation emitted by the human body: the detector collects the infrared radiation from the outside world and then gathers on the infrared sensor by the Fresnel lens.After the pyroelectric element receives infrared radiation, the temperature will change, then the charge will be released outward, and the alarm will be generated after the detection and processing[7]. The intrusion detector simulates the movement of the human body in the detection coverage area, and the device rotates the detector by using a fixed reference target[8]. The detector is fixed on the mounting bracket, the stepper motor rotates the bracket at a certain speed by changing gears, and the equivalent human body moves at a uniform speed in the detection coverage area. Moving speed as formula (1) and (2):

$$V=2\pi r/\omega \quad (1)$$

$$\omega=2\pi n \quad (2)$$

ω: stepping motor rotation angular velocity
r: reference target distance from the detector center
n: step motor rotation speed
formula (2) the equivalent moving velocity can be calculated by bringing in formula (1). The equivalent speed and distance can be controlled by changing the rotation speed and angle of the stepping motor.

Also calculated with reference to the target background temperature is as follows:
The total heat dissipation area of the human body simulator as formula (3):

$$A=0.3*(0.235+1.50)=0.52m^2 \quad (3)$$

The natural convection heat transfer coefficient of air as formula 4:

$$h=5\sim25\ W/(m^2*K) \quad (4)$$

Maximum temperature difference Δt=9K(ambient temperature 15°C, setting temperature 24°C)Heat dissipation power as formula (5)

$$Q = h*A*\Delta t=25*0.52*9=117W \quad (5)$$

If the power removal margin is n=2, then the heating power density as formula (6)

$$p=n*Q/A=450W/m^2 \quad (6)$$

The reference target is designed according to the standard size and consists of silicone rubber heater,6mm thick aluminum plate and insulation material. The silicone rubber heater is mainly composed of Ni-Cr alloy electric heating wire and high temperature insulation layer of silicone rubber.It has the advantages of fast heating, uniform temperature, high thermal efficiency, convenient use and not easy aging.The silicone rubber heater has a good softness and can be in close contact with the heated object, and the heating element can be processed with nickel alloy gold foil, the heating power can reach 1.2 W/cm^2, and the heating is more uniform. Aluminum sheet has good heat transfer characteristics, so that the heat distribution more evenly up to the standard requirements [9].

The control system adopts PT1000 (precision better than 0.1°C), 16 bit ΣΔ ADC, intelligent PID control algorithm,modularization design, distributed temperature control, heating power density 1000W/m2.

The PID control consists of the proportional unit P, integral unit I and the differential unit D. The relationship between the input e (t) and the output u (t) is as formula (7):

$$u(t)=k_p[e(t)+1/T_I\int e(t)dt+T_D*de(t)/dt] \quad (7)$$

The upper and lower bounds of the integral in the formula are 0 and t respectively so its transfer function is formula (8):

$$G(s)=U(s)/E(s)=k_p[1+1/(T_I*s)+T_D*s] \quad (8)$$

where k_p is a proportional coefficient,T_I is an integral time constant,and TD is a differential time constant.

2.4. Anti-headlamp (light) test device

The anti-headlamp (light) test device is mainly composed of a lamp assembly device, a dark box, a control system, etc. The detector is illuminated by light equivalent to the headlamp through the glass. At the beginning of the test, the halogen lamp is first turned off and the detector is electrified. Wait for detector to enter alert state,then turn on halogen lamp, turn on 2s, break 2s, repeat 5 times, check detector whether alarm state [10]. Lamp assembly is the plane mirror before the installation of vehicle H4 halogen lamp, halogen lamp installation and replacement convenient. The industrial Siemens PLC is used to control the power supply of lamps and lanterns, which has the advantages of simple operation, stable performance and high control precision.

2.5. Turbulent flow resistance test device

The anti-turbulent flow test device is used to detect the detector should be able to work normally when it is disturbed by hot air flow in the alert state. A 1000W hot air machine with a cross section area of 55mm×180mm is required to be placed at the 1m below the detector and 1m in front of the detector. The flow velocity of the hot air engine is 12° angle toward the detector, the flow velocity from the hot air engine is 2.2m/s ±0.2 m/s, and the flow velocity is 0.7m/s±0.1 m/s when passing through the surface of the detector. Fan power for 5 minutes, the test results should not produce alarm state. The test device is shown in figure 3.

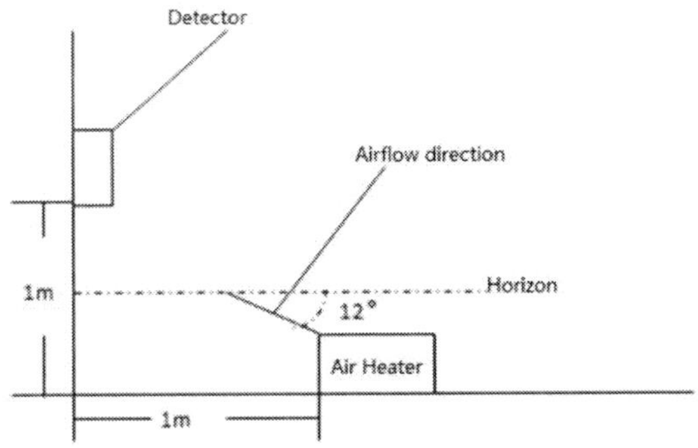

Figure 3. Anti-turbulent airflow test apparatus diagram.

The device consists of a speed regulating fan,a heating element, an air duct,a wind speed sensor, a temperature sensor and a control system. The speed regulating fan forms the PID closed loop control through the heating element installed in the air pipe, the wind speed sensor,the temperature sensor and the PLC, SCR in the control system, and the flow velocity and temperature are automatically adjusted according to the set value.

2.6. Anti-fluorescence interference test device

The anti-fluorescence interference test device is used to detect the influence of the interference of fluorescent lamp at the prescribed distance from the detector to the alarm state of the detector. The technical requirements are as follows: the lamp tube is placed at 0.5m above the detector and 2.0m ahead. For the detector installed with suction top, the lamp tube should be placed at 1.0m below the detector. The fluorescent lamp should be turned off first, then the detector should be electrified, waiting for the detector to be on alert, then the fluorescent lamp should be turned on for 60s, turned off for 30s,and repeated for 5 times. The detector does not alarm each time the light is on, and then turns the fluorescent lamp 90 degrees again.

The device consists of detector mounting bracket and fluorescent lamp mounting bracket, the height of the support is adjustable, the outer surface is blackened,and the light reflection effect test is reduced. Supports have top mounting and suspension mounting alternative, fluorescent lamp mounting bracket has rotation limit. The on-off of fluorescent lamp power supply is controlled by Siemens PLC in the control box, the control is stable and the on-off time is accurate. The whole control needs only one button to start the test, and the alarm state of the detector triggers the buzzer alarm in the control box.

3. Results

The intrusion detector test device is mainly used to simulate the performance parameters of the intrusion detector. The equipment is divided into five sets of equipment. The test device is mainly composed of an active infrared intrusion detector performance test device, an intrusion detector analog reference target device, an anti-vehicle headlamp (light) test device, an anti-turbulent flow test device, an anti-fluorescence interference test device.Remote monitoring terminal. The device has the advantages of complete function, simple operation, stable performance, high control precision, accurate collection and good security. In addition, the wireless transmission technology is adopted, and the remote operation can be carried out.

4. Conclusion

The system has passed the Anhui Provincial Department of Science and Technology. The findings of the expert review are as follows:

1. Meet the performance test and anti-jamming test of intrusion detector products, according to the relevant requirements of GB 10408, we can effectively judge whether the key performance indexes of intrusion detector such as false alarm, missed report, interference and so on are qualified.
2. The system can be used for wireless transmission and remote operation to simulate the testing of performance parameters in the operating state of the intrusion detector.
3.The system is divided into 7 equipment, 7 devices do not interfere with each other, can carry on the sample test at the same time, can carry on the single performance test.
4.The system uses PLC automatic control, the function is complete, the operation is simple, the performance is stable, the control precision is high, the collection is accurate, the security is good.

Main technical performance index as follows:

1. Detection category: active infrared, passive infrared, ultrasonic wave, microwave, vibration
2. Rotation angle: 0 °-360 °adjustable
3. Cylinder velocity: 5 m/s, 10m /s adjustable
4. Temperature range: 20°C-45°C adjustable

Acknowledgments

Supported by Anhui Key Research and Development Project(201904d07020002); Supported by National Natural Science Fund(61672204); Supported by Anhui Science and Technology Major Project(17030901026).

References

[1] Chen Suping, "Pyroelectric Infrared Sensor and Alarm Circuit Thereof [J]", Automatic Control, vol. **10** (2007).
[2] Shang Zhongfeng, Guo Jinghua, Qi Mingfeng, Fan Ziliang, "Pyroelectric Infrared Sensor and its Application", Research on Urban Construction Theory[J], vol. **21** (2012).
[3] Wang Xiaomin, Wang Zhimin, Sensor Detection Technology and Application, Peking University Press[P], Jan (2011).
[4] Caixia Liu, Jin Huang, "Technical analysis on the pyroelectric infrared sensor in intrusion detection devices". 2017 32nd Youth Academic Annual Conference of Chinese Association of Automation (YAC) (2017) [C], May 2017, Hefei, China.
[5] GB 10408.1-2000, Detectors for intruder alarm system Part1: Requirements for detectors-General [S].
[6] GB 10408.4-2000, Detectors for intruder alarm systems-Part 4: Active infrared intrusion detectors [S].
[7] Wang Xiaoli, A survey of the mainstream technology and principle of intrusion tectors[J]. Microcomputer & Its Applications, 33 (18): 1-3 (2014).
[8] Chen ZHAN, Research on three-dimensional Detection Model of passive Infrared detector (in Chinese) [J]. China Security & Protection, 03:60-84 (2014).
[9] GB 10408.5-2000, Detectors for intruder alarm systems-Part 5: Passive infrared detectors for use in buildings [S].
[10] Guo Li, Analysis of Problems and Standards of Invade Detector [J]. China Public Security (ComprehensiveVersion), (6): 181-82 (2003).

IOP Conf. Series: Materials Science and Engineering **715** (2020) 012099 doi:10.1088/1757-899X/715/1/012099

Attitude control of fixed-wing UCAV based on adaptive fuzzy PID

KU Shuo[1,*] DING Da-li[1], YANG Wei-li[2] and ZHU Wen-qiang[1]

[1]Aeronautics Engineering College, Air Force Engineering University, 710038 Xi'an, China

[2]The Chinese People's Liberation Army 93617 Unit, 101400 Beijing, China

*E-mail: kushuono1@163.com

Abstract. Aiming at the nonlinear and strong coupling characteristics of the fixed-wing UCAV's rigid body six-degree-of-freedom model, an adaptive fuzzy PID attitude control method was designed based on expert experience. The nonlinear model of UCAV was linearized with the principle of small disturbance, thus the state space expressions of longitudinal and lateral channels were obtained and the decoupling of the longitudinal and lateral channels was achieved. To verify the designed adaptive fuzzy PID controller, the simulation was carried out aiming at the pitch angle and velocity of the longitudinal channel and the roll angle of the lateral channel. The result showed that the designed controller had faster responsive speed and higher stability, and it could control the UCAV to achieve the desired attitude effectively.

1. Introduction

With the rapid development of UAV technology, the traditional man-in-the-loop way for controlling will no longer be applied to the sharply changing air combat situation[1]. Due to the absence of pilots on the aircraft, the stability, robustness and control accuracy of the control system are critical to ensure UCAV's own safety and successful completion of attack occupancy[2]. In literature[3], the root locus method in classical control theory is used to analyze and design the longitudinal control law of a fixed-wing UAV. The literature[4] introduced the LQR control method of the quadrotor UAV. Since the quadrotor UAV is usually in the disturbed state, the anti-interference ability is poor. In literature[5], a fuzzy PID controller is designed for the four-rotor aircraft's hovering control. The controller is used to perform inner-ring's attitude control and outer-ring's position control on the aircraft to achieve stable hovering flight. In this paper, an adaptive fuzzy PID attitude controller is designed for a fixed-wing UCAV to realize fast and stable attitude control of UCAV.

2. Six degree of freedom model of UCAV

The UCAV's six-degree-of-freedom model describes the relationship between external forces, moments, and motion parameters of UCAV. It mainly includes dynamic models and kinematic models[6]. The UCAV dynamics model is as follows:

$$\begin{cases} \dot{u} = rv - qw - g\sin\theta + \frac{1}{m}(\bar{X} + P) \\ \dot{v} = pw - ru + g\sin\phi\cos\theta + \frac{1}{m}\bar{Y} \\ \dot{w} = qu - pv + g\cos\phi\cos\theta + \frac{1}{m}\bar{Z} \end{cases} \tag{1}$$

In the above formula, P represents the UCAV's engine thrust and θ,ϕ are the pitch angle and roll angle of the UCAV. $\bar{X},\bar{Y},\bar{Z}$ respectively represent the resistance, lateral force and lift of the aircraft. m is the body mass of the aircraft and u,v,w are the sub-speeds of the UCAV along the x, y, and z axes of the aircraft. p,q,r are the angular velocities of the UCAV around the x, y, and z axes of the aircraft. The dynamic model of UCAV's rotation around the centroid is as follows[7]:

$$\begin{cases} \dot{p} = (c_1 r + c_2 p)q + c_3\bar{L} + c_4(\bar{N} + qheng) \\ \dot{q} = c_5 pq - c_6(p^2 - r^2) + c_7(\bar{M} - rheng) \\ \dot{r} = (c_8 p - c_2 r)q + c_4\bar{L} + c_9(\bar{N} + qheng) \end{cases} \tag{2}$$

In formula (2), $\Gamma = I_x I_z - I_{xz}^2$, $\Gamma c_1 = (I_y - I_z)I_z - I_{xz}^2$, $\Gamma c_2 = (I_x - I_y + I_z)I_{xz}$, $\Gamma c_3 = I_z$, $\Gamma c_4 = I_{xz}$, $c_5 = \frac{I_z - I_x}{I_y}$, $c_6 = \frac{I_{xz}}{I_y}$, $c_7 = \frac{1}{I_y}$, $\Gamma c_9 = I_x$, $\Gamma c_8 = I_x(I_x - I_y) + I_{xz}^2$, where I_x, I_y, I_z, I_{xz} are the rolling moment of inertia, the pitching moment of inertia, the yaw moment of inertia and the inertial product[8]. $\bar{L},\bar{M},\bar{N},heng$ are respectively UCAV's roll moment, pitch moment, yaw moment and engine angular momentum.

The UCAV's kinematics model is as follows:

$$\begin{bmatrix} \dot{x} \\ \dot{y} \\ \dot{z} \end{bmatrix} = \begin{bmatrix} \cos\psi\cos\theta & \cos\psi\sin\theta\sin\phi - \sin\psi\cos\phi & \cos\psi\sin\theta\cos\phi + \sin\psi\sin\phi \\ \sin\psi\cos\theta & \sin\psi\sin\theta\sin\phi + \cos\psi\cos\phi & \sin\psi\sin\theta\cos\phi - \cos\psi\sin\phi \\ -\sin\theta & \cos\theta\sin\phi & \cos\theta\cos\phi \end{bmatrix} \begin{bmatrix} u \\ v \\ w \end{bmatrix} \tag{3}$$

$$\begin{cases} \dot{\theta} = q\cos\phi - r\sin\phi \\ \dot{\phi} = p + \tan\theta(q\sin\phi + r\cos\phi) \\ \dot{\psi} = \frac{q\sin\phi + r\cos\phi}{\cos\theta} \end{cases} \tag{4}$$

In the above formula, $\dot{x},\dot{y},\dot{z}$ represent the velocity of the UCAV along each axis in the inertial coordinate system.

3. Stability analysis of UCAV's model

3.1. Linearization of UCAV's model

The above model is a nonlinear function of the UCAV's motion parameters. There are many parameters and the coupling is serious. Therefore, it is very difficult to find the analytical solution[9]. But it can be linearized with a small perturbation equation. And since the change of aerodynamic force and moment acting on UCAV is basically a linear function of motion parameters, the linear differential equations obtained by small perturbation equations can be used to analyze the stability of UCAV[10].

The UCAV model is trimmed at a height of $6000m$ and a flight speed of $158m/s$ to find the angle of attack required to balance the gravity of the model and the elevator yaw angle required to balance the pitch moment of the aircraft.

$$\begin{bmatrix} \dot{h} \\ \dot{\theta} \\ \dot{V}_T \\ \dot{\alpha} \\ \dot{q} \end{bmatrix} = A_{lo} \begin{bmatrix} h \\ \theta \\ V_T \\ \alpha \\ q \end{bmatrix} + B_{lo} \begin{bmatrix} \delta_t \\ \delta_e \end{bmatrix} \tag{5}$$

$$\begin{bmatrix} h \\ \theta \\ V_T \\ \alpha \\ q \end{bmatrix} = C_{lo} \begin{bmatrix} h \\ \theta \\ V_T \\ \alpha \\ q \end{bmatrix} + D_{lo} \begin{bmatrix} \delta_t \\ \delta_e \end{bmatrix} \tag{6}$$

$$\begin{bmatrix} \dot{\phi} \\ \dot{\psi} \\ \dot{\beta} \\ \dot{p} \\ \dot{r} \end{bmatrix} = A_{la} \begin{bmatrix} \phi \\ \psi \\ \beta \\ p \\ r \end{bmatrix} + B_{la} \begin{bmatrix} \delta_a \\ \delta_r \end{bmatrix} \tag{7}$$

$$\begin{bmatrix} \phi \\ \psi \\ \beta \\ p \\ r \end{bmatrix} = C_{la} \begin{bmatrix} \phi \\ \psi \\ \beta \\ p \\ r \end{bmatrix} + D_{la} \begin{bmatrix} \delta_a \\ \delta_r \end{bmatrix} \tag{8}$$

Equations (5) and (6) constitute the longitudinal state space expression of UCAV, and equations (7) and (8) constitute the lateral state space expression of UCAV.

3.2. Stability analysis of UCAV's model

The motion mode of the UCAV can be analyzed based on the state space expression of the UCAV and the inherent characteristics of UCAV's small perturbation motion can be analyzed by the distribution of eigen roots of UCAV's linear motion equation.

3.2.1. Longitudinal motion mode analysis of UCAV

In the equilibrium state, the distribution of longitudinal characteristic root of UCAV is as shown in the figure below. It can be seen that the longitudinal eigenvalue contains a pair of pure virtual roots, a pair of conjugate complex roots with negative real parts and a pair of unequal negative real roots, and it has no eigen roots with positive real parts. Therefore, the system is stable and controllable.

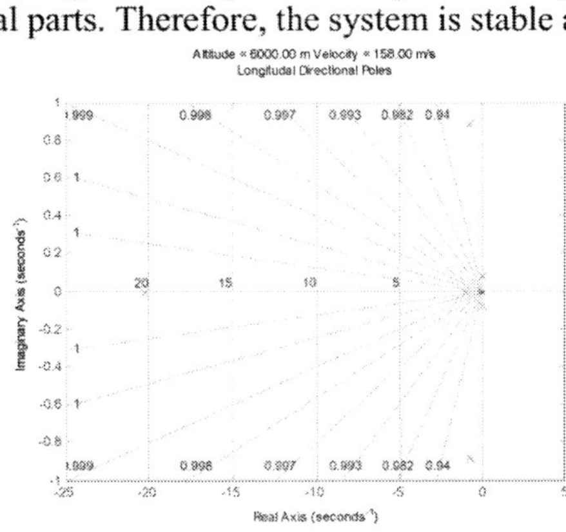

Figure 1. Distribution of longitudinal characteristic roots of UCAV.

3.2.2. Longitudinal frequency domain characteristics analysis of UCAV

When the elevator yaw angle is input, the amplitude-frequency characteristics and phase-frequency characteristics of the impulse response of the longitudinal state quantities such as the height, pitch angle, angle of attack, and pitch rate corresponding to the UCAV are as shown in the following figure.

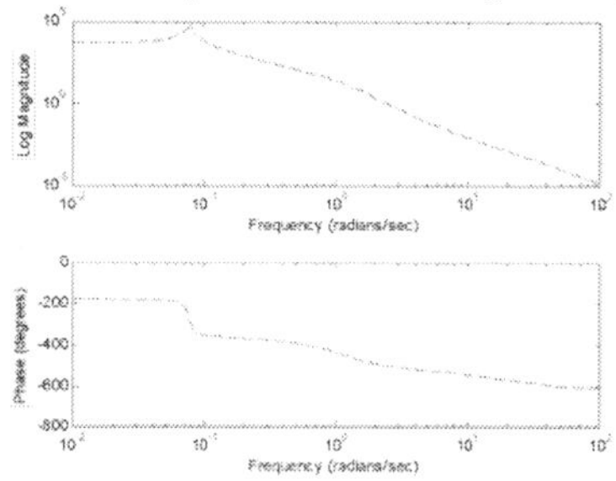

(a) Impulse response of altitude

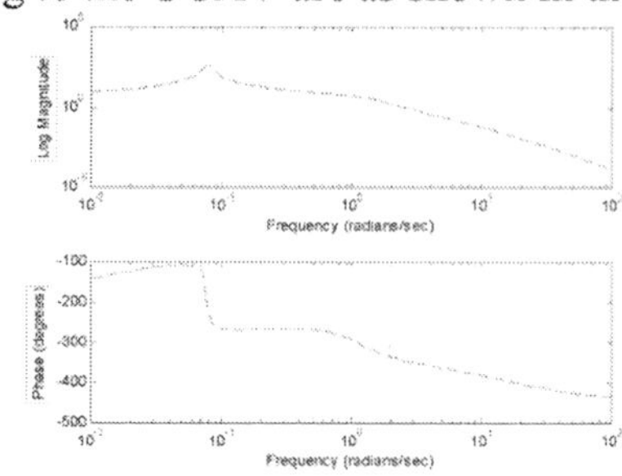

(b) Impulse response of pitch angle

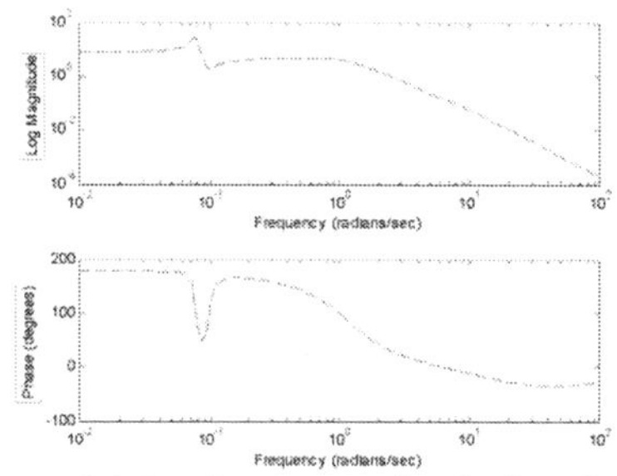

(c) Impulse response of angle of attack

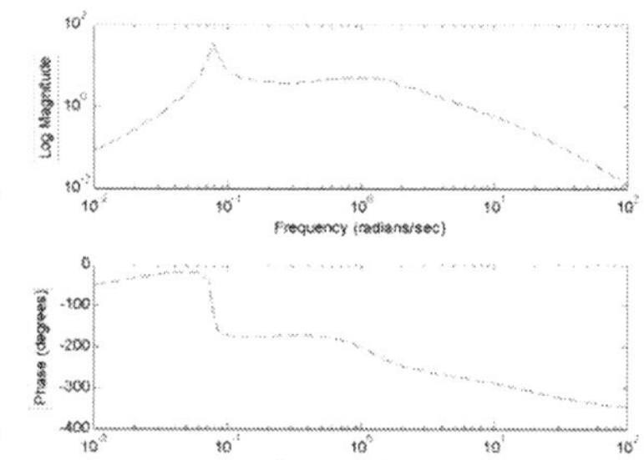

(d) Impulse response of pitch rate

Figure 2. Impulse response with elevator as input.

It can be seen from the above figure that when the elevator is a pulse input, the transient process of the impulse response of the longitudinal state quantities such as UCAV's height, pitch angle, angle of attack and pitch rate is similar to the steady state process. Therefore, a typical trimming point can be selected to design the UCAV's longitudinal control law.

3.2.3. Lateral motion mode analysis of UCAV

In the equilibrium state, the distribution of lateral characteristic root of UCAV is as shown in the figure below. It can be seen that the lateral characteristic root contains a pair of approximately pure virtual roots and two pairs of unequal negative solid roots, and it has no characteristic roots with positive real parts. Therefore, the system tends to be stable and controllable.

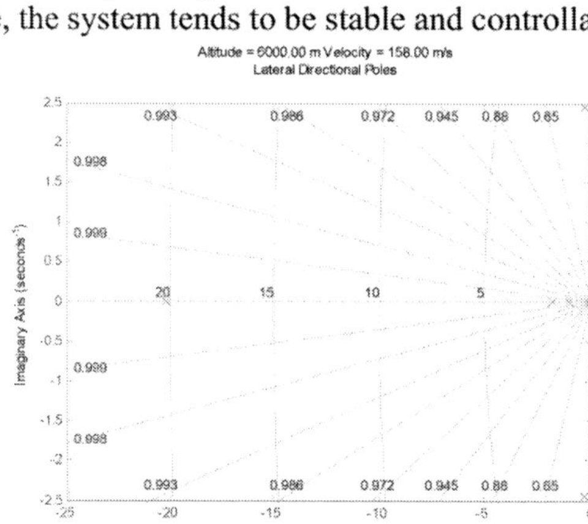

Figure 3. Distribution of lateral characteristic roots of UCAV.

3.2.4. Lateral frequency domain characteristics analysis of UCAV

The UCAV's lateral control inputs primarily have ailerons and rudders. When the aileron declination is used as the input, the amplitude-frequency characteristics and phase-frequency characteristics of the impulse response of the lateral state quantities such as the roll angle, yaw angle, side slip angle, roll angular rate, and yaw rate corresponding to the UCAV are as shown in the following figure:

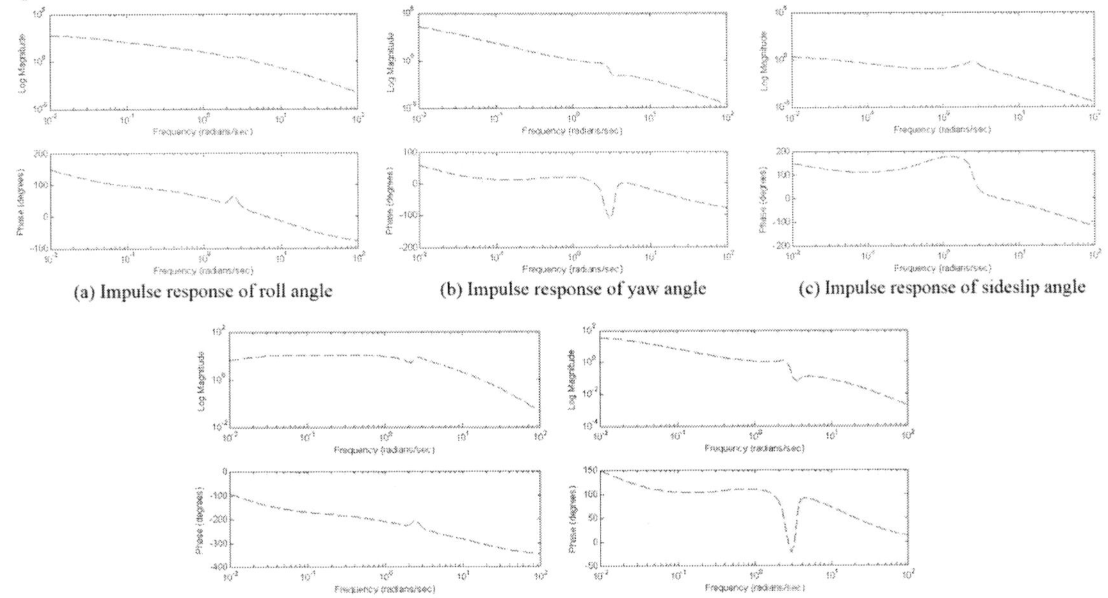

(a) Impulse response of roll angle (b) Impulse response of yaw angle (c) Impulse response of sideslip angle

(d) Impulse response of roll rate (e) Impulse response of yaw rate

Figure 4. Impulse response with aileron as input.

When the rudder declination is used as the input, the amplitude-frequency characteristics and phase-frequency characteristics of the impulse response of the lateral state quantities such as the roll angle, yaw angle, side slip angle, roll angular rate and yaw rate corresponding to the UCAV are as shown in the following figure:

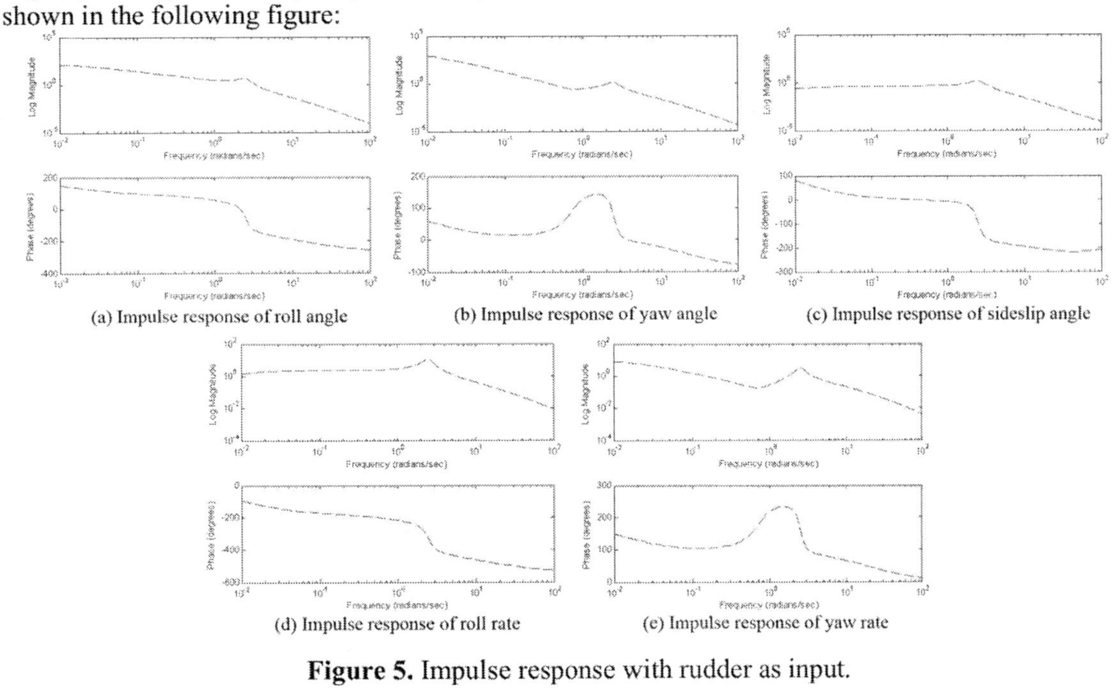

(a) Impulse response of roll angle (b) Impulse response of yaw angle (c) Impulse response of sideslip angle

(d) Impulse response of roll rate (e) Impulse response of yaw rate

Figure 5. Impulse response with rudder as input.

IOP Conf. Series: Materials Science and Engineering **715** (2020) 012099 doi:10.1088/1757-899X/715/1/012099

It can be seen from the above figure that when the aileron or rudder is a pulse input, the transient process of the impulse response of the lateral state quantities such as UCAV's roll angle, yaw angle, side slip angle, roll angular rate and yaw rate is similar to the steady state process. Therefore, a typical trimming point can be selected to design the UCAV's lateral control law.

4. Design of adaptive fuzzy PID attitude controller

The traditional PID controller cannot meet different expected output values because of the fixed parameters, and the parameter adjustment is also time consuming, so it can not meet the design requirements of the UCAV's attitude controller[11]. In order to truly reflect the pilot's operational characteristics, the pilot's operating experience can be learned from the design of the UCAV's attitude controller. Considering that the pilot's operating experience in controlling the attitude of the aircraft is ambiguous, the fuzzy control method can be used to construct the controller[12]. The specific attitude control process is shown in the figure below:

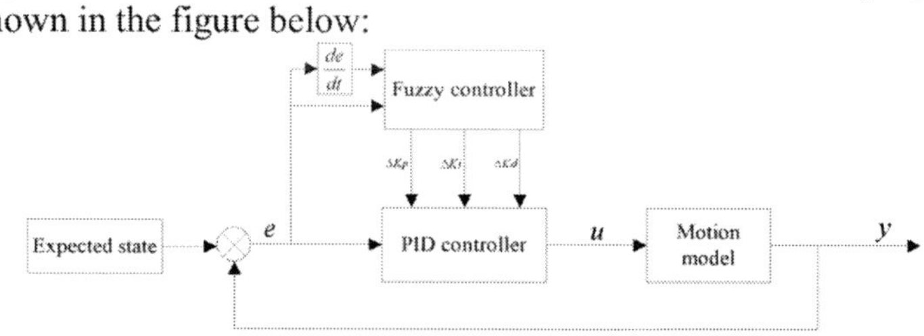

Figure 6. Flow diagram of attitude control of UCAV.

The input variable of the fuzzy PID attitude controller is the error of the expected state and the actual state e and the change rate of error $\frac{de}{dt}$. Inside the controller, the output of the fuzzy controller is the amount of change of Kp, Ki, Kd, which is used to adjust the three parameters of the PID controller. The output of the PID controller is the joystick offset $\Delta joystick$. The domain of discourse of input and output of the fuzzy controller is as follows:

$$\begin{cases} e \in [-\pi, \pi] \\ \dot{e} \in [-2\pi, 2\pi] \\ \Delta joystick \in [-0.1, 0.1] \end{cases} \tag{9}$$

The domain of discourse of input and output is divided into 7 linguistic variables, respectively NB, NM, NS, ZO, PS, PM, PB, and they correspond to 7 fuzzy subsets: negative big, negative medium, negative small, zero, positive small, positive medium and positive big. The membership function is a Gaussian membership function with the following expression:

$$U(x) = \exp(-(\frac{x-\overline{x}}{\sigma})^2) \tag{10}$$

The key to fuzzy control design is to summarize the experience of experts, construct appropriate fuzzy rules, and obtain fuzzy rule tables for adjusting Kp, Ki, Kd, as follows:

Table 1. Fuzzy rule list of ΔKp , ΔKi and ΔKd .

ΔKp ΔKi ΔKd		de/dt NB	NM	NS	ZO	PS	PM	PB
		PB	*PB*	*PM*	*PM*	*PS*	*ZO*	*ZO*
e	*NB*	*NB*	*NB*	*NM*	*NM*	*NS*	*ZO*	*ZO*
		PS	*NS*	*NB*	*NB*	*NB*	*NM*	*PS*
		PB	*PB*	*PM*	*PS*	*PS*	*ZO*	*NS*
	NM	*NB*	*NB*	*NM*	*NS*	*NS*	*ZO*	*ZO*
		PS	*NS*	*NB*	*NM*	*NM*	*NS*	*ZO*
		PM	*PM*	*PM*	*PS*	*ZO*	*NS*	*NS*
	NS	*NB*	*NM*	*NS*	*NS*	*ZO*	*PS*	*PS*
		ZO	*NS*	*NM*	*NM*	*NS*	*NS*	*ZO*
		PM	*PM*	*PS*	*ZO*	*NS*	*NM*	*NM*
	ZO	*NM*	*NM*	*NS*	*ZO*	*PS*	*PM*	*PM*
		ZO	*NS*	*NS*	*NS*	*NS*	*NS*	*ZO*
		PS	*PS*	*ZO*	*NS*	*NS*	*NM*	*NM*
	PS	*NM*	*NS*	*ZO*	*PS*	*PS*	*PM*	*PB*
		ZO	*ZO*	*ZO*	*ZO*	*ZO*	*ZO*	*ZO*
		PS	*ZO*	*NS*	*NM*	*NM*	*NM*	*NB*
	PM	*ZO*	*ZO*	*PS*	*PS*	*PM*	*PB*	*PB*
		PB	*NS*	*PS*	*PS*	*PS*	*PS*	*PB*
		ZO	*ZO*	*NM*	*NM*	*NM*	*NB*	*NB*
	PB	*ZO*	*ZO*	*PS*	*PM*	*PM*	*PB*	*PB*
		PB	*PM*	*PM*	*PM*	*PS*	*PS*	*PB*

5. Simulation results and analysis

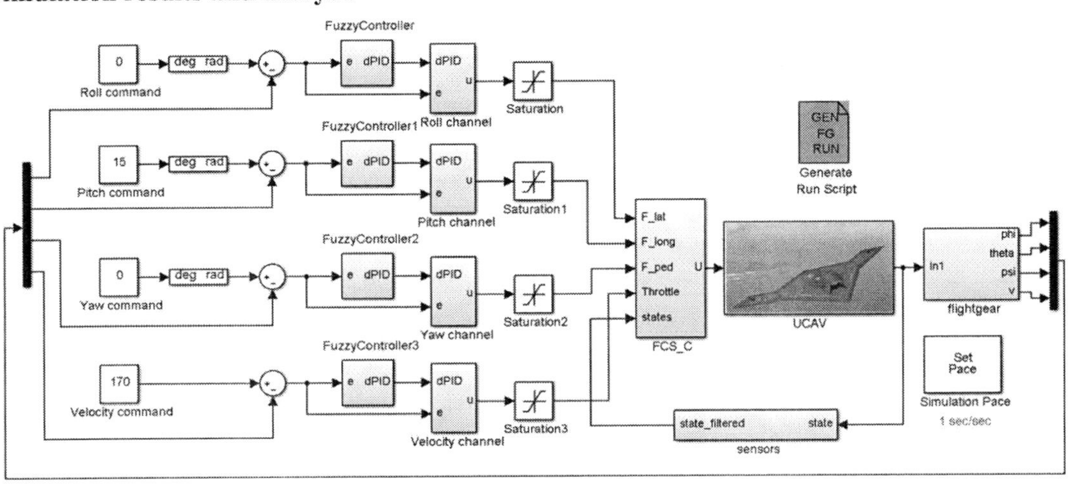

Figure 7. Nonlinear motion model of UCAV.

The constructed UCAV's nonlinear motion model is shown in the figure above. The trimming point is selected at a height of $6000m$ and a flight speed of $158m/s$. The pitch angle and velocity are selected to simulate the longitudinal control law and the roll angle is selected to simulate the lateral control law. The simulation results are compared with those of the traditional PID controller and they are shown in the figure below:

IOP Conf. Series: Materials Science and Engineering **715** (2020) 012099 doi:10.1088/1757-899X/715/1/012099

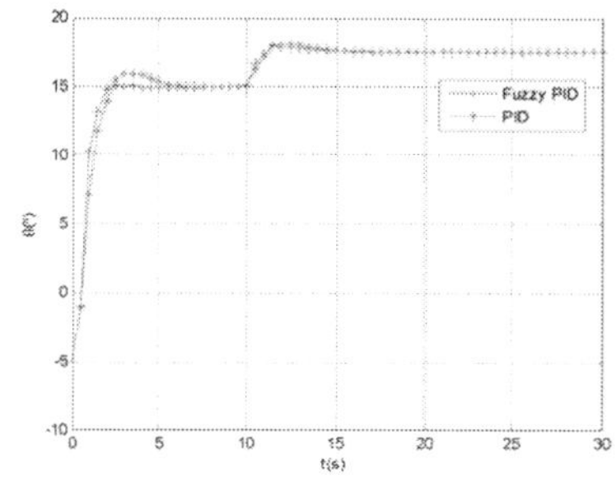

Figure 8. Response curve of pitch angle.

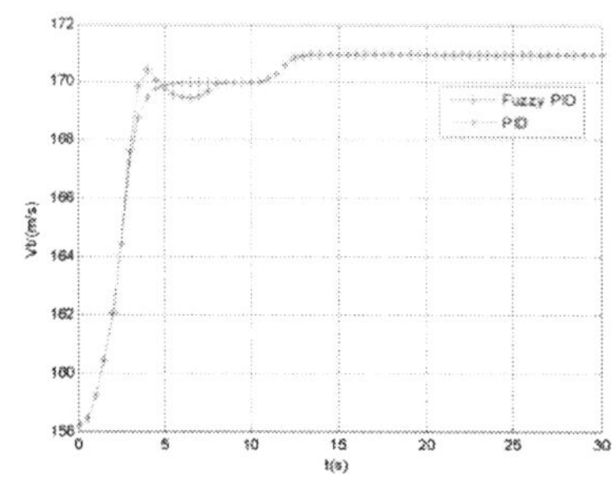

Figure 9. Response curve of velocity.

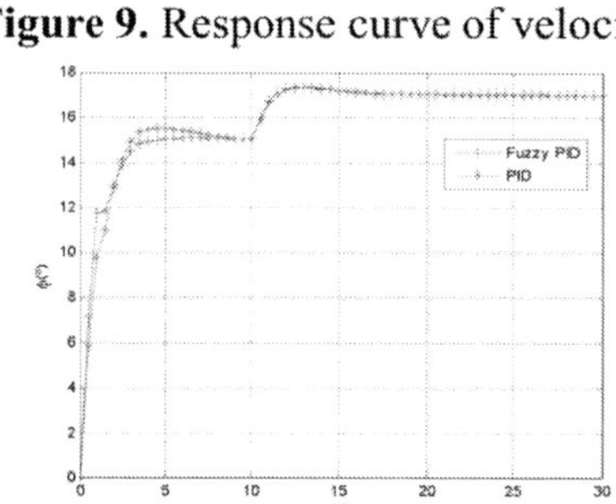

Figure 10. Response curve of roll angle.

It can be seen that compared with the traditional PID controller, the adaptive fuzzy PID controller designed in this paper has faster response speed, higher steady-state accuracy, and the overshoot is almost zero, that is, when approaching the desired state, the oscillation is smaller. In the 10th second of the simulation, step signal interference is added, and compared with the traditional PID controller, it can be seen that the adaptive fuzzy PID controller can track the step signal more quickly and accurately than the traditional PID controller. It also verifies that the adaptive fuzzy PID controller has better anti-interference ability.

6. Conclusion

In this paper, an adaptive fuzzy PID attitude controller is designed for UCAV's nonlinear motion model combined with expert experience. Using the small perturbation principle to linearize the UCAV's nonlinear motion model, the longitudinal and lateral state space expressions of the UCAV's model are obtained, and the decoupling control of the longitudinal and lateral channels is realized. The SIMULINK platform is used to build the UCAV's nonlinear motion model for simulation verification. The simulation results show that the designed adaptive fuzzy PID attitude controller can effectively control the UCAV to achieve the desired attitude and remain stable. The tracking performance can be maintained after the introduction of step signal interference, which verifies the effectiveness of the adaptive fuzzy PID attitude controller.

References

[1] DONG S T, YANG L H, KANG C J, et al. Fixed wing UAV attitude control and simulation[J]. Journal of Northeast Agricultural University, **46(9)**: 87-92, (2015)

[2] Garcia G A K S. Nonlinear model predictive controller for navigation,guidance and control for a fixed-wing UAV[C]. AIAA Guidance, Navigation, and Control Conference, 1-14, (2011)

[3] HE X Z, WANG R C, LUO Q Q. Portrait Control Law's Design and Simulation of a Fixed-wing UAV[J].Science Technology and Engineering, **10(9)**: 2134-2138,(2010)

[4] RINALDI F, GARGIOLI A, QUAGLIOTTI F. PID and LQ regulation of a multirotor attitude: Mathematical modelling, simulations and experimental results[J]. Journal of Intelligent & Robotic Systems, **73(1)**: 33-50, (2014)

[5] WANG J, WANG C L, LI Z, et al. Hovering control of four-rotor aircraft based on adaptive fuzzy proportion integration differentiation[J]. Journal of Guilin University of Electronic Technology, **36(5)**:406-411, (2016)

[6] SONNEVELDT L. Nonlinear F-16 Model Description [M]. The Netherlands: Control & Simulation Division, Faculty of Aerospace Engineering, Delft University of Technology, 5-6,(2010)

[7] RICHARD S R. Non-linear F-16 Simulation using Simulink and Matlab[M].The United States of America: University of Minnesota, 2, (2003)

[8] ANDRA S, VAITHEESWARAN S M, SHENDGE P D. Uncertainity Estimation Based Approach To Attitude Control of Fixed Wing UAV[C]. International Federation of Automatic Control, 278-283, (2016)

[9] WEI Q T, CHEN M, WU Q X. Backstepping-based attitude control for a quadrotor UAV with input saturation and attitude constraints[J]. Control Theory &Applications, **32(10)**:1361-1369, (2015)

[10] SUN X Y, FANG Y C, SUN N. Backstepping-based adaptive attitude and height control of a small-scale unmanned helicopter[J]. Control Theory & Applications, **29 (3)**: 381-388,(2012)

[11] SU J Y, FAN P H, CAI K Y. Attitude control of quadrotor aircraft via nonlinear PID[J].Journal of Beijing University of Aeronautics and Astronautics, **37(9)**:1054-1058, (2011)

[12] XIAO L B, LUO M K, YUAN Y. Fuzzy PID control based on behaviors[J].Journal of Sichuan University, **48(1)**: 27-32, (2011)

Action sports science with sensor technology and statistical analysis methods

Kazunari Ishida*

Hiroshima Institute of Technology, 2-1-1 Miyake, Saeki-ku, Hiroshima, 731-5143, Japan

*Email: drkishida@hotmail.com

Abstract. This paper describe how to apply sensor technology and statistical analysis methods to formalize action sports science. Action sports are composed of quick motions to achieve difficult tricks. Inertial sensors are suitable equipment to record quick motions. Statistical analysis methods are required to identify key factors to make successful tricks. To show how to apply these technology and methodologies for analysing action sports, three types of sports and three types of terrains are selected. These examples show us that the technology and methodologies can detect different sports and skill levels, while they can also identify key factors for successful tricks.

1. Introduction

This paper describe how to apply sensor technology and statistical methods to formalize action sports science. Sensor technology is widely accepted because of wide spread of smartphone use in our society. Basic activities, e.g. jogging and walking, are analysed as a task of human activity recognition [1]. Inertial sensors are also effective tools for sports performance evaluation [2] [3]. According to the review, there are many researches on team sports, individual sports, cyclic sports, winter sports, and outdoor sports. The last category contains skateboarding, which is analysed in only one paper, although it is one of the popular action sports and it will be one of the official competition in 2020 Tokyo Olympic games. This paper focuses on three types of action sports, i.e., skateboard, inline skate, and BMX. To show how to apply these technology and methodology for analysing action sports, three types of sports and three types of terrains are employed. These examples show us that the technology and methodologies can detect different skill levels and can identify key factors for successful tricks.

2. Development of wearable sensor system

Original wearable sensor system has been developed to record motion with precise time stamp to synchronize multiple time series data. Motion in action sport is very quick and high impact. Due to these very specific conditions, we developed our own recording device.

2.1. Hardware

The developed wearable sensor is composed of a microcomputer (Arduino), accelerometer and gyroscope (MPU6050), digital compass (HMC5883L), and GPS (MTK3339). The motion and location data is stored on a micro SD-card. The device's sampling rate is approximately 50 Hz. The data contains a precise time stamp and position from the GPS on each record at each sampling time. The record also contains sensing data from the three-axis accelerometer, three-axis gyroscope, and three-axis digital compass [4].

2.2. Software
The sensor data is processed by pre-processing software. The software converts date and time into timestamp in the data. It also synchronizes and normalizes the timestamp on the data collected from multiple sensors on body and equipment. After normalization, it detects the time series of each action automatically [4].

3. Statistical analysis
Motion data with inertial sensors is multi-dimensional time series data. Auto-correlation and cross-correlation are suitable indexes to evaluate periodicity of body part's motion and linkage between two different body parts' motions. Clustering method is effective method to separate different sports and skill levels, respectively. Regression and discriminant analyses are suitable methods to identify key factors for successful tricks or actions. Statistical hypothesis testing is also effective method to evaluate statistical significance concerning identified key factors.

3.1. Correlation
Auto-correlation is suitable indexes to evaluate periodicity of body part's motion [4]. Formulas (1), (2), and (3) are definitions of sample average $\mu_{i,j}$, sample covariance function $C_{i,j,k}$, and sample correlation function $R_{i,j,k}$. based on a time sequence $\{y_{i,j,1}, y_{i,j,2}, \ldots, y_{i,j,n}\}$, where i is sequential run number, j is sensor number, and t is time period between 1 and n. Autocorrelations on all sensors' axes are calculated to define the distance function between each pair of time series data based on the squared difference between them in formula (4), where S is a set of sensors and K is the maximum time lag for calculating distance.

$$\mu_{i,j} = \frac{1}{n}\sum_{t=1}^{n} y_{i,j,t} \quad \ldots. \tag{1}$$

$$C_{i,j,k} = \frac{1}{n}\sum_{t=k+1}^{n}\left(y_{i,j,t} - \mu_{i,j}\right)\left(y_{i,j,t-k} - \mu_{i,j}\right) \tag{2}$$

$$R_{i,j,k} = \frac{C_{i,j,k}}{C_{i,j,0}} \quad \ldots. \tag{3}$$

$$dist(a,b,S,K) = \sum_{s \in S}\sum_{k=0}^{K}\left(R_{a,s,k} - R_{b,s,k}\right)^2 \tag{4}$$

Cross-correlation is suitable indexes to evaluate periodicity of body part's motion and linkage between two different body parts' motions. Especially, local cross-correlation is effective similarity index to analyze quick motion [5]. To define local cross-correlation, local average is defined by formula (5) with respect to a point of time series trajectory represented as $y_{c,i,j,p+t}$, where the parameters are run (c), body part of sensor (i), sensor axis (j), window size (w), start time of window (p), and time in window (t). The range of p is [0, l-w] where l is length of time series. A local variance is defineded by formula (6). Local cross-variance between run c1 and c2 is defined by formula (7). Local cross-correlation coefficient is derived by formula (8), (9), and (10).

$$Y(p,w,c,i,j) = \frac{1}{w}\sum_{t=1}^{w} y_{c,i,j,p+t} \quad \ldots. \tag{5}$$

$$V(p,w,c,i,j) = \frac{1}{w-1}\sum_{t=1}^{w}\left(y_{c,i,j,p+t} - Y(p,w,c,i,j)\right)^2 \quad \ldots. \tag{6}$$

$$D(p,w,k,c1,c2,i,j) = \frac{1}{w-1}\sum_{t=1}^{w}\left(y_{c1,i,j,p+t} - Y(p,w,c1,i,j)\right)\left(y_{c2,i,j,p-k+t} - Y(p-k,w,c2,i,j)\right) \tag{7}$$

$$C_{c1,c2,i,j,k,w,p} = \frac{D(p,w,k,c1,c2,i,j)}{s_{c1,i,j,w,,p} \cdot s_{c2,i,j,w,p-k}} \quad ..... \tag{8}$$

$$s_{c1,i,j,w,p} = \sqrt{V(p,w,c1,i,j)} \quad ..... \tag{9}$$

$$s_{c2,i,j,w,p-k} = \sqrt{V(p-k,w,c2,i,j)} \quad ... \tag{10}$$

3.2. Clustering

To detect and separate different types of sports and skill levels, clustering is an effective method with certain similarity measuerments as described in 3.1. To evaluate these distance measurements in 3.1, hierachical clustering with dendrogram is suitable method for visual confermation of measurment appropriateness. For example, good measurment can generate well separated sub-clusters, howeve, poor measurment is going to result bad sub-clusters which contain different types of sports and skill levels [4][5][6][7][8].

3.3. Multivariable analysis

To identify successful factors concerning sports skill and trick, multivariable analisys is an effective method, becasue multiple inartial sensors generate multiple variable time seriase data. We can apply regression analysis to indentify relation between running speed and effective motion. We can also emply discriminant analysis to find effective motions in terms of each axis of an ineartial sensor [9].

3.4. Statistical hypothesis testing

To identify difference between success and failure in terms of timing of mortion, statistical hypothesis testing is an effective method, because the difference is evaluated in terms of statisitcal significance. We can employ t-test or Mann–Whitney U test to decide statisitical significance of difference of motions between success and failure. A t-test is most commonly applied when the test statistic would follow a normal distribution. On the other hand, U-test does not require the assumption of normal distributions [10].

4. Type of sports and terrains

Skateboard, BMX, and inline skate are selected target action sports, because they were official competitions in X-games and the former two sports will be official competitions in 2020 Tokyo Olympic Games. Halfpipe, big air ramp, and flat ground are selected target terrains.

4.1. Periodic motion

Turn maneuver on halfpipe is periodical motion (figure 1). Auto correlation described in 3.1 is effective to calculate similarity between pair of runs on halfpipe. Hierarchical clustering based on the auto-correlation illustrates properly separated sub clusters in terms of type of sports, i.e. skateboard, inline skate, BMX, and level of skills, beginner, intermediate, and advanced (figure 2) [4].

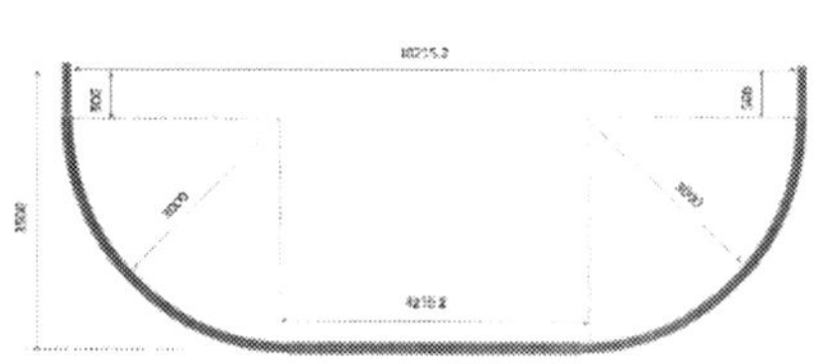

Figure 1. Halfpipe.

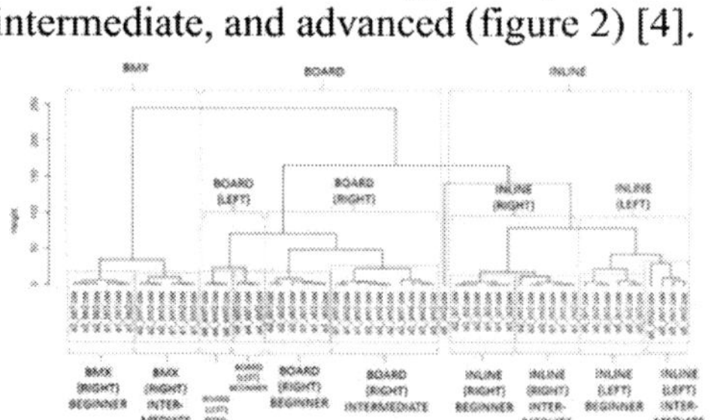

Figure 2. Cluster with Auto-Correlation.

4.2. Aperiodic motion

Set of jumps on big air ramp is aperiodic motion (figure 3). Local cross correlation described in 3.1 is effective to calculate similarity between pair of different runs on big air ramp, because it focuses on important motions and ignores the other unimportant motions. Hierarchical clustering based on the local cross-correlation illustrates properly separated sub clusters in terms of type of sports and level of skills (figure 4) [5] [6] [7].

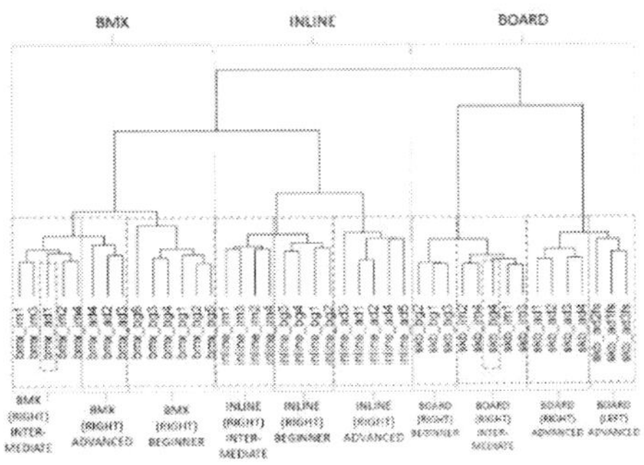

Figure 3. Big air ramp.

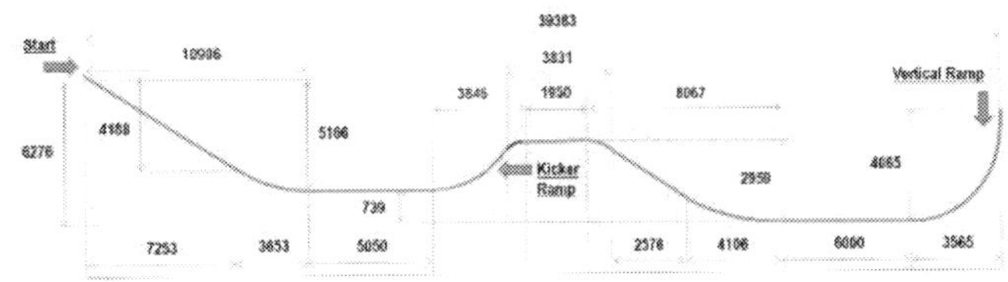

Figure 4. Cluster with local cross- correlation.

4.3. Repetitive balance

Flatland BMX is repetitive balancing sport on flat ground (figure 5).

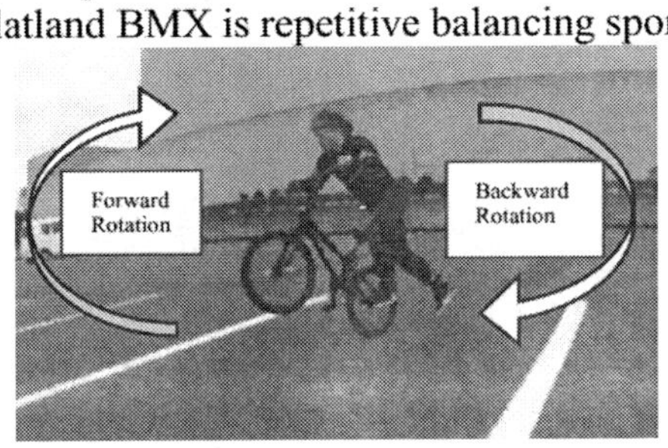

Figure 5. Example trick - peg wheelie turbine.

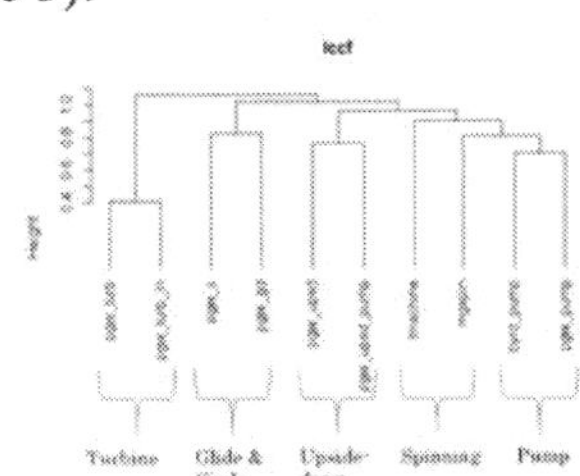

Figure 6. Cluster with local cross-correlation.

The local cross correlation and hierarchical clustering are also effective to classify ten tricks into similar subgroups, i.e. Turbine, Glide, Upside-down, Spinning, and Pump (figure 6) [8].

4.4. Repetitive drive

Moving forward with skateboard on flat ground is basic trick, i.e. Tic-Tac, while it is difficult for beginner. Tic-Tac is repetitive drive motion trick. Reglession and discriminant analyses reveals that huge periodic swing motion of skateboard and counter motion of upper body are key factors to get fast speed (figure 7) [9].

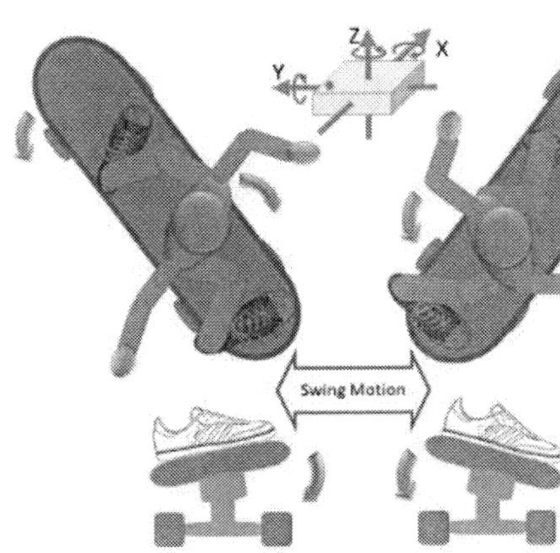

Figure 7. Swing and lean motions for Tic-Tac.

4.5. Trick sequence

Flipping skateboard is an advanced trick, which is composed of five steps (e) beginning, (d) preparation, (c) execution, (b) catch, and (a) landing (figure 8) [10].

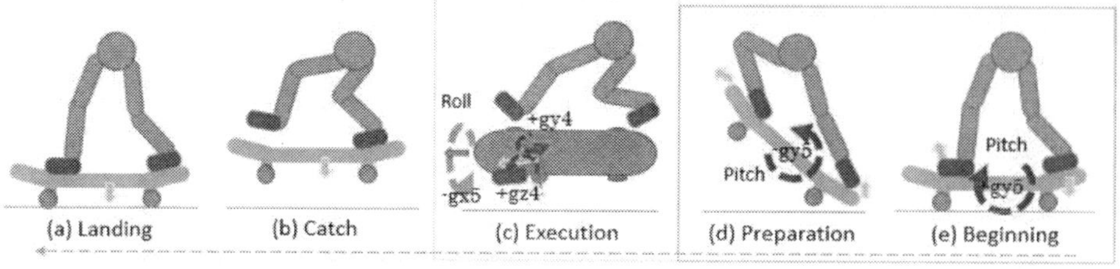

Figure 8. Kickflip sequence.

The sequence is completed in approxmetely 200 milli-seconds. Hence, timing is the most inportant factor to make successful trick. Especially, the duration between (e) beginning and (d) preparation is the most important period to do successful execution or kicking skateboard in step (c). Hence, lifting front side of skateboard and preparing kick foot are statistically analyzed to find difference between success and failura cases with Mann–Whitney U test. According to the statistical hypothesis testing, longer lifting time of front side of skateboard leads to successful kickflip, because steep angled skateboard is easy to kick and flip it. Moreover, early preparation or counter rotation of kick foot is desireble to do successful kickflip [10].

5. Conclusion

This paper summarized examples of sensor technology and statistical analysis methods to formalize action sports science. Original wearable sensor system has been developed to to analyse quick motions of action sports. To separate different sports and skill levels and to identify key factors to make successful action sports tricks, sensor data is analysed with statistical methods, i.e. auto-correlation, cross-correlation, hierarchical clustering, regression, discriminant analyses, and statistical hypothesis testing. This paper explained five examples, periodic motion on halfpipe, aperiodic motion on big air ramp, repetitive balance with BMX, repetitive drive with skateboard, and trick sequence with skateboard. These examples show us that the technology and methodologies can detect different type of sports and skill levels and can identify key factors for successful tricks. These findings are first step toward formalizing action sports science. In future work, wearable devices will be proposed to support action sports training based on the findings with IoT and edge computing technologies.

References

[1] Y. Chen, C. Shen, "Performance Analysis of Smartphone-Sensor Behavior for Human Activity Recognition," IEEE Access, vol. 5, pp. 3095 – 3110 (2017)

[2] V. Camomilla, E. Bergamini, S. Fantozzi, and G. Vannozzi ,"Trends Supporting the In-Field Use of Wearable Inertial Sensors for Sport Performance Evaluation: A Systematic Review," Sensors, 18, 873; doi:10.3390/s18030873 (2018)

[3] J. W. Harding, C. G. Mackintosh, A. G. Hahn, D. A. James, "Classification of Aerial Acrobatics," Proceedings of the 7th Snowboarding Using Inertial Sensors ISEA, Biarritz, France, June 2-6 (2008)

[4] K. Ishida, "Classification of Turn Maneuvers on Halfpipe with Multiple Sensors," International Journal of Electronics and Electrical Engineering, vol. 4, no. 5, pp. 420-425 (2016)

[5] K. Ishida, "Action Sports Analysis Based on Local Cross Correlation and Action Measurement Units with GPS timestamp," Proceedings of the 3rd International Workshop on Skill Science, JSAI International Symposia on AI, Kanagawa, November 16 (2016)

[6] K. Ishida, "Analyzing Action Sports Based on Jumping and Aerial Movements with Inertial Sensors and GPS Timestamp," The 8th Asia-Pacific Congress on Sports Technology (APCST 2017), October 16-19, Tel-Aviv, Israel (2017)

[7] K. Ishida, " Application of Computer and Sensor Technology for Analyzing Action Sports," Asian Conference on Engineering and Natural Sciences (2018 ACENS), Osaka, Japan, February 6-8 (2018)

[8] K. Ishida, " Action Sports Learning Based on Expert Instruction Revised by Wearable Sensor Data," 14th International Conference on Literature, Education, Humanities and Social Sciences (LEHSS-18), Cebu, Philippines, June 14-15 (2018)

[9] K. Ishida, "Sport Skill Evaluation with Wearable Sensors and Statistical Analysis," Proceedings of the 3rd International Conference on Sport Science and Physical Education, Xi`an, China, March 23-25 (2019)

[10] K. Ishida, "Statistical Analysis of Critical Factors in Action Sports Performance with Wearable Sensors," Proceedings of Asia-Singapore Conference on Sport Science (ACSS 2018), Singapore, July 17-18 (2018)

IOP Conf. Series: Materials Science and Engineering **715** (2020) 012101 doi:10.1088/1757-899X/715/1/012101

Microstructure and mechanical property of aluminium based die-cast alloy (ALDC6) for spider arm component

Min Heo, J.S.S.Babu, Seongmo Bae, Jun Park and Chung Gil Kan*

School of Mechanical Engineering, (ERC/NSDM), Pusan National University, Busan 46241, S.Korea

*Email: cgkang@pusan.ac.kr

Abstract. In this paper aims on the development of aluminum alloy (ALDC6) using high pressure die casting (HPDC) for spider arm component in washer. Presently, ALDC3 alloy is commonly used for the fabrication of spider arm but it has poor corrosion resistance. ALDC6 alloy which has good corrosion resistance can be used as an alternative material for spider arm. However due to high Si (Silicon) content of 10wt% in ALDC3 alloy it has good castability compared to ALDC6 alloy with 1wt% of Si. Hence there is a need to modify the ALDC6 alloys by increasing the percentage of silicon and magnesium elements so as to improve the castability, strength and corrosion resistance. The optical microstructure of the developed ALDC6 alloy was investigated and its phases were detected by XRD analysis. It was found out that presence of Mg_2Si in ALDC6 alloys is higher than that of the ALDC3 alloy. The microstructure and mechanical properties of ALDC6 alloys were evaluated and compared with existed ALDC3 alloys. The tensile strength and yield strength of the developed ALDC6 alloys has improved but elongation decreased. This implies that the spheroidal primary alpha grains and decrease in grain size due to the presence of Si and Mg within the alloy enhance the strength and yield stress.

1. Introduction

Al-Mg alloys (ALDC-6) are widely used in die casting in different industries because of the high strength to weight ratio, good ductility and good corrosion resistance [1]. The heavier shape component such as spider arm in washer which is coupled to the outer surface of the drum is typically mounted in a tub configured to make contact with water and hence likely to be corrosive. As the washers are fabricated by mass production, the spiders should have good corrosion resistance and castability. Presently, spider arm components are manufactured using ALDC3 aluminum alloy due to its good castability and strength properties. However, in the environments with water having relatively high sodium results in higher corrosion of spider arm made of ALDC3 alloy. Hence ALDC6 aluminium alloy is considered for the fabrication of spider arm component in washer. The ALDC6 alloy has good corrosion resistance due to the protective coating of oxide film on the alloy surface [2]. However, the ALDC6 alloy has low castability due to which such aluminium alloys has limited applications involving simple shaped components with low corrosion resistance. Hence there is a need for weight reduction, improved mechanical properties, wear and corrosion resistance of these aluminium alloys. The high pressure die casting (HPDC) is highly efficient in near net shape manufacturing whereby molten metal is injected into a metallic die cavity under high speed and high intensification. This paper describes the fabrication of ALDC6 alloy by using die casting and study the microstructure and mechanical properties of ALDC6 alloys.

2. Results and discussion

Figure 1(a) shows the curves of solid fraction with temperature of ALDC6 alloys measured by DSC test. From the profiles it can be clearly seen that near 0.9 solid fractions the slope of curve was narrow due to the high percentage of Si, 10wt%. The figure also indicates the low temperatures with sensitivity ranging from 200°C to 590°C and higher temperature from 590°C to 600°C. The working temperature was fixed at 550-590°C with solid fraction ranging from 0.5-0.1 to avoid changes of void fraction due to the variation in solidification temperatures. In figure 1(b), the results show that the density changes with increase in temperature of ALDC6 alloys. Upon melting, the increase in melt density decreases when the temperature reaches the solidus temperature, indicating the formation of precipitates which cause for the decrease in the density.

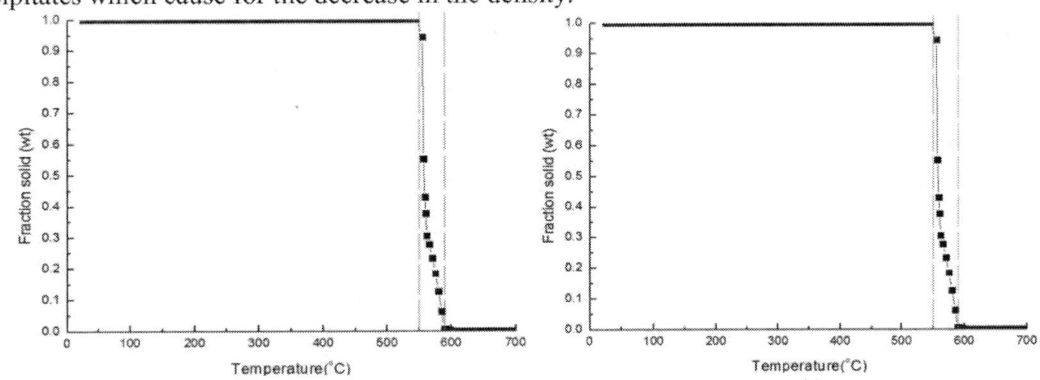

Figure 1. (a)Solid fraction vs temperature curves of the ALDC6 obtained by DSC analysis, (b) Density vs temperature of ALDC6 alloy.

3. Optical Microstructural and EDX analysis

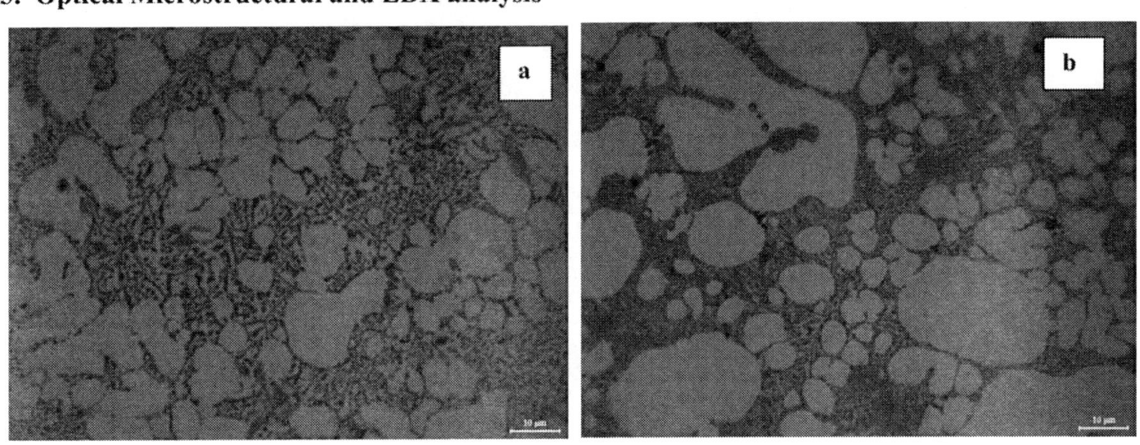

Figure 2. Optical microstructure of die casted (a)ALDC3, (b) ALDC6.

The figure 2(a) shows the optical microstructures of ALDC3 alloys, large α-Al grains and needle shaped Si eutectic structures were observed which results in the deterioration of the mechanical properties [3]. The optical microstructure of ALDC6 alloy with compounds such as $Al_6(MnFe)$ and Mg_2Si appears in the grain boundaries (Figure 2(b)). From the EDX analysis by SEM ofALDC6 alloy it indicates that a small fraction of Fe-rich intermetallic compound of β-$Al_{13}Fe_4$ phase and α-$Al_8Si_6Mg_3Fe$ with Chinese script were observed (Figure 3). The presence of high Fe content and low Si/Fe ratio cause the formation of β-$Al_{13}Fe_4$ compound in the aluminum alloy. These formations of intermetallic phases were the result of binary eutectic reactions. Moreover, the formations of different types of morphological changes from plate to Chinese script were reported to enhance the mechanical properties [4]. The presence of higher content of silicon can prevent the formation of hot tearing defects and improve the castability of aluminum alloys [5].

4. Mechanical properties of ALDC6 and ALDC3 alloys

The mechanical properties of tensile strength, yield strength and elongation will be evaluated with five tensile samples. Figure 4 shows the tensile strength and yield strength of ALDC3 andALDC6 alloys. The tensile strength of ALDC6 alloy is higher than that of the ALDC3 alloy with an elongation of 1-2%. However, the elongation decreased significantly from 5% to 2% for ALDC6 alloy. The results reveal that the addition of Si and Mg enhances the mechanical properties of the developed ALDC6 alloy. Therefore, the increase in tensile strength, yield strength and elongation of the sample is attributed to the presence of phases such as Mg_2Si and $Al_8Si_6Mg_3Fe$ in the alloys. In the case of ALDC6 base alloy, the Si content is 1wt% and the developed ALDC6 alloy has 10wt%. The presence of Si with 10wt% could significantly improve the castability and reduce hot tearing defects in the ALDC6 alloys. This implies that the addition of Si over 1wt% is required to develop die castings with reduced hot tearing defects.

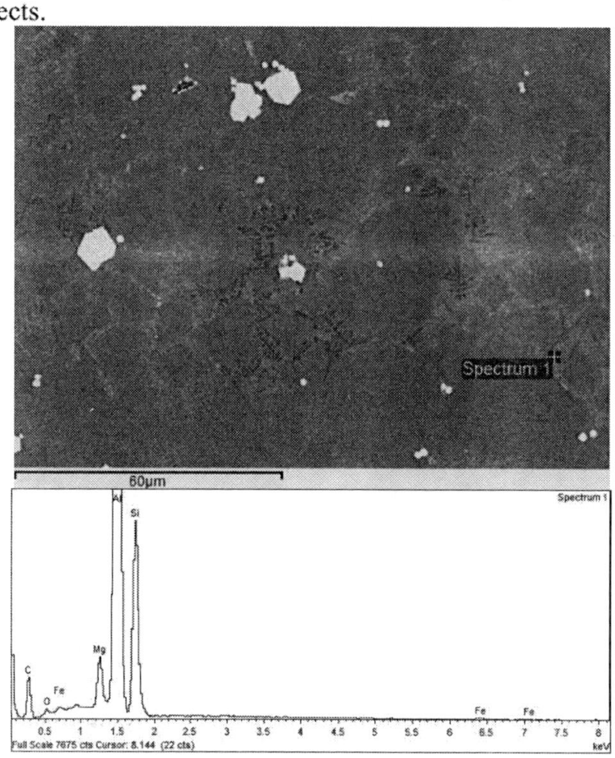

Figure 3. EDX analysis of ALDC6 alloy with presence of Mg_2Si.

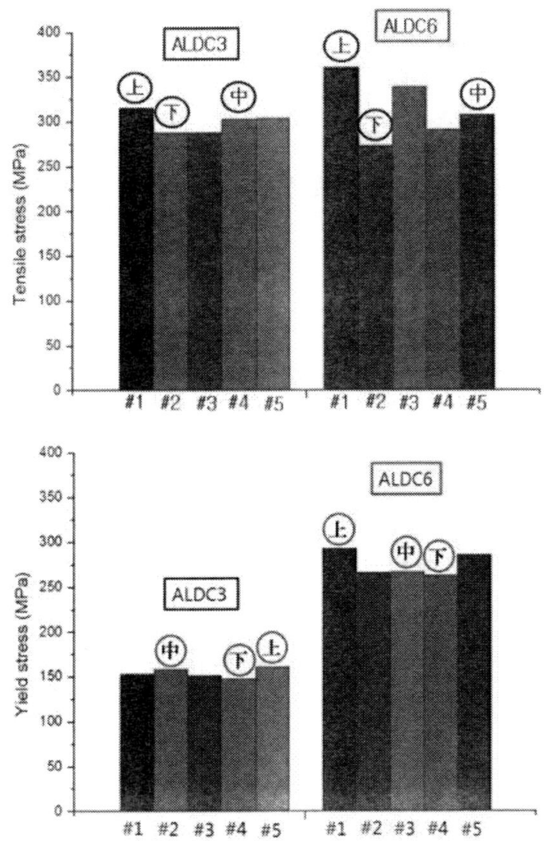

Figure 4. (a) Tensile strength of ALDC3 alloy and ALDC6 alloy, (b) Yield strength of ALDC3 alloy and ALDC6 alloy; 1, 2, 3, 4, 5-number of samples.

5. Conclusions

The following conclusions were drawn from this study:

1. The ALDC6 alloys for spider arm component have been developed by using HPDC. The addition of elements such as Si and Mg increase the fraction of intermetallic phases Mg_2Si and $Al_8Si_6Mg_3Fe$, thereby enhancing the mechanical properties.
2. The mechanical properties of the developed ALDC6 alloys are higher than that of ALDC3 alloy. The tensile strength and the yield strength of ALDC6 alloy are 361MPa and 280MPa respectively, with 2% elongation in die cast condition.
3. The results of the study confirm that the developed ALDC6 alloy is suitable for fabrication of spider arm component in washer parts.

Acknowledgments

This work was supported by the National Research Foundation of Korea (NRF) grant funded by the Korean government (MSIT) through GCRC-SOP (No.2011-0030013) and by the National Research Foundation of Korea (NRF) grant funded by the Korean government (MSIT) (No.2017-R1A2B4007884).

References

[1] Chung Gil Kang et al., (2011). Fabrication process analysis and experimental verification for aluminium bipolar plate in fuel cells by vacuum die-casting, J. Power Sources, vol.196, 8241-49.

[2] M.Okayasu et al., (2013). Corrosion and mechanical properties of cast aluminium alloys, Inter. J of cast Metals Res., vol.26, 2013, 319-29.

[3] Zhenghua Li and Hong Yan, (2015). Modification of primary α-Al, eutectic silicon and β-Al 5 FeSi phases in as cast AlSi 10 Cu_3 alloy with (La+Yb) addition, J of rare earths, vol.33, 2015, 995-1003.

[4] Misuhiro Okayasu et al.,(2015). Mechanical properties of various cast aluminum alloys made using a heated mold continuous casting technique with and without ultrasonic vibration, Metals, vol.5, 1440-53.

[5] S.Li and D.Apelian, Hot Tearing of Aluminum Alloys: A Critical Literature Review, Inter. J. of Metal casting. vol.5, 2011, 23-40.

IOP Conf. Series: Materials Science and Engineering **715** (2020) 012102 doi:10.1088/1757-899X/715/1/012102

Effect of interface bonding process on bending strength in CFRP/metal hybrid composites

Sung-Jin Kim[1], Jun Park[1], J. S. S. Babu[2], Hyeong Yun Seo[3] and Chung-Gil Kang[4,*]

[1]School of Mechanical and Precision Engineering, Pusan National University, San 30 Chang, Jun-dong, Geum Jung-Gu, Busan, 609-735, South Korea
[2]Convergence Research Center of 3D Laser Innovative Manufacturing Technology, Pusan National University, Busan-46241, Korea
[3]Department of Computer Software Engineering, ChangShin University, Masanhoiwon-gu 51352, Changwon-si, Gyeongsangnam-do, South Korea
[4]School of Mechanical Engineering, Engineering Research Center for Net Shape and Die Manufacturing(ERC/NSDM) Pusan National University, San 30 Chang Jun-dong, Geum Jung-Gu,Busan 609-735, South Korea

*E-mail: cgkang@pusan.ac.kr

Abstract. Recently, vehicle parts such as center pillars and impact beams used in automobiles have been requiring materials that satisfy both high strength and impact absorption. Therefore, in this study, the applicability of the forming process of CR340 (cold rolled steel sheet)/CFRP Hybrid Composites which are a bonded material of cold rolled steel sheet having excellent formability and carbon fiber prepreg having high strength and high collision absorption in the vehicle parts was investigated. Hybrid composites are fabricated by stacking carbon fiber reinforced plastic (CFRP) prepreg on a cold rolled steel sheet (CR340) with Zn coating. Bonding materials were used at the interface between the CFRP prepreg and steel sheet. In addition, in the case of CR340/CFRP Hybrid composites which are made by bonding CFRP prepreg to the surface of press-formed CR340 material, it is impossible to manufacture complicated shapes due to an interfacial bonding separation phenomenon between the reinforcement and steel sheet. An adhesive agent developed by this research team was applied on the cold rolled sheet between CFRP and CR340 materials to enable the blank production of CR340/CFRP Hybrid composites. The heating time and temperature required for bending test were set similar to the curing curve of the CFRP prepreg. The surface roughness of steel materials is important in terms of their tensile strength and bending strength In this study, the steel sheet used was cold rolled 340 with Zn coating without surface roughness treatment for application to mass production in industry. This study investigates the properties of a CFRP/metal hybrid composite material. The hybrid was fabricated after spreading a developed binder on high-strength steel sheets (CR980) prepared by cold rolling and pressing. During the preparation of the CR980/CFRP hybrid composite, and its mechanical characteristics were investigated through bending, tensile, and lap shear adhesion test experiments.

1. Introduction

There is now a demand for high-fuel-efficiency and lightweight vehicle in the automotive industry. As gas mileage has received increasing attention because of environmental regulations and high oil prices, weight reduction has become a necessity rather than an option for automobile manufacturers [1]. Recently, the weight reduction of automobiles has become a hot topic as environmental regulations and gas mileage regulations have been reinforced. Thus, there are many studies on weight reduction using aluminium alloy, magnesium alloy, titanium alloy, and high strength steel (HSS) in place of steel materials [2]. In addition, advanced high strength steel (AHSS) and composite materials are used to reduce weight. One of these composite materials is carbon fibre reinforced plastic (CFRP), which has a higher strength-to-weight ratio and rigidity than steel materials while being lighter than aluminium. Therefore, CFRP must be studied to reduce the weight of automobiles.

Currently, there are many studies being conducted on fibre reinforced plastic (FRP) [3] and its machining [4]. There are also many ongoing studies on the use of pre-impregnated carbon fibre reinforced plastic (CFRP prepreg) [5]. Prepreg is intermediate material of FRP, wherein resin, which is a matrix, is pre-impregnated into fibres.

Yanagimoto et al. researched formability using a dummy sheet for embossing (shallow drawing), hat bending, and V-bending experiments. In their experiments, unidirectional CFRP prepreg, which is one of the CFRP prepregs, was hardened using various laminating methods. Uriya et al. conducted a V-bending experiment and studied the subsequent elastic recovery [6].

In addition to the above, studies using carbon fibre reinforced thermoplastic (CFRTP) are also in progress. Isogawa et al. used a hemispherical punch in their research [7], while Behrens et al. used deep drawing [8]. CR340LA is a material used in automobile pillars, which play a very important role in the safety and protection of the passengers by absorbing the shock during a car crash. However, CRFP has mechanical properties that vary according to the fibre orientation and forming processes. Because of its low elongation and fracture toughness, there are many limitations to its application in automobile parts. Thus, many studies are being performed to overcome such problems using hybrid CFRP-metal composite materials.

2. Experiment procedure

2.1. Tensile and bending test

2.1.1. Preparation of specimen

Tables 1 and 2[9] show the properties of the carbon fiber composite material and steel sheet(CR340) used in the test.

In this study, the fabric style with epoxy in a liquid state is plain weave-type CFRP prepreg using automobile part fabrication. This weave-type was used as it has higher strength as well as low thickness compared to stain and twill weaves.

Table 1. Properties of plain weave carbon fabric prepreg (CF3327EPC, HANKUK CARBON).

Construction	Carbon fibre weight (gr/m2)	Resin weight (gr/m2)	Resin content (%)	Total weight (gr/m2)	Fabric thickness (mm)
Plain	205	150	42±2	352	0.27±0.050

Table 2. Properties of carbon fibre composites material (Fiber/Epoxy resin 120°C cure) (HANKUK CARBON Co. Ltd) [9] and steel (CR340).

Type	Tensile strength (MPa)	Young modulus (GPa)	Elongation (%)	Density (g/cc)	n
3K Carbon fibre	1500	135	1.05	1.6	-
CR340LA	396	496	27	7.6	0.18

A 25-ton material testing system (MTS) device was used in this experiment for the preparation of the specimen. The specimen was prepared by varying the number of CFRP laminations with the compression moulding method and curing for 30 min while maintaining a pressure of 0.5 MPa and temperature of 140°C. There are many methods for producing vehicle parts using CFRP. It is recommended that production methods appropriate for each vehicle part be based on KS M 3713:2012 [10].

In addition, the standard guide for CFRP testing methods is ASTM D4762 [11].

2.1.2. Tensile and bending test of CFRP

ASTM D638 and D3039/D3039M were referred to for the CFRP tensile testing method [12]. A 10-ton MTS device was used for the tensile test, with a displacement rate of 2 mm/min applied in the test. Caution is required when testing a composite material because it is very sensitive to the strain rates [13].

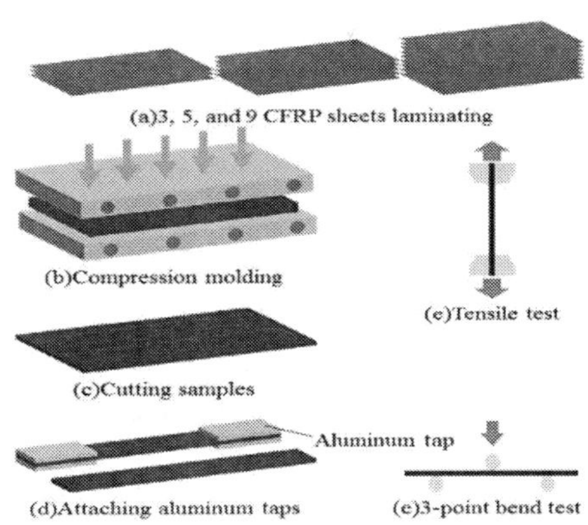

Figure 1. Schematic diagram of tensile and 3-point bend tests.

Figure 2. The tensile and 3-point bend tests of CFRP.

The use of an extensometer is not recommended because the high strength of the composite material compared to its low elongation causes a great impact and can overload the testing equipment. ASTM D790, D7264/D7264M, and KS M ISO 14125:2012 were referred to for the CFRP bending testing method [14, 15, 16].

The 25-ton MTS device was used for the bending test. In the case of the bending test, the span-to-thickness ratio was used for the experiment, with a displacement rate of 1 mm/min. Figure 1 briefly describes the specimen preparation and testing methods for the tensile and bending tests, while figure 2 shows the equipment used.

2.2. Lap shear adhesion test

For bending strength, the CR340LA/CFRP hybrid composite, the interfacial shear strength of the bi-material is also important. CR340LA is one of the materials used for automobile body parts [17]. It is also used for the automobile pillar. Galvanized CR340LA (GI CR340LA) is used as the steel in automobiles to prevent corrosion. It is reported that galvanizing above a certain thickness is considerably effective at preventing corrosion [18]. Preventive effects against the galvanic corrosion that may be present with hybrid composite materials can also be obtained [19]. There are many studies being conducted on the bonding relationship between CFRP and steel [20, 21].

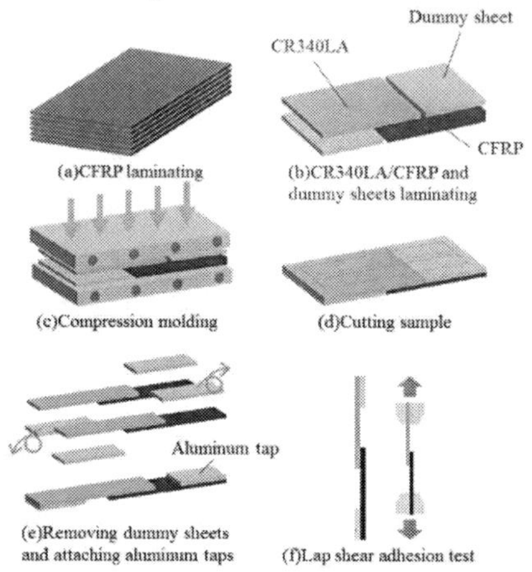

Figure 3. Schematic diagram of lap shear adhesion test.

Here, a lap shear adhesion test was conducted for the interfacial shear strength, which is briefly described in figure 3. Four sheets of CFRP were laminated to prepare a specimen with the same thickness as the CR340LA. Then, a dummy sheet material was added to the laminated CFRP to prevent the separation of the bi-material and deformation of the carbon fibre during the compression moulding. After it was cut to an appropriate specimen size, aluminium tabs were attached to prevent any moment that could be present in the specimens due to the misalignment of the pulling axis during the lap shear adhesion test. ASTM D5868 and D3168 were referred to for the test method [22, 23].

3. Experiment results and discussion

3.1. Results of tensile and bending test of CFRP

3.1.1. Results tensile and bending test of CFRP

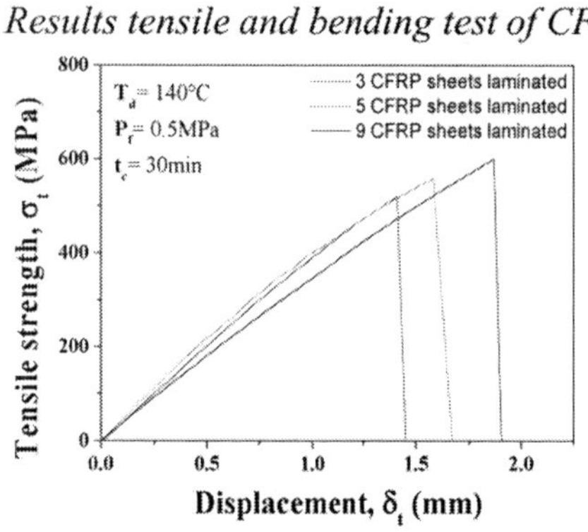

Figure 4. Tensile strength and displacement of CFRP according to number of CFRP prepreg laminating.

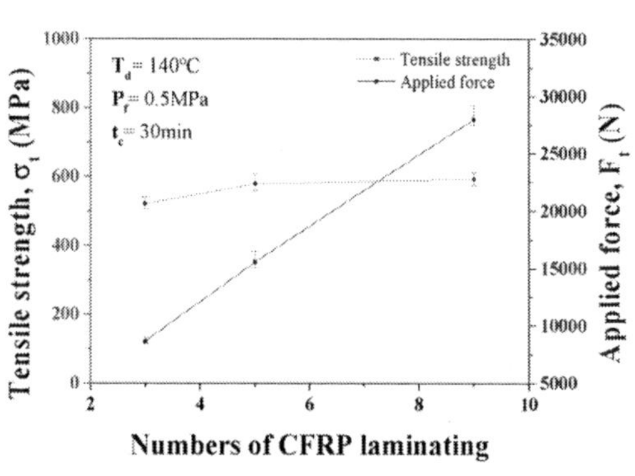

Figure 5. Ultimate tensile strength and applied force of CFRP according to numbers of CFRP prepreg laminating.

Figure 4 and 5 show the results of the tensile test. As presented, it can be observed that the tensile strength increases as the number of laminations increases. The average tensile strengths were 520.717, 578.511, and 592.317 MPa. For every condition, 10 specimens were prepared, for a total of 30 specimens used in the experiment. However, because of the characteristics of composite materials, many of the tests failed, and such failed tests were excluded from the results. The types of failures in the tensile tests are shown in detail in the previously noted ASTM D3039/D3039M [12].

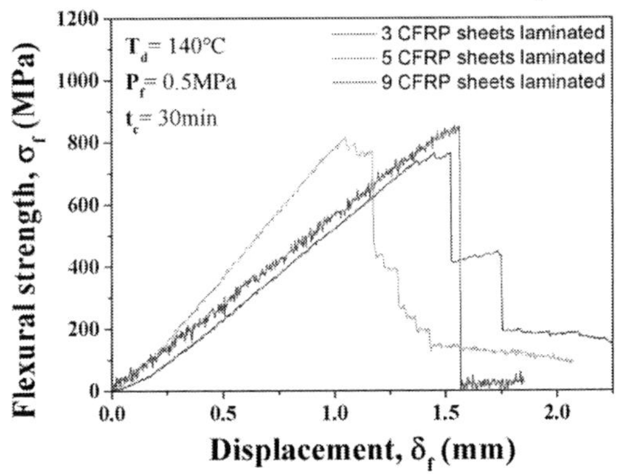

Figure 6. Flexural strength and displacement of CFRP according to number of CFRP prepreg laminating.

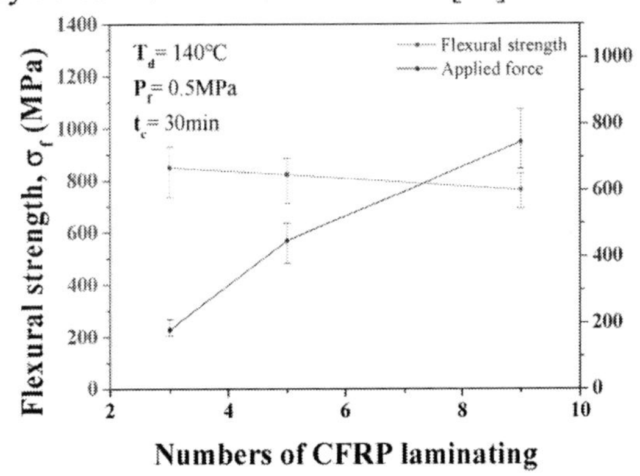

Figure 7. Ultimate flexural strength and Applied force of CFRP according to numbers of CFRP prepreg laminating.

The results of the bending test are shown in figure 6 and 7. In contrast to the tensile test results, it can be observed that as the number of laminations increases, the results decrease. In addition, the average flexural strengths were 850.476, 823.432, and 764.780 MPa, according to the number of laminations. In the case of the bending test, 12 specimens were prepared for every condition for a total of 36 specimens used in the experiment. Unlike the tensile test, all the bending tests were successfully conducted without failures.

3.1.2. Results of lap shear adhesion test

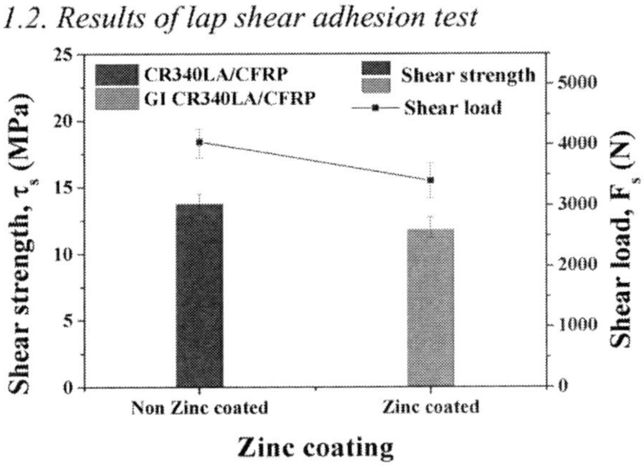

Figure 8. Interfacial shear strength (τ_S) and load (F_S) of CFRP to CR340LA with or without zinc coating.

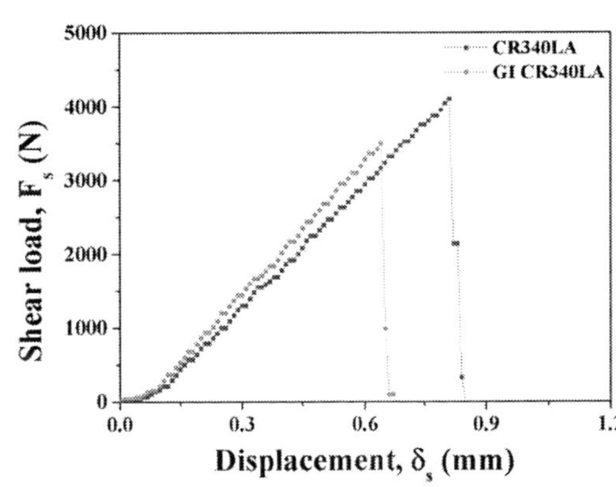

Figure 9. Shear load (F_S) and displacement (δ_S) according to surface conditions of CR340LA with or without zinc coating.

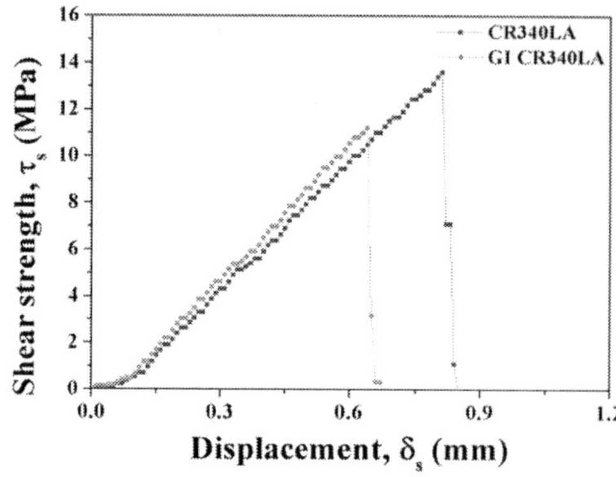

Figure 10. Shear strength (τ_S) and displacement (δ_S) according to surface conditions of CR340LA with or without zinc coating.

Figure 8, 9, and 10 show the shear strengths from the lap shear adhesion test of the CR340LA according to the presence or absence of a galvanized coating. When an automobile pillar is produced from the CR340LA/CFRP hybrid composite, the CR340LA corrodes at a faster rate than the galvanized CR340LA because of the galvanic corrosion between the two materials. Therefore, a galvanized material is sometimes used.

4. Conclusion

In this study, Effect of Interface Bonding Process on Bending Strength in CR340LA/CFRP hybrid composite material. Based on the changes in various mechanical properties in relation to the number of laminated layers of CFRP, and the lap shear adhesion test results for CR340LA and CFRP, the following conclusions were reached in this study

(1) When producing a CFRP part using the compression moulding method, the drawn part possesses different mechanical properties according to the number of laminations. The tensile strength increases with the number of laminations, whereas the bending strength decreases.

(2) Because of the increased outflow of epoxy as the number of laminations increases, the epoxy volume fraction decreases, and the carbon fibre volume fraction increases. The factor that most affects the tensile test is the carbon fibre. The results obtained are the results with an increased carbon fibre volume fraction. In the case of the bending test, the results obtained are the results for the decreased epoxy volume fraction that transfers stresses between the carbon fibres.

(3) The shear strength of the CR340LA and CFRP was 13.73 MPa, and that for the CR340LA with the zinc coating and CFRP was 11.79 MPa. Thus, the shear strength of the CR340LA and CFRP was higher than that of the CR340LA with the zinc coating and CFRP

Acknowledgments

This work was supported by the National Research Foundation of Korea (NRF) grant funded by the Korean government (MSIT) through GCRC-SOP (No.2011-0030013) and by the National Research Foundation of Korea (NRF) grant funded by the Korean government (MSIT) (No.2017R1A2B4007884).

References

[1] IRS Global, 2013, "Development trends of materials and components and their business strategy of domestic and foreign industry participation for light-weighting technologies of automobile", IRS Global

[2] M Kleiner, M. Geiger, A. Klaus, 2003, "Manufacturing of Lightweight Components by Metal Forming", CIRP Annals - Manufacturing Technology, 52, 521-542

[3] R Teti, 2002, "Machining of Composite Materials", CIRP Annals - Manufacturing Technology,

IOP Conf. Series: Materials Science and Engineering **715** (2020) 012102 doi:10.1088/1757-899X/715/1/012102

51, 611-634

[4] B C Shin, K B Kim, S J Ha, M W Cho, 2011, "A study on the machinability of Carbon Fiber Reinforced Plastics on tool shape", The Korean Society of Manufacturing Technology Engineers, 20, 799-804.

[5] J Yanagimoto, K Ikeuchi, 2012, "Sheet forming process of carbon fiber reinforced plastics for lightweight parts, "CIRP Annals - Manufacturing Technology, 61, 247-250

[6] Y Uriya, K Ikeuch, J Yanagimoto, 2014, "Cold and warm V-bending test for carbon-fiber reinforced plastic sheet", Procedia Engineering, 81, 1633-1638

[7] S Isogawa, H Aoki, M Tejima, 2014, "Isothermal forming of CFRTP sheet by penetration of hemispherical punch", Procedia Engineering, 81, 1620-1626

[8] B Behrens, S Hübner, A Neumann, 2014, "Forming sheets of metal and fibre-reinforced plastics to hybrid parts in one deep drawing process", Procedia Engineering, 81, 1608-1613

[9] http://www.performance-mposites.com/carbonfibre/mechanicalproperties_2.asp

[10] Korean Agency for Technology and Standards, 2012, "Preparation of carbon fibre reinforced plastic panels for test purpose", KS M 3713:2012

[11] ASTM International, 2011, "Standard Guide for Testing Polymer Matrix Composite Materials", ASTM D4762

[12] ASTM International, 2014, "Standard Test Method for Tensile Properties of Polymer Matrix Composite Materials", ASTM D3039/D3039M

[13] L.M. Welsh, J. Harding, 1985, "Effect of strain rate on the tensile failure of woven reinforced polyester resin composites", J. Phys. Colloques, 46, C5-405-C5-414

[14] ASTM International, 2010, "Standard Test Methods for Flexural Properties of Unreinforced and Reinforced Plastics and Electrical Insulating Materials", ASTM D790

[15] ASTM International, 2007, "Standard Test Method for Flexural Properties of Polymer Matrix Composite Materials", ASTM D7264/D7264M

[16] Korean Agency for Technology and Standards, 2012, "Fibre-reinforced plastic composites– Determination of flexural properties", KS M ISO 14125:2012

[17] H Gong, J Wang, Z Zhao, 2011, "Study on the Springback Characteristics of CR340LA Steel during the Typical Auto Part Stamping Process", Advanced Materials Research, 322, 98-101

[18] American Galvanizers Association, 2011, "Zinc coatings", American Galvanizers Association

[19] W Wang, Y Takao, T Matsubara, 2007, "Galvanic corrosion-resistant carbon fiber metal laminates", 16th International Conference on Composite Materials

[20] T Yu, D Fernando, J G Teng, X L Zhao, 2012, "Experimental study on CFRP-to-steel bonded interfaces", Composites: Part B, 43, 2279-2289

[21] J G Teng, D Fernando, T Yu, X L Zhao, 2012, "Debonding failures in CFRP-strengthened steel structures", The Third Asia-Pacific Conference on FRP in Structures

[22] ASTM International, 2014, "Standard Test Method for Lap Shear Adhesion for Fiber Reinforced Plastic (FRP) Bonding", ASTM D5868

[23] ASTM International, 2014, "Standard Test Method for Determining Strength of Adhesively Bonded Rigid Plastic Lap-Shear Joints in Shear by Tension Loading", ASTM D3163

IOP Conf. Series: Materials Science and Engineering **715** (2020) 012103 doi:10.1088/1757-899X/715/1/012103

Effects of extrusion temperatures on mechanical properties of B4C/Al composites

ShuangP Wu, ShangY Feng, Jia Wang and Qiu-Lin Li*

Graduate School at Shenzhen, Tsinghua University, Shenzhen 518055, China

*Email: liql@sz.tsinghua.edu.cn

Abstract. This article studied the effects of extrusion at different temperatures on the microstructures and mechanical properties of B_4C particle reinforced Al neutron shielding composite produced by powder metallurgy method. The results indicated that, the interfacial bonding between the B_4C particles and the matrix was greatly improved by hot extrusion at studied temperatures. In addition, the mechanical properties of the composites were obviously improved after hot extrusion. However, by further increased the extrusion temperature, yield strength (YS) and ultimate tensile strength (UTS) of the composites decreases. Meanwhile, the fractography of extruded composite revealed that fracture mechanism changed from particle brittle cracking to matrix plastic damage as extrusion temperature increased.

1. Introduction

In the industrial field, the common reinforcements of aluminum matrix composites include silicon carbide (SiC)、aluminum oxide (Al_2O_3)、boron carbide (B_4C) and so on. Among above reinforcements, B4C as a potential reinforcement is widely used in neutron absorption industry with excellent capacity for neutron absorption [1-3]. In recent years, the study of B_4C/Al composites has made much progress. Many conventional fabrication methods have been used to produce the composites, such as liquid-stirring casting [4], pressure infiltration [5], powder metallurgy (PM) [6], and so on. Compared with other manufacturing techniques, the powder metallurgy (PM) method is attractive, because of the easy controlling of microstructures. However, PM processing technique normally employs lower temperature sintering, resulting in the lower diffusion rates and weak interfacial bonding strength. Another problem of the PM method is that the Al_2O_3 film on most commercially atomized powders can act as the barrier for the atom diffusion during sintering, especially solid phase sintering. Under that condition, it is commonly accepted secondary hot deformation like extrusion, which can significantly improve interfacial bonding strength and uniformity of added particles, and finally improve its density and mechanical properties [7-9]. As for hot extrusion, extrusion temperatures can significantly affect grain size, distribution of large particles, dynamic recrystallization (DRX) in the material, thus affect the mechanical properties of composites. However, few research exported the details of above influences.

Thus, this paper further analyzed the extrusion temperatures and their influence on the mechanical property of the composites. 12 vol.% B_4C/Al composite was successfully prepared by using fine atomized Al powders and B_4C particles and then followed hot extrusion process. The study of the microstructure and tension property of the composites were carried out.

2. Experimental

2.1. Raw materials

The raw materials used in this study were B_4C particles (99% in purity) and fine atomized Al powders (99.9% in purity). The average size of B_4C particles and Al powders were 4.3μm and 1.5μm, respectively. Figure 1 shows scanning electron microscopy (SEM) of B_4C particles and Al powders.

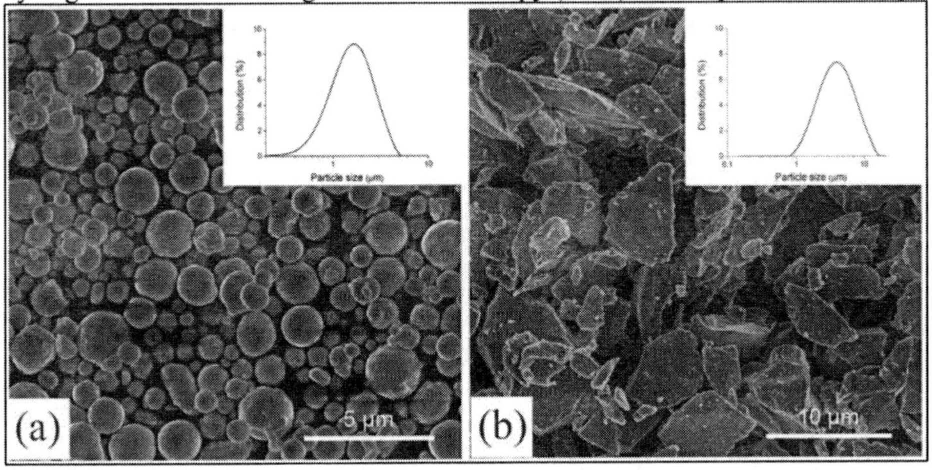

Figure 1. The SEM micrographs of Al and B_4C particles with an insert of particle size distribution (a) Al-1.5μm, (b) B_4C-4.3μm.

2.2. Processing

The aluminum powders were mixed with about 12 vol.% B_4C particulates in alcohol medium for 2 hours by using an attritor at 150 rpm/min. The ball to powder weight ratio was 6:1. The mixed powders were dried in air and compacted by cold isostatic pressing (CIP) at 400 MPa. The specimens were heated in a vacuum furnace with a heating rate of 10°C per minute to 450°C for 12 hours. The B_4C/Al billets with dimension of Φ 75 mm and thickness of 100 mm were machined from the as-sintered samples. And then conducted the preparation for the extrusion experiments. The surfaces of the billets, extrusion die and dummy block were coated with a graphite-base lubricant. The billets were put into the container and the extrusion container was heated to given temperatures in a pit-type furnace. Soaking time of 1 hour was set after the container attained the predetermined temperature. Billets were extruded with an extrusion ratio of 22:1 as well as a ram speed of 2 mm/s and then cooled in air. Three extrusion temperatures (540, 570 and 600°C) were employed.

2.3. Testing

Microstructural examination was carried out by scanning electron microscopy (SEM) with energy dispersive spectrometer (EDS) and electron back-scatter diffraction (EBSD) system. Specimens were cut parallel to extrusion direction with a size of 2×4×8 mm^3. The specimens for SEM and EBSD were polished by mechanical polishing and followed by ion beam (IB-09020CP, JEOL) polishing. Room temperature tensile test samples were machined into rectangular specimens, conforming to ASTM E8M standard. Tensile tests were carried out at a strain rate of 1×10^{-3} s^{-1} on an MTS-LANDMARK 370.25 test system. The density of the as-sintered and the as-extruded samples were measured by the Method of Archimedes.

3. Results and discussion

3.1. Microstructures

Figure 2 shows the SEM micrographs of B_4C/Al composites before and after extrusion at different temperatures. Clusters of B_4C particles (red circles) and micro-pores (red arrow) are observed in the samples before extrusion (Figure 2a and e). Meanwhile, adjacent spherical aluminum powders and the poor interface between the boron carbide and matrix can be clearly observed. By comparing with SEM micrographs of B_4C/Al composites before and after extrusion, the B_4C particles distribution is significantly improved by extrusion. For all extruded samples, it is hard to see the clusters of B_4C particles (Figure 2b, c and d). And the interface between B_4C particles and matrix is clean and no cracks which can be seen in figure 2f, g and h. As can be seen from the longitudinal section, the number of pores in the as-extruded composite were completely eliminated in all set temperatures.

The distribution of reinforcement particles depended on the blending and the consolidation procedures as well as the matrix-to-reinforcement particle size ratios [10]. In this study, micro-pores cannot be completely eliminated during sintering due to the Al_2O_3 film on the surface of fine atomized Al powders. During the hot extrusion, shearing enlarges the surface of the powders, and breaks up the Al_2O_3 film. Besides, the severe plastic deformation of the matrix, leading to the densification of aluminum matrix composites and rearrangement of the reinforcements [11-12].

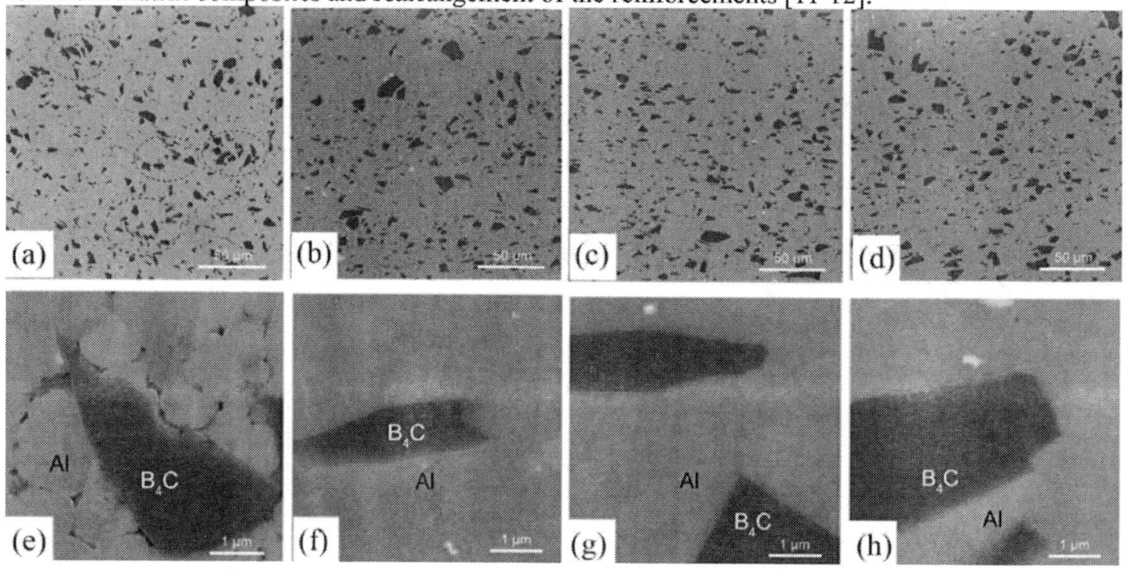

Figure 2. SEM micrographs for the composites before extrusion and being extruded at different extrusion temperatures: (a and e) before extrusion, (b and f) 540 °C, (c and g) 570 °C, (d and h) 600°C.

Figure 3 shows the EBSD maps and grain statistical distribution of the composites extruded at different temperatures. The average grain size of composites increases as extrusion temperature increased. The average grain sizes of the composites extruded at temperatures of 540, 570 and 600 °C are 0.7, 0.9 and 2.2 μm, respectively. In the E540 sample, only fine grains are observed, indicating that the full dynamic recrystallization (DRX) occurred during deformation. In the E570 and E600 samples, the grains became coarser, caused by higher extrusion temperature.

In a word, for 12vol.% B_4C/Al composites, when the temperature of hot extrusion reaches or exceeds to 540°C, the micro-pores in the matrix can be completely eliminated and the interface between B_4C particles and matrix is smooth. Meanwhile, hot extrusion at this temperature could result in a uniform particle distribution. And the density of extruded samples increases as the extrusion temperature increases. In addition, extrusion temperature greatly affects the size of matrix grains, the average grain size of composites increases as extrusion temperature increased. So that the hardness of being extruded samples decrease as the extrusion temperature increased.

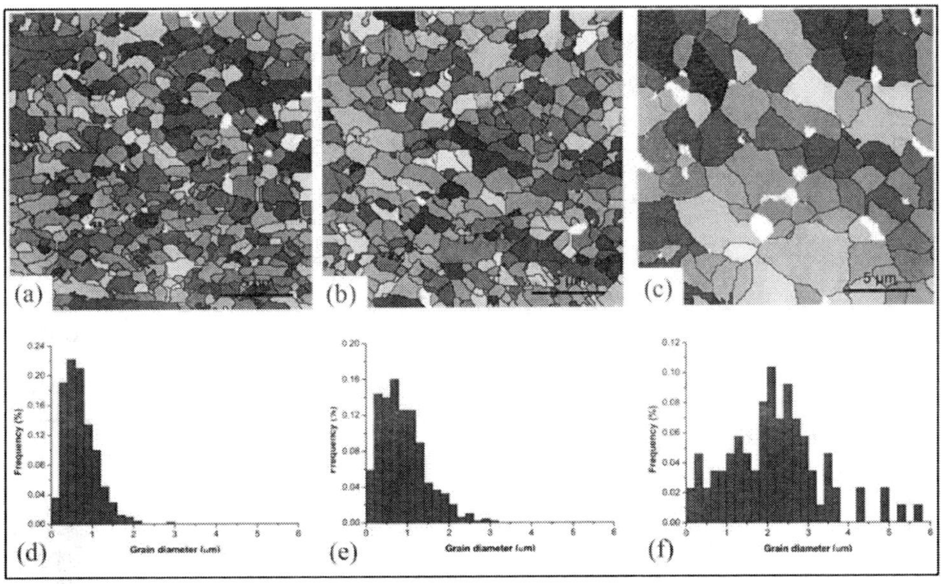

Figure 3. EBSD maps of the composites extruded at different extrusion temperatures: (a) 540°C, (b) 570°C, (c) 600°C. The grain size distribution obtained from EBSD measurements (d_{EBSD}) are shown in (d), (e), (f), respectively.

3.2. *Mechanical properties of the extruded composites*

Figure 4 shows the tensile mechanical properties of the composites extruded at different extrusion temperatures. The ultimate tensile strength of the E540 sample is 256 MPa, which is almost three times compared with sample before extrusion in this study (90 MPa). That is mainly attributed to the fine grain, uniform B_4C distribution and improved interface bonding strength between B_4C and matrix. As the extrusion temperature increasing, the yield strength (YS) and ultimate tensile strength (UTS) of the composites decrease, while the elongation of the composites increases.

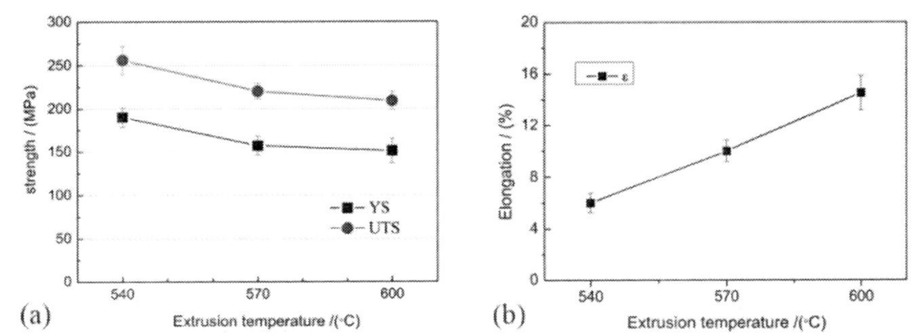

Figure 4. Mechanical properties of as-extruded composites: (a) YS and UTS, (b) elongation.

The mechanical properties of composites depend on interface bonding strength, particle dispersion and matrix grain size. A strong interfacial bonding helps to improve composites strength by transferring external load from soft matrix to hard reinforcements[13,14] Besides, the grain boundaries and reinforced particles in the composites act as the obstacle for dislocation movement to improve the composites' strength [15].

3.3. *Fracture behavior of the composites*

Figure 5 shows the fractograph of the composites extruded at different temperatures. In figure 5a, lots of fractured B_4C particles exist on the fracture surface (marked by red arrows) and shallow dimples in the matrix can be observed. As the extrusion temperature increased, the number of the fractured B_4C particles reduce (Figure 5b) and even disappear (Figure 5d), while the dimples are deeper (Figure 5e and 5f). At the same time, in the E540 sample (Figure 5d), large number of B_4C particles, especially with high aspect ratio, fractured and induced micro-cracks in the area near to tensile fracture. But, in the E600 sample (Figure 5f), there exists lots of micro-pores in the aluminum matrix and adjacent reinforcing phases without any fractured particles. There is a large mismatch between pure aluminum matrix and B_4C particles, and the plastic deformation of the soft matrix would be limited by the hard and brittle particles. Therefore, the fracture of composites is a combination of ductile fracture and brittle fracture.

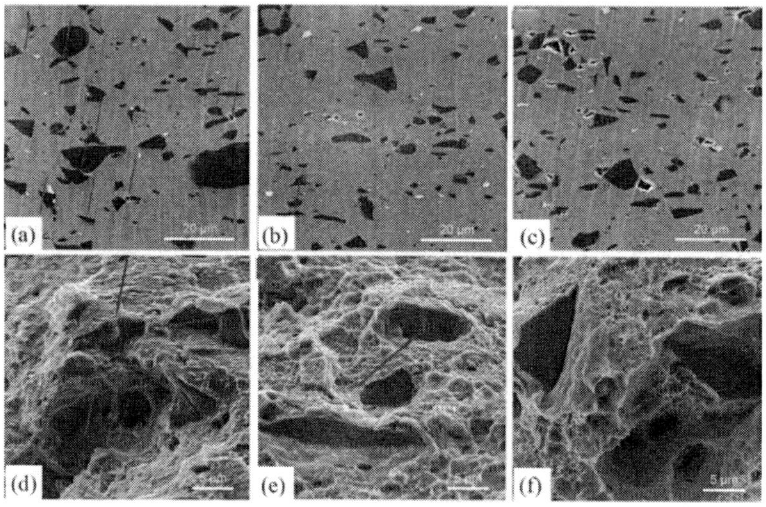

Figure 5. Fractographs of the composites and morphologies of fracture sides of composites extruded at different extrusion temperatures: (a and d) 540°C, (b and e) 570°C, (c and f) 600°C.

4. Conclusions

In this work, the effects of extrusion temperature on the microstructures and mechanical properties of 12 vol.% B_4C/Al composite were studied. Several conclusions have been drawn:

(1) For as-extruded 12 vol.% B_4C/Al composites, When the temperature of hot extrusion reaches or exceeds 540°C, can effectively reduce the number of micro-pores, increase the bonding strength of the interface between the B_4C particles and matrix, and uniform distribution of B_4C particles.

(2) The 12 vol.% B_4C/Al composites extruded at 540°C had the highest strength (UTS=256 MPa, YS=190 MPa), which is about three times compared with sample before extrusion (90 MPa). The average grain size of composites increases as extrusion temperature increasing, resulting in the decrease of the yield strength and tensile strength.

(3) The 12 vol.% B_4C/Al composites matrix strength decreases with the increased extrusion temperature, the fracture mechanism changes from particle brittle cracking to matrix plastic damage.

Acknowledgments

This work was founded by Shenzhen Engineering Laboratory of Nuclear Materials and Service Safety.

References

[1] Yu Li, Qiu-lin Li, Dong Li, Wei Liu, and Guo-gang Shu, T NONFERR METAL SOC **26**, (2016).
[2] D. B. Miracle, COMPOS SCI TECHNOL **65**, 2526-40 (2005).
[3] Jiyun Zheng, Qiulin Li, Wei Liu, and Guogang Shu, J COMPOS MATER **50**, (2016).
[4] Yu Li, Qiulin Li, Wei Liu, and Guogang Shu, J ALLOY COMPD **684**, (2016).
[5] Jin-Ju Park, Sung-Mo Hong, Min-Ku Lee, Chang-Kyu Rhee, and Won-Hyuk Rhee, NUCL

ENG DES **282**, 1-7 (2015).
[6] Peng Zhang, Yuli Li, Wenxian Wang, Zhanping Gao, and Baodong Wang, J NUCL MATER **437**, 350-58 (2013).
[7] G. A. ROZAK, J. J. LEWANDOWSKI, J. F. WALLACE, and A. ALTMISOGLU, J COMPOS MATER **26**, 2076-2106 (1992).
[8] Sajjad Amirkhanlou, Mohammad Reza Rezaei, Behzad Niroumand, and Mohammad Reza Toroghinejad, MATER DESIGN **32**, 2085-90 (2011).
[9] Chang-Soo Kim, Kyu Cho, Mohsen H. Manjili, and Marjan Nezafati, J MATER SCI **52**, 13319-349 (2017).
[10] Zhangwei Wang, Min Song, Chao Sun, Daihong Xiao, and Yuehui He, MAT SCI ENG A-STRUCT **527**, 6537-42 (2010).
[11] R. Dasgupta, S. Das, S. Chaturvedi, and A. K. Jha, T NONFERR METAL SOC **20**, 2229-33 (2010).
[12] A. M. Davidson and D. Regener, COMPOS SCI TECHNOL **60**, 865-69 (2000).
[13] J. Segurado and J. Llorca, ACTA MATER 53, 4931-42 (2005).
[14] Jie Chen, Chonggao Bao and Fulei Chen, MAT SCI ENG A-STRUCT **667**, 426-34 (2016).
[15] S. Sankaranarayanan, M. K. Habibi, S. Jayalakshmi, K. Jia Ai, A. Almajid, and M. Gupta, MATER SCI TECH-LOND **31**, 1122-30 (2015).

IOP Conf. Series: Materials Science and Engineering **715** (2020) 012104 doi:10.1088/1757-899X/715/1/012104

Research on EEMD for vibration extreme feature preserving of rotating machinery

Sanhua Li[1], Chunhua Tian[1], Chen Wang[1] and Zaichao Ma[2,*]

[1] Innovation Center for Industrial Big Data co. LTD, Beijing, 100083, China
[2]School of Software, Tsinghua University, Beijing, 100084, China

*Email: mazaichao@mail.tsinghua.edu.cn

Abstract. Ensemble Empirical Mode Decomposition (EEMD) has been widely used due to the more advanced processing of mode mixing than Empirical Mode Decomposition (EMD). However, EEMD still suffers mode mixing in the vibration analysis of rotating machinery. It proposes Extremum Feature Preserving Ensemble Empirical Mode Decomposition (EFPEEMD) for the feature extraction of rotor systems in rotating machinery. In the method, a fluctuant variation (FV) index is developed to quantitatively recognize the amplitude of added noise on the basis of the empirical rules in EEMD. The results of a misalignment vibration signal of a rotor system and an inner race defect signal of a rolling bearing show that, the primary vibration modes can be extracted rapidly and independently based on the elimination of the non-stationary and non-linear interference. Moreover, the fluctuant variation index removes the originally manual and empirical steps, thus enhancing the accuracy and automation of EEMD on the basis of preserving the extreme information

1. Introduction

The industrial rotating machinery have been widely used not only in the aerospace, national defense construction and other high-tech industries, but also in the petroleum chemical industry, metallurgy and other national pillar industry, such as steam turbine, generator and different types of compressors. The rotating equipment usually has a compact structure with big size and big weight, and long runs in a high speed and high-power way with strong vibration and noise interference. As the core part of rotary machinery, a rotor-bearing system is prone to be failure under vibration condition, which not only decreases its normal service life, but also reduces production efficiency. Therefore, it is significant to ensure the safety and stability of the production field and quickly and accurately identify the running state of rotating machinery, according to the vibration condition of the rotor-bearing system.

Correct analysis and effective extraction of the state parameters and dynamic response information of rotating machinery are the basis of subsequent fault pattern recognition [1]. However, vibration signals of rotating machinery often show significant harmonic frequency characteristics with non-linearity, non-stationarity and noise interference, which have been concerned for a long period [2,3]. Huang et al. proposed empirical mode decomposition [4], and initially established the time-frequency analysis system that measures the signal alternating with instantaneous frequency and expresses the signal components with basic mode components [5]. In this method, the application research in the field of mechanical equipment fault diagnosis was quickly obtained. Yu et al. [6] combined EMD and envelope analysis for fault diagnosis of rolling bearings. Gai [7] used EMD to analyze the starting signal of the rotor and draw a Bode diagram. Liu et al. [8] used the B-spline curve EMD and Hilbert spectrum for gear box fault diagnosis. However, mode aliasing is easily caused by the abnormal events such as noise interference [9-12]. Wu et al. proposed ensemble empirical mode decomposition [13], which assists EMD with the white noise and effectively decrease the mode aliasing phenomenon. This method is also quickly applied in the field of fault diagnosis of rotating machinery. Cao et al. [14] studied the noise reduction of non-stationary signals during rotor start-up using EEMD. Yu et al. [15] studied the rotor local rubbing using EEMD and compared the effect with EMD. Lei et al. [16,17] used EEMD for rotor impact rub fault diagnosis and studied the noise adding method of EEMD. Chen et al. [18] studied the component selection of EEMD and the noise-adding times for noise reduction of rotor vibration signals. Zhang et al. [19] combined morphological filtering, EEMD, sample entropy and grey correlation analysis to study the classification of common rotor faults. Though the achievements in the above research on rotor vibration pattern extraction, EEMD is still disturbed by mode mixing with the nonlinearity, non-stationarity and noise interference, which affect the rapid and accurate fault identification. It lies in the problem of quantitative selection of noise amplitude parameter. Therefore, an indicator is needed to be constructed to quantitatively select the parameters of noise amplitude and further solve the mode aliasing of EEMD, which is expected to accurately extract fault features of rotating machinery.

In this study, the nonstationary information of the rotor-bearing transmission system of the rotating machinery was investigated. EEMD method was proposed with extreme feature preserving, which improves the application ability of traditional EEMD in vibration analysis of rotating machinery.

2. Ensemble empirical mode decomposition (EEMD)

2.1. *Flow of the algorithm*

EEMD uses the uniform statistical distribution of the white noise spectrum to provide a uniform distribution for the decomposition scale of the investigated signals. Furthermore, mode mixing is eliminated by average effect due to the zero-mean property of the white noise, which is equal to the average of EMDs. The steps of EEMD algorithm can be summarized as follows:

Step 1: The original signal $x(t)$ plus the gaussian white noise $n_i(t)$ whose mean value is 0 and whose amplitude and standard deviation are constant for N times, as shown in equation (1).

$$x_i(t) = x(t) + n_i(t) \tag{1}$$

where, i=1, 2, ..., N.

Step 2: Performing EMD on each $x_i(t)$, K_i IMFs and a residual $r_i(t)$ are obtained, as shown in equation (2).

$$x_i(t) = \sum_{j=1}^{K_i} C_{ij}(t) + r_i(t) \tag{2}$$

where, $C_{ij}(t)$ denotes the jth IMF with the ith noise assisted signal, j=1, 2, ..., K_i

Step 3: Carrying out the ensemble process on the obtained IMFs to suppress the effect of the added gaussian white noise for several times, the real IMF and a residual $r(t)$ were obtained, as shown in equations (3) and (4).

$$C_j(t) = \frac{1}{N}\sum_{i=1}^{N} C_{ij}(t) \tag{3}$$

$$r(t) = \frac{1}{N}\sum_{i=1}^{N} r_i(t) \tag{4}$$

where, $C_j(t)$ is the jth IMF of the decomposed original signal.

Step 4: Obtain the K IMFs and a residual $r(t)$, as shown in equations (5).

$$x(t)=\sum_{j=1}^{K}C_j(t)+r(t) \quad (5)$$

2.2. *Noise amplitude selection*

Reference [13] points out that the effect of white noise on the investigated signal follows the statistical rule as shown in equation (6).

$$\varepsilon_n=\frac{\varepsilon}{\sqrt{N}} \text{ or } \ln\varepsilon_n+\frac{\varepsilon}{2}\ln N=0 \quad (6)$$

where, ε_n is the deviation between the original signal and the reconstructed signal, ε is the amplitude of white noise, and N is the number of the added noise.

According to the formula (6), the signal decomposition precision is directly proportional to the noise amplitude level, and inversely proportional to the open square of the noise-adding number N. A small ε is difficult to modify the local extreme point of the signal. A large N will increase the complexity of the algorithm. Therefore, ε is taken to be 0.2 times of the standard deviation of the investigated signal according to the noise adding experiment, as recommended in literature [16]. ε can be smaller if the signal is mainly characterized by higher frequency. On the other hand, ε can be higher if the signal is mainly characterized by lower frequency.

3. EFPEEMD for fault signal of rotating machinery

3.1. *Vibration mode extraction*

The goal of EEMD is to decompose the signal into IMFs and a residual with frequency ranging from high to low based on noise reduction. From the engineering experience of the rotor vibration analysis, IMFs with high frequency can be categorized to random noise, and the low frequency area of the IMFs may be classified as trends, false or residuals. If the invalid IMFs can be removed, the effective IMFs are obtained to describe the vibration mode of the rotating machinery. These IMFs can be analyzed separately, and also can be used for signal reconstruction. For the special conditions of the rotating machinery, the IMF component obtained by EEMD still has a certain degree of frequency aliasing. Therefore, the component selection is aimed at preserving the main vibration mode.

3.2. *Parameters of EEMD*

The typical vibration band of an investigated signal from the rotating machinery is mainly concentrated in the range of integer frequency within the base frequency, dividing frequency and octuple base frequency. It is generally classified as a lower frequency signal. Therefore, ε can be set as 0.2 times of the standard deviation of the analysis signal. However, the qualitative process needs artificial assistance. It is necessary to ensure a small distortion between the reconstructed signal and the original signal when we need a high orthogonality between IMFs due to noise inducing the increase of the signal disorder degree and the change of extreme point. Therefore, a Fluctuant Variation indicator is developed to measure the extreme fluctuation between the original signal and reconstructed signal, and further quantitatively determine the noise amplitude parameters, as shown in the equation (7). If FV has a small value, it is considered that the waveform distortion is small between the original signal and the reconstructed signal, and the corresponding noise amplitude parameter is optimal.

$$FV=\left|\frac{y_p}{y_{rms}}-\frac{x_p}{x_{rms}}\right| \quad (7)$$

where, x_p is the peak of the original signal, x_{rms} is the mean square amplitude of the original signal, y_p is the peak of reconstructed signal, and y_{rms} is the mean square amplitude of the reconstructed signal.

Additionally, the number of added noise N can be set as 100 according to the literature [13]. In [18], N can be a small value on the base of the vibration property of the rotating machinery.

3.3. Parameters of EEMD

The overall process framework is shown in figure 1.

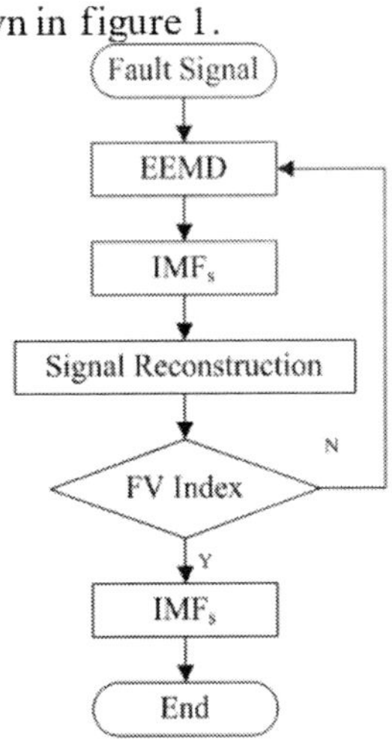

Figure 1. FEPEEMD for fault signal of rotating machinery.

Step 1: Set the initial noise amplitude of EEMD according to equation (6);
Step 2: Analyze the investigated signal by EEMD and obtain IMFs;
Step 3: Choose IMFs and reconstruct the signal;
Step 4: calculate the FV index and step to adjust the noise amplitude;
Step 5: Repeat Step1 to Step4 until satisfying the termination condition and output the optimal EEMD decomposition results.

4. Apparatus and application

4.1. The analysis of the rotor misalignment signal

This platform, as shown in figure 2, is mainly consisted of a rotor system and a vibration testing system. The rotor system includes a rotor, a motor, a pair of bearing and a foundation. The vibration testing system includes six groups of eddy current sensors and a set of data acquisition instrument connecting computer. Observing in detail, sensors 1–4 are divided into two groups for capturing the vibration signals located in the cross section with the directions of 45° and 135°. Sensor 5 is used to measure the phase and sensor 6 is applied to capture the rotating speed.

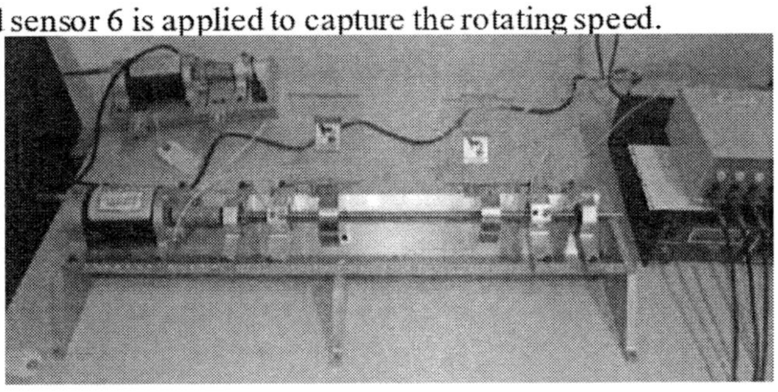

Figure 2. Structure of the rotor test bench.

The rotor test bench is used to simulate the misalignment fault. The sampling rate is set to be 2048Hz, the sampling number is set as 2048 points, and the rotating speed is set as 4000 rpm. The vibration signal both in time and frequency domain are illustrated in figure 3 and figure 4.

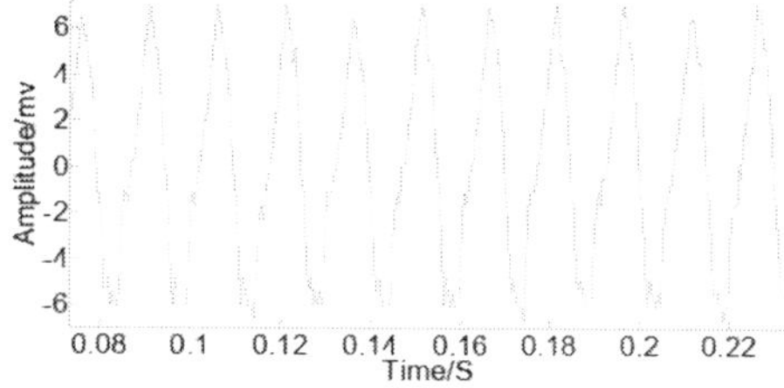

Figure 3. Misalignment signal of the rotor in time domain.

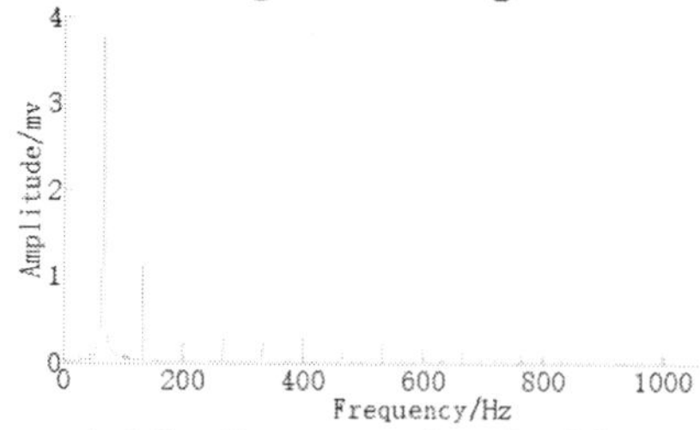

Figure 4. Misalignment signal of the rotor in frequency domain.

The rotor vibration signal and spectrum show that the dominant fundamental frequency component accompanies with the second harmonic component, indicating that the rotor is running with misalignment fault. According to the typical fault recognition process of the rotor system summarized in 2.4, the standard deviation of the calculated signal is 4.26. Therefore, set the initial parameter ε=0.85, N=100, and conduct EEMD on the signal. IMFs are obtained and shown in figure 5.

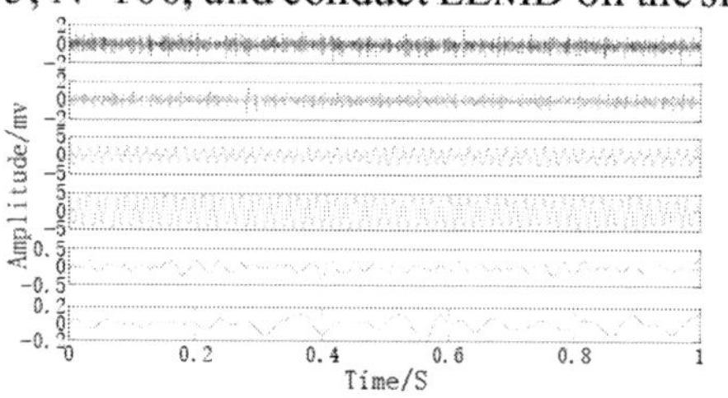

Figure 5. IMFs of misalignment signal in time domain(ε=0.85).

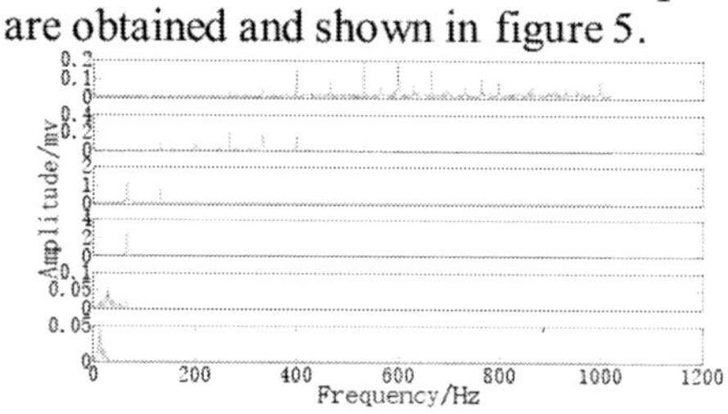

Figure 6. IMFs of misalignment signal in frequency domain (ε=0.85).

In figure 5, the main vibration modes of the signal are concentrated in IMF3 and IMF4. Thus, FFT is applied to the first six IMFs, and the spectrum of each IMF is shown in figure 6.

As shown in figure 6, IMF1 and IMF2 have broadband characteristics with small amplitude and are mainly concentrated in high frequency area. Therefore, IMF1 and IMF2, which mainly represent noise, can be deleted. Both IMF3 and IMF4 contain the fundamental frequency components of the signal, which shows significant mode aliasing phenomenon. Hence, IMF3 and IMF4 should be selected. However, IMF5 and IMF6 display the low-frequency noise characteristics with weak amplitude, implying the deletion of IMF5 and IMF6. The result demonstrates that there is a strong correlation between IMF3 and IMF4 with mode aliasing. IMF3 and IMF4 are then used for signal reconstruction. The reconstructed signal and its spectrum are shown in figure 7 and figure 8.

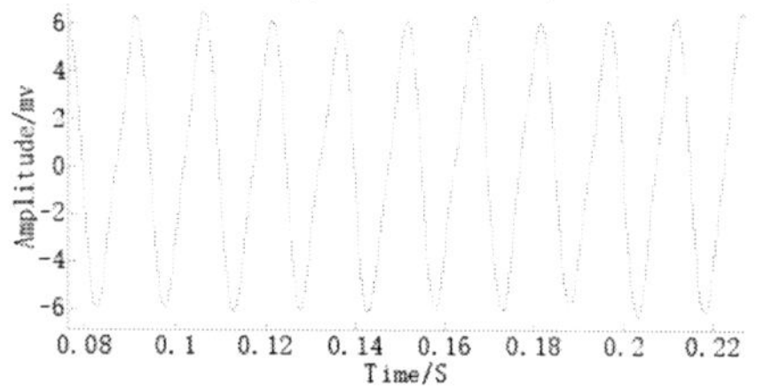

Figure 7. Reconstruction of misalignment signal of the rotor in time domain(ε=0.85).

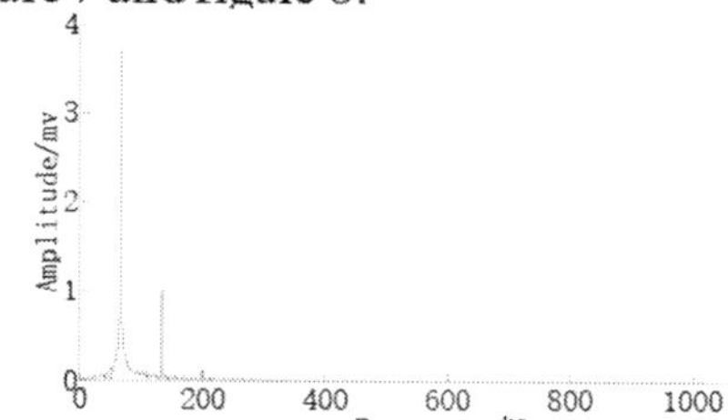

Figure 8. Reconstruction of misalignment signal of the rotor in frequency domain(ε=0.85).

Figure 7 and figure 8 show that the misalignment vibration mode has been extracted, and the high frequency noise is suppressed accompanying with weak low frequency noise. However, it is obvious that the mode aliasing is significant. Therefore, ε can be larger according to that the rotor vibration frequency mainly manifests low frequency property. Tentatively, it is adjusted to be 2.0. EEMD with a new noise amplitude is carried out on the investigated signals. IMFs and its spectrum can be observed in figure 9 and figure 10.

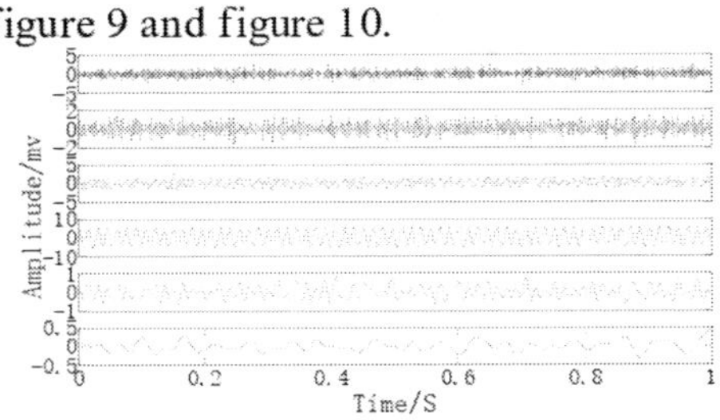

Figure 9. IMFs of misalignment signal in time domain(ε=2.0).

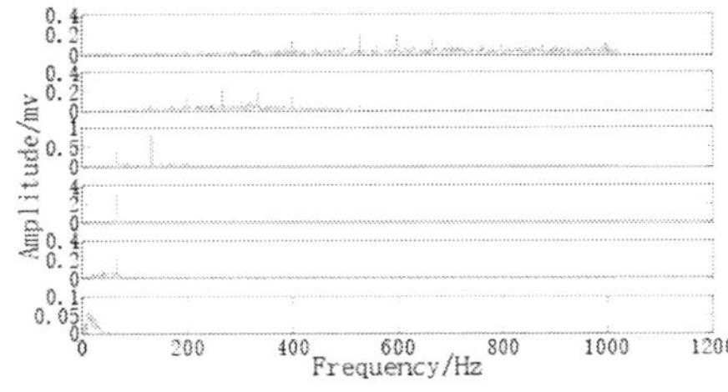

Figure 10. IMFs of misalignment signal in frequency domain(ε=2.0).

The comparison between figure 5 and figure 9, figure 6 and figure 10 show that IMF3 has the second harmonic component accompanied with a slight basic frequency, while the basic frequency is dominant in IMF4 without the interference from the second harmonic component. IMF5 gives a slight basic frequency component with a weak low-frequency noise. The mode aliasing phenomenon has been improved. Therefore, IMF3, IMF4 and IMF5 are used for signal reconstruction, and the reconstructed signal and its spectrum are shown in figure 11 and figure 12.

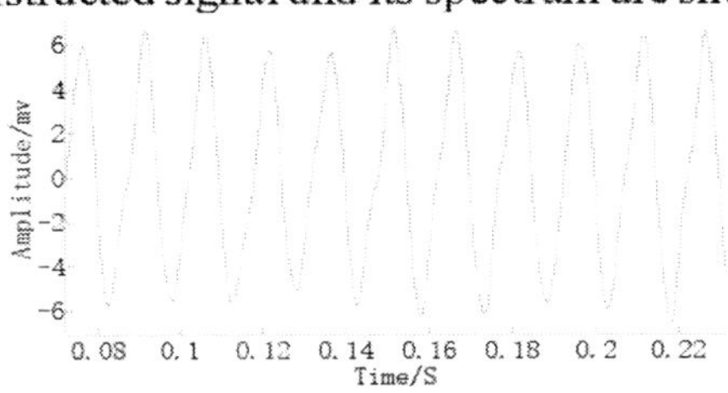

Figure 11. Reconstruction of misalignment signal of the rotor in time domain(ε=2.0).

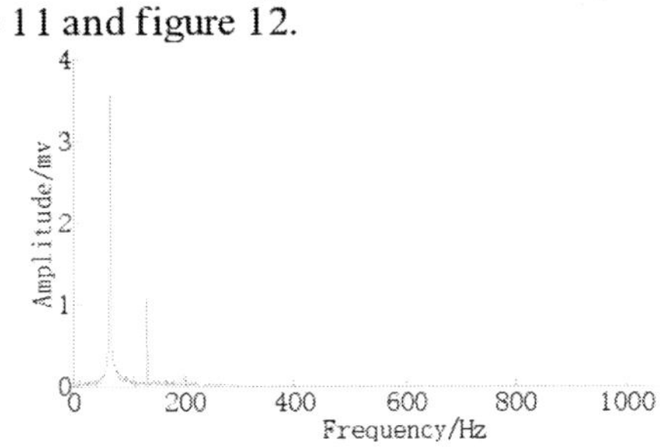

Figure 12. Reconstruction of misalignment signalof the rotor in frequency domain(ε=2.0).

Compared with figure 7, the fluctuation of the reconstructed signal is relatively higher when ε is equal to 2.0. The main reason is that the noise amplitude is too large and there is no obvious difference between the spectrums. Therefore, the fluctuation and change rule of reconstructed signal can be calculated iteratively, and appropriate noise amplitude can be selected. On the base of eliminating mode mixing, it ensures the small change of the reconstructed signal, and the IMF component has a clearer meaning of single vibration mode. At this point, the iterative search is conducted between the range of 0.85~2.0. The orthogonal change of IMF3 and IMF4, and the fluctuation of the reconstructed signals are shown in figure 13 and 14.

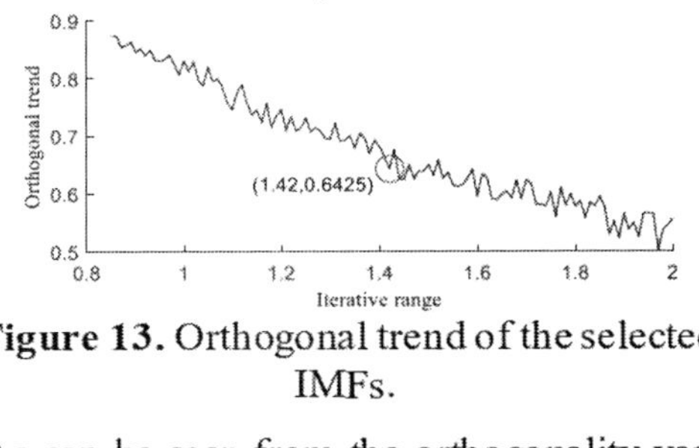

Figure 13. Orthogonal trend of the selected IMFs.

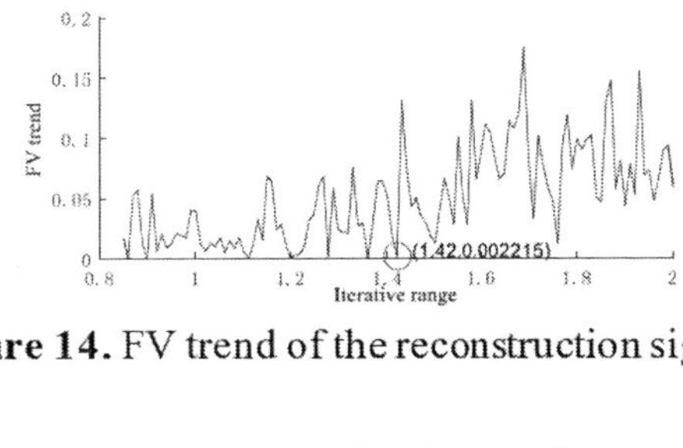

Figure 14. FV trend of the reconstruction signal.

As can be seen from the orthogonality variation rule of IMF, the value of orthogonality generally shows a trend of fluctuation linear decline with the increase of noise amplitude, which is consistent with the rule that the noise enhancement causes the orthogonality between the signals improved. The fluctuation variation rule of the reconstructed signal was observed further. Taking ε of 1.40 as the dividing line, the fluctuation on the left has a convergence point close to zero. On the right side, except the 1.42 marked red circle, there is no convergence point close to zero, and there is a tendency of increasing fluctuation. The noise amplitude of 2.0 is inconsistent with the selection condition given by the formula (6). Therefore, 1.42 marked red circle is the critical value reflected in the figure, and this point also corresponds to the 1.42 extreme value point of the marked red circle in the graph of the orthogonal fluctuation change rule. At this point, it can be judged that when ε is equal to 1.42, the orthogonality of components is guaranteed and the signal fluctuation of reconstruction changes little. The obtained IMFs and spectrum are shown in figure 15 and 16.

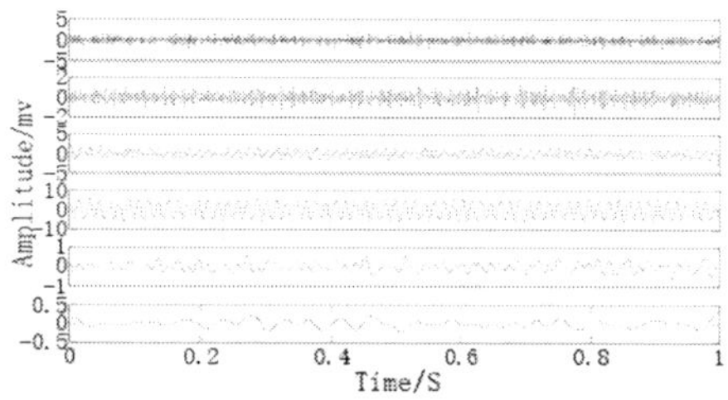

Figure 15. IMFs in time domain(ε=1.42).

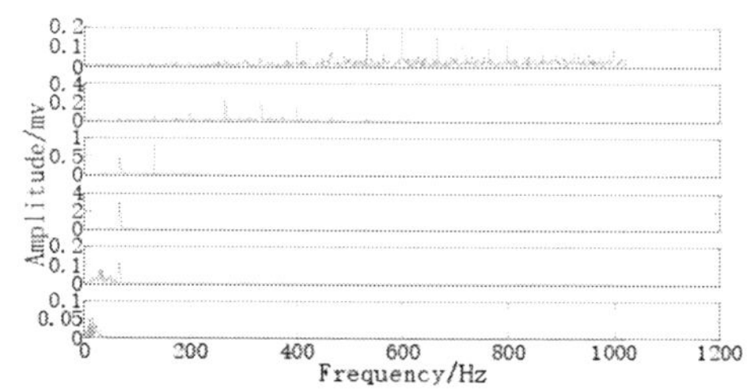

Figure 16. IMFs in frequency domain(ε=1.42).

Figure 15 and 16 show that the mode aliasing of basic frequency has been suppressed to half of the result whose ε is 0.85, which ensures the independent extraction of vibration mode. Moreover, the basic frequency component of IMF5 is extremely small and can be ignored. The reconstructed signal and spectrum with the selected IMF3 and IMF4 are shown in figure 17 and figure 18.

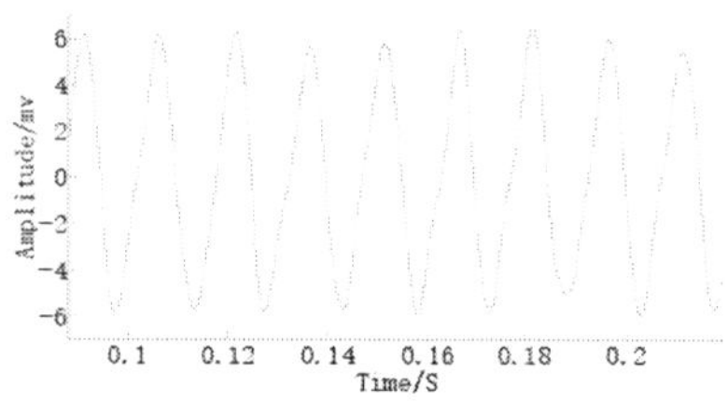

Figure 17. Reconstruction of misalignment signal of the rotor in time domain (ε=1.42).

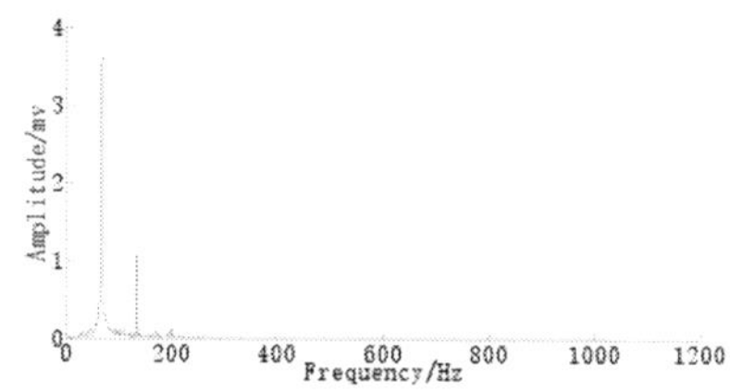

Figure 18. Reconstruction of misalignment signal of the rotor in frequency domain (ε=1.42).

Figure 17 and figure 18 show the extracted main misalignment mode. The high frequency noise is eliminated, and the fluctuation of the time-domain signal is less than that of the case of 2.0. The results show that by introducing the FV index, the qualitative learning process of traditional EEMD noise amplitude can be transformed into the quantitative learning process, leading to a clear physical significance of the signal components.

4.2. Fault signal analysis of bearing inner ring

EFPEEMD is verified by the fault signal of the bearing inner ring simulated on a bearing test bench from Case Western Reserve University as shown in figure 19.

Figure 19. Structure of the bearing test bench.

The bearing test bench consists of a 2 hp motor (left), a torque transducer/encoder (center), a dynamometer (right), and control electronics. The test bearings support the motor shaft. Single point faults were introduced to the test bearings using electro-discharge machining with fault diameters of 7 mils, 14 mils, 21 mils, 28 mils, and 40 mils (1 mil=0.001 inches). Here, the 6205-2RS JEM SKF, deep groove ball bearing whose inner race has fault diameters of 21 mils is employed. The corresponding defect frequency is 5.41 Hz. Moreover, the sampling frequency is 12 kHz, the sampling length is 122136 and the rotating frequency is 1797 rpm. As a result, the feature frequency of the inner ring is 162.03 Hz. The vibration signal both in time and frequency domain are respectively shown in figure 20 and figure 21.

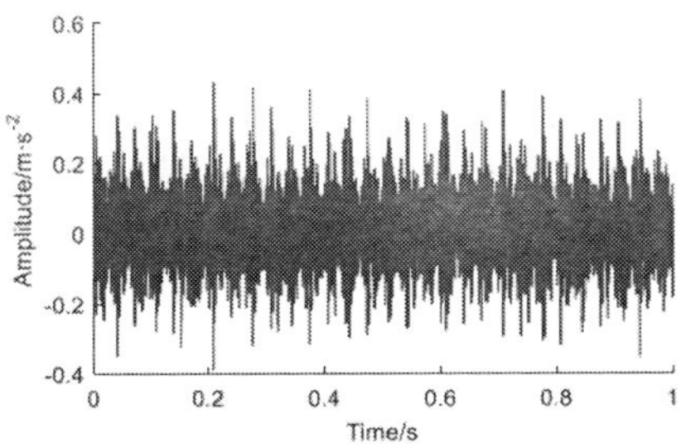

Figure 20. Inner ring defect signal of the bearing in time domain.

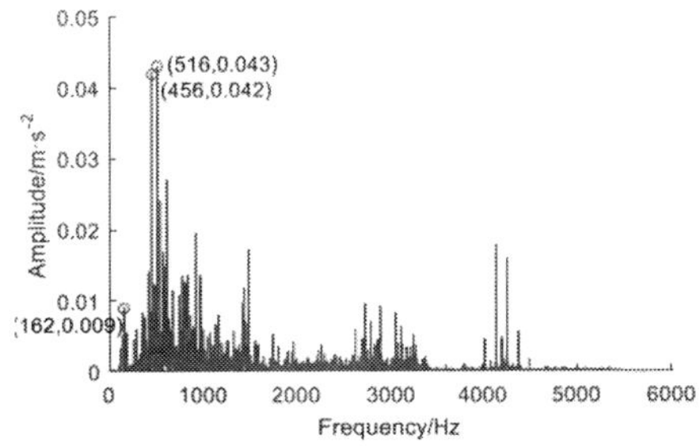

Figure 21. Inner ring defect signal of the bearing in frequency domain.

The signal of bearing inner ring and its spectrum show the frequency of 162 Hz of bearing vibration and the frequency modulation of high harmonics. However, the rotating frequency is not obvious. The signal envelope and EFPEEMD are used to decompose the signal, and the effects of IMF in time domain and frequency domain are obtained in figure 22 and figure 23.

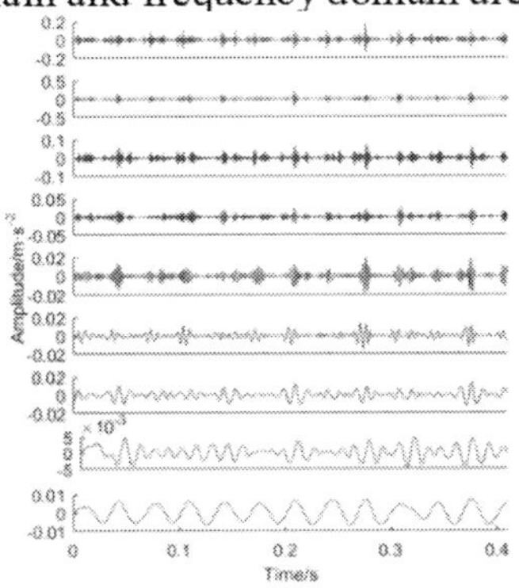

Figure 22. EFPEEMD IMFs of Inner ring defect signal of the bearing in time domain.

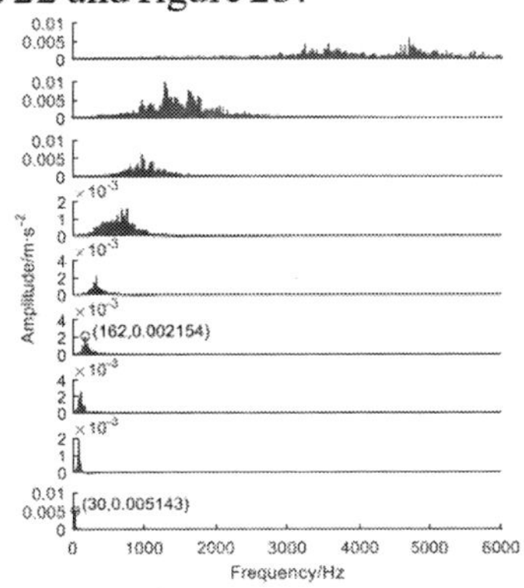

Figure 23. EFPEEMD IMFs of Inner ring defect signal of the bearing in frequency domain.

In figure 22 and figure 23, we can obtain the defect feature frequency of the inner ring and the rotating frequency in IMF6 and IMF9, respectively. The amplitude of 162 Hz component is 0.002154 m . s-2, and the amplitude of 30 Hz component 0.005143 m . s-2. Here, we use the two variables to compare the effect between EFPEEMD and EEMD.

In figure 24 and figure 25, the defect feature frequency of the inner ring and the rotating frequency are also located in IMF6 and IMF9. Moreover, the amplitude of 162 Hz component is 0.002122 m . s-2, and the amplitude of 30 Hz component 0.005137 m . s-2. As a result, the feature extraction capability of EFPEEMD is higher than that of EEMD, while the difference is small. It may attribute to the much more complexity of the bearing signal than that of the rotor signal.

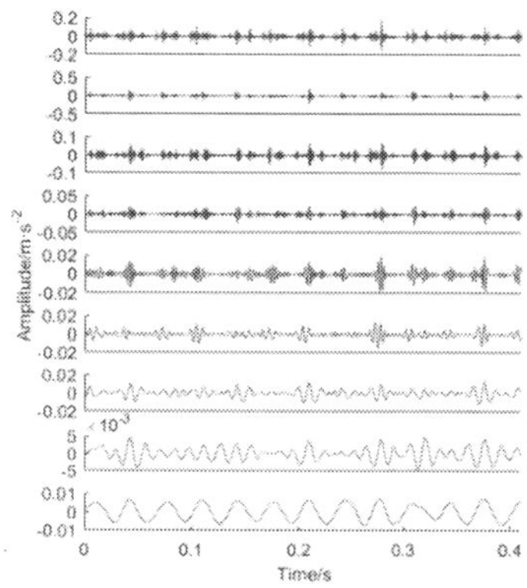

Figure 24. EEMD IMFs of Inner ring defect signal of the bearing in time domain.

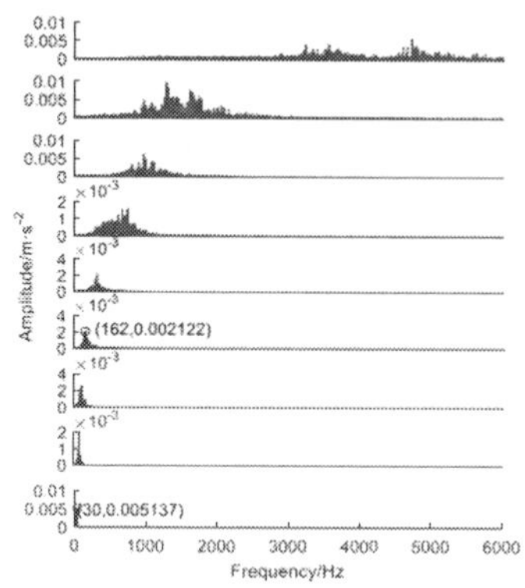

Figure 25. EEMD IMFs of Inner ring defect signal of the bearing in frequency domain.

5. Conclusions

FV index was developed to realize the quantitative calculation of the EEMD noise amplitude, which provides an efficient method to deal with problem of mode aliasing. EFPEEMD is significantly better than EEMD on the analysis of the rotor misalignment fault. The bearing signal is much more complex than the rotor signal, which causes the result of EFPEEMD close to that of EEMD. There are still several aspects that we can pursue to further improve the diagnostic performance. 1) Intelligent optimization algorithm can be attempted to search a suitable noise amplitude by iterating FV indicator. The flow should be investigated. 2) With the well-separated IMFs, the running condition of rotating machinery can be identified by adding deep learning to establish the mapping relation.

References

[1] Chen BQ. Overcomplete Frame Expansion Based on Wavelet Tight Frame and its Applications to Machinery Fault Detection and Diagnosis[D]. Xi'an: Xi'an Jiaotong University Press, 2013.

[2] Feng ZP, Chen XW. Adaptive iterative generalized demodulation for nonstationary complex signal analysis: Principle and application in rotating machinery fault diagnosis[J]. Mechanical Systems and Signal Processing, 2018, 110.

[3] Zhao HY, Wang JD, Jay L, et al. A compound interpolation envelope local mean decomposition and its application for fault diagnosis of reciprocating compressors[J]. Mechanical Systems and Signal Processing, 2018, 110.

[4] Huang NE, Shen Z, Long SR, et al. The empirical mode decomposition and the Hilbert spectrum for nonlinear and non-stationary time series analysis[J]. Proceedings of the Royal Society of London Series A: Mathematical, Physical and Engineering Sciences, 1998, 454 (1971): 903-995.

[5] He ZJ, Zi YY, Zhang XN. Modern signal processing and engineering application[M]. Xi'an: Xi'an Jiaotong University Press, 2007.

[6] Yu DJ, Cheng JS, Yang Y. Application of EMD method and Hilbert spectrum to the fault diagnosis of roller bearings[J]. Mechanical Systems and Signal Processing, 2005, 19 (2): 259-270.

[7] Gai GH. The processing of rotor startup signals based on empirical mode decomposition[J]. Mechanical Systems and Signal Processing, 2006, 20 (1): 222-235.

[8] Liu B, Riemenschneider S, Xu Y. Gearbox fault diagnosis using empirical mode decomposition and Hilbert spectrum[J]. Mechanical Systems and Signal Processing, 2006, 20 (3): 718-734.

[9] Huang N, Shen Z, Long SR. A new view of non-linear water waves: The Hilbert Spectrum[M], 2003.

[10] Zhao JP. Study on the Effects of Abnormal Events to Empirical Mode Decomposition Method and the Removal Method for Abnormal Signal[J]. Journal of Ocean University of Qingdao, 2001, 31 (6): 805-814.
[11] Rato RT, Ortigueira MD, Batista AG. On the HHT, its problems, and some solutions[J]. Mechanical Systems and Signal Processing, 2008, 22 (6): 1374-1394.
[12] Hu AJ, Sun JJ, Xiang l. Mode Mixing in Empirical Mode Decomposition[J]. Journal of Vibration, Measurement &Diagnosis, 2011, 31 (4): 429-434.
[13] Wu Z, Huang NE. Ensemble Empirical Mode Decomposition: A Noise-Assisted Data Aanlysis Method[J]. Advances in Adaptive Data Analysis, 2009, 01 (01): 1-41.
[14] Cao CF, Yang SX, Jiang XY. De-noising method for non-stationary vibration signals of large rotating machineries based on ensemble empirical mode decomposition[J]. Journal of Vibration and Shock, 2009, 28 (9): 33-38.
[15] Yu Y, Lang H. Fault diagnosis of rotor rub based on ensemble EMD[C], 2009: 2-144-142-148.
[16] Lei YG, He ZJ, Zi YY. Application of the EEMD method to rotor fault diagnosis of rotating machinery[J]. Mechanical Systems and Signal Processing, 2009, 23 (4): 1327-1338.
[17] Lei YG, Li NP, Lin J, et al. Fault Diagnosis of Rotating Machinery Based on an Adaptive Ensemble Empirical Mode Decomposition[J]. Sensors, 2013, 13 (12): 16950.
[18] Chen RX, Tang BP, Lv ZL. Ensemble Empirical Mode Decomposition De-noising Method Based on Correlation Coefficients for Vibration Signal of Rotor System[J]. Journal of Vibration, Measurement &Diagnosis, 2012, 32 (4): 542-546.
[19] Zhang WB, Zhou YJ, Zhu JX, et al. A New Rotor Fault Diagnosis Method Based on EEMD Sample Entropy and Grey Relation Degree[J]. Applied Mechanics and Materials, 2013, 347-350: 426-429.

Acquisition design and implement of Beidou B3I signal based on GPU

Gezheng Li [1,*], Qing Chang[1], Chuang Xi[1] and Xu Ma[2]

[1]School of electronic information engineering, Beihang University, 100191, Beijing, China

[2]China satellite navigation center, 100191, Beijing, China

*E-mail: ligezheng@foxmail.com

Abstract. China's Beidou satellite navigation system (BDS) has developed rapidly in recent years. In 2018, China recently publicized the latest Beidou system space signal interface control file (ICD) of the Beidou B3I. Since the Beidou B3I signal has only been released for a short time, there are no references for its implementation. Based on the research on the development status of signal acquisition at home and abroad, this paper studies the real-time problem of software receivers and optimize the module in the capture process on graphic processing unit (GPU) according to the characteristics of B3I signal, design and implement a Beidou B3I software receiver capture architecture and algorithm based on GPU. By comparing with acquisition on CPU, each computing module achieves a GPU/CPU acceleration ratio of 4-50 times, the overall signal acquisition speed has been improved by 2.5 times, and the real-time performance of Beidou B3I software receiver has been effectively improved. At the same time, aiming at the problem of Beidou signal acquisition accuracy, a secondary capture method is designed to improve the capture accuracy. In addition, an optimization idea based on compute unified device architecture (CUDA) programming is proposed to further parallel computing performance on GPU.

1. Introduction

The BDS-2 includes B1, B2 and B3 bands, of which B1 and B2 are for civilian use and B3 is for military use[1]. With the rapid development of GNSS systems and continuous improvement of signal processing algorithms, the software receiver based on PC only needs to replace the functional modules in the software, which has the characteristics of low cost, good openness and expandability, has become a research hotspot in receiver design. Aiming at the shortcomings of serial acquisition of the algorithm on the CPU, the implementation of the software receiver using the GPU with strong parallel processing capability has been widely used in recent years[2-3].

Beidou B3I is the latest publicly available frequency band in 2018 and with few references about it. When using B3I signal with a sampling rate of 62MHz, the huge amount of data affects the Beidou B3I software receiver's performance because parallel FFT acquisition costs the large amount of computation and time-consuming and there is few relevant information to introduce it.

Based on the FFT parallel code phase search method and the parallel processing ability of GPU, this paper transplanted Beidou B3I acquisition algorithm to GPU high-performance platform, designed and implemented Beidou B3I signal parallel acquisition algorithm based on CUDA, accelerated and optimized the software receiver of B3I frequency point with 62MHz sampling rate. Finally make it possible to meet real-time requirements. Meanwhile, for the Beidou system, due to the modulation of the Newman-Huffman code (NH code) on navigation message, its capture performance would be affected. Aiming this problem, we optimized the signal capture process of MEO/IGSO satellite which broadcasting D1 messages to improve capture performance.

2. Parallel-acquisition algorithm of beidou B3I signal

According to the ICD, the spread spectrum sequence of Beidou new signal system adopts the hierarchical structure of the combination of primary code and secondary code. The primary code uses a pseudo-random sequence code, and the secondary code uses Newman-Huffman code[4]. The C/A code of the Beidou B3I signal has a chip rate of 10.23Mcps, a period of 1ms, which is equal to the length of one chip of the secondary code, and the period of secondary code is 20ms equal to the length of one navigation message data bit[5].

The B3I signal is composed of a "ranging code + navigation message" modulated on the carrier, and its signal expression is:

$$S^j_{B3I}(t) = A_{B3I}C^j_{B3I}(t)D^j_{B3I}(t)\cos 2\pi f_3 t + \varphi^j_{B3I} \tag{1}$$

In the above formula, where the upper corner j is the satellite number, A is the amplitude of the B3I signal, C is the B3I signal ranging code, D is the data code modulated on the B3I signal ranging code, f_3 represents the B3I signal carrier frequency, φ is the initial phase of the B3I signal carrier[6].

Beidou B3I adopts code division multiple access (CDMA) spread spectrum communication technology. All satellite signals use the same center frequency and are modulated by different spreading codes. Signal acquisition is a two-dimensional search process. Currently, commonly used searching strategies include serial search, parallel frequency search, and parallel code phase search strategy. Parallel code phase search strategy replaces the correlation operation of digital correlator by using FFT[7], which greatly improves the efficiency of calculation, and is often used in the design and implementation of software receivers. Formula (2) is the specific transformation process.

$$z(n) = \frac{1}{N}\sum_{m=0}^{N-1} x(m)y(m-n) = IFFT[FFT(x(n))\cdot FFT^*(y(n))] \tag{2}$$

Figure 1 is the search principle based on parallel FFT code phase.

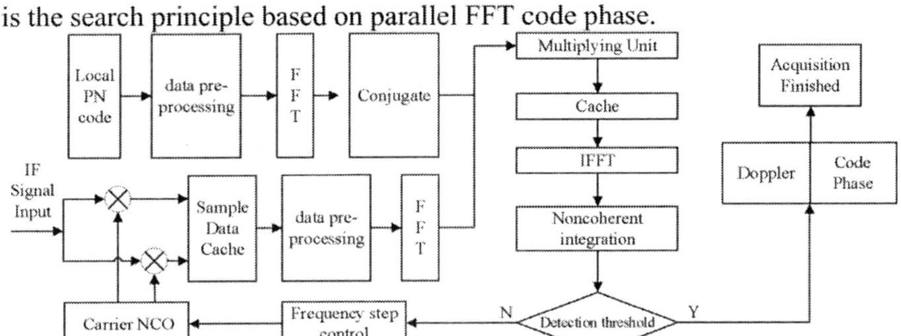

Figure 1. Parallel code phase search schematic.

According to figure 1, the implementation process of the parallel code phase search acquisition algorithm is as follows: First, the received IF signal is multiplied by the local carrier signal to obtain I and Q branch signals. Then, the FFT is performed on the complex signal composed of the signals. Next, Performing FFT transformation on the satellite pseudo random noise code (PRN) signal generated by the local PRN code generator to the frequency domain and then taking the conjugate. And multiply the frequency domain results in the above two steps and perform IFFT, take the modulus as the correlation result. Finally, perform non-coherent accumulation and determine the preset detection threshold to determine whether the acquisition is successful[8]. During the whole operation process, due to the large-point FFT operation and the circular search, there is a case where the calculation resource consumption is too large and the calculation takes too long.

3. B3I signal parallel capture scheme based on GPU

In 2006, NVIDIA released the GPU's general programming structure CUDA[9]. CUDA allows developers to develop without the need to learn complex graphical programming languages, greatly simplify the GPU development process. Moreover, CUDA is developed based on C language and is compatible with software receivers based on PC platform. Meanwhile, it can process large amounts of data with simple calculations in parallel, which is very conducive to the development of software receivers.

3.1. Design of acquisition algorithm based on CUDA

First, we need to transfer the IF data from the CPU to the GPU. But pay attention when programming with CUDA, if there is a dependent calculation between data and data, that is, the result of the current data is required when calculating the following data, at this point, the GPU will not be able to perform parallel calculations on the data[10]. In this paper, the blocks or functions that can be parallelized are allocated to the GPU, and the parts that cannot be paralleled will be run on the CPU, so that the respective advantages of the CPU and the GPU can be fully used. Although the GPU has strong ability in parallel computing, all the data calculated on the GPU needs to be copied from the CPU, which is the most time-consuming part. Therefore, the data transmission should be reduced as much as possible during optimization. The mapping between the portion of the signal capture on the GPU and the portion of the signal acquisition implemented on the CPU is illustrated from the following tables. The function mapping in the CPU is shown in table 1.

Beidou B3I is the latest publicly available frequency band in 2018 and with few references about it. When using B3I signal with a sampling rate of 62MHz, the huge amount of data affects the Beidou B3I software receiver's performance because parallel FFT acquisition costs the large amount of computation and time-consuming and there is few relevant information to introduce it.

Based on the FFT parallel code phase search method and the parallel processing ability of GPU, this paper transplanted Beidou B3I acquisition algorithm to GPU high-performance platform, designed and implemented Beidou B3I signal parallel acquisition algorithm based on CUDA, accelerated and optimized the software receiver of B3I frequency point with 62MHz sampling rate. Finally make it possible to meet real-time requirements. Meanwhile, for the Beidou system, due to the modulation of the Newman-Huffman code (NH code) on navigation message, its capture performance would be affected. Aiming this problem, we optimized the signal capture process of MEO/IGSO satellite which broadcasting D1 messages to improve capture performance.

Table 1. Function mapping in the CPU.

Map	Function
Mapping in the CPU	Read IF data Dependent calculation Store acquisition results

The function mapping in the GPU is shown in table 2.

Table 2. Function mapping in the GPU

Map	Function
Mapping in the GPU	Mixer Correlator Coherent/Incoherent integration C/A code generator Carrier NCO

In this paper, the length of data used for capture is 10ms, the sampling frequency of the processed signal is 62MHz, and so there are 62000 sample points per 1ms of data transferred to the GPU. And 3D search range of satellite signals is: (1) 12 Beidou satellites. (2) The doppler frequency searching range is ±10kHz, the frequency searching step is 500Hz, So the searching times is 41. (3) Search

10230 chips, the search step is half a chip, and a total of 20460 code phases need to be searched.The algorithm flow is shown in figure 2.

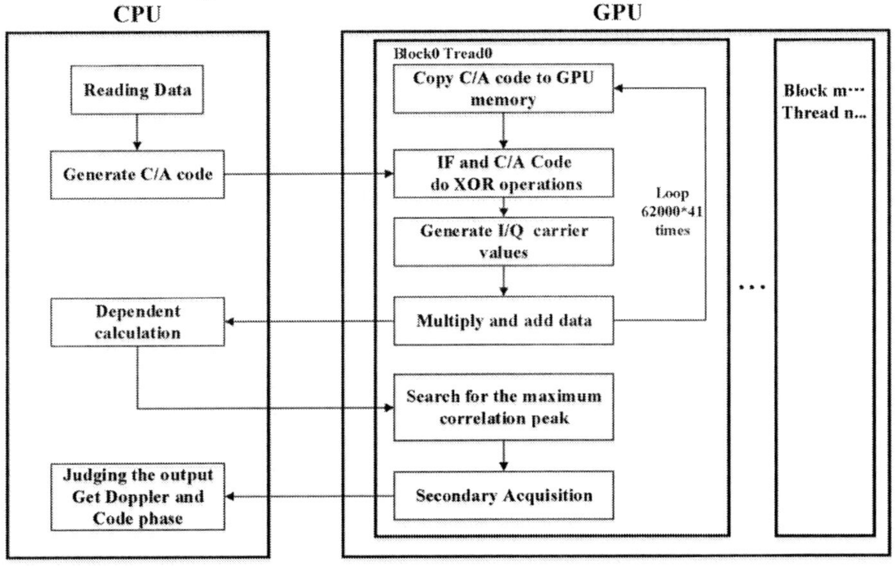

Figure 2. Schematic diagram based on GPU capture algorithm.

In the figure 2, we can see that in order to reduce the time consumption in the copy process, the locally generated C/A code need to be copied to the GPU calling function cudaMemcopy first, so that each time the data is used, it is not necessary to spend time copying again. Then copy the data that needs to perform FFT and IFFT calculation to the GPU, use the CUFFT library in CUDA and call the cufftExecC2C function for parallel computing. And other modules like exclusive-OR operation on the IF signal and the C/A code, multiply and add the I/Q two-way carrier value data are written in the GPU as corresponding kernel functions, which will minimize the copy frequency. Finally, copy the calculated result from the memory on GPU back to the CPU for non-coherent accumulation.

Next step we need to do is searching for the maximum correlation peak, which is a 2D search range of 41*31000. Copy the non-coherent accumulation data from the CPU back to the memory, calling the cublasIsamax function in the CUBLAS library in CUDA. The function it has to achieve is to get the index number of the first largest modulus in a vector. The first time using cublasIsamax function to calculate 31,000 points in parallel, which only needs to count 41 times, the second time using the function is to calculate the maximum value of 41 points, this method can further optimize the calculation speed.

3.2. Secondary capture of B3I signals

The Beidou satellite of type MEO/IGSO broadcasts the D1 navigation message. Since the NH code is modulated on the D1 message, it will have a certain impact on the accuracy of signal acquisition. In order to improve the accuracy of B3I signal acquisition, this paper will capture the signal that reaches the threshold value for the first time in the satellite of MEO/IGSO for a second time, as shown in figure 2. The software receiver will again perform the capture procedure at the position of each half step before and after the currently captured frequency, and find the largest peak among them as the new maximum correlation peak. This method will improve the accuracy of the capture, and provide relatively more accurate doppler value and code phase for the next tracking process.

4. Simulation verification and result analysis

4.1. Hardware environment of the simulation

1)Platform environment: This paper uses Visual Studio 2015 platform, CUDA8.0 for simulation testing, the CPU is an Intel Core i7-6700 chip with a frequency of 3.40GHz. The GPU is GeForce GT 730 graphics card of NVIDIA with a memory capacity of 1024MB and a memory frequency of 5012MHz.

2)Experimental data: The Beidou B3I IF data used in this paper is generated from an analog signal source generator, which sampling rate is 62MHz, and the center frequency is 15.52MHz.

4.2. Simulation results and analysis

The simulation completed the search of 12 channels of the Beidou B3I frequency point, and tested the running time of 1 non-coherent integration of single satellite on CPU and GPU. Choose the parts with relatively large calculation like FFT on local C/A code, generate co-directional and orthogonal carriers, FFT on carrier, Code multiplication, IFFT and search for the maximum correlation peak are selected for efficiency comparison, and the simulation results are shown in table 3.

Table 3. Calculation time of each function module.

Function module	accumulation time	CPU Time(ms)	GPU Time(ms)
C/A code FFT	1	0.469	0.1227
Generate co-directional and orthogonal carriers	1	4.923	0.3675
Carrier FFT	1	0.467	0.1289
Code multiplication	1	2.377	0.0431
IFFT	1	0.593	0.1413
Maximum peak search	1	3.039	0.2364

As shown in table 3, the same method is used to implement the acquisition function on the CPU and GPU, the CPU cannot meet the real-time requirements for processing 1ms of data. But the calculation speed of each function module on the GPU is significantly higher than that on the CPU. Then satellite signal acquisition is performed on all 12 channels, and the acquisition time of each channel shown in figure 3.

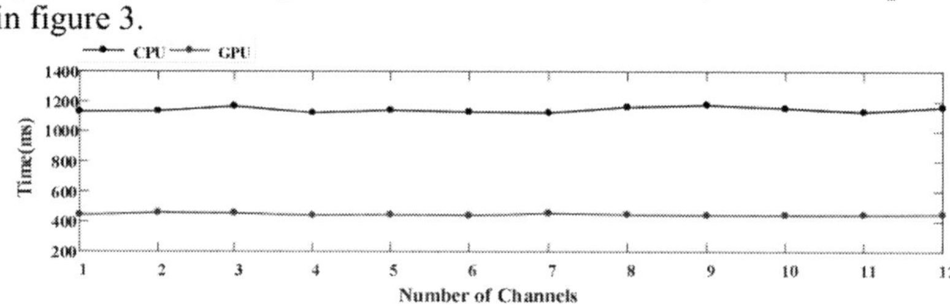

Figure 3. Time of each Channel.

The correlation peak results calculated when using the GPU for signal acquisition are plotted as shown in figure 4. A significant peak can be found, indicating that the signal was successfully captured and the result is correct.

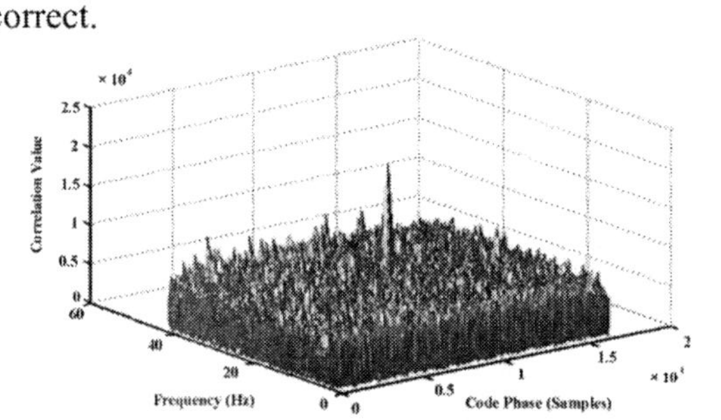

Figure 4. 2D parallel search related peak results.

From figure 3, we can see that the Beidou B3I signal capture speed based on GPU is much higher than that on the CPU, but the overall acceleration effect is not as good as the acceleration effect of each module. The reason for this result is that when executing a program on the GPU, the data needs to be copied from the memory on CPU to the memory on GPU. And after the calculation is finished, the data needs to be copied back to the memory on CPU. This part has a great impact on the efficiency of the program. Therefore, it is necessary to reduce the data transmission between the CPU and the GPU as much as possible. Meanwhile, it is also necessary to reduce the small-scale data transmission, and integrate the fragmented data into a large amount of data transmission as much as possible, maximizing the advantage of parallel computing on GPU, allowing the GPU to process more threads at once and allocate thread blocks and number of threads reasonably[11].

5. Summary

According to the newly disclosed Beidou signal frequency band of B3I, this paper designs and implements the Beidou B3I signal parallel acquisition algorithm based on CUDA, using the parallel architecture of GPU for the problem of poor real-time performance of software receivers based on CPU, and improves the real-time performance of software receivers at high sampling rates. The simulation results show that the software receiver optimized by GPU has an obvious acceleration effect. Under the condition of 62MHz sampling rate and 4bit quantization, the calculation speed of the function modules in the acquisition process has increased by 4-50 times and the entire acquisition process runs 2.5 times faster, enhancing the real-time capture performance of B3I signals. In order to solve the problem that the NH code is modulated on the D1 message of the B3I signal broadcasting by the MEO/IGSO satellite, which acquisition accuracy would be affected, the maximum value is adjusted by a second acquisition method near the maximum correlation peak of the captured result to further improve the accuracy of the acquisition.

As there are many kinds of memory on the GPU, such as registers, shared memory, constant memory, etc. and have their specific characteristics. Storing different types and sizes of data in the appropriate memory in the GPU, will make memories perform their optimal performance, which can further improve the capability and speed of GPU parallel computing, and maximize the real-time performance of the software receiver. In the future research, we will continue to study the software receiver based on GPU, which is conducive to the establishment of a high-performance simulation platform for technology development, algorithm improvement and innovation in the field of satellite navigation, which will greatly improves the efficiency of developers.

References

[1] Stirling-Gallacher R A, Hulbert A P. A fast acquisition technique for a direct sequence spread spectrum signal in the presence of a large doppler shift[C]. Spread Spectrum Techniques and Applications Proceedings, 4th International Symposium on IEEE, **1**: p. 156-160. (1996)

[2] Park K W, Yang J S, Park C, et al. Implementation of GPGPU based real-time signal acquisition and tracking modulefor multi-constellation GNSS software receiver. Proceedings of the 27th International Technical Meeting of the Satellite Division of the Institute of Navigation, Tampa, p. 1410-1416. (2014)

[3] Pany T, Riedl B, Winkel J. Efficient GNSS signal acquisition with massive parallel algorithms using GPUs. Proceedings of the 23rd International Technical Meeting of the Satellite Division of the Institute of Navigation, Portland, p.1889-1895. (2010)

[4] Niu, H, W. Research on the key technologies of signal acquisition for BEIDOU satellite [D]. China:Taiyuan university of technology, p. 6-7. (2016)

[5] China satellite navigation system management office. Beidou satellite navigation system Spatial signal interface control file Public service signal B3I (version 1.0). (2018.2.)

[6] Tsui J B Y. Fundamentals of Global Positioning System Receivers [M]. 2nd Ed. New York: Wiley-Interscience Publication, p.133-159. (2005)

[7] Gao, J, L, Liu, P, L, Chen, X. Acceleration of signal processing in broadband GNSS receiver

based on CUDA [J]. Information technology, 1: p.122-122. (2017)
[8] VAN NEE S J R, GOENEN A J R M. New fast GPS code-acquisition technique using FFT [J]. Electronics Letters, **17**: p.158-160. (1991)
[9] Sanders J, Kandrot E. CUDA by Example: An Introduction to General-Purpose GPU Programming [M]. Boston: Addison-Wesley Professional, (2011)
[10] Kandrot E. CUDA by Example [M]. Amsterdam: Addison-Wesley Longman, (2010)
[11] Gu, Q, Gao, N, Bao, Z, Z. Implementation and optimization of high-speed AES algorithm based on GPGPU and CUDA[J]. Journal of the Chinese Academy of Sciences, **28**: p.776-785. (2011)

Theoretical study of wakefield acceleration of electrons in capillary Z-pinch plasma waveguide

A A Shapolov[1,*], B Fekete[1], M Kiss[1], S Szatmari[2] and S V Kukhlevsky[1]

[1]Department of Physics, University of Pecs, Pecs, Hungary
[2]Department of Experimental Physics, University of Szeged, Szeged, Hungary

*Email: sapolov@gamma.ttk.pte.hu

Abstract. The theoretical investigation of wakefield acceleration of electrons by CO_2-laser pulse with central wavelength of 10.6 μm and input peak intensity of ~10^{17} W/cm^2 in transient hydrogen-plasma waveguide has been conducted. The plasma waveguide is produced by the fast Z-pinch discharge inside a 3 mm inner diameter and 50 mm long capillary. The waveguide properties of the capillary Z-pinch plasma were obtained from the space and frequency dependent wave equation that combines the attenuated charged particle inertia and the light wave effects in the plasma for the ideal Gaussian laser beam. For simulation of temporal and spatial evolution of electron density and other plasma variables during the capillary discharge, we used a standard one-fluid, two-temperature one-dimensional magnetohydrodynamic (MHD) model complemented with atomic data of hydrogen. Simulations showed that the guiding channel occurs far from capillary wall and exists for a few nanoseconds. In terms of laser driven plasma wakefield accelerators (LWFA), the Z-pinch waveguide is able to extend the acceleration length over the whole capillary, assuming that a single-mode transmission takes place. To quantify such ability of the channel, a correlation coefficient was introduced and computed at different input beam spot sizes. An optimal beam spot size was determined by taking maximum of time average of the coefficient over the channel lifetime. At the end of the guiding channel existence, the repetitive focusing and defocusing patterns were observed with intensity increase at the focal points. For simulation of the wakefield acceleration of electrons in the different waveguiding regimes we used the particle-in-cell (PIC) combined with the aforementioned MHD model. Influence of the waveguiding regimes on the electron acceleration was demonstrated.

1. Introduction

The waveguide effect of capillary discharge and its applications attract interest of many researchers [1-23]. The guiding of high intensity laser pulse has been widely used in laser driven plasma wakefield accelerators [1]. In the absence of guiding channel, the interaction length is basically limited by diffraction to the Rayleigh range [9]. In partially ionized plasmas, further limitation of the interaction length can be caused by ionization-induced refractive defocusing. Therefore, most of the researchers deal with hydrogen gas filled micro-capillary discharge waveguides [5-14], [16-22] in which the guiding core is extended to the capillary wall, i.e., the effects of the wall are inevitable. For example, the study [6] reported the variation of the pulse energy transmission during the ablating current and the temporal evolution of the transverse spatial profile of the exiting beam, as a result of many physical effects such as inverse bremsstrahlung, stimulated Raman scattering, stimulated Brillouin scattering, atomic modulation instabilities and the ionization of the discharge plasma loss mechanisms. Transient Z-pinch guiding channel was experimentally demonstrated in [5], where fully ionized He-plasma column generated in a capillary was used for channelling high-intensity (>10^{17} W/cm^2) Ti:sapphire laser pulse. The Ar-plasma capillary waveguide was demonstrated in [15]. The Ar-plasma capillary waveguide excited by Z-pinch was modelled in [15, 23]. The investigation of wakefield acceleration of electrons by laser pulse in plasma waveguides has been conducted in [16-21].

In this paper, we present the theoretical investigation of wakefield acceleration of electrons by CO_2-laser pulse with central wavelength of 10.6 μm and input peak intensity of $>10^{15}$ W/cm^2 in transient H-plasma waveguide. The plasma waveguide is produced by the fast Z-pinch discharge inside a 3 mm inner diameter and 50 mm long capillary. In Sec. 2, the waveguide properties of the capillary Z-pinch plasma were obtained from the space and frequency dependent wave equation that combines the attenuated charged particle inertia and the light wave effects in the plasma for the ideal Gaussian laser beam. For simulation of temporal and spatial evolution of plasma variables, during the capillary discharge, we used a standard one-fluid two-temperature one-dimensional MHD model [22, 23] complemented with atomic data of hydrogen (Sec. 2). Influence of the different waveguiding regimes on the wakefield acceleration of electrons was demonstrated in Sec. 3. For simulation of the wakefield acceleration of electrons we used the particle-in-cell (PIC) model [11-14] combined with the aforementioned MHD model (Sec. 3).

2. Waveguiding properties of the capillary Z-pinch plasma

Transverse electric field perturbations $\delta\vec{E}_t$ in plasma become light waves in the vacuum limit when the plasma effects are negligible. The inhomogeneous wave equation governing such perturbations can be written as

$$\nabla^2\delta\vec{E}_t - \frac{1}{c^2}\frac{\partial^2\delta\vec{E}_t}{\partial t^2} = \mu_0\frac{\partial\delta\vec{J}_e}{\partial t} \tag{1}$$

The inhomogeneous term $\delta\vec{J}_e$ represents the plasma effects. Because the EM waves in plasma are fast phenomena, we expect that the plasma response will be inertial. Thus, the current density perturbation in plasma induced by the electric field $\delta\vec{E}_t$ is given by $\delta\vec{J}_e = -en_e\delta\vec{v}_e$, where e is the electron charge, n_e is the electron density and $\delta\vec{v}_e$ is the electron velocity perturbation, respectively. The right-hand side of (1) can be rewritten as follows:

$$\frac{\partial\delta\vec{J}}{\partial t} \approx -en_e\frac{\partial\vec{v}_e}{\partial t}, \tag{2}$$

Since the capillary Z-pinch dynamic is much slower compared to the perturbation caused by the electric field. The motion of an electron with charge e and mass m_e in response to the electric field is governed by

$$m_e\frac{\partial\delta\vec{v}_e}{\partial t} = -e\delta\vec{E}_t - m_e f_{ei}\delta\vec{v}_e \tag{3}$$

where the last term on the right-hand side represents the impulse exchange between electrons and ions with a frequency f_{ei} in which the velocity of ions is neglected. Notice, we do not deal with the contribution of magnetic force. In that regard note, that the magnetic field of EM waves does not influence the magnitude of $\delta\vec{v}_e$ just change its direction. For electric field perturbations, we consider the wave solutions for ideal input Gaussian-beam $\delta\vec{E}_t(r,z,t) = \delta\tilde{\vec{E}}_t(r,z)\exp(i\omega t)$ in which $\delta\tilde{\vec{E}}_t(r,z)$ is space-dependent complex-valued electric field amplitude. For such perturbation, we expect a phase-shifted time harmonic inertial plasma response $\delta\vec{v}_e = \delta\tilde{\vec{v}}_e(r,z)\exp(i\omega t)$. Solving (3) basing on this form, we get the wave equation that combines the attenuated charged particle inertia and the light wave effects in a plasma:

$$\nabla^2\delta\vec{E}_t - \frac{1}{c^2}\frac{\partial^2\delta\vec{E}_t}{\partial t^2} - \frac{1}{c^2}\frac{\omega\omega_{pe}}{\omega - if_{ei}}\delta\vec{E}_t = 0 \tag{4}$$

where $\omega_{pe} = (e^2n_e/\varepsilon_0 m_e)^{1/2}$ is the electron plasma frequency. Using the proposed ideal input Gaussian-beam, (4) can be rewritten as the space- and frequency-dependent wave equation:

$$\left(c^2\nabla^2 + \omega^2\tilde{\varepsilon}_r\right)\delta\tilde{\vec{E}}_t(r,z) = 0 \tag{5}$$

where $\tilde{\varepsilon}_r = 1 - \omega^2 /(\omega^2 - i\omega f_{ei})$ is the relative permittivity of plasma, from which the refractive index is $\tilde{n}_r = \tilde{\varepsilon}_r^{1/2}$. Thus, the electric field of light wave in capillary Z-pinch determined by (4) and (5) depends on the refractive index of dynamically changing plasma channel via ω_{pe} and f_{ei}.

For the simulation of temporal and spatial evolution of the electron density and other plasma variables during the capillary Z-pinch discharge, we used a standard one-fluid, two-temperature, 1-D MHD model [22, 23]. In a cylindrical coordinate system. Equations of the model are as follows:

$$\frac{\partial n_i}{\partial t} + \frac{1}{r}\frac{\partial}{\partial r}(r v_r n_i) = 0 \tag{6}$$

$$m_i n_i\left(\frac{\partial v_r}{\partial t} + v_r\frac{\partial v_r}{\partial r}\right) = J_z\frac{\partial A_z}{\partial r} - \frac{\partial p}{\partial r} - \frac{1}{r}\frac{\partial}{\partial r}(r\pi_{rr}) + \frac{\pi_{\varphi\varphi}}{r} \tag{7}$$

$$\frac{3}{2}n_e\left(\frac{\partial T_e}{\partial t} + v_r\frac{\partial T_e}{\partial r}\right) = -\frac{p_e}{r}\frac{\partial}{\partial r}(r v_r) - \frac{1}{r}\frac{\partial}{\partial r}(r q_e^r) - \pi_e^{rr}\frac{\partial v_r}{\partial r} - \pi_e^{\varphi\varphi}\frac{v_r}{r} - \pi_e^{rz}\frac{\partial v_e^r}{\partial r} + Q_e \tag{8}$$

$$\frac{3}{2}n_i\left(\frac{\partial T_i}{\partial t} + v_r\frac{\partial T_i}{\partial r}\right) = -\frac{p_i}{r}\frac{\partial}{\partial r}(r v_r) - \frac{1}{r}\frac{\partial}{\partial r}(r q_i^r) - \pi_i^{rr}\frac{\partial v_r}{\partial r} - \pi_i^{\varphi\varphi}\frac{v_r}{r} + Q_i \tag{9}$$

$$-\frac{1}{r}\frac{\partial}{\partial r}\left(r\frac{\partial A_z}{\partial r}\right) = \mu_0 J_z \tag{10}$$

$$\frac{\partial J_z}{\partial t} + \frac{1}{r}\frac{\partial}{\partial r}(r v_r J_z) = \frac{e}{r m_e}\frac{\partial}{\partial r}(r\pi_{rz}) \tag{11}$$

$$n_e - \langle Z\rangle n_i = 0 \tag{12}$$

Here, n_e - electron density, n_i - ion density, v_r - plasma radial velocity, J_z - axial current density, T_e - electron temperature, T_i - ion temperature, $\langle Z\rangle$ - the average ion charge, $p = p_e + p_i$ - plasma pressure, Q_e - electron heating, Q_i - ion heating, A_z - the axial vector potential, $\pi_{\alpha\beta} = \pi_e^{\alpha\beta} + \pi_i^{\alpha\beta}$ - stress tensor, q_e^r - radial electron heat flow and q_i^r - radial ion heat flow. Spatial evolution of plasma variables are treated in Euler's specification of the flow field, whereas the total number of ions in capillary is constrained to be unchanged, i.e.

$$n_0 = \frac{p_0}{T_0} \equiv \frac{2}{R^2}\int_0^R r n_i dr \tag{13}$$

where R is the capillary radius, and n_0, p_0, T_0 are the initial neutral particle density, gas pressure and temperature, respectively. The electric current is assumed to be sinusoidal:

$$I(t) = I_{\max}\sin\left(2\pi\frac{t}{T_{cur}}\right) \equiv 2\pi\int_0^R r J_z dr, \tag{14}$$

where $I_{\max}$ and T_{cur} are the peak and cycle duration of the electric current, respectively. The system of equations (6)-(12) was solved for the boundary conditions corresponding to no-diffusive wall $\left(\partial n_i/\partial r|_{r=R} = 0\right)$, no-viscous wall $\left(\partial v_r/\partial r|_{r=R} = 0\right)$, in case of ions thermally insulated wall $\left(\partial T_i/\partial r|_{r=R} = 0\right)$ and in regards of electrons thermally conductive wall caused by magnetic pressure acting on electrons $\left(q_e^r \propto T_e J_z (\partial A_z/\partial r)|_{r=R} \neq 0\right)$. Nernst effect was also taken into consideration.

The general waveguiding properties of the capillary Z-pinch plasma have been derived for different gases. We have used plasma variables provided by the MHD model. The peak value of the current pulse, half-cycle duration, and the initial gas pressure were set to 30 kA, 100 ns and 0.6 mbar, respectively. The studies were conducted for CO_2-laser pulse with an input peak intensity of 10^{15} W/cm^2 and different spot sizes of $w_0 = 80 \ldots 200$ µm. As an example, figure 1-4 illustrate evolution of the spatial and spectral intensity distributions of guided laser beam with optimal spot size of $w_0^{opt} = 134$ µm in Ne and Kr-plasma. The spatial distribution was computed at the central wavelength ($\lambda = 10.6$ µm) of CO_2-laser pulse. In case of spectral distribution, the evaluation was studied at the capillary axis. We found that higher frequency components of the guiding pulse propagate a bit faster resulting in a slight shape-distortion of pulse after a longer distance. The studies showed, that in the first few nanosecond of the guiding channel lifetime its single-mode transmission ability is fairly low, and this time range varies with the applied beam spot size. A transient guiding channel is always emerging prior the maximum Z-pinch compression, independently of the used gases.

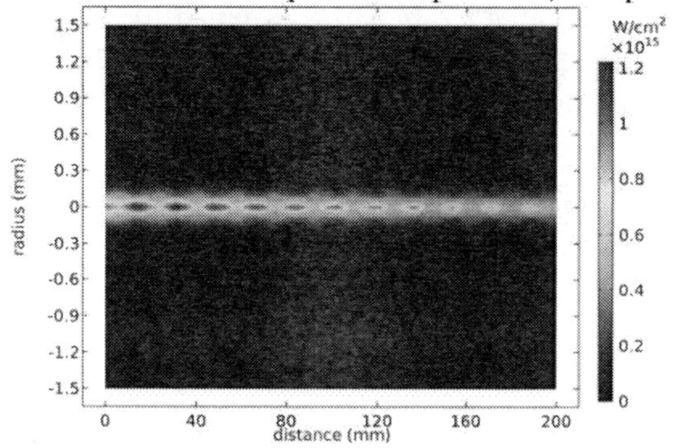

Figure 1. The intensity distribution of guided laser beam in Ne-plasma.

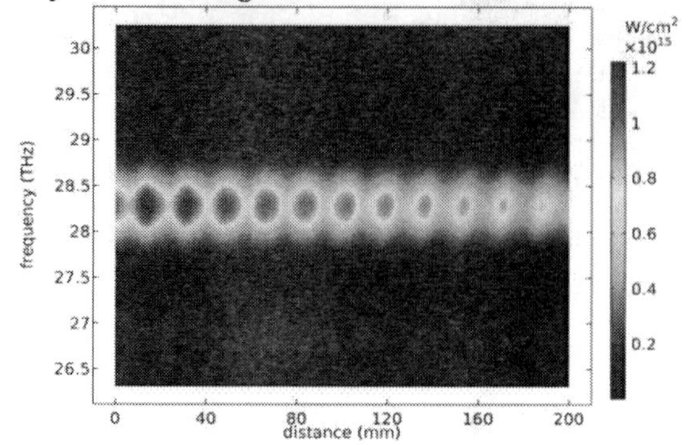

Figure 2. The spectral distribution of guided laser beam in Ne-plasma.

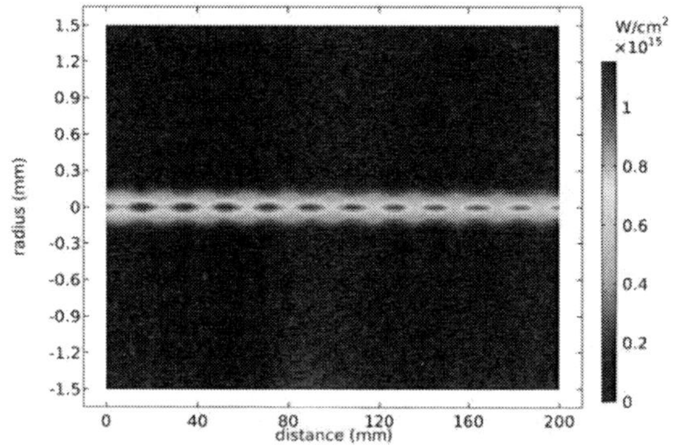

Figure 3. The intensity distribution of guided laser beam in Kr-plasma.

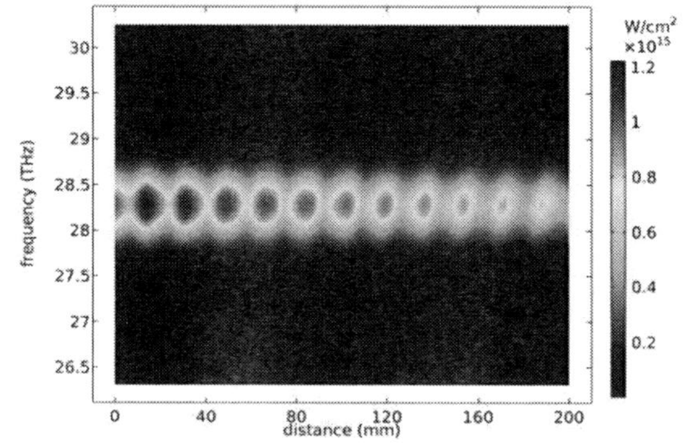

Figure 4. The spectral distribution of guided laser beam in Kr-plasma.

In single-mode regime, the guiding channel is able to extend the interaction length (Rayleigh-range) over the whole capillary length. In this regime the shape distortion of propagated pulse is insignificant even after longer distances (>100 mm). In order to quantify ability of the channel to produce the single mode regime, a correlation coefficient [23] was computed at different input beam spot sizes. The optimal beam spot size w_0^{opt} = 134 μm was determined by taking maximum of time average of the coefficient over the channel lifetime. The ability of plasma channel to guide the high intensity laser pulse is a crucial tool in the LWFA. Dynamic of the electron density distribution in the capillary Z-pinch involves changes in channel transmission. Therefore, the relatively short existence of the guiding channel necessitates low timing jitter between the laser system and the Z-pinch discharge. In order to improve the transmission ability, two options may come up, namely, the timing jitter should be reduced further or the time-dependent channel radius should be followed by the matched spot size. When the guiding channel significantly shrinks, the input beam spot size exceeds the matched spot size, resulting the repetitive focusing and defocusing patterns. Intensity at the focal points can be higher as at the input (figure 1-4). Formation of the guiding channel strongly depends on rising time of the excitation current pulse amplitude of which can be decreased by using lighter gases.

For hydrogen, the waveguiding properties of plasma channel produced by the capillary Z-pinch are shown in figure 5 and 6. The computations were performed for I_{max} = 9.9 kA, T_{cur} / 2 = 47 ns and the initial pressure of hydrogen p_0 = 1.1 mbar. The studies were conducted for CO_2-laser pulse with duration of 0.5 ps and input peak intensity of 10^{17} W/cm² at the different spot sizes. As an example, figure 5 and 6 show evolution of the spatial and spectral intensity distributions of guided laser beam for w_0 = 150 μm.

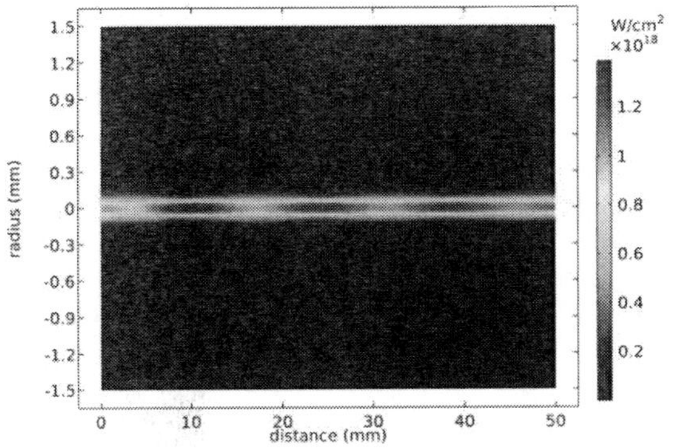

Figure 5. The intensity distribution of guided laser beam in the H-plasma.

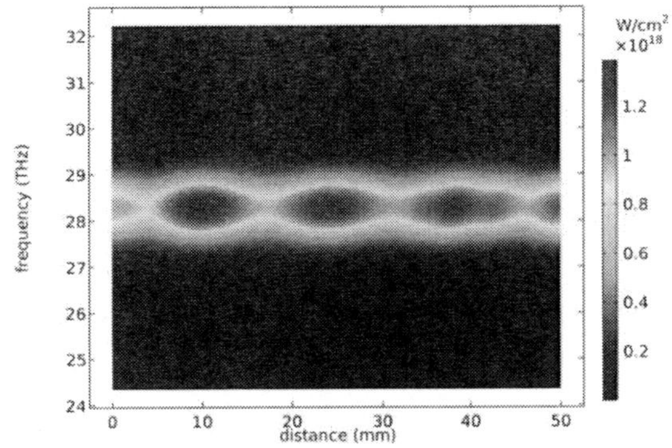

Figure 6. The spectral distribution of guided laser beam in the H-plasma.

3. Acceleration of electrons in capillary Z-pinch plasma waveguide

For simulation of the wakefield acceleration of electrons in hydrogen plasma channel produced by capillary Z-pinch, we used the particle-in-cell (PIC) model combined with the MHD model presented in Sec. 2. Newton's 2nd law for j-th electron with relativistic correction is given by

$$\frac{d}{dt}\left(m_e\gamma^j\vec{v}_e^j\right)=\vec{F}, \qquad (15)$$

where m_e is the rest mass of electron, $\gamma^j=\left(1-\vec{v}_e^{j2}/c^2\right)^{-1/2}$ is the relativistic factor and $\vec{F}=\vec{F}_e+\vec{F}_P$ is the force acting on electron. It is a superposition of two most dominant electric and ponderomotive forces:

$$\vec{F}=-e\vec{E}-\frac{e^2\nabla\vec{E}_L^2}{4m_e\omega_0^2}. \qquad (16)$$

Here, $\vec{E}=-\nabla U$ is the electric field caused by charge separation, $\vec{E}_L$ is the electric field amplitude of laser and ω_0 is the central angular frequency of laser pulse. Electric field caused by charge separation can be obtained from Gauss's law or Poisson's equation

$$\nabla\cdot\vec{E}=\rho/\varepsilon_0 \quad \text{or} \quad -\nabla^2U=\rho/\varepsilon_0, \qquad (17)$$

where $\rho = \langle Z\rangle en_i - en_e$ is the charge density. In case of hydrogen plasma, $\langle Z\rangle = 1$, i.e. $\rho = e(n_i - n_e)$. In the discrete case, the expression in bracket can be written as follows

$$\rho^d=e(N_i-N_e)=e\sum_{j=1}^{N}n_i\delta(\vec{r}-\vec{q}_i^j)-e\sum_{j=1}^{N}n_e\delta(\vec{r}-\vec{q}_e^j) \qquad (18)$$

where $\vec{q}_e^j$ and $\vec{q}_i^j$ are the positions of j-th electron and ion. Correspondently, $\vec{v}_e^j=d\vec{q}_e^j/dt$. All of equations above describing electron movement can be used for ions by interchanging m_e with m_i. Since $m_e \ll m_i$, ponderomotive force doesn't affect ions. Thus, they would "standing still" during whole process, i.e. $\vec{v}_i^j=0$ and $\vec{q}_i^j=\vec{r}^j$. On the other hand, in hydrogen plasma $n_i = n_e = n$. Number of particles in a mesh volume at $\vec{r}$ position is $V_{mesh}(\vec{r})=G_c(\vec{r})/n$, where $G_c(\vec{r})$ is the coupling constant which makes equality between discrete and continuous plasma frequency:

$$\omega_{pe}(\vec{r})\approx\omega_{pe}^D(\vec{r})=\sqrt{\frac{e^2}{\varepsilon_0m_e}\sum_{j=1}^{N}\frac{G_c(\vec{r})}{V_{mesh}(\vec{r})}\delta(\vec{r}-\vec{q}_e^j)} \qquad (19)$$

Now (18) can be rewritten as follows

$$\rho^d=e\sum_{j=1}^{N}\frac{G_c(\vec{r})}{V_{mesh}(\vec{r})}\left[\delta(\vec{r}-\vec{r}^j)-\delta(\vec{r}-\vec{q}_e^j)\right] \qquad (20)$$

Combining all of results above we get system of equations of the PIC model:

$$\vec{v}_e^j=d\vec{q}_e^j/dt \qquad (21)$$

$$\gamma^j=\left(1-\vec{v}_e^{j2}/c^2\right)^{-1/2} \qquad (22)$$

$$\frac{d}{dt}\left(m_e\gamma^j\vec{v}_e^j\right) = e\nabla U - \frac{e^2\nabla\vec{E}_L^2}{4m_e\omega_0^2} \tag{23}$$

$$-\nabla^2 U = \frac{e}{\varepsilon_0}\sum_{j=1}^{N}\frac{G_c(\vec{r})}{V_{mesh}(\vec{r})}\left[\delta(\vec{r}-\vec{r}^j)-\delta(\vec{r}-\vec{q}_e^j)\right] \tag{24}$$

The electron density evolution computed by using the above presented PIC model was compared with the MHD-based density perturbation model which is formulated as follows. Continuity equation of mass density of electrons is given by

$$\frac{\partial\rho_e}{\partial t} + \nabla\cdot\rho_e\vec{v}_e = 0 \tag{25}$$

Introduction of perturbed term of mass density $_e = _{e0} + \delta_e$ yields

$$\frac{\partial\rho_{e0}}{\partial t} + \frac{\partial\delta\rho_e}{\partial t} + \nabla\cdot\rho_{e0}\vec{v}_e + \nabla\cdot\delta\rho_e\vec{v}_e = 0 \tag{26}$$

Since $\delta\rho e << \rho e0$ and $\delta\rho e$ varies much faster in time than ρ_{e0}, the first and the last terms of equation above can be neglected. Taking relativistic factor γ into account we get

$$\frac{\partial}{\partial t}\left(m_e\gamma\delta n_e\right) + n_{e0}\nabla\left(e\nabla U - \frac{e^2\nabla\vec{E}_L^2}{4m_e\omega_0^2}\right) + \left(\frac{e}{m_e}\nabla U - \frac{e^2\nabla\vec{E}_L^2}{4m_e\omega_0^2}\right)\cdot\nabla\cdot n_{e0} = 0 \tag{27}$$

In case of hydrogen plasma, the charge density on the right-hand side of Poisson's equation (17) can be expressed as follows

$$\rho = e(n_i - n_e) = e(n_i - n_{e0} + \delta n_e) = e\delta n_e. \tag{28}$$

Here, $n_i - n_{e0} = 0$ due to plasma quasi-neutrality. Thus, we can do the following transformations on the 2nd term of (27):

$$\frac{n_{e0}}{m_e}\nabla\left(e\nabla U - \frac{e^2\nabla\vec{E}_L^2}{4m_e\omega_0^2}\right) = \frac{e^2 n_{e0}}{\varepsilon_0 m_e\gamma}\gamma\delta n_e - \frac{e^2 n_{e0}}{4m_e^2\omega_0^2}\nabla^2\vec{E}_L^2 = \omega_{pe}^2\gamma\delta n_e - \frac{e^2 n_{e0}}{4m_e^2\omega_0^2}\nabla^2\vec{E}_L^2 \tag{29}$$

where ω_{pe} is the relativistic plasma frequency, and is the relativistic factor for Gaussian laser beam, in which f_{ei} is the collisional frequency between electrons and ions. Combining all of results above we get the system of equations of density perturbation model that includes Eq. (30) and the following equations.

$$\gamma = \left(1 + \frac{e^2\vec{E}_L^2}{2m_e^2c^2\left(\omega_0^2 + f_{ei}^2\right)}\right)^{1/2} \tag{30}$$

$$\left(\frac{\partial^2}{\partial t^2} + \omega_{pe}^2\right)\gamma\delta n_e = \nabla\cdot\left(\frac{e^2 n_{e0}}{4m_e^2\omega_0^2}\nabla\vec{E}_L^2\right) - \frac{e}{m_e}\vec{E}\cdot\nabla n_{e0} \tag{31}$$

$$\nabla\cdot\vec{E} = -\frac{e}{\varepsilon_0}\delta n_e \tag{32}$$

Evolution of the electron density and energy gain during the acceleration was computed by using the above presented PIC for different waveguiding regimes, which occur at the propagation time of 0.60, 2.55, 7.05 and 8.80 ps (figure 7-10). Evolution of the electron density and energy gain during the acceleration was computed by using the above presented PIC for different wave-guiding regimes, which occur at the propagation time of 0.60, 2.55, 7.05 and 8.80 ps (figure 7-10).

Figure 11 shows evolution of the electron energy maximum, where the vertical lines indicate the local maxima, which are related to four different regimes. The respective spectra of electrons are shown in figure 12. We note that that acceleration of electrons is nonlinear versus the acceleration time (figure 11). We investigated correspondence of the electron energy maxima at the four regimes (figure 11) with the peaks of axial intensity of guided laser beam (figure 13). The respective transverse intensity distribution of guided laser beam is shown in figure 14. Comparison of figure 11 and 13 indicates quasi-synchronization of the local electron energy maxima with the inflection points between peaks of intensity of guided laser beam. The above presented PIC model (figure 7-10) was compared with the MHD-based density perturbation model (figure 15-18) in case of the same waveguiding regimes (0.60, 2.55, 7.05 and 8.80 ps). The PIC model provides information about energy gain during the acceleration (figure 11 and 12), while the perturbation model gives more exact information about plasma frequency (figure 19). Plasma wavelength of the PIC model is about 200 μm (figure 9 and 10), while the exact value is 144 μm that corresponds to the plasma frequency $\omega_{pe} = (e^2 n_e/\varepsilon_0 m_e \gamma)^{1/2}$.

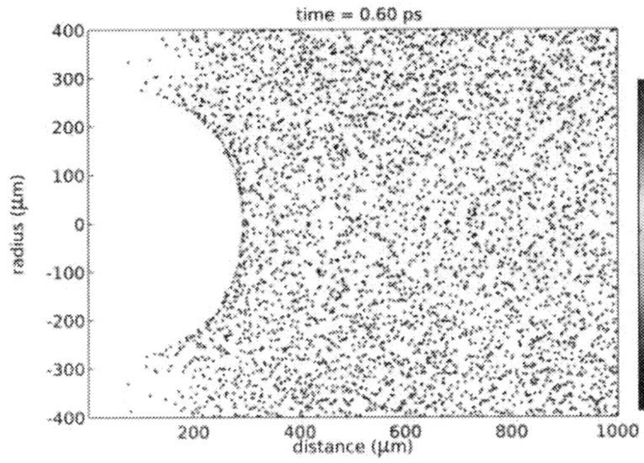

Figure 7. Electron distribution at 0.60 ps.

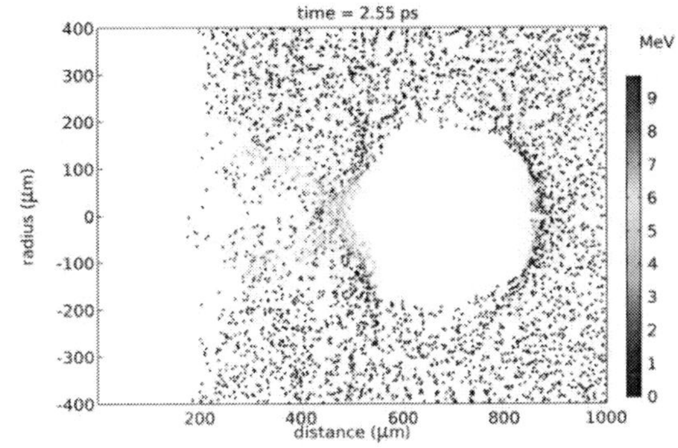

Figure 8. Electron distribution at 2.55 ps.

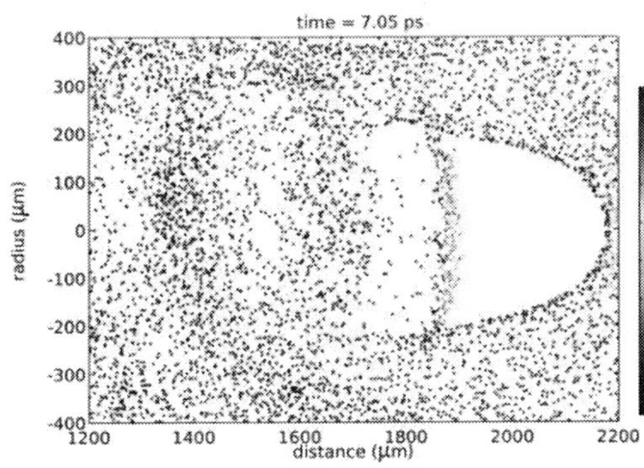

Figure 9. Electron distribution at 7.05 ps.

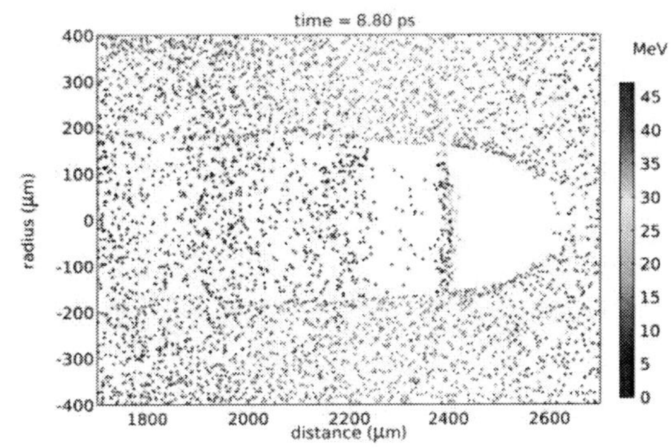

Figure 10. Electron distribution at 8.08 ps.

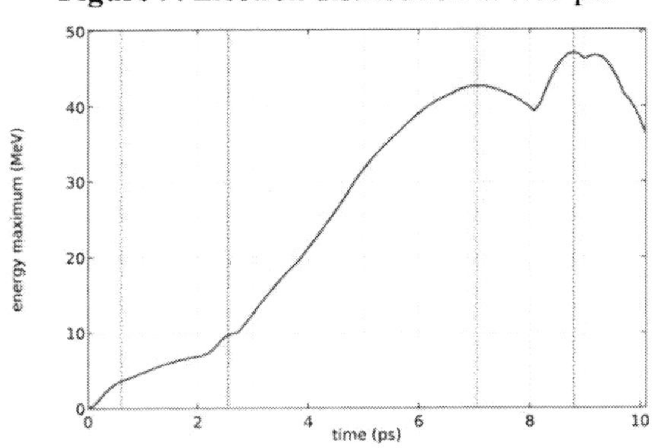

Figure 11. Electron energy maximum versus time.

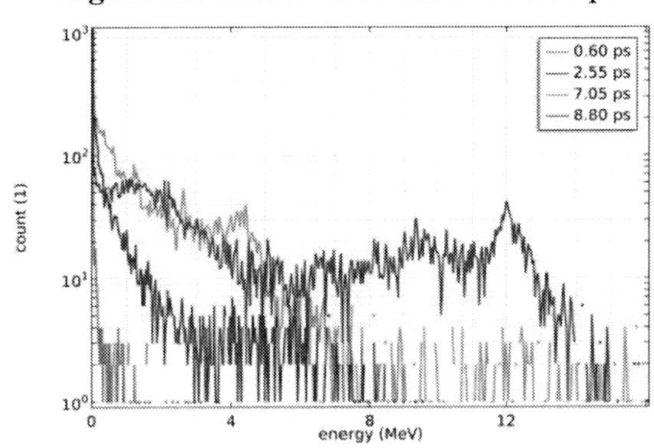

Figure 12. Energy spectra of electrons.

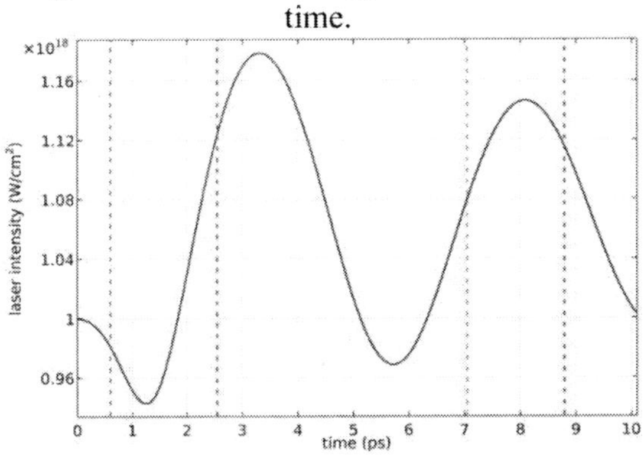

Figure 13. Axial intensity distribution of guided laser beam.

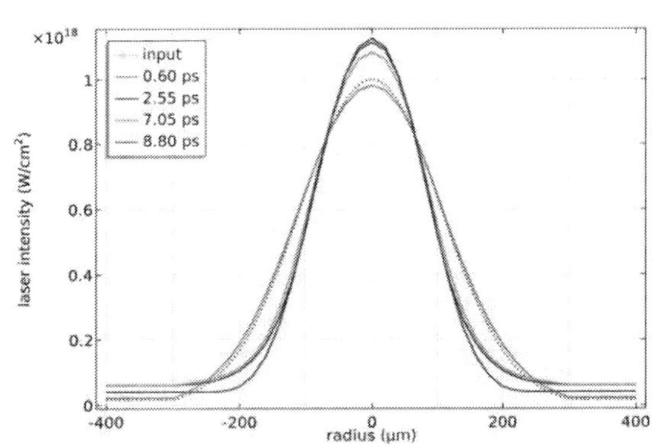

Figure 14. Transverse intensity distribution of guided laser beam.

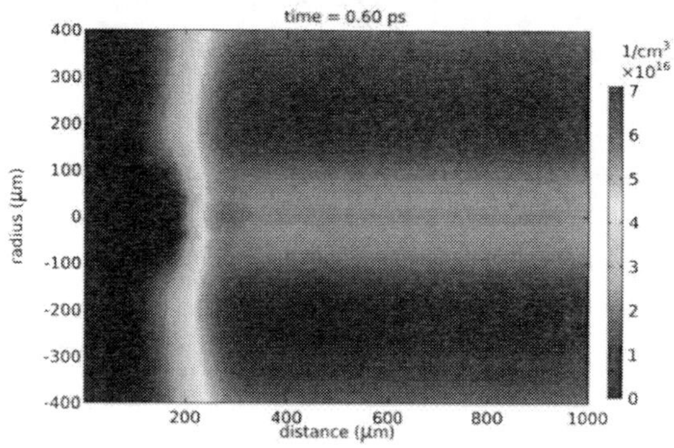

Figure 15. Electron density distribution at 0.6 ps.

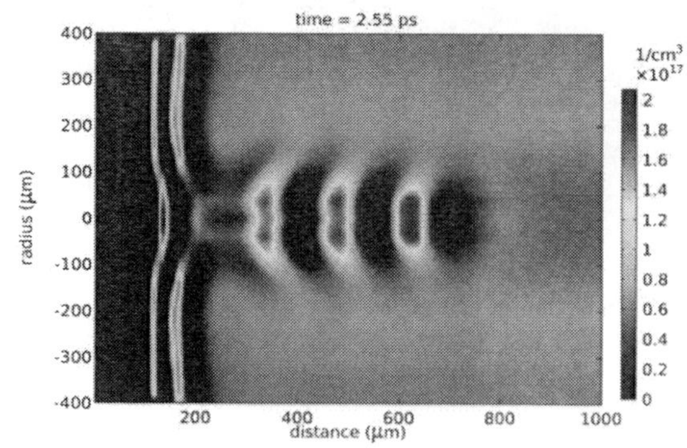

Figure 16. Electron density distribution at 2.55 ps.

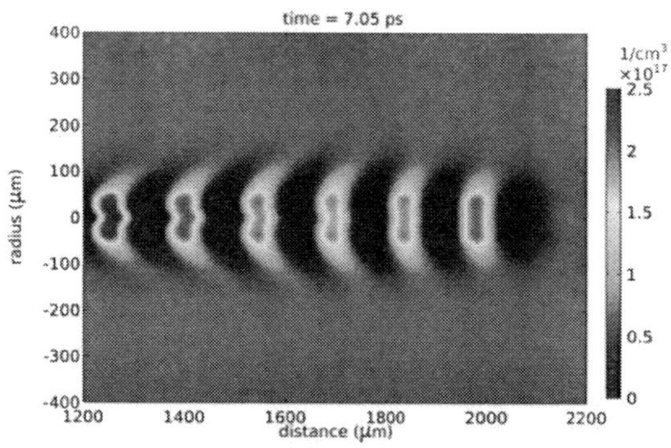

Figure 17. Electron density distribution at 7.05 ps.

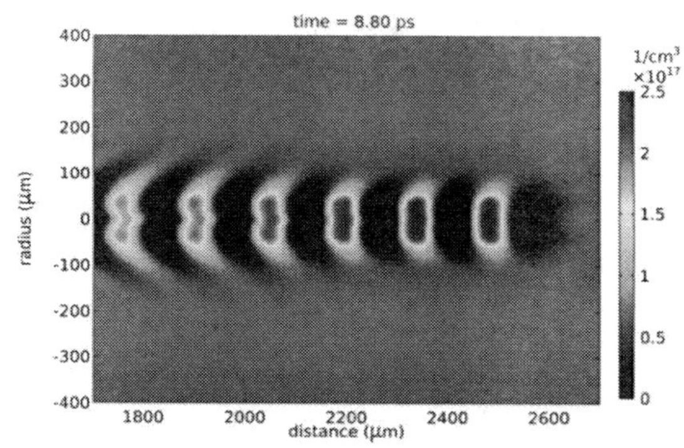

Figure 18. Electron density distribution at 8.80 ps.

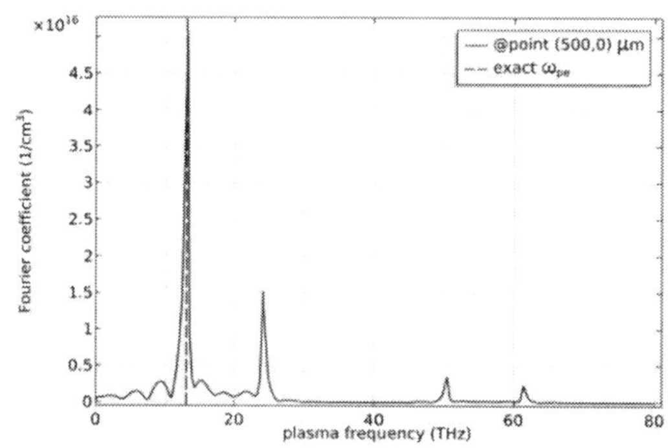

Figure 19. The plasma density spectrum derived from plasma oscillations at point (x = 500 μm, y = 0 μm), where $\omega_{pe} = (e^2 n_e/\varepsilon_0 m_e \gamma)^{1/2}$ is the exact plasma frequency.

4. Summary and conclusion

The theoretical examination of the Z-pinch H-plasma waveguide inside a 3-mm inner diameter and 50-mm long capillary was performed on CO_2-laser pulse with input peak intensity of >10^{15} W/cm^2. The main plasma variables of the fast Z-pinch discharge were computed by the 1-D MHD model. The waveguide properties of the capillary Z-pinch plasma were obtained from the space and frequency dependent wave equation that combines the attenuated charged particle inertia and the light wave effects in the plasma. Simulations showed that the guiding channel occurs far from capillary wall and exists for a few nanoseconds. In terms of laser driven plasma wakefield accelerators (LWFA), the Z-pinch waveguide is able to extend the acceleration length over the whole capillary. At the end of the guiding channel existence, the repetitive focusing and defocusing patterns were observed with intensity increase at the focal points. For simulation of the wakefield acceleration of electrons in the different waveguiding regimes, we used a PIC model. The influence of waveguiding regimes on the electron acceleration was demonstrated: quasi-synchronization of the electron energy maxima with the inflection points between peaks of intensity of guided laser beam was demonstrated. We showed that the acceleration of electrons is nonlinear versus the acceleration time.

Acknowledgments
This study was supported by the Human Resource Development Operational Program (contract EFOP-3.6.2-16-2017-00005).

References

[1] Tajima T and Dawson J M 1979 Phys. Rev. Lett. **43** 267
[2] Sprangle P and Esarey E 1992 Phys. Fluids B Plasma Phys. **4** 2241
[3] Ehrlich Y et al 1996 Phys. Rev. Lett. **77** 4186–4189
[4] Kukhlevsky S V and Kozma L 1998 Contrib. Plasma Phys. **38** 583
[5] Hosokai T et al. 2000 Opt. Lett. **25** 10
[6] Hooker S M and Spence D J 2000 J. Opt. Soc. Amer. B Opt. Phys. **17** 90
[7] Bobrova N A et al. 2001 Phys. Rev. E **65** 016407
[8] Spence D J Butler A and Hooker S M 2003 J. Opt. Soc. Amer. B, Opt. Phys. **20** 138
[9] Kumarappan V Kim K Y and Milchberg H M 2005 Phys. Rev. Lett. **94** 205004
[10] Leemans W P et al. 2006 Nature Phys. **2** 696
[11] Gordon D F 2007 IEEE Trans. Olasma Sci. **35** 1486
[12] Gedde C G R et al. 2007 J. Phys., Conf. Ser. **78** 012021
[13] Karsch S et al. 2007 New J. Phys. 9 415
[14] A. J. Gonsalves et al. 2007 Phys. Rev. Lett. **98** 025002
[15] Higashiguchi T et al. 2012 J. Appl. Phys. **111** 093302
[16] Leemans W P et al. 2014 Phys. Rev. Lett. **113** 245002
[17] Nakajima Ket al. 2015 High Power Laser Sci. Eng. **3** 11
[18] Gonsalves A J et al. 2016 J. Appl. Phys. **119** 033302
[19] Steinke S et al. 2016 Nature **530** 190
[20] Luo J et al. 2018 Phys. Rev. Lett. **120** 154801
[21] Zigler A et al. 2018 Appl. Phy. Lett. **113** 183505
[22] Sasorov P V Bobrova N A and Olkhovskaya O G 2015 The two-temperature equations of magnetic hydrodynamics of the plasma (Moscow, Keldysh Institute Preprints) pp 1–18
[23] Shapolov A A Kiss M and Kukhlevsky S V 2018 IEEE Transactions on Plasma Sci. **46** 3886

IOP Conf. Series: Materials Science and Engineering **715** (2020) 012107 doi:10.1088/1757-899X/715/1/012107

Preparation and characterization of flexible PVDF-HFP film for piezoelectric applications

Jureeporn Yuennan[1,*], and Nantakan Muensit[2]

[1]Department of Physics, Faculty of Science and Technology, Nakhon Si Thammarat Rajabhat University, Nakhon Si Thammarat 80280, Thailand
[2]Center of Excellence in Nanotechnology for Energy (CENE), Department of Physics, Prince of Songkla University, Songkhla 90112, Thailand

*E-mail: jureeporn_yue@nstru.ac.th

Abstract. We report on the preparation of electroactive β-phase in polyvinylidene fluoride hexafluoropropylene (PVDF-HFP) without any electrical poling treatment. The PVDF-HFP film with various mass fractions (wt%) of magnesium chloride hexahydrate salt ($MgCl_2{\cdot}6H_2O$) was fabricated by a solution casting technique. The morphological, mechanical, structural and dielectric properties were investigated. Results founded that the PVDF-HFP composite films show a large number of micro-pores on the top surface as a sponge-like structure. The Young's modulus and crystallinity values tend to be linearly decreased with the $MgCl_2{\cdot}6H_2O$ concentration leading to enhance flexibility. The largest β-phase fraction of 87.2% is found in the composite with 1 wt% $MgCl_2{\cdot}6H_2O$ filler. The obtained value is 4 times higher than the value of unfilled PVDF-HFP. Moreover, the maximum value of dielectric constant is about 19.2 at 10 Hz for 4 wt% $MgCl_2{\cdot}6H_2O$ filler. Therefore, the modified composites with the excellent flexibility, dielectric constant and β-phase may be a promising material for applications in the field of piezoelectric energy conversion.

1. Introduction

Nowadays, the development of flexible polymer composites with high dielectric properties has been paid attention of researchers worldwide. These can be used in wide range applications such as low-power portable electronics, actuator and different types of sensors [1]. Poly(vinylidene fluoride-co-hexafluoropropylene) (PVDF-HFP) have been designed as a polymer matrix due to their light weight, flexibility, environmental compatibility and brilliant electroactive responses (including piezo-, pyro- and ferroelectric effects) [2]. P(VDF-HFP) exhibits four different crystalline phases namely α-, β-, δ-, and γ-phases depending on molecular conformation [3]. However, the α- and β-phases are widely considered for investigation. The most stable nonpolar phase of α polymorph is found inactive with electroactive responses. Another is the largest spontaneous polarization of β-phase[4, 5]. Consequently, a variety of strategies have been focused on the enhancement of α- to β- transformation with reorientation of the dipole moment such as mechanical stretching, electrical poling, electrospinning, spin coating and adding of different fillers, etc. [6]. In this work, magnesium chloride hexahydrate ($MgCl_2{\cdot}6H_2O$) was employed as the active filler due to its easy process. It was found that the β-phase structure can be formed directly in the crystallization process by hydrogen bonding interaction between PVDF-HFP molecules and water in the $MgCl_2$ hydrated salts [7, 8]. The crystalline phase transformation, dielectric property, surface morphology and mechanical behavior of the composite films were investigated using Fourier transform infrared spectroscopy, scanning electron microscopy, atomic force microscopy and tensile testing, respectively.

2. Materials and methods

PVDF-HFP (Solef 11010/1001) powder was chosen as the polymer matrix purchased from Solvay Solexis, Inc. $MgCl_2 \cdot 6H_2O$ (M2393) filler and N-N, dimethylformamide (DMF, D158550) solvent were obtained from Sigma-Aldrich. PVDF-HFP films with and without $MgCl_2 \cdot 6H_2O$ filler were prepared by a simple solution casting method. Briefly, 20 mg PVDF-HFP was initially dissolved in 100 ml DMF (i.e., the polymer to the solvent ratio of 1:5) at room temperature by magnetic stirring for 6 h. Thereafter, the $MgCl_2 \cdot 6H_2O$ filler was dispersed in 10 ml DMF at four concentrations (1, 2, 3, and 4 wt%) by using an ultrasonic bath for 10 min. The prepared $MgCl_2 \cdot 6H_2O$ solution with various concentrations was added to the previous PVDF-HFP solution under continuous stirring for 12 h. Afterward, the mixture with different salt contents were cast on a clean glass by adjustable film applicator (SH1102; Sheen) and dried at 80 °C for 12 h in an oven to evaporate completely the DMF solvent. Finally, the PVDF-HFP samples were obtained with an average thickness of 100 μm.

Scanning electron microscope (SEM; Quanta 400) and atomic force microscope (AFM; Nanosurf Easyscan2) were used to study the surface morphology of prepared PVDF-HFP samples. An average pore size of PVDF-HFP film can be analyzed from the SEM images via the Java image processing program. Moreover, this software can be used to determine a percentage of porosity of the films proposed by Bilad and coworkers [9]. Each PVDF-HFP film was investigated in a non-contact mode in order to determine the surface roughness in term of root-mean-square surface roughness (*Sq*). Mechanical property of all samples was investigated using a tensile testing machine (z010, Zwick Roell). Fourier transform infrared spectrometer (FTIR; Vertex70, Bruker) was recorded from 400 to 1000 cm^{-1} in the Attenuated total reflectance (ATR) mode. The dielectric measurement was carried out on the samples by an LCR meter (IM 3533, HIOKI).

3. Results and discussion

3.1. Surface morphology

Figure 1 shows the SEM images of the PVDF-HFP films added with different concentrations of $MgCl_2 \cdot 6H_2O$ filler. It is shown a smooth surface in the bare PVDF-HFP sample. With loading $MgCl_2$ samples, a large number of micropores/cavities on the top surface is revealed as a sponge struture. As can be seen from table 1, both average pore size and porosity of uniform microporous film become increase with the $MgCl_2$ concentrations. This phenomenon may be the result of a low evaporation of the DMF solvent and water molecules of the $MgCl_2 \cdot 6H_2O$ filler in drying process, as was reported by Adhikary et al. [8]. The consistent distribution of micropores is attributed to the good compatibility between PVDF-HFP matrix and $MgCl_2 \cdot 6H_2O$ dopant resulting in a presence of the homogenous structure which is suitable for the fabrication of energy conversion devices. In addition, representative height images are also used to further investigate the surface topography of the samples. Figure 2 and table 1 are presented the 3D-AFM diagrams with a scanned area of 70 x 70 μm^2 and surface roughness values of as-prepared samples. The darkest and brightest regions are respectively assigned to the pores and film surface zones. It is confirmed that the pristine PVDF-HFP film represents a smoother surface compared to the others. Upon the addition of $MgCl_2 \cdot 6H_2O$ salt, the surface roughness of the samples are obviously increased with the filler contents agreed with the SEM results. Therefore, it can be indicated that the addition of $MgCl_2 \cdot 6H_2O$ salt causes a significant enhancement of pore size as well as roughness on the surface of PVDF-HFP films.

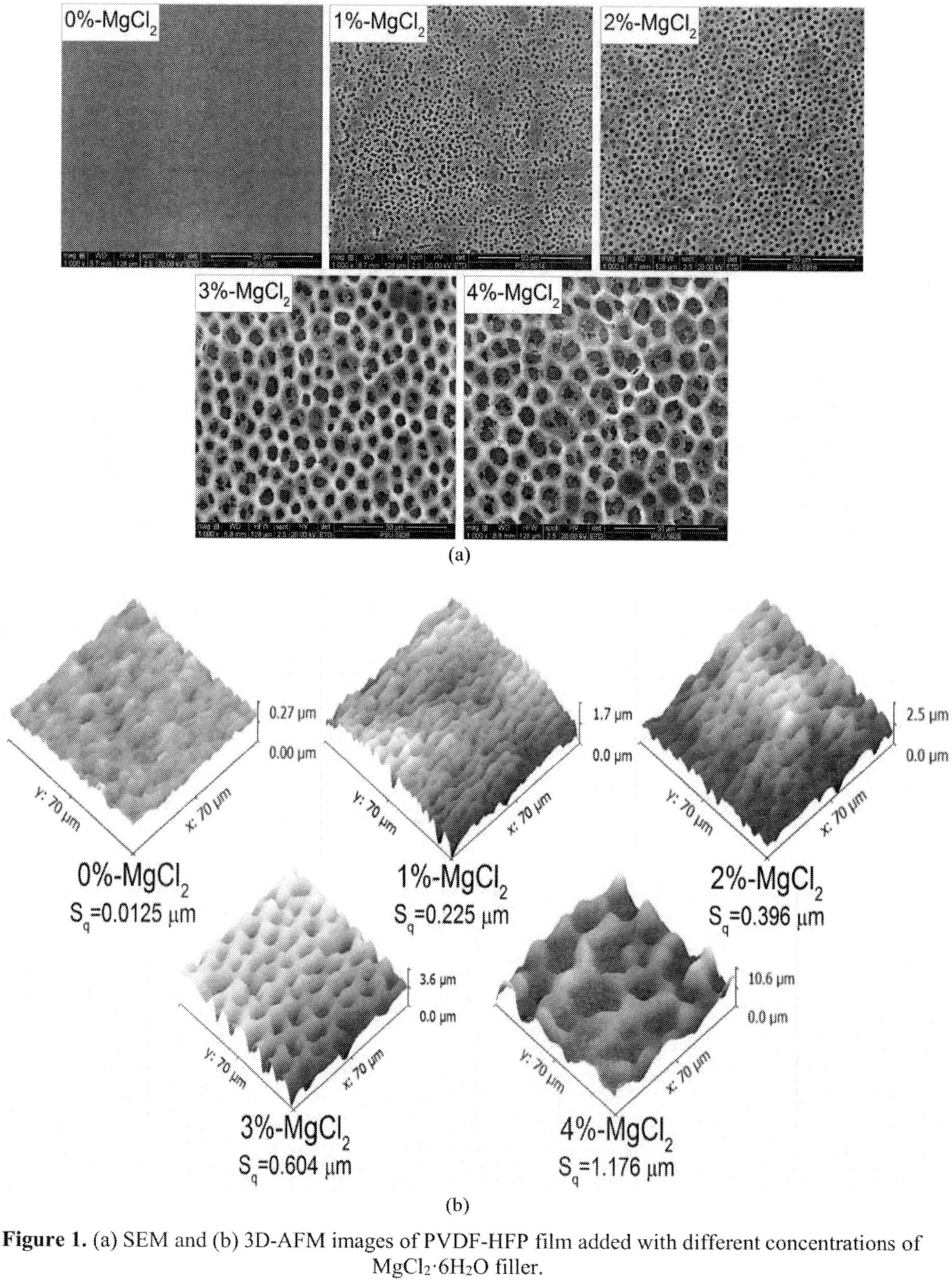

Figure 1. (a) SEM and (b) 3D-AFM images of PVDF-HFP film added with different concentrations of $MgCl_2 \cdot 6H_2O$ filler.

Table 1. Average pore size, surface roughness and porosity of PVDF-HFP film added with different concentrations of $MgCl_2 \cdot 6H_2O$ filler.

PVDF-HFP sample	Pore size (μm)	Surface roughness (μm)	Porosity (%)
0%-$MgCl_2$	N/A	0.0125	0.54 ± 0.1
1%-$MgCl_2$	1.25 ± 0.12	0.225	10.65 ± 0.5
2%-$MgCl_2$	2.02 ± 0.20	0.396	21.39 ± 0.9
3%-$MgCl_2$	3.56 ± 0.25	0.604	34.66 ± 1.5
4%-$MgCl_2$	5.68 ± 0.28	1.176	40.22 ± 1.8

3.2. Young's modulus

In order to study the mechanical property of PVDF-HFP film improved with $MgCl_2 \cdot 6H_2O$, the stress-strain results of all prepared PVDF-HFP films are listed in table 2. The tensile strength, elongation at break of the samples steadily reduce with increasing of the $MgCl_2 \cdot 6H_2O$ concentrations. The potential reason may be due to plasticizing effect of $MgCl_2 \cdot 6H_2O$ filler. A large amount of $MgCl_2$ molecule significantly suppresses of PVDF-HFP polymer chain movement in crystallization process resulting in low crystallinity [10]. Another reason is the formation of the spongy structure with higher porosity (SEM images) of the samples resulting in a weak linkage between the polymer and filler [11]. In addition, Young's modulus value tends to be linearly decreased with the $MgCl_2 \cdot 6H_2O$ concentrations. This leads to significantly develop in a greatly flexible film and easy deformation under applied force, which suggests that the composite sample is more favorable for the improvement in piezoelectric applications.

Table 2. Tensile properties performed on the pristine PVDF-HFP and composite films.

sample	Tensile strength (Mpa)	Elongation at break (%)	Young's modulus (Gpa)
0%-$MgCl_2$	23.7	408.6	1.56
1%-$MgCl_2$	22.8	347.0	1.22
2%-$MgCl_2$	20.8	300.5	0.87
3%-$MgCl_2$	19.9	236.4	0.78
4%-$MgCl_2$	15.2	154.9	0.33

3.3. Phase structure

The FTIR spectra of the PVDF-HFP samples assigned to crystalline α- and β-phases are given in figure 2a. According to the literature, the vibrational bands at 431, 531, 614 and 763 cm^{-1} are identified as the characteristic nonpolar α-phase, while the polar β-phase is represented at 837 cm^{-1} [12]. It is revealed that the pure PVDF-HFP film appears all characteristic peaks of α-phase. After addition of the $MgCl_2 \cdot 6H_2O$ filler, these α-peaks become disappear and the β-peak is gradually increased with $MgCl_2$ contents. The all-trans conformation of β-phase may be formed by hydrogen bond interactions between $-CH_2$ dipole of PVDF-HFP and water molecules in the filler [13]. Moreover, these specific bands can be used for determining the β-phase fraction ($\%F(\beta)$) and total β-phase fraction in crystalline region ($\%\beta$) using equation (1).

$$\%F(\beta) = (\frac{A_\beta}{1.26A_\alpha + A_\beta})x100\% \quad (1)$$

where A_β and A_α are the intensity peak at 837 and 763 cm^{-1}, respectively [14]. The $\%F(\beta)$ and $\%\beta$ ($\%\beta = [X_c \times F(\beta)] \times 100\%$) values of each sample are displayed in figure 2b. About 24.1% of $\%F(\beta)$ is obtained for the pure PVDF-HFP film. With the loading 1 wt% $MgCl_2 \cdot 6H_2O$ concentration in the PVDF-HFP matrix, maximum values of $\%F(\beta)$ and $\%\beta$ is about 87.2% and 38.7%. At higher doping concentrations, $\%F(\beta)$ almost saturates while $\%\beta$ gradually decreases. This is attributed to the reduction of crystalline degree with the increasing concentration of the filler.

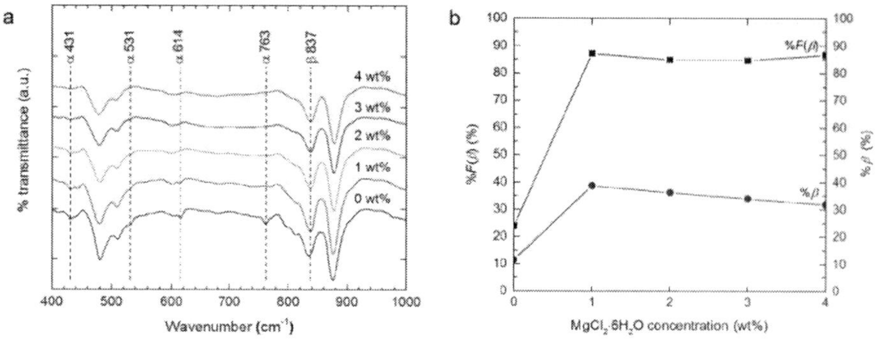

Figure 2. (a) FTIR spectra; and (b) β-phase fraction of PVDF-HFP film added with different contents of $MgCl_2{\cdot}6H_2O$ filler.

3.4. Dielectric behaviors

The variation of dielectric property as a function of frequency in the PVDF-HFP films is illustrated in figure 3. The values of dielectric constant (ε') and dielectric loss ($tan\ \delta$) are given as; $\varepsilon' = Ct / \varepsilon_0 A$ and $\tan\delta = \varepsilon'' / \varepsilon'$, where C is the sample capacitance measured by the LCR meter. A is an area of the sample, t is the sample thickness and ε_0 is a constant representing permittivity of free space. ε'' and ε' are the imaginary and real parts of permittivity, respectively [15]. From figure 3a, the dielectric constant of all samples slightly decreases with frequency. This behavior is caused by less enough time to arrange of dipole moment in electric field direction at high frequency [16]. It is also observed that dielectric constant linearly increases with $MgCl_2{\cdot}6H_2O$ concentration. The maximum value of dielectric constant is about 19.2 at 10 Hz for 4 wt% $MgCl_2{\cdot}6H_2O$ filler. This improvement of dielectric constant from hydrated $MgCl_2$ dopant results from several reasons. Firstly, the increase of hydrated $MgCl_2$ can induce large interfacial polarization formed by strong interaction between the water molecule of $MgCl_2$ filler and -CF_2 dipoles of polymer chains [8]. This interaction leads to transforming the β-phase structure described in the FTIR section. Secondly, the large accumulation of void charges at $MgCl_2$/PVDF-HFP interface may be formed as numerous micro-capacitors [17]. Moreover, a presence of micro-pores may be filled with air and separated by insulating dielectric polymer which is attributed to the property of a piezoelectric material [2]. From figure 3b, the dielectric loss is also gradually declined with increasing the applied frequency related to the interfacial polarization between the interface of hydrated $MgCl_2$ and polymer matrix [18]. The charge carriers accumulated on the interfaces have low mobility and need a long time interval to rotate in electric field direction [19]. Thus, maximum tangent loss of the samples presented at low frequencies (<1 kHz) is higher than that of ones at high frequencies. However, the dielectric loss value of all as-received samples remains relatively low (<0.5). The results suggest that the excellent piezoelectric properties of the samples with high dielectric constant and low dielectric loss may be used in modern electrical and electronic systems.

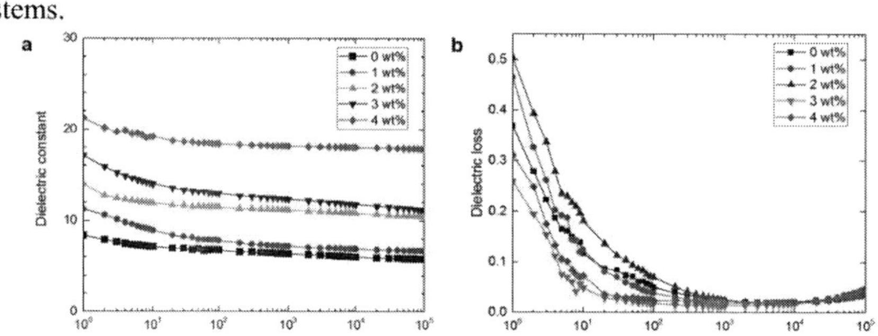

Figure 3. (a) Dielectric constant and (b) dielectric loss tangent of the PVDF-HFP film added with different concentrations of $MgCl_2{\cdot}6H_2O$ filler.

4. Conclusion

In this study, the $MgCl_2 \cdot 6H_2O$/PVDF-HFP films were prepared by the solution casting technique. The results pointed out that the composites showed the surface microstructure like sponge. The surface roughness of films increased thanks to the increase of size and density of pores with raising the $MgCl_2 \cdot 6H_2O$ filler. The PVDF-HFP composites with high porosity also exhibited the low Young's modulus. Furthermore, the $MgCl_2 \cdot 6H_2O$ filler played an important role in the β-phase transformation of films. The largest β-phase fraction of 87.2% was found in the composite with 1 wt% salt filler. The enhanced value was 3.6 times higher than the value of unfilled PVDF-HFP. The maximum dielectric constant and minimum Young's modulus of the film could achieve 19.2 and 0.34 GPa for 4 wt% salt, respectively.

Acknowledgements

The authors gratefully acknowledge the financial support from Faculty of Science and Technology, Nakhon Si Thammarat Rajabhat University and the instrumental supports from the Department of Physics and Center of Excellence in Nanotechnology for Energy at Prince of Songkla University.

References

[1] N.A. Hoque, P. Thakur, S. Roy, A. Kool, B. Bagchi, P. Biswas, M.M. Saikh, F. Khatun, S. Das, P.P. Ray, ACS Appl. Mater. Inter. **9** 23048-23059 (2017)
[2] S.R. Khaled, D. Sameoto, S. Evoy, Smart Mater. Struct. **23** 033001 (2014)
[3] A. Tamang, S.K. Ghosh, S. Garain, M.M. Alam, J. Haeberle, K. Henkel, D. Schmeisser, D. Mandal, ACS Appl. Mater. Inter. **7** 16143-16147 (2015)
[4] R. Song, G. Xia, X. Xing, L. He, Q. Zhao, Z. Ma, J. Colloid Interf. Sci. **401** 50-57 (2013)
[5] L. Wu, G. Huang, N. Hu, S. Fu, J. Qiu, Z. Wang, J. Ying, Z. Chen, W. Li, S. Tang, RSC Adv. **4** 35896-35903 (2014)
[6] P. Adhikary, S. Garain, D. Mandal, Phys. Chem. Chem. Phys. **17** 7275-7281 (2015)
[7] M.R. Bilad, J. Druyts, I.F.J. Vankelecom, IOP Conf. Ser. Mater. Sci. Eng. **180** 012272 (2017)
[8] X. Tang, R. Muchakayala, S. Song, Z. Zhang, A.R. Polu, J. Ind. Eng. Chem. **37** 67-74 (2016)
[9] P. Thakur, A. Kool, B. Bagchi, N.A. Hoque, S. Das, P. Nandy, RSC Adv. **5** 28487-28496 (2015)
[10] P. Martins, C.M. Costa, M. Benelmekki, G. Botelho, S. Lanceros-Mendez, Cryst . Eng. Comm. **14** 2807-2811 (2012)
[11] M. Sharma, G. Madras, S. Bose, Phys. Chem. Chem. Phys. **16** 14792-14799 (2014)H.-C. Eom, H. Park, H.-S. Yoon, Adv. Powder Technol. **21** 125-130 (2010)
[12] A.K. Nath, A. Kumar, J. Membrane Sci. **485** 30-41 (2015)
[13] V. Mamani, A. Gutiérrez, S. Ushak, Sol. Energ. Mat. Sol. C. **176** 346-356 (2018)
[14] J. Wu, Z.L. Wu, H. Yang, Q. Zheng, RSC Adv. **4** 44030-44038 (2014)
[15] A. Gasmi, M. Gouasmia, S. Etienne, Sol. St. Phen. **115** 151-156 (2006)
[16] Y.K.A. Low, L.Y. Tan, L.P. Tan, F.Y.C. Boey, K.W. Ng, J. Appl. Polym. Sci. **128** 2902-2910 (2013)
[17] J. Yang, P. Pan, L. Hua, X. Feng, J. Yue, Y. Ge, Y. Inoue, J. Phys. Chem. B **116** 1265-1272 (2012)
[18] Shalu, S.K. Chaurasia, R.K. Singh, S. Chandra, J. Appl. Polym. Sci. **132** 41456 (2015)
[19] T. Zhu, R. Lv, B. Wang, B. Na, M. Yin, Y. Zhu, J. Appl. Polym. Sci. **133** 43499 (2016)

IOP Conf. Series: Materials Science and Engineering **715** (2020) 012108 doi:10.1088/1757-899X/715/1/012108

Approach to structure management and customer-related configuration in the product lifecycle in mechanical engineering

Alexander Stekolschik

Faculty for Automotive Systems and Production, Cologne University of Applied Sciences, Betzdorfer Str. 2, 50679 Cologne, Germany

*E-mail: alexander.stekolschik@th-koeln.de

Abstract. Complexity in the product and production process in mechanical and aerospace engineering leads to the existence of a variety of structures like engineering and manufacturing bills of materials. The manual or partly automated creation of these structures leads to redundant data and quality issues. This paper describes the approach of an integrated structure management based on a reference structure. The development of an integrated reference structure management approach aims to enhance the quality of a product in combination with concurrent complexity reduction. The reference structure is customer- or order-neutral; it is utilized as a basis for other structures. This paper presents a software system-neutral approach for the customer-related configuration of this structure resulting in e- and mBOM. The implementation of the method framework (selection of use cases) in a standard product lifecycle management (PLM) system is part of the research.

1. Structures in the product lifecycle

The product and process complexity in machine and aerospace engineering is increasing daily, not least because of stronger customer orientation and individualization of customer-specific solution packages. In addition, rising quality standards and shortened delivery periods necessitate integrated processes in the product lifecycle combined with a high level of process reliability and stability. In the entire product lifecycle, different fractional structures are created by the involved product lifecycle parties. These multiple product structures [1] reflect the different degrees of product maturity as well as different views of the product (e.g. product development, industrial engineering, project execution, etc.). The different uses of these structures go along with the corresponding perceptions—this result in multiple requirements regarding the management of the structures. Different forms of appearance—requirement lists, bill of materials [2], service instructions, cost structures, among others—can be found in product requirements and functional structures, design and manufacturing structures, and eventually service structures (figure 1).

The origin and development of such structures mostly do not strictly follow a one-way progression. Instead, they arise in iterative processes and are partly dependent on each other, as is the case with spare-part structures based on the design engineering bill of materials. Basically, the degree of maturity and completeness increases with each step forward in the product lifecycle.

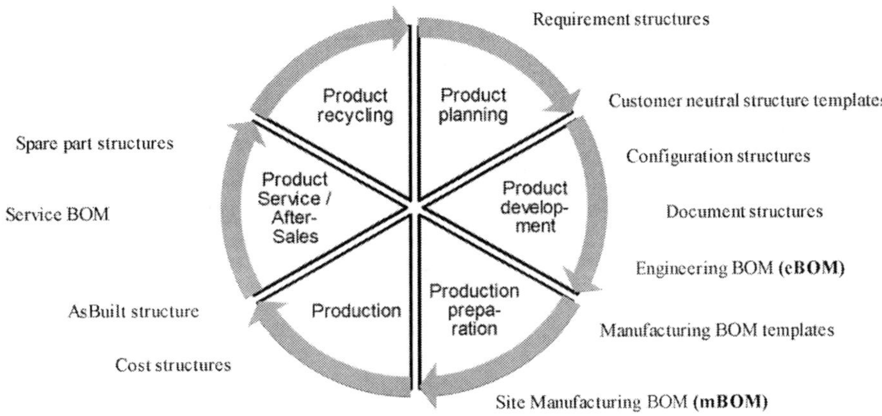

Figure 1. Different structures in the product lifecycle.

The engineering structure (eBOM) is created by product development groups. It represents the so-called 'as designed' structure and is mostly customer order-neutral. The eBOM composition represents the multidisciplinary view of the functional definition of the product; it is similar to the 3D CAD assembly structure. The eBOM is usually combined with related product specification documents, CAD models, non-CAD parts (e.g. oil), drawings, and annotations that should give a clear definition of the product. The mBOM [3] is a configuration of the product to show how it will be manufactured, assembled, serviced, and the like. An mBOM typically consists of a structured list of assemblies and sub-assemblies that describe the manufacturing process. This 'as shipped' structure is based on the eBOM (it consumes eBOM items). But most of the time, it is not the same as the eBOM because of the manufacturing process and localization.

The mBOM includes the equipment and tools required to build a product's components. It also includes details of packaging materials, user manuals, and brochures and thereby make the product shippable (figure 2). All these details help the manufacturing team to understand the design assembly and packaging procedures in a better way. The differences between the engineering and manufacturing structures often lead to a big effort in coordination along the creation process. The mBOM is mostly (logically) derived from the appropriate eBOM structure. But in most cases, it is created manually in the ERP or special BOM system—this results in time-consuming activities and errors. Moreover, the manual process makes efficient engineering change management [4] difficult or even impossible.

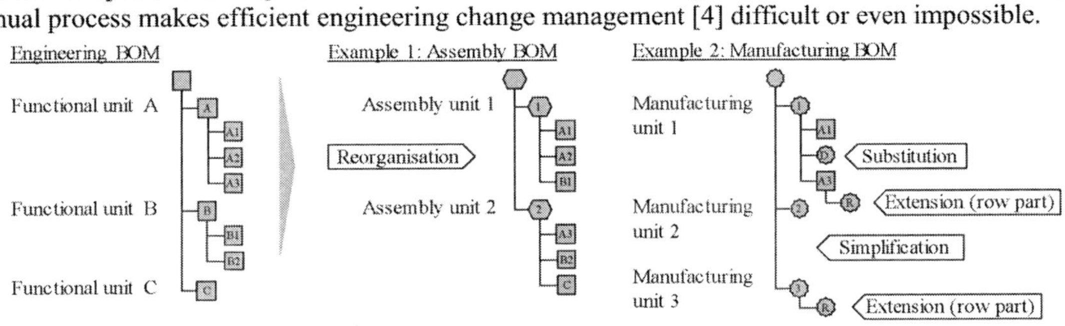

Figure 2. eBOM and derived structures.

BOM management [5] is an integrated part of different engineering tools, depending on the kind of the structure. The CAD BOM (design structure) and engineering BOM (eBOM) are often managed in the PDM system. The so-called CAD BOM or design BOM is generated based on the structure in the 3D-CAD system [6]. The CAD BOM can be different than the eBOM (partly due to the modelling-specific CAD structure and partly due to other engineering parts missing in the CAD BOM such as oil). Site-related manufacturing BOMs (mBOM) are mainly managed in ERP systems or in special

IOP Conf. Series: Materials Science and Engineering **715** (2020) 012108 doi:10.1088/1757-899X/715/1/012108

BOM management systems. They are mainly derived from the eBOM based on manufacturing and purchasing requirements like extension by semi-finished parts or re-ordering due to the assembly sequence.

2. Analysis of the creation processes of structures

The structure definition process was analysed in the first step of the research. Different types of logic have been identified for the creation of structures, depending on the process initiator.

2.1. Product structure-driven structure definition

This process is mostly used for manufacturing structures similar to or even identical to those of the product structure (eBOM). The definition of the eBOM or eBOM configuration [7] is given in the first step in engineering. Many companies copy their eBOMs and change the structure [8] in the next step, namely manufacturing. The change types are: reorganization of virtual assemblies (creation of new assembly nodes, movement of parts), structure simplification (flattening), structure extension (raw parts, blanks), and part substitution (global vs. local). In some cases, generic mBOM templates are used. These are reconfigured after the eBOM configuration. These processes are eBOM-driven and therefore the main part of the mBOM structure follows the eBOM composition, which is slightly adjusted for local site purposes.

2.2. Manufacturing process-driven structure definition

This process is applied if manufacturing requires structures heavily differing from the initial product structure. An example is the assembly structure, which consists of nodes that represent the local assembly process (for example, assembly steps, parts in the near vicinity). The industrial engineer defines the necessary mBOM nodes in the first step (a new mBOM or based on the mBOM template). For example, in the eBOM, the exhaust manifold is part of the engine compartment. During assembly, however, the exhaust manifold must be added after the body is built on the chassis, since the exhaust pipe must be put through the bumper. This type of regrouping and adding manufacturing-specific parts is most efficiently performed in a separate assembly mBOM structure. EBOM parts (items) are used for the filling of these mBOM nodes in the second step until all parts of the product structure are consumed in the mBOM. The eBOM is not the driver but a 'part supplier' for the manufacturing structures in such a way that mainly the parts, and not the structures, are reused. This type of creation mechanism works well for industries in which manufacturing procedures are standardized, such as the automotive or home appliances industries.

3. Approach to the structure management

The core of the new approach is the reference structure (RS), which represents the main framework and rules for communicating operative structures in a globally operating company (figure 3). It neither replaces former structures nor does it serve as an overall information carrier. It rather enables connecting miscellaneous heterogeneous structures and communicating changes faultlessly along the entire product lifecycle. The reference structure allows flexible management of all structures including engineering, logistical, financial, or customer-related structures. This approach is IT-system independent. The reference structure's appearance, composition, and information contents arise from a context-specific use case, process, and IT application.

The unique structure elements represent the actual reference structure. They serve as connection and interaction points to which operative structure contents or nodes (for example, from a CAD structure or BOM) are assigned to. They store the relations between different operative structure contents at different detail levels. From each unique RS element, it is possible to step into a specific operative structure area (nodes or directly contents) and vice-versa. Following this principle, it is possible to navigate to any operative structure element within the entire product lifecycle via the reference structure. This setup allows the derivation of use case-specific views on the reference structure, including all assigned operative structure contents, by filtering and sorting mechanisms.

Besides the related operative structure data, unique RS elements can directly contain any kind of pre-defined data describing the element, such as CAD data, documents, and cost information or guidelines. This data is usually project-independent and used for standardized product components. RS nodes serve as collection points in the reference structure hierarchy and do not have any relations with the operative structures or pre-defined data. They are not labelled by unique IDs and just combine specific groups and RS elements.

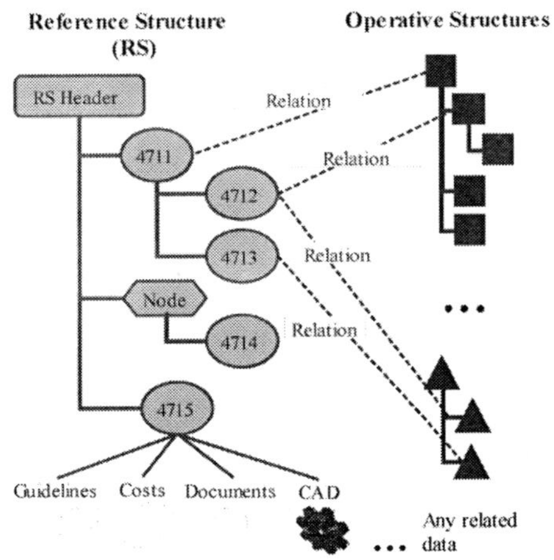

Figure 3. Reference structure approach.

For each specific use case (engineering design, manufacturing, project planning, etc.), specific views on the reference structure can be created. The reference structure even allows the derivation of operative structures like visualization structures, manufacturing structures, or CAD structures. This can be done by:

- direct usage: taking over the reference structure or sub structures without changes or with slight simplifications
- extensions/detailing: adding sub-structures (for example, CAD assemblies), additional description of properties, additional referenced data
- simplification: shortening or generalizing the structure composition (for example, for creating the project plan or cost accounting structures)

The reference structure approach can be used for the customer order-related configuration of structures, mainly at engineering (eBOM) and manufacturing (mBOM). The standardized RS elements may already contain pre-defined data, while empty placeholders will be filled with data and related with operative structures during the order-specific product lifecycle. For each product line or product standard in the company's product portfolio, one RS template should be defined.

4. Approach for the customer related structure configuration

In the next phase of the research, an approach to the structure configuration was developed. This approach is system-neutral and therefore can be implemented after the adjustment in data and process-management-engineering IT tools.

Figure 4 shows the approach to the product-structure-driven process. The main part of this process describes the so-called '150%' elements of the reference structure. This 'product configurator' contains all possible variants and manufacturing configurations so that a shippable product cannot be built using this structure. Moreover, it contains rules or constraints that determine when the variant items exist in the structures as well as the selections that may or may not be made in combination. This variant logic is combined in configuration set engineering (Ce). The manufacturing process is based on the mBOM logic, which is developed along the same line as that of eBOM and contains all possible configurations for the product parts and the required manufacturing items. MBOM-specific configuration tasks (e.g. blanks) are defined in the additional configuration set manufacturing (Cm). Both configuration sets are referenced to the reference product structure. In the first step, configuration set engineering is applied to the reference structure. The result is the customer-, project-, or order-related eBOM. The set Cm is applied to the derived eBOM after that so that the eBOM structure is considered. Both sets (Ce and Cm) completely control the mBOM and allow the order-related configuration. Minor adjustments can be done in the final step of the mBOM-creation process.

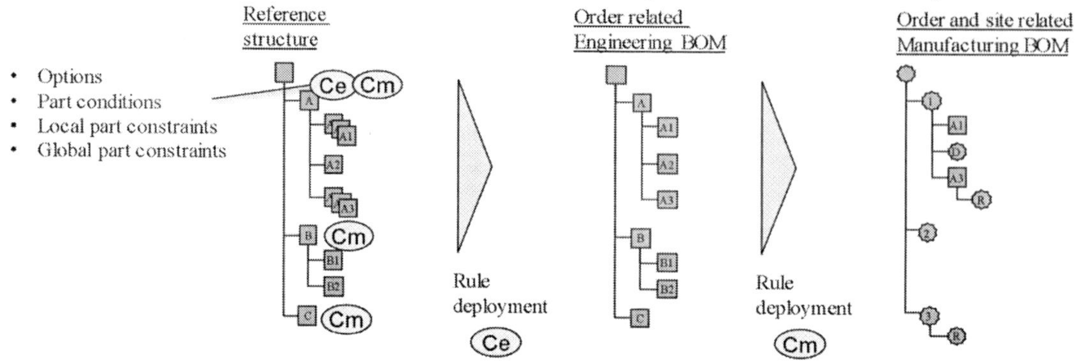

Figure 4. Configuration approach.

The developed configuration sets comprise different configuration rules.

- Options: The options contain a selection of alternative parts for the reference structure node depending on part conditions and constraints. This pre-developed components defer, for example, geometrically in materials or logistical processes.
- Part conditions: These are the main rules for the 'activation' of a component. They can contain rules relating to customer requirements, product lines, manufacturing specifications, or local site logistic chains.
- Local part constraints: Such constraints define part parameters based on their relations to other components or configuration parameter. For example, the geometric value of a component can be defined in relation to the size of another component.
- Global part constraints: Some constraints are defined for the overall company or organization. They are linked with the particular product or configuration. Typical global constraints are based on legal requirements and local customs guidelines or ISO standards, which can lead to the exclusion or inclusion of some components.

5. Industrial implementation and outlook

The developed concept was implemented as a prototype for different mechanical products. The engineering systems NX (for CAD modelling) and Teamcenter (product lifecycle management system for engineering and manufacturing) were used. The configuration logic was implemented using a customized configuration for all Teamcenter elements: options, variants, conditions, rule checks, and constraints. The implementation proved the practical relevance and usability of the framework. The primary benefits can be identified in the form of improved quality of product data, consistency of structures, and access to optimized product data .

The developed integrated reference structure management approach aims to enhance of the quality of the product in combination with concurrent complexity reduction. Globally operating businesses can only be built on integrated processes as well as on a common understanding of the business and product. To use the reference structure approach, it is essential to realign and harmonize the business processes. When a common process landscape is established and a product consolidation is carried out, the reference structure approach offers an optimal integration instrument. It allows connecting, assigning, and merging of information and processes along the entire product lifecycle. Following the real product lifecycle management (PLM) idea, the reference structure provides the fundamental basis

IOP Conf. Series: Materials Science and Engineering **715** (2020) 012108 doi:10.1088/1757-899X/715/1/012108

for an integration that is not only focused on engineering but on all lifecycle phases. In next phases of the research, advanced structures will be investigated. All requirements structures comprising both customer requirements and internal checklists are very important for the engineering-to-order business. The requirement could be linked with related structure nodes so that the tracking of its fulfilment can be realized. Another set of structures—maintenance service structure or recycling-related structures—emerge at the end of the product lifecycle. The extension will be based on the same basic principle of the central reference structure.

References

[1] Arnold V. 2011 *Product Lifecycle Management beherrschen* (Berlin: Springer)

[2] Jung A and Kim B-H 2012 *Korean Journal of Computational Design and Engineering* Vol **17/3**, *An integrated multi-BOM system for product data management*

[3] Eigner M and Koch W 2017 *Modellbasierter Entwicklungsprozess cybertronischer Systeme: Der PLM-unterstützte Referenzentwicklungsprozess für Produkte und Produktionssysteme* (Berlin: Springer)

[4] Stekolschik A 2016 *Second International Conference on Mechanical Engineering and Automation Science (ICMEAS 2016), Engineering Change Management Method Framework in Mechanical Engineering*

[5] Stark J 2015 *Product Lifecycle Management 21st Century Paradigm for Product Realisation* (Geneva: Springer)

[6] Giddaluru M P and James X. and Bhatti R 2017 International Conference on Innovative Design and Manufacturing, *Integrating product knowledge with modular product structures in PLM*

[7] Bruun H P L 2015 *Computers in Industry, PLM system support for modular product development*, pp. 97–111

[8] Malatesta M and Raffaelo R 2013, *International Conference on Engineering Design - ICED 2013, Supporting the modification process of products through a change management tool*

IOP Conf. Series: Materials Science and Engineering **715** (2020) 012109 doi:10.1088/1757-899X/715/1/012109

Study on a smart cross-polarization interference method to the polarized integration mono-pulse array radar

Haijun Wang*, Huanyao Dai, Xiaoliang Nie and Haiye Liu

Luoyang Electronic Equipment Test Centre of China, Luoyang, China

*Email: whjhra88@163.com

Abstract. In order to effectively interfere with radar in the complex electronic countermeasure environment, this paper studies the cross-polarization interference method to the polarized integration mono-pulse array radar based on the radar antenna's intrinsic property that it can receive the cross-polarization components. Firstly, the polarized integration mono-pulse angle measurement method of array radar is described. Then, the cross-polarization interference method for the polarized integration mono-pulse array radar is theoretically analysed, and it is proved mathematically that this interference method can bring the error term, leading to a decrease in measurement accuracy. Finally, a variant function of the singular function's derivative is proposed to simulate the cross-polarization pattern of the sub-array. On this basis, the cross-polarization interference on the polarized array radar angle measurement performance is simulated, and the performance curve is obtained. The simulation results show that the cross-polarization interference can lead to a decrease in the angle measurement performance, and when the cross-polarization interference component is close to the echo co-polarization component, the interference effect is good. The power requirement for implementing the interference is low, so it is conveniently implemented in various confrontational situations. This provides an effective and easy-to-use smart radar polarization interference method for electronic countermeasures.

1. Introduction

Electronic Countermeasures (ECM) [1,2] is a special combat mode and an indispensable combat force for modern warfare. The Polarization Array Radar (PAR) using the polarized integration mono-pulse angle measurement method can accurately measure the target angle [3] and plays an important role in the ECM. It is an important new system radar [4]. PAR makes full use of the advantages of strong anti-jamming capability and high multi-target resolution of the array radar, and can use the polarization information [5] of the target to further improve the accuracy. The polarized integration [6] mono-pulse angle measurement method is an improvement over the mono-pulse angle measurement method [7]. The measurement results of the H polarization channel and the V polarization channel are weighted and fused, then the final measurement angle is obtained. Because the polarized integration mono-pulse array radar adopts array antennas and polarization integration technology, it has strong anti-interference ability to the traditional single-polarized suppression interference and angle deception interference [8]. Seeking an effective angle deception jamming technique has been a hot issue in radar countermeasure research.

According to the inconsistency between the radar antenna main polarization and cross-polarization receiving vectors [9,10], cross-polarization interference[11] emits electromagnetic waves to illuminate the radar to achieve the purpose of angle deception. Because cross-polarization interference does not require spatially separated multiple interference sources, it has great application potential for important target protection or missile penetration [12].

In this paper, the feasibility of cross-polarization interference to the polarized integration mono-pulse array radar is analysed. It is proved that cross-polarization can effectively interfere with the angle measurement performance of the polarized integration mono-pulse array radar.

2. The polarized integration mono-pulse angle measurement method of the Array Radar

The polarized array antenna uses a uniform linear array composed of double orthogonal dipole pairs. It's shown in figure 1. Only the one-dimensional angle measurement of the pitch angle is considered here, and the azimuth angle $\phi = \pi/2$ and the pitch angle $\theta \in [-\pi/2, \pi/2]$.

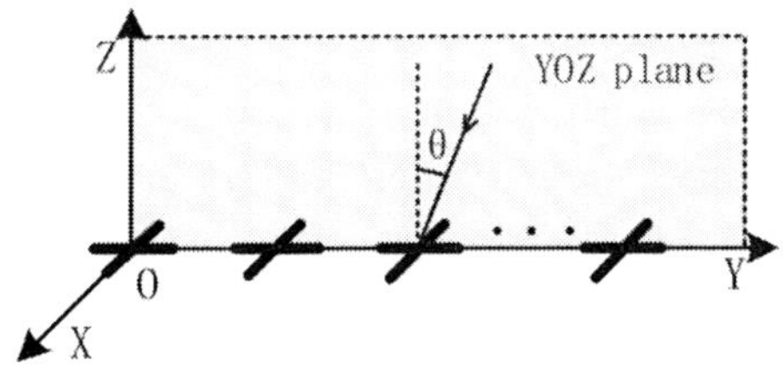

Figure 1. Polarized array antenna structure.

The array radar polarization integration mono-pulse angle measurement method is shown in figure 2.

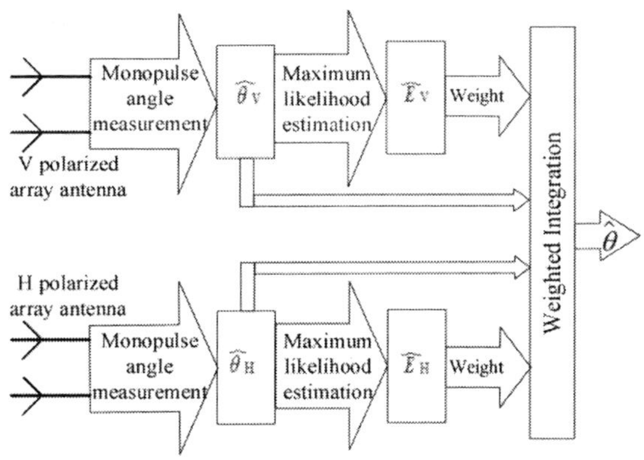

Figure 2. Array radar polarization integration mono-pulse angle measurement method.

The two sub-antenna arrays have the same beam direction and identical receive beams, only the phase center spacing $D = d \cdot N/2$. And the two beam receive signals have the same amplitude, and the phases are different $\Delta\varphi = 2\pi D \cdot \sin\theta/\lambda$. The two beam output signals are E_1, E_2, then $E_2 = E_1 \cdot e^{-i\cdot\Delta\varphi}$. The angle information of the target can be extracted by calculating the sum-difference ratio $\Delta/\Sigma = (E_1 - E_2)/(E_2 + E_1)$. The angle measurement formula is $\sin\hat{\theta} = \sin\theta_0 + k_1 \cdot \mathrm{Im}(\Delta/\Sigma)$, where θ_0 is the beam direction, $\hat{\theta}$ is the target angle estimation, and $k_1 = 2\lambda/(N\pi d)$.

The joint probability density function of the received signal and the complex amplitude is:

$$p(x;E) = \frac{1}{(2\pi\sigma^2)^{N/2}} \cdot \exp\left[-\frac{1}{2\sigma^2}\left(x - E\cdot s(\hat{\theta})\right)^{H} \cdot \left(x - E\cdot s(\hat{\theta})\right)\right] \quad (1)$$

$$x = E \cdot s(\hat{\theta}) + n \quad (2)$$

According to the maximum likelihood method, the estimated complex amplitude of the received signal is $\hat{E}_{\mathrm{H}} = \boldsymbol{s}^{\mathrm{H}}\left(\hat{\theta}_{\mathrm{H}}\right)\cdot \boldsymbol{x}_{\mathrm{H}} / N$, $\hat{E}_{\mathrm{V}} = \boldsymbol{s}^{\mathrm{H}}\left(\hat{\theta}_{\mathrm{V}}\right)\cdot \boldsymbol{x}_{\mathrm{V}} / N$.

The single-polarization angle measurements $\hat{\theta}_{\mathrm{H}}$ and $\hat{\theta}_{\mathrm{V}}$ are integrated and the angle estimation $\hat{\theta}$ under the polarization fusion method is obtained:

$$\hat{\theta} = \alpha_1\hat{\theta}_{\mathrm{H}} + \alpha_2\hat{\theta}_{\mathrm{V}},\ \alpha_1 = \sigma_{\mathrm{V}}^2 / (\sigma_{\mathrm{H}}^2 + \sigma_{\mathrm{V}}^2),\ \alpha_2 = \sigma_{\mathrm{H}}^2 / (\sigma_{\mathrm{H}}^2 + \sigma_{\mathrm{V}}^2) \tag{3}$$

Among them, σ_{H}^2 and σ_{V}^2 is the variance of $\hat{\theta}_{\mathrm{H}}$ and $\hat{\theta}_{\mathrm{V}}$:

$$\sigma_{\mathrm{H}}^2 = k_2\sigma_{\mathrm{n}}^2\theta_{\mathrm{3dB}}^2 / \left|\hat{E}_{\mathrm{H}}\right|^2, \sigma_{\mathrm{V}}^2 = k_2\sigma_{\mathrm{n}}^2\theta_{\mathrm{3dB}}^2 / \left|\hat{E}_{\mathrm{V}}\right|^2 \tag{4}$$

where $k_2 = 0.19N/\left(N^2-1\right)$ is a constant, and θ_{3dB} is the 3dB beamwidth. The final angle estimation is:

$$\hat{\theta} = (\hat{\theta}_{\mathrm{H}}\left|\hat{E}_{\mathrm{H}}\right|^2 + \hat{\theta}_{\mathrm{V}}\left|\hat{E}_{\mathrm{V}}\right|^2) / (\left|\hat{E}_{\mathrm{H}}\right|^2 + \left|\hat{E}_{\mathrm{V}}\right|^2) \tag{5}$$

3. Analysis of the cross-polarization interference method to the polarized integration mono-pulse array radar

The polarization array antenna can be seen as a combination of a horizontally polarized array antenna and a vertically polarized array antenna. When performing single-polarization angle measurement, each single-polarization array antenna is divided into two sub-arrays, so that there are four sub-arrays. Each sub-array corresponds to a main polarization beam and a cross-polarization beam.

It is assumed that the main polarization amplitude pattern and the phase pattern of the four sub-arrays are the same, respectively be $G_m(\theta)$ and $P_m(\theta) = 0°$ .But the cross-polarization amplitude patterns and the phase patterns of different antennas are difficult. Let the cross-polarization amplitude patterns of the horizontal polarization sub-array and the vertical polarization sub-array be respectively $G_{\mathrm{H1}c}(\theta)$, $G_{\mathrm{H2}c}(\theta)$, $G_{\mathrm{V1}c}(\theta)$ and $G_{\mathrm{V2}c}(\theta)$. Let the phase patterns are respectively $P_{\mathrm{H1}c}(\theta)$, $P_{\mathrm{H2}c}(\theta)$, $P_{\mathrm{V1}c}(\theta)$, $P_{\mathrm{V2}c}(\theta)$.

The difference signal obtained by the horizontally polarized array antenna is:

$$\begin{aligned}\Delta_{\mathrm{H}} = \Delta_{\mathrm{H}m} + \Delta_{\mathrm{H}c} &= G_m(\theta)\cdot E_{\mathrm{H}}\left(1-e^{-j\Delta\varphi}\right)\\ &+ G_{\mathrm{H1}c}(\theta)\cdot P_{\mathrm{H1}c}(\theta)\cdot\left(E_{\mathrm{V}}+E_{i\mathrm{V}}\right) - G_{\mathrm{H2}c}(\theta)\cdot P_{\mathrm{H2}c}(\theta)\cdot\left(E_{\mathrm{V}}+E_{i\mathrm{V}}\right)\cdot e^{-j\Delta\varphi}\end{aligned} \tag{6}$$

The sum signal is:

$$\begin{aligned}\Sigma_{\mathrm{H}} = \Sigma_{\mathrm{H}m} + \Sigma_{\mathrm{H}c} &= G_m(\theta)\cdot E_{\mathrm{H}}\left(1+e^{-j\Delta\varphi}\right)\\ &+ G_{\mathrm{H1}c}(\theta)\cdot P_{\mathrm{H1}c}(\theta)\cdot\left(E_{\mathrm{V}}+E_{i\mathrm{V}}\right) + G_{\mathrm{H2}c}(\theta)\cdot P_{\mathrm{H2}c}(\theta)\cdot\left(E_{\mathrm{V}}+E_{i\mathrm{V}}\right)\cdot e^{-j\Delta\varphi}\end{aligned} \tag{7}$$

Among them, E_{H} is the horizontal polarized wave reflected by the target, E_{V} is the vertically polarized echo reflected by the target, $E_{i\mathrm{V}}$ is the vertically polarized wave emitted by the jammer.

Similarly, the difference signal obtained by the vertically polarized array antenna is:

$$\begin{aligned}\Delta_{\mathrm{V}} = \Delta_{\mathrm{V}m} + \Delta_{\mathrm{V}c} &= G_m(\theta)\cdot\left(E_{\mathrm{V}}+E_{i\mathrm{V}}\right)\cdot\left(1-e^{-j\Delta\varphi}\right)\\ &+ G_{\mathrm{V1}c}(\theta)\cdot P_{\mathrm{V1}c}(\theta)\cdot E_{\mathrm{H}} - G_{\mathrm{V2}c}(\theta)\cdot P_{\mathrm{V2}c}(\theta)\cdot E_{\mathrm{H}}\cdot e^{-j\Delta\varphi}\end{aligned} \tag{8}$$

The sum signal is:

$$\begin{aligned}\Sigma_{\mathrm{V}} = \Sigma_{\mathrm{V}m} + \Sigma_{\mathrm{V}c} &= G_m(\theta)\cdot\left(E_{\mathrm{V}}+E_{i\mathrm{V}}\right)\cdot\left(1+e^{-j\Delta\varphi}\right)\\ &+ G_{\mathrm{V1}c}(\theta)\cdot P_{\mathrm{V1}c}(\theta)\cdot E_{\mathrm{H}} + G_{\mathrm{V2}c}(\theta)\cdot P_{\mathrm{V2}c}(\theta)\cdot E_{\mathrm{H}}\cdot e^{-j\Delta\varphi}\end{aligned} \tag{9}$$

When considering the existence of cross-polarization components, the signals received by horizontal polarization antenna and vertical polarization antenna are:

$$\begin{cases} \mathbf{x}_{\mathrm{H}} = G_m(\theta)\cdot E_{\mathrm{H}}\cdot \mathbf{s}(\hat{\theta}_{\mathrm{H}}) \\ +G_{\mathrm{H1}c}(\theta)\cdot P_{\mathrm{H1}c}(\theta)\cdot (E_{\mathrm{V}}+E_{i\mathrm{V}})\cdot \mathbf{s}_1(\hat{\theta}_{\mathrm{H}}) + G_{\mathrm{H2}c}(\theta)\cdot P_{\mathrm{H2}c}(\theta)\cdot (E_{\mathrm{V}}+E_{i\mathrm{V}})\cdot \mathbf{s}_2(\hat{\theta}_{\mathrm{H}}) + \mathbf{n}_{\mathrm{H}} \\ \mathbf{x}_{\mathrm{V}} = G_m(\theta)\cdot (E_{\mathrm{V}}+E_{i\mathrm{V}})\cdot \mathbf{s}(\hat{\theta}_{\mathrm{V}}) \\ +G_{\mathrm{V1}c}(\theta)\cdot P_{\mathrm{V1}c}(\theta)\cdot E_{\mathrm{H}}\cdot \mathbf{s}_1(\hat{\theta}_{\mathrm{V}}) + G_{\mathrm{V2}c}(\theta)\cdot P_{\mathrm{V2}c}(\theta)\cdot E_{\mathrm{H}}\cdot \mathbf{s}_2(\hat{\theta}_{\mathrm{V}}) + \mathbf{n}_{\mathrm{V}} \end{cases} \tag{10}$$

Among them, $\mathbf{s}(\theta) = \left[e^{j\phi_1}\ e^{j\phi_2} \cdots e^{j\phi_N}\right]$ is a phase-weighted vector.Due to the existence of interference, a large estimation error will be brought.

4. Simulation analysis

The number of array elements is set N=16, the radar frequency is f=5GHz, the array element spacing is set to half wavelength, the wavelength is 6cm, the beam is pointed $\theta_0 = 0°$, the angle between the target and the beam center is -2°, and the Monte Carlo times is set to M=1000. The cross-polarization component is set to be 20dB smaller than the co-polarization component, and the angle estimation root-mean-square error $\mathrm{RMSE} = \sqrt{E\left[(\hat{\theta}-\theta)^2\right]}$ is used to estimate the angle measurement performance.

The main polarization pattern of the sub-array is simulated using the singular function:

$$G_m(\theta) = \left[\sin(k_m\theta)/(k_m\theta)\right]^2 \tag{11}$$

The cross-polarization amplitude pattern of the sub-array is simulated using a variant function of the first derivative of the singular function, and has a plurality of control parameters to enable flexible simulation of cross-polarization patterns:

$$G_c = L\cdot\left[2k_c^2\theta\sin(k_c\theta)\cdot\cos(k_c\theta) - 2k_c\sin^{\alpha}(k_c\theta)\right]/\left((k_c\theta)^{\beta}+1\right) \tag{12}$$

Among them, the value of k is determined by the number of side-lobes, L is the attenuation adjustment amount, α and β are the beam shape parameters, and θ is the corresponding angle value whose unit is radian. In (10), set $k_m = 40$ and the beam width is $\theta_{3dB} = 4°$. The cross-polarization amplitude patterns of two sub-arrays of the horizontally polarized array antenna are set to $k_{\mathrm{H1}c} = 39$, $L_{\mathrm{H1}c} = 32$, $\alpha_{\mathrm{H1}c} = 1.8$, $\beta_{\mathrm{H1}c} = 1.5$, $k_{\mathrm{H2}c} = 40$, $L_{\mathrm{H2}c} = 32$, $\alpha_{\mathrm{H2}c} = 1.4$ and $\beta_{\mathrm{H2}c} = 2$. The cross-polarization amplitude patterns of two sub-arrays of the vertically-polarized array antenna are set to $k_{\mathrm{V1}c} = 35$, $L_{\mathrm{V1}c} = 33$, $\alpha_{\mathrm{V1}c} = 1.2$, $\beta_{\mathrm{V1}c} = 2$, $k_{\mathrm{V2}c} = 30$, $L_{\mathrm{V2}c} = 30$, $\alpha_{\mathrm{V2}c} = 2$ and $\beta_{\mathrm{V2}c} = 2$.

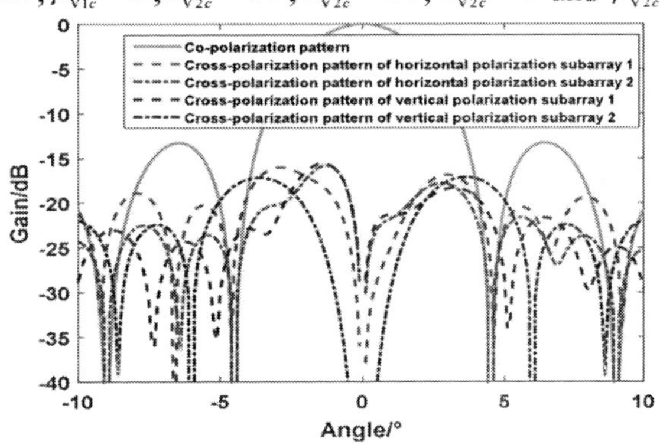

Figure 3. Main polarization and Cross-Polarization `of Sub-arrays.

Figure 3 shows the main polarization and cross-polarization amplitude pattern of each sub-array. Cross-polarization amplitude patterns of sub-arrays also differ greatly.

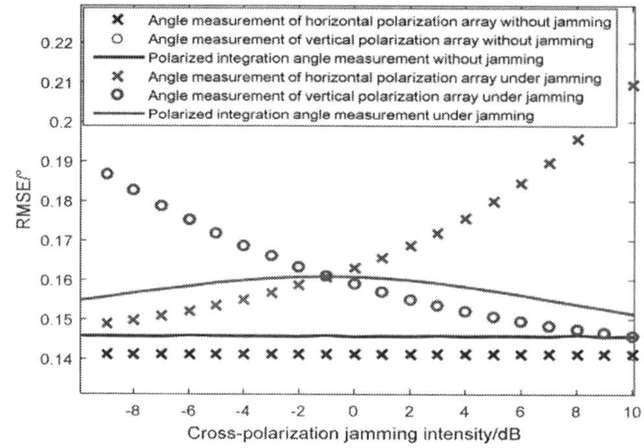

Figure 4. Relationship between measurement accuracy and cross-polarization interference intensity.

Figure 4 shows the relationship between the measurement accuracy and the interference intensity. When the cross-polarization interference intensity ranges from -10dB to 10dB, the angle measurement accuracy under interference is inferior to that without interference. As can be seen from figure 4, as the interference intensity increases, the horizontal polarization antenna measurement performance gradually decreases, and the vertical polarization antenna measurement performance gradually increases. When the cross-polarization interference is about 0 dB, the performance of the horizontal polarization array antenna is reduced large and the angle measurement performance of the vertical polarization array antenna is improved little. After polarization integration, the polarized integration angle measurement performance of the polarization array antenna is reduced.

5. Conclusion

In this paper, the cross-polarization interference method of polarized integration mono-pulse array radar is studied. The polarized integration mono-pulse angle measurement method of array radar is described. Then, the cross-polarization interference method of polarized integration mono-pulse array radar is theoretically analyzed, and it is proved mathematically that this interference method can bring the error term of angle measurement. Finally, the main polarization and cross-polarization antenna patterns of the four groups of sub-arrays are simulated. Based on this, the cross-polarization interference simulation of the angle measurement performance of the polarization array radar is performed, and the RMSE variation curve with cross-polarization intensity is obtained. The simulation results show that the cross-polarization interference will lead to the decline of the angle measurement performance of the array radar.

References

[1] Richard A. Poisel, Electronic Warfare Receivers and Receiving Systems (Artech House Press, 2014)
[2] Andrea De Martino, Introduction to Modern EW Systems (Artech House Press, 2012)
[3] Yangang Huang, Zhenhai Xu, Chong Dai, Ziyuan Xiong, Study on Monopulse Angle Measurement Method in Polarization Array Radar. Radar Science and Technology, **12**, (2), pp. 201-205(2014)
[4] Zhenhai Xu, Research on Signal Processing of Polarization Sensitive Array. PhD thesis, National University of Defence Technology, (2005)
[5] Haijun Wang, Jianbing Li, Tao Wang, Xuesong Wang.: 'Calculate the Polarized Scattering Properties of the Canted Rain Media with M-P DSD using Extended Mie Scattering Method', Int. Conf. 2015 IET, Hangzhou, China, October, pp. 1-5 (2016)
[6] Haijun Wang, Guangyong Zheng, Yonghu Zeng, Liandong Wang.: 'Experimental Analysis on

the Passive Bistatic Radar Based on Dual-polarized Integration Technology', Moder Radar, **39**, (10), pp. 11-20 (2017)

[7] Nichel U.: 'Overview of Generalized Monopulse Estimation', IEEE Aerospace and Electronic Systems Magazine, **21**, (6), pp. 27-56 (2006)

[8] Longfei Shi, Bo Ren, Jiazhi Ma, Yongzhen Li.: 'Recent Development of Radar Anti-interference Techniques with Polarimetry', Modern Radar, **38**, (4), pp. 1-7 (2016)

[9] Mianquan Li, Yongzhen Li, Jian Dong, Xuesong Wang.: 'Polarization characteristic analysis of crossed dipoles', Chinese Journal of Radio Science, **27**, (2), pp. 396-401 (2012)

[10] Anshi Zhu, Zili Chen, Xiaoqian Liu, Yunhui Zhen, Yong Yang.: 'Numerical Calculation of Radiation Characteristic of Monopole and Dipole Antenna', Journal of Microwaves, **31**, (2), pp. 33-38 (2015)

[11] David L. Adamy.: 'EW104: EW Against a New Generation of Threats' (Artech House Press, 2015)

[12] Huanyao Dai, Bo Zhou, Cuiqiong Mo, Jianlu Wang, Zheng Wang.: 'Electromagnetic interference mechanism and simulation in polarization for mono-pulse radar seeker', High Power Laser and Particle Beams, **27**, (10), pp. 103241 (2015)

IOP Conf. Series: Materials Science and Engineering **715** (2020) 012110 doi:10.1088/1757-899X/715/1/012110

Short text semantic feature extension and classification based on LDA

Liu-tong Xu, Zhao XUE* and Hai Huang

Beijing University of Posts and Telecommunications, Cloud Computing and Service Computing Center, 100876, Beijing, China

*Email: susannahermione@163.com

Abstract. To solve the problem of feature sparseness of short texts, we studied the application of LDA Topic Model on feature extension and classification of short texts. Training LDA on external long texts related to short texts, and achieving the inference and extension of short texts' topics based on LDA solves the feature sparseness of short texts and improves the accuracy of classification effectively. Latent semantic information in LDA can also effectively improve the interpretability of short texts.

1. Introduction

With the rapid development of information technology, we receive more and more information in daily life. At the same time, information technology has also brought some negative issues, such as "information overload" phenomenon-A lot of redundant information makes it inefficient in choosing valuable information. In order to better solve this problem, search engine technology is used to filter information, which makes the time required to obtain useful information greatly reduced. However, search engine is not perfect, it still has many problems, for example, one of the obvious ones is that users themselves are not very clear about what information they need in many situations, and they also can't accurately describe the information they need by keywords or phrases. At the same time, users' needs for information are constantly changing, even the same keyword may correspond to different information needs, so it is necessary to classify and identify the information needs which are described by texts. In this way, it can greatly narrow the scope of information retrieval. However, the efficiency of manual labelling is too low, and the accuracy often fails to meet usage requirements, therefore classifying texts by computer algorithms has been proposed.

With the rapid development of Internet and mobile terminal technology, the increasing number of varied types of information represented by short texts containing rich information have emerged, such as short messages. Its application has also gradually expanded to different types of information systems. Therefore, it is of great practical significance to study the short text classification methods which can lead to scientifically obtaining more and more valuable information and providing users with better information services.

Text classification mainly focused on the processing of long texts in the early stage of research. As the research further develops, some algorithms have achieved good results. However, compared with long texts, short texts are different from them, the differences specifically embody in the following aspects: firstly, short text has less characters, so its feature is sparse and it may contain more noise information; secondly, the dynamic characteristics of short texts are more obvious, which leads to

IOP Conf. Series: Materials Science and Engineering **715** (2020) 012110 doi:10.1088/1757-899X/715/1/012110

frequent update. Therefore, it is difficult to effectively process short texts using the classification methods which are currently applied to long texts. The technologies that have emerged are still not mature enough, which makes the application subject to certain restrictions.

2. Related work

Text classification has a long history. Back in the 1950s, Luhn H. P [1] proposed the concept of word frequency, which treats a text as a collection of words, then uses the word frequency to measure its importance. The earliest text classification methods mainly use customized classification rules, whose classification accuracy depends on the familiarity with the classification interval, and this makes classification difficult and its efficiency can't be guaranteed.

With the popularity of Internet applications, there have been massive text data, algorithms for classifying large-scale text data have also appeared and achieved good results. Among them, machine learning algorithms have been widely used because of high accuracy and intelligence. Salton [2] et al. designed a vector space model and applied it to text classification; Lewis [3] made a survey on text classification; Yang Yiming [4] studied the feature selection through experiments and compared different methods, such as Information Gain. At the end of the last century, Cortes [5] et al. proposed the Support Vector Machine (SVM) algorithm based on statistical theory, this method gradually attracted people's attention because of its good performance. On the basis of SVM, some experts have improved it so that its application has been further expanded. For example, the SVM-KNN proposed by Li Rong [6], successfully combined SVM with KNN; Liu Chunwei [7] made more improvements on SVM, and designed the PSO-SVM algorithm which used Particle Swarm Optimization to determine the parameters of SVM.

In the process of the rapid development of the Internet, there are more and more short text messages such as news, blogs, and chat records, how to process them and provide more valuable information has become a hot research in the field of text mining, experts in this field have conducted a lot of research work, and some algorithms or systems have also been applied to the processing of short text information. Rafiei [8] proposed a feature-based short text classification method, which makes information filtering effective; Sahami [9] proposed a search engine-based method to calculate the similarity between short texts; Xuan-Hieu Phan [10] et al. mainly adopted a method for obtaining hidden topics in the classification of short texts and achieved good results.

The current research and application of short text classification is mainly based on English texts, which can't be directly applied to Chinese language materials. Therefore, Chinese experts are also conducting research in this area. Fan Xinghua [11] proposed a short text feature extension method based on co-occurrence and achieved a certain degree of effect in short text classification; Wang Xiwei [12] proposed another short text feature extension method based on context semantics.

As experts are continuing to deepen research on short text classification, there are more and more new methods. The LDA topic model is the typical one among them. Blei [13] first proposed LDA in 2003 and applied it to text classification. It was originally mainly used in the field of mathematics, however, with the deepening of its research, its application has gradually expanded to more areas, and currently it has been applied to data mining processing. Lu Y [14] designed a topic- emotion mixed LDA model which divides words into ordinary words and words with emotional tendencies. The latter can be further subdivided into positive, negative, and middle categories, then it can generate the document's implicit semantic vector representation.

3. Our work

Compared with long text, short text has fewer characters and contains more noise information. Dynamic changes are obvious which leads short text to be updated frequently. All of these make short text classification still need more researches.

For long texts, after segmenting a large amount of text, the corpus will form a large dictionary. Then after filtering the stopwords, the dictionary and long texts still contain many words, and the text feature vector based on word frequency is still in high dimension. However, short text often contains only dozens of characters, even a dictionary generated by a large amount of short text, its scale is much smaller. After filtering the stopwords, only a few feature words remain, this will make fewer words in the dictionary and short texts which leads to "dimension disaster" phenomenon. Feature sparseness is a very obvious problem for short texts.

Feature extension can solve the sparseness problem. Text feature extension methods are divided into knowledge bases based [15] and LDA based [16-17]. The former mainly uses the currently widely used search engines such as Baidu to search and introduce relevant information of feature words. This method can increase short text features, and make its meaning clearer. However, its disadvantage is that the effect is too dependent on the rationality of the knowledge base. Currently, with the development of LDA, how to apply it to text feature extension has become a popular research. Our paper will conduct related research work in LDA, using the latent semantic information of long texts related with short texts to implement the feature extension and classification of short text.

3.1. LDA

In recent years, with the development and improvement of LDA, its application in text classification is also increasing. Parameter estimation of LDA usually adopts Gibbs sampling, then gets a text-topic probability distribution vector $\boldsymbol{\theta}$ and a topic-word probability distribution vector $\boldsymbol{\varphi}$, $\boldsymbol{\theta}_i$ represents the probability that the text implies a topic. The larger the value, the better the match between the text and the topic; $\boldsymbol{\varphi}_i$ represents the probability that a word appears on a topic. The larger the value, the greater the probability that the word appears on this topic, the top M words are usually selected according to the order of the probability to represent the topic. The generative graphical model of LDA is shown in figure 1.

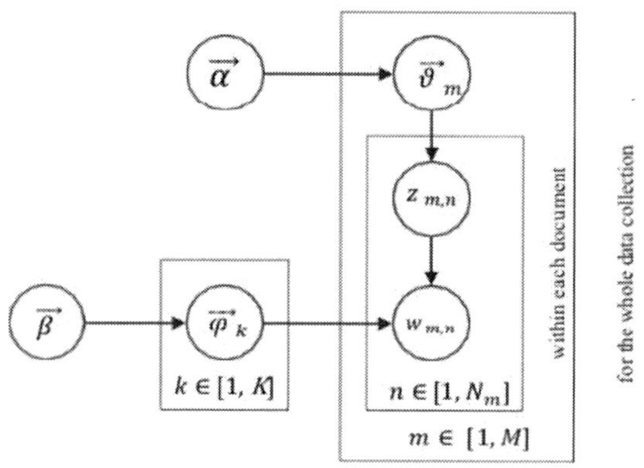

Figure 1. LDA: a generative graphical model.

From the generative graphical model depicted in figure 1, we can write the joint distribution of all known and hidden variables given the Dirichlet parameters which is shown in table 1.

Table 1. Generation process for LDA.

for all topics $\in [1, K]$ do
  sample mixture components $\vec{\varphi}_k \sim Dir(\vec{\beta})$
end for
for all documents $m \in [1, M]$ **do**
  sample mixture proportion $\vec{\vartheta}_m \sim Dir(\vec{\alpha})$
  sample document length $N_m \sim Poiss(\xi)$
  for all words $n \in [1, N_m]$ **do**
  sample topic index $z_{m,n} \sim Mult(\vec{\vartheta}_m)$
  sample term for word $w_{m,n} \sim Mult(\vec{\varphi}_{z_{m,n}})$
  end for

IOP Conf. Series: Materials Science and Engineering **715** (2020) 012110 doi:10.1088/1757-899X/715/1/012110

end for

Parameters and variables:

- M : the total number of documents
- K : the number of (hidden/latent) topics
- V : vocabulary size
- $\vec{\alpha}, \vec{\beta}$: Dirichlet parameters
- $\vec{\vartheta}_m$: topic distribution for document m
- $\Theta = \{\vec{\vartheta}_m\}_{m=1}^{M}$: a $M \times K$ matrix
- $\vec{\varphi}$: word distribution for topic
- $\Phi = \{\vec{\varphi}_k\}_{k=1}^{K}$: a $K \times V$ matrix
- N_m: the length of document m
- $z_{m,n}$: topic index of n th word in document m
- $w_{m,n}$: a particular word for word placeholder [m, n]

LDA uses the text-topic probability distribution vector $\boldsymbol{\theta}$ as the feature vector of the text, and implements classification based on it. But for short text, the length of the text itself is small, some of the topics could not reflect specific meaning. So, some methods [18] suggest that a large number of external related texts can be directly used to train LDA, then implement the feature extension of short text according to the topic information of this model. However, it is difficult to ensure that the extended added topic is completely consistent with the meaning of the original text. Some other methods [19-20] are applied to train LDA using a mixed topic model, but this makes the training time increase and the efficiency decline, and the number of topics K is manually set, corresponding optimal K of different corpus is often different, so it is difficult to determine the optimal number of mixed multi-topics, and these deficiencies have made the application limited. The above methods of short text classification based on LDA still have obvious deficiencies, summarized as the following aspects:

1) There is no filtering on unwanted topics. Redundant noise information affects classification accuracy.

2) Direct modeling of short texts at the level of implicit semantic information but ignoring the influence of word features on the statistical level, which causes the accuracy of classification to be excessively concentrated on the effects of LDA. Factors that influence the results are lopsided.

3) For multi-topic LDA models, more training time is required. Especially with the increase in the size of corpus, the time spent on the training will also increase significantly.

4) How to determine a reasonable number of topics is also an important issue. It takes a constant trial to determine the optimal value, which makes its application conditions too harsh.

3.2. Feature extension and classification

The improvement of the short text classification algorithm in our work is mainly to use the LDA built on the external related long texts to achieve the topic extension of short texts. The details are as follows: Infer the related topics of short texts using the LDA obtained on external related long texts, filter and remove unwanted topics, then calculate the text-topic probability distribution vector $\boldsymbol{\theta}$ and the topic-word probability distribution vector $\boldsymbol{\varphi}$ for short texts, and then introduce topics on external related texts that meet the condition by comparing with the threshold and introduce them to the short text representation vector. In this way, the semantic features of short texts are extended. As described above, the steps of the Chinese short text classification algorithm based on LDA Topic Model proposed in our work are as follows:

Step 1: A pre-processed external related long text set is used to obtain a LDA topic model;

Step 2: After the short text is preprocessed, infer the relevant topics of the short texts including training and testing set based on the model obtained in the first step, remove unwanted topics, and calculate the probability distribution $\boldsymbol{\theta}$ and $\boldsymbol{\varphi}$;

IOP Conf. Series: Materials Science and Engineering **715** (2020) 012110 doi:10.1088/1757-899X/715/1/012110

Step 3: Extend features based on the result of comparing $\boldsymbol{\theta}_i$ with the threshold, and determine the final text-topic probability distribution vector $\boldsymbol{\theta}$ of short text;

Step 4: Train the parameters of the classifier with the training short texts and obtain the classifier, then verify the performance of the classifier on the testing set.

The corresponding flow chart of the above steps is shown in figure 2.

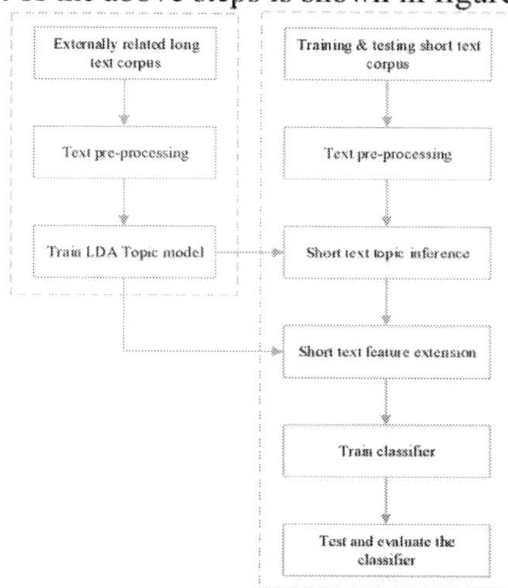

Figure 2. Flow chart of short text feature extension and classification algorithm based on LDA.

In our work, we select the classifier based on neural network [21], it uses short text representation vector $\boldsymbol{\theta}$ as input, and gets the probability that the short text belongs to a certain class. The class with the highest probability is the one which the short text is determined belong to.

4. Experiments

4.1. Data

In order to verify the performance of algorithm, we conducted related experiments. The short texts data used in our work is mainly a short news corpus crawled from China News Online[1], including eight classes of politics, sports, finance, culture, education, military, technology and entertainment from 2016.12-2017.12. The average number of texts per category in training and testing sets is 10,000, and the total amount of texts is about 160,000. The external related long texts set is Sogou network-wide news data (SogouCA[2]).

4.2. Evaluation

Evaluation metrics is an important basis for measuring the rationality of the algorithm which reflects the performance of the classification model. Through quantitative analysis of evaluation metrics, the advantages, disadvantages, and scope of application of the model can be directly derived, and establish a baseline for further improvement. At present, accuracy, recall, and precision are often used. By quantitatively analyzing the evaluation metrics, the experimental results can be accurately described. The definitions of each evaluation metrics are as follows.

(1) *Accuracy*

Accuracy is defined as the ratio of the number of correctly classified texts to the total number of texts, its calculation formula is as follows, *TP*, *TN*, *FP*, and *FN* respectively mean the number of correct positive samples, the number of correct negative samples, the number of wrong positive samples, and the number of wrong negative samples. For classification models, it should have a high accuracy.

[1] http://www.chinanews.com

[2] http://www.sogou.com/labs/resource/ca.php

$$Accuracy = \frac{TP+TN}{TP+FP+FN+TN} \quad (1)$$

(2) *Recall & precision*

Recall indicates the ratio of the number of documents that are classified into the class c_i by the classifier to the number of real documents within the class c_i. Its calculation formula is as follows.

$$R = \frac{TP}{TP+FN} \quad (2)$$

Precision indicates the ratio of the number of documents correctly classified into the class c_i and the number of documents determined to be c_i by the classifier. Its calculation formula is as follows.

$$P = \frac{TP}{TP+FP} \quad (3)$$

(3) *F1- Measure*

Recall and *precision* are complementary relationships. One increases, the other may decrease. So, using these two metrics to evaluate the overall performance of the classification model is often difficult to achieve good results. At this point, we can choose *F1- Measure* which comprehensively considers rate and precision. Its calculation formula is as follows.

$$F1 = \frac{2\times P\times R}{P+R} \quad (4)$$

(4) *Micro-R, Micro-P, Micro-F1*

The above three metrics mainly evaluate the performance of the classification model on each class in the corpus, but does not evaluate its overall classification effect. At this point, we can choose to use *Micro-R, Micro-P, Micro-F1* metrics to analyse the performance of the classification model.

The calculation formula for *Micro-R* is as follows.

$$Micro - R = \frac{\sum_{i=1}^{K} TP_i}{\sum_{i=1}^{K} (TP_i+FN_i)} \quad (5)$$

The calculation formula for *Micro-P* is as follows.

$$\mathrm{Micro} - \mathrm{P} = \frac{\sum_{i=1}^{K} \mathrm{TP_i}}{\sum_{i=1}^{K} (\mathrm{TP_i}+\mathrm{FP_i})} \quad (6)$$

The calculation formula for *Micro-F1* is as follows

$$Micro - F1 = \frac{2 \times (Micro-R)\times(Micro-P)}{(Micro-R) + (Micro-P)} \quad (7)$$

4.3. Results

Our work mainly extends topic information in the previous short text classification algorithm. It can better solve the problem of sparse semantic features of short texts and improve the interpretability of original texts. In order to verify the performance of the improved algorithm, the following experiment was conducted.

Previous research has confirmed that when LDA is not used, DF feature selection method has been able to achieve good results. We compare the performance of DF feature selection method, the document-topic probability distribution method, and the improved feature extension method in this paper. The specific experimental results are shown in figure 3.

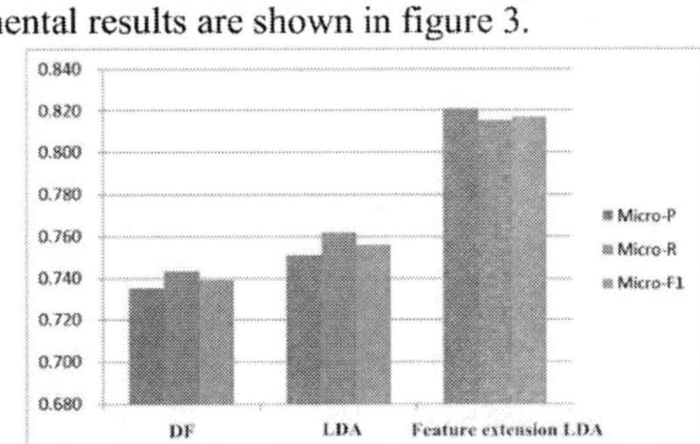

Figure 3. Comparison histogram of classification results of different algorithms.

5. Conclusion

In this paper, we improved the existing short text classification algorithm in feature extension by applying the LDA Topic Model to the feature extension and selection. First, LDA is trained on the external related long texts, and then the short text is represented as a text-topic probability distribution vector $\boldsymbol{\theta}$ based on the it. Then, by comparing $\boldsymbol{\theta}_i$ with the threshold, the topic obtained on the long texts satisfying the condition is introduced into the short text representation vector $\boldsymbol{\theta}$. In this way, the semantic information of short text is enriched, the feature extension of short text is realized, and it can get better results in Chinese short text classification.

References

[1] Luhn H P. The automatic creation of literature abstracts[M]. IBM Corp. (1958).
[2] Salton G. A vector space model for automatic indexing[J]. Communications of the ACM, (1974), **18**(11):613-620.
[3] Lewis D D. Feature selection and feature extraction for text categorization[C]// The Workshop on Speech & Natural Language. Association for Computational Linguistics, (2013):212--217.
[4] Yang Y, Pedersen J O. A Comparative Study on Feature Selection in Text Categorization[C]// Fourteenth International Conference on Machine Learning. Morgan Kaufmann Publishers Inc. (1997):412-420.
[5] Cortes C, Vapnik V. Support-Vector Networks. [J]. Machine Learning, (1995), **20**(3):273-297.
[6] Rong L I, Sun Y. Application of SVM-KNN Classifier into Web Page Classification[J]. Science Technology & Engineering, (2009).
[7] Liu C W, Luo J X. A PSO-SVM Classifier Based on Hybrid Kernel Function[J]. Journal of East China University of Science & Technology, (2014), **40**(1):96-101.
[8] Rafiei D, Bharat K, Shukla A. Diversifying web search results[C]// International Conference on World Wide Web, WWW 2010, Raleigh, North Carolina, USA, April. DBLP, (2010):781-790.
[9] Sahami M, Heilman T D. A web-based kernel function for measuring the similarity of short text snippets[C]// International Conference on World Wide Web, WWW 2006, Edinburgh, Scotland, UK, May. DBLP, (2006):377-386.
[10] Xuan-Hieu Phan, Le-Minh Nguyen, et al., Learning to classify short and sparse text & web with hidden topics from large-scale data collections [C]. In Proceedings of the 17th International Conference on World Wide Web, New York, ACM Press, (2008),91-100.
[11] Fan X, Hu H. A New Model for Chinese Short-text Classification Considering Feature Extension[C]// International Conference on Artificial Intelligence and Computational Intelligence. IEEE, (2010):7-11.
[12] Wang X W, Fan X H, Zhao J. Method for Chinese short text classification based on feature extension: Method for Chinese short text classification based on feature extension[J]. Journal of Computer Applications, (2009), **29**(3):843-845.
[13] Blei D M, Ng A Y, Jordan M I. Latent dirichlet allocation[J]. J Machine Learning Research Archive, (2003), 3:993-1022.
[14] Lu Y, Mei Q, Zhai C. Investigating task performance of probabilistic topic models: an empirical study of PLSA and LDA[J]. Information Retrieval, (2011), **14(2)**:178-203.
[15] Zhao H, Liu H. Classification Algorithm of Chinese Short Texts Based on Wikipedia[J]. Library & Information Service, (2013), **57(11)**:120-124.
[16] Hu Y. A New Method of Key Words Extraction for Chinese Short-text Classification[J]. New Technology of Library & Information Service, (2013).
[17] Zhang Peijing, Song Lei. Overview on Topic Modeling of Microblogs Text Based on LDA [J]. Library and Information Service, (2012), **56(24)**: 120-126.
[18] L Guo, Y Li, D Mu, T Yang, Z Li. A LDA Model Based Topic Detection Method[J]. Journal of Northwestern Polytechnical University, (2016),**34(4)**:698-702.
[19] Chen M, Shen D, Shen D. Short text classification improved by learning multi-granularity

topics[C]// International Joint Conference on Artificial Intelligence. AAAI Press, (2011):1776-1781.

[20] Guan P, Wang Y. Identifying Optimal Topic Numbers from Sci-Tech Information with LDA Model[J]. New Technology of Library & Information Service, (2016).

[21] H.K. Lam, Udeme Ekong, Hongbin Liu. A study of neural-network-based classifiers for material classification[J]. Neurocomputing, (2014), **144(1)**:367-377.

IOP Conf. Series: Materials Science and Engineering **715** (2020) 012111 doi:10.1088/1757-899X/715/1/012111

Design and application of a new type of transmitting mechanism for advanced manufacturing

Yanzhong He[1,*], Zhongliang Huo[1], Xingsong Wang[1], Jing Chen[2] and Chunlei Tu[1]

[1]School of Mechanical Engineering, Southeast University, 211189, China
[2]eTrust Power Group Ltd., CITIC Private Equity Funds Management Co., Ltd., 212132, China

*Email: 18136819996@163.com

Abstract. The speed change of the press slider in the tensile region should be slower and the full load operation time should be longer to adapt to the stretching process. By analyzing structure and motion characteristics of the existing mechanical press, changing the main drive gear of the press, a new press with a slide part added near the bottom dead centre (BDC) with low speed characteristic was designed, which is very suitable for advanced manufacturing. According to the working principle and main features of the new combined press, the parametric design of the mechanical model was established by using ADAMS simulation analysis software. Through the parametric design of the pendulum length, several groups of sliding motion curves were obtained, and the influence of the rod length on the motion characteristics of the slider was compared and analyzed.

1. Introduction

At present, the main configuration of mechanical press in domestic and foreign is the crank-slide mechanism. The working part (tool) of the crank press is a stamp, whose fixed part is mounted on the table of the press and whose movable part is mounted on the press slide. The slide is moved by the crank-slide mechanism. In one turn of the crank the connecting rod makes a complete pass; the stamping takes place during the forward motion of the slide. This structure is simple and easy to manufacture and maintain. It satisfies stamping process, cutting edge, shaping, and local forming without special kinematic requirement. Its motion characteristics are shown as sinusoidal curves. Moving speed of slider varies greatly near the bottom dead centre (BDC) and the working stroke is not uniform. When the mechanism works, it has great impact, vibration and noise. Especially in the drawing process, parts are prone to tear or wrinkle defects, which can not guarantee the accuracy and quality of the drawing parts. In order to meet some requirements of deep stamping and deep drawing, it is necessary to change the speed of slider slowly near the lower dead point, and the working area of full load is longer. In order to adapt to the above slider motion characteristics, the existing method is to transform the crank-slider press into six-bar, eight-bar and ten-bar press, which can meet the technological requirements of deep drawing and stretching, but the main transmission structure of this kind of mechanism is more complex, and the requirements for processing, assembly and debugging are higher. Moreover, the multi-link structure leads to poor rigidity and low precision of the whole machine. In order to ensure the quality of the whole machine, the processing and manufacturing requirements of the machine tool are further improved, thus the production cost is increased. However, the drawbacks of the hydraulic drawing machine are insufficient. In this paper, a new type of mechanical stretching pressure mechanism is designed for the deficiencies of deep drawing and stretching technology in the existing crank-slider press, and the kinematic and structural characteristics of the new structure are described in detail [1-2].

2. Working principle and structure scheme

2.1. Working principle

Figure 1 shows a schematic diagram of the slider motion of a new mechanical drawing press with new configuration. The working principle of the mechanical drawing press is as follows: The eccentric wheel is rotated by the transmission mechanism, and the pendulum rod moves up and down in a straight line while swinging with the eccentric wheel. The center of mass of the pendulum rod moves in the same straight line. The other end of the pendulum rod is connected with the slider, which moves up and down with the pendulum rod. Assuming that the eccentricity of eccentric wheel is e and the length of pendulum rod is L, its structural characteristic is that the relationship between eccentric wheel and pendulum rod should satisfy $e < L$. On the premise of satisfying the relationship between the length of the rod, different slider operating curves can be obtained according to the design of different lengths of L, and different slide technological motion characteristics can be obtained. In addition, the noise of crank press is lower than that of common crank press because of the lower working speed and the lower impact force. By changing the length of the pendulum rod, the speed of the slider at the end of the working stroke can be greatly slowed down, and it can stay at the limit working position for a period of time, so that the press can have the movement characteristics of tension and holding pressure.

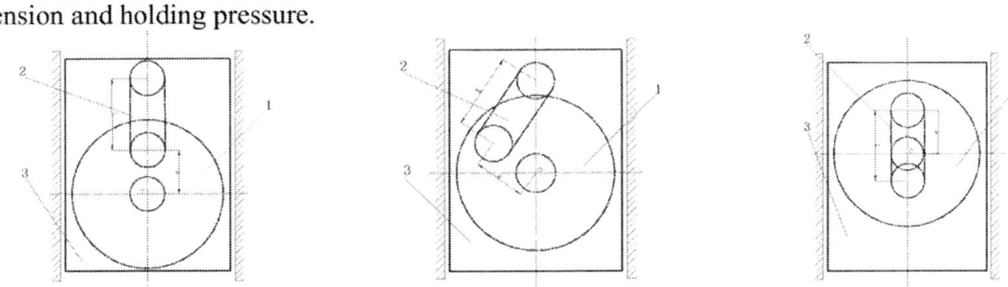

(a) Slider running to top dead centre (TDC) condition (b) Slider running process condition (c) Slider running to bottom dead centre (BDC) condition

Figure 1. New type mechanical press slide motion diagram.

The structure of the new mechanical tension press includes the fuselage and the transmission part. The eccentric wheel is mounted on the fuselage and driven by the transmission system. One end of the pendulum rod 2 is hinged with the eccentric wheel 1. With the rotation of the eccentric wheel 1, the pendulum rod 2 performs a compound motion of swing and linear reciprocating motion. The other end of the pendulum rod 2 is connected with the slider part 3, which follows the pendulum rod in reciprocating motion in the fuselage guide rail. The stroke of the press is H=2e, and the condition of $e < L$ must be satisfied [3-4].

2.2. Structure scheme

As shown in figure 2, the scheme structure of a new drawing mechanical press for the new configuration is presented. In order to ensure the balance of force and smooth transmission of the main transmission mechanism, the eccentric gears and eccentric wheels are symmetrically installed in the layout of the whole machine. The eccentric gear is rotated by the gear shaft, and the pendulum rod moves in a straight line while swinging with the eccentric wheel. The center of mass of the pendulum rod always moves in a straight line. The other end of the pendulum rod is connected with the middle axis of the slider. When the pendulum rod moves up and down, the slider moves up and down with the

IOP Conf. Series: Materials Science and Engineering **715** (2020) 012111 doi:10.1088/1757-899X/715/1/012111

pendulum rod. As shown in figure 3, the eccentric gear and the eccentric wheel in this mechanism are characterized in that the central axis of rotation of the eccentric portion and the central axis of rotation of the large gear are coaxial, which is different from the eccentric gear structure in eccentric gear press.

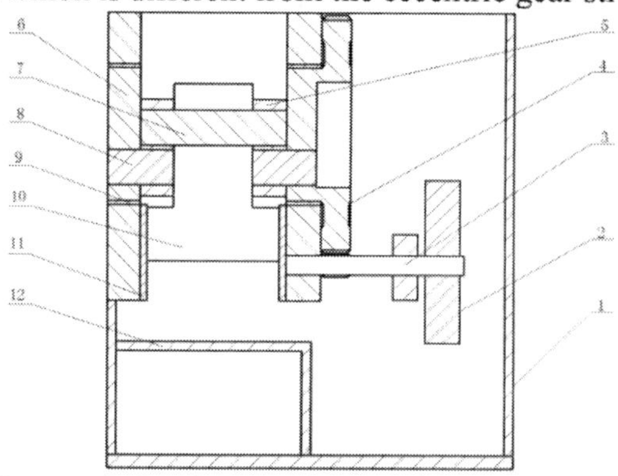

1 machine body, 2 flywheel, 3 gear shafts, 4 eccentric gear, 5 pendulum rod, 6 eccentric wheels, 7 axis, 8 connecting shaft, 9 sets of copper, 10 slider, 11 guide rails, 12 working table

Figure 2. Scheme structure of a new drawing mechanical press.

Figure 3. Eccentric gear structure.

The specific working process of the key parts installation in figure 2 is as follows: the eccentric wheel part of eccentric gear 4 is installed in the copper sleeve 9 of the fuselage 1, the large gear part of eccentric gear 4 meshes with the gear shaft 3, and the gear shaft 3 transmits power by the flywheel 2; the one end of the swing bar 5 is rotated and connected with the eccentric gear 4 through the copper sleeve through the connecting shaft 8. With the rotation of eccentric gear 4, the swing bar 5 oscillates and reciprocates with the upper and lower straight sports. The other end of the pendulum rod 5 is connected with the slider 10 through the rotation of the central axis 7. The slider 10 follows the pendulum rod 5 to reciprocate on the guide rail 11 of the fuselage 1. The structure of the eccentric wheel at the other end is the same as that of the side eccentric gear in the way of power transmission and connection of the slider.

3. Mechanism simulation and analysis

The rigid body simulation analysis model of the transmission mechanism of the new mechanical tension press is established by ADAMS simulation analysis software. Simplified model of main transmission mechanism is established. Simplified model of pendulum, eccentric wheel and slider of main transmission mechanism is created in ADAMS to analyse, which ensures that simulation analysis of main transmission mechanism is not distorted. As shown in figure 4a, the main characteristic parameters are eccentricity of eccentric wheel $e = 150mm$, and length of pendulum rod $L = 180\ mm$. figure 4b is the simulation operation condition of the main transmission mechanism slider when it goes down. It can be seen that when the slider goes down near the bottom dead centre (BDC), the pendulum rod gradually presents a vertical position, and the angle between the pendulum rod and the slider running speed decreases gradually. The smaller the angle is, the better the force condition of the pendulum rod is.

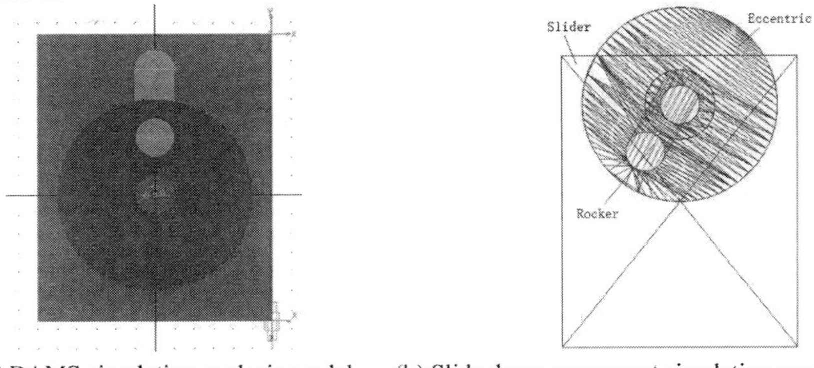

(a) ADAMS simulation analysis model (b) Slide down movement simulation condition diagram

Figure 4. ADAMS simulation experiment of slider motion.

Figure 5 is the motion characteristic curve of the main drive mechanism with new configuration obtained by ADAMS simulation. When the slider runs to the bottom dead centre (BDC), the speed of the slider changes slowly within a certain range of rotation angle, and the speed and acceleration are smaller, which is very suitable for sheets with low drawing rate.

Figure 6 is the displacement curve of the pendulum rod in X and Y directions, which can clearly show the position of the pendulum rod in a working cycle. In order to study the force variation law of the pendulum rod in the main transmission mechanism, the bottom of slider is assumed to bear the vertical force of F=1000N.

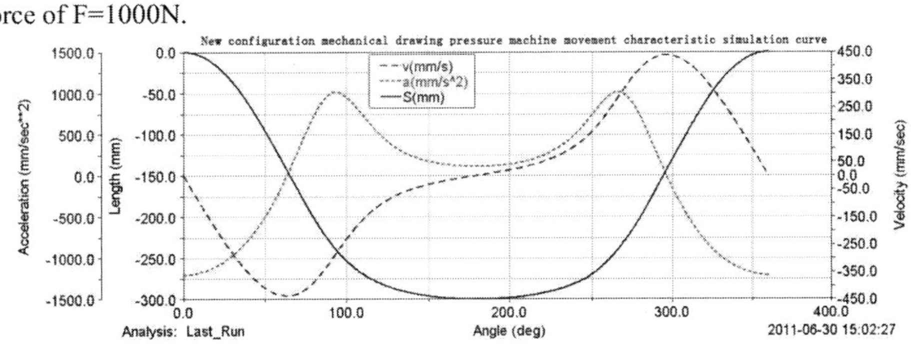

Figure 5. Main transmission mechanism movement simulation curve.

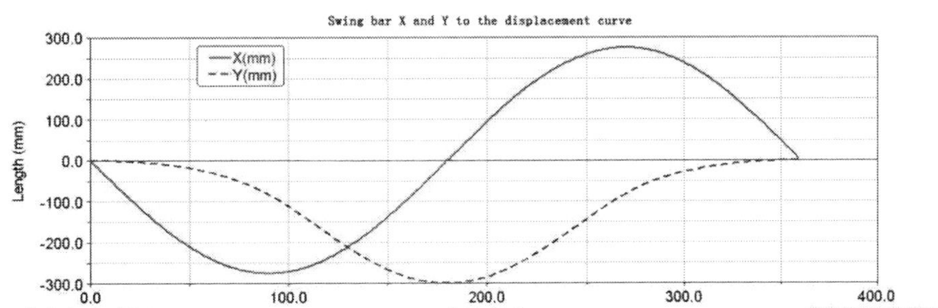

Figure 6. Displacement curve of the pendulum rod at the X and Y direction.

Figure 7 is the cure of the vertical direction force at the connecting point of the pendulum rod in the main transmission mechanism simulated by ADAMS. From figure 7a and figure 7b, it can be seen that the force on both sides of the pendulum rod is equal in size and opposite in direction when the slider goes down. The common point is that when the bottom of the slider is subjected to constant load, the force on the connecting position of the pendulum rod increases gradually as the slider goes down, but when the specific angle is reached, the force law changes, and the force on the connecting position of the pendulum rod decreases gradually with the slider approaching the bottom dead centre (BDC). Because the force of the pendulum rod has such characteristics, it is necessary to consider the angle position of the slider starting to stamp in order to optimize the force conditions at the pendulum rod connection, which is also the basis for checking the strength and stiffness of the pendulum rod.

4. Parametric design of the transmission mechanism

In order to more accurately analyse the influence of the relationship between rod length and eccentricity on the motion characteristics and force properties of the main transmission mechanism, the simulation analysis model in ADAMS is parameterized. Through the transmission simulation of parameterized rod length, the motion characteristics of slider and the force variation curve of pendulum rod under different rod length conditions are further investigated. The motion and force curves under various rod lengths are compared and analysed, and the rules are summarized, which provides a theoretical basis for the design of optimum parameters of the rod lengths of mechanisms.

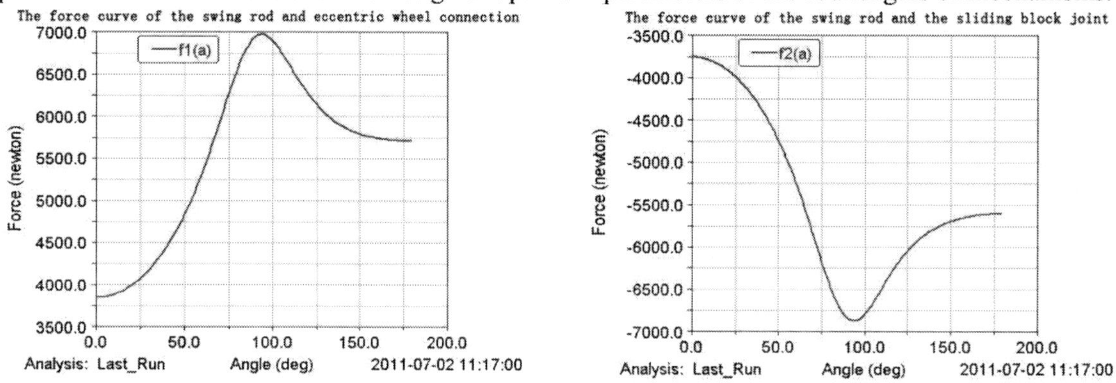

(a) Force variation curves of the pendulum rod is connected with the eccentric wheel

(b) Force variation curves of the pendulum rod is connected with the slide

Figure 7. Force variation curves of the pendulum rod.

4.1. Simulation analysis model parameterization

Based on the above simulation and analysis model created in ADAMS, its characteristic parameters are taken as the initial values of parametric analysis. The eccentricity of eccentric wheel e = 150 *mm*, the length of the pendulum rod L =180 *mm*. The length of the pendulum rod is parameterized and the value is L = [160, 170, 180, 190, 200] *mm*.

There are several parametric design models commonly used in ADAMS: parameterized POINT location, parameterized MARKER POINT location, parameterized geometry and parameterized drive. For the parameterization of the length of the pendulum rod, it can be transformed into the parameterization of the coordinates of the end point of the pendulum rod. When creating a connecting rod that needs to be parameterized, first create geometric points (two geometric points need to be parameterized at both ends, and one geometric point needs to be parameterized at one end), then select the two points when creating the connecting rod. Because the connecting rod is related to the geometric points, the parameterized connecting rod is indirectly realized by the position of the parameterized geometric points, the parameterized connecting rod is indirectly realized by the position of the parameterized geometric points.

The specific methods of parameterization are as follows: Open the simulation analysis model created by the ADAMS. A variable dialog box pops up after clicking on the menu [build] – [Design Variable] – [New]. According to the parametric analysis of the length of the pendulum rod, the design variables are 160, 170, 180, 190 and 200, and the standard value is 180. According to the coordinate position relationship of the simulation analysis model, the initial coordinate values of the parameterized points are (0, 330, 0). If the Y coordinate is parameterized instead, the coordinates are (0, (DV_1), 0), and the Y coordinate value (DV_1) is taken as [310, 320, 330, 350, 350]. After setting the parameterized coordinate point, the pendulum rod is correlated with the coordinate of the point. As shown in figure 8, the list of values of the parameterized design variable DV_1 is created. After defining the design variables, we need to define the objective function.

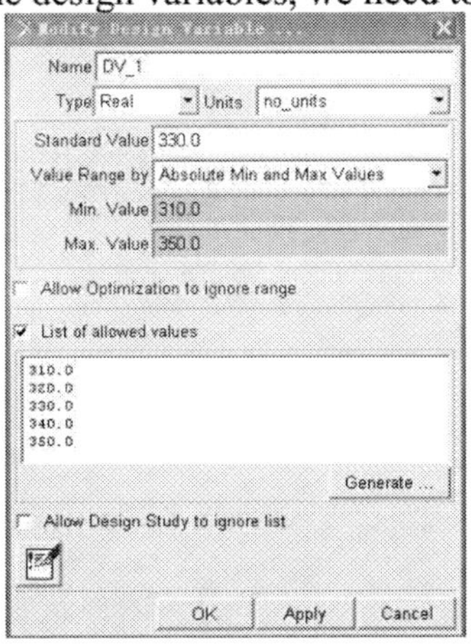

Figure 8. Settings dialog box about design variables of the parametric Y point coordinate values.

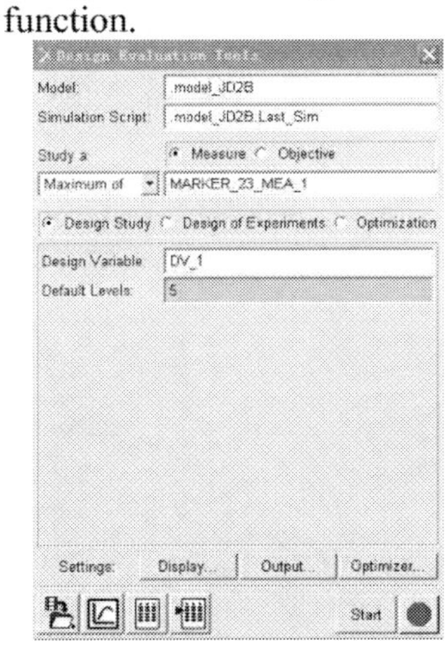

Figure 9. Settings dialog box about associated definition of the objective function and design variables.

Firstly, the stroke, velocity, acceleration curve, X, Y displacement curve, centroid velocity curve and force curve of the slider in the main transmission mechanism are measured as the optional values of the objective function. Then, the design variable length L of the pendulum rod is correlated with the corresponding objective function curve. As shown in figure 9, after clicking on the menu [simulate] – [Design Evaluation], the dialog box for correlation analysis of defining objective function and design variables is popped up. After defining the correlation function, clicking the [Start] button for parametric simulation analysis and calculation, the corresponding motion curve, displacement curve, centroid velocity curve and force curve of the slider under five different pendulum rod lengths can be obtained. Figure 10 shows the parameterized lines of the length of each pendulum rod.

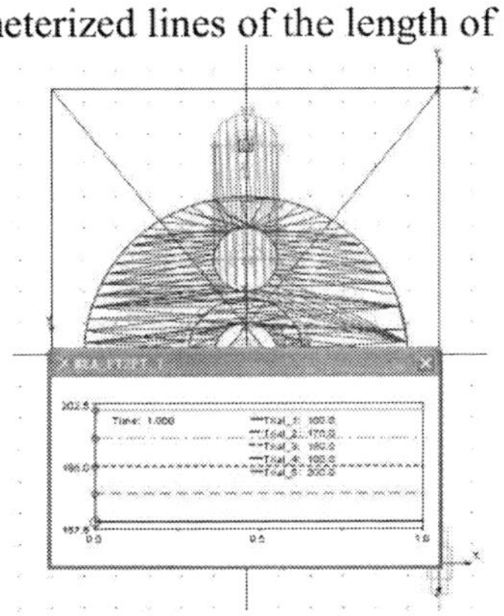

Figure 10. Parametric line of the pendulum rod.

4.2. Curve analysis

Figure 11 shows the parametric simulation curve of the transmission mechanism. As can be seen from the curves shown in figure 11a, all curves have a common point that the change is relatively gentle near the bottom dead centre. With the increase of the length of the rod, the smooth range increases, and different length of the pendulum will get different stroke curves of the slider. In this way, the required process stroke curve can be obtained by the length of the control rod.

As can be seen from the curves shown in figure 11b, the velocity curves of the pendulum rod are obviously different with the length of the pendulum rod. The common point is that the velocity near the bottom dead centre is small, but as the rod length increases, the speed decreases. In this way, the length of the control rod can be used to obtain a very suitable drawing speed for a certain material.

From the curve of figure 11c, it can be seen that the acceleration curves are also significantly different due to the length of the rod. The change of velocity near the bottom dead centre is slower, but with the increase of the length of the rod, the change of velocity is slower [5-9].

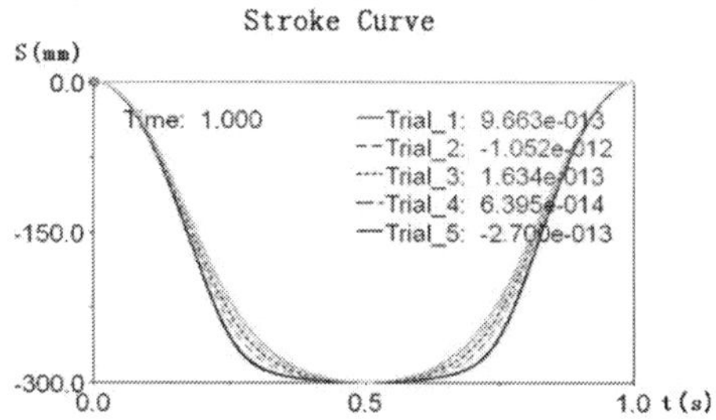

(a) Curve of the slide stroke

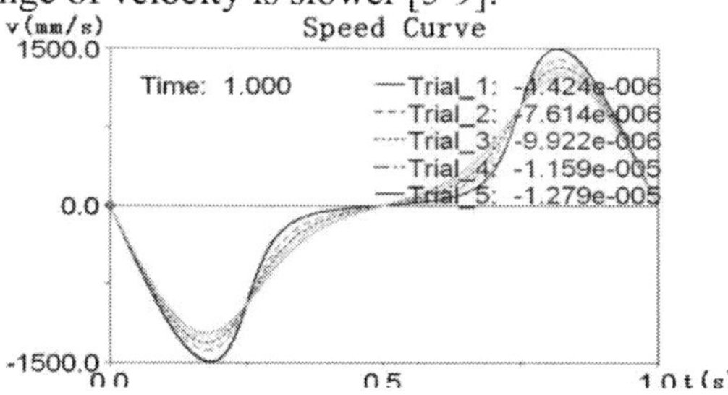

(b) Curve of the slide velocity

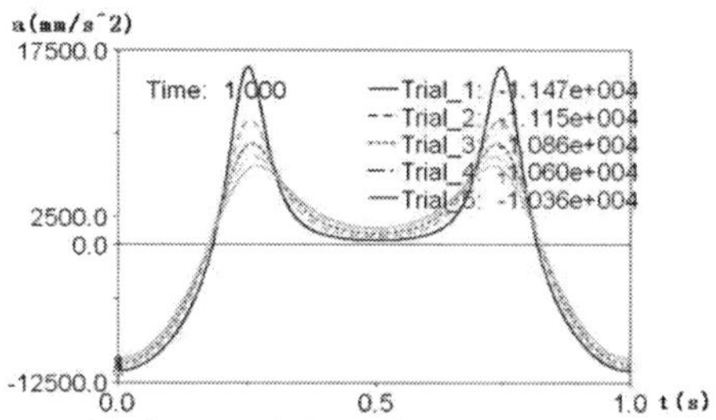

(c) Curve of the slide acceleration

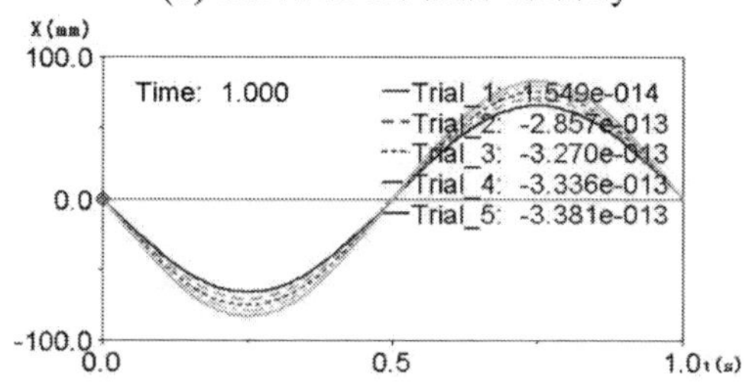

(d) Displacement curve of the pendulum rod at the X direction

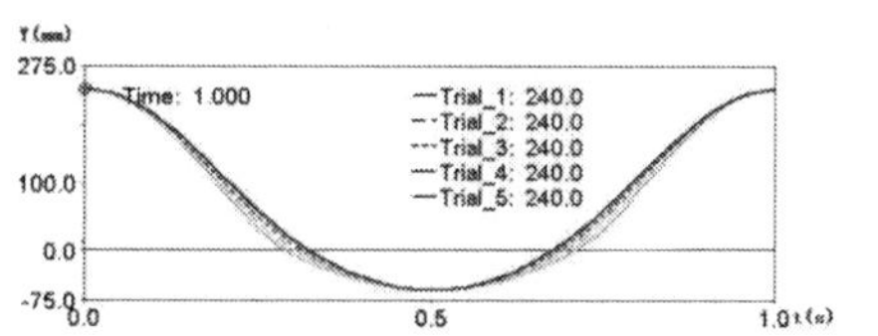

(e) Displacement curve of the pendulum rod at the Y direction

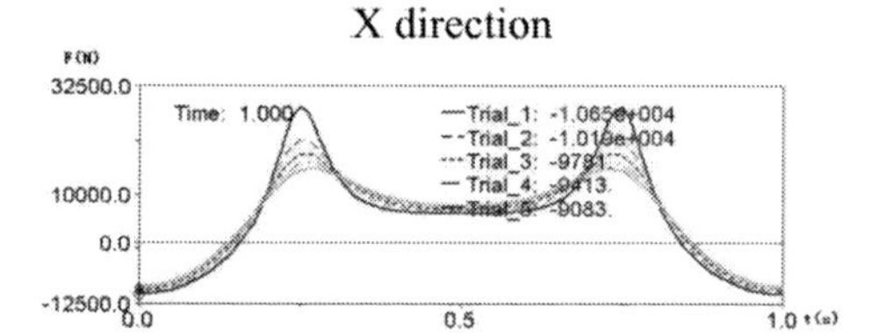

(f) Force variation curves of the pendulum rod is connected with the eccentric

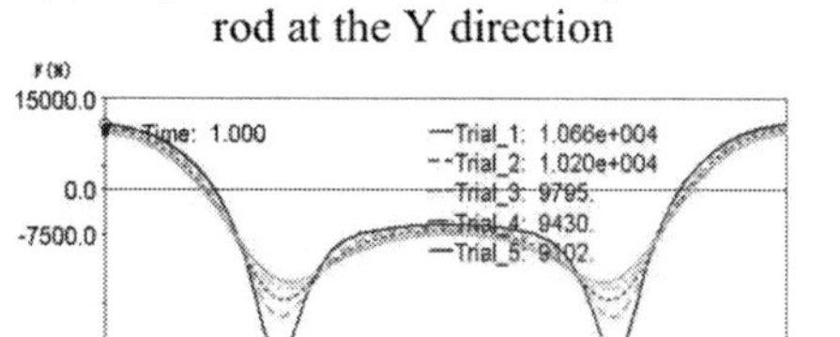

(g) Force variation curves of the pendulum rod is connected with the slide wheel

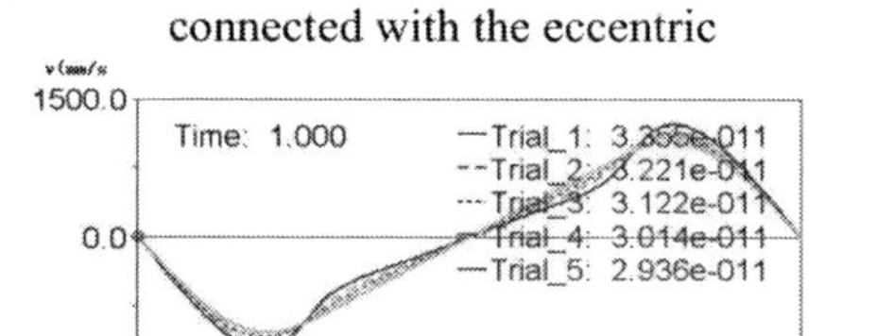

(h) Velocity curve of the pendulum rod mass center

Figure 11. Transmission mechanism parameterized simulation curve.

From the curves in figure 11d and 11e, it can be seen that with the increase of the length of the pendulum rod, the displacement space of the pendulum rod in the direction of X and Y has greater influence on the downward and upward process of the slider, but less influence near the bottom dead centre.

IOP Conf. Series: Materials Science and Engineering **715** (2020) 012111 doi:10.1088/1757-899X/715/1/012111

From the curves in figure 11f and 11g, it can be seen that with the increase of the length of the pendulum rod, the force at the junction of the two ends of the pendulum rod decreases, and the effect is most obvious near the maximum value of the force at a specific turning angle, when the effect is less near the bottom dead centre.

From the curves in figure 11h, the velocity characteristics of the centre of mass of the pendulum bar can be obtained. When the slider goes down, the speed increases gradually. After reaching a certain time, as the slider goes down to the bottom dead point, its speed decreases gradually, and the speed of reaching a certain time decreases gradually to zero. When the slider goes down and goes up, the sudden change of velocity at a specific time point is affected by the length of the pendulum rod. The longer the length of the pendulum rod is, the slower the sudden change of velocity is.

5. Conclusion

A new type of transmission mechanism of mechanical press suitable for deep stamping and drawing process is designed. The slider speed changes slowly near the bottom dead centre, and the working area of full load is longer. The mechanism can obtain different slider operating characteristic curves by changing the length of the pendulum rod, which has good versatility.

The working principle and main characteristics of mechanical drawing press are described in detail, and the scheme structure is designed. The key technology and working process of the drawing mechanical press are illustrated with an example.

With the help of ADAMS software, a simulation analysis model is created. The motion characteristics of the main transmission mechanism and the force characteristics of the pendulum rod are simulated and analysed, and the corresponding variation curve is obtained. By parameterizing the length of the pendulum rod in ADAMS, the corresponding slider motion curve, the displacement curve, the centroid velocity curve and the force curve of the pendulum rod under the condition of the length of the pendulum rod are obtained. The relationship curves are evaluated and analysed to provide theoretical basis for the design of the optimum rod length and the optimum technological motion curve of the main transmission mechanism.

This kind of mechanical stretching press has the advantage of sufficient stretching compared with hydraulic stretching press. In addition, the structure can greatly reduce the height of the press body and the weight of the whole machine. While the cost is reduced, the stiffness of the fuselage is increased and the precision of the whole machine is improved, which improves the precision of the parts and the life of the die. It provides an important reference for the design of mechanical press with this configuration.

References

[1] JIER Machine Tool Group Co., LTD. Design of Crank Press. [M]. Jinan: Publishing House of Jinan, 1973.

[2] D. Y. He. Crank Press [M]. Beijing: China Machine Press, 1989.

[3] Y. D. Lu, Y.Z.He, J.M.Huang, etc. Novel mechanical pressing machine [P]. China invention patent: CN201010562724.7.2010.11.25.

[4] Y. Z. He, J. M. Huang. Mechanical press with novel structure [P]. China patent for utility model: CN201020694554.3.2010.12.30.

[5] J. R. Zheng. Introduction and improvement of ADAMS virtual prototype technology [M], Beijing: mechanical industry press, 2002.

[6] MCS. Software. MSC. ADAMS/View advanced training course [M]. Beijing: tsinghua university press, 2004.

[7] L.G.Pu, M.G.Ji. Mechanical design [M]. Beijing: higher education press, 2004.

[8] "handbook of forging and pressing technology". Handbook of forging technology [M]. Beijing: defense industry press, 1988.

[9] Alexander Moser.Metalworking`s Mechanical Press Handbook [M].Boston: Metalworking Magazine, 1960.

An efficient method of logical expression constraint solving in path-oriented unit test

Yalan Wang[1], Songyi Liu[2], Yawen Wang[1,*] and Yunzhan Gong[1]

[1]State Key Laboratory of Networking and Switching Technology, Beijing university of Posts and Telecommunications, 100876, Beijing, China
[2]China Launch and Tracking Control, 100020, Beijing, China

*Email: wangyawen@bupt.edu.cn

Abstract. This paper extends the theory of constraint solving and presents a method to solve logical expression constraint problems existing in the process of generating test cases automatically. It defines the concept of the logical expression constraint on the path of C programs. In order to improve solving efficiency, it also presents an algorithm of pre-processing and heuristic backtracking. Our experiment results show that the method can narrow the domain of variables from meaningless infinite intervals to a finite range and achieve better coverage, and the results also show that the algorithm can improve the testing efficiency and halve the time of generating test cases.

1. Introduction

Path-oriented automatic generation of test cases is one of the typical ways in unit test, which depends on the establishment and solution of the constraint system. And the constraint system is mainly solved by the search method. The researchers have also done a lot of research on this method [1-3], from which it can be found that the problem of solving the logical expression constraint is less concerned. However, in a practical project, the logical expression constraints show relatively high proportion of the constrains (statistics from three large projects [4] - aa200c, qlib, and deco, which indicate that the number of logical expression constraints exceeds 30% of the total). ABased on the existing interval algorithm [5] in CTS (Code Testing System), a method of logical expression constraints solving in C-language programs is proposed in this paper, which can support logical expression constraints solving and improve the efficiency.

2. Path-oriented logical expression constraint solving algorithm

2.1. Problem definition

The Constraint Satisfaction Problem (CSP) consists of a set of variables $x_1,x_2,\cdots,x_n$, the domain $R_1,R_2,\cdots,R_n$ corresponding to the variable and a set of constraints. Solving the CSP problem is to find the value of the variables that in the variable domain so that the constraint is satisfied. [7] The logical expression constraint satisfaction problem (LECSP) is a special constraint satisfaction problem for C language. So the input of LECSP consists of three elements: variable set V, $V=\{v_1, v_2,\cdots,v_n\}$; the domain of the variable D, $D=\{d_1,d_2,\cdots,d_n\}$; and the constraint between the variables LC, $LC=\{Exp_1 \text{rop} Exp_2 \text{rop} \cdots rop Exp_n, \text{Lv}\}, \text{rop} \in \{\&\&, ||, !\}, Lv \in \{\text{T,F}\}$. Exp_i is a constraint composed of

IOP Conf. Series: Materials Science and Engineering **715** (2020) 012112 doi:10.1088/1757-899X/715/1/012112

mathematical expressions that connected by relational operators (=, <, >, <=, >=). We can also call it atomic expression constraint(AC). Lv is a logical value, when Lv is T, the true value of the constraint LC is True, and the logical value F represents False. The particularity of the logical expression constraint solving problem studied in this paper is not finding the value of variables, but making domain D reduction [6], so that when the variable V is searched in the domain D, it can reduce the search space in the constraint solving process, for the reason that can improve the efficiency of generating testing cases.

2.2. Introduction of algorithm

In this paper, a framework is established for logical expression constraints solving, and three algorithms are proposed in this framework. The whole algorithm is shown in Algorithm 1. The algorithm solves the non-LCS constraints on the path p first and takes the result as the initial variable domain. Secondly, we will call the "pruning" algorithm and the Variable Level Determination Algorithm for pre-processing, and heuristic backtracking algorithm will be called if there is a conflict in the solving process.

Algorithm 1 Path-oriented logical expression constraint solving algorithm

Input: The path to be covered, *p*;
Output: The variable domain that satisfies the logical expression constraint on path p, *variableDomain*;

```
procedure MyProcedure
    CS ← All non-LCS constraint sets on path p
    LCS ← All LCS constraint sets on path p
    result ← NULL
    initDomain ← intervalCompute(CS)
    for each lc ∈ LCS do
        ACS ← Get all atomic expression constraint sets that satisfy lc
        ACS ← getSimplification(ACS) //call "Pruning" algorithm
        ACS ← getOrder(ACS) //call variable level determination algorithm
        ac ← getFirstSet(ACS)
        conflict ← true.
        while conflict ≠ false do
            conflict ← intervalCompute(ac)
            if conflict = true then
                ac ← getAnotherSet(ACS) //call heuristic backtracking algorithm.
            end if
        end while
    end for
end procedure
```

2.2.1. "Pruning" algorithm for simplification

We can use the set of atomic expression constraints(ACS) to represent the logical expression constraint. In this case, the relationship among variables that inside the atomic expression is ignored and treated as a whole. In this way, we can find the logical value combination of atomic expressions satisfying the logical expression constraints by establishing truth tables, and the results will be expressed by union sets.

For example, a simple logical expression constraint, $LC = \{Exp_1\&\&Exp_2||Exp_3,\mathrm{F}\}$,and the truth table of it is as shown in table 1.

From table 1, we can see that there are diversity and complexity in the combination of atomic expressions, especially when a logical expression constraint consists of multiple atomic expression constraints, the solution space will grow exponentially. And from table 1,we can get ACS = $\{\{Exp_1,T\} \cap \{Exp_2,F\} \cap \{Exp_3,F\}\} \cup \{\{Exp_1,F\} \cap \{Exp_2,T\} \cap \{Exp_3,F\}\} \cup \{\{Exp_1,F\} \cap \{Exp_2,F\} \cap \{Exp_3,F\}\}$. In order to improve the efficiency of solving, the set of atomic expressions can be simplified. Simplification is based on the initial variable domain, and the atomic expression constraint that is always satisfied will be reduced by the "pruning" method, thus the unnecessary solving process is reduced. For example, in the ACS set, if it is certain that the initial variable domain of Exp_1 is constant T, then the ACS set can be simplified to ACS = $\{Exp_2,F\} \cap \{Exp_3,F\}$, it can be seen from this that the "pruning" algorithm can greatly reduce the solution space of the constraint.

Table 1. Truth table.

Exp_1	Exp_2	Exp_3	LC
T	T	T	T
T	T	F	T
T	F	T	T
T	F	F	F
F	T	T	T
F	T	F	F
F	F	T	T
F	F	F	F

2.2.2. Variable Level Determination Algorithm based on variable dependencies

According to the dependency between variables, the Variable Level Determination Algorithm(VLDA) sorts the ACS and solves the domain of the high-priority variable, so that the variable domain of the solution is more accurate, which is beneficial to the selection and generation of test cases.

Algorithm 2 Path-oriented logical expression constraint solving algorithm

Input: The set of atomic expression constraints, P;
Output: The set of variable; The set of atomic expression constraints, $SetP_n$;
1: Traverse all atomic expression constraints, set the highest priority p_0 to variables that directly depend on constants, and the corresponding constraint set is $SetP_0$;
2: Set the variable x that depends on the elements in p_0 to priority p_1, and the corresponding constraint set is $SetP_1$, and the constraint does not belong to $SetP_0$;
3: Continue recursively solving the set of priority P_n, until all elements are present in the constraint set $SetP_n$;
4: The variables at the same level is solved in order;
5: **return** P_n,$SetP_n$;

For example, suppose the logical expression constraint is $LC = \{a < b\&\&a > 5\&\&c > b,\mathrm{T}\}$, the variable a, b and c are integer variables, and the initial domain of the variables are $(-\infty, +\infty)$. We can quickly get ACS = $\{a < b,T\} \cap \{a > 5,T\} \cap \{c > b,T\}$. According to the initial variable domain and ACS, if we call the interval algorithm to calculate each atomic expression in order from left to right, the result of solving ACS will be $\{a{:}(5, +\infty),b{:}(-\infty, +\infty),c{:}(-\infty, +\infty)\}$. Looking at the logical expression constraint, we can get dependencies between variables, the variable c depends on the value of the variable b, the variable b depends on the value of the variable a, and the variable a depends on a constant. If we follow the dependency, we first solve $\{a > 5,T\}$, then solve $\{a < b,T\}$, and finally solve $\{c > b,T\}$, and the result will be $\{a{:}(5, +\infty),b{:}(6, +\infty),c{:}(7, +\infty)\}$. It can be seen that the algorithm can narrow the domain of variables, thus improving the efficiency of test case generation. The complete VLDA is defined as follows in Algorithm 2.

2.2.3. Heuristic backtracking algorithm for conflicts

After pre-processing, we can select an element of ACS and call the interval algorithm to solve the atomic expression to reduce the variable domain. However, there may be a conflict of solutions, that is, there is no value in the domain of variable that can satisfy the constraints. In this case, backtracking is required, and the new element of ACS will be re-selected to solve. One of the methods of selection is blind, that is, randomly choosing an element, which is simple and easy to implement, but it will easily lead to multiple conflicts; the other is heuristic, some heuristic information will be collected from the conflict. And according to this information, the results move in the direction most likely to reach the final state.

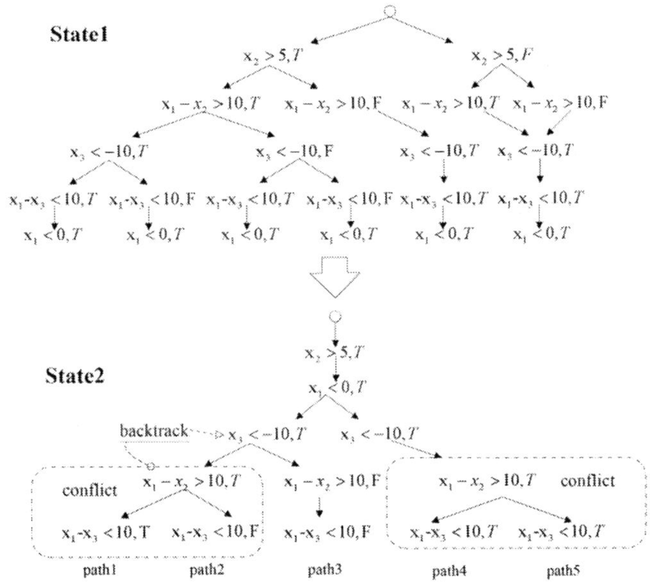

Figure 1. State at the time of constraint solving.

Suppose $\text{LCS} = \{(x_2 > 5\&\&x_1 - x_2 > 10||x_3 <- 10\&\&x_1 - x_3 < 10)\&\&x_1 < 0,\text{T}\}$, and the initial variable domain is $\{x_2{:}[8, +\infty]\}$. The ACS that satisfies LCS is represented by the path from the root node of the binary tree to the leaf node, as shown by state1 in figure 1. And after calling the pre-processing algorithm, the result will be shown in state2. As can be seen from the scale of the binary tree, the pre-processing algorithm greatly reduces the search space. Next, we can call the interval algorithm. Assuming that we select path1 when solving, start from the root node, when we calculate $\{x_1 - x_2 > 10,\text{T}\}$, we find conflict, then we need to backtrack and choose $\{x_1 - x_2 > 10,\text{F}\}$ in path2 to calculate. From the conflict, we can find that the relevant variable of the conflicting expression is $\{x_1,x_2\}$, so the relevant expression on the path1 is $\{x_2 > 5,T\}$ and $\{x_1 < 0,\text{T}\}$, the same as $\{x_1 - x_2 > 10,\text{T}\}$ in path4 and path5, so we can exclude path1, path2, path4 and path5 at the time of backtracking.

3. Experiments

The method RDP (Recursive Descent Parsing) solves logical expression constraints recursively and adds all possible values of variables to the set of solutions which will lead the result to be $(-\infty, +\infty)$. And the domain of $(-\infty, +\infty)$ is meaningless to the interval reduction. In the experiment, we compared the reduction of the variable domain between the strategy of this paper and the RDP method. Statistics from three large projects indicate that the more complex the logical expression constraints are, the less the number of constraints is. And the most complex logical expression constraints consist of six atomic expression constraints. The experiment in this paper is set as follows, six variables, and these variables are arbitrarily combined into a set of linear atomic expressions by using mathematical operators and relational operators. We will select $n(n \in [2,6])$ atomic expressions from the set and combine them into logical expressions by logical operators (&&,||,!), as a condition of the if statement in the program waiting for testing. [7]

The experiment tests 100 test programs and the test results of some functions are listed in table 2. As can be seen from experiment result, the strategy of this paper can support the logical expression constraint solving, reduce most of the variable domain to a limited range, and avoid the test case search failure.

IOP Conf. Series: Materials Science and Engineering **715** (2020) 012112 doi:10.1088/1757-899X/715/1/012112

Table 2. Comparison of the reduction of the variable domain.

function	LCS	reduction of variable domain		times of searching test cases	
		using the RDP method	using the strategy of this paper		
f01	$\{x_1 >- 5\&\&!(x_2 > 10),F\}$	$\{x_1:(-\infty, +\infty), x_2:(-\infty, +\infty)\}$	$\{x_1:[-4, +\infty), x_2:[11, +\infty)\}$	2	2
f02	$\{x_1 + x_2 < 10\&\&x_1 >- 5 \&\&x_2 + 2 * x_3 < 25,F\}$	$\{x_1:(-\infty, +\infty), x_2:(-\infty, +\infty) x_3:(-\infty, +\infty)\}$	$\{x_1:[-4, +\infty), x_2:(-\infty,13] x_3:[-6, +\infty)\}$	3	3
f03	$\{x_1 - x_2 > 10\&\&x_2 - x_3 > 5 \&\&x_3 > 5\&\&10*x_4 + x_3 > 20 \&\&x_5! = 0,F\}$	$\{x_1:(-\infty, +\infty), x_2:(-\infty, +\infty) x_3:[5, +\infty), x_4:(-\infty, +\infty), x_5:(-\infty, -1] \cup [1, +\infty)\}$	$\{x_1:[22, +\infty), x_2:[11, +\infty) x_3:[5, +\infty), x_4:(-\infty, +\infty), x_5:(-\infty, -1] \cup [1, +\infty)\}$	failed after searching 100 times	5
f04	$\{(x_2 > 5\&\&x_1 - x_2 > 10) \&\&(x_4 == 0\|\|x_3 + 2 * x_4 == 10) \&\&(!(x_5 * 2 < 4)\&\&x_6! = 0),F\}$	$\{x_1:(-\infty, +\infty), x_2:[6, +\infty) x_3:(-\infty, +\infty), x_4:(-\infty, +\infty), x_5:[2, +\infty), x_6:(-\infty, -1] \cup [1, +\infty)\}$	$\{x_1:[17, +\infty), x_2:[6, +\infty) x_3:[10,10], x_4:[0,0], x_5:[2, +\infty), x_6:(-\infty, -1] \cup [1, +\infty)\}$	failed after searching 100 times	6

The experiment also compares the time of generating test cases to illustrate the efficiency of the strategy. The experiment results show that 86 out of 100 functions can generate test cases more efficiently than the RDP method, and the time of solving the remaining functions is similar. The test results of some functions are listed in table 3. The result shows that the strategy of this paper has significantly improved the efficiency of generating test cases. This paper also analyses the cases that 14 of 100 functions where the test case generation time becomes longer after adding the logical expression constraint solving strategy. The main reasons are as follows. First, the test case generation radix is small, the difference is within a few milliseconds, and the machine state affects the efficiency. Second, when using the strategy of this paper, when there is a conflict that requires backtracking but still fails to solve, this will be accompanied by a large number of interval operations and waste more time.

IOP Conf. Series: Materials Science and Engineering **715** (2020) 012112 doi:10.1088/1757-899X/715/1/012112

Table 3. Comparison of test case generation efficiency.

function	Time of test case generation/ms		function	Time of test case generation/ms	
	using the RDP method	using the strategy of this paper		using the RDP method	using the strategy of this paper
f06	2,007	1,678	*f53*	6,640	3,536
f12	5,427	1,691	*f60*	5,936	4,711
f19	6,644	1,735	*f64*	7,542	4,120
f25	7,678	1,804	*f75*	5,501	1,822
f28	3,733	2,255	*f76*	7,418	4,309
f36	5,722	3,687	*f78*	6,793	3,359
f41	5,899	3,695	*f92*	8,2558	2,706
total	--	--	--	5,195	41,108

4. Conclusion

Based on the interval algorithm, this paper supports the logical expression constraint solving in path-oriented unit test and can successfully reduce the domain of the input variable. Meanwhile, the "pruning" and variable level determination algorithm improve the efficiency of test case generation. The experiment results show that the proposed logical expression constraint solving strategy can support complex logical expressions, and efficiency is also significantly improved compared with the RDP method. However, when there are many conflicts in the backtracking process, it will take a certain amount of time. How to make the conflict backtracking process faster and more efficient is a problem that needs to be studied in the future. In addition, the algorithm of this paper only supports linear constraints, and the solution to nonlinear constraints is also a problem to be solved in the future.

Acknowledgments

This paper is supported in part by the National Natural Science Foundation of China under Grant No.U1736110 and No.61702044.

References

[1] A. Gotlieb, B Botella, M Rueher, Automatic test data generation using constraint solving techniques[J], (ACM Press, 1998).

[2] R Dechter, D Frost, Backjump-based backtracking for constraint satisfaction problems[J], Artificial Intelligence, 147 (2002)

[3] R.A Demilli, A.J Offutt, Constraint-based automatic test data generation[J], Software Engineering IEEE Transactions on, 900 (1991).

[4] http://www.choco.solver,org/.

[5] G Alefeld, J Herzberger, Introduction to Interval Computations, (Academic Press, 1983).

[6] A.J Offutt, Z Jin, J Pan, The dynamic domain reduction procedure for test data generation[J], Software Practice & Experience (2015).

[7] https://github.com/byrcoderw/test/blob/master/experiment.c

AUTHOR INDEX

Institute of Physics
Dirac House, Temple Back
Bristol BS1 6BE UK

ISSN: 1757-8981
ISBN 978-1-7138-2574-6

CURRAN ASSOCIATES INC.
proceedings.com